T0139334

CONTROL ENGINEERING AND INFORMATION SYSTEMS

PROCEEDINGS OF THE 2014 INTERNATIONAL CONFERENCE ON CONTROL ENGINEERING AND INFORMATION SYSTEM (ICCEIS2014), YUEYANG, HUNAN, CHINA, 20–22 JUNE 2014

Control Engineering and Information Systems

Editor

Zhijing Liu

School of Computer Science and Technology, Xidian University, Xi'an, Shanxi, China

CRC Press
Taylor & Francis Group
Boca Raton London New York Leiden

CRC Press is an imprint of the
Taylor & Francis Group, an **informa** business

A BALKEMA BOOK

CRC Press/Balkema is an imprint of the Taylor & Francis Group, an informa business

© 2015 Taylor & Francis Group, London, UK

Typeset by V Publishing Solutions Pvt Ltd., Chennai, India
Printed and bound in the UK and the US

Published by: CRC Press/Balkema
     P.O. Box 11320, 2301 EH Leiden, The Netherlands
     e-mail: Pub.NL@taylorandfrancis.com
     www.crcpress.com – www.taylorandfrancis.com

ISBN: 978-1-138-02685-8 (Hbk)
ISBN: 978-1-315-72438-6 (eBook PDF)

Table of contents

Session 6: Algorithm and simulation

Preface

The 2014 International Conference on Control Engineering and Information System (ICCEIS2014) was successfully held in Yueyang, Hunan, China during June 20 to 22, 2014. ICCEIS2014 provided an international forum for the researchers to share knowledge and results in theory and applications of Control Engineering and Information System.

All papers including in this proceedings had undergone the strict peer-review by the experts before they are accepted for publications. This proceedings covers the subjects of Intelligent System, Teaching Cases, Pattern Recognition, Industry Application, Machine Learning, Systems Science and Systems Engineering, Data Mining, Optimization, Business Process Management, Evolution of Public Sector ICT, IS Economics, IS Security and Privacy, Personal Data Markets, Wireless Ad hoc and sensor network, Database and system security, Application of spatial information system and other related areas. We hope this proceedings will contribute in stimulating debate and research among scholars, researchers and academicians.

We would like to thank authors of this proceedings for their valuable contributions. We would also want to extend our thanks to all referees for their constructive comments on all papers. We are very grateful to have a strong organizing committee. They made lots of efforts before and during the conference. Finally, we would like to thank the Taylor & Francis publishing Inc. for publishing this proceedings.

We look forward to meeting you all in the next ICCEIS conference.

The Organizing Committee of ICCEIS2014

Session 1: Automation and control engineering

Control Engineering and Information Systems – Liu (Ed)
© 2015 Taylor & Francis Group, London, ISBN 978-1-138-02685-8

Research on performance test technology of industry control network

J. Luo, S.H. Chen, Y.Q. Zhu & L. Wu
Electric Power Research Institute of Guangdong Power Grid Corporation China Southern Power Grid, Guangzhou, China

ABSTRACT: The industrial Ethernet technology has been maturely applied in DCS real-time control network, so it is focused on how to evaluate the DCS network performance. This paper mainly studies the performance test of DCS network from the application view and gives the test method and partial actual test results of the transport media test, connectivity test, network throughput test, data communication traffic test, network health test and etc. The test method for DCS network data communication given in this paper is a good reference for the field test.

1 INTRODUCTION

Now the process industry automated control instrument and system is evolving to be digital, intelligent, networking and integrated, especially the application of the data communication network technology in the industry automation area accelerates plant and workshop automation progress including large power plants. The data collection, process control and production automated network system is transferring from analog to digitalization. The network system is evolving from 4–20 mA analog control to digital control and present fast Ethernet-based automated control (Luo, Pan, 2003). The DCS system of the power plant is fused with the information technology. As the main network architecture of the plant system, the Industrial Ethernet Technology including TCP/IP and etc. has become the main DCS real-time control network today. The DCS real-time communication network is the bridge connecting the controllers and operation stations and acts as the nerve center system of the DCS. A huge amount of monitoring and control data are transported via the real-time network, including process information and control commands to operators, which shared the information between all kinds of controllers.

2 CURRENT CONDITIONS OF PERFORMANCE TEST OF DCS DATA COMMUNICATION NETWORK

The performance of the data communication network will directly affect the effective transport of the control system, security and reliability of power production, so it is required to detect performance of the data communication network of the DCS system. The DCS system control is quickly fused with the Ethernet and TCP/IP data communication technology in recent years, but the knowledge and experiences on the data communication network test is limited (Lu, Deng, Zhang, 2007). In addition, the domestic control specialty is different form the data communication network specialty, it is lead to the data communication network test in DCS system lags behind the telecommunication test. Now no indicators and test specifications are for the data communication network test of the DCS system for power plant. The requirements for the data communication network indicators are simple in the current test acceptance specification (DL-T659-2006, Acceptance Test Specification of Distributed Control System for coal-fired power plant) and only the communication interface rate and load rate are required for test. in fact, the parameters and indicators for the control network performance are plentiful, for example: 1) cabling parameters at the physical layer; 2) link transport rate, transport delay, collision rate, error frames and token transport cycle at the link layer; 3) throughput rate, utilization rate, packet loss rate and broadcasting frame at network layer; 4) service performance of the DHCP, DNS and Web access at the application layer. These performance indicators of the real-time communication network affect operation of the DCS system.

3 TEST ITEMS AND TEST METHODS FOR DATA COMMUNICATION NETWORK OF ETHERNET-BASED DCS SYSTEM

3.1 Transport media test

Generally the non-shielded (shielded) pair-twisted cabling system such as Cat5, UltraCat5 or Cat6 is used. For the optical communication, the multi-mode or single-mode optical cabling system is used for LAN system. The transport indicators of the optical cabling system include the channel optical attenuation and length. The transport indicators, transport performance and test methods of the pair-twisted cable and optical cabling system should comply with the regulations in the ISO/IEC11801standard.

3.2 *System connectivity*

All networked terminals must be connected according to the requirements. It should be checked for load rate, communication rate and correctness of transport data at the communication interface between the DCS system and remote I/O and field bus. Connecting the online data network test instrument with double ports to the communication link in serial at the communication interface to analyze data streams in the DCS system, the real time load rate and communication rate are obtained, the error and error code are observed. The figure 1 shows the connectivity test results of the nodes at the field DCS system.

It should be checked for load rate, communication rate and correctness of transport data at the communication interface between the DCS system and other control systems. Connecting the online data network test instrument with double ports to the communication link in serial at the communication interface to analyze data streams in the DCS system, the real time load rate and communication rate are obtained, the error and error code are observed.

It should be test for the communication interface between the DCS system and SIS system. The integrity of the DCS data received by the communication interface should be checked by using the test instrument with the online data packet capturing capability to the redundant backup link of the SIS system, collecting data packets and parsing packets.

3.3 *Test of network throughput rate*

The throughput rate indicates the maximum data packet forwarding rate for the tested network link of the loadless network without packet loss. The throughput rate test is measured by frame lengths (including 64, 128, 256, 512, 1024, 1280 and 1518 and etc.). The special hardware equipment with application traffic simulation capability is used to test the TCP throughput according to the data frame format, TCP port number, VLAN ID, QOS and size of sliding window required by the DCS system in the bearing data network. Two instruments are used. One instrument simulates actual traffic of the DCS system and sends data at the

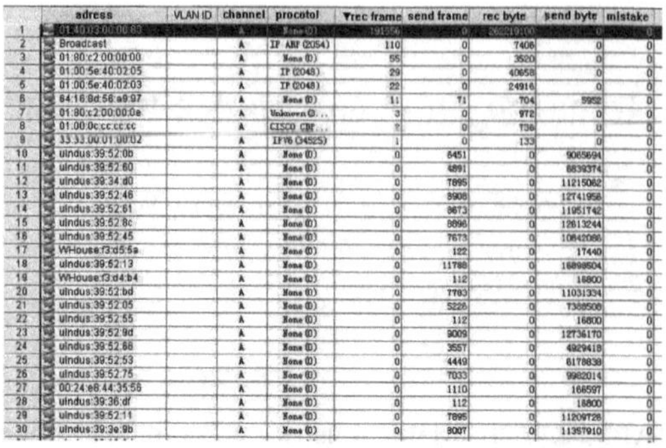

	adress	VLAN ID	channel	procotol	rec frame	send frame	rec byte	send byte	mistake
1	01:40:03:00:00:83		A	None (0)	191556	0	262219100	0	0
2	Broadcast		A	IP ARP (2054)	110	0	7406	0	0
3	01:80:c2:00:00:00		A	None (0)	55	0	3520	0	0
4	01:00:5e:40:02:05		A	IP (2048)	29	0	40658	0	0
5	01:00:5e:40:02:03		A	IP (2048)	22	0	24916	0	0
6	64:16:8d:56:a9:97		A	None (0)	11	71	704	5952	0
7	01:80:c2:00:00:0e		A	Unknown (0...	3	0	972	0	0
8	01:00:0c:cc:cc:cc		A	CISCO CDP...	2	0	736	0	0
9	33:33:00:01:00:02		A	IPV6 (34525)	1	0	133	0	0
10	uindus:39:52:0b		A	None (0)	0	6451	0	9065694	0
11	uindus:39:52:60		A	None (0)	0	4891	0	6839374	0
12	uindus:39:34:d0		A	None (0)	0	7895	0	11215062	0
13	uindus:39:52:46		A	None (0)	0	8908	0	12741958	0
14	uindus:39:52:61		A	None (0)	0	8673	0	11951742	0
15	uindus:39:52:8c		A	None (0)	0	8896	0	12613244	0
16	uindus:39:52:45		A	None (0)	0	7673	0	10842086	0
17	WHouse:f3:d5:5a		A	None (0)	0	122	0	17440	0
18	uindus:39:52:13		A	None (0)	0	11788	0	16898504	0
19	WHouse:f3:d4:b4		A	None (0)	0	112	0	16800	0
20	uindus:39:52:bd		A	None (0)	0	7783	0	11031334	0
21	uindus:39:52:05		A	None (0)	0	5226	0	7368506	0
22	uindus:39:52:55		A	None (0)	0	112	0	16800	0
23	uindus:39:52:9d		A	None (0)	0	9009	0	12736170	0
24	uindus:39:52:66		A	None (0)	0	3557	0	4929418	0
25	uindus:39:52:53		A	None (0)	0	4449	0	6178838	0
26	uindus:39:52:75		A	None (0)	0	7033	0	9982014	0
27	00:24:e8:44:35:56		A	None (0)	0	1110	0	166597	0
28	uindus:39:36:df		A	None (0)	0	112	0	16800	0
29	uindus:39:52:11		A	None (0)	0	7895	0	11209728	0
30	uindus:39:3e:9b		A	None (0)	0	8007	0	11357910	0

Figure 1. Test results of DCS connectivity.

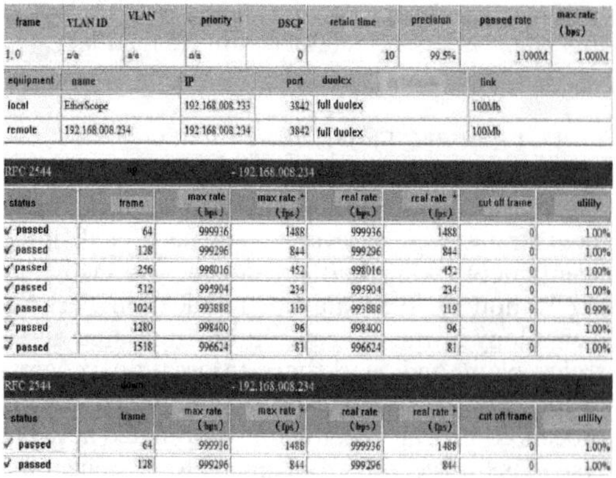

frame	VLAN ID	VLAN	priority	DSCP	retain time	precision	passed rate	max rate (bps)
1.0	n/a	n/a	n/a	0	10	99.5%	1.000M	1.000M

equipment	name	IP	port	duplex	link
local	EtherScope	192.168.008.233	3842	full duolex	100Mb
remote	192.168.008.234	192.168.008.234	3842	full duolex	100Mb

RFC 2544 up - 192.168.008.234

status	frame	max rate (bps)	max rate * (fps)	real rate (bps)	real rate * (fps)	cut off frame	utility
✓ passed	64	999936	1488	999936	1488	0	1.00%
✓ passed	128	999296	844	999296	844	0	1.00%
✓ passed	256	998016	452	998016	452	0	1.00%
✓ passed	512	995904	234	995904	234	0	1.00%
✓ passed	1024	993888	119	993888	119	0	0.99%
✓ passed	1280	998400	96	998400	96	0	1.00%
✓ passed	1518	996624	81	996624	81	0	1.00%

RFC 2544 down - 192.168.008.234

status	frame	max rate (bps)	max rate * (fps)	real rate (bps)	real rate * (fps)	cut off frame	utility
✓ passed	64	999936	1488	999936	1488	0	1.00%
✓ passed	128	999296	844	999296	844	0	1.00%

Figure 2. Throughput test results.

rated load rate. The peer instrument summarizes the received data and tests the loop. Figure 2 Field throughput test results of DCS System.

3.4 *Traffic test of data communication network of DCS system*

The test indicators include link transport rate, transport delay and packet loss rate. The link transport rate indicates the digital information transport rate in the network between equipment. The transport delay indicates the time from sending port (address) of data packets to the destination port (address). Generally the transport delay is related on transport distance, passing equipment and bandwidth utilization rate (Wang, Zhu, Huang, 2010). It is complicated to realize the accurate clock synchronization between test tools of sender and receiver. The transport delay is measured via the loop. One-way transport delay is one half of the round trip delay. The packet loss rate indicates the percent of non-forwarded data packets caused by network performance. The packet loss rate test is measured by frame lengths (including 64, 128, 256, 512,

frame	VLAN ID	VLAN	priority	DSCP	duration	step length	pass rate	rate(bps)
1,0	n/a	n/a	n/a	0	10	10%	100%	1.000M
equipment	name		IP		port	duplex		link
local	EtherScope		192.168.008.233		3842	total duplex		100Mb
remote	192.168.008.234		192.168.008.234		3842	total duplex		100Mb

192.168.008.234

status	frame	% target speed rate (real-bps)	% target speed rate (real-bps)	% lost pack rate	lost frame
pass	64	100% (999936)	100% (1483)	0%	0
pass	128	100% (999296)	100% (844)	0%	0
pass	256	100% (998016)	100% (452)	0%	0
pass	512	100% (995904)	100% (234)	0%	0
pass	1024	100% (993888)	100% (119)	0%	0
pass	1280	100% (998400)	100% (96)	0%	0
pass	1518	100% (996624)	100% (81)	0%	0

192.168.008.234

状态	frame	% target speed rate (real-bps)	% target speed rate (real-bps)	%lost pack rate	lost frame
pass	64	100% (999936)	100% (1483)	0%	0
pass	128	100% (999296)	100% (844)	0%	0
pass	256	100% (998016)	100% (452)	0%	0
pass	512	100% (995904)	100% (234)	0%	0
pass	1024	100% (993888)	100% (119)	0%	0
pass	1280	100% (998400)	100% (96)	0%	0
pass	1518	100% (996624)	100% (81)	0%	0

Figure 3.　Test results of network delay.

FRAME	VLAN ID	VALAN	priority	DSCP	retain time (sec)	repeat	delay time	rate (bps)
3, 0	n/a	n/a	n/a	0	10	1	1000.0 us	1.000M
equipment	name		IP		port	duplex		link
local	EtherScope		192.168.008.233		3842	total		100Mb
remote	192.168.008.234		192.168.008.234		3842	total		100Mb

RFC 2544　192.168.008.234

status	frame	repeat	bidirectional transmitting	unidirectional transmission
passed	64	1/1	24.0 us	12.0 us
passed	128	1/1	34.0 us	17.0 us
passed	256	1/1	55.0 us	27.5 us
passed	512	1/1	98.0 us	49.0 us
passed	1024	1/1	181.0 us	90.5 us
passed	1280	1/1	223.0 us	111.5 us
passed	1518	1/1	260.0 us	130.0 us

RFC 2544　192.168.008.234

frame	repeat	min	average	max
64	1	12.0 us	12.0 us	12.0 us
128	1	17.0 us	17.0 us	17.0 us
256	1	27.5 us	27.5 us	27.5 us
512	1	49.0 us	49.0 us	49.0 us
1024	1	90.5 us	90.5 us	90.5 us
1280	1	111.5 us	111.5 us	111.5 us
1518	1	130.0 us	130.0 us	130.0 us

Figure 4.　Test results of packet loss.

frame	VLAN ID	VLAN	priority	DSCP	restain time (sec)	error rate	rate (bps)
	n/a	n/a	n/a	0	2		1.000M
equipment	name		IP		port	duplex	link
local	EtherScope		192.168.008.233		3842	full duplex	100Mb
remote	192.168.008.234		192.168.008.234		3842	full duplex	100Mb

192.168.008.234

status	frame	send byte	error code	error rate
completed	64	1.714E+06	0	<5.804E-07
completed	128	1.837E+06	0	<5.445E-07
completed	256	1.909E+06	0	<5.231E-07
completed	512	1.947E+06	0	<5.136E-07
completed	1024	1.965E+06	0	<5.089E-07
completed	1280	1.978E+06	0	<5.055E-07
completed	1518	1.978E+06	0	<5.056E-07

192.168.008.234

status	frame	send byte	error code	error rate
completed	64	1.714E+06	0	<5.804E-07
completed	128	1.837E+06	0	<5.445E-07
completed	256	1.909E+06	0	<5.231E-07
completed	512	1.947E+06	0	<5.136E-07
completed	1024	1.965E+06	0	<5.089E-07
completed	1280	1.978E+06	0	<5.055E-07
completed	1518	1.978E+06	0	<5.056E-07

Figure 5.　Bit Error Rate test results.

frame	VLAN ID	VLAN	priority	DSCP	duration	target rate(bps)
	n/a	n/a	n/a	0	10	1.000M
equipment	name		IP		port	duplex	link
local	EtherScope		192.168.008.233		3842	total duplex	100Mb
remote	192.168.008.234		192.168.008.234		3842	total duplex	100Mb

192.168.008.234

status	frame	receive frame	lost frame	min interval	max interval	jithering
completed	64	14.88K	0	0.0 us	0.0 us	149.4 us
completed	128	8442	0	0.0 us	0.0 us	226.1 us
completed	256	4822	0	0.0 us	0.0 us	186.2 us
completed	512	2342	0	0.0 us	0.0 us	182.3 us
completed	1024	1191	0	0.9 us	0.0 us	174.8 us
completed	1280	961	0	0.0 us	0.0 us	233.0 us
completed	1518	811	0	0.0 us	0.0 us	155.1 us

192.168.008.234

status	frame	receive frame	lost frame	min interval	max interval	dithering
completed	64	14.88K	0	0.0 us	0.0 us	153.4 us
completed	128	8442	0	0.0 us	0.0 us	229.1 us
completed	256	4822	0	0.9 us	0.0 us	187.3 us
completed	512	2342	0	0.0 us	0.0 us	203.7 us
completed	1024	1191	0	0.0 us	0.0 us	169.3 us
completed	1280	961	0	0.0 us	0.0 us	190.4 us
completed	1518	811	0	0.0 us	0.0 us	149.4 us

Figure 6.　Jitter test results.

1024, 1280, 1518 and etc.). The special high-speed network traffic collection equipment is used to capture and store the DCS system traffic for a long term and the decoding analysis tool is configured (Mohammad, Kaiser, 1999). The figure 3 and 4 show the test results of the field measurement delay and packet loss.

3.5　Test of health condition of Ethernet link layer

The health condition test indicators of Ethernet link layer include link utilization rate, error rate, collision rate and jitter. The utilization rate indicates the ratio of the transport data throughput in the network link to the maximum physical bandwidth supported by this link. The link utilization rate includes maximum utilization rate and average utilization rate. The maximum utilization rate is related to the test statistics sampling interval. A shorter sampling interval can reflect network traffic burst, so the maximum utilization rate is bigger. The error rate indicates the percent of error frames in the network to total data frames (Wang, Luo,

Xue, 2008). The frequent Ethernet errors include long frame, short frame, FCS error frame, ultra-long error frame, under-long frame and frame alignment error frame. The collision rate indicates that two stations in one section may collide with each other when they simultaneously send Ethernet data frames (Wang, 2008). The collision is frequent under sharing Ethernet and half duplex switching Ethernet. Too many collisions will degrade the network transmission efficiency severely. The figure 5 and figure 6 show the test results of the nodes at the field DCS system.

4　CONCLUSIONS

Now the DCS performance test is one of the required route tests for new power plants pre-operation and old power plants after small and overhaul repair, but the related professionals involve DCS network performance test in reference and field test are little. This paper analyzes the test methods and meanings of the network performance indicators such as network transmission rate, transmission delay, collision rate, error rate, throughput rate, utilization rate and packet loss rate based on the field test results. It is expected to attract more professionals to study the bearing requirement and architecture design of the data communication network for the DCS system, summarize the requirements of the data network performance indicators for reliable and safe operation of the DCS system for the coal-fired power plants, further fill the gap in this industry.

REFERENCES

Luo Jia, Pan Xiao. "Network and communication of INFI-90 sequence of event system". Industrial Control Computer, 2003. 16(8):49–51.

Lu Huiming, Deng Hui, Zhang Zhiguang. "OPC scheme to realize openness of power plant DCS". Electric Power Automation Equipment, 2007, 27(7):95–97.

Mohammad Ali Livani, Kaiser and Jorg. "Scheduling hard and soft real-time communication in a controller area network" Control Engineering Practice. 1999. 7(12): 1515–1523.

Wang Jianfeng, Luo Zhenxin, Xue Peng. "Analysis of factors influencing DCS reliability". East China Electric Power,2008, 36(4): 109–111.

Wang Qi. "Discussion on anti-interference performance test of DCS". China Electric Power, 2008. 41(12):46–48.

Wang Qi, Zhu YA Qing, Hang Wei Jian. "Capabilities test test technique for distributed control systems". Science Press, 2010.

Control Engineering and Information Systems – Liu (Ed)
© 2015 Taylor & Francis Group, London, ISBN 978-1-138-02685-8

Design of antarctic astronomical telescope control system based on dsPIC30F

Z. Feng

National Astronomical Observatories, Nanjing Institute of Astronomical Optics and Technology,
Chinese Academy of Sciences, Nanjing, China
Key Laboratory of Astronomical Optics and Technology, Nanjing Institute of Astronomical Optics
and Technology, Chinese Academy of Sciences, Nanjing, China
University of Chinese Academy of Sciences, Beijing, China

H. Wang & X.L. Song

National Astronomical Observatories, Nanjing Institute of Astronomical Optics and Technology,
Chinese Academy of Sciences, Nanjing, China

ABSTRACT: This paper presents a dual-core backup motion control system based on dsPIC30F6010 A and LMD18200 to realize the precise control for the BDC's velocity and position of the Antarctic Astronomical Telescope. By using dual-core backup, digital PI regulating with correction of the integral term and other control strategies, the servo control of the BDC is implemented. The hardware and software design of the DC servo drives are introduced, and the experimental results are also given. This control system can be applied to the Antarctic Survey Telescope (AST3) control system.

1 INTRODUCTION

Many mechatronic products demand high control accuracy on speed and position control, especially the Antarctic Astronomical Telescope (AAT) that placed in a very bad environment, not only demands high control accuracy, but also requires a strong ability to resist disturbance, a very high reliability and environmental adaptability. Here we design a dual-core back up motion control system based on dsPIC30F and LMD18200 according to the actual situation. By using dual-core backup, digital PI regulating with correction of the integral term and other control strategies, the servo control of the BDC is implemented. The hardware and software design of the DC servo drives are introduced, and the experimental results are also given. This control system can be applied to the Antarctic Survey Telescope (AST3) control system.

2 OVERVIEW OF THE MOTION CONTROL SYSTEM

The AAT motion control system is composed of servo controller, photoelectric encoder, limit structure, driver, brush DC motor, gear box and PC.

In this paper, the brush DC motor is controlled by full digital method including the generation of PWM waveform, the tracking output of current, speed and position, and the PID control arithmetic

etc. The microcontroller dsPIC30F6010 A produces PWM signal that send to the power component LMD18200 to realize the fully digital motor drive, the efficiency equivalent to 100%.

To ensure the control system has a strong ability to resist disturbance, a very high reliability and environmental adaptability. The MCU unit and Drive Unit are configured with double redundancy. The MCU unit is switched by PC, and the drive unit is switched by the current work control unit. There is only one set of MCU unit work normally and another set work at the lowest power consumption state that is ready to replace the fault unit at any time.

In order to realize the precise control of DC motor, MCU must collect the position, speed and

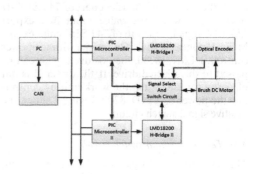

Figure 1. Structure diagram of motion control system.

current data, and the work conditions of some components on time, and adopts PID algorithm to realize the position, speed and current closed-loop control of the DC motor [1]. Figure 1 shows the whole control system block diagram.

3 HARDWARE DESIGN

3.1 Digital signal controller

In this paper we use the high performance digital signal controller dsPIC30F3010 A produced by Microchip Company as the control core. The MCU is easy to use and has processing capacity as strong as DSP but it costs lower, so it is the better solution in embedded system [2]. In the AAT motion control system, DSC mainly has the following functions:

- Collect relevant feedback data, complete the algorithm of the motor position, speed and current loop, and produce PWM, Direction and Brake signals;
- Collect the state of drivers and switch the fault driver;
- Read the temperature sensor value and control the temperature compensation unit;
- Communicate with PC through CAN Bus.

3.2 Positive signal switch circuit

In the control system, positive signals are composed of PWM, Direction and Brake in addition to the enable signals which are used on switching module. Due to double redundancy module, we design the positive signal switch circuit to ensure the positive signal only transfer to the current work drive unit, so that we can prevent the signal disturb the non-working device I/O ports.

Here we analysis basic principle of the circuit: First assume MCU-I and Driver-II are the normal work units. If MCU-I is out of order and system is switched to MCU-II by PC, the control signal of MCU-II export high electric value and MCU-I has no output, the select port of 74 LS157-I (U16) has high electric value input and export Chanel B signal, at the same time, the select port of 74 LS157-II (U19) has low electric value input and export Chanel A signal, and both of 74 LS157 chips export MCU-II control signals. Cooperate with the power control of LMD18200, the positive signals can be passed to the selected drive. If the driver II is out of order, we can switch the work unit by controlling the power of LMD18200. Figure 2 shows the positive signal switch circuit.

3.3 Feedback signal switch circuit

Through the feedback signal switch circuit, feedback signal can be reverse to the work unit and

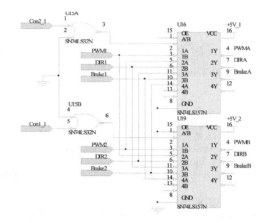

Figure 2. Positive signal switch circuit.

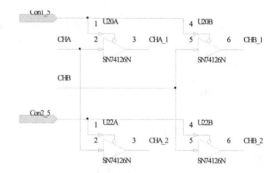

Figure 3. Digital feedback signal switching circuit.

prevent the signal to impact the non-working unit I/O ports. Here we use the three state output four bus buffer 74 LS126 for feedback signal switching, use the relay to switch the analog feedback signal. In the AAT motion control system, digital feedback signals are composed of photoelectric encoder pulse signal, LMD18200 temperature alarm signal, AAT rotary limit switch signal and so on. Analog feedback signal are composed of current signal and voltage detection signal. Digital feedback signal switch circuit is shown in Figure 3.

3.4 Drive and current feedback circuit

The drive part uses the professional H bridge component LMD18200 that can realize the PWM mode to control the AAT. The LMD18200 is a 3 A H-Bridge designed for motion control applications. The device is built using a multi-technology process which combines bipolar and CMOS control circuitry with DMOS power devices on the same monolithic structure. Ideal for driving DC and stepper motors; the LMD18200 accommodates

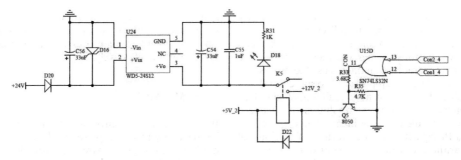

Figure 4.   The positive signal switch circuit.

peak output currents up to 6 A. An innovative circuit which facilitates low-loss sensing of the output current has been implemented [3]. The system adopts single polarity drive mode.

In order to get the current feedback value, we can use LMD18200 internal integrated current detection circuit. Then transform the current into voltage signal through a sample resistance, and send it to the ADC module of the MCU, after processing and the PWM duty will be updated.

But the voltage signal is too small to satisfy the MCU internal ADC module, therefore we need to amplify the value. Here we choose high precision op-amp LM358 to build amplify circuit, and then send the signal to ADC. Figure 4 shows the motor drive and current feedback circuit.

4 CONTROL ALGORITHM AND SOFTWARE DESIGN

4.1 Digtial PID controller

The incremental model of traditional digital PI regulator is

$$\Delta u(k) = u(k) - u(k-1) = Kp[e(k) - e(k-1)] + Kie(k) \tag{1}$$

In the formula: k is the sampling order, $u(k)$ is PI regulator output at k time, $e(k)$ is the error of input signal at k time, Kp is proportion gain, Ki is the integral gain, $Ki = KPT/Ti$, T is the sample period, Ti is the integration time constant [4].

To prevent PI regulator integral overflow and output saturation, the control system adopts digital PI regulating with correction of the integral term.

When the PI regulator in saturated zone, the integral term accumulation will be placed by weaken integral operation, so that can exit saturation quickly. The specific control algorithm is [5]

$$e(k) = r(k) - y(k)$$
$$u(k-1) = x(k-1) + Kpe(k) \tag{2}$$
$$x(k) = x(k-1) + Kie(k) + KcorEpi$$

In the formula: $Kcor$ is the correction gain factor, $Kcor = Ki/Kp$, $Epi = u0 - u(k)$, when $u(k) >$ umax, $u0 = $ umax; when $u(k) <$ umin, $u0 = $ umin, otherwise $u0 = uk$.

4.2 Software design

The motion control system is composed of an electric current feedback loop, a velocity feedback loop and a position feedback loop with a high precise optical encoder. When the motion control system is on the operating mode, by comparing the given position and encoder measured position, we will get the position error. After processing by anti-interference, the position PI regulator will output a given velocity value. In the velocity feedback loop, by comparing the given velocity and encoder measured velocity, we will get the velocity error. After processing by anti-interference, the velocity PI regulator will output a given current value. In the electric current feedback loop, by comparing the given current and the current measured by LMD18200, and after processing by anti-interference, the current PI regulator will output a control voltage and MCU will produce the PWM control signal to control the motor.

The motion control system software is programmed by modularized and digitally discretized method. All the related data are collected in interrupt mode and then processed. There is an on-chip

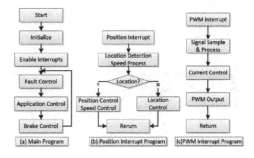

Figure 5.   System procedure flow chat.

9

timer in the dsPIC, so we can set the timer precisely through control program to collect the electric current, velocity and position data and then processed via servo control strategies. In order to improve the detection accuracy, mean filter is used to process the ADC sampling data [6]. In the motion control system, a quadruple frequency circuit is applied to take advantage of the inherent resolution of encoders, and improve the counting precision effectively. The digital PI regulating with correction of the integral term arithmetic is adopted to control the servo system that can assure the precision and responsiveness. Figure 5 shows the System Procedure Flow Chat.

5 EXPERIMENTAL RESULT

The motion control system is applied in the Antarctic Small Telescope which is driven by PITTMAN motor produced by AMETEK Inc. (USA), some related experiments are carried out, and the results convey that the control system works well. The motor parameters are the following: rated voltage is 6 V, torque constant is 0.97 and rated current is 0.49 A. In the motion control system, the encoder AEDR-8300-1P2 is adopted to sensor position and speed. The photoelectric encoder parameters are the following: rated voltage is 5 V, grating pulses are 512. Figure 6 shows the PWM and current detection waveform of LMD18200. This figure contains some parameters: the PWM frequency is 10 KHz, current sampling period is 125 μs and the position

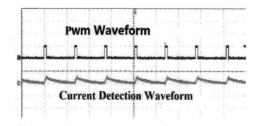

Figure 6. PWM waveform.

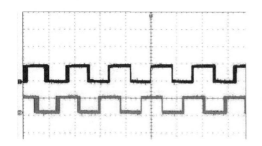

Figure 7. The encoder output.

and speed sampling period is 1 ms. Figure 7 shows the corresponding output of photoelectric encoder. The experiment result proves the motion control system can run effectively.

6 SUMMARY

In the light of harsh environment on Dome-A and low reliability, this paper presents a dual-core backup motion control system based on dsPIC30F6010 A and LMD18200 to realize the precise control for the BDC's velocity and position of the Antarctic Astronomical Telescope. The motion control system increases stability and reliability for further on the basis of the original microcomputer control system and it can be applied to the Antarctic Survey Telescope (AST3) control system. The running results show that this motion control system has perfect robustness, high control precision and can achieve satisfactory control effect.

In this paper the innovations are the following:

1. It uses dsPIC30F6010 A microcontroller as the core controller which supports abundant system resources and provides spatial extensibility for further;
2. Dual-core backup, digital PI regulating with correction of the integral term and other control strategies are applied in the system in order to increase the stability and reliability.

ACKNOWLEDGMENT

This work is supported by The National Basic Research Program of China (also called 973 Program) under Grant No. 2013CB834901 and the National Natural Science Foundation of China under Grant No. 10803013 which are gratefully acknowledged.

REFERENCES

[1] Wang Xiaoming. MCU Control of the Motor[M]. Beijing: Beihang University Press. 2002.
[2] Microchip Technology Inc. dsPIC30F6010 A Data Sheet. 2004.
[3] Wang Lei. Design of DC Motor Driver Based on LMD18200 [J]. Automation & Instrumentation, 2004(1).
[4] Liu Jinkun. Advanced PID control & MATLAB simulation [M]. Beijing: Beihang University Press. 2011.
[5] Zhu Pengcheng. Optimized Design of Boost Chopper Circuit PI and PID Regulator [J]. Power Electronics, 2001,35(4):28–31.
[6] Lucio Di Jasio. Programming 16-Bit PIC Microcontrollers in C-Learning to Fly the PIC24. Newnes, 2002.

Design of anti-windup using internal model control and conditioning techniques

H.X. Li

Xinzhou Teachers Institution, Xinzhou, Shanxi, China

ABSTRACT: It is well known the saturation can't be ignored in control system. A new anti-windup framework is presented in this paper. There are three controllers in the framework. The design of $Q_1(s)$ and $Q_2(s)$ takes advantage of the theory of internal model control. And the design of controller $Q_3(s)$ uses the method of conditioning techniques. The system's stability is proved by Popov's criterion. The results of simulations illustrate that the proposed framework has three functions: anti-windup, reference tracking and disturbance rejection.

1 INTRODUCTION

Most controllers, practically running in the industry, are tuned to operate in a linear range. They assume no controller and process limitations. Such assumption is valid if disturbance are small or there are only slight set point changes. However, real systems are always imposed to limitations. In such real systems, in the case of large set point change, control signal is limited what breaks a closed-loop path between controller and process. Brocken path can produce undesired controller's behavior (windup) if controller has the integral term or some other types of "memory" as in controllers with general rational transfer function or state-space controllers.

As a result of limitations and substitutions, the real plant input is temporarily different from the controller output. When this happens, if the controller is initially designed to operate in a linear range, the closed-loop performance will significantly deteriorate with respect to the expected linear performance. This performance deterioration is referred to as windup. A rational way to handle the problem of windup is to take into account, at the stage of control design, the input limitations. However, this approach is very involved and the resulting control law is very complicated. The nonlinearities of the actuator are not always known a priori. A more common approach in practice is to add an extra feedback compensation at the stage of control implementation. As this compensation aims to diminish the effect of windup, it is referred to as anti-windup.

The topic of anti-windup has been studied over a long period of time by many authors, and the

Fund Program: Scientific Research Project of Xinzhou Teachers institute.

most popular techniques are described in [2, 4, 7, 10, 11, 12]. However, many authors think that anti-windup is aimed at reducing the output overshoot and settling time in its step response. These thoughts need to be corrected.

Internal model control is a simple and direct vision technique. And when it is used in anti-windup, the saturation nonlinearity is considered in the object model together with the plant model. The design of the way of the internal model control is putting the object model and the real plant model in parallel connection. And the internal model controller can be designed using the object model. The reference [5] has illustrated the internal model control is useful for anti-windup, but the controller's output is still not the same as the plant input, which means the saturation is not completely disappeared. So this paper proposed a new framework of 3-DOF (degree of freedom) internal model control for anti-windup which increased a compensator based on 2-DOF internal model control structure. The compensator was designed using conditioning technique.

Conditioning technique was first presented in [Hanus, 1980]. To understand the conditioning technique, it is important to explain a concept of the "realizable reference". The realizable reference (w^r) is such that when applied to the controller instead of the reference (w), it results in the control variable (u) which is equal to the process input (u^r) and thus the limitation is not activated. Conditioning technique is a very simple and useful method for anti-windup. And it is suitable for all kinds of controllers. To illustrate the improved results, the third controller $Q_3(s)$ in the new framework which is proposed by this paper is designed by using the notion of the realizable reference.

2 DESIGN OF 2-DOF INTERNAL MODEL CONTROL STRUCTURE

Figure 1 shows 2-DOF internal model control structure for anti-windup. $G(s)$ represents real plant model. $G_0(s)$ represents object model. N represents saturation nonlinearity. According to the theory of internal model control, if the saturation nonlinearity is contained in the object model, the system's stability can't be changed. $Q_1(s)$ is a controller to make the system's output track the reference input. $Q_2(s)$ is another controller to resistant the disturbance. d represents disturbance coming from outside. u represents the controller's output. u^r is the real process input. $Q_1(s)$ and $Q_2(s)$ are designed on the propose that the saturation have not happened ($u = u^r$).

If the saturation have not happened, $N = 0$, $u = u^r$, we can get the system's closed-loop transfer functions:

$$\frac{y}{r} = \frac{Q_1(s)G(s)}{1 + Q_2(s)\big(G(s) - G_0(s)\big)} \tag{1}$$

$$\frac{y}{d} = \frac{1 - Q_2(s)G_0(s)}{1 + Q_2(G(s) - G_0(s))}$$

When $G(s) = G_0(s)$

$$y = G(s)Q_1(s)r + (1 - Q_2(s)G_0(s))d \tag{2}$$

In order to make the system's output y track the input reference r, and to make the system can resistant disturbance, we have:

$$G(s)Q_1(S) = 1 \tag{3}$$

And

$$(1 - Q_2(s))G_0(s) = 0 \tag{4}$$

Put the formulas (3) and (4) into (1), we can calculate: $y = r$ and $d = 0$. We can also have:

$$Q_1(s) = G^{-1}(S) \tag{5}$$

$$Q_2(s) = G_0^{-1}(s) \tag{6}$$

We can concluded that: the controller $Q_1(s)$ is used to make y track reference input and the controller $Q_2(s)$ is used to resistant the disturbance. When the models exactly matched with each other in the internal model structure ($G(s) = G_0(s)$), the ideal controllers can be designed as:

$$Q_1(s) = Q_2(s) = G_0^{-1}(s) \tag{7}$$

However, the equation (7) is calculated on the assumption that $G_0(s)^{-1}$ really existent and have no pure time-delay. As after sampling and holding, the plant's pulse transfer function $G(z^{-1})$ has a clap time delay and the controller's expression in equation (7) will have a priori factor z, the controller in equation (7) can't realized. We can changed it as:

$$Q_1(z^{-1}) = Q_2(z^{-1}) = z^{-1}G_0^{-1}(z^{-1}) \tag{8}$$

If the controlled plant has pure time delay or zeros beyond the unit circle, we can rewrite the plant model as:

$$G(z^{-1}) = G_+(z^{-1})G_-(z^{-1}) \tag{9}$$

where: $G_+(z^{-1})$ is the minimum phrase part of the plant model, $G_-(z^{-1})$ is non-minimum phrase part of the plant model. We can rewrite the controller's expressions by using $G_+(z^{-1})$:

$$Q(z^{-1}) = G_+(z^{-1})f(z^{-1}) \tag{10}$$

where: $f(z^{-1})$ is a filter and has the effect of anti-windup. When ignored the input reference and disturbance, we have:

$$u = -Q_2(s)(u^r G(s) - u G_0(s)) \tag{11}$$

$$u/u^r = Q_2(s)/(Q_2(s)G_0(s) - 1) \tag{12}$$

If put the equation (6) into (11), we have $u/u^r = \infty$. It is not reasonable absolutely. But if we put equation (10) into (11), we have:

$$u/u^r = f(z^{-1})/(f(z^{-1}) - 1) \tag{13}$$

We can also make u equal u^r approximately by adjusting the filter f for anti-windup.

3 DESIGN OF 3-DOF INTERNAL MODEL CONTROL STRUCTURE

Because choosing the filter is blindness, the above methods can not eliminate the windup completely. In fact, windup appears due to the fact that the

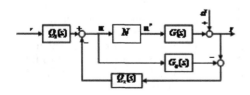

Figure 1. 2-DOF internal model control structure.

integral term increases too much. Thus, during saturation, the increase should be slowed down. It can be realized by a compensation which feeds back $u - u^r$ to the controller. Figure 2 is named 3-DOF internal model control structure which increases a controller $Q_3(s)$ on the basis of the 2-DOF internal model control structure. In Figure 2, $Q_3(s)$ is a compensator to eliminate $u - u^r$ exactly for anti-windup.

$Q_3(s)$ is designed by conditioning technique. According to the concept of conditioning technique, when applying the "realizable reference" w^r to the controller, the controller's output u would be the same as plant input u^r. So, conditioning technique can eliminate the windup completely.

See Figure 2, when ignoring the disturbance and $G(s) = G_0(s)$, we have:

$$u = wQ_1(s) - Q_3(s)(u^r - u) \tag{14}$$

Note that the realizable reference (w^r) can not be computed a priori, but only a posteriori. u^r should be computed a priori when using w. And u^r would stay the same as when using w^r. According to the definition of w^r, we have:

$$u^r = w^r Q_1(s) \tag{15}$$

Subtracting u^r (15) from u (14), we can calculate w^r as:

$$w^r = w + (1 - Q_3(s))/Q_1(s)(u^r - u)$$
$$= w + G_{w^r}(s)(u^r - u) \tag{16}$$

Originally, our goal is to make y track w, but y actually tracks w^r due to the input limitations. Thus, we would like to have such a realizable reference w^r that it will be as close as possible to w. Now, $w - w^r$ is expressed as a function of $u^r - u$. Normally, G_{w^r} is a dynamic transfer function. So w^r will not become the same as w at the moment when the controller leaves the limitation ($u^r = u$). We would like to have w^r as close as possible to w. That can be done by tuning G_{w^r} as a static gain:

$$G_{w^r}(s) = K^* \tag{17}$$

Here K^* is a constant. Put the equation (17) into (16), we can get:

$$K^* = (1 - Q_3(s))/Q_1(s) \tag{18}$$

What yields

$$Q_3(s) = 1 - K^* Q_1(s) \tag{19}$$

Put equation (19) into (15), we can get:

$$u = wQ_1(s) - (1 - K^* Q_1(s))(u^r - u) \tag{20}$$

Assume that $Q_1(s) = N_1(s)/D_1(s)$, according to the Hanus's methods of dealing with K^* in conditioning technique,

$$K^* = [D_1(s)/N_1(s)]_{s=\infty} \tag{21}$$

4 ANALYSIS OF THE SYSTEM'S STABILITY

The stability of 2-DOF internal model control structure can be guaranteed if the plant model and the internal model matched exactly with each other. But the analysis of the stability of 3-DOF internal model control structure is relatively complex. It can be guaranteed by Popov's criterion.

Figure 3(a) shows one kind of SISO nonlinearity system. $\Psi(.)$ which is at the feedback channel represents one-dimensional time invariant memoryless nonlinearity. $G(z^{-1})$ is a transfer function which satisfied the theory of Hurwitz strictly. Popov's stability criterion gives us a sufficient condition which can judge a nonlinearity system's progressive stability.

If $\Psi(.) \in (0,1]$ and $\text{Re}[(1 + jw)G(jw)] > -1, \forall w \geq 0$, the system in Figure 3(a) is progressive stable. The 3-DOF internal model control structure in Figure 2

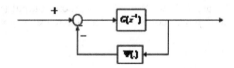

Figure 3(a). Nonlinear feedback connection.

Figure 3(b). Equivalent transforming of windup nonlinearity.

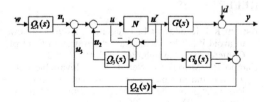

Figure 2. 3-DOF internal model control structure.

can be transformed into Figure 3(b). There are two subsystems in Figure 3(b). One is linear dynamic transfer function of the system and the other is the static saturation nonlinearity. The input and the output of the linear dynamic transfer function are u and u^r. Then we can express them as:

$$M = u/u^r \tag{22}$$

$$u = -Q_3(s)(u^r - u) \tag{23}$$

$$M = u/u^r = Q_3(s)/Q_3(s) - 1 \tag{24}$$

Substituting (19) in (24), we have

$$M = 1 - 1/k^* Q_1(s) \tag{25}$$

If $Q_1(s) = N_1(s)/D_1(s)$ and the degree of $D_1(s)$ is higher than $N_1(s)$'s, then $K^* = [D_1(s)/N_1(s)]_{s=\infty} = \infty$ and $M = 1$.

Because $0 < u/u^r \leq 1$, $\Psi(.) \in (0,1]$ and Re$(M(e^{-jwt}) = 1 > -1$. The system in Figure 2 satisfies Popov's criterion. So the system of 3-DOF internal model control structure is progressively stable.

5 SIMULATION RESULTS

To compare the mentioned anti-windup methods, we make a simulation with process:

$$G_0(s) = 0.2 \frac{s^2 + 2\xi_1 w_1 s + w_1^2}{s^2 + 2\xi_1 w_2 s + w_2^2} * \frac{s^2 + 2\xi_2 w_1 s + w_1^2}{s^2 + 2\xi_2 w_2 s + w_2^2}$$

where:

$$\xi_1 = 0.3827, \; \xi_2 = 0.9239, \; w_1 = 0.2115, \; w_2 = 0.0473.$$

The input is subjected to the following limits:

$$U_{max} = 1, U_{min} = -1$$

Figure 4 shows the different process output (y) with different controllers. The reference w goes

from 0 to 1 at the time origin. And the system encountered disturbance ($d = 0.8$) at the time 600 s.

In Figure 4, Curve 1 represents the output of Figure 1 when $Q_1(s) = Q_2(s) = [(s+1)/(18s+1)]$. $G_0^{-1}(s)$ and $N = 0$. Curve 2 represents the output of Figure 1 when the system encountered saturation and

$$Q_1(s) = Q_2(s) = [(s+1)/(18s+1)] \cdot G_0^{-1}(s).$$

Curve 3 represents the output of Figure 1 when

$$Q_1(s) = [(s+1)/(18s+1)] \cdot G_0^{-1}(s)$$

and

$$Q_2(s) = [(5s+1)/70s+1]G_0^{-1}(s).$$

Curve 4 represents the output of Figure 2 when

$$Q_1(s) = (s+1/18s+1)G_0^{-1}(s)$$

$$Q_2(s) = (5s+1/70s+1)G_0^{-1}(s)$$

and

$$Q_3(s) = 0.5s + 1/0.01s + 1$$

6 CONCLUSIONS

We have illustrated through simulations, 2-DOF internal model control structure has some effect on anti-windup. But the 3-DOF internal model control structure made by feedback compensator $Q_3(s)$ on the basis of 2-DOF internal model control does better on anti-windup. It can be seen that from Figure 4, using 3-DOF internal model control, the overshoot is smaller and the settling time is much shorter than in the absence of $Q_3(s)$. 3-DOF internal model control structure which designed using internal model control and conditioning technique can realize anti-windup, reference tracking and disturbance rejection. And the system is progressively stable guaranteed by Popov's criterion.

REFERENCES

Astrom K.J, Rundqwist L. Integrator windup and how to avoid it [J]. Proceedings of the American Control Conference, Pittsburgh, PA, 1989: 1693–1698.

Campo P.J, Morari M, Nett C.N. Multivariable antiwindup and bumpless transfer: A general theory [J]. Proceedings of the American Control Conference, Pittsburgh, PA, 1989: 1706–1711.

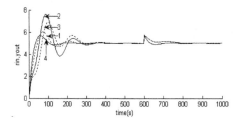

Figure 4. Process output (y) in closed loop.

Campo P.J, Morari M, Nett C.N. Rubust control of processes subject to saturation nonlinearities [J]. Computers & Chemical Engineering, 1990, 14(415): 343–358.

Chung Seng Ling, Michael D. Brown, Paul F. Weston et al. Gain Tuned Internal Model Control for Handling Saturation in Actuators [J]. American control conference, 2004:4692–4697.

Fertik H.A, Ross C W. Direct digital control algorithm with anti-windup feature [J]. ISA Transactions, 1967, 6(4): 317–328.

Funami Y, Yamada K. An Anti-windup Control Structure [C]. IEEE Int. Conf. on system, Man, and Cybernetics, Tokyo: IEEE Press, 1999(5):74–79.

Hanus R, Kinnaert M, Henrotte J.L. Conditioning Technique, a General Anti-windup and Bumpless Transfer Method [J]. Automatica, 1987, 23(6):729–739.

Hanus R, Kinnaert M. Control of constrained multivariable systems using the conditioning technique [J]. Proceedings of the American Control Conference, Pittsburgh, PA, 1989: 1711–1718.

Kothare M.V. Campo P.J, Morari M et al. A Unified Framework for the Study of Anti-windup Design [J]. Automatica, 1994, 30(12):1869–1883.

Wada N and Saeki M. Design of a static anti-windup compensator which guarantees robust stability [J]. Transactions of the Institute of Systems, Control and Information Engineers, 1999, 12(11):664–670.

Walgama K.S, Sternby J. Inherent observer property in a class of anti-windup compensators [J]. International Journal of Control, 1990, 52(3): 705–724.

Zheng A, Kothare M.V, Morari M. Anti-windup design for internal model control [J]. International Journal of Control, 1994, 60(5): 1015–1024.

The timing control technology for photoelectric theodolite based on dynamic compensation

Z. Jiang, H.Y. Liu, L.J. Zhang & J.C. Ma
Baichang Ordnance Test Center of China, Baicheng City, Jilin Province, China

ABSTRACT: In the paper a new kind of time control system based on microcontroller and peripheral interface circuit is introduced. The study in this paper focuses on the timing control technology for photoelectric theodolite. With the development of the ordnance, the measurement of the high-speed and multi-feature target is required. In order to obtain high-precision ballistic data, the smooth tracking, clear imaging, precision timing, and scientific algorithm is needed. Therefore, it is very important to keep strict alignment of the image with the encoder sampling time. External synchronization feature is an important indicator to determine whether the camera can meet test requirement. The frequency of the external synchronization signal determines the frequency of the camera sync shooting.

1 INTRODUCTION

Photoelectric theodolite is the most important equipment for acquisition the exterior ballistics data. A theodolite system consists of a central station and several measurement stations. The central station deal with information exchange and the final data processing; the measurement stations track the target, record the current frame time, angle and image. In order to obtain high-precision ballistic data, the smooth tracking, clear imaging, precision timing, and scientific algorithm is needed. With the quick development of the ordnance, the measurement of the high-speed and multi-feature target is required. Therefore, it is very important to keep strict alignment of the image with the encoder sampling time. This paper focuses on the study of the timing control technology for photoelectric theodolite based on dynamic compensation. To theodolite synchronization which take film as imaging medium, the synchronous refers to phase pulse leading edge coming from the motor is alignment with the pulse leading edge coming from the timing system when the control system is in stable operation. The true picture of the target movement can be achieved only when the high-speed camera gets synchronization in order to ensure the dot matrix on the frame corresponding to the information of the target. Nowadays most of new photoelectric theodolite equipment is digital measurement equipment, whose main measurement system is a CCD camera or CMOS camera [1,2,3]. The function of the synchronized control system is to ensure that the target image is corresponding to the data recorded by the host computer, which is crucial in providing a real basis for data processing after test.

2 THE SYNCHRONIZATION PERFORMANCE OF NEW HIGH-SPEED CAMERA

External synchronization feature is an important indicator to determine whether the camera can meet test requirement. The frequency of the external synchronization signal determines the frequency of the camera sync shooting. If the external sync signal frequency is 200 Hz, the camera can shoot 200 images per second. The highest shooting frequency of some new high-speed camera can reach hundreds millions of frames/sec controlled by its inner clock control. From the standpoint of both particle measurement and attitude measurement, as long as the real-time image download rate (bandwidth) and the rate of the memory capacity allows, the higher imaging frequency should be applied. A higher imaging frequency is more conducive to the use of inter-frame information on the characteristics of flight segments and characteristics of events for analysis, processing, resulting in higher accuracy of the test data.

External trigger signal is used to control the camera shooting or stop. The camera receives the rising edge of the signal to trigger the camera to start recording or stop recording. The trigger error of high-speed recorder is less than 1 μs.

The exposure time is the sensitive time of the imaging device in camera. The sensitivity of the CCD or CMOS device is very high, and some even reached 0.001 Lux.with the optical system, the A/D bias, the system gain and the image processing technology, the exposure time can be selected from μs to ms [4]. Imaging test results showed that the new high-speed recorder can exposure more than 10^5 times under natural light.

3 REFERENCE TIME INFORMATION

3.1 System components

Theodolite adopt international standards GPS/B code terminal. Its core device code card consists AC [B (AC)] code conversion circuit, DC [B (DC)] code conversion circuit, the divider chain logic control circuit, GPS receivers, PCI interface circuit and Single Chip Microcomputer [5]. The high-precision time data is provided with the computer standard PCI bus interface, as shown in Figure 1.

3.2 Working principle

The B input of the converter circuit (DC) code B (AC) code signal is converted to a TTL signal, and then supplied to the input selection circuit. The GPS receiver receives and demodulates the GPS signal, then the demodulated GPS information and the time data transmission through the serial port to the microcontroller. Then the data is transmitted to the FPGA latch after being processed by the microcontroller. Input selection circuit is controlled by the microcontroller. The microcontroller demodulates B code time information and second signal. The time information is then saved and sent to FPGA latch, synchronized with the internal clock signal, while generating the reference time pulse signal[6].

3.3 Main technical specifications

The frequency signal of time code crystal is 10 MHz, frequency accuracy better than 10^{-7} Hz, frequency stability less than 5×10^{-11} Hz/s. The accuracy of T0 is ±1 μs.

4 ENCODE INFORMATION ACQUISITION

4.1 System components

There are two kinds of encoders. One kind of encoder is incremental encoders and the other is absolute encoder. Incremental encoder is electromagnetic code disc, absolutely is usually photoelectric code disk. Currently, most of the theodolite adopts absolute optical shaft encoder. The system of encode information acquisition mainly consist of the encoder, the control circuit (data acquisition and processing circuit), interface circuit and display circuit. System components are shown as Figure 2.

4.2 Real-time problem

The encoder real-time problem can be attributed to the problem of real-time acquisition angle information. The key technical problem in encoder real-time problem is how to exactly correspond the sampling time (or media exposure) to the image integration time. To solve this technical problem, only strict control of the sampling signal (outer mining pulse) can be achieved. To make the sampling frequency and the camera shooting frequency is equal, the camera shooting frequency signal from timing system should be inputted into control circuit. With the rapid development of device and computer, a theodolite with high-speed cache (up to 512 G) has the function of synchronized camera frequency 1000 C/S above (1 k × 1 k resolution) superimposed. Furthermore, the angle information superimposed in real time sampling can reach more than 1000 Hz utilize the function of angle information real-time overlay. Theoretically, with the present

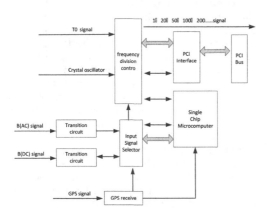

Figure 1.   The GPS/B code terminal.

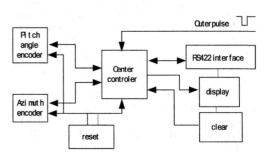

Figure 2.   The system of encode information acquisition.

technology we can realize high-precision real-time sampling of the encoder and overlay function at any camera frequency, only if the outside sampling pulse period is slightly longer than the angle information of the uplink control cycle[7]. But it is very difficult to make the exposure center time synchronize with the encoder sampling moment. On one hand, photography frequency and integration time imaging device timing relationship should be studied. On the other hand, the relationship of the synchronization correspondence between image interpretation point and encoder sampling time should be studied also.

5 THE DESIGN OF DYNAMIC COMPENSATION CONTROL SYSTEM

5.1 Detection principle

A stable adjustable frequency signal is provided with a signal generator and several light-emitting diodes were used as the target being photographed. When the external synchronization signal was inputted to the camera, we can find weather the frequency in the camera control software is consistent with the frequency of the generator. Converting diode lighting time t_1 and duration t_2, the exposure delay time of the camera can be measured. By detecting as above we found that with the integration time the timing of the image recording response is different. The time series relationship is shown in Figure 3.

5.2 Correction formula

As shown in Figure 3, the correction formula is as follow:

$$t_d = t_f + t_j + t_x$$

where t_d is a synchronous sampling amendment time; t_f is various photographic sampling frequency

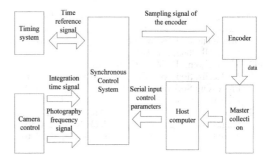

Figure 4.   System block diagram.

correction amount; t_j is different integration time sampling and interpretation of point correction amount; t_x is the exact amount of correction (statistics) based on a combination of factors. The precise amount of correction is based on data analysis of dynamic tracking shot, being processed with the method of least squares correction parameters.

5.3 Synchronous control system design

Synchronous sampling control system block diagram is shown in Figure 4.

Synchronous control system is consisting of the microcontroller and peripheral interface circuit. Accurate correction parameters are sent to the host computer through serial port. Synchronous control system determines the correction amount of the sampled signal output of the encoder according to photography frequency signal and the integration time of the signal. Microcontroller output encoder sampling control signal is accordance to the system time reference signal, photography frequency signal, the signal of the integration time and accurate correction parameter.

6 CONCLUSIONS

Timing synchronization control problem is of one of the key technologies in theodolite system, which ensure real-time or near real-time high-precision measurement. With the rapid development of weapons systems to the direction of the high-precision guided, multi-feature event time-high-precision measurement of the spatial information is particularly important. This requires new theodolite system must have a high frame rate recording capability to the time accurately determine the feature event.

REFERENCES

[1] Liu Lisheng. Data Processing for External Measurement [M]. Beijing: National Defence Industry Press.

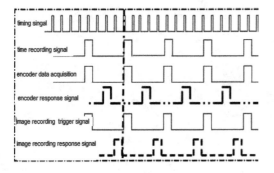

Figure 3.   Timing series relationships.

[2] Zhang Lingxia, Ma Caiwen, Liu Yi. Method Research Based On Data Fusion And Simulation Analysis of Multi-Theodolites by Intersection. Acta Photonica Sinica [J]. 2002, 31(12):1528–1531.

[3] Ma Caiwen, Li Yan, Chai Raojun et, al. Design & Realization of Complex Multi-target Image Processing Software System by TV Theodolite. Acta Photonica Sinica [J]. 2004, 33(8):994–998.

[4] T.G. Alley, M.A. Kramer, D.R. Martinez, et,al. Single-pass imaging through a thick dynamic distorter using four wave mixing [J]. OPT. Lett. 15, 81(1990).

[5] F. Alfonso. Target tracking with bearings-only measurements [J]. Signal Processing, 1999, 78(1):61–78.

[6] Hou Honglu, Zhou Deyun, Wang Wei. Angle accuracy assessment of photo-electric tracking system based on star calibration. Opto-Electronic Engineering [J]. 2006, 33(3):5–10.

[7] Che Shuangliang, Zhang Yaoming. Theory and Practice of Opticalelectronic Theodolite Calibration Method with Star in Range. Acta Photonica Sinica [J]. 2004, 33(10):255–1260.

Control Engineering and Information Systems – Liu (Ed)
© 2015 Taylor & Francis Group, London, ISBN 978-1-138-02685-8

Leader follower formation tracking control for multiple quadrotors

R. Abbas & Q.H. Wu
School of Automation, Beijing Institute of Technology, Beijing, China

ABSTRACT: In this paper, formation tracking control design for multiple quadrotor is discussed. Our approach is based on leader follower architecture for multiple quadrotor. The control approach used in this study is based on sliding mode and backstepping techniques. While the sliding mode is used to ensure the formation over z-axis and yaw angle (φ) and to compensate the external disturbance in the dynamic model of the quadrotor, the backstepping technique is used to ensure the formation tracking over x and y-axis. The system stability analysis is conducted through the Lyapunov theory where asymptotic formation tracking stability is demonstrated. Finally, simulation results demonstrate the effectiveness of the proposed controllers.

1 INTRODUCTION

Quadrotors have become the interest of many researches in the word. They present many advantages with respect to other helicopters such as its capability of vertical taking-off, landing and high maneuverability; it also can hover with a better environmental adaptability. Therefore, the use of quadrotors focuses on achieving the replacement of human being in dangerous missions, where the human life safety is not guaranteed.

Quadrotors can perform solo mission where they can achieve good performances, this characteristic will become more interesting when they operate in a coordinated fashion such as formation and trajectory tracking. Formation control of multiple quadrotors has been widely studied in the literature. In [10], Formation flight of group of UAV was considered. Cooperative control laws were based on integral backstepping. In [12], Formation control and trajectory tracking of mini rotorcraft was considered. A nonlinear controller based on separated saturations. A formation of multi-uavs tight formation flight controller based sliding mode is presented in [11].

In formation control there are many architecture and strategies have been developed. These include: leader follower [12, 15], behavior-based [14] and virtual structure [13]. In the first approach, one of the agents is designated as the leader, with the rest designated as followers. The followers need to position themselves relative to the leader and to maintain a desired relative position with respect to the leader. The second approach consists to design a formation control of agents as a weight average of the formation control of each agent participates in this team formation. The third approach considers that the formation as a single structure or rigid body.

In this work we present formation tracking for multiple quadrotor. The goal is to design formation tracking control for leader and follower quadrotors, where one quadrotor is designed to be the leader that tracks the desired trajectory and others are the followers. These followers track the leader's trajectory by keeping a fixed desired separation from it. The control approach used in this study is composed of two parts. In the first part due to the presence of external disturbances in the vertical motion (z-axis) of the fully actuated subsystem of the quadrotor dynamic model and the fact that it presents a good robustness to the external disturbances and provides a good tracking control [7, 8, 9], the sliding mode is used to compensate this effect and ensure the formation over z-axis and yaw angle (φ). In the second part, a second controller based on backstepping is introduced to ensure the formation tracking over x and y-axis. This controller is an effective technique based on lyapunov theory to design control algorithms for under-actuated systems [1, 2, 16].

The rest of this paper is organized as follows: In Section 2 dynamical model of a quadrotor based on Newton-Euler formalism is given. In Section 3 tracking formation control based on sliding mode and backstepping control is presented. Numerical simulations are provided to validate the proposed method in Section 4. Finally, the conclusions are presented in Section 5.

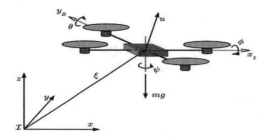

Figure 1. Quadrotor configuration.

2 QUADROTOR DYNAMIC MODEL

The quadrotor configuration is shown in Figure 1.

The motion of quadrotor is controlled by varying the rotation speed of the four rotors to change the thrust and the torque produced by each one.

The dynamic model is presented as [4, 5]:

$$
\begin{cases}
\ddot{x} = u_1(\cos\phi\sin\theta\cos\varphi + \sin\phi\sin\varphi) \\
\ddot{y} = u_1(\cos\phi\sin\theta\sin\varphi - \sin\phi\cos\varphi) \\
\ddot{z} = u_1(\cos\phi\cos\theta) - g + d(t) \\
\ddot{\phi} = u_2 l \\
\ddot{\theta} = u_3 l \\
\ddot{\varphi} = u_4
\end{cases}
\tag{1}
$$

$(x, y, z)^T$ correspond to the relative position of the mass centre of the quadrotor with respect to an inertial coordinate frame, $(\varphi, \theta, \psi)^T$ denotes the three Euler angles representing the attitude of the quadrotor, namely roll-pitch-yaw of the air vehicle. u_1 is the thrust force vector in the body system, u_2, u_3 and u_4 correspond to the control inputs of roll, pitch and yaw moments, respectively. g is the gravitational acceleration. l is the length from the mass centre to the rotor. $d(t)$ is the external disturbances in the vertical motion of the quadrotor.

3 FORMATION CONTROL DESIGN

The quadrotor can be considered as under-actuated system, this is due to the model has six outputs with only four independent inputs; it is not possible to control all of the states in the same time [3].

The state representation for (1) is divided in:

1. An under-actuated subsystem:

$$
S_1 : \begin{cases}
\dot{X}_{i1} = X_{i2} \\
\dot{X}_{i2} = u_{i1} G_{i1} \tau_i \\
\dot{X}_{i3} = X_{i4} \\
\dot{X}_{i4} = G_{i2} V_{i1}
\end{cases}
\tag{2}
$$

with:

$$
X_{i1} = \begin{bmatrix} x_i \\ y_i \end{bmatrix}, \ X_{i2} = \begin{bmatrix} \dot{x}_i \\ \dot{y}_i \end{bmatrix}, \ X_{i3} = \begin{bmatrix} \phi_i \\ \theta_i \end{bmatrix}, \ X_{i4} = \begin{bmatrix} \dot{\phi}_i \\ \dot{\theta}_i \end{bmatrix},
$$

$$
G_{i1} = \begin{bmatrix} \sin\varphi_i & \cos\varphi_i \\ -\cos\varphi_i & \sin\varphi_i \end{bmatrix}, \ G_{i2} = \begin{bmatrix} l & 0 \\ 0 & l \end{bmatrix},
$$

$$
\tau_i = \begin{bmatrix} \sin\phi_i \\ \cos\phi_i \sin\theta_i \end{bmatrix}, \ V_{i1} = \begin{bmatrix} u_{i2} \\ u_{i3} \end{bmatrix}
$$

2. A fully-actuated subsystem:

$$
S_2 : \begin{cases}
\dot{X}_{i5} = X_{i6} \\
\dot{X}_{i6} = G_3 V_{i2} + f
\end{cases}
\tag{3}
$$

With:

$$
X_{i5} = \begin{bmatrix} z_i \\ \varphi_i \end{bmatrix}, \ X_{i6} = \begin{bmatrix} \dot{z}_i \\ \dot{\varphi}_i \end{bmatrix}, \ G_3 = \begin{bmatrix} \cos\phi_i\cos\theta_i & 0 \\ 0 & 1 \end{bmatrix},
$$

$$
f_i = \begin{bmatrix} -g + d_i(t) \\ 0 \end{bmatrix}, \ V_{i2} = \begin{bmatrix} u_{i1} \\ u_{i4} \end{bmatrix}
$$

Note that: $f_i = f_{1i} + f_{2i}$, with $f_{1i} = \begin{bmatrix} -g \\ 0 \end{bmatrix}$, $f_{2i} = \begin{bmatrix} d_i(t) \\ 0 \end{bmatrix}$.

The following assumptions will be used.

Assumption 1: There exists a positive constant $C > 0$ such that the L-infinity norm of the external disturbances is bounded as: $\|d_i(t)\|_\infty \le C$.

Assumption 2: We assume that the roll, pitch and yaw angles are limited to

$$
-\frac{\pi}{2} < \phi < \frac{\pi}{2}, -\frac{\pi}{2} < \theta < \frac{\pi}{2} \ and \ -\pi < \varphi < \pi.
$$

The objective of formation control is to force a leader to track the desired trajectory (x^d, y^d, z^d and φ^d) and the group of followers to track the leader trajectory (x^l, y^l, z^l and φ^l) by keeping a fixed desired separation from it.

The design of the controllers is composed on two parts, The first controller is based on sliding mode to ensure the formation over z-axis and yaw angle φ, while the second is based on backstepping technique to ensure the formation over x and y-axis.

For the two subsystems, we propose the following controllers:

1. For fully-actuated subsystem S_2:

$$\begin{cases} V_{i2} = G_3^{-1}\{\ddot{X}_k^d - \alpha_i(X_{i6} - \dot{X}_k) - K_i tanh(\eta S_i) \\ \qquad - \beta_i S_i - f_1\} \\ S_i = (X_{i6} - \dot{X}_{iK}) + \alpha_i(X_{i5} - X_{ik}) \\ X_{ik} = \begin{cases} X_{i5}^d & \text{if } i \text{ is a leader } (i = 1) \\ X_{l5} + d_i & \text{if } i \text{ is a follower} \end{cases} \end{cases} \quad (4)$$

d_i is the fixed desired separation between followers and leader over z-axis and yaw angle (φ)

2. For under-actuated subsystem S_1:

$$\begin{cases} V_{i1} = (u_{i1}G_{i2})^{-1}[\dot{v}_{2i} + J_i^T Z_{i1} - A_2 Z_{i2}] \\ Z_{i2} = v_{2i} - X_{4i} \\ v_{i2} = J_i^{-1}(\dot{v}_{i1} - A_1 Z_{i1}) \\ Z_{i1} = v_{i1} - \tau_{i1} \\ v_{i1} = (u_{i1}G_{i1})^{-1}\{\ddot{X}_k - (\alpha_i + 1)(X_{i2} - \dot{X}_{ik}) \\ \qquad - \alpha_i(X_{i1} - X_{ik})\} \\ X_{ik} = \begin{cases} X_{i5}^d & \text{if } i \text{ is a leader}(i = 1) \\ X_{l5} + d_i & \text{if } i \text{ is a follower} \end{cases} \end{cases} \quad (5)$$

d_i is the fixed desired distance between followers and leader over x and y-axis.

α_i, β_i, K_i three positive constants.

Proof: First, we consider the fully-actuated subsystem (03). Defining a Lyapunov function as:

$$V(t) = \frac{1}{2}\sum_{i=1}^{n} S_i^T S_i \quad (6)$$

With S_i is the sliding surface given by:

$$S_i = (X_{i6} - \dot{X}_{iK}) + \alpha_i(X_{i5} - X_{ik}) \quad (7)$$

Differentiating V with respect to time, we have:

$$\dot{V} = \sum_{i=1}^{n} S_i^T \dot{S}_i$$
$$= \sum_{i=1}^{n} S_i^T[(\dot{X}_{i6} - \ddot{X}_K) + \alpha_i(X_{i6} - \dot{X}_k)] \quad (8)$$

We consider the second equation of (3)

$$\dot{X}_{i,6} = G_3 V_{i2} + f_{i1} + f_{i2} \quad (9)$$

If

$$V_{i2} = G_3^{-1}\{\ddot{X}_k - \alpha_i(X_{i6} - \dot{X}_k) \\ \qquad - K_i tanh(\eta S_i) - \beta_i S_i - f_1\}$$

Then (8) will be written as:

$$\dot{V} = \sum_{i=1}^{n} S_i^T[-K_i tanh(\eta S_i) - \beta_i S_i + f_{2i}] \quad (10)$$

where: $tanh(\eta x)$ is hyperbolic tangent function is used to approximate sign function.

$$\dot{V} = \sum_{i=1}^{n} S_i^T[-K_i sign(S_i) - \beta_i S_i + f_{2i}]$$
$$= -\sum_{i=1}^{n} S_i^T \beta_i S_i - \sum_{i=1}^{n} S_i^T(K_i sign(S_i) - f_{2i}) \quad (11)$$

$$\leq -\Sigma_{i=1}^{n}(K_i - C_i)\| S_i \|_2 \quad (12)$$

If we select $K > C$, then $\dot{V} \leq 0$ is negative definite. So $S \rightarrow 0$. which means that the leader state track the desired trajectory $X_{i5} \rightarrow X_{i5}^d$ and the followers track the leader with keeping a fixed distance $X_{i5} \rightarrow X_l + d$, so the formation is achieved over z-axis and yaw angle (φ).

Now we consider the under-actuated subsystem given in (02). For this system our approach is based on backstepping technique, and 3 steps are necessary to proof the formation tracking control for multiple quadrotor over x and y-axis.

First step: we consider the following system

$$\begin{cases} \dot{X}_{i,1} = X_{i,2} \\ \dot{X}_{i,2} = u_{i1} G_{i1} v_{i1} \end{cases} \quad (13)$$

where v_{i1} is the virtual control.

We remark that (13) takes the same form as (03) if $f = 0$ in the fully actuated subsystem (S_1). Thus the virtual v_{i1} takes the same form as V_{i2} and consequently the formation over x and y-axis is achieved. Based on this discussion, the virtual control v_{i1} is given by:

$$v_{i1} = (u_{i1}G_{i1})^{-1}\{\ddot{X}_k - (\alpha_i + 1)(X_{i2} - \dot{X}_k) \\ \qquad - \alpha_i(X_{i1} - X_k)\} \quad (14)$$

We select Lyapunov function candidate as:

$$V_1 = \frac{1}{2}\sum_{i=1}^{n}\left((X_{i2} - \dot{X}_k) + \alpha_i(X_{i1} - X_k)\right)^T \\ \left((X_{i2} - \dot{X}_k) + \alpha_i(X_{i1} - X_k)\right) \quad (15)$$

The time derivation of Lyapunov function is given by:

$$\dot{V} = \sum_{i=1}^{n}\left((X_{i2} - \dot{X}_k) + \alpha_i(X_{i1} - X_k)\right)^T \\ \left((u_{i1}G_{i1}v_{i1} - \ddot{X}_k) + \alpha_i(X_{i2} - \dot{X}_k)\right) \quad (16)$$

23

If

$$v_{i1} = (u_{i1}G_{i1})^{-1}\{\ddot{X}_k - (\alpha_i + 1)(X_{i2} - \ddot{X}_k) - \alpha_i(X_{i1} - X_k)\}$$

Then (16) will be writ as:

$$
\begin{aligned}
\dot{V} &= \sum_{i=1}^{n}\left((X_{i2} - \dot{X}_{ik}) + \alpha_i(X_{i1} - X_{ik})\right)^T \\
&\quad \left(\left(\ddot{X}_{ik} - (\alpha_i + 1)(X_{i2} - \dot{X}_{ik}) - \alpha_i(X_{i1} - X_{ik}) - \ddot{X}_{ik}\right)\right. \\
&\quad \left. + \alpha_i(X_{i2} - \dot{X}_{ik})\right) \\
&= -\sum_{i=1}^{n}\left((X_{i2} - \dot{X}_{ik}) + \alpha_i(X_{i1} - X_{ik})\right)^T((X_{i2} - \dot{X}_{ik}) \\
&\quad + \alpha_i(X_{i1} - X_{ik}))
\end{aligned}
$$

(17)

Then $\dot{V} \leq 0$ is negative definite, which implies that if v_{i1} is given by (14), then the formation over x and y-axis is asymptotically achieved.

In order to ensure $v_{i1} \rightarrow (u_{i1}G_{i1})^{-1}\{\ddot{X}_k - (\alpha_i + 1)(X_{i2} - \ddot{X}_k) - \alpha_i(X_{i1} - X_k)\}$, then the second and the third steps are necessary [4].

Second step:

Now we consider $\dot{X}_{i,3} = v_{i2}$ (18)

where v_{i2} is considered as virtual control.

Let define $Z_1 = v_1 - \tau \Rightarrow \dot{Z}_1 = \dot{v}_1 - \dot{\tau}$ (19)

with $\tau = [\tau_1 \ \tau_2 \ ... \ \tau_n]^T$, $v_1 = [v_{11}, v_{21} \ ... \ v_{n1}]^T$ and Z_1 $[z_{11} z_{21} ... z_{n1}]^T$.

For this step we consider the Lyapunov function given by : $V_2 = 0.5Z_1^T Z_1$ (20)

The time derivation of (20) is:

$$\dot{V}_2 = Z_1^T(\dot{v}_1 - \dot{\tau})$$ (21)

Note that

$$\dot{\tau} = \frac{\partial \tau}{\partial X_3}\frac{\partial X_3}{\partial t} = J\dot{X}_3 = Jv_2$$ (22)

with $v_2 = [v_{12} \ v_{22} \ ... v_{n2}]^T$ and $J = diag[J_1 \ J_2 \ ... \ J_n]$.
Then (21) will be written as:

$$\dot{V}_2 = Z_1^T(\dot{v}_1 - Jv_2)$$ (23)

In order to find $\dot{V}_2$ negative definite, we can choose v_2 as:

$$v_2 = J^{-1}(\dot{v}_1 - A_1Z_1)$$ (24)

with $A_1 = diag(a_1, ..., a_1)$, $a_1 > 0$.

Then $\dot{V}_2 = -Z_1^T A_1 Z_1 \leq 0$ (25)

$\dot{V}_2$ Negative definite.
Last step: we consider

$$\dot{X}_{i,4} = G_2 V_{i,1}$$ (26)

where $V_{i,1}$ is the final control of (S_1).

Let $Z_2 = v_2 - X_4 = J^{-1}(\dot{v}_1 + A_1Z_1) - X_4$ (27)

Then $JZ_2 = (\dot{v}_1 + A_1Z_1) - JX_4 = A_1Z_1 + \dot{Z}_1$ (28)

From (24) we can write

$$\dot{Z}_1 = JZ_2 - A_1Z_1$$ (29)

In this step, we select Lyapunov function as:

$$V_3 = \frac{1}{2}Z_1^T Z_1 + \frac{1}{2}Z_2^T Z_2$$ (30)

The derivative with respect time of (24) is given by:

$$
\begin{aligned}
\dot{V}_3 &= Z_1^T(JZ_2 - A_1Z_1) + Z_2^T(\dot{v}_2 - G_2)V_1 \\
&= -Z_1^T A_1 Z_1 + Z_2^T(\dot{v}_2 - G_2V_1 + J^T Z_1)
\end{aligned}
$$

(31)

In order to find $\dot{V}_3$ negative definite, we can choose V_1 as:

$$V_1 = (G_2)^{-1}[\dot{v}_2 + J^T Z_1 + A_2 Z_2]$$ (32)

with $A_2 = diag(a_2, ..., a_2)$, $a_2 > 0$

Then $\dot{V}_3 = -Z_1^T A_1 Z_1 - Z_2^T A_2 Z_2 \leq 0$. (33)

Finally we conclude by:
If $V_1 \rightarrow G_2^{-1}[\dot{v}_2 + J^T Z_1 + A_2 Z_2]$ then
$Z_2 \rightarrow 0$ *which implies that*
$v_2 \rightarrow J^{-1}(\dot{v}_1 - A_1Z_1)$ and $Z_1 \rightarrow 0$, thus
$v_{i1} \rightarrow (u_{i1} G_{i1})^{-1}\{\ddot{X}_k - (\alpha_i + 1)(X_{i2} - \ddot{X}_k) - \alpha_i(X_{i1} - X_k)\}$ Which implies that the tracking formation over x and y-axis is asymptotically achieved for multiple quadrotors.

4   SIMULATION

The proposed formation control has been simulated for the case of three quadrotors. The Euler angles, angular velocities for all quadrotors are all zero and the initial positions are given by:

$$(x_1, y_1, z_1) = (1.5, 3, 2)$$
$$(x_2, y_2, z_2) = (0.75, 0, 0.5)$$
$$(x_3, y_3, z_3) = (-1.5, 1, -1).$$

During this simulation, we assume that the external disturbance is fixed as $d_i(t) = 0.5\cos(t)$.

First we interest to the trajectory tracking, then we pass to the triangular formation of multiple quadrotor.

1. **Trajectory tracking:** In order to show the effectiveness of the control (04) under the external disturbances over z-axis. We consider trajectory tracking for three quadrotors.

   Figure 2 and Figure 3 illustrate the trajectory tracking for three quadrotors over z-axis with fixed $z_{1d} = 0$ and varying desired trajectory $z_{1d} = \sin(t) - \cos(t/3)$ respectively.

   It can be seen that the two followers (quadrotor 2 and 3) converge to the position of the leader (quadrotor 1) and the leader converges to the desired trajectory. It is noted that with a good choice of control parameters, the external disturbance is rejected.

2. **Formation tracking:** Now we propose a triangular formation of three quadrotor evolving in 2D (x-y) space. The relative distances between quadrotors over x-axis are given by [12]:

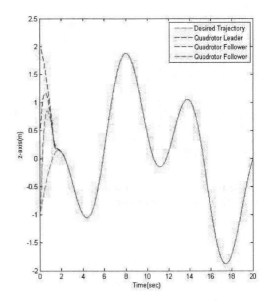

Figure 3. Trajectory tracking with time varying desired trajectory of three quadrotors.

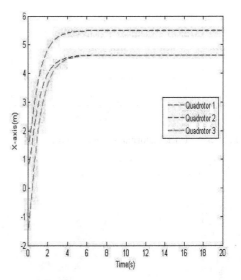

Figure 4. Tracking formation over x-axis.

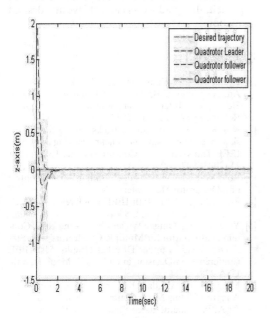

Figure 2. Trajectory tracking with constant desired trajectory of three quadrotors.

$$dx_{12} = x_1 - x_2 = \cos\left(\frac{\pi}{6}\right) \tag{34}$$

$$dx_{13} = x_1 - x_3 = \cos\left(\frac{\pi}{6}\right) \tag{35}$$

$$dx_{23} = x_2 - x_3 = \cos\left(\frac{\pi}{2}\right) \tag{36}$$

25

For the y-axis, the relative distances between quadrotors over y-axis are given by [12]:

$$dy_{12} = y_1 - y_2 = \sin\left(\frac{\pi}{6}\right) \qquad (37)$$

$$dy_{13} = y_1 - y_3 = -\sin\left(\frac{\pi}{6}\right) \qquad (38)$$

$$dy_{23} = y_2 - y_3 = -2\sin\left(\frac{\pi}{6}\right) \qquad (39)$$

Figure 4 and Figure 5 illustrate the formation control of three quadrotors over time for the

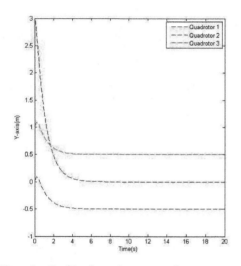

Figure 5.   Tracking formation over y-axis.

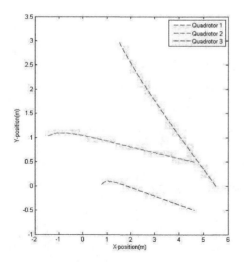

Figure 6.   Triangular formation of 3 quadrotors.

longitudinal and lateral subsystems respectively. It can be seen that the leader quadrotor converge to the desired trajectory and the two followers converge to the desired formation.

Figure 6 illustrates the formation control of three quadrotors over x-y space. It is shown that the triangular formation is asymptotically achieved.

5   CONCLUSION

In this work, we have presented a control approach to ensure trajectory and formation tracking for multiple quadrotors. The proposed approach is composed of two parts where the first is based on backstepping to ensure the trajectory and formation tracking over x and y-axis, the second is based on sliding mode in order to ensure the trajectory and formation tracking over z-axis and yaw angle. The external disturbance present in the quadrotor dynamical model is rejected by a good choice of parameters of controller presented.

Through the above results, the proposed control approach presented in this work ensures the stability of the system where it shows a strong robustness under the external disturbances and this is done by a good choice of the controller parameters. Simulation results based on leader follower approach of three quadrotors demonstrate the effectiveness of the method in trajectory tracking with both constant and varying time desired trajectory. With the proposed control approach, the quadrotors can achieve any desired formation.

REFERENCES

[1] Zheng Fang, Weinan Gao and Lei Zhang, Robust Adaptive Integral Backstepping Control of a 3-DOF Helicopter. International Journal of Advanced Robotic Systems. 20 Jun 2012.
[2] K.S. Kim, Y. Kim, Robust backstepping control for slew maneuver using nonlinear tracking function. IEEE Transactions on Control Systems Technology. 11 (2003) 822–829.
[3] RongXu and Umit Ozguner, Sliding Mode Control of a Quadrotor Helicopter.
[4] Proceedings of the 45th IEEE Conference on Decision and Control, CA, USA December 13–15. 2006.
[5] Yinqiu Wang, Qinghe Wu and Yao Wang 2012, Containment Control for Multiple Quadrotors with Stationary Leaders under Directed Graphs. 51st IEEE Conference on Decision and Control. Maui, Hawaii, USA:2781–2786.
[6] Mingu Kim, Youdan Kim and Jaiung Jun 2012, Adaptive Sliding Mode Control using Slack Variables for Affine underactuated Systems. 51st IEEE Conference on Decision and Control. Maui, Hawaii, USA: 6090–6095.

[7] Jieunelee, Hyeyongseokkim and Youdankim, Formation geometry center based formation controller design using Lyapunov stability theorem. KSAS International Journal. November 2008.

[8] Zhang. yanjiao, Yang Ying Zhao Yu and Wen Guanghui 2012, Finite time consensus tracking for nonlinear multi agent systems undirected and directed communication topologies. Proceedig of the 31st Chinese Control Conference Hafei, China: 6214–6219.

[9] Fahimi, F. 2008, 'Full formation control for autonomous helicopter groups', Robotica: 143–156.

[10] Jia-Jun Wang 2012, Stabilization and tracking control of X–Z inverted pendulum with sliding-mode control. ISA Transaction: 763–770.

[11] David, Nadav Berman, Shai Arogeti, Formation flight using multiple integral backstepping controllers 2011 IEEE 5th International conference on cybernetics and Intelligent Systems (CIS).

[12] Ji Ding and Qiongjian Fan, A multi UAV Tight Formation Flight Controller. International Conference computer Science and Automation (CSAE), IEEE 2012.

[13] Jose Alfredo Guerro, Pedro Castillo, Sergio Salazar, Rogelio Lozano Mini Rotorcraft Flight Formation Using Bounded Inputs J Intell Robots Syst (2012) 65:175–186.

[14] Norman H.M.Li and Hugh H.T. Liu, Formation UAV flight Control using virtual structure and motion synchronization. American Control Conference 2008.

[15] Seungkeun Kim and Youdankim, Three Dimensional Optimum for Multiple UAV Formation Flight using Behavior-based Decentralized Approach. International Conference on Control, Automation and Systems 2007.

[16] H. H. Tanner, G.J. Pappas and V. Kumar, Leader to Formation Stability, IEEE Trans. Robot. Autom., vol 20. Jun 2004.

[17] Tarek Madani and Abdelaziz Benallegue, Backstepping Control for a Quadrotor Helicopter Proceedings of the 2006 IEEE/RSJ international Conference on Intelligent Robots And Systems. China.

[18] Jun Li and Yuntang Li, Dynamic Analysis and PID control for a quadrotor Proceedings of the 2011 IEEE international Conference on Mechatronics and automation Beijing, China.

Control Engineering and Information Systems – Liu (Ed)
© 2015 Taylor & Francis Group, London, ISBN 978-1-138-02685-8

Least-square optimal tracking control of linear discrete systems with time-delays

J. Xie & Y.L. Liu

Ocean University of China, Qingdao, Shandong, China
University of Michigan College of Literature, Ann Arbor, MI, USA

ABSTRACT: The optimal tracking control problem of a linear discrete system with time-delays is mainly discussed in this paper. A new method named as the least-square optimal tracking control is presented in the terms of {1,3}-inverse and M-P inverse matrices. The least-square optimal tracking control law and the minimum-energy optimal tracking control law are designed for the discrete system with time-delays, and the estimation method of minimal tracking errors are given. In the end, the effectiveness and the feasibility of the new method are illustrated for the proposed approach in this paper by a numerical example.

1 INTRODUCTION

The optimal control theory is an important component of modern control theory, and the optimal control problem of a discrete control system is often found in practical applications. The optimal control theory has been widely applied to electric power system, AC drive, power electronics and so on.

At present, the optimal control theory of the linear discrete system is getting perfect and mature, and a lot of theories and methods to solve the optimal control problem. Xu & Qian (2006) transformed the quadratic optimal control problem into the dynamic programming problem for a linear discrete system, and it was researched by Hopfield Neural Network. Zhen et al. (2009) presented information fusion optimal estimation by proximity sensors. A generalized inverse matrix method of solving the optimal control problem of linear discrete time-varying systems was presented (Liu & Liu 1995). While, there are not enough conclusions for linear discrete systems with time-delay. The optimal sliding mode surface and optimal discrete variable structure control law were designed for a discrete system with delay (Zhou et al. 2010). Adopting a disturbance observer, the feed-forward and feedback optimal control law was designed for a linear discrete system with delay (Zhang 2007). The above two papers are both on the basis of sliding mode variable structure control theory. Then Karbassi (2005) gave the time-optimal controller of disturbance-rejection tracking systems with time-delays by state feedbacks. And saturating actuators based on heuristic dynamic programming

were given for the optimal control laws (Song et al. 2010).

In this study, we present a least-square control method to research the optimal tracking control problem of a linear discrete system with time-delays. Firstly, we transform this optimal tracking control problem into a least square dynamic programming problem. Secondly, according to the least square theory to solve differential equations, we get the least-square optimal tracking control law and minimum-energy optimal tracking control law by {1,3}-inverse and M-P inverse matrices. And then, the estimation method of minimal tracking errors are given. Lastly, a numerical example is given to illustrate the effectiveness and the feasibility of the method.

2 PRELIMINARIES

First we give the presentation of generalized inverse matrix. In Penrose (1955) it was shown that, for any matrix $A \in \Re^{n \times n}$, there is four equations

$$AXA = A, \tag{1}$$

$$XAX = X, \tag{2}$$

$$(AX)^{\mathrm{T}} = AX, \tag{3}$$

$$(XA)^{\mathrm{T}} = XA, \tag{4}$$

where $(\cdot)^{\mathrm{T}}$ denotes the transpose. X satisfying (1) is called {1}-inverse of A; and that satisfying (1) and (3) is called {1,3}-inverse. If X satisfies the above

(1–4), then X is called the M-P inverse of A, and usually denoted by A^+.

Lemma 1 (Israel & T.N.E. 1974) For any matrix $A \in \Re^{m \times n}$, $b \in \Re^m$, the problem

$$\min \| Ax - b \|, \tag{5}$$

has the least square solutions

$$x = A^{(1,3)}b + (I - A^{(1,3)}A)z,$$

where $z \in \Re^n$ is arbitrary. And the solution $x = A^+b$ is the minimum of all the least square solutions.

According to Lemma 1, now we give the definitions of least-square and minimum-energy optimal tracking control law for the linear discrete system.

Definition 1 For the linear discrete system

$$\begin{cases} x_{k+1} = Ax_k + Bu_k, \\ y_k = Cx_k, \end{cases} \quad k = 0,1,\dots,N. \tag{6}$$

Let the terminal time $N \in Z^+$ be fixed and the desired outputs $d_0, d_1, \dots, d_N \in \Re^{q \times m}$ be given, the control sequence $u_0, \dots, u_N$ is called the least-square optimal control of (6), if it satisfies

$$\begin{cases} \min \sum_{j=0}^{N} \| W_j(y_j - d_j) \|^2, \\ \text{s.t.} \quad (6), \end{cases} \tag{7}$$

where $W_0, W_1, \dots, W_N \in \Re^{q \times q}$ are weighting matrices.

In general, the optimal control of (7) is not unique, so there is the definition of minimum-energy optimal control.

Definition 2 Under the definition 1, the least-square optimal control $u_0^*, u_1^*, \dots, u_N^*$ of (7) is called the minimum-energy optimal control, if for any least-square optimal control $u_0, u_1, \dots, u_N$ of (7) satisfies

$$\sum_{j=0}^{N} \| T_j u_j^* \|^2 \le \sum_{j=0}^{N} \| T_j u_j \|^2, \tag{8}$$

where $T_0, T_1, \dots, T_N$ are nonsingular weighting matrices.

Lemma 2 (Campbell et al. 1979) There is

$$A^{(1,3)} = (A^T A)^{(1)} A^T.$$

Lemma 3 (Zlobel Formula) There is

$$A^+ = A^T (A^T A A^T)^{(1)} A^T.$$

Lemma 4 (Luo & Fang 2006) For $A \in \Re^{m \times n}$, then

$$A^+ = \begin{cases} A^T(AA^T)^{-1}, & \text{if } \operatorname{rank}(A) = m; \\ (AA^T)^{-1}A^T, & \text{if } \operatorname{rank}(A) = n. \end{cases}$$

Assume A1 The weighting matrices $W_0, W_1, \dots, W_N$ and $T_0, T_1, \dots, T_N$ are all identity matrices with corresponding orders.

If $W_0, W_1, \dots, W_N$ and $T_0, T_1, \dots, T_N$ are not all identity matrices, we have the following reversible transformations

$$u_k' = T_k u_k, \quad y_k' = W_k y_k, \quad d_k' = W_k d_k,$$

and then the system (5) is transformed into

$$\begin{cases} x_{k+1} = Ax_k + BT_k^{-1}u_k', \\ y_k' = W_k^{-1}Cx_k, \end{cases} \quad k = 0,1,\dots,N. \tag{9}$$

Obviously, all weighting matrices of the system (9) are identify matrices.

Assume A2 Let matrices $H_i, \bar{H}_{i-1}$ and H_i' have the same number of rows, and there exits

$$H_i = \begin{bmatrix} \bar{H}_{i-1} & H_i' \end{bmatrix}, \tag{10}$$

Under A2, there is Lemma 5 as follows.
Lemma 5 (Liu 2009) If $\bar{Q}_{i-1} = I - \bar{H}_{i-1}\bar{H}_{i-1}^{(1,3)}$, $\bar{Q}_{-1} = I$ and $\bar{A}_i = \bar{Q}_{i-1}H_i'$, then

$$H_i^{(1,3)} = \begin{bmatrix} \bar{H}_{i-1}^{(1,3)}M_i \\ \bar{A}_i^{(1,3)} \end{bmatrix}, \quad M_i = I - H_i'\bar{A}_i^{(1,3)} \tag{11}$$

Obviously, $\bar{Q}_{i-1}$ is symmetric and idempotent, that is $\bar{Q}_{i-1}^T\bar{Q}_{i-1} = \bar{Q}_{i-1}$. By Lemma 2, we yield

$$\bar{A}_i^{(1,3)} = (\bar{A}_i^T\bar{A}_i)^{(1)}H_i'^T\bar{Q}_{i-1}^T\bar{Q}_{i-1} = \bar{A}_i^{(1,3)}\bar{Q}_{i-1} \tag{12}$$

$$\bar{A}_i^{(1,3)}H_i' = \bar{A}_i^{(1,3)}\bar{Q}_{i-1}H_i' = \bar{A}_i^{(1,3)}\bar{A}_i \tag{13}$$

$$\bar{A}_i^{(1,3)}\bar{H}_{i-1} = \bar{A}_i^{(1,3)}\bar{Q}_{i-1}\bar{H}_{i-1} = 0. \tag{14}$$

3 MAIN RESULTS

Considering a linear discrete system with time-delays

$$\begin{cases} x_{k+1} = A_0 x_k + \sum_{j=0}^{r} B_j u_{k-j}, \\ y_k = C_0 x_k, \end{cases} \quad k = 0,1,\dots,N, \tag{15}$$

30

where $x_k \in \Re^n$ is a state vector, $u_{k-j} \in \Re^m$ is a input vector, $y(k) \in \Re^q$ is an output vector; A_0, B_j and C_0 are system matrices; N is a terminal time and $r \in Z^+$. Giving initial vectors $u_{-r}, u_{-r+1}, ..., u_{-1}$ and $d_0, d_1, ..., d_N$. By the linear property of (15), we know that $y_{N-i+1}, ..., y_N$ are linear combinations of x_{N-i} and $u_N, ..., u_{N-i+1}, u_{N-i}, u_{N-i-1}, u_{N-i-r}$.

For $i = 0, 1, ..., N$, let

$$u(i) = \begin{bmatrix} u_N \\ \vdots \\ u_{N-i+1} \\ u_{N-i+r} \\ \vdots \\ u_{N-i} \end{bmatrix}, \; y(i) = \begin{bmatrix} y_N \\ y_{N-1} \\ \vdots \\ y_{N-i} \end{bmatrix}, \; d(i) = \begin{bmatrix} d_N \\ d_{N-1} \\ \vdots \\ d_{N-i} \end{bmatrix}, \quad (16)$$

then we have

$$y(i-1) = R_{i-1} x_{N-i+1} + H_{i-1} u(i-1). \quad (17)$$

Denoting

$$B(i) = \begin{bmatrix} \underbrace{0, ..., 0}_{i}, B_r, B_{r-1}, ..., B_1 \end{bmatrix},$$

where $0 \in \Re^{m \times n}$ is the zero matrix, we have

$$x_{N-i+1} = A_0 x_{N-i} + B_0 u_{N-i} + B(i) u(i-1).$$

So the system (15) is equal to the following system

$$\begin{cases} x_{N-i+1} = A_0 x_{N-i} + B(i) u(i-1) + B_0 u_{N-i}, \\ y_{N-i} = C_0 x_{N-i}. \end{cases} \quad (18)$$

Now substituting (17) into (16), there exists

$$\begin{cases} y(i-1) = R_{i-1} A_0 x_{N-i} + \left(R_{i-1} B(i) + H_{i-1} \right) \\ \qquad \times u(i-1) + R_{i-1} B_0 u_{N-i}, \\ y_{N-i} = C_0 x_{N-i}. \end{cases} \quad (19)$$

As a result, the system (19) can be denoted by the following matrix equation

$$y(i) = R_i x_{N-i} + H_i u(i), \quad i = 0, 1, ..., N, \quad (20)$$

where $R_0 = C_0$, $H_0 = H_0' = 0$;

$$R_i = \begin{bmatrix} R_{i-1} A_0 \\ C_0 \end{bmatrix} = \begin{bmatrix} C_0 A_0^i & C_0 A_0^{i-1} & ... & C_0 A_0 & C_0 \end{bmatrix}^T,$$

$$H_i = \begin{bmatrix} R_{i-1} B(i) + H_{i-1} & \vdots & R_{i-1} B_0 \\ 0 & \vdots & 0 \end{bmatrix}, i = 1, ..., N.$$

3.1 The least-square optimal tracking control

Assume $d_0, d_1, ..., d_N$ are desired outputs, the tracking problem of (20) can be regarded as a dynamic programming problem

$$\min \| H_i u(i) - (d(i) - R_i x_{N-i}) \| \quad (21)$$

By Lemma 1 and Definition 2, we obtain the least-square optimal tracking control law of (20), that is

$$u^*(i) = H_i^{(1,3)}(d(i) - R_i x_{N-i}) + (I - H_i^{(1,3)} H_i) z(i), \quad (22)$$

where $z(i) = \begin{bmatrix} Z_N & ... & Z_{N-i} \end{bmatrix}^T$, $z_k \in \Re^n$ is arbitrary.

Let

$$H_i = \begin{bmatrix} R_{i-1} B(i) + H_{i-1} & \vdots & R_{i-1} B_0 \\ 0 & \vdots & 0 \end{bmatrix} \underset{=\!=\!=}{\text{def.}} \begin{bmatrix} \bar{H}_{i-1} & H_i' \end{bmatrix} \quad (23)$$

$$\bar{H}_{i-1} = \begin{bmatrix} R_{i-1} B(i) + H_{i-1} \\ 0 \end{bmatrix} \underset{=\!=\!=}{\text{def.}} \begin{bmatrix} \tilde{H}_{i-1} \\ 0 \end{bmatrix}. \quad (24)$$

According to Lemma 4, we derivative the $\{1,3\}$-inverse matrix of H_i, that is

$$H_i^{(1,3)} = \begin{bmatrix} \bar{H}_i^{(1,3)} M_i \\ \bar{A}_i^{(1,3)} \end{bmatrix} \quad (25)$$

where $\bar{Q}_{i-1} \in \Re^{(i+1)q \times (i+1)q}$, $\bar{A}_i^{(1,3)} \in \Re^{n \times (i+1)q}$, $M_i \in \Re^{(i+1)q \times (i+1)q}$.

$$M_i = I - H_i' \bar{A}_i^{(1,3)}, \quad \bar{Q}_{-1} = I, \quad H_i' = 0;$$

$$\bar{Q}_{i-1} = \begin{bmatrix} I - \tilde{H}_{i-1} \tilde{H}_{i-1}^{(1,3)} & 0 \\ 0 & I \end{bmatrix},$$

$$\bar{A}_i = \begin{bmatrix} (I - \tilde{H}_{i-1} \tilde{H}_{i-1}^{(1,3)}) R_{i-1} B_0 \\ 0 \end{bmatrix}$$

It is obvious that $H_i^{(1,3)}$ is responding to the last n rows of $\bar{A}_i^{(1,3)}$. And we also have

$$(I - H_i^{(1,3)} H_i) z(i)$$

$$= \begin{bmatrix} I - \bar{H}_{i-1}^{(1,3)} M_i H_{i-1} & -\bar{H}_{i-1}^{(1,3)} M_i H_i' \\ -\bar{A}_i^{(1,3)} \bar{H}_{i-1} & I - \bar{A}_i^{(1,3)} H' \end{bmatrix} \begin{bmatrix} z(i-1) \\ z_{N-i} \end{bmatrix} \quad (26)$$

From (13) and (14), there is $\bar{A}_i^{(1,3)} \bar{H}_i = 0$. Let $\bar{C}_i = I - \bar{A}_i^{(1,3)} H_i' = I - \bar{A}_i^{(1,3)} \bar{A}_i$, $\bar{C}_i \in \Re^{n \times n}$. Now we

31

study the last n rows u_{N-i}^* of the least-square optimal control $u^*(i)$, combined with (22), (25) and (26), we obtain

$$u_{N-i}^* = \bar{A}_i^{(1,3)}(d(i) - R_i x_{N-i}) + \bar{C}_i z_{N-i}, \ i = 0, \dots, N,$$

(27)

where $z_{N-i-r} \in \Re^n$ is arbitrary.

Further simplify the calculation. Firstly, by Lemma 2, we know

$$\bar{A}_i = \begin{bmatrix} (I - \tilde{H}_{i-1}\tilde{H}_{i-1}^{(1,3)})R_{i-1}B_0 \\ 0 \end{bmatrix},$$

where $I - \tilde{H}_{i-1}\tilde{H}_{i-1}^{(1,3)}$ is symmetric and idempotent, and

$$\bar{A}_i^{(1,3)} = \left[B_0^T R_{i-1}^T (I - \tilde{H}_{i-1}\tilde{H}_{i-1}^{(1,3)})R_{i-1}B_0 \right]^{(1)} \bar{A}_i^T.$$

Let $Q_{i-1} = R_{i-1}^T(I - \tilde{H}_{i-1}\tilde{H}_{i-1}^{(1,3)})R_{i-1}$, $Q_{-1} = I$, it is obvious that

$$\bar{A}_i^{(1,3)} = (B_0^T Q_{i-1}B_0)^{(1)} \bar{A}_i^T,$$

(28)

$$\bar{A}_i^T \bar{A}_i = B_0^T Q_{i-1}B_0.$$

(29)

Therefore, (27) is simplified as

$$\bar{A}_i^{(1,3)}d(i) = (B_0^T Q_{i-1}B_0)^{(1)}$$

$$B_0^T R_{i-1}^T(I - \tilde{H}_{i-1}\tilde{H}_{i-1}^{(1,3)})d(i-1)$$

(28)

$$\bar{A}_i^T R_i = B_0^T R_{i-1}^T(I - \tilde{H}_{i-1}\tilde{H}_{i-1}^{(1,3)})R_{i-1}A_0 = B_0^T Q_{i-1}A_0$$

$$\bar{A}_i^{(1,3)}R_i = \left(B_0^T Q_{i-1}B_0\right)^{(1)} B_0^T Q_{i-1}A_0$$

(29)

$$\bar{A}_i^{(1,3)}\bar{A}_i = \left(B_0^T Q_{i-1}B_0\right)^{(1)} B_0^T Q_{i-1}B_0$$

$$\bar{C}_i = I - \left(B_0^T Q_{i-1}B_0\right)^{(1)} B_0^T Q_{i-1}B_0.$$

(30)

And then substituting (30), (31) and (32) into (27), there exists

$$u_{N-i}^* = (B_0^T Q_{i-1}B_0)^{(1)} B_0^T R_{i-1}^T (I - \tilde{H}_{i-1}\tilde{H}_{i-1}^{(1,3)})d(i-1)$$

$$- \left(B_0^T Q_{i-1}B_0\right)^{(1)} B_0^T Q_{i-1}A_0 x_{N-i}$$

$$+ \left[I - \left(B_0^T Q_{i-1}B_0\right)^{(1)} B_0^T Q_{i-1}B_0\right]z_{N-i}.$$

(31)

Specially, when $i = 0$, there exists

$$u_N^* = \bar{A}_0^{(1,3)}(d(0) - R_0 x_N) + \bar{C}_0 z_N$$

(32)

where $\bar{A}_0 = \bar{Q}_{-1}H_0'$. Finally, we get the conclusion.

Theorem 1 Assuming desired outputs of system (14) are $d_0, d_1, \dots, d_N$ and initial inputs are $u_{-r}, u_{-r+1}, \dots, u_{-1}$. Let $d(i) = [d_N \ d_{N-1} \dots d_{N-i}]^T$, $i = 0, \cdots, N$. Then the least-square optimal tracking control sequence $u_0^*, u_1^*, \dots, u_N^*$ satisfies

$$\begin{cases} u_N^* = \bar{A}_0^{(1,3)}(d(0) - R_0 x_N) + \bar{C}_0 z_N, \\ u_{N-i}^* = (B_0^T Q_{i-1}B_0)^{(1)} B_0^T R_{i-1}^T(I - \tilde{H}_{i-1}\tilde{H}_{i-1}^{(1,3)})d(i-1) \\ \qquad - \left(B_0^T Q_{i-1}B_0\right)^{(1)} B_0^T Q_{i-1}A_0 x_{N-i} \\ \qquad + \left[I - \left(B_0^T Q_{i-1}B_0\right)^{(1)} B_0^T Q_{i-1}B_0\right]z, \end{cases}$$

where

$$B(i) = \left[\underbrace{0, \dots, 0}_{i}, B_r, B_{r-1}, \dots, B_1\right],$$

R_i and $\tilde{H}_i$ are denoted by (18); and

$$Q_{i-1} = R_{i-1}^T(I - \tilde{H}_{i-1}\tilde{H}_{i-1}^{(1,3)})^{(1)} R_{i-1}.$$

3.2 The minimum-energy optimal tracking control

Lemma 6 Assuming $H_i = \begin{bmatrix} \tilde{H}_{i-1} & H_i' \end{bmatrix}$, and $\bar{H}_{i-1}$ has linear relationship with H_i'. Let

$$\bar{Q}_{-1} = 0, \quad P_{-1} = 0, \quad \bar{A}_i = \bar{Q}_{i-1}H_i', \quad \bar{B}_i = P_{i-1}H_i',$$

$$\bar{C}_i = I - \bar{A}_i^+ \bar{A}_i, \quad D_i = I + \bar{C}_i \left(H_i'\right)^T \bar{B}_i \bar{C}_i,$$

$$\bar{C}_i = I - \bar{A}_i^+ \bar{A}_i, \quad D_i = I + \bar{C}_i \left(H_i'\right)^T \bar{B}_i \bar{C}_i.$$

If there exist $G_i = \bar{A}_i^+ + \bar{C}_i D_i^{-1} \bar{B}_i^T \left(I - H_i' \bar{A}_i^+\right)$, $Q_i = \bar{Q}_{i-1}M_i$, $P_i = M_i^T P_{i-1}M_i + G_i^T G_i$, $M_i = I - H_i' G_i$, then

$$H_i^+ = \begin{bmatrix} \bar{H}_{i-1}^+ M_i \\ G_i \end{bmatrix}, \quad i = 0, 1, \dots, N.$$

For the dynamic programming system (21), according to Lemma 1 and Definition 2, we get the minimum-energy optimal tracking control of (15) as

$$u^*(i) = H_i^+ \left[d(i) + R_i x_{N-i}\right], \ i = 0, 1, \dots, N$$

(35)

As we know, $H_i = \begin{bmatrix} \bar{H}_{i-1} & H_i' \end{bmatrix}$ and H_i' is the last n columns of H_i. By Lemma 6,

$$H_i^+ = \begin{bmatrix} \bar{H}_{i-1}^+ M_i \\ G_i \end{bmatrix}$$

32

and G_i is the last n rows of H_i^+. Now let substitute H_i^+ into (33) and take the last n rows of $u^*(i)$, there is the following conclusion.

Theorem 2 Assume $u_0^*, u_1^*, ..., u_N^*$ be the minimum-energy optimal tracking control sequence of system (15), then the sequence is denoted by

$$u_{N-i}^* = G_i[d(i) - R_i x_{N-i}], \quad i = 0,1, ..., N,$$

where the matrix G_i satisfies

$$\bar{Q}_{-1} = 0, \quad P_{-1} = 0, \quad H_0' = 0, \quad \bar{Q}_i = \bar{Q}_{i-1} M_i,$$

$$H_i' = \begin{bmatrix} R_{i-1} B_0 \\ 0 \end{bmatrix} = \begin{bmatrix} C_0 A_0^{i-1} B_0 \\ \vdots \\ C_0 B_0 \end{bmatrix}, \bar{A}_i = \bar{Q}_{i-1} H_i',$$

$$M_i = I - H_i' G_i,$$

$$\bar{B}_i = P_{i-1} H_i', \bar{C}_i = I - \bar{A}_i^+ \bar{A}_i, D_i$$

$$= I + \bar{C}_i (H_i')^T \bar{B}_i \bar{C}_i,$$

$$P_i = M_i^T P_{i-1} M_i + G_i^T G_i, G_i$$

$$= \bar{A}_i^+ + \bar{C}_i D_i^{-1} \bar{B}_i^T (I - H_i' \bar{A}_i^+).$$

Remark. When the matrix $\tilde{H}_i$ satisfies the conditions of Lemma 4, we will obtain the formula of $\tilde{H}_i^+$, and then by the theorem 2 the minimum-energy optimal control of (14) will be get directly.

3.3 *Minimal tracking errors*

The optimal tracking problem of a system is to get an optimal control law so that the output of the system could follow the tracks of the desired output and the tracking error is a minimal value. This section studies the estimation method of minimal tracking errors of system (15) on the basis of above theorems and conclusions.

When $i = N$, the formula (22) can be transformed into

$$u^*(N) = H_N^{(1,3)}[d(N) - R_N x_0]$$
$$+ (I - H_N^{(1,3)} H_N) z(N) \quad (36)$$

Now let (36) multiplied by H_N, we have

$$H_N u^*(N) = H_N H_N^{(1,3)}[d(N) - R_N x_0]. \quad (37)$$

Let $Q_i' = I - H_i H_i^{(1,3)}$, and $M_i = I - H_i' \bar{A}_i^{(1,3)}$, then

$$Q_i' = \bar{Q}_{i-1}(I - H_i' \bar{A}_i^{(1,3)}) = \bar{Q}_{i-1} M_i. \quad (38)$$

Theorem 3 Assume

$$\left(y^*(N) \right)^T = \left[\left(y_N^* \right)^T, \quad ..., \quad \left(y_0^* \right)^T \right]$$

be the optimal output of system (15) under the optimal control $u_0^*, u_1^*, ..., u_N^*$. Let $r_N = d(N) - R_N x_0$, then the estimation formula of the tracking error is

$$\left\| y^*(N) - d(N) \right\|^2 = \sum_{k=0}^{N} (y_k - d_k)^2 = r_N^T Q_N' r_N.$$

4 NUMERICAL EXAMPLE

This example illustrates the effectiveness and the feasibility of Theorem 2 and 3. Consider a simple linear discrete system with time-delays

$$\begin{cases} x_{k+1} = A_0 x_k + B_0 u_k + B_1 u_{k-1}, \\ y_k = C_0 x_k, \end{cases}$$

where

$$A_0 = \begin{bmatrix} 0.9512 & 0 \\ 0 & 0.9048 \end{bmatrix}, B_0 = \begin{bmatrix} 4.8770 & 4.8770 \\ 0 & 0 \end{bmatrix},$$

$$B_1 = \begin{bmatrix} 0 & 3.589 \\ -1.895 & 0 \end{bmatrix}, C_0 = \begin{bmatrix} 0.01 & 0 \\ 0 & 0.01 \end{bmatrix},$$

$N = 1, k = 0,1.$ Assume $x_0 = [1 \ 1]^T$ and the desired outputs

$$d_0 = \begin{bmatrix} 0.0001 \\ 0.0001 \end{bmatrix}, d_1 = \begin{bmatrix} 0.0002 \\ 0 \end{bmatrix}.$$

Then

$$H_0 = 0, H_0' = 0, \tilde{H}_0 = \begin{bmatrix} 0 & 0 & 0 & 0.03589 \\ 0 & 0 & -0.01895 & 0 \end{bmatrix},$$

$$H_1 = \begin{bmatrix} 0 & 0 & 0 & 0.03589 & 0.04877 & 0.04877 \\ 0 & 0 & -0.01895 & 0 & 0 & 0 \\ 0 & 0 & 0 & 0 & 0 & 0 \\ 0 & 0 & 0 & 0 & 0 & 0 \end{bmatrix}.$$

Obviously, rank$(\tilde{H}_0) = 2$ and the row number of $\tilde{H}_0$ is 2. By Lemma 4, we get

$$\tilde{H}_0 \tilde{H}_0^+ = \begin{bmatrix} 0.9999923 & 0 \\ 0 & 1.0000067 \end{bmatrix} \approx \begin{bmatrix} 1 & 0 \\ 0 & 1 \end{bmatrix},$$

$$H'_{-1} = 0, \; H'^{+}_{-1} = 0, \; \tilde{H}^{+}_{0} = \begin{bmatrix} 0 & 0 \\ 0 & 0 \\ 0 & -52.7708 \\ 27.8627 & 0 \end{bmatrix},$$

$$Q_{-1} = I, \; \bar{A}_0 = \bar{Q}_{-1} H'_0 = 0.$$

Above all, we have the following conclusions. For $i = 0$,

$$u^*_1 = \bar{A}^{+}_0 (d_1 - C_0 x_1) + \bar{C}_0 z_1 = z_1.$$

For $i = 1$, because

$$Q_0 = R^{\mathrm{T}}_0 (I - \tilde{H}_0 \tilde{H}^{+}_0)^{(1)} R_0 = R^{\mathrm{T}}_0 (I - I)^{(1)} R_0 = 0,$$

Then $u^*_0 = z_2$.
For the arbitrariness of $z_1, z_2 \in \Re^n$, we design the minimum-energy optimal tracking control as

$$u^*_0 = \begin{bmatrix} 0.001 \\ 0.001 \end{bmatrix}, u^*_2 = \begin{bmatrix} 0 \\ 0.002 \end{bmatrix}.$$

Consider the minimal errors. Let

$$r_1 = d(1) - R_1 x_0 = \begin{bmatrix} -0.009312 \\ 0.009048 \\ -0.0099 \\ -0.0099 \end{bmatrix},$$

then

$$\left\| y^*(1) - d(1) \right\|^2 = r^{\mathrm{T}}_1 Q'_1 r_1 = 2 \times (0.0099)^2$$
$$= 0.000196 \approx 0.$$

The above numerical example illustrates that the tracking error under the designed minimum-energy optimal control tends to zero. Therefore, the design of the least-square and minimum-energy optimal tracking control in this paper is effective and feasibility.

5 CONCLUSIONS

The optimal tracking control problem of a kind of linear discrete systems with time-delays is researched in this paper. And the main method is the least square theory to solve differential equations. We present a least-square control method to research the optimal tracking control problem of a linear discrete system with time-delays. Firstly, we transform this optimal tracking control problem into a least square dynamic programming problem. Secondly, according to the least square theory and the properties of {1,3}-inverse and M-P inverse to solve the dynamic programming system, we get the least-square optimal tracking control law and minimum-energy optimal tracking control law. And then the estimation method of minimal tracking errors are given. Lastly, we give a numerical example to illustrate the effectiveness and the feasibility of presented conclusions in this paper.

REFERENCES

Campbell, S.L. et al. 1979. *Generalized inverses of linear transformations.* London: Pitman publishing Ltd.
Israel, B. & A., T.N.E 1974. *Generalized inverses: theory and applications.* New York: Wiley.
Karbassi, S.M. 2005. Time-optimal control of disturbance-rejection tracking systems for discrete-time time-delayed systems by state feedback. *Computer and Mathematics with Applications* 50(8–9):1415–1424.
Liu, X.H. 2009. The algorithms for generalized inverses and least square estimations. *Journal of Jiangxi Vocational and Technical College of Electricity* 22(1):71–74.
Liu, D.Y. & Liu, X.H. 1995. Optimal and minimum-energy optimal tracking of discrete linear time-varying systems. *Automatica* 31(10):1407–1419.
Luo, J.H. & Fang, W.D. *Introduction to matrix analysis.* Guangzhou: South China University of Technology Press.
Song, R.Z. et al. 2010. Optimal control laws for time-delay systems with saturating actuators based on heuristic dynamic programming. *Neurocomputing* 73(16–18):3020–3027.
Xu, D.X. & Qian, F.C. 2006. Optimal control of discrete bilinear system based on the hopfield neural network. *Information and Control* 35(1):90–92.
Zhang, B.L. 2007. Optimal control design for linear discrete time-delay systems with external disturbances. *Journal of Ningxia University* 28(1):25–28.
Zhen, Z.Y. et al. 2009. Discrete system optimal tracking control based on information fusion estimation. *Control and Decision* 24(1):81–85.
Zhou, S.S. et al. 2010. Optimal sliding mode control for discrete time systems with time-delay. *Control and Decision* 25:300–306.

Control Engineering and Information Systems – Liu (Ed)
© *2015 Taylor & Francis Group, London, ISBN 978-1-138-02685-8*

The stability of a safety system containing two redundant robots

X. Qiao, D. Ma, Y.X. Sha, G.C. Zhao & R. Guo
Department of Mathematics, Daqing Normal University, Daqing, China

ABSTRACT: A repairable safety system model which is consisted of two redundant robots and one safety unit is established in this paper. We are devoted to discussing the unique existence of the solution and the stability of the studied system by analyzing the spectrum distribution of the system operator in detail. Thus we can obtain stability analysis of the repairable system.

1 INTRODUCTION

Today, robots are used to perform various types of tasks in the industrial sector. Even though one of the purposes of using robots was to replace humans performing unsafe and dangerous tasks, the use of robots has unfortunately resulted in serious accidents and other safety-related problems (see Altamuro 1985, Lauck 1986, Nicolaisen 1987). So, its reliability has also become a challenging issue.

Consequently, a robot not only has to be reliable, but also safe. Thus, the safety mechanism is an important element of the overall robot system. More specifically, an overall robot system is made up of a robot and its associated safety or mechanism. To improve the reliability of a system, the concept of redundancy is practiced, which can also be applied to robots. Therefore, this paper presents stability analysis of a system containing two redundant robots and one safety unit.

2 FORMULATION OF MODEL

The basic system (i.e. without considering the repair process) starts operating at time $t = 0$ and it may be degraded either due to a failure of a robot or the system. From these two degradation states when either a robot or the safety system fails (i.e. from state 1 only), the overall system is further degraded to a state having only one robot working with the failed safety system. The system fails whenever the single operating robot fails. The symbol u in Figure 1 and indicates the repair action from various system states.

The general assumptions associated with the analysis presented in this article include: (i) statistically independent failures; (ii) the overall system is composed of two identical and redundant active robots; (iii) all failure rates are constant; (iv) a repair robot or the safety system is as good as new

(when applicable); and (v) the overall system fails only when both the robots fail.

The physical meaning of relative signals in Figure 1 is as follows.

i—i th state of the overall robot system: $i = 0$ (both the robots and the safety system operating normally). $i = 1$—one robot and the safety system working normally, the other robot failed. $i = 2$—both the robots operating normally, the safety system failed. $i = 3$—one robot working normally and the other along with the safety system failed. $i = 4$—both the robots and the safety system failed. $i = 5$—both the robots failed, the safety system functioning. $P_i(t)$—probability that the overall robot system is in state i at time t, for $i = 0, 1, 2, 3, 4, 5$. λ—constant failure rate of a robot. λ_s—constant failure rate of the safety system associated with the robot system. μ_i—ith constant repair rate; $i = 1$ (of robot), $i = 2$ (of the safety system). $\mu_i(x)$—time-dependent repair rate when the overall robot system is in state i and has an elapsed repair time of x, satisfying $0 \leq \mu_i(x) < \infty$, $0 < G = \sup_{x \in [0,\infty)} \mu_i(x)$, $\int_0^x \mu_i(\tau)d\tau < \infty$, $\int_0^\infty \mu_i(\tau)d\tau = \infty$, for $i = 4,5$. $P_i(x,t)$—probability that at time t, the system is in state i and the elapsed repair time lies in the interval $[x, x + \Delta x]$, for $i = 4,5$.

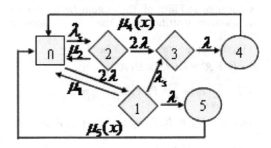

Figure 1. Total robot system transition diagram.

Taking into account the practical physical meaning, we can supposed that the mean of repair rates exists and does not equal to 0, namely

$$0 < \lim_{x \to \infty} \frac{1}{x} \int_0^x \mu_i(\tau)d\tau < \mu_i < \infty, \; i = 4,5,$$

$$\mu = \min\{\mu_4, \mu_5\}.$$

For simplicity, let $a_0 = 2\lambda + \lambda_s$, $a_1 = \lambda + \lambda_s + \mu_1$, $a_2 = 2\lambda + \mu_2$. For arbitrarily distributed failed over all robot system repair times, using the supplementary variables method (see Dhillon 1983, Gaver 1963), the system under consideration can be formulated as the following equations:

$$P_0'(t) = -a_0 P_0(t) + P_2(t)\mu_2 + P_1(t)\mu_1$$
$$+ \sum_{i=4}^{5} \int_0^x P_i(x,t)\mu_i(x)dx, \tag{1}$$

$$P_1'(t) = -a_1 P_1(t) + 2\lambda P_0(t), \tag{2}$$

$$P_2'(t) = -a_2 P_2(t) + \lambda_s P_0(t), \tag{3}$$

$$P_3'(t) = -\lambda P_3(t) + 2\lambda P_2(t) + \lambda_s P_1(t), \tag{4}$$

$$\frac{\partial P_i(x,t)}{\partial x} + \frac{\partial P_i(x,t)}{\partial t} = -\mu_i(x)P_i(x,t), \; i = 4,5. \tag{5}$$

The associated boundary and initial conditions are as follows:

$$P_4(0,t) = \lambda P_3(t), \tag{6}$$

$$P_5(0,t) = \lambda P_1(t), \tag{7}$$

$$P_0(0) = 1, \tag{8}$$

$$P_1(0) = P_2(0) = P_3(0) = P_4(x,0) = P_5(x,0) = 0. \tag{9}$$

3 STABILITY OF THE SYSTEM

In this section, in order to further study of the properties of the system, we will formulate the problem into a suitable Banach space and study the unique existence of its solution and explain the exponential stability of the system by analyzing the spectrum distribution of the system operator in detail.

Firstly, let the state space X be

$$X = \{P \in R^4 \times L^1(R^+) \times L^1(R^+) \, | \,$$

$$\| P \| = \sum_{j=0}^{3} |P_j| + \sum_{j=4}^{5} \| P_j(x) \|_{L(R^+)} < \infty\},$$

where $P = (P_0, P_1, P_2, P_3, P_4(x), P_5(x))$, it is obvious that

X is a Banach space.

Secondly, we will introduce some operators in X,

$$AP = \text{diag}\left(-a_0, -a_1, -a_2, -\lambda, -\frac{d}{dx} - \mu_4(x),\right.$$
$$\left. -\frac{d}{dx} - \mu_5(x)\right)P,$$

$$D(A) = \{P = (P_0, P_1, P_2, P_3, P_4(x), P_5(x))$$

$$\in X \left| \frac{d}{dx} P_j(x) \in L^1(R^+),\right.$$

and $P_j(x)(j = 4,5)$ are absolutely continuous functions and satisfying $P(0) = (P_0, \; P_1, \; P_2, \; P_3, \; \lambda P_3, \lambda P_1\}$,

$$EP = \begin{pmatrix} 0 & \mu_1 & \mu_2 & 0 & \int_0^\infty \mu_4(x)dx & \int_0^\infty \mu_5(x)dx \\ 2\lambda & 0 & 0 & 0 & 0 & 0 \\ \lambda_s & 0 & 0 & 0 & 0 & 0 \\ 0 & \lambda_s & 2\lambda & 0 & 0 & 0 \\ 0 & 0 & 0 & 0 & 0 & 0 \\ 0 & 0 & 0 & 0 & 0 & 0 \end{pmatrix}$$

$$\times \begin{pmatrix} P_0 \\ P_1 \\ P_2 \\ P_3 \\ P_4(x) \\ P_5(x) \end{pmatrix}, \; D(E) = X.$$

Then the above equations (1)–(9) can be formulated as an abstract Cauchy problem into the suitable Banach space X.

$$\begin{cases} dP(\cdot, t)/dt = (A + E)P(\cdot, t), t \geq 0 \\ P(\cdot, t) = (P_0(t), P_1(t), P_2(t), P_3(t), P_4(\cdot, t), P_5(\cdot, t))^T. \\ P(\cdot, 0) = P^0 = (1,0,0,0,0,0)^T \end{cases}$$
$$\tag{10}$$

Since the system (1)-(9) is rewritten as an abstract Cauchy problem, it is necessary to prove that the system (10) has a unique nonnegative solution by using C_0-semigroup theory. It presents the expression of the dynamic solution of the system equation.

For convenience, we will present a few of lemmas in the first place, which can be readily proved by (Cao & Cheng 2006).

Theorem 3.1 There exists constant $M > 0$ such that for any $t > 0$,

$$\int_0^\infty \exp[-\int_t^x \mu_j(\tau)d\tau]dx \leq M, \; j = 4,5. \tag{11}$$

Theorem 3.2 For any $\gamma \in \{\gamma \in \mathbb{C} \, | \, \text{Re}\,\gamma > 0, \text{ or } \gamma = ia, a \in R, a \neq 0\}$,

36

$|\int_0^\infty \mu_j(x)\exp[-\int_0^x (\gamma + \mu_j(\tau))d\tau]dx| < 1, j = 4,5.$ (12)

Theorem 3.3 The system operator $A + E$ is a dispersive operator (see Nagel 1986) with dense domain.

The proof of Theorem 3.3 can see (Geni etal. 2001, Zheng & Qiao 2009, Yuan & Xu 2011).

Theorem 3.4 $\{\gamma \in \mathbb{C} \mid \operatorname{Re}\gamma > 0$, or $\gamma = ia, a \in R, a \neq 0\} \subset \rho(A + E)$.

Proof. For any $G = (g_0, g_1, g_2, g_3, g_4(x), g_5(x)) \in X$, considering the equation $[\gamma I - (A + E)]P = G$, $\forall G \in X$,

namely,

$$(\gamma + a_0)P_0 - \mu_1 P_1 - \mu_2 P_2 - \sum_{j=4}^5 \int_0^\infty P_j(x)\mu_j(x)dx = g_0,$$
(13)

$(\gamma + a_1)P_1 - 2\lambda P_0 = g_1,$ (14)

$(\gamma + a_2)P_2 - \lambda_s P_0 = g_2,$ (15)

$(\gamma + \lambda)P_3 - 2\lambda P_2 - \lambda_s P_1 = g_3,$ (16)

$P_j'(x) + (\gamma + \mu_j(x))P_j(x) = g_j(x), j = 4,5.$ (17)

And we can assume

$P_4(0) = \lambda P_3,$ (18)

$P_5(0) = \lambda P_1.$ (19)

Solving the equation (17) with the help of (18)–(19) gets

$$P_j(x) = P_j(0)\exp[-\int_0^x (\gamma + \mu_j(\eta))d\eta]$$
$$+ \int_0^x \exp\left[-\int_\tau^x (\gamma + \mu_j(\eta))d\eta\right] g_j(\tau)d\tau, j = 4,5.$$
(20)

Noting that $g_j(x) \in L^1[0,\infty)$, $j = 4,5$, together with Theorem 3.1, we know $P_j(x) \in L^1[0,\infty)$, $j = 4,5$. This implies that $[\gamma I - (A + E)]$ is an onto mapping.

Secondly, we will prove that this operator is also an injective mapping. This is, the operator equation $[\gamma I - (A + E)]P = 0$ has a unique solution 0. Set $G = 0$ in the former discussion. Then we can gain the following matrix equation by combing (13)–(16) and (20),

$$\begin{pmatrix} \gamma + a_0 & -\mu_1 - \lambda w_5 & -\mu_2 & -\lambda w_4 \\ -2\lambda & \gamma + a_1 & 0 & 0 \\ -\lambda_s & 0 & \gamma + a_2 & 0 \\ 0 & -\lambda_s & -2\lambda & \gamma + \lambda \end{pmatrix} \begin{pmatrix} P_0 \\ P_1 \\ P_2 \\ P_3 \end{pmatrix} = \begin{pmatrix} 0 \\ 0 \\ 0 \\ 0 \end{pmatrix},$$
(21)

where $w_j = \int_0^\infty \mu_j(x)\exp\left[-\int_0^x (\gamma + \mu_j(\xi))d\xi\right]dx$, $j = 4,5.$

It is clear that

$|\gamma + a_0| = |\gamma + 2\lambda + \lambda_s| > |-2\lambda - \lambda_s| > 0,$

$|\gamma + a_1| = |\gamma + \lambda + \lambda_s + \mu_1| > |-\mu_1 - \lambda_s - \lambda w_5| > 0,$

$|\gamma + a_2| = |\gamma + 2\lambda + \mu_2| > |-\mu_2 - 2\lambda| > 0,$

$|\gamma + \lambda| > |-\lambda w_4| > 0,$ for $\gamma > 0$ or $\gamma = ia$, $a \in R/\{0\}$.

And from Theorem 3.2, we can obtain $|w_j| < 1$, $j = 4,5$. Thus the matrix of coefficients of the linear equation; (21) is a strictly diagonal-dominant matrix about column. Therefore, it is invertible, which manifests that operator $[\gamma I - (A + E)]$ is a one-to-one mapping.

Because $[\gamma I - (A + E)]$ is densely defined closed in X, we can derive that $[\gamma I - (A + E)]^{-1}$ exists and is bounded by recalling Inverse Operator Theorem and Closed Graph Theorem. That is, set $\{\gamma \in \mathbb{C} \mid \operatorname{Re}\gamma > 0$ or $\gamma = ia, a \in R/\{0\}\}$ belongs to the resolvent set of the system operator $A + E$. Thus we complete the proof of Theorem 3.4.

Theorem 3.5 The simple eigenvalue of the system operator $A + E$ is 0.

Proof. Firstly, we will prove that 0 is the eigenvalue of $A + E$ with positive eigenvector. Considering the equation $(A + E)P = 0$ and assuming that P satisfies the boundary conditions (18)–(19). Then repeating the proof process of the injective mapping in Theorem 3.4 with $\gamma = 0$, we can get the following solutions:

$$P_1 = \frac{2\lambda}{a_1}P_0,$$
(22)

$$P_2 = \frac{\lambda_s}{a_2}P_0,$$
(23)

$$P_3 = 2\lambda_s\left(\frac{1}{a_1} + \frac{1}{a_2}\right)P_0,$$
(24)

$$P_4(x) = 2\lambda\lambda_s\left(\frac{1}{a_1} + \frac{1}{a_2}\right)P_0 \cdot \exp\left[-\int_0^x \mu_4(\eta)d\eta\right],$$
(25)

$$P_5(x) = \frac{2\lambda^2}{a_1}P_0 \cdot \exp[-\int_0^x \mu_5(\eta)d\eta],$$
(26)

where P_0 is an arbitrary real number. Then it can be deduced that $P_j(x), j = 4,5, \forall x \in [0,+\infty)$ by taking $P_0 > 0$ without loss of generality. So the vector

$$P^* = (P_0^*, P_1^*, P_2^*, P_3^*, P_4^*(x), P_5^*(x)),$$
(27)

is the positive eigenvector corresponding to eigenvalue 0 of the system operator $A + E$ and it is also the positive steady-state solution of the system, here $P_j^*(j=0,1,2,3)$ and $P_j^*(x)(j=4,5)$ respectively signifies P_j and $P_j(x)$ showed in (22)-(26). In addition, it is easy to see that the geometric multiplicity of eigenvalue 0 in X is one.

Secondly, we will prove that the algebraic multiplicity of eigenvalue 0 is one. Taking vector $Q=(1,1,1,1,1,1)$, then $Q \in X^*$. For any $P \in D(A+E)$, it is easy to show that $<(A+E)P,Q>=0$ by noticing the boundary conditions. Therefore, we can deduce that $<P,(A+E)^*Q>=0, \forall P \in X$, for $D(A)$ is dense in X. This shows that Q is the eigenvector of $(A+E)^*$, the adjoint operator of $A+E$, corresponding to eigenvalue 0.

In the light of (Dundord & Schwartz 1958, Qiao et al. 2009), we only need to explain that the algebraic index of 0 in X is one. We use the reduction to absurdity. Suppose that the algebraic index of 0 is 3 without loss of generality. Thus there exists $U \in X$ such that $(A+E)U=P^*$. Therefore,

$$< P^*,Q > = <(A+E)U,Q> = <U,(A+E)^*Q>=0. \tag{28}$$

However,

$$< P^*, Q > = \sum_{j=0}^{3} P_j + \sum_{j=4}^{5} \int_0^\infty P_j(x)dx >0. \tag{29}$$

(28) contradicts (29). Thus the algebraic index of 0 in X is one. Then the algebraic multiplicity of 0 in X is one. The proof of Theorem 3.5 is complete.

Theorem 3.6 The system operator $A+E$ generates a positive contraction C_0-semigroup $T(t)$.

Proof. We can obtain the proof of Theorem 3.6 by Phillips Theorem (see Pazy 1986), Theorem 3.3 and Theorem 3.4.

Theorem 3.7 The system equation (10) has a unique $P(\cdot,t)$ nonnegative time-dependent solution which satisfies $\| P(\cdot,t) \|=1, \forall t \in [0,\infty)$.

Proof. In the light of Theorem 3.6 and (Pazy 1986), we have that the system (10) has a unique nonnegative solution $P(\cdot,t)$ and it can be expressed as $P(\cdot,t)=T(t)P^0, \forall t \in [0,\infty)$.

Because $P(\cdot,t)$ satisfies (1)–(9), it is easy to receive that $d/dt \| P(\cdot,t) \|=0$. Here $\| P(\cdot,t) \|=\| T(t)P^0 \|=1, \forall t \in [0,+\infty)$. It is just consistent with the physical intuition of the system.

Theorem 3.8 The time-dependent solution of the system (1)–(8) strongly converges to its steady state solution. That is $\lim_{t \to \infty} P(\cdot,t)=P^*$, where P^* is the eigenvector corresponding to 0 in X satisfying $\| P^* \|=1$.

Proof. In the light of Theorem 3.7, the nonnegative solution of the system (1)–(9) can be expressed

as $P(\cdot,t)=T(t)P^0$, $\forall t \in [0,+\infty)$. Thus combing Theorem 2.10 (see Nagel 1986) together with Theorem 12.3 in (Taylor & Lay 1980), we can deduce that

$$P(\cdot, t) = T(t)P^0 = < P^0,Q > P^* + R(t)P^0$$
$$= P^* + R(t)P^0, \tag{30}$$

where Q is the same as defined in Theorem 3.5, $R(t)=Ce^{-\varepsilon t}$ for suitable constants $\varepsilon > 0$ and $C > 0$. Hence we have $\lim_{t \to \infty} P(\cdot,t) = < P^0, Q > P^* = P^*$.

As a result, the exponential stability of the solution of the studied system was obtained.

Thus we explained that the system of interest is exponentially stable. Exponential stability is a very important property in reliability study. We can solve some problems readily and derive some better conclusions by using it. For example, by using the property, the governors can decide how to assign the repairman to do repair in his work time to increase the profit of the system benefit.

As far as such a problem is concerned, previous literatures such as (Liu et al. 2005) only pointed out that when the profit of the system benefit with repairman vacation in steady state is larger than that of the classical system benefit. But this is a less practical condition because they can not solve the following problems. Firstly, how long the system will go to the stability state. Secondly, whether the steady state indices such as steady state availability can substitute for the transient ones. Thirdly, what is the probability that the repairman can do repair.

However, by studying the exponential stability of the system, all these problems can be solved easily. Actually, for a given fault, the system can go to the steady state at a very fast speed and its steady state availability can substitute for the dynamic one by considering a safety factor. Moreover, $P_j(t)(j=0,1,2,3)$ means the probability that the system is operating normally after every repair and the repairman is on vacation at time $t \geq 0$ and $P_j(t) \to P_j > 0$, here $P_j(j=0,1,2,3)$ is the first four coordinate of the eigenvector P^* in Theorem 3.8. Then Theorem 3.8 indicates that after a certain time $t > 0$, the repairman can always be urged to do repair with a fixed probability to increase the total profit of the system benefit.

ACKNOWLEDGEMENTS

This work is supported by Natural Science Foundation of Heilongjiang Province, P.R. China (No. QC2010024) and Daqing Normal University Nature Science Foundation (No. 10ZR06).

REFERENCES

Altamuro, V.M. 1985. Working safely with the iron collar worker. In Working Safely with Industrial Robots, ed. P.P. Strubhar.

Cao, J.H. & Cheng, K. 2006. Introduction to Reliability Mathematics, Higher Education Press, Beijing.

Dhillon, B.S. 1983. Reliability Engineering in Systems Design and Operation, Van Nostrand Reinhold, New York.

Dundord, N. & Schwartz, J.T. 1958. Linear Operators, Part I. Wiley, New York.

Gaver, D.P. 1963. Time to failure and availability of paralleled systems with repair, IEEE Trans. Reliab., vol.12, pp. 30–38.

Geni, G. et al. 2001. Functional Analysis Method in Queueing Theory, Research Information Ltd, Hertfordshire.

Lauck, K.E. 1986. New standards for industrial robot safety, CIM Rev., pp. 60–68.

Liu, R.B. et al. 2005. A N-unit series repairable system with a repairman doing other work. Journal of Natural Science of Heilongjiang University, Journal of Natrual Science of Heilongjiang University, 22(4), pp. 493–496.

Nagel, R. 1986. One-parameter Semigroup of Positive Operator, Lecture Notes in Mathematics, New York: Springer Verlag.

Nicolaisen, P. 1987. Safety problems related to robots, Robotics, vol.3, pp. 205–211.

Pazy, A. 1986. Semigroup of linear operators and application to partial differential equations, Springer-Verlag, New York.

Qiao, X. et al. 2009. Exponential asymptotic stability of repairable system with randomly selected repairman, Chinese Control and Decision Conference, IEEE Press, pp. 4580-4584, doi:10.1109/CCDC.2009.5191527.

Taylor, A.E. & Lay, D.C. 1980. Introduction to functional analysis, John Wiley & Sons, New York Chichester Bresbane Toronto.

Yuan, W. & Xu, G. 2011. Spectral analysis of a two unit deteriorating standby system with repair, WSEAS Transactions on Mathematics, vol. 10, no. 4, pp. 125–138.

Zheng, F. & Qiao, X. 2009. Exponential stability of a repairable system with N failure modes and one standby unit, International Conference on Intelligent Human-Machine Systems and Cybernetics, IEEE Press, pp. 146–149, doi:10,1109/IHMSC.2009.160.

Control Engineering and Information Systems – Liu (Ed)
© 2015 Taylor & Francis Group, London, ISBN 978-1-138-02685-8

Application of stepping motor in cutting fixed-scale sheets

W.N. Han & L.W. Liu
North China Institute of Aerospace Engineering, Hebei Province, China

P. Yu
Dalian Ocean University, Dalian, China

ABSTRACT: Based on the PLC of SIEMENS S7-200 series, this article focuses on the application of the stepper motor in the system of the cutting sheet metal's controlling in the flame cutting machine. The stepper motor is driven and controlled by PLC and high power transistor. The sheet metal precision position is achieved and the fixed length cutting is finished by the flexible application of PLC instruction system. The control system has the advantages of simple structure, high reliability, so it has good versatility and utility value.

1 INTRODUCTION

Stepping motor is a kind of common open-loop control electric actuator that can turn electric pulse signal into mechanical angular displacement or linear displacement. Under non-overloaded conditions, the rotating speed and stopping position of motor only depend on the frequency of pulse signal and number of pulses, and meantime, it can transfer stably from one motion state into the other in the operating frequency so that realize high-precision position control without accumulated error. It is one of key products of mechatronics and widely used in varieties of automation equipment (Miao 2007).

Programmable logic controller, or PLC for short, is one of three pillars of modern control, which was first developed in the 1960's (Zhou 2009). With the continuous development of PLC, its functions are greatly extended and improved and more widely used for the control of complex and special systems. SIEMENS SIMATIC S7-200 PLC is well applied with simple interface, convenient configuration and strong realtime capability (Guo 2007, Long & Nie 2002), and it is equipped with the built-in High-speed pulse generator with related control instructions (Xu et al. 2010), which can well control the stepping motor.

Plates cutting machines have been widely applied with the development of modern machinery industry. According to the used resource, cutting machines can be divided into various types, of which flame cutting machines have advantages of cutting medium or large steel plates and welding grooves. Taking sheet metal flame cutting (Liu 2011) as an example and based on SIEMENS S7-200 PLC, the thesis made a study on the control of stepping motor, which realized the fixed-scale cutting of sheets (Kong et al. 2007), increased the antijamming capability of the system and the cutting precision, and greatly enhanced the level of automation of the cutting machines.

2 ANALYSIS OF CONTROL FUNCTION

The function of sheet cutting control system is mainly to accomplish the movement and cutting of sheets, including the tractor, clamping device, feed system, cutting gun, etc. The system structural representation is as shown in Figure 1.

Tractor is cart, comprised of front and back two groups of parallel rollers, moving with sheets driven by the stepping motor and controlling by PLC high-speed pulse output and direction output,

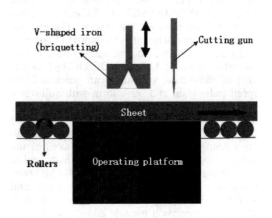

Figure 1. System structural representation.

thus ensuring both the operating speed and the setting length, which realizes precise control and sheet transport function. The clamping device is like that V-shaped iron moves driven by the cylinder which is controlled by the monocoil two-position solenoid valve under the control of PLC. Feed system can realized the movement up and down, left and right of cutting gun, which are controlled through two double-coil two-position solenoid valves driving pneumatic actuators under the control of PLC. The gas valve of cutting gun is also turned on and off under the control of PLC. The work procedure of this system includes the following actions.

1. At first, the cart moves with the work piece beneath the cutting gun for the test through limit switch.
2. As the cart arrives in place, start the clamping device, and after the time delay, the clamping ends.
3. After clamping ends, drop the cutting gun in place.
4. After the cutting gun is dropped in place, switch on the oxygen valve and acetylene valve, and light up the flame.
5. When the cutting gun is preheated for a certain time, switch on the cutting gun and turn right of the solenoid valve, and thus start the cutting process.
6. When the cutting is finished, switch off the acetylene valve and oxygen valve successively, turn the cutting gun left to the original position.
7. Loosen the clamping device, and the cart moves for a fixed length driven by the stepping motor.
8. Restart the clamping device and go on the next cutting process.

3 DESIGN OF SYSTEM HARDWARE

According to the functional requirement of the control system and taking into account the economy and reliability, the SIEMENS S7-200 PLC can be selected to control the sheet cutting system. The external hardware of control system is as shown in Figure 2.

There are totally 22 digital value inputs and 10 outputs. Choosing CPU 224DC/DC/DC as the host of the control system, it can generate high-speed pulse train and wave form with adjustable pulse width at Q0.0 and Q0.1, with a frequency up to 20 kHz. Use S7-200 PLC high-speed pulse output function to output pulse signals to the driving circuit composed of GTR, and control the stepping motor through the driver so as to realize positioning control. It adopts YKA2404MC two-phase hybrid stepping motor subdivided driver and 42BYG250B stepping motor with a step angle of 1. 8°. System expansion module adopt EM223, one type of input/output hybrid module, and use 24 V

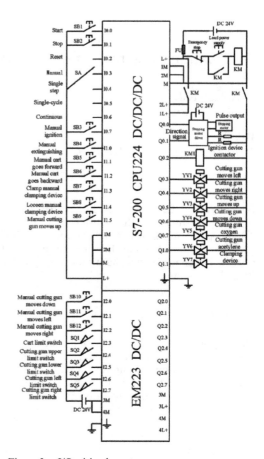

Figure 2. I/O wiring layout.

digital value combination with 8 inputs/8 outputs, which can both satisfy the requirements of system control and I/O points.

In order to satisfy the requirement of production, the control system is divided into five work modes, i.e. manual, stepping, single-cycle, continuous and reset operation. Manual operation is just to complete the manual control of each action, with no need for making actions as per the working procedure. Input circuit has the button, travel switch and hand switch, switching on and off reliably, input circuit is equipped with S7-200 24 V self-contained power that can meet the requirement of input signals. Output driver includes solenoid valve coil, contactor coil and stepping motor driver, and uses DC 24 V additional power supplies.

4 PLC CONTROL OF STEPPING MOTOR

As there are a number of work modes of the control system, each part of procedure can be made into relatively independent subprogram modules respectively, and execute function selection through call instructions. Under the continuous work mode, the work procedure of the control system is shown as in Figure 3.

When using the high-speed pulse output function of S7-200 PLC, the output terminal Q0.0 and Q0.1 are subject to the control of PTO/PWM generator. When the setting of corresponding bit of PTL control register (Liao 2007), it can be adjusted through executing pulse (PLS) output instruction. PLS instruction is that S7-200 PLC reads the corresponding bit of the special register and makes actions for the corresponding PTO generator. The thesis studies on using Q0.0 to generate high-speed pulse train, using Q0.1 to control the rotation direction of stepping motor, and applying multi-step Pulse Train Output (PTO) function which provides square wave output (Xie 2010). Users control the cycle period and pulse number.

Before each call of cart stepping subprogram SBR_0, the pulse output end Q0.0 and the direction control end Q0.1 shall be reset to zero. When the direction control end Q0.1 shows at state "1", the stepping motor moves forwards, which means that the cart goes forwards. When it shows at state "0", the stepping motor moves to the opposite direction, which means that the cart goes backwards.

The cart going backwards is only applicable for manual procedure. Cart stepping subprogram is shown as Figure 4.

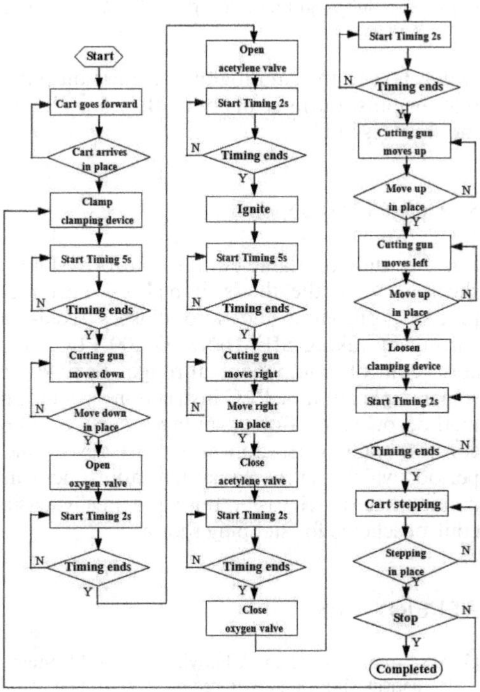

During multi-step PTO operations, executing one PLS instruction can realize multi-step pulse output, and CPU can automatically read the characters of each pulse train from the envelop table of V storage zone. When using multi-step operation mode, the special register SM67.5 must be set at state "1" and loaded into the initial address of envelop table in the V storage zone (SMW168 or SMW178). The time base can be made in microsecond or millisecond, but all cycle values in the envelop table must use the same base that must remain unchanged in the execution of envelop. When corresponding parameters are set, PLS instructions can be used to start multi-step PTO operation.

The operation of stepping motor should have four stages: speeding up startup, uniform motion, retarded motion and low-speed stop. The stepping motor shall start at low frequency and then gradually increase the frequency, so that it can reach a state of high-speed operation. Likewise, it needs decrease the pulse frequency so as to stop normally. Therefore, it needs multi-step output of pulse train. In the thesis, the stepping motor driver adopts 10 subdivisions, so the pulse number as per round that the motor rotates is 2000 pulses. If the roller driven by the stepping motor has a perimeter of 300 mm, the distance that the cart moves for one pulse is 0.15 mm. If the sheet cutting length

Figure 3. Work flow chart of cutting system.

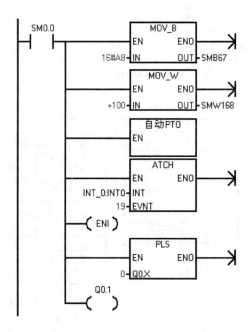

Figure 4. Carts tepping subprogram (SBR_0).

is 3 m, the number of pulses required for motor stepping each time is 20000 pulses. In the design of multi-step PTO process, the 20000 pulses shall be distributed into the four operation stages of stepping motor, so that the stepping motor can meet the work demand (Li & Li 2009).

The subprogram "auto PTO" (SBR_1) nested in the subprogram of cart going forward is the subprogram of envelope table for cart stepping, as shown in Figure 5. The number of segments of envelope is stored at the initial address of envelope table. As the operation of stepping motor has four stages, envelope segment is set to be 4 and stored in VB100 address. Under the initial address, every 8 bytes are set with certain attribute, and composed of 16 cycle values, 16 cycle increment values and 32 pulse count values. In programming, it shall set the pulse cycle value, cycle increment value and pulse number of four stages respectively.

After the pulse output terminal Q0.0 outputs pulses, it enters into the interrupt program and PLS0 pulse number is completed. During the interrupt program, set the state of M0.7 as "1", so that it can meet the condition of next step which becomes active step. And more, send the control word "16#0" to SMB67 in the interrupt program, which stops the output of high-speed pulse train. The interrupt program is shown as Figure 6.

The subprogram of cart going forward is shown as Figure 7. The pulse cycle in the subprogram of cart going forward shall not exceed the minimum start frequency of the stepping motor, or it neither starts normally nor ensures precise positioning. Moreover, the number of pulses shall be sufficient so as to ensure that the cart can move to the position where the cutting machine is accurately. In order to ensure sheets in right place, connect in series the normally-closed contacts of cart travel

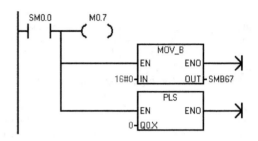

Figure 6. Interrupt subprogram (INT_0).

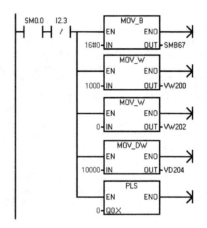

Figure 7. Subprogram of cart going forward (SBR_2).

switch I2.3 to the subprogram of pulse output, so that the pulse output can be cut off promptly when the sheets is in place.

5 CONCLUSION

For the study on flame cutting control system for sheet metal in the thesis, it realizes sheet metal precise positioning and accomplishes fixed-scale cutting by taking SIEMENS S7-200 PLC as the core control component through flexible and smart application of PLC instruction system. This method owns a lot of advantages, such as simple control, stable operation and short development period, which makes the system adjustment and maintenance much easier. It is a practically feasible control scheme for stepping motor.

REFERENCES

Guo Yanping and Zhang Chaoying, Control System of Industrial Manipulator Based on PLC, 9th ed., vol. 1. China: Instrument Technique and Sensor, 2007, pp. 31–33.

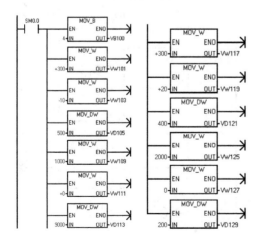

Figure 5. Subprogram of cart stepping envelope (SBR_1).

Kong Qinghua and Liu Guihang, Application of PLC in Cutting Machine's Control System, 1st ed., vol. 1. China: Shanghai Electric Appliance Technology, 2007, pp. 46–48.

Liao Changchu, PLC Programming and Applications of S7-200. Beijing, China: Mechanical Industry Press, 2007.

Li Ming and Li Quanqing, Design of the Hydraulic Conveying Manipulator Based on PLC, 37th ed., vol. 1. China: Machine Tool & Hydraulics, 2009, pp. 99–101.

Liu Fangxia, Application of PLC in the Automatic Cutting Machine, 30th ed., vol. 1. China: Techniques of Automation and Applications, 2011, pp. 61–64.

Long Wei and Nie Guanhong, The Realization of Real-time Communication Between S7-200 PLC and PC, 24th ed., vol. 5. China: Journal of Nanchang University (Engineering & Tchnology), 2002, pp. 81–83.

Miao jianxin, Application and Optimization of Flame Cutting Machine in Heavy Plate of Baogang, 11th ed., vol. 1. China: Electric Age, 2007, pp. 2–3.

Xie Liping, PLC Quick Start and Practice of Siemens S7-200 series. Beijing, China: Posts & Telecom Press, 2010.

Xu Zhi, Du Yiming, Xiong Tianzhong and Sun Chengzhi, Design and Application based on Siemens S7-200 PLC Control of Stepping Motor, 5th ed., vol. 1. China: Engineering Technology, 2010, pp. 113–114.

Zhou Xiwei, Key Techniques on Industrial Control System Based on PLC, 16th ed., vol. 5. China: Computer Knowledge and Technology, 2009, pp. 4340–4341.

Control Engineering and Information Systems – Liu (Ed)
© 2015 Taylor & Francis Group, London, ISBN 978-1-138-02685-8

Generalized Model Predictive Control

Y.C. Zhao
Department of Electronics, College of Engineering, University of Buenos Aires, Buenos Aires, Argentina

ABSTRACT: The Generalized Model Predictive Control (GMPC) based on ARMAX (auto-regressive moving average with exogenous signal) model is developed in present paper. The process model is used to predict the future process output. The difference between the predicted output and the future reference is minimized by using a sequence of future control signals calculated in such a way that it optimizes a multistage objective function. The obtained control law contains implicitly the integral action. GMPC controller is very robust when faced with uncertainties of process delays. And this, consequently, gives the perspective of its application at the case of poor estimation of delay times.

1 INTRODUCTION

1.1 Model predictive control

The source of basic concepts of Model Predictive Control (MPC) began around 1960. The Smith predictor (Smith 1957), Linear Quadratic Gaussian (LQG) algorithm (Kalman et al. 1960a,b) and receding horizon concept (Propoi 1963) all laid the foundation for the development of MPC.

Consequently, MPC emerged in the end of 1970s. Model Predictive Heuristic Control (MPHC) (Richarlet et al. 1976) and Dynamic Matrix Control (DMC) (Cutler & Ramaker 1980) were the first industrial implementations. Both methods use a quadratic objective function and the convolution model.

Many transfer function model based predictive controls arose from another field of research related to adaptive control. Two examples are Extended Horizon Adaptive Control (EHAC) (Ydstie 1984) and Extended Prediction Self Adaptive Control (EPSAC) (De Keyser & Van Guawenberghe 1985). One of the most popular methods, Generalized Predictive Control (GPC) (Clarke et al. 1987a,b) is also based on the transfer function model, in which the mathematic expectation of a quadratic objective function is the criterion to be optimized.

MPC formulated in the state-space context has also developed. Some of them are A State-space Description for GPC Controllers (Ordys & Clarke 1993), and Model Predictive Control with Linear Models (Muske & Rawlings 1993).

MPC based on other models such as neural or fuzzy model were also presented.

MPC presents many advantages such as the ability to control a process with long delay times, the feasibility to handle easily the multivariable case and the relative simplicity to deal with constrained control.

1.2 Goals of present paper

The above-mentioned GPC has become one of the most popular MPC in both industry and academia since it was proposed in 1987. It has been successfully implemented in many industrial applications (Clarke 1988), (Peng 2006), (Nishizaki et al. 2011). However, a solid delay estimation is fundamental to GPC, because errors of more than one unit the system can become unstable if the process delay is high (Camacho & Bordons 2007). Therefore, is it possible to make a controlled system less sensitive to those estimation errors of delays?

MPC as a whole and GPC as an individual calculate the increment of the control signal and obtain the integral action, which is required in a great deal of process control, through an explicit way:

$$u(k) = u(k-1) + \Delta u(k).$$

Is it possible that the optimization yields directly the control signal in which the integral action is incorporated implicitly?

2 PROCESS MODEL

The process model plays a decisive role in MPC. The model must be able to capture the process dynamics and to predict precisely the future process output, moreover it must be simple to analyze and implement.

2.1 Transfer function model

Assume the process dynamics are characterized by the linearized model

$$x(k) = \frac{B(z^{-1})}{A(z^{-1})}[u(k-d) + v(k-d)]$$

$$y(k) = x(k) + \frac{C(z^{-1})}{D(z^{-1})}e(k) \tag{1}$$

where $A(z^{-1})$, $B(z^{-1})$, $C(z^{-1})$ and $D(z^{-1})$ are polynomials in the backward shift operator without any common factors, among which $A(z^{-1})$ and $D(z^{-1})$ are monic, u is the control signal, d is the process delay, v is the disturbance acting on the process input, y is the output and $\{e(k)\}$ is a sequence of independent random variables (white noise) with zero mean and a certain value of standard deviation. Model (1) is called ARMAX (auto-regressive moving average with exogenous signal) model.

However, when taken into account neither the disturbance nor the noise, the model (1) becomes

$$y(k) = \frac{B(z^{-1})}{A(z^{-1})}u(k-d). \tag{2}$$

It will be the start point to derivate predictors in the next section.

3 OUTPUT PREDICTION

The prediction of the future output is essential to calculating the future control signal. The more precise the prediction is, the more efficient the control will be. One of the basic ideas of MPC is rewriting the model of the process in order to obtain an explicit expression for the output at a future instant. The prediction is usually separated into two parts: one depends on past data, another on assumed future control signals.

3.1 Predictor

It follows from (2) that

$$y(k+d) = \frac{B(z^{-1})}{A(z^{-1})}u(k). \tag{3}$$

From now on, the argument z^{-1} is dropped for simplifying the writing. The polynomial A is so factorized $A = A_0 A_2$ that A_0 contains only stable and well damped mode(s) or equals 1. Since polynomial A is monic, A_0 and A_2 are monic too, therefore A_2 can be expressed as $A_2 = 1 - z^{-1}A_1$. Thus (3) can be rewritten as

$$y(k+d) = \frac{B}{A_0}u(k) + A_1 y(k+d-1). \tag{4}$$

And $h-1$ step ahead

$$y(k+d+h-1) = \frac{B}{A_0}u(k+h-1) + A_1 y(k+d+h-2). \tag{5}$$

Nevertheless, as mentioned in Section 2, the prediction (5) does not take either disturbance or noise into account as a result of using the model (2). It is remedied by adding a correction term

$$\varepsilon(k+d+h-1) = y(k) - \frac{B}{A_0}u(k-d) - A_1 y(k-1) \quad \forall h$$

which is from (4) and supposed to be probably different from zero. The amount of the correction term is related to the difference between measured output and that according to the model (2). The last equation tells that a constant error is assumed over the prediction horizon. Its value, according to the model (1) is

$$\varepsilon(k+h) = \frac{B}{A_0}v(k-d) + \frac{A_2 C}{D}e(k) \quad \forall h$$

Introduce the identity

$$\frac{A_1^i B}{A_0} = F_h^i + z^{-(h-d+1)}\frac{G_h^i}{A_0} \tag{6}$$

where $h = 1, 2, ..., N_2$ and $i = 0, 1, ..., N_2 - 1$, moreover F_h and G_h are the quotient and remainder of the division expressed in the left hand side of (6), respectively. Notice that the superscript i of the terms F_h and G_h is no more than a notation. The numerator of the left side of (6) is a polynomial of degree $in_1 + n - 1$, where n_1 and n are the degrees of the polynomials A_1 and A respectively.

In digital control, normally, the output is measured first, it is used then to calculate the control signal; so the first output that is influenced by $u(k)$ is $y(k+d)$. This leads to $F_i = 0$, $i < d$. Iterating Equation (5) with the correction term for all values of h and using the identity (6) yield the following.

$$y(k+d) = F_d^0 u(k)$$

$$+ \frac{1}{A_0}\left[G_d^0 + B\sum_{i=0}^{d-1}(A_1^{i+1}z^{1-i} - z^{1-d}A_1^i)\right]u(k-1)$$

$$+ \left[1 + (1-z^{-1})\sum_{i=1}^{d-1}A_1^i\right]y(k)$$

$$y(k+d+1) = -F_d^1 u(k) + F_{d+1}^0 u(k+1)$$

$$+ \frac{1}{A_0}\left(G_d^1 + G_{d+1}^0 + B\sum_{i=2}^{d} A_1^i z^{2-i} - z^{1-d} B\sum_{i=0}^{d} A_1^i \right) u(k-1)$$

$$+ \left[1 + (1-z^{-1})\sum_{i=1}^{d} A_1^i \right] y(k)$$

$$\vdots$$

$$y(k+d+N-1) = \sum_{i=1}^{N} F_{d+i-1}^{N-i} u(k+i-1)$$

$$+ \frac{1}{A_0}\left(\begin{array}{c} \sum_{i=1}^{N} G_{d+i-1}^{N-i} + B\sum_{i=N}^{N+d-2} A_1^i z^{N-i} \\ -z^{1-d} B\sum_{i=0}^{N+d-2} A_1^i \end{array} \right) u(k-1)$$

$$+ \left[1 + (1-z^{-1})\sum_{i=1}^{N+d-1} A_1^i \right] y(k) \tag{7}$$

where $N = N_2 - N_1 + 1$. The prediction can be expressed in condensed form as

$$\mathbf{y} = \mathbf{F}u + \mathbf{H}u(k-1) + \mathbf{M}y(k) \tag{8}$$

where

$$\mathbf{y} = [y(k+d) \quad y(k+d+1) \quad \cdots \quad y(k+d+N-1)]^T$$
$$\mathbf{u} = [u(k) \quad u(k+1) \quad \cdots \quad u(k+N-1)]^T$$

$$\mathbf{F} = \begin{bmatrix} F_1^0 & 0 & \cdots & 0 \\ F_1^1 & F_2^0 & \cdots & 0 \\ \vdots & \vdots & \ddots & \vdots \\ F_1^N & F_2^{N-1} & \cdots & F_N^0 \end{bmatrix}$$

$$\mathbf{H} = \frac{1}{A_0}\begin{bmatrix} G_d^0 + B\sum_{i=1}^{d-1} A_1^{i+1} z^{1-i} - z^{1-d} B\sum_{i=0}^{d-1} A_1^i \\ G_d^1 + G_{d+1}^0 + B\sum_{i=2}^{d} A_1^i z^{2-i} - z^{1-d} B\sum_{i=0}^{d} A_1^i \\ \vdots \\ \sum_{i=1}^{N} G_{d+i-1}^{N-i} + B\sum_{i=N}^{N+d-2} A_1^i z^{N-i} - z^{1-d} B\sum_{i=0}^{N+d-2} A_1^i \end{bmatrix}$$

$$\mathbf{M} = \begin{bmatrix} 1 + (1-z^{-1})\sum_{i=1}^{d-1} A_1^i \\ 1 + (1-z^{-1})\sum_{i=1}^{d} A_1^i \\ \vdots \\ 1 + (1-z^{-1})\sum_{i=1}^{N+d-1} A_1^i \end{bmatrix}$$

Equation (8) can be expressed as

$$\mathbf{y} = \mathbf{F}u + \tilde{\mathbf{y}} \tag{9}$$

where $\tilde{\mathbf{y}} = \mathbf{H}u(k-1) + \mathbf{M}y(k)$ which is depends on the past information.

The output prediction can of course be expressed in the increment of control signal instead of control signal as in (9). Applying the difference operator $\Delta = 1 - z^{-1}$ on both sides of (2) yields

$$\Delta y(k) = \frac{B}{A}\Delta u(k-d)$$

Similarly, the following predictor yields

$$\mathbf{y} = \mathbf{F}\Delta \mathbf{u} + \mathbf{H}\Delta u(k-1) + \mathbf{M}y(k) \tag{10}$$

It can be expressed in a more condensed form

$$\mathbf{y} = \mathbf{F}\Delta \mathbf{u} + \tilde{\mathbf{y}} \tag{11}$$

where $\tilde{\mathbf{y}} = \mathbf{H}\Delta u(k-1) + \mathbf{M}y(k)$.

3.2 *Comment*

The predictor is the essential part of the controller and the behavior of a controller is closely related to the predictor.

No matter how the output prediction is calculated, with control signal or the increment of control signal, there is only one predictor for a process without any stable and well damped mode, and there are at least two candidates for predictor if the process to be controlled has at least one stable and well damped mode. The latter means there will be more resources to rely upon when designing a specified controller.

4 OBJECTIVE FUNCTION

Different objective functions have been proposed for different MPC algorithms, but the aim is basically the same. The distance between the future output and the reference and the control efforts are penalized. A general expression of such an objective function is

$$J = \sum_{i=N_1}^{N_2} \lambda(i)[y(k+i) - r(k+i)]^2$$
$$+ \sum_{j=1}^{N_u} \rho(j)[\Delta u(k+j-1)]^2 \tag{12}$$

where N_1 and N_2 are the minimum and maximum prediction horizons respectively, and N_u is the control horizon. $\lambda(i)$ and $\rho(j)$ are coefficients whose elections depend on pretension to future behavior of the controlled system, Δu is the increment of control signal and the signal r is the reference or set point.

Redefining the objective function is required in order to obtain a controller with double integral action. I introduce

$$J = \sum_{i=N_1}^{N_2} \lambda(i)[y(k+i) - r(k+i)]^2$$
$$+ \sum_{j=1}^{N_u} \rho(j)[\Delta^2 u(k+j-1)]^2 \qquad (13)$$

where $\Delta^2 = 1 - 2z^{-1} + z^{-2}$. Notice that the objective function (13) penalizes the accelerations of the effort as well as the discrepancy between the predicted output and the future reference.

In this paper $\rho(j)$ = constant for all j, and $\lambda(i) = 1$ for all i, are adopted for the sake of simplicity unless otherwise indicated.

5 CONTROL LAW

The aim of the MPC is that the future output on the considered horizon should follow a determined reference signal, at the same time the control effort is taken into account, thus the objective function should be minimized. The product of minimization is the control law.

5.1 *Optimization with control signal*

The objective function (12) is optimized and the following are introduced.

$$\mathbf{r} = [r(k+d) \quad r(k+d+1) \quad \cdots \quad r(k+N+d-1)]^T$$
$$\mathbf{u}(k) = [u(k|k) \quad u(k+1|k) \quad \cdots \quad u(k+N-1|k)]^T$$
$$\mathbf{u}(k-1) = [u(k-1|k-1) \quad \cdots \quad u(k+N-2|k-1)]^T$$

where $u(k + i | k)$ denotes $u(k + i)$ calculated at instant k, so

$$\Delta \mathbf{u} = [\mathbf{u}(k) - \mathbf{u}(k-1)]^T.$$

The problem of optimization is solved by the following theorem.

THEOREM 1 Optimal Control Law 1

Consider the process (1). Use the predictor (9) and criterion (12). Then the optimal sequence of control signals $\mathbf{u}$ is given by

$$\mathbf{u}(k) = (\mathbf{F}^T \mathbf{F} + \rho \mathbf{I})^{-1}[\mathbf{F}^T(\mathbf{r} - \tilde{\mathbf{y}}) + \rho \mathbf{u}(k-1)]. \qquad (14)$$

Proof: Assume $N_1 = d$ without any loss of generality. The cost function (12) can be written as

$$J(d, N, N_u) = (\mathbf{y} - \mathbf{r})^T(\mathbf{y} - \mathbf{r}) + \rho \Delta \mathbf{u}^T \Delta \mathbf{u}$$
$$= [\mathbf{F}\mathbf{u}(k) + \tilde{\mathbf{y}} - \mathbf{r}]^T[\mathbf{F}\mathbf{u}(k) + \tilde{\mathbf{y}} - \mathbf{r}]$$
$$+ \rho[\mathbf{u}(k) - \mathbf{u}(k-1)]^T[\mathbf{u}(k) - \mathbf{u}(k-1)]$$
$$= (\mathbf{r} - \tilde{\mathbf{y}})^T(\mathbf{r} - \tilde{\mathbf{y}}) + \mathbf{u}^T(\mathbf{F}^T\mathbf{F} + \rho \mathbf{I})\mathbf{u}^T - \mathbf{u}^T\mathbf{F}^T\mathbf{r}$$
$$+ \mathbf{u}^T\mathbf{F}^T\tilde{\mathbf{y}} - \rho \mathbf{u}^T \mathbf{u}_{k-1} - \mathbf{r}^T\mathbf{F}\mathbf{u}$$
$$+ \tilde{\mathbf{y}}\mathbf{F}\mathbf{u} - \rho \mathbf{u}_{k-1}^T\mathbf{u} + \rho \mathbf{u}_{k-1}^T\mathbf{u}_{k-1}$$

where $\mathbf{u}$ and $\mathbf{u}(k)$ are equivalence, so do $\mathbf{u}_{k-1}$ and $\mathbf{u}(k-1)$. Since $\mathbf{F}^T\mathbf{F} + \rho \mathbf{I}$ is always nonnegative definite, the function J has a minimum. It is, moreover, quadratic in $\mathbf{u}$. One way to find the minimum is to complete the square. Another way to minimize the cost function is to determine the gradient of (15) with respect to $\mathbf{u}(k)$.

$$\frac{\partial J}{2\partial \mathbf{u}} = (\mathbf{F}^T\mathbf{F} + \rho \mathbf{I})\mathbf{u} - \mathbf{F}^T(\mathbf{r} - \tilde{\mathbf{y}}) - \rho \mathbf{u}_{k-1}$$

Making the gradient to zero yields

$$(\mathbf{F}^T\mathbf{F} + \rho \mathbf{I})\mathbf{u}(k) = \mathbf{F}^T(\mathbf{r} - \tilde{\mathbf{y}}) + \rho \mathbf{u}(k-1).$$

The sequence of control signals is, if the matrix $\mathbf{F}^T\mathbf{F} + \rho \mathbf{I}$ is invertible,

$$\mathbf{u}(k) = (\mathbf{F}^T\mathbf{F} + \rho \mathbf{I})^{-1}[\mathbf{F}^T(\mathbf{r} - \tilde{\mathbf{y}}) + \rho \mathbf{u}(k-1)]$$

and the theorem is proven. $\square$

The first element $u(k)$ of vector $\mathbf{u}(k)$ is applied to the process in question. The control law is calculated again when a new measurement is obtained at the next sampling instant. The receding-horizon control concept is thus used.

The control law (14) incorporates implicitly the integral action. Refer to (Zhao & Zanini 2012) for a proof. A control law with integral action makes a controlled system able to track step set point or to reject load disturbances. Controllers with integral action are required in many process controls. They are more appropriate for many industrial applications in which disturbances are non-stationary.

The result of the last theorem shows one of the goals proposed in Section 1 is obtained: the optimization yields directly the control signal in which the integral action is implicitly incorporated.

A comparison simulation with an integrating plant

$$y(k) = \frac{0.0094 + 0.0088z^{-1}}{(1 - z^{-1})(1 - 0.8187z^{-1})} u(k - 10)$$

shows that the plant controlled by GMPC method (the control law (14) with $A_0 = 1 - 0.8187$ and $A_1 = -1$, and parameters: $N = 9$, $N_u = 1$, $\rho = 6$ and $\lambda(i) = 30$ for all i) can tolerate the delay mismatch ± 1 unit of time and that controlled by GPC method (with parameters: $N = 9$, $N_u = 1$, $\rho = 10$ and $\lambda(i) = 50$ for all i) can tolerate 0 unit of time. Notice that the two controlled system have the same response to step reference signal under those just mentioned parameters. So another goal proposed in Section 1 is also attained: making a controlled system less sensitive to delay mismatch.

5.2 Optimization with increment of control signal

The objective function (13) is optimized and the following are introduced.

$$\mathbf{r} = [r(k+d)\ r(k+d+1)\ \cdots\ r(k+d+N-1)]^T$$
$$\Delta\mathbf{u}(k) = [\Delta u(k\,|\,k)\ \Delta u(k+1\,|\,k)\ \cdots\ \Delta u(k+N\,|\,k)]^T$$
$$\Delta\mathbf{u}(k-1) = [\Delta u(k-1\,|\,k-1)\ \cdots\ \Delta u(k+N-2\,|\,k-1)]^T$$

where $\Delta u(k+i\,|\,k)$ denotes $\Delta u(k+i)$ calculated at the instant k, hence

$$\Delta^2\mathbf{u} = [\Delta\mathbf{u}(k) - \Delta\mathbf{u}(k-1)]^T.$$

Now another optimal control law is presented by the following theorem.

THEOREM 2 Optimal Control Law 2

Consider the process (1). Use the predictor (11) and the performance criterion (13). The optimal sequence of increments of control signals $\Delta\mathbf{u}$ is such a one that satisfies

$$(\mathbf{F}^T\mathbf{F} + \rho\mathbf{I})\Delta\mathbf{u}(k) = [\mathbf{F}^T(\mathbf{r} - \tilde{\mathbf{y}}) + \rho\Delta\mathbf{u}(k-1)].$$

If the matrix $\mathbf{F}^T\mathbf{F} + \rho\mathbf{I}$ is nonsingular, the sequence of increments of control signals is unique and given by

$$\Delta\mathbf{u}(k) = (\mathbf{F}^T\mathbf{F} + \rho\mathbf{I})^{-1}[\mathbf{F}^T(\mathbf{r} - \tilde{\mathbf{y}}) + \rho\Delta\mathbf{u}(k-1)]$$

The optimal control is

$$\Delta u(k) = \mathbf{L}[\mathbf{F}^T(\mathbf{r} - \tilde{\mathbf{y}}) + \rho\Delta\mathbf{u}(k-1)]$$
$$u(k) = u(k-1) + \Delta u(k) \qquad (16)$$

where $\mathbf{L}$ is the first row of the matrix $(\mathbf{F}^T\mathbf{F} + \rho\mathbf{I})^T$.

Proof: The proof is based on a similar way of the proof of Theorem 1 and taking $u(k) = u(k+1) + \Delta u(k)$ into account. $\square$

The control law (16) incorporates double integral action. A less rigorous proof can be found in (Zhao & Zanini 2012). A control law with double

integral actions is required for a controlled system to track ramp set point. Controllers with double integrator are appropriate to control many industrial processes such as pharmaceutical and food where piecewise references with ramps are frequently encountered. It is also of interest that the position or velocity follows evolutions of that type in the control of motors and in robotics applications. In general it would be desirable for the process output to follow a mixed trajectory composed of steps and ramps.

5.3 Comment

If the double integral action is not necessary for a given control problem, the criterion (12) can be used instead of (13) to obtain (16).

The optimal control laws (14) and (16) were obtained analytically without considering constraints on either manipulated or controlled variables. If that is not the case, the optimization problem is solved with Quadratic Programming (QP) techniques such as feasible direction methods, active set methods and initial feasible point methods.

6 SUMMARY AND CONCLUSION

6.1 Summary

An ARMAX model based generalized predictive control (GMPC) has been developed in present paper. New predictors, new controllers, and a new criterion of performance are presented. GMPC consists three essential parts:

- Predictor based on ARMAX model.
  There is more than one candidate for predictor if the process in question has at lest one stable and well damped mode, in consequence this provides a design with more freedom.
- Control law derived by means of minimizing an objective function.
  Two objective functions were introduced for different purpose. Which should be used depends mainly upon how many integrators the controller must possess.
- Receding horizon concept.
  This concept is used in GMPC as in all MPC methods to give a controller the desired feedback characteristic.

6.2 Conclusion

The presented GMPC method can, according to simulations, control a great deal of variety of processes from stable and unstable ones to those of nonminimum phase with certain degree of robustness. The integral action is an intrinsic property of all the controllers proposed. The proposed GMPC controller is dependent on prediction horizon and on delay as well if $d > 1$. All the matrixes, **H** and **M** as well as **F**, involved in the calculation of a control law can be obtained recursively. This will be a matter of great importance in adaptive applications. The highlight of GMPC is that it is very robust when faced with the uncertainty of delay times, thus there are potential applications to the cases of rough estimation of process delay or of time-variant process delays.

REFERENCES

Camacho, E.F. & Bordons, C. 2007. *Model Predictive Control.* London: Springer-Verlag, pp.111–113.

Clarke, D.W. 1988. Application of Generalized Predictive Control to Industrial Processes. *IEEE Control Systems Magazine,* 122:49–55.

Clarke, D.W., Mohtadi, C. & Tuffs, P.S. 1987. Generalized Predictive Control. Part I. The Basic Algorithm. *Automatica,* 23(2): 137–148.

Clarke, D.W., Mohtadi, C. & Tuffs, P.S. 1987. Generalized Predictive Control. Part II. Extensions and Interpretations. *Automatica,* 23(2): 149–160.

Cutler, C.R. & Ramaker, B.C. 1980. Dynamic Matrix Control—A Computer control Algorithm. In *Automatic Control Conference, San Francisco.*

De Keyser, R.M.C. & Van Cuawenberghe 1985. Extended Predictive Self-adaptive Control. In *IFAC Symposium on Identification and System Parameter Estimation,* York, UK, pages 1317–1322.

De Keyser, R.M.C. 1991. Basic Principles of Model Based Predictive Control. In 1st *European Control Conference, Grenoble,* pages 1753–1758.

Kalman, R.E. 1960a. Contributions to the theory of optimal control. *Bulletin de la Societe Mathematique de Mexicana,* 5, 102–119.

Muske, K.R. & Rawlings, J.B. 1993. Model Predictive Control with Linear Models. *AIChE Journal,* Vo. 39, No. 2.

Nishizaki, J., Okazaki, S., Yanou, A. & Minami, M. 2011. "Application of strongly stable generalized predictive control to temperature control of an aluminum plant" *IEEE Conference Publications,* pages 2602–2607.

Ordys, A.W. & Clarke, D.W. 1993. A State-space Description for GPC Controllers. *International Journal of System Science,* 24(9):1727–1744.

Peng Guo 2006. Research of Multivariable GPC and Its Application in Thermal Power Plant. ICICIC'06. *IEEE Conference Publications,* pages 35–38.

Propoi, A.I. 1963. Use of LP Methods for Synthesizing Sampled-data Automatic Systems. *Automatic Remote Control,* 24.

Richalet, J., Rault, A., Testud, J.L. & Papon, J. 1976. Algorithmic Control of Industrial Processes. In 4th IFAC *Symposium on Identification and System Parameter Estimation.* Tbilisi USSR.

Smith, O.J.M. 1957. Close Control of Loops with Dead Time. *Chemical Engineering Progress,* 53(5):217.

Ydstie, B.E. 1984. Extended Horizon Adaptive Control. In *Proc. 9th IFAC World Congress, Budapest, Hungrary.*

Zhao, Yuchen & Zanini, Aníbal. 2012. ARMA Model Based GPC. *Proceedings of the Int. Conf. on Integrated Modeling and Analysis in Applied Control and Automation. IMAACA. pp. 178–190, vienna, Austria.*

Control Engineering and Information Systems – Liu (Ed)
© 2015 Taylor & Francis Group, London, ISBN 978-1-138-02685-8

MIMO sliding mode for flightpath angle control during carrier landing

Q.D. Zhu, X. Meng & Z. Zhang
College of Automation, Harbin Engineering University, Harbin, Heilongjiang Province, China

ABSTRACT: The landing of high-performance aircraft on an aircraft carrier in rough seas is a demanding task, requiring precision control of the flightpath angle. In this paper, a Multi-Input Multi-Output (MIMO) sliding mode control method is advanced in order to hold the aircraft move along the ideal glide path in the complicated landing environment, and a flightpath angle controller is designed via MIMO sliding mode theory to replace the function of the autopilot and APCS equipped for the longitudinal flight control system. An improved longitudinal flight control system model is established based on the good performance of the controller. The simulation results show that the flightpath controller designed by this paper could make the variable quantity of flightpath angle track the variable quantity of pitch angle quickly and accurately, and the improved longitudinal flight control system possesses a better tracking performance and the ability of suppress airwake disturbances, and also has strong robustness.

1 INTRODUCTION

Precise control of the flightpath angle (γ) is required over the approach trajectory to land an aircraft on an aircraft carrier in heavy seas. The touchdown point is restricted to a small area on a deck that is pitching and heaving due to wave action. Furthermore, the glide path is continually disturbed by severe atmospheric turbulence generated in part by the carrier superstructure. Landing the aircraft is generally regarded as the most exacting of all routine airplane operations. In order to ensure the aircraft can land on the angle deck safely, we usually control γ by controlling the pitching angle (θ). Thus, the dynamic response characteristics of γ caused by θ affect the performance of the aircraft tracking ideal trajectory directly.

The landing stage of aircraft is in low dynamic pressure situation, lead to the strong coupling between γ and velocity and the handling characteristics degradation of aircraft, which increases the difficulty of controlling γ. The reason for this phenomenon is analyzed in [1], and it indicates that the Automatic Power Compensation System (APCS) with constant angle of attack (α_d) corresponds with speed stability, flightpath stability and maneuverability that is better than APCS with constant airspeed. Reference [2] indicates that the aircraft with variable structure APCS has a better tracking performance of the path angle to pitch angle and ability of anti airwake disturbance. Reference [3] describes that the carrier landing flightpath angle controller is designed based on total energy control theory and proves the robust performance is good.

The aircraft system is a typical a Multi-Input Multi-Output (MIMO) system. According to the characteristics, a new thinking and method of controlling γ is provided in this paper. A controller is designed via MIMO sliding mode theory to realize the precise control of γ. Simulation validates that the controller could make the variable quantity of flightpath angle ($\Delta\gamma$) follow the change of the variable quantity of pitch angle ($\Delta\theta$) very good and the improved longitudinal flight control system has better robustness and no high frequency chattering phenomenon.

2 PROBLEM FORMULATION

Generally, carrier-based aircraft flight control system uses conventional PID control strategy. The structure diagram is shown as follows. The system consist of two parts, one is the autopilot which controls θ, the other is APSC which control the engine thrust of aircraft automatically in adverse

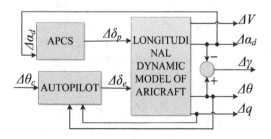

Figure 1. Longitudinal flight control system.

conditions. The autopilot automatically adjust the elevator deflection to change θ based on the guidance commands, which gives rise to the change of γ and thus achieve the correction of the glide path. Meanwhile, the APCS control the airspeed or α_d, $\Delta\gamma$ can track $\Delta\theta$ quickly and accurately. It ensures the stability of the aircraft movement.

At present, most articles use the intelligent control theory to design the autopilot or APCS in order to improve the performance of the flight control system. This paper presents an improved flight control system, as shown in Figure 2. It can make use of MIMO sliding mode controller to realize the features of the autopilot and APCS equipped for the flight control system in order to ensure precise γ control and suppression of airwake turbulence.

2.1 Description of longitudinal aircraft movement model

In this paper, we consider that the airwake affects the aircraft movement. In landing stage, the initial state variables are:

$$\begin{cases} v_{k0} = v_0, \ \alpha_{k0} = \alpha_0, \ \gamma_0 = 0, \ \theta_0 = \alpha_0 \\ q_0 = 0, \ \dot{v}_k = \dot{\alpha}_k = \dot{q} = \dot{r} = 0 \end{cases} \tag{1}$$

When the aircraft is flying in the reference state, the longitudinal perturbation model with wind can be described as follows:

$$
\begin{bmatrix} \Delta\dot{v} \\ \Delta\dot{\alpha}_d \\ \Delta q \\ \Delta\theta \end{bmatrix} =
\begin{bmatrix}
x^v & x^\alpha + g\cos\gamma_* & x^q & -g\cos\gamma_* \\
-y^v & -y^\alpha + g\sin\gamma_*/v_* & 1 & -g\sin\gamma_*/v_* \\
\mu_z^v - \mu_z^\alpha y^v & \mu_z^\alpha - \mu_z^\alpha(y^\alpha - g\sin\gamma_*/v_*) & \mu_z^q - \mu_z^\alpha - \mu_z^\alpha g\sin\gamma_*/v_* & \\
0 & 0 & 1 & 0
\end{bmatrix}
\begin{bmatrix} \Delta v \\ \Delta\alpha_d \\ \Delta q \\ \Delta\theta \end{bmatrix}
$$

$$
+ \begin{bmatrix}
x^{\delta_e} & x^p \\
-y^{\delta_e} & -y^p \\
\mu_z^{\delta_e} - \mu_z^\alpha y^{\delta_e} & \mu_z^p - \mu_z^\alpha y^p \\
0 & 0
\end{bmatrix}
\begin{bmatrix} \Delta\delta_e \\ \Delta\delta_p \end{bmatrix}
$$

$$
+ \begin{bmatrix}
-x^v & -x^\alpha/v_* \\
y^v & y^\alpha/v_* \\
\mu_z^\alpha y^v - \mu_z^v & (\mu_z^\alpha y^\alpha - \mu_z^\alpha)/v_* \\
0 & 0
\end{bmatrix}
\begin{bmatrix} w_{xg} \\ w_{zg} \end{bmatrix}
\tag{2}
$$

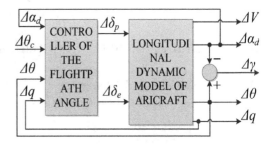

Figure 2. Improved longitudinal flight control system.

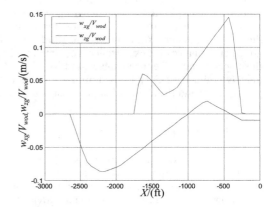

Figure 3. Steady wind ratios.

The meaning of the parameters and the process of formula derivation can be found in [4].

For convenient expound, we denote it as follow:

$$\dot{x} = \mathbf{A}x + \mathbf{B}u + \Gamma_w w \tag{3}$$

where $x = [\Delta v \ \Delta\alpha_d \ \Delta q \ \Delta\theta]^T$ is state vector, $u = [\Delta\delta_e \ \Delta\delta_p]^T$ is control vector, $w = [w_{xg} \ w_{zg}]^T$ is gust vector, ΔV is the variable quantity of velocity, $\Delta\alpha_d$ is the variable quantity of angle of attack, Δq is the variable quantity of pitching angular velocity, $\Delta\delta_e$ is the variable quantity of elevator deflection, $\Delta\delta_p$ is the variable quantity of throttle deflection, w is the airwake disturbance.

2.2 Airwake disturbance model

The airwake has a great influence on the glide path of the aircraft in the landing period. The influence is the largest source of landing error. If the aircraft trajectory can't be amended, only the steady component of airwake would enable the aircraft deviation from the desired touch point about 38.1 m. The airwake disturbance model should be established for the sake of assuring the safety of landing, and it is benefit to develop the technology of airwake attenuation.

Reference [5] specifies the model of atmospheric disturbances to be used for carrier landing operations. This paper only introduces the steady component of airwake model. The steady components of the airwake consist of a reduction in the steady wind and a predominant upwash aft of the ship which are functions of range. Figure 3 illustrates the steady wind functions w_{xg}/V_{wod} and w_{zg}/V_{wod} as functions of position aft of the ship center of pitch. V_{wod} is velocity of wind over the deck.

3 DESIGN OF FLIGHTPATH ANGLE CONTROLLER

3.1 Design of MIMO sliding mode controller

Generally, sliding mode control design involves two steps:

i. Selection of stable hyperplane (S) in the state/error space on which motion should be restricted, called the switching function, and
ii. Synthesis of a control law which makes the selected sliding surface attractive.

Let us consider a linear process, eventually a MIMO system, defined by

$$\dot{x} = \mathbf{A}x + \mathbf{B}u + \mathbf{\Gamma}_w w \qquad (4)$$

where $x \in R^n$, $u \in R^m$, $m \leq n$ and rank $\mathbf{B} = m$.

Let us also define the sliding surface as the intersection of m linear hyperplanes

$$S = \{x \in R^n : s(x) = \mathbf{C}x = 0\} \qquad (5)$$

where $\mathbf{C}$ is a full rank $(m \times n)$ matrix and let us assume that a sliding motion occurs on S.

In order to drive the state towards to the sliding switching surface, the reaching condition must be satisfied, which is

$$s\dot{s} < 0 \qquad (6)$$

But this condition can not reflect the detail that the movement is how to approach the hyperplanes. The quality of the normal movement requires a good approach process, such as fast. Thus we can use the reaching law to guarantee it.

This paper uses the exponential reaching law [6], which is expressed as

$$\dot{s} = -\varepsilon\text{sgn} - ks, \ \varepsilon > 0, \ k > 0 \qquad (7)$$

where s is the distance between the state and the hyperplanes, $\dot{s}$ is the reaching speed of the state ε and k are parameters.

The above equation can be calculated as

$$s(t) = \frac{\varepsilon}{k} + \left(s_0 - \frac{\varepsilon}{k}\right)e^{-ks} \qquad (8)$$

It can be seen that the parameter ε is used to ensure system reaching the hyperplane with a certain speed and the parameter k ensures the realization of sliding mode. Adjusting parameters properly ε and k are not only guarantee dynamic performances of sliding mode reaching phase, but also reduce the high frequency of controller output. Although the reaching law is a little complex, it is useful to the quality of the normal movement.

In sliding mode, $\dot{s} = \mathbf{C}\mathbf{A}x + \mathbf{C}\mathbf{B}u + \mathbf{C}\mathbf{\Gamma}_w w = 0$ and $s \equiv 0$. Assuming that $\mathbf{C}\mathbf{B}$ is invertible (which is reasonable since $\mathbf{B}$ is assumed to be full rank and s is a chosen function). The sliding motion is affected by the so-called equivalent control.

$$u = -(\mathbf{C}\mathbf{B})^{-1}\left[-(-\varepsilon\text{sgn}(s) - ks) + \mathbf{C}(\mathbf{A}x + \mathbf{\Gamma}_w w)\right] \quad (9)$$

This paper replaces the discontinuous term $sgn(s)$ by its continuous approximation $sat(s)$. Where $sat(.)$ is the standard saturation function, defined by:

$$sat(s) = \begin{cases} ks, & |s| \leq \Delta \\ \text{sgn}(s), & |s| > \Delta \end{cases}, \quad k = \frac{1}{\Delta} \qquad (10)$$

where Δ is the boundary layer.

This method can eliminate chattering but often at the cost of a non-zero steady-state error that is proportional to Δ. In order to obtain smaller errors, it is therefore necessary to make Δ smaller, which in turn, leads to chattering again [7]. So it is important to select the value of Δ. The steady-state error caused by this method can't be avoided, but the power of reducing the high frequency chattering makes it possible to use the sliding mode control in engineering application.

Then the above controller is replaced with its continuous approximation.

$$u = -(\mathbf{C}\mathbf{B})^{-1}\left[-(-\varepsilon sat(s) - ks) + \mathbf{C}(\mathbf{A}x + \mathbf{\Gamma}_w w)\right] \quad (11)$$

Proper sliding mode controller can be designed according to the boundary conditions of the system, as well as a reasonable value of Δ.

3.2 Design of flightpath angle controller

It can be easy proved that the longitudinal aircraft model is a MIMO system. Therefore, we can use the MIMO sliding mode controller to control γ. The good dynamic response characteristics of γ

caused by the change of θ can be guaranteed by using the above sliding mode controller. So, the hyperplanes are defined as

$$s = \begin{bmatrix} c1(\Delta\theta_c - \Delta\theta) + c2\Delta q \\ c3\Delta\alpha + c4\int\Delta\alpha \end{bmatrix} \quad (12)$$

where $c1$, $c2$, $c3$, $c4$ are constant, $\Delta\theta_c$ is the command signal of pitch angle. When the controller is applied to the improved flight control system, we just need to choose the proper value of the parameters to achieve the precise and rapid control of γ.

4 RESULTS AND DISCUSSIONS

In the MATLAB/simulink environment, we verify the feasibility and effectiveness of the improved longitudinal flight control system which is built in this paper. The step signal is considered to be the severe control input for systems. We usually use it to test the dynamic response and tracking performance of systems. At the beginning, $\theta_c = 0$, when $t = 1$, let $\theta_c = 1$. In addition, the parameters of MIMO sliding mode controller are set to $c1 = 20$, $c2 = -10$, $c3 = 1$, $c4 = 0.1$ and $\Delta = 2$. We can obtain these change curves of $\Delta\theta$, $\Delta\alpha_d$, $\Delta\gamma$, $\Delta\delta_e$ and $\Delta\delta_p$.

Figure 4 shows, $\Delta\theta$ reaches the steady state in 2.5 seconds with little steady state tracking error. On account of using the saturation function, we need to reach a compromise between chattering reduction and minimum steady state error. The change curve of $\Delta\theta$ arises a slight fluctuation because of suffering the airwake turbulence during 20–30 seconds.

From Figure 5, it's clear that the value of $\Delta\alpha_d$ changes very small and nearly remains in zero. From Figure 4 and Figure 5, as we can see that the

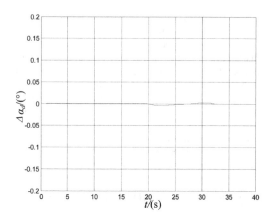

Figure 5.   Change curve of the angle of attack.

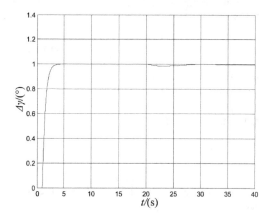

Figure 6.   Change curve of the flightpath angle.

controller designed in this paper is not only accomplish the function of the autopilot but also APCS. Contrast Figure 4 and Figure 6, we can notice that $\Delta\gamma$ can follow the change of $\Delta\theta$ exactly with negligible time delay. From the principle of flight dynamics, we know that the change of θ and γ are not synchronized. The aircraft's α and aerodynamic are changed after the change of θ, and then impact on γ. The theory confirms that the simulation results of this paper are reasonable and correct. Simulation results indicate that $\Delta\gamma$ can track the change of $\Delta\theta$ fast and accurately with MIMO sliding mode controller even under the influence of airwake disturbance.

The change curves of the elevator deflection and the throttle deflection are given respectively in Figure 7 and Figure 8. From the observation on the results, the phenomenon of high frequency chattering does not occur. Obviously, the controller designed in this paper can suppress the

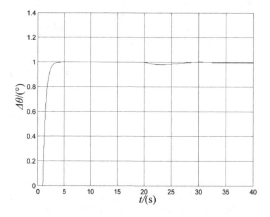

Figure 4.   Change curve of the pitch angle.

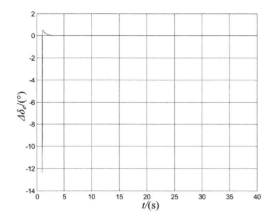

Figure 7. Change curve of the elevator deflection.

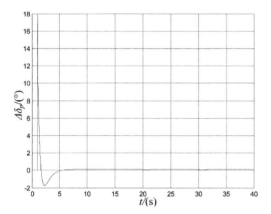

Figure 8. Change curve of the throttle deflection.

high frequency chattering of controller output effectively.

5 CONCLUSION

Be aimed at the characteristics of landing stage, the MIMO sliding mode control theory is applied to the design of the flightpath angle controller in this paper. The controller which realizes the precise control of the flightpath angle could replace the function of the autopilot and APSC equipped for the longitudinal flight control system. And it provides a new way to improve the control performance of the flight control system. The simulation results demonstrate that the MIMO sliding mode controller designed in this paper can eliminate the airwake disturbance effectively to make $\Delta\gamma$ follow $\Delta\theta$ quickly and exactly without the high frequency chattering phenomenon. Moreover, the controller has practical significance in engineering owing to relatively simple structure and easy running on the airborne computer.

REFERENCES

Department of Defense Interface Standard, 1997, *Flying qualities of pilot aircraft.*
Gao W.B, 1990, *Variable structure control theory*, Beijing: China Science and Technology.
Li J.X, Hou Z.Q, Xu Y.J. and Lin Z, 2010, Design of carrier landing flightpath angle controller, *Journal of Naval Aeronautical and Astronautical University*, 25(3): 303–306.
Robert C.N, 2008, *Flight stability and automatic control*, Beijing: National Defense Industry.
Sridhar S, Hassan K.K, 2005, Robust output feedback regulation of minimum-phase nonlinear systems using conditional integrators, *Automatic.* 41:43–54.
Wang X.H, Yang Y.D. and Zhu H, 2007. Research of handling characteristic of aircraft in low dynamic pressure situation, *FLIGHT DYNAMICS*, 25(4), 29–32 and 36.
Xiao Y.L, 1987, *Aircraft equations of motion*, Beijing: Aviation Industry.
Zhu Q.D, Wang T, Zhang W. and Zhou F, 2009, Variable structure approach power compensation system design of an automatic carrier landing system, *Proc. IEEE Chinese Control and Decision Conference*, Guiling: IEEE.

Control Engineering and Information Systems – Liu (Ed)
© 2015 Taylor & Francis Group, London, ISBN 978-1-138-02685-8

Pipeline robot and intelligent control method

Q.Y. Li, H.Q. Zhu & Z.F. Feng
College of Computer and Communication Engineering, China University of Petroleum (East China),
Qingdao, Shandong, China

ABSTRACT: Pipeline robot and the development of pipeline robot at home and abroad were introduced. The defects of traditional control theory in robotics applications were introduced, and the intelligent control of the pipeline robot was analyzed in detail. For the common and lack of fuzzy control and neural network control, the fusion of two kinds of intelligent control which are fuzzy control and neural network was discussed.

1 INTRODUCTION

With the rapid development of the computer technology, the microelectronic technology and the network technology in the late 20th century, the robot technology has been quickly developed (Cai 2009). The robot has been widely used in such fields as industrial production, ocean exploration, medical rehabilitation as well as military. With the continuous development of society and scientific technology, robot technology is no longer limited to the traditional industrial robots now. The robot which was used in the special environment (non-structural environment) has become a hot topic such as special robot. And as a branch of the special robot, pipeline robot has also been developed rapidly.

2 PIPELINE ROBOT

As we all know, the pipeline has been widely adopted in such fields as petroleum, chemical and nuclear industry. However, due to the relatively poor environment the pipeline which was always subject to heavy pressure and corrosion will crack or leak. In order to extend the life of the pipeline and to prevent the occurrence of the accident (such as the leakage), it is essential to do some regular inspection and maintenance work on the pipeline. However, pipeline is always located in complex environment where people are not easy to reach, which brought great difficulties to the repair and maintenance work. The appearance of the pipeline robot solves the difficulties of pipeline inspection and maintenance.

Pipeline robot (Wang et al. 2008) is a machine which can walk along the pipe wall. It can carry one or more sensors and operating devices and carry out a series of pipeline testing and maintenance operations under the control of the operator. The pipeline robot has many types: according to the driving mode, there are three types including motor drive, the fluid thrust (pneumatic and hydraulic) and external force on the elastic rod; according to its pattern of action, there are wheeled, legged, crawler, crawling, telescopic peristaltic type. The pipe robot can work in complex and closed pipes including straight horizontal pipe, pipe with different angle bends and pipe with slope tube as well as vertical tube.

2.1 The research progress of pipeline robot

In 1950s, pipeline running accidents occur frequently. Under the circumstances, people attached great importance to the pipeline inspection and maintenance, especially people in Europe and America and other developed countries. The early research on pipeline robot is about some clean-up and testing of long-distance pipeline which gain lots of achievements. The most representative is called "pipeline pig" which is non-powered equipment which can do some clean-up and testing work. It's generally believed that the pipeline pig is the origin of the pipeline robot.

In around 1970s, pipeline robot achieved unprecedented development. It mainly benefited from the development of microelectronic, computer technology, sensor technology and automation technology which provided the technology guarantee.

From the end of the 20th century to the beginning of this century, the research on micro-electromechanical systems, micro-driven machine and special materials gain great progress, then the research on tiny pipeline begin to boom. The tiny pipeline robot (Wang & Zuo 2010) can enter the important area of small tubes such as nuclear power stations and do some inspection, testing and maintenance work. Currently, the tiny pipeline robot adopted in medicine is mainly used for diagnosis and treatment and check of human intestine.

2.1.1 *The foreign pipeline robot's research status*

Study on the pipeline robot started earlier in foreign countries. The French man J.VERTUT was believed to do research on the theory and prototype of pipeline robot early. He proposed the model PRIV in1978 which is wheel-legged simple structure and can walk in the pipe. Japan has many research achievements in the area of pipeline robot. In 1987, the Japanese scholars like T. Morimitsu successfully developed a vibrating pipeline robot. The robot has relatively simple structure and is easy to achieve miniaturization. However, this robot has its own defects: its traveling speed is difficult to control; the vibration make the pipeline robot rotate along the circumferential direction; its movement posture is unstable and difficult to be effectively controlled because. Then, Shigeo Hirose, HidetakaOhno and others (Hirose et al. 1999) from the University of Tokyo successfully developed Thes-I, Thes-II, Thes-III pipeline robot early or late. Thes-I and Thes-II could run in the pipe with diameter of 50 mm, Thes-III was suitable for pipe with diameter of 150 mm. In 2008, Qinxue Pan and ShuxiangGuo and others (Pan et al. 2009) form the Japan University of Kagawa developed a spiral wireless tiny pipe robot which was designed for medical testing. This robot uses external magnetic field as driving force, so the rotational forward speed of the robot is related to the current frequency through the external magnetic field of the wire. In 2010, Osaka University and a gas company jointly developed a robot (Online 2010) that could penetrate into the internal and do inspection which was able to take photos in the pipeline and detect the aging and broken place. Its appearance was like a threaded shaft equipped with a motor.

In 1962, the world's first industrial robot was created in the United States. The United States is one of the robot powers. In 2001, Daphne DpZurko from New York Gas Group Company (NYGAS) and Hagen Schempf from Carnegie Mellon University developed the pipeline robot system Exlorer with funding from the US National Aeronautics and Space Administration (NASA). Exlorer can run in a long distance by cableless way which was dedicated to the detection of underground gas pipes. In 2010, Carnegie Mellon University robotics experts like HargentSemfen launched the wireless remote controlled robot "explore II"(Ling 2011) that was 2.4 meters long and 30 kg weight. The robot can measure slight variation of the magnetic field in real-time which

is created by dynamic electromagnetic coil in the pipeline. The variation of magnetic field reflects the thickness of the pipe wall that may be signs of corrosion or broken seam or even explosion, which is helpful for the testing of the pipeline. The position of the robot can be detected by electromagnetic induction, so if the robot system was failure the operator could retrieve the robot. In addition, Maria Feng of the University of California set up a research group which was dedicated to the manufacture of a robot (Online 2010) that could crawl in the pipe and repair the pipe. This robot can easily enter the water pipes, inspect pipe and find the leak points or weak points, then repair pipe by using some strengthening material.

South Korean also carried out the research work of the pipeline robot and has made great progress. Scholars like SegonRoh from Sungkyunkwan achieved a lot in the research of pipeline robot. The representative is the MRINSPECT series (Roh et al. 2009) (from MRINSPECT I to MRINSPECT V) of pipeline robot. MRINSPECT I was a single structure robot adopting non-independent driving mode which could achieve flexible support through the spring and was able to adapt to certain changes in the pipe diameter. MRINSPECT II was multi-section structure which was linked through controllable connecting sections and adopted spring-supported linkage. MRINSPECT III was also multi-section structure, but different from MRINSPECT II, it's train-style. There was a drive section and a steering knuckle at each end with the work section in the middle. The universal joints were controlled by dual-motor which could allow the robot to controllably turn and make it to have the ability of active steering. MRINSPECT IV improved on the basis of the MRINSPECT III's structure and smaller than it was one section driving mode. And different from MRINSPECT I, MRINSPECT IV adopted the independent driving mode that can control the driving mode according to the different speed, so it can run through the elbow and T type pipe easily. The defects of it are that it can only adapt a small range of pipe diameter and its capacity of the streamer is limited. MRINSPECT V was an upgraded version of the MRINSPECT IV by using train-style multi-section structures and different driving speed. It could run in L and T type pipe steadily and can be controlled in remote distance intelligently, so it is more intelligent than the other robots in the series.

2.1.2 *The domestic pipeline robot's research status*

The research of the pipeline robot in China starts late. Even though the domestic research has made remarkable achievements, it's still far behind the research of foreign countries. Professors from

Harbin Industrial University like Deng Zongquan carried out the research of pipeline robot early. They began the research on the walking mechanism of pipeline robot as early as 1987 developed a variety of inspection robot prototype which could apply to pipes with different diameter. In 2001, scholars like Qian Jinwu from Shanghai University developed tiny robot system Tubot II which could apply to the pipe with diameter of 20 mm and automatically detect pipeline defects. This robot system consists of a moving mechanism carrying the eddy current probe controlled by two levels of computer, and the eddy current analyzer as well as the data recording device and other components.

In 2008, Beijing University of Science and Technology and North China University jointly developed a pipeline robot (Zhao et al. 2009) using SMA (Shape Memory Alloy) as the driving element. The robot could run in the pipe with diameter from 50 mm to 80 mm. It applied to the U-shaped and T-shaped pipe with its minimum speed as 20 cm/min.

In 2009, teachers and students from Wuhan University of Science and Engineering invented a cleaning and inspection pipeline robot (Lu 2009) that can do cleaning on the pipe wall in 360 degrees and can adapt to the working environment of long pipeline with different diameters. The robot used the structure of umbrella and open-closed brush head and tail positioning system that enables the robot to achieve the stability of the vehicle body and the adaptability of the pipe diameter, and equipped with a camera for pipe inspection.

In 2011 students of Suzhou University invented a miniature robot (Online 2011) which was able to drive into the pipe of the nuclear power plant steam generator to check the pipeline safety situation. The movement of the robot is based on resonant principle that with the micro-motor the robot drives eccentric wheel and then produces some vibration, through asymmetric collision and friction of the burr and wall so as to drive robot. The robot has novel driving mode, agile controlling mode, and low power consumption, so 6V voltage can drive it.

3 INTELLIGENT CONTROL METHODS

With the rapid development of robot, both traditional and modern control technology get a different degree of application in the robot system. However, the traditional control theory has many defects in the application such as: requesting to establish a precise mathematical model; with single control task; without a good solution to the control problem of nonlinear systems; with low ability of controlling. Intelligent control technology is a new stage in the development of control theory. It's proposed for dealing with the controlling problem

of complex system which is difficult to solve by traditional control methods.

Intelligent control (Cai 2009) is able to independently drive intelligent machines to achieve its target of automatic control without human intervention, which is an advanced control technology simulating human intelligence.

With the development of fuzzy control, neural networks, evolutionary algorithms and expert system technology, intelligent control technology has got great development and contributes to the development automation and intellectualization of robot.

3.1 The pipeline robot's intelligent control

Pipeline robot is a system with complex structure and multi-degree of freedom. It has time-varying, strong coupling and nonlinear dynamics characteristics, and has characteristics of multi-information, multi-variable and multi-tasking. Pipeline robot mainly depends on the environmental conditions of the internal pipe motion in the pipeline, i.e. the roughness of the surface of the pipe wall, the pipeline obstructions and pipeline defects, etc (Xu 2003). Changing in environmental conditions within the pipe can change the state of the robot, the external force, the control parameters and control signals. There would not be accurate and complete robot motion model in practical applications. Problems of the complex pipeline environment, various unknown job, no precise mathematical model, etc. make the pipeline robot technologies not to use traditional control method, so the appearance of the intelligent control technology provides technical guarantee for the development of robot technology by dealing with the problems in robot technology.

The intelligent control system of the pipeline robot must have the capacities of simulating human learning, adaptive and self-organizing. This requires the control system of the pipeline robot have the following basic characteristics: (1) Effective global control of complex system, and strong fault tolerance; (2) Combination of qualitative decision-making and quantitative control to control the function of multi-modal combinations; (3) From the perspective of system function and overall optimization analysis and synthesis system to achieve the intended target, and have the ability of self-organization; (4) At the same time with mixed control process of a non-mathematical model with knowledge representation and accurate mathematical model.

3.2 Fuzzy neural network

Fuzzy control and neural network control are the two main control methods of intelligent control. They are often used in combination in practical applications. The integration of them is called fuzzy neural network.

Fuzzy control and neural network have many similarities: both of them imitate human intelligence; they all have the ability to parallel processing; no need to create precise mathematical model; able to solve a lot of problems of complexity, uncertainty, nonlinear which the traditional automation technology cannot solve; great fault-tolerant capability. At the same time they have their own inadequacies. On the one hand, in the process of training the neural network often turn into local extreme minimum and sometimes the network training time is over long. Existing empirical knowledge cannot be made good use of. The initial weights are always casually taken as zero or random number. On the other hand, the rules used in the fuzzy control system are stable, which lead to the lack of self-learning and adaptive ability. Therefore, it is possible to combine the two methods by using the respective advantages to compensate for each other's deficiencies in order to achieve the best control effect. The connections among the neurons of the neural network structure are casual and the results of the training are instable while the thinking reasoning function in fuzzy control can complement. Comparing with the fuzzy control, the neural network has powerful learning function which can use neural networks online learning membership function of the fuzzy set to achieve the reasoning process and fuzzy decision. These characteristics of fuzzy control and neural network are very suitable for intelligent control of the pipeline robot.

This literature (Xu 2003) will apply the learning algorithm based on enhanced fuzzy neural network to the pipeline robot's attitude control, and simulating results show that the enhanced fuzzy control neural network can improve the control effect to a certain extent.

For robot obstacle avoidance, Yang Juan, Sun hua and others (Yang et al. 2005) will apply the information fusion algorithm based on TS fuzzy neural network model to robot obstacle avoidance movement by using a plurality of ultrasonic ranging sensors to detect the distance and direction of the obstacles and after the information fusion of fuzzy neural network to achieve robot identification of obstacles and environment types as well as conflict-free movement. The experiments show that this method cannot only enable the robot recognize environment and avoid static obstacles.

4 CONCLUSION

Currently the pipeline robotic research has made a lot of achievements. Pipeline robot is developing toward miniaturization, multi-function and intelligent direction. Its technology has been rapidly developed with the development of scientific technology, but there are also many problems to achieve application universality of pipeline robot. There are following problems:

1. The energy supply problem. Energy supply is divided into cable way and no cable way according to different energy supplying way. In the cable way, the pipeline robot mainly face the problem of the shortening of the maximum walking distance of the robot operation and a series of reliability problems due to the friction between the cable and wall. For no cable way, the main problems are the driving capability and the walking distance of the robot. The energy supply depends mainly on the carried miniature battery, so driving capability and maximum walking distance of the robot is limited because the battery capacity is limited and t is also affected by the quality of the battery.

2. The communication problems. The pipeline robot needs to contact with the outside world at any time, if the transfer of information uses wire, problem of friction will occur. In addition, if pipeline uses wireless communication way, the signal will face shield and penetration fading.

3. Robot position, posture, and the recognition of pipeline environment. Now as the network technology develops rapidly, the combination of the pipeline robot and the network is a certain trend, so network-based remote control of the pipeline robot is also a problem needed to study. Comparing with traditional control technology, intelligent control technology can improve the speed and stability of the robot, but there are also limitations. For example, if the rule base of robot fuzzy control is too large, the time of reasoning process will last too long; if the rule base is very simple, the control accuracy is also restricted and the neural network is easy to fall into the local minimum and so on. All above problems are need to be solved in the design of intelligent control.

ACKNOWLEDGMENTS

R.B.G. thanks the support of the National Natural Science Fund with the grant number of 50674099, and the Fundamental Research Funds for the Central Universities (14CX02031 A).

REFERENCES

Cai, Zixing. Sep. 2009. *Robotics*, 2rd ed. Beijing: Tsinghua University Press.

Ling, Qiyu. Mar. 2011. Chinapipe homepage. [Online]. Available:http://www.chinapipe.net/national/2011/11185.html.

Lu, Shuiping. Oct. 2009. Teachers and students who are College of mechanical and electronic invented pipeline robot.[Online]. Available: http://newspaper.wutnews.net/ show.aspx? id = 10888.

Pan, Qinxue. Guo, Shuxiang & Li, Desheng, et al. Feb. 2009. Mechanism and control of a spiral type of microrobot in pipe. //Proceedings of the 2008 IEEE *International Conference on Robotics and Biomimetics*, pp. 43–48, doi: 10.1109/ROBIO.2009.4912977.

Roh, S. gon. D.W. Kim. Lee, J. sub. H. Moon & H.R. Choi. Feb. 2009. *In-pipe Robot Based on Selective Drive Mechanism*. International Journal, vol. 7, pp. 105–112, doi: 10.1007/s2555–009–0113-z.

Rototain. 2010. Japanese invented pipeline inspection robot. [Online]. Available: http://www.robotain.com/news/keji/201003/02429.html.

Rototain. Mar, 2010. Rototain homepage. [Online]. Available: http://www.robotain.com/news/keji/201003/10458.html.

Shigeo, Hirose. Hidetaka, Ohno. Takeo, Mitsui. Kiichi, Suyama. May 1999. Design of in-pipe inspection vehicles for Φ25, Φ50, Φ150 pipes. IEEE *International Conference on Robotics and Automation*, vol. 3, pp. 2309–2314, doi: 10.1109/ROBOT.1999.770450.

Unknown. Apr. 2011. Students which were in Soochow University invented the micro pipe robot. [Online]. Available: http://www.gusuedu.com/html/View9469.html.

Wang, Dianjun. Li, Runping & Huang, Guangning. Apr. 2008. Progresses in study of pipeline robot. *Machine tool & hydraulics*. vol. 36, pp. 185187, doi: 1001–3881(2008)4–185–3.

Wang, Yaohua & Zuo, Rengui. Dec. 2010. Research of micro in-pipe robot at home and abroad. *Journal of machine design*, vol. 27, pp. 1–5, doi:1001–2354 (2010)12–0001–05.

Xu, xiaoyun. 2003. Study on Pipe inspection robot system and based on fuzzy neural network control. Shanghai:Shanghai Jiaotong University.

Yang, Juan. Sun, Hua & Wu, Lin. 2005. Application of fuzzy neural networks in information fusion for obstacle avoidance. *Control theory and applications*, vol. 24, pp. 22–24, doi: 1003–7241(2005)02–0022–03.

Zhao, Yuxia. He, Guangping & Gao, Dewen, et al. Dec. 2009. Micro in-pipe robot mechanical structure design of shape memory alloy driving. //Proceedings of the 2009 IEEE *International Conferfence on Robotics and Biomimetics*. IEEE Press, pp. 360–365, doi: 10.1109/ROBO.2009.5420672.

Control Engineering and Information Systems – Liu (Ed)
© 2015 Taylor & Francis Group, London, ISBN 978-1-138-02685-8

Inverse optimal adaptive fault-tolerant attitude control for flexible spacecraft

H.H. Long & J.K. Zhao
Shanghai Key Laboratory of Navigation and Location Based Services, Shanghai Jiao Tong University, Shanghai, China

ABSTRACT: This paper studies adaptive fault-tolerant attitude control scheme based on inverse optimal technique for flexible spacecraft subject to actuator faults, uncertainty inertia matrix and external disturbances. Additive faults and the partial loss of actuator effectiveness are considered simultaneously to establish the dynamic model of flexible spacecraft. The attitude of spacecraft is represented by the unit quaternion, which is singularity-free. Two adaptive parameter update laws are introduced to estimate the unknown inertia matrix and the lower bound of the partial loss of actuator effectiveness fault, and an adaptive control Lyapunov function is constructed based on backstepping techniques. An inverse optimal adaptive fault-tolerant attitude control law is designed based on adaptive control Lyapunov function and inverse optimal methodology which ensures robustness, stabilization and H_∞ optimality with respect to a family of cost functional. The usefulness of the proposed algorithm is assessed through numerical simulations.

1 INTRODUCTION

Accuracy and stability are the significant considered factors in satellite attitude control. However, some disastrous faults or failures may be induced due to the aging or damage of actuators and sensors during the mission of a spacecraft, which can cause severe performance deterioration of control systems or even result in the failure of the mission. Hence, fault tolerance of the spacecraft attitude control system is one of the crucial issues that need to be studied. Fault-Tolerant Control (FTC) has been counted as one of the most promising control technologies for maintaining specified safety performance of a system in the presence of unexpected faults. Several FTC methods have been proposed for attitude control problem of a spacecraft in recent years (Jin 2008, Xiao 2011 & Hu 2009). However, the degrees of optimality of these fault-tolerant controllers were not stated explicitly.

A so-called inverse optimal control (Krstic et al. 1998) is an alternative method to solve the nonlinear optimal control problem by circumventing the need to solve the HJIPD equation directly. (Bharadwaj et al. 1998) and (Krstic et al. 1999) first addressed the inverse optimal method to solve the attitude control problem. They designed an inverse optimal feedback controller for the attitude regulation problem of a rigid spacecraft without external disturbances and uncertainties in the inertia matrix. (Luo et al. 2005) derived an adaptive inverse optimal approach to solve the rigid spacecraft with external disturbances and uncertainty inertial matrix by introducing an adaptive control Lyapunov function (aclf). However, these existing inverse optimal control approaches solving attitude problem are limit only to rigid spacecraft and can hardly be straightly extended to a flexible spacecraft system since the vibration in the flexible appendages induced by the orbiting attitude slewing operation may degrade the attitude pointing accuracy, and also no fault tolerant are considered in these controller designs.

To overcome the shortcomings of the preceding research for spacecraft attitude control systems, in this work a novel fault-tolerant attitude control strategy is addressed for flexible spacecraft in the presence of actuator faults, uncertainty inertia matrix and external disturbances. The proposed fault-tolerant attitude controller is designed by using the adaptive Lyapunov methodology and the inverse optimal technique for stabilizing systems, which is shown to be optimal with respect to a family of cost functional and achieves H_∞ disturbance attenuation without solving the associated HJIPD equation directly. Meanwhile, the derived control law can achieve the goal of fault-tolerant control without the need of any fault detection and isolation mechanism to determine the fault information.

2 MODEL DESCRIPTION AND PROBLEM FORMULATION

In this paper, we use the unit quaternion to represent the attitude of the spacecraft which is free of singularity. The unit quaternion Q is defined by

$$Q = \begin{bmatrix} q_0 \\ q \end{bmatrix} = \begin{bmatrix} \cos(\varphi/2) \\ e\sin(\varphi/2) \end{bmatrix} \quad (1)$$

where $e \in \Re^3$ and φ denote the Euler axis and Euler angle, respectively; $q = [q_1, q_2, q_3]^T \in \Re^3$ and $q_0 \in \Re^1$ are called the vector part and the scalar part of the unit quaternion, respectively. The quaternion Q satisfies the unit norm constraint $\|Q\| = 1$. Then the kinematic equations are described in terms of the attitude quaternion and are given by

$$\dot{Q} = \begin{bmatrix} \dot{q}_0 \\ \dot{q} \end{bmatrix} = \frac{1}{2} \begin{bmatrix} -q^T \\ q^\times + q_0 I_3 \end{bmatrix} \omega \quad (2)$$

where $\omega \in \Re^3$ is the angular velocity of the body frame with respect to a reference frame and expressed in the body frame, I_3 is unit matrix with order 3, and '$\times$' is an operator on the three dimensional vector q and $q^\times$ denotes

$$q^\times = \begin{bmatrix} 0 & -q_3 & q_2 \\ q_3 & 0 & -q_1 \\ -q_2 & q_1 & 0 \end{bmatrix} \quad (3)$$

which is a skew-symmetric matrix.

Next, consider the dynamic equations of a flexible spacecraft. Here, we consider the case where both partial loss of control effective and additive fault in actuators simultaneously. Then the general nonlinear spacecraft attitude dynamics model can be given by

$$J_s \dot{\omega} + \sigma^T \ddot{\eta} = -\omega \times (J_s \omega + \sigma^T \dot{\eta}) + (\delta u + f) + d_s \quad (4)$$

$$\ddot{\eta} + D\dot{\eta} + E\eta + \sigma \dot{\omega} = 0 \quad (5)$$

where $J_s \in \Re^{3\times 3}$ is the symmetric inertia matrix of the whole structure, $\eta \in \Re^N$ is the model coordinate vector, $\sigma \in \Re^{N\times 3}$ is the coupling matrix between the elastic and rigid structure, $u \in \Re^3$ is the control torque acting on the main body. $d_s \in \Re^3$ is a term taking into account disturbance torque. Whereas, $D = diag\{2\xi_i \Lambda_i^{1/2}, i = 1, 2, ..., N\}$ and $E = diag\{\Lambda_i, i = 1, 2, ..., N\}$ are the damping and stiffness matrices, respectively, in which N is the number of elastic modes considered, ξ_i is the corresponding damping ratio, and $\Lambda_i^{1/2}$ is the natural frequency. $\delta = diag\{\delta_{11}, \delta_{22}, \delta_{33}\}$ denotes the partial loss of actuator effectiveness fault with

$$0 < \mu \le \delta_{ii} \le 1, i = 1, 2, 3 \quad (6)$$

and $f \in \Re^3$ represents the additive actuator fault.

In fact, due to onboard payload motion, rotation of solar arrays, fuel consumption and out-gassing, ect., the inertial matrix J_s is time-varying. Here, we let $J_s = J + \Delta J$, where J represents the constant component and ΔJ denotes the time-varying component of inertial matrix J_s. If the terms $\Delta J \dot{\omega}$ and $\omega \times \Delta J \omega$ are considered as the disturbances, then (4) can be rewritten as

$$J\dot{\omega} + \sigma^T \ddot{\eta} = -\omega \times (J\omega + \sigma^T \dot{\eta}) + \delta u + (d + f) \quad (7)$$

where $d(t) = \Delta J \dot{\omega} - \omega \times \Delta J \omega + d_s(t)$ is considered as the lumped disturbances of the spacecraft.

In this paper, the control objective is to design a fault-tolerant controller for fault attitude system in Eqs. (2), (4) and (5) such that for all initial conditions the desired aims are achieved, that is $\lim_{t\to\infty} q = 0$, $\lim_{t\to\infty} \omega = 0$, $\lim_{t\to\infty} \eta = 0$, and $\lim_{t\to\infty} \dot{\eta} = 0$.

3 INVERSE OPTIMAL CONTROL FOR FLEXIBLE SPACECRAFT

3.1 Inverse optimal adaptive fault-tolerant controller design under actuator faults

In this section, an inverse optimal adaptive control method is employed to design a robust and optimal controller against actuator faults and external disturbances.

Note that kinematics equation given by (2) describes a system in cascade interconnection, which implies that the kinematics system in (2) is controlled indirectly through the angular velocity vector ω. Thus, in order to stabilize the kinematics system, ω is regarded as the virtual control input and design the following control law:

$$\omega_d = -Kq \quad (8)$$

with $K \in \Re^{3\times 3}$ and $K = K^T > 0$. It has been proved in [8] that the control law ω_d globally asymptotically stabilizes the kinematics system at the origin.

For the simplicity of the development, the following variable is introduced $v = \sigma\omega + \dot{\eta}$. Therefore, if we let $\xi = [\eta^T, v^T]^T$, and take into account the equation in (5), then the following equation can be yielded

$$\dot{\xi} = A\xi + B\omega \quad (9)$$

where

$$A = \begin{bmatrix} 0 & I \\ -E & -D \end{bmatrix}, \quad B = \begin{bmatrix} -\sigma \\ D\sigma \end{bmatrix} \quad (10)$$

In view of Eq. (7) and (9), one has

$$(J - \sigma^T\sigma)\dot{\omega} = -\omega^\times J\omega + [\sigma^T E \ \sigma^T D - \omega^\times \sigma^T]\xi$$
$$- (\sigma^T D - \omega^\times \sigma^T)\sigma\omega + \delta u + (d + f).$$
$$\tag{11}$$

we define a linear operator $L : \Re^3 \to \Re^{3\times 6}$ acting on $a = [a_1, a_2, a_3]^T$ by

$$L(a) = \begin{bmatrix} a_1 & 0 & 0 & 0 & a_3 & a_2 \\ 0 & a_2 & 0 & a_3 & 0 & a_1 \\ 0 & 0 & a_3 & a_2 & a_1 & 0 \end{bmatrix}$$

Let $\vartheta = [J_{11}, J_{22}, J_{33}, J_{23}, J_{13}, J_{12}]$, it follows that $Ja = L(a)\vartheta$. Let $\hat{\vartheta}$ represent the parameter estimate of ϑ and $\tilde{\vartheta}$ be the estimation error defined by $\tilde{\vartheta} = \vartheta - \hat{\vartheta}$, and define the expect error as following

$$x = \omega - \omega_d = \omega + Kq \tag{12}$$

As a result, the subsystem (11) becomes

$$J_\sigma \dot{x} = \big(\phi(\omega) + \varphi(q, q_0, \omega)\big)\vartheta + r_1(\omega)\xi$$
$$+ r_2(q, q_0, \omega)\omega + \delta u + (d + f) \tag{13}$$

where $J_\sigma \in \Re^{3\times 3}$, $\phi(\omega) \in \Re^{3\times 6}$, $\varphi(q, q_0, \omega) \in \Re^{3\times 6}$, $r_1(\omega) \in \Re^{3\times 1}$ and $r_2(q, q_0, \omega) \in \Re^{3\times 1}$ are expressed by

$$J_\sigma = J - \sigma^T\sigma \tag{14}$$

$$\phi(\omega) = -\omega^\times L(\omega) \tag{15}$$

$$\varphi(q, q_0, \omega) = L\left(\frac{1}{2}K(q_0 I_3 + q^\times)\omega\right) \tag{16}$$

$$r_1(\omega) = [\sigma^T E \ \ \sigma^T D - \omega^\times \sigma^T] \tag{17}$$

$$r_2(q, q_0, \omega) = -(\sigma^T D - \omega^\times \sigma^T)\sigma$$
$$- \frac{1}{2}\sigma^T \sigma K(q_0 I_3 + q^\times) \tag{18}$$

3.2 Inverse optimal adaptive fault-tolerant controller design under actuator faults

In this section, an inverse optimal adaptive control method is employed to design a robust and optimal controller against actuator faults and external disturbances.

Theorem 1. Consider an auxiliary system that consists of (2), (20) and the following equation:

$$J_\sigma \dot{x} = \big(\phi(\omega) + \varphi(q, q_0, \omega)\big)\vartheta + r_1(\omega)\varsigma$$
$$+ r_2(q, q_0, \omega)\omega + \delta u + \frac{2x}{\gamma^2} \tag{19}$$

under the dynamic feedback control law

$$u = \alpha(q, q_0, \omega)$$
$$= -R_1(q, q_0, \omega)^{-1}x$$
$$= -\tilde{\mu}\beta(q, q_0, \omega) \tag{20}$$

together with the adaptive parameter update laws

$$\dot{\hat{\mu}} = \psi(q, q_0, \omega, x, \hat{\mu})$$
$$= x^T \hat{\mu}^3 \beta(q, q_0, \omega) \tag{21}$$

and

$$\dot{\hat{\vartheta}}(q, q_0, \omega) = \Pi\pi(q, q_0, \omega) \tag{22}$$

where

$$\beta(q, q_0, \omega) = \Big[\Omega(q, q_0, \omega)^T \Omega(q, q_0, \omega) + 2K_1$$
$$+ \Delta(q, q_0, \omega)^T K_1^{-1}\Delta(q, q_0, \omega) + \frac{4x^T}{\gamma^2}\Big]x \tag{23}$$

$$\pi(q, q_0, \omega) = \big(\phi(\omega) + \varphi(q, q_0, \omega)\big)^T x \tag{24}$$

$$\Omega(q, q_0, \omega) = \frac{1}{\sqrt{c}}K^{1/2}$$
$$\times \Big[cK^{-1} - \omega^\times \hat{J} + \frac{1}{2}\hat{J}K(q_0 I_3 + q^\times) + r_2(q, q_0, \omega)\Big]^T \tag{25}$$

$$\Delta(q, q_0, \omega)$$
$$= \Big[-\omega^\times \hat{J} + \frac{1}{2}\hat{J}K(q_0 I_3 + q^\times) + r_2(q, q_0, \omega)\Big] \tag{26}$$

$\hat{\mu}$ is the parameter estimate of μ in (6) and $\tilde{\mu}$ is defined by $\tilde{\mu} = 1/\hat{\mu} - \mu$, $\Pi \in \Re^{3\times 3}$ is a positive definite matrix, c is an adjustable parameters satisfying $c > 0$, both $K \in \Re^{3\times 3}$ and $K_1 \in \Re^{3\times 3}$ are positive definite symmetric matrices, and the matrices or vectors J_σ, $\phi(\omega)$, $\varphi(q, q_0, \omega)$, $r_1(\omega, \eta, \dot{\eta})$ and $r_2(\omega)$ are as defined in (14)-(18), respectively. Then if α, β, ρ, c, K, K_1 and λ_ξ satisfy the following conditions

$$\xi^T PBKq \leq 2\alpha\|\xi\|\|q\| \tag{27}$$

$$x^T[\sigma^T E \ \sigma^T D - Kq^\times \sigma^T]\xi$$
$$+ \xi^T PBx \leq 2\rho\|x\|\|q\| + 2\beta\|q\|\|\xi\| \tag{28}$$

$$\Lambda = \begin{bmatrix} \frac{c}{4}\lambda_K^{\min} & 0 & -\alpha \\ 0 & \frac{1}{4}\lambda_{K_1}^{\min} & -(\beta + \rho) \\ -\alpha & -(\beta + \rho) & \frac{\lambda_\xi}{2} \end{bmatrix} \geq 0 \tag{29}$$

67

where $\lambda_K^{\min}$ and $\lambda_{K_1}^{\min}$ are the minimum eigenvalue of matrices K and K_1, respectively, the dynamic feedback control u in (20) together with the adaptive parameter update laws in (21) and (22) adaptively stabilize the auxiliary system in (2), (10) and (13), that is, $\lim_{t\to\infty} q = 0$, $\lim_{t\to 0}\omega = 0$, $\lim_{t\to 0}\eta = 0$ and $\lim_{t\to 0}\dot{\eta} = 0$ for all initial conditions.

Proof. Since A defined in (10) has all its eigenvalues in the left-hand plane, for every fixed $\lambda_\xi > 0$ there exist a symmetric and positive-definite solution $P \in \Re^{2N\times 2N}$ of the Sylvester equation

$$(PA + A^T P)/2 = -\lambda_\xi I \qquad (30)$$

Consider the smooth positive-definite radially unbounded Lyapunov function as follows

$$V_1 = cq^T q + c(1-q_0)^2 + \frac{1}{2}x^T J_\sigma x + \frac{1}{2}\tilde{\vartheta}^T \Pi^{-1}\tilde{\vartheta} + \frac{1}{2}\tilde{\mu}^2 \qquad (31)$$

The time derivate V_1 along Eqs. (2), (9) and (19), we have

$$\dot{V}_1 = -cq^T Kq + x^T \left[cq + \big(\phi(\omega)+\varphi(q,q_0,\omega)\big)\vartheta + \Gamma_1(\omega)\varsigma \right.$$
$$\left. + \Gamma_2(q,q_0,\omega)\omega + \alpha u + \frac{2x}{\gamma^2} \right] + \xi^T P(A\xi + B\omega)$$

$$+ \tilde{\vartheta}^T \left(-\big(\phi(\omega)+\varphi(q,q_0,\omega)\big)x \right) - \frac{\tilde{\mu}\dot{\hat{\mu}}}{\hat{\mu}^2}$$

$$= -cq^T Kq + x^T \left[cq + \big(\phi(\omega)+\varphi(q,q_0,\omega)\big)\hat{\vartheta} + \Gamma_1(\omega)\varsigma \right.$$
$$\left. + \Gamma_2(q,q_0,\omega)\omega + \alpha u + \frac{2x}{\gamma^2} \right] + \xi^T P(A\xi + B\omega) - \frac{\tilde{\mu}\dot{\hat{\mu}}}{\hat{\mu}^2}$$

$$= -cq^T Kq - \lambda_\xi \xi^T \xi + x^T \alpha u + \sqrt{c}x^T \Omega^T K^{1/2}q + x^T \Delta x$$

$$+ \xi^T \Gamma_1(\omega)x + \xi^T PBx + \xi^T PBKq + \frac{2}{\gamma^2}x^T x - \frac{\tilde{\mu}\dot{\hat{\mu}}}{\hat{\mu}^2}$$

$$\leq -cq^T Kq - \lambda_\xi \xi^T \xi + \sqrt{c}x^T \Omega^T K^{1/2}q + x^T \Delta x$$

$$+ \xi^T \Gamma_1(\omega)x + \xi^T PBx + \xi^T PBKq + \frac{2}{\gamma^2}x^T x$$

$$- x^T \mu\hat{\mu}\beta(q,q_0,\omega) - \frac{1-\hat{\mu}\mu}{\hat{\mu}}x^T \hat{\mu}\beta(q,q_0,\omega)$$

$$= -\frac{c}{2}q^T Kq - \lambda_\xi \xi^T \xi - \frac{1}{2}x^T K_1 x \quad \frac{1}{2}\left\| \sqrt{c}K^{1/2}q - \Omega x \right\|^2$$

$$+ x^T \Gamma_1(\omega)\xi.$$

If inequalities (27), (28) and (29) hold, we have

$$\dot{V} \leq -\frac{c}{2}\lambda_K^{\min}\|q\|^2 - \lambda_\xi\|\xi\|^2 - \frac{1}{2}\lambda_{K_1}^{\min}\|x\|^2$$

$$-\frac{1}{2}\left\| \sqrt{c}K^{1/2}q - \Omega(q,q_0,\omega)x \right\|^2$$

$$-\frac{1}{2}x^T (K_1 - \Delta(q,q_0,\omega))^T K_1^{-1}(K_1 - \Delta(q,q_0,\omega))x$$

$$+ 2\rho\|x\|\|\xi\| + 2\alpha\|\xi\|\|q\| + 2\beta\|q\|\|\xi\|$$

$$\leq -\frac{c}{4}\lambda_K^{\min}\|q\|^2 - \frac{1}{4}\lambda_{K_1}^{\min}\|x\|^2 - \frac{\lambda_\xi}{2}\|\xi\|^2$$

$$-\big[\|q\|,\|x\|,\|\xi\|\big]\Lambda \begin{bmatrix} \|q\| \\ \|x\| \\ \|\xi\| \end{bmatrix}$$

$$\leq -\frac{c}{4}\lambda_K^{\min}\|q\|^2 - \frac{1}{4}\lambda_{K_1}^{\min}\|x\|^2 - \frac{1}{2}\lambda_\xi\|\xi\|^2$$

It follows from Barbalat's theorem that $q \to 0$, $x \to 0$ and $\xi \to 0$, and therefore $\omega \to 0$, and consequently $\eta \to 0$, $\dot{\eta} \to 0$ as $t \to \infty$.

Theorem 2 The dynamic feedback control law

$$u = \alpha^*(q,q_0,\omega)$$
$$= -2R_1(q,q_0,\omega)^{-1}x \qquad (32)$$

together with the adaptive parameter update laws

$$\dot{\hat{\mu}} = \psi(q,q_0,\omega,x,\hat{\mu}) \qquad (33)$$

$$\dot{\hat{\vartheta}}(q,q_0,\omega) = \Pi\pi(q,q_0,\omega) \qquad (34)$$

where $R_1(q,q_0,\omega)$, $\psi(q,q_0,\omega,x,\hat{\mu})$ and $\pi(q,q_0,\omega)$ are defined in (20), (21) and (22), respectively, $\Pi \in \Re^{3\times 3}$ is a positive definite matrix, solve the H_∞ inverse optimal control problem for the attitude control system (2), (9) and (13) by minimizing the cost functional

$$L_a = \lim_{t\to\infty}\left[4V_1(T) + \int_0^t \left(\begin{array}{c} l(q,q_0,\omega) \\ +u^T Ru - \gamma^2\|(f+d)\|^2 \end{array} \right) d\tau \right] \qquad (35)$$

for each $\vartheta \in R^6$, where

$$l(q,q_0,\omega) = -4\left\{ cq^T\omega + \xi^T P(A\xi + B\omega) \right.$$
$$+ x^T \left[\big(\phi(\omega)+\varphi(q,q_0,\omega)\big)\hat{\vartheta} + \Gamma_1(\omega)\xi \right.$$
$$\left. + \Gamma_2(q,q_0,\omega)\omega \right] \Big\} + 4x^T R_1^{-1}x - \frac{4}{\gamma^2}x^T x - \frac{4\tilde{\mu}\dot{\hat{\mu}}}{\hat{\mu}^2},$$

c is an adjustable parameters satisfying $c > 0$, and the matrices or vectors J_σ, $\phi(\omega)$, $\varphi(q,q_0,\omega)$, $\Gamma_1(\omega,\eta,\dot{\eta})$, $\Gamma_2(\omega)$, A and B are as defined in (14)–(18) and (10) respectively.

68

4 A NUMERICAL EXAMPLE

To illustrate and evaluate the effectiveness of the proposed control schemes, numerical simulations have been conducted for a flexible spacecraft system given in Eqs. (2), (4) and (5) with the developed attitude control laws given by Eqs (20), (21) and (22). The numerical simulation parameters are given as in Table 1. In the simulation analysis, only the first three elastic modes have been considered in the controller design, so taking into account possible spillover effects. The rest-to-rest maneuver is considered in the simulation.

We considered the conditions that both additive faults and the partial loss of actuator effectiveness occur. The actuator effectiveness matrix α and additive fault vector f are given by, respectively,

$$\alpha_i(t) = \begin{cases} 1, & t < 5\ s \\ 0.2 + 0.4\sin(2\pi t), & t \geq 5\ s \end{cases}$$

$$f_i(t) = \begin{cases} 1, & t < 5\ s \\ 0.35 + 0.05\sin(2\pi t), & t \geq 5\ s \end{cases}$$

The performance of the controller in (51) is investigated in this section. Figures 1–2 shows the

Table 1. Simulation parameters.

Parameter name	Value
Inertia matrix	$J_0 = \begin{bmatrix} 350 & 3 & 4 \\ 3 & 280 & 10 \\ 4 & 10 & 190 \end{bmatrix}$ kg/m^2
Natural frequency	$\omega_{n1} = 0.768,\ \omega_{n2} = 1.103,$ $\omega_{n3} = 1.873,\ \omega_{n4} = 2.549$ rad/s
Damping ration	$\xi_1 = 0.0056,\ \xi_2 = 0.0086,\ \xi_3 = 0.013,\ \xi_4 = 0.025$
Coupling matries	$\sigma = \begin{bmatrix} 6.45637 & 1.27814 & 2.15629 \\ -1.25619 & 0.91756 & -1.67264 \\ 1.11687 & 2.48901 & -0.83674 \\ 1.23637 & -2.6581 & -1.12503 \end{bmatrix}$
Initial conditions	$q(0) = [-0.2 \quad 0.7 \quad -0.35]^T$ $\omega(0) = [0 \quad 0 \quad 0]^T$ $\eta_i(0) = \dot{\eta}_i(0) = 0,\ i = 1,2,3,4$ $\hat{\vartheta}(0) = [340, 290, 175, 1, 2, 10]$ $\hat{\mu}(0) = 1$
External disturbance Control parameters Maximum allowable torque input	$d(t) = \begin{bmatrix} 0.3\cos(0.01t) + 0.1 \\ 0.15\sin(0.002t) + 0.3\cos(0.025t) \\ 0.3\sin(0.001t) + 0.1 \end{bmatrix}$ $\lambda_\xi = 1000,\ c = 1000,\ K = I_3,$ $K_1 = 4000I_3,\ \Pi = 200I_6,\ \gamma = 2,$ $u_{max} = 5\ N \cdot m$

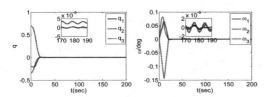

Figure 1. Time responses of the quaternion and angular velocity.

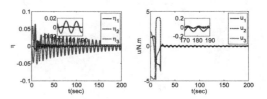

Figure 2. Time responses of vibration displacement and control torques.

69

simulated results obtained by the designed inverse optimal adaptive fault-tolerant attitude controller in (51). The quaternion and angular velocity responses of the attitude system are shown in Figure 1. One can observe that high control precision of attitude and angular velocity are obtained within 10 s even if in the presence of actuator faults, system parametric uncertainties and external disturbances. The responses of the modal displacements η_1, η_2, η_3 are presented in Figure 2, in which a low vibration level is illustrated. This illustrates that the designed controller is capable of suppressing the system vibration while controlling the attitude of the spacecraft. The applied control torques on the flexible spacecraft with actuator faults is shown in Figure 2. From the simulation results we can clearly see that the proposed fault-tolerant controller can accommodate the actuator fault.

REFERENCES

Bharadwaj S., K.M. Qsipchuk, K.D. Mease, F.C. Park, 'Geometry and inverse optimality in global attitude stabilization', J. Guid., Control Dyn., 21, (6), 930–939, 1998.

Hu Q.L., 'Robust adaptive sliding-mode fault-tolerant control with L_2-gain performance for flexible spacecraft using redundant reaction wheels', IET control Theory Appl., 4, (6), pp. 1055–1070, 2009.

Jin J.H., S.H. Ko, C.K. Ryoo, 'Fault tolerant control for satellites with four reaction wheels', Control Eng. Pract., 16, (10), pp. 1250–1258, 2008.

Krstic M., P. Tsiotras, 'Inverse optimal stabilization of a rigid spacecraft', IEEE Trans. Auto. Control, 44, (5), pp. 1042–1049, 1999.

Krstic M., Z.H. Li, 'Inverse optimal design of Input-to-State stabilizing nonlinear controllers', IEEE Trans. Auto. Control., 43, (3), pp. 336–350, 1998.

Luo W.C., Y.C. Chu, K.V. Ling, 'Inverse optimal adaptive control for attitude tracking of spacecraft', IEEE Trans. Auto. Control., 50, (11), pp. 1639–1654, 2005.

Xiao B., Q.L. Hu, 'Fault-tolerant attitude control for flexible spacecraft without angular velocity magnitude measurement', J. Guid. Control Dyn., 34, (5), pp. 1556–1561, 2011.

Control Engineering and Information Systems – Liu (Ed)
© *2015 Taylor & Francis Group, London, ISBN 978-1-138-02685-8*

Robust fuzzy-neural tracking control for robot manipulators

H.J. Rong
State Key Laboratory for Strength and Vibration of Mechanical Structures, School of Aerospace,
Xi'an Jiaotong University, Xi'an, Shaanxi, China

J. Yang
Institute of Control Engineering, School of Electronic and Information Engineering,
Xi'an Jiaotong University, Xi'an, Shaanxi, China

J.M. Bai
Optical Direction and Pointing Technique Research Department, Xi'an Institute of Optics and Precision
Mechanics of CAS, Xi'an, Shaanxi, China

ABSTRACT: The paper presents a robust fuzzy-neural control scheme for an n-link robot manipulator with uncertainties. In the proposed scheme a fuzzy neural network is utilized to construct the control input by approximating the unknown nonlinearities of dynamic systems. The parameters of the fuzzy neural network approximator are modified using the recently proposed fuzzy-neural algorithm named Online Sequential Fuzzy Extreme Learning Machine (OS-Fuzzy-ELM), where the parameters of the membership functions characterizing the linguistic terms in the if-then rules are assigned by random values independent from the training data. Different from the original OS-Fuzzy-ELM algorithm, the consequent parameters of if-then rules are updated based on the stable laws derived based on Lyapunov stability theorem and Barbalat's lemma so that the asymptotical stability of the system can be guaranteed. Also a sliding mode controller is incorporated to compensate for the modelling error of fuzzy neural network. Finally the proposed robust fuzzy-neural controller is applied to control a two-link robot manipulator and the simulation results verify the effectiveness of the proposed control scheme.

1 INTRODUCTION

In the past decade, considerable research development has been achieved in the use of fuzzy inference system for the tracking control of robot manipulators when they suffer from structured and unstructured uncertainties such as load variation, friction, and external disturbances etc. In the earlier methods [1,2] the fuzzy rules for designing the fuzzy controller are built according to the designer experience and can not be varied once they are determined. By combing the learning ability of neural networks, fuzzy neural networks can overcome the shortcomings of the conventional fuzzy inference systems. Many adaptive fuzzy control schemes based on fuzzy neural networks have been developed for the robot manipulators to achieve accurate trajectory tracking and good control performance [3,4].

In the paper, a robust fuzzy-neural control scheme for the robot manipulator is presented. In the scheme, fuzzy neural network is used as function approximator to estimate the control input. Different from the existing method, the parameters of the fuzzy neural networks are adjusted based on the recently proposed fuzzy-neural algorithm named Online Sequential Fuzzy Extreme Learning Machine (OS-Fuzzy-ELM) [5] where the parameters of the fuzzy membership functions need not be adjusted during training and one can randomly assign the values to them. This results in a linearly parameterized model thus achieving a less computation and simple design process.

2 DYNAMICS AND PROPERTIES OF ROBOT MANIPULATORS

The dynamics of a general n-link robotic manipulator is expressed in the Lagrange-Euler formation as

$$\mathbf{M}(\mathbf{q})\ddot{\mathbf{q}} + \mathbf{C}(\mathbf{q},\dot{\mathbf{q}})\dot{\mathbf{q}} + \mathbf{G}(\mathbf{q}) + \tau_d = \tau \qquad (1)$$

where $\mathbf{q} = [q_1, q_2, ..., q_n]^T \in R^n$, $\dot{\mathbf{q}} \in R^n$, and $\ddot{\mathbf{q}} \in R^n$ are the vectors of joint position, velocity, and acceleration respectively. $\mathbf{M}(\mathbf{q}) \in R^{n \times n}$ represents the inertia matrix; $\mathbf{C}(\mathbf{q},\dot{\mathbf{q}}) \in R^{n \times n}$ denotes the matrix of centripetal and Coriolis forces; $\mathbf{G(q)} \in R^n$ is the gravity vector. $\tau_d \in R^n$ is the lumped uncertainty denoting external disturbances and unmodeled dynamics. $\tau \in R^n$ represents joint torque supplied by the actuators. The above dynamic equation has the following properties:

Property 1: The inertial matrix $\mathbf{M(q)}$ is symmetric and positive definite.

Property 2: The matrix $\dot{\mathbf{M}}(\mathbf{q}) - 2\mathbf{C(q,\dot{q})}$ is skew symmetric and satisfies $\mathbf{x}^T(\dot{\mathbf{M}} - 2\mathbf{C})\mathbf{x} = 0, \forall \mathbf{x} \in R^n$.

The tracking error vector and its filtered error vector are expressed as,

$$
\begin{aligned}
\mathbf{e} &= \mathbf{q}_d - \mathbf{q} \\
\mathbf{s} &= \left(\frac{d}{dt} + \lambda\right)\mathbf{e} = \dot{\mathbf{e}} + \lambda\mathbf{e} = \dot{\mathbf{q}}_d - \dot{\mathbf{q}} + \lambda\mathbf{e}
\end{aligned} \tag{2}
$$

where $\mathbf{q}_d$ and $\dot{\mathbf{q}}_d$ are the desired position and velocity, λ is the diagonal positive-definite matrix to be designed by the user.

The control objective is to find a suitable control law τ so that the filtered error vector $\mathbf{s}$ will approach to zeros as $t \to \infty$. Under this condition, the joint position vector $\mathbf{q}$ can follow the desired signals $\mathbf{q}_d$ as closely as possible.

Introduce

$$
\dot{\mathbf{q}} = \dot{\mathbf{q}}_r - \mathbf{s} \tag{3}
$$

where $\dot{\mathbf{q}}_r = \dot{\mathbf{q}}_d + \lambda\mathbf{e}$.

The dynamics of the n-joint robot arm is rearranged as,

$$
\begin{aligned}
\mathbf{M(q)}\dot{\mathbf{s}} &= \mathbf{M(q)}\ddot{\mathbf{q}}_r + \mathbf{C(q,\dot{q})}\dot{\mathbf{q}}_r - \mathbf{C(q,\dot{q})}\mathbf{s} \\
&\quad + \mathbf{G(q)} + \tau_d - \tau
\end{aligned} \tag{4}
$$

To guarantee the convergence of tracking error and the stability of the whole control system, the control law is defined as,

$$
\begin{aligned}
\tau &= \mathbf{M(q)}\ddot{\mathbf{q}}_r + \mathbf{C(q,\dot{q})}\dot{\mathbf{q}}_r + \mathbf{G(q)} + \tau_d + \mathbf{Ks} \\
&= \mathbf{f(z)} + \mathbf{Ks}
\end{aligned} \tag{5}
$$

where $\mathbf{z} = [\mathbf{q}, \dot{\mathbf{q}}, \dot{\mathbf{q}}_r, \ddot{\mathbf{q}}_r] \in R^{4n}$, $\mathbf{K} = \mathrm{diag}[k_1, k_2, ..., k_n]$ $\in R^{n \times n}$ is a diagonal positive constant matrix and chosen by the user. Substituting Eq. (5) into Eq. (4) will result in the closed-loop dynamics $\mathbf{M(q)}\dot{\mathbf{s}} + (\mathbf{K} + \mathbf{C(q,\dot{q})})\mathbf{s} = 0$, which will ensure the stability of the system and also guarantee the exponential convergence of the tracking error $\mathbf{e}$.

However, in practical applications the system dynamics ($\mathbf{M(q)}$, $\mathbf{C(z)}$, $\mathbf{G(q)}$, τ_d) are perturbed or unknown, thus it is difficult to implement this controller. To solve the problem, Fuzzy Neural Network (FNN) is used to approximate $\mathbf{f(z)}$ and based on this the control law is rewritten as,

$$
\tau_f = \hat{\mathbf{f}}(\mathbf{z}) + \mathbf{Ks} \tag{6}
$$

where $\hat{\mathbf{f}}(\mathbf{z})$ is the estimated value of the function $\mathbf{f(z)}$ from FNN.

The approximation error between the ideal model and the approximated model will arise in Eqn. (6) when the FNN is utilized to approximate $\mathbf{f(z)}$. To compensate for the approximation error, a sliding mode controller τ_s is used to augment the fuzzy-neural controller above and thus the overall control law is considered as,

$$
\tau = \tau_f + \tau_s \tag{7}
$$

In the following section, we will describe the design procedure of the adaptive fuzzy-neural controlle.

3 DESIGN PROCEDURE OF THE ROBUST FUZZY-NEURAL CONTROLLER

3.1 *Fuzzy Neural Network*

OS-Fuzzy-ELM can handle the fuzzy inference systems including the most commonly used Mamdani type of fuzzy models where the antecedent (if) part and the consequent (then) part are both described by the fuzzy sets, and Takagi-Sugeno-Kang (TSK) type of fuzzy models where only the antecedent part is described by fuzzy sets whereas the consequent part is described by linear functions about input variables. TSK fuzzy model can achieve better performance with fewer rules and has been widely applied into the nonlinear system control problem. Here in the paper a FNN with TSK fuzzy model is used to construct the control law.

The TSK fuzzy model is given by the following rules [6]: Rule i: if $(x_1 \text{ is } A_{1i})$ AND $(x_2 \text{ is } A_{2i})$ AND ... AND $(x_n \text{ is } A_{ni})$, then y_1 is β_{1i} ... y_m is β_{mi} where $\mathbf{x}(\mathbf{x} = [x_1, ..., x_n]^T)$ and $\mathbf{y}(\mathbf{y} = [y_1, ..., y_m]^T)$ are the input and output of the system, respectively, $A_{ji}(j = 1, 2, ..., n; i = 1, 2, ..., \tilde{N})$ is the fuzzy set of the jth input variable x_j in rule i, n is the dimension of the input vector, m is the dimension of the output vector, and $\tilde{N}$ is the number of fuzzy rules. $\beta_{ki}(i = 1, 2, ..., \tilde{N}; k = 1, 2, ..., m)$ is the crisp value and a linear combination of input variables, that is $\beta_{ki} = q_{ki0} + q_{ki1}x_1 + \cdots + q_{kin}x_n$. The system output y for given input $\mathbf{x}$ is calculated by

$$
\mathbf{y} = \frac{\sum_{i=1}^{\tilde{N}} \beta_i R_i(\mathbf{x}; \mathbf{c}_i, a_i)}{\sum_{i=1}^{\tilde{N}} R_i(\mathbf{x}; \mathbf{c}_i, a_i)} = \sum_{i=1}^{\tilde{N}} \beta_i G(\mathbf{x}; \mathbf{c}_i, a_i) = \boldsymbol{\beta}^T \mathbf{G}(\mathbf{x}; \mathbf{c}, \mathbf{a})
$$

$$
\tag{8}
$$

For TSK fuzzy model, the consequence is the linear equation of the input variables and given by

$$\beta_i = \mathbf{x}_e^T \mathbf{q}_i \qquad (9)$$

where $\mathbf{x}_e$ is the extended input vector by appending the input vector $\mathbf{x}$ with 1, that is $[1, \mathbf{x}^T]^T$; $\mathbf{q}_i$ is the parameter matrix existing in the TSK model for the ith fuzzy rule and given by

$$\mathbf{q}_i = \begin{bmatrix} q_{1i0} & \cdots & q_{mi0} \\ \vdots & \cdots & \vdots \\ q_{1in} & \cdots & q_{min} \end{bmatrix}_{(n+1) \times m} \qquad (10)$$

Thus the output equation (8) for the TSK model becomes as:

$$y = \sum_{i=1}^{\tilde{N}} \mathbf{x}_e^T \mathbf{q}_i G(\mathbf{x}; \mathbf{c}_i, a_i) \qquad (11)$$

The equation further is written in the following compact form,

$$y = \mathbf{Q}^T \mathbf{H} \qquad (12)$$

where $\mathbf{H}$ is the hidden matrix weighted by the normalized firing strength of fuzzy rules and $\mathbf{Q}$ is the parameter matrix for TSK model, respectively. The details can refer to [5].

3.2 *Fuzzy-neural controller*

The function $\hat{\mathbf{f}}(\mathbf{z})$ in the control law τ_f will be modelled using equation (12) and given as,

$$\hat{f}(\mathbf{z}) = \mathbf{Q}^T \mathbf{H}(\mathbf{z}; \mathbf{c}, \mathbf{a}) \qquad (13)$$

The parameters $\{(\mathbf{c}, \mathbf{a})\}$ in Eqn. (13) are updated using OS-Fuzzy-ELM algorithm developed based on the Extreme Learning Machine [7] for adjusting the parameters of the fuzzy neural network. In OS-Fuzzy-ELM, the parameters of the fuzzy membership functions need not be adjusted during training and one can randomly assign the values to them. And then the consequent parameters are analytically determined. Performance of OS-Fuzzy-ELM has been evaluated on some benchmark problems in the areas of function approximation and compared with other existing fuzzy algorithms. The results show that the proposed OS-Fuzzy-ELM has superior performance in terms of the generalization ability and training speed.

Here, OS-Fuzzy-ELM is utilized to determine the parameters $\{(\mathbf{c}, \mathbf{a})\}$. In OS-Fuzzy-ELM, all these parameters could randomly be generated according to any given continuous probability distribution without any prior knowledge about the target function. When the antecedent parameters ($\{\mathbf{c}, \mathbf{a}\}$) are determined by assigning the random values, the fuzzy neural network becomes a linear approximator in the consequent parameters. But different from the original OS-Fuzzy-ELM algorithm, the consequent parameter ($\mathbf{Q}$) is adjusted based on the adaptive law derived using the Lyapunov second method, in which the stability of the entire control system is satisfied.

According to the universal approximation theorem of neural fuzzy systems, there exists optimal FNN estimator to approximate the nonlinear dynamic function $\mathbf{f}(\mathbf{z})$ in equation (5) such that

$$\mathbf{f}(\mathbf{z}) = \mathbf{Q}^{*T} \mathbf{H}(\mathbf{z}) + \varepsilon(\mathbf{z}) \qquad (14)$$

where $\varepsilon(\mathbf{z}) \in \mathbf{R}^n$ is the approximation error. Using the approximation theory, the inherent approximation error $\varepsilon(\mathbf{z})$ can be reduced arbitrarily by increasing the number of fuzzy rules and thus it is reasonable to assume that $\varepsilon(\mathbf{z})$ is bounded, i.e., $\|\varepsilon(\mathbf{z})\| \leq \bar{\varepsilon}$. $\mathbf{Q}^*$ is the optimal parameter and defined as follows:

$$\mathbf{Q}^* \triangleq \operatorname{argmin} \left[\sup_{z \in M_z} \left\| \mathbf{f}(\mathbf{z}) - \hat{\mathbf{f}}(\mathbf{z}) \right\| \right] \qquad (15)$$

where $\|\cdot\|$ represents the two-norm of a vector, M_z is the predefined compact set of input vector $\mathbf{z}$.

On the basis of equation (7) and equation (14), the tracking error of equation (4) becomes as,

$$\mathbf{M}(\mathbf{q})\dot{\mathbf{s}} = \tilde{\mathbf{Q}}^T \mathbf{H}(\mathbf{z}) - \mathbf{C}(\mathbf{q}, \dot{\mathbf{q}})\mathbf{s} - \mathbf{Ks} - \tau_s + \varepsilon \qquad (16)$$

where $\tilde{\mathbf{Q}}(= \mathbf{Q}^* - \mathbf{Q})$ is the parameter error. The control objective of asymptotic tracking now becomes the problem of adapting the parameters ($\mathbf{Q}$) in order to make the approximation error as small as possible. The parameter ($\mathbf{Q}$) is updated using the stable adaptive law for satisfying the stability of the entire system, which is illustrated in the following.

3.3 *Stable adaptive laws*

To derive the stable tuning law, consider the following Lyapunov function candidate,

$$V = \frac{1}{2} \mathbf{s}^T \mathbf{M}(\mathbf{q})\mathbf{s} + \frac{1}{2\eta_1} \operatorname{tr}(\tilde{\mathbf{Q}}\tilde{\mathbf{Q}}^T) \qquad (17)$$

where η_1 is a positive constant representing the learning rate.

73

Using equation (16), the derivative of the Lyapunov function is given as,

$$
\begin{aligned}
\dot{V} &= \mathbf{s}^T \mathbf{M}(\mathbf{q})\dot{\mathbf{s}} + \frac{1}{2}\mathbf{s}^T \dot{\mathbf{M}}(\mathbf{q})\mathbf{s} + \frac{1}{\eta_1}\mathrm{tr}(\tilde{\mathbf{Q}}\dot{\tilde{\mathbf{Q}}}^T) \\
&= \mathbf{s}^T \tilde{\mathbf{Q}}\mathbf{H}(\mathbf{z}) - \mathbf{s}^T \mathbf{C}(\mathbf{q},\dot{\mathbf{q}})\mathbf{s} + \mathbf{s}^T(-\mathbf{Ks} - \tau_s + \varepsilon) \\
&\quad + \frac{1}{2}\mathbf{s}^T \dot{\mathbf{M}}(\mathbf{q})\mathbf{s} + \frac{1}{\eta_1}\mathrm{tr}(\tilde{\mathbf{Q}}^T \dot{\tilde{\mathbf{Q}}})
\end{aligned}
\tag{18}
$$

Using property 2, $1/2\, \mathbf{s}^T \dot{\mathbf{M}}(\mathbf{q})\mathbf{s} - \mathbf{s}^T \mathbf{C}(\mathbf{q},\dot{\mathbf{q}})\mathbf{s} = 0$, $\forall\, \mathbf{s} \in R^n$ Eqn. (18) becomes

$$
\dot{V} = \mathrm{tr}(\tilde{\mathbf{Q}}\mathbf{H}(\mathbf{z})\mathbf{s}^T) + \mathbf{s}^T(-\mathbf{Ks} - \tau_s + \varepsilon) + \frac{1}{\eta_1}\mathrm{tr}(\tilde{\mathbf{Q}}\dot{\tilde{\mathbf{Q}}}^T)
\tag{19}
$$

where $\mathbf{s}^T \tilde{\mathbf{Q}}\mathbf{H}(\mathbf{z}) = \mathrm{tr}(\tilde{\mathbf{Q}}\mathbf{H}(\mathbf{z})\mathbf{s}^T)$ is utilized.
If $\tilde{\mathbf{Q}}$ is selected as,

$$
\dot{\tilde{\mathbf{Q}}}^T = -\eta_1 \mathbf{H}(\mathbf{z})\mathbf{s}^T
\tag{20}
$$

Equivalently, since $\dot{\tilde{\mathbf{Q}}} = -\dot{\mathbf{Q}}$, the tuning rule of the consequent parameters $\mathbf{Q}$ is given by,

$$
\dot{\mathbf{Q}}^T = \eta_1 \mathbf{H}(\mathbf{z})\mathbf{s}^T
\tag{21}
$$

Equation (19) becomes,

$$
\dot{V} = -\mathbf{s}^T \mathbf{Ks} - \mathbf{s}^T \tau_s + \mathbf{s}^T \varepsilon
\tag{22}
$$

To make Eqn. (22) less than or equal to zero, the sliding mode control term τ_s is chosen as

$$
\tau_s = \bar{\varepsilon}\,\mathrm{sgn}(\mathbf{s})
\tag{23}
$$

Since $\|\varepsilon\| \le \bar{\varepsilon}$, Eqn. (22) can be written as follows:

$$
\dot{V} \le -\mathbf{s}^T \mathbf{Ks} - \bar{\varepsilon}\|\mathbf{s}\| + \|\mathbf{s}\|\|\varepsilon\| \le -\mathbf{s}^T \mathbf{Ks} \le 0
\tag{24}
$$

The above equation illustrates that $\dot{V}$ is negative semidefinite and thus the overall control scheme is guaranteed to be stable. Using Barbalat's lemma, it can be seen that as $t \to \infty$, $\mathbf{s} \to 0$. In turn, this implies that the tracking error of the system will converge to zero.

In the following section, a two-link robot arm will be performed to evaluate the proposed fuzzy-neural controller.

4 SIMULATIONS

In this section, the performance of the proposed fuzzy-neural controller is evaluated on a two-link

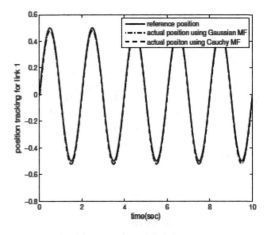

Figure 1.   Position tracking of link 1.

robot manipulator used widely in the literature. Moreover OS-Fuzzy-ELM can be applied into any bounded nonconstant piecewise continuous membership function and thus both Gaussian and Cauchy fuzzy membership functions are studied in the problem to verify the effectiveness of the proposed controller.

For brevity, the details about the dynamic equation of the two-link manipulator is referred to [?]. The initial states are $\mathbf{q}(0) = 0$ and $\dot{\mathbf{q}}(0) = 0$. The parameters associated with the controller design are as follows: $\lambda = \mathrm{diag}(5,5)$, $\mathbf{K} = \mathrm{diag}(100,100)$, $\eta_1 = 10$, the modelling error magnitude $\bar{\varepsilon} = 0.01$, the number of fuzzy rules $\tilde{N} = 5$, the membership function parameter $(\mathbf{c}, a)$ are chosen randomly in the intervals $[-1,1]$ and $[0,1]$ respectively.

The simulation results of the position tracking are illustrated in Figure 1, where solid line represents the reference signals, the dash dot line represents the results using Gaussian fuzzy membership function and the dash line represents the results using Cauchy fuzzy membership function. From the figure, it can be found that the actual outputs of the link can follow the reference outputs very well. In addition, the simulation results depict that the controllers constructed using Gaussian fuzzy membership function and Cauchy fuzzy membership function have quite similar tracking performance.

5 CONCLUSIONS

In this paper, a robust fuzzy-neural control scheme for an n-link robot manipulator is proposed. In the scheme, the fuzzy neural network is used to approximate the control input and their parameters are modified based on the OS-Fuzzy-ELM algorithm where the antecedent parameters are assigned randomly. The consequent parameters are adjusted based on the stable laws derived from a Lyapunov function for ensuring the asymptotic stability of the proposed control system. Besides, to attenuate the modelling error of fuzzy neural network, a sliding mode controller is augmented with the fuzzy-neural controller. Computer simulation studies of a two-link arm verify the adaptation and tracking performance of the proposed control approach. Also, two different membership functions named Gaussian and Cauchy membership functions are utilized to evaluate the proposed scheme and obtain similar results.

ACKNOWLEDGMENTS

This work is funded in part by National Natural Science Foundation of China (Grant No. 61004055), National Science Council of ShaanXi Province (Grant No. 2012 JQ8032 and 2014 JM8337) and the Fundamental Research Funds for the Central Universities.

REFERENCES

Chen, C.S. Robust self-organizing neural-fuzzy control with uncertainty observer for mimo nonlinear systems, IEEE Transactions on Fuzzy Systems, 19(4): 694–706.

Er, M.J. & Gao, Y. 2003, Robust adaptive control of robot manipulators using generalized fuzzy neural networks, IEEE Transactions on Industrial Electronics, 50, (3): 620–628.

Huang, G.-B., Zhu, Q.-Y., Mao, K.~Z., Siew, C.-K., Saratchandran, P. & Sundararajan, N. 2006, Can threshold networks be trained directly? IEEE Transactions on Circuits and Systems II, 53(3):187–191.

Lim, C.M. & Hiyama, T. 1991, Application of fuzzy logic control to a manipulator, IEEE Transactions on Robtics and Automation, 70(5): 688–691.

Rong, H.-J., Huang, G.-B., Sundararajan, N. & Saratchandran, P. 2009, Online sequential fuzzy extreme learning machine for function approximation and classification problems, IEEE Transactions on Systems, Man, and Cybernetics, Part B: Cybernetics, 39(4): 1067–1072.

Tagaki, T. & Sugeno, M. 1985, Fuzzy identification of systems and its application to modelling and control, IEEE Transactions on Systems, Man, and Cybernetic, 15(1): 116–132.

Yi, S.Y. & Chung, M.J. 1997, A robust fuzzy logic controller for robot manipulators with uncertainties, IEEE Transactions on Systems, Man, and Cybernetics, Part B: Cybernetics, 27(4): pp.706–713.

Control Engineering and Information Systems – Liu (Ed)
© 2015 Taylor & Francis Group, London, ISBN 978-1-138-02685-8

The statistics method of key parameters in network control system

L. Dong
College of Information Science and Engineering, Northeastern University, Shenyang, Liaoning, China
College of Information Engineering, Shenyang University, Shenyang, Liaoning, China

J.H. Wang
College of Information Science and Engineering, Northeastern University, Shenyang, Liaoning, China

H. Zhou
College of Information Science and Engineering, Northeastern University, Shenyang, Liaoning, China
T-Fine Automation Co., Ltd., Shenyang, Liaoning, China

S.S. Gu
College of Information Science and Engineering, Northeastern University, Shenyang, Liaoning, China

ABSTRACT: A random processing method is studied for the key parameters in network control system. The network time delay and packet loss are the most important problem in network control system. There are both the random variable, the time delay accords to Poisson distribution and the packet loss accords to the binomial distribution. The method that measures the time delay and packet loss is present in this paper. The arithmetic mean of the observed values can use of the mathematical expectation of the network time delay. At last, two dimension random variable probability distribution of network control system is given for more studying.

1 INTRODUCTION

The literature on modeling, analysis and controller design of Networked Control Systems (NCSs) expanded rapidly over the last decade [1–3]. The use of networks offers many advantages such as low installation and maintenance costs, reduced system wiring (in the case of wireless networks) and increased flexibility of the system. However, from a control theory point of view, the presence of the network also introduces several disadvantages such as time-varying network-induced delays, aperiodic sampling or packet dropouts. To understand the impact of these network effects on control performance several models have been developed. Roughly speaking, these NCS models can be categorized into continuous-time and discrete-time models. A further discrimination can be given on the basis of which network phenomena they include.

In the continuous-time domain, [4–5] applied a descriptor system approach to model the sampled-data dynamics of systems with varying sampling intervals in terms of (infinite-dimensional) Delay Differential Equations (DDEs) and study their stability based on the Lyapunov-Krasovskii functional method.

In [6–7], this approach is used for the stability analysis of NCSs with time-varying delays and constant sampling intervals, using linear matrix inequality-based techniques. The recent results in [8] also involve H_∞ controller designs based on Linear Matrix Inequalities (LMIs). However, Mirkin [9] showed that the use of such an approach for digital control systems neglects the piecewise constant nature of the control signal due to the zero-order-hold and delay jointly introduce a particular linearly increasing time-varying delay within each control update interval (sometimes indicated by the sawtooth behaviour of the delay), whereas in the modeling approach mentioned above it is replaced by an arbitrary bounded time-varying delay.

The majority of NCS models are discrete-time formulations based on the exact discretization of the continuous-time linear plant over a sample interval [10–14]. Such models avoid the problem of an infinite-dimensional state that is encountered in the continuous-time (DDE) models due to delays. Moreover, in these discrete-time models the piecewise constant nature of the control signal due to the zero-order hold is taking into account exactly.

Additionally, it has been shown in [15], that for systems with aperiodic sampling and time-varying delay less than the sampling interval the use of discrete-time models for stability analysis gives less conservative characterization of stability than the use of (impulsive) delay differential equations. Under simplified assumptions, such that the delay is a multiple of the sampling interval or it takes values in a finite set, the obtained models lead to switched linear systems and corresponding stability conditions can be applied. [16–17].

However, these models are not so realistic as in practice one typically encounters an infinite number of possible values for the delay. Moreover, more realistic models should take into account that the sampling periods might be aperiodic.

In the literature, two ways of modeling network-induced uncertainties (such as time-varying delays and sampling intervals and packet dropouts) can be distinguished. Firstly, in [18–19], bounds are imposed on the delays, sampling intervals and the maximum number of subsequent dropouts. Secondly, in [20–21], a stochastic modeling approach is adopted. In this paper, we will adopt the first approach. Given bounds on delays, sampling intervals and subsequent dropouts, we will formulate stability conditions and constructive controller synthesis results independent of the probability distribution of the uncertain variables. So, such robust results also apply in the stochastic setting and can be seen as 'probability distribution-free' results for the stochastic case if the domain of the probability distribution function is bounded.

This paper is structured as follows: In Section 2 we present a method to measure the time delay in network control system. Section 3 is to discuss the probability distribution and the digital features of the network time delay. Section 4 is research the random variable time delay and packet loss probability distribution of network control system. Section 5 closes with concluding remarks.

2 THE MEASURE OF NETWORK TIME DELAY

According to a specific network control system, such as TCP/IP network, when the network parameter is configured, we can test the performance of the network control system. The network time delay and packet loss are the most important problem in network control system. We will measure the time delay with the following method, and get the rule of the time delay and packet dropout distribution in network control system.

A controller or computer control multiple nodes, and each node is a single loop feedback control system in Figure 1. If the object in network control system are a complex industrial process and have multiple sensors and actuators, we can resolve to multiple single loop control system with DCS control system. So the above structure has the universal significance. The method of

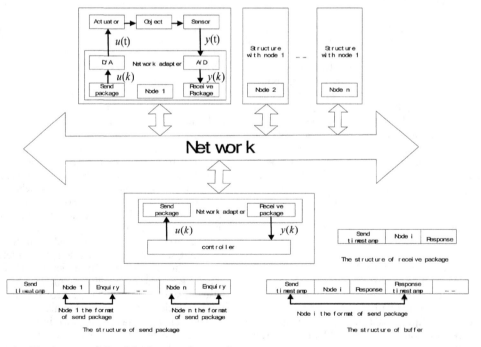

Figure 1. The measure of time delay in network control system.

real-time measurement process of network delay τ is this:

1. The controller (or computer) sends a contract awarding program to network by interface in the form of radio. The program data has the information of time delay measurement packets, its structure as shown in the bottom left corner in Figure 1. The time delay measurement packets include send timestamp and node information.
2. All the network node will receive the time measurement packets by network adapter. When the NO. i node receive the time measurement packets and confirm the response information of the packets, this node will produce a response information packet. This packet is time delay measurement receive packet, its structure as shown in the bottom right corner in Figure 1. This packet is send to network with the contract awarding program by network interface.
3. The controller (or computer) receives the response information packet of the NO. i node with the receiver function package by network adapter. This information packet append the receive timestamp and deposit to the receive information cache area.
4. The calculation of the ith node in the network time delay: τ_i = receive timestamp (the ith node)—send timestamp (the ith node).
5. Repeat the above (1)–(4) process, we will real-time measure the network delay of the NO. i node. We are going to get the rule of the network time delay distribution.

3 THE PROBABILITY DISTRIBUTION AND THE DIGITAL FEATURES OF THE NETWORK TIME DELAY

After testing the network experiment, we find that the time delay is an uncertain random variable in network control system. The rule of the time delay in network control system, similar to the density problem, is in line with poisson distribution.

The network time delay random variable τ can be zero and all the positive integer $m = 0,1,2,...$, while the probability $P(\tau = m)$ is that

$$P_\lambda(\tau) = \frac{\lambda^\tau}{\tau!} e^{-\lambda} \qquad (1)$$

where $\lambda > 0$ is constant. The time delay in network control system accord with the Poisson distribution.

All possible values of the discrete random variable τ is τ_i. The product of τ_i and the probability $P(\tau = \tau_i)$ is the mathematical expectation of the random variable τ, and write for $E[\tau]$.

If the measure values of the random variable τ are finitude,

$$\tau_1, \tau_2, ..., \tau_l$$

While the probability of these values are

$$p(\tau_1), p(\tau_2), ..., p(\tau_n)$$

So the mathematical expectation of the random variable τ is

$$E[\tau] = \tau_1 p(\tau_1) + \tau_2 p(\tau_2) + ... + \tau_n p(\tau_n) = \sum_{i=1}^{n} \tau_i p(\tau_i)$$

$$(2)$$

If the measure values of the random variable τ are infinity,

$$\tau_1, \tau_2, ..., \tau_n, ...$$

While the probability of these values are

$$p(\tau_1), p(\tau_2), ..., p(\tau_n), ...$$

So the mathematical expectation of the random variable τ is the sum of the following series.

$$E[\tau] = \tau_1 p(\tau_1) + \tau_2 p(\tau_2) + ... + \tau_n p(\tau_n) + ...$$
$$= \sum_{i=1}^{\infty} \tau_i p(\tau_i) \qquad (3)$$

We suppose that these series are absolute convergence, so the sum of series have nothing to do with the order of the various.

From the above experiment, we can get that the random variable τ accord with the Poisson distribution. From the formula (1) and (3), we can get

$$E[\tau] = \sum_{\tau=0}^{\infty} \tau \frac{\lambda^\tau}{\tau!} e^{-\lambda} = \lambda e^{-\lambda} \sum_{\tau=1}^{\infty} \frac{\lambda^{\tau-1}}{(\tau-1)!} \qquad (4)$$

using $k = \tau - 1$, we can get

$$E[\tau] = \lambda e^{-\lambda} \sum_{k=0}^{\infty} \frac{\lambda^k}{k!} = \lambda e^{-\lambda} e^{\lambda} = \lambda \qquad (5)$$

It follows that the parameter of the Poisson distribution λ is the mathematical expectation of the variable τ. When we compute the probability of the Poisson distribution, the λ is equal to the mean value of the each period of network time delay. It is in fact that we can use the arithmetic mean of the observed values $\bar{\tau}$ to substitute the mathematical expectation $E[\tau]$.

4 TWO DIMENSION RANDOM VARIABLE PROBABILITY DISTRIBUTION OF NETWORK CONTROL SYSTEM

The time delay and packet loss are the most important problem in network control system. When we design the controller in network control system, these factors should be fully taken into account. The time delay submits to Poisson distribution in most network protocol. And the loss packet accord with the binomial distribution.

From the Figure 1, we can measure the case of the network packet loss, and get the packet loss rate. The network time delay τ and the packet loss l are both random variable. In some network protocol, the τ is zero or positive integer value and l is zero or one. When $l = 0$, no packet loss occurs, $l = 1$, the packet loss occurs.

Generally speaking, the rate of packet loss is very small for the sake of reliability in the network control system. The packet loss l accord with the binomial distribution. The network packet loss random variable l can be zero or one, while the probability $P(l = k)$ is that

$$P_k = P(l = k) = \binom{n}{k} p^k (1-p)^{n-k}, k = 0, 1, ..., n \quad (6)$$

where k is the frequency of the packet loss, n is the total number of the experiment, p is the rate of the packet loss.

$$\frac{P_{k-1}}{P_k} = \frac{(1-p)k}{p(n-k-1)} \leq 1 \quad (7)$$

$$\frac{P_k}{P_{k+1}} = \frac{(1-p)(k+1)}{p(n-k)} \geq 1 \quad (8)$$

We can obtain the most probable frequency of the packet loss from the formula (6), (7) and (8). We can get the formula (9).

$$(n+1)p - 1 \leq k \leq (n+1)p \quad (9)$$

If the rate of the packet loss p is 0.001, and the total number of the experiment n is 5000. We can get the most probable number of the packet loss and the corresponding probability form formula (7), (8) and (9).

$$k - [(n+1)p] = [(5000+1) \times 0.001] = 5 \quad (10)$$

$$P_{5000}(5) = \binom{5000}{5}(0.001)^5(0.999)^{4995} \approx 0.1756 \quad (11)$$

Table 1. Time delay and packet loss probability distribution.

l	τ			
---	τ_1	τ_2	...	τ_n	...
0	$p(\tau_1, 0)$	$p(\tau_2, 0)$		$p(\tau_n, 0)$	
1	$p(\tau_1, 1)$	$p(\tau_2, 1)$		$p(\tau_n, 1)$	

We can gain probability of the packet loss at least in the network control system from the formula (6).

$$P(k \geq 1) = 1 - P(k < 1) = 1 - P(0) = 0.9934 \quad (12)$$

From the formula (12), we can get the conclusion that after a long period running, no matter how small packet loss rate, the network packet loss phenomenon must have occurred.

So we can get two dimension random variable probability distribution of network control system. The probability of two dimension random (τ, l) is that the random variable τ gets the value of τ_i and the random variable l gets the value of l_j.

$$P(\tau_i, l_j) = P(\tau = \tau_i, l = l_j), i = 1, 2, ..., n, ...; j = 0, 1 \quad (13)$$

From the formula (13), we can get the probability distribution table.

The mathematical expectation of the random variable τ and l can be gotten form the formula (14).

$$Ef(\tau, l) = \sum_i \sum_j f(\tau_i, l_j) p(\tau_i, l_j) \quad (14)$$

where $f(\tau_i, l_j)$ is the function of the random variable τ and l. The form of the $f(\tau_i, l_j)$ is different in the various network control systems. We can get the value of the $p(\tau_i, l_j)$ from the Table 1, so we can get the mathematical expectation of the random variable τ and l.

5 CONCLUSION

In this paper, we present a method to measure the time delay in network control system. The time delay accords to Poisson distribution and the packet loss accords to the binomial distribution. The arithmetic mean of the observed values can use of the mathematical expectation of the network time delay. The probability distribution and the digital features of the key parameters are researched by the random processing method in network control system.

REFERENCES

Cloosterman M.B.G, Hetel L, van de Wouw N, Heemels W.P.M.H, Daafouz J. and Nijmeijer H. Controller synthesis for networked control systems [J]. Automatica, 2010, 46(10), 1584–1594.

Cloosterman, M.B.G., van de Wouw, N., Heemels, W.P.M.H. and Nijmeijer, H., Robust stability of networked control systems with time-varying networkinduced delays [C], In Proc. of the 45th IEEE conference on decision and control, San Diego, CA, USA, 2006, 4980–4985.

Felicioni F.E. and Junco S.J. A Lie algebraic approach to design of stable feedback control systems with varying sampling rate [C]. In Proceedings of the 17th IFAC world congress, Seoul, Korea 2008, 4881–4886.

Fridman E, Seuret A. and Richard J.P. Robust sampleddata stabilization of linear systems: an input delay approach [J]. Automatica, 2004, 40, 1441–1446.

Fridman E, Shaked U. Stability and guaranteed cost control of uncertain discrete delay systems [J]. International Journal of Control, 2005, 78(4), 235–246.

Gao H, Chen T. and Lam J. A new delay system approach to network-based control [J]. Automatica, 2008, 44(1), 39–52.

Gu K, Kharitonov V.L. and Chen J. Stability of time delay systems [M]. Boston: Birkhauser, 2003.

Han C.Y. and Zhang H.S, Linear optimal filtering for discrete time systems with random jump delays, Signal Processing, 2009, 89(6), 1121–1128.

Mirkin, L. Some remarks on the use of time-varying delay to model sampleand-hold circuits [J]. IEEE Transactions on Automatic Control, 2007, 52(6), 1109–1112.

Tipsuwan Y. and Chow M.Y. Gain scheduler middleware: a methodology to enable existing controllers for networked control and teleoperation-Part 1: networked control [J]. IEEE Transactions on Industrial Electronics, 2004, 51(6): 1218–1227.

Tipsuwan, Y. and Chow, M.Y. Control methodologies in networked control systems [J]. Control Engineering Practice, 2003, 11(10), 1099–1111.

van de Wouw N, Naghshtabrizi P, Cloosterman M. and Hespanha J.P. Tracking control for networked control systems [C]. In Proc. of the 46th IEEE conference on decision and control, New Orleans, LA, USA, 2007, 4441–4446.

van de Wouw N., Naghshtabrizi P, Cloosterman M. and Hespanha J.P. Tracking control for sampled-data systems with uncertain time-varying sampling intervals and delays [J]s. International Journal of Robust and Nonlinear Control, 2010, 20(4), 387–411.

Wang Z.D, Ho D.W.C. and Liu X.H, Robust filtering under randomly varying sensor delay with variance constraints [J], IEEE Transactions on Circuits and Systems, 2004, 51(6), 320–326.

Yang Y.H, Fu M.Y. and Zhang H.S, State Estimation Subject to Random Communication Delays [J], Acta Automatica Sinica, 2013, 3, 39(3), 237–243.

Yu M, Wang L. and Chu T.G, et al. Stabilization of networked control systems with data packet dropout and transmission delays: continuous-time case [J]. European Journal of Control, 2005, 11: 40–49.

Yue D, Han Q.L. and Lam J. Network-based robust H_∞ control of systems with uncertainty [J]. Automatica, 2005, 41: 999–1007.

Yue D, Tian E.G. and Wang Z.D, et al. Stabilization of systems with probabilistic interval input delays and its application to networked control systems [J]. IEEE Transactions on Systems, Man, and Cybernetics, Part A: System and Humans, 2009, 39(4): 939–945.

Zhang, W, Branicky, M.S. and Phillips, S.M. Stability of networked control systems [J]. IEEE Control Systems Magazine, 2001, 21(1), 84_99.

Zhu X.L. and Yang G.H. Stability analysis and state feedback control of networked control systems with multi-packet transmission [C]. American Control Conference. Seattle, 2008: 3133–3138.

Zhu X.L. and Yang G.H. State feedback controller design of networked control systems with multi-packet transmission [J]. International Journal of Control, 2009, 82(1): 86–94.

Control Engineering and Information Systems – Liu (Ed)
© 2015 Taylor & Francis Group, London, ISBN 978-1-138-02685-8

Network based dead-beat response control strategy of the heating stove control system

X.S. Huang
Zhejiang Institute of Mechanical and Electrical Engineering, Hangzhou, Zhejiang Province, P.R. China

J.P. Jiang
College of Electrical Engineering, Zhejiang University, Hangzhou, Zhejiang Province, P.R. China

ABSTRACT: Network with random time delay based dead-beat response control strategy of the heating stove control system has been studied in this paper. At first, the mathematical models of the networked based heating stove control system both in frequency domain and discrete state equation have been established, then using time-stamped BP neural network method to predict the delay time of network is presented. The digital controller with dead-beat response design method has been given, and the control algorithms has been derived. Finally, the realizable conditions of control system is analysed. The simulation results illustrate that the accurate temperature control without steady-state error can be achieved. The system has good dynamic and static performance.

1 INTRODUCTION

Heating stove networked control system is which refers to the formation of the closed-loop heating stove control system by the Internet. There are series of advantages of the networked control system. First, it can realize remote control. When heating stove operate in extreme environments, or not be involved in the distance, it needs using the Internet to remote control and monitor. Second, any node on the Internet can be operated and enhance system functions. Third, the wealth of information resources can be obtained from internet. Forth, especially with connecting to the Internet by a variety of mobile communications equipment, many special functions can be realized in the networked control systems. In addition, it can simplify hardware structure of the control systems and reduce the cost, etc. Therefore, study on the networked control system is to be very cost effective and also is a common concern research topic in nowadays.

Although the networked control system has many advantages, but also has the obvious lack, which affect its widespread use. A main drawback of the networked control system is existence of information transmission delay, which is related with the communication protocols, transmission capacity, transmission distance, transmission time and other factors. So delay time is an uncertain,

time-varying, random variable. Besides, there is existence of information transmission packet loss or leakage in the networked control systems. Those above bring more difficulties to the networked control system modeling, control, and optimization, and even make system unstable.

This paper focuses on the networked heating stove control system with uncertain information transmission delay. The main content of the paper may be divided into four parts:

1. An algorithm by the time stamp BP neural network for uncertain information transmission delay time predictive was developed.
2. Using the predicted value of time delay, the mathematical model of the heating stove system networked based both in frequency domain and discrete-time domain representations had been established.
3. In order to speed up the temperature dynamic response and achieve precise temperature control, the dead-beat strategy was applied, and the controller design method was given.
4. The physical conditions can be realized of the dead-beat digital controller were discussed. A design method of dead-beat digital controller was presented when the system has unstable pole. Finally, the simulation result shows the effectiveness of the proposed control algorithms.

2 MATHEMATIC MODEL OF THE HEATING STOVE SYSTEM BASED ON NETWORKED CONTROL

2.1 Simplified structure diagram of the networked-based heating stove control system

The simplified structure diagram in frequency domain for the networked heating stove control system as shown in Figure 1.

Where τ_{sc} is the network information transmission delay time of feedback loop; τ_{CA} is the network information transmission delay time of feed-forward loop; τ_h is delay time of the heating stove itself; Total delay time of the networked control system of heating stove is

$$\tau_N = \tau_{sc} + \tau_{cA} + \tau_h \tag{1}$$

$Ke^{-\tau_h s}/(T_h s+1)$ is the open-loop transfer function of the heating stove itself; $G_D^*(s)$ is z transfer function of the dead-beat digital controller; $Z.O.H$ is z transfer function of the zero order holder. Considering the network information transmission time-delay, open-loop transfer function of the networked heating stove control system can be expressed as

$$\frac{Y(s)}{R(s)} = \frac{K}{(T_h s+1)}e^{-\tau_N s} \tag{2}$$

Due to delay time is a uncertainty, time-varying, random variable, so the time-stamped BP neural network method to predict delay time is used in this paper. If the actual delay value $\tau_N(k)$ of system is replaced with predictive value $\hat{\tau}_N(k)$ in real time, the equation (2) can be rewritten as

$$\frac{Y(s)}{R(s)} = \frac{K}{(T_h s+1)}e^{-\hat{\tau}_N s} \tag{3}$$

2.2 Time-stamped BP neural network prediction of the network transmission delay time [3][4]

Work process of the Time-Stamped BP Neural Network (TSBPNN) is that: In current time $(k-1)$, a

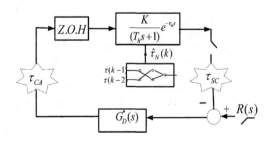

Figure 1. Simplified structure diagram of the networked heating stove control system.

data packet is sent out from the controller, and the send time is marked as τ_{TS}^{CA} by time-stamped. When the data packet reaches the server through the Internet and implements the control of the controlled object, then returns back to controller through the Internet, the arrival time is marked as τ_{TS}^{SC}. The information transmission delay time is obtained:

$$\hat{\tau}(k) = \tau_{TS}^{CA}(k-1) - \tau_{TS}^{SC}(k-1) \tag{4}$$

After a period of time, the available delay time data series can be obtained in controller side:

$$\tau_i\{\tau(i):i=1,2,\ldots k-1\}$$

According to those data series, the one step ahead delay time predicted value can be obtained by BP neural network. This article uses 2-2-1 structure of the BP neural network, as shown in Figure 2. There are 2 neurons in input layer; 2 neurons in hidden layer, and one neuron in output layer.

Using the former two times delay values $\tau(k-1), \tau(k-2)$ as the input of the BP neural network, the delay time predicted value $\hat{\tau}_N(k)$ can be created after the neural network learning and training. Assuming

$$e(k) = \tau_N(k) - \hat{\tau}_N(k) \tag{5}$$

The last time error $e(k)$ can be used to train the neural network in next time $(k+1)$.

Define energy function or performance index

$$E(k) = \frac{1}{2}e(k)^2 = \frac{1}{2}[\tau(k) - \hat{\tau}(k)]^2 \tag{6}$$

The question focuses on how to set up the connection weight coefficients to minimize the cost function $E(k)$. The first-order gradient optimization method is used. In the $(k+1)$ time, the connection weights are adjusted as

$$\left.\begin{array}{l} w_{1i}^{(2)}(k+1) = w_{1i}^{(2)}(k) + \eta^{(2)}\delta_1^{(2)}x_i^{(1)} \\ \quad + \lambda(w_{1i}^{(2)}(k) - w_{1i}^{(2)}(k-1)) \\ \\ w_{ji}^{(1)}(k+1) = w_{ji}^{(1)}(k) + \eta^{(1)}\delta_j^{(1)}x_i^{(0)} \\ \quad + \lambda(w_{ji}^{(1)}(k) - w_{ji}^{(1)}(k-1)) \\ \quad\quad\quad\quad i = 1\sim 2; j = 1\sim 2 \end{array}\right\} \tag{7}$$

where $\eta^{(1)}$ is learning rate for [0, 1].
In output layer

$$\delta_1^{(2)} = [\tau(k) - \hat{\tau}(k)]f'(s_1^{(2)}) \tag{8}$$

where $f'(s_1^{(2)})$ is derivative of the activation function. If linear activation function is used in this output layer, then $f'(s_1^{(2)}) = 1$ and

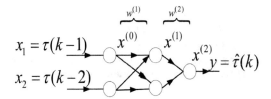

$$x_1 = \tau(k-1)$$
$$x_2 = \tau(k-2)$$

Figure 2. BP neural network structure.

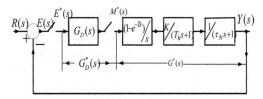

Figure 3. The frequency domain transfer function diagram of the networked heating stove control system.

$$\delta_1^{(2)} = [\tau(k) - \hat{\tau}(k)] = e(k) \qquad (9)$$

In the hidden layer and input layer

$$\delta_j^{(l)} = f'(s_j^{(l)}) \sum_{k=1}^{2} \delta_k^{(2)} w_{kj}^{(2)} \, (j = 1 \sim 2) \qquad (10)$$

In order to make convergence faster and inhibit oscillates in the learning process, a momentum factor λ is introduced ($0 \le \lambda < 1$).

2.3 Mathematical model of the networked heating stove control system

Equation (3) shows the open-loop transfer function of the networked heating stove control system, and strictly speaking, it includes the infinite dimensional factor[5], namely

$$e^{-\hat{\tau}_N s} = \frac{1}{\displaystyle\sum_{i=0}^{n} l_i s^i} \qquad i \in N; n \in N; l_i = \frac{1}{i!}\tau_k^i;$$

If those higher order terms in the factor are ignored, then, n = 1 and the equation can be expressed as

$$e^{-\hat{\tau}_N s} = \frac{1}{(\tau_N s + 1)} \qquad (11)$$

Substitute equation (11) into equation (3) yields open-loop transfer function of the networked heating stove control system. It can be expressed as

$$\frac{Y(s)}{R(s)} = \frac{K}{(T_h s + 1)} \frac{1}{(\tau_N s + 1)} \qquad (12)$$

Now Figure 1 (simplified structure diagram of the networked heating stove control system) can be rewritten as Figure 3.

In Figure 3, $(1 - e^{-Ts})/s$ is zero order hold transfer function; T is sampling period; $G_D(s)$ is digital controller transfer function with dead-beat response; $G_D^*(s)$ is digital controller transfer function; $G^*(s)$ is combinatorial z transfer function of the zero order holder and controlled object.

3 THE DIGITAL CONTROLLER WITH DEAD-BEAT RESPONSE

3.1 Design principle of dead-beat response digital controller

The so-called dead-beat response is: when the control system of the typical input (unit step, speed and acceleration input), the system output of the steady-state error is zero, and the response of the system is the fastest. Take the "star" transformation for system shown in Figure 3, and obtain

$$Y(s) = E^*(s)G_D^*(s)G(s) \qquad (13)$$

where $G(s) = \dfrac{(1 - e^{-Ts})}{s} \dfrac{K}{(T_h s + 1)(\tau_N s + 1)}$,

Take "star" transformation for equation (13)

$$Y^*(s) = E^*(s)G_D^*(s)G^*(s) \qquad (14)$$

Due to

$$E^*(s) = R^*(s) - Y^*(s) \qquad (15)$$

Substitute Eq. (15) into Eq. (14), and then yield

$$Y^*(s) = (R^*(s) - Y^*(s))G_D^*(s)G^*(s)$$

$$\frac{Y^*(s)}{R^*(s)} = \frac{G_D^*(s)G^*(s)}{1 + G_D^*(s)G^*(s)} = K^*(s) \qquad (16)$$

Rewriting equation (16) into z transfer function

$$\frac{Y(z)}{R(z)} = \frac{G_D(z)G(z)}{1 + G_D(z)G(z)} = K(z) \qquad (17)$$

where $G(z)$ is combination z transfer function of the zero order holder and the networked heating stove control system; $K(z)$ is z transfer function of the closed-loop control system; $G_D(z)$ is z transfer function of the digital controller with dead-beat response.

In order to make the networked heating stove system can obtain the accurate temperature control, and the temperature steady-state error of system output is zero, so a digital controller with dead-beat response is used. According to the final-value theorem, the steady-state error of the system can be obtained from the error of z transfer function.

From equations (15) and (17), obtain

$$E(z) = R(z) - Y(z) = R(z)[1 - K(z)] \qquad (18)$$

According to the final-value theorem, the steady-state error of the system can be expressed as

$$\lim_{k \to \infty} e(k) = \lim_{z \to 1}(1 - z^{-1})E(z)$$
$$= \lim_{z \to 1}(1 - z^{-1})R(z)[1 - K(z)] \qquad (19)$$

The input signal of the heating stove is unit step function

$$R(z) = \frac{1}{(1 - z^{-1})} \qquad (20)$$

From equations (19) and (20), to make the steady-state error of the system is zero, as long as meet the following equations

$$[1 - K(z)] = (1 - z^{-1})F(z) \qquad (21)$$

where $F(z)$ is an optional item. Choose $F(z) = 1$

$$K(z) = z^{-1} \qquad (22)$$

Substitute Eq. (22) into Eq. (17):

$$\frac{Y(z)}{R(z)} = \frac{G_D(z)G(z)}{1 + G_D(z)G(z)} = z^{-1} \qquad (23)$$

By equation (23), we receive z transfer function of the digital controller for the networked heating stove control system with dead-beat response, namely

$$G_D(z) = \frac{z^{-1}}{G(z)(1 - z^{-1})} \qquad (24)$$

According to the type (24) designed digital controller, from equations (19), (22) and (23), see not hard, the steady-state error of the networked heating stove system is zero, and the temperature accurate control of the system is realized. It will not produce temperature deviation. It can be seen from equation (23) that the output response of the system lag behind the input signal only one sampling period. It is undoubtedly the fastest response. So, the system designed has achieved dead-beat response.

3.2 The algorithms of digital controller with dead-beat response

1. Find the combination z transfer function of the zero order holder and the networked heating stove control system. Obtain from Figure 3

$$G(z) = K(1 - z^{-1})Z\left[\frac{1}{s(T_h s + 1)(\tau_N s + 1)}\right]$$
$$= \frac{Az + B}{z^2 + Ez + D} \qquad (25)$$

where

$$A = -K\left[\left(e^{-\frac{T}{T_h}} + e^{-\frac{T}{\tau_N}}\right) + \frac{T_h}{(\tau_N - T_h)}\left(e^{-\frac{T}{\tau_N}} + 1\right)\right.$$
$$\left. + \frac{\tau_N}{(T_h - \tau_N)}\left(e^{-\frac{T}{T_h}} + 1\right)\right]$$

$$B = K\left[e^{-\frac{T}{T_h}}e^{-\frac{T}{\tau_N}} + \frac{T_h}{(\tau_N - T_h)}e^{-\frac{T}{\tau_N}}\right.$$
$$\left. + \frac{\tau_N}{(T_h - \tau_N)}e^{-\frac{T}{T_h}}\right]$$

$$E = -\left(e^{-\frac{T}{T_h}} + e^{-\frac{T}{\tau_N}}\right)$$

$$D = e^{-\frac{T}{T_h}}e^{-\frac{T}{\tau_N}} \qquad (26)$$

2. Find the digital controller z transfer function. From equations (24), (25), obtain

$$G_D(z) = \frac{z^{-1}}{G(z)(1 - z^{-1})} = \frac{1 + Ez^{-1} + Dz^{-2}}{A + (B - A)z^{-1} - Bz^{-2}} \qquad (27)$$

3. Find the control algorithm of the digital controller. From Figure 3, obtain

$$\frac{M(z)}{E(z)} = G_D(z) \qquad (28)$$

Substitute Eq. (27) into Eq. (28), take the inverse z transfer function of $M(z)$, the control algorithm of the digital controller can be expressed:

$$m(k) = \frac{1}{A}e(k) + \frac{E}{A}e(k - 1) + \frac{D}{A}e(k - 2)$$
$$- \frac{(B - A)}{A}m(k - 1) + \frac{B}{A}m(k - 2) \qquad (29)$$

4. Realizable conditions analysis for the digital controller. The design of the digital controller must be physical is achieved, the conditions are:

a. For the digital controller $G_D(z)$, the order of molecular must be less than or equal to the denominator order;

b. When the composite z transfer function of the zero order holder and the networked heating stove control system is spread out into z^{-1} form, then in the closed-loop z transfer function, the minimum order of $K(z)$ must be equal to or higher than that of $G(z)$. For example, if $G(z)$ start from z^{-1} order, the first item of the $K(z)$ must be equal to zero, that is

$$K(z) = \alpha_0 + \alpha_1 z^{-1} \ldots + \alpha_n z^{-n} = \alpha_1 z^{-1} \ldots + \alpha_n z^{-n} \tag{30}$$

where α_0 must be equal to zero.

c. The $G_D(z)$ can be realized in physical conditions, but also must include stability conditions. If the $G(z)$ has one zero of the unit circle outside, such as

$$G(z) = \frac{z + 1.5}{z^2 + 1.4z + 0.4} \tag{31}$$

In order to make the digital controller does not contain unstable pole, it must take the z transfer function of closed-loop system:

$$K(z) = k(z + 1.5)z^{-2} \tag{32}$$

where k is undetermined coefficient. Setting the system input with the unit step function, the k value can be determined. By equations (21), (32), obtain

$$1 - k(z + 1.5)z^{-2} = (1 - z^{-1})F(z) \tag{33}$$

Choose

$$F(z) = a_0 + a_1 z^{-1} \tag{34}$$

Substitute Eq. (34) into Eq. (33), then

$$K(z) = 0.4(z + 1.5)z^{-1} \tag{35}$$

From equations (35), (23) and (31)

$$G_D(z) = \frac{K(z)}{G(z)[1 - K(z)]} = \frac{0.4(1 + 1.4z^{-1} + 0.4z^{-2})}{(1 - 0.4z^{-1} - 0.6z^{-2})} \tag{36}$$

4 SIMULATION

4.1 The technical data of tested stove

Type: SX2–4-10; Rated power: 4 kw; Rated voltage: 220v; Rated temperature: 1000°C; Heating stove time constant: $T_h = 300$ s; Heating stove delay time: 100 s; Heating stove system magnification: K = 7.5.

4.2 Algorithm

1. Setting the input signal $R(s)$ as the unit step representation; choose the sampling period: T (in seconds); setting the number of calculation value K_N; setting $k = 0$; Calculating coefficient: A; B; C; D;

2. Setting $m(k)$ and error initial value $e(k)$;

3. $k = k+1$, $e(k) = r(k)[1 - z^{-1}] = r(k) - r(k-1)$;

4. Calculation control algorithm of the digital controller

$$m(k) = \frac{1}{A}e(k) + \frac{E}{A}e(k-1) + \frac{D}{A}e(k-2)$$
$$- \frac{(B-A)}{A}m(k-1) + \frac{B}{A}m(k-2);$$

5. Calculation $y(k)$ from Figure 3 and equation (25)

$$\frac{Y(z)}{M(z)} = G(z) = \frac{Az^{-1} + Bz^{-2}}{1 + Ez^{-1} + Dz^{-2}}$$

namely,

$$y(k) = Am(k-1) + Bm(k-2)$$
$$- Ey(k-1) - Dy(k-2);$$

6. setting

$$\begin{cases} e(k-2) = e(k-1); e(k-1) = e(k); \\ m(k-2) = m(k-1); m(k-1) = m(k); \end{cases}$$

7. If $k < K_N$, return to step (4), otherwise stop the calculation.

4.3 The simulation results and conclusions

The simulation results show in Figure 4 and Figure 5.

The unit step response of the system is shown in Figure 4. The curve 1 corresponds to the unit step dynamic response of system with sampling period 60 seconds; Curve 2 corresponds to the unit step dynamic response of system with sampling period 120 seconds. The square wave input response of the system is shown in Figure 5. The curve 1 corresponds to the square wave input pulse dynamic response of system with sampling period 60 seconds. The curve 2 corresponds to the square wave input response with sampling period 120 seconds.

Figure 4 and Figure 5 show that whether with an unit step input or a square wave input, the system has dead-beat response characteristics and the shortest transition process, that is, the steady-state error of the system is zero, and the system output lags behind the input signal only one sampling period. It's clearly in Figure 4 and Figure 5 when the sampling period is shorter, the system transition time also reduce. But it must be pointed out that when sampling cycle reduced to a certain value, because the amplitude of system input signal is limited (namely the energy of the system is limited), the system output cannot meet the dead-beat response rules. It must also be pointed out that, the input signal of this paper studying on the heating stove networked control system is the unit step representation. For the servo system, the input signal is a speed function; For some other control systems, the input signal may be acceleration functions. Speed function and acceleration function are respectively

$$R(z) = \frac{1}{(1-z^{-1})^2} \text{ and } R(z) = \frac{1}{(1-z^{-1})^3}$$

By equations (21) and (23), is not hard to find out the corresponding closed-loop transfer function and digital controller transfer function.

The deficiency of the dead-beat response control strategy is: the mathematical model accuracy of controlled object must be ensured. System dynamic and static performance will be significantly affected by the model error. Therefore, the user must pay attention to modeling of the controlled object.

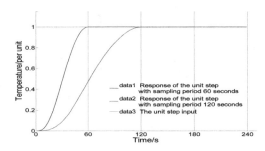

Figure 4. Dynamic response of the system with unit step input.

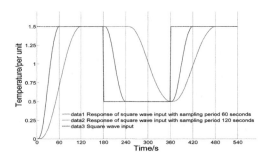

Figure 5. Dynamic response of the system with square wave signal input.

REFERENCES

Deepak Srinivasagupta & Heinz Schattler. 2004. Time-Stamped model predictive control: An algorithm for control of processes with random delays [J]. *Computers and Chemical Engineering*, 28(8):1337–1346.

Guobo, Xiang. 2008. *Optimization control of time-delayed systems* [M]. Beijing: China Electric Power Press.

Hu, S & Weiyong, Yan. 2009. Stability of networked control systems subject to input and output packet loss [C]// *The 28th Chinese Control Conference CDC/CCC*: 15–18.

Lining, Sun et al. 2004. Research on neural network based prediction algorithm of time delay [J]. *Robot*, 26(3): 237–240.

Qzh, Zhang & Wd, Zhang. 2006. Time-stamped predictive functional control in networked control systems [J]. *Journal of Control Theory and Applications*, 23(1): 126–130.

Rees, D. 2006. Networked Predictive Control of Internet Based Systems [C]// *The 25th Chinese Control Conference*: 516–521.

Xinlan, Wang et al. 2011. Adaptive predictive functional control for networked control systems with random delays [J]. *International Journal of Automation and Computing*, 8(1):62–68.

Yi, Liu & Hexu, Sun. 2010. Time-stamped multivariable generalized predictive control of network control systems [C]// *Computer Design and Applications (ICCDA) International Conference*: 211–215.

Control Engineering and Information Systems – Liu (Ed)
© 2015 Taylor & Francis Group, London, ISBN 978-1-138-02685-8

The design of plant combustion control system

L.J. Hu, Y.J. Yang & B. Hai

Electric Power College, Inner Mongolia University of Technology, Hohhot, Inner Mongolia, China

ABSTRACT: Combustion control system is a cascade-type control system that is composed of the main steam pressure and combustion rate. The system is based on load instructions that the main steam pressure regulator offers to control the amount of fuel, send-air volume and lead-air volume to coordinate changes. By analyzing the combustion control system and on the basis of setting principle of cascade-type control system, three subsystems and the main steam pressure regulator parameters are respectively set, and the control system is simulated integrated in MATLAB. The simulation shows the system can complete the task of control perfectly, maintain the steam pressure and the furnace pressure stable and guarantee the combustion process efficiency.

1 INTRODUCTION

In the operation of thermal power plant boiler, steam load is often changes, the amount of fuel that must be timely adjust the supply boiler load change, to adapt to the need of load. Automatic regulation mainly includes the automatic feed water regulator, steam temperature automatic regulation and automatic adjustment of boiler combustion process. The combustion process of automatic regulation requires not only the heat provided by burning fuels adapt to the need of load, but also ensures economical combustion and safety operation.

The boiler heat balance of input and output is used to characterizing the steam pressure, so load regulation is the pressure regulator, while pressure control is through the adjustment of fuel to maintain a regulated pressure for a certain value. As an important parameter of the boiler steam pressure state, not only directly related to the safe operation of boiler equipment, but also the stability reflects the relationship between supply and demand of energy in combustion process. So to maintain the stability of main steam pressure is an important task, and then to maintain the boiler hearth negative pressure stability, Maintain the normal operation of the fuel system, and Ensure the economy of the combustion process.

2 THE COMPOSITION OF THE COMBUSTION CONTROL SYSTEM

Control system of the combustion process is composed of main steam pressure control and combustion rate control of cascade system. Main steam pressure control loop can be regarded as the system of the primary loop. Combustion rate control can be seen as inner loop of the system. Burning rate control circuit is made up of multiple parallel subsystem parameter ratio control system. In accordance with the pressure regulator (The master regulator) is given of the load instruction (Combustion rate) LD (Li zhunji 1997). To control fuel quantity, send air volume, ventilation rate into the appropriate ratio of coordinate change, to ensure furnace calorific value corresponding to the load instruction (Bian lixiu, 2001). Boiler combustion control system for the use of direct blowing pulverizing system, the combustion rate control also includes other subsystem. Combustion structure as shown in Figure 1 of control system.

In Figure 1, pressure control task is to maintain the stability of main steam pressure, the main controller W_{PII} according the change of the main steam pressure PM, to load instructions issued to each subsystem of LD combustion rate

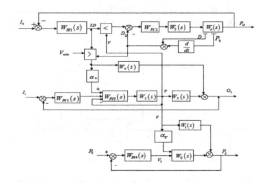

Figure 1. Combustion control system block diagram.

control. Amount of fuel quantity circuit to load instructions LD as given value. When load changes, compared with the small value selector and total quantity signal by LD, to ensure that the total wind-coal ratio in the process of combustion. The output of small value selector as a given signal, and compared with the comparator and heat signal, by the regulator W_{PI2} to adjust the difference for coal. W_s as fuel quantity controlled object, W_0 as steam pressure controlled object. Steam pressure controlled object W_0 can lead to the steam flow D and steam drum pressure Pb, steam drum pressure P_b by the differentiator and D together constitute D_Q, fuel quantum system and the main steam pressure system composed of cascade system.

The air supply control subsystem according to the combustion rate adjust supply air flow in the furnace, so as to ensure the thermal load of furnace to meet the needs of the change of the outside load, while ensuring the combustion economy (HU shousong, 2001). Flue gas oxygen content correction circuit is based on a deviation of flue gas oxygen content and the given value, correction of two air volume, so as to ensure optimum excess air.

As the furnace negative pressure can rapidly reflect the disturbance of air supply and air draught, according to chamber pressure PS regulation to ensure coordinated variation of suction fan and blower, so as to maintain the stability of the furnace pressure PS. (Wen qunying 2006).

3 REGULATOR SETTING AND SIMULATION

3.1 The furnace pressure control system engineering tuning

The task of furnace pressure control system is to maintain coordination of air quantity and air volume change, for a given value of negative pressure of furnace hearth (Zhang liangming, Xia dujuan, 2006). The equivalent diagram of furnace pressure control system is shown in Figure 2. The given value is given by a given device, W_{PI4} is the furnace pressure regulator, $W_1(s)$ is the air flow disturbance

transfer function of the furnace pressure P_s. $W_0(s)$ is the transfer function of generalized controlled object control channel. Regulator feedforward branch on behalf of α_V, the general design of the differential device (Zhang zidong, Wang huaibin, LI bingxi, 1994).

The furnace pressure control system tuning parameters including differential feedforward branch α_V and furnace pressure regulator W_{PI4}. W_{PI4} is a proportion differential link, so tuning parameters for δ_s and T_{is}. (Cehng laijiu, 1984).

When the input is 0.5; step signal of 1 for input of air volume V. α_V setting result is 0.6. The results of the controller parameter W_{PI4} setting for the $T_{is} = 5s$, $\delta_s = 100\%$. If the design meets the system requirements, the chamber pressure P_s should not be affected by the wind disturbance, can eventually restored to the given value 0.5. Figure 3 is the furnace pressure P_s response curve.

From the curve can be seen in the furnace pressure P_s there is a first step to rise, reaching a peak of 0.85, show that the addition of air flow disturbance effect on the system, but by regulating the regulator, and quickly restored to the given value 0.5. In the more than 40 seconds have reached the given value, the regulator parameter selection well enough, it can be quickly eliminate the disturbance of air, to keep up with the given value (Huang zhonglin, 2001).

3.2 The fuel control system engineering tuning

The fuel control system using heat as feedback signal, to reflect changes in the amount of fuel, so control system of fuel heat setting should be first setting signal D_Q (Weng siyi, 1987). Setting heat signal D_Q is choosing the differentiator parameters T_D, K_D. The principle of differentiator parameters T_D, K_D is that the heat signal can accurately reflect the changes in the amount of fuel, and outside interference (i.e. steam turbine regulating valve changes) heat signal should remain unchanged.

Desired value LD = 0.5, Steam turbine valve opening μ_T is 0.5, Fuel quantity adjusting parameters tuning $\delta = 22\%$, $T_i = 50s$. The differentiator

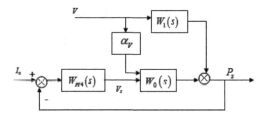

Figure 2. Furnace pressure control system structure diagram.

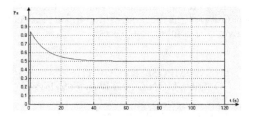

Figure 3. Furnace pressure control system response curve.

parameter $W_0(s)$ setting results as shown in Figure 4 and 5, figure below response curves can be obtained by simulation.

It is shown that variation of response curve of heat signal can accurately follow the fuel quantity, in 18 seconds on the follow up fuel signal, therefore, selection of parameters for the obtained heat signal is well enough.

3.3 Air control system engineering tuning

Air control system is the air traffic control system. Regulation of air flow is regulated by two times the total air volume, ensure that the combustion process of total air-coal ratio. Air conditioning subsystem consists of two air conditioning circuit, signal measuring circuit of total air and flue gas oxygen content correction circuit. Air control system structure diagram as shown in Figure 6.

LD by α_D as the air flow rate of a given value, W_{P15} as a Oxygen correction regulator, W_{P13} as a furnace pressure controller. The furnace pressure controller according to the total air volume and the given value, the deviation between the adjusted secondary air volume. Flue gas oxygen content correction circuit is based on a deviation of flue

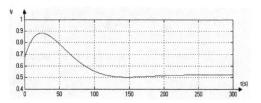

Figure 7. Flue gas oxygen response curve.

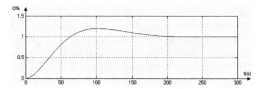

Figure 8. Supply air response curve.

gas oxygen content and the given value, correction of the secondary air volume, to ensure The optimum excess air. The content of setting includes Oxygen correction regulator W_{P15}, furnace pressure controller W_{P13} and feedforward branch α_D. (Owens, D.H., 1978).

Results α_D tuning is 0.7, Oxygen correction parameter W_{P15} tuning is carried out according to the dynamic characteristics of $W_3(s)$. The air supply control loop system diagram in approximate proportion link. By flue gas oxygen content in air output disturbance under the response curve of available oxygen correction regulator parameters. The simulation results of air control system.

Diagram of oxygen content in flue gas is given a value of 1, LD as a air control system of a given value, the signal input LD = 0.5.

3.4 Simulation of the combustion control system

Through the front face of the subsystems of the combustion control system parameter setting, based on the system structure diagram will be setting a good parameter generation into the system, as shown in Figure 9. On the system simulation, get the response of the main steam pressure curve, flue gas oxygen content of the response curve and hearth negative pressure response curve.

Main steam pressure is given in the Figure is 10, flue gas oxygen content of the given value is 30%, the furnace negative pressure set point of −50 Pa. Air control system of a given value is The load instruction LD. Air flow at the same time as the perturbation of the furnace negative pressure control system. Respectively setting the furnace pressure regulator W_{P14} and furnace pressure feedforward and branch of the system α_v; The fuel quantity

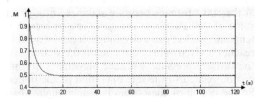

Figure 4. Fuel signal response curve.

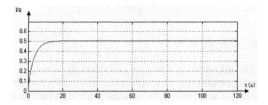

Figure 5. Heat signal response curve.

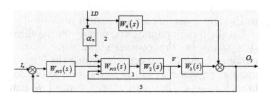

Figure 6. Supply air control system structure diagram.

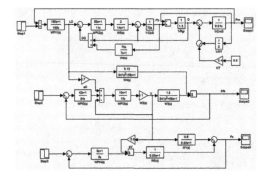

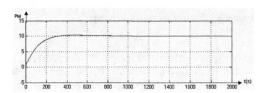

Figure 9. Combustion control system simulation diagram.

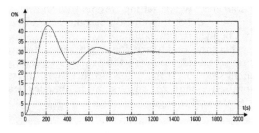

Figure 11. Flue gas oxygen response curve.

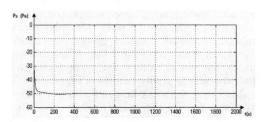

Figure 12. Furnace pressure control system response curve.

PM
15
10
5
0
-5
0 200 400 600 800 1000 1200 1400 1600 1800 2000 t(s)

Figure 10. Main steam pressure response curve.

control system of the heat signal D_Q and fuel regulator W_{PI2}; The furnace pressure controller W_{PI3} and feed forward branch α_p and The content of setting includes Oxygen correction regulator W_{PI5}. At last, Main steam pressure setting of the main steam pressure control loop of the regulator W_{PI1}. The results as shown in Figure 10. The main steam pressure response curve, in the form of flue gas oxygen content of the response curves and hearth negative pressure response curve is shown below.

As can be seen from the figure, the main steam pressure reaches a given value in 850 s, maintain the stability of main steam pressure, also is the completion of the combustion control system of the main task.

The flue gas oxygen content in a given value is 30%, the Figure 11 can be seen the given value reaching 30% in 1400 s, In the curve can also know that some of its dynamic performance index, the rise time t_r is 120 s, so The response of the system so quickly. The peak time is 230 s, the overshoot is 43.3%. Therefore, through the analysis of response curves, setting the parameters of the system is well enough, so that the oxygen content in flue gas can reach the given value as soon as possible, to ensure that the economy of the combustion process. And the final, response curve of furnace pressure.

It can be seen from the figure, the chamber pressure is 0 Pa at the initial time, then decreased rapidly, the first time reach to a given value at

150 s. Since then has been in a given value of about −50 Pa to do a small change, until the time arrives at the 800 s furnace pressure stabilized at a given value of −50 Pa. From the curve can be seen system tuning effect can well enough, it can be quickly reaches a given value, to maintain the stability of the furnace pressure. Furnace pressure is normal or not, related to the safe operation of the boiler. So the design of the system is well enough, when there is a disturbance, after the pressure regulator control, the system can recover soon for a given pressure, thereby restoring normal running very quickly (Norman Anderson, 1989).

4 CONCLUSION

According to the design of combustion control system mission requirements, the combustion process control system for the overall analysis. And then According to the parameter tuning principle of the system of three subsystems respectively amount of fuel combustion control system, the air supply control system, the furnace negative pressure control system controller are set, system simulation.

On the timing parameters, mainly on the three subsystems of the combustion control system parameter setting, respectively, setting the furnace pressure regulator and furnace system feedforward branch of α_V; The fuel quantity control system of the heat signal and fuel regulator; The furnace pressure controller and feedforward branch and

The content of setting includes Oxygen correction regulator, at last, Main steam pressure setting of the main steam pressure control loop of the regulator. Each subsystem parameters setting of system simulation is carried out after completion of all, the main steam pressure can be seen in the simulation results follow the system for a given pressure value, which completed the stable main steam pressure is the main task; Delivery follow the load instruction LD given value, Flue gas oxygen content in the end also remains at a given value; Hearth negative pressure in hearth negative pressure control system follow the given value, and is not affected by disturbance of air output.

So, the whole system tuning results of combustion control system can achieve the task requirements, to ensure that the system in a safe, economic conditions.

ACKNOWLEDGMENTS

Author introduction: Hu Linjing, female, master tutor, the associate professor at the Electric Power College Inner Mongolia University of Technology, engaged in teaching, and research of power system and thermal automatic control. The research funded by the Education Department Science Project of Inner Mongolia (NJZY13103) and the Natural Science Foundation of Inner Mongolia (2013MS0919).

REFERENCES

Bian lixiu. Thermal Control System [M]. Beijing: China Electric Power Press. 2001. 53–145.

Chen laijiu. Thermal processes and application of automatic control theory [M]. Beijing: Beijing Water Power Press. 1982.

Hu shousong. Principles of Automatic Control [M]. Beijing: Science Press. 2001.

Huang Zhonglin. MATLAB computational and control system simulation [M]. Beijing: National Defence Industry Press. 2001, 11.

Li zunji. Thermal Control System [M]. Beijing: China Electric Power Press. 1997. 54–106.

Norman Anderson. Instrumentation for Process Measurement and Control [M]. U.S.A. Published in Radnor. Pennsylvania. 1989.

Owens, D.H. Feedback and Multivariable Control System [M]. Peter Peregrines, London 1978.

Wen qunying. Thermal control system [M]. Beijing: China Electric Power Press. 2006:89–94.

Weng Siyi. Automatic control system computer aided design and simulation [M]. Xi'an: Xi'an Jiaotong University Press. 1987.

Zhang liangming, Xia dujuan. Industrial boiler thermal detection and process control [M]. Tianjin: Tianjin University Press. 2000.

Zhang zidong, Wang huaibin, Li bingxi. Boiler automatically adjust [M]. Harbin: Harbin Institute of Technology Press. 1994.

Design of the specific analog module and the control logic based on FPGA

Z. Feng & M. Yang
Department of Computer Application, Shijiazhuang Information Engineering Vocational College, Shijiazhuang, China

X.F. Di
Electronic Information Specialty Group, Shijiazhuang Information Engineering Vocational College, Shijiazhuang, China

ABSTRACT: The specific analog module which takes the FPGA as the control core is designed. Then the control scheme is given and the structure of the logic is designed. In order to validate the correctness of the work, an experiment is done. The result shows that the specific analog module can realize the two functions: the A/D conversion with analog current input in eight channels and the D/A conversion with analog currentoutput in four channels.

1 INTRODUCTION

Special analog module can realize mutual conversion between digital and analog. The module can be widely used in medical equipment, and it has a very broad application prospects.

At present in the control method of A/D transformation and D/A transformation in two kinds of methods are commonly used: one is based on MCU or CPU as the control core, auxiliary control with the peripheral interface chip; another kind is FPGA as the core control. Two methods have different characteristics, control module, single-chip microcomputer as the core of the writing process is relatively simple, but now many models of SCM has integrated the A/D, D/A module, the control is very flexible. But because the SCM processing speed is relatively low, so this kind of control method is generally suitable for low speed sampling occasions. In contrast, using FPGA as the control core module has great advantages in speed, has the advantages of short development cycle, the general capacity is good, easy to develop, expand, not only reduces the design difficulty, but also accelerate the product development cycle. Thus in the high-speed data acquisition is widely used in control systems.

2 HARDWARE CIRCUIT DESIGN

The special analog module is mainly composed of FPGA chip EP1C6T144I7, A/D chip ADS8327 and D/A chip DAC7634, complemented by the necessary peripheral circuits, such as power supply circuit, grounding circuit, input circuit, output circuit, isolation circuit, analog switches, oscillation circuit. According to the hardware structure of special analog module and the function module, which can be divided into three parts: FPGA and drive circuit, A/D conversion circuit, D/A conversion circuit. Hardware block diagram shown in Figure 1.

We can see from Figure 1, the FPGA in the whole working process of analog module to PC data and command receiving and control signal generation effect. According to the working principle of FPGA input and output signal module can be divided into the following five groups: the first group is the read/write signal, the main role is to realize the communication of FPGA and PC; the second group is the drive control signal, the main role is to control the direction of data transmission between FPGA and PC; the third group is the A/D control signal, the main role is to achieve the control of A/D

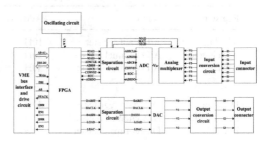

Figure 1. Hardware principle diagram.

conversion; the fourth group is the D/A control signal, the main role is to control the D/A conversion; the fifth group is the analog multiplexer address signal, the main role is to select the channel adc.

3 CONTROL LOGIC DESIGN

Partition corresponding to hardware, the special simulation control logic module can also be divided into three parts. The logic structure diagram is shown in Figure 2.

Through the VME bus communication between host computer and FPGA in the diagram, PC to send commands and data to the FPGA to realize the control of the whole module, the VME bus logic can be used as a top-level logic module. And because the module can realize A/D conversion and D/A conversion of two kinds of function, no connection between these two kinds of function, so the two part of the logic can be used as two parallel of the underlying logic module.

VME bus interface control logic is responsible for the communication between FPGA and PC in Figure 2, data and command receives the control signal is given, and the underlying logic, in addition to and the module for data transmission; A/D control signal and the data logic and D/A control module receives the logic is given, and a given control signal ADC and DAC the ADC and DAC, control.

Special analog module using register control mode, so in the FPGA internal definition of the six registers, as listed in Table 1. The host computer through the register read and write operations to achieve the control of A/D conversion and D/A conversion.

3.1 Implementation of VME bus logic state machine

This module main function is to realize communication between FPGA and PC, PC and control for the A/D control logic control design and D/A logic design to produce the control signal. In order to realize PC to read and write FPGA, this module uses the state machine implementation of receiving and sending data. Design of state machine as shown in Figure 3, a total of four states: idle, read, write, end state state. State machine three state to complete a read/write operation.

3.2 Realization of A/D control logic state machine

In order to realize the control of the A/D chip, a finite state eight state machine design, it is the main body of the A/D conversion control logic. The eight state of the state machine can be divided into two groups: S1 to S5 to A/D converter during normal working state, S6 to S8 a reset state. Figure 4 is the state machine, the machine is in the S1 state after power on.

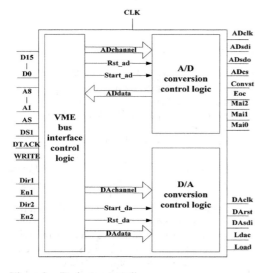

Figure 2. Logic structure diagram.

Table 1. FPGA internal register definition.

Register	Name	Address	Effect
AD data	A/D data register	10H	Storage conversion results.
AD channel	A/D channel register	20H	The lower three bits are analog switches.
AD ctrl	A/D control register	30H	Fifteenth high given the A/D control logic starting signal Start_ad, fourteenth bit high given the A/D control logic reset signal Rst_ad.
DA data	D/A data register	40H	Stored data to be converted.
DA channel	D/A channel register	50H	High eight D/A channel selection code.
DA ctrl	D/A control register	60H	Fifteenth high given the D/A control logic starting signal Start_da, fourteenth bit high given the D/A control logic reset signal Rst_da.

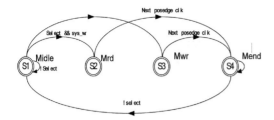

Figure 3. VME bus read/write state machine.

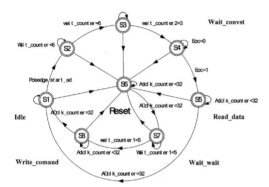

Figure 4. A/D control logic state machine.

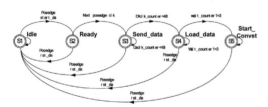

Figure 5. D/A control logic state machine.

3.3 *Realization of D/A control logic state machine*

The D/A control logic state machine and A/D control logic state machine is similar to a total of five states, as shown in Figure 5.

4 EXPERIMENT

An experiment platform is collaborated in order to proving the accuracy of the hardware and control logic. It includes the following parts: Panel, Embedded computer (VME), the voltage source, DVM, a general purpose computer. The data collected is displayed in the form of waves using Quartus II and SignalTap II. The result is showed in Figure 6.

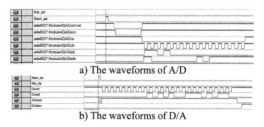

a) The waveforms of A/D

b) The waveforms of D/A

Figure 6. The result of experiment.

5 CONCLUSION

It includes a complete design procedure of module to special analog signal, such as design of hardware circuit, established of the control system, the preparation and implementation of specific logic, simulation and testing of the logic.

1. Control logic includes three modules, VME Bus control logic, D/A and A/D convertion control logic. VME Bus control logic is top—level of the whole logic, D/A and A/D convertion control logic are parallel bottom level of the whole logic.
2. VME Bus control logic is mainly composed of two major sections, controler of read—write state and logic of internal control signals generate. It can achieve communication between FPGA and the upper computer. It generates internal logic control signals and controls the work of bottom level modules.
3. A/D convertion control logic includes controler of A/D state convert and SPI Bus control logic. It controls proper operation of ADC and realizes A/D convertion of eight analog input currents.
4. D/A convertion control logic includes controler of D/A state convert and SPI Bus control logic. It controls proper operation of DAC and realizes D/A convertion of four analog output currents.

REFERENCES

Huang Ronglan, Wan De. FPGA control module design based on A/D conversion and acquisition [J]. China sciencepaper Online, 2009.

Li Jinghu. Research of high speed and low power A/D converter and related technology [D]. Haerbin: Doctoral Dissertation of Harbin Institute of Technology. 2009:3–4.

Li XiuRong, Li zhuoxuan. Application and development of A/D conversion technology [J]. Application of technology of photoelectric. 2010, 25(3): 45–48.

Ramniceanu, Spiridon, Eynde. A 6 bitresolution 1 G samples/sec analog to digital converter. Semiconductor Conference, 2004 [J]. CAS2004 Proceedings, 2004, 13(6): 191–194.

Control Engineering and Information Systems – Liu (Ed)
© 2015 Taylor & Francis Group, London, ISBN 978-1-138-02685-8

A defects classification method for aerospace measurement and control software

C. Ma, J.R. Li, C.L. Sun, J.D. Yang & Z.F. Zhou
China Satellite Maritime Tracking and Control Department, Jiangyin, China

ABSTRACT: With the development of Aeronautics and Space undertakings, various civil and military satellite, manned spacecraft and deep space probes will enter orbit, this raises many challenging requirements for the aerospace measurement and control software. If the defects in software can not be discovered and exposed adequately in testing process, serious consequences and great loss may be caused once the software is delivered in orbit. In order to establish a systemic method to analyze the introduction, discovering, positioning, harm and root-cause of aerospace measurement and control software defects, in the paper we analyze the aerospace measurement and control software development process and defect types found in practical tecsting, then propose a classification method for aerospace measurement and control software defects, based on the results, we try to improve the testing effectiveness and stability of aerospace measurement and control software.

1 INTRODUCTION

The development of aerospace industry has incomparable importance to the position of a country in the world; software as the critical high-tech products of aerospace industry, its quality plays a decisive role in aerospace task. However, in recent years, aerospace measurement and control software defects have become the dominant cause of system unavailability; improvements in software reliability and quality have become highly imperative. As we know software defect is a natural state of the software itself, different software development stages may introduce different types of software defects [1][2], when the defects contained software code is executed, the implementation results will not be consistent with the anticipated results, consequently the system or function of the system can not meet the task requirements. That is to say software defect is the cause of software failure; in addition different software defects will have different consequences. For aerospace measurement and control software, if the defects in software can not be discovered and exposed adequately in testing process, serious consequences and great loss may be caused once the software is delivered in orbit.

The purpose of doing research on classification method of software defects [6][7], especailly of the aerospace measurement and control software, is to establish a systemic method to analyze the introduction, discovering, positioning, harm and root-cause of aerospace measurement and control software defects, afterwards in the process of software defect

management targeted testing effort can be focused on several specified types of defects, or sepcial testing tools can be studied and developed to find defects as soon as possible and repair defects completely.

2 COMMON DEFECT CLASSIFICATION METHODS

There are many software defect classification methods with different purpose, different viewpoint and complexity. Several typical software defect classification methods are as follows:

2.1 Putnam classification method

Putnam and Myers divide software defects into six categories [1]: requirement defect, design defect, algorithm defect, interface defect, performance defent and documentation defent; In the mean time, according to the consequences of the software defects and the action to be taken, the severity level of software defect is also graded as follows: fatal, serious, moderate and insignificant. This relatively simple method mainly covers three stages of software development including requirement analysis, software design and software implementation, but it's function to analyze and eliminate the software defects is limited.

2.2 Thayer classification method

Using the defect information obtained from problem reports filled by testing personnel in the testing process and fed back by software users, Thayer, according to the property of software defects, divide software defects into sixteen categories including calculation error, logical error, I/O error, data processing error, operating system and supporting software errors, configuration error, interface error,etc. This mechod mainly use the problem reports of the testing process to guide the development personnel to locate and eliminate the defects, however it doesn't consider other factors which may introduce defects in the whole software development process, besides it's analysis of defects' introduction stage and cause is inadequate.

2.3 *Orthogonal defects classification method*

Orthogonal Defects Classification (ODC) method [8] 978-1-4799-0641-3/13/$31.00:copyright: 2013 IEEE proposed by IBM company classify the defects into seven categories: assignment, inspection, algorithm, timing, interface, function and relevance.

The classification process consists of two steps. In the first step, when the defect is found, according to the environment causing defects to be found and the possible inpact on users, the following three attributes of defect can be identified: defect-detection activity, defect trigger and impace of defect. In the second step, when the defect is fixed and closed, five attributes of defect can be identified: defect carrier, defect type, defect determiner, defect age and defect source. The above eight attributes are critical to the prevention and elimination of defects.

This detailed classification method is applicable to the defects' location, elimination, cause analysis and prevention, the information provided by defect attributes is helpful to create the condition for preventing defects, eliminating defects and improving the software development process [2]. However, with the disadvantages of complexity to classify the defect and the difficulty of using the classification standards, this method is difficult to use in practice.

2.4 *IEEE classification method for anomaly*

IEEE classification method [3] holds that classification process consists of four procedures: recogniton, investigation, plan of action and implementation of treatment and the investigation procedure divideds the anomalies into eight categories: logical problem, calculation problem, interface error, timing errror, data processing problem, documentation problem, documentation quality problem, and enhancemen problem.

IEEE classification method classifies the software anomalies comprehensively and provieds a complete property framework for the whole defect classification process, but it's classification process is too diffcult, normally it needs to be refined or adapted to the specific domain and project environment. As a result, it is not helpful to classify and measure software defects quickly.

3 AEROSPACE MEASUREMENT AND CONTROL SOFTWARE DEFECTS CLASSIFICATION METHOD

We summarize the above traditional software defects classification methods. Because of containing insufficient defect information, some methods are too simple to help efficiently fixing the software defects. Without considering the reason and happening process of the defects, some methods is not appy to improve software development process. Some methods' classification standards is difficult to use in practice because of it's complexity. Therefore, combining the advantages and disadvantages of the above mentioned software defects classification methods, at the same time considering the development process of the aerospace measurement and control software especially the process of the defect's detection, location, elimination, and prevention, we propose a classification method of aerospace measurement and control software defects, it can be used to classify aerospace measurement and control software defects, meanwhile it provides accurate and complete help information to locate, fix and prevent aerospace measurement and control software defects.

The objective of classifying defects of aerospace measurement and control software is to measure the defects and analyze their occurred process and cause, improve, and then improve software development process, prevent the defect and enhance organizational capability maturity. Therefore, classification method for aerospace measurement and

Table 1. Defect's attributes and definition.

Attributes	Detailed meaning of attributes
Intruduction phase	Software development stage when defects are introduced
Detection phase	Software development stage when defects are discovered
Defect type	Defect type classified according to it's attributes
Programming language	Software programming language
Root cause	The fundamental cause of software defects
Severity	The severity of defect caused adverse impact on aerospace task

control software defects should satisfy the following requirements:

1. Classify the detected defects accurately
2. No overlap between categories, the classification system should cover all types of defects
3. The classification standard should combine with the software life cycle
4. The method should satisfy the developing process of aerospace measurement and control software and requirements of defects' detection, location, elimination, prevention.

We can define the following six demensions to describe attributes of aerospace measurement and control software defects, namely intruduction phase, detection phase, defect type, programming language, root cause of defects and severity, as shown concretely in Table 1.

According to the definition of software development phase, aerospace measurement and control software life circle contains ten phases, hence we define the following ten demensions to describe the intruduction phase and detection phase, namely system requirement analysis and design, software requirement analysis, architectural design, detailed design, software implementation, integration testing, validation testing, system joint testing, acceptance of delivery, operational maintenance. Generally speaking, introduction phase doesn't contain development phase operational maintenance beginning of software testing, but for the convenience of expression, intruduction phase and detection phase are defined consistently.

Defect type define the attribute values of defects' type, and the attribute values can be ascertained by means of describing the type of the actual repair

Table 2.   Defects types and forms.

Type number	Type	The main representation of defect
1	Requirement analysis defect	1. Missing requirement definition 2. Incomplete requirement 3. Requirement logical error 4. Requirement is not updated
2	Defect of design phase	1. Missing design 2. Incomplete design 3. Design error 4. Design is not updated
3	Functional defect	1. Function is not implemented 2. Incomplete function 3. Function error 4. Function is not updated
4	Performance defect	1. Performance is not implemented 2. Incomplete performance 3. Security defect 4. Reliability defect [4]
5	Data error	1. Wrong data definition 2. Wrong data access 3. Wrong data operation
6	Structure error	1. Wrong software structure 2. Wrong control flow
7	Interface error	1. Wrong interface definition 2. Wrong interface access 3. Wrong interface fault trentment
8	Software integration defect	1. Wrong intraface 2. Wrong external interface
9	Implementation and coding errors	Seen in the below detailed description
10	Test equipment defect	1. Equipment function defect 2. Equipment performance defect
11	Test software defect	1. Function is not implemented 2. Functon error
12	Test execution defect	1. Wrong test design 2. Wrong test data 3. Wrong test execution 4. Inadequate test case
13	Other defects	The other defects that are difficult to classify

work done on the defects. The traditional software defect classification methods differentiate software defects only considering existing problems of the software itself. But in the actual software testing, many of the problems and anomalies exposed are not software defects, in many cases, they are caused by the misunderstanding and misoperation of tester, or the defects of testing equipment and software. Given all that, in order to entriely reflect the real information of the aerospace measurement and control software testing process, we should incorporate the abnormal phenomena exposed in software testing process into the result of software defects classification, as can be seen in Table 2.

Requirement analysis defects are mainly introduced in the requirement analysis phase, so they can be eliminated by the perfection of requirement analysis, similarly, defect intoduced in the design phase can be eliminated by the perfection of design documents. Errors of function, performance, data, structure, interface, integration mainly reflect the real software defects caused by imperfect software requirement analysis or design, we can lilewise eliminate these errors by the perfection of requirement analysis, software design or by the modification of software code. Testing equipment, software and execution defects mainly reflect the unexpected results of testing process, which are caused by the wrong testing hardware, software or activities of tester. These defects can be eliminated by redesigning the test software-hardware platform, test methods or test case.

Further more, based on the different phases of software code compilation, we continue to classify the implementation and coding errors into the following four categories: lexical defect, grammar defect, semantic defect and maintainability defect. In this way, we can determine the detect-found compilation phase, as a result during this compilation phase we can change the compiler to partly find similar defects, correspondingly tester can improve the efficiency and accuracy of defect elimination.

The objective of lexical analysis is to recognize all the words and then inspect possible problems through the scanning and decomposition of character strings that make up the source code. Grammatical analysis is based on the lexical analysis to decompose word strings into various grammatical unit according to the grammar rule of programming language. If the coding error is not the above two kinds of defects, but the program is still different from what the developer expected, then that is semantic defect. Defects like problems of program comments, redundant variables and statement are maintainability defects. As mentioned above, we try to further classify the defects of implementation and coding errors to provide the basis for the prevention of defects, as can be seen in Table 3.

The programming languages of implementing the aerospace measurement and control software are maily the VB, C++, C# and Delphi.

Through investigating the root causes of aerospace measurement and control software defects

Table 3. Implementation and coding error.

Defect type	The main representation of defect
Lexical defect	1. Problems of number representation system, like missing the character of 'H' in hexadecimal notation 2. Confusion of similar operator, like " = " and " = = "
Grammar defect	1. Wrong macro definition 2. Missing return value of the program branch 3. Wrong usage of operator priority level 4. Mismatch problem of push and pop instructions
Semantic defect	1. Conflict of data access 2. Problem of variable initialization 3. Problem of judging the equality of floating point numbers 4. Problem of losing accuracy 5. Out of rage problem 6. Port access without using volatile 7. Problem of suspending scene protection 8. Dimension problem 9. Problem of dividing by zero 10. Stack overflow problem
Maintainability defect	1. Wrong comments 2. Ambiguous comments 3. Redundant variables 4. Redundant statement

Table 4. Severity level of defects.

Level number	Severity level	Meaning
1-2	Negligible	Affect system efficiency but don't affect the schedule and quality
3-4	Minor	Cause the system difficult to operate and affect system Interface beauty, but don't affect function and performance
5-6	Medium	System damage (failure of nonessential function) cause the software to be degraded or to modify the parameters (repairable on the spot)
7-8	Fatal	Failure of essential function, (unrepairable on the spot) cause to update the version of software
9-10	Disastrous	Invalid critical system cause the Space Misstion not to be completed

collected in recent three years, according to the defect repairing method, we divide the root causes of aerospace measurement and control software defect into the following ten categories: incomplete understanding of user requirement, misunderstanding of user requirement, incomplete function design, wrong function design, incomplete performance design, coding omission, inaccurate understanding of programming language, operating environment and standard, operating errors.

The severity of aerospace measurement and control software defect is classifed according to the degree of adverse impact on Space Mission, as can be seen in Table 4.

4 CONCLUSION

In the paper, we first analyze several typical software defect classification methods, then according to the development process of aerospace measurement and defect types found in practical tecsting, then propose a classification method for aerospace measurement and control software defects, based on the results, we can establish a systemic method to analyze the introduction, discovering, positioning, harm and root-cause of aerospace measurement and control software defects. Besides, the method can be used to analysis the defect occurrence frenquency data of different software; we can accordingly make innovation measures of software process management to improve software quality and productivity of software organization.

In the development of future aerospace measurement and control software, we can utilize the defect classification method presented in this paper to make defect statiscal analysis, then we can according adopt different test methods and test tools [5] in different test phase to discover and expose certain type of defects, thus we can futher try to improve the testing effectiveness and stability of aerospace measurement and control software.

REFERENCES

Badgett, T. & Sandler, C. 2004. The Art of Software Testing, 2 st Edition, John Wiley & Sons.

Beizer, B. 1990. Software Testing Techniques. Thomson Learning, 2nd edition.

Freimut, B., Denger, C. & Ketterer, M. 2005. An Industrial Case Study of Implementing and Validating Defect Classification for Process Improvement and Quality Management. In Proc. 11th IEEE International Software Metrics Symposium (METRICS '05). IEEE Computer Society.

IEEE Std 1044-1993. 1993. IEEE standard classification for software anomalies [S] the institute of Electrical and Electronics Engineers.

Levendel, Y. 1989. Defects and reliability analysis of large software systems: Field. Proc. 19th Fault Tolerant Computing Symp, pp. 238–243.

Putnam, L.H. & Myers, W. 1992. Measures for excellence: reliability software on time, within budget [M]. New Jsesey: Prentice hall.

Ram, C., Inderpal, S.B. & Jarir, K.C. 1992. Orthogonal defect classification-a concept for in-process measurements [J] IEEE transaction on software enginnering, pp. 943–956.

Wagner, S., Koller C. & Trischberger, P. 2005. Comparing Bug Finding Tools with Reviews and Tests. In Proc. 17th International Conference on Testing of Communicating Systems (TestCom'05), volume 3502 of LNCS, Springer, pp. 40–55.

Control Engineering and Information Systems – Liu (Ed)
© *2015 Taylor & Francis Group, London, ISBN 978-1-138-02685-8*

An automatic control system of waiter robot

M.C. Meng, J.Y. Liu & Y.W. Liu
Qinggong College, Hebei United University, Tangshan, China

ABSTRACT: This design is a waiter robot system with S3C2410 microprocessor, and it was developed by Bochuang Company's "three musketeers" development board. This system contains robot intelligent control, infrared emission and receiving, wireless data transmission, QT interface, touch screen, wireless serial technology. It can achieve restaurant guest' order conveniently, quick and accurately. It saves man power.

Keywords: S3C2410; wireless data transmission; QT interface; touch screen

1 SUMMARY

The waiter robot is designed to solve the high cost of human resources problems in food and beverage department, and it also make the catering service industry more intelligent and work more efficient. And the system has high stability. Its development cost is not very high. Based on these advantages, the system has good development prospects.

The system is mainly used in hotels and other catering department. When the guests arrive, they can use the infrared emission module on each table to call the robot to order. The robot evades obstacles automatically to arrive the table, then the guest will order on the touch flat which is on the robot, every dish can be selected and removed. When a guest finish the order, click finish options, then press the infrared transmitting module, the robot will automatically return to the starting point for the next guest's call. At the same time, the order list has been sent to the kitchen through the wireless device, the kitchen can accurately receive the order list which is send by each table, and then the order list is display through the 12864 screen. The system's main function is to realize intelligent order, improve work efficiency, saving human resource and space resource.

2 THE STRUCTURE OF SYSTEM

The system is mainly composed of three parts: the robot action control section, order and transmission part of order list and part of displaying order list in the kitchen. The waiter robot uses triangle car simulation, with black and white box as the venue. The robot walks by square form to reach the target table and change their route to avoid touch

other obstacles in real time. The model diagram as shown in Figure 1. Now I will introduce this system through the three parts above.

2.1 *Robot action controlled section*

Complete the system on "the Three Musketeers" development board of Bochuang Company. It uses the S3C2410 as the core control chip to control the robot to walk. The control theory mainly includes: tracing principle, obstacle avoidance principle and principle of positioning [1,2].

Tracing principle is detected black and white line through infrared. It will return low voltage level when it detected black line and it will return high voltage level when it detected white line. It controls the robot always walking down the line by returning the voltage level of different.

Obstacle avoidance function is implemented through the infrared sensors. The sensor returns low voltage level when it detected obstacles. The robot judge road conditions according to the

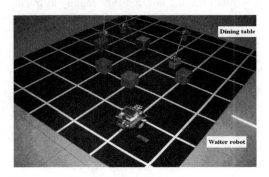

Figure 1. Model of the waiter robot.

voltage level, then select true path according to the terminal target.

The robot relies on memory and updating the coordinates to realize localization. When the initial coordinates determined later, robot update the current coordinates in order to determine its position when it walk by a grid, and then analysis the barricades and destination to judge the direction of walk, ultimately reaching the target.

Robot walking space is black and white squares on a grid line, the robot find the target point through walking around on the grid line.

Each table and robot communication system is achieved by infrared transmitting and receiving. When the guests arrived at the table they can emission signal to the robot by via infrared emission system above the table. Then the robot will automatically get to the table for the guest's order. When the guest finished the order, the system send a completion signal, the robot will automatically return to the starting points for the next call of other guest. The system adopts 555 chip to make 38 KHZ pulse, and then use the 1 KHZ, 2 KHZ, 3 KHZ, 4 KHZ pulse to its modulation. It will send signal through infrared tube. The robot determines the different target table through the receiving different pulse signal.

The robot use a group of infrared tube TCRT5000 as the sensor which can trace and record the coordinates. The circuit diagram as shown in Figure 2: TCRT5000 infrared tube is composed of transmitting tube and receiving tube. The infrared signal was absorbed by the black line when the transmitting infrared tube transmits signal to the black line. Now the pin 3 in Figure 2 output low voltage level. The infrared signal was reflected by the white line and was received by receiving tube when the transmitting infrared tube transmits signal to the white line. Now the pin 3 in Figure 2 output high voltage level. Then the high level and low level voltage were compared through the comparator to get the signal which can be identified by processor. Based on the infrared tube can identify black line and white line, the robot use the infrared tube mainly in tracing

and localization functions. Tracing function is the robot use a row of infrared tube in front of it to walk down the line, to ensure that the robot can adjust the position of its own in time. Localization function is that a horizontal single sensor was installed in the right front of the robot. The sensor can memory coordinates whenever the robot walk through a lattice. Then the processor will update the current coordinates according to their direction of travel to [3].

There are infrared sensors in the front and sides of the robot. The infrared sensors are used to avoid obstacle. Each infrared sensor consists of three lines, the yellow line, green line and the red line. The red line is connected to the power of 4.5~5.5 V high voltage level. The green line is connected to the GND. The yellow line is a signal line, it is connected to a pull-up resistor which resistance between 1 K~10 K. When there are no obstructions in front of robot the yellow line output high voltage level. When there are obstacles the output port will turn to low voltage level. The working principle is marked on the Figure 2. There is a potentiometer on the back of the infrared sensors which can be adjusted to change the detecting distance. In the effective distance the infrared sensors will output low voltage level once the potentiometer was adjusted. The control chip can identify the low voltage level. The robot can successfully avoid obstacles through the infrared sensor detect the obstacles in front of robot.

The waiter robot use driving motor L298. The L298 is a special driving integrated circuit which belongs to the H Bridge integrated circuit. It contains double H Bridge and it can realize two-wheel motor reversing. Its biggest characteristic is that its power will enhancement when the output current increases. Its output current is 2 A, and its maximum current is 4 A, its maximum working voltage is 50V. It can drive inductive loads such as DC motor for large power, stepper motor and battery valve. In particular its input end can be directly connected with the single-chip microcomputer then it can be controlled very easily by the single chip microcomputer. It only needs to change the input logic voltage level to realize the motor positive inversion. The robot relies on such a drive motor to achieve turning around flexibility and have a flexible action [4].

2.2 Order and the list of reder transmitting part

The waiter robot ordering system is developed based on QT4. It uses QT checkbox and pushbutton to add pictures and a plurality of windows switching. The touch screen can show menus for guests to order, guests can select their love of food on touch screen. There is no mouse displaying, the

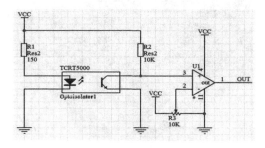

Figure 2. A part of the circuit diagram about sensor.

guests can select order directly. When the order finished, press the finish button, QT4 can response the pressing events. Then we call the serial driver to send a message to the serial port. The serial port was set to a baud rate of 9600, 8 bits of data. There is a parity bit with no odd-even, a stop bit, and a control bit with no data flow. Then the list of order was send to kitchen through the wireless serial. The wireless serial communication module is made of ZigBee. When it is inserted in the embedded development board, it can realize the data transmission through processor driving the serial port.

As shown in Figure 3, first select the number of table on the touch screen, then a variety of dishes is displayed. As shown in Figure 4, guests can choose different dishes according to personal likes to. If the guest does not want the dish which he choice, he can click the touch screen again to cancel the order. As the screen is very small, when food is selected over in this page, the guests can press touch screen to next page to select the next page of the menu. When the guests finish the order can press a finish options, then the orders will be send to the kitchen by serial.

We use APC220-43 multichannel micro power embedded wireless data transmission module to

Figure 5. Order list display interface.

menu data transmission. The APC220-43 module is a highly integrated half duplex wireless data transfer module. It is embedded microprocessor and high performance RF chip. The adoption of innovation and efficient loops interweave check-to-correct code. Anti interference and sensitivity are greatly improved. Maximum continuous burst error correcting 24 bits to achieve the industry leading level. The APC220-43 module provides a plurality of channel selection. It can be modified online serial rate, transmission power, radio frequency rate and other parameters. The APC220-43 module can transmit any size data. While the user without having to write complex settings and transmission procedures. At the same time, small volume, wide voltage operation of far transmission distance, rich convenient software programming setting function which can be applied to very wide field [5].

2.3 *The part of menu display in kitchen*

As shown in Figure 5. Menu display system for displaying each table has ordered dishes. The system receives data through the wireless module, and use single chip microcomputer to deal with these data. Then the guests point dishes achieved by driving 12864. At the same time we can switch different desktop view to see the dishes that each table has ordered. The corresponding table screen operation when the dishes have is neat.

3 COMPOSITION OF SYSTEM HARDWARE

As shown in Figure 6, this complete intelligent ordering service system constitutes based on S3C2410MCU core and counseling. When the infrared transmitting module transmits signals, the robot on the signal received by the infrared

Figure 3. Table selection interface.

Figure 4. Order interface.

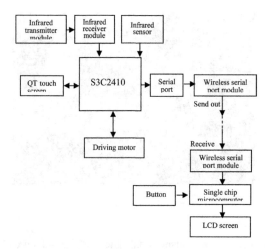

Figure 6. Order list display interface.

receiving head, the received signal is sent to the controller. Through the motor driving controller made robot walking. The robot update site coordinate through the two infrared tubes when it is walking, thereby defining its position. Through the infrared obstacle avoidance sensors for obstacle detection, the detection signal is sent to the controller, robot obstacle avoidance for controller controls the robot steering. When the robot arrives in front of the guest, touch QT interface to complete the order. Then every touch, QT interface and S3C2410 will conduct a information exchange. When the order is complete, the S3C2410driver serial send data. Then the display is in the kitchen will show this table all the dishes. Through the keys to view each table is ordered dishes in the display module. We can also use the button for screen update operation [6].

4   CONCLUSION

This paper describes the whole design process of the robot. The design includes robot controlled, QT interface, wireless serial data transmission and reception. Each module and process are basically achieved the expected target. Of course, there are many shortcomings because of the level and time caused by inadequate. In the future we will continue to improve the research.

REFERENCES

Haoqiang Tan, C Language Program Design. Qinghua University press, 2006, pp. 40–65.
Jasmin Blanchette, Mark Summerfirld, C++ GUI Programming with QT 4, Publish of electronics industry, 2008, pp. 52–63.
Linux DSP Gateway Specification Rev 3.3, 2010, pp. 60–72.
OMAP5910/5912 Multimedia Processor DSP Subsystem Reference Guide, Literature Number: SP-RU890 A, 2005.
Shi Yan, Fundamentals of Digital Electronic Technology, Higher Education Press, 2006, pp. 112–128.
Shibai Tong, Fundamentals of Analog Electronic Technology, Higher Education Press, 2009, pp. 98–110.

Control Engineering and Information Systems – Liu (Ed)
© 2015 Taylor & Francis Group, London, ISBN 978-1-138-02685-8

Fuzzy control and simulation of single inverted pendulum

Y.H. Huang, L. Zhang, P.P. Zhang & F.F. Ru
Henan University, Kaifeng, China

K.Q. Wang
Nanjing Artillery Academy, Nanjing, China

ABSTRACT: The single inverted pendulum simulation model with input disturbance is built by using SimMechanics toolbox, at the same time, the simplified fuzzy controller based on Single Input Rule Modules (SIRMs) dynamically connected fuzzy inference model is presented, which can achieve equilibrium of the inverted pendulum system. The number of fuzzy rules is greatly reduced and the "curse of dimensionality" problem in the fuzzy multi-variable control is overcame through transforming multi-dimension fuzzy control into single dimension fuzzy control by using SIRMs, and pendulum angle control takes priority over cart position control by introducing the dynamic importance degrees. The simulation and real-time control results show that the proposed method can effectively restrain input disturbance, realize stability control of the inverted pendulum system, and has some advantages such as simple structure, good control effects, easy realization and strong robustness.

1 INTRODUCTION

The inverted pendulum system is a typical multi-variable, nonlinear, strong coupling and non-minimum phase system with natural instability characteristics, which can effectively reflect many key problems such as stabilization, nonlinear, robustness and tracking in the controlling process. It is an ideal model to verify the different control theories and techniques in teaching and scientific research. In engineering applications, the inverted pendulum system has similarities with satellite attitude control, the joint motion of robot control and steady hook device of lifting machinery. Therefore, as typical experimental equipment, the inverted pendulum system becomes a bridge from theory to practice. The study has important engineering background and practical significance.

For single inverted pendulum system, the classical fuzzy control accomplish fuzzy inference for the four input variables at the same time, which can realize the stability control of inverted pendulum system, but the control effect is not effective, because the fuzzy rules are too much and the system is relatively complicated. Whereas the control system can greatly reduce the number of fuzzy rules, make a difference between each input variable and realize effective control by using SIRMs. Relative to other fuzzy controller, the simplified fuzzy controller which based on SIRMs can accomplish the control goals that pendulum upright and the cart back to the initial position in a short time. It also can effectively restrain input disturbance

on the system and has strong robustness. SIRMs algorithm is easy for hardware implementation in practical application for only a small number of multiplication but no division.

2 SINGLE INPUT RULE MODULES (SIRM)

The Single Input Rule Modules (SIRMs) dynamically connected fuzzy inference model transform multi-dimension fuzzy control into single-dimension fuzzy control which can greatly reduce the number of fuzzy rules. In the control system, the effects of all input variable on the system are different. For example, the angle and angular velocity of the pendulum directly affect the success in the stability control of inverted pendulum system, and the effects of the position and speed of the cart are relatively weak. That is to say, the pendulum control should take priority over the cart control before the pendulum reach equilibrium; the cart position control can only start when the pendulum is in equilibrium. As a result, each input variable corresponding to a SIRM and a dynamic importance degree, and the dynamic importance degrees will change with the control conditions. Each input variable in the SIRM has

$$\text{SIRM-i}: \left\{ R_i^j : \text{if } x_i = A_i^j \text{ then } f_i = C_i^j \right\}_{j=1}^{mi} \quad (1)$$

Here, SIRM-i expresses the ith input item, and R_i^j is the jth rule in the SIRM-i. x_i is the ith input variable, A_i^j is fuzzy sets of the fuzzy variable x_i of the jth rule in the SIRM-i, f_i is the consequent variable of the SIRM-i. C_i^j is output variable value of the jth rule in the SIRM-i, m_i is the index number of the rules in the SIRM-i.

To express the importance degree of each input variable, the SIRMs dynamically connected fuzzy inference model define a dynamic importance degree w_i^D

$$w_i^D = w_i + B_i \Delta w_i \qquad (2)$$

The base value w_i guarantees the necessary function of the corresponding input item in the control process, the dynamic value is defined as the product of the breadth B_i and the inference result Δw_i which can adjust the importance degree of the corresponding input item on system according to the changes of control conditions.

The SIRMs dynamically connected fuzzy inference model obtain the value of the output item f by

$$f = \sum_{i=1}^{n} w_i^D f_i \qquad (3)$$

It is the summation of the products of the output f_i of each SIRM and its dynamic importance degrees for all input items.

3 SIMULATION AND ANALYSIS

3.1 The mathematic model of inverted pendulum

After neglecting some secondary factors, such as air resistance, the system internal friction and regardless the deformation of the component, the linear single inverted pendulum system can be abstracted into the system that consists of the cart and the homogeneous quality rod, as shown in Figure 1, this is a typical rigid body system. The meaning and size of each parameter are shown in Table 1.

The inverted pendulum system which used in real-time control experiment is placed on the

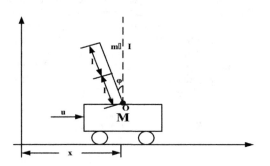

Figure 1. Configuration of the inverted pendulum.

Table 1. Physical parameters of inverted pendulum.

Symbols	Physical interpretation	Value (unit)
M	Mass of the cart	1.096 Kg
m	Mass of the pendulum	0.109 Kg
l	Length of rotation axis to center of mass for the pendulum	0.25 m
I	Rotation inertia of the pendulum	0.0034 Kg*m*m
b	Friction coefficient of the cart	0.1 N/m/sec
ϕ	The angle of pendulum and upright	rad

unstable desk which can produce interference on the inverted pendulum system, affecting the stability due to the oscillation of the table itself. Figure 2 is the structure block diagram, $R(t)$ is the input and $Y(t)$ is the output of the inverted pendulum system, represents the input disturbance on the system. In order to more close to the reality and verify effectiveness of the proposed controller.

The single inverted pendulum simulation model with input disturbance is established by using Sim-Mechanics toolbox, it is a fourth-order system which has one input and two outputs, the four state variables are cart position, cart speed, pendulum angle and pendulum angular velocity, respectively. Near the equilibrium point, $\phi = 0$, then, $\cos\phi = 0$, $\sin\phi = \phi$, $\dot{\phi}^2 = 0$. u represents the input acceleration and w represents the input disturbance on the inverted pendulum system, the state equation can be expressed as:

$$
\begin{bmatrix} \dot{x} \\ \ddot{x} \\ \dot{\phi} \\ \ddot{\phi} \end{bmatrix} =
\begin{bmatrix}
0 & 1 & 0 & 0 \\
0 & \dfrac{-b(I+ml^2)}{N} & \dfrac{m^2gl^2}{N} & 0 \\
0 & 0 & 0 & 1 \\
0 & \dfrac{-bml}{N} & \dfrac{(M+m)mgl}{N} & 0
\end{bmatrix}
\begin{bmatrix} x \\ \dot{x} \\ \phi \\ \dot{\phi} \end{bmatrix}
$$

$$
+ \begin{bmatrix} 0 \\ \dfrac{ml}{N} \\ 0 \\ \dfrac{M+m}{N} \end{bmatrix} w
+ \begin{bmatrix} 0 \\ \dfrac{-(I+ml^2)}{N} \\ 0 \\ \dfrac{ml}{N} \end{bmatrix} u
$$

$$
y = \begin{bmatrix} x \\ \phi \end{bmatrix} =
\begin{bmatrix} 1 & 0 & 0 & 0 \\ 0 & 0 & 1 & 0 \end{bmatrix}
\begin{bmatrix} x \\ \dot{x} \\ \phi \\ \dot{\phi} \end{bmatrix}
\qquad (4)
$$

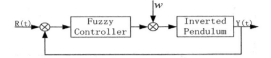

Figure 2. The system structure block diagram.

Here, $N = (I + ml^2)(M + m) - m^2l^2$, after bring the actual values in Table 1, the state equation of inverted pendulum system is obtained:

$$\begin{bmatrix} \dot{x} \\ \ddot{x} \\ \dot{\phi} \\ \ddot{\phi} \end{bmatrix} = \begin{bmatrix} 0 & 1 & 0 & 0 \\ 0 & -0.0890 & 0.7132 & 0 \\ 0 & 0 & 0 & 1 \\ 0 & -0.2671 & 31.5397 & 0 \end{bmatrix} \begin{bmatrix} x \\ \dot{x} \\ \phi \\ \dot{\phi} \end{bmatrix}$$

$$+ \begin{bmatrix} 0 \\ 2.6708 \\ 0 \\ 118.1042 \end{bmatrix} w + \begin{bmatrix} 0 \\ 0.8903 \\ 0 \\ 2.6708 \end{bmatrix} u$$

$$y = \begin{bmatrix} x \\ \phi \end{bmatrix} = \begin{bmatrix} 1 & 0 & 0 & 0 \\ 0 & 0 & 1 & 0 \end{bmatrix} \begin{bmatrix} x \\ \dot{x} \\ \phi \\ \dot{\phi} \end{bmatrix} \qquad (5)$$

3.2 The block diagram of SIRMs

The block diagram of SIRMs is shown in Figure 3. Each rule of the fuzzy control based on SIRMs has one antecedent variable and one consequent variable, the fuzzy rules are simple and have less number and easy to implement. Because there is only one output variable and each input variable on system performance is not the same, so the output result needs to be multiplied by corresponding dynamic important degree, and then add up them to control the inverted pendulum system.

According to the actual control experience we know that if you push the cart to one direction, due to inertia, the inverted pendulum will fall to the opposite direction. The SIRMs of each input variable are shown in Table 2. Here, membership functions NB, NM, Z, PM, PB are defined as triangles or trapezoids. It assumes that if the inverted pendulum system has a right driving force, the driving force is positive, whereas the driving force is negative. If the pendulum angle is positive, the system will produce a positive driving force which

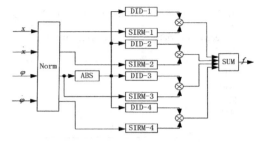

Figure 3. The block diagram of SIRMs.

Table 2. SIRM for each input item.

Antecedent variable $|x_1|$	Consequent variable $\Delta w_i(i = 1,2)$ | $\Delta w_i(i = 3,4)$
DS	0.0 | 1.0
DM	0.5 | 0.5
DB	1.0 | 0.0

can bring the cart to right, so that the pendulum would rotate counterclockwise. If the cart on the right side of the origin, the positive driving force will bring the cart to the right further away from the origin, then the pendulum will rotate to the left, indirectly realizing the position control of the cart by control the pendulum angle.

When the absolute value of the pendulum angle is large, the inverted pendulum system is not in equilibrium, the pendulum angle should be controlled preferential; when the absolute value of the pendulum angle is close to zero, the pendulum angle control has completed, and the cart position should be controlled in time to achieve the control target. That is to say, the pendulum angle control is primary and the cart position control is secondary. Therefore, the fuzzy rules of the dynamic variable Δw_1, Δw_2, Δw_3, Δw_4, which corresponding the angle and angular velocity of the pendulum, position and velocity of the cart can be set up in Table 3. The antecedent variables of all fuzzy rules are the absolute value of the pendulum angle. The membership functions DS, DM, DB are defined as triangles or trapezoids. In addition, the control parameters which through trial and error are shown in Table 4.

3.3 The simulation block diagram of simulink

The fuzzy control system of the inverted pendulum is established by using simulink toolbox. The system is composed of the simplified fuzzy controller and the model of the inverted pendulum. The block diagram of the simplified fuzzy controller is shown in Figure 4.

Table 3. Fuzzy rules for the dynamic variables.

Antecedent variable $x_i(i=1,2,3,4)$	Consequent variable $f_i(i=1,2,3,4)$
NB	−1
NM	−0.5
Z	0
PM	0.5
PB	1

Table 4. The base value and breadth of the input items.

Input item	Base value	Breadth
Pendulum angle	4	4
Angular velocity	4	4.5
Cart position	2.4	1.4
Cart velocity	2.4	1.4

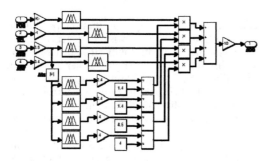

Figure 4. The SIRMs controller model.

3.4 The simulation and real-time control result analysis

Figure 5 denotes the simulation result without input disturbance. From Figure 5 we can see that the system will reach equilibrium about 3.5 s, and the process of dynamic equilibrium relatively steady, the maximum displacement is no more than 0.2 m. Guaranteeing the system parameters unchanged, Figure 6 denotes the simulation result with input disturbance. The stabilization time of system is shown in Figure 6 about 3.5 s; the maximum displacement of the cart is 0.2 m. Comparing with Figure 5, we can find that the stabilization time of the system and the maximum displacement are little changed, but the process of dynamic equilibrium shocks fiercely in contrast, illustrating the controller has strong robustness.

Figure 7 denotes the real-time control result with input disturbance. The figure shows that the

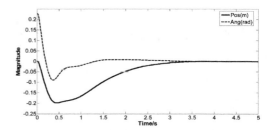

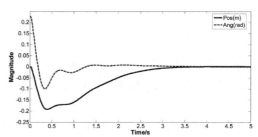

Figure 5. The simulation result without input disturbance.

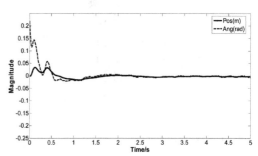

Figure 6. The simulation result with input disturbance.

Figure 7. The real-time control result with input disturbance.

stabilization time of the system is short, the overshoot is small and the steady-state error is close to zero. The control effect is relatively ideal and also verifies the effectiveness of the fuzzy controller based on SIRMs.

4 CONCLUSIONS

In this paper the single inverted pendulum system is controlled by using SIRMs dynamically connected fuzzy inference model, not only can greatly reduce the number of fuzzy rules and make the fuzzy rules become easy, but also adjust the weights of the input variables through the corresponding dynamic importance degree, make a difference between each input variable and control the impact on the system of each input variable. The simulation and real-time control results show that the fuzzy controller based on the SIRMs can implement the single inverted pendulum system stability control effectively restrain the input disturbance and have good dynamic and static characteristics and strong robustness.

REFERENCES

Guo-hui Li, "An Simulation Study of Single Inverted Pendulum Based on T-S Fuzzy Control Model," Journal of Dalian Jiaotong University, vol. 29, Feb. 2008, pp. 68–72.

Hongwei Li, "Reduced fuzzy control of single inverted pendulum and its simulation," Control Engineering of China, vol. 17, Nov. 2010, pp. 769–773.

Jianqiang Yi, "The single input rule modules dynamically connected fuzzy inference model," Contribution to science and academic symposium of Qian Xuesen, Beijing, 2001.

Xueming Ding, Peiren Zhang, Xingming Yang, Yongming Xu, "Motion control of two-wheel mobile inverted pendulum based on SIRMs. Journal of System Simulation," vol. 16, Nov. 2004, pp. 2618–2621.

Yi J., Yubazaki N., "Stabilization fuzzy control of inverted pendulum systems," Artificial Intelligence in Engineering, vol. 14, Mar. 2000, pp. 153–163.

Zhenfeng Chen, Peng Zhang, Yufeng Cai, Lifen Ding, "Fuzzy control technology based inverted pendulum," Journal of civil aviation university of China, vol. 22, Dec. 2004, pp. 11–14.

Zhong Luo, Zhiyong Shi, Peng Hu, Hongyi Liu, "Analysis and experimental investigation of factors affecting control characteristics of line inverted pendulum systems," Journal of Northeastern University (Natural Science), vol. 32, Apr. 2011, pp. 559–562.

Zifan Fang, Bingfei Xiang, Wenhui Shu, "Research on new fuzzy control method of semi-active suspension based on single input rule module," Journal of China Three Gorges University (Natural Sciences), vol. 33, Oct. 2011, pp. 58–62.

Control Engineering and Information Systems – Liu (Ed)
© 2015 Taylor & Francis Group, London, ISBN 978-1-138-02685-8

Design of in-wheel motor PID controller based on artificial immune algorithm

L.Y. Lei, H. Ren, L.P. Liu, J.J. Xu & S. Wu
Zhejiang Agriculture and Forestry University, Hangzhou, Zhejiang, China

ABSTRACT: Control system of In-wheel motor electric vehicles is core problem of research in auto industry, this paper proposes artificial immune algorithm PID (Proportional Integral Derivative) control strategy based on the structure of brushless DC hub motor, An in-wheel motor controller will be designed, including the hardware and software part, circuit board will be shown.

1 INTRODUCTION

In-wheel motor driving technology was invented by Robert in early 1950s, which was applied to large-sized mining trucks by the General Electric Company (GE) for the first time [1] [2]. In-wheel motor electric vehicles have many advantages, like reducing pollution to our environment, simplifying chassis structure, higher space and energy utilization rate, independent control for each driving motor.

Among many unsolved problems of In-wheel motor electric vehicles, the control system is the key to obtaining wonderful performance, controller with new driving structure and advanced control algorithm have become a hotspot for research in Auto industry.

PID (Proportional Integral Derivative) controller has made great progress in machine controlling, however, it can't be applied in a complicated control environment, therefore, many new methods in PID control are developed [3], like Neural Network PID control, Fuzzy PID control, Genetic Algorithm PID control, self-learning Intelligent PID control, Fuzzy Immune PID control, they can help to make the driving system simpler, getting faster response, higher anti-interference ability, higher performance of whole system [4–7]. In-wheel motor controller is designed based on Artificial Immunity Algorithm PID, an intelligent algorithm on the basis of biological immune principle.

2 STRUCTURE AND PRINCIPLE OF BRUSHLESS DC HUB MOTOR

Brush DC motor, AC motor, three-phase AC induction motor, permanent magnet brushless DC motor and switched reluctance motor can be used as driving motor in electric vehicles, among them, we choose brushless DC hub motor because of its reliable operation and long lifespan.

2.1 The basic structure of brushless DC hub motor

Based on the location of rotor in the motor, permanent magnet brushless DC hub motor can be divided into outer rotor, inner rotor and disc type. As for its rotor structure, magnetic pole is constructed by magnetic steel amounted on the yoke of rotor, air-gap magnetic field is generally square wave or trapezoidal wave. For stator structure, its three-phase wings are distributed evenly with an electrical angle of 120 degrees. We choose three-phase outer rotor permanent magnet brushless DC motor as In-wheel motor, which consists of motor, position sensor and controller.

Brushless DC hub motor generates signal by means of reflecting the position of rotor position sensor, drives power switch connected with armature windings using electronic commutation phase circuit, thereby generates a leap rotating magnetic field in the stator to make the permanent magnet rotate. With the rotation of rotor, the position sensor sends varied signal to change the charge state of armature windings to make sure that current direction of conductor in magnetic pole remains the same, this is the essence of without touch conversion process for a brushless DC motor.

2.2 Torque control strategy

Current hysteresis control with direct input torque (current) instruction, current control with integral action (PI control) as adjustment coefficient, space voltage vector control based on torque control, are used as torque control strategy. In an ideal power condition, the mechanical characteristics of the torque control strategy is a straight line parallel to the speed axis, which is different in actual power condition, due to the limited capacity, its mechanical characteristics are shown in Figure 1.

We can see from Figure 1, good acceleration feature is obtained if we choose this control strategy in electric vehicles with limited voltage, besides; the falling parts of mechanical characteristics are almost coincident with different input, which indicates stable traveling velocity of vehicle is nearly equal corresponding with input changes when vehicle runs in this section.

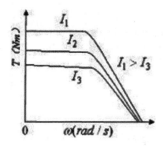

Figure 1. Mechanical characteristic output curves in torque control.

3 ARTIFICIAL IMMUNE ALGORITHM PID CONTROL STRATEGY

Artificial Immune Algorithm (AIA) proposed in this paper is to solve the problem of PID parameter optimization problem.

The design process for AIA PID controller is actually the determination for three gain coefficients K_p, K_i, K_d, therefore, to optimize the design of AIA algorithm PID controller, the objective function of the optimization problem is regarded as antigen, gain coefficients K_p, K_i, K_d are as antibodies, the algorithm flow is shown in Figure 2.

The concrete procedures are as followes:

a. Set the parameter values of const, β, θ, Nc and m(the size of antibodies); set A_b^* and $A_d^* = \{K_p^*, K_i^*, K_d^*\}$, which are used to store the elite antibody and three PID gained values, whose initial value is 0;

b. Generate randomly several antibodies, and form initial antibodies;

c. Make $t = 1$;

d. For each antibody k ($k = 1,2, ..., m$), input K_p^k, K_i^k, K_d^k to PID controller, perform a complete control operation to each control object, calculate fitness f_k of antibody. Then select one with maximum fitness among antibodies as elite antibody, copy it and save it to A_b^* and A_d^*, turn to step 8;

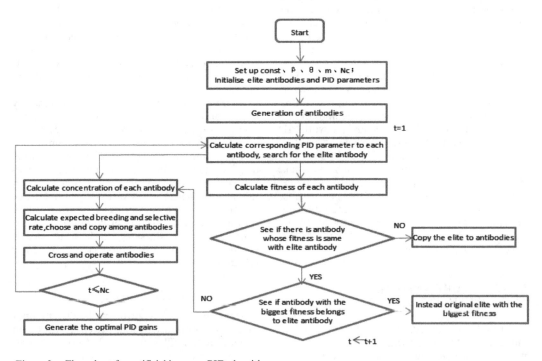

Figure 2. Flow chart for artificial immune PID algorithm.

116

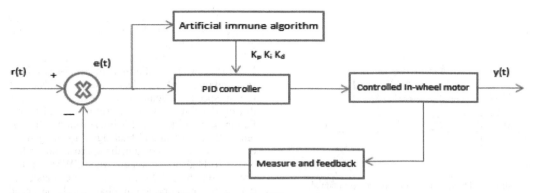

Figure 3. The flow chart for artificial immune algorithm PID control system.

e. For each antibody k ($k = 1,2, ..., m$), obtain its corresponding gain values K_p^k, K_i^k, K_d^k, perform a complete control operation to each control object, calculate fitness f_k of antibody;

f. Check if there is antibody whose fitness is the same with that of elite antibody, once there exists, the antibody will be stored in A_b^*, and have antibody with minimum fitness deleted; otherwise continue;

g. If the maximum fitness of antibody is bigger than that of the elite, then copy the antibody with the maximum fitness, make it be new elite antibody, and save it in A_b^* and A_d^*; otherwise continue;

h. Calculate the concentration of each antibody;

i. Calculate the expected rate of reproduction of each antibody; calculate the probability of selection for each antibody; choose and copy one among antibodies using roulette method;

j. Crossover of antibodies;

k. Variation of antibodies;

l. Make $t \leftarrow t + 1$, if $t \leq$ Nc, then back to step 5, otherwise, put output value in A_d^*, get the best result $\{K_p^*, K_i^*, K_d^*\}$, then whole flow is over.

In this paper, we choose the artificial immune PID control system to control motors, which can be of great help to enhance the control precision, the flow chart of control system is shown in Figure 3.

4 DESIGN OF IN-WHEEL MOTOR CONTROLLER

4.1 Hardware design for in-wheel motor controller

The PIC18F4480 (4580) is an enhanced FLASH 28/40/44-pin microcontroller with Nanowatt Technology, and 10-bit A/D conversion and, whose FLASH/Data EEPROM retention period can last more than 40 years, with interruption priority. Each microcontroller has 1, 2 or 4 channel PWM outputs, with advantages like PWM polarity selection, programmable dead time, automatic shutdown and restart function, besides, PWM resolution rates 1–10, the conversion rates 100 Ksps. The ECAN module information bit of this microcontroller rates 1Mbps, complying with available specification of CAN2.0, there are three modes that can use it: Legacy mode, enhanced Legacy mode and FIFO mode. In this paper, we choose PIC18F4480 as In-wheel motor controller due to its speciality.

The power system is composed of six 12 V (72 V) 120 Ah lead-acid batteries in series. It is the 72 V voltage that is used to charge motor windings. We get 24 V voltage from two of them, which is converted to +5 V, +15 V via DC-DC direct current power, respectively, as power supply for control

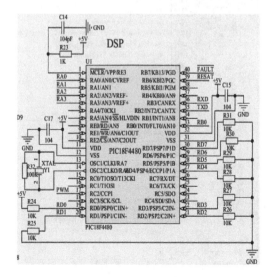

Figure 4. The outer circuit diagram of main control circuit.

117

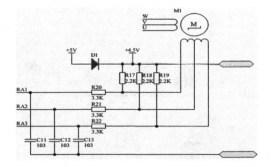

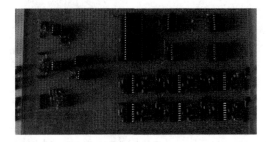

Figure 5.   Detection circuitry of position signal.

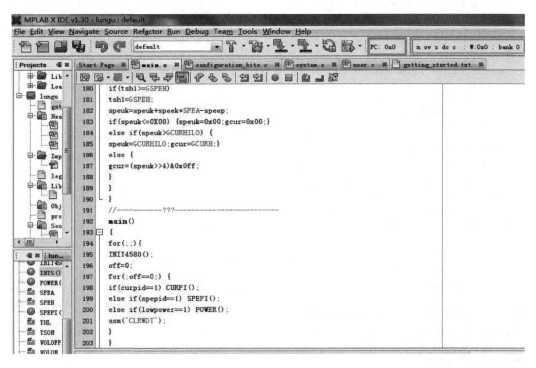

Figure 6.   The circuit board for in-wheel motor controller.

board, Hall position sensors and drive plate. The core control chip is PIC18F4480 for control circuit. The functions of main control circuit are as follows: CPU is initialized in the main program, I/O, timer, A/D is initialized, the PWM is initialized and interrupted to determine the stop/start of motor, speed is calculated and regulated by PID; during the time of ON interruption (Hall rising, falling part will trigger this interruption), the system will take 3-channel Hall signal for judgement of direction and calculate the speed count value; access to Hall position state when PWM is interrupted, set the PWM output, rewrite register to determine the conduction phase, and the speed PID values are sent to PWM register, generating the corresponding PWM signal. This PWM signal is isolated, amplified through the drive circuit, will drive the power tube, ultimately complete the adjustment of motor speed.

4.2   Design of main control circuit

The PIC18F4480 is employed as core control chip in the main control circuit, including clock circuit, reset circuit, analog input circuit and switch input and output circuit. At the same time, CPU gives out 6 channels PWM pulse signal to chip differentiated

Figure 7.   Main control programme of in-wheel motor controller.

by DS26F3IM as the input signal of power drive circuit. As we know, main control circuit is the command center for normal operation and servo function of brushless DC motor, its outer circuit is shown in Figure 4.

4.3 Design for drive circuit

Drive circuit amplifies pulse sent by control circuit, drives power devices; the switching characteristics of power devices are closely relative with the performance of driving circuit. The two same power switches can get different characteristics using different driving circuit. The excellent drive circuit can help improve switching characteristics of power devices, help reduce switching losses and improve reliability and efficiency of motor. Take voltage, current demands of motor into account, we choose HCPL-316 J as drive chip, which is the special drive chip for MOSFET and IGBT with Vce (sat) protected. In addition, we choose MOSFET as the power device.

Position signal detection not only detects the relative position of motor stator and rotor used to control phase conversion, but also generates speed values used to constitute feedback link of speed, we choose hall sensors to detect position signal of the motor, whose schematic diagram is shown in Figure 5.

4.4 The entity of in-wheel motor controller

Debug circuit board is built according to the design of each circuitry, in Figure 6.

4.5 The software design for in-wheel motor controller

The drive programme of In-wheel motor controller is compiled in MPLAB, which is shown in Figure 7.

REFERENCES

C.C. Chan An Overview of Electric Vehicle Technology. Proc. of IEEE, 1993, 8l(9): 1202–1213.

Chuanbao Li. One Control system of In-wheel Motor and its use in electric vehicles [D]. University of Science and Technology of China, 2008.

Guijiao Wang. The Movement Type and Energy Distribution of Drive System for In-wheel Motor Electric vehicles [D]. Wuhan University Of Technology, 2009.

H. Shimizu et al. Advanced Concepts in Electric Vehicle Design. IEEE onIE, 1997, 44(1): 14. 18.

Huihui Zhang. Simulation for Power Drive System for Wheel Drive Electric vehicles based on Electronic Differential [D]. Chang'an University, 2010.

Yongyue Yan, Qingzhou Li, Shuxin Yu. Overview of Intelligent PID Control [J]. PLC & FA, 2006, (12): 9–13.

Yuanyuan Zhang. Torque Coordination Control for In-wheel Motor Electric vehicles [D]. Jilin University, 2009. rol circuit.

Control Engineering and Information Systems – Liu (Ed)
© 2015 Taylor & Francis Group, London, ISBN 978-1-138-02685-8

UAV flight controller design method based on genetic algorithm

Y.H. Li, M. Li, N. Zhang & H.G. Jia
Changchun Institute of Optics, Fine Mechanics and Physics, Chinese Academy of Sciences, Changchun, China

ABSTRACT: A flight controller design method based on model matching and genetic algorithm optimization is proposed in this paper. A nonlinear model is used as the controlled object of this method. Compared with the classic flight controller design method, this method can access the needed controllers quickly and conveniently. A nonlinear mathematical six-degree-of-freedom model of a UAV is established, the aerodynamics, engine and environmental model are all included in this model. Using the above method, a UAV flight controller is designed. Finally, the six degrees of freedom nonlinear simulation is carried out to verify the effectiveness of the designed controller.

1 INTRODUCTION

For the classic flight controller design method, firstly, we have to select a series of trim points within the flight envelope, then the nonlinear dynamic model of the aircraft must be linearized on the trim point [1]. And the Flight controller is designed based on the linearized model in the end. The most important part of the work is the choice of the controller parameters. For the classic flight controller design method, the trial and error method is often used to enable the control system to achieve the appropriate targets. This control system design methods are inefficient, and mostly depends on the engineering experience of designer. Although the designed control system can be achieved in engineering, it is not necessarily the optimal result. With the improved performance of aircraft, flight control systems are complex more and more.

Using the trial and error method to select the controller parameters has become a bottleneck restricting the design of flight control system. To solve this problem, a flight controller design method based on model matching and genetic algorithm optimization is proposed in this paper [2].

Li Guangwen Adjusted the parameters of the flight control system by using Niche genetic algorithm optimization strategy based on uniform design [3]. Wang Yong Designed Lateral Deviation Correction Controller for UAV Taxiing by using genetic algorithms optimization method [4].

There are many successful applications in the system parameter identification and process control based on genetic algorithm optimization [5] [6].

Firstly, the nonlinear simulation model of six DOF of the UAV was built, then the flight control and controller structure of the UAV, as well as the concrete expression of control rule, were presented. The frequency model, which was used as reference model to represent the anticipated performance of the flight controller, can neatly meet the requirements of the flight controller, and the dynamic and steady state indicator proved to be fairly good. The global optimization employing the genetic algorithm was implemented in the design of flight controller pattern to make the characteristic of the controller approach the reference model. Finally, according to the application of the method mentioned above in some UAV project, it shows that the devised controller is valid and the employed design method of the control rule is highly effective.

2 GENETIC ALGORITHM

Genetic Algorithm (GA) is a search heuristic that mimics the process of natural evolution [7]. This heuristic (also sometimes called a metaheuristic) is routinely used to generate useful solutions to optimization and search problems. Genetic algorithms belong to the larger class of Evolutionary Algorithms (EA), which generate solutions to optimization problems using techniques inspired by natural evolution, such as inheritance, mutation, selection, and crossover. Genetic algorithms in particular became popular through the work of John Holland in the early 1970s. Genetic Algorithm steps: encoding, the initial population generation, fitness evaluation testing, selection, crossover and mutation.

Crossover probability and mutation probability is calculated as follows.

$$P_c = \begin{cases} \dfrac{k_1(f_{\max} - f)}{f_{\max} - f_{avg}}, & f \geq f_{avg} \\ k_2, & f < f_{avg} \end{cases} \tag{1}$$

$$P_m = \begin{cases} \dfrac{k_3(f_{max} - f')}{f_{max} - f_{avg}}, & f' \geq f_{avg} \\ k_4, & f' < f_{avg} \end{cases} \quad (2)$$

f_{max} is the maximum fitness value groups, f_{avg} is the Average fitness value, f is the larger the fitness value of the two individuals to cross, f' is variation in individual fitness value, k_1, k_2, k_3, k_4 is constant.

3 SIX DOF OF DYNAMICS MODEL

Figure 1 is the structure of the UAV Six Degrees of Freedom of Closed-Loop Flight Simulation.

3.1 UAV dynamics model

In this paper, the movement of the UAV in the air is mainly considered. The Aerodynamic force F_A, the thrust of the engine F_P, and the gravity of the UAV itself G are the main Forces acting on the UAV. The six degrees of freedom dynamic model of the UAV is established in the body coordinate system.

The velocity component is u,v,w and the Angular velocity component is p,q,r, Euler angles is ϕ, θ, ψ.

The translational dynamic equation of the UAV in the body coordinate system is as follows

$$m(\dot{u} - vr + wq) = F_{A_x} + F_{P_x} + mg_x$$
$$m(\dot{v} + ur - wp) = F_{A_y} + F_{P_y} + mg_y$$
$$m(\dot{w} - uq + vp) = F_{A_z} + F_{P_z} + mg_z \quad (3)$$

Translational kinematic equations of the UAV in the ground coordinate system is as follows

$$\begin{bmatrix} \dot{x}_e \\ \dot{y}_e \\ \dot{z}_e \end{bmatrix} = C_{gb} \begin{bmatrix} u \\ v \\ w \end{bmatrix} \quad (4)$$

C_{gb} is the transformation matrix from body coordinate system to the ground coordinate system.

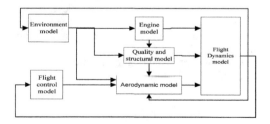

Figure 1. Flight simulation structure.

Because the UAV is symmetric, $I_{yz} = I_{xy} = 0$. Rotation dynamics equation of the UAV in the body system is as follows

$$I_x \dot{p} - I_{xz}\dot{r} - I_{xz}pq + (I_z - I_y)rq = M_{A_x} + M_{P_x}$$
$$I_y \dot{q} + I_{xz}p^2 - I_{xz}r^2 + (I_x - I_z)pr = M_{A_y} + M_{P_y}$$
$$I_z \dot{r} - I_{xz}\dot{p} + I_{xz}qr + (I_y - I_x)pq = M_{A_z} + M_{P_z} \quad (5)$$

Rotation motion equation

$$\dot{\phi} = p + \tan\theta(q\sin\phi + r\cos\phi)$$
$$\dot{\theta} = q\cos\phi - r\sin\phi$$
$$\dot{\psi} = (q\sin\phi + r\cos\phi)\sec\theta \quad (6)$$

3.2 Aerodynamic model

The aerodynamic forces acting on the UAV can be decomposed into drag D, lift L and lateral force C in the airflow coordinate system

$$D = C_D QS, L = C_L QS, C = C_C QS, Q = \frac{1}{2}\rho V^2 \quad (7)$$
$$C_D, C_L, C_C = f(Ma, Re, \alpha, \beta, \delta_e, \delta_r, \delta_a) \quad (8)$$

The aerodynamic moments acting on the UAV are as follows

$$MA_x = C_l QSb, MA_y = C_m QSc, MA_z = C_n QSb \quad (9)$$
$$C_l = C_{l\beta}\beta + (C_{lp}p + C_{lr}r)(b_{ref}/2V) + C_{la}\delta_a + C_{lr}\delta_r$$
$$C_m = C_{m0} + C_{m\alpha}\alpha + (C_{mq}q + C_{m\dot{\alpha}}\dot{\alpha})(c_{bar}/2V) + C_{me}\delta_e$$
$$C_n = C_{n\beta}\beta + (C_{np}p + C_{nr}r)(b_{ref}/2V) + C_{na}\delta_a + C_{nr}\delta_r$$

$$(10)$$

3.3 Engine model

The thrust of the engine is depending on the Flight speed V, flight height H, the Angle of attack α, the throttle δ_p.

$$F_P = f(V, H, \alpha, \delta_p) \quad (11)$$

During the simulation, According to the real-time flight status, thrust is calculated by the interpolation method based on the experimental data table of thrust.

The simulation also takes into account the impact of the fuel consumption of the engine on the weight of the aircraft.

The UAV engine fuel consumption is depend on the flight altitude, the flight Mach number, the Angle of attack and the throttle.

$$Sfc = f(V, H, \alpha, \delta_p) \qquad (12)$$

3.4 Quality and structural model

The quality of the UAV is changing over time

$$\frac{dm}{dt} = -sfc \qquad (13)$$

The change of UAV quality characteristic is mainly caused by the fuel consumption.

Because the Fuel tank is close to the center of mass, the change of center of mass can be ignored approximately, at the same time ignore the UAV moment of inertia and moment of inertia changes.

3.5 Environment model

Because the aerodynamic forces and moments, as well as engine thrust and Atmospheric physical quantities, the settlement of the aircraft equations of motion need to create the atmosphere model.

The ISA International Standard Atmosphere model is used in the simulation.

$0\ m < h \le 11000\ m$:

$$T = 288.15 - 0.0065 * h,$$

$$a = \sqrt{kRT}$$

$$p_h = 101325 * (1 - 0.0000225577 * h)^{5.25588}$$

$$\rho = 1.225 * (1 - 0.0000225577 * h)^{4.25588} \qquad (14)$$

$k = 1.4,\ R = 287.053\ \mathbf{J/(kg \cdot K)}.$

4 CONTROLLER DESIGN

Flight controller is designed on a trim point for the six degrees of freedom dynamics model.

The design process is as follows:

First we need to specify the flight control law configuration.

Then Select a reference model to meet the flight quality requirements.

In the end we just need the computer automatically optimize the control parameters of the closed-loop system to make the response characteristics as much as possible consistent with the reference model.

The selected reference model in this paper is an ideal single-input single-output second-order system model.

$$G(s) = \frac{w_n^2}{s^2 + 2\xi w_n s + w_n^2} \qquad (15)$$

$w_n = 7.5,\ \xi = 0.707.$

Then the control law configuration is established.

The control system uses flight status parameters obtained by solving the six-degree-of-freedom model of the aircraft as its input, and its output is passed to the six-degree-of-freedom model of the aircraft by the servo model. Thereby a closed-loop control system is formed.

The controller structure is the selected configuration with the control parameters (P, I, etc.) as its variables; it is also the optimization variables that optimization process required for.

Under the same input, aircraft dynamics model receives the rudder deflection angle generated by the control system.

It is according to a certain adaptive evaluation function, the response generated by the dynamic model is processed with the response of the reference model under the same input.

Thus the objective function value which optimization required for is obtained.

In the longitudinal we designed two modal controllers: the pitch angle stability control and the highly stable control.

Using these two controllers, the UAV can accomplish the mission profile with climbing, declining and cruising.

4.1 Pitch angle stability control

The outer loop of the controller uses the pitch angle as feedback. Meanwhile, in order to improve the damping characteristics of the aircraft angular movement, a rate of change of the angle of the pitch is introduced. The Angular rate feedback is used to adjust the aircraft pitching motion damping ratio. So as to improve the flying qualities of the aircraft.

$$\delta_e = [(Kp + Ki/s)(\theta_{ref} - \theta)] - Kq * \dot{\theta} \qquad (16)$$

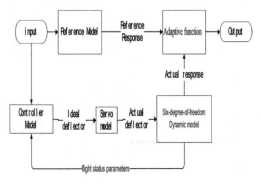

Figure 2. Controller structure.

The optimization model is established based on this controller structure as shown in Figure 3.

Optimization variables: *ki, kp, kq*

Optimization goals:

$$\min\left(sqrt\left(\sum_{i=1}^{n}\left(y1(iT)^2 - y2(iT)^2\right)\right)\right)$$

Optimization results:

$ki = 0.1683,\ kp = 1.3257,\ kq = 0.1956.$

4.2 Altitude hold/control system

Altitude Hold/control system use the signal output by INS/DGPS to determine the deviation of the height relative to a given value, then the height differential signal is passed to the pitch angle control system. The track inclination angle is changed until the height difference is zero to control aircraft Increase and Decrease. In Altitude Hold/control process, pitch angle deviates signal is the damping

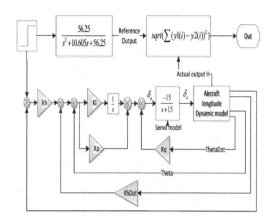

Figure 5.   Height controller design model.

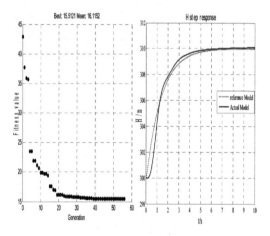

Figure 6.   Best fitness with generation.

signal of the height stable system. Altitude Hold/control system is usually formed on the basis of the pitch angle control system.

$$\delta_e = \left(Kp + \frac{Ki}{s}\right)\left[Kh \times (h_{ref} - h) - Khdot \times \dot{h} - \theta\right] - Kq \times \dot{\theta}$$

(17)

Optimization results:

$ki = 0.0732,\ kp = 3.04,\ kq = 2.34,\ kh = 5.0,$

$khdot = 3.3.$

5   NONLINEAR SIMULATION

The simulation process to achieve the mission profile: climbing with given pitch angle, cruise with

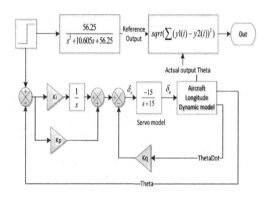

Figure 3.   Pitch angle controller design model.

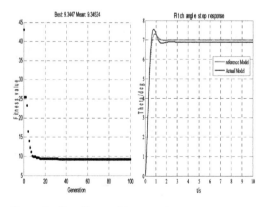

Figure 4.   Best fitness with generation.

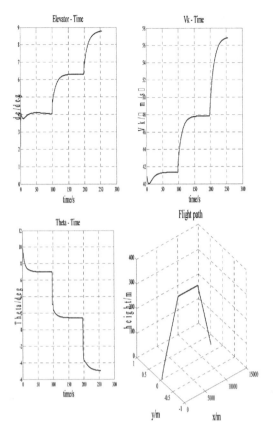

Figure 7. Simulation results.

6 CONCLUSION

A flight controller design method based on model matching and genetic algorithm optimization is proposed in this paper. A UAV flight controller is designed using the above method. Through non-linear simulation verification, we prove that the controller can well meet the relevant requirements, the pitch angle of the aircraft climb, high cruising, such as pitch angle decreased control. Compared with the classic controller design method, the method is more intelligent, can save a lot of energy and time.

REFERENCES

Kristinsson K, Dumont G.A. System identification and control using genetic algorithms. IEEE Transactionson System, Man, and Cybernetics, 1992, 22(5): 1033–1046.

Lansberry J.E, Wozniak L, Goldberg D.E. Optimal hydrogenerator governor tuning with a genetic algorithm. IEEE Transactions on Energy Conversion, 1992, 7(4): 623–630.

Li Guangwen, Zhang Weiguo, Liu Xiaoxiong, Li Dong. NGA Based on Uniform Design and Its Application in Optimization of FCS. Acta Aeronautica Et Astronautica Sinica Vol. 29 Sup May 2008.

Li Renhou. Theory and method of intelligent control. Xi'an: Xidian University Press, 2002. (in Chinese).

Liu Bo, Li Ping, A Stateflow Based Simulation of UAV Multi-mode Flight Control. Proceedings of the 8th World Congress on Intelligent Control and Automation, July 6–9 2010, Jinan, China.

Pan Wei, Wang Xueyong, Jing Yuanwei. Mixed $H2/H\infty$ state feedback controllers based on genetic algorithm. Control and Decision, 2005, 20(2): 132–136. (in Chinese).

Peter H. Zipfel. Modeling and simulation of aerospace vehicle dynamics. AIAA Education Series. AIAA inc, 2007.

Robert C. Nelson. Flight stability and automatic control. McGraw-Hill Book Company, 1998.

Wang Xiaoping, Cao Liming. Genetic algorithm. Xi'an: Xi'an Jiaotong University Press, 2005. (in Chinese).

Wang Yong, Wang Yingxun. Lateral Deviation Correction Control for UAV Taxiing. Acta Aeronautica Et Astronautica Sinica. Vol. 29 Sup. May 2008.

given height, decreased with given pitch angle. The mission profile is achieved using the Simulink finite state machine module: Stateflow. The sequence and switching conditions of the mission profile as shown below.

First the UAV climbs with the given pitch angle of 7 degree.

After climbing to the height of 200 m, the UAV begins to cruise with the given height of 200 m.

When the fight distance is 10000 m, the UAV begins to dive with given angle −3 degree.

When the height is 30 m, the simulation stops.

The simulation results is shown in Figure 7.

Control Engineering and Information Systems – Liu (Ed)
© 2015 Taylor & Francis Group, London, ISBN 978-1-138-02685-8

HIL test platform for unmanned aerial vehicle flight control system

M. Li, L. Zhou, Y.H. Li & Z.Y. Luo

Changchun Institute of Optics, Fine Mechanics and Physics, Chinese Academy of Sciences, Changchun, Jilin, China

ABSTRACT: The avionics systems, especially the flight management and control systems, are some of the most critical elements of any aircraft, more so for an autonomous aircraft. Testing these systems in a realistic synthetic environment is an important process. Flight testing full scale aircraft in the real world environment has always been expensive. Additionally, new emerging technologies require extensive testing and doing so in the full scale environment is cost prohibitive. It is logical therefore, to develop a suitable simulation environment in which the avionics systems can be tested, just as they would be on the actual aircraft. This paper discusses the development of such a simulation system, and conduct a series of flight tests to compare the data between the hardware in the loop test with digital simulation.

1 INTRODUCTION

Presently there is vast interest in UAV (Unmanned Aerial Vehicle) development given its civilian and military applications. One of the main UAV components is the flight control system. Its development invariably demands several simulations and tests. Thus, before embedding a flight control system, it has to be exhaustively lab tested. Based on this necessity, with educational and research purposes in autopilot control systems development area, a test platform is herein proposed. The test platform provides an environment where the designed autopilot system can be applied into models very similar to real aircraft. Parameters of flight as well as aircraft responses can be monitored and easily analyzed. An essential component of the simulation system is the incorporation of the actual flight hardware into the simulation loop. Incorporating the flight hardware makes the simulation system more complex and potentially more expensive however the benefits from this inclusion are significant. Most of the aircraft systems, such as avionics and sensors, are directly built into the simulation just as they would be on the actual aircraft that will identify all of the potential failures.

It employs Matlab/Simulink to run the dynamic flight model, sensor models and control algorithms, a microcontroller with the autopilot controller under test to command model aircraft flight control surfaces, a servo to drive these control surfaces, and the flight simulator X-Plane with the aircraft to be commanded for the flight visualization. The mission management computer and other systems including the flight control system are connected to the Simulink/X-Plane systems through data communication buses. The test platform provides the means for designers to construct mission and control algorithms and to test the avionics hardware and software systems in real time, by simulating full missions, with realistic visual feedback of the aircraft's response.

2 FLIGHT AND SENSOR MODELS

According to Newton's 2nd Law, the equations of motion written in the local aircraft frame for a rigid body, are given by: $F = mdV/dt + m\omega \times V$. Similarly, the angular equation is expressed by: $M = dh/dt + \omega \times h$. Where F is the force applied, M is the moment, V is the velocity vector, and h is the angular moment. It is possible to write h as a product between the inertia tensor and the angular velocity: $h = [I]\omega$. Where ω is the angular velocity and $[I]$ is the inertia tensor. Or, using the vector components and neglecting terms usually small in the matrix:

$$x = m[du/dt + qw - rv]$$
$$y = m[dv/dt + ru - pw]$$
$$z = m[dw/dt + pv - qu]$$

and

$$L = I_{xx}(dp/dt) - I_{xz}(dr/dt + pq) + qr(I_{zz} - I_{yy})$$
$$M = I_{yy}(dq/dt) + I_{xz}(p^2 - r^2) + pr(I_{xx} - I_{zz})$$
$$N = I_{zz}(dr/dt) + I_{xz}(qr - dp/dt) + pq(I_{yy} - I_{xx}).$$

An airplane in free flight has three translational motions (vertical, horizontal and transverse), three rotational motions (pitch, yaw and roll) and numerous elastic degrees of freedom [1]. The model used to simulate the dynamics of the aircraft is a general aviation UAV. It has a capacity for 40 kg payloads, six meters wing span and has a 40 hp engine. The model is developed through a Simulink library, named AeroSim. It was used as test model to simulate the dynamics of an aircraft. This model consists of several blocks, modeling the different equations of the aircraft dynamics [2] [3]. The outer Block Diagram is shown in Figure 1.

In control, it is possible to define the aircraft as a system where the entrance is the control input and the output is the movement or the path. In order to reduce the complexity of this mathematical modeling problem, some simplifying assumptions may be applied. To analyze and design the flight control system, it is usual to define the equations of the aircraft motion, linearized around a straight level flight trim condition, and expressed as a LTI state space system. The chosen controller, according to the theory of optimal control, was the Linear Quadratic Regulator, LQR. This controller is characterized by a straightforward design, based on the definition of a cost function.

For the longitudinal control case, the states to control in servo were the speed and altitude. The aircraft follows the reference values by controlling the elevator and the Throttle.

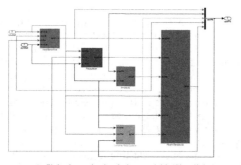

a) flight dynamic simulation model in Simulink

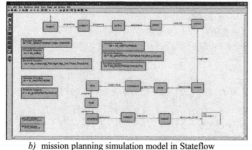

b) mission planning simulation model in Stateflow

Figure 1. Matlab simulink block diagram of the UAV dynamic model.

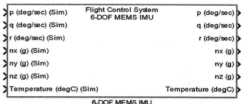

a) the IMU block

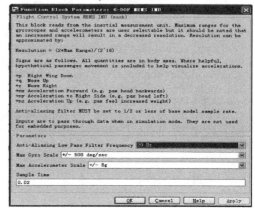

b) the IMU mask

Figure 2. The IMU block and mask.

The Inertial Measurement Unit is a MEMS device that incorporates a 3-axis accelerometer with a 3-axis gyroscope together. Figure 2 shows the IMU block and mask. The IMU block has three options that the user must select in the mask of the IMU block. The IMU has a built in digital low pass filters with user selectable cut-off frequencies to prevent anti-aliasing. The low pass filter frequency must be set to half or less of the base sample rate. The range of the gyros and accelerometers must also be set, and the sample time should match the overall sample time of the autopilot. The inputs to the block are simply the simulated values of each of the accelerations and angular rates as well as temperature. These are passed through the block in simulation, and not used when the model is embedded to the hardware. The outputs of the block are the measured values from the inertial measurement unit.

3 INCORPORATION OF FLIGHT HARDWARE

The architecture of the test platform is shown in Figure 3. The system is comprised of three separate systems all linked together via the real-time simulation system. The real-time simulation system, general control system and the visualization

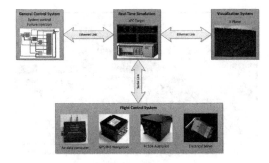

Figure 3. The architecture of the test platform.

system run on three separate PC based systems. An Ethernet link is used for communications between the real-time simulation system, general control system and the flight visualization system. The interface between the real-time simulation system and the flight hardware is currently handled through serial communications however this is easily modifiable. The general control system provides a central point for controlling the simulation. All elements of the system can be monitored and controlled.

The real-time simulation system implements the dynamic models on an embedded system using xPC Target. This implementation ensures that the dynamic models are run in real time and provides a means to synchronize all elements of the system. The model can fully control every element of the simulated aircraft. A wide range of failures can be simulated and any repair actions can be simulated. The system implements its own flight and sensor models providing complete control over every aspect of the system. Overall, although the shift of the dynamic models increased the complexity and computing requirements of the real-time simulation system, the shift has made it possible to more fully simulate missions. Currently the dynamic models of the aircraft and sensors are implemented in Matlab Simulink. The model provides a full nonlinear, six degrees of freedom aircraft model based upon the stability coefficients for an aircraft. It also provides standard sensor models. These models are implemented in such a manner that they can easily be replaced with models based upon analyses of the sensors to be used on the actual aircraft.

The intelligent flight control system will be tested while running on the flight hardware. In order to achieve this, the flight hardware must be linked into the simulation system in as transparent a manner as possible. By this we mean that the flight hardware should ideally not be capable of detecting any difference between the simulation and the real world [4] [5].

The system uses RS422 serial communications to link the flight hardware to the dynamic models. This communications method was retained as serial communications are present on the majority of PC's and microprocessor based systems. All communications through this link are packet based and are either commands or reports. Reports are used by the real-time simulation system to provide sensor data to the flight hardware. Commands are sent from the flight hardware to provide new input to the dynamic model or to reconfigure aspects of the dynamic model. The disadvantage of the RS422 communication links are that they create potential bottlenecks in the simulation environment. The RS422 links can handle a maximum data throughput of 115, 200 bits per second (bps). Due to the large quantity of data which will be being transferred through these links (e.g. aircraft state information) a slowdown may occur due to the inability of the link to handle the full data throughput. This problem can be removed by changing the link between the real-time simulation system and the flight hardware to CAN (1 Mbps link).

4 VISUALIZATION SYSTEM

The system uses the X-Plane flight simulation system for flight visualization. X-Plane is a commercially available flight simulation package produced by Laminar Research. Although there exist several simulators like Microsoft's Flight Simulator, and FlightGear, X-Plane provides extremely accurate flight models and allows for external communication as well as airfoil design. It is accurate enough to be used to train pilots. Unlike Microsoft Flight Simulator, however, X-Plane also allows for input and output from external sources. While FlightGear has I/O capabilities similar to X-Plane, it is not quite as stable as X-Plane and doesn't provide the level of support. X-Plane provides future capabilities that unmanned aerial vehicles will need, including navigation markers, changing weather conditions, and air traffic control communication. While X-Plane's plane maker provides an interface to design a vehicle based on physical dimensions, power, weight, and a host of other specifications, it produces a specialized file that allows the simulator's built in physics to properly interact with the model. Thus, the equations for the simulator model are not available.

The advanced visualization system is a software package which enables Simulink to communicate with the flight simulator X-Plane. The software enables testing of custom control systems, observers, guidance and navigation systems.

Figure 4.   The X-Plane visualization system.

It is also possible to use the software to conduct Hardware In the Loop (HIL) testing. The software package consists of two parts. A Simulink block, which send control input to the simulator and receive state information. The other part of the package is an X-plane plugin, which receive control input, and set the data references of X-plane accordingly. Further the plugin send the state information. The communication is done by UDP protocol, thus enable the system to run on separate computers [6].

Data packets are sent and received through the microcomputer Ethernet port. UDP protocol is used to establish the communication between Matlab/Simulink (Controller and Aircraft Dynamics) and X-Plane (Visualization). Using standard ethernet communications it is possible to both send commands to X-Plane's flight model and to completely override the flight model and directly set the aircraft state. The X-Plane visualization system is shown in Figure 4. Aircraft state commands (position and orientation) are sent to X-Plane and no data is received back. Commands are sent to X-Plane using a custom Simulink block. This Simulink block was written specifically to communicate to X-Plane, however it could easily be adapted to communicate with any ethernet capable simulation system. The block is provided with the desired aircraft position and orientation which are then sent to X-Plane.

5   APPLICATIONS AND RESULTS

The test platform has been designed with hardware in the loop testing a priority. This system has been developed in order to test intelligent avionics systems for usage on unmanned aircraft. The platform is a critical component of the development and testing of these systems and will be used extensively in the design process.

a)   the avionics systems hardware in the loop experiment

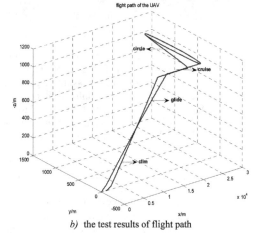

b)   the test results of flight path

Figure 5.   The experiment process and test results.

The intelligent avionics systems being developed will perform onboard mission planning and execution and failure detection and compensation. The flight model will be used to simulate all aspects of the mission from takeoff through to landing. Faults will be injected into the system and then repaired by the intelligent avionics. This requires a level of realism in the simulation.

In order to demonstrate how the test platform could be applied in laboratory classes and also help autopilot systems development, the autopilot system was designed and integrated into the platform. The experiment process and the test results are shown in Figure 5.

6   CONCLUSION

The development of this test platform resulted in a valuable tool to aid flight control systems study and design. It allows monitoring the aircraft responses for a designed autopilot system with high degree of realism. It is possible to change control system parameters very easily and check the results out in a friendly environment. This possibility easies design task as well minimizes risks of embedding the system for field tests.

The usage of the test platform for simulation of vehicles other than aircraft (e.g. satellites) has also been considered. With the current architecture there is no limitation on the situations to which system could be applied. Providing a capable visualization system, then it could be used to simulate satellite, ground based or underwater based operations. Minor modifications would be required to elements of the general control system and the dynamic models would need to be replaced. This extensive and varied usage has led to the most recent developments on the system which has made it capable of simulating highly complex situations. Overall the development and usage of the test platform has provided significant benefit for relatively little cost. The low cost and incorporation of hardware in the loop makes system testing simple and effective. The visualization and real time nature provides a powerful design and teaching tool for control systems.

REFERENCES

Adang Suwandi Ahmad1, Jaka Sembiring, "Hardware in the Loop Simulation for Simple Low Cost Autonomous UAV (Unmanned Aerial Vehicle) Autopilot System Research and Development" Institut Teknologi Bandung, Indonesia, 2007.

Barnard, P., "Graphical Techniques for Aircraft Dynamic Model Development," AIAA Modeling and Simulation Technologies Conference and Exhibit [CD-ROM], Providence, RI, 2004.

Daniel Ernst, "Development of Research Platform for Unmanned Vehicle Controller Design, Evaluation, and Implementation System: From MATLAB to Hardware Based Embedded System", Department of Computer Science and Engineering College of Engineering University of South Florida, Florida, 2007.

Kimberlin, R.D., Flight Testing of Fixed-Wing Aircraft, AIAA, Virginia, 2003, Chaps 4, 15, 21.

Laminar Research, "X-Plane Description", X-Plane Manual, 2009.

Nelson, R.C. "Flight Stability and Automatic Control" McGraw-Hill Second Edition, 1998.

Control Engineering and Information Systems – Liu (Ed)
© 2015 Taylor & Francis Group, London, ISBN 978-1-138-02685-8

Studying for stabilization of singular large-scale linear control systems

S.L. Sun

Guangdong Polytechnic Normal University, Guangzhou, China

ABSTRACT: This paper studies the state feedback stabilization of singular linear continuous large-scale control systems by using singular Lyapunov matrix equation and singular vector Lyapunov function method. There gives the parameter regions of asymptotic stability and instability. At last, an example is given to show the application of main result.

1 INTRODUCTION

With the development of modern control theory and its permeation into other application area, one kind of systems with extensive form has appeared which form follows as:

$$E\dot{x}(t) = f(x(t), t, u(t))$$

where $x(t) \in R^n$ is a n−state vector, $u(t) \in R^m$ is a m−control input vector, E is a $n \times n$ matrix, it is usually singular. This kind of systems generally is called as the singular control systems. It appeared large in many areas such as the economy management, the electronic network, robot, bio-engineering, aerospace industry and navigation and so forth. Singular large-scale control systems have a more practical background. At present, the some research results of the problem above are obtained in the references. The asymptotical stability and stabilization of the discrete singular large-scale control systems has been studied by using scalar sum Lyapunov function method in the references. This paper consider the state feedback stabilization of linear continuous singular large-scale control systems by introduce singular vector Lyapunov function method, and give a more large interconnecting parameters regions of stability.

2 DEFINITIONS AND PROBLEM FORMULATION

Consider the linear continuous singular large-scale control systems with m subsystems:

$$E_i \dot{x}_i(t) = A_{ii} x_i(t) + \sum_{\substack{j=1 \\ j \neq i}}^{k} A_{ij} x_j(t) + B_i u_i(t) \qquad (1)$$

where $x_i(t) \in R^{n_i}, u_i(t) \in R^{m_i}, \sum_{i=1}^{k} n_i = n, \sum_{i=1}^{k} m_i = m, x(t) = [x_1^T(t), x_2^T(t), ..., x_k^T(t)]^T \in R^n$ is a semi-state vector, $u(t) = [u_1^T(t), u_2^T(t), ..., u_k^T(t)]^T \in R^m$ is a control input vector; $A_{ii}, E_i \in R^{n_i \times n_i}, A_{ij} \in R^{n_i \times n_j}, B_i \in R^{n_i \times m_i}$, are constant matrices; $E = Block - diag(E_1, E_2, ..., E_k), N = \{1, 2, ..., k\}, rank(E_i) = r_i \leq n_i, \sum_{i=1}^{k} r_i = r, rank(E) = r \leq n$.

The isolated subsystems of the systems (1) are

$$E_i \dot{x}_i(t) = A_{ii} x_i(t) + B_i u_i(t), i \in N \qquad (2)$$

We have following definitions:

Definition 1 [Dong-Mei Yang]: The every singular subsystem of systems (2) be said to be stabilization if and only if exist a matrix $K_i \in R^{m_i \times n_i}$, and the linear state feedback law:

$$u_i(t) = -K_i x_i(t) \qquad (3)$$

subject to the closed-loop system of the system

$$E_i \dot{x}_i(t) = (A_{ii} - B_i K_i) x_i(t) \qquad (4)$$

become stable.

The closed-loop systems of the singular large-scale control systems (1) are

$$E_i \dot{x}_i(t) = [A_{ii} - B_i K_i] x_i(t) + \sum_{\substack{j=1 \\ j \neq i}}^{k} A_{ij} x_j(t), i \in N \qquad (5)$$

The isolated subsystems of (6) are

$$E_i \dot{x}_i(t) = (A_{ii} - B_i K_i) x_i(t), i \in N \qquad (6)$$

Definition 2 [Dong-Mei Yang]: The singular systems (4) is said to be regular if for some one $s \in C$

$\det(sE_i - A_{ii} - B_iK_i) \neq 0$.

Definition 3 [Dong-Mei Yang]: The singular systems (5) is said to be impulsive-free if for any $s \in C$

$\deg\{\det(sE_i - A_{ii} - B_iK_i)\} = rank(E_i)$.

In order to investigate the stability of singular large-scale control systems (1), we give the following lemmas:

Lemma 1 [Dong-Mei Yang]: Assume that x, $y \in R^n$, $B \in R^{n \times n}$ is a positive semi-definite matrix, then

$$2x^T By \leq \varepsilon x^T Bx + \varepsilon^{-1} y^T By \tag{7}$$

holds for any real numbers $\varepsilon > 0$.

Lemma 2 [Dong-Mei Yang]: If the singular system (4) is regular, impulsive-free, and stable, then for the any positive definite matrix M_i, there exists positive definite matrix W_i which satisfies

$$(A_{ii} - B_iK_I)^T W_iE_i - E_i^T W_i(A_{ii} - B_iK_i) = -E_i^T M_iE_i \tag{8}$$

take singular Lyapunov function as follows:

$$V_i[E_ix_i(t)] = [E_ix_i(t)]^T W_i[E_ix_i(t)] \tag{9}$$

then

$$\lambda_m(W_i)\|E_ix_i(k)\|^2 \leq V_i[E_ix_i(k)] \leq \lambda_M(W_i)\|E_ix_i(k)\|^2 \tag{10}$$

where $\lambda_m(W_i)$ and $\lambda_M(W_i)$ represent respectively minimum and maximum eigenvalue of the matrix W_i.

3 MAIN RESULTS

Theorem 1: The singular large-scale control system (1) is stabilizable. If the following conditions are satisfied.

a. The every isolated subsystems (2) of the singular large-scale control systems (1) are stablilizable; and the every isolated closed-loop subsystem (6) of the isolated subsystems (5) are regular and impulsive-free, that is, for any a positive definite matrix M_i there exists positive definite matrix W_i which satisfies (8);
b. For singular large-scale control systems (1), there exist real numbers $\delta_{ij} > 0$, which satisfies

$$[A_{ij}x_j(t)]^T W_i[A_{ij}x_j(t)]$$
$$\leq \delta_{ij}[E_jx_j(t)]^T W_j[E_jx_j(t)] \quad (i,j \in N, i \neq j) \tag{11}$$

c. Assume the $k \times k$ matrix $-D = (-d_{ij})_{k \times k}$ is a M-matrix. Where

$$d_{ij} = \begin{cases} 1 - \dfrac{\lambda_m(M_i)}{\lambda_M(W_i)}, & j = i \\ \delta_{ij}(k-1), & j \neq i \end{cases} \tag{12}$$

Proof: Construct singular vector Lyapunov function as follows

$$V[EX(t)] = \{V_1[E_1X_1(t)], V_2[E_2X_2(t)],$$
$$..., V_k[E_kX_k(t)]\}^T$$

where $V_i[E_ix_i(t)]$ is given by (9). By the using (7)~(9), we have

$$\dot{V}_i[E_ix_i(t)]\big|_{(1)} = \{[E_i\dot{x}_i(t)]^T W_i[E_ix_i(t)]$$
$$+ [E_ix_i(t)]^T W_i[E_i\dot{x}_i(t)]\}\big|_{(1)}$$

$$= \left[A_{ii}x_i(t) + \sum_{\substack{j=1 \\ j \neq i}}^{k} A_{ij}x_j(t) + B_iu_i(t)\right]^T W_i[E_ix_i(t)]$$

$$+ [E_ix_i(t)]^T W_i\left[A_{ii}x_i(t) + \sum_{\substack{j=1 \\ j \neq i}}^{k} A_{ij}x_j(t) + B_iu_i(t)\right]$$

$$= [(A_{ii} - B_iK_i)x_i(t)]^T W_i[E_ix_i(t)]$$
$$+ [E_ix_i(t)]^T W_i[(A_{ii} - B_iK_i)x_i(t)]$$
$$+ 2[E_ix_i(t)]^T W_i\sum_{\substack{j=1 \\ j \neq i}}^{k} A_{ij}x_j(t)$$

$$\leq -[E_ix_i(t)]^T M_i[E_ix_i(t)]$$
$$+ [E_ix_i(T)]^T W_i[E_ix_i(t)]$$
$$+ \left[\sum_{\substack{j=1 \\ j \neq i}}^{k} A_{ij}x_j(t)\right]^T W_i\left[\sum_{\substack{j=1 \\ j \neq i}}^{k} A_{ij}x_j(t)\right]$$

$$\leq -\frac{\lambda_m(M_i)}{\lambda_M(W_i)}[E_ix_i(t)]^T W_i[E_ix_i(t)]$$
$$+ [E_ix_i(t)]^T W_i[E_ix_i(t)]$$
$$+ \sum_{\substack{j=1 \\ j \neq i}}^{k}\sum_{\substack{s=1 \\ s \neq i}}^{k} [A_{ij}x_j(t)]^T W_i[A_{is}x_s(t)]$$

$$\leq \left(1 - \frac{\lambda_m(M_i)}{\lambda_M(W_i)}\right)[E_i x_i(t)]^T W_i [E_i x_i(t)]$$

$$+ \frac{1}{2} \sum_{\substack{j=1 \\ j \neq i}}^{k} \sum_{\substack{s=1 \\ s \neq i}}^{k} \left\{ [A_{ij} x_j(t)]^T W_i [A_{ij} x_j(t)] \right.$$

$$+ [A_{is} x_s(t)]^T W_i + [A_{is} x_s(t)] \}$$

$$= \left(1 - \frac{\lambda_m(M_i)}{\lambda_M(W_i)}\right)[E_i x_i(t)]^T W_i [E_i x_i(t)]$$

$$+ (k-1) \sum_{\substack{j=1 \\ j \neq i}}^{k} [A_{ij} x_j(t)]^T W_i [A_{ij} x_j(t)]$$

By using (10), we have

$$\dot{V}_i[E_i x_i(t)]_{(1)} \leq \left(1 - \frac{\lambda_m(M_i)}{\lambda_M(W_i)}\right) V_i[E_i x_i(t)]$$

$$+ \delta_{ij}(m-1) \sum_{\substack{j=1 \\ j \neq i}}^{k} V_j[E_j x_j(t)]$$

Thus we have

$$\dot{V}_i[E_i x_i(t)] \leq \sum_{j=1}^{k} d_{ij} V_j[E_j x_j(t)]$$

That is

$$\dot{V}[Ex(t)] \leq DV[Ex(t)]$$

Noticing the condition (c) of the theorem 1, and using the Comparison Principle, we have $\lim_{t \to \infty} V[Ex(t)] = 0$, thus $\lim_{t \to \infty} V_i[E_i x_i(t)] = 0$, $i \in N$, therefore $\lim_{t \to \infty}[E_i x_i(t)] = 0, i \in N$.

Since the every subsystem of systems (5) are regular and impulsive-free, so there exist nonsingular matrices P_i, Q_i, such that

$$P_i E_i Q_i = \begin{bmatrix} I_i^{(1)} & o \\ o & o \end{bmatrix}, \; P_i(A_{ii} - B_i K_i)Q_i = \begin{bmatrix} N_i & o \\ o & I_i^{(2)} \end{bmatrix}.$$

Let

$$P_i A_{ij} Q_j = \begin{bmatrix} A_{ij}^{(1)} & A_{ij}^{(2)} \\ A_{ij}^{(3)} & A_{ij}^{(4)} \end{bmatrix}, \; z_i(t) = Q_i^{-1} x_i(t) = \begin{pmatrix} z_i^{(1)}(t) \\ z_i^{(2)}(t) \end{pmatrix},$$

where $I^{(1)}, I^{(2)}$ is $r_i \times r_i$ and $(n_i - r_i) \times (n_i - r_i)$ identity matrix,^{i} $P_i, Q_i, N_i, A_{ij}^{(1)}, A_{ij}^{(2)}, A_{ij}^{(3)}, A_{ij}^{(4)}$ are respectively corresponding dimension constant matrices. We shall be to prove

$$z_i(t) = Q_i^{-1} x_i(t) = \begin{pmatrix} z_i^{(1)}(t) \\ z_i^{(2)} t) \end{pmatrix} \to 0(t \to \infty).$$

The singular closed-loop system (6) is equivalent to

$$\begin{cases} \dot{z}_i^{(1)}(t) = N_i z_i^{(1)}(t) + \sum_{\substack{j=1 \\ j \neq i}}^{k} [A_{ij}^{(1)} z_j^{(1)}(t) + A_{ij}^{(2)} z_j^{(2)}(t)] \\ \\ 0 = z_i^{(2)}(t) + \sum_{\substack{j=1 \\ j \neq i}}^{k} [A_{ij}^{(3)} z_j^{(1)}(t) + A_{ij}^{(4)} z_j^{(2)}(t)] \end{cases}$$

Noticing that

$$P_i E_i x_i(t) = P_i E_i Q_i Q_i^{-1} x_i(t) = \begin{pmatrix} I_i^{(1)} & 0 \\ 0 & 0 \end{pmatrix}\begin{pmatrix} z_i^{(1)}(t) \\ z_i^{(2)}(t) \end{pmatrix}$$

That is

$$P_i E_i x_i(t) = \begin{pmatrix} z_i^{(1)}(t) \\ 0 \end{pmatrix},$$

We have

$$z_i^{(1)}(t) = \begin{pmatrix} I_i^{(1)} & 0 \end{pmatrix} P_i E_i x_i(t)$$

Then from

$$E_i x_i(t) \to 0(t \to \infty)$$

We have

$$z_i^{(1)}(t) \to 0(t \to \infty)$$

Noticing that

$$P_i A_{ij} x_j(t) = P_i A_{ij} Q_j Q_j^{-1} x_j(t) = \begin{pmatrix} A_{ij}^{(1)} & A_{ij}^{(2)} \\ A_{ij}^{(3)} & A_{ij}^{(4)} \end{pmatrix}\begin{pmatrix} z_j^{(1)}(t) \\ z_j^{(2)}(t) \end{pmatrix}$$

We have

$$A_{ij}^{(3)} z_j^{(1)}(t) + A_{ij}^{(4)} z_j^{(2)} = \begin{pmatrix} 0 & I_i^{(2)} \end{pmatrix} P_i A_{ij} x_j(t)$$

Noticing that $A_{ij} x_j(t) \to 0(t \to \infty)$ holds from (9), that is

$$A_{ij}^{(3)} z_j^{(1)}(t) + A_{ij}^{(4)} z_j^{(2)} \to 0(t \to \infty)\;(i, j \in N, i \neq j)$$

We know

$$-z_i^{(2)}(t) = \sum_{\substack{j=1 \\ j\neq i}}^{k} A_{ij}^{(3)} z_j^{(1)}(t) + A_{ij}^{(4)} z_j^{(2)}$$

Then

$$z_i^{(2)}(t) \to 0 (t \to \infty)$$

Hence

$$z_i(t) \to 0 (t \to \infty)$$

And

$$x_i(t) \to 0 (t \to \infty)$$

That is

$$x(t) \to 0 (t \to \infty). \ \square$$

Theorem 2: (a) Assume all closed-loop isolated systems (6) of the systems (2) are regular, impulsive-free, and given positive definite matrix M_i there exists positive definite matrix W_i which satisfies

$$(A_{ii} - B_i K_I)^T W_i E_i - E_i^T W_i (A_{ii} - B_i K_i) = E_i^T M_i E_i$$

(b) Assume there exists a real number $\delta_{ij} > 0$, which satisfies

$$\left[A_{ij}x_j(t)\right]^T W_i \left[A_{ij}x_j(t)\right] \le \delta_{ij} \left[E_j x_j(t)\right]^T$$
$$W_j \left[E_j x_j(t)\right] (i, j \in N, i \neq j)$$

(c) Assume the $k \times k$ matrix $D = (d_{ij})_{m \times m}$ is a M-matrix. Where

$$d_{ij} = \begin{cases} \dfrac{\lambda_m(M_i)}{\lambda_M(W_i)} - 1, & j = i \\ \delta_{ij}(1-k), & j \neq i \end{cases}$$

Then singular large-scale control systems (1) are not stabilizable.

Proof is similar to the theorem 1, here it be omitted.

4 EXAMPLE

Consider the following 5-order discrete singular large-scale control system which consists of two sub-systems

$$E_i \dot{x}_i(t) = A_{ii} x_i(t) + \sum_{\substack{j=1 \\ j\neq i}}^{2} A_{ij} x_j(t) + B_i u_i(t), i = 1,2 \quad (13)$$

where

$$E_1 = \begin{pmatrix} 1 & 0 \\ 0 & 1 \end{pmatrix}, E_2 = \begin{pmatrix} 1 & 0 & 0 \\ 0 & 1 & 0 \\ 0 & 0 & 0 \end{pmatrix}, A_{11} = \begin{pmatrix} 1 & 0 \\ 1 & -1 \end{pmatrix},$$

$$A_{12} = \begin{pmatrix} \frac{1}{2} & \frac{1}{2} & 0 \\ -\frac{1}{2} & \frac{1}{2} & 0 \end{pmatrix}, A_{21} = \begin{pmatrix} \frac{1}{2} & \frac{2}{3} \\ -\frac{2}{3} & \frac{1}{2} \\ 0 & 0 \end{pmatrix}$$

$$A_{22} = \begin{pmatrix} 1 & 0 & 0 \\ 1 & 0 & 1 \\ 2 & 1 & 1 \end{pmatrix}, B_1 = \begin{pmatrix} 1 \\ 0 \end{pmatrix}, B_2^T = \begin{pmatrix} 1 & 0 & 1 \\ 0 & 1 & 0 \end{pmatrix},$$

we choose the control law:

$$u_i(t) = K_i x_i(t), (i = 1,2)$$

Where

$$K_1 = (2 \ \ 1), K_2 = \begin{pmatrix} 2 & 1 & 0 \\ 0 & 1 & 1 \end{pmatrix}, M_1 = \begin{pmatrix} 1 & 0 \\ 0 & 1 \end{pmatrix},$$

$$M_2 = \begin{pmatrix} 1 & 0 & 0 \\ 0 & 1 & 0 \\ 0 & 0 & 1 \end{pmatrix}$$

Then

$$W_1 = \begin{pmatrix} \frac{1}{2} & 0 \\ 0 & \frac{1}{2} \end{pmatrix}, W_2 = \begin{pmatrix} \frac{1}{2} & 0 & 0 \\ 0 & \frac{1}{2} & 0 \\ 0 & 0 & \frac{2}{3} \end{pmatrix}.$$

it is easy to test that (8) is hold, and we have obtained matrix

$$D = \begin{pmatrix} -1 & \delta_{12} \\ \delta_{21} & -\frac{1}{2} \end{pmatrix}$$

then when $\delta_{12} = \delta_{21} = \delta < \sqrt{2}/2$, the matrix $-D$ is a M-matrix, then we know this system (13) is satbilizable from Theorem 1.

REFERENCES

Chao-Tian Chen & Jian-Chuan Zhao.2007. Stability of a Kind of Continuous Singular Large-Scale Dynamical Systems. Proceedings of the Sixth Conference on Machine Learning and Cybernetics, Hong Kong. pp. 2319–2321.

Dong-Mei Yang & Qing-ling Zhang. 2004. Singular Systems. Science Press, Beijing.

Shui-Ling Sun & Ping Peng. 2009. Connective Stability of Singular Linear Time-Invariant Large-Scale Dynamical Systems with Sel-Intraction. Advances in Diffrential Equations and Conntrol Processes. vol 3(1), pp. 63–76.

Shui-Ling Sun & Yun-Yun Chen 2011. State Feedback Stabilization of Discrete Singular Large-scale Control Systems, Studies in Mathematical Sciences. vol 2(2), pp. 36–42.

Song-Lin Wo. 2004. Stability and Decentralized Control for the Singular Large-Scale Systems. Nanjing University of Science and Technology May. Vol, 35(2), pp. 47–52.

Control Engineering and Information Systems – Liu (Ed)
© 2015 Taylor & Francis Group, London, ISBN 978-1-138-02685-8

Research on role-based access control model of fundamental spatial database system

X.Q. Yu & T. Zhang
China School of Geomatics and Prospecting Engineering, Jilin Jianzhu University, Changchun, China

P. Xu
School of Traffic and Transportation Engineering, Changsha University of Science and Technology, Changsha, China

L. Jiang
School of Resources and Environment, University of Electronic Science and Technology of China, Chengdu, China

ABSTRACT: In order to access the information in the database for multi-user hierarchically, the author discusses the problem of user role and privilege management and the role-based access control model, makes an intensive study on the flow of system access control, and implements a role-based access control model of fundamental spatial database system of Jilin water resources finally. The experiment result indicates that this model can radically reduce the complexity of system management, assign the privilege of the user reasonably, and insure the system run safely on the net.

The conventional access control strategy includes Discretionary Access Control (DAC), and Mandatory Access Control (MAC). DAC provide uses with flexible, user-friendly data access control mode, but its security is lower, so the invalid user may gain the access privileges when he avoids its security protection. MAC provides the entire subject and its object a specified sensitive label to control the access privileges, so the security is high. The implementation of the MAC mode is so complex that it is unfavorable for the large-scale application. Role-Based Access Control (RBAC) is a hot topic of the field of information security access control in recent years. The major advantage of RBAC lies in its support for the privilege administration. In the RBAC, role is a bridge that transforms the access privilege for users into for roles. Subsequently, users have been combined with the specific roles.

Fundamental Spatial Database System of Jilin Water Resources (FSDSJWR) is an information platform that is multi-user and multi-department oriented. The system integrates heterogeneous and cross-industry water resources spatial data of each department in Jilin to meet the various service needs of users. In the FSDSJWR, we want to select and analyze the water resources information, and obtain the decision function. Because of the difference in data, users, application program, and spatial location, a strict access control system which can realize separation of duty means everything.

In this paper, the author use Microsoft.NET as a development platform, taking example for the database system of FSDSJWR. The development and experiment is rose-based, which is superior to traditional frame. The system structure consists four layers such as representation layer, role and privilege administration layer, business logic layer, and data access layer. This model can radically reduce the complexity of system management, meet the need of the multi-user, assign the privilege of the user reasonably, and insure the system run safely on the net.

1 ROLE-BASED ACCESS CONTROL

Figure 1 shows the construct of the RBAC model.

User sets (U) is the subject to deal with the data.

Role sets (R) is the position within the organization which is endowed with rights and duties.

Privilege sets (P) is the privilege of the role which can make specific operation for the specific data object.

Say sets (S) creates for the user when he login in the system, and activate the corresponding role the user possesses.

Administration Role sets (AR) in which the role finishes the management and maintenance of the system itself.

Administration Privilege sets (AP) is utilized to manage and maintain the privilege sets of the access control subsystem. The administration

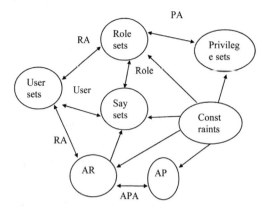

Figure 1.   The construct of the RBAC model.

spatial database, and global spatial database. The ordinary user can use the water resources fundamental spatial database and data market. Department spatial database is exclusively belongs to the department heads. Global spatial database is built for the highest department to make decision.

Figure 2 shows the grades of the fundamental spatial database of water resources in Jilin province.

Ordinary user: According to the authorization, they have the privilege of fundamental spatial water resources data browse and selection, multimedia video on demand, and general download service.

Special user: According to the authorization, they have the privilege of fundamental spatial water resources data download, data computation service, data transform service, and cross-database retrieval service. At the same time, they also have the privileges that the ordinary users have.

Advanced privilege user: They mainly refer to the provider of the fundamental spatial water resources data. Besides the privilege of the special users, they have the privilege of data application for registration, service definition, service publishing, and data updating and backup.

Administrator: They have all the privilege of FSDSJWR manipulation. Besides the privilege of the advanced privilege user, they also can add, delete, modify, suspend, and activate the privilege of the roles and the users, and maintain the system.

Among of the four roles, the ordinary user has the lowest role; the special user can inherit the privilege of the ordinary user; the advanced user can inherit the privilege of the special user; the administrator has the highest role which can inherit the privilege of the advanced user. Because of the hierarchy inheritance of role, the privilege need not be assigned to each role. The role can inherit the privilege of its lower role to improve the reusability of the system.

privilege includes the privilege to manage the P sets and the R sets.

Role Assignment (RA) refers that role can be assigned to user. One user can gain many roles, whereas, a role can be assigned to several specific users.

Privilege Assignment (PA) refers that privilege can be assigned to role. One role can have many privileges, whereas, a privilege can be assigned to many roles.

Administration Privilege Assignment (APA) refers that administration privilege can be assigned to the administration role which includes the assignment of the privilege and role.

User: S-U is a map function between the say and the user.

Role: S-R is a map function between the say and the role.

Constraints refer the value constraint of the user, the role, the privilege, and the say.

2   DESIGN OF ROLE-BASED ACCESS CONTROL MODEL OF FSDSJWR

2.1   *User role and privilege administration*

The user role and privilege administration of FSDSJWR are divided into four types according to the requirement analysis.

The water resources are utilized to serve for all the water resources departments, even for the different departments. The knowledge background of the users and the application purpose is different each. In order to satisfy all the different level of the user application and response it quickly, the database of the water resources of Jilin is divided into four grades.

The four grades are water resources fundamental spatial database, data market, department

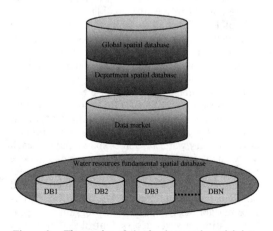

Figure 2.   The grades of the fundamental spatial database of water resources in Jilin province.

2.2 The flow of the FSDSJWR access control

The flow of the FSDSJWR access control is shown in Figure 3.

The user login: The user logins in the interface of the FSDSJWR and enters the user name and password.

Login authentication: The system accepts the user application for the login authentication and identifies the user according to the user information in the user database. If the login authentication is successful, the access controller provides the user with role information. Otherwise, notify the login authentication failure and login again.

Privilege access: The legal user will be given corresponding privilege and privilege access interface by the controller.

Server: The user gets into the server of FSDSJWR through the access interface and does all kinds of operation service at the same time.

Administrator login: The administrator can login in the system from the background directly, which can access the user database, role database, privilege database directly without the login authentication. The administrator can define the privilege decision and criterion rule for the access controller.

2.3 The design of role and privilege database

Role and privilege database is built in the background of the FSDSJWR which store the user, the role, the privilege, and their relationship. The role and privilege database of FSDSJWR is composed of the following tables shown in the Figure 4.

The table of user information: The table of user information which consists of user ID, user name, and user password.

The he table of role information: The table of role information which consists of role ID, role name, role set-up time, and role remarks.

The table of privilege information: The table of privilege information which consists of privilege ID, privilege name, function ID related to the privilege and system database object ID related to the privilege. With regard to the advanced requirement, a fine-grained access control unit can be defined, which will be subfunction and database object accessed by the subfunction.

The relation table of user and role: The relation table of user and role which consists of user ID, Role ID, and valid flag. The assignment of the role by the user is implemented by the relation table.

The relation table of role and privilege: The relation table of role and privilege which consists of role ID, privilege ID, valid flag, and privilege assignment name. The assignment of the privilege by the role is implemented by the relation table.

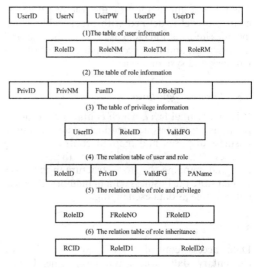

UserID	UserN	UserPW	UserDP	UserDT	

(1)The table of user information

RoleID	RoleNM	RoleTM	RoleRM

(2) The table of role information

PrivID	PrivNM	FunID	DBobjID	

(3) The table of privilege information

UserID	RoleID	ValidFG

(4) The relation table of user and role

RoleID	PrivID	ValidFG	PAName

(5) The relation table of role and privilege

RoleID	FRoleNO	FRoleID

(6) The relation table of role inheritance

RCID	RoleID1	RoleID2

Figure 4. The main table of the role and privilege database.

Figure 3. The flow of the FSDSJWR access control.

The relation table of role inheritance: The relation table of role inheritance which consists of role ID, father role number, and father role ID. The table describes the inheritance relation between the roles.

The conflict limitation table of role: The conflict limitation table of role which consists of role conflict ID and other role ID. The function of the table is the limitation of the role conflict when assigns the roles by the user.

3 IMPLEMENTATION OF ROLE-BASED ACCESS CONTROL MODEL OF FSDSJWR

In this paper, the author use Microsoft.NET as a development platform, taking example for the database system of FSDSJWR. The development and experiment is rose-based, which is superior to traditional frame. The system structure consists four layers such as representation layer, role and privilege administration layer, business logic layer, and data access layer.

3.1 Hierarchy representation layer

It is representation layer that supply interface for terminal user to have a direct access to program platform. Representation layer is not only a business component view and control role to identify the validity and hierarchy of user, but also a role which provide service for user. Therefore, a user can have an access to database.

3.2 Role and privilege administration layer

The relation information of the role-based access control model is defined with oracle table which includes information table of the user, the role, and the privilege, and the access privilege table of system task and system service.

3.3 Business logic layer

It is an essential layer, which contains conditional relations among business modules, data validation, business disposal, and method control. Defining business data entity in accordance with actual business and mapping the store structure in database, we can store the business data in memory. Business entity is role as data exchanging.

3.4 Data access control layer

Data access control layer mainly consists of elementary data access interface, general data access interface, and abstract data access interface. Elementary data access interface provide each business object elementary service. The service of business logic layer absolutely depends on the data access layer. General data access interface afford general data access service which has no relation to business. Abstract data access interface afford access interface which has no relation to data resources so as to access varies data.

4 CONCLUSION

The author proposes a role-based access control model of fundamental spatial database system of Jilin water resources and implements it in the. net. FSDSJWR can set different privileges according to the user role, so the administrator can avoid complex modification of the privilege because of the change of the personnel department, position function, and security requirement. The model adopts role inheritance to realize the automatic assignment of the privilege, therefore, reduces the complexity of the management of the system. The privilege assignment for the system submodule and database object access is set with finer-grained user control, so the security management of the system can be better and the FSDSJWR can run in security on the net.

The work was supported by the National Natural Science Foundation for young. (No.41201388). The work was also supported by Youth Science Technology Foundation of Jilin Jianzhu University. (No. J20111007).

REFERENCES

Al-Kahtani M.A, Sandhu R.S. 2002. A Model for Attribute-based User-roleAssignment.*18th Annual Computer Security ApplicationsConference.* 353–364.

Axel Kern. 2002. Advanced features for enterprise-wide role-based access control. Proceedings of 18th Annual Computer Security Application Conference, Washington, US:IEEE Computer Society. 333–342.

Cai qiong, Han hongmu, Zuo cuihua. 2007. Research on the Role Hierarchy and Permission Management of the Role-Based Access Control Model. *Computer Engineering & Science.*(4):36–50.

Sandhu R.S, Coyne E, FeiMtein H.L. 1996. Role-based access control model. *IEEE Computer*(2):38–47.

Sandhu R.S, Ferraiolo D, Kuhn R. 2000. The NIST model for role-based access control: towards a unified standard. *Proceedings of the Fifth ACM Workshop on Role Based Access Control, Berlin, Germany: ACM.* 47–63.

Zhang tao. 2010. The research and realization of service-oriented urban fundamental geographical information metadata catalog system. *Science of Surveying and Mapping*(4):75–79.

Control Engineering and Information Systems – Liu (Ed)
© 2015 Taylor & Francis Group, London, ISBN 978-1-138-02685-8

Decentralized adaptive tracking of nonlinear interconnected systems with quantized inputs

X.S. Xiao
School of Information and Electrical Engineering, Hunan University of Science and Technology, Xiangtan, China

ABSTRACT: This paper addresses the problem of decentralized tracking control for a class of uncertain nonlinear interconnected systems with quantized inputs. A simplified adaptive fuzzy controller, which contains only one adaptive parameter, is synthesized. The proposed controller guarantees that the output of the closed-loop system converges to a small neighborhood of the reference signal and all signals in the closed-loop system remain bounded.

1 INTRODUCTION

Due to the widespread use of network and digital computer in control systems, quantization in feedback control systems have received much attention in recent years [1]–[3]. The primary feature of network-based control systems is that signals between controllers and plants are transmitted through a network medium with limited bandwidth. Consequently, the control signal can not be directly transmitted to the plant before being quantized, which indicates that the data are only available with finite precision [4]. In the existing literature, there are mainly two approaches for handling quantized feedback control systems. The first one considers the dynamic quantizers [1], which scales the quantization levels dynamically in order to improve the steady state performance. The second one uses static quantizers such as logarithmic and uniform quantizers [2]. Compared with dynamic quantizers, the static quantizers have much simpler structures since they assume that data quantization is dependent on the data at each instant of time only. In the framework of the two approaches mentioned above, almost all existing literature considered the problem of stability or stabilization for certain linear systems (continuous-time or discrete-time), only a few literature considered the problem of stabilization of uncertain linear system [5] or nonlinear systems [4], [6]. To the best of our knowledge, the problem of quantized tracking control of nonlinear system (certain or uncertain) has never been investigated.

On another research front, adaptive control design of nonlinear systems has been being a hot topic in control community, see e.g. [7]–[9] and reference therein. However, these works are based on the assumption that system uncertain nonlinearities are either known functions with unknown linear parameters or bounded by known nonlinear functions. Thus, if such a prior knowledge of the structure or the upper bounds of these unknown nonlinearities is not available, these approaches become infeasible. Such restrictions have been removed by using adaptive fuzzy control, see e.g. [10], [11]. Nevertheless, due to the unknown weight vector is used as the estimated vector, these control schemes require a lot of adaptive parameters when high-order systems are considered, which maybe results in the learning time to be unacceptably large. This fact severely prevents the adaptive fuzzy control from extensive practical application. To solve this problem, [12] and [13] firstly considered the norm of the ideal weighting vector in fuzzy logic systems as the estimation parameter instead of the elements of weighting vector, where the number of adaptation laws is reduced considerably.

Motivated by the aforementioned background, this paper considers the problem of decentralized tracking control for a class of uncertain nonlinear interconnected systems with quantized inputs. Base on uniform quantizers, a simplified adaptive fuzzy controller, in which only one parameter is needed to be estimated on-line, is synthesized. During the controller design process, Mamdani type fuzzy systems are used to directly approximate the desired control input signals instead of the unknown system nonlinearities, which not only circumvent the control sigularity problem but also has less adaptive parameters. The proposed controller guarantees the boundedness of all the signals in the closed-loop system and achieves good tracking performance.

2 PROBLEM FORMULATION AND PRELIMINARIES

Consider a class of interconnected systems composed of N subsystems governed by

$$\begin{cases} x_i^{n_i} = f_i(\bar{x}_i) + g_i(\bar{x}_i)Q_i(u_i) + d_i(t) + \sum_{\substack{j=1 \\ j\neq i}}^{N} h_{ij}(\bar{x}_j) \\ y_i = x_i, \qquad i = 1, 2, \ldots, N \end{cases} \quad (1)$$

where $\bar{x}_i = [x_i \; x_i^{(1)} \; \cdots \; x_i^{(n_i-1)}]^T$ and $y_i \in R$ are the ith subsystem's state vector and output, respectively. $f_i(\bar{x}_i)$ and $g_i(\bar{x}_i)$ are unknown nonlinear smooth functions. $d_i(t)$ is the unknown external disturbance and satisfies $|d_i(t)| \leq \sigma_i$ with σ_i being a positive constant (unnecessarily known). $h_{ij}(\bar{x}_j)$ is an unknown continuous nonlinear function which denotes the interconnection between subsystems i and j. $Q_i(u_i) \in R$ is the uniform quantized input defined as

$$Q_i(u_i) = \rho_i \cdot \text{round}(u_i/\rho_i) \quad (2)$$

where $\rho_i > 0$ is called the quantization level and round($\cdot$) is a function that rounds toward the nearest integer. The quantization error is defined as $\Delta u_i = Q_i(u_i) - u_i$. Based on the condition (2) and the definition of Δu_i, each component of Δu_i at time t is bounded by the half of the quantization level ρ_i, i.e.,

$$\|\Delta u_i\|_\infty \leq \rho_i/2 \quad (3)$$

The control objective is to make the output $y_i(t)$ follow the reference signal $y_{mi}(t)$. To this end, the tracking error and error vector are defined as $e_i = y_i - y_{mi}$ and $\bar{e}_i = [e_i \; e_i^{(1)} \; \ldots \; e_i^{(n_i-1)}]^T$, respectively. Also, the following assumptions are imposed.

Assumption 1: The sign of $g_i(\bar{x}_i)$ does not change and there exist constants b_{mi} and b_{Mi} (unnecessarily known) such that

$$0 < b_{mi} \leq \left|g_i(\bar{x}_i)\right| \leq b_{Mi}$$

Without loss of generality, we also assume that $g_i(\bar{x}_i)$ is positive in this study.

Assumption 2: The reference signal $y_{mi}(t)$ and its time derivatives up to the n_i-th order are continuous, bounded and available.

Assumption 3: For interconnected system (1), there exist unknown smooth functions $\alpha_i(\bar{x}_i)$ $(i = 1, \ldots, N)$ such that

$$2\bar{e}_i^T P_i b \alpha_i(\bar{x}_i) - \sum_{\substack{j=1 \\ j\neq i}}^{N} h_{ji}^2(\bar{x}_i) \geq 0 \quad (\text{when } \bar{e}_i^T P_i b \neq 0)$$

where the matrix $b = [0 \; \cdots \; 1]^T$ and the matrix P_i is a symmetric positive-definite matrix that will be specified latter.

In this paper, a fuzzy logic system will be used to approximate a continuous function $q(s)$ defined on some compact set. Adopt the singleton fuzzifier, the product inference and the center-average defuzzifier to deduce the following fuzzy rules:

$$R_i : \text{IF } s_1 \text{ is } F_1^i \text{ and } \ldots \text{ and } s_n \text{ is } F_n^i,$$
$$\text{Then } y \text{ is } B^i \; (i = 1, 2, \ldots, M)$$

where $s = [s_1 \; \ldots \; s_n]^T \in R^n$ and $y \in R$ are the input and output of the fuzzy system, respectively, F_j^i and B^i are fuzzy sets in R, M is the number of fuzzy rules. Since the use of singleton fuzzification, center-average defuzzification and product inference, the output of the fuzzy system can be formulated as

$$y(s) = \frac{\sum_{i=1}^{M} \varphi_i \prod_{j=1}^{n} \mu_{F_j^i}(s_j)}{\sum_{i=1}^{M} \prod_{j=1}^{n} \mu_{F_j^i}(s_j)}$$

where φ_i is the point at which the fuzzy membership function $\mu_{B^i}(y)$ achieves its maximum value (assumed to be 1). Define

$$\xi_i(s) = \prod_{j=1}^{n} \mu_{F_j^i}(s_j) \Big/ \left(\sum_{i=1}^{M} \prod_{j=1}^{n} \mu_{F_j^i}(s_j)\right),$$

$$\xi(s) = [\xi_1(s) \; \xi_2(s) \; \cdots \; \xi_M(s)]^T$$

and $\Phi = [\varphi_1 \; \varphi_2 \; \cdots \; \varphi_M]^T$, then the fuzzy logic system can be rewritten as

$$y(s) = \Phi^T \xi(s) \quad (4)$$

where Φ and $\xi(s)$ are usually called weight vector and fuzzy basis function vector, respectively.

Lemma 1 ([14]): Let $\mathcal{L}(s)$ be a continuous function defined on a compact set Ω. Then for any given constant $\epsilon > 0$, there exists a fuzzy logic system (4) such that

$$\sup_{s \in \Omega} |\mathcal{L}(s) - \Phi^T \xi(s)| \leq \epsilon$$

In the following discussion, variables of functions will be omitted when no confusion can arise.

144

3 SIMPLIFIED ADAPTIVE FUZZY CONTROLLER DESIGN

For system (1), if the following control

$$u_i^* = g_i^{-1}(-\bar{f}_i + y_{mi}^{(n_i)} + K_i^T \bar{e}_i) \tag{5}$$

where $\bar{f}_i = f_i + \frac{3}{2} g_i^2 b^T P_i \bar{e}_i + \alpha_i(\bar{x}_i) + (N-1)\bar{e}_i^T P_i b/2$ and $K_i = [k_{i1} \ .. \ k_{in}]^T$, is adopted, then the closed-loop interconnected system is

$$\dot{\bar{e}}_i = A_{ci}\bar{e}_i + b\Big[g_i(u_i - u_i^*) + g_i \Delta u_i$$
$$+ d_i(t) - \frac{3}{2} g_i^2 b^T P_i \bar{e}_i$$
$$- \alpha_i(\bar{x}_i) - \frac{N-1}{2}\bar{e}_i^T P_i b + \sum_{\substack{j=1 \\ j \neq i}}^N h_{ij}(\bar{x}_j) \Big] \tag{6}$$

where

$$A_{ci} = \begin{bmatrix} 0 & 1 & 0 & \cdots & 0 \\ 0 & 0 & 1 & \cdots & 0 \\ \vdots & \vdots & \vdots & \ddots & \vdots \\ 0 & 0 & 0 & \cdots & 1 \\ k_{i1} & k_{i2} & k_{i3} & \cdots & k_{in} \end{bmatrix}$$

By appropriately choosing the feedback matrix K_i, A_{ci} can be a Hurwitz matrix. In other words, for any symmetric positive-definite matrix Q_i, the following Lyapunov equation

$$A_{ci}^T P_i + P_i A_{ci} = -Q_i \tag{7}$$

Has a matrix solution P_i which is symmetric and positive-definite.

Since the unknown functions f_i and g_i are involved in u_i^*, u_i^* can not be implemented actually. However, according to Lemma 1, there exists fuzzy logic system $\Phi_i^T \xi(z_i)$ such that

$$u_i^* = \Phi_i^T \xi_i(z_i) + \delta_i(z_i), |\delta_i(z_i)| \leq \epsilon_i \tag{8}$$

where $z_i = [\bar{x}_i^T, y_{mi}, \cdots, y_{mi}^{(n_i)}]^T$ and $\delta_i(z_i)$ is the approximation error bounded by any given positive constant ϵ_i.

Theorem 1: If Assumption 1, 2 and 3 are satisfied, then there exists a decentralized simplified adaptive fuzzy control

$$u_i = -\frac{b^T P_i \bar{e}_i}{2a_i^2} \hat{\theta}_i \xi_i^T(z_i)\xi_i(z_i) \tag{9}$$

With adaption law

$$\dot{\hat{\theta}}_i = \frac{r_i}{a_i^2}\bar{e}_i^T P_i bb^T P_i \bar{e}_i \xi_i^T(z_i)\xi_i(z_i) - k_{i0}\hat{\theta}_i \tag{10}$$

where a_i, r_i, k_{i0} are positive design parameters and P_i is the solution of Lyapunov equation (7) such that all signals in the overall closed-loop interconnected system remain bounded. Furthermore, for any given scalar $\varepsilon > 0$, the tracking error satisfies

$$\lim_{t \to \infty} \sum_{i=1}^N \|y_i - y_{mi}\| \leq \varepsilon$$

Proof: For the overall closed-loop interconnected system, choose the Lyapunov function as

$$V(t) = \sum_{i=1}^N V_i(t) = \sum_{i=1}^N \left\{ \bar{e}_i^T P_i \bar{e}_i + \frac{b_{mi}}{2r_i}\tilde{\theta}_i^2 \right\} \tag{11}$$

where $\tilde{\theta}_i = \theta_i - \hat{\theta}_i$ in which $\theta_i = \|\Phi_i\|^2/b_{mi}$ and $\hat{\theta}_i$ is the estimate of θ_i.

Then the time derivative of $V(t)$ along the trajectories of (6) is given by

$$\dot{V}(t) = \sum_{i=1}^N \Big\{ 2\bar{e}_i^T P_i A_{ci}\bar{e}_i + 2\bar{e}_i^T P_i b\Big[g_i(u_i - u_i^*)$$
$$- \frac{3}{2} g_i^2 b^T P_i \bar{e}_i + g_i \Delta u_i + d_i(t) - \alpha_i(\bar{x}_i)$$
$$- \frac{N-1}{2}\bar{e}_i^T P_i b + \sum_{\substack{j=1 \\ j \neq i}}^N h_{ij}(\bar{x}_j) \Big] - \frac{b_{mi}}{r_i}\tilde{\theta}_i\dot{\hat{\theta}}_i \Big\} \tag{12}$$

Note the fact that

$$-2\bar{e}_i^T P_i b g_i u_i^* = -2\bar{e}_i^T P_i b g_i\Big[\Phi_i^T \xi(z_i) + \delta_i(z_i)\Big]$$

$$= -2\bar{e}_i^T P_i b g_i \frac{\Phi_i^T \xi(z_i)}{\|\Phi_i\|}\|\Phi_i\| - 2\bar{e}_i^T P_i b g_i \delta(z_i)$$

$$\leq \frac{b_{mi}}{a_i^2}\bar{e}_i^T P_i bb^T P_i \bar{e}_i \theta_i \xi_i^T(z_i)\xi_i(z_i) + a_i^2 b_{Mi}^2$$
$$+ g_i^2 \bar{e}_i^T P_i bb^T P_i \bar{e}_i + \epsilon_i^2 \tag{13}$$

and

$$2\bar{e}_i^T P_i b d_i(t) \leq g_i^2 \bar{e}_i^T P_i bb^T P_i \bar{e}_i + \frac{\sigma_i^2}{g_i^2}$$

$$\leq g_i^2 \bar{e}_i^T P_i bb^T P_i \bar{e}_i + \frac{\sigma_i^2}{b_{mi}^2} \tag{14}$$

$$2\bar{e}_i^{\mathrm{T}} P_i b g_i \Delta u_i \le g_i^2 \bar{e}_i^{\mathrm{T}} P b b^{\mathrm{T}} P \bar{e}_i + \frac{\varrho_i^2}{4} \tag{15}$$

we have

$$\dot{V}(t) \le \sum_{i=1}^{N} \left\{ 2\bar{e}_i^{\mathrm{T}} P_i A_{ci}\bar{e}_i + a_i^2 b_{Mi}^2 + \epsilon_i^2 + \frac{\sigma_i^2}{b_{mi}^2} + \frac{\varrho_i^2}{4} \right.$$
$$+ \frac{b_{mi}}{a_i^2} \bar{e}_i^{\mathrm{T}} P_i b b^{\mathrm{T}} P_i \bar{e}_i \theta_i \xi_i^{\mathrm{T}}(z_i)\xi(z) + 2\bar{e}_i^{\mathrm{T}} P_i b g_i u_i$$
$$- \frac{b_{mi}}{r_i} \tilde{\theta}_i \dot{\hat{\theta}}_i - 2\bar{e}_i^{\mathrm{T}} P_i b \alpha_i(\bar{x}_i) - (N-1)(\bar{e}_i^{\mathrm{T}} P_i b)^2$$
$$\left. + 2\bar{e}_i^{\mathrm{T}} P_i b \sum_{\substack{j=1 \\ j \ne i}}^{N} h_{ij}(\bar{x}_j) \right\} \tag{16}$$

Substituting (9) into (16) gives

$$\dot{V}(t) \le \sum_{i=1}^{N} \left\{ 2\bar{e}_i^{\mathrm{T}} P_i A_{ci}\bar{e}_i + a_i^2 b_{Mi}^2 + \epsilon_i^2 + \frac{\sigma_i^2}{b_{mi}^2} + \frac{\varrho_i^2}{4} \right.$$
$$+ \frac{b_{mi}}{r_i} \tilde{\theta}_i \left[\frac{r_i}{a_i^2} \bar{e}_i^{\mathrm{T}} P_i b b^{\mathrm{T}} P_i \bar{e}_i \theta_i \xi_i^{\mathrm{T}}(z_i)\xi(z) - \dot{\hat{\theta}}_i \right]$$
$$- 2\bar{e}_i^{\mathrm{T}} P_i b \alpha_i(\bar{x}_i) - (N-1)(\bar{e}_i^{\mathrm{T}} P_i b)^2$$
$$\left. + 2\bar{e}_i^{\mathrm{T}} P_i b \sum_{\substack{j=1 \\ j \ne i}}^{N} h_{ij}(\bar{x}_j) \right\} \tag{17}$$

By the definition of $\tilde{\theta}_i$, we have

$$\frac{b_{mi} k_{i0}}{r_i} \tilde{\theta}_i \hat{\theta}_i \le \frac{b_{mi} k_{i0}}{2r_i} (\theta_i^2 - \tilde{\theta}_i^2) \tag{18}$$

It follows from (10), (17) and (18) that

$$\dot{V}(t) \le \sum_{i=1}^{N} \left\{ 2\bar{e}_i^{\mathrm{T}} P_i A_{ci}\bar{e}_i + a_i^2 b_{Mi}^2 + \epsilon_i^2 + \frac{\sigma_i^2}{b_{mi}^2} + \frac{\varrho_i^2}{4} \right.$$
$$+ \frac{b_{mi} k_{i0}}{2r_i} \theta_i^2 - \frac{b_{mi} k_{i0}}{2r_i} \tilde{\theta}_i^2 - 2\bar{e}_i^{\mathrm{T}} P_i b \alpha_i(\bar{x}_i)$$
$$\left. - (N-1)(\bar{e}_i^{\mathrm{T}} P_i b)^2 + 2\bar{e}_i^{\mathrm{T}} P_i b \sum_{\substack{j=1 \\ j \ne i}}^{N} h_{ij}(\bar{x}_j) \right\} \tag{19}$$

It is clear that when $\bar{e}_i^{\mathrm{T}} P_i b = 0$, then the last three terms of (19) equal to zero. When $\bar{e}_i^{\mathrm{T}} P_i b \ne 0$, then we have

$$\sum_{i=1}^{N} 2\bar{e}_i^{\mathrm{T}} P_i b \sum_{\substack{j=1 \\ j \ne i}}^{N} h_{ij}(\bar{x}_j)$$
$$\le \sum_{i=1}^{N} \left\{ (N-1)(\bar{e}_i^{\mathrm{T}} P_i b)^2 + \sum_{\substack{j=1 \\ j \ne i}}^{N} h_{ji}^2(\bar{x}_i) \right\} \tag{20}$$

Combing (19) and (20) yields

$$\dot{V}(t) \le \sum_{i=1}^{N} \left\{ 2\bar{e}_i^{\mathrm{T}} P_i A_{ci}\bar{e}_i + a_i^2 b_{Mi}^2 + \epsilon_i^2 + \frac{\sigma_i^2}{b_{mi}^2} + \frac{\varrho_i^2}{4} \right.$$
$$+ \frac{b_{mi} k_{i0}}{2r_i} \theta_i^2 - \frac{b_{mi} k_{i0}}{2r_i} \tilde{\theta}_i^2 - 2\bar{e}_i^{\mathrm{T}} P_i b \alpha_i(\bar{x}_i)$$
$$\left. + \sum_{\substack{j=1 \\ j \ne i}}^{N} h_{ji}^2(\bar{x}_i) \right\} \tag{21}$$

From Assumption 3 we know that in this case the last two term of (19) are negative. Thus we have

$$\dot{V}(t) \le \sum_{i=1}^{N} \left\{ 2\bar{e}_i^{\mathrm{T}} P_i A_{ci}\bar{e}_i + a_i^2 b_{Mi}^2 + \epsilon_i^2 + \frac{\varrho_i^2}{4} + \frac{\sigma_i^2}{b_{mi}^2} \right.$$
$$\left. + \frac{b_{mi} k_{i0}}{2r_i} \theta_i^2 - \frac{b_{mi} k_{i0}}{2r_i} \tilde{\theta}_i^2 \right\}$$
$$\le \sum_{i=1}^{N} \left\{ -a_{i0} V_i(t) + b_{i0} \right\} \le -\gamma W(t) + \eta \tag{22}$$

where $a_{i0} = \min\{k_{i0}, \lambda_{\min}(Q_i)/\lambda_{\max}(P_i)\}$, $b_{i0} = a_i^2 b_{Mi}^2 + \epsilon_i^2 + \frac{\varrho_i^2}{4} + \frac{\sigma_i^2}{b_{mi}^2} + \frac{b_{mi} k_{i0}}{2r_i} \theta_i^2$, $\gamma = \min\{a_{i0} : i = 1, \cdots, N\}$ and $\eta = \sum_{i=1}^{N} b_{i0}$. From (22), it is clear that $\bar{e}_i$ and $\tilde{\theta}_i$ are bounded. Moreover, it follows from (11) and (22) that

$$\sum_{i=1}^{N} \left\| \bar{e}_i^2 \right\| \le 2 \left(V(0) - \frac{\eta}{\gamma} \right) e^{-\gamma t} + 2\frac{\eta}{\gamma} \tag{23}$$

which implies

$$\sum_{i=1}^{N} e_i^2 \le 2 \frac{\eta}{N \lambda \gamma} \le \varepsilon^2 \tag{24}$$

where $\lambda = \min\{\lambda_{\min}(P_i) : i = 1, \ldots, N\}$. This completes the proof. $\square$

146

4 CONCLUSION

In this paper, the decentralized tracking problem has been investigated for a class of uncertain nonlinear interconnected systems with quantized inputs. A simplified adaptive fuzzy control scheme is proposed, which can guarantee that system's output follows the reference signal and all signals in the closed-loop system remain bounded.

ACKNOWLEDGMENTS

This work was supported by University Scientific Research Program of Hunan Province (12C0865) and Hunan Provincial Natural Science Foundation of China (14JJ3108).

REFERENCES

[1] D. liberzon, "Hybrid feedback stabilization of systems with quantized signals," Automatica, vol. 39, no. 11, pp. 1543–1554, 2003.

[2] T. liu, Z.P. Jiang and D.J. Hill, "A sector bounded approach to feedback control of nonlinear systems with state quantization," Automatica, vol. 48, no. 1, pp. 145–152, 2012.

[3] G. Prior and M. Krstic, "Qunatized-input control lyapunov approach for permanent magnet synchronous motor drives," IEEE Trans. Control Syst. Tech., vol. 21, no. 5, pp. 1784–1794, 2013.

[4] C.D. Persis, "Robust stabilization of nonlinear systems by quantized and ternary control," Syst. Control Lett., vol. 59, no. 2, pp. 602–608, 2009.

[5] S.W. Yun, Y.J. Choi and P. Park, "h2 control of continuous-time uncertain linear systems with input quantization and matched disturbances," Automatica, vol. 45, no. 12, pp. 2435–2439, 2009.

[6] T. Hayakawa, H. Ishii and K. Tsumura, "Adaptive quantized control for nonlinear uncertain systems," Syst. Control Lett., vol. 58, no. 3, pp. 625–632, 2009.

[7] M.-W. Koo, H.-L. Choi and J.-T. Lim, "Universal control of nonlinear systems with unknown nonlinearity and growth rate by adaptive output feedback," Automatica, vol. 47, no. 3, pp. 2211–2217, 2011.

[8] W.X. Shi, M. Zhang, W.C. Guo and L.J. Guo, "Stable adaptive fuzzy control for mimo nonlinear systems," Comput. Math. Appl., vol. 62, pp. 2843–2853, 2011.

[9] C.Y. Wen, J. Zhou, Z.T. Liu and H.Y. Su, "Robust adaptive control of uncertain nonlinear systems in the presence of input saturation and external disturbanc," IEEE Trans. Automatic Control, vol. 56, no. 7, pp. 1672–1678, July 2011.

[10] C.W. Tao, J. Taur, J.H. Chang and S.-F. Su, "Adaptive fuzzy switched swing-up and sliding control for the double pendulum-and-cart system," IEEE Trans. Syst. Man Cybern. Part B: Cybern., vol. 40, no. 1, pp. 241–252, Feb. 2010.

[11] S.C. Tong, C.Y. Li and Y.M. Li, "Fuzzy adaptive observer backstepping control for mimo nonlinear systems," Fuzzy Sets Syst., vol. 160, pp. 2755–3775, 2009.

[12] Y.S. Yang, G. Feng and C.B. Feng, "A combined backstepping and small-gain approach to robust adaptive fuzzy control for strict-feedback nonlinear systems," IEEE Trans. Syst. Man Cybern. Part A: Syst., vol. 34, pp. 406–420, 2004.

[13] Y.S. Yang and C.J. Zhou, "Robust adaptive fuzzy tracking control for a class of perturbed strict nonlinear systems via small-gain approach," Inform. Sci., vol. 170, pp. 211–234, 2005.

[14] L.X. Wang and J.M. Mendel, "Fuzzy basis functions, universal approximation and orthogonal least squarses learning," IEEE Trans. Fuzzy syst., vol. 3, no. 5, pp. 807–814, May, 1992.

Control Engineering and Information Systems – Liu (Ed)
© 2015 Taylor & Francis Group, London, ISBN 978-1-138-02685-8

Two guards min-sum algorithm in a scanning simple polygon

J. Zhang

School of Information Science and Technology, Dalian Maritime University, Dalian, China
Dalian Educational Technology and Computing Center, Dalian Ocean University, Dalian, China

Y.J. Li

School of Information Science and Technology, Dalian Maritime University, Dalian, China

ABSTRACT: In 2010, Xuehou Tan and Bo Jiang put forward an optimal scanning algorithm for solving the min-sum problem in two-guard simple polygon problem. The min-sum problem is the minimized problem that the two guards walk the total distance. This paper explores in-depth the detailed process through choosing the dominated shot, constructing the ray-shooting segment diagram, classified counting arc set, and at last citing dijkstra to settle shortest path algorithm to compute step by step, whose time complexity of the algorithm is o (n²). Finally, to use program to realize the algorithm and then input the test data, analyze the results to verify the feasibility and validity of the algorithm.

1 INTRODUCTION

The two-guard problem is one of the classical problems in computational geometry, which mainly discusses: for a given simple polygon P, there is an entrance s and an exit t at its edge, this polygon P is called corridor. Two-guard problem is starting from the entrance s, to exile t target (refer to the illegal invader). This kind of guard problem is stimulated by the famous Art Gallery and Watchman Route Problems, which is committed to forming a group of mobile guards P on an n edge shape to detect an unpredictable and mobile target.

This paper aims at studying the min-sum problem in a simple polygon, which is the subproblem of two-guard problem. There is a basic agreement to this problem, namely the two guards moving on the edge in the simple polygon, and they can always be seen each other, and the segment which connects the two guards to separate polygon P into a "clear" region (already searched region) and an "unclear" region (unsearched region). In this case, the first problem to be solved is whether the simple polygon P is clean, that is, the target can be detected, if it can, the simple polygon can be completely searched. This paper proposes the algorithm to solve this problem, which is through the control point selection, the structure ray segment, the classification of calculation arc set, and the dijkstra algorithm for solving the shortest path to complete the calculation steps. The time complexity of the algorithm is o (n2). Finally, applying for program to realize the algorithm, to analyze the results and to verify the feasibility and validity of the algorithm.

2 PRELIMINARY

This paper discusses that the two guards are limited in a simple polygon, that is, there doesn't exist a given simple polygon crossing edge and no empty hole.

Given a simple polygon P, set ∂P as the edge of the polygon. Two points guard are moving at the segment g1 and g2, g1(t), g2(t) are the position on the edge of P at the time of t. It needs to consider the following two sweep instructions [2].

In the whole moving process, all the line segment g1(t) g2(t) are not crossing, as shown in Figure 1(a).

In the whole moving process, all the line segment g1(t) g2(t) are crossing, as shown in Figure 1(b).

Definition 2.1: for a vertex v on the simple polygon P, Succ(v) shows the next adjacent v vertex on

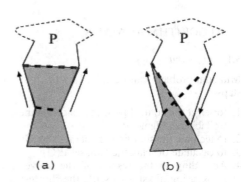

Figure 1. Sweep instructions.

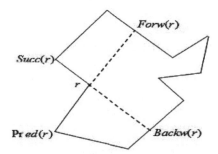

Figure 2.   Ray shoot and ray segment.

the clockwise direction, Pred(v) shows a vertex on the clockwise direction, as shown in Figure 2.

If a vertex of P interior angle is greater than 180 degrees, so this vertex is concave (also known as the reflex [3]), this kind of vertex is the reflex vertex. The reflex vertex has two rays shoot both in the front and the back.

Definition 2.2: back shot point of reflex vertex r, denoted as Backw(r), it is the first point of P along the direction between Succ(r) to r, that is the first crossing point on the edge of the Succ(r) r line and P edge. Front shot point of reflex vertex r, denoted as Forw(r), it is the first point of P along the direction between Pred(r) to r, that is the first crossing point on the edge of the Pred(r) line and P edge, as shown in Figure 2. Vertex r is the source point of both shot point Backw(r) and Forw(r), line segment r Forw(r) and r Backw(r) are called the ray segments.

Definition 2.3: to set L and R is the chain of edge sequence in the simple polygon P, L and R at most intersect only two vertices (respectively is the starting point and end point of the two chains). If there exists a point q ∈ R to each point p ∈ L, make the q from p visible, then L from R is weakly visible [4]. If L and R are weakly visible each other, than P is scanned. This paper studies only a scanning simple polygon.

3   ALGORITHM ANALYSIS

3.1   *Algorithm*

Min-sum problem can be solved by the following steps:

1. to select the control point, to calculate the controllable ray segment;
2. to tectonic the shot line segment graph G;
3. to calculate the classified arc set E(G);
4. According to the results of the above three steps, using dijkstra to settle the shortest path from s to t in picture G.

The main algorithm is as follows:

Input: set node of ray segment G V(G) = {v0,v1, ..., vm-1}, m-1<4n, in which contains the output of starting point s and end point t, the shortest path ρ between s and t, and the shortest path length.

1.   to set the starting point *s* and the end point *t*, to select the ray segment set V1(G)
2.   for(int i = 0;i< V1(G).size();i++)
2.1.   if(in V1(G), the source point and the starting point of the node is not on the same chain)
2.2.   to put the node into V2(G), there exists the useful node, that is, the ray segment.
3.   for(i = 0;i < V2(G).size();i++)
3.1.   for(int j = i+1;j< V2(G).size();j++)
3.1.1.   bool temp_flag = Equal_point()
3.1.2.   if(temp_flag = = false)
3.1.3.   if(Intersect() = = true)
3.1.4.   Straightsweep()
3.1.5.   else
3.1.6.   Coutersweep()
4.   to output the controllable force ray segment set, and the starting point and the end point, this node set is Vc(G)
5.   to calculate arc set E(G) by the set Vc(G)
6.   to tectonic the ray segment picture G
7.   to calculate the shortest path ρ and the shortest path length by dijkstra algorithm

The main function of algorithm is to initialize the node set, the starting point and the end point, and then to select ray segment set V1(G) in the node set to further select effective ray segment set V2(G). To transfer the direct scanning straight sweep subroutine and counter sweep subroutine to get the controllable ray segment Vc(G), and then to calculate arc set E(G) to get the ray segment picture G, at last to calculate the shortest path and the shortest path length of ray segment in picture G by dijkstra algorithm.

Straight sweep subroutine algorithm is described as follows:

8.   if (the source point of ray node a, b are on the same chain)
9.   if (node end point are all Forw or Backw)
10.   if(Forw(a)<Forw(b))
11.   then to add the scanning controllable (b,Forw(b)) node to Vc(G)
12.   else
13.   if(Backw(a)<Backw(b))
14.   then to add directly the scanning controllable node (a,Backw(a)) to Vc(G)
15.   else
16.   Mostnear()
17.   else
18.   the situation of inverse scanning intersection, to add Mostnear subroutine to clockwise starting point which is the nearest ray segment node Vc(G)

19. else
20. if node end point are all Forw or Backw
21. if (end point is Backw)
22. if((a<Backw(b))∪(Backw(a)<b))
23. then to add direct scanning controllable node (a,Backw(a)) to Vc(G)
24. else
25. if((Backw(b)<a)∪(b<Backw(a)))
26. then to add direct scanning controllable node (b,Backw(b)) to Vc(G)
27. else
28. Mostnear()
29. else
30. if((Forw(b)<a)∪(b<Forw(a)))
31. then to add direct scanning controllable node (a,Forw(a)) to Vc(G)
32. else
33. if((a<Forw(b))∪(Forw(a)<b))
34. then to add direct scanning controllable node (b,Forw(b)) to Vc(G)
35. else
36. Mostnear()
37. else
38. the situation of inverse scanning intersection, to add Mostnear subroutine to clockwise starting point which is the nearest ray segment node Vc(G).

Subroutine straightsweep through comparative to the direct scanning controllable ray segment in Figure 3, to add the direct controllable ray segment to the new ray segment set Vc(G).

Both subroutine Countersweep and Straightsweep are similar, to add controllable inverse scanning ray segment to ray segment Vc(G), here omit it.

After structuring the ray segment picture G, it needs to calculate arc set E(G), that is, each arc is assigned a weight of E(G). A weight of an arc is defined as the sum of two guards walking distance. Here the weight set GS is the weight G;

the following three kinds of situations are respectively calculated.

①Two arcs are from the initial vertex to the first action (control line segment) and the last action (control line segment) to end point.

The weight of this kind of arc is the vertex (terminate vertex) to one of the point on the ray segment (action), till the sum of another point line segment.

②When the two actions (control line segment) are directly scanning, the weight of this arc is both the sum of two left points plus the sum of two right points.

③When two movements (control line segment) are inverse scanning, the situation is complex. As shown in Figure 4(a), a guard moves along clockwise L, and stop at every concave vertex to continue searching to find higher back shooting point, here as utp. First Utp is q4, and then is q6. Once searching the front shot point is less than utp, as shown in Figure 4(a), q2 s is the front shot point, and it is marked by ltp. At this time, it can detect a wedge, and utp is a defined larger turning point, while ltp is still smaller.

Guard continues walk on the strand of L, and there are two situations that can happen. Firstly, when it meets vertex p4, a former front shot point is lower than the current front shot point ltp, as shown in Figure 4(a), at this time, utp is equal to q6, and ltp is equal to q2. And then q1 has become the lower turning point ltp. Secondly, a back shooting point of a vertex is higher than ltp, as shown in Figure 4(a), the point q5 is. Further, in this situation, the front shooting point must be higher than ltp, then the guard on R must turn at ltp, at least at this time to the back shooting point q5. Moreover, it draws the conclusion of continuous turning point (utp, ltp). The back shooting point q5 has become a higher steering candidate point. If there exists a corresponding lower turning point, then this point is to be detected. Then ltp is set to t. After the above calculating, the turning point on R chain can be obtained, the walking path of guard on R chain as shown in Figure 4(b).

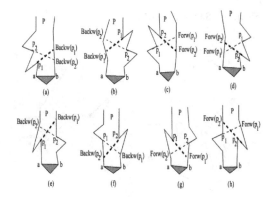

Figure 3. A direct scanning controllable point case study.

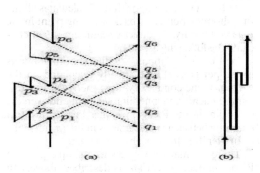

Figure 4. Weight calculation of inverse scanning arc.

To calculate the turning point on L chain can apply to another similar algorithm, here omit it.

The above studies shows, any scanning of P have to give the turning point of the more ordered turning points and the lower turning point.

For inverse scanning fragment, from s to t to scan L chain, and to give the inverse turning point of R. Algorithm turning point is described as follows:

1. utp: = s
2. ltp: = g
3. For all the concave vertexes in the direction of clockwise p∈ L Do
4. If ltp > utp
5. Then If Backw(p) > utp
6. Then utp: = Backw(p)
7. Else If Forw(p) < utp
8. Then ltp: = Forw(p)
9. End If
10. End If
11. Else If Forw(p) < ltp
12. Then ltp: = Forw(p)
13. Else If Backw(p) > ltp
14. Then output turning point (utp,ltp)
15. utp: = Backw(p)
16. ltp: = g
17. End If
18. End If
19. End If
20. End For
21. If ltp < utp
22. Then output a pair of turning point (utp,ltp)
23. End If

The rest part of arc calculation algorithm is similar to the above, here omit it.

Finally, to input the above data into dijkstra algorithm, then to obtain the solution of the min-sum problem.

3.2 Time complex degree analysis

The following complex degree algorithm is analyzed.

At first, using the algorithm [5] calculates all the first shooting point and back shooting point in the time complexity. To directly scanning, it can conclude the following aspects:

Lemma 3.1: Suppose all the shooting points in the polygons P have already calculated on the ○ (nlogn) time complexity, then the given two disjoint segments and inner line segment and even the end point in the P edge, whether there is a direct scanning between two segments in linear time [6].

In literature [6], it is given its lemma.

There are many n vertices in a simple polygon P, to a initial segment **ab**, as described above, can find two concave vertices a1 and b1 in the linear

time. But to judge whether there is a direct scanning from **ab**, at most, needs twice calculations, so all the direct scanning in P can be calculated with time complexity.

Inverse scanning is similar to it, here omit it.

Through direct scanning and reverse scanning to select ray segment with control forces as the node in ray segment picture G. Whether there exists a direct or inverse scanning between two control line as the arc in the picture G. Apparently V(G) and E(G) are of the same size ○ (n), so the ray segment G can structure in the time complexity.

Finally, owing to the time complexity of dijkstra algorithm ○ (n2), to sum up, the time complexity of the algorithm is ○ (n2).

4   ALGORITHM VERIFICATION

Given a simple polygon P, as shown in Figure 5.

Through the shooting point, to calculate the shooting point of polygon P. The ray segment can be obtained respectively: bForw(b), bBackw(b), dForw(d), dBackw(d), hForw(h), hBackw(h), iForw(i), iBackw(i), mForw(m), mBackw(m), as shown in Figure 6(a).

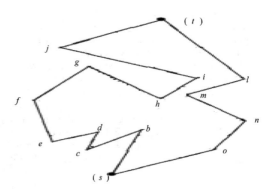

Figure 5.   Simple polygon P.

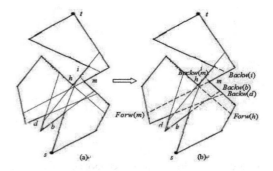

Figure 6.   Ray segment and control point.

Through Countersweep and Straightsweep, to give the shooting point with controllable force, as shown in Figure 6(b). The ray segment with red wide dotted line marking, in this case, these lines are: bBackw(b), dBackw(d), hForw(h), iBackw(i), mForw(m), mBackw(m). With the starting point s and the end point t, a total of 8 nodes, these nodes are the vertex in the ray segment G.

According to the algorithm to calculate the weight of each arc of arc set E(G), as the weights among each vertex in the ray segment G, thus it completes the structure of the ray segment G. Then the corresponding operating results as shown in Figure 7.

At last, the application MFC interface to realize dijkstra algorithm, to comparatively give ray segment picture G, as shown in Figure 8. v0,v1, ..., v7 is the 8 vertices of ray segment picture G, which respectively represent the starting node, the end node and 6 ray segment nodes with controllable force. Arc set E(G) = { e0,e1,...,e8} is the directing weight arc set, each arc weight can be calculated through the program classification, the structure ray segment picture G as shown in Figure 8.

In the picture, the red path is the shortest path of the ray segment G according to dijkstra algorithm, v0, v7 are respectively the starting point s

```
This program imports divide idea.
The Dominated Segments:
Segment Source Point   Segment Target Point   Vertex of graph G
    <280,750>               <450,590>               v1
    <410,410>               <480,450>               v6
    <180,760>               <470,610>               v3
    <330,540>               <495,765>               v4
    <390,510>               <60,760>               v2
    <390,510>               <360,490>               v5

Origin v0: <210,1030>    Destination v7: <330,30>

Arc   Path     Weight
e1    v0->v1   885.2350
e2    v1->v2   575.115
e3    v2->v4   1170.47
e4    v1->v3   292.6175
e5    v3->v4   1285.540
e6    v4->v5   351.324
e7    v5->v6   236.438
e8    v6->v7   1463.0875

The length of the shortest path is 4514.24.
The shortest path is v0  v1  v3  v4  v5  v6  v7  .
Press any key to continue
```

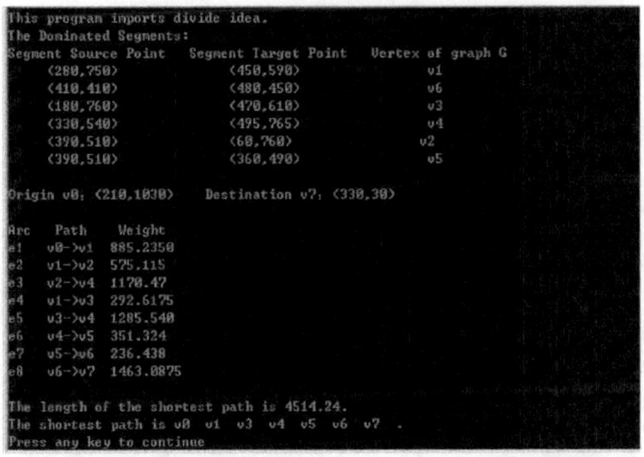

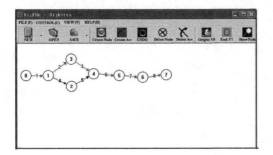

Figure 7. Operating result.

Figure 8. Shortest path of ray segment picture G.

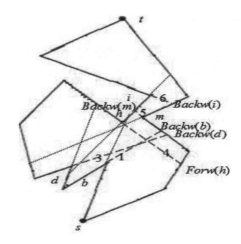

Figure 9. Optimal scanning of simple polygon P.

and the end point t, arc 1 is arc e1, followed by analogy.

Through using dijkstra algorithm to calculate the shortest path $\rho = \{v_0, v_1, v_2. v_3, v_4, v_5, v_6, v_7\}$, to obtain the optimal scanning scheme of polygon P.

Two guards a and b respectively start from initial vertex s, a along the left chain, and b along the right chain.

The first scanning rule is: a, b respectively from s scan directly to ray segment b Backw (b);

The second scanning rule is: a, b respectively from b Backw (b) inversely scan to ray segment d Backw (d);

The third scanning rule is: a, b respectively from d Backw (d) inversely scan to ray segment h Forw (h);

The forth scanning rule is: a, b respectively from h Forw (h) directly scan to ray segment m Backw (m);

The fifth scanning rule is: a, b respectively from m Backw (m) directly scan to ray segment i Backw (i);

The sixth scanning rule is: a, b respectively from i Backw (i) directly scan to termination vertex point t.

5 CONCLUSION

This paper in-depth explores the two-guard problem of simple polygon, on the basis of Xuehou Tan and Bo Jiang's subproblem of two-guard problem, to give the realized algorithm, the time complexity of algorithm is o (n2), and then specifically realizes and input test data to analyze the results to verify the validity and feasibility of the algorithm.

The research has made the preliminary results; further research can start in 3D space. In addition, this paper gives the classified arc set; each arc weight calculation of the whole arc set still needs further study.

REFERENCES

[1] C.L. Lawson. 1972. *Transforming triangulations. Discrete Math.*1972, 3:365–372.

[2] C. Icking, R. Klein. 1992. *The two guards' problem. IJCGA.*1992, 2:257–285.

[3] D. Avis, G.T. Toussaint. 1981. *An optimal algorithm for determining the visibility of a polygon from an edge. IEEE Transaction on Computers.*1981, 30:910–914.

[4] M. De. Berg, M. van Kreveld, M. Overmars. 2005. *Translated by Deng Junhui. Computational Geometry—Algorithm and Application (Second Editon). Tsinghua University Press.* 2005.

[5] P.J. Heffernan. *An optimal algorithm for the two-guard problem.* IJCGA.1996, 6:15–44.

[6] S. Fisk. *A short proof of Chvatals's watchman theorem. Journal of Combinatorial Theory.* 1978, B 24:374.

[7] Tan X., Jiang B. 2010. *Optimum sweeps of simple polygons with two guards. J. Comput. Sci. Tech.* 2010. Beijing:304–315.

Control Engineering and Information Systems – Liu (Ed)
© *2015 Taylor & Francis Group, London, ISBN 978-1-138-02685-8*

Virtual instrument and neural network in application of grain moisture detection

J.M. Liu, Z.Z. Xu & D.H. Sun
North China Institute of Astronautic Engineering, Langfang, China

ABSTRACT: Moisture content is affected by temperature during moisture detection process. In this paper, neural network algorithm is used in virtual instrument to reduce the influence of temperature on the measurement of grain moisture, so that temperature will compensate for moisture measurement. Standard signal which was sampled by a capacitive sensor and a temperature sensor and processed by conditioning circuit output is directly input into the NI data acquisition. The voltage value of moisture and temperature is sent to the computer through the data collector, and is optimized by the neural network algorithm. The effect of temperature on the moisture measurement is reduced, so the virtual instrument temperature compensation function is realized.

1 INTRODUCTION

This paper uses virtual instrument technology to achieve intelligent detection grain moisture, as compared to the traditional instruments, virtual instruments with short development cycle, easy to maintain, CorelDraw, and high price/performance advantages [Zhu, 2007]. More important is the use of virtual instrument technology to develop intelligent instruments, making the user more participatory and ease of applying the new theories, new technologies achieve the flexibility of instrument and want to upgrade, this is defined by the vendor, feature a fixed, independent of traditional instruments are not up to.

Virtual instrument technology is building on the combine of computer technology and test instrument. It makes full use of computer hardware and software resources. Virtual instrument which replaced appropriate electronic circuits can complete the traditional instruments parts as well as all of the features of the hardware.

Capacitance sensor input-output is nonlinear, which is vulnerable to the influence of environment temperature, so that it brings to the actual measure error. The artificial neural network applies for temperature compensation for virtual instrument environment. Grain moisture detection based on neural network and virtual instrument which complete main calculation, control and display functions. Computer software instead of the hardware circuit can realize temperature compensation so as to reduce cost.

2 VIRTUAL INSTRUMENT LABVIEW

Virtual instrument is based on computer hardware and software platform, used in industrial automation and other aspects. It is dominate by software technology, adding a small number of hardware for data processing, analysis, control, displays and other functions. Development languages of virtual instrument has not only genera programming languages such as Basic, standard C, Visual C++, also has their own proprietary software, such as NI company LabVIEW, LabWindows/CVI.

A testing application named virtual instrument was developed using the LabVIEW software. LabVIEW (Laboratory Virtual Instrument Engineering Workbench) is based on program development using the graphical editing have a block diagram language application in the form of a built in powerful data acquisition and instrument control functions, such as libraries and development tools. LabVIEW programs can mimic the appearance of the actual instrument and manipulation capabilities.

3 SYSTEM COMPONENTS

The front-end data acquisition main used Capacitance type sensor and AD590 temperature sensor, which sensor output standard of signals directly entered NI data acquisition Manager, through data acquisition devices which can achieve for analog to digital conversion of water and temperature. Digital voltage value can directly

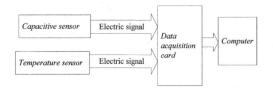

Figure 1. Block diagram of system design of temperature compensation.

input computer, so that LabVIEW software achieved acquisition and displayed of water and temperature, and MatLAB achieved corresponding of algorithm in the temperature compensation of moisture. LabVIEW software displays temperature compensation before and after of moisture values [Plexus, 2009]. Figure 1 shows block diagram of system design of temperature compensation.

4 SENSOR DESIGN

4.1 Capacitive sensor

Capacitive sensor is an instrument which can converse the non-physical quantity to the capacitance change. It has high resolution, non-contact measurement, and can work in the environment, such as high temperature, low temperature, strong radiation. Full drying of food dielectric constant General is about 2~3, but water of dielectric constant is 80. Because drying food of dielectric constant far is less than water, capacitance detection law was measuring food into Capacitance type sensor poles plate between of media empty cavity in the and because food containing water rate of different, will makes sensor of relative dielectric constant occurs changes, turn caused capacitance occurs change to measurement food containing water rate of. According to the different electrode structure, capacitive sensors have parallel plate, cylinder-shaped and so on [Lu, 2007]. Because cylindrical capacitors under the influence of edge effect is small, and has a large effective area of the plate, so the system uses a cylindrical sensor whose capacitance equation for:

$$C = \frac{2\pi\varepsilon_x L}{\ln(R/r)} \quad (1)$$

In the equation, L is the height of the electrode. R and r is the radius of the cylindrical inner and outer electrodes respectively. ε_x is relative permittivity of the medium.

Measuring food will be placed in media space between the sensor polarization plate cavity, because of differences in moisture content of food, so that the relative permittivity change capacitive

sensors, capacitance change caused, plate thickness in the design to be fully small, to reduce edge effects, can also be loaded in plate plus the metal shielding enclosures to reduce the parasitic voltage of external interference.

Capacitance detection circuit using pulse width modulation circuit, this circuit is to use the sensor principle of charging and discharging of the capacitor, so that the circuit output pulse duty cycle change depending on changes in capacitance of capacitance sensor, and then through a low-pass filter are corresponds to measure DC signal, through the load resistor for standard voltage signals sent NI data acquisition card.

4.2 Temperature sensor

AD590 direct current signal output proportional to the thermodynamic temperature, at the output series resistance converts the voltage signal. In addition, AD590 temperature does not need to reference point, the advantages of strong anti-jamming ability and good interchangeability.

Real-time measured AD590 temperature T0, the ambient temperature is −25~25 °C, AD590 linear current output sensitivity is 1 μg A/K, the output current between 273.2 μA~323.2 μA changes. Selected sampling standard resistance of 1k, then cold-side voltages between 273.2 mV~323.2 mV, changes in temperature every 1 °C, the output voltage changes 1 mV, amplifier gain set to 15. Selection of United States PMI produces voltage operational amplifier OP-07. when the temperature changes 0.5 °C, and AD590 0.5 μA, V590 = 0.5 mV output, the OP-07 output voltage is 7.5 mV, quantitative units greater than acquisition card, meet the demands of cold-junction temperature measurement resolution. Secondly, according to the AD590 temperature measurement principle, determine the relationship end resistance measuring temperature and sensor output voltage, which can be provided under the temperature measuring range, end resistance the output voltage of the

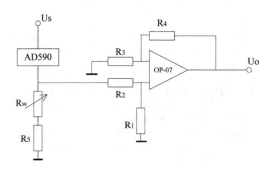

Figure 2. Temperature sensor circuit.

sensor ranges, assuming 10–40 mV, data acquisition card is (0–10 V interface), then conditioning circuit magnification (10 V/40 mV = 250). Figure 2 shows temperature sensor circuit.

5 BP NEURAL NETWORK

Using BP neural network back-propagation algorithm is a simple feed-forward neural network, is most commonly used in artificial neural networks, and one of the most sophisticated models. BP artificial neural networks are made up of several layers of neurons, in addition to outside input, and output layers, there are one or more layers of hidden layer, layer with no connections between nodes in the network, without feedback between inputs and outputs (Lv, 2006).

Algorithm of BP Network belongs to δ, is a mentor-learning algorithm. It is composed of two parts: from the information being passed to the input layer with calculation of hidden layer-by-layer to output layer, each layer under the influence of the State of neurons a layer of neurons in the State. Didn't get expected output if the output layer, then calculate the error signal, and then bring the signals along the original path of error back-propagation and modified back propagation neuron's weights and thresholds, reduce network errors output tend to expect goals.

This design using MatLAB Script nodes in LabVIEW software, called from MatLAB neural

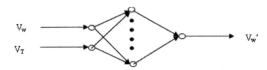

Figure 3. Structure diagram of neural network algorithm.

network Toolbox corresponding function, realization of neural networks algorithm calculations.

The temperature compensation system consists of three layers, first layer for the input layer, there are two input points, including moisture voltage and temperature voltage values; the second layer is hidden; third layer to output layer has an output point, for temperature compensation of moisture voltage value. Figure 3 shows structure diagram of neural network algorithm.

According to the calibration experiments of the learning sample, use of learning remediation network BP algorithm weights and thresholds, until such time as meet accuracy requirements. It can not be used immediately after the neural network is trained. The additional samples should be used to verify the performance. If it is failure to meet the requirements, the neural network should be re-trained. Therefore, the neural network training is an iterative process. Table 1 shows the relationship between output voltages at different temperature and moisture.

6 BASED ON THE LABVIEW DATA ACQUISITION PROCESSING AND STORAGE

DAQ Assistant is an interactive tool used to quickly create measurement applications. Use DAQ Assistant to the user graphically selects the desired measurement type, and used directly in LabVIEW block diagram of the program that generated the code.

LabVIEW Database Connectivity Toolkit with full SQL functionality, interactive operations directly with your local or remote databases, and the need for structured query language programming. It can more easily connect a variety of common databases such as Microsoft Access, SQL Server, Oracle and so on.

This design calls DAQ Assistant in LabVIEW block diagram of the program, read the data

Table 1. Relationship between output voltages at different temperature and moisture.

10 °C		15 °C		20 °C		25 °C	
Water content (%)	Voltage (mV)	Water content (%)	Voltage (mV)	Water content (%)	Voltage (mV)	Water content (%)	Voltage (mV)
9.1	45.2	8.7	52.4	8.3	49.1	8.2	58.0
10.1	62.3	9.9	76.4	10	79.7	10	96.7
11	77.2	11.1	96.6	11.5	107.6	12	113.3
12.8	97.4	13	109.3	13.8	121.6	14.4	128.2
14.6	104.1	14.9	120.6	15.2	132.4	15.8	143.6
15.3	112.3	15.8	137.4	16.9	181.5	17.6	182.3
16.5	140.2	16.5	160.3	18.4	232.2	19.8	250.4
17.8	177.5	18.3	212	19.5	266.4	20.4	296.1

collected in the NI data acquisition device. Data is saved in the database in Access by using packet connection tool.

7 CONCLUSION

Study on virtual instrument and neural network algorithm based on the principle of using LabVIEW software to complete the virtual temperature compensation of moisture, through the connection with the database, and further improved the design of temperature compensation system. It is able to realization of software instead of hardware circuit, algorithm used for temperature compensation function. Self learning function based on the neural network algorithm can continuously improve the precision of the sensor, making the system design of practical significance.

ACKNOWLEDGMENTS

Number The article comes from Hebei Provincial Key Disciplines of detection technology and automatic equipment. The work was supported by the Project of the Science and Technology Bureau of Hebei Provincial. Project ID is 13210332. The work was supported by the Project of the Science and Technology Bureau of Langfang city. Project ID is 2013011029.

REFERENCES

Lu Liangjun. "Automatic compensation of pressure sensor calibration system based on virtual instruments." *Wuhan Huazhong University, 2007.*

Lv Wei, guanglin Li. "LabVIEW and BP neural network-based research on temperature compensation". Information technology, 2006 (31): 66–67.

Plexus cool. "Theory and application of MATLAB neural network Toolbox-oriented." Beijing: China University of science and technology press, 2009.

Zhu Binfeng, Xu Guiyun, Li Junmin, imperative June yan. "Virtual instrument design of pressure sensor based on BP neural network temperature nonlinear correction". Instrument standardization and metrology, 2007 (1): 39–49.

Structure thermal design checking of solid rocket motor

L. Zhou, X.F. Han, W.H. Dong & H.G. Jia
Department of New Technology, Changchun Institute of Optics, Fine Mechanics and Physics,
Chinese Academy of Sciences, Changchun, China

ABSTRACT: The temperature of the gas of solid rocket motor is high, and combustion chamber structure endure high gas pressure, so it is necessary to perform structure thermal protecting and strength analysis of solid rocket motor. In this paper, structure thermal protecting design is performed for combustion chamber and nozzle, and structure strength check of solid rocket motor is performed by using experience formula and FEM. The results show structure thermal protecting and strength check achieve demands.

1 INTRODUCTION

Combustion chamber and nozzle are important parts of solid rocket motor. Combustion chamber is the place of propellant grain burning. Heat energy of the gas is converted into kinetic energy and thrust is supplied by nozzle [1]. Generally temperature of engine gas is high and the gas pressure in combustion chamber is large. To ensure the safety of the combustion chamber of solid rocket motor, it is necessary to perform structure thermal protecting and strength check of combustion chamber which expose to the heated gas [2]. Usually erosion at the nozzle is serious, and there is important equipment near the nozzle. So it is also necessary to design erosion layer and heat insulating layer reasonably and check structure strength to ensure reliability of motor itself and other important equipment [3].

The gas temperature of a solid rocket motor is up to 4000 K and the gas pressure is up to 10 Mpa. To ensure motor structure reliability, structure thermal protecting and strength check are studied in this paper. Firstly, thermal protecting of former head, cylindrical section and nozzle is performed. Secondly strength check of overall structure of solid rocket motor is performed. Finally, the results of thermal protecting design and strength check are presented.

2 THERMAL PROTECTING OF SOLID ROCKET MOTOR

Now erosion protection method is often used in thermal protecting of solid rocket motor. Erosion protection method is that a series of physical and chemical changes of materials or components, which are coated or installed on the surface of structure of solid rocket motor, produce heat dissipation for thermal protecting. The method is simple, reliable, high efficiency, and the material can be adaptable with the changes of environment strongly.

2.1 Thermal protection layer thickness of front head and front of cylindrical shell

According to thermal protection layer model with fixed thickness [4], the expression of the carbonization layer thickness is as follows:

$$x = \sqrt{2kt} \tag{1}$$

$$k = \frac{\lambda\left(T_c - T_{df}\right)}{\rho_v H + \rho_v \bar{c}_v\left(T_{di} - T_i\right)} \tag{2}$$

where, x is thickness of carbonization layer, t is operating time of solid rocket motor, λ is thermal conductivity, T_c is gas temperature, T_{df} is temperature at decompasition district, T_{di} is the initial temperature at decomposition district, T_i is initial temperature, ρ_v is density of thermal protecting material, H is heat of decomposition, c_v is average specific heat of thermal protecting material.

The thermal protection input parameters are as follows: $T_c = 4000$ K, $t = 0.1$ s, $T_{df} = 700$ K, $T_{di} = 600$ K, $\lambda = 0.00038$ J/(cm s K), $c_v = 1.6$ J/(g K), $T_i = 293$ K, $\rho_v = 1.2$ g/cm^3, $H = 900$ J/g.

Carbonization layer thickness can be calculated by formula (1) and formula (2), namely $x = 0.123$ mm. The total thickness of thermal protecting layer is calculated by the following formula:

$$\frac{T^* - T_i}{T_{df} - T_i} = erfc\left(\frac{x_{T^*}}{2\sqrt{\frac{\lambda}{\bar{c}_v \rho} t}}\right)\left(erfc\left(\frac{x}{2\sqrt{\frac{\lambda}{\bar{c}_v \rho} t}}\right)\right)^{-1} \tag{3}$$

where, $T^* = 300$ K is maximum allowable wall temperature, x_{T^*} is the total thickness of thermal protecting layer. The total thickness of thermal protecting layer can be calculated by formula (3) and safety factor 1.5, namely $x_{T^*} = 0.314$ mm.

2.2 Thermal protecting layer thickness of rear of cylindrical shell and bottom cover

According to thermal protection layer model with variation thickness, the expression of carbonization layer thickness is

$$t = \frac{k}{v_e^2}\left(e^{\frac{v_e x}{k}} - 1\right) + \frac{x}{v_e} \qquad (4)$$

where, $v_e = 0.05$ cm/s is erosion rate. Carbonization layer thickness can be calculated by formula (4), namely $x = 0.142$ mm. The total thickness of thermal protecting layer can be calculated by formula (3) and safety factor 1.5, namely $x_{T^*} = 0.333$ mm.

From the calculation results, it can meet thermal protecting requirement that the design value of total thermal protecting thickness equal 0.35 mm.

2.3 Thermal protecting of nozzle

Thermal protection material thickness of Nozzle can be calculated by the following formula

$$\delta = \delta_e + \delta_c + \delta_d \qquad (5)$$

where, δ_e is erosion thickness of thermal protecting layer, δ_c is carbonation thickness of thermal protecting layer, δ_d is design margin of thermal protecting layer thickness, namely $\delta_d = (0.2\sim0.5)(\delta_e + \delta_c)$.

The calculation formula of δ_e is as follows:

$$\delta_e = v_e t \qquad (6)$$

The calculation formula of δ_c is as follows:

$$\delta_c = At^m e^{(-B/q)} \qquad (7)$$

where, q is heat flux, A, B and m are empirical constants. Carbon cloth–phenolic: $A = 0.9144$, $B = 755000$, $m = 0.68$. High silica oxygen–phenolic: $A = 0.7874$, $B = 1026630$, $m = 0.68$.

The calculation formula of q is as follows [5]:

$$q = \alpha \left(\frac{h_e}{0.5(h_w + h_e) + 0.22(h_r - h_e)}\right)^{0.5} (T_r - T_w) \qquad (8)$$

where, α is convective heat transfer coefficients, h is specific enthalpy, the subscript w represents the wall surface, the subscript e represents the free flow, the subscript r represents the recovery specific enthalpy. T is temperature, the subscript w represents wall temperature, the subscript r represents the recovery temperature.

The calculation formula of α is as follows

$$\alpha = \frac{0.026}{d_t^{0.2}}\left(\frac{\mu^{0.2}c_{pg}}{Pr^{0.6}}\right)\left(\frac{p_c g}{c^*}\right)^{0.8}\left(\frac{d_t}{r_c}\right)^{0.1}\left(\frac{A_t}{A}\right)^{0.9}\sigma_1 \qquad (9)$$

where, d_t is throat diameter of nozzle, μ is gas dynamic viscosity coefficient, c_{pg} is gas specific heat capacity at constant pressure, Pr is Prandtl number, p_c is combustion chamber pressure, g is gravity acceleration, c^* is gas characteristic velocity, r_c is curvature radius of throat, A_t is throat area, A is cross-section area at nozzle calculation.

σ_1 is coefficient and its calculation formula is as follows:

$$\sigma_1 = \left[0.5\frac{T_w}{T_r}\left(1 + \frac{\gamma - 1}{2}Ma_e\right) + 0.5\right]^{-0.68}$$

$$\left[1 + \frac{\gamma - 1}{2}Ma_e\right]^{-0.12} \qquad (10)$$

where, γ is ratio of specific heats, Ma_e is Mach number of free stream.

The thermal protecting material is high silica oxygen–phenolic and thermal protection layer thickness of nozzle throat is calculated. The input parameters as follows: $T_c = 4000$ K, $t = 0.1$ s, $Ma = 1$, $d_t = 8$ mm, $c^* = 1600$ m/s, $\gamma = 1.2$, $T_w = 300$ K, $c_{pg} = 1863$ J/(kg K), $v_e = 0.4$ mm/s.

The results is $\delta_c = 0.19$ mm, $\delta_e = 0.04$ mm, $\delta_d = 0.115$ mm. The total thickness of thermal protecting layer $\delta = 0.35$ mm.

3 STRUCTURE STRENGTH CHECK OF SOLID ROCKET MOTOR

The wall temperature is 300 K through thermal protecting technology, so structure strength check is performed according to room temperature. The required input parameters are shown in Table 1.

The material of front head and cylindrical section is TC4, and its parameters are shown in Table 2.

The material of nozzle is 2A12, and its parameters are shown in Table 3.

Table 1. Input parameters.

Parameters	Value
Maximal operation pressure	10 MPa
Inner diameter of cylindrical shell	73 mm

Table 2. Material parameters of TC4.

Parameters	Value
Modulus of elasticity	1.08×10^{11} Pa
Poisson's ratio	0.34
Yield strength at room temperature	860×10^6 Pa

Table 3. Material parameters of 2A12.

Parameters	Value
Modulus of elasticity	7.0×10^{10} Pa
Poisson's ratio	0.31
Yield strength at room temperature	255×10^6 Pa

3.1 Strength check of former head

Former head is elliptical head. Its long is inside diameter of cylindrical section $D_{i1} = 73$ mm, and short side is $h_{i1} = 15$ mm, thickness $\delta_i = 1$ mm. The yield pressure is obtained by the following formula:

$$P_s = \frac{2\sigma_s \delta_1}{\frac{1}{6}\left[2 + \left(\frac{D_{i1}}{2h_{i1}}\right)^2\right]D_{i1} + 0.5\delta_1} = 20.465 \text{ MPa} \quad (11)$$

And safety factor is obtained as follows:

$$n_s = \frac{P_s}{\xi P_{max}} = 2.0465 \quad (12)$$

3.2 Strength check of the cylindrical section

The thickness of cylindrical section $\delta_2 = 3$ mm, the yield pressure is obtained by the following formula:

$$P_s = \frac{2\sigma_s \delta_2}{D_{1i} + \delta_2} = 20.132 \text{ MPa} \quad (13)$$

And safety factor is obtained as follows:

$$n_s = \frac{P_s}{\xi P_{max}} = 2.0132 \quad (14)$$

3.3 Finite element check

Mises stress cloud graph of solid rocket motor is shown in Figure 1. Safety factor cloud graph is shown in Figure 2. It can be seen from Figure 1 and Figure 2, maximum Mises stress is 440.74 MPa, and minimum safety factor is 1.1264. Safety factor

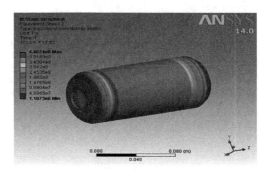

Figure 1. Mises stress cloud graph.

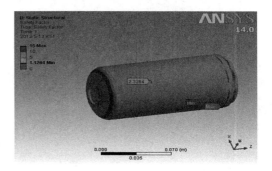

Figure 2. Safety factor cloud graph.

is 2.0132, which is obtained by empirical formula, and it is slightly smaller than 2.3264, which is obtained by FEM. The reason is that there are holes at former head and nozzle in finite element model which cause larger safety factor at cylindrical section than fully-closed pressure vessel with less pressure.

4 CONCLUSION

Through thermal protecting design, the wall temperature of solid rocket motor is 300 K. Structure strength check is performed by empirical formula and FEA, and safety factors at former head and cylindrical section are about 2. The minimum safety factor is 1.1264. Therefore, thermal protecting design and structure strength design meet operating requirements.

REFERENCES

Fu Peng, Jian Ze-qun, Zhang Gang-chui, Gao Bo, "Engineering calculation for erosion and thermal structure of throat-insert of a SRM nozzle," Journal of Solid Rocket Technology. Beijing, vol. 28, pp. 15–19, January 2005.

Sun Bing, Liu Xiao-yong, Lin Xiao-shu, Cai Guo-biao, "Computation of ablation of thermal protection layer in solid rocket ramjet combustor," Journal of Propulsion Technology. Beijing, vol. 23, pp. 375–378, May 2002.

Wang Wei, "Ablation Calculation of Nozzle Divergence Cone Insulation," Journal of Solid Rocket Technology. Beijing, vol. 22, pp. 16–19, March 1999.

Zhang Bin, Liu Yu, Wang Chang-hui, Ren Jun-xue, "Computation of ablation of thermal-protection layer in long-time working solid rocket motors," Journal of Solid Rocket Technology. Beijing, vol. 34, pp. 189–201, February 2011.

Zheng Ya et al. Heat Transfer of Solid Rocket Motor, 3rd ed. Beijing: Beijing Aeronautics and Astronautics university Press, 2006, pp. 15~25.

Session 2: Signal and image processing

Control Engineering and Information Systems – Liu (Ed)
© 2015 Taylor & Francis Group, London, ISBN 978-1-138-02685-8

Study on multi-layered classification of image

Y. Wang, M. Yu, J.C. Chen & S.Z. Han
School of Information, Hebei University of Technology, Tianjin, China

ABSTRACT: Considering the single feature can not classify different images quickly and accurately, a multi-stage classification method is proposed in this paper. After analyzing different features of images, several typical methods of feature extraction are selected. Features which effect different kinds of image are analyzed through the classifier of SVM. Then, a multi-stage classification method is put forward, which conducts coarse classification according to remarkable features of an image, and fulfills the detailed classification through extracting other features of the image. The experiments show that the range of image classification is reduced and the interference of irrelevant images are avoided by the method which based on coarse classification. Furthermore, the method improves the recognition rate effectively compared with the traditional single-layered classification.

1 INTRODUCTION

The large amounts of images on the internet make the user confused. These images are in urgent need of reasonable management. Image classification can solve this problem with the advantage of fast speed of classification and saving in manpower.

In image classification, different users have various needs in image classification or even one user has different needs at different times. So we should put forward an adaptable image classification to meet these needs. Considering the single remarkable characteristics cannot classify different images quickly and accurately, this paper puts forward a multi-stage classification method, which conducts image coarse classification according to remarkable single feature, and achieves detailed classification through reintegration the low-level features of the images.

The experiments show that by multi-layered classification which based on rough classification, can reduce the range of image classification, and thus avoid the interference of other irrelevant images. This method improves the recognition rate effectively compared to traditional classification algorithms. Besides, the users can choose different features to satisfy the classification requirement of various images, which can make the classification more adaptable and applicable.

2 FEATURE EXTRACTION

In this section we extract low-level features of the images, such as color features, texture features and edge features. The methods of extracting color, texture and edge features are selected, through comparing the classification results and the number of dimensions of feature vectors brought about by different feature extraction methods.

2.1 Feature extraction of color moment based on HSV space

Spatial color histogram, color set, color polymerization vector and color correlation chart et al. are all color feature extraction methods. But these algorithms may lead to misclassification due to the process of quantization on image. Additionally, important relations between various pixels were always overlooked in these methods, and the features they extracted have high dimensions.

In 1996, on the basis of the mathematical thoughts that moment can represent the distributions of color features well, Stricker and Orengo proposed an improved algorithm—color moments [1] to extract color features of image. The color information of an image mainly distribute in low order moments. So mean, variance and skewness

Figure 1. The experimental images.

Table 1. The results of color moment.

Image	Mean			Variance			Skewness		
	H	S	V	H	S	V	H	S	V
Office	−0.0235	−0.3299	0.0752	0.0119	−0.1425	−0.0002	−0.0033	−0.0703	0.0009
Tree	0.0979	0.0250	0.1564	−0.0007	0.0111	−0.0306	0.0061	−0.0011	0.0181
Building	0.2976	0.3813	−0.0031	−0.1080	−0.1316	−0.0352	0.0483	0.0695	0.0024
Bus	−0.0382	−0.0407	0.2870	−0.0630	−0.0442	−0.0343	−0.0164	−0.0380	0.0355
Mountain	0.1103	0.4913	0.1303	−0.0181	−0.3523	−0.0780	0.0037	0.2849	0.0171
Flower	−0.2946	−0.1837	−0.0390	−0.1284	0.0036	0.0390	−0.1091	−0.0397	0.0179

can fully represent color distribution of the whole image. Figure 1 shows the 6 experimental images. We extract the 9 color moments of H, S, and V channel, shows in Table 1.

2.2 Feature extraction of texture

As Gabor filter can easily get valid texture information which relate to space situation and space frequency, we choose Gabor filter to extract text features of an image. Before extract texture features, texture images need to be quantified. Every image is divided into 16 sub-images and the size of every sub-image is 4×4 [2]. Then all pixels in one sub-image are averaged. So the dimensions of feature vectors are $4 \times 4 \times 4 = 64$.

2.3 Feature extractions of edge direction histogram based on Canny

Sobel, Prewitt and Laplacian et al. are the common edge detection algorithms. Canny is proved an excellent algorithm of edge detection. It include filter, strengthen and detect stages. As a kind of statistical methods, edge direction histogram can reflect shape and direction information of an image. It was widely used in the field of image processing by the advantage of its simple calculation, fewer dimensions of feature vector and valid classification. So we choose edge direction histogram to represent edge features.

The process of edge direction histogram is [3]:

1. Obtain grayscale images though gray processing.
2. Calculating dx and dy of pixcel (x,y) by canny algorithm.
3. Calculating the edge ditection $\theta(x,y) =$ argta (dx/dy) of all pixcels.
4. The angles of edge direction are quantifyed. The angle range [−180°, 180°] is quantified to 36 level by the interval of 10°.
5. Obtain the histogram of edge direction θ.

In Figure 2, image named Lena with 256*256 pixels is laid on the upper left, and image with edge

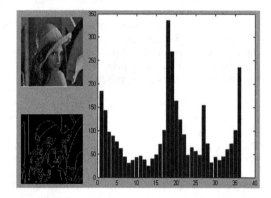

Figure 2. Edge direction histogram.

features extracted by Canny is laid on the lower left. Image on the right is the histogram which is got in step 5. From the histogram, we can see that different angles have different features. So the features got from Canny are useful in edge feature extraction.

The above low-level features play an important role in representation of an image. The features of color, texture and edge provide the necessary data for the following auto classification of images.

3 THE ANALYSIS OF IMAGE CLASSIFICATION

3.1 Classification based on single feature

In the experiment, we choose different kinds of features respectively as input of SVM [4] [5] [6]. Six groups of training data and test data are selected randomly. Recognition rates of six groups are averaged as the final classification results. The results are showed in Table 2.

From Table 2, we can see: The classification rate especially in experiments of flower and tree, which we choose texture features as input extracted is higher than color features extracted by moment and edge features. We can draw the conclusion that

Table 2. The classification results of single feature.

Feature	Feature dimension	Recognition rate (%)						
		Office	Tree	Building	Bus	Mountain	Flower	Sum
Color	9	81.333	72	48.667	73.333	68	53.333	66.111
Texture	64	72.667	93.333	45.333	76	66.667	83.333	72.889
Edge	36	73.333	52	58.667	95.333	49.333	44.667	62.222

Table 3. The classification results of combined features.

Feature	Feature dimension	Recognition rate (%)						
		Office	Tree	Building	Bus	Mountain	Flower	Sum
Color + Texture	73	84.667	93.333	73.333	70.667	83.333	87.333	82.111
Color + Edge	45	84	62.667	84.667	91.333	76.667	60.667	76.667
Texture + Edge	100	75.333	89.333	72	87.333	76.667	83.333	80.667
Color + Texture + Edge	109	84.667	96	80.667	94.667	85.333	75.333	86.111

Gabor features can well classify this kind of natural images.

The features extracted by edge direction histogram have the worst classification rate. However, it has the best result when recognizing the image of Bus. It is owing to the fact that bus has remarkable edge features. Besides, the rate is improved when recognizing the image of Building. So edge direction histogram is fit to classify the images such as artificial scene and other such kind of scenes.

The classification rate achieved by color moment features is better than edge features and worse than texture features. Whereas it is still an useful methods for its simple calculation and low feature dimension.

3.2 Classification based on combined features

In this section, several kinds of combined features are selected as the input of SVM. Experiment is the same as section 3.1. The result is showed in Table 3. Comparing Table 2 with Table 3, we can see that the recognition rates obtained by combined features are obviously better than which by single feature, especially in images of Building and mountain. The Bus image which has obvious edge features and tree image which has clear texture features get worse recognition rate. The classification rates of other images are improved.

4 MULTI-LAYERED CLASSIFICATION

We can draw the conclusion from Table 2, that image Tree get an ideal recognition rate when texture features are inputs of SVM, and image Bus get a high rate when edge direction histogram is input of SVM. Single-layered classification can classify simple images of remarkable features quickly and exactly [7]. The Office, Building and Bus belong to man-made object, whereas Tree, Mountain and Flower belong to natural object. The edge features of man-made images are different from natural images. That the former has regular edge, yet the latter has complex and irregular edge. Therefore, edge features can easily distinguish man-made and natural image. So we propose the multi-layered classification. The method include two stages: First, we conduct image coarse classification according to the remarkable features of image. Second, we proceed detailed classification as section 3.2 to reach a high recognition rate.

In order to avoid the influence of the pattern and size of different images, we should convert the format of image into JPEG. Then extract the features of color, texture and edge as we state at section 2. Users can choose obvious feature to conduct coarse classification according to their own needs. They should reclassify through choosing other features if the results of coarse classification unsatisfied. The result may be saved if the coarse recognition rate ideal. If not, we should go on with detailed classification. In this stage, users can combine the various kinds of features optionally.

4.1 Experimental results

First, the image database is divided into natural image and man-made image according to the remarkable difference of edge features. Man-made image contain images of Office, Building and Bus.

Table 4. The classification results of two methods.

| Classify | Recognition rate (%) | | | | | | |
	Office	Tree	Building	Bus	Hill	Flower	Sum
Single-layer	84.7	96	80.7	94.7	85.3	75.3	86.1
Multi-layer	89.3	96.7	88	93.3	85.3	86.7	89.9

Natural image contain images of Tree, Mountain and Flower. The process of the experiment is:

1. Low-layered coarse claasification: distinguish man-made and natural images.

 The image is divided into two calss as we mentioned above, man-made and natural image. The training data is obtained from 3/4 of the number of every class. The rest of data is applied as testing data. General classification rate of the two classes is 91.333%.

2. Two-layered detailed classification:

 From Table 3, we can conclude that the features with three classes are better than those with two classes and single class. Next, we conduct experiment with the other two kind of features. Table 4 shows the experimental results of single-layered and two-layered classification.

From the experimental results, we can see that multi-layered classification outperforms single-layered classification, especially in images of Office, Building and Flower. Only the recognition rate of Bus is lowered slightly. In general, the recognition rate is improved to a certain extent by multi-layered classification. It is due to the fact that the influence of irrevant images on classification result is avoided by coarse classification, which contributes to the increase of detailed classification rate.

5 CONCLUSIONS

The multi-layered classification we proposed can be widely used in image classification, which can get higher recognition rate. We can also divide the image on the internet into day and night, or sunrise and sunset according to color features, or texture image and non-texture image according to texture features and so on. The multi-layered classification can also be used in fast image classification. The features in every layer can be choosing freely in accordance with the users' needs.

REFERENCES

Flickner. M. Query by image and video content: the QBIC project: querying images by content using color, texture, and shape [C]. In Proc. of the SPIE Conf. on Storage and Retrieval for Image and Video Databases, 1993.

Hartmann. A. & Lienhar R.W. Automantic classification of images on the web [C]. In Proc. of Storage and Retrieval for Media Databases, 2002.

Huiqiang. X. & Guoyu. W. Partition-Based Image Classification Using SVM [J]. Microcomputer Information, 2006, 22(5): 210–212.

Jin. G. Application of SVM in Image Classification [D]. Northwest University, 2010.

Stricker. M. & Orengo. M. Similarity of color images. SPIE Storage and Retrieval for Image and Video Databases III, 1995, 381–392.

Xianglin. M., Zhengzhi. W. & Lizhen. W. Building global image features for scene recognition [J]. Patter Recognition, 2012, 45: 373–380.

Yinghui. K. & Liang. S. Image classification method based on hierarchy semantics [J]. Journal of Computer Applications, 2011, 31(10): 3045–3047.

Control Engineering and Information Systems – Liu (Ed)
© 2015 Taylor & Francis Group, London, ISBN 978-1-138-02685-8

Improved differential coherent based bit synchronization algorithm for high-dynamic and weak GPS signal

X.S. Li & W. Guo

National Key Laboratory of Science and Technology on Communications, University of Electronic Science and Technology of China, Chengdu, Sichuan, China

ABSTRACT: Previous work proposed Differential Coherent Accumulation Algorithm (DCAA) to achieve bit synchronization of high-dynamic and weak GPS signal, which just uses the product of complex sample after coherent integration times the complex conjugate of another complex sample that the interval between them is an integer number of navigation data bit period. In this paper, in order to obtain the processing gain more, we proposed an Improved Differential Coherent Accumulation Algorithm (IDCAA). The proposed algorithm makes full use of the conjugate product of coherent integration outputs between them separated by arbitrary C/A code length and shows a 1.5~2 dB improvement in processing gain over DCAA. After introducing the complex signal model, we present the theoretical expressions of the Edge Detection Rate (EDR) and false-alarm probability, as well as the simulation results of comparing the proposed IDCAA with DCAA under large frequency deviation, large chirping rate, and large frequency deviation and chirping rate condition.

1 INTRODUCTION

In the GPS system, each satellite broadcasts Coarse Acquisition (C/A) code at the L1 frequency. The C/A code has a 1ms period while the data rate of the GPS navigation data is 50 Hz. It means that there are 20 C/A code periods in every navigation data bit period. After satellite signals acquisition, the navigation data bit boundary position must be determined in order to detect navigation data bits. This process is called bit synchronization.

There are multiple algorithms available to achieve bit synchronization. The histogram method is widely used in inexpensive commercial GPS receivers (Van Dierendonck 1996, Krumvieda et al. 2001). This approach partitions the assumed navigation data bit period into 20 1ms C/A-code periods and senses sign changes between successive correlator outputs. For each sensed sign change, a corresponding histogram cell count is incremented. The navigation data bit boundary position can be determined from a peak in the histogram that exceeds a pre-specified upper threshold. With this approach, the sign of correlator output is based on a hard decision (i.e., a two value decision, where +1 corresponds to a positive sign while −1 corresponds to a negative sign). So it is easily affected by frequency deviation between the local carrier replica and incoming signal. Therefore, it is not suitable for bit synchronization under high dynamic conditions. Moreover, it is difficult to

detect bit boundary from weak signal on account of not making full use of the data bit energy.

Kokkonen & Pietilä (2002) proposed a bit synchronization algorithm, which basically computes time averages of bit energies for all twenty possible bit boundary positions and selects the position which maximizes the bit energy. This method uses bit energy efficiently, so it can operate with weak signal. But it similarly can't be applied to bit synchronization under high dynamic conditions.

GPS receivers are now found on cellular phones, automobiles, ballistic and cruise missiles, and innumerable other fast-moving devices. It is necessary that researchers make a special study of bit synchronization under high dynamic and weak signal conditions (Lashley et al. 2009, Kamel 2010). In order to solve the problem, based on Differential Coherent Accumulation Algorithm (DCAA) in (Li et al. 2011), we further propose an Improved Differential Coherent Accumulation Algorithm (IDCAA). With this algorithm, every one of complex samples after coherent integration within a navigation data bit period is multiplied by the complex conjugate of another sample that has been delayed of n C/A code periods rather than n navigation data bit periods and the products are accumulated coherently, and then the sums are accumulated non-coherently as a statistic. By calculating the value of statistic of 20 bit boundary candidates, the starting position of differential coherent accumulation corresponds with the largest value is determined as navigation data bit boundary position to achieve bit synchronization.

2 COMPLEX SIGNAL MODEL

After satellite signals acquisition, complex output signal of correlator can be written as (Yang et al. 2008, Schmid & Neubauer 2004):

$$X_n = b_n R(\Delta \tau)\mathrm{sinc}(f_0 T_c + \alpha n T_c^2)A_n$$
$$\times \exp\left\{ j\left[2\pi\left(f_0 n T_c + \frac{1}{2}\alpha n^2 T_c^2\right) + \phi_0\right]\right\} + w_n \quad (1)$$

where X_n is the complex output signal with the sample index n, $b_n = \pm 1$ is the unknown sign of navigation data bit, $R(\cdot)$ stands for the correlation function with an ideally triangular shape, $\Delta \tau$ is code offset between received code and local replica, $\mathrm{sinc}(\cdot)$ is the sinc function, f_0 is the frequency deviation between received carrier and local replica, T_c is coherent integration time, α is the chirping rate, A_n is the amplitude, ϕ_0 is the initial phase, and w_n is the complex white Gaussian noise that real and imaginary parts are uncorrelated noises with zero mean and unity variance.

According to (1), the frequency deviation between received carrier signal and local replica can be expressed by

$$\Delta f_n = f_0 + \alpha n T_c \quad (2)$$

Since the change of signal amplitude is rather small, so the signal amplitude is assumed to be constant ($A_n = A$) for simplicity. Formula (2) can be simplified as:

$$X_n = b_n A \exp\left\{ j\left[2\pi\left(f_0 n T_c + \frac{1}{2}\alpha n^2 T_c^2\right) + \phi_0\right]\right\} + w_n \quad (3)$$

Let Θ_n denote the phase angle of signal:

$$\Theta_n = 2\pi\left(f_0 n T_c + \frac{1}{2}\alpha n^2 T_c^2\right) + \phi_0 \quad (4)$$

Then, (4) can be further simplified as:

$$X_n = b_n A \exp(j\Theta_n) + w_n \quad (5)$$

3 ALGORITHM STATEMENT AND PERFORMANCE ANALYSIS

3.1 Description of IDCAA

The receiver operations for improved differential coherent accumulation algorithm are illustrated in Figure 1.

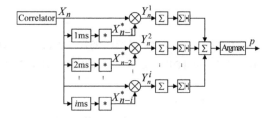

Figure 1. Baseband architecture with improved differential coherent accumulation algorithm.

where i ms denotes a delay of i milliseconds, $*$ denotes complex conjugation, Σ denotes coherent accumulation, $\Sigma|\cdot|$ denotes non-coherent accumulation, p denotes navigation data bit boundary position.

The differential coherent value of the nth correlator output and the previous ith correlator output can be written as:

$$Y_n^i = X_n X_{n-i}^* = b_n b_{n-i} A^2 e^{j(\Theta_n - \Theta_{n-i})} + b_n A e^{j\Theta_n} w_{n-i}^*$$
$$+ b_{n-i} A e^{-j\Theta_{n-i}} w_n + w_n w_{n-i}^*$$
$$\triangleq S_n^i + N1_n^i + N2_n^i + N3_n^i \quad (6)$$

In (7), the first summand represents the deterministic signal component; the other three summands represent the noise component.

If the sign of navigation data bit in (7) doesn't change, the phase of signal component can be written as:

$$\Delta\Theta_n^i = \Theta_n - \Theta_{n-i} = 2\pi\left[f_0 i T_c + \frac{1}{2}\alpha i(2n - i)T_c^2\right] \quad (7)$$

Otherwise, the phase of signal component can be written as:

$$\Delta\Theta_n^i = \Theta_n - \Theta_{n-i} = 2\pi\left[f_0 i T_c + \frac{1}{2}\alpha i(2n - i)T_c^2\right] + \pi \quad (8)$$

It is easy to know that the amplitude of signal component due to coherent accumulation will be maximum when all of summands have the same sign. This amplitude has relation to the second order phase difference of two corresponding complex coherent integration values, so it is insensitive to frequency deviation derived from fast-moving receiver and clock drift.

According to Figure 1, assume the length of navigation data to be processed is N bits, we get the statistic as follows:

$$Z_{iklp} = \sum_{i=1}^{M} \sum_{k=fix(i/20)}^{N-2} \left| \sum_{l=1}^{20-(i)\bmod(20)} Y_{20k+l+p}^{i} \right| \quad (9)$$

where $i = 1, 2, \ldots, M$ is delay time between correlator outputs taken differential coherent in units of milliseconds, l is times of coherent accumulation, k is times of non-coherent accumulation, $p = 0, 1, 2, \ldots, 19$ is bit boundary candidate, fix is truncated function, mod denotes modulo operation. The estimated navigation data bit boundary position is obtained as:

$$\hat{p} = \arg\max_{p} Z_{iklp} \quad (10)$$

3.2 Performance analysis

In (7), the deterministic signal component rewritten as:

$$S_n^i = A^2 b_n b_{n-i} e^{j(\Theta_n - \Theta_{n-i})}$$
$$= A^2 b_n b_{n-i} e^{j\left\{2\pi\left[f_0 i T_c + \frac{1}{2}\alpha i(2n-i)T_c^2\right]\right\}} \quad (11)$$

The noise component rewritten as:

$$N1_n^i = Ab_n e^{j\Theta_n} w_{n-i}^* = Ab_n e^{j\left[2\pi\left(f_0 n T_c + \frac{1}{2}\alpha n^2 T_c^2\right)\right]} w_{n-i}^* \quad (12)$$

$$N2_n^i = Ab_{n-i} e^{j\Theta_{n-i}} w_n^*$$
$$= Ab_{n-i} e^{j\left\{2\pi\left[f_0(n-i)T_c + \frac{1}{2}\alpha(n-i)^2 T_c^2\right]\right\}} w_n^* \quad (13)$$

$$N3_n^i = w_n w_{n-i}^* = (w_{I,n} + jw_{Q,n})(w_{I,n-i} - jw_{Q,n-i})$$
$$= w_{I,n} w_{I,n-i} + w_{Q,n} w_{Q,n-i}$$
$$+ j(w_{Q,n} w_{I,n-i} - w_{I,n} w_{Q,n-i}) \quad (14)$$

$N1$ and $N2$ obey a zero-mean complex Gaussian distribution with variance

$$\sigma_{N1}^2 = E\left[|N1 - E(N1)|^2\right] = E\left[|N1|^2\right] = A^2 \sigma_w^2 \quad (15)$$

$$\sigma_{N2}^2 = E\left[|N2 - E(N2)|^2\right] = E\left[|N2|^2\right] = A^2 \sigma_w^2 \quad (16)$$

where σ_w^2 is the variance of complex white Gaussian noise w_n

$N3$ obeys a normal product distribution with variance (Schmid & Neubauer 2004)

$$\sigma_{N3}^2 = E\left[|N3|^2\right] = \sigma_w^4 \quad (17)$$

According to (Schmid & Neubauer 2004), the normal product sum distribution nearly obeys

Gaussian distribution if only the times of coherent accumulation is greater than 9.

We assume that the cumulant of differential coherent accumulation in a navigation data bit length is expressed as follows:

$$Z_{lp}^i = \sum_{l=1}^{20-(i)\bmod(20)} Y_{n+l+p}^i \quad (18)$$

Through the above analysis, the real and imaginary parts of the cumulant $Z_{lp}{}^i$ have a Gaussian distribution, thus the envelope $|Z_{lp}{}^i|$ is a Ricean distribution defined by (Kaplan & Hegarty 2006):

$$p(z) = \begin{cases} \dfrac{z}{\sigma_n^2} e^{-\left(\frac{z^2+S^2}{2\sigma_n^2}\right)} I_0\left(\dfrac{zS}{\sigma_n^2}\right), & z \geq 0 \\ 0, & z < 0 \end{cases} \quad (19)$$

where z is the value of the random variable, σ_n^2 is the RMS noise power, S is RMS signal amplitude, $I_0(zS/\sigma_n^2)$ is modified Bessel function of zero order.

If p is navigation data bit boundary position p_0, i.e. $p = p_0$, then RMS signal amplitude is expressed as

$$S_{lp}^i = \left| \sum_{l=1}^{20-(i)\bmod(20)} S_{n+l+p}^i \right| = A^2 \left| \sum_{l=1}^{20-(i)\bmod(20)} e^{j2\pi\alpha i l T_c^2} \right| \quad (20)$$

The RMS noise power is

$$\sigma_n^2 = \frac{1}{2} \sum_{l=1}^{20-(i)\bmod(20)} (\sigma_{N1}^2 + \sigma_{N2}^2 + \sigma_{N3}^2)$$
$$= \frac{1}{2} \sum_{l=1}^{20-(i)\bmod(20)} (2A^2 \sigma_w^2 + \sigma_w^4) \quad (21)$$

If $p \neq p_0$, then RMS signal amplitude is expressed as

$$S_{lp}^{i\,\prime} = \left| \sum_{l=1}^{20-(i)\bmod(20)} S_{n+l+p}^i \right|$$
$$= A^2 \left| \sum_{l=1}^{20-(i)\bmod(20)} b_{n+l+p} b_{n-i+l+p} e^{j2\pi\alpha i l T_c^2} \right| \quad (22)$$

The RMS noise power has the same expression as (22).

When the value of i is small, the amplitude of the statistic Z_{iklp} obeys Ricean sum distributions. Cumulative Density Function (CDF) and Probability Density Function (PDF) closed-form approximations have been proposed for the sum of multiple independent Ricean variables in (L'opez-Salcedo 2009). Under weak signal conditions, it may take hundreds of bits to achieve bit synchronization, so the number of accumulation is very large. According to the Central Limit Theorem (CLT) for the sum of independent and identically distributed random variables, the distribution of the statistic Z_{iklp} obeys a Gaussian distribution with mean and variance as follows:

$$E(Z_{iklp}) = \begin{cases} \displaystyle\sum_{i=1}^{M} \sum_{k=fix(i/20)}^{N-2} S_{lp}^i & p = p_0 \\ \displaystyle\frac{1}{2}\sum_{i=1}^{M} \sum_{k=fix(i/20)}^{N-2} (S_{lp}^i + S_{lp}^{i\prime}) & p \neq p_0 \end{cases} \quad (23)$$

$$D(Z_{iklp}) = \sum_{i=1}^{M} \sum_{k=fix(i/20)}^{N-2} \sigma_n^2 \quad (24)$$

We assume that the probability of sign change is 0.5 in (24). The synchronization performances have been evaluated with an index called Edge Detection Rate (EDR) (Schmid & Neubauer 2004) which represents the percentage of successful synchronizations over a certain number of attempts and is given as a function of the carrier-to-noise ratio (C/N_0). We may assume that $p_0 = 0$ is the correct bit boundary position, the Edge Detection Rate can be written as:

$$P_d = P\left\{ Z_{iklp}\big|_{p=0} > \max\left(Z_{iklp}\big|_{p=1,2,\cdots,19}\right) \right\} \quad (25)$$

Let U denote the maximum of the other 19 statistic, i.e.

$$U = \max(Z_{iklp}\big|_{p=1,2,\cdots,19}) \quad (26)$$

Then, (26) can be expressed as:

$$P_d = \int_{-\infty}^{+\infty} f_{Z0}(v) \int_{-\infty}^{v} f_{Zp}(u) \, du dv$$

$$= \int_{-\infty}^{+\infty} f_{Z0}(v) \prod_{p=1}^{19} F_{Zp}(v) \, dv \quad (27)$$

where f_{Z0} is the PDF of the statistic of the correct bit boundary position, f_{Zp} and F_{Zp} are the PDF and CDF of the pth statistic respectively, $p = 1, 2, 3, \ldots, 19$.

The corresponding probability of false alarm can be expressed as:

$$P_{fa} = P\left\{ Z_{iklp}\big|_{p=0} < \max\left(Z_{iklp}\big|_{p=1,2,\cdots,19}\right) \right\} = 1 - P_d \quad (28)$$

4 SIMULATION AND RESULTS

Our previous studies have shown that Differential Coherent Accumulation Algorithm (DCAA) has excellent bit synchronization performance compared to the histogram method and the bit energy method under high-dynamic and large frequency deviation conditions, thus we only compare bit synchronization performance of IDCAA with DCAA's in this section, meanwhile, we give the value of Theory Analysis (TA) as well. Complex signal model is given by (4), all EDR curves have been obtained processing a data sequence of 50 symbols which equals a time interval of 1 second, and repeating the simulation 10000 times to obtain a good statistics. The integration interval $T_c = 1$ ms, and w_n is the complex white Gaussian noise that real and imaginary parts are uncorrelated noises with zero mean and unity variance, the initial phase $\phi_0 = 0°$. In addition, for different initial frequency deviation and chirping rate, we have simulated three cases. We let $M = 59$ in IDCAA and let $M = 6$ in DCAA in All of these cases.

Case 1 we compare the synchronization performance of two algorithms under large frequency deviation condition. Figure 2 shows the simulation result with $f_0 = -250$ Hz and $\alpha = 0$ Hz/s. It can be seen from Figure 3 that the IDCAA algorithm offers a sensitivity gain over the DCAA algorithm around 2dB when the Edge Detection Rate is 0.9.

Case 2 we compare the synchronization performance of two algorithms under large chirping rate. Figure 3 shows the simulation result with $f_0 = 0$ Hz and $\alpha = 150$ Hz/s. Similarly, the IDCAA algorithm offers a sensitivity gain over the DCAA algorithm around 2 dB.

Case 3 we compare the synchronization performance of two algorithms under large frequency

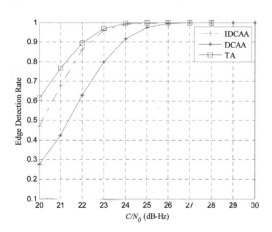

Figure 2. Synchronization performance of IDCAA and DCAA with $f_0 = -250$ Hz and $\alpha = 0$ Hz/s.

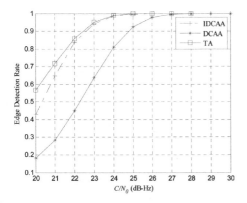

Figure 3. Synchronization performance of IDCAA and DCAA with $f_0 = 0$ Hz and $\alpha = 150$ Hz/s.

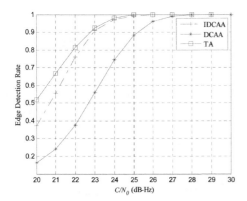

Figure 4. Synchronization performance of IDCAA and DCAA with $f_0 = -250$ Hz and $\alpha = 150$ Hz/s.

deviation and chirping rate condition. Figure 4 shows the simulation result with $f_0 = -250$ Hz and $\alpha = 150$ Hz/s. Like the previous two cases, when the Edge Detection Rate is 0.9, the sensitivity gain of the IDCAA algorithm is improved by about 1.5 dB than the DCAA algorithm.

5 CONCLUSION

An Improved Differential Coherent Accumulation Algorithm (IDCAA) is proposed in this paper for GPS bit synchronization. The IDCAA algorithm offers a sensitivity gain of 1.5~2 dB over the DCAA algorithm that depends on the respective scenario. The improvement of sensitivity gain benefits from making full use of the differential coherent value

with different interval time. The simulation results show that the IDCAA algorithm can detect weak GPS signal with carrier to noise ratio of 22~23 dB-Hz, when the Edge Detection Rate is 0.9 and the length of processed data is 1 second. The IDCAA algorithm is more suitable for bit synchronization of high-dynamic, weak GPS signal.

ACKNOWLEDGMENTS

Research is supported by the 863 project (Grant No. 2014AA01A701); The National Natural Science Foundation of China (Grant No. 61271168); Foundation of National Key Laboratory of Science and Technology on Communications (Grant No. 9140C020302130C02007).

REFERENCES

Kamel, A.M. Design and testing of an intelligent GPS tracking loop for noise reduction and high dynamics applications. *Proc. of ION GNSS 2010*, Portland, OR, 2010:3235–3243.

Kaplan, E.D. & Hegarty, C.J. *Understanding GPS: principles and applications*. 2nd edition. Norwood, MA: Artech House, 2006:221–222.

Kokkonen, M. & Pietilä, S. A new bit synchronization method for a GPS receiver. *2002 IEEE Position Location and Navigation Symposium*, Palm Springs, CA, 2002:85–90.

Krumvieda, K., Madhani, P. & Cloman, C. et al. A complete IF software GPS receiver: a tutorial about the details. *Proc. of ION GPS 2001*, Salt Lake City, UT, 2001:789–829.

L'opez-Salcedo, J.A. Simple closed-form approximation to Ricean sum distributions. *IEEE Signal Processing Letter*, 2009, 16(3):153–5.

Lashley, M., Bevly, D.M. & Hung, J.Y. Performance analysis of vector tracking algorithms for weak GPS Signals in high dynamics. *IEEE Journal of Selected Topics in Signal Processing*, 2009, 3(4):661–673.

Li, X.S., Guo, W. & Xie, X.B. A GPS bit synchronization method for high-dynamic and weak signal. *Journal of Electronics & Information Technology*, 2011, 33(10):2521–2525.

Schmid, A. & Neubauer, A. Performance evaluation of differential correlation for single shot measurement positioning. *Proc. of ION GNSS 2004*, Long Beach, CA, 2004:1998–2009.

Van Dierendonck, A.J. GPS receivers. In: Parkinson BW, Spilker JJ, Axelrad P, et al., editors. *Global positioning system: theory and applications*. Washington, DC: AIAA, 1996.

Yang, C. & Nguyen, T. Blasch, E. et al. Post-correlation semi-coherent integration for high-dynamic and weak GPS signal acquisition. *Proc. of IEEE/ION PLANS 2008*, Monterey, CA, 2008:1341–1349.

Control Engineering and Information Systems – Liu (Ed)
© 2015 Taylor & Francis Group, London, ISBN 978-1-138-02685-8

Image retrieval using morphology based Scale Invariant Feature Transform

W. Han & W.S. Shao
NJUE, Nanjing, Jiangsu, China

Z.Y. Yu
Nanjing Certus Net Incorporation, Nanjing, Jiangsu, China

ABSTRACT: We propose a new feature extraction method for image retrieval. The method describes the morphology relationship between the SIFT key points and the edge of an image. Our algorithm uses such morphology relationship to describe the key points whose position is calculated by the SIFT algorithm, and we call the proposed algorithm as morphology-based SIFT. The simulation results demonstrate that the proposed algorithm can describe image feature well and is efficient for image retrieval.

1 INTRODUCTION

Searching a digital library having large number of digital images or video sequences has become more and more important in this visual age. People search and browse multimedia database through Internet every day. To make such searching practical, effective image retrieval becomes more and more important. The technology of Content Based Image Retrieval (CBIR) grows fast these years. CBIR aims to find images with the required characteristics in an image database. A successful CBIR system is based on distinctive feature detection and efficient indexing, and our proposed algorithm can extract distinctive image feature for image retrieval.

An image can be described by different visual features to retrieval images. The features include color, texture, shape, edge and so on. Color, texture, and shape are often used as image features. In these low-level visual features, color feature is a simple way to represent an image, and it is common to use color histogram to represent image feature. However, using color histogram to retrieve image is not reliable because same kind of object can have totally different color information. So in some image retrieval circumstances, we need to character an image in other ways.

T. Ojala et al. characterized the image spatial configuration of local image texture in a rotation invariant way [9] (Ojala et al. 2002), and the measure output of [9] (Ojala et al. 2002) depicts objects distinct texture feature. M. Safar et al. introduced some shape-based object retrieval techniques [8] (Safar et al. 2000), which extract shape

information as features. Edge information can also be extracted and used in image retrieval, which is illustrated in [10(Zhou et al. 2001), 7(Banerjee et al. 2003), 3(Ohashi et al. 2003)]. J.Z. Wang applied a Daubechies wavelet transform for each of the three opponent color components in an image [5] (Wang et al. 1998), the wavelet coefficients and their variances in the lowest few frequency bands are stored as feature vectors, and these feature vectors are further used for image retrieval and indexing.

Similar to this paper, some papers use edge information as feature for image retrieval. In [10] (Zhou et al. 2001), some information embedded in the structure of edges is extracted. In [7] (Banerjee et al. 2003), compactness of edge for image retrieval is used. In [3] (Ohashi et al. 2003), the edge based feature extraction method functioned by representing the relative position relationship between edge pixels. However, the extracted edge features in [10] (Zhou et al. 2001), [7] (Banerjee et al. 2003), [3] (Ohashi et al. 2003) do not contain the edge spatial position information which relative to key points, and they are not effective for image retrieval.

In CBIR system, an image query uses one or more visual features as well as techniques such as object centric and relevance feedback. However, the first step for image retrieval is still feature extraction. For the extracted features, the invariance with respect to imaging conditions is a big challenge. Specifically, the extracted features should be invariant with respect to geometrical variations, such as translation, rotation, scaling, and affine transformations. At the same time, these features should be invariant with respect to photometric variations such as illumination, direction, intensity, colors, and highlights.

Therefore, we propose Morphology-based SIFT algorithm, this algorithm extracts robust feature from key points for image retrieval.

Like SIFT, [6] (Zhao et al. 2011) and [11] (Ke et al. 2004) calculate the key point positions as the SIFT algorithm, and encode the salient aspects of the image gradient in the key point's neighborhood. However, [6] (Zhao et al. 2011) used a circular region to describe every key point's feature, while [11] (Ke et al. 2004) applied Principal Component Analysis (PCA) to the normalized gradient patch and form new feature vectors.

This paper also calculates the key point positions as the SIFT algorithm, the relationship between the edge information and the SIFT key point information is used to describe the key points. Such relationship describes the image morphology and is used as feature for image retrieval. Experiment results show that the feature using the proposed algorithm can effectively retrieve different image classes.

The remainder of this paper is organized as follows: The second section introduces the background of SIFT algorithm, while the third section introduces the proposed algorithm. The forth section uses simulation to compare the image retrieval results of the proposed algorithm and other algorithms. The final section gives a conclusion.

2 SIFT ALGORITHM

The SIFT algorithm [1] (Lowe. 2004) mainly contains the following four steps: (1) scale-space extreme detection, (2) a model is used to accurately determine the critical point, then remove low-contrast and edge points to get the key points, (3) key points have been accurately identified the direction specified parameter, (4) description of feature points generated feature vector.

In the first stage of SIFT algorithm, potential interest points are identified by scanning the image over location and scale. The SIFT algorithm is implemented by constructing a Gaussian pyramid and searching for interest points at the extreme of difference-of-Gaussian images.

Figure 1 represents the difference-of-Gaussian scale space which is built by the difference-of-Gaussian function, the difference-of-Gaussian is defined by a convolution function as follows:

$$D(x,y,\sigma) = (G(x,y,k\sigma) - G(x,y,\sigma)) * I(x,y) \quad (1)$$

where $G(x,y,\sigma)$ is a variable-scale Gaussian, $I(x,y)$ is an input image.

In order to detect the local extreme points, we need to compare each point in the difference-of-Gaussian space with its adjacent 26 pixels which is

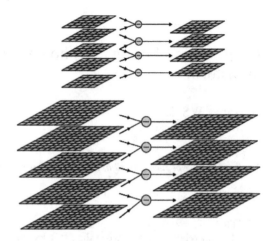

Figure 1. Building the Gaussian and difference of Gaussian pyramid.

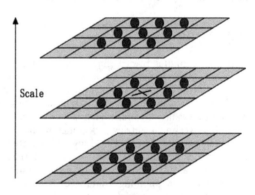

Figure 2. Detect the extreme of the scale space.

shown in Figure 2. After comparing with all of these 26 points, if the point is the local extreme point, we can define this point as the local extreme point of the image. And all of these local extreme points constitute the candidate set of key points of the image. After eliminating the low-contrast points and the edge points from such candidate set in the second stage of SIFT algorithm, we get the key points sets.

In the second stage of SIFT algorithm, SIFT algorithm removes low-contrast and edge points. So, it is highly possible that the key points are not on the edge of the image which we get using canny method [4] (Canny. 1986) in our proposed Morphology-based SIFT algorithm. In the Morphology-based algorithm, we use the distance between edge and key points in different direction as feature vector. If the key points are not on the edge, the key points can describe the image morphology more effectively.

In the process of removing edge points, the Hessinn matrix is calculated using the method which is similar to the Harris edge detection:

$$H = \begin{bmatrix} D_{xx} & D_{xy} \\ D_{xy} & D_{yy} \end{bmatrix} \qquad (2)$$

define $Tr(H) = D_{xx} + D_{xy}, Det(H) = D_{xx}D_{yy} - D_{xy}^2$ if the $Tr(H)^2/Det(H) < (\gamma+1)^2/\gamma$, we can get the conclusion that this point is the edge point. Then we eliminate it. γ is the ratio between the largest eigenvalue and the smaller one of H.

The third and forth stages of SIFT algorithm build a representation for each key point based on the gradient of a patch of pixels in its local neighborhood. However, if two objects of same kind have total different texture and gradient, which is the usual circumstances, the SIFT algorithm cannot detect the corresponding key point matches between two objects.

3 MORPHOLOGY-BASED SIFT ALGORITHM

3.1 *Algorithm description*

Morphology-based SIFT calculates the key point positions as the original SIFT algorithm, and expresses the relation between the key point positions and the edge information of an image as feature for image retrieval.

The proposed algorithm can be described as follows: (1) Pre-detect the edge information using canny edge [4] (Canny. 1986) detection method. (2) Calculate key point positions as the original SIFT algorithm. (3) For each key point, search in different directions, find whether there is an edge point in different directions. If there is an edge point in one direction, find the distance between the key point and the edge point in that direction. Otherwise, find the distance between the key point and the corner of the image in that direction. (4) Use the relationship between the edge and key points to derive feature vectors. (5) Use the derived feature vectors for image retrieval.

The edges in our algorithm are calculated using Canny [4] (Canny. 1986) edge operator and only the strong edges are retained, so the relationship between the edge and key points depicts the morphology of the strong edge which can describe the character of an image well.

3.2 *Feature extraction*

Similar to SIFT, our algorithm uses one feature vector to describe a key point, and we call the feature vector as descriptor for that key point.

The descriptors in our algorithm are significantly smaller than the standard SIFT feature vector, but perform better in the process of image retrieval.

For each key point, we search if there is an edge point in different directions as shown in Figure 3, if there is an edge point in one direction, we define that key point is closed in that direction. Otherwise, we define that key point is open in that direction.

We also record the distance between the key point and edge or image corner in different directions. Whether a key point is closed or open and the distance between the key point and edge or image corner represents their spatial relationship. Such representation can well signature the morphology of the objects in an image. And we use such representation as a description of an image.

In our algorithm, we use 8 features as the descriptor of each key point. The 8 features are whether the key point is closed or open and the length of the key point in upper, below, left and right directions. If a key point is closed in one direction, the length of the key point in that direction is the distance between the key point and the edge point in that direction. If a key point is open in one direction, the length of the key point is the distance between the key point and the most outside pixel of the image in that direction.

We use parameter l, r, u, b to represent the length of the key point at left, right, upper, below directions respectively, as shown in Figure 3. And we use value lc, rc, uc, bc to represent whether the key point is open at left, right, upper, below directions respectively. For a key point, we value lc, rc, uc, or bc as M or zero corresponding to whether it is open in four directions. If it is open in one direction, we set the corresponding value of that direction as M. Otherwise, We set it zero. We set the value M as one forth length of the image. In our algorithm, the descriptor of a key point consists the elements

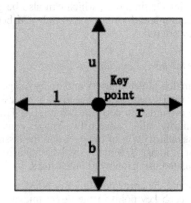

Figure 3.   The morphology relationship between a key point and the edge.

of l, r, u, b, lc, rc, uc, bc. And we use these elements of the key points in an image to depict the edge morphology of that image.

Considering a white circle on a black background, which will have a single scale space maximum where the circular positive central region of the different-of Gaussian function matches the size and location of the circle. So a key point will generate for that circle. And the descriptors of our Morphology-based SIFT algorithm can effectively describe the morphology of that circle.

3.3 Dissimilarity between two key points

When we compare a key point in the query image and a key point in the target image, we compare both key point positions and descriptors between two key points. For each key point, we form histogram p to describe that key point. Bins of the histogram consist of the key point position information and descriptor vectors, i.e. p consists of μ x, μy, l, r, u, b, lc, rc, uc, bc. Where x, y are the horizontal and vertical position of a key point, and the parameter μ is a regularization parameter governing the tradeoff between the position difference and the descriptor vector difference.

And the histogram difference is used to describe the similarity between two key points. Key Point Dissimilarity between two images is measured using following distance.

$$D_k = \mu(\Delta x + \Delta y) + \Delta l + \Delta r + \Delta u + \Delta b$$
$$\Delta lc + \Delta rc + \Delta uc + \Delta bc \qquad (3)$$

where D_k represents Key Point Dissimilarity, Δ represents the absolute difference between the target image and query image.

When we compare two key points, the difference between open and closed in one direction for a key point will generate a difference value of M for the Key Point Dissimilarity, which can also be generated from a length difference of value M between two key points.

3.4 Matching algorithm

If we match target image A and query image B, we use the total dissimilarity value in (3) to represent the degree of similarity between two images. The total dissimilarity value between two images consists of the dissimilarity cost of image A and the dissimilarity cost of image B. And the total dissimilarity value is calculated using following three steps.

1. Finding the corresponding key point
For each key point in the target image A, find the most similar key point in query image B, and then calculate the\Key Point Dissimilarity

value. The process of finding the most similar key point is depicted as follows:

Let PA = { pA1, pA2, ..., pAm } be m local features in image A, and PB = { pB1, pB2, ..., pBn } be n local features in image B. Each feature corresponds to one key point, and is represented as a vector of weighted pixel position and Morphology-based SIFT descriptor of that key point. For the local feature pAs in image A, the corresponding feature pBt, in image B should meet the following condition (nearest neighbor).
$$\arg \min_{p_B^t \in P_B} \left| p_A^s - p_B^t \right|$$

2. Calculate the dissimilarity cost of two images
We add the Key Point Dissimilarity value of all the key points in image A, and normalize by $\sqrt{n+1}$. The result is the dissimilarity cost of image A. n is the total number of key points of that image. And then we use same method to calculate the dissimilarity cost of image B.

3. Add the dissimilarity cost of two images as the total dissimilarity value
We add two dissimilarity cost values together as the total dissimilarity value between image A and image B. The total dissimilarity value we have got indicates the dissimilarity between the target image and the query image. We use total dissimilarity value as the criterion to retrieve images.

When two similar objects have different textures, the key point number between two objects may be greatly different, though the key points between two images may have similar descriptor information. Through normalizing by $\sqrt{n+1}$, the effective of texture difference resulting large total dissimilarity value can be greatly reduced.

Normally, the key point number in image A and image B is different, so instead of using histogram difference of descriptors of the whole image to compare, we search the nearest key point and comparing each key point separately, and then sum the dissimilarity.

3.5 Summation of our algorithm

Same kind objects often have totally different color and texture information and totally different geometric information. But their morphology has some similarity, and our Morphology-based SIFT features can effectively describe such similarity.

Step 1: Calculate the edge of the target image and the query image.
Step 2: Using the method in SIFT calculate the key points in the target image and query image.
Step 3: Get the descriptors of each key point, which contains elements of l, r, u, b, lc, rc, uc, bc.
Step 4: For each key point in the query image, calculate the Key Point Dissimilarity of all the

matches of each key point in the target image, and choose the match with least dissimilarity. For each key point in the target image, calculate the Key Point Dissimilarity of all the matches of each key point in the query image, and choose the match with least dissimilarity.

Step 5: Calculate the dissimilarity cost of the query image and target image separately, and add the two costs as the total dissimilarity value between two images.

Step 6: Use total dissimilarity value for image retrieval.

4 SIMULATION RESULTS

When two images of same kind of object have total different texture or total different geometric information, the SIFT algorithm is hard to match the key points, while our algorithm can effectively match the key points, as shown in Figure 4.

We use two images of elephant for illustration in Figure 4. We manually set threshold to have our algorithm and the original SIFT algorithm each return 14 matches. In the original SIFT algorithm, 10 key points on the elephant match 7 key points on the elephant and 3 key point at the background; 4 key points at the background all match on the elephant. In our algorithm, 12 key points on the elephant all match on the elephant. 2 key points at the background match 1 key point on the elephant, and 1 key point at background.

(a) Key point matches using the original SIFT algorithm

(b) Key point matches using Morphology-based SIFT algorithm

Figure 4. A comparison of image matches found using the original SIFT algorithm (a), with the ones found using Morphology-based SIFT algorithm (b) for a same kind of object with total different texture.

From Figure 4, we can see that when two same kind objects have totally different texture, color information, our proposed SIFT matching is more accurate than the original SIFT matching algorithm for the key points matching.

For image retrieval, 1080 images of the Caltech 101 dataset [2] (Li et al. 2004) are used to evaluate the performance of the proposed methods. There are 27 kinds of objects of the choosing images, each kind of objects has 40 images.

To prove the feature of our algorithm is better than the feature of previous algorithms, the proposed scheme compares the precision rate with the algorithm of original SIFT and the algorithm of [3] (Ohashi et al. 2003). For each algorithm, we choose the best 5, 10, 20 images, and we calculate the precision rate in each choosing method.

$$\Pr ecision = n/N$$

where N is the number of retrieved images after comparing the degree of similarity, n is the number of relevant images in the retrieved image.

For using the original SIFT method, image retrieval is formulated as follows. Given two images, we first extract their corresponding feature vectors. For each feature vector in an image, we compare it against all feature vectors in the other image and find whether there is a match for which the ratio comparing the distance of the closest neighbor to that of the second-closed neighbor is less than a threshold. We treat the number of matches as the similarity between images.

The detailed results of the average precision of the top ranked results for different algorithms are presented in Table 1.

From Table 1, we can see that the proposed algorithm can describe the image feature better and perform better in the image retrieval process. Figure 5 and Figure 6 show some image retrieval results for target images. The retrieval results include the top 5 retrieval results using Morphology-based SIFT algorithm, the original SIFT algorithm, and the algorithm in [3] (Ohashi et al. 2003). We can see that the propose algorithm has better image retrieval results. For the original SIFT algorithm, it is hard to find key points matches which are nearly same, even when

Table 1. Average image retrieve precision of top ranked results for different algorithms.

Algorithm	Top 5	Top 10	Top 20
Proposed algorithm	68.9%	63.9%	52.5%
SIFT algorithm	7.1%	6.3%	5.4%
Algorithm in paper [3]	17.8%	13.3%	11.7%

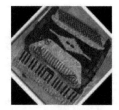

(a) The target image

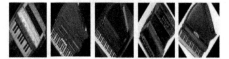

(b) Image retrieval result using Morphology-based SIFT algorithm

(c) Image retrieval result using original SIFT algorithm

(d) Image retrieval result using the algorithm in [3]

Figure 5.   For an image of an accordion, top 5 image retrieval results of our algorithm, SIFT algorithm and the algorithm in [3].

(c) Image retrieval result using original SIFT algorithm

(d) Image retrieval result using the algorithm in [3]

Figure 6.   (*Continued*)

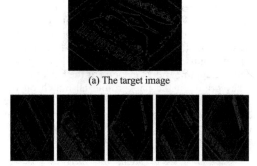

(a) The target image

(a) The target image

(b) Image retrieval result using Morphology-based SIFT algorithm

Figure 7.   Top 5 image retrieval results of our algorithm, SIFT algorithm and the algorithm in [3].

the objects are same kind. Because the details of the objects are different. But our algorithm does not focus on the details, our algorithm focus on the morphology of objects, which can depicts the kind of the objects well. Figure 7 shows the edge information of the target image and the retrieval images using Morphology-based SIFT algorithm. In Figure 7, we can see that their edge information have some similarity, and our algorithm can effectively use such similarity.

5   CONCLUSION

We propose a new algorithm which can extract a new image feature for image retrieval. This feature is about the relationship between the edge and key points, and can describe the image morphology well.

(b) Image retrieval result using Morphology-based SIFT algorithm

Figure 6.   For an image of an accordion, top 5 image retrieval results of our algorithm, SIFT algorithm and the algorithm in [3].

The feature we introduce is an alternate representation for image descriptors of the SIFT algorithm. Compared to the standard SIFT, our algorithm is more robust to texture and illumination change. The simulation results show that the image feature in our algorithm can be effectively used in the field of image retrieval, and our algorithm has a better performance than the original SIFT algorithm in the process of image retrieval.

REFERENCES

[1] D. Lowe; "Distinctive Image Features form Scale-Invarient Key points"; International Journal of Computer Vision, IJCV; Volume:60, Issue:2, 2004, pp:91–110.

[2] F.F. Li, R. Fergus and P. Perona; "Learning generative visual models from few training examples: an incremental Bayesian approach tested on 101 object categories"; Computer Vision and Pattern Recognition Workshop, CVPR, 2004, pp:178.

[3] G. Ohashi and Y. Shimodaira; "Edge-Based feature extraction method and its application to image retrieval"; Journal of Systemics, Cyberneticsand Informatics; Volume:1, Number: 5, 2003, pp:25–28.

[4] J. Canny; "A Computational Approach to Edge Detection"; IEEE Transaction on Pattern Analysis and Machine Intelligence, PAMI; Volume:8; Number:6,1986, pp:679–698.

[5] J.Z. Wang, G. Wiederhold, O. Firschein and S.X. Wei; "Content-based Image Indexing and Searching Using Daubechies Wavelets"; International Journal on Digital Libraries; Volume:1, Number:4, 1998.

[6] J. Zhao, T. Xin, G. Men; "Improved SIFT feature in image retrieval using"; International Conference on Computer Research and Development, ICCRD; Volume:2, 2011, pp:393–397.

[7] M. Banerjee, M.K. Kundu; "Edge based features for content based image retrieval"; The journal of the Pattern Recognition Society; Volume:36, Issue:11,:2003, pp:2649–2661.

[8] M. Safar, C. Shahabi and X. Sun; "Image Retrieval By Shape: A Comparative Study"; IEEE International Conference on Multimedia and Expo, ICME; Volume:1, 2000, pp:141–144.

[9] T. Ojala, M. Pietikainen and T. Maenpaa; "Multiresolution Gray-Scale and Rotation Invariant Texture Classification with Local Binary Patterns"; IEEE Transctions on Pattern Analysis and machine intelligence; Volume:24, Issue:7, 2002, pp:971–987.

[10] X.S. Zhou, T.S. Huang; "Edge-Based Structural Features for Content-Based Image Retrieval"; \ Pattern Recognition Letters; Volume:22, Issue:5, 2001, pp:457–468.

[11] Y. Ke, R. Sukthankar; "PCA-SIFT: A More Distinctive Representation for Local Image Descriptors"; Proceedings of the IEEE Computer Society Conference on Computer Vision and Pattern Recognition, CVPR; Volume:2, 2004, pp:506–51.

Control Engineering and Information Systems – Liu (Ed)
© 2015 Taylor & Francis Group, London, ISBN 978-1-138-02685-8

The design of video images segmentation system based on NiosII core

Z.P. Cui, Y.X. Qin & R. Hu
Guangxi University of Science and Technology, Liuzhou, China

ABSTRACT: Realized a system of real-time video image sampling, segmenting and processing based on NiosII core. The system achieves its main function by configuring NiosII soft core processor and related functional modules within a FPGA chip. By adopting the pure hardware structure, this design takes advantage of FPGA's high-speed parallel processing ability to realize the video sampling, segmenting and display. Experimental tests show that, the system of video segmentation has a small volume, low power consumption, strong real-time performance and strong stability.

1 INTRODUCTION

Taking advantage of current computer network platform, video monitoring system puts the dates of video and audio of monitoring area into digital format, at the same time, transmitting, storing and sharing those dates over the network, it makes safety monitoring department obtain the real-time, accurate, specific, intuitive data of monitoring site in a timely manner, and provides a new technical means for safe production supervision. In recent years, with integration rising while prices falling, power consumption lowering, FPGA, the field programmable device, has become a hot technology in the field of video signal processing because of its high-performance, high reliability and other advantages. Especially due to the high-speed parallel processing ability of FPGA, make it has an obvious advantages in the processing speed, reliability, cost, development cycle and the ability to expansion. The system of Real-time video data acquisition designed in his paper, adopts CMOS camera[1–2] and image sensor OV7620 to capture analog video data and output the digital video data stream, while processing data in the FPGA, displaying. video surveillance shows with LCD video monitor. The NiosII processor of Altera corporation[3–5] has many features, such as customizability, high reliability, low cost, usability, wide applicability etc., Designing a system of video image collection and segmentation based on Nios soft II core, has strong real-time performance, can realize multi-point and real-time video monitoring, collection and segmentation.

2 PRINCIPLE AND STRUCTURE OF THE SEGMENTATION SYSTEM

2.1 System working principle

System realizes the real-time image acquisition, segmentation, display, and other functions[6–7] by embedding NiosII CPU soft core within the FPGA chip[8]. In this design, the FPGA device selects the EP2C35F484C8, one of Cyclone II series chips of Altera Company. Video decoding chip OV7620 (CMOS color/black & white picture transducer) was selected as the analog-to-digital conversion device of video data, which supports both continuous and interlaced scanning mode, outputs VGA and QVGA two image formats with the highest pixels of 664×492 and the frame rate of 30fp8, can satisfy the requirements of the general image acquisition system; 32-bit SDRAM chip K4S641632 with high speed, low power consumption was select as image data buffer. E1709Wc, 17 inches liquid crystal display of DELL was selected to show the segmentation image. System adopted NiosII soft core processor technology can solve the clock change and many other issues well, also can be simplified the structure by putting NiosII CPU and LCD driver controller into one piece of FPGA, system flow chart as shown in Figure 1.

2.2 System structure

System structure design is divided into two parts: the first part is a multi-channel real-time image data flow input, the second part is the sorting and distribution of multi-channel image and data flow.

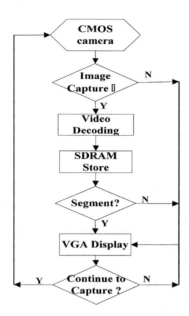

Figure 1. Flow chart of system structure.

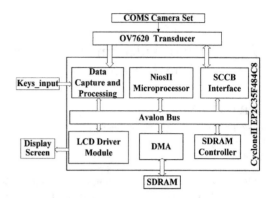

Figure 2. System structure diagram.

The real time data input in the first part adopts VHDL hardware description language to describe, for the real-time transmission of data flow. The second part is established based on NiosII. In this data processing system, chose LCD as display component, SDRAM as collecting image data storage, switches and keys as input of image analysis and control. After CMOS camera group sampling video data, the video data flow was transformed from YUV format to RGB format through image sensor OV7620. When need to load image data of a certain channel, the NiosII soft core will intercept appropriate data signal from RGB frame format by controlling switch keys, then primary color signal is intercepted separately to buffer through the FIFO, and read into SDRAM from DMA in twice, complete storage of each channel image data, after this restore the primary colors data signal via another FIFO and showed by the LCD display. The system composition block diagram is shown as Figure 2.

2.3 System hardware design

Hardware of this system consists of memory, FPGA, control keys and the LCD interface[9–10]. The memory portion includes two pieces of external SDRAM memory, image data cache in the SDRAM; The FPGA chip control to other functional modules by adding NiosII CPU cores through SOPC Builder toolbox; Video segmentation image can be displayed on the LCD through the LCD driver interface. NiosII CPU core contains CPU, SDRAM controller, DMA, FIFO, PIO, flash and

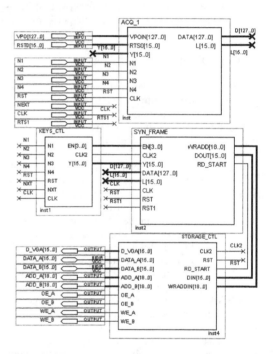

Figure 3. Image segmentation system internal structure.

other main IP core. Its structure mainly contains four modules: pressed key control module KEYS_CTL, data acquisition and extraction module ACQ_1, frame synthesis module SYN_FRAME and frame storage control module STORAGE_CTL. Hardware structure diagram is shown in Figure 3.

Keys control module KEYS_CTL sends enable signal and clock signal of different segmentation requirement for a data processing module according to different pressed keys, to realize the image segmentation number controllable and picture switching in turn. Data acquisition module ACQ_1 completes data acquisition and processing coming from 16 video channels. Each channel receive the line effective signals RTS0, pixel clock signal CLK and 8-bit video data VPO [7.. 0] from a piece of VO7620. This module is to solve three problems: data cache, data collection and extraction. Because of each channel of video data field synchronization signal is separate, a LPM_FIFO is needed to be set for each video channel to cache the video flow data, and after the valid data from each video channel being distributed to 4, 9, 16 frames under the button control and after the data from each sub-frame completely entering into cache LPM_FIFO, those data is written into external SDRAM to be caching.

The function of frame synthetic module SYN_FRAME is splicing the split-screen video frame to make a complete frame of the video signal; this process is divided into read out of FIFO and writes into SDRAM two steps. After a line effective video data of each video channel write into respective buffer FIFO, starting to display the read enable signal of FIFO from left to right sequentially. The address of external two data storage is generated by frame storage control module STORAGE_CTL, each pixel of the store address and the position of the screen are one-to-one correspondence. In order manage to process data sequentially, this system use the two SDRAM chips to work alternately, when one is inputting frame video data, another one have read out the date that synthesized from frame, after read up or write full, the system will generate an interrupt, then exchange the date of the two SDRAM chips and deal with the next frame data.

2.4 The design of system software

System software is mainly the construction of the SOPC, include the designs such as CPU, SDRAM, DMA and FIFO. First of all, add CPU, TJAG UART, SDRAM controller, Flash and DMA into the SOPC Builder, and then adding the custom-FIFO and LCD Driver, after Assigning address and interrupt, the System software is generated. NiosII Integrated Development Environment (IDE) is a basic software development tool of NiosII series embedded processors. All of the software development tasks can be completed in the NiosII IDE, involve editing, compiling, debugging and downloading. After system start, through the SDRAM to control the data storage space applied by IP core, and then initialize the system, which mainly complete the initialization of CMOS sensor chip, to ensure the position and the size of images open-window, and setup the Color/black & white working mode. While the control key of the split screen is on, startup the data acquisition

and processing module, began to collect the data flow of each video channels. When SDRAM1 is wrote full with a complete frame video image data, the system will shift to write the data of the next frame into SDRAM2, then the video data began to be read out line by line from SDRAM1 and sent to LCD interface module, after insert a timing signal, the encoded video signal will be transferred to the LCD monitor. System software is designed in the NiosII IDE development environment, based on the software generated by SOPC Builder, edit source program with C/C++ language, and then compile and debug, generate an executable file, finally Debug and run the executable program via the download cable.

3 VIDEO IMAGE SEGMENTATION SYSTEM PERFORMANCE ANALYSIS

In order to ensure the real-time performance of the system, a line of effective storage of video data must be finished before the advent of the next line. The CMOS camera adopted in this system choose the 320×320 resolution black & white mode to output, and the liquid crystal display adopted in this system is DELL LCD monitor E1709Wc with the resolution of 1280×1024. Set four split-screen (As shown in Fig. 4) display systems as an example, storing or displaying a line of effective video data needs 1280 clocks. But due to the camera output Y component of video data requires format adjustment, and the data written into the dual port RAM need to be stable to read. External SDRAM read or write address need to shift, All these factors have led to several more clocks paid in line efficient video data storage. Draw from the simulation waveform; it takes 1288 system clock cycles to store a line of effective data. The highest clock frequency of SDRAM can be up to 125 MHZ, so time needed in storing one line of effective data is: T1 = 1288/125 = 10.304 ns,

Figure 4.   Four images segmentation image.

The biggest line of video effective data output from CMOS camera sensor OV7620 is 664 bytes, invalid data byte is: 664-320 = 344 bytes, clock of output is 27 MHz, then the time needed in invalid data is: $T2 = 344/27 = 12.74$ ns. As a result of time taken by invalid data $T2$ is greater than the time needed for a line of video data storage $T1$, so a line of real-time video can store completely during the blanking period, it is fully meet the requirements of real-time video data processing.

4 SUMMARY

The processing of video image segmentation involves large amounts of data exchanging. NiosII core rely on FPGA powerful and flexible logic control function, advanced internal channel interconnection, which make it with an ultra small time delay in real-time video data processing. AS introduced in this paper, due to its high processing speed, stable and reliable performance, easy installation and flexibility features, the image segmentation system implementation scheme, which based on NiosII IP core, will be more and more widely applied in the field of image processing.

REFERENCES

A. Luk'yanitsa Image Synthesis for Auto stereoscopic systems [J]. Computational Mathematics and Modeling, 2012, 23(2):195–207.

Altera Corp., NiosII Software Developer's Handbook [M]. Altera 2005.

Altera Corp., SOPC Builder Data Sheet [M]. Alt era 2005.

Fugang Duan, Zhan Shi. The Implement of MPEG-4 Video Encoding Based on NiosII Embedded Platform [J]. Modern Applied Science, 2009, 3(9):1913–1852.

Ignacio Bravo, Javier Baliñas, Alfredo Gardel, José L Lázaro, Felipe Espinosa, Jorge García. Efficient Smart CMOS Camera Based on FPGAs Oriented to Embed Image Processing [J]. Sensors, 2011, 11(3):2282–2303.

Johan Lie, Marius Lysaker, Xue-Cheng Tai. A variant of the level set method and applications to image segmentation [J]. Mathematics of Computation, 2006, 75(255): 1155–1174.

Ning Guan, Xu Zhang, Bo Liu, Zan Dong, Bei Ju Huang, Yun Gui, Jian Qiang Han, Yuan Wang, Zan Yun Zhang, Hong Da Chen. CMOS Image Sensor with Optimal Video Sampling Scheme [J]. Science China Information Sciences, 2011, 55(6):1429–1435.

Rui Wang, Zengshuai Mi, Haihang Yu, Wei Yuan. The Design of Image Processing System Based on SOPC and OV7670 [J]. Procedia Engineering, 2011, 24:237–241.

Takashi Saegusa, Tsutomu Maruyama. An FPGA implementation of real-time K-means clustering for color images [J]. Journal of Real-Time Image Processing, 2007, 2(4):309–318.

Xie, Benju. Modeling Abstract NiosII Multiprocessor System [J]. International Journal of Advancements in Computing Technology, 2012, 4(13):294–299.

Control Engineering and Information Systems – Liu (Ed)
© 2015 Taylor & Francis Group, London, ISBN 978-1-138-02685-8

Lossless and progressive coding method for stereo images

S.G. Li
Wuhan Polytechnic University, Wuhan, Hubei, China

ABSTRACT: As an expansion of lifting scheme, Vector Lifting Scheme (VLS) provides a flexible decorrelation approach for stereo image coding by directly incorporating information of a reference image into the lifting procedure of a target image. In this paper, a distinctive VLS is proposed for lossless and progressive coding of stereo image. A feature of the proposed method is that no side information needs to transmit to receivers for reconstruction. This feature can be ascribed to two estimation procedures which take advantage of lossless coding. First, to eliminate transmission of disparity vectors, level by level disparity estimation is applied. Second, adaptive prediction filters are used to estimate parameters on the fly for VLS. The experimental results demonstrate that this approach leads to a competitive performance.

1 INTRODUCTION

Stereoscopic imaging systems are extensively applied in photogrammetry, entertainment and machine vision. In the field of digital photogrammetry, stereo image pairs are used to generate DEM or DTM. However, a mass of image data bring a challenge to image storage and transmission. Especially, how to cater to the capacity of wireless channel for those stereo sensors set on satellites is a strenuous task. Compression techniques are usually used to solve the problem. And stereo image compression technique has received many researchers' attention.

In last two decades, wavelet transform applied to image and video compression has achieved a great success. Accordingly, many research works for stereo image compression were done by this means (Jiang et al. 1999; Boulgouris & Strintzis 2002). In general, these methods can be briefed as follows. First, left image of image pair is used to predict the right image. As a result, a residue image is generated. Second, wavelet transform is applied to the left image and the residue image independently. Finally, an encoder such as EBCOT as illustrated by Taubman (1998) is used to encode those subband coefficients generated by wavelet transform. Although an improved performance is reported compared with DCT-based approaches as introduced by Perkins (1992), spatial disparity compensation violates resolution scalability. Disparity estimation and disparity compensation in wavelet domain (or subbands) have also been considered (Edirisinghe et al. 2004).

Accordingly, a novel Vector Lifting Scheme (VLS) for lossless coding and progressive archival of multispectral images is proposed by Benazza-Benyahia & Pesquet (2002). Novelty of this approach is that it combined intra-image and inter-image prediction. Following Benazza-Benyahia & Pesquet (2002), Mounir Kaaniche applied VLS to stereo image compression (Kaaniche et al. 2009). In the VLS for stereo image coding, the Disparity Compensation (DC) is incorporated into the decomposition procedure of target images. Because the information of the reference image directly joins the lifting process of target images, no residue images are generated. Experiment results indicated that VLS achieves a coding gain up to 1dB compared with the conventional lifting scheme.

Considering of the good compression performance and multiresolution of VLS, this paper focus on lossless coding for stereo images. Although the method (Kaaniche et al. 2009) can provide a lossless and progressive code-stream, a lot of side information composed of disparity vectors and prediction weights must send to receivers synchronously. This impairs the performance of VLS applied to stereo image compression. Therefore, how to eliminate the information in the case of lossless coding is a problem on which this paper focuses. A main contribution is that the proposed coding scheme does not need any side information. Two estimation procedures which take advantage of lossless coding are devoted into this feature. The first one is level by level disparity estimation. It carries out in the encoder and decoder to eliminate disparity vectors. Secondly, to eliminate the compensation parameters of VLS, the procedure of recursive least square is used to estimate them on the fly. Experimental results indicate a better performance than other methods like JPEG2000. Maybe, those lossless coding methods based on spatial prediction such as JPEG-LS lead to a better performance. However, they can't provide a progressive code-stream. Therefore, the comparison between the proposed method and them isn't made in this paper.

The rest of this paper is organized as follows. Section 2 reviews VLS for stereo image compression. In section 3, we give the detail of our distinctive VLS for lossless coding of stereo images. Finally, in section 4, experimental results are given and some conclusions are drawn in section 5.

2 VLS FOR STEREO IMAGE

Sweldens proposed a lifting scheme as an efficient implementation of Discrete Wavelet Transform (DWT) (Sweldens 1998; Daubechies & Sweldens 1998). In the lifting scheme, DWT is decomposed into a series of prediction operator P and update operator U. P is applied to even samples of a signal and the predicted values of odd samples are produced. The predicted errors are just the high frequency coefficients of DWT. And then the high-frequency coefficients are employed to update those corresponding even samples by using operator U to generate the low-frequency coefficients of wavelet transform. The lifting scheme gives DWT greater flexibility and wider range of applications.

As illustrated by Kaaniche et al. (2009), an expansion of lifting scheme VLS is proposed for stereo image coding. The VLS can be depicted as Figure 1. Like other stereo image decomposition, the conventional lifting scheme is applied to the reference image, this process can be expressed as equation (1) (2).

$$\tilde{d}_{j+1}^{(l)}\left(m_x,m_y\right)=I_j^{(l)}\left(m_x,2m_y+1\right)-\left(\mathbf{P}_j^{(l)}\right)^T\mathbf{I}_j^{(l)} \qquad (1)$$

$$\tilde{I}_{j+1}^{(l)}\left(m_x,m_y\right)=I_j^{(l)}\left(m_x,2m_y\right)+\left(\mathbf{U}_j^{(l)}\right)^T\tilde{\mathbf{d}}_{j+1}^{(l)} \qquad (2)$$

where **P** denotes the prediction vector, **U** denotes the update vector, $\mathbf{I}_j$ denotes the horizontal even samples and $\mathbf{d}_{j+1}$ denotes the produced high frequency coefficients.

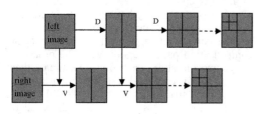

Figure 1. One level VLS for stereo image.

The difference between the traditional decomposition and VLS is the decomposition applied to the target image. In VLS, the even samples of target image and the corresponding samples of reference image are jointly used to predict the odd samples, but the conventional lifting scheme only use the even samples itself. Obviously, VLS will achieve a better performance because of the strong correlation between the left image and the right image. The better prediction performance is conducted, the smaller high-frequency coefficients are obtained. The prediction procedure in VLS can be formulated as (3).

$$\tilde{d}_{j+1}^{(r)}\left(m_x,m_y\right)=I_j^{(r)}\left(m_x,2m_y+1\right)$$
$$-\left(\left(\mathbf{P}_j^{(r)}\right)^T\mathbf{I}_j^{(r)}+\left(\mathbf{P}_j^{(r,l)}\right)^T\mathbf{I}_j^{(c)}\right) \qquad (3)$$

where $\mathbf{I}_j^c$ denotes the corresponding wavelet coefficients of the reference image resulted from Disparity Estimation (DE), $\mathbf{P}_j^r$ denotes the predict vector for the even samples itself and $\mathbf{P}_j^{r,l}$ for the reference image. $\mathbf{P}_j^{r,l}$ must be estimated by linear-regression mode or the solution of Yule-Walker equation for stable image data.

In vector lifting scheme, the update process is an issue worthy of discussion. Because the reference image information is introduced into the prediction procedure, the conventional update procedure will destroy the low-frequency components. This will definitely affect the results to the next level decomposition. This phenomenon is called update leakage effect (Kaaniche et al. 2009). An improved method is thus proposed, that is Predict-Update-Predict (P-U-P) lifting structure. First a conventional lifting process is applied and a normal low-frequency subband and a high frequency subband are obtained. And then the normal low-frequency information and the reference image information are jointly employed to predict acquired high frequency subband. As a result, smaller high frequency coefficients are obtained. Kaaniche et al. (2009) have proven that the P-U-P structure can achieve a better performance. The rest of this paper will focus on this structure. This lifting structure can be expressed as follows.

$$\tilde{d}_{j+1}^{(r)}\left(m_x,m_y\right)=I_j^{(r)}\left(m_x,2m_y+1\right)-\left(\mathbf{P}_j^{(r)}\right)^T\mathbf{I}_j^{(r)} \qquad (4)$$

$$\tilde{I}_{j+1}^{(r)}\left(m_x,m_y\right)=I_j^{(r)}\left(m_x,2m_y\right)+\left(\mathbf{U}_j^{(r)}\right)^T\tilde{\mathbf{d}}_{j+1}^{(r)} \qquad (5)$$

$$\overset{\mathrm{v}}{d}_{j+1}^{(r)}\left(m_x,m_y\right)=\tilde{d}_{j+1}^{(r)}\left(m_x,m_y\right)$$
$$-\left(\mathbf{q}_j^T\tilde{\mathbf{I}}_{j+1}^{(r)}+\left(\mathbf{P}_j^{(r,l)}\right)^T\mathbf{I}_j^{(c)}\right) \qquad (6)$$

188

3 VLS WITHOUT SIDE INFORMATION

VLS above-mentioned provides a good performance code method, of course it can be used as a lossless coding method for stereo images. However, a lot of side information composed of disparity vectors and prediction weights must send to receivers with the code-stream. This section will introduce a modified VLS without side information.

The modified VLS without side information is depicted as Figure 2. At first, to eliminate transmission of disparity vectors, level by level disparity estimation carries out during encoding. A low-frequency approximation is firstly generated for disparity estimation of current resolution. And then, adaptive prediction filters is used to estimate the filter parameters for VLS instantly. When decoding for reconstruction at the side of decoder, the same operations can repeat because no distortion exists.

3.1 Level by level disparity estimation

Because disparity vectors are not available, we should estimate disparity vectors by existing information at the side of decoder. And also, the disparity vectors generated at the side of decoder should be identical to that used in encoding processing. Thanks to lossless compression, the synchronization between encoder and decoder is possible. In addition, to get a progressive (resolution scalable) compression code-stream, we should estimate disparity vectors of the current resolution by using the previous resolution. As shown in Figure 2, to decompose $\mathbf{I}_j^{(r)}$, a conventional DWT is used to get $\mathbf{I}_{j+1}^{(r)}$ and $\mathbf{I}_{j+1}^{(l)}$. And then disparity estimation carries on between $\mathbf{I}_{j+1}^{(r)}$ and $\mathbf{I}_{j+1}^{(l)}$. The disparity vector

generated by the estimation procedure is used to decompose $\mathbf{I}_j^{(r)}$. Similarly, $\mathbf{I}_{j+1}^{(r)}$ can be decomposed. We call this iterative process Level by Level Disparity Estimation (LbLDE).

At the side decoder, we can get $\mathbf{I}_{j+1}^{(r)}$ and $\mathbf{I}_{j+1}^{(l)}$ prior to $\mathbf{I}_j^{(r)}$ because of progressive code-stream. The disparity estimation can repeat and an iterative synthesis process can implement.

3.2 Adaptive prediction filters

As introduced by Kaaniche et al. (2009), the solution of Yule-Walker equation is used as the prediction coefficients in (4)(5)(6). It's obvious that least squares should be used to get the solution for the nonstationary case. Accordingly, Recursive Least Square (RLS) is employed to estimate those prediction coefficients adaptively. An alternative is Least Mean Squared Error (LMS) adaptive algorithm. It's well-know that RLS has faster convergence speed than LMS.

RLS tries to seek prediction weight vectors to minimize the cost function.

$$J(n) = \sum_{i=0}^{n} \lambda^{n-i} \, | \, d(i) - \mathbf{w}^H(n)\mathbf{u}(i) \, |^2 \qquad (7)$$

The basis RLS algorithm can be described as follows.

1. Initialization: $\mathbf{w}(0) = 0$, $\mathbf{P}(0) = \delta\mathbf{I}$, where δ is a very small real number.
2. Update and prediction:
   $n = 1, 2 \ldots$

   $e(n) = d(n) - w^H(n-1)u(n)$

   $$k(n) = \frac{\mathbf{P}(n-1)\mathbf{u}(n)}{\lambda + \mathbf{u}^H(n)\mathbf{P}(n-1)\mathbf{u}(n)}$$

   $$\mathbf{P}(n) = \frac{1}{\lambda}[\mathbf{P}(n-1) - \mathbf{k}(n)\mathbf{u}^H(n)\mathbf{P}(n-1)]$$

   $\mathbf{w}(n) = \mathbf{w}(n-1) + \mathbf{k}(n)e * (n)$

where $\mathbf{w}(n)$ is the prediction weight vector in (6) at time instant n, namely $(\mathbf{q}_j, \mathbf{P}_j^{(r,l)})$. Correspondingly, $\mathbf{u}(n)$ is the prediction signal vector in (6), namely $(\mathbf{I}_{j+1}^{(r)}, \mathbf{I}_j^{(c)})$. The expected response d(n) is thus equivalent to $\tilde{d}_{j+1}^{(r)}$ and the prediction error e(n) is identical with $\overset{\vee}{d}_{j+1}^{(r)}$.

The scalar λ is called forgetting factor. This factor heavily influences the convergence speed and precision. Large number of experiments show that 0.999 is an appropriate value for λ.

The perfect restructure property is preserved because the same adaptation algorithm is used at the encoding and the decoding stage. Since no

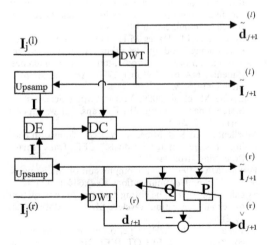

Figure 2. Modified VLS.

quantization distortion existing in $\mathbf{u}(n)$, the prediction weight $\mathbf{w}(n)$ get at the side of encoder and decoder take same values.

4 EXPERIMENTS AND RESULTS

This section gives several stereo image compression examples using the mention-above algorithm. Following Kaaniche et al. (2009), we use 5/3 wavelet and the order of adaptive filter is also 5. In the step of LbLDE, we use different block sizes for different decomposition levels. The block size of disparity estimation for the first level of decomposition is 8, the block size for the second is 4, and the block size is reduced by half in turn. The forgetting factor λ is set to 0.999.

In the following experiments, we used the MQ-coder to encode the transform coefficients, which is used in EBCOT as a part of JPEG2000. We have compared the proposed method with other two compression methods which also support progressive code-stream. One is independent encoding method using JPEG2000. We use OpenJPEG software (version 1.3) that is an open source JPEG2000 codec. Another is VLS proposed by Kaaniche et al. (2009). The average bitrate are computed as (8).

$$R = \frac{R_{total}}{2} \qquad (8)$$

where R_{total} is equal to $R^{(l)}+R^{\circledR}$, $R^{(l)}+R^{(e)}+R^{(side\,information)}$ and $R^{(l)}+R^{(e)}$ for JPEG2000, VLS proposed by Kaaniche et al. (2009) and method proposed in this paper respectively.

We have used four image pairs ("pentagon", "houseof", "fruit" and "wdc1r") as test images, which are downloaded from http://vasc.ri.cmu. edu / idb / html / stereo / index.html. Table 1 shows the lossless compression performance of three methods. It can be noticed that the average lossless compression performance of VLS is close to that of independent encoding using JPEG2000. This phenomenon may result from the side information generated by VLS consumes a little bitrate. Our proposed method results in an average gain of about 0.15 bpp over independent

encoding using JPEG2000. Compared with VLS, our proposed method also leads to an improvement of about 0.1bpp. An exception is the image of 'fruit'. Both VLS and the proposed method are inferior to the independent encoding method. This may result from few correlations between the left view image and the right view image.

5 CONCLUSION

This paper proposed a modified vector lifting schemes for lossless compression of stereo images. This modified scheme makes use of the lossless property and does not need any side information. At first, level by level disparity estimation carries out in encoder or decoder. And then, adaptive prediction filters is used to estimate the filter parameters for VLS instantly. With the adaptive methods, this method leads to a good lossless compression performance. Foremost, the code stream is progressive.

REFERENCES

Benazza-Benyahia A. & Pesquet J.-C. 2002. Vector-lifting schemes for lossless coding and progressive archival of multispectral images. IEEE Trans. Image Process 40(9):2011–2024.

Boulgouris N.V. & Strintzis M.G. 2002. A Family of Wavelet-Based Stereo Image Coders. IEEE Trans. Circuits Syst. Video Technol. 12(10):898–904.

Daubechies I. Sweldens W. 1998. Factoring wavelet transforms into lifting steps. J. Fourier Anal. Appl. 4(3):247–269.

Edirisinghe E.A., Nayan M.Y., Bez H.E. 2004. A wavelet implementation of the pioneering block-based disparity compensated predictive coding algorithm for stereo image pair compression. Signal Processing: Image Communication 19(37–46).

Jiang Q., Lee J.J., Hayes M.H. 1999. A wavelet based stereo image coding algorithm. Acoustics, speech, and signal processing; Proc. International Conference, Phoenix, 684–696 October 1999. Arizona: Arizona State University.

Kaaniche M. et al. 2009. Vector Lifting Schemes for Stereo Image Coding. IEEE Trans. Image Process 18(11):2463–2475.

Moellenhoff M.S. & Maier M.W. 1998. Transform coding of stereo image residuals. IEEE Trans. Image Processing 7(6):804–812.

Perkins M.G. 1992. Data Compression of Stereo pairs. IEEE Trans. Communications 40(4):684–696.

Sweldens W. 1998. The lifting scheme: A construction of second generation wavelets. SIAM J. Math. Anal. 29(2):511–546.

Taubman D.S. 2000. High performance scalable image compression with EBCOT. IEEE Trans. Image Proc. 9:(1158–1170).

Table 1. Performance of lossless compression in terms of average bitrate (in bpp).

Image	VLS	JPEG 2000	Proposed
Pentagon	5.04	5.05	4.90
Houseof	5.24	5.24	5.11
Fruit	3.71	3.67	3.68
wdc1r	3.49	3.54	3.38

Target detection and decision fusion of multi-band SAR images

X.J. Liu & X.D. Wang
Radar Teaching and Research Department, Navy Submarine Academy, China

B. Chen
Naval Academy of Armament, China

ABSTRACT: CFAR detection and decision fusion is proposed in this paper to improve multi-band SAR images target detection performance Experiments on real images prove the advantage of this method.

1 INTRODUCTION

Different band SARs (Synthetic aperture radar) show different characters (Chen & Zhang. 2011). Low band SAR is of highly penetrability and can detect hidden targets in the wood or under the ground. High band SAR can get clear profile and more detail. Ghasem and Luo zhi-quan propose self-adapted decision fusion of the distributed detecting system (Mirjalily & Luo. 2003). Martin and Mark propose decision fusion of the multi image based on Boole fusion criterion (E. Liggins II & Nebrich. 2000). But all of them don't apply the decision fusion to the real multi-band SAR images. How to use multi-band SAR images to get as much information as possible of the targets and improve the detection performance is an important problem to be solved.

2 ALGORITHM ANALYSIS

This paper proposes decision fusion based on Neyman-Pearson criterion of multi-band SAR image target detection. Theoretic analysis and experiment on real images all shows that this method not only can fuse the information of multi-band SAR image to get accurate detection result, but also can improve the target detection probability greatly.

The flow chart of decision fusion of multi-band SAR target detection result based on Neyman-Pearson criterion is as Figure 1, detailed steps are as following, firstly, apply CFAR (Constant false alarm rate) detection on multi-band SAR images (Lv. 2010), then co-registrate the detection result, finally use Neyman-Pearson criterion to fuse the detection result on decision level (Xiang & Zhao. 2001).

In the flow chart, u_i(i = 1,2, ... N) denotes the binary hypothesis testing result of the detectors of every band SAR images, $u_i = 1$ denotes that the

detection judgment is H_1 (target present); $u_i = 0$ denotes that the detection judgment is H_0 (target absent). Use Neyman-Pearson criterion to fuse the result of the detectors on the decision level, get the fused detection result u_0 (Xu & Jian. 2011).

Let $\delta(u)$ denote the probability of the fusion judgment to be H_1, $\Lambda(u)$ denote the likelihood ratio, the decision rule is as following,

$$\delta(u) = \begin{cases} 1 & \Lambda(u) > t \\ \gamma & \Lambda(u) = t, \quad \Lambda(u) = \dfrac{p(u \mid H_1)}{p(u \mid H_0)} \\ 0 & \Lambda(u) < t \end{cases} \tag{1}$$

Randomizing constant γ and fusion threshold t are determined by the false alarm probability. Let P_{di} and P_{fi} denote the detection probability and false alarm probability of the i-th detector, P_d^f and P_f^f denote the detection probability and false alarm probability after the fusion, so,

$$p(u \mid H_1) = \prod_{i=1}^{n} (P_{di})^{u_i} (1 - P_{di})^{(1-u_i)} \tag{2}$$

$$p(u \mid H_0) = \prod_{i=1}^{n} (P_{fi})^{u_i} (1 - P_{fi})^{(1-u_i)} \tag{3}$$

$$\Lambda(u) = \frac{p(u \mid H_1)}{p(u \mid H_0)} = \prod_{i=1}^{n} \left(\frac{P_{di}}{P_{fi}} \right)^{u_i} \left(\frac{1 - P_{di}}{1 - P_{fi}} \right)^{(1-u_i)} \tag{4}$$

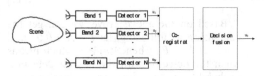

Figure 1. Flow chart of decision fusion of multi-band SAR.

So P_d^f and P_f^f can be written as following,

$$P_d^f = \sum_{\Lambda(u)>t} p(u|H_1) + \gamma \sum_{\Lambda(u)=t} p(u|H_1) \quad (5)$$

$$P_f^f = \sum_{\Lambda(u)>t} p(u|H_0) + \gamma \sum_{\Lambda(u)=t} p(u|H_0) \quad (6)$$

Here the false alarm probability of every band SAR images and false alarm probability after the fusion are all the same, thus $P_f^f = P_{fi} = P_f$, so,

$$P_f^f = \sum_{\Lambda(u)>t} p(u|H_0) + \gamma \sum_{\Lambda(u)=t} p(u|H_0) = P_f \quad (7)$$

In the following, decision fusion based on Neyman-Pearson criterion is applied to three band SAR images. Let $P_{d1}>P_{d2}>P_{d3}$ and $P_{di}>P_{fi}$. The γ and t are computed on three conditions, at the same time proves the detection probability is improved.

1. When $P_{d1}/1 - P_{d1}<P_{d2} P_{d3}/(1 - P_{d2})(1 - P_{d3})$, we can get the sequence of $\Lambda(u)$ as following, $\Lambda(0,0,0)<\Lambda(0,0,1)<\Lambda(0,1,0)<\Lambda(1,0,0)<\Lambda(0,1,1)$ $<\Lambda(1,0,1)<\Lambda(1,1,0)<\Lambda(1,1,1)$, let $t = \Lambda(1,0,0)$, by computing the following equation (7), we can get,

$$\gamma = \frac{1-2P_f}{1-P_f} \quad (8)$$

So,

$$P_d^f = \sum_{\Lambda(u)>t} p(u|H_1) + \gamma \sum_{\Lambda(u)=t} p(u|H_1)$$
$$= P_{d1} + [(1-P_{d1})P_{d2}P_{d3} - (1-\gamma)$$
$$P_{d1}(1-P_{d2})(1-P_{d3})] \quad (9)$$

Since $P_{d1}/1 - P_{d1} < P_{d2} P_{d3}/(1 - P_{d2})(1 - P_{d3})$, so,

$$(1-P_{d1})P_{d2}P_{d3} > P_{d1}(1-P_{d2})(1-P_{d3})$$
$$> (1-\gamma)P_{d1}(1-P_{d2})(1-P_{d3}) \quad (10)$$

Then,

$$P_d^f = P_{d1} + \begin{bmatrix} (1-P_{d1})P_{d2}P_{d3} \\ -(1-\gamma)P_{d1}(1-P_{d2})(1-P_{d3}) \end{bmatrix} > P_{d1}$$
$$(11)$$

Based on the above prove, we can get the conclusion as following, let $t = \Lambda(1,0,0)$ and $\gamma = 1-2P_f/1 - P_f$, when $\Lambda(u) \geq t$, which means when the detector with the highest detection probability or two detectors at least decides target present, the fusion result decides target present and $P_d^f > \max (Pd_1,Pd_2,Pd_3)$.

2. When $P_{d1}/1-P_{d1}>P_{d2} \quad P_{d3}/(1-P_{d2})(1-P_{d3})$, we can get the sequence of $\Lambda(u)$ as following, $\Lambda(0,0,0)<\Lambda(0,0,1)<\Lambda(0,1,0)<\Lambda(0,1,1)<\Lambda(1,0,0)$ $<\Lambda(1,0,1)<\Lambda(1,1,0)<\Lambda(1,1,1)$, let $t = \Lambda(1,0,0)$, by computing the equation (7), we can get:

$$\gamma = 1 \quad (12)$$

Submit equation (12) into the equation (5), we can get:

$$P_d^f = \sum_{\Lambda(u)>t} p(u|H_1) + \gamma \sum_{\Lambda(u)=t} p(u|H_1) = P_{d1} \quad (13)$$

Based on the above prove, we can get the conclusion as following, let $t = \Lambda(1,0,0)$ and $\gamma = 1$, when $\Lambda(u) \geq t$, which means when the detector with the highest detection probability or two detectors (the two detectors with lower detection probability except) at least decides target present, the fusion result decides target present and $P_d^f = \max(P_{d1},P_{d2},P_{d3})$.

3. When $P_{d1}/1 - P_{d1} = P_{d2} P_{d3}/(1 - P_{d2})(1 - P_{d3})$, we can get the sequence of $\Lambda(u)$ as following, $\Lambda(0,0,0)<\Lambda(0,0,1)<\Lambda(0,1,0)<\Lambda(0,1,1) = \Lambda(1,0,0)<$ $\Lambda(1,0,1)<\Lambda(1,1,0)<\Lambda(1,1,1)$, let $t = \Lambda(1,0,0)$, by computing the equation (7), we can get:

$$\gamma = 1 \quad (14)$$

Submit equation (14) into the equation (5), we can get:

$$P_d^f = \sum_{\Lambda(u)>t} p(u|H_1) + \gamma \sum_{\Lambda(u)=t} p(u|H_1) = P_{d1} \quad (15)$$

Based on the above prove, we can get the conclusion as following, let $t = \Lambda(1,0,0)$ and $\gamma = 1$, when $\Lambda(u) \geq t$, which means when the detector with the highest detection probability or two detectors at least decides target present, the fusion result decides target present and $P_d^f = \max(P_{d1},P_{d2},P_{d3})$.

3 EXPERIMENT ON REAL IMAGES

To verify the validity of the decision fusion based on Neyman-Pearson criterion, we choose three (P, X and L) band SAR images of the same scene to experiment, as Figures 2, 3 and 4. The CFAR detection result of them when the $P_f = 10^{-6}$ is shown by Figures 5, 6 and 7. Apply decision fusion based on Neyman-Pearson criterion to Figures 5, 6 and 7 when the $P_f = 10^{-6}$, we get the fusion image Figure 8.

Figure 2.　P band SAR image.

Figure 3.　L band SAR image.

Figure 4.　X band SAR image.

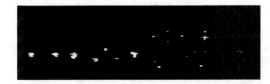

Figure 5.　P band SAR image detection result.

Figure 6.　L band SAR image detection result.

Figure 7.　X band SAR image detection result.

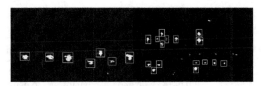

Figure 8.　Decision fusion result.

Figure 9.　One part of P band SAR image detection result.

Figure 10.　One part of L band SAR image detection result.

Figure 11.　One part of X band SAR image detection result.

Figure 12.　One part of decision fusion result.

In Figure 8, not only all the targets but also the posture and distribution are shown in one image by the decision fusion, so we can say that we can get more comprehensive and practical information by decision fusion.

The improvement of the detection probability is proved by comparison of the detection probability before and after the fusion. Figures 9, 10 and 11 is one part of P, X and L band SAR image detection result of the same area respectively; Figure 12 is one part of the decision fusion result of the same area.

Table 1 shows the detection probability of corresponding 3 pixels (gotten by co-registration) in the P, X and L band SAR image detection result in row 2 to 4, and the decision fusion result in row 5.

193

Table 1. Comparison of the detection probability before and after the fusion ($P_f = 10^{-6}$).

Pixel num	P band	L band	X band	Fusion
1	0.9659	0.9617	0.9548	0.9659
2	0.9610	0.9603	0.9401	0.9603
3	0.9716	0.9722	0.9670	0.9801

By comparison of the detection probability before and after the fusion, we can get the conclusion that the detection probability is improved greatly by the decision fusion based on Neyman-Pearson criterion.

4 CONCLUSION

The decision fusion of detection based on Neyman-Pearson criterion is proposed and experimented on 3 band SAR images. The comparison before and after the fusion proves that this method not only can comprehensively fuse the information of every source images and get one image concluding all of the targets with posture and distribution, but also can improve the detection performance greatly. Just 3 band SAR images are study here, the experiment on SAR images of more than 3 bands is carried on in progress.

REFERENCES

Cong-an Xu, Tao Jian. 2011. An improved VI-CFAR detector, *Signal Processing*(27): 926–931.

Ghasem Mirjalily, Zhi-quan Luo. 2003. Blind adaptive decision fusion for distributed detection. *IEEE Trans. on Aerospace and Electronic Systems*(39):34–52.

Hua-jie Chen, Yu Zhang. 2011. Target coverd by camouflage netting detection based on multi-band and polarization SAR image fusion. *Opto-Electronic Engineering*(38): 106–118.

Jiu-ming Lv. 2010. Study of N-P principle and CA-CFAR detection technology. *Ship Electronic Engeering*(30)18–121.

Martin E. Liggins II, Mark A, Nebrich. 2000. Adaptive Multi-Image Decision Fusion, In Signal Processing, Sensor Fusion, and Target Recognition. *Proceedings of SPIE*(4052): 218–228.

Ming Xiang, junwei Zhao. 2001, On the Performance of Distributed Neyman-Pearson Detection Systems. *IEEE Trans. on Systems, Man And Cybernetics*(31): 2001.

Control Engineering and Information Systems – Liu (Ed)
© 2015 Taylor & Francis Group, London, ISBN 978-1-138-02685-8

Fast SPIHT for satellite spectral image compression

C.X. Zhong
Software College, Jiangxi Normal University, Nanchang, China

ABSTRACT: With the fast development of aviation and spaceflight technology, the image coding method based on wavelet transform has been widely used in spectral image compression. A successful example is an improvement to SPIHT was used to compress multi-spectral images. Although this improvement decreases the requirement for space and time, it still contains a great deal of redundant search, greatly reducing encoding speed. Hence, the authors of this paper proposed an optimized fast search algorithm, and applied it to the improved SPIHT algorithm for multi-spectral image compression, greatly reducing redundant search and improving coding speed.

1 INTRODUCTION

With the fast development of aviation and spaceflight technology, spectral imaging technology has achieved wide application. Spectral image can obtain simultaneously the spatial and spectral information of the observed object. But with the increase of the spatial resolution of imaging spectrometer and the number of wave bands, the amount of produced data has also become increasingly huge, bringing great difficulty for the image storage and transmission of the satellite system. Hence, it is necessary to compress multispectral image, and more and more attention from home and abroad are being paid to the compression of multispectral image [6]. The image coding method based on wavelet transform has become the first choice of coding methods for multispectral image compression. SPIHT [1,5] as a state-of-the-art algorithm for wavelet-based image compression has especially wide applications in spectral image compression. For example, Chen [4] improved SPIHT and applied the improved algorithm to multispectral image compression. Although his improvement decreases SPIHT's requirement for space and time, and makes SPIHT a more appropriate algorithm for hardware implementation, the improved SPIHT still contains a great deal of redundant search, having not fully improved coding speed. Hence, the authors of this paper present an optimized fast search algorithm, which only requires scanning backward the wavelet decomposition matrix once to decide the significance of all D(i,j) and L(i,j), making hereafter coding operations relying on the significance of D(i,j) and L(i,j) need only to look up the table storing the significance of all D(i, j) and L(i, j), thus greatly reducing redundant search and improving coding speed.

Using this optimized fast search algorithm, we can improve SPIHT to get faster coding speed. Hence it is meaningful to introduce this optimized fast search algorithm and the corresponding improved SPIHT algorithm.

2 AN IMPROVEMENT OF SPIHT FOR MULTISPECTRAL IMAGE COMPRESSION

An analysis on SPIHT algorithm shows that during the coding of SPIHT, each time when a significant coefficient is scanned, it is moved into LSP, then in each refinement pass, a most significant bit of the coefficient is output. After several scans, the coding is terminated, but usually only several previous bits of each significant coefficient had been output. For a given compression ratio, there are always some bits at the back abandoned, these abandoned bits are called scan error [2]. In fact, the bit number of scan error (calling it *ebnum*) can be determined before scan, so during the coding, when a significant coefficient is scanned, all of it previous (n+1-ebnum) bits are output once and for all. Thus it isn't necessary to move the coefficient's coordinate into LSP, saving an array (LSP) and reducing time and space consumption.

Generally, for a specific image, given a compression ratio *Rate*, a corresponding optimal bit number of scan error can be determined. The detail algorithm is as follows:

1. Set an initial threshold $T_0 = 2$, and the number of wavelet coefficients less than T_0 be N = 0.
2. Estimate the bit rate using the following formula:
   $rates = 8*(nRow*nColumn-N)/(nRow*nColumn)$

Comparing *rates* with the given bit rate *Rate*:
if *rates* ≥ *Rate* then go to (3) else go to (4).
3. Double the thresholds: $T_0 = 2*T_0$, and calculate the number (N) of wavelet coefficients less than T_0, go to (2).
4. Terminate the loop, and obtain the bit-number of scan error: $ebnum = [\log_2(T_0)]$.

In the original SPIHT algorithm for image coding, after a decomposition to the image, a majority of energy are distributed in lower subbands, but the coefficients in the higher subbands are usually smaller than the threshold and the number of these coefficients is very large, so in the last stages of scan, both array LIP and LIS expands very quickly and can not be reduced efficiently, causing much waste of time and space. Hence Chen [4] proposed an improved SPIHT algorithm for multispectral image, calling it Algorithm 1. In his improved algorithm, a flag matrix *Flag* of the same size as wavelet coefficient matrix is defined as follows before coding scan:

If the absolute value of a wavelet coefficient is smaller than 2^{ebnum}, set $Flag(i, j) = 0$, otherwise set $Flag(i, j) = 1$. Consequently, the encoder can determine whether the coefficient should be added to LIP by judging the corresponding flag bit.

Through an analysis of the above improved algorithm, we can see that although the algorithm has reduced SPIHT's requirement for space and time, making SPIHT a more appropriate algorithm for hardware implementation, the improved SPIHT still contains a great deal of redundant search, having not fully improved coding speed. Hence, the authors of this paper find a fast search algorithm, which only requires scanning backward the wavelet decomposition matrix once to decide the significance of all D(i, j) and L(i, j) needed by SPIHT's coding passes. This fast search algorithm can be described as follows:

Algorithm 2 Fast search algorithm for SPIHT
Input: A *K*-level wavelet decomposition matrix $(c_{i,j})_{0 \le i,j < N}$ of an image ($N = 2^m$, $K \le m$)
Output: $DM(N, N)$ as a table storing the maximal absolute value among descendants of every node (x, y), and $LM(N, N)$ as a table storing the maximal absolute value among indirect descendants of every node (x, y)
Begin
    for $l \leftarrow 1$ to K do   /*process level 1 to level K iteratively */
    for $orient \leftarrow 1$ to 3 do   /*the orientation number of HL_l, LH_l, HH_l are 1,2,3 respectively*/
/*process each group of four coefficients in the subband with orientation number *orient* on level l, this subband has $2^{l-1}N \times 2^{l-1}N$ such groups */

$u = 2^{-l}N$; $v = u/2$;
$c_1 = u \times [orient/2]$; $c_2 = u \times (orient-2 \times [orient/2])$;
for $x \leftarrow 1$ to v do
for $y \leftarrow 1$ to v do
1. Compute the coordinates of the four coefficients of each group:
$x_1 = 2x-1$; $y_1 = 2y-1$; $i_1 = x_1+c_1$; $j_1 = y_1+c_2$;
$x_2 = 2x-1$; $y_2 = 2y$; $i_2 = x_2+c_1$; $j_2 = y_2+c_2$;
$x_3 = 2x$; $y_3 = 2y-1$; $i_3 = x_3+c_1$; $j_3 = y_3+c_2$;
$x_4 = 2x$; $y_4 = 2y$; $i_4 = x_4+c_1$; $j_4 = y_4+c_2$;
Compute the coordinates of their parent coefficient: $ip = x+ c_1/2$; $jp = y+ c_2/2$;
2. Compute the maximal absolute value among the four coefficients:
$Mx = \max\{|c_{i1, j1}|, |c_{i2, j2}|, |c_{i3, j3}|, |c_{i4, j4}|\}$;
3. If $l > 1$ then
    3.1. find also the maximal absolute value among the descendants of the four coefficients:
        $My = \max\{D(i1, j1), D(i2, j2), D(i3, j3), D(i4, j4)\}$;
    3.2. set My as the maximal absolute value among indirect descendants of the parent of the four coefficients:
        $LM(ip, jp) = My$;
    3.3. find the maximal value between Mx and My:
        $Mx = \max\{Mx, My\}$
4. Set Mx as the maximal absolute value among descendants of the parent of the four coefficients:
$DM(ip, jp) = Mx$
end for y end for x
end for *orient* end for l
End

If the Algorithm 2 is applied to Algorithm 1, we can obtain a faster SPIHT for multispectral image compression, which is an improvement of a SPIHT for multispectral image compression. This improved algorithm can be described as follows:
Algorithm 3 Fast SPIHT for multispectral image compression

1. Initialization
    1.1. Call Algorithm 2 (i.e. fast search algorithm) to scan the wavelet matrix to determine the maximal absolute value among descendants and the maximal absolute value among indirect descendants of every node (x, y) in the matrix, and store these maximal values in two tables DM(N, N) and L(N, N).
    1.2. Compute the number of passes
        $n = \log_2(\max_{0 \le i,j < N}\{|c_{i,j}|\})$
    Set a list of insignificant coefficients as follows:
    LIP = $\{(i,j)|(i,j) \in LL_K \cup HL_K \cup LH_K \cup HH_K\}$
    Set a list of insignificant descendant set as follows:
    LIS = $\{(i,j)D|(i,j) \in HL_K \cup LH_K \cup HH_K\}$

1.3. Compute a flag matrix as follows
   If $|c_{i,j}| \geq 2^{ebnum}$ then $Flag(i,j) = 1$
   else $Flag(i,j) = 0$
2. Sorting pass
   2.1. for each entry (i,j) in the LIP do:
      (2.1.1) output $S_n(i,j)$;
      (2.1.2) If $S_n(i,j) = 1$ then output the first $(n+1\text{-}ebnum)$ bits of $c_{i,j}$ once and for all, and also output the sign bit of $c_{i,j}$.
   2.2. for each entry (i,j) in the LIS do:
      (2.2.1) if the entry is of type D then
      a. If $DM(i, j) \geq 2^n$ then output "1" else output "0";
      b. If $DM(i, j) \geq 2^n$ then for each $(k,l) \in O(i,j)$ do:
         • If $|c_{k,l}| \geq 2^n$ then output the first $(n+1\text{-}ebnum)$bits of $c_{k,l}$ once and for all, and output the sign bit of $c_{k,l}$;
         • If $|c_{k,l}| < 2^n$ then judge whether $Flag(k,l)$ is equal to 1,if yes then add (k,l) to the end of LIP.
      c. If $L(i,j) \neq \Phi$ then move (i,j) to the end of LIS, as an entry of type L, and go to step (2.2.2); else, remove entry (i,j) from LIS.
      (2.2.2) if the entry is of type L then
      a. If $LM(i,j) \geq 2^n$ then output "1" else output "0";
      b. If $LM(i,j) \geq 2^n$ then add each $(k,l) \in O(i,j)$ to the end of LIS as an entry of type D; else, remove (i,j) from LIS.
3. Preparation for the next encoding pass
   n is reduced by 1 and go to step (2).

3 EXPERIMENT RESULT

In order to compare Algorithm 1 with Algorithm 3, we used MatLab to implement the two algorithms. We used some images transmitted back by Changebenyue 2 (Danieel_512 × 512) to do the compression experiment, and select the number of wavelet decomposition to be 6 and wavelet filter to be bior4.4. The coding time ratios (defined as

Table 1. Performance comparison between Algorithm 1 and Algorithm 3.

Image size	Bit rate	Coding time ratios	PSNR0 (db)	PSNR1 (db)
512 × 512	0.1	2.73	24.4233	24.4233
512 × 512	0.2	2.66	26.6386	26.6386
512 × 51212	0.3	2.44	28.6964	28.6964
512 × 512	0.4	2.17	30.1545	30.1545
512 × 512	0.5	2.33	30.5172	30.5172

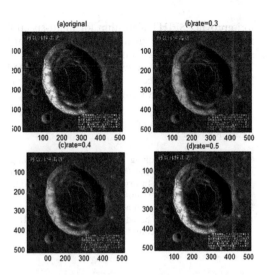

Figure 1. The quality of reconstructed images of the improved SPIHT.

"the execution time of the original algorithm/the execution time of the improved algorithm") and PSNRs of reconstructed images of the original algorithm (denoted as PSNR0) as well as PSNRs of reconstructed images of the improved algorithm (denoted as PSNR1) are listed in Table 1.

Table 1 shows that the improved algorithm's encoding speed is twice more than the original algorithm's encoding speed, especially for lower bit rates. Figure 1 shows that the quality of reconstructed images of the improved algorithm is completely the same as that of the original algorithm. Hence, the improved algorithm improves the encoding speed greatly without loss of reconstruction quality.

4 CONCLUSIONS

Since SPIHT has many advantages, wavelet-based image coding method has become the first choice of coding methods for multispectral image compression. As a state-of-the-art algorithm for wavelet-based image compression, SPIHT also has wider application in the compression of spectral images. A typical successful example is that an improvement to SPIHT was used to compress multi-spectral images. Although this improvement decreases the requirement for space and time, and makes SPIHT a more appropriate algorithm for hardware implementation, it still contains a great deal of redundant search, having not fully improved coding speed. In order to solve this problem, the authors of this paper proposed an optimized fast search algorithm, and applied it to improving SPIHT algorithm for multi-spectral image compression, greatly reducing redundant search and improving coding speed.

Experiment data show that the optimized SPIHT for multi-spectral image compression is really faster than the original SPIHT.

REFERENCES

Cuixiang Zhong, Guoqiang Han and Minghe Huang. An Improvement to Set Partitioning in Hierarchical Trees [J]. Journal of Information and Computational Science, 2007, 4(1):397–404.

Ren-Jie Chen, Xue-bin Liu and Bin Hu. Multi-spectral image compression based on an improved SPIHT [J]. Journal of computer applications, 25(12) (2005):255–259.

Said A. and W.A. Pearlman. A new, fast and efficient image code based on set partitioning in hierarchical trees [J]. IEEE Transactions on Circuits and Systems for Video technology, 1996, 6(6):243–250.

Said A. and W.A. Pearlman. An Image multiresolution representation for lossless and lossy compression [J]. IEEE Transaction on Image Processing, 1996, 5(9):1303–1310.

Yan-kui Sun. *Wavelet Analysis and its Applications* [M]. Tsinghua University Press, Beijing, 2005 (In Chinese).

Zhou Youxi, Li Yunsong, Wu Chengke. Environmental Satellite Multispectral Images Compression Algorithm [J]. Acta Optica Sinica, 2006,26(3):336–340.

Control Engineering and Information Systems – Liu (Ed)
© 2015 Taylor & Francis Group, London, ISBN 978-1-138-02685-8

An advanced human motion image tracking algorithm of Gaussian model

H. Long & N. Huo
Xuzhou College of Industrial Technology, Xuzhou, Jiangsu, China

ABSTRACT: The paper studied the tracking optimization of human image sequence and improves the accuracy of the track. In the image sequence frames of the human motion, because the background image frame change faster and the differences between the pixels frame are large, the traditional method of tracking can't adapt to the situation that the extraction of dynamic pixel is too few which is caused by rapid, fast image background frame changes. It will lead to misjudgment of the tracking foreground, lag of the tracking effect and inaccuracy. In order to solve this problem, this paper puts forward a Gaussian mixture model based on the human motion tracking algorithm. It is performed by establishing a Gaussian mixture model simulation to remove interference, and eliminate the impact caused by rapid changes by updating the iterative dynamic pixel parameters to solve the tracking lag problems. The experimental results show that the method can greatly improve the human motion accuracy in complex background.

1 INTRODUCTION

Traditional tracking method based on image frame difference has certain defects. When applied to the tracking scenario with rapid background pixels changes, the algorithm is to detect tracking feature based on image pixel difference. Once the background differs greatly or change too fast, it can lead to the result that the variance calculated according to frame differential is bigger, the dynamic pixel is missing and distortion of tracking model are serious. Then tracking always results in the misjudgment of the foreground and inaccuracy of the human motion tracking[1][2][3]. Therefore, accurate tracking, under the condition of rapidly changing scene or lack of dynamic pixel, has always been a difficult problem in its field.

In order to solve tracking problem caused by the rapid changing background, an improved sequence tracking method, Gaussian mixture model, is proposed. It is performed by establishing a Gaussian mixture model simulation to remove interference, and eliminate the foreground misjudgment by updating the iterative dynamic pixel parameters, solving the tracking lag problem and greatly improving the accuracy rate.

2 THE TRACKING PRINCIPLE OF HUMAN MOTION IMAGE

2.1 Human motion tracking based on frame difference

The current popular tracking method of human image motion is to extract human dynamic features based on pixel frame difference to realize the effective tracking. The specific methods are as follows[4][5]:

First of all, calculate the frames difference of the background image:

$$p = \sum_{j=1}^{n} T_j \Big/ \sum_{j=1}^{n} g_j - \sum_{i=1}^{n} T_i \Big/ \sum_{i=1}^{n} g_i \qquad (1)$$

In the above formula, p is pixels calculated based on the effective frame difference between the adjacent frames, T is gray-scale feature, and g is the sum of pixels features.

Step 2: Build a dynamic tracking model according to the pixels of interframe features. The method is as follows:

$$h = \sum_{j=2}^{n} \left(p_j, \dots p_0 \right)^2 \Big/ n \qquad (2)$$

Thereinto, H is dynamic feature model of human motion tracking.

Step 3: Utilize the dynamic feature model to perform the dynamic tracking of image sequence as follows:

$$s(x,y) = n\frac{P_{k-1}}{h(t)} + \bar{c}_t \times h(t) \qquad (3)$$

In the above formula, $\bar{c}_t$ is switching pixel ratio of the background, symbolizing the background changing speed. $s(x,y)$ is the actual calculated tracking result, expressed with two-dimensional coordinate.

In order to measure the accuracy of tracking, the formula of lag rate of the human motion tracking is given as follows:

$$v = \frac{s_i(x,y)}{s_j(x,y)} = \frac{n\frac{P_{i-1}}{h(t_i)} + h(t_i)}{n\frac{P_{j-1}}{h(t_j)} + h(t_j)} \cdot \overline{c_t} \qquad (4)$$

Thereinto, $s_i(x,y)$ is the actually calculated result while $s_{ji}(x,y)$ is the standard tracking results. V is to symbolize the accuracy of the tracking. The larger gap between V value and 1, the worse the tracking effect is.

2.2 Defects in traditional method

It is known from the above principle that there is a drawback in the traditional moving target tracking method. When the method is applied under the condition of relatively complex background or faster image frames switch, it is not guaranteed every frame difference contains complete human motion features. In the event of feature missing, dynamic pixel number in formula 1 frame will reduce quickly, which leads to the results inaccuracy of tracking model built on basis of dynamic pixel in formula 2. Then the existing calculation error in formula 3 will be serious, and the pixel is mixed together with that of the previous frame, causing foreground misjudgment. In formula 4, the deviation of numerator and denominator is bigger, leading to a tracking lag and larger tracking deviation[6][7].

3 IMPROVEMENT OF HUMAN MOTION TRACKING OF GAUSSIAN MODEL

3.1 Build the Gaussian mixture model

Traditional tracking model can not avoid the above-mentioned weak points. In order to remove the tracking interference under fast changing background, a Gaussian mixture model is built to deal with the rapid changing tracking background. The regular pattern of background distribution is shown bellow:

$$f(x) = \sum_{i=1}^{k} \omega_{i,t} \eta(x,\mu_{i,t},\Sigma_{i,t})$$

Within it, $\mu_{i,t}$ is mean value, $\Sigma_{i,t}$ is covariance matrix, $\omega_{i,t}$ is the weight of each Gaussian distribution and $\sum_{i=1}^{k} \omega_{i,t} = 1$. $\eta(x,\mu_{i,t},\Sigma_{i,t})$ is to express the i th mixture Gaussian distribution at the time of t, then:

$$\eta(x,\mu_{i,t},\Sigma_{i,t}) = \frac{1}{(2\pi)^{\frac{n}{2}}\left|\Sigma_{i,t}\right|^{\frac{1}{2}}} e^{-\frac{1}{2}(x_t-\mu_{i,t})'\Sigma_{i,t}^{-1}(x_t-\mu_{i,t})}$$

In order to reduce the amount of calculation and improve real-time performance of the system, the value of covariance matrix is simplified to:

$$\Sigma_{i,t} = \sigma_k^2 I$$

In the above formula, σ is the standard deviation of the Gaussian mixture distribution and I is unit matrix.

Through the above assumption, it can reduce the amount of calculation and improve the system's real-time performance, but meanwhile, it also reduces the accuracy of the algorithm. New observed value x_t and $k(1 \le k \le K)$ Gaussian mixture distribution sorted by priority ($p = \omega/\sigma$, where is gaussian weight distribution, σ is standard deviation) are matched and the matching formula is $|x_t - \mu_{i,t-1}| < 2.5\sigma_i$. Under the condition of unmatched mixture Gaussian distribution, when $k < K$, add a new Gaussian distribution; when $k = K$, use the new Gaussian distribution instead of the one with the minimum priority. New Gaussian distribution sets x_t as the mean value, initializing a larger variance and a low weight. Gaussian model weight is updated as follows:

$$\omega_{i,t} = (1-\alpha)\omega_{i,t-1} + \alpha M_{i,t}$$

When matching, $M_{i,t}$ is 1, otherwise $M_{i,t}$ is 0. The matching Gaussian distribution updates the mean value and variance with the following formula. The unmatched remain unchanged.

$$\mu_{i,t} = (1-\beta)\mu_{i,t-1} + \beta x_t \qquad (5)$$

$$\sigma_{i,t}^2 = (1-\beta)\sigma_{i,t-1}^2 + \beta(x_t - \mu_{i,t})^T (x_t - \mu_{i,t}) \qquad (6)$$

α is the learning rate, a fixed value. The relationship between α and β meets the following formula:

$$\beta = \alpha\eta(x_t | \mu_{i,t},\sigma_{i,t}) \qquad (7)$$

α is empirical value ranging from 0.04 to 0.05. Then the background model is generated. In mixed Gaussian distribution, not all models correspond with the background. To determine whether it belongs to the background or detected target needs to further process these Gaussian models. The greater the priority $p = \omega/\sigma$ is, the greater possibility of being background. The smaller the priority, the greater possibility of being detected target. In multi-Gaussian model, each model is ranked from the largest priority to the smallest one. Then, B Gaussian model are selected as background. The B value is obtained from the following formula:

$$B = \arg\min_b \left(\sum_{k=1}^{b} p > T \right) \quad (8)$$

The choice of T is very important; if T is smaller, the background may only be performed by a Gaussian mixture distribution, then the Gaussian mixture distribution degrades to single Gaussian distribution; if T is larger, the detected target will be mistakenly considered as background, impacting the detection result; when the T value is appropriate with a plurality of background models, it is possible to deal with the interference caused by image flicker and rapid changes.

3.2 Phenomenon of foreground misjudgment

Traditional Gaussian mixture model algorithm can not avoid the pixels missing caused by rapid changes, so it will cause foreground misjudgment and tracking lag. The foreground misjudgment means the foreground pixels are considered as background pixels. Along with the motion of the object, this phenomenon is becoming more serious. From formula 6, it is known that the increase and decrease of variance are mainly determined by the absolute value of the difference between pixel value and the average value. The variance in the Gaussian distribution has been increasing until covering the entire color domain. Pixels are classified as foreground or background according to the Gaussian distribution weight. Foreground misjudgment will not be studied too much in the paper, and is deemed to be arising to adapt to the Gaussian distribution changes itself. But it can be seen from formula 5 that this assumption is not convincing because the minimum time $t_{\min}$ Gaussian distribution spent on becoming background model can be expressed by the following formula:

$$t_{\min} \geq \frac{1}{\ln(1-\alpha)} \ln \left[\frac{K + \lambda - 3}{(K-2)(\alpha-1)} \right] \quad (9)$$

Suppose in formula 9, $\lambda = 0.7$, $\alpha = 0.005$ and the number of Gaussian distribution $K = 3$. Under this condition, $t_{\min} \geq 70$, then foreground misjudgment can not be explained by the quick background updating. The experiments show that Lee method greatly increased foreground misjudgment.

3.3 Update of iterative dynamic pixel parameter

After analysis, it is known that because the traditional methods uses the fixed dynamic number of pixels, it can not handle the situation with rapid changed background, then it leads to greater changes of the variance, resulting in the loss of dynamic pixel and causing foreground misjudgment. Under such condition, a new algorithm is proposed to compensate for the characteristics of the missing pixels by updating iterative dynamic pixel parameter, reducing foreground misjudgment. So, two learning rates are used to deal with dynamic pixel mean value and variance respectively. First, set an adaptive learning rate γ_k related to the mean μ_k to satisfy the following formula:

$$\gamma_{k,t} = \gamma_{k,t-1} + q_k - \frac{1}{K} \sum_{i=1}^{K} q_i \quad (10)$$

where $q_k = \eta(x, \mu_k, \sigma_k)$ is the probability of current dynamic pixel belonging to the kth Gaussian distribution. The experiments show that when initial value of γ_k is $\gamma_{k,0} = 0.05$, the experimental results are very successful. As it is seen from the above formula that when the Kth Gaussian probability distribution is greater than that of the average, γ_k increases. Otherwise, it decreases. The adaptive learning rate can quickly update the average value of the Gaussian mixture model and adapt to changes of light intensity, so as to solve the problem of lack of dynamic pixel in traditional method.

The changes of learning rate is of no relationship with the average value, thereby reducing the foreground misjudgment. But the problem of excessive variance update is not solved. So the Iterative dynamic pixel parameter model is used to control the change of the variance, and when the mean value changes are small, semi-linear is used to adapt to the changes; when the mean value changes rapidly, stop changing. The purpose can be realized by the following s-curve function.

$$f_{a,b}(x, \mu_k) = a + \frac{b-a}{1 + e^{-s\varepsilon(x,\mu_k)}}$$

where in, $\varepsilon(x, \mu_k) = (x - \mu_k)^T (x - \mu_k)$ and s is the slope of the curve. In the experiment, $s = 0.005$. Then, the formula 5 becomes:

$$\sigma_{i,t}^2 = (1 - \eta)\sigma_{i,t-1}^2 + \eta f_{a,b}(x, \mu_{i,t-1}) \quad (11)$$

$\eta = 0.6$ and is a fixed learning rate. $f_{a,b}(x, \mu_k)$ limits the range of the dynamic pixel variance value to $[a + b/2]$, b, and the choice of a and b should make sure the variance value range covers at least a Gaussian distribution. Then the dynamic pixel within the region can be used as reference pixels, compensating for insufficience and eliminating the drawbacks.

The above method can eliminate the drawback caused by rapid background changes and increase the dynamic pixel, eliminating foreground misjudgment and tracking lag. Finally, the tracking will be done accurately.

4 EXPERIMENTAL RESULTS

In order to verify the effectiveness of the method, a set of comparative experiments are performed. Choose two experimental sites with quite different lighting, background and use the two different methods respectively to do the comparative tracking experiments.

Figure 1 showed the tracking results of the two algorithms. The left image is about the traditional frame difference tracking algorithm and the right one is about the one researched in the paper. As it is seen from Figure 1, the new algorithm reduced the probability of foreground misjudgment and

Figure 1. Comparison of two methods in complex background.

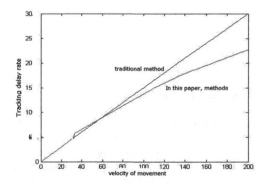

Figure 2. Tracking lag rate comparison of two algorithms.

significantly inhibited the tracking lag. In order to measure the accuracy of the two methods, the tracking lag rate formula in Chapter 4 is use to collect the intuitive trend results of the two different methods and the results are shown in Figure 2.

From the above figure, it is known that with the increased moving velocity, the tracking lag rate is much smaller than that of the traditional method. It can greatly eliminate the error and achieve a satisfactory result.

5 CONCLUSION

The traditional frame difference tracking method of human motion is difficult to deal with the problem of dynamic pixels reduction caused by rapid changes of the background. A Gaussian mixture model-based tracking algorithm is put forward to solve the above mentioned problem by building a Gaussian mixture model simulation and updating the iterative dynamic pixel parameters. The experimental results show that this method can significantly improve the accuracy of human motion in complex background, and achieve good results.

REFERENCES

Alin C. Popescu. Statistical Tools for Digital Image Forensics. Ph.D Dissertation, Department of Computer Science, Dartmouth College, 2005.
Friedman N. and Russell S. Image segmentation in video sequences: A probabilistic approach[J], In Proceedings 13. Conf. on Uncertainty in Articial Intelligence, 1997:1–3.
Grimson W., Stauffer C., Romano R. Using adaptive tracking to classify and monitor activities in a site[C] Proceedings of IEEE Conference on Computer Vision and Pattern Recognition. Washington, DC: IEEE Computer Society, 1998: 22–31.
Kaew Tra Kul Pong P., Bowden R. An improved adaptive background mixture model for real-time tracking with shadow detection [C] The 2nd European Workshop on Advanced Video-based Surveillance Systems. Kingston: Kluwer Academic Publishers, 2001:149–158.
Lee, D.-S. Effective gaussian mixture learning for video background subtraction[J] IEEE Trans. on Pattern Analysis and Machine Intelligence, 2005, vol. 27, no. 5, 827–832.
Micah K. Johnson. Lighting and Optical Tools for Image Forensics. Ph.D. Dissertation, Department of Computer Science, Dartmouth College, 2007.
Stahl M., Aach T., Buzu T.M., et al. Noise-resistant weak structure enhancement for digital radiography [A]. SPIE Conference on Image Processing [C], San Diego, 2009, 2:1406–1417.

Control Engineering and Information Systems – Liu (Ed)
© 2015 Taylor & Francis Group, London, ISBN 978-1-138-02685-8

Image segmentation by cue integration and polar space transformation

M. Liu

College of Physics, Mechanical and Electrical Engineering, Jishou University, Jishou, China

X. Luo & W.S. Zheng

School of Information Science and Technology, Sun Yat-sen University, Guangzhou, China

ABSTRACT: This paper proposes a polar-space based method to segment the image automatically. It aims at segmenting the object of interest by integrating all the visual cues and finding the "optimal" closed contour in the polar space. Experimental results further verify and demonstrate the efficacy of the proposed polar-space based method on the challenging datasets.

1 INTRODUCTION

In computer vision literature, segmentation is the process of partitioning an image into disjoint, homogeneous and compact regions where each part constitutes connected pixels with similar properties, such as brightness, color, texture. Over the years, many algorithms (Tu & Zhu, 2002, Shi & Malik, 2000, Felzenszwalb & Huttenlocher, 2004) have been proposed for segmentation. Generally speaking, all the algorithms could be classified into three categories: (1) feature-space based techniques; (2) image-domain based techniques; (3) physics based techniques (Luccheseyz & Mitray, 2001).

This paper addresses the problem of separating the object of interest from a static image. We propose a segmentation framework that takes a point as its input and outputs the region containing that point, shown in Figure 1. Essentially, segmenting this region is equivalent to find the enclosing contour, which is a connected set of boundary edge fragments in the edge map, around the input point. The edge map is generated by using all available visual cues.

The proposed algorithm framework is a four step process: First, the edge map of the image is generated using low-level cues; second, the edge map is transformed from Cartesian space to polar space with the input point as the pole; third, the "optimal" path through this transformed edge map is found; fourth, the path is mapped back to a enclosing contour, Figure 1b shows the final result.

2 GENERATING THE EDGE MAP BY CUE INTEGRATION

In this section, we explicate the first step of the proposed segmentation algorithm: generating the edge map where the boundary edges (the actual boundary) are much brighter than the internal edges.

In most scenes, the static visual cues such as color, intensity, or texture can precisely locate the edges. In our framework, we use the Berkeley edge detector (Fowlkes & Malik, 2004), which learns the color and texture properties of the boundary pixels versus the internal pixels from a dataset containing human-labeled segmentations of 300 images, to generate our initial boundary edge map, shown in Figure 2b. This edge detector handles texture much better than any intensity-based edge detectors. We can see that the unauthentic texture edges have been basically removed while the boundary edges are bright.

However, some internal edges (see BC, CD, DF in Fig. 2b) still remain bright. In order to separate the boundary edges from internal edges, we can use the motion cue sufficiently. It is known that the optical flow value changes significantly at the boundary of an object while substantially remain unchanged inside an object. Based on this concept, we can modify the edge map so that the edge pixels with strong gradient of optical flow values are stronger than the ones with weak gradient.

Figure 1. (a) The image with an input point. (b) The final segmentation given by the proposed approach.

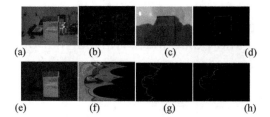

(a)   (b)   (c)   (d)

(e)   (f)   (g)   (h)

Figure 2. (a) An example of the input image. (b) The Berkeley edge map. (c) The magnitude of optical flow field. (d) The boundary edge map combining the static visual cues with motion cue. (e) The final output by mapping back the optimal path to the Cartesian space. (f) The optimal cut dividing the polar image into two parts: left (inside the object) and right (outside the object). (g) The optimal contour overlapped on the polar edge map. (h) The corresponding polar edge map.

We break the initial edge map into straight line segments and select rectangular regions of width α at a distance β on its both sides (see FC and FA in Fig. 2c). Then, we calculate the average flow inside these rectangles. The difference in the magnitude of the average flow on both sides is the standard measurement of the probability of the segment to be boundary edge. The brightness of an edge pixel on the segment is changed as

$$I'(x,y) = \lambda I(x,y) + (1-\lambda)\Delta f / \max(\Delta f), \qquad (1)$$

Δf represents the change in optical flow; λ is the weight related to the relative importance of the static visual cues based boundary estimate. Figure 2d shows the final boundary edge map where the internal edges are clearly fainter (low probability) and the boundary edges are brighter (high probability).

3 POLAR SPACE TRANSFORMATION

Let us say $I^p(\cdot)$ is the corresponding polar edge map of the edge map $I^c(\cdot)$ in the Cartesian space, and $Q(x_0, y_0)$ is chosen as the input point. We can see that a pixel $I^p(r, \theta)$ in the polar coordinates corresponds to a pixel location $\{I^c(x, y): x = r\cos\theta + x_0, y = r\sin\theta + y_0\}$ in the Cartesian space. Adopting the method of bilinear interpolation, which only considers 4 immediate neighbors, we can figure out $I^c(x, y)$ directly.

We propose to generate a continuous 2D function $F(\cdot)$ by putting 2D Gaussian kernel functions on every edge pixel and align the major axis of those Gaussian kernel functions with the orientation of the edge pixel. Let S be the set of all edge pixels. The intensity at any pixel location (x, y) in the Cartesian coordinates is defined as

$$F(x,y) = \sum_{i \in S} \exp\left(-\frac{x_i^t{}^2}{\sigma_{x_i}^2} - \frac{y_i^t{}^2}{\sigma_{y_i}^2}\right) * I^c(x_i, y_i), \qquad (2)$$

where

$$\begin{bmatrix} x_i^t \\ y_i^t \end{bmatrix} = \begin{bmatrix} \cos\theta_i & \sin\theta_i \\ -\sin\theta_i & \cos\theta_i \end{bmatrix} \begin{bmatrix} x_i - x \\ y_i - y \end{bmatrix}, \qquad (3)$$

$$\sigma_{x_i}^2 = \frac{A_1}{\sqrt{(x_i - x_0)^2 + (y_i - y_0)^2}}, \qquad (4)$$

$$\sigma_{y_i}^2 = A_2, \qquad (5)$$

θ_i is the orientation of the edge pixel i, A_1 and A_2 are constant. On account of keeping the gray values of the edge pixels in the polar edge map the same as the corresponding edge pixels in the Cartesian edge map, we set the square of variance along the major axis, $\sigma_{x_i}^2$, to be inversely proportional to the distance between the edge pixel i and the pole Q.

The polar edge map $I^p(r, \theta)$ is obtained by sampling $F(x, y)$. The intensity values of $I^p(r, \theta)$ are scaled to the interval ranging from 0 to 1, which can depict the probability of an edge pixel being at boundary. Figure 2h displays the polar edge map corresponding to Figure 2d. What's more, we come to an agreement that the angle $\theta \in [0°, 360°]$ is represented along the vertical axis and increases from top to bottom while the radius $r \in [0, r_{max}]$ varies along the horizontal axis and increases from left to right. r_{max} stands for the maximum Euclidean distance between two arbitrary pixels in the image.

4 FINDING THE OPTIMAL CUT

Let us regard every pixel $p \in P$ of I^p as a node in a graph.

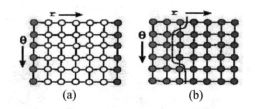

(a)   (b)

Figure 3. (a) The green nodes in the first column are initialized to be inside the object while the red nodes in the last column are initialized to be outside the object. (b) The binary labeling output after minimizing the energy function by means of graph cut algorithm.

Every node (pixel) is connected with their 4 immediate neighbors (see Fig. 3). A row of the graph represents the radiation from the input point at an angle (θ) equal to their row number. The first and the last rows are the rays $\theta = 0°$ and $\theta = 360°$, which are exactly the same in the polar space. So, the pairs of nodes $\{(0°, r), (360°, r)\}, \forall r \in [0, r_{max}]$ could be connected by edges in the graph. Let us say $l = \{0, 1\}$ are the two possible labels for each pixel, where $l_p = 0$ indicates inside the object and $lp = 1$ indicates outside the object. γ denotes the set of all the edges between neighboring nodes in the graph. Evidently, we aim at assigning a label l_p to every pixel p (i.e., finding the mapping $M(p) \rightarrow l$), which corresponds to the minimum energy. The energy function is

$$G(M) = \sum_{p \in P} \phi_p(l_p) + \mu \sum_{(p,q) \in \gamma} \varphi_{p,q} \cdot \delta(l_p, l_q), \quad (6)$$

$$\varphi_{p,q} = \begin{cases} \exp(-\tau I_{pq}^p) & \text{if } I_{pq}^p \neq 0 \\ C & \text{otherwise,} \end{cases} \quad (7)$$

$$\delta(l_p, l_q) = \begin{cases} 1 & \text{if } l_p \neq l_q \\ 0 & \text{otherwise,} \end{cases} \quad (8)$$

where $\mu = 50$, $\tau = 5$, $C = 20$, $I_{pq}^p = (I^p(r_p, \theta_p) + I^p(r_q, \theta_q))/2$, the cost of assigning a label l_p to the pixel p is denoted by $\phi_p(l_p)$ and the cost of assigning different labels to the neighboring pixels p and q is denoted by $\varphi_{p,q}$.

At the beginning of this process, we assign value 0 to the data term $\phi_p(l_p)$ for all the nodes except those in the first column and the last column:

$$\phi_p(l_p) = 0, \forall p \in (r, \theta), r \in (0, r_{max}), \theta \in [0°, 360°], \quad (9)$$

mean while, the nodes in the first column (corresponding to the input point in the Cartesian space) must be inside the object and are assigned label 0:

$$\phi_p(l_p = 1) = Z, \phi_p(l_p = 0)$$
$$= 0, p \in (0, \theta), \theta \in [0°, 360°], \quad (10)$$

similarly, the nodes in the last column must be outside the object and are assigned label 1:

$$\phi_p(l_p = 0) = Z, \phi_p(l_p = 1)$$
$$= 0, p \in (\gamma_{max}, \theta), \theta \in [0°, 360°], \quad (11)$$

shown in Figure 3. $Z = 1000$ must be a high value in order to make sure the initial labels to the first and the last columns could not change in the minimization step. To find the optimal path in the polar edge map, we apply the planar graph cut algorithm (Boykov & Kolmogorov, 2004) to minimize the energy function $G(M)$, given in equation (6).

The global optimization process divides the polar edge map into two parts: left side (inside the object) and right side (outside the object), see Figure 2f and Figure 2g. Finally, we map the resulting binary segmentation back to the Cartesian space to obtain the desired result. The boundary between the left (label 0) and the right (label 1) parts in the polar space corresponds to the closed contour around the input point in the Cartesian space, see Figure 2f and Figure 2e.

5 EXPERIMENTAL RESULTS

The purpose of our framework is to provide stable and accurate segmentation. We evaluated the performance of the proposed algorithm on 50 image pairs along with their ground-truth segmentation, which is created by identifying and segmenting the most outstanding object of interest for each image pair manually.

The final segmentation of our algorithm is compared with the ground-truth segmentation according to the empirical F-measure:

$$\Omega = \frac{2PR}{P + R}, \quad (12)$$

where P denotes the precision which calculates the score of our segmentation overlapping with the ground-truth segmentation and R symbolizes the recall which calculates the score of the ground-truth segmentation overlapping with our segmentation.

See Table 1, there has been considerable improvement in the proposed algorithm after combining static visual cues with motion cue. With the visual cues only, the actual boundary contour is not drawn precisely owing to the strong internal edges (see the row 2 of Fig. 4). Nevertheless, the affiliation of optical flow cue makes the internal edges fade away and the correct contour is found (see the row 3 of Fig. 4). On the other hand, we evaluate the performance of the proposed algorithm by comparing it with state of the art methods (Bagon et al., 2008, Alpertetal., 2007, Tu & Zhu, 2002, Shi & Malik, 2000). The public Alpert image database (Alpert et al., 2007) is used for our comparative experiments. Table 2 indicates we perform better than (Tu & Zhu, 2002, Shi & Malik,

Table 1. The performance of the proposed method.

For test images	F-measure
Without motion	0.65 ± 0.02
With motion	0.94 ± 0.01

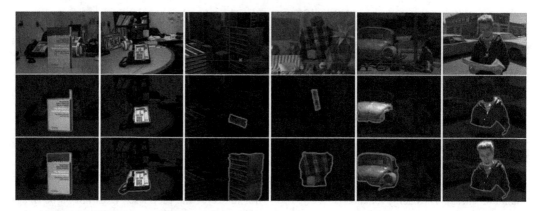

Figure 4. Row 1: the original images with an input point. Row 2: the segmentation using the static visual cues only. Row 3: the segmentation for the same input point after combining static visual cues with motion cue.

Table 2. The performance of the proposed method.

Algorithm	F-measure
Bagon [7]	0.87 ± 0.010
Alpert [8]	0.86 ± 0.012
Ours	0.84 ± 0.018
NCut [2]	0.72 ± 0.012
MeanShift [1]	0.57 ± 0.023

2000) and extraordinarily close to (Alpert et al., 2007, Bagon et al., 2008).

6 DISCUSSION AND CONCLUSIONS

In this paper, we have presented an efficient image segmentation algorithm. The framework cleverly combines static visual cues with motion cue to distinguish the internal contours from the actual boundary contours and then transforms the edge map from Cartesian space to polar space in order to find the optimal segmentation. The main contribution of the paper is to solve the old problem (segmentation) from a unique perspective, which leads us to segmenting the object of interest automatically and effectively. As the future extension of this work, the algorithm can be used to segment large number of images and the extracted regions can be studied for the high-level processes, such as object detection, tracking and recognition.

REFERENCES

Jianbo Shi and Jitendra Malik. 2000. Normalized cuts and image segmentation. T-PAMI, 22(8):888–905.

Luccheseyz, L. and S.K. Mitray. 2001. Color image segmentation: A state-of-the-art survey. Proceedings of the Indian National Science Academy (INSA-A). Delhi, Indian: Natl Sci Acad 67(2):207–221.

Martin D., C. Fowlkes and J. Malik. 2004. Learning to detect natural image boundaries using local brightness, color and texture cues. T-PAMI, 26(5):530–549.

Pedro F. Felzenszwalb and Daniel P. 2004. Huttenlocher. Efficient graph-based image segmentation. IJCV, 59(2):167–181.

Shai Bagon, Oren Boiman and Michal Irani. 2008. What is a good image segment? a unified approach to segment extraction. In ECCV, volume 5305, pages 30–44.

Sharon Alpert, Meirav Galun, Ronen Basri and Achi Brandt. 2007. Image segmentation by probabilistic bottom-up aggregation and cue integration. IEEE Conference on (pp.1–8).

Tu Z.W. and S.C. Zhu. 2002. Mean shift: a robust approach toward feature space analysis. T-PAMI, 24(5):603–619.

Yuri Boykov and Vladimir Kolmogorov. 2004. An experimental comparison of min-cut/max-flow algorithms for energy minimization in vision. T-PAMI, 26:359–374.

Control Engineering and Information Systems – Liu (Ed)
© 2015 Taylor & Francis Group, London, ISBN 978-1-138-02685-8

2-D maximum entropy threshold segmentation-based SIFT for image target matching

M. Zhou
Chongqing Key Lab of Mobile Communications Technology, Chongqing University of Posts and Telecommunications, Chongqing, China

X. Hong
Chongqing Key Lab of Mobile Communications Technology, Chongqing University of Posts and Telecommunications, Chongqing, China
Chongqing Lab of Material Physics and Information Display, Chongqing University of Posts and Telecommunications, Chongqing, China

Z.S. Tian
Chongqing Key Lab of Mobile Communications Technology, Chongqing University of Posts and Telecommunications, Chongqing, China

H.N. Dong
Chongqing Lab of Material Physics and Information Display, Chongqing University of Posts and Telecommunications, Chongqing, China

ABSTRACT: An improved Scale Invariant Feature Transform (SIFT) by the 2-Dimensional Maximum Entropy Threshold Segmentation (2DMETS) is proposed. First, using the pixel gray levels and the corresponding neighborhood space information, we can construct a 2-D gray histogram from the original images. Then, the image segmentation is conducted with the help of the maximum entropy of gray histogram. Finally, the local extreme value points in both of the 2-D image plane space and difference of Gaussian (DoG) scale space are selected as feature matching points. Experimental results conducted in a real environment show that the 2DMETS-based SIFT effectively reduce the interference of background noise and edge pixels and thereby improve correct matching rate.

1 INTRODUCTION

In recent decade, image target matching plays a significant role in the computer vision and digital image processing, like the autonomous navigation, 3-dimensional map reconstruction, target and scene recognition, visual tracking and positioning. In general, the target matching depends on the extraction of stable image characteristics. Although the typical Harris algorithm [1] can quickly extract the image characteristics, the performance cannot be guaranteed in the condition of the image rotation or scaling invariance. Compared with Harris algorithm, the target matching results by the Forstner algorithm [2] will significantly rely on the contrast of images. Moreover, the feature extraction threshold involved in the Forstner algorithm is also difficult to be selected. As another representative matching algorithm, the SIFT algorithm [3] has been demonstrated to perform well in the situations of image rotation, transformation and zooming.

The most significant contribution from this paper is that we propose a new 2DMETS-based SIFT algorithm for the image target matching. By conducting the 2DMETS on the raw color images, the SIFT can more effectively and reliably identify the image targets. Moreover, in different conditions of the signal-to-noise ratio and image size, the 2DMETS-based SIFT can not only save a large amount of calculation cost, but also guarantee the stable performance on the image segmentation without any requirement of the targets' prior feature information. Based on the 2DMETS processing, both the pixel gray and neighborhood spatial information will be used to construct a 2-D image gray histogram for the image threshold segmentation. Furthermore, the performance of the feature point searching and matching can be guaranteed by 2DMETS-based SIFT algorithm because the background noise and interference of edge pixel points are suppressed.

2 RELATED WORKS

As one of the most representative algorithms to detect the local invariant features in images, the SIFT algorithm [4] was proposed by Lowe in 1999, he improved it in 2004. The SIFT algorithm simulates the scale and normalizes the rotation and translation in images to compute the affine invariant. Because of the low calculation cost and well matching performance, the SIFT algorithm is widely used for the real-time feature matching in large database situation.

In order to more effectively save the calculation cost of the SIFT algorithm, a large number of institutes and universities have proposed a variety of improved SIFT algorithms in recent decade, such as the Principle Component Analysis (PCA)-based SIFT, Speeded Up Robust Features (SURF), Harris-SIFT, Shape SIFT (SSIFT) and Affine SIFT (ASIFT) algorithms. Compared to the conventional SIFT, the descriptors of the PCA-based SIFT can effectively reduce the feature points and decrease the dimensions of feature vectors. The PCA-based SIFT not only effectively improves the matching speed, but also significantly saves the storage space [5]. By applying the integral images to the image convolution computation, the SURF only relies on a small number of histogram to quantize gradient orientations [6]. To improve the calculation efficiency of the conventional SIFT, the Harris-SIFT extracts the feature points and calculates the SIFT descriptor of each feature point by the Harris operator [7]. By the consideration of the global shape context, the significant application of the SSIFT is that it can effectively recognize the Chinese characters from images which have been contaminated by the complex circumstance [8]. Different to the conventional SIFT, the ASIFT simulates the two camera axis parameters: latitude and longitude angles [9].

The above algorithms fail to consider the influence of the background noise and edge pixels in the image target matching. To solve this problem, this paper gives the idea of integrating the 2DMETS and SIFT. The main purpose of the 2DMETS is to divide the gray levels in image into segments to achieve the maximum entropy and cluster the same pixel gray values into the same target for the SIFT matching [10].

3 2DMETS-BASED SIFT FOR TARGET MATCHING

The 2DMETS-based SIFT algorithm depends on the pixel and average gray of the neighborhood information to process image and obtain the ideal information from images. First of all, the 2DMETS-based SIFT construct a 2-D gray level histogram by the distribution of the grayscale values

of image pixels and the corresponding neighbor pixels. Then, by setting the optimal threshold of the maximum entropy in gray histogram, the segmentation removes the background noise and edge pixel points. Last but only least, the SIFT is applied to match and identify the image targets based on the extraction of the features from the divided images.

3.1 Construction of 2-D grayscale histogram

By dividing the gray level into L levels, the gray level at each pixel and the corresponding average gray level of the neighborhood form a level pair. Because each pair belongs to a 2-D bin, the pairs can generate a 2-D grayscale histogram. The numbers of histogram bins and pixels are $m \times n$ and $M \times N$. The probability of (i, j) is calculated by (1).

$$p(i,j) = \frac{r(i,j)}{M \times N}, \quad \sum_{i=0}^{L-1}\sum_{j=0}^{L-1} p(i,j) = 1 \tag{1}$$

where $0 \le p(i,j) \le 1$, $r(i,j)$ is the occurrence number of level pair (i,j).

3.2 Selection of 2-D maximum entropy threshold

If the 2-D maximum entropy threshold is at the pair (s,t), $r(i,j)(i=1,\cdots,s; j=1,\cdots,t)$ is recognized as the conditional probability function. Then, by the normalization processing, the information entropy of the target and background in images can be calculated in (2) and (3).

$$\begin{cases} H_1 = \sum_{i=0}^{s}\sum_{j=0}^{t} \frac{p(i,j)}{P_1(s,t)} \lg \frac{p(i,j)}{P_1(s,t)} \\ \quad = \lg P_1(s,t) + \frac{h_1(s,t)}{P_1(s,t)} \\ P_1(s,t) = \sum_{i=0}^{s}\sum_{j=0}^{t} p(i,j) \\ h_1(s,t) = -\sum_{i=0}^{s}\sum_{j=0}^{t} p(i,j)\lg p(i,j) \end{cases} \tag{2}$$

$$\begin{cases} H_2 = \sum_{i=s+1}^{L-1}\sum_{j=t+1}^{L-1} \frac{p(i,j)}{P_2(s,t)} \lg \frac{p(i,j)}{P_2(s,t)} \\ \quad = \lg P_2(s,t) + \frac{h_2(s,t)}{P_2(s,t)} \\ P_2(s,t) = \sum_{i=s+1}^{L-1}\sum_{j=t+1}^{L-1} p(i,j) \\ h_2(s,t) = -\sum_{i=s+1}^{L-1}\sum_{j=t+1}^{L-1} p(i,j)\lg p(i,j) \end{cases} \tag{3}$$

Then, we define the total information entropy $H(s,t)$ as the summation of H_1 and H_2, as illustrateds in (4).

$$H(t,s) = \lg p_1(t,s) p_2(t,s) + \frac{h_1(t,s)}{p_1(t,s)} + \frac{h_2(t,s)}{p_2(t,s)} \quad (4)$$

Therefore, the 2-D maximum entropy threshold (s^*,t^*) is selected to maximize the value $H(s,t)$, such as

$$(s^*,t^*) = \arg\max_{1<s<L-1;\ 1<t<L-1} H(s,t) \quad (5)$$

3.3 Generation of SIFT feature vector

The SIFT recognizes the local extreme points in both the 2-D image flat space and DoG scale space as the image matching feature points. During the recognition, each sample pixel will be compared with its 26 neighbor pixels to find the maximal or minimal DoG operator response value. The homologous pixel point is defined as the key point [3]. The DoG operator will be named as the Gaussian kernel difference of two different scales, as shown in (6).

$$D(x,y,\sigma) = \left[G(x,y,k\sigma) - G(x,y,\sigma) \right] * I(x,y)$$
$$= \Gamma(x,y,k\sigma) - \Gamma(x,y,\sigma) \quad (6)$$

where $I(x,y)$ is the raw color image. (x,y) is the coordinates in DoG space. σ is the scale factor. k is the scale space nuclear. $G(x,y,\sigma)$ is the variable Gaussian function. $\Gamma(x,y,\sigma) = G(x,y,\sigma) * I(x,y)$ stands for the Gaussian convolution nuclear. "$*$" means the convolution.

In the SIFT matching, once a key point has been detected, the next concern is to accurately locate its position and discard the other unstable ones which are much sensitive to the noise in images. By assigning a consistent orientation to each key point, it can be guaranteed that the SIFT operator will be invariant to the image rotation. The distribution characteristics of neighbor pixel gradient directions are used to specify the parameters of the main direction of each key point. Then, the descriptor for the key point which is highly distinctive and invariant will be computed. Finally, we select the 16×16 pixel point window around the key point, draw the 8-direction gradient histogram on each 4×4 pixel block, and calculate the accumulated value of each gradient direction to form a seed point. At this point, a key point will be composed by 4×4 seed points and each seed point has the vector information in eight directions. Therefore, the 128 gradient data can be generated as the 128-dimensional SIFT feature vector for each key point.

3.4 SIFT feature matching

After the SIFT, the feature key points of retrieval images are generated. Then, the best matching for each key point can be obtained by identifying its nearest neighbor in the database from the training images. The nearest neighbor is defined as the key point with the minimal Euclidean distance to the invariant descriptor vector. In order to avoid the error matching which is caused by the background noise, we use nearest distance ratio to match the features, as shown in (7).

$$\text{Dist-ratio} = \frac{\text{Distance(1st } NN)}{\text{Distance(2nd } NN)} \quad (7)$$

where Distance (1st NN) and Distance (2nd NN) are the first and second nearest Euclidean distances respectively. Based on (7), if a feature is not matched by the other features, the values of the first and second nearest distances are much similar. Then, The Dist-ratio will approximately equal to one. In the contrary, if a matching occurs, the Dist-ratio could be close to zero.

4 EXPERIMENTAL RESULT

4.1 Image description

There are three groups of images with the same pixel size of which are selected for our experiments: 1) group 1 (see Fig. 1): two indoor short-distance shooting images with slight angle rotation; 2) group 2 (see Fig. 2): two outdoor long-distance shooting images with slight angle rotation; 3) group 3 (see Fig. 3): two outdoor long-distance shooting images with significant angle rotation. Obviously, the targets in group 1 are static and slightly influenced by the illumination changes. The targets in group 2 are highly influenced by the background noise (e.g., the moving vehicles and

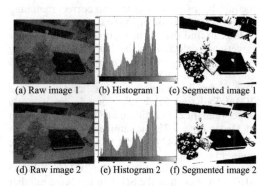

(a) Raw image 1   (b) Histogram 1   (c) Segmented image 1

(d) Raw image 2   (e) Histogram 2   (f) Segmented image 2

Figure 1.   2DMETS processing on images in group 1.

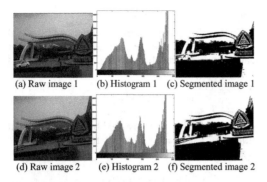

(a) Raw image 1   (b) Histogram 1   (c) Segmented image 1

(d) Raw image 2   (e) Histogram 2   (f) Segmented image 2

Figure 2.   2DMETS processing on images in group 2.

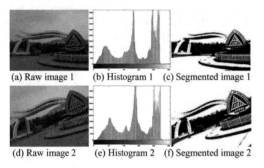

(a) Raw image 1   (b) Histogram 1   (c) Segmented image 1

(d) Raw image 2   (e) Histogram 2   (f) Segmented image 2

Figure 3.   2DMETS processing on images in group 3.

pedestrians). Compared to the group 2, the targets in group 3 are in different angles.

4.2 *Matching results*

As illustrated in Figures 1–3, by the 2DMETS, the raw color images will be transformed into the black-and-white images in uniform grayscale. The background and edge noise in images have been effectively eliminated from images.

Figures 4–6 show the results of image target matching. Figures 7 and 8 give the variations of the number of total matches and correct matching rate with the increase of Dist-ratio.

We can find that: 1) in Figures 4–6, several new targets are detected by the 2DMETS processing on the SIFT algorithm. For instance, targets ⑥ and ⑦ in Figure 4(b) and ④ and ⑤ in Figure 6(b); 2) in Figure 7, the number of total matches by the 2DMETS-based SIFT algorithm is significantly larger than the conventional SIFT. In the condition of Dist-ratio = 0.75, the number of total matches is increased by 142.86%, 59.15% and 75% for the images in groups 1–3 respectively; 2) In Figure 8, the correct matching rate of the 2DMETS-based SIFT is higher than the SIFT in the large number of total matches condition (i.e., the images in

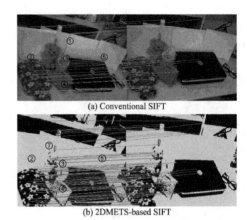

(a) Conventional SIFT

(b) 2DMETS-based SIFT

Figure 4.   Comparisons of target matching for the images in group 1.

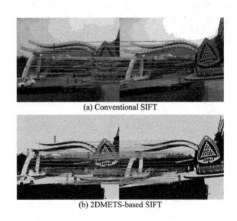

(a) Conventional SIFT

(b) 2DMETS-based SIFT

Figure 5.   Comparisons of target matching for the images in group 2.

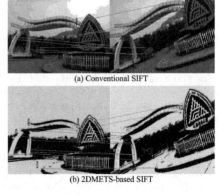

(a) Conventional SIFT

(b) 2DMETS-based SIFT

Figure 6.   Comparisons of target matching for the images in group 3.

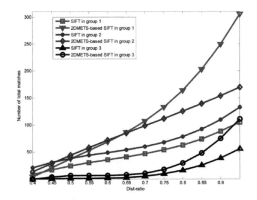

Figure 7. Variations of number of total matches in different Dist-ratio.

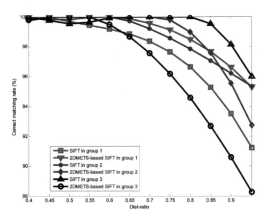

Figure 8. Variations of correct matching rate in different Dist-ratio.

groups 1 and 2). However, if the number of total matches is small (i.e., the images in groups 3), the correct matching rate cannot be guaranteed. Therefore, the increase of the number of total matches should be suggested as an effective way to improve matching performance of 2DMETS-based SIFT algorithm.

5 CONCLUSION

By the analytical and experimental analysis of the image target matching in the indoor and outdoor environment in the different conditions of illumination intensity, background noise and rotation angles, the robustness and environmental adaptability of SIFT algorithm can be effectively improved by 2DMETS processing on the raw color images. However, a more effective way to increase the number of total matches and improve the correct matching rate of the 2DMETS-based SIFT forms another interesting work in future.

ACKNOWLEDGMENTS

This work was supported in part by the Program for Changjiang Scholars and Innovative Research Team in University (IRT1299), National Natural Science Foundation of China (61301126), Special Fund of Chongqing Key Laboratory (CSTC), Fundamental and Frontier Research Project of Chongqing (cstc2013jcyjA40041, cstc2013jcyjA40032, cstc2013jcyjA40034), Scientific and Technological Research Program of Chongqing Municipal Education Commission (KJ130528), Startup Foundation for Doctors of CQUPT (A2012-33), Science Foundation for Young Scientists of CQUPT (A2012-77), and Student Research Training Program of CQUPT (A2013-64).

REFERENCES

Bay, H. Ess, A. Tuytelaars, T. & Cool, L.V. 2008. Speeded-up robust features (surf). *Compuer Vision and Image Understanding* 110(3): 346–359.

Durrant, A. & Dunbabin, M. 2011. Real-time image classification for adaptive mission planning using an autonomous underwater vehicle. *Oceans*: 1–6.

Kapur, J.N. Sajoo, P.K. & Wong, A.K.C. 1985. A new method for gray-level picture thresholding using the entropy of the histogram. *Computer Vision, Graphics, and Image Processing* 20(3): 273–285.

Lowe, D.G. 1999. Object recognition from local scale-invariant features. *International Conference on Computer Vision*: 1150–1157.

Lowe, D.G. 2004. Distinctive image features from scale-invariant keypoints. *Computer Vision* 60(2): 91–110.

Ryu, J.B. & Park, H.H. 2010. Log-log scaled Harris corner detector. *Electronics Letters* 46(24): 1602–1604.

Yu, G.S. & Morel, J.M. 2009. A fully affine invariant image comparison method. *IEEE International Conference on Acoustics, Speech and Signal Processing*: 1597–1600.

Zhang, Y. & Wei, K.B. 2010. Research on wide baseline stereo matching based on PCA-SIFT. *IEEE 3rd International Conference on Advanced Computer Theory and Engineering*: V5 137–140.

Zhang, Q. Rui, T. & Fang, H.S. 2012. Particle filter object tracking based on Harris-SIFT feature matching. *International Workshop on Information and Electronics Engineering*: 924–929.

Z. Jin, Qi, K.Y. & Zhou, Y. 2009. SSIFT: an improved SIFT descriptor for Chinese character recognition in complex image. *International Symposium on Computer Network and Multimedia Technology*: 62–64.

Shape detection based on Hough transform in images

W. Zhang

Baicheng Normal University, Jilin, China

ABSTRACT: The identification of objects' profile is the most important means to recognize objects by people. It is considerable to detect objects' profile in image processing and pattern recognition. Hough transform as an effective method of shape feature extraction is widely used. The paper analyzed the theory of Hough transform, pointed out the disadvantages, discussed the improved algorithm and applied it to detect shape in images.

1 INTRODUCTION

Technology of Image target detection extends people's vision cognition. It is very important for image analysis. It also plays important roles in target auto identification system. With the development of image processing, target detection has been the basic technology in artificial intelligence and is applied in many fields widely. The basic analytical means develop deeply with the development of mathematical tools. Today the application of image detection has transcended the field of vision. It has more features of machine intelligence and digital technology.

The detection means mainly includes image segmentation and feature extraction and analysis. Image segmentation partitions the image into some blocks. In one block, the features are similar, but very different in different blocks. The function of segmentation is to extract useful information from image. The feature of image mainly includes grayscale, veins and shape. The identification of objects' profile is the most important method to recognize objects by people. It is considerable to detect objects' profile in image processing and pattern recognition. Hough transform as an effective method of shape feature extraction is widely used. The advantage of this method is not sensitive to noise, but it also consumes great deal of times and storage space when parameter space has two or more dimensions. Many scholars have put forward many improved algorithms based on Hough transform. For example, the method based on gradient computes gradient according to original image and segments image with some threshold. It calculates the gradient angles which has non-zero value in segmented image. It obtains circles parameters through calculations. Another similar method computes edge gradient by Sobel operator to reduce invalid calculations. Although the

methods reduce time, they are still very complex. Someone presented the Point Hough Transform (PHT) implementation of circle detection, by using the geometric characteristics of circle. The method selects three points in the curve to calculate. It changes the 3D parametric statistics into one-dimensional parametric statistics and greatly reduces the complexity of the algorithm, but there are still a large number of invalid points involved in operation. It affects the speed and accuracy of detection.

Hough transform is mainly applied to binary image which is also edge image. Therefore before the Hough transform, grayscale images require pre-treatment including image filtering and edge detection. Image pre-processing is an essential preparatory task in the target detection course using Hough transform. Its results will directly affect the quality of the detection. Because Hough transform is easy to complete parallel processing and detects the edge information correctly in the case of inference, it has been used widely in computer vision and auto target recognition system.

2 TRADITIONAL HOUGH TRANSFORM

Paul Hough put forward Hough transform in 1962 and applied for patent. The basic idea is that one point in image space is converted into one curve or curved surface in parameter space. The ports, which have the same parameter feature, cross after transformation in parameter space. It completes detection by accumulation extent at the cross. Hough transform as an effective image target detection method can detect straight lines, circles, ellipses, parabolas and many other analytical graphics. The generalized Hough transform has to do some extend to the traditional method, which can detect any graphics using pre-set look-up table and is no longer restricted by graphic analytical expressions. The generalized Hough transform has resolved many problems in method recognition and computer vision. Now, Hough transform not only is used simply to detect shape in image but also is applied widely to SAR/ISAR image processing, office documents image processing, aviation image auto processing and biomedical engineering. If the detected image is not very complex, traditional Hough transform can get satisfied results. But because it is a map matching algorithm based on one to many, every feature point in image may match parameters which belong to all curves crossing this point and it accumulates these parameter units vote. Another disadvantage is in its requirement of a large storage space for the 3-D cells needed to cover the radius and the curve center position (x,y), its computational complexity, and its low processing speed. Even the worse, the accuracy of the extracted parameters of the detected circle is poor, particularly in presence of noise. All these disadvantages result in unaccepted shortcomings:

1. Time-consuming. It needs large computation.
2. Space-consuming. It needs a great deal of spaces to store parameters.
3. Low precision. It disperses the image space and parameter space.
4. Confusion of noise and feature points. It casts vote equally for all feature points by accumulating ones. So it couldn't distinguish noise and feature points.

Based on these disadvantages it has to do something to improve traditional Hough transform. Recent years, some valuable improved algorithms have been put forward. Such as probability Hough, random Hough, fuzzy Hough, level iteration and so on.

3 LINE DETECTION

The basic idea of line detection is to utilize duality of point-line in image and parameter space. All points which are in one line must correspond to the lines crossed at one point in parameter space. Vice versa.

In image space, all lines crossed the point (x,y) can be expressed by the equation 1:

$$y = px + q \qquad (1)$$

In equation, p is gradient and q is intercept. Equation 1 can be expressed in another form:

$$q = -px + y \qquad (2)$$

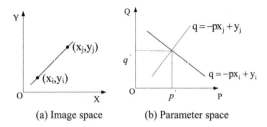

(a) Image space (b) Parameter space

Figure 1. Duality of point-line in image and parameter space.

Equation 2 describes one line crossed at the point (p,q) in parameter space. The relationship is shown in Figure 1.

In image space, equation 1 describes all lines crossed at the point (x_i, y_i). It can be changed into $q = -px_i + y_i$. It is same to $y_j = px_j + q$ and $q = -px_j + y_j$. The latter expresses another line in parameter space. In parameter space, two lines cross at the point (p', q'). In fact, they correspond to one line through the point (x_i, y_i) and (x_j, y_j) in image space. Because it satisfies equation $y_i = p'x_i + q'$ and $y_j = p'x_j + q'$. Based on discussion above, we can know that in image space every point in the line through the point (x_i, y_i) and (x_j, y_j) corresponds to one line in parameter space and these lines intersect on one point (p', q').

The points through one line in image space correspond to the crossed lines in parameter. On the contrary, in parameter space all lines intersected on one point match to the points through one line. That is the point-line duality. Corresponding to the duality, we can work out the line equation connected all these points by Hough transform from the edge points in image space. Hough transform converts the problem of line detection in image space into the problem of point detection in parameter space. It can be completed by simple adding.

4 CIRCLE DETECTION

Circle detection plays important roles in digital image pattern recognition. The problem of detecting circular features arises in many areas of image detection and is of particular relevance in industrial applications such as automatic inspection of manufactured products and components, aided vectorization of drawings, target detection, etc. Traditionally, the circle detection in digital images is realized by means of Hough transform. In Circle Hough Transform (CHT) the center point and radius are basic parameters. CHT makes use of accumulating 3D space parameters to extract circle.

The set, $\{(x_i, y_i) | i = 1, 2 \ldots n\}$ includes the points on the circle which is to detected. The center point is (a, b), radius is r, the circle equation in image space is:

$$(x_i - a)^2 + (y_i - b)^2 = r^2 \tag{3}$$

In the same case, if the point (x, y) is in the image set, its equation in parameter coordinate system (a, b, r) is:

$$(a - x)^2 + (b - y)^2 = r^2 \tag{4}$$

Obviously, the equation is 3D cone. Any point in image corresponds to one 3D cone in parameter space. For any point in the set $\{(x_i, y_i) | i = 1, 2 \ldots n\}$, all these 3D cones consist of cones cluster according to the parameters in equation (4). If the points are on the same circle, the cones must cross on one point (a_0, b_0, r_0) in the parameter space. This point just corresponds to the center and radius of the circle. Their relationship is shown in Figure 2.

Generally, after transformation the parameter space is 3D. It needs to create a 3D accumulating array, $A(a, b, r)$. Every edge point is computed and accumulated into A. The concrete steps are shown as following:

1. Compute the maximum and minimum of a, b, r and create the parameter space. The size of parameter space depends on the maximum and minimum.
2. Create the accumulating array $A(a, b, r)$ and set 0 for every element.
3. Perform Hough transform on the edge point in the image. According to equation 2, work out the relative curve on 3D grid for the point (a, b, r). At the same time, plus 1 to the corresponding array. That is, $A(a, b, r) = A(a, b, r) + 1$.
4. Find the local maximum in the array which is corresponding to the points on the same circle. The value provides the center and radius on the same circle in the image.

Because of the dispersed error, the equation 4 can be changed into equation 5 in the computing course.

$$\left| (a_0 - x_i)^2 + (b_0 - y_i)^2 - r^2 \right| \leq \xi \tag{5}$$

The parameter is the compensation for image digitization and quantization. During the computing, r is a progressive increment parameter. On every step, r is fixed. Work out the points on the circle which center is (x_i, y_i) on the plan vertical to r and accumulate the array, which is corresponding to this plan. R increases progressively from 0 to the maximum which can be accepted by image plan. In every increasing, there is always a relative plan. For the point (x_i, y_i) in the image, the varying range of a, b is 2r.

In the Hough transform, if the dimension of parameter space is less than or equal to 2, the transform result is satisfied. However, if the dimension is greater than 2, the transform will take very long time to complete and consume a great deal of store space. So Hough transform has more theory value. In practice, few people use this means directly. We must try to decrease the dimension of parameter space. If we can get the radius or the varying rang of radium before transform, we can save some time and space.

For the circle in the image, shown in Figure 3, (a, b) is the center, θ is the direction of gradient of edge point (x_i, y_i). And the gradients direct to the center. The information of the gradient is:

$$\begin{cases} g = \sqrt{g_x^2 + g_y^2} \\ \theta = \arctan\left(\dfrac{g_x}{g_{\partial y}}\right) \end{cases} \tag{6}$$

The equation used to compute the center is:

$$\begin{cases} a = x_i - r\cos\theta \\ b = y_i - r\sin\theta \end{cases} \tag{7}$$

In equation 7, $\sin\theta, \cos\theta$ can be computed directly, $\cos\theta = g_x/g$, $\sin\theta = g_y/g$.

We can make use of the local gradient of edge point to decrease calculation and improve the precision of the center location. At first we can extract

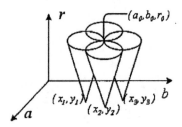

Figure 2. Parameter space of the circle, (a, b, r).

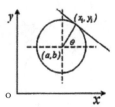

Figure 3. Circle in the image space.

the edge from image, calculate the gray varying gradient of pixel. Then change the gradient image into binary image according the threshold. The gradient value which is greater than the threshold is set into 1. At the same time, record its direction. On the contrary, the value is set into 0. In the binary image, we add 1 to the 2D array relative to every 1. If all pixels, which value is 1, have been processed, the local maximum in the array can be regarded as the location of center. The essence of this means is that the dimension is decreased from 3 to 2 during the accumulating and changes the mapping from one to many to one to one. This means has been regarded as the standard Hough transform.

5 CONCLUSION

Hough transform has been developed greatly since it was put forward in 1962. It has been applied widely in many fields of information processing. The paper discussed how to use improved algorithm to detect circle in images. To overcome the shortcomings, many researchers have put forward may improved algorithms. But because of uncertainty and inaccuracy, all kinds of improved algorithms still have some disadvantages. How to improve Hough transform to overcome shortcomings is a crucial task for everyone who engages in this field.

REFERENCES

CHU Guangli Video Object Segmentation Algorithm Based on Background Reconstruction ICCDA 2010.
Hough P.V.C. Method and Means for Recognizing Complex Patterns. U.S. Patent. No.3069654, 1962. 2.
Jung C.R., R. Schramm. Rectangle detection based on a windowed Hough transform. 17th Brazilian Symposium on Computer Graphics and Image Processing, 2004, 113–120.
Lan-Mark Geusebroek, Arnold W.M. Smeulders, Joost van de Weijer. Fastanisotropic gauss filtering. IEEE Trans. Image Processing, 2003,12(8):93 8–943.
Thuy Tuong Nguyen K., Xuan Dai Pham, JaeWook Jeon. An improvement of the Standard Hough Transform to detect line segments. IEEE International Conference on Industrial Technology, 2008, 1–6.
Thuy Tuong Nguyen, Xuan Dai Pham, Dongkyun Kim, framework for the accuracy of line detection by Hough Jae Wook Jeon. A test Transforms. 6th IEEE International Conference on Industrial formatics, 2008,1528–1533.

Control Engineering and Information Systems – Liu (Ed)
© 2015 Taylor & Francis Group, London, ISBN 978-1-138-02685-8

Application and research of vehicle image segmentation based on the improved snake model

L. Xu & X.C. Li

School of Optical-Electrical and Computer Engineering, University of Shanghai for Science and Technology, Shanghai, China

ABSTRACT: Although the gradient vector flow deformable model is a useful tool for the image segmentation, but when the concave region of the image is deep and fine, deformable model can not enter the deep region, so the iterative gradient vector flow method is proposed. Experimental results show that: the method is effective to solve the vehicle image segmentation.

1 INTRODUCTION

Image segmentation (Song 2010) is the foundation of image understanding, and it occupies the important position in image field, especially in medical (Mahmoud 2001, Jayadevappa & Demirkaya 2009), traffic, military field. In the recently twenty years, Image segmentation based on snake model attracts more attention of researchers. So it is very promising to study image segmentation technology further.

2 GRADIENT VECTOR FLOW

Since Kass (Kass 1987) proposed the snake model theory, the active contour model, attracted more attention in the field of the image research. In the process of finding appointed object boundary, the control points in the elastic curve move from an initial state to another state that has lesser image energy. The target contour is defined as a parameter curve $V(s) = (x(s), y(s))$, A way is given by the minimum total energy of the snake, which is the result of the equation:

$$E_{total} = \int E_{Internal} + E_{External}$$

$$= \int E_{Elastic} + E_{Bending} + E_{External}$$

$$E_{Elastic} = \frac{1}{2}\int \alpha(s)|v_s|^2 \, ds$$

$$E_{Bending} = \frac{1}{2}\int \beta(s)|v_{ss}|^2 \, ds$$

$$E_{External} = -\gamma(s)|\nabla I[v(s)]|^2 \qquad (1)$$

$E_{Elastic}$ is Elastic energy, $E_{Bending}$ is the bending energy. α and β are weighting parameters that con-

trol the snake's tension and rigidity respectively, V_s and V_{ss} denote the first and second derivatives of $V(s)$ with respect to s. $E_{External}$ is external energy. The external energy is composed of the following formula:

$$E_{ext}(V(s)) = E_{img}(V(s)) + E_{con}(V(s)) \qquad (2)$$

In the formula, $E_{img}(V(s))$ *is* the image energy, $E_{con}(V(s))$ is the external binding energy.

Given a gray level image $I(x, y)$ which viewed as a function of continuous position variables (x, y), typical external energies designed to lead an active contour toward step edges are

$$E_{img}(V(s)) = |\nabla I(x, y)|^2 \qquad (3)$$

$$E_{img}(V(s)) = -|\nabla[G_\sigma(x, y) * I(x, y)]|^2 \qquad (4)$$

where $G_\sigma(x, y)$ is a two-dimensional Gaussian function with standard deviation σ and ∇ is the gradient operator. A snake that minimizes E must satisfy the Euler equation:

$$\alpha X''(s) - \beta X''''(s) - \nabla E_{ext} = 0 \qquad (5)$$

This equation can be used as an energy balance equation, i.e.

$$F_{int} - F_{ext} = 0 \qquad (6)$$

$X''''(s)$ shows four order differential

$$F_{int} = \alpha X''(s) - \beta X''''(s), \quad F_{ext} = -\nabla E_{ext} \qquad (7)$$

Internal force stops the curve bending and extension, External force pushes the curve to the desired boundary, when the active contour and the target contour overlap, internal and external force and is zero, achieving force balance, and energy function gets to the minimum.

However, traditional snake model has two problems: one is the initialization of deformable model, that is initial contour is needed to set close to the target curve as far as possible; the other question is that the deformable model is difficult to move into the edge of the concave region. To solve these problems, Xu (PrinceJ 1996, Xu 1998, Xu 2000) proposed gradient vector flow (Gradient Vector Flow, GVF) model. The basic idea of the GVF snake is to extend influence range of image force to a larger area by generating a GVF field. The GVF field is computed from the image. In detail, a GVF field (Cheng 2004, Tang 2006) is defined as a vector $V(x,y) = (\mu(x,y), v(x,y))$ that minimizes the energy function:

$$\varepsilon = \iint \mu\left(\mu_x^2 + \mu_y^2 + v_x^2 + v_y^2\right) + |\nabla f|^2 |V - \nabla f|^2 \, dxdy \tag{8}$$

In particular, when $|\nabla f|$ is small, energy function is mainly decided by the partial derivative of vector field that is in the homogeneous region. Energy is dominated by the square of the partial derivative, producing a slowly varying field. In addition, when $|\nabla f|$ is large, the function is decided by the second part at this time, that is in the edge area of the image, when $v = |\nabla f|$, this function is minimum. Parameter μ is regularization parameter, equalizing the first and the second in the integrand.

3 ITERATIVE GRADIENT VECTOR FLOW SNAKE MODEL

Because the set of the GVF initial contour is roughly described, it is often far away from the bottom of the concave region. So when the GVF model curse is convergencing, the both walls of the concave region will produce the strong tensile force. In addition, the gradient vector flow field force from the bottom of the concave region is weak, it is not strong enough to lead the curve to the bottom of the concave region, so GVF model curve can only move limited distance and not reach the expected goal. To solve the above problems, it is to design the initial contour (Zhang 2012). And the following iterative GVF algorithm can make it.

3.1 Algorithm description

- Set the initial contour of a GVF deformable model (composed of discrete points).
- Depending on the generalized force balance principle, by much iteration, GVF deformable

model curve can get a closed contour curve that is composed of the evenly spaced discrete points.
- Whether the closed contour curve above meets the requirement? If so, the algorithm terminates; if it is not satisfied, the next step.
- Find out the failed discrete points that can not reach the edge of the image, the way is to calculating the average gray value between every discrete point and other surrounding eight points. If the gray value < D (preset value), this point is not convergent point. Failed point and its surrounding points should be moved along the direction of the inner Normal, The moved contour curve will be regarded as the initial contour curve of the next GVF deformable model, then turn to the second step, go on iterating.

3.2 Explanatory notes

- The algorithm is iteration of the GVF algorithm, GVF itself is a iterative algorithm. Each iteration is the evolution process that the deformable model curve approaches the generalized force balance.
- Because the depth of concave region is not the same, only the distance from the initial curve to the bottom of the concave region is less than a certain value, the curve will converge to the target contour. Otherwise the curve will be pulled back by the internal force. If the target contour still can not be achieved after (n-1) time's iteration, we must increase ingression distance.
- Each time the adjacent discrete points also should be moved in order to increase the bearing width. Because in some cases, the non-convergence discrete points may be only one, this moment, the V shape will be here, in this case, the internal force is very strong and have a strong resistance force.
- In order to ensure average interval, inserting new discrete points is necessary.

3.3 Experiment

1. One time GVF (Fig. 1)
   From the picture, we can see that the curve doesn't achieve convergence of the target contour by the 1000 iterations.

a   b   c

Figure 1. (a) an image with the concave regions (b) Initial border (c) final result.

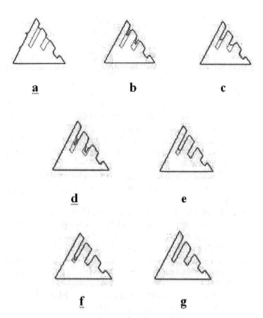

a   **b**   **c**

d   **e**

f   **g**

Figure 2. (a)(c)(e)(g) show the result of one time GVF, two times GVF, three times GVF, four times GVF(50 iterations each time). (b)(d)(f) show the result that non-convergence points are moved 10, 20, 30 pixels in the direction of the inner Normal.

2. The iterative GVF (Fig. 2)

From the experimental result, it is very clear that the efficiency of IGVF was very higher than GVF. By 200 iterations totally, IGVF can converge to the target contour, and GVF still can not make it by 1000 iterations, in fact, no matter how many times GVF iterates, the deformable model can not converge to the target contour. IGVF not only inherits the advantages of GVF, but also has unique advantages on image containing fine and deep concave region. The IGVF algorithm is simple and easy to achieve. If the initial contour of the deformable model is inside the image, IGVF algorithm only needs to correct the moving direction of the discrete points from the inner Normal to the outward Normal.

4 APPLICATION IN VEHICLE IMAGE SEGMENTATION

A. First application (Fig. 3)

Application in vehicle image segmentation is based on the iterative GVF snake model method. This experiment adopts MATLAB as the programming tool

B. Another application (Fig. 4)

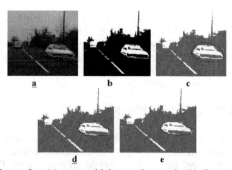

a   **b**   **c**

d   **e**

Figure 3. (a) one vehicle on the road. (b) Segment image by image enhancement and Shuangfeng method (Liu 2010). (c) The convergence of the traditional snake image segmentation. (d) The convergence of the GVF snake image segmentation. (e) The convergence of the iterative GVF Snake image segmentation.

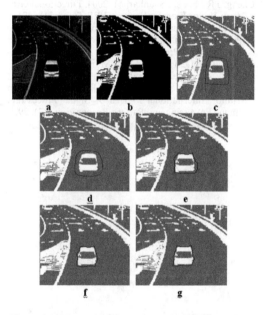

a   **b**   **c**

d   **e**

f   **g**

Figure 4. (a) one vehicle on the road (b) Segment image by image enhancement and Shuangfeng method (c) the initial border (d) convergence of the traditional Snake image segmentation (e) convergence of the GVF Snake image segmentation. (f) After revising the points of both sides, that is the result of the improved iterative GVF Snake image segmentation. (g) After revising the points of the bottom, that is the result of the improved iterative GVF Snake image segmentation.

5 CONCLUSION

From this paper, we can see clearly that the iterative GVF snake method has all the advantages of the GVF algorithm, and it has the better convergence effect for the deep concave region. The method is also simple and easy to achieve and it is well used in the vehicle image segmentation. In addition, I believe non-convergence points will be revised automatically by software, not man-made in the future research.

ACKNOWLEDGEMENT

This work is supported by graduate student education innovation projects of Shanghai education committee, the authors would like to thank the anonymous reviewers for providing suggestions to improve the paper.

REFERENCES

Cheng, J.R., Foosw, Shankar, M. 2004. Directional Gradient Vector Flow for Snakes. IEEE Trans: 318–321.
Demirkaya, O., Asyali, M.H., Sahoo, P.K. 2009. Image Processing with Matlab: Applications in medicine and biology: CRC Press.
Jayadevappa, D., 1, Kumar, S.S., Murty, D.S., A Hybrid. 2009. Segmentation Model based on Watershed and Gradient: Vector Flow for the Detection of Brain Tumor. International Journal of Signal Processing, Image Processing and Pattern Recognition 2(3): 29.
Kass, M., Witkin. A., Terzopouls. D. 1987. Snake: active contour models. Proceeding of International Journal of Computer Vision, Jan: 321–331.
Liu Zhiqun, Yang Wanting, Zhu Qiang. 2010. Several image enhancement algorithm research. Hefei normal College Journal: 60–63.
Mahmoud, M.K.A., Al-Jumaily, A. 2001. Segmentation of skin cancer images based on gradient vector flow (GVF) snake. International Conference on Mechatronics and Automation: 216–220.
Prince, J.L., Xu, C. 1996. A new external force model for snakes. InProc. 1996 Image and Multidimensional Signal Processing Workshop: 66–71.
Song Yinmao and Liu Lei. 2010. The methods and progress of image segmentation. Computer learning: 1–3.
Tang, J., et al. 2006. Surface extraction and thickness measurement of the articular cartilage from MR images using directional gradient vector low snake. IEEE Trans. Biomed. Eng. 52 (5): 896–907.
Xu, C., Prince, J. 1998. Snakes, shapes, and gradient vector flow. IEEE Transl. on Images Processing: 359–369.
Xu, C., Prince, J.L. 2000. Gradient Vector Flow Deformable Models. BankmanIsaac. Handbook of Medical Imaging. SanDiego: Academic Press.
Zhang Rongguo, Liu Kun. 2012. Edge shape curve and surface active contour model. National Defense Industry Press.

Control Engineering and Information Systems – Liu (Ed)
© *2015 Taylor & Francis Group, London, ISBN 978-1-138-02685-8*

Digital color image encryption based on Chaos map

Y. Huang

Department of Physics, Science and Technology of Kunming University, Kunming, China

S.L. He

Department of Asset and Equipment Management of Kunming University, Kunming, China

ABSTRACT: The new encryption scheme for digital color image has been proposed. The image can be encrypted through encryption matrixes generated with Hénon map chaotic sequences to XOR color matrixes many times. The R, G and B components of the color image can be treated randomly and cipher image becomes much uniform. Because the chaotic sequences are extremely sensitive to the initial values, even if there is a little wrong in keys, any useful information of the original image can not be obtained. The specific experiments confirmed the validity of this scheme.

1 INTRODUCTION

With the rapid development and extensive application of the computer technology, communication technology, network technology, the security of digital images becomes more and more important. To fulfill such a task, many image encryption methods have been proposed to protect the content of digital images. Because the chaotic systems are usually ergodic and are sensitive to system parameters and initial conditions, they are very suitable for information encryption. In recent years, many of the digital image encryption schemes which are based on chaos have been proposed [1–3]. The two-dimensional chaotic map is usually used to confuse the pixels of a gray-level image in almost of these methods. But, with the improvement of color CCD technology, the more and more digital color images occur in every aspects of social life, which includes e-Commerce of course. Digital color images encryption becomes key problem to be solved urgently. In this paper, the scheme to encrypt digital color images with two-dimension chaotic Hénon map has been proposed.

2 CHOOSE A HÉNON MAP TO GENERATE CHAOTIC SEQUENCE

Mathematically, chaos refers to a very specific kind of unpredictability: deterministic behavior that is very sensitive to its initial conditions. In other words, infinitesimal variations in initial conditions for a chaotic dynamic system lead to large variations in behavior. Because Logistic chaotic map is well-known by everyone, the random sequences used in most of chaotic encryption methods are generated by it. But, there are still other maps, such as Hénon map, circle map and so on, can generate chaotic sequences. They can be also used to encrypt image information.

The Hénon map is a simple two-dimension discrete chaotic system, which is mapping as

$$\begin{cases} x_{i+1} = 1 + y_i - ax_i^2 \\ y_{i+1} = bx_i \end{cases},$$

(1)

where a and b are system parameters. $\{x_i\}$ and $\{y_i\}$ are iterative sequences.

When $b = 0.3$, the bifurcation diagram of x changing with a is shown as Figure 1. It is found that chaos will yield while $a \geq 1.1$, but there are still some periodic windows. While $a \in [1.390, 1.395]$, the bifurcation diagram becomes uniform and there are hardly cycle solutions. Figure 2 shows this situation.

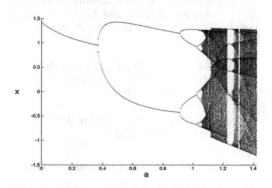

Figure 1. The full bifurcation diagram ($b = 0.3$).

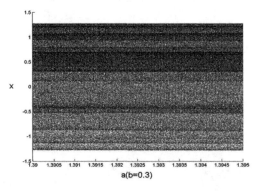

x-axis label: a(b=0.3)

Figure 2. The bifurcation diagram of $a \in [1.390, 1.395]$.

3 THE SCHEME TO ENCRYPT A DIGITAL COLOR IMAGE

A color image corresponds intuitively to the perceived representation of our colored environment. Computationally, a color image is treated as a vector function. For a three-channel digital color image, three vector components are given for one image pixel. Each of the vectors characterizes a color in the basic color space [4]. The RGB digital color image is composed of three color matrixes. The number of elements of each matrix is equal to the number of the pixels of the image. The value of each element is determined by the level numbers of each basic color. For common 24 bit RGB digital color images, each basic color has been allocated 8 bits, and every basic color has been divided into 256 levels. As the results, the value of an element of color matrix is the integer which is among 0 to 255.

If the values of elements of the three color matrixes were confused by a known way, the digital color image should be encrypted. We will use the Hénon map to produce encrypting matrixes which values of elements are randomly. Then make these encrypting matrixes XOR the color matrixes many times to randomize them and the cipher image could not be seen any useful information.

4 THE DESIGN AND REALIZATION OF THE ENCRYPTION ALGORITHM BASED ON MATLAB

In order to encrypt a digital color image effectively, the following steps will be taken.

4.1 Separating the R, G, B of a digital color image

As mentioned above, the digital color image can view as a combination of three color matrixes. To cipher a digital color image, the R,G,B components of it must be separated and the three color matrixes f_R, f_G, f_B can be gotten.

4.2 Yielding random encryption matrixes

Supposing each color dividing into m levels and there are n elements for every color matrix, the encryption matrixes $Ma, Mb, Mc, \ldots$ will be generated by following procedures. According to the initial values which inputted by user severed as keys, the three chaotic sequences have been generated by Hénon map in which the system parameters a and b have been selected three times with the program. Each value of the three sequences has been multiplied by the number of color levels m, and has been changed into integer to sever as the one of n elements. Then the encryption matrixes Ma, Mb and Mc have been yielded with same way. Because the elements of these matrixes generated by chaotic Hénon map sequences, they are very randomly. For 24 bit RGB digital color image, all elements of encryption matrixes are 8 bit integers which are less than or equal to 255.

4.3 Forming cipher image by making encryption matrix XOR color matrix

Using encryption matrixes Ma, Mb, Mc to XOR color matrixes f_R, f_G, f_B bit-by-bit, the encryption R, G and B components of a color digital image have been obtained respectively. To make cipher color matrixes more randomly, much times of XOR have been processed with many encryption matrixes. Combination of cipher color components, the cipher image will be given.

4.4 The implement of algorithm with Matlab

The Matlab [5] has been selected to implement the algorithm to encrypt a digital color image proposed above. In following program, there are 2 keys which will input by user. The keys are the initial values of three Hénon maps, and 3 chaotic encryption matrixes Ma, Mb, Mc are yielded. Every color matrix of f_R, f_G and f_B XOR all of 3 encryption matrixes Ma, Mb, Mc in a different order. The special codes saved in the file htxjm.m are:

```
flnm = input('Please input filename of a data
image:'); %when decryption, this statement replace
by 'load shy1.mat y'
x - imread(flnm);
image(x);
s = input('Please input key1(0~1):');
t = input('Please input key2(0~1):');
fR = x(:,:,1);
fG = x(:,:,2);
```

```
fB = x(:,:,3);
k1 = size(fR);
l1 = numel(fR);
xn = ones(l1,1);
Ma = uint8(ones(k1));
Mb = uint8(ones(k1));
Mc = uint8(ones(k1));
for k = 1:800;
xm = s;
ym = t;
s = ym + 1 – 1.39 * xm. * xm;
t = 0.3 * xm;
end
for n = 1:11
xm = s;
ym = t;
s = ym + 1 – 1.39 * xm. * xm;
t = 0.3 * xm;
xn(n) = s;
end
zn = abs(xn. * 100) – floor(abs(xn. * 100));
for i = 1:11;
Ma(i) = uint8(zn(i) * 255);
end
for k = 1:800;
xm = s;
ym = t;
s = ym + 1 – 1.392 * xm. * xm;
t = 0.3 * xm;
end
for n = 1:11
xm = s;
ym = t;
s = ym + 1 – 1.392 * xm. * xm;
t = 0.3 * xm;
xn(n) = s;
end
zn = abs(xn. * 100) – floor(abs(xn. * 100));
for i = 1:11;
Mb(i) = uint8(zn(i) * 255);
end
for k = 1:800;
xm = s;
ym = t;
s = ym + 1 – 1.395 * xm. * xm;
t = 0.3 * xm;
end
for n = 1:11
xm = s;
ym = t;
s = ym + 1 – 1.395 * xm. * xm;
t = 0.3 * xm;
xn(n) = s;
end
zn = abs(xn. * 100) – floor(abs(xn. * 100));
for i = 1:11;
Mc(i) = uint8(zn(i) * 255);
end
```

```
fR = bitxor(bitxor(bitxor(fR,Ma),Mb),Mc);
fG = bitxor(bitxor(bitxor(fG,Mb),Mc),Ma);
fB = bitxor(bitxor(bitxor(fB,Mc),Ma),Mb);
y = cat(3,fR,fG,fB);
save shy1.mat y; %when decryption, deleted this
statement
image(y);
imwrite(y,'cream2.jpg');
```

4.5 The decryption of cipher image

If the keys of getting a cipher image are known, the decryption image can be obtained by running again the encryption program. In other words, the decryption method is same as the encryption method.

5 RESULT OF EXPERIMENT

Running htxjm.m to encrypt a 1201×900 JPG color image 'fruit' gotten from Internet. The keys are key1 = 0.1, key2 = 0.5. In Figure 3, ① is the original image, and ② is the cipher image. None of the useful information of the original image can be seen in the ③. This demonstrated the digital image encryption algorithm proposed above is much effective. The ④ of Figure 3 is result decrypted with

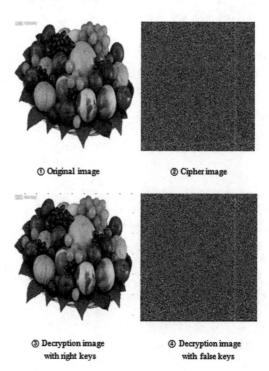

① Original image   ② Cipher image

③ Decryption image   ④ Decryption image
with right keys   with false keys

Figure 3.   The result of encryption and decryption of 'fruit'.

correct keys 0.1 and 0.5; it is same as the original image. The ⑤ of Figure 3 is the result decrypted with keys 0.1 and 0.5000000000000001. Although only the second key is wrong with one of sixteen numbers, the decrypted image appears no useful information. This result reveals that selecting the initial values of the chaotic Hénon map as keys is very effective.

6 CONCLUSION

In this paper, new scheme to encrypt the digital color image has been proposed. The image can be encrypted through encryption matrixes generated by the chaotic sequences of Hénon maps to XOR color matrixes many times. The experiment results fully proved the new algorithm is much effective. This work enriched the research of image encryption.

REFERENCES

Bao G., S. Ji & J. Shen, 2002. "Magic Cube Transformation and Its Application in Digital Image Encryption", Computer Applications, Vol 22, No. 11:23–25.

H. Liu, 2008. Essentials of MATLAB R2007, Tsinghua University Press, Beijing (in Chinese).

Koschan & M. Abidi "DIGITAL COLOR IMAGE PROCESSING".

S. Yang, 2010 "Novel Chaotic crambling and Encryption Method of Image", Computer Engineering, Vol 35, No. 10: 135–137 (in Chinese).

Yen & J. Guo, 2000. "A New Chaotic Key-Based Design for Image Encryption and Decryption", IEEE International Symposium on ISCAS 2000:49–52.

Control Engineering and Information Systems – Liu (Ed)
© 2015 Taylor & Francis Group, London, ISBN 978-1-138-02685-8

Analysis of land use changes based on Landsat TM/ETM+ images

H.S. Chen
School of Information Technology, Beijing Normal University, Zhuhai, Guangdong, China

B. Xia
School of Marine Sciences, Sun Yat-sen University, Guangzhou, Guangdong, China

ABSTRACT: Taking Dongguan City as the study area and applying Landsat TM/ETM+ images, authors studied land use changes and its land surface temperature effect in Dongguan City from 1995 to 2005. Conclusions are: (1) land use changes is characterized by the rapid expansion of built-up and the huge loss of arable land; (2) the impact of those changes on the regional land surface temperature is mainly indicated by a rising proportion of built-up in the high land surface temperature range.

1 INTRODUCTION

Since 1980s, Pearl River Delta (PRD) region of China has witnessed rapid urbanization and industrialization. Driven by the rapid urbanization and industrialization, the traditional land use pattern dominated by agricultural land has changed significantly by a rapid expansion of urban land and a huge loss of arable land Those changes inevitably affect the regional land surface environmental conditions. Therefore, many researchers took the PRD region to study its land use changes (Li 2004, Yan 2006, Zhang 2003 and Zhou 2005) and their environmental effects (Liu 2006 and Qian 2005). Results implied that the PRD region has been undergoing significant changes, but there were interior differences for the PRD region due to different natural and socio-economic conditions between cities.

In our study, Dongguan City, a typical city of rapid urbanization in the PRD region, was taken as our study area. The purpose of our study was to examine land use changes in Dongguan from 1995 to 2005 and explore its land surface temperature effect. We hope that our study can provide scientific basis for the study area's urban planning, living environment improvement and sustainable development and a meaningful reference for other regions of China which is undergoing rapid urbanization.

2 DATA AND METHODS

2.1 Study area and data

Dongguan City, located on the northeast of the Pearl River Delta (Fig. 1), has been selected as the study area considering its rapid urbanization since

1980s. It is located between 113°31′E–114°15′E and 22°39′N–23°09′N. It is mostly hilly to the east and flat in the west, and has a subtropical monsoon climate with abundant sunshine and rainfall over the year. Dongguan City has 32 towns, with a total area of 2,465 square kilometers and a total population of 8.22 million.

To quantitatively examine land use changes in the study area, Landsat 5 TM images (30 December, 1995; 22 December, 1998 and 23 November, 2005) and Landsat 7 ETM+ image (7 November, 2002) were selected. All images bands 1–5 have a spatial resolution of 30 meters and band 6 have a spatial resolution of 120 meters for Landsat 5 TM images and 60 meters for Landsat 7 ETM+ image. In addition, digital land use maps with a scale 1:10,000 for the study area in 1995 and 2000 were used to test the accuracy of land use classification.

Figure 1. Location of Pearl River Delta and the study area in Guangdong Province, China.

2.2 Image pre-processing

As having been geometrically corrected when acquired, Landsat 7 ETM+ images in 2002 is selected as a reference image. We select 25~30 evenly distributed control points for each image and use binary quadratic polynomial method to register other images with the reference image. It is guaranteed that the Root-Mean-Square (RMS) error after registration is less than 0.5 pixels. Then registered images are re-sampled by bi-linear interpolation at a spatial resolution of 30 m.

2.3 Land use classification

According to characteristics of land use in the study area, land use types in Dongguan City are divided into seven types: water, woodland, grassland, orchard, arable land, built-up and bare land. Firstly, interpretation marks are separately established based on false-color composite images with band 5,4,3 and band 4,3,2. Then classification templates for different times are established with a reference to digital land use maps in 1995 and 2000. The error matrix is used for evaluating classification accuracy. It is ensured that error matrix between different types in classification templates must be greater than 85%. Finally, all images are classified by Maximum Likelihood Classification (MLC) and clusters less than 5 pixels are eliminated by merged into adjacent ones. Meanwhile, the classification accuracy assessed by digital land use maps and filed survey data. Results are shown in Table 1.

2.4 Derivation of land surface temperature

There are three steps to derivate the Land Surface Temperature (LST) from Landsat TM/ETM+ images band 6. Firstly, we derivate pixel Digital Number (DN) values into radiance (L_λ) by the following equation (Irish 2001):

$$L_\lambda = gain \times DN + offset \tag{1}$$

where L_λ is the radiance of band 6 in the unit W/($m^2 \cdot sr \cdot \mu m$), *gain* and *offset* respectively represents

Table 1. Accuracy of land use classification from 1995 to 2005.

Classification accuracy	1995	1998	2002	2005
Overall accuracy	87.03%	88.42%	86.94%	87.92%
Kappa coefficient	0.821	0.832	0.865	0.845

Table 2. Emissivity of different land use types.

Land use type	Emissivity
Water	0.995
Woodland	0.985
Grassland	0.982
Orchard	0.985
Arable land	0.975
Built-up	0.97
Bare land	0.97215

gains and biases value of band 6, which can be found in the header file of images.

Then, land surface brightness temperature (T_6) is deviated from L_λ by the following equation (Irish 2001):

$$T_6 = \frac{K_1}{\ln\left(K_2/L_\lambda + 1\right)} \tag{2}$$

where T_6 represents land surface brightness temperature in the unit K, K_1 and K_2 are constants, which $K_1 = 607.76 W/\left(m^2 \cdot sr \cdot \mu m\right)$ and $K_2 = 1260.56\ K$ for Landsat 5 TM images and $K_1 = 666.09\ W/(m^2 \cdot ster \cdot \mu m)$ and $K_2 = 1282.71\ K$ for Landsat 7 ETM+ images.

Finally, land surface brightness temperature (T_6) must converted into Land Surface Temperature (LST). There are some common algorithms such as split window algorithm (Price 1984), mono-window algorithm (Qin and Karnieli 2001) and Artis & Carnahan algorithm (Artis and Carnahan 1982). Because split window algorithm and mono-window algorithm both complex and need some atmospheric parameters, we chose Artis & Carnahan alorithm to retrieve land surface temperature by the following equation (Artis and Carnahan 1982):

$$T_s = T_6/\left[1 + \left(\lambda \times T_6/\alpha\right)\ln \varepsilon\right] \tag{3}$$

where T_S is the land surface temperature, λ and α are constant with values $\lambda = 11.5\ \mu m$ and $\alpha = 0.01438\ mK$, ε represents the emissivity of land surface. The value of ε is determined by land use type mainly referencing Qin and Li (2004) and actual land surface characteristics of study area, and results are listed in Table 2.

3 RESULTS AND DISCUSSION

3.1 Land use changes from 1995 to 2005

As can be seen from Figure 2, land use pattern of Dongguan City had undergone significant changes

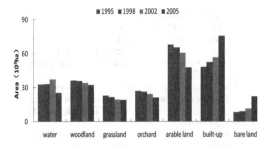

Figure 2. Land use change of Dongguan City from 1995 to 2005.

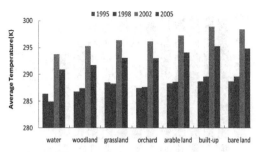

Figure 3. Average temperature of different land use types from 1995 to 2005.

from 1995 to 2005. Among all land use types, arable land reduced most rapidly in area, which declined from 68021.01 ha in 1995 to 47728.08 ha in 2005, and built-up increased most rapidly in area, which rose from 48145.86 ha in 1995 to 75823.56 ha in 2005. As for other land use types, woodland, grassland and orchard had a tendency to decrease in area and bare land had a tendency to increase in area.

Those changes led to significant changes to the land use pattern in Dongguan City. In 1995, the land use pattern is dominated by agricultural land, including arable land, woodland, grassland and orchard, which took up 63.4% of the study area, while built-up only took up 19.7%. In 2005, the proportion of built-up rose to 31.1%, and the agricultural land declined to 49.4%.

3.2 Temperature variation over different land use types

Because of differences in heat capacity, emissivity and other factors, different land use types are very different in land surface temperature. We calculated the average land surface temperature of each land use type and results are shown in Figure 3.

Among all land use types, built-up and bare land have highest average temperature, because they have smaller heat capacity, which built-up is mainly impervious surface composed of cement and other building materials, and bare land is lack of vegetation. On the other hand, water and woodland have lowest average temperature, because water has a larger heat capacity and woodland is with good vegetation cover. As can be seen from above, built-up and bare land maybe have a significant role in rising region's land surface temperature, water and woodland have a significant role in decreasing region's land surface temperature.

3.3 Land use types in different temperature ranges

It is widely applied in studying regional thermal environment that temperatures usually are divided

Table 3. The definition of different LST ranges.

LST ranges	Criteria
Low LST range	$T^* <$ Tmean**-S***
Normal LST range	Tmean $-$ S $\leq$ T $\leq$ Tmean $+$ S
High LST range	T $>$ Tmean $+$ S

*T represents the pixel's land surface temperature.
**Tmean represents the average LST of all pixels in the study area.
***S represents the standard deviation of all pixels' land surface temperature.

into different temperature ranges (Xu 2003, Zhang 2006). In our study, we divided Land Surface Temperature (LST) into three different temperature ranges: low LST, normal LST and high LST, and criteria for division are shown in Table 3.

In order to quantitatively examine composition of different land use types in each temperature range, we calculated the proportion of each land use type in different temperature range by the following equation:

$$PRO_{ij} = \frac{S_{ij}}{S_j} \times 100\% \qquad (4)$$

where PRO_{ij} represents the proportion of the ith land use type in the jth temperature range, S_{ij} represents the area of the ith land use type in the jth temperature range, and S_{ij} represents the total area of the jth temperature range.

Figure 4 shows proportions of different land use types in each temperature range from 1995 to 2005. Among all land use types, built-up contributed most to the high LST range, while water and woodland contributed most to the low LST range. As for changes over time, the proportion of built-up in high LST range rose from 1995 to 2005, the proportion of water in low temperature area rose from 1995 to 1998 and then declined from

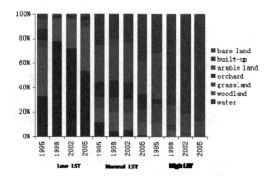

Figure 4. Proportions of different land use in each LST ranges from 1995 to 2005.

1998 to 2005, while the proportion of woodland in low temperature area declined from 1995 to 1998 and then rose from 1998 to 2005.

4 CONCLUSIONS

We studied land use changes and their land surface temperature effect in Dongguan City based on four different Landsat TM/ETM+ images from 1995 to 2005, and conclusions were: (1) land use changes is characterized by the rapid expansion of built-up and the huge loss of arable land; (2) the impact of those changes on the regional land surface temperature is mainly indicated by a rising proportion of built-up in high LST ranges. But, driving factors and other ecological environment effects of those changes aren't explored in this paper, and we will do works on those aspects in future.

REFERENCES

Artis, D.A. & Carnahan, W.H. 1982. Survey of emissivity variability in thermography of urban areas. Remote Sensing of Environment 12(4):313–329.

Irish, R. 2001. Landsat 7 Science Data Users Handbook. In: http://landsathandbook.gsfc.nasa.gov/.
Li, X. 2004. Spatio-temporal analysis of land use patterns in the development corridor of the Pearl River Delta in 1988~1997. Journal of Natural Resources 19(3):307–315.
Liu, Y. & Kuang, Y.Q. 2006. Impact of land use on urban land surface temperature-a case study of Dongguan, Guangdong Province. Scientia Geographica Sinica 26(5):597–601.
Price, J.C. 1984. Land surface temperature measurements from the split window channels of NOAA7 advanced very high resolution radiometer. Journal of Geophysical Research 89(D5):7231–7237.
Qian, L.X. & Ding, S.Y. 2005. Influence of land cover change on land surface temperature in Zhujiang Delta. Acta Geographica Sinica 60(5):761–769.
Qin, Z.H. & Karnieli, A. 2001. A mono-window algorithm for retrieving land surface temperature from Landsat TM data and its application to the Israel-Egypt border region. International Journal of Remote Sensing 22(18):3719–3746.
Qin, Z.H. & Li, W.J. 2004.The estimation of land surface emissivity for Landsat TM6. Remote Sensing For Land & Resources 16(3):28–32, 36, 41.
Xu, H.Q. & Chen, B.Q. 2003. An image processing technique for the study of urban heat island changes using different seasonal remote sensing data. Remote Sensing Technology and Application 18(3):129–133.
Yan, X.P. & Mao, J.X. 2006. Research on the human dimensions of land use changes in the mega-urban region: a case study of the Pearl River Delta. Acta Geographica Sinica 61(6):613–621.
Zhang, W.Z. & Wang, C.S. 2003. Coupling relationship between land use change and industrialization & urbanization in the Zhujiang River Delta. Acta Geographica Sinica 58(5):677–685.
Zhang, Y. & Yu, T. 2006. Land surface temperature retrieval from cbers-02 irmss thermal infrared data and its applications in quantitative analysis of urban heat island effect. Journal of Remote Sensing 10(5): 789–797.
Zhou, J. & Wu, Z.F. 2005. A quantitative analysis on the fragmentation of cultivated land on the two sides of the Zhujiang estuary. Tropical Geography 25(2):107–110.

Control Engineering and Information Systems – Liu (Ed)
© 2015 Taylor & Francis Group, London, ISBN 978-1-138-02685-8

Iterative estimator of instantaneous frequency of multiple Linear Frequency Modulation signals

P.P. Yu & M.S. Ai

Sergeant College of the Second Artillery Engineering University, Qingzhou, China

ABSTRACT: An iterative algorithm is proposed aiming at the problem of instantaneous frequency estimation of multiple linear frequency modulation signals. The algorithm is based on subspace tracking, and the matrix linear transformations as well as polynomial rooting are adopted to gain the parameter estimation. The advantages of the proposed algorithm are as following: lower computational complexity; higher frequency resolution; without cross terms problem in the multiple signals environment. However, since the matrix inverse is adopted, the performance of the novel algorithm will be lost in a low Signal-Noise-Ratio (SNR) environment. Simulation verified that the proposed algorithm possesses obvious superiority when the SNR is not lower than 6 dB.

1 INTRODUCTION

The Linear Frequency Modulation (LFM) signal has been used widely in the fields such as Radar, Sonar and radio communication. Since the LMF signal is a kind of typical non-stationary signal, the joint Time-Frequency Analysis (TFA) is usually adopted to estimate the parameters of the signal. The Short-Time Fourier Transformation (STFT) is a simple TFA method, but it has lower time-frequency concentration. The Wigner-Ville distribution [1] (WVD) is better in the time-frequency concentration, but the cross terms are unavoidable in the multiple signals situation because that the quadratic transformation is adopted. Some kernel functions have been designed to reduce the cross terms in the Cohen time-frequency distribution [2], but the calculation load of that is always a problem.

All the TFA methods mentioned above emphasize on the power-spectrum in the time-frequency plane, and the frequency resolution has to been limited by the length of the time window, which is called "Rayleigh Limit".

A sound way to breakthrough the Rayleigh Limit is the super-resolution estimation, which can obtain a pseudo frequency spectrum by taking advantage of the signal's eigensubspaces. The calculation complexities of the traditional super-resolution estimation methods are big, so that few methods of this kind have been used to estimate the parameters of the non-stationary signal. A novel algorithm has been proposed to obtain super-resolution instantaneous frequency estimation with lower calculation complexity. A data projecting method was adopted to trace the signal's eigensubspaces, and the instantaneous frequency estimation has been obtained with a polynomial roots finding method. The pseudo frequency spectrum possesses good performance in the time-frequency concentration, and the point is that the pseudo spectrum has no cross terms in the multiple LFM signals environment.

2 SIGNAL MODEL

Given that a time series $S(t)$ is comprised of M LFM signals, which can been written as

$$S(t) = \sum_{m=1}^{M} A_m s_m(t) + n(t), t = 1, 2, ..., T \qquad (1)$$

Here $s_m(t) = \exp(-j2\pi(f_m t + 1/2 \, k_m t^2))$, $m = 1, 2, ..., M$, A_m, f_m and k_m indicate the amplitude, the original frequency and the frequency modulation rate of the signals respectively, and T indicates sampling length. $n(t)$ indicates the channel noise, which is thought as zero mean white Gaussian noise. The signal is interval sampled, and N successive sampled points make up a vector that is called the snapshot. When the noise is ignored, the snapshot $y(t_0)$ is written as following

$$y(t_0) = [S(t), S(t-1), ..., S(t-N+1)]_{t=t_0}$$

$$= \left[\sum_{m=1}^{M} A_m s_m(t), \sum_{m=1}^{M} A_m s_m(t) \cdot e^{-j2\pi f_{m0} \Delta t}, \right.$$

$$\left. ..., \sum_{m=1}^{M} A_m s_m(t) \cdot e^{-j2\pi f_{m0}(N-1)\Delta t} \right]_{t=t_0}$$

$$= [A_1 s_1(t), A_2 s_2(t), ..., A_M s_M(t)] \cdot \mathbf{F}_{t=t_0} \qquad (2)$$

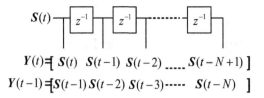

$$Y(t) = [\ S(t)\ \ S(t-1)\ S(t-2)\ \text{.....}\ S(t-N+1)\ ]$$

$$Y(t-1) = [S(t-1)\ S(t-2)\ S(t-3)\text{-----}\ S(t-N)\ ]$$

Figure 1. The schematic of the reconstructed snapshot vectors.

The letter F indicates the matrix which includes the information of instantaneous frequency, and the expression of F is as follow

$$F = \begin{bmatrix} F_1 & F_2 & \cdots & F_M \end{bmatrix}^T \tag{3}$$

$$F_m = [1, e^{-j2\pi f_{m0}\Delta t}, ..., e^{-j2\pi f_{m0}(N-1)\Delta t}]^T, \tag{4}$$
$$m = 1, 2, ..., M$$

The time series $S(t)$ is realigned to be the snapshots $y(t)$, and the schematic is showed in Figure 1. The covariance matrix $R(t) = y^H(t) \cdot y(t)$ is a singular matrix, and the eigen-values of the matrix are not helpful to estimate the instantaneous frequency, but it's useful in the subspace tracking that adopts an iterative mode.

3 THE SUBSPACE TRACKING ALGORITHM

Given that the $G \in C^{N \times N}$ is a nonnegative definite matrix, and the eigen values are aligned as $\lambda_1 \geq \lambda_2 \geq \cdots \lambda_N > 0$. The $u_1, u_2, ..., u_N$ correspond to the eigen values. $\{U_n\}$ is a orthogonal matrix sequence, which is defined by the expression following

$$U(t) = \text{orthnorm}(G \cdot U(t-1)) \quad t = 1, 2, ... \tag{5}$$

The "orthnorm" indicates ortho-normalization, which will be realized by the Gram-Schmidt method usually. If the original matrix U_0 meets that the $U_0^T[u_1, u_2, ..., u_L]$ is nonsingular, we can get the equation as follows

$$\lim_{t \to \infty} U(t) = [u_1, u_2, ..., u_L]$$

The theorem above had been proofed in article [3], based on which the Data Projection Method (DPM) has been proposed in article [4], and the mathematic expression is as follows.

$$U(t) = \text{orthnorm}\{(I - \mu \cdot R(t)) \cdot U(t-1)\} \quad t = 1, 2, ... \tag{6}$$

Given that the number of the signals M has been estimated beforehand and the original stochastic matrix $U(0)$ is a $N \times (N-M)$ dimension matrix, the $U(t)$ will approximate the noise eigen subspace of the sampled signals. Here the μ is called step factor. The iteration expression (6) is stable when the μ meets $0 < \mu \ll 1/\lambda_{max}$, and the λ_{max} indicates the maximum eigen value of the covariance matrix $R(t)$.

According to the MUSIC algorithm [5], the instantaneous frequency vectors F_m and the noise eigen subspace U_n are orthogonal, so expression (7) can be obtained as follows.

$$F_m \cdot U_n = 0 \tag{7}$$

The F_m is unknown value and the U_n as the coefficients can be obtained by the iteration (6), so a group of linear equations can be gotten according to the expression (7). The complex roots Z_m of the equations are keeping updated since the coefficients U_n are successive changed with the new sampled data. The M complex roots which are nearest to the unit circle in the complex plane will be chosen to estimate the instantaneous frequency f_m with the following expression.

$$f_m = -\arctan\left(\frac{img(Z_m)}{real(Z_m)}\right) / 2\pi, m = 1, 2, ..., M \tag{8}$$

With the continuous iteration, the instantaneous frequencies of the signals can be tracked, and the whole procedure of the tracking algorithm is described as follows.

a. Reconstruct the sampled data $S(t)$ to be the vector $y(t)$, which has N dimension, and the number N has to be bigger than M.
b. Calculate the matrix $R(t) = y^H(t) \cdot y(t)$ and compute the iteration (6).
c. Solve the equation (7), and choose M roots which are nearest to the unit circle in the complex plane.
d. Calculate the instantaneous frequencies with the expression (8).
e. If the sampling is end, the procedure would be finished, other else the new sampled data are taken in and return the step b).

4 NUMERICAL SIMULATIONS

Given that two LFM signals with the parameters as $(A_1, f_1, k_1) = (1, 260 \text{ MHz}, -7.5 \times 10^{13} \text{ Hz/s}^2)$, $(A_2, f_2, k_2) = (0.5, 180 \text{ MHz}, 4 \times 10^{13} \text{ Hz/s}^2)$. Set that the sampling frequency $f_s = 10^9$ Hz, the sampling time $T = 1u$ s, the length of the snapshot $N = 8$, the step factor $\mu = 0.5 \times 10^{-2}$. Take the STFT method as a comparison to the proposed method, and the length of time window is set to be $50n$ s.

The performances of the methods are showed in the Figure 2 and Figure 3 respectively.

The simulations show that both of the two methods can avoid cross term problem and the proposed method possesses good performances on the time-frequency concentration. Beside that, when the power of the two signals are different markedly, the small one will be submerged by the big one and hard to be detected. However, the proposed method is not sensitive to the power difference between the signals. The simulations below show that the proposed method is still effective when the power difference between the two signals is up to 14 dB, as a comparison, the small signal is nearly submerged when the STFT method is adopted. The performances of the methods are showed in the Figure 4 and Figure 5 respectively.

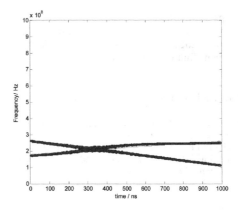

Figure 4. The pseudo spectra obtained with proposed method (SNR = 12 dB).

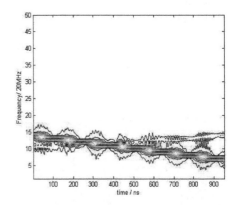

Figure 5. The power spectra obtained with STFT method (SNR = 12 dB).

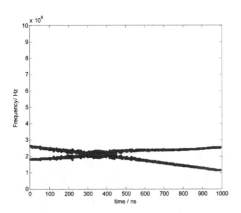

Figure 2. The pseudo spectra obtained with proposed method (SNR = 6 dB).

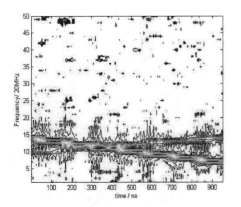

Figure 3. The pseudo spectra obtained with STFT method (SNR = 6 dB).

As an iterative algorithm, the proposed method possesses lower computational complexity, and it possesses good time-frequency concentration as well as avoids the cross term problem. But the proposed method has its shortcoming that it performs badly in the low Signal Noise Ratio (SNR) environment. When the SNR is lower than 0 dB, the proposed method will be fail, so it is suitable in a medium or higher SNR environment only.

5 CONCLUSIONS

An iterative algorithm is proposed in this paper for instantaneous frequency estimation, which is suitable for multiple LFM signals. The algorithm possesses lower computational complexity and higher time-frequency concentration than the classical TFA method. What's more important, the proposed method avoids the cross term problem. However, because of that the matrix inverse is adopted, the novel method is not suitable the environment that the SNR is lower than 0 dB.

REFERENCES

Cohen L. Time-frequency distributions a review[C] *Proc. of the IEEE*, 1989, 77 (7):941–981.

G.H. Golub, C.F. Van Loan, Matrix Computations (3rd edition)[M]. Baltimore and London, *MD: The John Hopkins Univ. Press,* 1996,

J.F. Yang, M. Kaveh. Adaptive Eigensubspace Algorithms for Direction or Frequency Estimation and Tracking[J]. *IEEE Trans. on Acoust., Speech, Signal Process.* 1988, 36(2):241–251.

R.O. Schmidt. Multiple Emitter Location and Signal Parameter Estimation[J]. *IEEE Trans. on antennas and propagation.* 1986; 3:278–80.

Zhao P.H., Ping D.F., Deng B. New method to suppress cross terms of the Winger-Ville distribution[J]. *Journal of computer*, 2010, 30(8):2218–2220.

Control Engineering and Information Systems – Liu (Ed)
© *2015 Taylor & Francis Group, London, ISBN 978-1-138-02685-8*

The digital signal calibration techniques for spectral reconstruction

B. Li & Q. Chen

Shenzhen Polytechnic, Shenzhen, Guangdong, China

ABSTRACT: The spectral reconstruction of digital camera as a starting point, and presents a linear digital signal acquisition and calibration methods which can be used in spectral reconstruction. By comparing the root mean square error and color difference, proved that experiment with digital signal correction method is feasible, this correction method used to reconstruct spectral data pre-processing can effectively improve the accuracy of spectral reconstruction.

1 INTRODUCTION

The traditional copy of color based on the metamerism phenomenon, which depends on the specific observer and observation condition, the metamerism phenomenon will be destroyed once these conditions are changed. With the improvement of the importance of color reproduction technology in many areas [1,2], multi-spectral imaging technology has become a major research tools to copy the color image. However, different color device has its own characteristics, which makes color copying difficult. In order to record and process the color image more faithfully, to minimize the impact of equipment limitations and differences, the date which used to obtain a digital image can not be directly applied to the spectral reconstruction, RGB signals must be corrected firstly.

The RGB signals of the camera is the integral value obtained from the optical radiation energy on the photoreceptor and camera spectral response function interaction, which is a set of linear data. But in the practical use of digital camera, the signal by the CCD or CMOS photoelectric conversion must experience a series of transmission and compression, and be output to the display device finally, such the RGB signals has become non-linear data. So in order to get an accurate spectral reconstruction effect, making the spectrum from the reconstruction of digital signal and from the spectrum instrument measuring as accurately as possible, it is necessary to convert the digital camera signal acquired from the image back to linear RGB values, that is making linear correction of the camera digital signals. in the Page Setup dialog box (File menu) see Table 1.

2 LINEAR CORRECTION

For any pixel of the digital image, the signal from linear camera and the radiation energy which incident to the CCD also satisfy the linear relationship for a certain color channels, when the incident radiation energy is changed, the signal of linear camera in conjunction with the proportional change, meanwhile there is also a linear relationship between the response value RGB digital signal digital and the photoelectric signal from camera, so that we can achieve the linearity correction of the RGB data through establishing conversion equation between the RGB digital signal value of camera in each channel and the radiation energy which CCD suffered.

The energy of light radiation reflected by the surface color under different light sources can not be consistent, linear correction of RGB data must

Table 1. Radiation energy T and RGB signals ON the SG card.

n	T	t	R	G	B	r	g	b
14	11.098	0.013	7	9	7	0.030	0.031	0.039
13	21.721	0.026	24	27	24	0.105	0.108	0.117
12	55.513	0.068	53	55	49	0.233	0.221	0.239
11	74.351	0.092	67	69	63	0.295	0.285	0.3
10	97.831	0.121	81	85	77	0.356	0.348	0.369
9	122.355	0.151	92	98	89	0.405	0.402	0.426
8	159.052	0.197	116	119	108	0.511	0.488	0.517
7	245.805	0.304	148	152	141	0.651	0.638	0.66
6	314.655	0.389	167	170	159	0.735	0.719	0.739
5	369.415	0.457	178	182	171	0.784	0.773	0.791
4	439.929	0.544	189	194	184	0.832	0.832	0.843
3	524.088	0.649	202	207	197	0.889	0.891	0.9
2	691.671	0.856	219	222	214	0.964	0.968	0.965
1	807.347	1		227	230	221	1	1	1

be done in a certain light conditions. Before linear correction, date of spectral power distribution $S(\lambda)$ of the light source and date of the spectral reflectance $R(\lambda)$ of the object surface must be got in a darkroom, then the date of light radiation incident to the CCD will be obtained multiplied $S(\lambda)$ by $R(\lambda)$.

$$t_{(i)} = \int S(\lambda) \cdot R(\lambda) \cdot d\lambda \qquad (1)$$

The optical radiation energy $t(i)$ can be obtained by calculation, and the camera digital signal $D(k,i)$ of the three RGB channels can be obtained by shoot. Thus A linear conversion equation between $D(k,i)$ and $t(i)$ can be built, that be expressed as:

$$t_{(i)} = f(x) \cdot D_{(k,i)} \qquad (2)$$

In order to improve the accuracy of linear conversion equation, fifth power polynomial fitting for each channel be used in this experiment to get the linearity correction.

$$f(x) = c_1 x^5 + c_2 x^4 + c_3 x^3 + c_4 x^2 + c_5 x + d \qquad (3)$$

In formula (3), x represents each camera digital signal, and d is a constant.

3 DATA ACQUISITION

3.1 Digital signal acquisition

The digital signal is taken by shooting the 14 gray-scale color samples of Color Checker SG standard color card using a Nikon D700 under a uniform light source lighting. Due to the optical effect of camera's internal convex lens through the digital signal is obtained, the central portion of CMOS received the strongest light radiation energy, and decreasing along a radial direction. So that the sample area should be placed in the central area of the camera field of view as much as possible. In addition, due to the visual design for human eye of the camera, the digital response value of part of the color will be higher than the white of the same gray level, in order to express more level of tone of the image, and prevent the digital camera signal from reaching or exceeding 255 that is the maximum RGB signal of the channel, and meanwhile be able to identify the dark blocks of color, light radiation energy reaches the surface of the CCD must be controlled during the experiment, so that the RGB signal is within a reasonable value range.

Several experiments proved that when the maximum digital signal value of standard white block channel of the experimental color card is controlled to 227, the phenomenon of bright tone overflow and dark tone does not appear. And when the aperture is set to 2.8 and the exposure time is 1/25 sec, the distribution of the RGB values of the captured images are more reasonable in darkroom scene of this experiment.

The Nikon D700 used in this experiment is 12.1 million pixels, and supports storage for RAW format, so it's more convenient and accurate to set the white balance. In order to make the picture data have better uniformity and rationality, some photographing norms must be made, all experiment follow the following conditions and methods.

1. Arrange shooting scene, make the test samples in the central of the two light sources and into a 45 ° angle, distance with 1.5 meters; adjust the balance of the sample holder and the various parts of the knob, to ensure that the optical axis of the lens of the digital camera parallel to the surface of camera platform, perpendicular to the vertical platform surface of the bracket with the experimental samples.
2. Place Color Checker SG standard color card on the vertical platform surface of the bracket with the experimental samples, adjust the height of the bracket or horizontal move the bracket, so that all the color samples of the color card is within the scope of the visual field of the digital camera, and the center of camera's field of view central quasi-color card. In this case, the distance between the camera lens and the color card plane is 1.3 meters. Mount camera on the tripod, and adjust the camera in a horizontal position.
3. After turn on the light source for three minutes, record the scene color temperature that is 4895 K using a illuminance meter (Konica Minolta Color Meter CL-200).
4. Set parameters of the camera, ISO200, RAW format, DX-closed format, and full frame format. Turn off all other optimization features.
5. Adjust the camera to M shooting mode, photograph according to the parameters of the fourth step. Tripod and cable release must be used, and the reflectors lockup function should be opened when shooting, to ensure that pixels of photographs completely overlap.
6. Open the images of Raw format in Camera Raw, adjust the color temperature value of all images according to the measured value, then save image with the white balance.

The spectral radiation energy of the imaging light source is measured in the scene as arranged above (1), take a uniform whiteboard on the vertical stage of experimental sample holder, and then obtain the spectral radiation energy distribution curve by measure the whiteboard using Eyeone spectrophotometer. In the distribution of spectral radiation energy of the light source do limited, and the maximum energy value is limited to 100, Figure 1 is the visible band energy distribution curve of the limited light source.

Through the above conditions and methods, the illumination of gray block area in these images is very uniform, and the obtained data is more reasonable, which ensure the accuracy of the linear calibration data, so that the accuracy of the prediction results of spectral reconstruction model is improved. The obtained data can be used to the calculation of linear correction model.

3.2 Sample spectral reflectivity measurement

In order to obtain the spectral reflectance of the gray color sample, the Spectrolino with high stability and accuracy is used in this experiment. Figure 2 shows the spectral reflectance curve of each gray sample, we can see that the reflection curve of the respective color blocks in the range of 400–700 nm remains substantially, and the reflection coefficient level is similar in each gray color blocks, we can also assume relative spectral reflectance degree of

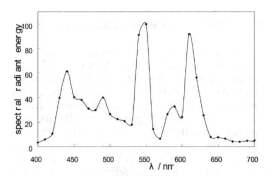

Figure 1.   Light radiation energy distribution curve.

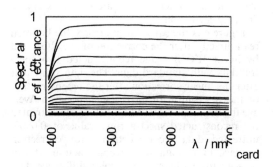

Figure 2.   14 gray-scale spectral reflectance on the SG standard color.

various color samples is the same, and the optical radiation energy T is divided into 14 according to the difference of grayscale.

4   DATA PROCESSING

4.1   Data normalization

According to the formula (1), the respective spectral radiation energy T of 14 gray block can be calculated using the measured spectral distribution of light source and spectral reflectance of 14 gray block, and normalize reference to the maximum value, obtained normalized optical radiation energy t; then process the RGB digital signal acquired by the shooting, select the lightest sample of the RGB values (white) as a standard, normalize each channel of the gray color RGB signal, obtain the normalized digital signals r, g, b values. The processed data will be used in the matching linear equation eventually. The normalized data of gray patches is shown in Table 1.

4.2   Fitting the linear transformation equation

Curve fitting refers to a complex function of $p(x)$, to obtaining a simple function $f(x)$ facilitate calculation, and it requires that $f(x)$ with the minimum error for $p(x)$ [3], that is to make the difference between approximation and the measured value smallest. Obviously, the difference is an important symbol to measure the quality of the fitting curve.

$$\sum_{i=1}^{n}\delta_i^2 \approx 0 \qquad (4)$$

As measured standard, minimum sum of squares between approximation and the measured value is used in this experiment, that means the method of least squares is used to curve fitting, it is described as follows: according to the respective optical radiation energy t normalized by the known data of 14 gray block and normalized digital signals r, g, b, obtain an approximate linear fit function $f(r), f(g), f(B)$ of the three channels of the digital camera by fitting, respectively make the following difference sum of squares minimum.

$$\sum_{i}^{n}\delta_i^2 = \sum_{i=1}^{n}[t_i - f(r_i)]^2$$

$$\sum_{i}^{n}\delta_i^2 = \sum_{i=1}^{n}[t_i - f(g_i)]^2$$

$$\sum_{i}^{n}\delta_i^2 = \sum_{i=1}^{n}[t_i - f(b_i)]^2$$

The method of the least squares is achieved by Matlab. Get the final linear conversion equation [4] calculated by Matlab, and then make the normalized digital signals r, g, b as the abscissa, the normalized energy of optical radiation t as the ordinate, conversion relationship of the respective channels is painted in Figure 3.

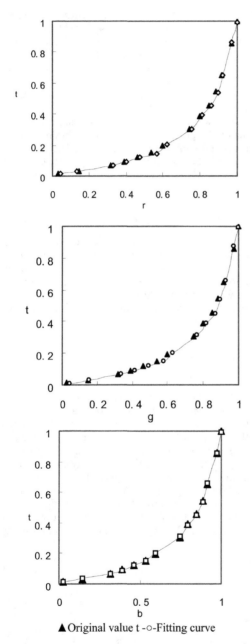

▲ Original value t -○-Fitting curve

Figure 3. Digital camera r, g, b three-channel linear equation fitting.

Simulate the trend line of the test points for each channel by polynomial diagram, get three-channel linear conversion equation of digital cameras under the illumination of the light source obtained as follows:

$$
\begin{cases}
f(r) = 2.234r^5 - 2.9436r^4 + 1.2976r^3 \\
\quad + 0.229r^2 + 0.169r + 0.0072 \\
f(g) = 4.949g^5 - 9.6831g^4 + 7.0481g^3 \\
\quad - 1.768g^2 + 0.4451g - 0.0017 \\
f(b) = 4.9478b^5 - 9.4389b^4 + 6.7928b^3 \\
\quad - 1.7343b^2 + 0.4319b - 0.0034
\end{cases}
\tag{5}
$$

In the formula (5), $f(r)$, $f(g)$, and $f(b)$ respectively stand for the normalized digital signals after correcting, r, g, b respectively stand for the to normalized digital signals directly obtained from a digital camera. When processing RGB signals by linear transformation equations, the same standard white sample must be used with that when the equation is constructed to normalized the data, so it can be substituted into the linear transformation equations.

5   LINEAR PROCESSING RESULTS

In order to express the advantage of digital signal linearity correction directly, a Color Checker 24 color card was photographed, and then use the pseudo-inverse method [5,6,7,8] to reconstruct the spectrum with the digital signal directly readed camera and the digital signals after the linearity correction. Then the root mean square error was used to evaluate the errors between reconstruct spectrum and original standard spectrum, also write program to make spectral reflectance converted to the XYZ values of the color, then converted to Lab values, and finally calculate the CIE1976L * a * b * color value to evaluate the reconstruction color aberration, in order to illustrate the necessary of linearity correction.

Figure 4 is the square root error value between read directly after the correction of the RGB and by linear R'G'B' value reconstructed two spectral data with standard data. The data show that after the linear correction after the reconstruction of the spectrum with the original spectrum Average the rmse value is 0.0512, whereas the untreated the reconstructed spectra rmse value of 0.0803, untreated data reconstruction error correction is twice more data than reconstruction. It can be seen from Figure 5, the color difference data ΔE between the color blocks and shooting color card reconstructed using correction significantly

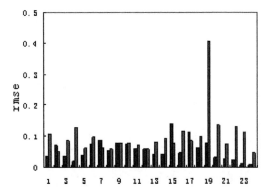

Figure 4. Spectral reconstruction accuracy compared.

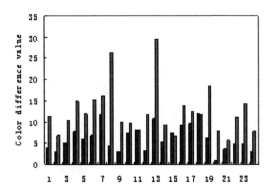

Figure 5. Color differencel accuracy compared.

reduced, the average color difference value down to 5.11 from 12.86.

Experiments prove that this acquisition image method and linear correction method for spectrum reconstruction and the camera's colour char-acterization provides important data guarantee. These work can effectively improve and enhance spectrum reconstruction accurate degree.

ACKNOWLEDGMENTS

The study is financially supported by Science and Technology Program of Shenzhen (No. JC2009 03180754A) and National Natural Science Foundation of China (61108087). Project team thanks the financial support of Science and Technology Bureau of Shenzhen and Shenzhen Polytechnic.

REFERENCES

Friedhelm K., Kenro O., Masahiro Y., et al 2000 Amultiprimary display: Discunting observer metamerism[C]//The 9th Congress of the International Color Association, Rochester, NY, 4421:898–901.

Ke Wang. 2005. Based on Matlab least squares curve fitting[J]. Beijing Broadcasting Institute Journal, 12(2):52~56.

Marimont D.H., Wandell B.A. 1992, Linear Models of surface and illuminant spectra[J]. Journal of the Optical Socity of America, 9:1905–1913.

Murakami Y., Obi T. 2001. Ymaguchi M., et al. Spectral reflectance estimation from multi-band imaging using color chart[J]. Optics Communi cations, 188:47–54.

Pratt W.K., Mancill C.E. 1976, Spectral estimation techniques for the spectral calibration of a color image scanner[J]. Applied Optics, 15:73–75.

Vrhel M.J., Gershon R., Iwan L.S. 1994, Measurement and analysis of object reflectance spectra[J]. Color Research and Application, 19:4–9.

Yangw P. 2005. Study on Procedure Simplification of Cross-Media Colour Reproduction[D]. Beijing: Beijing Institute of Technology.

Zhimin Liu. 1981. Error and data processing[M]. Beijing: Atomic Energy Publishing House.

Control Engineering and Information Systems – Liu (Ed)
© 2015 Taylor & Francis Group, London, ISBN 978-1-138-02685-8

Properties evaluation of signal synthesis using Fourier series and Walsh series

M. Sun

Department of Electronic Information Engineering, Foshan University, Guangdong, P.R. China

ABSTRACT: Fourier series, whose basic idea is that a time signal can be decomposed into a sum of an orthogonal set of sines and cosines, plays a fundamental role in signal s and systems analysis. In this paper, waveform synthesis is discussed based another orthogonal set–the Walsh functions. For discontinuous waveform synthesis, the conclusion shows that Walsh series obviously has potential advantages comparing to the Fourier series.

1 INTRODUCTION

Rectangular pulse waves, triangular waves, saw-tooth waves and trapezoidal waves have been widely used in the field of production and scientific research, which become the common waveform in electronics.[1] The square wave particularly become common waveforms of signals and systems analysis because it's easy to produce and process by digital circuit.

There are many different ways to represent a signal mathematically.[2] Usually a signal can be described as a linear combination of elementary functions, which are called basis functions.

The set of basis functions are defined by $\phi_0(t)$, $\phi_1(t)$, $\phi_2(t)$, ..., $\phi_N(t)$,

Any function $f(x)$ can be described as

$$f(t) = \sum_{n=0}^{\infty} \hat{a}_n \phi_n(t) \tag{1}$$

In orthogonal basis functions, trigonometric function plays a fundamental and important role in signal and system analysis. In the wide application of Fourier analysis, any periodic signal often can be regarded as an infinite sum of different frequencies sine and cosine functions. In engineering practice, many common signals are nonsinusoidal signals.

In this paper, signal synthesis characteristics of the Walsh series and Fourier series is analyzed discussed respectively. At the same time the relationship is discussed between Fourier series expansion and Walsh series expansion. The classical Fourier series based on triangular functions is promoted to a digital Fourier series.

2 INTEGRITY ORTHOGONAL BASIS FUNCTIONS

Condition of orthogonality for real basis functions is:

$$\int_{t_1}^{t_2} \phi_n(t)\phi_m(t)\,dt = \begin{cases} 0, n \neq m \\ \lambda_m, n = m \end{cases} \tag{2}$$

Orthogonal conditions of complex basis functions is:

$$\int_{t_1}^{t_2} \phi_n(t)\phi_m^*(t)\,dt = \begin{cases} 0, n \neq m \\ \lambda_m, n = m \end{cases} \tag{3}$$

where λ_m is a constant, if $\lambda_m = 1$ therefore orthonomal.

In practice, N is finite number terms. A linear combination of these orthogonal basis functions $\phi_0(t)$, $\phi_1(t)$, $\phi_2(t)$, ..., $\phi_N(t)$ will approximatively represent the desired signal $f(t)$ in the interval $[t_1, t_2]$.

$$f_N(t) \approx \sum_{n=0}^{N-1} a_n \phi_n(t) \tag{4}$$

If $f_N(t)$ is used to approximate represent $f(t)$ (the partial sum of the first N terms of the series), the mean-square *error* in the approximation is the square difference of $f(t) - f_N(t)$. The series converges *in the mean* to $f(t)$ in the interval (t_1, t_2). The square difference is

$$MSE = \varepsilon^2 = \frac{1}{t_2 - t_1}\int_{t_1}^{t_2}\left[f(t) - \sum_{n=0}^{N-1} a_n \phi_n(t)\right]^2 dt \tag{5}$$

If the N approach to infinite, the limit of the epsilon 2 is equal to zero, that is $\lim_{n\to\infty}\varepsilon^2 = 0$, then this function set is called integrity orthogonal functions set.

3 FOURIER SERIES

The trigonometric function set, such as {1, cos(1x), sin(1x), cos (2x), sin(2x), …, cos(nx), sin(nx), …}, $n = 0, 1, 2, …$, has very good properties—orthogonality.

When satisfying Dirichlet's condition, any periodic time-varying function $f(t)$ can be decomposed into an infinite sum of sines and cosines of different frequencies and amplitudes in terms of orthogonal functions. As shown in Figure 1, series expansion is

$$f(t) = \frac{a_0}{2} + \sum_{k=1}^{\infty}\left(a_k \cos\frac{2\pi kt}{T} + b_k \sin\frac{2\pi kt}{T}\right) \quad (6)$$

where a_0, a_k, b_k are the Fourier coefficients, which determine the amplitude of the different components.

$$a_0 = \frac{2}{T}\int_{-\frac{T}{2}}^{\frac{T}{2}} f(t)dt,$$

$$a_k = \frac{2}{T}\int_{-\frac{T}{2}}^{\frac{T}{2}} f(t)\cos\frac{2\pi kt}{T}dt,$$

$$b_k = \frac{2}{T}\int_{-\frac{T}{2}}^{\frac{T}{2}} f(t)\sin\frac{2\pi kt}{T}dt,$$

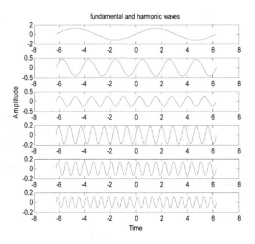

Figure 1.   Fundamental and harmonic wave of Fourier series.

Fourier series can be written as an exponential Fourier expansion

$$f(t) = \sum_{n=1}^{\infty} F_n e^{jn\omega t} \quad (7)$$

where

$$F_n = \frac{1}{T}\int_{-\frac{T}{2}}^{\frac{T}{2}} f(t)e^{-jn\omega t}dt \quad (8)$$

When a periodic signal $f(t)$ with discontinuity points is expressed by Fourier series with a finite number terms, if fulfilling Dirichlet's conditions, the greater N increases, the closer it gets to the signal. And Fourier series will converge uniformly as $N\to\infty$, it completely closes to the signal. And then

$$\int|f_N(t) - f(t)|^2 dt \to 0, N \to \infty \quad (9)$$

The truncated series approximates closely the function $f(t)$ as N increases, which unavoidable exhibits an oscillatory behavior and an overshoot. No matter how many terms are added, an overshoot remains at a value of 8.95% in the vicinity of the discontinuity at the nearest peak of oscillation. That is what is called the Gibbs phenomenon[3] This shows that nonsinusoidal periodic signal can not be synthesized with a finite number terms sinusoidal function, whose harmonic function is unable to fit discontinuous point. And at the same time the high-frequency interference signals are added.

4 WALSH SERIES

The Walsh function is a set of non-sinusoidal integrity orthogonal functions, each of which only takes on only two amplitude values, +1 and −1. All of them constitute a set complete orthogonal functions over the interval [0,1).[4][5][6] As shown in Figure 2.

Walsh functions can also be expressed in terms of even and odd waveform symmetry

$$\begin{cases} Wal(2n,t) = Cal(n,t) \\ Wal(2n-1,t) = Sal(n,t) \end{cases}, n = 1,2,…,\frac{N}{2} \quad (10)$$

Walsh functions Sal and Cal can be visualised as the respective sine and cosine basis functions in Fourier Series.

Therefore, every function $f(t)$ that is absolute integrable can be presented by a series of Walsh functions over the interval [0,1].

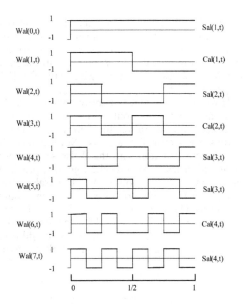

Figure 2. Continuous Walsh function of Walsh index.

$$f(t) = a_0 + \sum_{n=1}[a_n Cal(n,t) + b_n Sal(n,t)] \qquad (11)$$

where a_0, a_n, b_n are the Walsh coefficients, which determine the amplitude or weighting of each Walsh function

$$a_0 = \int_0^1 f(t) Cal(0,t) dt = \int_0^1 f(t) dt,$$

$$a_n = \int_0^1 f(t) Cal(n,t) dt,$$

$$b_n = \int_0^1 f(t) Sal(n,t) dt$$

The trigonometric functions in the Fourier series are replaced by a series of orderly rectangular function in the Walsh series.[7] Walsh series is digital Fourier series.[8]

Walsh series can also be expressed as:

$$f(t) = \sum_{n=1}^{N-1} d_n W_{wal}(n,t) \qquad (12)$$

where the coefficients d_n is: $d_n = \int_0^1 f(t) W_{wal}(n,t) dt$, there $n = 0, 1, 2, \ldots, N-1$.

$$\int_0^1 W(t)_n W_m(t) dt = \begin{cases} 0 & m \neq n \\ 1 & m = n \end{cases}$$

The size of steps and the number N will be determined by how many harmonics are combined

in synthetic signal. The more number of terms is used, the smaller the step size is. Therefore, the better will be of closeness of $f_N(t)$ to $f(t)$.

5 ROOT-MEAN-SQUARE ERROR ANALYSIS

The closeness of the approximation is measured by Mean Square Error. If $f(t)$ is periodic signal, MSE is defined as follows:

$$MSE = \varepsilon^2 = \frac{1}{T}\int_0^T \left[f(t) - f_N(t)\right]^2 dt$$

$$= \frac{1}{T}\int_0^T \left[f(t) - \sum_{n=0}^{N-1} a_n \phi_n(t)\right]^2 dt$$

$$= \frac{1}{T}\left[\int_0^T f^2(t) dt - \sum_{n=0}^{N-1} a^2{}_n + \sum_{n=0}^{N-1} (a_n - \hat{a}_n)^2\right]$$

$$\qquad (13)$$

MSE is minimum when $a_n - \hat{a}_n = 0$. Where $\hat{a}_n$ is coefficient of $f(t)$, a_n is coefficient of $f_N(t)$. The smaller MSE, the better approximation.
And Root-mean-square error is:

$$Error = \left(\frac{MSE}{Signal\ Energy}\right)^{1/2}$$

$$= \sqrt{\frac{1}{T}\left[\frac{\int_0^T f^2(t) dt - \sum_{n=0}^{N-1} a^2{}_n}{\int_0^T f^2(t) dt}\right]}$$

$$= \sqrt{\frac{1}{T}\left[1 - \frac{1}{E}\sum_{n=0}^{N-1} a^2{}_n\right]} \qquad (14)$$

RMSE of Fourier series and Walsh series are shown as follows respectively:

$$Error_F = \sqrt{\frac{1}{T}\left[1 - \frac{1}{E}\sum_{n=0}^{N-1} F^2{}_n\right]} \qquad (15)$$

$$Error_W = \sqrt{\frac{1}{T}\left[1 - \frac{1}{E}\sum_{n=0}^{N-1} d^2{}_n\right]} \qquad (16)$$

6 COMPARISON ANALYSIS OF SIGNAL SYNTHESIS

In this section the signal synthesis performance of the WS and FS is analyzed. One sample of rectangular waves with T = 1 and triangular waves T = 6 is shown in Figure 3. The signals are represented by linear combination of base functions.

Figure 4 shows that a square wave is fitted by finite terms of N = 512 using the Walsh series and Fourier series respectively. The Walsh method had better reconstruction reliability than Fourier method in fitting of in the vicinity of the discontinuity. The waveforms be synthesized are of closeness to the original signal waveforms. While by using Fourier method, the discontinuities can't be fitted by harmonic functions because of an infinite number of high-frequency oscillation, aliasing.

Figure 5 shows that a triangular waves is fitted by the Walsh series and Fourier series respectively. Original signal can be fitted well by means of the Walsh series with a finite number terms. While with the Fourier series the reconstructed signal cannot be fitted well at the sharp peak of original one, and the great root mean square error is generated. A comparison of the root mean square error is shown in Figure 6 that the signal is synthesized by the Walsh series and Fourier series with N = 1024 respectively.

It can be seen from the above analysis that FS does not achieve is continuities well and reach high accuracy. Moreover the Gibbs phenomenon always exists even for large N. Nevertheless, the signal reconstruction with the Walsh series is efficient and of less error, especially good approximation degree for unsmooth signal. The Walsh series is superior in having no Gibbs phenomenon comparing to The Fourier series. And not only that, since the amplitude of a Walsh function is either +1 or −1. The Walsh series simply with additions and subtractions, rather than multiplications as

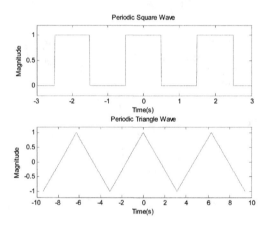

Figure 3. Periodic square wave with $T = 2$ and triangular wave with $T = 6$.

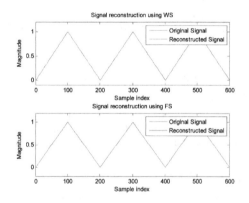

Figure 5. Synthesis of periodic triangular wave by Walsh series and Fourier series.

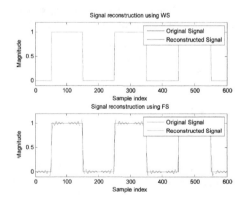

Figure 4. Synthesis of periodic square wave by Walsh series and Fourier series.

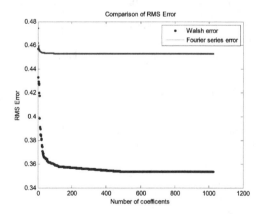

Figure 6. Root Mean Square Error of square wave synthesis by Walsh series and Fourier series.

242

in the Fourier series. The savings in computation time makes it possible to compute the Walsh series. Consequently, the simpler implementation is in its hardware.

7 CONCLUSION

A comparison of Walsh and Fourier series properties had been discussed in this paper. In engineering practices, most of the common signals are nonsinusoidal form. Oscillating and overshoting behavior in the discontinuous points arise from the truncated Fourier series. Accordingly, signal reconstruction has poor effect, low accuracy. The results obtained show that the Walsh series are more suitable in the representation of non-sinusoidal signals for its low error, high accuracy. Under the same condition of the Mean Square Error (RMSE), a better signal fitting performance is achieved.

REFERENCES

Abiyev A.N. and K. Dimililer. Reactive Power Measurement in Sinusoidal and Nonsinusoidal Conditions by Use of the Walsh Functions. IEEE Conference Proceedings on Instrumentation and Measurement Technology, IMTC 2008.

B. Golubov, A. Efimov and V. Skvortsov. Walsh Series and Transforms. Kluwer Academic Publishers. 1991.

B. Lin, B. Nguyen and E.T. Olsen. Orthogonal Wavelets and Signal Processing. Signal Processing Methods for Audio, Images and Telecommunications.

Benjamin Jacoby. Walsh Functions: A Digital Fourier Series. The BYTE Book of Computer Music, McGraw-Hill (March 1979).

Maqusi, M. Applied Walsh Analysis, Heyden & Sons Ltd., London, 1981.

Schipp F, Wade W.R., Simon P. and Pál J. Walsh series, Introduction to dyadic harmonic analysis, Adam Hilger, Bristol and New York, 1990.

Xiong Yuan_xin, Liu Di_chen. Convergence of Fourier serier and Gibbs phenomenon. Engineering Journal of Wuhan University. 2001, Vol. 34 No. 1:69–71, 85.

Zhang Qi shan, Yan Guang-wen, Wei Yu-chuan. Common Waveform Series. Journal of Beijing University of Aeronautics and Astronautics, 2001, Vol. 27, No 15:540–543.

Control Engineering and Information Systems – Liu (Ed)
© 2015 Taylor & Francis Group, London, ISBN 978-1-138-02685-8

Air valve fault diagnosis of diesel engine based on acoustic signal analysis

S.H. Cao, D.Y. Ning & J.J. Xu
Transportation Equipments and Ocean Engineering College, Dalian Maritime University, Dalian, China

ABSTRACT: Acoustic fault diagnosis system of diesel engine is based on the analysis of the measured sound signals and try to extract the fault information contained in the signals which can be able to characterize the fault, and the contents of the research relates to the separation and feature extraction of the sound signal. In this paper, the research object is the diesel engine air valve abnormal sound signal. In order to identify the fault type, some digital signal analysis methods such as digital filtering, wavelet transform and power spectral analysis method are adopted to extract and separate the abnormal sound signal. The results show that the methods owning some good analysis effects expressed as small amount of calculation and easy-to-real-time processing, which will contribute to the development of online sound fault detection device.

1 INTRODUCTION

Some parts of the diesel engine work in the harsh conditions as high temperature, high pressure and possess a higher probability of failure. In particular, the failure rate of the air valve is relatively high up to 11.9% [1]. The diesel engine sound contains information on the running state. Experienced staff can identify abnormal sound from the noisy sound and there from determine the type of failure. The key identifying the fault from abnormal sound lies in the operator familiar with these sound, but even for experienced operators, when the abnormal sound is weak or background noise is too loud, the discrimination of the fault will not be easy.

In this study, a real-time on-line diesel engine fault diagnosis system based on acoustic signal analysis is designed. The feasibility of the method is verified in the basis of the air valve abnormal sound signal. Its main task is to extract the frequency characteristics of the sound signal of air valve failure. When the diesel engine is running, a variety of acoustic signals mix together. Because these various failures is provided with different characteristic frequencies, it is possible to extract the characteristic frequency of sound signal by time-frequency analysis. Existing separation and extraction methods aim at sound characteristics involve spectrum analysis, short-time Fourier transform, wavelet transform, and so on. And these methods have a successful application [2].

2 THE IMPLEMENTATION OF THE SOUND SIGNAL DETECTION SYSTEM

In this study, the fault that air valve clearance is too large is taken as the research object. The sound signals are collected during normal and abnormal operation of diesel engine. Applying the digital filtering, windowing function, wavelet transform and power spectral analysis, the collected sound signals are processed, and finally extracted the characteristic frequency of the abnormal noise signal.

3 SIMULATION EXPERIMENTS AND DATA ANALYSIS

3.1 *The fault sound signal acquisition*

In this paper, based on the composition of the signal acquisition system, a test stand has been built which can achieve real-time acquisition of the sound signal and data processing, using a microphone as a sound collection device, the sampling frequency being 44.1 kHz. The sound is collected as normal sound signal in the first experimental procedure when the the diesel engine is in the normal operation. And then the exhausting air valve clearance of the first cylinder is increased 0.4 mm and collected the sound as fault signal.

The normal sound signal is shown in Figure 1, and the abnormal sound signal is shown in Figure 2.

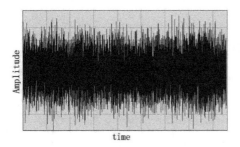

Figure 1. Normal sound signal.

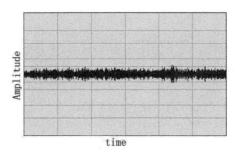

Figure 3. Normal sound signal.

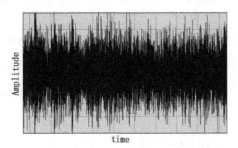

Figure 2. Fault sound signal.

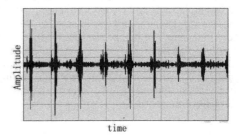

Figure 4. Fault sound signal.

3.2 *The data processing*

In the collected sound signal, it may be mixed with the ambient noise or other signal which must be removed. During the experiment, multiple sets of signal datas are measured in order to improve the accuracy and pertinence of the study. And the filter was applied to suppress noise [3, 4].

3.2.1 *Signal filtering*
Signal filtering is one of the primary means of suppression and elimination of noise. And the signal can be frequency-selective through filter. So that the specific frequency component of the signal can be passed, while the other frequency component can be greatly attenuated. In the present experiment, the same filtering operation was performed on the normal sound signal and the failure sound signal. By manually adjusting the filter bandwidth repeatedly, the maximum different frequency band was found between the normal signal and the fault signal, which was identified as the characteristic frequency band of the fault sound.

The time domain curve of signal filtered is shown in Figure 3 and Figure 4.

3.2.2 *Applying window function*
The math tools commonly used in digital signal processing is the Fourier transform, which only is applicable to the analysis for the entire time domain and frequency domain. However, regarding to the computer signal processing, it is impossible to infinitely long signal processing, while analysis can only be taken a limited time segments. In order to short-term analysis, window function is applied to the signal processing [5].

Once the signal is applied to the window function as well as the energy leakage will be produced. Therefore, when the window function is selected, we should consider the nature of the signal and processing requirements. The window function caused the energy leakage should have as little as possible, while not affect the frequency resolution. Taking into account the Hanning window is more suitable for the analysis of non-periodic continuous signal; the selected window function in this paper is the Hanning window.

Time-domain form of the Hanning window function can be expressed as

$$w(k) = 0.5\left(1 - \cos\left(2\pi\frac{k}{n+1}\right)\right) k = 1, 2, \ldots, n \quad (1)$$

The frequency domain characteristics expressed as

$$W(\omega) = \begin{cases} 0.5W_R(\omega) \\ +0.25\left[\begin{matrix} W_R\left(\omega - \dfrac{2\pi}{N-1}\right) \\ +W_R\left(\omega + \dfrac{2\pi}{N-1}\right) \end{matrix}\right] \end{cases} e^{-j\omega\left(\frac{N-1}{2}\right)} \quad (2)$$

246

where $W_R(\omega)$ is the amplitude frequency character-istic function of the rectangular window:

$$W_R(\omega) = \frac{2\sin \omega T}{\omega T} \qquad (3)$$

where ω is angular frequency, T is the length of the window.

3.2.3 Wavelet transform

The wavelet transform is an analysis method in the time domain at the same time in the frequency domain, which has a good localization character-istic in the time-frequency domain. According to the different frequency components of the signal, wavelet transform can automatically adjust the density of the sampling [6, 7].

In this paper, the filtered sound signal by time was splitted to several sections, whose each length is 1 second. Three sections signal each of from the nor-mal signal and fault signal has been studied. Differ-ent wavelet bases were selected to analyze the sound signal wavelet decomposition, through comparative to select the proper wavelet. The final choice is db10 orthogonal wavelet bases, and then wavelet decom-position of four layers is performed to the signal. The results are shown in Figure 5 and Figure 6.

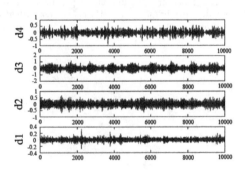

Figure 5. Db10 wavelet decomposition of the normal sound signal after filtering.

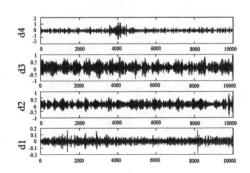

Figure 6. Db10 wavelet decomposition of the failure sound signal after filtering.

3.2.4 Power spectrum analysis

By comparison of Figures 5 and 6, it is difficult to distinguish between the normal signal and the fault signal. Thus, Power spectrum of the signal is obtained through further analytic signal, which can be expressed some statistical characteristics of the signal. In order to extract the characteristic frequency of the fault acoustic signal after filter-ing, the Hilbert envelope and the power spectrum analysis are executed for the detailed signal of the first layer d1 after the wavelet decomposition. As a result, shown by Figure 7 (a), (b), and (c).

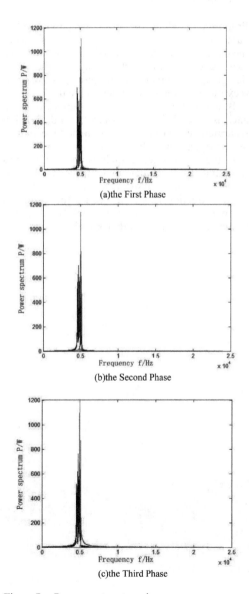

Figure 7. Power spectrum envelope.

Black as a normal signal power spectrum envelope line, gray as failure acoustic signal envelope as shown in Figure 7. Curve comparison of the two colors, it is easy to see that a clear distinction of power spectrum between the fault sound signals with normal sound signal. And three sections of the fault signal peak all appear at the location of 0.5×10^4 Hz, which fully explains the characteristic frequency of sound signal in trouble is about 5 KHz.

4 CONCLUSIONS

Proposed diesel engine fault detection method based on the sound signal. The working status of the diesel engine can be judged based on the unusual sound. After the spectral analysis of the acoustic signal, the characteristic frequency of the signal is extracted as the basis to determine the fault.

Based on the Sound signal of "air valve clearance is too big", the collected sound signal is separated and extracted, and obtained the characteristic frequency is about 5 kHz.

In the paper, wavelet analysis and power spectrum analysis were integrated efficiently, since the power spectrum analysis can achieve the pure fault signal extraction as well as reduce the frequency components of the analyzed signal, filter off frequency interference of other signals. And wavelet analysis has the some advantages of acquiring mathematical details of signal. The experimental results show that the method is feasible and effective.

REFERENCES

Gang, CHEN. Mingfu, LIAO. 2007. Fault Diagnosis for a Rolling Element Bearing Based on Wavelet Analysis. *Science Technology and Engineering* 7(12):2810–2814.

Ghamry, M.El. Reuben, R.L. Steel, J.A. 2003. The development of automated pattern recognition and statistical feature isolation techniques for the diagnosis of reciprocating machinery faults using acoustic emission. *Mechanical Systems and Signal processing* 17:805–823.

Heji, YU. Qingda, HAN. Shen, LI. 2001. Equipments Fault Diagnosis Engineering. BeiJing. Metallurgical Industry Press.

Nianxi, XUE. 2008. Application of MATLAB in digital signal processing. BeiJing. Tsinghua University Press.

Ping, QIN. Heping, FU. Bing, YAN. 2006. Diagnosing of contact friction fault of a plain bearing under dead load based on AE monitoring. Tribology 26(6):585–589.

Shuangxi, JING. Zhanxu, TIE. Yingqi, ZHANG. 2000. Study for fault diagnosis techniques based on wavelet analysis. *Journal of China Coal Society* 25:143–146.

Zengyong, LIU. Baojie, SHI. Hongwen, MA. 2006. Analysized diesel air valve unusual inspection based on cylinder cap vibration signal. *Movable Power Station & Vehicle* 2:31–34.

Control Engineering and Information Systems – Liu (Ed)
© 2015 Taylor & Francis Group, London, ISBN 978-1-138-02685-8

Study on acquisition algorithm of GNSS signal

F.Q. Gao & X.W. Gu
School of Information Science, Zhejiang Sci-Tech University, Hangzhou, China

K. Zhang
Key Laboratory of Integrated Electronic System Technology of Ministry of Education,
University of Electronic Science and Technology, Chengdu, China

ABSTRACT: Soft-radio technology based GNSS receiver is the trend of GNSS receiver product. Now, there are lots of achievements about GNSS signal acquisition algorithm, but those achievements have big limitations such as the parameters of acquisition algorithm can't be adjusted according to the running conditions adaptively and it can't give attention to two or more things such as speed and sensitivity. Therefore, a algorithm that can adjust parameters of GNSS signal acquisition algorithm adaptively according to running conditions is put out in this paper. Finally, The simulation shows that the algorithm this paper can give attention to two or more things such as speed and sensitivity.

1 INTRODUCTION

Globe satellite navigation system "GNSS" can provide information such as the location of the receiver, time for every kind of carrier that running on the land or flying in the shy or sailing on the sea, etc. In present, GNSS has been used into various fields of national economic and daily life such as urban traffic, commercial product transport, telling the exact time and geodetic surveying. Global Position System "GPS" of America, Global Navigation Of Satellite System "GLNOSS" of Russia, Galileo of Europe, COMPASS of China all belong to GNSS.

As He et al. (2011), Yang & Tian (2011) mentioned, GNSS receiver must acquire the signal of GNSS first so as to estimate the doppler frequency in carrier and the rude phase of Pseudo Random Code (PRN) such as C/A code, because the receiver's capacity of tracking signal is limited.

According to the strategy of searching satellite signal. Shen (2012) mention that Acquisition algorithm is divided into serial one and parallel one. serial acquisition algorithm search unit by unit in the arrange of the doppler frequency in carrier and the rude phase of PRN. While parallel acquisition algorithm search parallelly in the domain of carrier's doppler frequency or in the domain of PRN's rude phase. Acquisition algorithm can be divided into four kind that is carrier serial and PRN parallel, carrier serial and PRN serial, carrier parallel and PRN parallel, carrier parallel and PRN parallel.

According to the method of signal processing. Acquisition algorithm is divided into time domain one and frequency domain one. serial acquisition algorithm search unit by unit in the arrange of the doppler frequency in carrier and the rude phase of PRN. While parallel acquisition algorithm.

The most commonly used method in the time domain is the sliding correlation method. Shen (2012) mention that Circular correlation algorithm, The earliest frequency domain acquisition algorithm, is proposed in 1991, which can acquire GNSS signal quickly, but it requires large amount of computation and a high performance of CPU. Therefore, sun (2011) and Borio et al. (2010) put forward some kind of improvement to it. For example, a paper provided a kind of C/A code acquisition algorithms that combines FFT and circular product. patent No. CN 200910072325.X opened a kind of differential correlation cumulation acquisition algorithm that can acquire feeble signal. Which solve the problems of long acquisition time and square consuming which accompany the semibit method and fullbit method in the acquisition of feeble GNSS signal.

We can see from above that there are lots of achievements about the study of GNSS signal acquisition algorithm in the world. But these achievements have their limitation such as that parameters can not be adjusted automatically or performance optimization when situation and other external facts changed as Borio et al. (2010) referred. What's more, these achievements can not solve the problem that speed and sensitivity of acquisition can not be made in the same time. In this paper, we will dicuss this problem.

GNSS signal acquisition is one of the most important function in the mid-frequency processing of GNSS receiver and its capability will impact the performance of GNSS receiver directly. The mainly method of boost speed of signal acquisition is by reducing the computation complexity. The relation of the time of coherent integration, the times of non-coherent integration, and bandwidth of tracking loop is studied to improve acquisition sensibility. For instance, if the time of coherent integration is extended, the acquisition sensibility will be improved. But the change range of tracking signal is lower.

The relation of the time of coherent integration, The times of non-coherent integration, and mobile state of receiver and the input signal to noise ratio is investigated in this paper in order to improve both the acquisition sensibility and the acquisition speed.

2 THE ADAPTIVE ADJUST STRATEGY OF ACQUISITION

Parameter table and control logical are set in order to let different GNSS signal get through the same physical channel by time divide multiplex access, which are used to configure acquisition channel and control the process of acquisition. Submodule of doppler frequency estimation is designed to enhance the acquisition speed. Fuzzy logic module is designed to adjust the time of coherent integration and the times of non-coherent integration according to situation and receiver's mobile state so that the acquisition sensibility and the acquisition speed can be improve simultaneously and the acquisition algorithm is more robust. The configuration of our acquisition algorithm is told in Figure 1.

It can be saw from the above figure that the acquisition algorithm provided in this paper is made some improvement based on the thought of

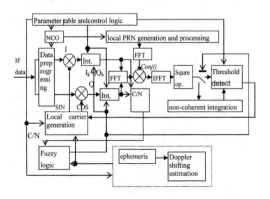

Figure 1. Configuration of acquisition algorithm.

parallel searching in frequency domain. such as follows:

a. data preprocess

The acquisition speed can be enhanced through cutting down the FFT computation data amount by sampling the input intermediate frequency data again.

b. parameter table and control logic

Parameter table and submodule of control logical are used to store the controlling and state parameters in order to let different logic channel running on the same physical channel by time divide multiplex access. Which is also required FIFO data storage to buffer the mid-frequency data.

c. fuzzy logic algorithm—adjusting strategy of acquisition. Its mainly usage is to enhance the ability of the acquisition algorithm to stand all kinds of change of situation and mobile state.

Its input includes speed of the receiver which is output by submodule of ephemeris and doppler frequency shift estimation, carrier to noise ratio (abrre. C/N) which reflect the intensity of the GNSS signal, the type of GNSS signal. The output of fuzzy logic algorithm includes adjusting strategy of acquisition that the GNSS receiver should adapt, such as the time of coherent integration, the times of non-coherent integration.

d. ephemeris and dopplor frequency shift estimation: it calculate the value of doppler frequency shift in PRN and carrier and control acquisition algorithm to work efficiently.

e. threshold judgement.

The maximum of result of square calculation result output by the threshold judgement function is searched (if non-coherent integration is required, non-coherent integration is done before threshold judgement). If The maximum is bigger than the threshold, then the GNSS signal is acquired. And then, This maximum will be compared with the nearest two data and the maximum of these three data will be take out. Accordingly, we can get the phrase shift of PRN, so that we can improve precision.

As the most important part, fuzzy logic algorithm that used to adjust strategy of acquisition automatically is studied in order to enhance the adaptivity of acquisition algorithm and improve both the acquisition sensibility and the acquisition speed simultaneously.

3 MODELING OF THE FUZZY LOGIC ALGORITHM

Receiver maneuvering state estimation information such as velocity and acceleration are used to determine the GNSS receiver's maneuverability. And the signal to noise ratio of correlator output signal is used to estimate the signal strength. Based on these two parameters, the fuzzy logic algorithm that adjust the strategy of acquisition is designed. When the signal is strong enough to be acquired, such as stronger than 38 Db/Hz, the time of coherent integration is set as the periods of PRN so as to acquire the signal as quickly as possible, whereas, when the signal is very weak, such as weaker than 35 Db/Hz, the time of coherent integration is set as several times of the periods of PRN so as to enhance the acquisition sensibility.

The time of coherent integration and the times of non-coherent integration under different C/N are got by computer simulating and analyse, as listed in Table 1 and Table 2. In addition, the up-limit of coherent integration is various according to different situation, for example, it is about 3 ms if the speed is not more than 150 m/s and it is about 1 ms if simultaneously the acceleration is more than 30 m/s².

When the C/N is smaller than usually, it should enhance the time of coherent integration firstly. Only if the time of coherent integration has been up to its maximum, then augmenting the times of non-coherent integration should be considered. In this paper, the maximum of the times of non-coherent integration is set 10. if the GNSS signal can be acquired when maximum is got, then the acquisition of this channel is failure. acquisition algorithm for next channel will be run consequently.

Fuzzy logic algorithm designed in this paper includes modules as follows.

Unit of membership degree simulating. It is used to calculate the function value of member-

ship degree of input/output parameters of the algorithm. The input fuzzy data got in this paper are listed in Table 3, Table 4 and The output fuzzy data can be seen in Figure 2.

Unit of fuzzy logic inference rule contains rule that used to deducing calculation, which are listed in Table 5.

Unit of reasoning and de-fuzzy get strategy of acquisition which includes the time of coherent integration and the times of non-coherent integration.

In this paper, fuzzy logic algorithm designed above is used to adjust strategy of acquisition of the receiver so that it can adapt to the situation and enhance its adaptivity to environment.

Table 3. Fuzzy partition of C/N (units: Db/Hz).

C/N(Db/Hz)	[~35]	[20~45]	[35~]
Fuzzy set	Small	Middle	Big

Table 4. Fuzzy partition of velocity (units: m/s).

Velocity (m/s)	[0~10]	[0~150]	[10~]
Fuzzy set	Zero	Small	Big

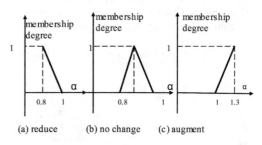

(a) reduce   (b) no change   (c) augment

Figure 2. Membership degree function of fuzzy logic algorithm's output.

Table 1. The times of non-coherent integration under different C/N.

C/N(Db/Hz)	>28	28	27	26	25	24	23	22–21
The times of non-coherent integration	1	2	3	4	4	5	7	9

Table 2. The time of coherent integration under different C/N.

C/N(Db/Hz)	>38	37–35	34	33–32	31	30–29	28
The time of coherent integration (ms)	1	2	3	4	6	7	10

Table 5. Inference rule.

		Velocity		
Output		Zero	Small	Big
C/N	Small	Augment	Augment	Augment (1–3 ms)
	Middle	No change	No change	No change
	Big	Reduce at least 1 ms	Reduce at least 1 ms	Reduce at least 1 ms

The time of coherent integration should be enhanced firstly if the C/N is smaller than usually. Only if the time of coherent integration has been up to its maximum, then the times of non-coherent integration should be augmented. In this paper, the maximum of the times of non-coherent integration is set 10. If the GNSS signal can be acquired when maximum is got, then the acquisition of this channel is over and the acquisition of next channel will be run consequently.

4 SIMULATION RESULT

First, satellite signals in different situation is simulated by soft-simulator, then the performance of the acquisition algorithm is tested by computer simulation.

Signal of one GPS satellite is simulated. Its mid-frequency is at 4.124 MHz with a sampling rate of 16.368 MHz, its C/N is 38 dB/Hz with carrier doppler frequency shift of 2760 Hz and code phase shift of 2972. The result of test the acquisition algorithm are list in Figure3.

Result of original acquisition algorithm with the time of coherent integration being 1 ms is listed in Figure 3(a). if the C/N is reduced by 0.2 dB, it can not get the acquisition result. Whe256 the improved acquisition algorithm can acquire the signal. What's more, when the C/N is reduced by 1 dB the improved acquisition algorithm can still acquire the signal and the acquisition result is listed in Figure 3(b). It can be seen that the performance of improved acquisition algorithm is better than original one's.

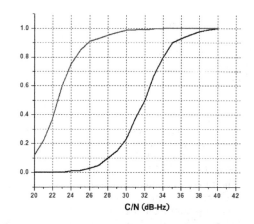

Figure 4. The graph of acquisition detect probability and C/N.

Then, the graph of acquisition detect probability of acquisition algorithm is got by computer simulation, which is listed in Figure 4. It can be seen that the improved acquisition algorithm have a high acquisition detect probability and its acquisition sensibility increase by about 8 dB.

Adjusting C/N till the original acquisition algorithm only can acquire the signal when the time of coherent integration is bigger that 2 ms. The running time of the improved acquisition algorithm proposed by this paper is reduced by 4.2 seconds calculated by tic-toc command (work frequency of CPU is 1.86 GHz and memory storage is 1 G).

It can be proved by computer simulation that the improved acquisition algorithm proposed by this paper can adjust its strategy automatically according to different situation. And it can improve both the acquisition sensibility and the acquisition speed simultaneously.

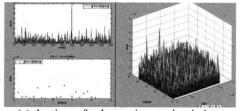

(a) the time of coherent integration is 1ms

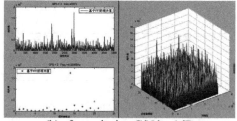

(b) after reducing C/N by 1dB
(the time of coherent integration is 2ms)

Figure 3. Results of test.

ACKNOWLEDGMENT

This work was financially Supported by The open fund of Key Lab of Integrated Electronic System Technology of Ministry of Education, China, and an innovation project of Zhejiang province university students' science and technology.

REFERENCES

Borio, D., O'Driscoll, C., et al. Composite GNSS Signal Acquisition over Multiple Code Periods[J]. IEEE Transactions on Aerospace and Electronic Systems, 2010, 46(1):193–206.

He xiaofeng, Nie zuguo, et al. Performance analysis on high dynamic signal acquisition aided by SINS for GNSS satellites[J], Journal of Chinese Inertial Technology, 2011, 19(4):447–451. (In Chinese).

Ma yongkui, Zhang yi, et al. Modified method of high dynamic & high sensitivity GPS signal acquisition[J]. Systems Engineering and Electronics. 2009, 31(2): 265–269. (In Chinese).

Shen yingjie. Research on Acquisition and Code Tracking Technology of Weak GNSS Signal[D]. Hangzhou: Zhejiang University, 2012. (In Chinese).

Sun, K. Differential channels combining strategies for composite GNSS signals acquisition[C]. Institute of Navigation-International Technical Meeting 2011,2: 1218–1231.

Yang, L., Tian, J.A new threshold setting method of GNSS signal acquisition under near-far situation[J]. IEICE Transactions on Communications, 2011, E94-B(7):2082–2091.

Control Engineering and Information Systems – Liu (Ed)
© 2015 Taylor & Francis Group, London, ISBN 978-1-138-02685-8

The design and implement of adaptive speech enhancer

B.Y. Huang & P. Yang
Naval Marine Academy, Guangzhou, China

ABSTRACT: The traditional hospital information system could be developed into telemedicine system. The information of text, speech, videos must be transmitted. And the speech signal is mix with much of environment noise in the trail telemedicine system, which can greatly restrict the use of the telemedicine system. In order for improving the definition of speech, an adaptive speech enhancer based on autocorrelation is proposed. The frequency of the enhancer and the principle of speech enhancing are analyzed in theory. And a speech enhancer device using the method is designed and implemented. The hardware structure is analyzed. The result from tests demonstrated that the speech signal in the white noise can be enhanced effectively using the method, and performance of the device meet the demand of speech enhancing in the telemedicine system.

1   INTRODUCTION

To take good care of soldiers' stationed remote area or island, hospitals of PLA are constructing long-range medical treatment system. The telemedicine system is a new type of medical treatment. Applying communication network and computer multimedia, we can transmit multimedia medical information to carry out remote medical treatment. This transfer medical information includes text, picture, video and audio information. We found in practical interference occurs during speech signal collection because of limit of remote island and characteristic of medical. So the speech signal is hardly understand. This is the reason why we design filter of speech signal.

This paper presents an adaptive filter of speech based on autocorrelation. The filter can improve definition of voice. The hardware of filter is tight and low power consumption. Power supply is provided by USB. The performance of this hardware is excellent and stable in practical use.

2   ADAPTIVE SPEECH ENHANCER BASED ON ASCE

In 1982, Nasir Ahmed and S. Vijayendra put forward the algorithm of adaptive short time autocorrelation filter (Ahmed 1982). The structure of method and performance of frequency is as following. The structure of ASCE method is described as Figure 1. This algorithm is a typical algorithm of open loop adaptive filtering. At present, we cannot found document about apply the algorithm to adaptive enhance.

2.1   ASCE algorithm

Assumed input signal $x(n)$ is sums of speech signal $s(n)$ and white noise $v(n)$. Δ denote time delay and used to inverse correlation. The result of correlation $R_{x,x-\Delta}(n,l)$ is used as infinite impulse response of adaptive filter. So the output of filter is

$$y(n) = \sum_{i=0}^{2L} h(i,n)x(n-i) \tag{1}$$

$$h(i,n) = R_{x,x-\Delta}(n,-L+i) \tag{2}$$

Cross correlation $R_{x,x-\Delta}(n,l)$ can be calculated as:

$$R_{x,x-\Delta}(n,l) = \beta R_{x,x-\Delta}(n-1,l)$$
$$+ (1-\beta)x(n)x(n-\Delta-l) \tag{3}$$

where $l \leq |L|$, $0 < \beta < 1$ denote smooth parameter.

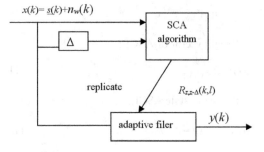

Figure 1.   ASCE algorithm.

2.2 Steady characteristic of ASCE algorithm

We assumed input signal $x(n)$ is sums of brand band white noise and sine signal to describe steady characteristic of ALE algorithm as follow:

$$x(n) = v(n) + s(n) \tag{4}$$

where $v(n)$ is brand band white noise of zero mean, power is δ_n^2.

$s(n)$ is discrete sine signal and defined as

$$s(n) = Ae^{jw_0 nT} \tag{5}$$

where A is amplitude, ω_0 is frequency, T is discrete sampling interval.

To describe ASCE algorithm, we define these vectors, where in $M = 2L$.

Input vector:

$$\mathbf{X}(n) = [x(n), x(n+1), \ldots, x(n+M-1)]^T \tag{6}$$

Weight coefficient vector:

$$\mathbf{W}(n) = [w_1(n), w_2(n), \ldots, w_M(n)]^T \tag{7}$$

Delay vector:

$$\mathbf{q} = [1, e^{-jw}, e^{-j2w}, \ldots, e^{-j(M-1)w}]^T \tag{8}$$

Input sine signal vector:

$$\mathbf{S}(n) = Ae^{jw_0 nT}[1, e^{-jw}, e^{-j2w}, \ldots, e^{-j(M-1)w}]^T$$
$$= Ae^{jw_0 nT}\mathbf{q} \tag{9}$$

Cross correlation matrix between input signal and delayed signal:

$$R_{x,x-\Delta} = E[x(n)\mathbf{X}^T(n-\Delta)]$$
$$= E[x(n) \cdot (Ae^{jw_0(n-\Delta)T}\mathbf{q})^T]$$
$$= A^2 e^{-jw_0 \Delta T}\mathbf{q} \tag{10}$$

Weight vector of ASCE:

$$h(l) = R_{x,\,x-\Delta} \tag{11}$$

So steady-state ASCE:

$$y(n) = X^T(n-\Delta)H^* \tag{12}$$

Impulse response function $H(w)$ of filter as follows equation:

$$H(w) = e^{-j\Delta wT}H(w)$$

$$= A^2 e^{-j(w-w_0)\Delta T}\sum_{m=0}^{M-1} e^{-jm(w-w_0)T}$$

$$= A^2 \cdot \frac{1-e^{jM(w-w_0)T}}{1-e^{j(w-w_0)T}} \cdot e^{-j(w-w_0)\Delta T}$$

$$= A^2 \cdot \frac{\sin\left(\dfrac{M}{2}(w_0-w)T\right)}{\sin\left(\dfrac{1}{2}(w_0-w)T\right)} \cdot e^{j\frac{M-1}{2}(w_0-w)T}$$

$$\cdot e^{j(w_0-w)\Delta T}$$

$$= A^2 \cdot \frac{\sin\left(\dfrac{M}{2}(w_0-w)T\right)}{\sin\left(\dfrac{1}{2}(w_0-w)T\right)} \cdot e^{j\left(\frac{M-1}{2}+\Delta\right)(w_0-w)T} \tag{13}$$

where $\left|\sin\left(\dfrac{M}{2}(w_0-w)T\right)\middle/\sin\left(\dfrac{1}{2}(w_0-w)T\right)\right| \leq M$

Frequency response of filter is

$$|H(w)| = A^2 \cdot \frac{\sin\left(\dfrac{M}{2}(w_0-w)T\right)}{\sin\left(\dfrac{1}{2}(w_0-w)T\right)} \leq A^2 M \tag{14}$$

If $w = w_0$, $|H(w)| = A^2 M$.

If noise is gauss noise, delayed Cross correlation between signal of delay time Δ and original signal can restrain effect of gauss noise. This method makes use of this property.

3 DESIGN AND IMPLEMENT OF ADAPTIVE SPEECH ENHANCER

3.1 Requirement of system

To fulfill requirement of remote medical speech enhancer, adaptive speech enhancer should have function as follow: the hardware of filter is tight and low power consumption; the hardware can enhance speech signal in real time (Li 2009).

In adaptive enhancer, received speech signal is sampled and converted to discrete signal by AD. The discrete signal is filtered by adaptive filter. The result is exported by DA.

We can online DSP program by JTAG. The design can ensure simply apply and is convenient to upgrade.

3.2 Structure of system

In telemedicine system probationer ship, there is a problem of much of environment noise contaminating speech signal. We design an adaptive speech enhancer based on autocorrelation. The result from tests demonstrated that the speech signal in the white noise can be enhanced effectively using the method, and performance of the device meet the demand of speech enhancing in the telemedicine system.

An adaptive speech enhancer based on autocorrelation includes five parts: input signal adjustment, signal collection, DSP signal processor, DA converter, output signal adjustor. The key parts are A/D converter, D/A output, DSP signal processor (Yang 2008). The part of input is DC coupling amplifier and provide credible analog signals to postposition circuit. The D/A converters are using AD5445 of Analog Device.

The sampling frequency is 44.1 kHz. The DSP system is TMS320C5402 produced by TI. We use 16 bits fixed point DSP which has the calculation performance over 10MIPS. It combines an advanced modified Harvard architecture(with one program memory bus, three data memory buses, and four address buses), a CPU with application specific hardware logic, on-chip memory, on-chip peripherals, and a highly specialized instruction set. On-chip memory is 128 KB. D/A is ADS831 produced by Burr-Brown. Read-Write clock, operation state, A/D and D/A are control by DSP.

3.3 Design examples

In order to validate the presented adaptive speech enhancer, some simulation results are presented in this section. The speech signal is shown in Figure 2(a). Figure 2(b) shows is noise-free speech signal. In Figure 2(c), autocorrelation enhancer perform well.

Figure 2(b) illustrate how the shape of the signal contaminated by additive noise is blurred. Figure 2(c) shows the enhanced signal using adaptive speech enhancer. We can easily understand what speech said because enhanced signal can preserve signature of original noise-free signal.

In Figure 2, we show adaptive speech enhancer based on autocorrelation perform well in Gaussian white noise.

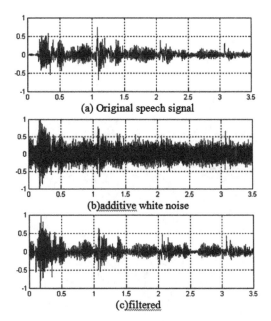

Figure 2. Plot of adaptive enhancer.

4 CONCLUSION

We design an adaptive speech enhancer using adaptive autocorrelation algorithm. The result demonstrated that the speech signal in the white noise can be enhanced effectively using the method, and performance of the device meet the demand of speech enhancing in the telemedicine system. This hardware of enhancer easily handles and is applicable to many cases. The adaptive algorithm based on correlation presented is simple structure and operates in real-time.

REFERENCES

Ahmed N. & S. Vijayendra. 1982. An algorithm for line enhancement, *Proc. IEEE,* 70(12): 1459–1460.
Bo Li & Guang Zhao. 2009. Study on video image processing system based on DSP, *Journal of Shenyang Institute of Engineering (Natural Science),* 3(5): 272–274.
Peng Yang, Sheng zhou & Bingcheng Yuan. 2008. Study on Real-time Simulation of Underwater Target Radiating Noise, *Audio Engineering,* 32(8): 37–40.

Control Engineering and Information Systems – Liu (Ed)
© 2015 Taylor & Francis Group, London, ISBN 978-1-138-02685-8

A novel method of calculating the weight of Adaptive Side-Lobe Cancellation

N. Pang
Department of Information and Electronics, Beijing Institute of Technology, Beijing, China

P.J. Xu
Beijing Electric Power Design Institute, Beijing, China

H.Q. Mu
Beijing Radio Measurement Institute, Beijing, China

ABSTRACT: This paper first introduces the basic principle of adaptive side-lobe cancellation. Then a novel method of calculating cancellation weight is presented. The modified method can eliminate the decline of the performance due to the diseased self-interrelated matrix. Compared with traditional methods of computing side-lobe canceling weight, this method has more effective results of ECCM.

1 INTRODUCTION

With the development of radar jamming technique, radar anti-jamming measurement becomes more and more important. Interference includes active interference and passive interference. Active interference is more difficult to overcome. Because the main lobe is so narrow that interference is difficult to come through but easy to come through side-lobe. Lower the antenna side-lobe from a certain degree can repress the source interference, but in general, interference signal is stronger than useful signal and usually drowns the useful signal, causes the difficulty aggrandizement to examine signal. Usually the valid method of dealing with this kind of interference from the side-lobe is Adaptive Side-lobe Cancellation (ASLC) technique. As a kind of space filter technique, ASLC has already been adopted at home and abroad. The key step to use this technique is to compute weight value.

This paper introduces a novel method of calculating cancellation weight which can eliminate the decline of the performance due to the diseased self-interrelated matrix.

2 THE PRINCIPLE OF ASLC

2.1 *Open loop adaptive ASLC*

ASLC system usually consists of primary element and auxiliary element. The primary element includes antennas with high gain and explicit orientation. The auxiliary element includes one or several omni-

directional antennas whose gain corresponds to the first side-lobe gain of the main antenna. The main beam of the primary antenna always controls alignment signal arrival direction. In most cases, there will be a lot of interference from side-lobe antenna into the primary element and at the same time into the auxiliary element. The ASLC system calculates the weight coefficient according to the receiving signal from the primary and auxiliary antenna. Then the auxiliary element is determined adaptively in an unwanted interference in the primary element. The resulting output, or the residual, is formed by subtracting the auxiliary element output from the primary element. The output of the interference cancellation tends to zero, so as to achieve the purpose of interference suppression (Ding 1989). Its principle diagram is shown in Figure 1.

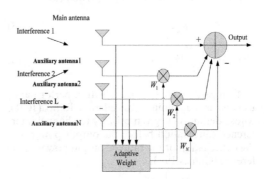

Figure 1. Open loop adaptive sidelobe cancellation principle block diagram.

Figure 1 is for N sub-channel adaptive side-lobe cancellation principle block diagram. $X = [X_1, X_2, \ldots X_N]$ said N auxiliary antenna receiving signal. $W = [W_1, W_2, \ldots W_N]$ said weighted coefficient. The formula is below in (1).

$$V = Y - W^T \cdot X \tag{1}$$

where V = cancellation output; T = transposition; and Y = primary antenna receiving signal.

2.2 *The critical method of calculating weights*

Cancellation residual is the main antenna signal minus the weights vector and auxiliary antenna signal of the inner product. Our aim is to make cancellation residual power minimum. This is called the Least Mean Square (LMS) criterion, with statistical expressed as that $E[|V|^2]$ reaches a minimum:

$$
\begin{aligned}
P &= E\left\{|V|^2\right\} \\
&= E\left\{\left(Y - W^H \times X\right) \times \left(Y^* - W^T \times X^*\right)\right\} \\
&= E|Y|^2 - R_{XY}^H \times W - W^H \times R_{XY} \\
&\quad + W^H \times R_{XX} \times W
\end{aligned} \tag{2}
$$

In (2), * = taken conjugate; $E[\cdot]$ = taken mean; R_{XY} = the main and auxiliary channel cross-correlation function matrix; R_{XX} = Auxiliary channel autocorrelation function matrix. To type (2) weights vector for derivation can get the equation (3).

$$\frac{\partial P}{\partial W} = -2R_{XY} + 2(R_{XX} \times W) = 0 \tag{3}$$

Finishing (3), we can get (4).

$$R_{XX}W = R_{XY} \tag{4}$$

When (4) of the autocorrelation matrix is nonsingular matrix, we can get (5) and (6).

$$W_{opt} = R_{XX}^{-1} \times R_{XY} \tag{5}$$

$$r(t) = y(t) - \sum_{i=1}^{N} W_n x_n(t) \tag{6}$$

To measure side-lobe cancellation performance usually uses Jamming Cancellation Ratio (JCR), whose definition is given priority to channel interference cancellation before the output power and the cancellation of the ratio of output power interference (Games 1993).

$$JCR = \frac{E\{X^2\}}{E\{V^2\}} \tag{7}$$

In the calculation of the JCR we can take main antenna in the direction of interference gain and side lobe cancellation after processing interference in the direction of the ratio of the gain.

3 THE NOVEL METHOD OF CALCULATING THE OPTIMAL WEIGHTS

3.1 *The principle of the novel method*

The process of calculating the optimal power value of is actually solution of Wiener-hopf equation. We can use multiple coefficient equations to solve the equation, or can check all primary gauss elimination method to calculate directly, also can use transformation method to calculate the matrix (Richard 1993). But in one or some directions there is no interference. Once it is appeared, side-lobe cancellation performance will be serious deteriorated. To avoid this kind of situation, the sick matrix and the distribution of eigenvalues is closely related. When the characteristic value size difference between the appearance of morbid R_{XX} is for singular matrix or morbid matrix, with direct inversion method or direct solution of equations to calculate the weight, the method will influence the cancellation performance, through the characteristic value remove does not exist in the direction of interference elements, can eliminate the bad influence for the cancellation of the results. In this paper based on the ideas in the matrix transformation method process, the AIC criterion can remove the interference into the characteristic value, and finally calculate the optimal weights.

As R_{XX} is for Hermite complex matrix particularly, it can be transformed from complex matrix into real symmetric matrix (Li 2005). Then we can do real symmetric matrix transformation to calculate the eigenvalues and eigenvectors, thus obtain the optimal weights. There are several methods to calculate the eigenvectors and eigenvalues of real symmetric matrix. First, we can use Hessenberg-QR, but this method can only find out characteristic value, not directly get feature vector. Secondly, Jacob method, this method for low order matrix (<5) solving accuracy is higher, for high order accuracy variation. Thirdly, it is Householder-QR method (Lu 2007), this kind of method for computation moderate, not morbid matrix precision, for morbid matrix are also iterative error, often lead to get eigenvectors and eigenvalues of results are inaccurate. Fourth, Jacob for improved Jacob pass method, this method operation efficiency is relatively slow, for any type of torque operation accuracy is higher, this paper uses the method.

3.2 The steps of the novel method

The specific steps of the novel method of calculating optimal weights of are as follows:

- Transform the N Hermite matrix into 2 * N real symmetric matrix S.

$$R_{XX} = Z * D * Z^T \qquad (8)$$

where, R_{XX} = complex matrix,

$$R_{XX} = A + j * B, Z = U + j * V, D = \Lambda \qquad (9)$$

Here, A and B, U, V, D are for real matrix of diagonal matrix. Requirements take eigenvectors and eigenvalues of, after transformation can be transformed into realistic symmetric matrix S eigenvectors and eigenvalues of the problem. That is below in (10):

$$S = \begin{bmatrix} A \ ..... -B \\ B \ ...... \ A \end{bmatrix} \qquad (10)$$

- Use of real symmetric matrix S Jacob-pass method obtains its 2 * N eigenvalues and corresponding eigenvectors.
- Transform eigenvalue and eigenvector of the matrix S into eigenvalues and corresponding eigenvectors of the matrix R, and form a diagonal matrix and orthogonal matrix.
- After the characteristic value diagonal matrix inversion, the AIC criterion (M Wax & T Kailath 1985) removes the items in the direction which does not exist interference, and forms a new diagonal matrix DN_{N*N}.
- The inversion of matrix R is obtained by matrix multiplication.

$$R^{-1} = Z * DN_{N*N} * Z^T \qquad (11)$$

- Weights can be obtained by matrix multiplication.

$$W = R^{-1} * P \qquad (12)$$

4 THE SIMULATION AND ANALYSIS OF THE RESULTS

Building model and simulating accords to the five auxiliary antennas, assuming that the JNR = 36 dB, signal to interference ratio = −10 dB, the main antenna pattern main lobe width of 3 dB for 6°, main lobe to the first side lobe are −13 dB. Five auxiliary antennas is for wide side lobe antennas. The gain is slightly higher than the first one side lobe main antenna gain.

Main antenna in target tracking, has been directed towards the target direction. Interference signal is for broadband white noise interference. Receiver noise is set to white Gaussian noise. It is assumed that the main lobe of the main antenna receives the target signal, and the side lobe receives

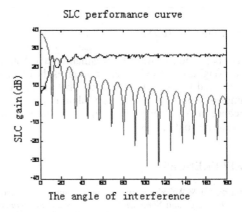

Figure 2. 0–180 degrees JCR.

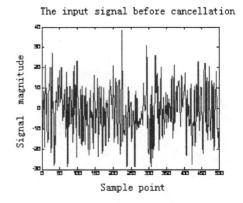

Figure 3. Echo signal before the cancellation.

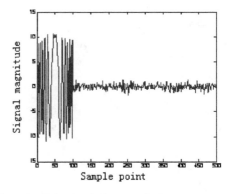

Figure 4. Echo signal after cancellation.

261

the interference signal. Five auxiliary antennas receive four interference signals of the environment. JCR and echo before and after the cancellation is shown in Figures 2–4. In the range of 0 to 180 degrees of the main antenna patern, the JCR is more than 25 dB. It is a more satisfactory result in ECCM system.

5 CONCLUSIONS

The paper puts forward a novel method to calculate the weights of ASLC system, and gives the specific steps and the simulation results. This method can improve the interference cancellation ratio, and reduce or avoid bad effect on the cancelling results of the morbid matrix.

REFERENCES

Jia-qi, Lu & Wei-xiong, Bai & Jiang-chuan, Teng. 2007 *The design and Simulation of adaptive sidelobe canceling system based on digital signal processing*. Aeronautical computing technique, vol. 37, June: 10–12.

Jun-ping, Zhang & Wang-jie, Song & Zi-jing, Zhang & Min Hu. 2008. *Performance analysis and Simulation of adaptive sidelobe canceller. Radar science and technology*. vol. 6, June: 486–491.

Lu-Fei, Ding. 1989. *The Principle of Radar*. Beijing: Science Press.

M Wax & T Kailath. 1985. *Detection of Signals by Information Theoretic criteia*. IEEE Transactions on Acoustics, Speech and Signal Processing, 33(2):387–392.

Richard, A. Games & Willard, L. Eastman & Michael, J. Sousa. 1993. *Fast Algorithm and Architecture for Constrained Adaptive Side-Lobe Cancellation*. IEEE Trans. Antennas and Propagat. vol. AP-41, NO. 5.

Shu-Jie, Zhao. 2010. *Radar Signal Processing. Beijing*: Tsinghua University Press.

Xiao-bo, Li & Wang-wei, Xue & Zhi-yong Sun. *A method for solving complex eigenvalues of Hermite matrix*. Data acquisition and processing, vol. 20, pp. 403–406, April, 2005.

Control Engineering and Information Systems – Liu (Ed)
© 2015 Taylor & Francis Group, London, ISBN 978-1-138-02685-8

Key frame matching and dual direction Monte Carlo particle filter based video tooning

J.B. Song
School of Information Engineering, Communication University of China, Beijing, China

J.L. Wan
College of Computer and Information Engineering, Henan University of Economics and Law, Zhengzhou, China

L. Ye & Q. Zhang
Key Laboratory of Media Audio and Video, Ministry of Education, Communication University of China, Beijing, China

ABSTRACT: This paper aims at the difficulty that lack of observation model and high-dimensional sampling in video tooning, proposes a method based on key frame matching and dual-directional Markov chain Monte Carlo sampling of video motion redirection. At first, after extracting the key frame of a given video, By affine transformation and linear superposition, the subject initializes the video's space-time parameters and forms the observation model; Secondly, in each space-time, based on the bi-directional Markov property of each frame, This paper proposed a dual-directional Markov chain Monte Carlo sampling particle filter structure and takes full advantage of the relationship of the front and back frame of the parameters to estimate motion redirection parameters. At the same time, for high-dimensional sampling problem, the subject according to the directional parameters' correlation implements classification of skeleton parameters-morphological parameters-physical parameters, proposes a hierarchical genetic strategy to optimize the output parameters and improves the efficiency of the algorithm. The research of this paper will produce an efficient and prominent animation expressive video motion redirection method and play an important role on video animation of the development.

1 INTRODUCTION

Nowadays with the blooming development of new media and cultural creative industries, the video animation has drawn so many researchers' attention due to animation production efficiency and animation expression. In order to reduce the number of key frames painted workload, "performance animation" gradually grows based on motion capture. But motion capture only extracts the information of articulation point demarcated ahead of time and discards a large number of extremely rich animation material elements contained in the video information. While making full use of these elements can help people enhance the efficiency of animation production.

The existing research can be divided into two categories for the draw of the role movement in the animation. One is for running water (Yu 2008), smoke (Selle 2004), rain (Zhang 2009), snow (Yu 2007), such natural phenomena as the typical irregular fuzzy objects, emphasizes the whole effect, emphasizes the visual style, and uses the shape change of the same object to form a motion. Second

is for parametric modeling to the render objects in the video, and then the model parameters are estimated and corrected through the optimized theory and the filtering technology, and finally formed animation characters motion by the mapping. This stylized method of video motion is called motion redirection. (Li 2010) proposes an animation generation algorithm of stylized redirection of falling leaves which is used to produce the stylized animation of the flying and falling leaves. On the basis of the falling leaves gesture redirection algorithm and path synthesis algorithm, the algorithm uses the movement details of different falling leaf as a unit, achieves stylized animation effects of a variety of types of leaves flying and falling through the dynamic control, as shown in Figure 1.

This kind of animation generation method is simple, but the effect expressive force is not strong. Root cause is for the target characterization only limited to the location in space and the outer envelope, skeleton structure and physical characteristics (such as illumination change, topology structure, color information, etc.) do not show up, and the skeleton motion of the target role based on the

Figure 1. Achievement of stylized animation of falling leaves.

joints just is the most stylized motion information. After extending characterization parameters from the shape point to the multi-type parameter set of skeleton points, shapes, and physical characteristics, the corresponding parameter estimation and redirection are extended to the high dimension and nonlinear Bayesian estimation problem. This paper makes in-depth study in order to solve this problem, proposes a method based on key frame matching and dual-directional Markov chain Monte Carlo sampling of video motion redirection.

2 RESEARCH TARGETS

Video motion redirection is in the case of the given video sequence I and parameterized animation object, the estimate of animation role model related parameters based on video sequence I. Each redirection parameter should include U points of skeleton structure characterization (S_i), V points of the target outside contour characterized by R_j, and the target morphological parameters set F = [T, D, C], T stands for the possible topology transformation of the object, D stands for the illumination changes, C stands for color changes, as shown in Figure 3.

Define g = [S, R, F], video motion redirection is to obtain the Maximum A Posteriori (MAP) in the statistical learning, namely:

$$(g_0, \ldots g_\tau)_{opt} = \arg\max p(g_0, \ldots g_\tau \mid I_0, \ldots I_\tau) \quad (1)$$

$$p(g_\tau \mid g_1, \ldots, g_{\tau-1}) = p(g_\tau \mid g_{\tau-1}) \quad (2)$$

$$p(I_1, \ldots, I_\tau \mid g_1, \ldots, g_\tau) - \prod_{t=1}^{\tau} p(I_t \mid g_t) \quad (3)$$

Among them, $(I_0, \ldots, I_\tau)$ stands for video sequence, $(g_0, \ldots, g_\tau)$ stands for the parameters set of the animation characters corresponding to the characterization of video sequences.

Because there is difference between the animation objects and the referred object in the video in shape, color and texture, and many other physical characterizations, the observation model is missing. Two assumptions in the sense of time domain recursive estimation (Eq. (2) the Markov assumption in the state transition and the observation conditional independence assumption in the Eq. (3)) can't be established. So the traditional particle filter process can't be directly applied to the project to solve the problem of video motion redirection. Some information can be used lies in that animation model will form some key gestures generated from the animation, accordingly, for a given video sequence, some key gestures in the key frames are extracted as the a priori of the data processing. We define $(t_1, t_2, \ldots, t_T)$ as the moments of key frames of the video sequence, each key frame contains key gestures $\{KP^* (m)\}_{m=1:M}$ remain to match the target, M is the number of key gestures, accordingly, define the animation model key gestures as $\{KP (m)\}_{m=1:M}$, then KP (m)~$p(g_{ti}|I_{ti})$. Through $\{KP^* (m)\}_{m=1:M}$ and $\{KP (m)\}_{m=1:M}$, each key gesture redirection mapping h_m can be obtained, namely the $KP^* (m) = h_m(KP (m))$.

Through the above definition, Eq. (1) can be extended as follows:

$$
\begin{aligned}
& p(g_1, g_2, \ldots g_\tau \mid I_1, I_2, \ldots, I_\tau) \\
& = p(g_1, g_2, \ldots g_\tau \mid I_1, I_2, \ldots, I_\tau, g_{t_1}, g_{t_2}, \ldots g_T) \\
& \quad \cdot p(g_{t_1}, g_{t_2}, \ldots g_T \mid I_{t_1}, I_{t_2}, \ldots I_T) \\
& = p(g_1, g_2, \ldots g_\tau \mid I_1, I_2, \ldots, I_\tau, h) \\
& = \prod_{t=t_1}^{T-1} p(g_t, g_{t+1}, \ldots g_{t+\delta} \mid I_t, I_{t+1}, \ldots I_{t+\delta}, h) \\
& = \prod_{t=t_1}^{T-1} p(I_t, I_{t+1}, \ldots I_{t+\delta}, h(g_t), h(g_{t+1}), \ldots h(g_{t+\delta})) \\
& \quad \cdot p(g_t, g_{t+1}, \ldots g_{t+\delta}) \quad (4)
\end{aligned}
$$

The objective of the research can be summarized as:

a. particle filter structure design that is suitable for video stylized motion, namely the solution of the Eq. (4). It specifically includes: the redirection mapping method of the components [S, R, F] of the video motion redirection parameter g and the matching of the key frames; the generate way of every moment parameters status based on h_m, inside the time and space body formed from the key frames; optimization strategy of state estimation after having generated the time-space body internal parameters status.

b. based on the above key technology to implement a hierarchy video motion redirection structure of the key frames—time-space body—redirection video sequence, as shown in

264

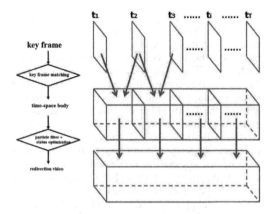

Figure 2. The hierarchy video motion redirection structure of the key frames—time-space body—redirection video sequence.

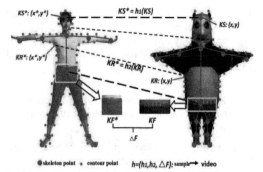

Figure 3. Key frame matching in the video motion redirection.

Figure 2, at the same time build video redirection analysis and implementation platform.

3 THE RESEARCH CONTENT

3.1 Affine transformation and linear superposition based gesture redirection and key frame matching

Each redirection parameter should include U points of skeleton structure characterization (S_i), V points of the target outside contour characterized by R_j, and the target morphological parameters set $F = [T, D, C]$, T stands for the possible topology transformation of the object, D stands for the illumination changes, C stands for color changes, as shown in Figure 3. Gesture redirection contains three aspects of content: skeleton point redirection mapping, namely the mapping h_1 in Figure 3, contour point redirection mapping, namely the mapping h_2 in Figure 3, and the statistical matching of the key frame physical features, namely ΔF shown in Figure 3. Gesture redirection and key frame matching forms into calculation of $h = [h_1, h_2, \Delta F]$.

Because S and R are point feature, we define $L = [S, R]$ represents the target object gesture, $KL \in L$ for the key gesture of the target. This paper uses several key gestures to cover the whole gesture space, then uses the stack of the key gestures to show the actual gesture of the target. The gesture change is described as a combination of two types of deformation: 1) affine deformation, including translation, rotation, scaling and shearing deformation; 2) key gesture deformation, achieved by interpolation to a corresponding set of key gestures, these key gestures need to be manually selected in advance. While ΔF is obtained by the linear superposition of F information in each key gesture.

3.2 Dual-directional Markov chain Monte Carlo sampling based video time-space body filter

After defining the video key frames and the key frames matching, in each time-space body formed by the key frames, the calculation of $h = [h_1, h_2, \Delta F]$ is able to provide the necessary observation model for the implementation of the particle filter and make the constraint relations of the previous and next frames inside the time-space body, namely Bidirectional Markov Chain Monte Carlo Sampling.

Due to the property of the Markov Chain (the unidirectional Markov property is equivalent to the bidirectional Markov property), we can get the following bidirectional Markov property:

$$p(g_t \mid g_1, ..., g_{t-1}, g_{t+1}, ..., g_N) = p(g_t \mid I_{t-1}, I_{t+1}) \quad (5)$$

While observing independence still holds:

$$p(I_1, ..., I_\tau \mid g_1, ..., g_\tau) = \prod_{t=1}^{\tau} p(I_t \mid g_t) \quad (6)$$

So we can get:

$$p(g_t \mid g_1, ..., g_{t-1}, g_{t+1}, ..., g_N, I_1, ... I_N)$$
$$= c_t p(g_t \mid g_{t-1}, g_{t+1}) p(I_t \mid g_t) \quad (7)$$

When we have video time-space body, we can relax the requirement:

$$p(g_t \mid g_1, ..., g_{t-1}, g_{t+1}, ..., g_N, I_1, ... I_N)$$
$$= c_t p(g_t \mid g_{t-1}, g_{t+1}, I_{t-1}, I_t, I_{t+1}) \quad (8)$$

More generally:

$$p(g_t \mid g_1, ..., g_{t-1}, g_{t+1}, ..., g_N, I_1, ... I_N)$$
$$= c_t p(g_t \mid g_{\partial t}, I_{\partial t}, I_t) \quad (9)$$

Among them, ∂t represents the neighboring frame. Dynamics model and observation model couple at this time, which forms dual-directional Markov chain Monte Carlo sampling based video time-space body filter. And the redirection parameter estimate of each frame video image is formed within each time-space body.

3.3 *Video redirection parameters optimization based on hierarchical genetic optimization strategy*

After getting the posterior probability density distribution of the video motion redirection parameters in the use of particle filter, because of high dimension of the parameters, and because the used particle number is limited, we can't guarantee that we can produce the optimal solution of the results after filtering. So we need to take a certain optimization strategy to find the optimality.

Hierarchical genetic optimization strategy is proposed as a tree optimization model, this model has three necessary assumptions:

a. the parameters in the same level of are independent of each other;
b. on the same branch, high level parameter distribution is independent of the low level parameters;
c. all parameters of the same branch eventually has a same measurement model.

After analyzing the optimal status variables parameter of in this paper, if it can be divided into the levels of skeleton structure points (S)—contour points (R)—physical characteristics (F), then the level hierarchy is in accordance with the above three necessary assumptions. So the parameters optimization of the video redirection will be realized by using hierarchical genetic optimization strategy.

4 SUMMARY

Nowadays with the blooming development of new media and cultural creative industries, the video animation has drawn so many researchers' attention due to animation production efficiency and animation expression. This paper aims at the difficulty that lack of observation model and high-dimensional sampling in video tooning, propose a method based on key frame matching and dual-directional Markov chain Monte Carlo sampling of video motion redirection. At first, after extracting the key frame of a given video, By affine transformation and linear superposition, the subject initializes the video's space-time parameters and forms the observation model; Secondly, in each space-time, based on the bi-directional Markov property of each frame, This paper proposed a dual-directional Markov chain Monte Carlo sampling particle filter structure and takes full advantage of the relationship of the front and back frame of the parameters to estimate motion redirection parameters. At the same time, for high-dimensional sampling problem, the subject according to the directional parameters' correlation implements classification of skeleton parameters-morphological parameters-physical parameters, proposes a hierarchical genetic strategy to optimize the output parameters and improves the efficiency of the algorithm. The research of this paper produces an efficient and prominent animation expressive video motion redirection method and plays an important role on video animation of the development.

ACKNOWLEDGMENT

This paper is supported by National Natural Science Foundation of China (61201236) and Engineering Planning Project of Communication University of China (XNG1230).

REFERENCES

Li 2010. Haiyan Li, Qingjie Sun, Xiaodan Liu, Qin Zhang, Example-based Stylized Trajectory Synthesis of Falling Leaf, 2010 International Conference on Computer Engineering and Technology, Chengdu, China.

Selle 2004. Andrew Selle, Alex Mohr, Stephen Chenney, Cartoon Rendering of Smoke Animations, Proceedings of the 3rd international symposium on Non-photorealistic animation and rendering, Annecy, France, 2004, pp. 57–60.

Yu 2007. Jinhui Yu, Xinan Jiang, Haiying Chen, Cheng Yao, Real-time Cartoon Water Animation, The Journal of Computer Animation and Social Agent, vol. 18, 2007, pp. 405–414.

Yu 2008. Yu Jinhui, Patterson John, Liao Jing, Modeling the interaction between objects and cartoon water, Computer Animation and Virtual Worlds, Volume 19, Numbers 3–4, July 2008, pp. 375–385.

Zhang 2009. Zhang S, Chen T, Zhang Y, Hu S, Martin R, Video-based Running Water Animation in Chinese Painting Style. In Science in China Series F: Information Sciences 52, February 2009, pp. 162–71.

Control Engineering and Information Systems – Liu (Ed)
© 2015 Taylor & Francis Group, London, ISBN 978-1-138-02685-8

PCA-SC based object recognition

W.G. Huang, C. Gu & Z.K. Zhu
School of Urban Rail Transportation, Soochow University, Suzhou, Jiangsu, China

ABSTRACT: Shape matching and object recognition are very critical problems in the field of computer vision. In order to improve the matching efficiency and anti-noise performance, various methods have been studied in recent years. In this paper, we propose a new descriptor, PCA-SC descriptor, which applies Principal Components Analysis (PCA) algorithm to reduce the dimensions of feature matrix formed by Shape Context. In the proposed PCA-SC algorithm, we build a covariance matrix, and reduce its dimensions according to the size of eigen value. PCA-SC descriptor can not only remove noise interference and improve the recognition accuracy, but also enhance the matching efficiency for real-time application. The experimental results of MNIST database indicate that the PCA-SC descriptor outperforms previous SC algorithm: the recognition speed is double and the accuracy reaches to 96.15% by 0.5%. Furthermore, the anti-noise performance becomes stronger. Hence, this novel algorithm shows better performance for shape matching and object recognition in efficiency, accuracy and anti-noise.

1 INTRODUCTION

The shape of an object is very important in object recognition. Using the shape of an object for object recognition and image understanding is a growing topic in computer vision and multi-media processing, and finding good shape descriptors is the central issue in these applications. In order to realize this purpose, the shape characteristics are put forward, and widely used in engineering application, such as wide-baseline matching, object class recognition, image and video retrieval, robot navigation, scene classification, texture recognition and data mining technology [1].

According to the source of feature, Zhang D [2] divided into two categories method based on shape characteristics description: silhouette method and transform domain method. The shape matching and object recognition based on the transform domain method have the ability of strong distinguishment and stability, which means that it is not easy to affect by environmental interference. Moreover, it can keep the invariance in the process of geometric transformation, gray transformation, convolution transformation and perspective transformation [3]. The main method is to select some invariance local features, such as feature points [4], object contours [5], feature regions [6]. Banerjee A [7] put forward Fourier Descriptors method to shape matching, which is simple and efficient, but the ability of capturing local characteristics is not strong. Then GCH Chuang [8] put forward Wavelet Descriptors method. It can capture global and local features of the image exactly

because wavelet transform has the characteristic of multi-resolution analysis. This method has a high matching precision, but the mount of calculation is very large and it is sensitive to translation and rotation. Serge Belongie [9] proposed Shape Context method, which is based on the concept of statistical information. This method has high precision of matching, and local feature can be captured efficiently as well. Meanwhile translation invariance and rotation invariance are available, which will overcome some weaknesses of Fourier descriptor and Wavelet descriptor. However, its computational complexity is high, and the ability of restraining noise is limited as well.

In this paper, we propose a new descriptor, PCA-SC descriptor, which applies PCA algorithm to overcome the shortcomings of SC method in the field of high dimensional feature matrix, large computational complexity and weak anti-noise performance. As a result, PCA-SC algorithm can reduce the dimension of feature matrix, simplify the computational complexity, improve matching speed, suppress noise interference, and enhance the rate of recognition effectively.

2 MODELING BY SC [11] ALGORITHM

2.1 *Log-polar translation*

This algorithm makes use of Canny edge detection operator to extract edge feature from of target image. Then get coordinate information of target silhouette points based on contour extraction algorithm. Next contour point positions are exchanges from Cartesian coordinate to polar coordinate via the transformation equation:

$$m(r,\theta) = \ln r,\; n(r,\theta) = \theta \qquad (1)$$

The method of forming Shape Context feature descriptor is shown in Figure 1.

Each contour point corresponding to the log-polar histogram can be obtained by the above steps, which will generate a $n \times 60$ dimensional characteristic matrix to describe the features of this image.

2.2 Compute matching cost degree

Consider a point P_i on the target image and a point Q_j on the template image. Let denote the cost of matching these two points. As shape contexts are distributions represented as histograms, it is natural to use the test statistic:

$$C_{ij} = C(P_i, Q_j) = \frac{1}{2}\sum_{k=1}^{K}\frac{[h_i(k)-h_j(k)]^2}{h_i(k)+h_j(k)} \qquad (2)$$

where $h_i(k)$ and $h_j(k)$ denote the K-bin normalized histogram at P_i and Q_j respectively and $K = 60$.

The matching cost degree C_{ij} is between 0 to 1. The similar of two images is greater when the size of C_{ij} is smaller.

3 RECOGNITION WITH PCA-SC DESCRIPTOR

PCA is one of the classic methods for linear transformation. The basic idea of PCA is to attempt to find a group of base vectors which can best reflect the distribution characteristics. Hence, with PCA method applied, it will simplify the problem, improve the operation speed and promote the ability of noise immunity performance.

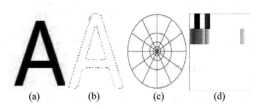

(a)　　　(b)　　　(c)　　　(d)

Figure 1. Shape Context computation and matching. (a) Original image. (b) Sampled edge points of letter A. (c) diagram of log-polar histogram bins used in computing the shape contexts. We use five bins for logr and 12 bins for θ. (d) Example shape contexts for reference samples marked by ° in (b). Each shape context is a log-polar histogram of the coordinates of the rest of the point set measures using the reference point as the origin.

3.1 PCA-based SC descriptors

PCA enables us to linearly-project high-dimensional samples onto a low-dimensional feature space. For our application, this projection (encoded by the patch eigenspace) can be pre-computed once and stored. The input feature matrix formed by SC has n × 60 elements. However, this matrix in the high-dimensional space is not random or scattered distribution, but contains a certain rule in a relatively low-dimensional space. Hence, we can use base vectors to present the image feature subspace and realize the purpose of reducing the dimension of matrix. The method of applying PCA to reduce the dimensions of feature matrix formed by SC is as follows:

1. Get a $n \times 60$ feature matrix formed by SC of target image, and generate a n dimensional feature descriptor:

$$X = [x_1, x_2, \ldots, x_n]^T \qquad (3)$$

where x_i denote the 60-bin normalized histogram as $x_i = [x_{i,1}, x_{i,2}, \ldots, x_{i,60}]$, and n is the number of the contour points in the matching image.

2. Calculate the average feature vector $\bar{x}$:

$$\bar{x} = \frac{1}{n}\sum_{i=1}^{n} x_i \qquad (4)$$

Then we can get residual vector between sample feature vector and average feature vector:

$$d_i = x_i - \bar{x},\; i = 1, 2, \ldots, n \qquad (5)$$

3. Generate a covariance matrix C:

$$C = DD^T \qquad (6)$$

where

$$D = \begin{bmatrix} d_{1,1} & d_{1,2} \cdots & \cdots & d_{1,60} \\ d_{2,1} & d_{2,2} \cdots & \cdots & d_{2,60} \\ \vdots & \vdots & & \vdots \\ d_{n,1} & d_{n,2} \cdots & \cdots & d_{n,60} \end{bmatrix}_{n\times 60}$$

Then calculate the eigenvalues λ_i and corresponding eigenvectors e_i of this covariance matrix.

4. Arrange eigenvalues from largest to smallest $\lambda_1 \geq \lambda_2 \geq \cdots \geq \lambda_n$, and compute corresponding eigenvectors $[e_1, e_2, \ldots, e_n]$. Extract the first k largest eigenvalues as principle component

(the size of k is based on the shape complexity of the image). Meanwhile, generate a projection matrix E, which is composed of k largest eigenvalues corresponding eigenvector $[e_1, e_2, \dots, e_k]$.

5. Based on the formula $Y = E' * X$, exchange n dimensional feature descriptor of SC into k dimensional feature descriptor of PCA-SC, which is $y_1, y_2, \dots, y_k$.

6. Calculate k dimensional feature descriptor of PCA-SC in template image, and compute the cost degree between template image and object image to judge whether the two images are matched or not.

3.2 PCA-SC algorithm computational complexity analysis

The amount of calculation with SC algorithm is:

$$N_1 = 60 \times \sum_{i=1}^{l} nm_i \tag{7}$$

where n is the number of target image contour points, l is the number of template images, and respectively, each number of contour points of template image is.

The amount of calculation with PCA-SC algorithm is:

$$N_2 = 60 \times lk^2 \tag{8}$$

where we use PCA method to decrease the dimension of feature matrix to k, and $k \le n$, $k \le m_i$.

Furthermore, when $k \ll n$, $k \ll m_i$, namely, the shape of target image and template image is complex. Hence, the matching speed can be greatly improved, which is efficient and can expand engineering application.

Meanwhile, we just extract the principle component of feature vectors, and ignore some minor interferential information. Hence, the novel feature matrix has certain noise rejection capability and higher accuracy, especially in the background of weak noise [12].

4 EXPERIMENTAL EVALUATION

This paper sets up three different experiments to make a comparison between SC algorithm and PCA-SC algorithm. Here, we present results on the MNIST [13] data set of handwritten digits, which consists of 10,000 test digits. Test environment: 64 Windows7 operating system, Pentium(R) Dual-Core CPU E6600, 4G memory. The test software is Matlab R2011b. Error rate is used to test the performance between SC algorithm and PCA-SC algorithm in this experiment. It can be shown as:

$$\text{Test error rate} = \frac{\text{Number of false matches}}{\text{Total number of matches}} \tag{9}$$

4.1 Recognition speed test

The first experiment tests the recognition speed of number 4 between SC algorithm and PCA-SC algorithm. With PCA method applied, the operation speed increases greatly and the recognition speed is double than SC method, which is shown in Figure 2. When the number of test digits is 1000, SC algorithm spends 7825s, while PCA-SC algorithm spends 3652s.

4.2 The recognition accuracy test

As we can see from Table 1, the first ten order eigenvalues are larger than others, which is represented the principle components of this matrix. After the 14th, the eigenvalue becomes smaller and smaller, which is represented the minor component. What is more, maybe it is generated by interferential information. This experiment tests the recognition error rate of SC method and PCA-SC method when $k = 5, 10, 20$. When PCA-SC method is fused, the error rate is smaller than SC method. Furthermore, the accuracy rate is the highest when $k = 10$ in Figure 3. So we can draw a conclusion that PCA-SC method has a stronger robustness and higher recognition accuracy rate.

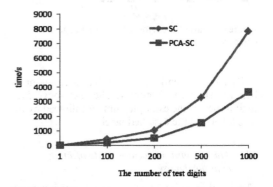

Figure 2. Speed test of recognition.

Table 1. Eigenvalue of covariance matrix.

2.8930	1.9000	0.6119	0.2596	0.1615
0.0877	0.0667	0.0490	0.0360	0.0317
0.0187	0.0162	0.0121	0.0094	0.0084
0.0066	0.0059	0.0050	0.0048	0.0032

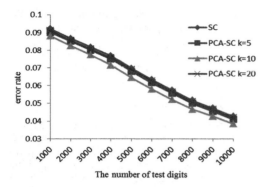

Figure 3. Recognition error rate of object uncontained noise.

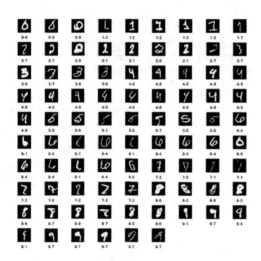

Figure 4. Part of the error recognition image in MNIST database.

Part error identification of handwritten numbers are shown in Figure 4. The text above each digit indicates the example number followed by the true label and the assigned label.

4.3 The recognition accuracy rate of image contained noise

The third experiment tests the recognition accuracy rate of image contained gaussian noise with average of 0 and variance of 1. In this experiment, Canny edge detection operator is adopted to extract target edge because it has a certain anti-noise performance and can successfully detect the edge of image. The first twenty eigenvalues of covariance matrix can be seen in Table 2.

The recognition error rate between SC and PCA-SC algorithm is shown in Figure 5. After

Table 2. Eigenvalue of covariance matrix.

3.1103	2.1612	1.6385	0.5951	0.4084
0.3452	0.1519	0.1051	0.0824	0.0766
0.0695	0.0625	0.0384	0.0321	0.0297
0.0259	0.0220	0.0209	0.0153	0.0141

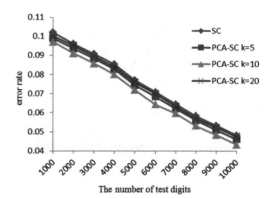

Figure 5. Recognition error rate of object contained noise.

fusing PCA-SC algorithm, the first eight eigenvalues are larger than others in Table 2, so the corresponding eigenvectors are the main ingredients, which play the leading role in shape matching and object recognition. The recognition accuracy rate of PCA-SC is higher than SC algorithm. Meanwhile, the recognition effect is best when the dimension reduction to 10 in Figure 5. As a result, we can come to the conclusion that the recognition accuracy and robustness are improved with PCA method, because some minor interferential constituents are ignored.

5 CONCLUSIONS AND FUTURE WORK

In this paper, PCA-SC algorithm is proposed to satisfy the requirement of high precision and fast speed for shape matching and object recognition in modern engineering application. Experiments shows that PCA-SC descriptor can simplify the computational complexity, improve the speed of recognition, suppress noise effectively, and improve the rate of recognition, which reaches to 96.15% by 0.5%.

In the future, we can use PCA-SC algorithm to match and recognize some more complex shapes and objects because of the improvement of efficiency and accuracy. It can be used in the engineering application of face recognition, the train bogie fault diagnosis, etc.

ACKNOWLEDGEMENT

The authors gratefully acknowledge the support from the Perspective Research Foundation of Production Study and Research Alliance of Jiangsu Province of China (No. BY2012112) and a Natural Science Foundation of Jiangsu Province (No. BK2010225).

REFERENCES

Banerjee A, Dutta A, "Fuzzy matching scheme on fourier descriptors for retrieval of 2 dimensional shapes," National Conference on Computing and Communication Systems, pp. 1–5, November 2012.

Chuang G.C.H, Kuo C.C.J, "Wavelet descriptor of planar curves: Theory and applications," IEEE Transactions on Image Processing, vol. 5, no. 1, pp. 56–70, January 1996.

Chunjing Xu, Jianzhuang Liu, Xiaoou Tang, "2D Shape Matching by Contour Flexibility," IEEE Transactions on Pattern Analysis and Machine Intelligence, vol. 21, no. 1, pp. 180–186, January 2009.

Edward Hsiao, Alvaro Collet, Martial Hebert, "Making specific features less discriminative to improve point-based 3D object recognition," IEEE International Conference on Compurter Vision and pattern Recognition, pp. 2653–2660, June 2010.

Iasonas Kokkinos, Michael M. Bronstein, Roee Litman, Alex M. Bronstein, "Intrinsic shape context descriptors for deformable shapes," IEEE Conference on Computer Vision and Pattern Recognition, pp. 159–166, June 2012.

Kunttu I, Lepisto L, Rauhamaa J, "Multiscale Fourier descriptor for shape-based image retrieval," Proceedings of the 17th International Conference on Pattern Recognition, pp. 765–768, 2004.

Serge Belongie, Jitendra Malik, Jan Puzicha, "Shape Matching and Object Recognition Using Shape Contexts," IEEE Transactions on Pattern Analysis and Machine Intelligence, vol. 24, no. 24, pp. 509–522, April 2002.

Sun Hao, Wang Cheng, Wang Runsheng, "A Rewiew of Local Invariant Features," Journal of Image and Graphics, vol. 16, no. 3, pp. 141–151, February 2011.

Xiao Q, Liu X, Liu M, "Object Tracking Based on Local Feature Matching," IEEE International Conference on Computational Intelligence and Design, pp. 399–402, October, 2012.

Y. Le Cun, The MNIST database of handwritten digits. http://yann.lecun.com/exdb/mnist.

YKe Y, Sukthankar R, "PCA-SIFT: A more distinctive representation for local image descriptors," IEEE Computer Society Conference on Computer Vision and Pattern Recognition, vol. 2, July 2004.

Zhang D, Lu G, "Review of Shape Representation and Description Techniques," Pattern Recognition, vol. 37, no. 1, pp. 1–19, July 2004.

Zhang S, Du J, Zhang L, "Circular Cone: A novel approach for protein ligand shape matching using modified PCA," IEEE International Conference on Computer Methods and Programs in Biomedicine, pp. 168–175, February, 2012.

Control Engineering and Information Systems – Liu (Ed)
© 2015 Taylor & Francis Group, London, ISBN 978-1-138-02685-8

A line spectrum estimation method of underwater target radiated noise

P. Yang & S.H. Chen
Naval Marine Academy, Guangzhou, China

ABSTRACT: A line-spectrum estimation method of underwater target radiated noise is studied, and the method is based on fourth-order cumulant slice spectrum. The characteristic of fourth-order cumulant slice spectrum is analyzed. Using fourth-order cumulant slice spectrum, the gauss-white noise and symmetry distribution noise can be reduced, and the spectrum structure is similar with power spectrum. The characteristic and calculating process of Lofargram is introduced. The calculating method of fourth-order cumulant slice spectrum diagram is introduced with analyzing the characteristic of Lofargram and fourth-order cumulant slice spectrum diagram. The line in the fourth-order cumulant slice spectrum diagram can be transformed into peak value in the Radon-domain using Radon transform. After the peak value is gained, the frequency and frequency modulation can be calculated. At last, Simulation results show that this algorithm is effective.

1 INTRODUCTION

The characteristic of radiated noise is very important in the detector and recognition of underwater targets. Among many characteristic parameters of radiated noise, the line spectrum in low frequency that is produced by a series of circumrotating machines is more important. The analyzing of Lofargram is one of representative passive hydro acoustic signal process methods. Using the method, the non-stable characteristic of a signal can be figured with a 3-D figure which is gained by the short-time FFT transform of the continuing sample data. Lofagram analyzing is based on the signal's power spectrum. The high-order cumulant slice spectrum is based on the high-order spectrum. The high-order cumulant slice spectrum can reserve the excellent characteristic of reducing the additional gauss-white noise. So, the dynamic high-order cumulant slice spectrum can be gained by the high-order spectrum of the continuing sample data. And the line spectrum characteristic can be analyzed using the dynamic high-order cumulant slice spectrum.

2 DEFINITION AND CHARACTERISTIC OF FOURTH-ORDER CUMULANT SLICE SPECTRUM

Let $x(t)$ be a zero-mean complex random variables. The 4th-order cumulant slice spectrum of this process, denoted $C(\omega)$, is defined as the fourier transform of the 4th-order cumulant diagonal slice. Let $c4x(\tau,\tau)$ denote the diagonal slice of the 4rd-order cumulant.

$$C(\omega) = \int_{-\infty}^{\infty} \int_{-\infty}^{\infty} x(t)x^3(t+\tau)e^{-j\omega\tau}dtd\tau$$

$$C(\omega)$$
$$= \int_{-\infty}^{\infty} x(t)e^{-j(-\omega)t}dt\int_{-\infty}^{\infty} x^3(t+\tau)e^{-j\omega(t+\tau)}d(t+\tau)$$
$$= X^*(\omega)[X(\omega)*X(\omega)*X(\omega)]$$

$$(1)$$

$X(\omega)$ is the Foutier transform of $x(t)$, and $X^*(\omega)$ is the complex conjugate of $X(\omega)$.

Following are some important properties of 4th-order cumulants slice spectrum, which are used in theoretical developments (Fan 2002).

1. Let $x(t) = \sum_{i=1}^{P} a_i \cos(\omega_i n + \phi_i)$ be a zero-mean real nth-order harmonic wav, and ϕ_i be a random variables with uniform distribution in $[-\pi, +\pi)$. Then, the 4th-order cumulant diagonal slice spectrum is equivalent.

2. Let $n(t)$ be a zero-mean gauss noise, then

$$C_{4n}(\omega) \equiv 0$$

3. Let $n(t)$ be a zero-mean random noise with symmetry distribution probability density, and be irrelevance at any different time, then

$$C_{3n}(\omega) \equiv 0$$

4. Let $x(t) = \sum_{i=1}^{P} a_i \cos(\omega_i n + \phi_i)$ be a zero-mean real nth-order harmonic wav, and ϕ_i be a random variables with uniform distribution in $[-\pi, +\pi)$. Then, the 4th-order cumulant diagonal slice spectrum is the same spectrum structure with the power spectrum.

3 LOFARGRAM

In order for improving the performance of line-spectrum detecting and tracking and extracting, the Lofar process may be used. Using the passive sonar, the signal that received from a certain direction can be time-frequency analyzed, and the result can be displayed to the operators. If x-axis represents the signal's frequency, and y-axis represents the time, the lighteness represents the amplitude, then the two-dimension image in the time-frequency plane is a Lofargram.

The Lofargram analyzing is one of representative passive hydro acoustic signal process methods in the past decade. The method of calculating the Lofargram is introduced in the paper (Xiong 2007).

1. The original sampling data of the signal may be separated into a sequence of segments. And each segment contains N sampling points. The segments can overlap partly.
2. Each segment of the sampling data, denoted $L(n)$, can be processed according the following steps.

$$u(n) = \frac{L(n)}{\max_{1 \le i \le N}[L(i)]}$$

$$x(n) = u(n) - \frac{1}{N}\sum_{i=1}^{N}u(i)$$

3. $X(\omega) = \text{FFT}[x(n)]$

4 PHOTOGRAPHS AND FIGURES

Lofargram is gained by the FFT transform from a segment of signal. And 4th-order cumulant diagonal slice spectrum has more better performance of reducing the additional gauss-white noise than the Fourier transform.

4.1 Dynamic 4th-order cumulant diagonal slice spectrum

If the sample rate of the original signal, denoted $s(t)$, is fs, then the calculating process of the dynamic 4th-order cumulant diagonal slice spectrum is shown as following.

1. The original sampling data of the signal may be separated into a sequence of segments. And each segment contains N sampling points. The segments can overlap partly, and for example the number of overlap point is M.
2. Each segment of the sampling data, denoted $L(n)$, can be processed according the following steps.

$$u(n) = \frac{L(n)}{\max_{1 \le i \le N}[L(i)]}$$

$$x(n) = u(n) - \frac{1}{N}\sum_{i=1}^{N}u(i)$$

3. According formula (1), the dynamic 4th-order cumulant diagonal slice spectrum can be gained by calculating 4th-order cumulant diagonal slice spectrum of the signal, denoted $x(n)$.

4.2 Line-spectrum detecting method

A narrow-band signal is shown as a line-spectrum in a Lofargram or a dynamic 4th-order cumulant diagonal slice spectrum. On the Lofargram or dynamic 4th-order cumulant diagonal slice spectrum, a clear line that generates from many points can be displayed, and the points represent the position of line-spectrum. That is spectrum-line. The problem of frequency detecting and tracking can be resolved with line detecting and extracting on the Lofargram or dynamic 4th-order cumulant diagonal slice spectrum.

The line-spectrum extractor is used image process arithmetic. The line-spectrum can be detected using Radon transform. If the spectrum-line is a beeline, the method can obtain good performance of line-spectrum extracting in the low SNR.

Let a beeline in a image plane, denoted $\rho = x\cos\theta + y\sin\theta$, is mapped into a point in the Radon plane, denoted (ρ, θ). The Radon transform of a image is given by (Wood 1996 & Xin 2001)

$$R(\rho,\theta) = \iint_D f(x,y)\delta(\rho - x\cos\theta - y\sin\theta)dxdy$$

where D is the whole image plane, and $f(x,y)$ is the value of the point (x,y).

ρ is the distance of a line and the origin. θ is the angle of the origin and the vertical line of the line.

For the point (ρ, θ) in the Radon-space, the reconstruct formula of the original image space is given by

$$y = \rho/\sin\theta - x\cot\theta$$

Thus, the frequency and frequency modulation rate of the line spectrum can be gained by the parameters of the line in the Radon-space as follows.

$$f = \frac{\rho}{\sin\theta} \cdot \frac{fs}{K} \tag{2}$$

$$\mu = \cot\theta \cdot \frac{fs/K}{ts \cdot M} \tag{3}$$

274

5 SIMULATION RESULT

The following simulations were performed with double precision floating-point arithmetic on a Matlab software. The SNR for each line-spectrum is defined as the power of the line-spectrum over the total noise power.

The original signal is given by

$$s(t) = \sin(2\pi f_0 t) + \sin\left[2\pi(f_1 + \frac{\mu_1}{2}t)t\right]$$

$$+ \sin\left[2\pi(f_2 + \frac{\mu_2}{2}t)t\right]$$

where the sampling frequency is 1500 Hz, and the SNR is −10 dB.

$f_0 = 20$ Hz, $f_1 = 120$ Hz, $\mu_1 = 5$ Hz/s, $f_2 = 500$ Hz, $\mu_2 = -8$ Hz/s.

Referring to Figure 1, the simulation compares the performance of a Lofargram and a 4th-order cumulant diagonal slice spectrum. In the Figure 1(a), the line-spectrums are difficult to distinguish in the signal's Lofargram. And three line-spectrums can be gained in the 4th-order cumulant diagonal slice spectrum. The 4th-order cumulant diagonal slice spectrum has better performance in the reducing the additional gauss-white noise than the Lofargram.

The Radon-transform of the Lofargram and 4th-order cumulant diagonal slice spectrum is shown as Figure 2.

Referring to Figure 2, only one peek-value was gained in the Lofargram using Radon-transform, and three peek-values were gained in the 1½D spectrum.

A line in the Lofargram or a 4th-order cumulant diagonal slice spectrum can be transformed to a peek-value. If the point in the Radon-plane, denoted (ρ, θ), is gained, the parameters of the

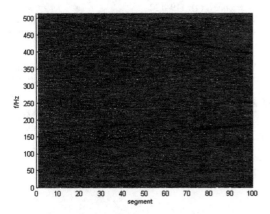

Figure 1(a). The simulation signal's Lofargram.

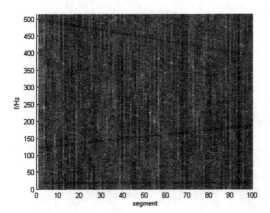

Figure 1(b). The dynamic 4th-order cumulant diagonal slice spectrum.

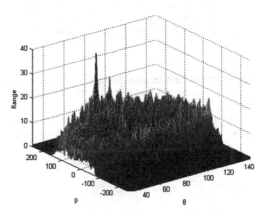

Figure 2(a). Radon-transform of Lofargram.

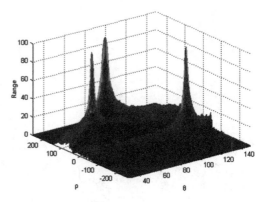

Figure 2(b). Radon-transform of 4th-order cumulant diagonal slice spectrum.

Table 1. Line-spectrum parameters using Lofagram.

| No. | Parameter | Lofargram | |
		Radon-plane	Detecting para.
1	$f_0 = 20$ Hz $\mu_0 = 0$ Hz/s	$\rho = 241$ $\theta = 89.8$	$f_0 = 21.98$ Hz $\mu_0 = 0.04$ Hz/s
2	$f_1 = 120$ Hz $\mu_1 = 5$ Hz/s	–	–
3	$f_2 = 500$ Hz $\mu_2 = -8$ Hz/s	–	–

Table 2. Line-spectrum parameters using new method.

| No. | Parameter | Dynamic 1½D spectrum | |
		Radon-plane	Detecting para.
1	$f_0 = 20$ Hz $\mu_0 = 0$ Hz/s	$\rho = 241$ $\theta = 90.4$	$f_0 = 20.47$ Hz $\mu_0 = 0.07$ Hz/s
2	$f_1 = 120$ Hz $\mu_1 = 5$ Hz/s	$\rho = 135$ $\theta = 65.0$	$f_1 = 122.90$ Hz $\mu_1 = 5.12$ Hz/s
3	$f_2 = 500$ Hz $\mu_2 = -8$ Hz/s	$\rho = -39$ $\theta = 125.6$	$f_2 = 497.17$ Hz $\mu_2 = -7.87$ Hz/s

line-spectrum can be calculated using formula (3) and (4). The results were listed in the Table 1 and Table 2.

The results have shown that the method in the paper has better performance in the line-spectrum detecting and parameter extractor.

6 CONCLUSIONS

In the paper, the calculating method of 4th-order cumulant diagonal slice spectrum diagram is introduced with analyzing the characteristic of Lofargram and 4th-order cumulant diagonal slice spectrum diagram. The line in the 4th-order cumulant diagonal slice spectrum diagram can be transformed into peak-value in the Radon-domain using Radon transform. The frequency and frequency modulation can be calculated with the parameters of the peak-value in the Radon-domain. And Simulation results show that this algorithm is effective.

REFERENCES

Ran Xin, Ma Shiwei & Cao Jialin. 2001. Analysis of Multicomponent LFM Signal Based on Radon-Wigner Transform, *Journal of Shanghai University (Natural science)*, (4): 119–122.

Wood J.C. & Barry D.T. 1996. Radon transformation of time-frequency distributions for analysis of multicomponent signal, *IEEE Transactions on signal processing*, (6): 3166–3177.

Yangyu Fan, Baoqi Tao & Ke Xiong. 2002. Feature extraction of ship-radiated nose by 1½-spectrum, *Acta Acustica*, (1): 71–76.

Ziying Xiong & Xiqing Zhu. 2007. Ship radiated-noise research based on the LOFAR spectrum and DEMON spectrum characteristics, *Journal of Ship Mechanics*, (4): 300–306.

An algorithm for distinguishing the semantic objects in video

L. Zhang & C.S. Zhou
Computer Center, Beijing Information Science and Technology University, Beijing, China

ABSTRACT: Recognition of The Semantic Objects in Video is important in video analysis. It has great significance for the classification of video management and video retrieval. In this paper, we design an algorithm for recognition of the semantic objects in video. We analyzed problems of the current video object recognition algorithm, and propose a video object recognition algorithm based on the KNN algorithm. This paper selected some low-level features of video, such as grayscale, color and two-dimensional wavelet feature. Some similarity determination methods were analyzed in the paper. The experimental results show that the algorithm is effective and feasible in monitor video.

1 INTRODUCTION

With the continuous development of multimedia technology, digital video in recent years has gradually become an important way of information documentation and dissemination of Digital video with its record of events is accurate and intuitive, informative features a wide range of applications. Every day the vast amounts of video data are generated, such as surveillance video, sports video, film and television works. With the rapid growth of applications and processing of video data, video data management, classification, retrieval, also had a new problem. General video category management manually marked, using the video title, keywords and other methods of video text description and classification of management. However, manual annotation methods cannot adapt to the rapid growth of video data classification and retrieval needs. In this case, people began to study content-based video semantic analysis, can be extracted directly from the video low-level features high-level semantic information, and automatic annotation of video semantic content.

Stored in the video file is some low-level features, such as color information, but the people watching the video information is the cognitive layer. This information was mainly shown in the video objects (such as characters, animals, plants) as well as object movement and mutual relations. If you want to analyze and organize video, or for further video retrieval, an important step is to effectively identify the semantic objects in video.

At present, the object recognition using template matching method, made of the feature vector value of the object and its relative position to a fixed template, use the template to traverse the search for a matching object to be identified, calculate the distance of the object feature vector to be identified with the template or degree of correlation exceeds a certain threshold, reports to identify success. Fixed template during recognition is relatively simple. However, due to the establishment of a fixed template according to specific application areas of a priori knowledge. Types of video there are several specific applications, you need to establish an appropriate set of templates for each field, and each field of each video object scale change is difficult to use a template description of the need to use a variety of templates, resulting in a lot of calculation. Fixed template-based approach is more suitable for video object is relatively simple and fixed video object detection, such as lights in the traffic video detection.

In the paper, we propose a video object recognition algorithm based on the KNN algorithm. This paper selected some low-level features of video, such as grayscale, color and two-dimensional wavelet feature. Some similarity determination methods were analyzed in the paper. The experimental results show that the algorithm is effective and feasible in monitor video.

2 THE BASIC OF OBJECT RECOGNITION

Before the video object recognition, the need to solve three problems, including image preprocessing, low-level feature selection and similarity determination method selected.

2.1 *Image preprocessing*

Video image data may introduce bias because of changes in light, lens zoom, and other reasons, thus affecting the extraction and video vector calculation, so the image analysis than the former, the image gray scale adjustment and geometric normalization, reduction pattern-matching errors. Common image pre-processing methods to convert color mode, gray-scale transformation, geometric correction, and filtering.

2.1.1 Gray-scale transformation

From the intuitive feeling of the people in the video analysis process, if a color video is converted to the corresponding gray-scale video, people apart from the sensory level of comfort, the amount of information loss is very small. General video images converted to grayscale for analysis, so as to achieve the purpose of dimensionality reduction and noise reduction.

Set of f(x, y) for the gray-scale transformation (x, y) point gray value, g(x, y) is the gray value of the gray-scale transformation (x, y) point. The gray-scale transformation methods are the following:

Linear transformation: set the range of gray-scale image f (x, y) [a, b], linear transformation, the range of the image g (x, y) is [a', b'], the relationship between the g(x, y) and f(x, y) can be expressed as:

$$g(x,y) = a' + \frac{b'-a'}{b-a}(f(x,y)-a) \tag{1}$$

Binarization: Though gray image binarization, the image showing a clear black and white, better division of the outline of an image region. Image binarization threshold method. The threshold T used in the binarization process, the gray value of each pixel of the image compared with T, if greater than T—, then take the white, or else take the black.

$$g(x,y) = \begin{cases} 255 & f(x,y) \geq T \\ 0 & f(x,y) < T \end{cases} \tag{2}$$

2.1.2 Image normalization

In the video, because of the distance adjustment of the camera or object movement, the same object in different frames or different video sizes phenomenon, resulting in geometric distortion, to address these issues way is to use a normalized way geometric correction of the input object, unified to the same size.

2.1.3 Image filtering

In the image acquisition process, the inevitable subject to various disturbances, in order to improve the quality of the image, to avoid the interference noise caused by image analysis requires the elimination of these noise images filtering operator.

Mean filter: mean filtering the image data to the local average, average gray value. At pixel (x, y) to obtain the N × N neighborhood. Is calculated using the following formula:

$$g(x,y) = \frac{1}{N \times N} \sum_{k=i-\lfloor N/2 \rfloor}^{i+\lfloor N/2 \rfloor} \sum_{l=j-\lfloor N/2 \rfloor}^{j+\lfloor N/2 \rfloor} f(k,l) \tag{3}$$

Mean filter to remove high-frequency information in the image, the image becomes smooth.

Gaussian filter: Gaussian filter is used to smooth the performance of the Gaussian function in the frequency domain construct low-pass filter can effectively remove obey the noise is too distributed. In image processing, usually two-dimensional zero mean Gaussian function for smoothing filter:

$$g(x,y) = e^{-\frac{(x+y^2)}{2\sigma^2}} \tag{4}$$

Two-dimensional Gaussian functions with rotational symmetry, so the smoothness of the filter in all directions is the same. The parameter σ determines the degree of smoothing, σ greater the Gaussian filter of the band is wider, the better the smoothness of. By adjusting the σ parameter, you can make the image to get a better treatment effect.

2.2 Low-level feature extraction

In algorithm design, we can see that the low-level features of the sample is the main basis for matching samples and data entry, low-level feature selection is successful, the correct rate to match the results of low-level features of improper selection can cause the error of matching results.

2.2.1 Gray histogram

Image gray distribution function, it represents the proportion of each pixel of the grayscale image, reflecting the frequency of occurrence of each gray-scale image.

The gray histogram is defined as follows:

$$H(k) = \frac{n_k}{N} \tag{5}$$

where k denotes the image intensity values, n_k that the gray value of k, the number of pixels, N is the total number of image pixels. The gray histogram contains all color values in the image frequency, while the loss of a pixel where the spatial location information. Any object can be calculated which corresponds to the histogram, the gray histogram is only with the gray of the object itself, has nothing to do with location. Gray histogram features to some extent, translational stability and deformation stability. However, different objects may have the same color distribution, in order to have the same histogram. If a single gray level histogram features, will result in the identification of the false detection rate.

2.2.2 *Image contour*

Contour of the image is an important feature of the video object edge detection algorithm in 3.2, we described in detail, this section will be adopted to detect the contour of characteristics, one of the characteristics of the video object recognition. The main use of profile characteristic shape parameters

$$F = \frac{\|B\|^2}{4\pi A} \qquad (6)$$

where B is the contour of the perimeter, A is the contour contains the size of the area. Here for perimeter and area to calculate the number of pixels.

2.3 *The determination of similarity*

In pattern recognition, in order to determine the similarity of two objects, you need to calculate the distance between the feature vectors of these two objects. Commonly used distance calculated methods include Euclidean distance and Mahalanobis distance. Here, we selected Euclidean distance:

$$d(x,y) = \|x - y\| = \left[\sum_{i=1}^{n}(x_i - y_i)^2\right]^{1/2} \qquad (7)$$

Euclidean distance with translation invariance and rotation invariance, dimensionless determined case, two vectors are more similar, the smaller the distance.

3 DESIGN OF RECOGNITION ALGORITHMS

For universal Video, more appropriate is based on learning algorithm, KNN (k-Nearest Neighbor) classification algorithm, is a theoretically more mature approach is relatively simple machine learning algorithms. The idea of the method: To investigate the mode to be recognition of the k-nearest neighbor samples, the k nearest neighbor samples of what kind of samples up to, it will be identified samples normalized where a class. KNN algorithm, the choice of neighbors are correctly classified objects. The method in the given class of decision-making only on the basis of the category of the nearest one or several samples to determine the category to be sub-sample.

3.1 *Create a video sample training set*

Step1: object segmentation algorithm to obtain the sample of objects to form a candidate sample set.

Step 2: treat selected sample set of sample data pre-processing, these include normalization and Gaussian filter.

Step 3: to be selected sample set of data sorting, filtering out which characteristics, easy to distinguish the samples as training samples. The principle of the principle of the selected first, select the training data set is broadly consistent with the number of various types of samples is selected should be representative of the historical data.

Step 4: on the training samples in the category label, label the contents of the object name and category.

Step 5: training sample set of training samples to obtain sample characteristics values to form a feature vector.

3.2 *Parameter settings*

To determine the distance function and the values of K.

The distance function determines which samples of the K nearest neighbors to be classified in this selection depends on the actual data and decision-making. If the sample space midpoint, the most commonly used is the Euclidean distance. The other commonly used distance function is the absolute distance, the sum of squared deviations and standard deviation.

Number of neighbors have a certain impact on the classification results, usually to determine an initial value (set to 5 in this article), and then adjust until you find the right value.

3.3 *Sample recognition*

Similar determination to get to be classified samples from the nearest sample of N labeled sample in the sample characteristics of the object with the video memory.

Set up the categories set $V = \{v_1, v_2, ... v_n\}$;

Input: the training data set $D = \{(X_i, Y_i), 1 \leq i \leq N\}$, where X_i is the condition attributes of the sample-i, Y_i is the category;

New sample: X_q;

Output: X_q is the type of Y_q.

Calculation process:

Step 1: Calculate X_q with the sample in the training set of training samples to calculate the distance.

Step 2: training distance sort, access to the nearest k training samples, denoted as $x_1, x_2, ..., x_k$.

Step 3: Calculate the k samples, the largest proportion of label semantics is the sample to be to identify semantic information, the formula:

$$Y_q = \arg\max_{v \in V} \sum_{i=1}^{k} \delta(v, Y_i) \qquad (8)$$

In which, when a = b, δ (a, b) = 1, otherwise δ (a, b) = 0.

The KNN algorithm, all the neighbors the right to vote the same, without taking into accounts the factors of distance. This article uses an improved KNN algorithm:

Improved KNN algorithm in the training samples marked with the same process as the traditional KNN algorithm will not repeat narrative, major improvements in the stage of a new sample of object identifiers from the weights.

$$Y_q = \arg\max_{v \in V} \sum_{i=1}^{k} \omega_i \delta(v, Y_i) \qquad (9)$$

4 EXPERIMENTAL RESULT AND ANALYSIS

In this study, we use the Vs 2008 and OpenCV library v2.1, and select three kind of video (news, sports, monitor and movie) clips as test video, experimental steps flow the section III. Analysis of overall performance

From statistical data, the video object recognition algorithm using the KNN algorithm to the video object identification, and monitoring class video recognition rate, reaching 79%, the class of the press and sports video also reached a good recognition rate. Improved KNN algorithm is

Table 1. Performance of algorithm.

Video type	KNN		Improved KNN	
	Precision	Recall	Precision	Recall
News	74	76	82	83
Sports	79	80	82	85
Monitor	81	82	84	83
Movie	62	65	68	70

In Table 1, the precision an recall are computed by (1) and (2)

Precision = Correct/(Correct + False)   (1)
Recall = Correct/(Correct + Miss)   (2)

that the recognition rate to higher than the KNN algorithm.

4.1 K value and the correct rate of recognition

With the changes in the value of K, the recognition rate of the algorithm will be some changes, this trend of the response of the following figure:

By changing the K value, the recognition rate of the video changes, can be obtained from the above table, when K = 7, the overall recognition rate of the video.

4.2 Number of training samples and the correct rate of recognition

The number of training sample size will affect the video object recognition correct rate. Through experiments, we get the following data.

From the recognition accuracy of the Chart can be seen, when the number of training samples, the recognition rate will increase with the increase in the number of samples, to a certain amount (in this experiment is about 80) when the number of samples, the recognition rate is no longer significantly increased slight fluctuations in the vicinity of the stable value.

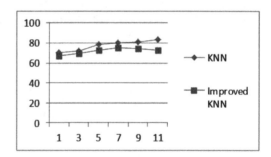

Figure 1. K value and the correct rate of recognition.

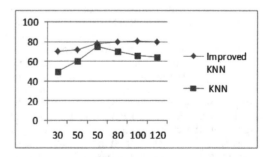

Figure 2. Number of training samples and the correct rate of recognition.

5 CONCLUSION

We analyzed problems of the current video object recognition algorithm. Video object recognition is an important foundation of the entire video semantic analysis, video analysis plays an important role in this chapter, the object recognition algorithm based on the KNN algorithm, the improved KNN algorithm, the system and summarized in the video object recognition need to use the image preprocessing, feature extraction and similarity determination method. Commonly used algorithms in pattern recognition and machine recognition in other fields have achieved some success, this paper these several algorithms applications in the video semantic object recognition, the experiment proved the three algorithms can be good for video identification, the effective realization of the automatic annotation of video semantics and video category judgment. The experimental results show that our algorithm is effective and feasible in sports and monitor video.

ACKNOWLEDGEMENT

The project is sponsored by the Scientific Research Common Program of Beijing Municipal Commission of Education. (SQKM201211232019).

REFERENCES

Alan Hanjalic. Content-Based Analysis of Digital Video [M], Springer, 2004. Chengcui Zhang, Xin Chen, Liping Zhou, Wei-Bang Chen, Semantic retrieval of events from indoor surveillance video databases [J], *Pattern Recognition Letters* 30 (2009) 1067–1076.

Chengcui Zhang, Xin Chen, Liping Zhou, Wei-Bang Chen, Semantic retrieval of events from indoor surveillance video databases [J], *Pattern Recognition Letters* 30 (2009) 1067–1076.

Hanjalic. A. "Content-based analysis of digital video", Springer, 2004.

Murugappan, M. Human emotion classification using wavelet transform and KNN: *Proceedings of the 2011 International Conference* on 276, 2010.

Patel, B.V.; Deorankar, A.V.; Meshram, B.B. Content based video retrieval using entropy, edge detection, black and white color features, *ICCET 2010–2010 International Conference on Computer Engineering and Technology, Proceeding*s, v 6, p V6272–V6.

Pattern Analysis and Intelligent Robotics, ICPAIR 2011, v 1, p 148–153, 2011.

Zhu, Liuzhang (Anhui Electric Power Jiyuan Software Co. Ltd. Hefei, China); Li, Zimian; Cao, Zheng.An efficient and effective video similarity search method, *Proceedings of SPIE—The International Society for Optical Engineering*, v 8285, 2011.

Control Engineering and Information Systems – Liu (Ed)
© *2015 Taylor & Francis Group, London, ISBN 978-1-138-02685-8*

The video segmentation algorithm based on region selection

F.Q. Meng
Baicheng Normal University, Jilin, China

ABSTRACT: The video object segmentation algorithm comes forward based on region selection. This algorithm starts with a simple and practical spatial region segmentation method, then performs multiple selections on received regions via the motion information and space-time energy model, finally obtains the accurate segmentation object by certain post-processing technology. Experiments prove that this algorithm has preferable robustness.

1 INTRODUCTION

In the real world, the human visual system can distinguish different target region relying on single frame image gray-level information alone. This ability of accurate target segmentation is inseparable with both developed human nervous system and long-term experiential learning. Another important factor is the visible boundary of spatial domain between target and background. If there is no such kind of boundary, human beings can't distinguish any more; for example, in a dark room, people can't make a distinction between objects.

Each object region is made up with patchwork of different gray subdomains in a same subdomain pixels possess equal or very similar grayscale or unitive texture. In the process of video segmentation, if regional information and time domain motion information can be fully used at the same time, the accuracy of segmentation can be further improved.

Collection of pixels with color or texture coherence and adjacency in space is often called as region, region attributes can make up for shortage of morphology filtering method which neglect consistency of color or texture, play better restriction role on the local place where object motion information is inadequate. Method based on region selection provides better segmentation effect while under static background and when moving object is simple.

2 SELF-ADAPTIVE FILTERING ALGORITHM OF IMPULSE NOISE

For reasons of its own noise problem or image follow-up algorithms, figures collected by image system usually need preprocessing. During the process of imaging, digitization and transmission,

figures will be affected by various kinds of noise interference, and figure quality degrades disappointingly, and will influence the image visual effect. The noise interference creates image degradation and quality reduction. These behave as the image blurring, characteristics submerging, so that handicap the image analysis, and obtain poor-quality images. Image noise can be understood as various factors that interfering people's visual organ or system sensors to understand or accept analysis of image source information. There are several reasons for causing noise of an image: internal noise from sensors or electronic components due to the random movement of loading particles, noise from electric current or electromagnetic field changes generated by electrical motion of electrical equipments' internal parts, external noise generated when natural magnetoelectricity or project magnetoelectricity entering into internal system through the atmosphere or power lines, noise made by particles of photosensitive materials on photo negative film or by flaws on disk surfaces, interference or quantizing noise of transmission channel and decoding error noise, etc. All these factors determine the characteristics of noise distribution, as well as the relationship between it and the image signal. General image noise are unpredictable random signals, it can only be cognized by probability and statistics method. Noise act on overall process of image entering, collecting and exporting, especially the incoming noise during inputting and collecting will affect the whole image processing and the output results. Image noise has a variety of types, such as Gaussian noise, Rayleigh noise, gamma and impulse noise. Among them impulse noise (also named as salt and pepper noise or bipolar noise) is most common. During image production and transmission the impulse is always produced, mainly reflected in the whistle stop of imaging when major influence on image quality occurs, then image filtering method can be used to remove it. Noise and random disturbance must be reduced and filtered, in order to strengthen useful information, increase effectiveness and reliability of follow up processing. Filtering image noise turns into a most important step of image processing, it is very important for image processing.

Usually pixels of an image always possess high correlations between their neighboring pixels. That is to say the most energy of an image is concentrated in the low frequency region. But the energy of the image detail (especially information of boundaries) is mainly concentrated in the high frequency region. And interference introduced from the process of imaging, and digital transmission is generally also concentrated in the high frequency region, how to filter out the noise and preserve the details of image becomes the main problem when filter is designed. That is how to protect the image detail information during the process of eliminating or weakening noise.

The typical representative of traditional nonlinear filter, median filtering algorithm, is mainly used to suppress impulse noise. This algorithm is widely applied to various actual image preprocessing. But when the filter window is small, although you can protect image details very well, but when the number of the pixel points polluted by impulse noise is greater than half of the number of the whole pixels in the filtering window, the median filtering will come to a complete failure; If the filter window is increased, the calculated amount will increase significantly, required time increase, then. One way to resolve this problem is to filter the impulse with self-adopting filter windows. The filter window changes size in line with different pollution degrees of impulse noise in different regions of the image. Set the initial size of filtering window to be $n \times n$ (n is an odd number), and the number of noise within the window to be m, contaminated degree of filtering window to be γ:

$$\gamma = \frac{m}{n \times n} \times 100\% \qquad (1)$$

Self-adoptive median filter divides into two layers of a and b, adaptive algorithms is showed as follows:

A. Judgment on pollution level: $\gamma < T$, then

$$s_{ij}^n = s_{ij}^M = Median(s_{ij}^n, \tau);$$

$\gamma \geq T$, then comes to layer B.

B. To enlarge the filter window: filter window is enlarged to $(\hat{n}+2) \times (\hat{n}+2)$, where, $(\hat{n}+2) < \nu$. Recalculate γ, then calls layer A.

In formula (1), $\hat{n} \times \hat{n}$ is the size of the last filter window, τ and S_{ij}^M represents respectively the number and the median of non-noise pixels in the filter window. ν is the maximum filter window, T is threshold value for pollution level in the filter window. When the filter window is with the definite median filter, and noise sum reaches 0.3 of the filter window, filtering effect becomes unsatisfied, so generally the threshold value T is taken as 0.3, then good filtering effect shows.

Adopting filtering windows self-adapting method mainly possesses following superiorities:

1. Self-adaptive filtering algorithm can change filter size according to different noise pollution levels, so not only median filtering algorithmic failure can be avoided effectively when pollution is serious, but also better filter effect can be achieved when different filter window can be chosen according to different contamination levels.
2. Signals that are not polluted by noise is protected effectively, due to filtering merely noise signals. Image filtering allows only non-noise points partake filtering, and forecloses noise points, thus influence of filtering results by noise points be reduced.
3. Due to merely filtering detective impulse noise and the adaptivity of filtering windows, required time is greatly reduced, algorithmic practical applicability is increased.
4. The setting of the maximum filtering window ν is to confirm the maximum filtering window in accordance with physical truth, and prevent a few images come out with its own contents constituted with parts of extremum grayscale.

3 SEGMENTATION ALGORITHM BASED ON REGION SELECTION

3.1 Image segmentation in the current frame

First threshold value T_1 of region segmentation must be determined. Take the first pixel $f_k(1,1)$ of the kth frame as the first region. Then for each pixel $f_k(x,y)$ find the minimum value of the absolute difference with confirmable pixels in its 8 neighborhoods. The pixels in 8 neighborhoods can be indicated as $f_k(x \pm m, y \pm m)$, while $m = 0$ or 1, then:

$$d_{\min} = \min[\text{abs}[f_k(x,y) - f_k(x \pm m, y \pm m)]] \qquad (2)$$

If $d_{\min} < T_1$, $f_k(x,y)$ will be classified into the same region with $d_{\min}$, or else a new region be started. All pixels being disposed as above mentioned, will all belong to a certain region that is to segment current frame into multiple small homogenous regions. This algorithm adopts the standard test sequence mother & daughter as the experimental subject. Figure 1 is the first frame of the original figure, which is divided into multiple small homogeneous regions after preliminary segmentation, as is shown in Figure 2. We can find that each region can basically reflect their features. The threshold value T_1 is 20 in this experiment.

3.2 Initial classification of region

Change detection can be used for initial classification of region. Set the frame difference image as:

$$d_k(x,y) = |f_{k+n}(x,y) - f_k(x,y)| \tag{3}$$

If the adjacent frame difference method is adopted, the n value is 1, when the object moves slowly, frame skip difference can be used, set the n value to be 3–5. The next step is binarization processing of frame difference image, set the threshold as T_2:

$$M_1 = \begin{cases} 1 & \text{if } d_k(x,y) > T_2 \\ 0 & \text{if } d_k(x,y) \le T_2 \end{cases} \tag{4}$$

The difference figure is as shown in Figure 3 and Figure 4. The threshold T_2 is 15. In Figure 3 is the difference image when $n = 1$, Figure 4 is the image for $n = 3$, as can be seen from the experimental results, for slowly moving video sequence, taking frame skip difference method can obtain more sufficient motion information, more complete 0–1 value mask.

Figure 1. The first frame of the original figure.

Figure 2. Segmentation image in space region.

Figure 3. Direct difference figure.

Figure 4. Frame skip difference figure.

Calculate each region pixels, then carry out normalization processing.

$$\rho_i = \frac{n_i}{m_i} \tag{5}$$

n_i is pixels moving in region i, that is homologous pixels as 1 in M_1. m_i is the total number of pixels in this region, ρ_i is the percentage of the moving pixel numbers in the total, if $\rho_i > 0.8$, it means more than 80% pixels in this region are changing movement, judge this to be moving regions; If $\rho_i < 0.2$, it means less than 20% pixels are moving, judge this to be stationary regions. Regions between the two need further judgment to be moving objects or static background.

3.3 Re-classification for intermediate regions

For intermediate regions with $0.2 \le \rho_i \le 0.8$ need to be reclassified, to identify whether it is the motion region or the static background region. This kind of region can be classified according to its space-time characters, and by adopting time-space energy model method.

Time energy of intermediate region i is defined as:

$$E_{time}^i = \sum_{(x,y) \in M_i} M_1^i(x,y) \tag{6}$$

where, M_i is the domain of region i.

285

Space energy of intermediate region i can be defined by the following steps:

Set $V(x, y)$ to be space energy of pixel (x, y), defined as:

$$V(x, y) = \begin{cases} -1 & n(x, y) \in B \\ 1 & n(x, y) \in M \\ 0 & n(x, y) \in C \end{cases} \quad (7)$$

$n(x, y)$ is 3 neighborhood of pixel (x, y), B is the background region, M is moving region, C is the intermediate region. Based on the above regional space energy can be defined as:

$$E_{space}^i = \sum_{(x, y) \in M_i} V_i(x, y) \quad (8)$$

According to the two components of space-time energy the total energy can be calculated as:

$$E_{total}^i = \frac{1}{2N}(E_{time}^i + E_{space}^i) \quad (9)$$

N is the total number of pixels in the region.

Set energy threshold as T_3, when $E_{total}^i > T_3$, the region is the motion object region, or else it is the background region. Further classification can be carried out according to space-time energy model, threshold $T_3 = 0.2$. The result is shown in Figure 5.

From the experiment results, space-time energy model can accurately carry out reclassification, and the segmentation object is basicly complete, only except some leakage points and irregular phenomena on several edges.

Background region and moving object region mask can be obtained through the above steps.

$$M_2 = \begin{cases} 1 & (x, y) \in M \\ 0 & (x, y) \in B \end{cases} \quad (10)$$

Motion objects are made up in the mask region corresponding to 1, rather complete video segmentation object can be obtained, but there are still internal holes and minority edges loss, these need

Figure 5. Regional space-time energy option.

Figure 6. The final segmentation result.

post-processing by methods such as morphology. The final segmentation result is shown in Figure 6.

4 CONCLUSIONS

This paper carries out the relevant research aiming at impulse noise on image filtering algorithm, provides an adaptive image filter algorithm. The relatively commonly used image filtering algorithm of this algorithm can filter out image noise more effectively on the basis of protecting the image details effectively. At the same time puts forward a video segmentation algorithm based on region selection, this algorithm not only considers algorithmic accuracy but also reduces the complexity and improves the robustness as far as possible. It can segment obvious accurate video objects from the complex video sequence. But the algorithm also has some shortcomings, that is, it did not solve the occlusion problems in segmentation, this is a problem to be solved in future research.

REFERENCES

Bao Hongqiang. Based on the content video motion object segmentation technology research. [Doctoral Dissertation]. Shanghai: University of Shanghai. 2005.3 (3):55–56.

Chu Guangli Video Object Segmentation Algorithm Based on Background Reconstruction ICCDA 2010.

Guowen CHE, Lijun MA. An Excel model of analogue orthotropic plate method to calculate load transverse distribution coefficients [J]. Proceedings of 2011 International conference on Electronic and Mechanical Engineering and Information Technology vol. 7: 3796–3799.

Lijun Ma the Integrity Rules and Constraints of Database [J]. Proceedings of 2011 International conference on Electronic and Mechanical Engineering and Information Technology vol.7:3747–3749.

Wang Jun, Cha Yun, Wu Yushu. Automatic video segmentation method based on background model. Computer engineering and applications 2003.9.

Zhu Hui. Object-oriented generated video segmentation technology research. [Doctoral Dissertation] University of electronic science and technology. 2002.10 (3)25–37 (8)100–110.

Control Engineering and Information Systems – Liu (Ed)
© 2015 Taylor & Francis Group, London, ISBN 978-1-138-02685-8

A method for distinguishing and counting the persons from video

L. Zhang
Computer Center, Beijing Information Science and Technology University, Beijing, China

J. Wang
Office of the Principal, Beijing Information Science and Technology University, Beijing, China

ABSTRACT: Video is currently being applied to many places. Extracting people objects automatically from the video and count them is an important application. This paper analyzes the shortcomings of the current video object extraction algorithm, and we propose an improved algorithm which based on the adjacent frame the contours of moving objects and get the head regions in accordance with the method of calculation of threshold, thus obtain the number of people in the video. We also designed a statistical strategy to reduce the computational error. The experimental results show the algorithm achieved very good results.

1 INTRODUCTION

At present, the video is applied to many places. The video recorded all the information within the scope of sight in real time. In video analysis, Object segmentation is largely dependent on the difference between the target shapes in the image and matching. Video objects in the general motion of folding, scaling, occlusion, and other changes, there will be, as the brightness changes, the corresponding feature will be a corresponding change. Has yet to find and feel consistent with the object of cognitive mathematical model can only be more accurate to identify some simple shape, a smaller change in the object.

Video semantic object segmentation process can use the video in a variety of prior knowledge as a guide, the knowledge of the object can be seen as narrow, specific high-level semantic information, as a top-down task-oriented information in a specific application to find and split certain types of objects that meet the specific concept. Video human face object segmentation can take advantage of the color of the face, facial organ distribution constraints a priori knowledge to define the similarity criteria for the face, and then split in regions with high similarity values to represent the human face.

This paper analyzes the shortcomings of the current video object extraction algorithm, and we propose an improved algorithm which based on the adjacent frame the contours of moving objects and get the head regions in accordance with the method of calculation of threshold, thus obtain the number of people in the video. We also designed a statistical strategy to reduce the computational error.

2 THE CURRENT EXTRACTION TECHNIQUES

Segmentation of video object extraction is the bases of video analysis, video object segmentation generally use two algorithms:

2.1 Edge detection

Some algorithm for video object segmentation using edge detection methods, namely the operator in the candidate region to extract edge pixels to get the target contour. Commonly used edge detection is the use of operator images calculus. Operator can be viewed as an array, its essence is to act as a fixed size, by numerical convolution of the nuclear operator's reference point is usually located in the center of the array. Non-zero part of the array size is known as the nuclear support.

To calculate the value of a pixel point of the convolution, the first nuclear reference point for positioning the pixel array of the remaining points to cover the pixels around, and then multiplying and summing these values, the specific calculation formula is as follows:

$$H(x,y) = \sum_{i=0}^{M_i} \sum_{j=0}^{M_j} I(x+i-a_i, y+j-a_j) G(i,j) \quad (1)$$

In formula (1), $H(x,y)$ is convolution value, $I(x,y)$ is a specific pixel value, the value of $G(i,j)$ is the nuclear.

If the pixel brightness value of a certain threshold, the pixels are boundary pixels of the original

Figure 1. Edge detection operator.

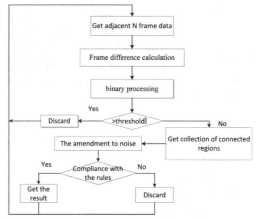

Figure 2. Edge detection based on fixed background framer.

image. The figure (Fig. 1) is the results of the canny operator to detect sub-image.

From the picture we can see that the boundary image obtained by the ordinary integral edge detection algorithm is very vague, and highlight the border does not function well.

2.2 Edge detection based on fixed background frame

In the video, still camera to take video grayscale image as a background frame, the grayscale image of the video follow-up with the frame of the background subtraction, and then the binary operation, you can get the outline of the movement of moving objects.

From the figure (Fig. 2) we can see that in the case of camera stability, light constant, this method can get better test results. But during our video semantic analysis, cannot always keep the camera stable and light constant. Most used in intelligent video of target detection.

3 IMPROVED VIDEO OBJECT SEGMENTATION ALGORITHM

In view of the ordinary based segmentation algorithm based on the edge of the video, and video emotional semantic analysis features. In this paper, two types of moving object-based video object segmentation algorithm, and through experiments, the performance of the two algorithms were compared.

First of all, we made two assumptions:

Most of the video frames object movement is gradual and video the same lens in a certain period of time relative stability.

Attention and emotional response of the main factors caused the moving objects in video.

Moving objects based on the background of the video can be stable in the camera, the light is constant to obtain good segmentation results, and we can constantly update the background frame of video object segmentation.

The algorithm process is as follows:

Step 1: Color mode conversion

The RGB color space in the most basic image, but the disadvantage of the RGB space is not intuitive, difficult to obtain from the RGB value of color perception attributes. In the Human Visual System (HVS), the sensitivity of the brightness is higher than the degree of sensitivity of the color, the HSV color space commonly used in video analysis, change space including brightness, color saturation.

RGB space into HSV space using the following formula:

$$V = \frac{1}{\sqrt{3}}\sqrt{R+B+G} \tag{2}$$

$$S = 1 - \frac{\sqrt{3}}{V}\min(R,G,B) \tag{3}$$

$$H = \begin{cases} \theta & G \geq B \\ 2\pi\theta & G < B \end{cases} \tag{4}$$

In formula (4),

$$\theta = \cos^{-1}\left[\frac{\frac{1}{2}[(R-B)+R-G]}{\sqrt{(R-B)^2+(R-B)(G-B)}}\right] \tag{5}$$

$H \in [0°, 360°]$ $S \in [0,1]$ $V \in [0,1]$

Step 2: Frame difference calculation
Using the formula (6) calculate the difference between the adjacent three frames.

$$D_{k,k+1}(x,y) = |I_k(x,y) - I_{k-1}(x,y)|$$
$$+ |I_{k-2}(x,y) - I_{k-1}(x,y)| \qquad (6)$$

$I_k(x,y)$ represents the k-th frame in the (x,y) point pixel brightness values, $D_{k,k+1}(x,y)$ is the k-th frame k+l frame corresponding to the location of the pixel difference.

Step 3: Binary processing
Binarization of the difference matrix of the frame difference.

$$D_{k,k+1}(x,y) = \begin{cases} 255 & if \ |I_k(x,y) - I_{k+1}(x,y)| > T \\ 0 \end{cases} \qquad (7)$$

Step 4: Correction to obtain the separation area.
General external segmented binary image, the outline of the candidate region there may be prominent within the hole. Mathematical morphology can be used to remove a small piece of interference and noise.

Erosion after dilate is called close, CLOSE(X) = E(D(X)). The closing operation can eliminate internal voids.

Dilate after erosion known as the open (open), OPEN(X) = D(E(X)). The opening operation can eliminate the external small protrusion, the contours of the object becomes smooth.

Step 5: Get the candidate region
By constructing a color probability in detecting skin color pixels, is given a probability value for each grid in the color space, the probability of the pixels belonging to color by the rules of the lookup table, or the Bayesian classifier.

In image processing, the different objects tend to have different chromaticity. For example, in the recognition of human face, the skin color is an important feature to distinguish between face and non-face.

Set pixels than the color value of pixel (x,y) M × N video frames, $f(x,y)$, the color values in the color mode, color three-dimensional, need to be converted to one-dimensional representation. The formula is as follows:

$$f(i,j) = 16H(i,j) + 4S(i,j) + V(i,j) \qquad (8)$$

set $p(f(i,j),c)$ as probability of the pix $f(x,y)$ is the point on face.

The probability of the hole image can be expressed as:

$$P = \frac{1}{M \times N} \sum_{i=0}^{M-1} \sum_{j=0}^{N-1} p(f(i,j),c) \qquad (9)$$

If the probability exceeds a threshold T, we identified the region as a candidate region.

$$R = \begin{cases} True & if(P > T) \\ false & if(P < T) \end{cases} \qquad (10)$$

Step 6: Eliminate interference area
Positioning in order to reduce the amount of computation, improve accuracy, the need for the candidate region to interference unless the head. From the aspects of the candidate region for processing:

– Extracted to face pixel may also be extracted to the arms, legs, skin color pixels. According to the principle of hair in the upper part of the face within a certain range, you can rule out non-face skin regions. Finally, the merger of regional and hair color region to determine the candidate region.
– The image contour is an important feature of the video object, the main use of the profile characteristic shape parameters.

Removal area is too small area or too big area. As the region is too small or too big, noise will not be complete in the head area.
Set A be the area of the contour:
if A > T1 and A < T2 then {Keep the area}
else Discard
T1 and T2 is the threshold of the area.

– The centroid of the head should be in the connected region. Depending on the shape of the head, the contour into convex. Determine the location of the centroid can remove the concave head area.
– The aspect ratio of the head area within limits. Calculation of the candidate region of the external rectangle of aspect ratio, excluding the non-conforming area.

Set B be the circumference of the contour. We use the following formula contour feature:

$$F = \frac{\|B\|^2}{4\pi A} \qquad (11)$$

if F > T1 and F < T2 then {Keep the area}
else Discard

– Remove shadows, saturation and brightness of the pixel, the brightness of the shaded basic black or gray, the basic color characteristics, you can exclude the shaded suppression of false alarms generated by the light to produce the figure. (You can use stereo vision to expand the viewing angle, to reduce false negatives)

- Reduce light interference, light; impact on the brightness, chrome, the use of mixed-color model can to some extent to avoid the interference of light.
- Variable threshold, anti-interference. Calculate the mean brightness of the entire video, the formation of the waveform, if the waveform transition and the new value is more stable, strong light changes. Adjust the threshold to adapt to the new environment.

4 CALCULATE NUMBER OF PEOPLE

Though extract people's head regions, we can get the number of people in a video frame. However, due to the people in the video is in motion. It is likely caused by the overlap between the objects and occlusion, so that the number of statistics in the video is not accurate. Here, we designed a statistical strategy to reduce the computational error.

When collecting head count to sudden changes. Set up a new poll number N', the original head count as N.

If N'> N calculation of the video head's center of mass displacement to determine the head position in the video edge of the area. If the new head centroid edge in the video, that newcomers to enter. If there is no head centroid in the video edge is tentative, the person in doubt, follow-up testing 2–3 times, the head count is still N' shows the number of one, otherwise, discard the test results, the same number.

If N'<N computing center of mass displacement, and scan the first X-frame of the video, and calculate the centroid distance, observing the smallest centroid distance is smaller than a certain threshold, if it is, that appeared to block the same number. X frame and then observing whether anyone in the video edge and determine the direction of movement. If not and outward movement to reduce the number of people.

5 EXPERIMENTAL RESULT AND ANALYSIS

In this study, we use the Vs 2008 and OpenCV library v2.1, and select 100 video clips as test video, experimental steps flow the section III and section IV.

Examples of experimental results are shown in Figure 3.

The precision an recall are computed by (12) and (13)

Figure 3.   Example of experimental results.

Table 1.   Performance of algorithm.

Video type	Based on fixed background frame		Improved algorithm	
	Precision	Recall	Precision	Recall
News	0.56	0.58	0.78	0.76
Sports	0.35	0.45	0.65	0.62
Monitor	0.71	0.75	0.84	0.88
Movie	0.23	0.25	0.53	0.43

$$\text{Precision} = \text{Correct}/(\text{Correct} + \text{False}) \qquad (12)$$

$$\text{Recall} = \text{Correct}/(\text{Correct} + \text{Miss}) \qquad (13)$$

We did a survey of various types of video detection results obtained by e proposed algorithm in this paper, and re compared them with results obtained based on the background of the frame algorithm. We can see test results on all kinds of video algorithms have greatly improved. In particular, sports video upgrade more obvious. Algorithm for monitoring video gets the highest detection rate, more than 80%, and get lowest detection rate for movie video, less than 60%.

6 CONCLUSION

In this paper, we analyze the problems of video edge detection and object segmentation algorithm, proposed an improved algorithm based on the background frame object segmentation algorithm, and though experimental verification, the proposed object segmentation algorithm can basically meet the video the needs of object segmentation. Higher than the performance of edge detection algorithms and object segmentation algorithm based on a fixed background frame. Video people counting strategy to solve the problem of the detection error caused due to camera movement and background motion. However, from the experimental results, the proposed algorithm is more suitable for sports video, and monitoring the number of video statistics, but is not suitable for film video character extraction. The reason may be the movie video structure is more complex and requires further study.

ACKNOWLEDGMENT

The project is sponsored by the Scientific Research Common Program of Beijing Municipal Commission of Education. (SQKM201211232019).

REFERENCES

Alan Hanjalic. Content-Based Analysis of DigitalVideo[M], Springer, 2004. Chengcui Zhang, Xin Chen, Liping Zhou, Wei-Bang Chen, Semantic retrieval of events from indoor surveillance video databases[J], Pattern Recognition Letters 30 (2009) 1067–1076.

Chen, Duan-Yu; Chu, Kuei-Cheng; Liu, Yu-Chien; Chen, Yung-Sheng Human subject-based video browsing and summarization,2010 International Conference on Machine Learning and Cybernetics, ICMLC 2010, v 6, p 2796–2801, 2010.

Hanjalic. A. "Content-based analysis of digital video", Springer, 2004.

Jamil, Akhtar; Siddiqi, Imran; Arif, Fahim; Raza, Ahsen. Edge-based features for localization of artificial Urdu text in video images, Proceedings of the International Conference on Document Analysis and Recognition, ICDAR, p 1120–1124, 2011.

Patel, B.V.; Deorankar, A.V.; Meshram, B.B. Content based video retrieval using entropy, edge detection, black and white color features, ICCET 2010–2010 International Conference on Computer Engineering and Technology, Proceedings, v 6, p V6272–V6276, 2010.

Zhang, Hui; Zhao, Zhicheng; Cai, Anni; Xie, Xiaohui. A novel framework for content-based video copy detection, Proceedings—2010 2nd IEEE International Conference on Network Infrastructure and Digital Content, IC-NIDC 2010, p 753–757, 2010.

Session 3: Communication and networking

Control Engineering and Information Systems – Liu (Ed)
© 2015 Taylor & Francis Group, London, ISBN 978-1-138-02685-8

Scalable QoS multicast provisioning in Wide-Sense Circuit tree

K.X. Huang, Y. Chen, C.L. Li & X.B. Zhang
State Key Laboratory of Mathematical Engineering and Advanced Computing, Zhengzhou, Henan, China

ABSTRACT: It is still a hot topic that how to provide dynamic reconfigurable QoS multicast service in the network. The MPLS network combined with the DiffServ can provide end-to-end QoS guarantee for differentiated services, and aggregated multicast can reduce the cost of maintaining multicast state. In this paper, based on the technology of the Wide-sense circuit Tree technique, we propose an architecture to provide the dynamic QoS multicast service, in the reconfigurable network environment. With the idea of aggregated multicast to reduce multicast state, and using the Wide-sense circuit technology to guarantee the quality of the transport service.

1 INTRODUCTION

IP Multicast technology is one of the key technologies of the next generation of the Internet. It increases the efficiency of Internet resources utilization and the bandwidth. However, due to the characteristics of the multicast and the architecture of the existing network, it is very hard to implement in practical, even providing the QoS. In order to solve scalability of the multicast, UCLA proposed a scheme to reduce the multicast forwarding state called as aggregated multicast [Fei 2001], by forcing multiple multicast groups to share one distribution tree, then they proposed AQoSM algorithm [Cui 2006], the algorithm based on MPLS and DiffServ technology, designed to solve scalability and efficiency issues of QoS multicast. MMT [Boudani 2002] technology proposed to resolve the establishment and management of the MPLS multicast tree. Aggregated multicast can effectively reduce multicast state but brought bandwidth consumption, Jiang Y. et al. [Yong 2010] proposed aggregation multicast strategy RMM, to avoid waste of bandwidth, increase the adaptability of dynamic changes.

MPLS can be used to control network resources, avoid network congestion and improve resource utilization. Through the rational allocation of bandwidth resources and effective control of the routing process, it can optimize the resources utilization, but it is based on the same strategy to process each stream. DiffServ model can provide the appropriate QoS requirements through the data stream classification and control, and configuration network resources, but it lacks effective end-to-end routing policy. Through combining DiffServ and MPLS, we can get a very attractive strategy which can provide a scalable QoS service in the backbone network.

Never use letter spacing and never use more than one space after each other.

To meet basic needs of the future telecommunications network, Flexible Architecture of Reconfigure able Infrastructure (FARI) [Julong 2011] builds a basic network whose function is dynamically reconfigurable and scalable, and based on this network can provide basic network services which can be customized and meet different business needs. The Wide-Sense Circuit (WSC) is the key technology. In this paper, though the combining the idea of aggregated multicast and the WSC, we propose a provisioning scalable QoS Multicast scheme. In this scheme, to reduce the routing state and the cost of label, the same class services are aggregated. And it guarantees QoS of the data transmission with the WSC.

2 WSTC

2.1 *Definitions and function*

WSCT is a tree structure WSC established for uniformly transmitting the traffic of a multicast traffic class. WSC is a kind of adaptive virtual circuit dynamically established for the traffic flows with the same class transmitting along the same path. As a basic service unit in reconfigurable network, WSC can be established, deleted and adjusted according to applications' requirements and network conditions. The requirements are consisted of two kinds of parameters and index: a) the QoS requirements including bandwidth, delay, packet loss rate, throughput and so on. b) The non-Qos requirements for example the security.

Wide-Sense Circuit Domain (WSCD) is constituted by the WSC server and Reconfigurable Router (RR). The WSC server is the "brain" of all WSCD, responsible for managing the WSC in the domain and the state of the network. As WSCT be a special form of WSC, it also is managed and built by this server. In the Domain it is centralized model, but it is distributed model cross domain.

In one WSCD, the number of the WSCT depends on the service requirements and the state of the network. It can provide the dynamic QoS multicast service. Generally, one kind of multicast service is corresponding to one multicast tree. According the classification of the business, the multicast groups are aggregated. Then we got the distribution tree. At last, we got WSCT based on this tree.

Shown in Figure 1, in this WSCR, one WSCT (A, B, C, D, E, F) be built to transmit multicast data, where path (C-F) still transmit data with IP packet switching, path (A-B, B-C, C-D, D-E) transmit data with WSC. WSC server managers the dynamic changes of the group members to leave and join by collecting the state of link of the RR, calculates and managers the multicast tree, and managers WSCT by exchanges the interactive messages between the server and the RR. When a RR join, quit or fail, Server should to recalculate the WSCT and distribute the new tree structure information to the member nodes.

2.2 WSCT protocol structure

WSCT is logically composed of managing layer, controlling layer and data layer. RR collects the current network status by the managing layer, calculates the location of WSC, and sends the establishment, adjustment, and removal command to the controlling layer. The controlling layer receives and executes the command form the managing layer. The functions of data routing and forwarding, and guarantee the traffic QoS are completed by the data layer.

The protocol between centralized WSC servers across domain: It is used to manage the WSCT across domain, by exchanging the messages between the WSC servers in different domains. The messages including the requirement of the multicast traffic, the class of the multicast traffic, the view of traffic across domain, and the state of WSCT across domain and so on.

The protocol between WSC server and RR: It is used to send the WSCT establishment, adjustment, and removal commands to the controlling layer. The command of the traffic statistics request is sent to the controlling layer in a cycle, and the WSC server will analyze the statistics of the messages about the traffic. In some case, such as the outbreak of the traffic, the WSC server will directly get the request form the controlling layer to adjust the WSC.

The protocol between RR: It is used to establish, adjust or remove the path, under the commands of the managing layer.

As shown in Figure 2, the managing layer is responsible for maintaining the deployment strategy. The controlling layer is responsible for the executing the deployment. The data layer is responsible for implementing the data forwarding. By communicating between three layers, the function of the WSCT is accomplished. The centralized management is easy to maintain the consistency and efficiency of the multicast tree. But it is also easy to overload or be single-point-of-failure. We can enhance the reliability by cluster, distributed,

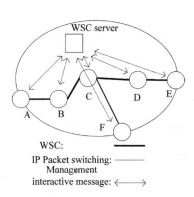

Figure 1.   An example of WSCT.

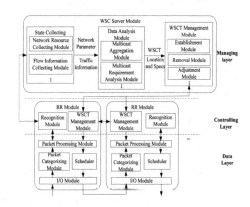

Figure 2.   Logical model of WSCT.

and redundant backup and so on. The WSC server only participates in deciding the policy, not in forwarding the packet. In practice those problems are limited to influence the network.

3 WSCT MANAGEMENT

3.1 Establishing policies

In the Flexible architecture of reconfigureable infrustructure, though the function of cognitive restructuring, the real-time running state, the resources distribution, and the application state of the network are extracted, analyzed, and integrated comprehensive multi-dimensional. Under detecting the data link by controlling layer and analyzing by WSC sever, according to the kind of the multicast service, the multicast groups are gathered category. Assuming that multicast groups have the same QoS requirements are noted as G_{c_i}, where G is the multicast group, c_i is the class of the QoS. Then managing layer calls the establishment algorithm according the state of the network, and sends the commands to the controlling layer. The WSCT is established in the data layer under the control of controlling layer.

The WSCT establishing algorithm can be described as follows.

We assume that $G_{c_i} = (c_i, \{g_1, g_2, ..., g_n\})$ represents a king of multicast groups, $T_{G_{c_i}}$ represents the distribution tree, which all groups share the same class service c_i, we get the $T_{G_{c_i}}$ according to the traffic statistics and the c_i.

It is assumed that $U_{c_i} = (u_{c_i, ij})_{n \times n}$ represents the distribution of the proportion of the bandwidth occupancy of the class c_i of the traffic in the $T_{G_{c_i}}$, Where n is the number of the nodes, $u_{c_i, ij} = Q_{c_i, ij}/(b_{ij} \times T)$ represents the traffic of the class c_i in the link l_{ij}, b_{ij} is the bandwidth of the link l_{ij}, T is the time interval to statistics.

In this step, we need to build the path of WSCT, according to the criterion and relevant factor. $W_{c_i} = (w_{c_i, ij})_{n \times n}$ Is the distribution of the link of WSC, if $w_{c_i, ij} = 1$, we established the path, otherwise, not. $U_{c_i, th}$ Is the threshold of the proportion of the bandwidth occupancy of the link. If $u_{c_i, th} \geq U_{c_i, th}$, $w_{c_i, ij} = 1$.

In this paper, we need to build different WSCTs according to the QoS requirement. We assume that B is the threshold of percentage bandwidth overhead. When we get the multicast groups G_{c_i}, we should try to divide the G_{c_i} to some subsets G_l, and the members of each subsets transmit the data through the corresponding aggregated tree T_l. We need minimum the number of l (the number of the aggregated trees). In other words, we can define the problem of optimizing the aggregated

tree as the minimum group model [Fangjin 2011]. The goal is to get the minimum L. The constraints are described as follows.

$$
\begin{cases}
\delta(T_l, g_i) \leq B \quad g_i \in G_l, l = 1, 2, ..., L \\
\bigcup_{l=1}^{L} G_l = G \\
G_m \cap G_n = \phi \qquad 1 \leq m \neq n \leq L \\
G_l \neq \phi \qquad l = 1, 2, ..., L
\end{cases}
$$

where $\delta(T_l, g_i)$ is the additional percentage bandwidth overhead, noted as:

$$
\frac{\cos t(T_l) - \cos t(T_{g_i})}{\cos t(T_{g_i})}.
$$

We assume that $\cos t(T_l)$ is the overhead of transmitting the data through the tree T_l, $\cos t(T_{g_i})$ is the overhead of transmitting the data through the native tree T_{g_i}.

3.2 Adjustment policies

In practice, the members of the multicast group are always changing, for example: join or exit. So when a new kind of QoS multicast applies to join, there are three cases, a) the existing tree cover new multicast group members; b) We should expand the existing tree; c) we should set up a new one for the new multicast tree. When the new multicast group wants to join the network, to calculate the additional percentage bandwidth overhead, and the update the aggregated tree, according to the constraint conditions. When a group left, however, generally it will not damage the constraints, the appropriate corresponding multicast group tree can reach a smaller percentage bandwidth overhead.

Aggregation tree after the update, the network will detect each link of the tree a) When the proportion of occupied bandwidth of a link does not meet the appropriate parameters, the removal of WSC command sent to the corresponding nodes; b) when the occupied bandwidth of a link to the appropriate parameters, an establishment WSC command sent to the corresponding nodes. RRs in turn to accept and implement the WSC removed and establishment commands from the WSC server. The entire process of adjusting WSCT was finished. Only when the entire network does not exist in a certain class of QoS multicast service, it will perform the WSCT removal commands.

4 PERFORMANCE ANALYSIS

In the simulation experiment, as shown in Figure 3 the network topology, the number of WSCT in the domain is directly related to the number of kinds of QoS. Here, we consider only one kind of QoS requirement. We should analysis the number of WSCT, the forwarding state in the RRs and the percentage bandwidth overhead. R_1, R_2, R_3, R_4, R_5, R_6 are the border routers. The multicast groups are randomly generated, and the aggregated algorithm is the Jun-Hong Cui greedy algorithm.

In order to ensure efficient data transmission services, WSCT uses the method of the Label Switching to transmit the data. Most of the research on aggregated multicast, it usually aggregates the tree by the similarity. In this paper, we want to provide the QoS. So we aggregate the tree by the same class of the QoS requirement. Though the experiment, we find that when there are a lot of kinds of QoS requirement, the bandwidth overhead and the number of the tree are not better than the method by tree similarity, but we can guarantee the traffic QoS. As shown in Figure 4, the setting of the threshold B percentage bandwidth overhead will affect the number of final WSCT.

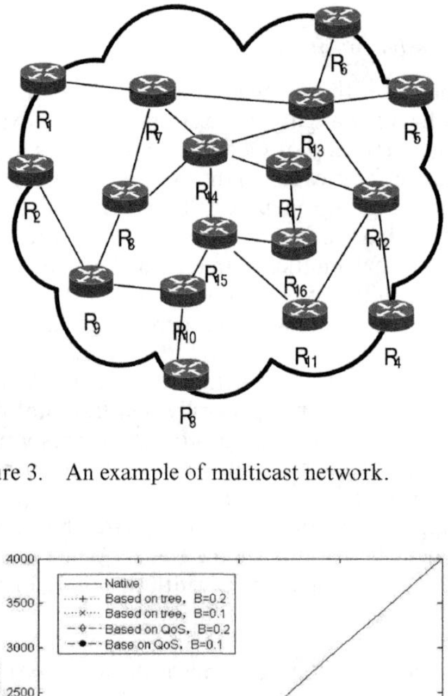

5 CONCLUSIONS

With the rapid expanding of different kinds of services and applications in the network, the ability of the Internet to meet the demand for all kinds of business requirements is also increasing. But the current structure of the Internet only provide a single "best effort" service, it has greatly restricted the ability. In this paper, based on the technology of the Wide-sense circuit Tree technique, we propose an architecture to provide the dynamic QoS multicast service, in the reconfigurable network environment. We can reduce multicast states effectively with the idea of the aggregated multicast, and guarantee the quality of the transport service by the Wide-sense circuit technology.

ACKNOWLEDGEMENT

Sponsored by the National Basic Research Program of China (973 Program) under Grant No. 2012CB315901.

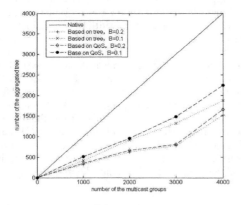

Figure 3. An example of multicast network.

Figure 4. Number of the multicast groups.

REFERENCES

Boudani A, Cousin B. "A new approach to construct multicast trees in MPLS networks". Proc. of Seventh International Symposium on Computers and Communications, ISCC 2002, July, 2002:913–919.

Cui, J.H. L. Li, M. Faloutsos and M. Gerla, "AQoSM: Scalable QoS Multicast Provisioning in Diff-Serv Networks," Computer Networks, vol. 50, pp. 80–105, Jan. 2006.

Fei, A. J.H. Cui, M. Gerla and M. Faloutsos, "Aggregated multicast: an approach to reduce multicast state," in Proceedings of Sixth Global Internet Symposium (GI), 2001.

Jiang Yong, Hu Song-Hua "Rendezvous Multicast: A Novel Architecture for MPLS QoS Multicast." Hournal of Software, 2010, 21(4):827–837.

Lan Julong, "Research on the Architecture of the Reconfigurable Fundamental Information Communication Network," Report of the National Basic Research Program of China (973 Program), 2011.

Zhu Fangjin, "Research on optization model and algorithm for aggragted multicast" Shandong Universtity, Jinan, Qctober, 2011.

298

Design of a wideband circularly polarization RFID reader antenna

B. Wang, X.J. Kong & L.J. Huang
College of Electronic Engineering, Chongqing, China

ABSTRACT: A novel wideband circularly polarization microstrip antenna is proposed for application of UHF Radio Frequency Identification (RFID) in China. Excellent performance is obtained based on an easy fabrication of three-layer configuration, including two corner cut patches, an S shaped horizontally meandered strip, three FR4 substrates, and a ground plane. Especially, the upper side of FR4 substrate is printed with an S shaped horizontally meandered strip. It obviously improves the impedance bandwidth, axial ratio bandwidth, and radiation pattern symmetry. Systematic optimization achieved impedance bandwidth of 20.1% (760–930 MHz) with VSWR approximately below 1.4, the 3 dB circular polarization AR bandwidth about 11% (840–938 MHz), and the corresponding gain of 8.9 dB. Compared to the technique standard for UHF RFID, the proposed antenna is a good candidate for UHF RFID readers/writers antenna operating between 840 MHz and 960 MHz.

1 INTRODUCTION

Radio Frequency Identification (RFID) is a new automatic identification technology, deriving from 80s last century and gaining a fast progress at the starting of present century. Utilizing RF signals, it automatically identifies target object and obtains relevant data, characterized by long distance, Non Line of Sight (NLOS) and meanwhile identifying several tags. Generally the whole RFID system includes reader, electronic tag and processor, and it is extensively applied to vehicle automatically management, the charging system of highway, tracking goods, production line and so on. Compared to other passive RFID, the passive UHF RFID has longer transmission distance and faster speed, so it has a broad application prospect. There are diverse frequency bands in different countries, as shown in Table 1. It is shown that the frequency band in most countries is 865–928 MHz, except China and Japan. With the development of RFID technology in China, we need to manufacture wideband reader/writer

Table 1. UHF RFID frequency bands in different countries.

Country	Frequency (MHz)
China	840–845 and 920–925
Japan	952–955
Europe	865–868
North America	902–928

antenna which is appropriate for Chinese UHF RFID frequency range (840–928 MHz, relative bandwidth is 10% approximately).

In fact, the basic operation principle of passive UHF RFID is not complex. First of all, the reader antenna sends RF modulating signal. Secondly when the tag enters the reader district, it transmits the product code depending on the induction current energy. Thirdly the reader antenna obtains the product code and decode. Finally the message is disposed in the centralized information system. Generally the UHF RFID tag utilizes deforming symmetrical structure (Rao et al. 2005) which is linear polarization while the reader antenna demands for identifying the tag at different orientation, so the reader antenna should be circular polarization.

The circular polarization of microstrip antenna with single feed or several feeds is usually generated by two stimulated orthorhombic linear polarization waves whose amplitude is uniform and whose phase is 90 degree discrepancy. Single feed circular polarization microstrip antenna has the characteristics of simple structure, lower cost, easily fabrication. Although it only needs to ensure the appropriate feed position so as to engender the circular polarization wave, there is a disappointed AR bandwidth which is about 0.5~1% of the centre frequency (Wang et al. 2009, Lee et al. 2005). If we adopt the lower dielectric constant and the thicker substrate for the microstrip antenna of single feed and single patch, the relative AR bandwidth will be improved to 6~14%, but there is a problem that the height of the designed antenna will exceed 50 mm which is the maximum size for UHF RFID readers/writers antenna (Yang et al. 2008).

Assuming utilizing stacked patch structure, the relative AR bandwidth of microstrip antenna based on the single coaxial feed can be enhanced to 18.8% (Sudha et al. 2004). But the input impedance can not be satisfied for the requirement of communication system of VSWR < 1.5. In order to resolve the impedance matching for the stacked microstrip antenna, there were several researchers who presented the aperture coupling feed technology (Lee et al. 2001) and L shaped strip feed technology (Meshram 2007). However, the aperture coupling feed technology needed the transmitting board decrease back scattering, consequently resulting in the height of the designed antenna larger. Furthermore, the horizontal and perpendicular length of L shaped strip influenced both the input impedance and reactance, finally leading to the designed antenna more complex. Therefore, Chen Zhi Ning with others proposed a novel feed technology based on half wave strip (Chen & Chia 2003), which can easily obtain the applicable input impedance and reactance by adjusting the structure parameters, and soon afterwards, this technology was used to design a stacked microstrip antenna, whose impedance bandwidth is improved to 22% in conditions of VSWR < 1.2 (Wang et al. 2010), but the radiation pattern was not symmetrical.

In consideration of the above problems, this paper proposes a certain feed technology based on S shaped horizontally meandered strip, one end of which is connected to the main radiating patch by a probe, and the other end of which is connected to SMA connector. Through making use of S shaped strip technology, we obtain an approving impedance matching and symmetrical radiation pattern, meanwhile as the air layer between the stacked patch is applied, we acquire an improved impedance and AR bandwidth.

2 STRUCTURE AND DESIGN OF ANTENNA

The operating frequency of the designed reader/writer antenna in the paper is 888 MHz, which is based on the centre frequency of 840–928 MHz for Chinese UHF RFID frequency range. The antenna structure is shown in Figure 1. There is a copper board on the bottom of the designed antenna, whose size is $L \times L = 250$ mm × 250 mm; then successively upward, the first layer is a rectangle substrate etched S shaped horizontally meandered strip, whose size is $L_1 \times W_1 = 105$ mm × 70 mm; the second layer is a square substrate whose size is $F_1 \times F_1 = 180$ mm × 180 mm, etched a square patch whose size and corner cut size are $L_2 \times L_2 = 138$ mm × 138 mm and $\Delta L_2 = 41$ mm respecttively; the third layer is a square substrate whose size is $F_2 \times F_2 = 180$ mm × 180 mm,

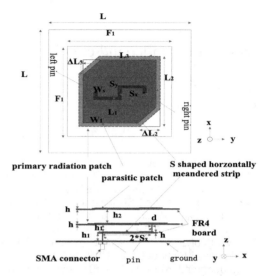

Figure 1. Structure and dimensions of the proposed UHF RFID antenna.

etched a square patch whose size and corner cut size are $L_3 \times L_3 = 123$ mm × 123 mm and $\Delta L_3 = 33$ mm respectively. The above substrate is FR4 board (dielectric constant $\varepsilon = 4.4$ and loss tangent $\tan\delta = 0.02$), and the second layer patch is the primary radiation patch, compared to the third parasitic patch which improves the AR bandwidth. In addition, between the stacked substrates, there are three layer airs, whose height are $h_1 = 6$ mm, $h_2 = 12.7$ mm respectively. In fact, the air layer improves the bandwidth and gain. At last, utilizing two probes whose diameter d is equal to 1.27 mm, one end of S shaped horizontally meandered strip is connected to the primary radiation patch and the other end of it is connected to the SMA connector. The connected position is also shown in Figure 1. The width W_s of S shaped horizontally meandered strip is 6 mm.

The horizontal size S_x and vertical size S_y of the S shaped horizontally meandered strip mainly determines the matching of the antenna. Changing S_x influences both the input impedance and reactance of the antenna. For instance, with S_x increasing, the input impedance gradually becomes larger, and the changing trend is similar to the status that the coaxial probe position influences the input impedance of the common microstrip antenna (Richards et al. 1981). Similarly changing S_y only affects the input reactance. Therefore, we need to firstly adjust S_x in order to change input impedance to 50 Ω, then make the S_x fixed and adjust the S_y so as to change input reactance to zero. Through analyzing the simulation results shown in Figure 2 and Figure 3, we can acquire the exact length of S_x

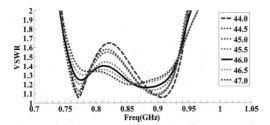

Figure 2. VSWR curve with S_x changed from 44 mm to 47 mm.

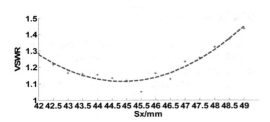

Figure 3. VSWR of centre frequency point with the different S_x.

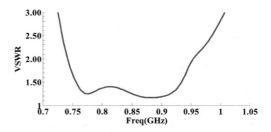

Figure 4. Simulated results: VSWR.

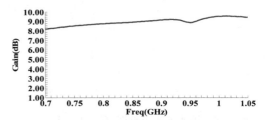

Figure 5. Simulated results: gain.

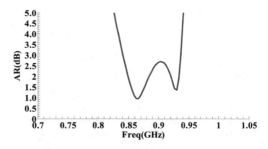

Figure 6. Simulated results: AR.

and S_y. It is shown in Figure 2 that the VSWR varies with S_x gradually changing 0.5 mm from 44 mm to 47 mm, and the VSWR of center frequency point based on the different S_x in Figure 2 is reflected in Figure 3. Through analysis, we ultimately ascertain the length of S_x is 46 mm and S_y is 27 mm. In essence, the feed technology of S shaped horizontally meandered strip is similar to half wave strip, but utilizing S shaped horizontally meandered strip technology, we can get symmetrical directional pattern, increase the half power beam width, decrease the VSWR and improve the gain.

3 RESULT AND DISCUSSION

Through simulation and optimization with HFSS, we acquire good simulation results for the UHF RFID reader/writer antenna.

The VSWR curve of UHF RFID reader/writer antenna is showed in Figure 4. From the curve, we can see the impedance bandwidth is 170 MHz (from 760 MHz to 930 MHz, and relative bandwidth is 20.1%) while the VSWR < 1.4. The result attributes the success to make use of the S shaped horizontally meandered strip and thick air layer between substrates.

The gain curve from 700 MHz to 1050 MHz is revealed in Figure 5, from which we acquire that the gain is greater than 8.9 dB from 840 MHz to 928 MHz which is equal to Chinese operating frequency, so the gain has been improved obviously.

The AR curve is showed in Figure 6, and the AR bandwidth is 88 MHz (from 840 MHz to 938 MHz, and relative bandwidth is 11%) while the AR < 3 dB. It is showed that Chinese UHF RFID frequency range has been absolutely covered.

In Figure 7, Figure 8 and Figure 9, three different frequency points 840 MHz, 900 MHz and 930 MHz have been respectively selected for observing their directional pattern. There are two principal planes E and H respectively displayed in Figure 7, Figure 8 and Figure 9. From the E and H plane, we obtain the radiation pattern is relatively symmetrical.

Through the analysis of above results, the half power beam width based on the three different frequency point is revealed in Table 2. In fact, we can summarize there is only 4 degree range discrepancy between 840 MHz and 930 MHz, thus it can be seen that there are good directionality and gain within the above frequency range.

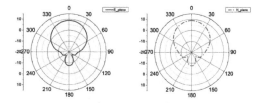

Figure 7. Simulated results: radiation patterns at 840 MHz.

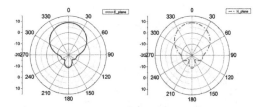

Figure 8. Simulated results: radiation patterns at 900 MHz.

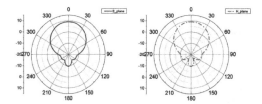

Figure 9. Simulated results: radiation patterns at 930 MHz.

Table 2. Half power beam width.

Frequency	E plane (Degree)	H plane (Degree)
840	80	82
900	80	79
930	80	78

4 CONCLUSIONS

Through a certain feed technology based on S shaped horizontally meandered strip, one type novel reader/writer antenna has been proposed and designed, which is suitable for Chinese UHF RFID frequency range. The simulation results ultimately prove the designed antenna with characteristics of the well impedance matching and the symmetrical radiation pattern, and meanwhile, because of the parasitic patch on the top of the antenna and the air medium between the substrates, the impedance

and AR bandwidth are broadened obviously. The simulation results show that within the frequency range 840–928 MHz, the VSWR is smaller than 1.4, the AR is smaller than 3 dB and the gain is larger than 8.9 dB. By utilizing the single feed structure and the low-cost FR4 board, the antenna is with the characteristics of simple structure, easy fabrication and low cost. All in all, the designed antenna is fit for Chinese UHF RFID application. It is the next work for the designed antenna which will be manufactured.

ACKNOWLEDGEMENT

This work was supported by the Natural Science Foundation of CQ CSTC under Grant cstcjjA40022 and the Science & Technology Research Project of Chongqing Municipal Educational Commission of China under Grant KJ120505.

REFERENCES

Chen, Z.N. & Chia, M.Y.W. 2003. Broad-band suspended probe-fed plate antenna with low cross-polarization levels. *IEEE Transactions on Antennas and Propagation* 51(2): 345–346.

Lee, B. Kwon, S. Choi, J. 2001. Polarisation diversity microstrip base station antenna at 2 GHz using T-shaped aperture coupled feeds. *IEEE Proceedings Microwave, Antenna and Propagation* 148(5): 334–338.

Lee, J.M. Kim, N.S. Pyo, C.S. 2005. A circular polarized metallic patch antenna for RFID reader. Proceeding of 2005 Asia Pacific Conference on communications: 116–118. Perth: Western Australia.

Meshram, M.K. 2007. Analysis of L-strip proximity fed rectangular microstrip antenna for mobile base station. *Microwave and Optical Technology Letters* 49(8): 1817–1824.

Rao, K.V.S. Nikitin, P.V. Lam, S.M. 2005. Antenna design for UHF RFID tags: a review and a practical application. *IEEE Transactions on Antennas and Propagation* 53(12): 3870–3876.

Richards, W.F. Lo, Y.T. Harrison, D.D. 1981. An improved theory for microstrip antennas and applications. *IEEE Transactions on Antennas and Propagation* 29(1): 38–46.

Sudha, T. Vedavathy, T.S. Bhat, N. 2004. Wideband single-fed circularly polarised patch antenna. *Electronics Letters* 40(11): 648–649.

Wang, Z.B. Fang, S.J. Fu, S.Q. 2009. A low cost miniaturized CP antenna for UHF radio frequency identification reader applications. *Microwave and Optical Technology Letters* 51(10): 2382–2384.

Wang, Z.B. Fang, S.J. Fu, S.Q. 2010. Broadband stacked patch antenna with low VSWR and low cross-polarization. *ETRI Journal* 32(4): 618–621.

Yang, S.L.S Lee, K.F. Kishk, A.A. Luk, K.M. 2008. Design and study of wideband single feed circularly polarized microstrip antenna. Progress in Electromagnetics Research 80: 45–61.

Control Engineering and Information Systems – Liu (Ed)
© 2015 Taylor & Francis Group, London, ISBN 978-1-138-02685-8

Improved BP of 2-D PSD nonlinear error correction

H.H. Xiao & Y.M. Duan
Department of Computer and Information Science, Hechi University, Yizhou, Guangxi, China

ABSTRACT: When PSD (Position Sensitive Detector) applied, how to overcome the nonlinear problem is an important research direction. In order to solve this critical problem, LM algorithm which based on the improved BP neural network algorithm to advantaged the PSD non-linear, established experimental platform, analyzed using BP neural network algorithm and LM algorithm which based on improved BP neural network algorithm to corrected PSD non-linear error; when training data, using two hidden layer neural network training, achieved the PSD nonlinear compensation.

1 INTRODUCTION

Position sensitive sensor PSD (Position Sensitive Detector) leap in the development in recent years as the representative of the position sensitive device (K 2000 a). PSD itself has time to respond to fast and high position resolution, PSD quickly get a wide range of applications (Q & W 2011b), but due to the non-linear effects of PSD, causing the entire position sensitive sensor measurement data confidence (WJ, Mcgraw-Hill 1988 c). Although the structure and use of the material of the PSD to be improved in a way to improve the linearity of the PSD (M & J & T 2012 d), it is still difficult to overcome the non-linearity of the PSD. PSD having a good linearity in the A region (central region), but the nonlinearity of the PSD of its linearity of the B area (edge area) have a substantial impact and constraints (K & S 2011 e), thus limiting the PSD measurement range and accuracy. The papers from the research impact of the non-linear factors of the two-dimensional PSD, followed by a two-dimensional PSD non-linear error correction algorithm, derived on the basis of the two-dimensional PSD nonlinear error correction algorithm deficiencies and shortcomings of the PSD, and proposed a new algorithm-LM algorithm PSD nonlinear error correction.

2 2-D PSD

2-D PSD of the four electrodes is the most obvious difference between the 1-D PSD. Duplex structure, pillow-type structure, the quadrilateral structure is the most common 2-D PSD structure.

Figure 1 shows the gradual increase in the square structure forming long strip electrodes, the transfer impedance gradually becomes zero, which is characterized by relatively easy to bias, fast response, small dark current and other advantages.

PSD sectional structure is shown in Figure 2, when the PSD photosensitive surface when a certain intensity of the incident light, and light energy is proportional to the charge will be generated in the incident position, the output of such charge through the resistor layer. Ideal symmetry due to the resistance of the resistive layer, the light spot is inversely proportional to the distance and the output current to the respective electrodes. Set the light spot from the photosensitive surface center distance is L, the output current I_1 and I_2, the current on the electrode is set to the total current I_0, therefore: $I_0 = I_1 + I_2$.

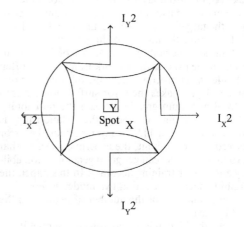

Figure 1. 2-D PSD electro-optical position sensor basic schematic diagram.

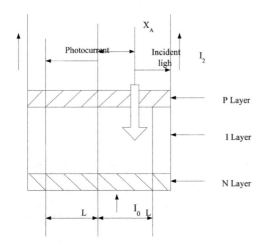

Figure 2.   2-D PSD schematic structure.

3   SIMULATION ANALYSIS

According to the above described experiments and BP neural network algorithm and LM algorithm, Matlab7.1 programming and debugging, the experimental data and select the hidden layers and the number of neurons. This article uses two nodes entered layer, and physical parameters of the output layer into two nodes: a position coordinate points in the 2-D PSD ideal output x represents a position coordinate point (x_0, y_0) in the 2-D PSD(x, y). The select two hidden layers of the middle layer, two hidden layers take appropriate number of neurons, respectively, 38 and 34, in the first hidden layer activation function using 'tansig' second hidden layer activation function The take tanlm output layer to take activation function purelin. The maximum number of training set to 500, the target convergence accuracy of 1.0×10^{-5}.

Determine the number of hidden layers:

Gradient instability, the number of local minimum in the process of network training, it is often a hidden layer processing, the use of more than one hidden layer makes the error surface formed into a local minimum greatly increase the probability. Data processing, mapping the relationship is not continuous, a hidden layer so that the error exceeds a certain limit, the experiments show that two hidden layers can get good generalization ability and higher training accuracy. In this paper, the training network selection two hidden layers.

Determination of the number of neurons in the hidden layer:

Take the hidden layer neurons is essential if the little number of neurons in the right amount of information in the network, often during training

can not match; neurons number too much may cause the training time particularly long, cause confusion match is very important to determine the number of neurons in the hidden layer.

BP neural network training process is as follows.

Open MATLAB7.1 input window in the program are as follows:

The run training simulation is as follows.

Wherein Figure 3, the abscissa indicates the training step; the ordinate represents the training accuracy.

The LM algorithm training process is as follows:

Open MATLAB7.1 input window in the program are as follows:

The run training simulation is as follows:

3,4 simulation diagram can be drawn, compared to traditional BP neural network algorithm, LM algorithm of network convergence speed significantly faster convergence and high precision, in the case of the same error accuracy, the conventional BP neural network algorithm by 200 times in order to achieve the required accuracy, the LM algorithm is reached after 7 times the required accuracy.

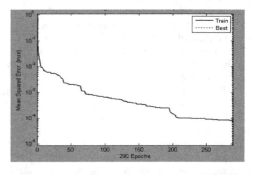

Figure 3.   Non-linearity correction of 2-D PSD based on BP Neural Network.

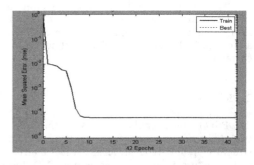

Figure 4.   Non-linearity correction of 2-D PSD based on LM.

4 CONCLUSIONS

1. Improved BP neural network algorithm optimization results are better than BP neural network algorithm in the optimization process, BP neural network algorithm is limited to local optimal solution, and improved BP neural network method has excellent global optimization, get is a global optimal solution.

2. Improved BP neural network algorithm optimization is better than BP neural network algorithm improved, but improve marginally improved BP neural network algorithm based on BP neural network algorithm However, BP neural network algorithm relies searching for the best local information at the point of initial design, can only converge to the optimal solution in the vicinity of the initial point: improved BP neural network algorithm for solving the number of iterations is less than BP neural network algorithm solving iterations, thereby improving the convergence rate of the BP neural network algorithm is faster and more efficient.

3. Improved BP neural network algorithm is a global convergence and fast convergence of modern optimization algorithm. Has optimized high efficiency, optimization effect at the same time, it applied to the micro electric wheel hub motor optimal design is feasible, and has a broad engineering application value.

REFERENCES

Cx51 compiler user's guide. Optimizing C compiler and library reference for classic and extended 8051 micro-controllers [z]. Keil Software, Inc. 2000. 60–66.

Dahong Qian et al. A method for measurement of multiple position on one position-sensitive detector (PSD) [J]. IEE Trans. Instrum. Meas. 2011, 42(1), pp. 14–18.

Goodman W.J. Introduction to Fourier Optics [M]. New Seok H. Jang-Mok K. et al. Decopling Control of Bridle Rolls for System [J]. IEEE Trans. On Industry Applications, 2011, 35(1), pp. 119–125 York: Mcgraw-Hill, 1988, pp. 96–97.

Tomizuka M. et al. Synchronization of Two Motion Control Axes under Adaptive Freeforward Control [J]. ASME Journal of Dynamic System, Measurement and Control, 2012, 114(6), pp. 183–203.

Control Engineering and Information Systems – Liu (Ed)
© 2015 Taylor & Francis Group, London, ISBN 978-1-138-02685-8

Novel public key cryptosystem based on Chebyshev polynomials

P. Zhen & L.Q. Min
University of Science and Technology Beijing, Beijing, China

G. Zhao
Beijing Electronic Science and Technology Institute, Beijing, China

ABSTRACT: Different from previously proposed Chebyshev polynomials public key cryptosystem with Elgamal structure, this paper describes a novel public key algorithm based on Chebyshev polynomials over finite field and modular operation. Its intractability is based on integer factorization and discrete Chebyshev problem, which provides double security insurance. Experimental results and performance analyses show that the improved algorithm has much higher security and practical value. The algorithm is also easy for software implementation because of its simple structure.

1 INTRODUCTION

Chaotic systems are characterized by complicated behaviors which are sensitive to initial conditions or system parameters. Meanwhile, chaotic dynamics is also deterministic in the sense that it has an apparent function or equation. These properties, as required by several cryptographic primitives, render chaotic systems a potential candidate for constructing cryptosystem.

The application of chaotic maps in cryptography has been studied for more than twenty years (Kocarev 2001), but most of work concentrate on symmetric key cryptosystems (Schmitz 2001, Liu et al. 2010, Pareek et al. 2010 & He et al. 2010). There are also some public key cryptosystem based on chaos (Tenny et al. 2003 & Ruanjan 2005), but they are analyzed to be unpractical or unsecure. Kocarev et al. proposed a public key cryptosystem based on the commutative property of Chebyshev polynomials over real numbers (Kocarev et al. 2003), but it was soon broken by a triangle substitution attack (Bergamo et al. 2005). To resist the attack, a modified cryptosystem was presented (Kocarev et al. 2005), which expands the definition of Chebyshev polynomials from real field to finite field. Lima et al. proved that to break the cryptosystem through triangle substitution in finite field is equivalent to solving discrete logarithm problem (Lima et al. 2008). Recently, some new developments have been made. In Ref. (Liao et al. 2010, Chen et al. 2011 & Li et al. 2011), the authors analyzed the detailed properties of Chebyshev polynomials and pointed out some problems about its security. Cheong (Cheong 2010) constructed a one-way function

based on Chebyshev polynomials. Recently, Sun et al. proposed an improved public key encryption algorithm (Sun et al. 2013), which import the alternative multiply coefficient to forge the cipher text tactfully.

In this paper, we proposed a public key encryption algorithm, which is a combination of Chebyshev polynomials and modular operation and provided a higher level of security. It has simple structure and is easy for software implementation.

This paper is organized as follows. Section 2 introduces the Preliminaries about Chebyshev polynomials. Then the proposed novel public key cryptosystem is described in section 3. Performance analyses are given in section 4. Section 5 presents the conclusions.

2 PRELIMINARIES

2.1 Chebyshev polynomials

Definition 1. Let $n \in Z^+$ and $x \in R$, then Chebyshev polynomial of order n, $T_n(x): R \to R$ is recursively defined using the following recurrent relation:

$$T_n(x) = 2xT_{n-1}(x) - T_{n-2}(x), \ n \geq 2 \qquad (1)$$

where $T_0(x) = 1$ and $T_1(x) = x$.
The first few Chebyshev polynomials are

$$T_2(x) = 2x^2 - 1$$
$$T_3(x) = 4x^3 - 3x$$
$$T_4(x) = 8x^4 - 8x^2 + 1$$
$$\cdots \qquad (2)$$

When x is a real number, $T_n(x)$ always has the following explicit algebraic expression:

$$\begin{cases} T_n(x) = \cos(n\cos^{-1}(x)), & x \in [-1,1] \\ T_n(x) = \cosh(n\cosh^{-1}(x)), & x \in [1,+\infty) \end{cases} \tag{3}$$

Some important properties of Chebyshev polynomials are as follows.

1. $T_r(T_s(x)) = T_s(T_r(x))$ (4)

2. $T_n\left(\dfrac{x+x^{-1}}{2}\right) = \dfrac{x^n + x^{-n}}{2}$ (5)

Proposition 2 can be easily deduced from the explicit algebraic Equation 2 and it is this commutative property that is employed by Kocarev et al. to construct a novel public key algorithm (Kocarev et al. 2003 & Kocarve et al. 2005).

When $x \in R$, the explicit algebraic expression of $T_n(x)$ to a security loophole in the public-key cryptosystem based on Chebyshev polynomials defined over the real number field (Bergamo et al. 2005). Therefore, Kocarev et al. extended the definition of $T_n(x)$ to the finite field Z_n (Kocarve et al. 2005).

Definition 2. Let $n \geq 0$ be an integer, a variable $x \in Z_N$ and N be a positive integer. Chebyshev poynomial of order n is recursively defined by

$$T_n(x) \equiv (2xT_{n-1}(x) - T_{n-2}(x)) \bmod N \tag{6}$$

where $T_0(x) \equiv 1 \bmod N$ and $T_1(x) \equiv x \bmod N$.

It is easy to verify that the above propositions of $T_n(x)$ also holds over Z_N.

2.2 Elgamal cryptosystem based on Chebyshev polynomials

Utilizing the commutative semi-group property, the Elgamal public key cryptosystem (Kocarve et al. 2005) is constructed. Its process can be described as follows:

1. Key pair generation
   a. Randomly generates a large integers s, selects a random number $x \in Z_N$, $s \neq 1$, $x \neq 1$ and then computes $T_s(x) \bmod N$.
   b. Then s is private key and $(x, T_s(x))$ is the public key.
2. Message encryption
   Assume Alice wants to send a message $M \in Z_N$ to Bob. The encryption process is:
   a. Alice randomly selects an integer r and computes $C_1 = T_r(x) \bmod N$, $C_2 = M \times T_r(T_s(x)) \bmod N$.
   b. Then she sends the cipher text $C = (C_1, C_2)$ to Bob.

3. Message decryption
   Having received the cipher, Bob can decrypt it as follows:
   a. Bob uses his private key s to compute $T_s(C_1) = T_s(T_r(x)) = T_{sr}(x) = T_r(T_s(x))$,
   b. He recovers the plaintext by computing $M = X / T_s(T_r(x))$.

3 THE NOVEL PUBLIC KEY CRYPTOSYSTEM

3.1 The cryptosystem

The detailed procedure of the proposed public key cryptosystem is described as follows:

1. Key generation
   Bob chooses two large prime p, q, $n = pq$ and select an integer $x \in Z_n$. Then he computes

$$e = T_r(x)(\bmod n) \tag{7}$$

$$d = e(\bmod p) = T_r(x)(\bmod n)(\bmod p)$$
$$= T_r(x)(\bmod p) \tag{8}$$

Bob's public key is (x, e, n) and private key is (r, d, p).
2. Message encryption
   Alice wants to send message m (m is an positive integer and $m < d$) to Bob. She selects a random number k and encrypts the message using Bob's public key (x, e, n) to obtain the cipher text c:

$$c = (kn + ex + mx) \tag{9}$$

Then she sends cipher text c to Bob.
3. Message decryption
   Upon receiving the cipher text c, Bob can recover the message m' with his private key.

$$m' = x^{-1}c(\bmod p)(\bmod d) \tag{10}$$

This cryptosystem has very simple structure which is different from original algorithm. Let's proof its correctness. Firstly, a proposition will be introduced.

Lemma Let $n = pq$ and p, q are two large primes, then for any $a \in Z^+$,

$$a(\bmod n)(\bmod p) = a(\bmod p) \tag{11}$$

According to the lemma, the explicit proof is as follow:

$$m' = x^{-1}c(\bmod p)(\bmod d)$$
$$= x^{-1}(kn + ex + mx)(\bmod p)(\bmod d)$$
$$= x^{-1}(ex + mx)(\bmod p)(\bmod d)$$
$$= (d + m)(\bmod d)$$
$$= m$$

This shows the proposed cryptosystem is correct.

3.2 *Example demonstration*

Let's demonstrate an example of numerical simulation. We implement this algorithm by utilizing Visual C++ 6.0 and NTL number theory library. Let p, q, r, x be 512-bits and m be 500-bits integers respectively.

a. Key generation
Bob chooses the following parameters:
$p =$ 10114792273660656874618568712406420
34417622045779056317809222229293377869163
74923318745284718351487926620784410619571
58788753119587936299054539196971556855507
$q =$ 72489478148005407389881420097557548
32175437334699700212220795530383889529953
75687737294838451490977753632450792945
6694482323164238709048440202838708497091
$x =$ 57543407607326151964820304905295653
3297587507204818260902937813368494604065
47762556697379537715451183030579215453921
18395838609115357277660796621551185471661
$r =$ 51241046293815974215351694592782659
02790275633960702595035634808487215959390
22250844832581229302231487395838913450776
1942653403349052128427459623672988197667
Then he computes:
$n = p * q =$ 733216013493138117105381971313
13626030462995997154899899493027836398778
66646483487862939529074355312220396271499
71881659511784140850825781167646303448408
15372680491258955657214795171355395155611
02793517365904202512209763013248940460491
62195913709111131781980009908061118288632
695192942010776399927302182036013726
$e = T_r(x)(\bmod n) =$ 614372841594614153753
15315920963494037275591419999299996137244
14057437284712204114811549013829429513731
53581581813141353393928275077353766249893
33978629353545223905864650280694512972649
52611592627765003964462407371530749610410
91430839906087317539802432248607433645262
4498332573355629869087647104247196425851616
$d = T_r(x)(\bmod p) =$ 423980596098549728389
08207321266336828164441167482151785651458
55587049808560942251792742743027336686951
51518775429747014160426753382865898627052
7017468299683
b. Message encryption
Alice wants to send the message:
$m =$ 3189311446577289447661641312991302
70806241899143557854634394964208104553281
66887241839592041613965516298362772384045
58946719184722717578261334697172899
She chooses a random integer:
$k =$ 73310130821754254084728326598436157
86510146724341560502748413740487837302660
37941916502634774601407093111007315998789
26097433302335057716778206613705065567

Then she encrypts the cipher text:
$c = (kn + ex + mx) =$ 8910526871653719273
87099678729153154898914222026331249680572
61392112207977414272881348671091514127108
45589451069343033279177961511292719702049
15000240929212215726852628668043971556073
76636933649438777433927654819948561225512
48328847387845068275868374889635509825511
49154144188394048086786168350023850977293
13410831947346517642961294327825304801531
30275451057398874715622323927337579523478
38068737220358773872518567325842368533720
39594139467196595381515211786131217

c. Message decryption
Upon receiving the cipher text c, Bob can use his private key to recover the message m':
$m' = x^{-1}c(\bmod p)(\bmod d) =$ 318931144657728
94447661641312991302708062418991435578546
34394964208104553281668872418395920416139
65516298362772384045589467191847227175782
61334697172899
So this completes the whole encryption process and the decryption result is correct.

4 PERFORMANCE ANALYSIS

4.1 *Security analysis*

Two difficulties will be confronted if attacker want to obtain the private key (r, d, p) from the public key (x, e, n) in order to break the cryptosystem. (1) To compute p from n. This involves large integer factorization problem. Even though there is no strict theory to proof that integer factorization is a difficult problem, no efficient method is found to solve it so far. So it's still believed to be a well-known intractability. (2) The attacker needs to get $T_r(x)$. Since r is kept secret, it's almost impossible to compute the private key d. For the above two reasons, the designed cryptosystem can hide the private key efficiently and strengthen its security.

4.2 *Parameter selection*

1. To increase the difficulty of decomposing n, the larger value of n, the better. So the two primes p, q should be large enough, but they shouldn't be too close.
2. x should also be large enough. When $x = 0$, $T_r(0)(\bmod n) = 1, 0, -1, 0$, with the period of 4, when $x = 1$, $T_r(1)(\bmod n) = 1$, When $x = n-1$, $T_r(n-1) \ (\bmod n) = 1$, $p-1$. So the selected x should avoid these particular values.
3. In the cryptosystem, encrypted message m should satisfy the condition $m < d$. Since d is secret key, the sender may be ambiguous about

its value. To solve the problem, a threshold T can be set, which means d has the number of bits larger than T while the bits of m smaller than it. In order to increase the plaintext and cipher text space, a large enough T should be chosen.

4.3 Software implementation

The main software implementation issues of this algorithm is how to evaluate Chebyshev polynomials so that the computation time of $T_n(x)$ could be reduced. There are several kinds of measures. The first is to assume the large number s (r is the same) is written as

$$s = s_1^{k_1} s_2^{k_2} \dots s_i^{k_i} \tag{12}$$

then

$$T_s(x) \bmod P = \underbrace{T_{s_1}(\dots T_{s_1}}_{k_1} \dots \underbrace{T_{s_i}(\dots T_{s_i}}_{k_i}(x))) \bmod P$$

To compute $T_s(x)$, one needs only $(s_1-1) \times k_1 + (s_2-1) \times k_2 + \dots + (s_i-1) \times k_i$ iterations of the Chebyshev map instead of s iterations (Kocarve et al. 2003). Because choosing s and then factorizing s to get $k_i (i=1,2,3\dots)$ may cost a lot of time, but a reverse order can be adopted easily, i.e., k_i is chosen randomly and then s is constructed by k_i. The second is the fast algorithm of Chebyshev polynomials (Sun et al. 2013). Rewrite the Chebyshev polynomials as:

$$\begin{bmatrix} T_n(x) \\ T_{n+1}(x) \end{bmatrix} = \begin{bmatrix} 0 & 1 \\ -1 & 2x \end{bmatrix} \begin{bmatrix} T_{n-1}(x) \\ T_n(x) \end{bmatrix} = \begin{bmatrix} 0 & 1 \\ -1 & 2x \end{bmatrix}^n \begin{bmatrix} T_0 \\ T_1 \end{bmatrix} \tag{13}$$

From equation 13, we can find out that the key point of computing $T_n(x)$ is to compute the value of matrix $\begin{bmatrix} 0 & 1 \\ -1 & 2x \end{bmatrix}^n$. Sun et al. has verified the high efficiency (Sun et al. 2013) of the fast algorithm of Chebyshev polynomials.

4.4 Speed test

This test is operated on AMD Athlon Dual Core 2.10 GHz processor, 2.00 GB RAM. We record the whole running time, including the three process: key generation, message encryption and decryption, according to the different bits of p. As is shown in the following Figure 1.

The results show that the cryptosystem can reach high efficiency when the number of bits is relatively small. The consumed time may be a little longer when p has the bits larger than 512 bits.

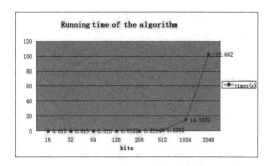

Figure 1. Running time of the algorithm.

For further test, we can see that the process of key generation takes up much of the spent time, while the time for message encryption and decryption is quite small, even if p has very large bits. This may lead to little influence in practical use since the key generation process are not needed to be frequently executed.

5 CONCLUSIONS

In this paper, we utilize the Chebyshev polynomials and modular operation to construct a novel public key cryptosystem. Its structure is quite different from the original Elgamal public key Chebyshev algorithm. It involves integer factorization and discrete chebyshev problem. Experimental results and performance analyses show that it is secure and practical. However, further performance analysis are necessary for our future work.

ACKNOWLEDGEMENTS

The authors would like to thank the anonymous reviewers for helpful comments and suggestions. This research is supported by the National Natural Science Foundation of China (Grant Nos. 61170037, 61074192) and the specialized Research Fund for Doctoral Program of Higher Education of China (No. 11280102).

REFERENCES

Bergamo, P. D'Arco, P., De Santis, A. & Kocarev, L. 2005. Security of public-key encryption based on Chebyshev polynomials. *IEEE Trans. Circuits and Systems I: Regular Papers*, 52(7): 1382–1393.

Chen, F., Liao, X., Xiang, T. & Zheng, H. 2011. Security analysis of the public key algorithm based on Cheyshev polynomials over the integer ring Z_N, *Information Science*, 181(22): 5110–5118.

Cheong, K.Y. 2012. One-way functions from Chebyshev polynomials. *Cryptololy ePrint Archive*, Report 2012/263, http://eprint.iacr.org.

He, B., Luo, L.Y. & Xiao, D. 2010. A method for generating S-box based on iterating chaotic maps. *Journal of Chongqing University of Posts and Telecommunications (Natural Science Edition)*, 22(1): 89–93.

Kocarev, L. 2001. Chaos-based cryptography: A brief overview. *IEEE Circuits and Systems Magazine*, 1(3): 6–21.

Kocarev, L. & Tasev, Z. 2003. Public-key encryption based on Chebyshev maps. *In: Proceedings of the IEEE international symposium on circuits systems*, 25–28 May 2003, vol. 3, pp. III-28, III–31.

Kocarev, L., Makraduli, J. & Amato, P. 2005. Public-Key Encryption Based on Chebyshev Polynomials. *Circuits, Systems and Signal Processing*, 24(5):497–517.

Lima. J.B., Campello, R.M. & Panario, D. 2008. Security of public key cryptosystems based on Chebyshev polynomials over prime finite fields. *Proceedings of the IEEE International Symposium on Information Theory (ISIT'08)*, 6–11 July, 2008, Toronto, Canada. Piscataway, NJ, USA: IEEE, 2008: 1843–1847.

Liao, X., Chen, F. & Wong, K.W. 2010. On the security of public-key algorithms based on Cheyshev polynomials over the finite field Z_N, *IEEE Transactions on Computer*, 59(10): 1392–1401.

Liu, H. & Wang, X. 2010. Color image encryption based on one-time keys and robust chaotic maps. *Computers and Mathematics with Applications*, 59(10): 3320–3327.

Li, Z., Cui, Y., Jin, Y. & Xu, H. 2011. Parameter Selection in Public Key Cryptosystem based on Chebyshev Polynomials over Finite Field. *Journal of Communications*, 6(5): 400–408.

Pareek, N.K., Patidar, V. & Sud, K.K. 2010. A random bit generator using chaotic maps. *International Journal of Network Security*, 10(1): 32–38.

Ruanjan, B. 2005. Novel public key encryption technique based on multiple chaotic systems. *Phys. Rev. Lett*, 26:098702.

Schmitz, R. 2001. Use of chaotic dynamical systems in cryptography. *Journal of the Franklin Institute*, 338(4): 429–441.

Sun, J., Zhao, G. & Li, X. 2013. An improved public key encryption algorithm based on Chebyshev polynomials. *Telkomnika*, 11(2): 864–870.

Tenny, R., Tsimring, L. & Abrabanel, H. 2003. Using distributed nonlinear dynamics for public key encryption. *Physical Review Letters*, 90(4): 47903:1–4.

Control Engineering and Information Systems – Liu (Ed)
© 2015 Taylor & Francis Group, London, ISBN 978-1-138-02685-8

Benchmarking openVPN bundle with private encryption algorithm

J. Chu, M. Chen, Z.J. Dai & Z.P. Shao
Research Institute of Information Technology and Communication, China Electric Power Research Institute, Nanjing, China

ABSTRACT: This paper shows how to benchmark our OpenVPN implementation. There are three problems in this benchmarking. First, according to the complexity of SSL/TLS and OpenVPN's protocol, packets been actually transported on the wire are different every new connection, so we cannot just record these packets and replay them to simulate many concurrent clients. Second, the OpenVPN client will open a virtual NIC to catch application data packets, which is usually too few to use for the testing purpose. And finally, our OpenVPN implementation uses a private encryption algorithm, which are only provided by the specific hardware accelerator, so we must find a way to simulate the algorithm, to avoid running many clients with the same number of accelerators, which is unrealistic and inefficient. This paper shows our way to solve these two problems.

1 INTRODUCTION

OpenVPN is an open source software application that implements Virtual Private Network (VPN) techniques for creating secure point-to-point or site-to-site connections in routed or bridged configurations and remote access facilities. It uses a custom security protocol that utilizes SSL/TLS for key exchange. It is capable of traversing Network Address Translators (NATs) and firewalls (Yonan 2008).

To improve OpenVPN's efficiency and security, we modified the implementation of the original version, replace its symmetric encryption algorithms with SM1, which is standardized by State Cryptography Administration. There are no pure software implementation of this algorithm, and the users can use it through specific hardware accelerator only (Manavski 2007).

The usual approach to benchmark a network service is to record packets generated from a real session of a client, and then simulate many concurrent clients by replaying these packets with multi-thread or multi-process programming. There are several such benchmarking software, such as HP LoadRunner (a general purpose network protocol benchmarking tool) (Yang & Li 2007) and Apache HTTP server benchmarking tool (ab, a tool for benchmarking HTTP servers specifically).

There are three problems to benchmark our OpenVPN implementation using this approach. First, according to the complexity of SSL/TLS and OpenVPN's protocol, packets been actually transported on the wire are different every new connection, so we cannot just record these packets and

replay them to simulate many concurrent clients. Second, the OpenVPN client will open a virtual NIC (Krasnyansky & Maksim 2007) to catch application data packets, which is usually too few to use for the testing purpose. And Finally, our OpenVPN implementation uses the SM1 private encryption algorithm, which are only provided by the specific hardware accelerator, so we must find a way to simulate the algorithm, to avoid running many clients with the same number of accelerators, which is unrealistic and inefficient.

Here are how we solve these problems.

2 PROBLEMS WITH THE PROTOCOL AND VIRTUAL NIC

The protocol of OpenVPN is a little complicated. It exchanges session keys and handshake other information with OpenSSL's SSL/TLS protocol, but encrypt and transport the real application data with its own protocol. The whole process is as shown in Figure 1.

OpenVPN's protocol is based on UDP. At the handshake stage, OpenVPN implements a TCP over UDP protocol, to retransmit OpenSSL's TCP packets with UDP. With this, OpenVPN exchanges its own encryption and integration keys through OpenSSL. At the application data stage, it uses these keys to encrypt and transmit other applications' IP packets with its own format.

When OpenVPN generates its own keys, it uses a random number generator to generate some random data as the source of the keys. To record and replay packets, we must fix these keys to some

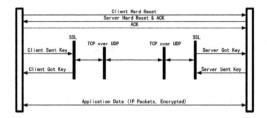

Figure 1.   The process of OpenVPN protocol.

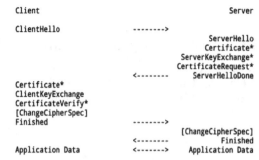

Figure 2.   The process of OpenSSL protocol.

literal value. We can just remove the invoking of the random number generator functions. Modify the function "prng_init" in the file "crypto.c", and replace the function call "RAND_bytes" with a call to function "memset".

When the client of OpenVPN finished the handshake with the server, it will open a virtual NIC, and set its address and routes. This is useful when in real scenario, but not for testing purpose, as there are no more than one virtual NIC can be opened usually, so we must disable this operation too. Modify functions "open_tun" and "close_tun" in the file "tun.c", and delete their whole function body to prevent the client to open the virtual NIC. Modify the function "check_add_routes_dowork" in the file "forward.c", and replace its function body with a call to the function "initialization_sequence_completed" to prevent the client to setup virtual network card's routing table entries.

Without a virtual NIC, there will be no application data to transmit, and our testing will be useless, so we must generate some application data by ourselves. Modify the function "read_tun_buffered" in the file "tun.h", instead of calling function "tun_finalize", fill the reading buffer "buf" with a fixed ICMP/IP package's content, as if it is read from the virtual NIC. Then modify the function "io_wait_dowork" in the file "forward.c" to triggle the virtual NIC's reading event every second or so.

Because OpenVPN implements its handshaking process through OpenSSL, we also need to modify OpenSSL to fix its randomness. The process of the OpenSSL protocol is as shown in Figure 2 (Dierks 2008).

There is similar problem in this process as in OpenVPN's own protocol. When the OpenSSL client sends the "ClientHello" message to the server, there is a field called "random", which is used to generate encryption and integration keys for OpenSSL itself. As in OpenVPN, we must also fix this field to some literal value to prevent it using different keys for each session. Modify the function "ssl3_client_hello" in the file "ssl/s3_clnt.c", replace "time" and "RAND_pseudo_bytes" function calls with a "memset" function call, preventing the client to generate random

bytes for keys. In the file "ssl/s3_srvr.c", the modification of functions "ssl3_get_client_hello" and "ssl3_send_server_hello" are similar to the modification of the above "ssl3_client_hello" function.

3   PROBLEM WITH THE PRIVATE ENCRYPTION ALGORITHM

Instead of standard encryption algorithms such as DES, AES and so on, we are using SM1 encryption algorithm. This algorithm is prescribed by State Cryptography Administration, and its private. There is no pure software implementation available. We must add a specific hardware accelerator to use this algorithm.

To avoid using too many accelerators to simulate OpenVPN clients (on the other hand, if we want to use the method of record and replay, there is no chance to use any external hardware), we must simulate this algorithm in the software. Fortunately, after we solving the first two problems of our benchmarking, the application data to be encrypted and transmitted are fixed, so we can develop a small program to generate these application data beforehand, encrypt them with a real accelerator, and then store the results in some file. In the OpenVPN client to be test, we can then modify the encryption function to read the file, find the corresponding result of a packet, instead of invoking the hardware at all. Functions to be modified are "openvpn_encrypt" and "openvpn_decrypt" in the file "crypto.c".

4   BENCHMARKING RESULTS

Now we can start our benchmarking. Start a modified client to connect to the server, and record its network flow using LoadRunner. When this connection is over, we can specify the number of the clients to be run concurrently by LoadRunner, to

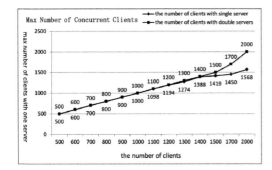

Figure 3. Max number of concurrent clients.

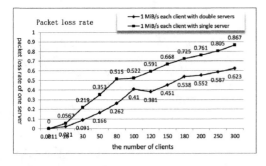

Figure 4. Throughput.

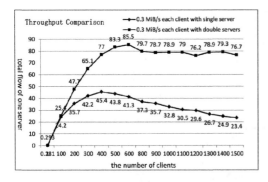

Figure 5. Packet loss rate.

analyzes the server's performance under different loads and conditions.

We've tested two use cases, one is a single server and the other is two servers with an LVS system. LVS (Linux Virtual System) is a system to balance the network loads among many servers, to improve the whole system's load performance. Among these, we have three measures to test: the max number of concurrent clients, throughput (Padhye 1998), and packet loss rate (Perkins & Orion & Vicky 1998).

1. The max number of concurrent clients (Fig. 3)
2. Throughput (Fig. 4)
3. Packet loss rate (Fig. 5).

5 CONCLUSION

This paper solves several problems when benchmarking OpenVPN, and can finally test its performance with the method of record and replay. This method provides us a very straight feedback to improve the performance of the OpenVPN server.

ACKNOWLEDGMENT

This work was financially supported by the China Electric Power Research Institute Nanjing Branch.

REFERENCES

Dierks T., "The transport layer security (TLS) protocol version 1.2." (2008).
Krasnyansky M., Y. Maksim, "Universal TUN/TAP device driver." URL: http://www.kernel.org/pub/linux/ kernel/, FILE: people/marcelo/linux-2.4/Documentation/networking/tuntap.txt (2007).
Manavski S.A., "CUDA compatible GPU as an efficient hardware accelerator for AES cryptography." Signal Processing and Communications, 2007. ICSPC 2007. IEEE International Conference on. IEEE, 2007.
Padhye J., "Modeling TCP throughput: A simple model and its empirical validation." ACM SIGCOMM Computer Communication Review. Vol. 28. No. 4. ACM, 1998.
Perkins C., H. Orion, H. Vicky, "A survey of packet loss recovery techniques for streaming audio." Network, IEEE 12.5 (1998): 40–48.
Yang P., J. Li, "Using Load Runner to Test Web's Load Automatically [J]." Computer Technology and Development 1 (2007): 080.
Yonan J., "OpenVPN–an open source SSL VPN solution." URL http://openvpn.net (2008).

315

Control Engineering and Information Systems – Liu (Ed)
© 2015 Taylor & Francis Group, London, ISBN 978-1-138-02685-8

Research on combined modulation technique for Current Source Inverter

J. Bai, S.Q. Lu & J. Liu

Electrical and Information Engineering Department, Beihua University, Jilin, China

ABSTRACT: This paper presents a new MSM technology with many advantages such as the better output waveforms and the high utilization ratio of DC voltage based on the study of the CSI-SPWM technology and CSI-TPWM technology. In order to make CSI get a bigger extended speed range in the high-voltage and high-power inverter applications field, this paper provides a new MSM-SHE piecewise modulation scheme. Finally, the simulation results validate the feasibility and effectiveness of this scheme.

1 INTRODUCTION

1.1 Source of the paper

With the development of the superconducting magnetic energy storage technology, current source inverter has become one of the most widely used inverter topology in high-voltage transmission system due to its advantages such as simple topology structure and reliable short-circuit protection (D. Xu & B. Wu 2005, H. Komurcugil 2010, Dan Wang et al. 2012). At present, the mature modulation methods are Selective Harmonic Elimination (SHE) Technology, Trapezoidal Pulse Width Modulation (TPWM) technology and so on (Haixian Yu 2010). However, Sinusoidal Pulse Width Modulation (SPWM) technology can produce a better output waveform, but requires a higher sampling frequency, so that the switching frequency of inverter is higher. Generally, the switching frequency is low for high-power inverters, and SPWM Inverter with low switching frequency will produce huge amounts of low-order harmonics and a narrow pass-band, greatly reduce the performance of the control system (Changyong Wang et al. 1999). Compared with SPWM technology, the output waveform of TPWM technology is not good, but the utilization ratio of the DC voltage is high and the switching frequency is low. In order to make CSI has better output waveform; this paper presents a new modulation method which is MSM technology. MSM technology has the advantage of SPWM and PWM. Besides, in order to realize the greater speed range in the premise of good output waveform, the paper that combined with the advantages of SHE technology and MSM technology proposes a method of piecewise modulation which is MSM-SHE method in the high-power variable frequency speed control system.

2 CURRENT SOURCE INVERTER CIRCUIT

2.1 The topology of three-phase CSI

Figure 1 shows the topology of three-phase CSI, Which consists of a current source, Impedance load for Y connection, six IGCT power switching elements and filter capacitor.

Three-phase CSI Principles: the state of the power switches S_1~S_6 is expressed with bi-logic variables 0 and 1, which 0 for off and 1 represents conduction. Compared to the Voltage Source Inverter (VSI). The working mode of CSI requires one and only one switch is turned on at the same time in Upper and lower half-bridge, so that the current of the inductance is continuous and the energy of the inductance is balance. The switch state variables satisfy the following constraints: $S_1 + S_3 + S_5 = 1$, $S_4 + S_6 + S_2 = 1$. This paper introduces the trio-logic phase switch status variables for describing the working state of CSI. The upper bridge arm switch turn-on and the lower bridge arm switch turn-off, $S_j = 1$; the upper and lower bridge arm switches turn-on or turn-off, $S_j = 0$; the

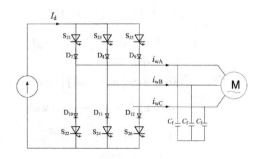

Figure 1. Circuit topology of current source inverter.

lower bridge arm switch turn-on and the upper bridge arm switch turn-off, $S_j = -1$, Where $j = a$, b, c. The relationship of phase switch status variables and a single switch state meet: $S_a = S_1 - S_4$, $S_b = S_3 - S_6$, $S_c = S_5 - S_2$.

3 CURRENT SOURCE INVERTER MODULATION TECHNIQUE

3.1 The outline of modulation technique

The requirements of three-phase CSI for three-phase PWM signal is only two switches turn-on, which one lies in the upper bridge arm and the other lies in the lower bridge arm at any time (except during commutation). If only one switch is turned on, the current will lose the continuity and the DC inductance will produce an extremely high voltage which will damage to the switches. If more than two switches are simultaneously turned on, PWM current will no longer meet the defined switching waveform. This is the important reason that the three-phase CSI-PWM control becomes more complex than voltage-type (Zhaokai Yuan 2008).

3.2 Selective harmonic elimination

Selective Harmonic Elimination (SHE) technology is an off-line modulation method and can eliminate some main low-order harmonics in the inverter PWM current. The switch working angles of power devices are precalculation and then stored in the digital controller. Figure 2 shows the elimination of 5th, 7th and 11th harmonic modulation schematic.

The expression for the current i_w:

$$i_w(\text{wt}) = \sum_{n=1}^{\infty} a_n \sin(nwt) \qquad (1)$$

$$a_n = \frac{4}{\pi} \int_0^{\frac{\pi}{2}} i_w(wt) \sin(nwt) d(wt)$$

$$= \frac{4I_{dc}}{\pi n} \times \begin{cases} \cos(n\theta_1) + \cos\left[n\left(\frac{\pi}{3} - \theta_1 \right) \right] - \cos(n\theta_2) \\ \quad - \cos\left[n\left(\frac{\pi}{3} - \theta_2 \right) \right] + \cos(n\theta_k) \\ \quad + \cos\left[n\left(\frac{\pi}{3} - \theta_k \right) \right] - \cos\left(\frac{n\pi}{6} \right) \\ k = \text{odd number} \\ \cos(n\theta_1) + \cos\left[n\left(\frac{\pi}{3} - \theta_1 \right) \right] - \cos(n\theta_2) \\ \quad - \cos\left[n\left(\frac{\pi}{3} - \theta_2 \right) \right] + \cdots - \cos(n\theta_k) \\ \quad - \cos\left[n\left(\frac{\pi}{3} - \theta_k \right) \right] + \cos\left(\frac{n\pi}{6} \right) \\ k = \text{even number} \end{cases}$$

$$(2)$$

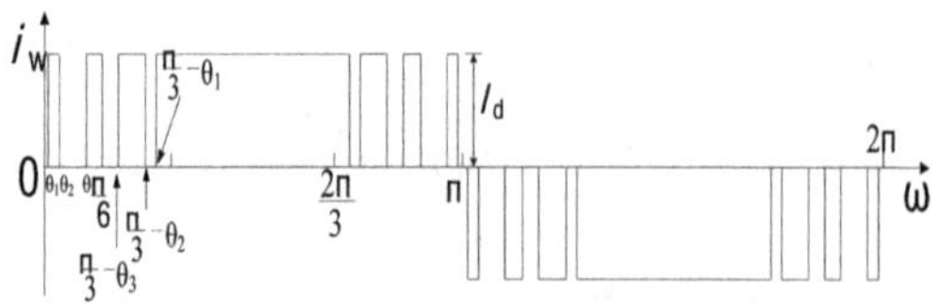

Figure 2. Selective harmonic elimination modulation schematic.

For $a_5 = a_7 = a_{11} = 0$,

$$\begin{cases} \cos(5\theta_1) + \cos\left[5\left(\frac{\pi}{3} - \theta_1 \right) \right] - \cos(5\theta_2) \\ \quad - \cos\left[5\left(\frac{\pi}{3} - \theta_2 \right) \right] + \cos(5\theta_k) \\ \quad + \cos\left[5\left(\frac{\pi}{3} - \theta_k \right) \right] - \cos\left(\frac{5\pi}{6} \right) = 0 \\ \cos(7\theta_1) + \cos\left[7\left(\frac{\pi}{3} - \theta_1 \right) \right] - \cos(7\theta_2) \\ \quad - \cos\left[7\left(\frac{\pi}{3} - \theta_2 \right) \right] + \cos(7\theta_k) \\ \quad + \cos\left[7\left(\frac{\pi}{3} - \theta_k \right) \right] - \cos\left(\frac{7\pi}{6} \right) = 0 \\ \cos(11\theta_1) + \cos\left[11\left(\frac{\pi}{3} - \theta_1 \right) \right] - \cos(11\theta_2) \\ \quad - \cos\left[11\left(\frac{\pi}{3} - \theta_2 \right) \right] + \cos(11\theta_k) \\ \quad + \cos\left[11\left(\frac{\pi}{3} - \theta_k \right) \right] - \cos\left(\frac{11\pi}{6} \right) = 0 \end{cases} \qquad (3)$$

The solutions of the above equations are $\theta_1 = 2.24^0$, $\theta_2 = 5.60^0$, $\theta_3 = 21.16^0$.

3.3 Trapezoidal PWM

TPWM technology is a relatively common modulation method of CSI. Figure 3 shows TPWM schematic. TPWM technology obtains the Symmetrical PWM waveform of the gate driving signal V_{g1} of the power switch S_1 in the first half cycle by comparing the trapezoidal modulating reference waveform V_m with the triangle carrier waveform V_{cr} in the oblique section of $0^0 \sim 60^0$ and $120^0 \sim 180^0$. The PWM waveforms of the power switch $S_1 \sim S_6$ gate driving signal are queued in order delay 60^0. Then the PWM waveform of the current i_w is obtained by comparing the PWM waveforms of the gate driving signal $V_{g1} \sim V_{g6}$.

3.4 Sinusoidal PWM

As is shown in Figure 4, CSI-SPWM technology adopts the control strategy of dynamic trio-logic SPWM, which transforms the bi-logic SPWM

318

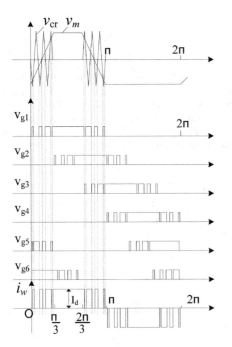

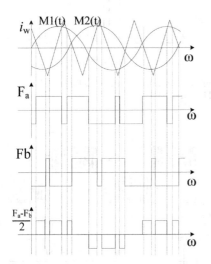

Figure 3. Trapezoidal PWM modulation schematic.

Figure 4. Sinusoidal PWM modulation schematic.

signals F_a, F_b, F_c into the trio-logic signals Y_a, Y_b, Y_c. Transform relations as follows (Yu Xiong et al. 2003):

$$\begin{bmatrix} Y_a \\ Y_b \\ Y_c \end{bmatrix} = \frac{1}{2} \begin{bmatrix} 1 & -1 & 0 \\ 0 & 1 & -1 \\ -1 & 0 & 1 \end{bmatrix} \begin{bmatrix} F_a \\ F_b \\ F_c \end{bmatrix} \quad (4)$$

The control strategy meets the requirements of three-phase CSI for three-phase PWM control signal, so the relationship between the output current and the input modulation signal is a linear, which to ensure the normal operation of three-phase CSI.

3.5 MSM technology

The comparative analysis of TPWM and SPWM shows that the output current waveform of CSI using SPWM is better, but the switching frequency is higher and the DC voltage utilization is lower; while the DC voltage utilization of CSI using TPWM is higher and the switching frequency is lower, but the output current waveform is worse. A new MSM modulation method is proposed for achieving better waveform output and higher utilization ratio of DC voltage based on the above analysis. As is shown in Figure 5, SPWM is used within $0^0 \sim 60^0$, $120^0 \sim 240^0$, $300^0 \sim 360^0$ and Similar TPWM is used within $60^0 \sim 120^0$, $240^0 \sim 300^0$ in each cycle of the reference waveform.

3.6 Combined piecewise modulation

In AC Adjustable-speed System, the formula of the motor speed is given by

$$n = \frac{60f}{p}(1-s) \quad (5)$$

where f is defined as power input frequency, p is defined as motor pole number, s is defined as slip.

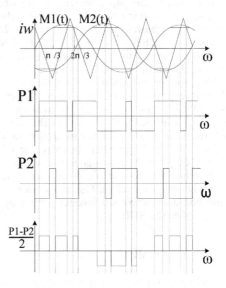

Figure 5. MSM modulation schematic.

319

By the formula, the motor speed can be adjusted by changing the frequency of power supply. Assume that the pole pair number and slip is constant, achieving greater speed range requires that the conversion device can output larger frequency range. However, the switching device of high power density has a greater resistance, and it will generate a lot of heat which will likely lead to damage to the device if not promptly discharge when running in a higher switching frequency in the high-voltage and high-power applications. Therefore low switching frequency control scheme is generally used in high-voltage and high-power CSI, which limits the inverter output frequency range to a certain extent.

Further, the output voltage waveforms of the inverter using different modulation techniques also have their own advantages and disadvantages. SHE technology can output close to the sine wave voltage waveform in the lower switching frequency, so it is widely used in high-power inverter (Fanghua Zhang et al. 2007). Although SHE-PWM technology has many advantages, the biggest obstacle is to solve the harmonic elimination model in practical applying. The SHE-PWM switch-angle equations are nonlinear and transcendental equation, so the solution of more than four harmonic components is still very difficult, or even impossible to achieve (Eryong Guan et al. 2005). Because of the influence of TPWM technology, The MSM technology will produce two large harmonics in the $3 (N_p - 1) \pm 1$ and $n = 3 (N_p - 1) \pm 5$ nearby. If the half-cycle pulse number N_p is less, these harmonics will be difficult to filter, and if N_p is larger, these harmonics through a filter have little effect on the motor, so N_p is usually greater than 7 in practical applications.

In summary, this paper presents a broadband piecewise modulation scheme (MSM-SHE technology) for CSI achieves greater speed range problem with lower switching frequency. The MSM-SHE technology can effectively suppress harmonics and improve the output waveform in the high-voltage and high-power applications. When the frequency of the inverter output voltage is lower, this scheme uses MSM technology which ensures that the sampling frequency is the higher, so that the output waveform is good and the utilization ratio of DC voltage is high; yet when the frequency of the inverter output voltage is higher, the scheme uses SHE technology, so that the output waveform is relatively good. Therefore, CSI can obtain wideband and good-waveform inverter output by the scheme of MSM-SHE.

4 MATLAB SIMULATION

In the paper, simulation experiments are carried out in accordance with three-phase CSI topology

of Figure 1. Because the CSI current and the load doesn't matter, for the sake of discussion, the load in the simulation is purely resistive load. Parameters of simulation system are set (except for SHE) as follows: $I_d = 1000$ A, $R_1 = R_2 = R_3 = 1\Omega$, output current frequency is 20 Hz, Triangular carrier frequency is 600 Hz, M = 1. The simulation waveforms (A phase) and the spectrum (A phase) of TPWM, SPWM, MSM, SHE are respectively shown in Figure 6~Figure 9, where SHE is the

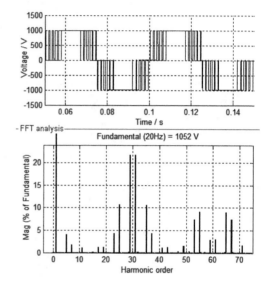

Figure 6. Output current waveform and the spectrum of TPWM at 20 Hz.

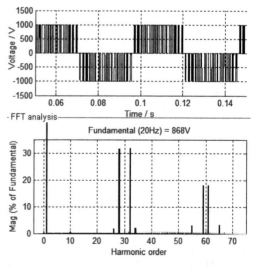

Figure 7. Output current waveform and the spectrum of SPWM at 20 Hz.

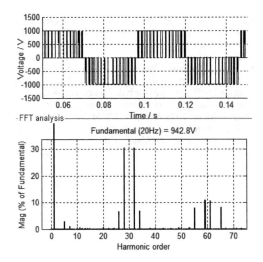

Figure 8. Output current waveform and the spectrum of MSM at 20 Hz.

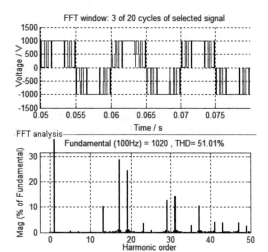

Figure 9. Output current waveform and the spectrum of SHE at 100 Hz.

Table 1. The data sheet of simulation experiments.

Modulation technique	Fundamental amplitude (V)	% of fundamental									
		5	7	11	13	17	25	29	31	35	37
TPWM	1054	0.39	0.2	0.03	0	0.12	1.06	2.33	2.34	1.06	0.43
SPWM	867.3	0.03	0.13	0.03	0.02	0.02	0.06	0.05	0.14	0.07	0.06
MSM	943	0.28	0.09	0.1	0.02	0.04	0.01	0.07	2.9	0.04	0.02
SHE	1020	0.06	0.04	0.01	1.06	2.94	0.1	1.31	0.16	0.28	1.09

elimination of 5, 7, 11 harmonic simulation. The experimental data of TPWM, SPWM, MSM and SHE are shown in Table 1.

By the comparative analysis of the experimental data shows that output voltage waveform of MSM technology is better because the relative content of 25th harmonic and the following harmonics are small and the utilization ratio of DC voltage is high, yet the content of 5th, 7th and 11th harmonics in the output voltage waveform of SHE modulation is substantially zero and the DC voltage utilization ratio is relatively high. Therefore, the MSM-SHE Piecewise Modulation method can obtain wider speed regulation frequency band to achieve greater speed range within the finite switching frequency in high-voltage and high-power AC drive applications.

5 CONCLUSIONS

MSM technology achieves the output of the high utilization ratio of DC voltage and the good waveform in a relatively lower switching frequency, and provides a better method for CSI. In addition, the piecewise modulation technology combined by MSM technology and SHE technology solves the problem that the lower sampling frequency of high-power switching device to a certain extent, so that CSI can be used more widely in the field of high-voltage and large-capacity variable frequency drive.

REFERENCES

Changyong Wang, Mao Liu, Zhongchao Zhang. 1999. Research on phase-shifted SPWM technique in current-source multi-converter. *Power Electronic Technology* (4):43–45.

Dan Wang, Zhen-hui Wu, Gang Xu, Da-da Wang, Meng Song, Xiao-tao Peng. 2012. Real-time power control of superconducting magnetic energy storage. *IEEE International Conference on Power System Technology*: 1–5.

Eryong Guan, Pinggang Song, Manyuan Ye, Shun-pao Liu. 2005. A method of solution to SHE-PWM swithing angles for CSI. *Proceedings of the CSEE* 29(17):62–65.

Fanghua Zhang, Yong Ding, Huizhen Wang, Yang-guang Yan. 2007. The SHE control strategy on three-phase four-leg inverter. *Proceedings of the CSEE* 27(7):82–87.

Haixian Yu. 2010. Study of current source high-voltage converter using IGCT. *HeFei University of Technology, MA thesis*: 31–40.

Komurcugil H. 2010. Steady-state analysis and passivity-based control of single-phase PWM current-source inverters. *IEEE Trans. Ind. Electron.* 57(3): 1026–1030.

Xu D. & Wu B. 2005. Multilevel current source inverters with phase shifted trapezoidal PWM. *IEEE 36th Conference on Power Electronics Specialists*: 2540–2546.

Yu Xiong, Danjiang Chen, Changsheng Hu, Zhongchao. Zhang. 2003. Analysis and experiment of carrier phase-shifted SPWM technique based on current-source multi-converter. *power electronic technology* 37(4):39–41.

Zhaokai Yuan. 2008. Study on PWM control method of three-phase current source inverter. *China University of Petroleum (East China), MA thesis*: 11–13.

Control Engineering and Information Systems – Liu (Ed)
© 2015 Taylor & Francis Group, London, ISBN 978-1-138-02685-8

Bi-directional detection MMSE algorithm in the LTE system

D.D. Yuan & R.H. Qiu
College of Information Sciences and Technology, Engineering Research Center of Digitized Textile and Fashion Technology, Ministry of Education, Donghua University, Shanghai, P.R. China

ABSTRACT: MIMO and OFDM are the key technologies used by LTE which can provide high-speed data stream, so the Signal detection at the receiving end of the LTE system is essential. Based on the analysis of the traditional detection algorithm, this paper presents a Bi-directional detection MMSE algorithm. In traditional MMSE algorithm for interference cancellation detection, different layers have different reliability, the first detection layer has the lowest reliability; the last detection layer has the highest reliability. In this paper, the Bi-directional detection MMSE algorithm detects from back to front, retain the value of the front layers, then detect from front to back, the value of the following layers should be reserved, at last, to consolidate the obtained results. The simulation results show that this algorithm has better bit error rate performance and system capacity than interference cancellation MMSE detection algorithm.

1 INTRODUCTION

In order to adapt to the development of modern communication system, in 2004, 3GPP launched a mobile communication Long Term Evolution (LTE). LTE adopt MIMO and OFDM technologies which can provide high spectrum efficiency, high channel capacity and flexible spectrum allocation [1]. It will play an important role of mobile communications in the future. VBLAST (Vertical layered space-time code) encoding which is presented by Bell Labs can provide a high transfer rate [2], so it is very suitable for LTE system. The traditional LTE detection algorithms are mainly Maximum Likelihood (ML) algorithm, Zero-Forcing algorithm (ZF) and Minimum Mean Square Error (MMSE) algorithm. Among the three algorithms, simulation shows that the ML algorithm is optimal, but the shortcomings of the ML algorithm is that the complexity of it is exponential growth which cannot be practical as in [3]. Interference progressive cancellation of ZF and MMSE detection algorithm can obtain a better BER performance in [4,5]. An improved MMSE detection algorithm is put forward in [6] can obtains a low error rate, but also increases complexity. The various detection algorithms based on MMSE is introduced and showed their system capacity in [7,8]. Analysis of existing detection algorithm, this paper presents a bi-directional detection minimum mean square error algorithm of VBLAST in LTE system, simulation results show that this algorithm has better bit error rate performance and system capacity than the interference cancellation MMSE algorithm.

2 LTE SYSTEM DETECTION MODEL

In order to support different applications such as multimedia and office needs, a high rate of reliable communication in a wireless communication link becomes more and more necessary. However, in a wireless communication system, there are many undesirable factors, the task becomes very challenging. These non-ideal factors are including various forms of electromagnetic wave propagation, multipath fading, signal attenuation, and a variety of bandwidth and power restrictions.

The combination of MIMO and OFDM signal processing technologies are used in the LTE system. The LTE system's transmit side carries Serial-to-parallel conversion, layer mapping, pre-coding, etc. to limit the non-ideal factors and generates an OFDM baseband signal. In the receiving side the procession is its Inverse process. The receiving end of the structure is shown in Figure 1.

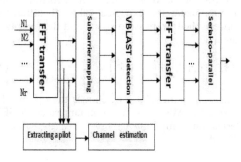

Figure 1. The receiving end of VBLAST encoding in LTE system.

Signal detection uses the channel impulse response which derived from the channel estimation and the received signal to recover the sending signal at the receiving end. In the LTE system, the complex-valued signal of the transmitting side is mapped to the Nt transmit antenna ports and converted into the OFDM baseband signal then it is emitted in each antenna port. Emission signal through the wireless channel to arrive the Nr receive antennas and all the receiving antennas receive the transmitted data, and process the OFDM baseband signal, then judge the emission signal by the signal detection algorithm.

If there is no inter-symbol interference, the MIMO system input and output relationship [9] is

$$Y = HX + N \qquad (1)$$

In the formula, H is a matrix of size $Nt \times Nr$, its element (i, j) means that the fading coefficient between the transmit antennas i and receiving antennas j. N is the noise matrix of size $N \times Nr$; The elements of H and N obey Complex independent and identically distributed with zero mean and variance σ^2 of the Gaussian random variable. Let $x_i(k)$ indicates the transmitted symbols at the time point k from the antennas i, then all the sub-stream of the transmitting antenna can be expressed as

$$X = \begin{bmatrix} x_1(1) & \cdots & x_{Nt}(1) \\ \vdots & \ddots & \vdots \\ x_1(N) & \cdots & x_{Nt}(N) \end{bmatrix}$$

N is the length of the transmit sequence from each transmit antenna. The average energy of each transmit antenna is $1/Nt$, its trace $\{X^H X\} = 1$, $[\cdot]^H$ is the conjugate transpose of $[\cdot]$.

3 BI-DIRECTION DETECTION MMSE ALGORITHM

The traditional MMSE detection algorithm aims at minimizing the expected value of difference between the transmission signal and the reception signal, i.e.

$$MSE = \min E\left[\|YW - X\|^2\right] \qquad (2)$$

is the minimum value, i.e., choose the optimal W so that the MSE has the minimum value. MSE's derivation computing can obtained the optimal W [10]

$$W_{opt} = H^H \left[HH^H + \sigma^2 I_{Nt}\right]^{-1} \qquad (3)$$

In the Formula, I_{Nt} is the $Nt \times Nt$ unit matrix.

During reception, while the right on both sides of the equation (2) multiplies the W_{opt}, if we detect the i layer, the other layers has been detected and those who are undetected are interference. Just as the following Formula (4).

$$\tilde{y}_i(k) = \sum_{j=1}^{Nr} w_{j,i}\left[x_i(k)h_{i,j} + \sum_{m=1,m\neq i}^{Nt} x_m(k)h_{m,j}\right] + \tilde{n}_i(k)$$

$$(4)$$

In order to eliminate the interference from the other layers, we can detect the layer by sentence and one by one, to eliminate interference from other detection layers. When is detecting the i layer, the receiving matrix layer in i–1 can be expressed as in [11]:

$$Y^{i-1} = Y^i - \hat{x}_i h_i \qquad (5)$$

The interference cancellation minimum mean square error detection algorithm is achieved. From the above discussion on the MMSE algorithm, we can know that interference cancellation MMSE algorithm can improve the performance of the MMSE detection algorithm. At the same time we can see that each layer in the process of eliminating interference has different reliability. The first layer to be detected out has the lowest reliability, and the last layer to be detected out with the highest reliability. This is because that with the increasing number of layers detected, the rank of the channel matrix H is also gradually decreases. The interference between the respective transmission signals also decreases. Based on the above analysis, this paper proposes a bi-directional detection MMSE algorithm. First we can detect the signal itemized from the first layer to the Nt layer, then reserve the latter detected values; at same time, detect the signal itemized from the Nt layer to the first layer, reserve the last layer's values. Then consolidate both sides of the obtained detection values to obtain high reliability detection value.

In Matlab simulation platform, we can be achieved with the following cycle:

Set $i = Nt$
Use formula (3) to calculated the W_{opt}

The following is a cycle:

While $i \geq 1$
Calculate $y_i(k)$
according to the formula (4), k = 1, 2 ... N.
Obtain Judgment of $\hat{y}_i(k)$.
Using formula (5) calculate the $\tilde{y}_i(k)$.
Removal of the i-th row H_i and update H matrix.
Calculate the new W_{opt} using the updated H.
$i = i - 1$

The end of the cycle, to obtain $y_i(k)$, Reserved $y_1(k)$, $y_2(k)$ ... $y_{Nt/2}(k)$

At the same time, we can set another detector to do reverse detection.

Set $i = 1$

Use formula (3) to calculated the W_{opt}

The following is a cycle:
While $i \le Nt$
Calculate $\tilde{y}_i(k)$, k = 1, 2 ... N.
Obtain Judgment of $\tilde{y}_i(k)$.
Calculate Y^{i+1}
Removal of the i-th row H_i and update H matrix.
Calculate the new W_{opt} using the updated H.
$i = i + 1$
The end of the cycle, to obtain $y_i(k)$, Reserved $y_{Nt/2+1}(k), y_{Nt/2+2}(k)...y_{Nt}(k)$

Merge the resulting value and output the final result.

4 THE SIMULATION RESULTS AND ANALYSIS

In order to obtain the performance of bi-directional MMSE detection algorithm, this paper has carried out a simulation of the BER and the system capacity, the transmission data is 10000, it use of the QPSK modulation scheme.

Figure 2 is a 4 × 4 antenna case BER curves between the layers when using QPSK modulation scheme. We can see from the curve in Figure 2, the fourth layer which is first detected have the worst BER performance. As the detection going on, bit error rate performance gradually increased.

Figure 3 is the BER performance of three detection algorithms with 4 × 4 antenna, the figure is QPSK modulation. From the figure, we can see that when interference cancellation is performed,

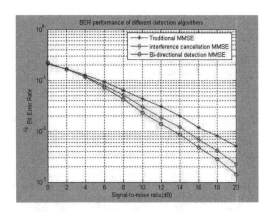

Figure 3. 4 × 4 antenna BER performance of different detection algorithms.

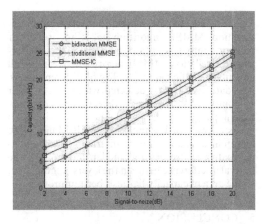

Figure 4. 4 × 4 antenna detection algorithm system capacity.

the system can obtain a certain improvement of BER performance, the bi-directional detection MMSE algorithm can obtain further BER improvement than the interference cancellation MMSE algorithm.

Figures 4 and 5 are system capacities of 3 different detection algorithms under 4 × 4 antennas mode and 6 × 6 antenna mode.

As can be seen from the figure, in the 4 × 4 antennas mode bi-directional detection MMSE algorithm improves about 3 bit/s/Hz than the MMSE algorithm and improves about 1bit/s/Hz than interference cancellation MMSE algorithm. In 6 × 6 antennas mode, bi-directional detection MMSE algorithm improves 5 bit/s/Hz than the MMSE algorithm and improves 1 bit/s/Hz than the interference cancellation interference cancellation MMSE algorithm.

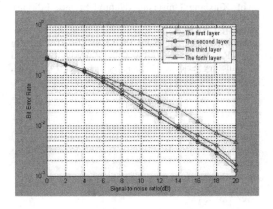

Figure 2. 4 × 4 antenna interference cancellation MMSE algorithm layers BER curve.

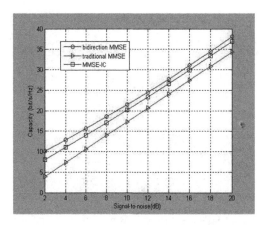

Figure 5. 6 × 6 antenna system capacity of the various detection algorithms.

According to the simulation results, the bidirection MMSE algorithm takes the order of reception into account on the basis of the interference cancellation MMSE algorithm. It takes full advantage of the data reliability of different layers. The data that is first detected which has a lower reliability is discarded in the detection process, only leaving part of the data with a higher reliability, so it has a lower BER simulation results. Due to the hierarchical elimination of interference, the channel matrix is also getting smaller and smaller; the mutual interference between the layers is eliminated, so the system capacity has also been improved.

5 CONCLUSION

In the modern communication, reducing the BER and receiver complexity has been the pursuit of the goal of engineering applications. This paper discusses the MMSE algorithm in the LTE system, the interference cancellation MMSE algorithm and bi-directional detection MMSE detection algorithm proposed on this basis, the algorithm reduce the interlayer interference between different layers of the transmission signal data and it have a better BER performance and system capacity than the interference canceller MMSE algorithm. In the actual LTE communication system Bi-directional detection MMSE algorithm has a certain value.

REFERENCES

Chandrasekaran M, Srikanth Subramanian, "Performance of precoding techniques in LTE." International Conference on Recent Trends in Information Technology, ICRTIT 2012, pp. 367–371, April 2012.

Chao Chen. BLAST structure MMSE detection methods analysis, International Conference on Information and Communications Security, CCICS 2011, pp. 253–256; 2011.

Gong Bing, Deng Fei qi, Research on MIMO Detection Algorithm in LTE System, Journal of Chongqing Institute of Technology Vol. 23, No. 9, pp. 115–117, Sep. 2009.

Guo Ge, Li Xiao-wen, Chen Fa-tang. Improved V-BLAST Algorithm in LTE System[J]. Wide band network. Vol. 34, No. 12, pp. 87–90. 2010.

Li Ying, Li Jia, Wang Xin-mei. An improved detection algorithm of layered space-time codes. Journal of China institute of communications. Vol. 24, No. 3. pp. 113–118; March 2003.

Lin Xu-feng, Yang Hui Performance analysis of detection algorithms in LTE. Telecom engineering technics and standardization, No. 1, pp. 84–88, 2012.

Shengbing Cai, Zhemin Duan, Jin Gao. Comparison of Different Virtual MIMO Detection Schemes for 3GPP LTE. Proceedings-5th International Conference on Wireless Communications, Networking and Mobile Computing, WiCOM 2009, pp. 1–5 Sept. 2009.

Wang Yu, Lin Yun. Impact of MMSE law-based Algorithm on System Capacity in MIMO System, Communications Technology, Vol. 40 No. 12, pp. 118–120, 2007.

Zhao Chen, Liu Ying-zhuang, Zhu Guang-xi, "Research on ZF Algorithm for VLST Architecture" Radio Communications Technology, Vol. 33, No. 2, pp. 37–39 2007.

Zhu Lin-lu, Channel Capacity of MIMO Communication System and MATLAB Simulation. Journal of Xinxiang University (Natural Science Edition). Vol. 29, No. 2, pp. 144–146 Apr. 2012.

Zijian Baia, Christoph Spiegela, etc. Novel block decision feedback equalizers for MIMO-OFDM used in UTRA LTE for UTRA LTE. Procedia Earth and Planetary Science, Vol. 1, No. 1, pp. 1425–1434, September 2009.

Control Engineering and Information Systems – Liu (Ed)
© *2015 Taylor & Francis Group, London, ISBN 978-1-138-02685-8*

Virtual channel based aggregated multi-source multicast across backbone

Y. Chen, C.L. Li, K.X. Huang & X.B. Zhang
State Key Laboratory of Mathematical Engineering and Advanced Computing, Zhengzhou, Henan, China

ABSTRACT: To achieve the tradeoff between state aggregation and traffic redundancy, different branches of the tree are formed in backbone and receivers' access networks. In receivers' access network, each branch corresponds to a multicast group; while in the backbone and on the paths of senders to backbone, each branch corresponds to a multicast traffic class, which achieves multicast states aggregation. Simulation results show that VC-AMSM will result in some traffic redundancy, but it can greatly reduce the number of multicast states and control messages in the network. And the one-tree-one-traffic-class characteristic provides a preferable basis to reconfigure the routing to label switching.

1 INTRODUCTION

Since it is proposed, IP multicast has been widely used in group communication applications, such as video broadcast, online video conference, and so on. However, because of the protocol complexity, poor scalability and no QoS assurance problems of traditional multicast, the deployment of large-scale multi-source multicast is difficult. In traditional inter-domain multi-source multicast protocols, such as MBGP/PIM-SM/MSDP [Fenner 2006, Bates 2000, McBride 2006], receivers can get the addresses of sources after the source discovering process. Then, receivers can join these sources by using specific source multicast protocol, and form multicast trees rooted at these sources. Routers on the multicast trees generate and maintain a <s, g> state for each source-group pair. When the backbone carries a large of multicast groups' traffic sourced from different accessed networks, the multicast states in backbone routers will increase rapidly. This will result in poor scalability.

If the group traffic with the same QoS requirement flows along the same path, it is much easier to aggregate them and provide QoS assurance to them. However, the aggregation will result in redundancy of some multicast traffic on the shared path. Because the bottleneck of carrying capacity of backbone is message processing rather than bandwidth provision, while, the access networks usually have fewer multicast states and but their bandwidth is limited, multicast aggregation is usually implemented in backbone.

To solve the problems of traditional multicast and achieve the tradeoff between state aggregation and traffic redundancy, we proposed VC-AMSM, a virtual channel based aggregated multi-source multicast scheme across backbone. By leveraging the mechanism of specific source multicast, VC-AMSM forms aggregated multicast tree based on virtual channel, to carry multi-source multicast traffic across backbone. By assigning a specific virtual channel to a class of multicast applications, it aggregates the traffic of a class of multicast applications to a shared tree in the backbone and provides a preferable basis to provide QoS assurance to multicast applications.

Section 2 introduces the main contents of VC-AMSM. Then, the results of simulating experiment are presented in section 3. At last, section 4 concludes work in this paper.

2 VC-AMSM

2.1 Definitions

Definition 1 Multicast Traffic Class Set (MTCS) C: Multicast traffic can be categorized into m classes $C = \{c1, ci, cm\}$ according to their QoS requirements, such as bandwidth, delay, packet loss rate. To aggregate multicast states in backbone by using extended PIM-SSM [Holbrook 2006], a multicast class is needed to be mapped to a preserved aggregated multicast address. We suggest using 224.0.2.0-224.0.2.255 as the preserved aggregated multicast address. The last 8 bits of the preserved address identify the multicast class and the extended PIM-SSM is assumed to be able to run on the addresses.

Definition 2 Same Class Multicast Group Set (SCMGS) G_{ci} It is assumed that multicast groups $g_1, g_2, ..., g_n$ ($n \geq 1$) are groups in the same multicast traffic class ci $(1 \le i \le m)$, then $G_{ci} = (ci, \{g_1, g_2, ..., g_n\})$ is called as a SCMGS with traffic class ci.

Definition 3 Set of Virtual Channels (VC): A virtual channel $vci <s_{ci}, g_{ci}>$ is a multicast channel established for forming the shared tree of SCMGS G_{ci}. Here, s_{ci} is an addressable IP address called virtual source, which is not the real multicast source and only tells the routers what direction they should send extended PIM-SSM messages to. It is a function of **ci**. VC-AMSM defines no particular function, but it requires that s_{ci} is an address in the backbone equipments, near the center of the participants. In addition, AMSM suggests that G_{ci} for G_{cj} and $(i \ne j)$, $s_{ci} \ne s_{cj}$ to decentralize multicast traffic in the backbone. g_{ci} is a preserved aggregated multicast address for ci. All virtual channels compose the VC.

2.2 Initial shared aggregated tree setup and source discovery

In VC-AMSM, the sources and receivers tell the DR (the router directly connected to the sender and/or receiver) their JOIN/REQS/LEAVE requirements by sending extended IGMPv3 [Cain 2002] (or MLDv2 [Vida 2004], for IPv6) messages. Routers need to generate corresponding $(*, g)$ states to form aggregated trees by using extended PIM-SSM protocol. The $(*, g_c)$ state formed by extended PIM-SSM includes four fields $<*, g_c, ilist, olist>$, where * means it can match any source, gc is a preserved multicast address corresponding to c, ilist is the incoming interface list, and olist is the outgoing interface list.

IGMP/MLD message extension and DR's processing: In VC-AMSM, a host needs to tell its DR that it applies to send/receive the groups traffic or leave a group g. The value of "Type" field in IGMP/MLD message can be extended to fulfill this. Take MLDv2 message for example. Its "Type" field includes 8 bits. Its value is 130, 131 or 132 represents the message is a Multicast Listener Query, a Multicast Listener Report or a Multicast Listener Done message. The field value can be extended as: when its value is 133, the message is a "Multicast Sender Report" message which represents the host applies to send the group traffic. In addition, the value of application type c (8 bits) is included in the extended IGMP/MLD messages.

When a DR receives an extended IGMP/MLD $(*, g)$ message from a host, where * means that the host doesn't know the real source of g, it creates or maintains its $(*, g_c)$ state. Diffident with standard PIM-SSM, when receiving a Multicast Sender Report message from a host, it adds the interface receiving the message to ilist of its $(*, g_c)$ state; when receiving a Multicast Listener Done message from a host, it removes the interface receiving the message from olist and ilist of its $(*, g_c)$ until the timer of the interface become expired to avoid other type of messages of groups belong to c appear come from the interface.

Then, according the different requests from the host, the DR sends JOIN/REQS/LEAVE message towards s_c, where s_c is the virtual source of the virtual channel corresponding to c. When sending JOIN message, it adds the sending interface to ilist of its $(*, g_c)$ state. When sending REQS message, it adds the sending interface to olist of its $(*, g_c)$ state. When sending LEAVE message, it removes the sending interface from ilist and olist of its $(*, g_c)$ state.

JOIN/REQS/LEAVE message processing by extended PIM-SSM: In control plane, VC-AMSM uses extended PIM-SSM JOIN/REQS/LEAVE messages to create and maintain $(*, g_c)$ states. The type of extended PIM-SSM message includes JOIN (for applying receiving the traffic of g_c), REQS (for applying sending the traffic of g_c) and LEAVE (for applying leaving g_c). JOIN and REQS can be combined in one message, and LEAVE must be used separately. The highest 3 bits in "reversed1" field of SSM message specifies the type of the messages. In the 3 bits, the bit from the highest to the lowest represents whether or not ('1' or '0') expecting to receive the group traffic, whether or not ('1' or '0') expecting to send the group traffic and whether or not ('1' or '0') expecting to leave the group respectively.

When a router r receives a JOIN/REQS/LEAVE message sent from a downstream router on interface I, it processes the message as follows.

If the message type is '100', '010' or '110', and r does not has $(*, g_c)$ state, r generates a $(*, g_c)$ state, with its ilist and olist be set {}.

If the message type is '1?0' (where ? represents '1' or '0'), r adds I to the olist of its $(*, g_c)$ state.

If the message type is '?10', r adds I to the ilist of its $(*, g_c)$ state.

If the message type is '??1', r removes I from the ilist and the olist of its $(*, g_c)$ state until the timer of the interface become expired to avoid JOIN/REQS messages of g_c appear come from the interface.

After that, suppose the shortest path interface from r to s_c is I', r processes as follows.

If $ilist \cup olist - \{I'\}$ is null, indicating there is neither sender nor receiver downstream, r immediately sends LEAVE message (whose type is '??1') to I', and eliminates its $(*, g_c)$ state.

If $ilist \cup olist - \{I'\}$ is null and r's $(*, g_c)$ state is a newly generated one, r immediately sends JOIN/REQS message to I', or else r sends JOIN/REQS message to I' when JOIN/REQS timer expires.

If $ilist - \{I'\}$ is null and $olist - \{I'\}$ is not null, indicating there is receiver and no sender downstream, *r* sends the message with type '100' and adds *I'* to the *ilist* of its (*, *g_c*) state.

If *olist*-{*I'*} is null and *ilist*-{*I'*} is not null, indicating there is sender and no receiver downstream, *r* sends the message with type '010' and adds *I'* to the *olist* of its (*, *g_c*) state.

If *olist*-{*I'*} is not null and *ilist*-{*I'*} is also not null, indicating there is sender and receiver downstream, *r* sends the message with type '110' and adds *I'* to the *ilist* and *olist* of its (*, *g_c*) state.

The (*, *g_c*) states are soft states. The expirations of the interface timer and the state timer will eliminate corresponding interface in *ilist* or *olist*, or the whole state.

Then, the initial shared aggregated tree is formed. Some branches of it are uni-directional and some are bidirectional. The tree is just act as the initial communication route way among the group participants with the same traffic class c and the sender's traffic will flow alone the tree to arrive the receivers. And then, the receivers can sense the real source set $S = \{s_1, s_2, \ldots\ldots, s_m\}$ from the receiving multicast packets' heads and send join messages to the direction of the real sources to form the source specific trees.

2.3 Forming the source specific trees in receivers' access networks

Being aware of a source address s_i from the multicast packet header, the receiving hosts apply to join <s_i, g> channel using extended IGMPv3 or MLDv2 messages, and the routers send extended JOIN (s_i, g) messages to the shortest path interface towards s_i. The IGMPv3/MLDv2 and JOIN messages are extended to include the 8-bit traffic class fields. To maintain multicast state aggregation in backbone, when a JOIN (s_i, g) arrive at an edge router in backbone, the router transform the JOIN (s_i, g) message to JOIN (s_c, g_c) message and send it to the shortest path toward s_c, where (s_c, g_c) is the virtual channel of (s_i, g) who's traffic class is c. Thus, the source specific trees are formed in receivers' access networks while maintaining the aggregated shared tree in backbone and on the paths of senders to backbone.

2.4 Packet forwarding rule

When a router *r* receives a multicast packet *P*, whose source address is s_i, destination address is g, and traffic class is *c*, on a interface *iif*, it will forward the packet according to the following rule (related functions can be found in [1,4], only PIM-SSM functions are considered and the assert mechanics is ignored).

O-list = NULL;
If ((s_i,g) state exists) and (RPF check succeeds)

o-list = *olist*(s_i, g);
else if ((*, *g_c*) state exists) and (*iif* is in *ilist* of (*, *g_c*))
{*o-list* = *olist*(*, *g_c*);
o-list = *o-list* (-) *iif*};
Forwards *P* on each interface in *o-list*;

Above process shows that if a router is the intersection of the shared aggregated tree and the source tree, it will firstly forward the packet along the source tree.

3 COMPARISON AND SIMULATION

To compare the performances of PIM-SM and VC-AMSM, we implement an example of multi-source multicast across backbone, of which topology is shown in Figure 1. We assume there are six access networks which are inter-connected by the backbone network. We also assume there is a SCMGS $G_c = (c, \{g_1, g_2, g_3, g_4, g_5, g_6, g_7\})$. The sources, receivers and the Rendezvous Points (RPs) of groups $G_c = (c, \{g_1, g_2, g_3, g_4, g_5, g_6, g_7\})$ are shown in Table 1. R_{14} is the virtual source of G_c. The time on which the sources and receivers take part in the group communication is shown in Table 2. And all sources send their multicast packets at 10 Kbps in our experiment.

We compare the performance of PIM-SM and VC-AMSM on number of control messages, number of multicast states and traffic flow. The comparison of number of control messages is shown in Figure 2, where the number of control messages is accumulated from time 0 to t_i. It can be seen that VC-AMSM can efficiently reduce the number of control messages in the network.

Routers in backbone and the access networks maintain multicast states in PIM-SM and VC-AMSM. The number of states is accumulated from

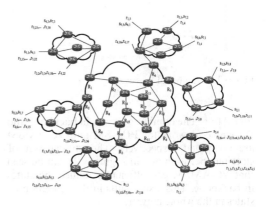

Figure 1. An example of multi-source multicast across backbone.

Table 1. Sources and receivers of groups.

Group	Sources	Receivers	RP
g_1	$s_{1,1}, s_{1,2}, s_{1,3}$	$r_{1,1}, r_{1,2}, ..., r_{1,8}$	R_{15}
g_2	$s_{2,1}, s_{2,2}$	$r_{2,1}, r_{2,2}, ..., r_{2,13}$	R_{10}
g_3	$s_{3,1}$	$r_{3,1}, r_{3,2}, ..., r_{3,9}$	R_{14}
g_4	$s_{4,1}, s_{4,2}, s_{4,3}, s_{4,4}$	$r_{4,1}, r_{4,2}, ..., r_{4,22}$	R_8
g_5	$s_{5,1}, s_{5,2}, ..., s_{5,8}$	$r_{5,1}, r_{5,2}, ..., r_{5,36}$	R_{13}
g_6	$s_{6,1}, s_{6,2}, s_{6,3}$	$r_{6,1}, r_{6,2}, ..., r_{6,9}$	R_{12}
g_7	$s_{7,1}, s_{7,2}, s_{7,3}, s_{7,4}$	$r_{7,1}, r_{7,2}, ..., r_{7,18}$	R_{17}

Table 2. The time of the hosts taking part in.

Time	Source	Receiver
t_1 (0 min)	$s_{1,1}, s_{1,2}, s_{1,3},$ $s_{2,1}, s_{4,1}, s_{4,2}$	$r_{1,1}, r_{1,2}, r_{1,5}, r_{1,6}, r_{2,1}, ...,$ $r_{2,7}, r_{4,1}, r_{4,5}, r_{4,6}, r_{4,8}, r_{4,9}$
t_2 (2 min)	$s_{2,2}, s_{3,1}$	$r_{2,8}, ..., r_{2,13}, r_{3,1}, ..., r_{3,5},$ $r_{4,21}, r_{4,22}$
t_3 (4 min)	$s_{5,1}, s_{5,2}, s_{5,3},$ $s_{5,4}, s_{5,5}$	$r_{4,2}, r_{4,3}, r_{4,4}, r_{4,7}, r_{5,1}, ...,$ $r_{5,20}$
t_4 (6 min)	$s_{4,3}, s_{5,6}, s_{5,7}$	$r_{3,6}, ..., r_{3,9}, r_{4,10}, ..., r_{4,20}$
t_5 (8 min)	$s_{4,4}, s_{5,8}$	$r_{5,21}, ..., r_{5,36}$
t_6 (10 min)		$r_{1,3}, r_{1,4}, r_{1,7}, r_{1,8}$
t_7 (12 min)	$s_{6,1}, s_{6,2}, s_{6,3}$	$r_{6,1}, r_{6,2}, ..., r_{6,9}$
t_8 (14 min)	$s_{7,1}, s_{7,2}, s_{7,3}, s_{7,4}$	$r_{7,1}, r_{7,2}, ..., r_{7,18}$

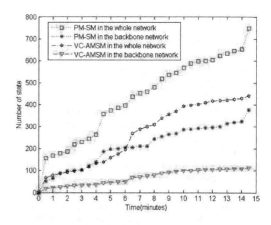

Figure 3. Number of states.

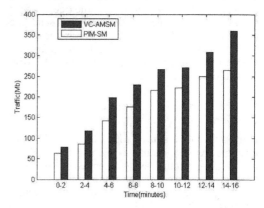

Figure 4. The accumulated traffic in backbone.

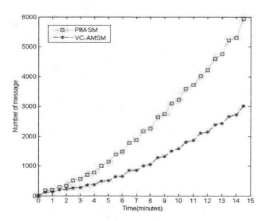

Figure 2. Number of messages.

time 0 to t_i. The number of states in backbone and the whole network for PIM-SM and VC-AMSM are computed respectively. The comparison of state number is shown in Figure 3. It can be seen that VC-AMSM can efficiently reduce the states in backbone. This also results in the reduction of states in the whole network.

Figure 4 shows the comparison of the accumulated traffic in backbone in each interval of t_i to

$t_i + 1$. It can be seen that VC-AMSM will result in the increase of flow because of the traffic redundancy. This is because the group's traffic belong to the same traffic class will flow along the same path in backbone and from the sender to backbone. Unlike PIM-SM source discovery process, the virtual channel mechanism does not necessarily rendezvous all traffic to the virtual source R_{14} (for example, traffic from $s_{1,2}$ will flows along R_5, R_{13}, R_{12}, R_4 in backbone to $r_{1,1}$, $r_{1,2}$, $r_{1,3}$). So, the traffic redundancy is not very high.

4 CONCLUSIONS

We propose a virtual channel based aggregated multi-source multicast scheme across backbone. It provides better scalability by aggregating multicast states in the backbone network and on the paths of senders to backbone. And this one-tree-one-traffic-class characteristic provides a preferable basis to further QoS assurance mechanism such as QoS routing and converting to label switching. For different SCMGSs, we manage to assign different virtual sources to disperse the multicast traffic and avoid causing unbalanced excessive burden on some routers. We will study how to reconfigure the VC-AMSM tree to label switching tree to provide better QoS assurance in further research.

ACKNOWLEDGMENT

Sponsored by the National Basic Research Program of China (973 Program) under Grant No. 2012CB315901.

REFERENCES

Bates T., Y. Rekhter, R. Chandra and D. katz, "Multiprotocol Extensions for BGP-4[S]," Internet Engineering Task Force (IETF), RFC 2858. June 2000.

Cain B., S. Deering, "Internet Group Management Protocol, Version 3," Internet Engineering Task Force (IETF), RFC 3376, October 2002.

Fenner B., M. Handley, H. Holbrook and I. Kouvelas, "Protocol Independent Multicast—Sparse Mode (PIM-SM): Protocol Specification (Revised)," Internet Engineering Task Force (IETF), RFC 4601, August 2006.

Holbrook H. and B. Cain, "Source-specific multicast for IP," Internet Engineering Task Force (IETF), RFC4607, August 2006.

McBride M., J. Meylor, "Multicast Source Discovery Protocol (MSDP) Deployment," Internet Engineering Task Force (IETF), RFC 4611, August 2006.

Vida R., Ed.r, "Multicast Listerner Discovery Version 2(MLDV2) for IPv6," Internet Engineering Task Force (IETF), RFC 3810, June 2004.

Control Engineering and Information Systems – Liu (Ed)
© 2015 Taylor & Francis Group, London, ISBN 978-1-138-02685-8

Optimal design of QR code encoding

Y.M. Qin, L. Chen, B.F. Chen & N. Lei
Department of Mathematics, Jilin University, Changchun, Jilin, China

ABSTRACT: As a type of matrix two-dimensional barcode, QR code is quick and omnibearingly readable. The system is playing a more and more important role in data storage and data transfer in all walks of life. To improve the quality of a QR code pattern and the accuracy of its decoding, we made some revisions of its encoding. First we elaborated the algorithm of partitioning masking. Then we replaced the Penalty Law with the Variance Method which was suggested by Zhang Li *et al.* to evaluate masking. The programming result showed that the modifications in both aspects help in improving the uniformity of the dark and light modules so these modifications are reasonable and effective. Last we rearranged the array of codeword series and realized the arbitrary distribution of error correction codes and data codes in the figure to avoid the loss of information caused by the figure edge damages. Finally, we applied these improvements in the design of the QR-code encoding software and got the revised version.

1 BACKGROUND

Barcode is a representation of data by varying the widths and spaces of parallel lines. It is a graphical identifier and can be automatically recognized by photoelectric scanning and reading equipment. QR code is a type of matrix 2D barcode developed by Denso Corporation in Japan in September 1994 (Pavlidis, 1990 & 1992; Reiley, 1998).

QR code can store more information than other normal barcode. It doesn't need to be pointed straight at the scanner when scanning. QR code is widely used in storage management in different industries in recent years. QR decoding softwares in cellphones lead more customers to learn and to enjoy the service. The popularities of coding and reading 2D barcode with cellphones truly make QR code walk into our daily life.

The process of QR's application is to code the data and information into a code pattern firstly and then it would be printed or pasted onto an object. The last step is to be read and decoded by a recognizer (Yang, 2003). The quantity of the code pattern directly affects the accuracy of recognizing.

Generally, there are 7 steps to generate a QR code pattern. The National Standards GB/T18284-2000 Quick Response Code, GB for short, makes detailed description and requirements of this process. Comprehensively speaking, the design program in GB meets the basic design requirements. But with the expansion of its application range and higher standards of recognition the quantity of the code pattern needs further improvement.

Considering the actual need, we improved the design of QR code in masking schemes and codewords arrangement.

2 IMPROVEMENT OF MASKING SCHEMES

2.1 Masking

The effects of scanning the code pattern by a QR code recognizer are greatly affected by the brightness of the code image. Large light modules have a strong reflection and may submerge a bit of dark modules among them. This would lead to information reading errors. So the more uniformly the dark and light modules distribute, the more efficient of QR code recognizing. Masking is to carry out XOR operations between special patterns and the code patterns with dark and light modules, so as to achieve the purpose of a balanced distribution of the modules in the entire code pattern.

Eight Mask Pattern References (000~111) are given in GB. Penalty Laws are proposed to pick out the best Mask Pattern Reference with the lowest penalty.

2.2 Improvements of masking schemes

Follow the aims and ways of masking, if we subdivide the code pattern into sub-regions and select out the best Masking Pattern Reference for each region, the combined pattern should be better than the entire code pattern masking. Considering that subdividing the code pattern would add edges, a small number of dark or light modules would happen to meet at edges when best-masked sub-regions are combined. So we took the combined-best Mask Pattern Reference as the ninth masking scheme and masked the entire code pattern together with other eight Mask Pattern References to select out the best masking scheme for the entire code pattern. Based on this thinking, we proposed the following algorithms of partitioning masking.

Algorithm 1: Algorithm of partitioning masking
S1. Divide the entire code pattern into N*N sub-regions.
S2. Mask each sub-region with eight Mask Pattern References, calculate the penalties and select out the reference with lowest penalty as the best masking scheme of the sub-region.
S3. Combine the best Mask Pattern References of all N*N sub-regions together as the ninth masking scheme.
S4. Mask the entire code pattern with these nine masking schemes and calculate the penalties and select the masking scheme with the lowest penalty as the best masking scheme of the entire code pattern.

In the experiment, we let N be 2 and divide the entire code pattern into 4 sub-regions. The best masking scheme were selected with the above algorithm.

If the best masking scheme is the combined scheme, then we need 12 bits to record 4 Mask Pattern References. The Format Information is 15 bits long in GB. It usually contains 2 bits of correction level, 3 bits of Mask Pattern Reference and 10 bits of correction codes. It appears twice in the code pattern to be deciphered correctly since that is crucial for the deciphering of the whole pattern. However, the improved masking scheme takes 14 bits to record the correction level and Mask Pattern References. It is unreasonable to keep correction code in only one bit. According to BCH coding principle (Feng, 2011; Huang, 2003; Wallace, 2001), BCH (31, 16) was the best correction algorithm to get the correction codes of 14-bits-long data of correction level and Mask Pattern Reference. That is to add '00' to 14 bits of information data to form 16 bits of information, and generate 15 bits of correction codes with BCH (31, 16). Such that the Format Information is 31 bits long. This is 16 bits longer than it (15 bits) in GB. Since the Format Information should appear twice in the pattern, so it is 31 bits longer in total.

According to GB, the lengths of each kind of characters are listed in Table 1.

So 32 bits can store 9.6 digits or 2.9 letters or 4.0 bytes or 2.4 Japanese characters or 2.4 Chinese characters.

These numbers are not so large. Even if the original data exceeds original storage limit, since the Format Information occupies more bits, and the code pattern has to take a higher version, the area of new code pattern will not take 1 mm^2 more than the original code pattern. So it is reasonable to take these bits to store Format Information.

2.3 *Revision of the capacity of codewords in each version*

Every 8 bits are called a codeword. Since Format Information takes additional 32 bits (4 codewords), data codewords in each version must be reduced 4 in Table 7–11 (Numbers of Symbol Characters and Data Capacities of Version 1-40). Error correction features in Table 13–22 (Error Correction Features of Version 1-40) in GB should make corresponding changes. To get revised Table 13-2, we summarize following revision rules.

1. Reduce 4 total codewords of each version.
2. Keep error correction codewords of each error correction level in each version.
3. For each error correction level of each version, if the codewords contain two kinds of error correction blocks, data of these two kinds of error correction blocks can be determined by rule 4).
4. Suppose the total number of codewords is D, total number of error correction codewords is C, error correction level is 1, total numbers of two kinds of error correction blocks are x and y, block numbers are n_1 and n_2, numbers of data codewords are d_1 and d_2 and error correction capacity is r, they should meet the followings.
   a. $D = n_1 * x + n_2 * y$

   b. $y = x + 1 d_2 = d_1 - 1$

   c. $C = n_1 * (x - d_1) + n_2 * (y - d_2)$
      $= (n_1 + n_2) * (x - d_1)$

   d. $x - d_1 = 2r$

   e. $\left| \dfrac{r}{x} - l \right| < 5\%$

Table 1. Lengths of five kinds of characters.

Modes	Digits	Letters	Bytes	Japanese characters	Chinese characters
Lengths (characters/bits)	3/10	2/22	1/8	1/13	1/13

f. $8 \le r \le 15$ If there are several solutions meet (1)~(7), select the solution such $|r/x - l|$ least.

5. If codewords cannot be divided into two kinds of error correction blocks that meet the requirements in 4), reduce data codewords in each block with equal rate based on the information in GB.

2.4 Arrangement of format information

For the position of Format Information, considering the convenience of storing and searching and least changes of other data codewords, we proposed an arrangement plan as Figure 1.

2.5 Results comparison

Generate the QR code pattern of test data '111111 11111111111111111111111111111111111' (input with random) separately with original masking scheme and revised masking scheme.

Figure 2 presents the code pattern masked with Mask Pattern Reference 110. Figure 3 presents the code pattern masked with combined Mask Pattern Reference showed in Figure 4. It is clear that dark and light modules in Figure 3 distribute more balanced than that in Figure 2.

On the other hand, we made a statistic analysis. We took 1,000 data coding experiments with revised masking scheme and recorded the numbers that the best masking scheme was the combined scheme

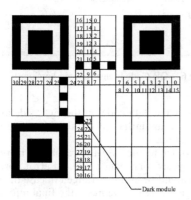

Figure 1. Arrangement of format information.

Figures 2–3. Masked code patterns and combined Mask Pattern Reference.

(i.e. the selected best Masking Pattern References of all 4 sub blocks are not completely the same). The result showed the rate that combined schemes were the best scheme was up to 78%. What's more, when the data are Chinese characters, the rate was even higher.

3 IMPROVEMENTS OF MASKING EVALUATION METHODS

3.1 Evaluation method of QR code masking result

The way to evaluate the results of masking scheme is clearly marked in GB.

In this Penalty Law, we have no idea on how those certain four features are chosen to get high penalties. There is no explanation on why weight of penalty score (N1 = 3, N2 = 3, N3 = 40, N4 = 10) is so determined. So this is highly subjective and it lacks convincing theory evidence. While this method is suitable for QR code only and it is difficult to promote and transplant.

Therefore, we adopted a more objective and more general evaluation method with more convincing theory evidence.

3.2 Objective evaluation method of masking results—variance method

3.2.1 Principle of variance method
The ideal situation of masking is that dark modules and light modules are equally distributed in any part of the pattern. Statistically, this is known as expectation. Zhang et al. (2009) proposed a more math-based Variance Method. The degree of the differences between masking results and expectations are used as criteria to evaluate the results.

3.2.2 Algorithm of variance method
Algorithm 2: Algorithm of Variance Method.
S1. Read modules in rows.
S2. Expectation of the distribution of two adjacent modules $E(x) = 1/2$. If two adjacent modules are dark and light separately, then the variance $D(x_i) = 0$. If the two modules are the same, the variance $D(x_i) = 1/4$.
S3. After scanning modules in a row, a sum of variance can be obtained.
S4. Scan the module block in columns, same as in rows, and calculate the sum of variance scanned in columns.
S5. Calculate mean square deviations of the whole code pattern masked with different masking schemes and select the optimal masking scheme with the least mean square deviation.

Table 2. 8 masking scheme scores under Penalty-Algorithm and Variance-Algorithm.

Masking scheme	Penalty law	Variance method
000	1069	0.483333*
001	1107	0.507142
010	1054	0.538095
011	1127	0.523809
100	1044*	0.533333
101	1057	0.526190
110	1064	0.519047
111	1085	0.516666

Penalty Law Variance method

Figure 4. Optimal masking method comparisons.

3.2.3 *Realization of variance method and result comparison*

Enter "qqqqqq" as test data and use two methods to score 8 masking schemes of QR code respectively. Result is shown in Table 2.

Table 2 suggests that the lowest score of two algorithms are masking scheme 100 for Penalty Law and masking scheme 000 for Variance method, respectively. The results are shown in Figure 3. As you can see in the figure, the optimal masking scheme selected through Variance Method is, to a large extent, better than the optimal masking scheme selected by Penalty Law.

4 IMPROVEMENTS OF CODEWORDS ARRANGEMENT

Take Version 5—H as an example. Its final codewords series are:

D1, D12, D23, D35, D2, D13, D24, D36, … D11, D22, D33, D45, D34, D46, E1, E23, E45, E67, E2, E24, E46, E68, … E22, E44, E66, E88. If necessary, code followed by remaining bit (0) in the end.

The entire code series is arranged from lower-right corner with data code on the right side and error correction code on the left side. The risk of this arrangement is that once left side of the code pattern is damaged, it is impossible to correct

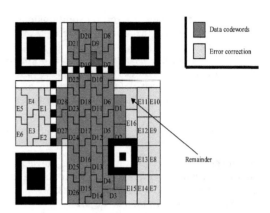

Figure 5. Improved codewords arrangement pattern.

Figure 6. QR Code Image Revised Version.

errors of the data code on the right side and this may have an influence on the identification of information.

In order to increase the reliability in reading, we proposed that data code and error correction code should be evenly arranged in case of pattern damage. Therefore we mark the final information flow array as N array. The only thing we need to do is to rearrange the elements in N array in any order. New array is named as N*, then arrange the elements in N* array by the original rule into the pattern. So we implement the cross arrangement of code data and correction data. For example, mark N = (N1, N2, N3), and then rearrange it as (N3, N1, N2), this will cause error correction codes distributed on both sides of the data code, as shown in Figure 4. This could increase the information security and reliability.

336

5 IMPROVEMENTS ON SOFTWARE

We implemented the algorithm of partitioning masking in QR code coding and valued the results with Variance Method. Rearranged the data code and correction code at last. By revising relevant functions based on open source codes of QR code generator, we got the following Revised QR code generator.

6 CONCLUSIONS

As a type of Matrix 2D barcode, QR code is featured as coding Chinese and Japanese characters effectively. The popularity of coding and reading 2D barcode with cellphones truly makes QR code walking into our daily life. It also improves the requirements of the accuracy of coding and reading data.

We improved the encoding algorithms of QR code in three aspects and achieved our aims of improving code pattern qualities in experiments.

Firstly we proposed the partitioning masking. Select out the best Mask Pattern References for each region, and then combine them together as a combined scheme. Compare the penalties of masking results separately with combined scheme and original entire masking scheme and select out the best masking scheme. Statistics data showed about 78% of the best masking schemes are combined schemes. So partitioning masking is reasonable.

Secondly we replaced the Penalty Law with the Variance Method which was suggested by Zhang Li *et al.* to evaluate masking. Variance Method has a stronger mathematical dependence, more detailed value standards and briefer algorithm realization. Coding Experiments also showed the final code patterns were with more balanced dark and light modules with Variance Method.

Finally we rearranged the codeword series and realized the distribution of error correction codes and data codes with arbitrary combination in the figure to avoid the loss of information by the damage of the figure edges.

REFERENCES

Daniel J. Reiley, Polarization of barcode readers[J]. Opt. Eng, 1998, 37(2): 688–695.

GB/T18284-2000, National Standards, Quick Response Code (QR code) [S].

Hanlu F., Yingwei H., Xiaojiao N., Yinchao Q. Principle and implementation of error correction coding of QR code[J]. Journal of Computer Applications, 2011, 31(1): 39–42.

Hank Wallace. Error Detection and Correction Using the BCH Code, [EB/OL], http://www.aqdi.com/bch.pdf, 2001.

Hongbo H., Junling X., Lijuan T. Error-correcting coding of QR code based on Reed-Solomon Algorithm[J]. Computer Engineering. 2003, 29(1): 93–95.

Pavlidis T., Swartz J., Wang Y.P. Fundamentals of bar code information theory[J]. IEEE. Computer, 1990, 23(4): 74–86.

Pavlidis T., Swartz J., Wang Y.P. Information encoding with two-dimensional bar codes[J]. IEEE Computer, 1992, 27(6): 18–27.

Qiuying Y. Research and Application of QR code, a kind of 2D barcode[D]. North University of China, 2003.

Li Z., Min D., Baifeng W. Sequence Two-dimension symbologies design and mask solution[J]. Journal of Image and Graphics. 2009, 14(7): 1426–1431.

The task-assigning model of network attack and defense test based on human role

F. Zhou
The Information System Important Laboratory of the 28th Research Institute of China Electronics Technology Group Corporation, Nanjing, China

ABSTRACT: In this paper, the human role set of network attack and defense process are proposed based on the description of dynamic adaptive OODA loop process model, and defining the function of various type roles. Then considering the ability evaluation test of military information system under network confrontation conditions, analyzing the configuration requirement of human roles during the test process, and building the task-assigning model based on the role mapping mechanism, realizing the association between the sub test task and human roles. At last, the role ability level requirement and the role undertaker are proposed according as the characteristic of test task.

1 INTRODUCTION

At present, the military information system has entered a new stage of integrated construction and development, transforming from hierarchical structural pattern based on platform-centric to flatten networking pattern based on network-centric. The scale of networking becomes more and more bigger, and the degree of network-centric becomes more and more higher, also those changes increase the security risk from network space[1–5], such as attacking the military sensor, communication center, command and control system, weapon control system and other military infrastructure, or injecting some false targets into the military sensor, or altering the communication link data, or forging the false intelligence data into the intelligence processing system and so on. In addition, the security risk also includes internal threat from the personnel misoperation and enemy counterespionage and penetration behavior.

Facing the above security threats, in order to strengthen the security of military information system has become an urgent demand for development and construction of system. In this paper, according to analyze the countermeasure process within cyberspace, the human roles of attacking side and defense side are presented. Then considering the capability evaluation test of military information system, the test task allocation mechanism based on human roles is built, in order to realize the flexible and efficient configuration against human roles during the process of system's capability evaluation test.

2 THE CONCEPT OF HUMAN ROLE

The role's definition can be understood as an object structure, including the task, the target, behavior, function, obligation, capability and relationship, etc[6–8]. And the role can receive information, process information and send information. The concept of human role has been used to research the automatic management of the information system, in which the role is denoted as the basic unit of system.

The network confrontation process can be described by dynamic adaptive OODA loop operation process model, which has four stages: observe, orient, decide and act.

Based on the above process model, according to analyze the activities of attacking side and defending side during the process of network confrontation, we present eight human roles against on both sides in this paper. In the attacking side, we present four attacker roles, that is: detecting controller, attack situation analyzer, attack decision maker, attack performer. And in the defending side, we present four defender roles, that is defense supervisor, defense situation analyzer, defense decision maker, defense performer. The above eight human role's function is defined as shown in Table 1.

3 TASK-ASSIGNING MECHANISM BASED ON ROLES

In the traditional evaluation tests of military information system, the test task-assigning mechanism can be described as follows: firstly the test task

Table 1. The human role's function of network attacking and defending.

Stage of OODA loop	Attacking		Defending	
	Role's type	Role's function	Role's type	Role's function
Observe	Detecting controller	✓ Detecting objective system and gathering network topology, the port, protocol, service, OS type and version, vulnerabilities and defense measures information using reconnaissance and detection tool. ✓ Adjusting detection way based on information gathered.	Defense supervisor	✓ Utilizing system security tools to monitor system running state, and collect system or system security equipment log file information.
Orient	Attack situation analyzer	✓ Analyzing and evaluating the vulnerabilities of objective system based on information gathered. ✓ Mining the key node of objective system and business relations among nodes.	Defense situation analyzer	✓ Generating system security situation, conducting threat estimation and prediction, mining the potential attack or abnormal behavior.
Decide	Attack decision maker	✓ Making the attack plan or strategy, includes that determining attack targets, setting attack way and choosing attack weapon. ✓ Adjusting dynamically the attack plan or strategy based on attack effect.	Defense decision maker	✓ Making defense plan according to the analysis results. ✓ Adjusting dynamically security defense strategy or plan.
Act	Attack performer	✓ Manipulating attack tool to implement attack activities, and providing feedback information about attack effect.	Defense performer	✓ Preventing illegal operation and intrusion behavior, scanning and remedying system vulnerability, data backup and fault recovery.

is decomposed into multiple test subtasks, then assigning test executing entity for each subtask. But this task-assigning mechanism is lack of the flexibility and adaptability. When the pre-assigned executing entity cannot accomplish test task, since the pre-specified task-assigning relationship, leading to other entities cannot replace the preassigned entity to complete the assigned task.

In order to overcome the above problem, we present a task-assigning mechanism based on human role. The basic idea is that adding a role layer between test task layer and executing entity layer, the subtasks will be associated with the human roles. And the subtask will be assigned to human roles rather than executing entity, then each role is taken on by different entity.

Considering that though the computer can simulate human's simple behavior in a certain extent, but cannot simulate some complex analysis behaviors, such as reasoning, planning and decision-making behavior. These complex behaviors must be fulfilled by people, even controlled by people. So the execute entity includes well-trained actual people and simulated people, in which the simulated people is defined as the agent program that

can imitate actual people to fulfill execution task, such as the information collection, information processing and simple decision-making task.

The merit of this mechanism shows that the test task is associated with the role, and separated from execution entity. So we only need to configure different role for specific tests rather than the executor or participant, thereby improving the flexibility of task-assigning. The detail principle and model can be shown as Figure 1.

In order to efficiently evaluate the capability and performance of military information system under the countermeasure condition, we need to build the high fidelity attack threat system within the test environment, and devise countermeasure experiment with back to back test mode. So we classify the test personals into the attacker and defender, respectively present the human role configuration requirement and the undertaken task of each role.

Firstly, we consider the attacker of military information system, whose tasks is to simulate the network attacks, mainly includes the following tasks:

The first mission (M1): monitoring the communication link, scanning network, detecting network

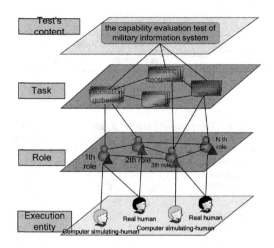

Figure 1. The task-assigning model based on role mapping.

topology and scanning vulnerability, etc, in order to gather system information, mainly include the link characteristic, communication protocol, topology structure, the server and host address, the OS type and version, the open port and system service, system vulnerability information etc.

The second mission (M2): analyzing and processing the gathered information. Identifying the key node of system through the traffic analysis method or the complex network theory, then judging the business relations among nodes, such as the intelligence support relation, or command and control relation, Collaborative operational relation etc.

The third mission (M3): determining attack target (such as communication network backbone node, communication link, key application node and system service software) and attack effect based on above analysis result and the computation ability, communication ability, serve ability of system and attacking entry node information. Then making attack plan through the expert knowledge base and historical experience, including that choosing attack way and attack tools, selecting attack time and making attack strength. In which, the choice of attack way and tool mainly depend on the attack effect, such as in order to block or destroy the intelligence processing center of radar intelligence group system, we can choose the DDOS attack way. The choice of attack strength mainly depends on communication network bandwidth, system computation ability. Attacking behavior will be easy detected by firewall or intrusion detection system when the strength value is set too large, otherwise difficult to achieve the attack purpose.

The attack time includes the start time and duration time. The fourth mission (M4): implementing attack activities against on the specified target, and monitoring the attack process. The fifth mission (M5): evaluating attack result based on the target running state and performance, and dynamic adjusting the making plan based on the evaluation results.

According to the role mapping mechanism proposed above, combined with each role's function. We build assigning relations between the above five missions and attacker roles, as shown in Table 2.

From the Table 2, we can conclude that different test mission can be understood by the same role with different capability level. In this paper, we respectively establish each role's capability level based on mission's difficulty.

In the first mission (M1), the detecting controller's capability level can be evaluated by three indexes, that is, the time of code breaking, the quantity of insensitive information gathered within a certain time, the time of system-detecting. Table 3 shows the detail measurement means.

In the second mission (M2), the attack situation analyzer's capability level can be evaluated by time rate and success rate of mining the key node of objective system and business relationships among nodes. Table 4 shows the detail measurement means.

In the third and fifth mission (M3 and M5), the attack decision maker's capability level can be evaluated by two factors. Table 5 shows the detail measurement means.

In the third and forth mission (M4), the attack performer's capability level can be evaluated by the access privilege obtained, the duration time of attack and the concealment of attack. Table 6 shows the detail measurement means.

Then considering the defender of military information system, whose tasks is to protect system from attack, mainly includes the following tasks:

In the first mission (M1): deploying the security defense measures in order to improve the security performance of system. We establish integrative protection architecture with defence in depth principles, which includes four levels: communication network protection, server and host protection, application protection and data protection.

In the second mission (M2): monitoring the system running state, communication network traffic, data from system security protection equipment and system business behavior.

In the third mission (M3): Comprehensive analyzing and processing all kinds of gathered information, and generating the system security situation. Then determining whether the system has been suffered from possible attack behavior.

Table 2. The assigning relations between mission and role of attacker.

Mission	M1	M2	M3	M4	M5
Role	Detecting controller	Attack situation analyzer	Attack decision maker	Attack performer	Attack decision maker

Table 3. The capability level of detecting controller.

Role's capability	Role's measurement rule	
The time of code breaking	$<=T_{m1}$	1 level
	(T_{m1}, T_{m2})	2 level
	$>=T_{m2}$	3 level
The quantity of insensitive information gathered within a certain time	$>=N_{info2}$	1 level
	(N_{info1}, N_{info2})	2 level
	$<=N_{info1}$	3 level
The time of system-detecting	$<=S_{d1}$	1 level
	(S_{d1}, S_{d2})	2 level
	$>=S_{d2}$	3 level

Table 5. The capability level of attack decision maker.

The role's capability	Role's measurement rule	
The time of making attack plan	$<=T_{d1}$	1 level
	(T_{d1}, T_{d2})	2 level
	$>=T_{d2}$	3 level
The effectiveness of making attack plan	$>=M_{d2}\%$	1 level
	$(M_{d1}\%, M_{d2}\%)$	2 level
	$<=M_{d1}\%$	3 level

Table 4. The capability level of attack situation analyzer.

The role's capability	Role's measurement rule	
The time rate of mining the key node of objective system and business relationships among nodes	$<=T_{i1}$	1 level
	(T_{i1}, T_{i2})	2 level
	$>=T_{i2}$	3 level
The success rate of mining the key node of objective system and business relation among nodes	$>=P_{s2}$	1 level
	(P_{s1}, P_{s2})	2 level
	$<=P_{s1}$	3 level

Table 6. The capability level of attack performer.

Role's capability	Role's measurement rule	
The access privilege obtained	Root privilege	1 level
	Normal user permissions privilege	2 level
The duration time of attack	$<=T_{a1}$	1 level
	(T_{a1}, T_{a2})	2 level
	$>=T_{a2}$	3 level
The concealment of attack	Successful	1 level
	Detected	2 level

Table 7. The assigning relations between mission and role of defender.

Mission	M1	M2	M3	M4	M5
Role	Defense performer	Defense supervisor	Defense situation analyzer	Defense decision maker	Defense performer

In the forth mission (M4): making the security defense plan and attack plan based on the system situation, and gathered information.

In the fifth mission (M5): responding rapidly to attack behavior according the defense plan. At the same time, recovering the damaged system node, function, service, data. For example, repairing the server or host vulnerabilities; strengthening the system password strength; changing the service port; attack deception; constructing high-interaction Honeynet; recoving the damaged data.

Table 8. The capability level of defense supervisor.

The role's capability	Role's measurement rule	
The sensitive information count	$>=N_{Dinfo2}$	1 level
	(N_{Dinfo1}, N_{Dinfo2})	2 level
	$<=N_{Dinfo1}$	3 level
The detection efficiency	$>=V_{detec2}$	1 level
	(V_{detec1}, V_{detec2})	2 level
	$<=V_{detec1}$	3 level

Table 9. The capability level of defense situation analyzer.

The role's capability	Role's measurement rule	
The time of identifying attack threat behavior or event	$<=T_{Di1}$	1 level
	(T_{Di1}, T_{Di2})	2 level
	$>=T_{Di2}$	3 level
The success ratio of identifying attack threat behavior or event	$>=S_{Di2}\%$	1 level
	$(S_{Di1}\%, S_{Di2}\%)$	2 level
	$<=S_{Di1}\%$	3 level
The efficiency of identifying attack threat behavior or event	$>=V_{iden2}$	1 level
	(V_{iden1}, V_{iden2})	2 level
	$<=V_{iden1}$	3 level

Table 10. The capability level of defense decision maker.

The role's capability	Role's measurement rule	
The time of making defense plan	$<=T_{Dd1}$	1 level
	(T_{Dd1}, T_{Dd2})	2 level
	$>=T_{Dd2}$	3 level
The effectiveness of making defense plan	$>=M_{Dd2}\%$	1 level
	$(M_{Dd1}\%, M_{Dd2}\%)$	2 level
	$<=M_{Dd1}\%$	3 level

Table 11. The capability level of defense performer.

The role's capability	Role's measurement rule	
The completeness of security strategy set	$<=P_{Ds1}$	1 level
	(P_{Ds1}, P_{Ds2})	2 level
	$>=P_{Ds2}$	3 level
The time of responsing attack	$<=T_{Da1}$	1 level
	(T_{Da1}, T_{Da2})	2 level
	$>=T_{Da2}$	3 level
The time of tracking attack threat	$<=T_{Dt1}$	1 level
	(T_{Dt1}, T_{Dt2})	2 level
	$>=T_{Dt2}$	3 level
The number of decoying attack threat source	$>N_{Dt2}$	1 level
	(N_{Dt1}, N_{Dt2})	2 level
	$<=N_{Dt1}$	3 level
The time of restoring system function and data	$<=T_{Dr1}$	1 level
	(T_{Dr1}, T_{Dr2})	2 level
	$>=T_{Dr2}$	3 level
The way of restoring system	$>N_{way}$	1 level
	$<N_{way}$	2 level
The degree of restoring system	$>=D_{sd2}\%$	1 level
	$(D_{sd1}\%, D_{sd2}\%)$	2 level
	$<=D_{sd1}\%$	3 level

Similar with the above attack's role mapping mechanism, we establish the relations between the defense missions and defender roles, as shown in Table 7.

In the second mission (M2), the capability level of defense supervisor is evaluated by two factors, as shown in Table 8.

In the third mission (M3), the capability level of defense situation analyzer is evaluated by three factors. Table 9 shows the detail measurement means.

In the fourth mission (M4), the capability level of defense decision maker is evaluated by two factors. Table 10 shows the detail measurement means.

In the first and fifth missions (M1 and M5), the capability level of defense performer is valuated by seven indexes, Table 11 shows the detail measurement means.

4 CONCLUSION

In this paper, we build the test task assigning model of military information system based on human-roles mapping mechanism. This model can realize the flexible and efficient configuration against human roles during the system test process. In the further research, we will focus on the research of the automatically role assignment method and how to verify the effectiveness of the above model under the simulation test environment.

REFERENCES

[1] Xu bo-quan, Wang heng Zhou guang-xia. Understanding and Studying Cyberspace[J]. Command information system and technology, 2010, 1(1):23–26.

[2] United States Army Training and Doctrine Command. The United States Army's Cyberspace operations concept capability plan 2016–2028[R]. Washington, D.C.:USDOD, 2010.

[3] Zhou guang-xia, Sun xin. Study on Cyberspace Operations[J]. Command information system and technology, 2012, 3(2):6–10.

[4] Neil C. Rowe. A Model of Deception during Cyber-Attacks on Information Systems[J]. IEEE first symposium on computing & Processing, 2004: 21–30.

[5] Igor Kotenko. Multi-agent Modelling and Simulation of Cyber-Attacks and Cyber-Defense for Home-land Security[J]. IEEE International Workshop on Intelligent Data Acquisition and Advanced Computing Systems: Technology and Applications, 2007:614–620. [6]SUN Qiang, WANG Chong-jun. Simulation model of network instruction and defense using multi-Agent[J]. Computer Engineering and Applications, 2010, 46(20):122–125.

[6] Song Yi-bing, Yang Yong-tian. Role-based Agent-CGF Research[J]. Computer simulation, 2006, 23(3): 12–15.

[7] Uzing P., Mors A., Valk J., et al. Coordinating self-interested planning agents[J]. Autonomous Agents and Multi-Agent Systems, 2006, 12(2):199–218.

[8] Thomas J. Parsist, Warrn H. Debanym, Jr. Advanced cyber attack modeling analysis and visualization. AFRL-RI-RS- TR-2010-078 Final Technical Report. 2010.

[9] Huang Jian-ming, Gao Da-peng. Combat Systems Dynamics Model with OODA Loop[J]. Journal of system simulation. 2012, 24(3):561–566.

Control Engineering and Information Systems – Liu (Ed)
© 2015 Taylor & Francis Group, London, ISBN 978-1-138-02685-8

Design of Android security reinforcement system for electric power mobile terminal

M. Chen, Z.P. Shao, J. Chu & Z.J. Dai
China Electric Power Research Institute, Nanjing, China

ABSTRACT: With the development of information technology, the mobile communication technology and a variety of mobile terminals have been widely used in the Electric power corporations. As for the security access requirements of electricity power mobile terminal, this paper analyzes of the security risk and the research status and designs a security reinforcement system for electric power mobile terminal based on Android platform. This system solves the risks alone with the access of electric power mobile terminal, such as data eavesdropping, data corruption and sensitive information leakage, and improves the overall security of the electric power mobile application.

1 INTRODUCTION

STATE GRID Corporation of China (SGCC) is related to national energy security and national economy. It is an important role of economic responsibility, political responsibility and social responsibility. With the construction of smart grid and SG-ERP, the SGCC has higher requirements of confidentiality, integrity and availability for information security(Yu, 2005, Zhang 2009).

With the development of information technology, the mobile communication technology and a variety of mobile terminals have been widely used in the construction of smart grid. It brings convenience to the application system of smart grid, but also caused serious security problems. Especially in some applications, mobile terminals need to do some information exchange with application system in the internal network of SGCC. Most of this information is so sensitive and classified, that it would cause serious damage to the social order and public interests if it was destroyed or compromised. Therefore, how to ensure the safe and reliable of the access of mobile terminal into the internal network, and how to prevent the disclosure of sensitive information have become an urgent problem to be solved in the construction of smart grid.

2 BACKGROUND

2.1 Risk analysis of mobile access

Mobile technology and mobile terminals have a natural disadvantage of security.

- Difficulty for terminal regulatory. Mobile terminal has the characteristics of portability and mobility. Compared with the fixed terminal, it is easier to be lost and harder to monitor. Meanwhile, the complex environment of the use of mobile terminal Increases the risk of virus and malware.
- Public telecommunications transport network with opening and serious security risks. Mobile terminal is using the public telecommunications transport network to access the internal network of SGCC. These public telecommunications transport networks such as GSM/GPRS/EDGE/ TD-SCDMA, GSM/GPRS/EDGE/WCDMA, CDMA1x/CDMA2000, connected into Internet, are weakened by a general lack of security. The data transferred by these networks is easy to be stolen.

2.2 Current research

Currently, there are server mainstream secure access systems including Cisco's NAC (Network Admission Control), Microsoft's NAP (Network Access Protection), and TNC (Trusted Network Connect) invented by TGG (Trusted Computing Group) (DAI 2008).

According to actual needs, the SGCC has conducted a series of research in view of the access technology and the terminal protection technology. It has set server norms and standards for the terminal access, and implemented a prototype of information security access platform for mobile terminal with Windows Mobile operating system.

2.3 New requirement

With the development of mobile technology, the operating system used by the mobile terminal is no longer only Microsoft's Windows Mobile or Nokia's Symbian. Google Inc.'s Android operating system has become the most popular mobile operating system in the world. Particularly in China's smart mobile terminal market, Android has a 90 percent market share. Android is well known for its openness and customizable, that will certainly lead Android devices to be widely used in enterprises. So reinforcing the Android devices to ensure the security of Android devices access into internal network of SGCC is realistic.

3 UNDERSTANDING ANDROID SECURITY LAYOUT OF TEXT

3.1 Analyzing security mechanism of Android

Android is a Linux-based operating system. Based on the security mechanisms in Linux, android adds some system-specific security mechanisms, including application permissions, sandboxes and application signing,

- Permissions are an Android platform security mechanism to allow or restrict application access to restricted APIs and resources.
- Android uses the concept of a sandbox to enforce inter application separation and permissions to allow or deny an application access to the device's resources such as files and directories, the network, the sensors, and APIs in general.
- The Android system requires that all installed applications are digitally signed—the system will not install or run an application that is not signed appropriately (Enck 2009, Liao 2009, Fu 2011).

3.2 Main threats of Android

While the features of Android such as open-source, openness and free, have brought convenience to the developer, they have also brought a lot of security risks. Now Android is facing two main risks.

- Malware. Android's open ecosystem is more conducive to malware. The malwares will install into the devices and get the permission by defrauding the user's trust, and then they will steal the user's data and control the device without the user knowing.
- Network attacks. Those attacks often trigger a rise in scam texts and mails intended to fool users into downloading malware, so that they can destroy or seal the user's sensitive data (Wang 2011).

4 DESIGN OF ANDROID SECURITY REINFORCEMENT SYSTEM

4.1 Protection strategies

To ensure security of enterprise applications, enterprise network and enterprise data, we must make our protection strategies though a different set of eyes, or different dimension, such as management, technology, operation processes. All of physical security, network security, data security and system security should be considered.

- Physical security strategies include man-machine certification based on hardware, location services on mobile devices and hardware management. These strategies protect the mobile terminals from illegal using after been lost or stolen.
- Network security strategies include authenticate before access, network access control, protection of data confidentiality and data integrity. These strategies protect the mobile terminals from the risks while using the network.
- Data security strategies include data access control, encrypted storage and data destruction. These strategies protect the sensitive data in the mobile terminal and prevent data breaches.
- System security strategies include process monitoring and version management. These strategies protect the operating system and the application software in the mobile terminal.

4.2 System structure

The reinforcement system is designed based on the protection strategies. The main structure is shown in "Figure 1".

Configure module configures strategies for the reinforcement system. Once in a while, it can synch strategies from the Terminal Control & Monitor Center and apply the Remote strategies to the other modules such as Access Module, process manager Module and Hardware manager Module.

Authentication module collects the terminal feature information by native method calls provided via the TelephoneManager object. All those information such as ICCID, IMEI, TF number, will be stored encrypted and used as the key elements of identity authentication when the terminal is turning on and accessing into the internal network. Any Unauthorized changes of SD card or SIM card will cause the terminal not available to prevent any unsafe use.

Network Access Module uses the transparent proxy technology to monitor the network data flow of the terminal and filter packets according to the strategies. It will communicate with the Secure Access Gateway using technology of mutual authentication and key agreement protocol and

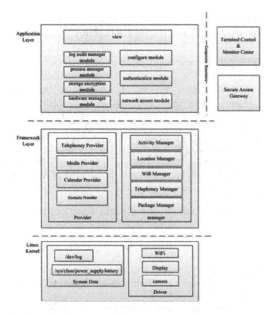

Figure 1. System structure.

then build a private data channel using technology of tunnel and encryption between the terminal and the enterprise internal servers.

The process module realizes the access control of system processes by adding mandatory access control mechanism. By combining users and processes to access control rights, it will real-time monitor files and achieve the file security access.

Storage Encryption Module will encrypt the data produced by the particular program or the files included sensitive words by Using encryption interface provided by the special SD card. It will Store those encrypted data in the special storage area that it will not lose any even if the terminal or SD card is lost.

Hardware manager Module will close specified hardware equipment such as camera, Bluetooth, WIFI according to the strategy by using the bottom hardware interface.

Log audit manager Module will collect the terminal's log and send those log data to the enterprise audit server for terminal action analyses and audit.

4.3 Operating mechanism

The reinforcement system has been running rely on the SIM card and the power special encryption safety SD card. The SIM card provides communication services for the terminal, so that the terminal can communicate with the internal servers. The SIM card is a key elements of identity

authentication as same as the safety SD card which provides Encryption Service for the sensitive data in the terminal. They will be bound so that any losing of them will not compromise security.

The full process of this system is shown in "Figure 2".

- System initialization. The system will do local identity authentication during the system initialization process and make sure make sure everything is in order. Any losing of the SIM card or the power special encryption safety SD card or the certificate password error will lead the system initialization to failure.
- Connection and identity authentication. When the system initialization is done, the system will communicate with the secure access gateway and do remote identity authentication using APN provided by the SIM card. All the communication data will be encrypted by the safety SD card. The reinforcement system will send the terminal's signature including the card number of SIM card and safety SD card, DN of the certificate, IMEI and hash of the terminal information for remote identity authentication. Only the terminal that has passed the remote identity authentication can gain access.
- Security access. When the terminal has gained access, the reinforcement system will build a private data channel using technology of tunnel

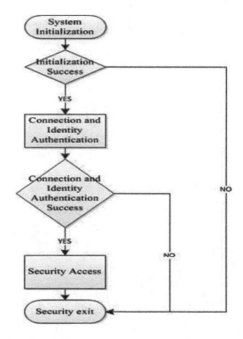

Figure 2. System process.

347

and encryption between the terminal and the enterprise internal servers. The terminal is only allowed to using this tunnel to access the internal servers. Outside access will be banned. Meanwhile, the latest security strategies will send to the terminal for synchronization. Depending on the strategies, the terminal can only access specified services, store the data in specified area and limit the usage of peripherals so that it won't reveal any sensitive information.

- Security exit. When the reinforcement system is ready to exit, it will turn off the connection with the internal servers, erase the operation records and the temporary data and hide the local encrypted data. After the system exit, the terminal can not process sensitive information.

5 CONCLUSIONS

This paper presents a design of Android Security Reinforcement System for Electric Power Mobile Terminal which will solve the risks alone with the access of electric power mobile terminal, such as data eavesdropping, data corruption and sensitive information leakage, and improves the overall security of the electric power mobile application.

REFERENCES

Dai Fei-jun, Feng Qi, Yao Li-ning, Wang Zhen-yu. Comparative Analysis of New Techniques for Secure Endpoint Access. Journal of Information Engineering University, 2008: Vol. 9 NO. 3 244–247.

Fu Yi-yang Zhou Dan-ping. Android's Security Mechanism Analysis. Netinfo Security, 2011, 10(9):19–20.

Liao Ming-hua, Zheng Li-ming. The Security Mechanism Analysis and Probe into the Solution of Android OS. Science Technology and Engineering, 2009, 26(11): 6351–6354.

Wang Kun. Android System-based Security Methods. Software, 2011, 11:68–69.

William Enck, Machigar Ongtang, Patrick McDaniel. Understanding Android Security. IEEE Security & Privacy, 2009, 7(1):53–54.

Yu Yong, Lin Wei-ming. Classified Protection of Information Security in Electric Power Industry. Netinfo Security, 2005, 2:52–52.

Zhang Wen-liang, Liu Zhuang-zhi, Wang Ming-jun, Yang Xu-sheng. Research Status and Development Trend of Smart Grid. Power System Technology, 2009, 33(13):1–11.

Control Engineering and Information Systems – Liu (Ed)
© 2015 Taylor & Francis Group, London, ISBN 978-1-138-02685-8

Deep-insight and fine-grained evaluation method of TCP performance in mobile networks

J.T. Luo, C.C. Xiang & R.R. Hu
*School of Communication and Information Engineering, Chongqing University of Posts and Telecom,
Chongqing, P.R. China*

ABSTRACT: In this paper, a fine revised evaluation method on TCP performance in mobile networks was proposed. The new method evaluates the TCP performance in its typical three stages with smaller time granularity in order to enhance the accuracy and efficiency of TCP performance evaluation. The method was tested in a Mobile Network testbed, and the results showed that the proposed method makes it possible to obtain more accurate and finer-grained analysis results than traditional methods.

1 INTRODUCTION

Currently, mobile networks have covered broadly, which provides easy access to Internet. Although the performance of Internet in the fixed-line is fairly well, in the mobile network, performance of Transmission Control Protocol (TCP) is affected due to the instability of the wireless link.

At the same time, the increasing proportion of mobile Internet data services leads to requirements for high quality of data service. Response speed of the web page, text, images and video quality are main factors to assess the Quality Of Service (QoS). Additionally, QoS is related with TCP performance. So, TCP performance monitoring and analysis could lay a foundation for improving the efficiency of packet data transmission and helping Operators to locate the cause of problems in packet network. As for TCP performance assessment methods, traditional methods are focused on overall assessment, which is not accurate enough to reflect the characteristics of data services, which has great impact on the service experience of mobile users.

In this paper, a fine revised evaluation method for TCP performance was proposed based on TCP transmission characteristics. The method aims at reducing the statistical time granularity to show features of the use of service data. In addition, a monitoring program for CDMA2000 1x EV-DO (EV-DO) was presented based on the data transmission flow of EV-DO.

2 THE ANALYSIS OF TCP PERFORMANCE INDICATORS

Before introducing our approach, the symbols used in the analysis are first listed in the following:

Key Index value of hash;
H_i Location of associated information in hash;
S_1 Previous sequence number;
S_2 Current sequence number;
S_{max} Max sequence number;
A_1 Previous acknowledge number;
A_2 Current acknowledge number;
T_1 Delay of TCP connection establishment;
T_{A1} Arrive time of previous TCP package;
T_{A2} Arrive time of current TCP package;
IP_{src} Source IP;
IP_{dst} Destination IP;
P_{src} Source Port;
P_{dst} Destination Port;
F Flag value of TCP header;
C_R Requested number of TCP connection;
C_A Acknowledge number of TCP connection;
D Direction;
R_{con} Success rate of TCP connection;
R_1 Data transmission rate of TCP in bps;
L_1 Size of previous package in Byte;
L_2 Size of current package in Byte;
Pkt_0 Up total number of packages;
Pkt_1 Down total number of packages;

2.1 Design of TCP performance evaluation

For a better statistical classification, the statistics of TCP performance are divided into the following three stages: TCP connection establishment, TCP data transfer and TCP connection release. In order to match the same TCP information, hash indexing algorithm is involved in Figure 1. Besides, the indexing key is consist of IP_{src}, IP_{dst}, P_{src} and P_{dst}.

Statistical data relies on the decoding results of raw network data and real-time TCP performance statistics depend on TCP information stored in the hash tables.

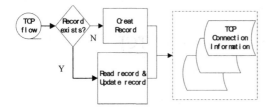

Figure 1. Hash-table query process.

2.2 An algorithm for TCP performance assessment

2.2.1 TCP connection establishment

TCP is a connection-oriented protocol, before data transmission, a TCP connection needs to establish. The process is called "three-way handshake"[4].

There are some important indicators during the connection establishment: C_R, C_A, T_1, R_{con}. The detailed procedure is listed in the following.

STEP 1: $Key = \{IP_{src}, IP_{dst}, P_{src}, P_{dst}\}$
//Get information of the TCP connection
IF $\{F = 0x02\}$ THEN
$H_i = GetTcpInfo(Key, TcpInfo)$;
$UpdateInfo(TcpInfo)$;
//Get and update TCP information in the hash-tables
STEP 2: IF $\{H_i\ ! = NULL\}$ THEN
$C_R = C_R + 1$;
ELSE $H_i = AddTcpInfo(Key, TcpInfo)$;
$C_R = 1$
//if TCP connection does not exist then build a new record
STEP 3: IF $\{F = 0x10\}$ THEN
$H_i = GetTcpInfo(Key, TcpInfo)$;
STEP 4: IF $\{H_i\ ! = Null\ \&\&\ A_1 = S_1 + 1\ \&\&$
$S_2 = A_1\}$ THEN
$C_A = C_A + 1$;
$Update(TcpInfo)$;
$T_1 = T_{c2} - T_{c1}$;
STEP 5: $R_{con} = C_A \div C_R$;
STEP 6: {Get the next TCP packet and go to STEP 1};

During STEP4, If previous acknowledge number, sequence number and current sequence number match the conditions: $A_1 = S_1 + 1$ and $S_2 = A_1$, then a TCP connection is established.

2.2.2 TCP data transfer

Due to the complexity of application protocols and large numbers of data over TCP, communication parties need to consult Maximum Transmission Unit (MTU). Because of this, there are so many fragmented packets during data transmission. In order to ensure the accuracy and completeness of data transmission, TCP uses sequences number and acknowledge number to determine the data packets is continuous or not.

Besides, during data transmission TCP window control is applied to increase the transmission speed and lower the retransmission rate.

The statistical approach is refined for a better representation of data usage. For example, excluding silence period during data transmission could reduce the mistake caused by long-term statistics. The detailed procedure of data transmission period is listed in the following:

STEP 1: $Key = \{IP_{src}, IP_{dst}, P_{src}, P_{dst}\}$
$H_i = GetTcpInfo(Key, TcpInfo)$;
//take upstream process as an example
STEP 2: IF $\{H_i\ ! = Null\ \&\&\ IP_{src} = Key.\ IP_{src}\ \&\&$
$P_{src} = Key.\ P_{src}\}$
THEN $D = 0$; //0: Up packet
$PKT_0 = PKT_0 + 1$;
$Update(UpInfo)$;
ELSE $D = 1$ //1: Down packet;
$PKT_1 = PKT_1 + 1$;
$Update(DownInfo)$;
STEP 3: IF $\{S_1 < S_2\}$ THEN
$S_{max} = S_2$;
ELSE $PKT_{retr0} = PKT_{retr0} + 1$;
STEP 4: $R_{retr0} = PKT_{retr0} \div PKT_0$;
STEP 5: IF $\{D = 1\ \&\&\ L_2 > 0\}$ THEN
$R_1 = L_2 * 8 \div (T_2 - T_1)$;
//refining the analysis of transmission rate
ELSE IF $\{D = 1\}$ THEN
$R_1 = 0$;
STEP 6: {Get the next TCP packet and go to STEP 1};

Transmission packet is judged by sequence number in every packets, if sequence number of current packet matches the conditions: $S_2 \geq S_{max}$, the packet must be a transmission packet. In addition, interval between two packets is used to calculate transmission rate which improve the accuracy of rate at the cost of computer memory.

2.2.3 TCP connection release

Similar to the process of TCP connection establishment, connection release phase takes 4 steps, which can be called "Four-way handshake". The statistical method goes the same way with TCP connection establishment phase, so the specific process needs no further elaboration.

3 EXPERIMENTAL ENVIRONMENT AND RESULT

3.1 EV-DO testbed

The experimental testbed used for the EV-DO monitoring system is illustrated in Figure 2. In the EV-DO test network Packet Control Function (PCF) and Packet Data Serving Node (PDSN) are necessary network elements for packet service.

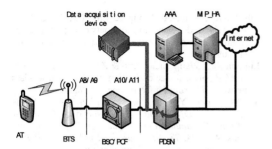

Figure 2.　EV-DO testbed.

Table 1.　Information of TCP connection.

Statistics field	Connection results
Source IP & Port	192.168.3.87: 41134
Destination IP & Port	58.217.200.14:80
Num of up packets	5618
Num of down packets	9402
Num of up bytes	335.9 KByte
Num of down bytes	13.430 MByte

Table 2.　Time statistics.

Statistics field	Connection results
Connection start time	16:48:37.00
Connection end time	16:54:57.04
Duration of connection	06:20.03
Duration of establishment	1.002s
Duration of release	1.295s

Table 3.　Retransmission rate statistics.

Statistics field	Connection results
Up retrans-packets	0
Down retrans-packets	583
Up retrans-rate	0
Down retrans-rate	0.0620

TCP performance monitoring is based on the data of A10 and A11 interface (R-P interface).

3.2　Experimental results

During experimental test, we use a smartphone to simulate the process of surfing the Internet with the aid of WiFi hotspot. The exchanged data is captured for further analysis, and the results are summarized in Table 1 to Table 3.

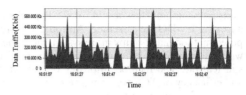

Figure 3.　Fine traffic statistics.

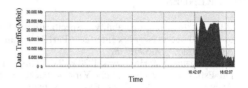

Figure 4.　Overall traffic statistics.

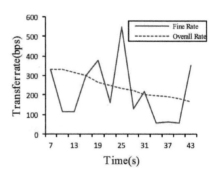

Figure 5.　TCP download rate statistics.

As it showed in Table 1, the statistics is a connection to a download website including the downloading process. The detailed connection information was summarized in Table 2. Then in order to simulate the unstable wireless environment, The phone was taken far away from the hotspot to test the statistic of retransmission packet. The results are listed in Table 3.

The accuracy of transfer rate is particularly important. In this paper, interval of every packet transmission is regarded as the minimum statistical time granularity.

As it showed in Figure 5, result of fine performance evaluation is more accurate. Especially, when the user holds the TCP connection without any data interaction, the statistic rate should be close to zero. So the overall rate assessment may ignore it and make the statistical rate inaccurate.

4　SUMMARY

In this paper the fine revised evaluation method on TCP performance is illustrated in detail. And then measurements and test results of TCP performance over EV-DO are presented. Although the proposed method may occupy more system memories, with the help of 64-bit computers, the method is still feasible. The method improves the accuracy and efficiency of TCP performance monitoring at the cost of system memories.

REFERENCES

3GPP2. 2004. Interoperability Specification for cdma2000 Access Network Interfaces-Part 7 (A10 and A11 Interfaces). *A.S0017-B*.

Goto, T., Tagami, A. & Hasegawa, T. 2009. TCP Throughput Estimation by Lightweight Variable Packet Size Probing in CDMA2000 1x EV-DO Network. *9th Annual International Symposium on Applications and the Internet. Bellevue, 1-8, July 2009*. USA.

Liu Yu-Jun, Fang Ai-Zhong & Liu Xiao-Bing. 2005. Analysis and realization of network performance monitoring system. *Computer Measurement & Control* (15):692–694.

Liu Xiao-bo, Wang Sheng-Kun & Zhang Jun-Wei. 2007. A TCP performance evaluation of MIPv6 in wireless network. *Information Technology* (7):98–99.

Ma Wen-Feng, Yi-Qiang & Ma Yi-Fei. 2002. Mobile IP Based CDMA2000 Packet Network. *Journal of Militart Communications Technology* (23):31–34.

Stevens, W. 2004. *TCP/IP Illustrated Volume 1: The Protocols, 2nd ed*. BeiJing:China Machine Press.

Wang Shi-Long & Luo Jiang-Tao. 2009. IP session statistic over time intervals based on Hash algorithms. *Digital Communication* (1):72–75.

Wang Huan, Yao Yuan-Cheng & Zhang Ying. 2007. Active-measurement and analysis of end-to-end TCP-throughput for different background stream. *Computer Engineering* (33):92–93.

Control Engineering and Information Systems – Liu (Ed)
© 2015 Taylor & Francis Group, London, ISBN 978-1-138-02685-8

Research and integration of a web-based distributed system

Y.G. Dai, T. Xu, J.S. Xiao & L.Q. Ou
Key Laboratory of China's National Linguistic Information Technology,
Northwest University for Nationalities, Lanzhou, China

ABSTRACT: Web technology can gather information of characters, pictures, audios and videos together. And it can be also used in navigation, which is unrelated to the platform. The characteristics of distributed technology include the following: It can satisfy the needs of lots of users at the same time; distributed web servers can store a lot of resource while a single web server cannot. The problem that a system integration has to solve is to better integrate the achievements that a project has made. A system integration should be widely used, should be in great need, and can store a great number of resource. And the Web based distributed system is a good one.

1 INTRODUCTION

World Wide Web (Web) is a technology that develops with the popularization of the Internet. Its invention facilitates people's organization and access to the source of the Internet [1]. The use of the Internet/Intranet based WEB has been more and more perfect and popular, from the initial HTML static text page to CGI ISAPI/NSAPI ASP/JSP technology. One of the important reasons that why Web is very popular is that it can display pictures and texts of all colors on the same page. The information of the Internet was only in texts before Web came into use. Web can gather information of characters, pictures, audios and videos together. And Web can be used in navigation easily. Just by changing from one link to another, you can browse on all pages and websites. In addition, Web is distributed. A number of pictures, audios and videos take up much disk space. We cannot even foresee the amount of the information. It's unnecessary to place all information together. Information can be place on different websites. But it's necessary to display the websites on the browser. Thus information that is physically different is unified logically. And it is unified to the users.

A distributed system is an assemblage of several independent computers. These computers are like single related systems for users. On the one hand, as for the hardware, the computer itself is independent. On the other hand, as for the software, users want to work on a single system. In a word, if the parts of a system are limited on the same place, the system is a centralized one. If its parts lie on different places, it is a dividual one.

2 BACKGROUND INFORMATION

2.1 The development of the web based distributed system

With the development of Web, the distributed system has been the main choice to the users. For example, the Electronic Commerce systems, the grid system, P2P share system, the intelligence system of enterprises' dynamic alliance are distributed systems. And the appearing of loud computing has highlighted the advantage of distributed systems. Cloud computing is the development of distributed computing, parallel computing and grid computing. In other words, cloud computing is the commercialization of these computer concepts. The basic principle of cloud computing is to distribute computing on lots of distributed computers, not on local computers or remote servers. The operation of enterprises' data centers is similar with the Internet. Thus enterprises can switch its resource to the application in need and access computers and storage system accordingly. The distributed Web system is main stream of system integration technology and also the orientation of system integration and its solution.

2.2 The research background

The research background of this thesis is Minhan distance education project. To better integrate the achievements that the project has made and to popularize the research achievement of this thesis, the system will build an integrated and jointly operated platform. The platform will take in the achievements of all the branch systems and operates jointly. And it will commercialize the service and the management, live and develop independently, and display the latest development of Minwen technology and the platform.

During the research, lots of research data is obtained as well as visual research achievements including characters, pictures, sounds and videos. As a result, a vivid display platform is needed to display the research achievements, thus the achievements can facilitate national education. However, the capacity of the research achievements is huge, especially video resource which takes up lots of storage space. But video resource can vividly show the achievements to people, especially to minority people. So it's impossible to store all the resource in a server, and a distributed system is needed. During the process of building the platform, the distribution includes Web server's distribution problem and multi-user distrtion.

3 DISTRIBUTED SYSTEM AND INTEGRATION RESEARCH

3.1 *Using B/S mode at clients*

Compared to C/S exploiting mode, B/S has more advantage. For example, exploit distributed client system by using C/S exploiting mode. Resolution is the first problem to be solved, because the sizes and the resolutions of the screens used by the clients could be different. The screen will display abnormally if we use the same software interface with the same resolution. However, there will not be such a problem if B/S exploiting mode is used. The problem can be simply solved just by setting the properties of the webpage on different computers.

However, another problem will occur if the B/S mode is used. That is the compatibilities of browsers of different versions are different. The distributed clients have to be compatible in IE Firefox opera safari browsers as well as IE7 IE8 IE9 IE10, different versions of the same web browsers. Therefore, this thesis tries to find out ways to solve the compatibility problem. For example, the following codes are some of the codes in the project to solve the problem.

function is IE (){return navigator. app Name. index Of ("Microsoft Internet Explorer")! = -1 && document. all;}
function is IE 6 () {return navigator. User Agent. Split (";") [1]. To Lower Case ().index Of ("msie 6.0") = = "-1"? false: true;}
function is IE 7 (){return navigator. User Agent. Split (";") [1]. To Lower Case (). Index Of ("msie 7.0") = = "-1"? false: true;}
function is IE 8 (){ return navigator. User Agent. Split (";") [1]. To Lower Case (). Index Of ("msie 8.0") = = "-1"? false: true;

By judging and disposing the codes above, we can make the system be compatible in different browsers. But the versions of different browsers are updated constantly. When the browser is updated and the codes are not compatible in the browser of new version, the distributed clients will not work. So we have to issue plugins to solve the compatibility problem during the process of exploiting the new system.

3.2 *The advantage of the distributed web*

A distributed system is a system that organizes the computers on network and communicates and coordinates through message passing. It is of high scalability [6].

The purpose of organizing a distributed system should be definite. It's foolish to organize a distributed system just because we are able to do that. The main purpose of a distributed system is to enable users to access long-distance resource and share the resource with other users in a controlled way. The resource could be everything. It can be a text, a picture, a voice, a video and so on.

The advantage of using the distributed Web is mainly its layered structure. Its layers are on different machines. Thus the business of a single computer is reduced. And the computer works in a high speed. The layered structure of a distributed system is as shown in Figure 1.

The Web distributed system is made of three parts: one is the user interaction; the second is the operating system or the data base; the third part is the middleware that includes the core function of application systems. The middleware belongs to the processing layer. It is different from user interface and data base. And it has little in common with other parts on the processing layer. If the special effects of the pictures and pages are

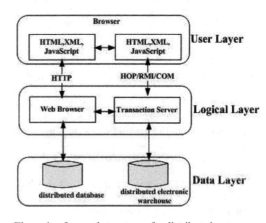

Figure 1.　Layered structure of a distributed system.

354

neglected, the user interface of the search engine is very simple: users can get the information that they want by inputting keywords. A database stores Web pages, and these Web pages can be taken out and indexed in advance. The core of the search engine is to convert the keywords to one or more data query. The data query lists the query results after finishing inquiring, and then convert the list to a group of HTML pages. Query requests can be from different clients, and the browsers that answer the requests can be on different machines.

From the above discussion, we can draw that a distributed system is the result of dividing the application programs to user interface, processing part and data layer. Each layer corresponds to the logic structure of the application programs. In many business environments, distributing processing is to organize the client browser application programs to a multi-layer structure, which is called longitudinal distribution. But as for the using distributed system, the client browser application programs that uses longitudinal distribution can not make a large scale system, and transverse distribution is needed to distribute the clients and browsers physically and to copy them on different computers. World-Wide-Web is a successful example of using transverse distribution.

4 THE CASE ANALYSIS OF WEB-BASED DISTRIBUTED SYSTEM

The case root in National science and technology support project, "Minority language information processing of common key technology research and demonstration application", and its task "Tibetan/Chinese distance education resources in the key technology research and demonstration application platform". The case is to exploit a Tibetan-Chinese bilingual remote education platform, based on the present situation of the remote education technology in Tibetan areas and the Chinese remote education technology system. Figure 2 is the first page of the platform.

There is lots of education resource of different forms like characters, pictures, voice and videos on the platform. Of all the resource on the platform, the Hd adult education video is of high capacity and great content. It was recorded by 15 teachers from Institute of Tibetan in Northwest University for Nationalities. And it took them 2 years to finish the video. Based on this situation, the case uses a Web-based distributed project to publish server. The Infrastructure Service of the platform is made of three parts: domain name resolution service, Web service and data storage service. And domain name resolution service is provided by domain name resolution service provider. And Web service

Figure 2. The first page of the platform.

and data storage service is set up by the project group. Web service is provided by an IBM tower-type browser. Its number is IBM 39Y8321/100A. The project group thinks that such a Web browser can meet the query needs at the begging. Data base service is provided by two minicomputers. Their model numbers are IBM P55A. When the platform is visited, the working process of the service is as following: domain name resolution browser gets the IP address of the Web browser by analyzing de domain name; Web browser answers users' inquiries through the TOMCAT browser, helps users to create processes, and goes back to the Web page. If a user submits a request again after browsing a Website, for example, the user wants to watch a video course; the Web browser will visit data base storage and call the video course back for the user.

In a word, this case uses a Web-based distributed system to a great extent. Compared with some previously deployed projects, its access speed is distinctly increased.

5 CONCLUSION

The distributed system can be found everywhere. Internet enables users all over the world access the service of the Internet. It is of profound importance for people in the remote minority areas. Distance education has to work out the problem of large visiting quantity at the same time and the problem of storing a large number of high definition video. The distributed Web and distributed data storage are the solutions. The Web-based distributed system has to get used to different hard wares, operating systems, different browsers and browsers of different versions. The Web-based distributed system can be the most favorable solution to solve the problem of system integration. With the development of cloud computing, the Web-based distribution will enjoy a greater development.

ACKNOWLEDGMENTS

The authors would like to thank Key Lab of China's National Linguistic Information Technology, Northwest University for Nationalities sharing their Plateform of Distributed on Computer.

This work is supported by National science and technology support project under Grant NO. 2009BAH41B07, the central college funds under Grant NO. ycx13014.

REFERENCES

[1] Gao Xiaoqin, Liu Guoxin. "Foreign Research Status of Enterprise Distributed Innovation"[J]. Journal of Wuhan University of Technology (Information & Management Engineering), 2009:253~255.

[2] Andrew S. Tanenbaum Maarten van Steen. "Distributed SystemsPrinciples and Paradigms"[M]. Tsinghua University Press. 2012:12~44.

[3] Tom White. Hadoop: The Definitive Guide. O'Reilly Media, Inc. 2009.

[4] Michael Miller. "Cloud Computation"[M]. China Machine Press.

[5] George Coulouris Tim Kindberg. "Mechanical industry publishing house design concept and distributed system"[M]. China Machine Press. 2004:28~40.

[6] Luo Jun-zhou, Jin Jia-hui, Song Ai-bo, Dong Fang. "Cloud computing: architecture and key technologies" [j]. Journal on Communications. 2011(07):3~20.

Control Engineering and Information Systems – Liu (Ed)
© 2015 Taylor & Francis Group, London, ISBN 978-1-138-02685-8

Research on Wireless Power Transmission for fractionated spacecraft system—*a review and some perspectives*

Y. Cheng, X.Q. Cheng & Y. Zhao
College of Aerospace Science and Engineering, National University of Defense Technology, Changsha, China

ABSTRACT: Wireless Power Transmission (WPT) technology is a new power transmission strategy which has an important application prospect in space missions. Taking fractionated spacecraft as its application object, this paper investigates and elaborates the latest advancements and key technologies on wireless power transmission including microwave method, laser method, magneto-inductive resonantly coupled mode, concentrated sunlight method, and so on. Principles, characteristics, difficulties and feasibility of these methods have been discussed in details. Furthermore, available approaches of wireless power transmission have been proposed for fractionated spacecraft and the application prospect has been described. It is supposed that this paper will benefit for future research and development in the field of wireless power transmission.

1 INTRODUCTION

The traditional space power system powers every single satellite by solar arrays and storage battery independently. Wireless power transmission technology, as a new power transmission strategy, has unique advantages for space missions that can avoid using large solar arrays, storage battery and shadow among satellites, extending the satellite's service life (Patel 2005). In 2005, a new concept of fractionated modular spacecraft system was proposed by the American scientist Owen Brown. The fractionated spacecraft system decomposes monolithic spacecraft into a set of separate heterogeneous free flying modules which incorporate various payload and infrastructure functions. They interact with each other through wireless communication and wireless power transmission, performing as a virtual holistic spacecraft (Mathieu, C. 2006). In 2007, the Defense Advanced Research Projects Agency (DARPA) embarked on a program, entitled System F6 (Future Fast, Flexible, Fractionated, Free-Flying Spacecraft united by Information exchange spacecraft), to develop, mature and demonstrate on orbit the key enablers of spacecraft fractionation (Mathieu 2006, Maciuca & Chow 2009). Wireless power transmission, as one of the key technologies in F6, has low technology maturity but is of great value, which can reduce the complexity of every single module, enhance their viability and extend the system agility. All of these determine the success of F6 Program to some extent. Aiming at enhancing the nation's space fast response ability and accelerating the breakthrough of key technologies, this paper investigates several methods about wireless power transmission then puts forward some technical solutions, laying foundation for the development of fractionated spacecraft system.

2 THE REALIZATION OF SPACE WIRELESS POWER TRANSMISSION

Wireless power transmission has been widely investigated for different applications in recent years. Methods such as RF, magnetic induction and concentrated sunlight for WPT have appeared in different fields, but totally immature for applications (Shinohara 2011). As a heterogenous spacecraft system, modules in the fractionated spacecraft clusters get necessary energy by wireless power transmission to maintain their own need. For different missions and safety consideration, modules need to keep distance at ten to ten thousand meters from each other with the average power consumption on hundreds to thousands watt per module. That's to say, wireless power transmission technology for space application must be able to transmit hundreds to thousands watt power in the range of 10 m~10 km. In the following paragraphs, several methods of WPT for fractionated spacecraft system have been discussed in detail.

2.1 *Microwave Power Transmission*

Microwave Power Transmission (MPT) is the transmitting and receiving of microwave energy in space, essentially. For space to space applications, system of MPT comprises three parts: the transmitter, the transmission in space and the receiver. A series of issues were involved in the system, including transmission technology, frequencies selection, transmitting/receiving antenna aperture sizes, transmission range, transmitting source technology, rectifying antenna configuration and microwave beam shaping (Lafleur & Saleh 2008).

In MPT, the original energy is from the sun. Solar energy is converted to DC which in turn is converted to microwave energy by microwave generator. The transmitting antenna, either omni antenna or directional antenna, is used to beam the microwave energy through free space to rectifying antenna located on the mission spacecrafts. Phased-array antenna and parabolic reflecting antenna are suitable options for different applications but the aperture size is constrained by the modules. The antenna size is about 10–50 m in diameter for large central resource module and ~1 m for small mission modules (Jamnejad & Silva 2008). The transmission frequency is determined by multiple factors, including available spectrum, range, antenna size constraints, the power level and the component maturity. The main frequencies of interest for fractionation spacecraft are S-band (2.45 GHz), C-band (5.8 GHz), X-band (8.51 GHz), Ka-band (35 GHz) and so on. The higher the frequency, the smaller the antenna aperture sizes. The selection of transmitter tube must be conditioned on high efficiency, high output power, low noise, acceptable voltage, adequate gain to minimize driver requirements and long life (Little 2000).

The microwave power beam transmission efficiency in free space is defined by Friss transmission equation as follows (Jamnejad & Silva 2008):

$$\eta = \frac{P_r}{P_t} = \frac{G_t G_r}{L_{fs}} = \frac{\lambda^2 G_t G_r}{(4\pi D)^2} = \frac{A_t A_r}{(\lambda D)^2} \qquad (1)$$

where P_r, P_t, G_r, G_t, A_r, A_t, λ and D are receiving power, transmitting power, the gain of transmitting antenna, the gain of receiving antenna, aperture area of transmitting antenna, aperture area of receiving antenna, wavelength and distance between transmitting antenna and receiving antenna, respectively. L_{fs} is the path loss in free space, determined by the transmission frequency and the range.

The relationship in eq. (1) can be used to estimate proper antenna sizes for different transmission efficiencies at the given transmission distance and efficiency. Table 1 show power beaming parameters for 5.8 GHz and 35 GHz power beaming systems at 30% and 80% collection efficiency for the transmission ranges both 300 m and 1000 m, with a 3 m diameter transmitter antenna (Little 2000). From the left tables, we can conclude that high transmission frequency is very essential for the system to improve the transmission efficiency and minify the rectenna size.

Furthermore, with given distance and aperture size, the amplitude and phase distribution on the transmitting antenna aperture can be optimized to achieve maximum transmission efficiency. In reflector antennas this can be achieved by defocusing the feed. In phased arrays this can be achieved by changing the phase of the individual elements of the array. Rectifying antenna is the key component of the MPT system which efficiently converts microwave energy to DC power. The core of the rectifying antenna is schottky barrier diode. To obtain the high efficiency across the entire collecting area, it is important to keep the energy at each diode sufficiently high by concentrating the incident flux. The patch rectenna is considered to be a near optimal solution for efficient capture of incident microwave energy. The imagination of future MPT subsystem for fractionated spacecraft system is shown in Figure1.

Table 1. 5.8 GHz and 35 GHz power beaming system*.

Collection efficiency (%)	Transmission frequency (GHz)	Rectenna diameter (m)	Output DC power (W)	DC to DC efficiency (%)
Range = 300 m				
30	5.8	4.1	461	15
	35	0.68	237	7
80	5.8	8.5	1196	40
	35	1.42	600	18
Range = 1000 m				
30	5.8	13.7	338	11
	35	2.26	150	5
80	5.8	28.4	865	28
	35	4.7	335	11

* Transmitted RF Power = 1.9 kW for 5.8 GHz and 1.1 kW for 35 GHz, 10 dB Gaussian Taper.

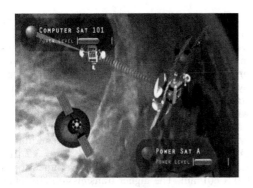

Figure 1. Imagination of Wireless Power Transfer by MPT.

Lafleur & Saleh (2008) put forward a method of utilizing microwave power beam for wireless power transfer inside the fractionated spacecraft systems. Despite some ideal assumptions such as the continuous microwave power transfer, no pointing losses, no space losses and so on, only 6% of the design space satisfy the 250 W need for small satellite power constraint. The great majority of available designs lie in high transmission frequency (>33 GHz), large antenna diameter (>0.93 m) and close range between satellites (<740 m).

Vahraz Jamnejad and Arnold Silva in Jet Propulsion laboratory also did broad scale researches on microwave power beam in a fractionated spacecraft environment. Various issues have been discussed such as the type of both transmitting antennas and rectifying antennas, the mechanisms of redirecting and refocusing the power beam, antenna aperture sizes, separation distances, transmitting frequency, and power levels. According to their study, beam shaping can be used to optimize the transmission efficiency and minimize the interference and the power losses through sidelobes (Jamnejad & Silva 2008). At last, they proposed a series of available technologies such as defocusing phase arrays, reflector systems, reflect arrays and multi-frequency to improve the transmission efficiency and increase the transmission distance.

2.2 Laser method

Laser is famous for its directing property, energy concentration and low losses in vacuum. Thus it can be used in space wireless power transmission. Laser WPT system is composed of laser pulse modules and laser receiving modules. The former is used to generate and transmit laser by DC power conversion or solar-pumped lasers. While, the latter is used to convert high-intensity laser to DC power, generally by photovoltaic cells. Solar-pumped laser

becomes a hot research in recent years because it doesn't need large and complex structures. Power system under this approach will be of simple composition, low mass, low cost and high efficiency. At present, many laser medium have been demonstrated in solar-pumped laser including Nd:YAG, Cr/Nd:GSGG, Cr/Nd:YAG ceramic, alexandrite, ruby, and so on (Hwang 1991, Raible et al. 2011). The highest laser output power can reach ~kW level and the direct conversion efficiency of solar energy to laser is nearly 20% when using solar-pumped lasers.

At receiving terminal, the wavelength and power intensity of laser beam have great influence on the conversion efficiency of photovoltaic cells. The highest receiver power density achieved to date is using a single vertical multi-junction photovoltaic cell with spectrum in the vicinity of 800~1000 nm, which can provide an exceptionally high electrical output of 13.6 W/cm^2 with 24% optical-to-electrical conversion efficiency (Lu 2002, Raible et al. 2011). An imagination of using laser for wireless power transmission in space is as follows: collect and focus the solar energy with large reflector arrays, send it to the laser oscillators to generate coherent laser beam from the terminal coupling mirror, then transmit it to the photovoltaic cells on the receiving spacecraft. Necessary components for the system are large solar energy collecting structures with high mass-area ratio, laser oscillator, low loss reflector and large cold spectral filter. The transmitting power is determined by the areas of solar energy collecting structures. The temperature and efficiency of the laser system are determined by thermal radiation rate and the whole conversions. Compared to other large space lasers, this design of laser system for wireless power transmission can work in low temperature with high efficiency constantly (Lu & Fan 2002). Also using a laser-based power transfer system can increase the power available to ion or electrical propulsion system.

Zhao & Zhu (2012) proposed an external optical feedback laser power transmission system, which can be used for wireless power transmission in space. The key design in this system is that besides photovoltaic cells, it place a partially reflecting mirror at the power receiver that return a portion of the laser beam to the power transmitting terminal. By choosing the proper reflectivities of the front and the external mirrors, it can reduce the threshold current, raise output power and improve power transmission efficiency. In addition, under the external optical feedback the system can automatically adjust the laser intensity according the pointing accuracy between transmitting and receiving terminal, maintaining safe and efficient power transmission among sub-modules in fractionated spacecraft systems.

2.3 Magneto-inductive resonantly coupled mode

Previous investigations about wireless power transmission using microwave or laser are far-field technologies. They are suitable for long distance transmission but the transmission efficiency is very low at present. Recently some experts restart researches on mid-range magneto-inductive resonantly coupled wireless power transmission, which was a potential approach for close range fractionated spacecraft. In MRC, power transferred among resonator of the same resonance frequency in the form of magnetic or electric coupling. Thus it won't be interrupted by non-magnetic barriers inside the clusters and has considerable transmission efficiency in mid-range.

The primary circuits of MRC system, in its simplest form, can be treated as the inductively coupled air transformer model as shown in Figure 2.

In Figure 2, U_1, U_2, L_1, L_2, R_1, R_2 and L are the equivalent voltage of driving source, inductance and resistance of the primary windings, secondary windings and load, respectively.

When the circuits is in resonance, the transmission efficiency is defined as in eq. (2), which can be used to choose proper parameters to improve the transmission efficiency (Fu et al. 2009, Masayoshi et al. 2010).

$$\eta = \frac{P_{OUT}}{P_{IN}} = \frac{I_2^2 R_L}{I_1 U_1}$$
$$= \frac{(\omega M)^2 R_L}{(R_2 + R_L)[(R_2 + R_L)R_1 + (\omega M)^2]} \quad (2)$$

where, $M = \pi \mu_0 r^4 n / 2D$ is mutual-inductance. And μ_0, r, n and D are vacuum magnetic conductivity, radius of windings, number of windings and range between primary windings and secondary windings.

Obviously, for mid-range power transmission by this method, range is the main factor in determining the transmission efficiency but some other factors also can be regulated to improve it, such as frequency and winding parameters.

To research this WPT technology and verify the above analysis, we design a MRC system which works at different frequencies with different windings. The prototype of the windings is made of printed plate (Fig. 3). The input source is instead by class C power amplifying circuit. The resonance state is regulated by matching capacitance with windings.

The maximum efficiency obtained in the above system is about 20% on the following conditions: D = 65 mm, f = 2.0 MHz, n = 6, L_1 = 3.2 μH, L_2 = 3.2 μH, R_L = 10 Ω. The winding dimension is (40 × 60) mm. The higher transmission efficiency relies on high frequency and large winding structure as well as optimal winding parameters.

In literatures (Simon 2009, Tan et al. 2010), wireless power transmission based on Mid-range magneto-inductive resonantly coupled technology for space application had been studied. Several prototype systems were designed, fabricated and tested. Characteristics of four solenoid coils and two planar spiral coils such as self-resonance frequency, quality factor and efficiency at different distance had been modeled and analyzed. An empirical law for MRC efficiency vs. approximately distance as $D^{-4.2}$ was concluded according to experiment results (Simon & Langer 2009). According to the authors, care must be taken to maximum efficiency with respect to magnetic field axis angular alignment among resonant structures to avoid geometry where the mutual inductance is low. They also pointed out that large resonant structures would have longer transmitting range. At last, they proposed one possible approach applying MRC to response space systems. In this approach, inter-bus and intra-bus architectures were put forward and three kinds of power control ways were analyzed, including centralized control, De-centralized control and hybrid control. Because the fractionated spacecraft system requires instantaneously high

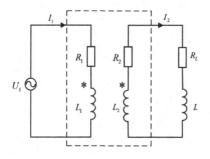

Figure 2. Circuit model of coupling windings.

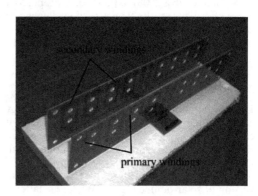

Figure 3. Prototype of the windings for experimental device.

power at long distance over ten meters, the MRC is more suitable to provide supplementary power for spacecraft at close range using a hybrid control scheme, especially providing emergency power or resurrect a non-operational spacecraft through orbital maneuvering. The imagination of future MRC systems using inter-bus hybrid control is shown in Figure 4.

2.4 Concentrated, unconverted sunlight

So far, a monolithic spacecraft makes its own energy by the use of photovoltaic power generation which need deploy expensive solar arrays, typically more than $750/W. For fractionated spacecraft system, previously investigated about power distribution schemes using lasers or microwaves have low efficiency under 10% due to multiple energy conversions. Under the contract of DARPA, researchers in Space Systems/Loral (SSL) proposed a new method for the power subsystems of F6, named concentrated, unconverted sunlight (Turner 2006). In fractionated spacecraft system, one spacecraft module named "mother ship" or resource ship directing to the sun collected sunlight centrally then reflected and distributed high-intensity beams of concentrated sunlight to other spacecraft modules which named "daughter ship" or mission ship through reflecting mirror on the resource ship.

Wireless power transmission for fractionated spacecraft system through unconverted, concentrated sunlight is shown in Figure 5. The RV collects solar energy by a big primary mirror then the sunlight is concentrated into 1~3 m diameter optical beam with intensity of ~100 suns. In order to reduce the beam divergency to provide a narrower beam or beam of the same width at longer distance, it would require a secondary mirror abroad the RV significantly larger than the primary. The secondary curved mirror must be able to move

Figure 4. Imagination of Inter-bus Hybrid Control WPT.

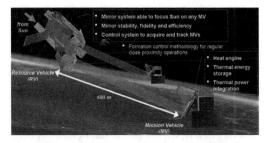

Figure 5. The RV supplies power to multiple MVs.

on the order of centimeters to maintain optical beam focusing on the multiple target MVs (Turner 2009).

Mission ships received and converted high-intensity sunlight using a Stirling cycle heat engine and Thermal Energy Storage (TES), including a Phase Change Material (PCM) and a working fluid. The phase materials which provides heat energy to heat engine in the thermal energy storage can store and release heat from the TES without a significant change in temperature. Considering the heat losses, the energy storage capacity of PCM is twice more than a rechargeable battery. The working fluid kept in liquid form to assure that heat is distributed evenly throughout the energy storge system to increase the efficiency of PCMs.

Before reaching to the MVs, the sunlight required 3 or 4 reflections, thus high reflectivity is very essential for the mirrors. Multiple layer coating mirrors can be used to reduce the effect of the atmosphere and resolve the tarnishing problem, providing 98~99% reflectivity across the entire solar spectrum (Turner 2009). The usual coating materials are sliver and aluminum. Because of no multiple energy conversions, the efficiency of sunlight transmission is more than 90% from RV to MVs. At the receiving end aboard the MV, the efficiency of the Stirling cycle dynamic heat engine for energy conversion is about 30%. Thus, the overall efficiency of concentrated, unconverted sunlight power transmission is 25% or much higher, more than five times as high as that for space-qualified microwave or laser systems.

In fractionated spacecraft system, all the MVs get concentrated sunlight from RV only a few percent of orbital period to support the peak power need for their payload at different time. During eclipses or lack of concentrated sunlight, MV gets power from its own thermal energy storage or reorientating itself to the sun with the receiving mirror to support a minimum survival mode. In case of emergency, the MV can obtain a small amount of energy from the heat trap to preserve itself if its body directed at the sun.

In this way, the RV which concentrated the high-cost power equipment, collected and distributed the solar energy in the form of concentrated, unconverted sunlight then transmitted it to MVs of low-cost which absorbed the sunlight with TES, heat engines and heat traps. This new method avoided high-cost, low efficiency and complex deployable solar arrays aboard all the spacecraft modules. Moreover, it enchance the stability and flexibility of the fractionated spacercraft system, reduce their cost and can support multiple MVs with one RV. More importantly, efficiency of this type WPT was hardly affected by transmission range in space. Thus it can realize wireless power transmission over a long distance with high efficiency. So this new technology got the funding from DARPA and would be an available WPT for the future fractionated spacecraft systems.

2.5 Other methods

Besides all the ways mentioned above, there are some other methods to realize wireless power transmission in space such as ultrasonic and space solar energy station. Ultrasonic is famous for its strong direction and large power transmitting capacity that can be used in mid-rang WPT. Ultrasonic wireless power transmission is based on piezo-electric effect and inverse piezo-electric effect of piezo-electric material, which realize the reversible conversion of electrical energy to mechanism energy in turn (Bai 2011). In contrast to MRC, ultrasonic wave has no Electromagnetic Interference (EMI) to equipment. And the output voltage of the receive transducer attenuating with transmitting range can be optimized by redesigning the directivity of the transducer. But this method has just started and need to be further studied. Solar energy satellite firstly introduced by Dr Peter Glaser which can be used to compose Space Solar Energy Station (SSPS) to collect solar energy in high orbit (For example, we can use the net of Galileo satellites). By multiple energy conversions, Power can be transferred to spacecraft or ground using lasers or microwaves. Due to the high solar radiation intensity, stable energy and enough sunlight (more than 99%) in high orbit, the energy utilization ratio is considerable and the scale of transmitted power can be very high (Henderson 2009). But before it coming to the truth, some key technologies need to be resolved at first, such as low-cost, high-efficiency and radiation-resistant photovoltaic arrays, lightweight, high-strength and modularized structures and components, low-cost space launching and transportation as well as on-orbit installation and module replacement.

3 FEASIBILITY ANALYSING

Different from the wireless communication which focuses on the signal-to-noise ratio, wireless power transmission mainly concerns the transmission efficiency. Especially for the spacecraft system, WPT is required to achieve high efficiency in long distance. Also the technology maturity must be taken into considerations. In section II, we mainly discuss four approaches for space wireless power transmission, including microwaves, lasers, Magneto-inductive resonantly coupled mode and concentrated, unconverted sunlight. Their overall WPT efficiency are 3%, 4%, 8%, 25%, respectively (Turner 2006).

In comparisons, research on WPT through microwaves is sufficient at home and aboard and the technology maturity is very high. But the end to end efficiency of microwave method is very low in space due to multiple energy conversions, less than 5%. To transmit large power over a long distance with microwave, such critical issues as very high transmitting frequencies, large antennas, microwave beams shaping and retrodirective system should be designed and configured appropriately which were hard to realize in practice. Theory and experiment results do suggest that microwave power transmission may be more suitable as a supplementary power mode for future small satellite clusters within short distance.

MRC is a mid-range technology which can transfer large power with high efficiency in near distance. But the efficiency attenuates rapidly with transmitting range. The significant transmitting range is on the order of ten meters or less. In some cases, the range and direction of modules in fractionated system vary in time under different missions which change the relative location of power provider and receiver spacecrafts, affecting the resonance structure, resulting poor transmission efficiency. High transmission efficiency also needs large resonance structure size. Thus this method may be taken as supplementary power to feed payloads with short-term peak power, provide necessary power in case of emergency or resurrect a non-operational spacecraft.

Laser is famous for its better directional property, larger power carrying capability, and so on. Wireless power transfer through laser belongs to far-field technology. In space application, the transmission efficiency has nothing to do with the transmission range because of the moderate atmospheres of space environment. Thus small power can be transferred over a long distance. And the number of transmitting and receiving equipments using in laser transmission is fewer than in microwave. So it is a promising technology that can be used for WPT in fractionated spacecraft systems.

Wireless power transmission using concentrated, unconverted sunlight has high efficiency because of no multiple energy conversions, high reflectivity mirrors and low attenuation in space. In addition, power equipment is centralized on the RV so that considerations can be given to other subsystems to further reduce cost, simplify operations and increase flexibility. MVs in fractionated spacecraft systems can get high-intensity sunlight from RV in a few percent of orbit period and survive during eclipses or lack of concentrated sunlight by their own power subsystems such as TES, heat engines and heat traps. Thus it can be taken as one of the major WPT technologies for fractionated spacecraft systems.

4 TECHNICAL CHALLENGES AND APPLICATION PROSPECT

From the discussion above, we can conclude that lasers and concentrated, unconverted sunlight can be taken as the main approaches, microwaves and MRC as the assistant approaches in wireless power transmission for fractionated spacecraft system. Although much effort has been made in theory and experiments, WPT for practice application in space still has a long way to go.

Challenge of lasers:

- Developing new crystal materials to increase the efficiency of direct solar-pumped laser.
- Seeking light-weight, high-strength material to build large area sunlight collecting structure to decrease space transportation cost.
- Spectrum multiplexing technology to reduce the system temperature, increase the efficiency.
- High-intensity laser power beaming receiving and converting system such as vertical multi-junction photovoltaic cells.

Challenge of concentrated, unconverted sunlight:

- Developing low-loss, high-convergence reflecting mirror to improve the utilization of spectrum.
- High-intensity light beaming concentrating and control technology to ensure that sunlight wouldn't diverge when transmitting over a long distance.
- Researching high efficiency heat engine and thermal energy storage system to ensure that high-intensity sunlight can convert into heat and power energy efficiently.

The main Problems faced with microwaves are the directional control technology for microwave power beaming and miniaturization of sending and receiving devices. Furthermore, the RF-interference to satellites communication system must be considered carefully in practice. The technical challenge

in MRC is developing isotropic coverage resonance structures for both RV and MVs to enhance their adaptability to attitude and relative position of each other. And impedance match must be designed carefully to ensure reliable resonance, in order to improve transmission efficiency.

It is expected that once wireless power transmission has been applied to the fractionated spacecraft systems successfully, it will greatly enhance the flexibility, improve the fractionated degree and reduce the global cost and structure complexity of the systems. In addition, these WPT approaches will have significant, broad application and alluring prospects in other fields because this new power transmission mode can provide flexible, reconfigurable and removable power, satisfying the needs of safe manufacturing in formidable working environment. Thus it will have a good market in mining industry and intelligent high-rise buildings. In transportation, this technology can provide convenient power source for electrical bus, electrical locomotive and maglev train. At the same time, this technology can be promoted into household appliances, biomedicine and man-machine integration devices. Moreover, such applications may extend into national defense and military areas in the future.

5 CONCLUSIONS

This paper introduces the concept of fractionated spacecraft system which may bring about a revolution in future space missions. Wireless power transmission, as one of the key technologies in fractionated spacecraft system, has been discussed detailedly throughout the whole paper. Principles, characteristics, and feasibility have been studied for various methods, respectively. Net system efficiency and transmitting range are the main concern for wireless power transmission. Taking account of various factors above, we suggest that lasers and concentrated, unconverted sunlight should be taken as the primary choices and microwaves and MRC should be secondary choices for WPT in fractionated spacecraft system. At last, difficulties and technical challenges for these methods as well as future applications have been described.

REFERENCES

Bai, Y et al. 2011. Study of Wireless Power Transfer System Based on Ultrasonic. *Trics & Acoustooptics* 33(2): 324–327.
Fu, W.Z et al. 2009. Maximum Efficiency Analysis and Design of Self-resonance Coupling Coils for Wireless Power Transmission System. *Proceedings of the CSEE* 29(18):21–26.

Henderson, E.M. 2009. Space Based Solar Power Flight Demonstration Concept. *IEEE Aerospace conference, 7–14 March 2009.* Big Sky, MT.

Hwang, I.H & Lee, J.H. 1991. Efficiency and Threshold Pump Intensity of CW Solar-Pumped Solid-state Lasers. *IEEE Journal of Quantum Electronics* 27(9):2129–2134.

Jamnejad, V & Silva, A. 2008. Microwave Power Beaming Strategies for Fractionated Spacecraft Systems. *IEEE Aerospace Conference, 1–8 March 2008.* Big Sky, MT.

Lafleur, J.M & Saleh, J.H. 2008. Feasibility Assessment of Microwave Power Beaming for Small Satellites. *6th International Energy Conversion Engineering Conference, 28–30 July 2008.* Cleveland, Ohio.

Little, F.E. 2000. A Wireless Power Transmission Power System For Microgravity Crystal Processing Satellites. *IEEE Aerospace Conference Proceedings, 18–25 March 2000.* Big Sky, MT.

Lu, X.Q & Fan D.Y. 2002. Progress of high-power Ti:sapphire laser-amplifier system. *Laser & Optoelectronics Progress* 39(9):15~22.

Maciuca, D.B & Chow, J.K. 2009. A Modular, High-Fidelity Tool to Model the Utility of Fractionated Space Systems. *AIAA SPACE 2009 Conference & Exposition, 14–17 September 2009.* Pasadena, California.

Masayoshi, K. et al. 2010. Mid-range Wireless Power Transmission in Space. *8th Annual International Energy Conversion Engineering Conference, 25–29 July 2010.* Nashville: TN.

Mathieu, C. 2006. Assessing the Fractionated Spacecraft Concept. *Master thesis: Massachusetts Institute of Technology, Engineering Systems Division, Technology and Policy Program, 2006.*

Patel, M.R. 2005. *Spacecraft Power Systems.* BeiJing: China Aerospace Press.

Raible, D.E, Dinca, D, Nayfeh, T.H. 2011. Optical Frequency Optimization of a High Intensity Laser Power Beaming System Utilizing VMJ Photovoltaic Cells. *2011 International Conference on Space Optical Systems and Applications, 11–13 May 2011.* Santa Monica, CA.

Shinohara, N. 2011. Power Without Wires. *IEEE microwave magazine* 12(7):S64–S73.

Simon, M.J & Langer, C. 2009. Wireless Power Transfer for Responsive Space Applications. *AIAA SPACE 2009 Conference & Exposition, 14–19 September 2009.* Pasadena: California.

Tan, L.L. et al. 2010. Study of Wireless Power Transfer System Through Strongly Coupled Resonances. *2010 International Conference on Electrical and Control Engineering, 25–27 June 2010.* Wuhan.

Turner, A. 2006. In-Space Power Transfer Bouncing the Light-Fanstastic! *Fractionated Spacecraft Workshop, Space Systems/Loral, 3–4 August 2006.* Colorado Springs.

Turner, A.E. 2006. Power Transfer for Formation Flying Spacecraft. *AIAA Space 2006, 19–21 September 2006.* San Jose, California.

Turner, A.E. 2009. In-Space Power Transfer using Concentrated, Unconverted Sunlight. *7th International Energy Conversion Engineering Conference, 2–5 August 2009.* Denver, Colorado.

Zhao, J.S & Zhu, J.F. 2012. Design and simulation of an efficient and safe laser power transmission system. *Journal of Tsinghua University (Sci & Tech)* 52(4):.

Control Engineering and Information Systems – Liu (Ed)
© 2015 Taylor & Francis Group, London, ISBN 978-1-138-02685-8

Intelligent data acquisition system based on wireless PDA

X.W. Tao & Z.T. Fu
School of Land Sciences and Technology, China University of Geosciences, Beijing, China

ABSTRACT: With the characteristics of high speed and short cycle in mineral exploration, the airborne geophysical technologies play an important role in our geological exploration and prospecting. Especially in the land resources survey, airborne geophysical technologies provide a wealth of geophysical information in basic geological research and potential prediction of metallic ores, oil and gas. However, there are still some gaps in the effective use of airborne geophysical data in geological prospecting when compared with international advanced levels. Ground verification work for airborne geophysical anomalies is lagging behind other. In order to solve the problem the paper applies the advanced techniques of the current sophisticated digital equipment and communications networks to develop an all-digital airborne geophysical ground verification system on the principle of combining field work with inside work. Through the analytical study of internal exceptions, the system designs the inspection route of field anomaly. Adopting the ground anomaly detection equipment in the field, staff in the field work carries out accurate verification work.

1 INTRODUCTION

With the increasing demand for mineral resources and the recovery of the international mining in recent years, China aviation and geophysical techniques have been developed rapidly. Especially in the field of direct prospecting, the work of finding insidious mineral has achieved good results. Aero geophysical surveying has the advantages of high efficiency and wide region. In many areas of China large-scale high-precision airborne geophysical surveying work as in [1] has been carried out. It sets an even higher demand on the ground verification work for the airborne geophysical anomalies. In order to achieve the modernization and informatization of the basic geological surveying work, many countries have established a set of system and standardized work processes. The field system such as the "Map IT" software in Tablet PC platform and the ESRI ArcPad software in PDA-based platform has been developed successfully. Compared with high efficiency of the airborne geophysical surveying, the anomaly inspection of the ground work still remains in the stage of depending on paper records, paper drawings and paper analysis, which results in a relatively low efficiency of the verification work. With the rapid development and wide application of surveying & mapping information technology and wireless communication technology, the digitization of field records, mapping and analysis in the ground verification work for the airborne geophysical anomalies becomes more and more urgent.

On the base of PDA platform, the hardware system integrates GPS, GIS and wireless communications network technology into a digital ground verification system. A seamless connection of geophysical survey equipment has been established through the SD card, USB and Bluetooth. Especially the application of GPS, GIS and wireless communications network technology, greatly improves the work efficiency of the ground verification. Furthermore the system has made great strides in standardization and normalization of the data, which provides a basis for field geophysical data sharing and gradually forms a unique set of verification work of the technical system. With the constant improvement of the technical system, ground verification is lagging behind this constraint is the bottleneck of the geological prospecting will be the initial solution.

2 HARDWARE SYSTEM DESIGN

2.1 Hardware system

PDA module is a kernel module in hardware system (see Fig. 1), integrated ground verification module and GPS module. PDA Module completes the storage, management, computing, analysis, display and data transmission of the field geophysical data and positioning data. Ground verification module completes the ground geophysical data acquisition of the ground verification equipment and real-time data transfer to the PDA Module. GPS positioning module completes the global positioning satellite

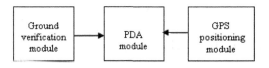

Figure 1. Field data acquisition system hardware components.

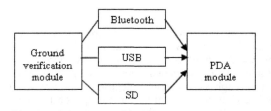

Figure 2. Integration of PDA and ground verification equipment.

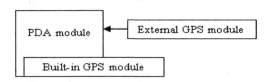

Figure 3. Integration of GPS module and the PDA module.

signal reception and sends the received position data to the PDA Module. It couples with the ground verification data to format spatial data.

2.2 Hardware integration

Hardware integration achieves the data communication between the different and stand-alone device modules, including ground verification module and the PDA module, GPS positioning module and PDA Module. Taking into account the different functions of the ground verification device as well as diversification of equipment, ground verification device uses the respective data transmission that includes Bluetooth, USB interface and SD card (see Fig. 2). Ground verification data is synchronized with the PDA Module. PDA Module developed data management system that will verify the standardized GPS match storage data of different times, different places and different types.

The integration between GPS positioning module and PDA module has internal way and external way (see Fig. 3).

Built-in GPS is integrated into the internal components of the PDA Module. This integration is relatively high, compact, lightweight and

measuring simple, while an external GPS Module builds a connection between the PDA internal components without interference by external COM interface or Bluetooth. After repeated tests, and in order to ensure the quality of reception of GPS signals in harsh environments, the system uses an external GPS.

3 SYSTEM FUNCTION MODULE AND KEY TECHNOLOGY RESEARCH

3.1 Functional design

Embedded devices systems are widely used in small electronic devices. The core design is relatively small, and has limited system resources. System functional design adopts component-based system design (see Fig. 4). DMS (Data Management System) module as a system Center module, integrates two component modules that are TMS (Task Management System) module and GIS module. Based on Center module, TMS module and GIS module are separate, and have relatively independent operation. If one of the two component modules is closed, it will not affect the normal operation of another component module, in order to save system resources and improve operational efficiency.

DMS module is responsible for data management and analysis of GIS data, GPS monitoring data, geomagnetic data and task data. TMS module is responsible for task management of ground verification. GIS module is responsible for geographic data loading, route designing, GPS navigating, geomagnetic data location matching with the geological description entry in task surveying region.

TMS module and DMS module was developed in Visual C development environment, based on the SQL Server CE, embedded database systems development. GIS module used ArcPad Application Builder development environment for secondary development of ArcPad.

For embedded devices such a special low-memory storage system environments, high consumption and high storage process are not stable running. After the system simulated testing, analyzing and summary of similar functions, the paper comes to the stability of the three functional modules and resources consumed by the three functional modules, running the program (see Table 1).

Figure 4. Division of the PDA management module.

366

Table 1. Module resources and stability.

Module	Resources consumption	Stability
TMS	Low	Very stable
GIS	Medium	Stable
DMS	High	Unstable

By comparison as shown in Table 1, it can be found in three modules that the resources consumption of TMS module is the least and the most stable. GIS module is worse than TMS module, mainly due to the large amount of geographic data query. In the late development, the method that establishes index to query the data will be used. DMS module, whether the consumption of resources or stability are the worst in the three modules, mainly due to the high operation frequency of the data, as well as a large amount of data synchronization and PC operation. The module in the system development will be as key part in order to improve the performance and stability of DMS module. Database technology is the focus of the system development.

3.2 Key technologies

It is necessary to establish management processes about anomaly inspection planning data collection. First, ground verification worker will be conveyed by the system planning routes and related information. When reaching the location by mobile terminal, the client program captures the current location coordinates automatically by the GPS module. Then, the client program compares these data with the central database of information and planning routes and related information and data display on the terminal screen. PDA can display electronic maps and navigation information in accordance with the planning directions route. Finally, the system will automatically activate the GPRS wireless network. The geographic location information and the current time together with the live line condition records sent to the central database.

In support of Real-time information collection system and GIS, the database data was processed, combined with abnormal verification principle. First, based on abnormal screening and delineation of prospecting targets for field verification, the route was designed. Second, to fully tap the GIS spatial analysis capabilities, the underlying data has been combined with collecting data. System dynamically planned and simulated verification route, and optimized program to make reasonable scheduling of scientific decision-making and personnel.

Ground verification data is stored in database as the spatial data form. For recording complete and accurate data, you need to regulate the structure of the attribute data. In addition to record ground verification data, it should record the necessary information, such as data integrity, data download status, the status of data transmission, the progress of work, work units and operator information.

In order to achieve the integration of the PDA terminal and ground verification work and timely information delivery, it must be completed that field PDA collection data of ground verification and central server data is synchronous and consistent. Taking into account the diversity of field ground verification hardware devices that have different functions and equipment, different driver software, different transmission method and complicated work situation of ground verification, data collected by ground verification equipment is transmitted to SQL CE database in PDA Module, using their own data transmission mode (Bluetooth, USB interface and SD card).

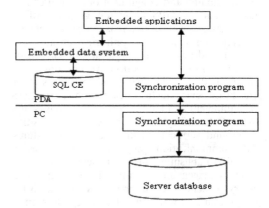

Figure 5. Data transmission.

Table 2. Data processing.

Data processing	Requirement
Standardization	Data is formatted in accordance with the geophysical anomaly data file.
GPS matching	Data collected of the different types at different times in different locations matches the GPS location.
Data storage	Data that have completed standardization and GPS matching is stored in the database, using a uniform format.

Database management program of PDA module formats data of different times and different places in accordance with Table 5. Data processing includes standardization, GPS matching and data storage. In the end of the field work data in the SQL CE database of PDA module is transmitted to the remote server database of ground verification by PDA synchronization program.

4 CONCLUSION

The system uses mobile geographic information system, global positioning satellite, network communication and database technology. It adopts the method that field data acquisition integrates with inside management. Relying on a central server to establish two-way communication between field data acquisition and inside management, real-time operating information of ground verification team is transmitted to remote server management system. At the same time management decision-making and task scheduling return back. There is a set of operating technical processes that includes route planning and design of ground verification, task scheduling, GPS navigation, ground verification data acquisition, database storage, network transmission, field GPS monitoring, ground verification and aviation geophysical anomaly analysis. Field ground verification system consists of two parts that are hardware and software system. Hardware devices include field handheld geophysical equipment and mobile communications equipment. Software systems include data acquisition of ground verification, management procedures, data communications program, the central server graphics publishing and task scheduler within inside management program. Therefore, it is necessary to achieve these objectives that hardware integration, software and data integration must be done, while the three aspects are different from each other and mutual unity. In order to establish an open space integration mechanism of anomaly inspection to speed up the process of mineral exploration, we must speed up the modern digital process of ground verification. Air-ground verification team must be mobilized to achieve the ground anomaly inspection. Aerial survey data is provided for ground verification team, and the results of ground geophysical team measurement feedback return to the aviation sector, thereby improving the efficiency of the geophysical work.

ACKNOWLEDGMENT

This work supported by "the Fundamental Research Funds for the Central Universities".

REFERENCES

Allen J., Anderson S., Browne M., Jones P., A Framework for Considering Policies to Encourage Sustainable Urban Freight Traffic and Goods/Service Flows. Summary Report. Transport Studies Group, University of Westminster, London; 2000.

Andrew Dursch, David C. Yen, Dong-Her Shih, Bluetooth Technology: an Exploratory Study of the Analysis and Implementation Frameworks, Computer Standards & Interfaces, 2004, pp. 263–277.

Briner, A.P., Kronenberg, H., Mazurek, M., Horn, H., Engi, M., Peters, T., FieldBook and GeoDatabase—Tools for Field Data Acquisition and Analysis. Computers & Geosciences, 1999, pp. 1101–1111.

Burnham, J.P., 2001. The Essential Guide to the Business of US Mobile Wireless Communications. Prentice-Hall PTR, Upper Saddle River, NJ, 2001.

Chen Liping, Yao Guangqiang, Zhao Chunjiang, Design and Development of Feed-formula System Based on Pocket PC [J]. Transactions of the Chinese Society of Agricultural Engineering, 2008, pp. 179–183.

De Donatis, M., Bruciatelli, L., MAP IT: the GIS Software for Field Mapping with a Tablet PC, Computers & Geosciences, 2006, pp. 673–680.

Dey, A.K., Salber, D., Abowd, G.D., Futakawa, M., 1999. An Architecture to Support Context-Aware Applications. Technical Report GIT-GVU, Georgia Institute of Technology, College of Computing.

Fang Hui, He Yong, Field Information Collection and Processing System Based on Palm-sized Personal Computer, Transactions of the Chinese Society of Agricultural Engineering, 2004, pp. 124–128.

Guo Wei, Ye Haijian, Ji Ronghua, Communication Protocol in Mobile Field Information Intelligent Service System Based on GPRS, Transaction of the CSAE, 2009, pp. 175–178.

Lindgren, M., Svensson, E., Jedbratt, J., 2002. Beyond Mobile: People, Communications and Marketing in a Mobilized World. Palgrave Global Publishing.

McCaffrey, K.J.W., Holdsworth, R.E., Clegg, P., Jones, R.R., Wilson, R., Using Digital Mapping Tools and 3-Dvisualization to Improve Undergraduate Fieldwork, Planet Special Edition 5, 2003, pp. 34–36.

Murakami E., Wagner DP. Can Using Global Positioning System (GPS) Improve Trip Reporting? Transportation Research Part C 1999; 7: 149–165.

Raul Morais, A ZigBee Multi-powered Wireless Acquisition Device for Remote Sensing Applications in Precision Viticulture [J]. Computer and electronics in agriculture, 2008, pp. 94–106.

ShiFeng, Virtual Reference Station and Mobile GPS PDA Platform Positioning System [D]. Shanghai: Tongji University, 2007, 3.

TanAnning, WuCaicong and ZhengLihua, Wireless Data Transmission Technology for Mobile Agriculture Terminal, 2009, pp. 244–247.

Visser J, van Binsbergen A, Nemoto T. Urban Freight Transport Policy and Planning. First International Symposium on City Logistics, Cairns, Australia; 1999.

Wilson, R.W., 2006. Digital Fault Mapping and Spatial Attributes of Basement Influenced Oblique Extension in Passive Margin Settings. Ph.D. Dissertation, University of Durham, Durham, England.

Control Engineering and Information Systems – Liu (Ed)
© 2015 Taylor & Francis Group, London, ISBN 978-1-138-02685-8

SDdwMDS: An improved distributed weighted multidimensional scaling algorithm in Wireless Sensor Network

M. Zhao

Department of Computer Science and Technology, Yanshan University, Qinhuangdao, China

ABSTRACT: With the development of the wireless sensor network, the position information of sensor node plays a crucial role, and the sensor node localization has become a crucial research topic. By researching on the distributed weighted multidimensional scaling algorithm in WSN, especially the dwMDS algorithm, the SDdwMDS algorithm is proposed in this paper by applying the adaptive neighbor selection mechanism and approved Gauss kernel weighted mechanism, and introducing the steepest descent method to optimize the local cost function. Through experimental analysis, the SDdwMDS algorithm is more precise and more suitable than the dwMDS algorithm for the sensor node localization in sparse and irregular topology structure network.

1 INTRODUCTION

Wireless Sensor Network (WSN)[1] is composed of a large number of low-cost, perceptive, sensor nodes in the network, which have wireless communication and calculation capabilities. The sensor network system is widely used in national security, military, medical, environmental monitoring, manufacturing industry, traffic management, disaster rescue, space exploration and other fields. So the sensor node position information plays a crucial role in WSN, and the sensor node localization has become a crucial research topic.

There is no doubt, the direct method of accessing sensor node position information is to use global positioning system GPS (Global Positioning System). However, due to the net construction cost and energy consumption, positioning all of the sensor nodes in WSN can not only use the GPS system. But through obtaining a few nodes' location information by GPS, and using other positioning technology to get the position of the other nodes.

In recent years, since multidimensional scaling techniques applied to wireless sensor network node localization technology. So far, in order to improve positioning accuracy, different types of network have been researched. For example, the classic MDS-MAP algorithm is only suitable for small size, uniform distribution of sensor network node. To meet the requirement of large scale, irregular topology structure network, MDS-MAP (P) algorithm, Local MDS algorithm and dwMDS algorithm have been proposed[2–5].

In this paper, the following chapters elaborated through the following aspects: firstly, introduce the dwMDS algorithm and analyze its shortcomings, secondly propose the SDdwMDS algorithm, then compare the SDdwMDS algorithm with dwMDS algorithm through experimental analysis, and finally conclude.

2 DISTRIBUTED WEIGHTED MULTIDIMENSIONAL SCALING ALGORITHM dwMDS

In 2005, JOSE A. COSTA of the Michigan University in USA proposed distributed multi-dimensional scaling localization algorithm dwMDS, which apply the Gauss kernel weighted mechanism on distributed multi-dimensional scaling localization. According to the distance weights of the unknown nodes, the algorithm uses the adaptive 2 hop neighbor selection mechanism to avoid the negative bias effect of the neighbor selection.

The positioning process of dwMDS algorithm is divided into the following stages:

(1) Using to the classic MDS algorithm to each nodes to construct the initial local coordinate system. (2) By using the incremental greedy method, convert all nodes of the initial local coordinates into global relative coordinates. (3) According to the coordinate information of the beacon nodes in sensor network, initialize the global relative coordinates to be global absolute coordinates. (4) According to the local properties of the network, make the unknown node to select the real 1 hop neighbor nodes, and participate in positioning solving. (5) Using the Gauss kernel weighted mechanism to calculate weights:

$$w_{ij} = \exp\{(\hat{d}_{ij})^2/h_i^2\} \qquad (1)$$

$\hat{d}_{ij}$ is the distance between node i and node j;

$$h_i = \max_j(\hat{d}_{ij})$$

(6) Optimizing the initial global absolute coordinates by iterative calculation.

3 SDdwMDS ALGORITHM DESCRIPTION

According to the shortcoming of the distributed weighted multidimensional scaling localization algorithm dwMDS, this paper proposes an improved algorithm SDdwMDS, which improves the dwMDS algorithm from the following 3 aspects.

3.1 *The adaptive neighbor selection mechanism*

The adaptive neighborhood selection is described as the following steps:

Step 1: Judge whether the number of 1 hop neighbor node of the unknown node is less than the neighbor node number involved in the positioning, if less, then select 2 hop neighbor nodes, or select only 1 hop neighbor nodes;
Step 2: Calculate localization relative errors of the neighbor nodes selected by the first step, and then sort by the ascending order. For the 2 hop neighbors, weight the shortest path length as the unknown distance between nodes;

The calculation of node localization relative error is described as the following process:

1. Before each iterative calculation, calculate the average location error by formula 2.

$$Average\ Error(i) = \sum_{j=1}^{M} |d_{ij} - \hat{d}_{ij}|/M \qquad (2)$$

M is the number of the neighbor nodes.
2. Do normalization processing to the position error by formula 3.

$$Norm(i) = \left\{\sum_{j=1}^{M} |d_{ij} - \hat{d}_{ij}|\right\} \bigg/ \left\{\sum_{j=1}^{M} \hat{d}_{ij}\right\} \qquad (3)$$

3. Considering the error introduced by the self position error between the unknown node i and the neighbor node j, and the shortest path length $\hat{d}_{ij}$ between node i and the neighbor node j, calculate the relative error of node j by formula 4.

$$Res\ Error_i(j) = Average\ Error(j) + Norm(i) \times \hat{d}_{ij} \qquad (4)$$

Step 3: Delete the neighbor nodes whose relative error is larger than preset threshold, and put the rest of the neighbor nodes into the positioning calculation.

3.2 *The approved Gauss kernel weighted mechanism*

Affected by the range error and the irregular topological structure, the localization error of the neighbor nodes near the unknown node might be very large. Therefore, the weight calculations need to consider the range error and position error of the neighbor nodes, to make the performance of neighbor nodes in the unknown node localization have a greater contribution. Based on this point, we can see that weights and the relative error are inversely proportional, so this paper put the normalized reciprocal of the relative error as the weight by formula 5.

$$w_{ij} = resError_i(j)^{-1} \bigg/ \sum_{j=1}^{M} resError_i(j)^{-1} \qquad (5)$$

3.3 *The steepest descent method to optimize the local cost function*

According to the classical metric multidimensional scaling technique of algorithm dwMDS is positioning the unknown node by minimizing the stress coefficient; this paper will convert the stress coefficient into the objective cost function shown as formula 6, by solving the objective cost function minimized condition on the unknown nodes in positioning.

$$S = 2 \sum_{1 \le i \le n} \sum_{i < j \le N} w_{ij}\left(\hat{d}_{ij} - \|z_i - z_j\|\right)^2;$$

$$\|z_i - z_j\| = \sqrt{(x_i - x_j)^2 - (y_i - y_j)^2} \qquad (6)$$

Through simple calculation, cost function is

$$S = \sum_{i=1}^{n} S_i + c$$

The constant c is not associated with the node coordinates, and S_i is the local cost function of each unknown node, shown as formula 7:

$$S_i = \sum_{j=1, j \ne i}^{n} w_{ij}\left(\hat{d}_{ij} - \|z_i - z_j\|\right)^2$$
$$+ 2 \sum_{j=n+1}^{N} w_{ij}\left(\hat{d}_{ij} - \|z_i - z_j\|\right)^2 \qquad (7)$$

If there is no beacon node (N = n), the relationship between S and S_i calculated by $\partial S/\partial z_i = 2\partial S/\partial z_i$.

By these relations we can see that, as each S_i takes minimum value, S will obtain the minimum value, and the unknown node position coordinates can be achieved by minimizing each unknown node local cost function. Then as the most simple and rapid method of the unconstrained optimization method, the steepest descent method is introduced to optimize the local cost function.

In the numerical analysis, the basic idea of the iterative optimization method is: given initial node $z^{(0)}$, get a series of nodes $\{z^{(k)}\}$, when $k \to \infty$, $z^{(0)}$ is the optimal solution when the formula 7 gets the minimum value, the iterative formula is $z^{(k+1)} = z^{(k)} + \alpha_k d_k$.

Because the cost function should be minimized, the descent direction of the optimization algorithm should be the negative gradient direction, shown as formula 8.

$$\frac{\partial S_i^{(k)}}{\partial z_i^{(k)}} = -\left[\sum_{j=1, j\neq i}^{n} 2w_{ij}\left(1 - \frac{\hat{d}_{ij}}{\left\| z_i^{(k)} - z_j^{(k)} \right\|}\right)(z_j^{(k)} - z_j^{(k)}) \right.$$

$$\left. + \sum_{j=n+1}^{N} 4w_{ij}\left(1 - \frac{\hat{d}_{ij}}{\left\| z_i^{(k)} - z_j^{(k)} \right\|}\right)(z_i^{(k)} - z_j^{(k)}) \right]$$

$$= -(z_i^{(k)} - z_1^{(k)}, z_i^{(k)} - z_2^{(k)}, ..., z_i^{(k)} - z_N^{(k)})$$

$$\times (b_1^{(k)}, b_2^{(k)}, ..., b_N^{(k)})^T$$

$$(8)$$

Make $D^{(k)} = (z_i^{(k)} - z_1^{(k)}, z_i^{(k)} - z_2^{(k)}, L, z_i^{(k)} - z_n^{(k)})$, $B^{(k)} = (b_1^{(k)}, b_2^{(k)}, L, b_n^{(k)})^T$, the coordinates Z_i of node i fit formula 9.

$$b_j^{(k)} = \begin{cases} 2w_{ij}\left(1 - \dfrac{\hat{d}_{ij}}{\left\| z_i^{(k)} - z_j^{(k)} \right\|}\right), & j \leq n, j \neq i \\ 0, & j = i \\ 4w_{ij}\left(1 - \dfrac{\hat{d}_{ij}}{\left\| z_i^{(k)} - z_j^{(k)} \right\|}\right), & j > n \end{cases}$$

$$\zeta t(\kappa+1) = \zeta t(\kappa) - \langle\langle \kappa \rangle \cdot \Delta(\kappa) \cdot B(\kappa). \qquad (9)$$

The iterative step length factor $\alpha^{(k)}$ is put into the linear search, shown as formula 10. And the multidimensional objective function is converted into a target function, shown as formula 11.

$$S_i\left(z_i^{(k)} - \alpha^{(k)} \times D^{(k)} \times B^{(k)}\right)$$

$$= \min_{\alpha>0} S_i\left(z_i^{(k)} - \alpha^{(k)} \times D^{(k)} \times B^{(k)}\right) \qquad (10)$$

$$\sqrt{(\langle \rangle)} = \zeta t(\kappa) - \langle \cdot \Delta(\kappa) \cdot B(\kappa). \qquad (11)$$

This paper uses the quadratic interpolation to do the linear search, set the function value $\Phi(\alpha)$ at α_i, α_{i-1} and α_{i-2} as $\Phi(\alpha_i)$, $\Phi(\alpha_{i-1})$ and $\Phi(\alpha_{i-2})$. The formula 12 can be got by the quadratic interpolation.

$$\langle_{t+1} = 1/2(\langle_{t} + \langle_{t-2} - \chi_1/\chi_2); \chi_1$$

$$= (\sqrt_{t-2} - \sqrt_{t})/(\langle_{t-2} - \langle_{t});$$

$$\chi_2 = ((\sqrt_{t-1} - \sqrt_{t})/(\langle_{t-1} - \langle_{t}) - \chi_1)/(\langle_{t-1} - \langle_{t-2}). \qquad (12)$$

Get $\Phi_{i+1} = \Phi(\alpha_{i+1})$, and remove the corresponding maximum function values among α_i, α_{i-1} and α_{i-2}. Put the new point α_{i+1} into the three new points, iterative calculate until get the demanded precision.

The SDdwMDS algorithm is proposed in this paper and shown as the following:

Input: $\{\hat{d}_{ij}\}$, m, ε;/* $\hat{d}_{ij}$ is the distance between node i and node j, m is the number of the beacon nodes, ε is the accuracy. */
Output: x_i, y_i/*The Coordinates of the node i.*/

Begin

1. Inputs: $\{\hat{d}_{ij}\}$, m, ε;
2. Bulid local map with MDS;
3. Change local maps into the global relative map;
4. Change relative map into absolute map to get the initial coordinates $X^{(0)}$;
5. Initialize: k = 0, $S^{(0)}$, compute $-\partial S_i^{(0)}/\partial z_i^{(0)}$ by formula 8;
6. Repeat:
7. $k \leftarrow k + 1$;
8. For i = 1 to n
9. Compute the relative location Error;
10. Select neighborhood to participate location;
11. Compute w_{ij} from formula 5;
12. Search step factor α_k and calculate $z^{(k+1)}$ by formula 9;
13. Compute $S_i^{(k)}$;
14. $S^{(k)} \leftarrow S^{(k)} - S_i^{(k-1)} + S_i^{(k)}$;
15. End for
16. Until $S^{(k-1)} - S^{(k)} < \varepsilon$.
End

4 EXPERIMENTAL ANALYSIS

In order to verify that the proposed SDdwMDS algorithm still has high positioning precision in sparse and irregular topology structure network. This paper will do simulation experiment in these two kinds of topology structure network. In sparse network, there are 50 nodes randomly deployed in

371

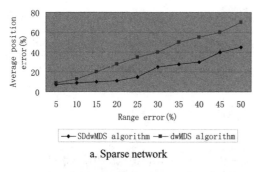

a. Sparse network

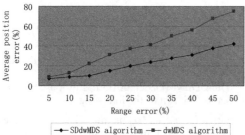

b. C type irregular topology structure network

Figure 1. The positioning comparison result between SDdwMDS and dwMDS in sparse network and C type irregular topology structure network.

the 200 m * 200 m square area. In irregular topology structure network, there are 160 nodes randomly deployed in C type area.

Compared the SDdwMDS algorithm with the dwMDS algorithm with the change of position error in sparse network and C type irregular topology structure network, the positioning comparison result is shown as Figure 1. Obviously, with the range error increases, the position error of the SDdwMDS algorithm is smaller than the dwMDS algorithm.

5 CONCLUSION

Researching on the dwMDS algorithm, the SDdwMDS algorithm is proposed in this paper by applying the adaptive neighbor selection mechanism and approved Gauss kernel weighted mechanism, and introducing the steepest descent method to optimize the local cost function. Through experimental analysis, the SDdwMDS algorithm is more precise and more suitable than the dwMDS algorithm for the sensor node localization in sparse and irregular topology structure network.

REFERENCES

Costa J.A., Patwari N, Hero A.O, et al. Distributed Weighted-Multidimensional Scaling for Node Localization in Sensor Networks [J]. ACM Journal, 2006, 2(1): 39–64.

Doherty L. Algorithms for Position and Data Recovery in Wireless Sensor Networks [D]. University of California Berkeley, 2009.

Luo H.Y, Li J.T, Zhu Z.M, et al. Distributed Multidimensional Scaling with Relative Error-based Neighborhood Selection for Node Localization in Sensor Networks[C]. Proceedings of the 2007 IEEE International Conference on Integration Technology, Shenzhen, 2007: 735–739.

Niculescu D, Nath B. Ad-Hoc Positioning Systems (APS) [C]. Proceedings of 2006 IEEE Global Telecommunications Conference, Washington, 2006: 2926–2931.

Shang Y, Zhang Y, Fromherz M. Localization from Connectivity in Sensor Networks [J]. IEEE Transactions on Parallel and Distributed Systems, 2007, 15: 961–974.

Shang Y. Improved MDS-Based Localization[C]. Proceedings of IEEE INFOCOM'04, Hong Kong, 2004, 4: 2640–2651.

Tubaishat M, Madria S. Sensor Networks: An Overview [J]. IEEE Potential, 2008, 22(2): 20–23.

Vivekanandan V, Wong V W S. Ordinal MDS-based Localization for Wireless Sensor Networks [J]. International Journal of Sensor Networks, 2006(I):169–178.

Control Engineering and Information Systems – Liu (Ed)
© 2015 Taylor & Francis Group, London, ISBN 978-1-138-02685-8

A new kind of wireless sensor network model using cloud computing platform for information awareness

F. Zhang, C. Chen, H.B. Yang, H.T. Cai & J.L. Zhang
Hefei New Star Institute of Applied Technology, Hefei, P.R. China

ABSTRACT: The application of huge advantage of cloud computing into wireless sensor network is a relatively new subject worth exploring. Based on technology characteristics and service models of cloud computing, combined with the node structure and system composition of wireless sensor network, a kind of wireless sensor network model combined with cloud computing is put forward by the authors, who have carried out beneficial discussion on the system model, the main principle and protocol conversion, and so on. So the model can be served as a new kind of way for information awareness.

Keywords: component; cloud computing; wireless sensor network; network model; communication protocol; routing node

1 INTRODUCTION

Cloud computing is a kind of calculation pattern based on the Internet. By this way, the sharing of software and hardware resources and information can be offered to the computers and other equipment according to the demand. Cloud computing is another upheaval after the large transformation from large-scale computer to client-server. Users no longer need to know the details of the infrastructure in the *cloud*, needn't to have relevant professional knowledge, also needn't to control directly.

With the rapid development of Micro-Electro-Mechanism System (MEMS), System on Chip (SOC), wireless communication and low-power embedded technology, Wireless Sensor Network (WSN) arises at the historic moment. With its characteristics of low-power consumption, low cost, distributed and self-organization, it has brought a profound change of information awareness.

With the progress of technology, industry put forward higher requirements for the computing ability of WSN. Therefore, how to apply the huge advantage of cloud computing into WSN is a relatively new subject worth exploring. In this paper, a kind of WSN model combined with cloud computing is put forward by the authors, who have carried out some beneficial discussion on the structure and function.

2 TECHNOLOGY BACKGROUND OF CLOUD COMPUTING AND WIRELESS SENSOR NETWORK

2.1 Overview of cloud computing

Since 1983, when the concept of **Network Style Computer** was put forward by Sun Microsystems Corporation, the development of cloud computing has passed 30 years. Including giant company such as Amazon, Google, IBM, Yahoo, HP, Dell, Novell, AMD, Intel, Cisco, Dell, etc., together with research institutions such as Carnegie Mellon University, MIT, Stanford University, University of California, Berkeley and University of Maryland, etc., have been respectively entering the technology fields of cloud computing.

2.1.1 Technical characteristics of cloud computing
Cloud computing describes a new mode of IT service increase, using and delivery based on the Internet, which usually provides dynamic and easy extension and often virtualized resources through the Internet. Service characteristics of cloud computing on the Internet have certain similarities with cloud and water circulation in nature. As a result, cloud is a pretty apt metaphor.

According to the definition of National Institute of Standard and Technology, cloud computing service should possess the following main features: on-demand self-service, access by any network device anywhere and anytime, multi-user shared resource pool, rapid deployment flexibility, and service can be monitored and measured [1]. In addition, it is generally believed it has the following characteristics: quick deployment of resources or access services based on virtualization technology, reduction the processing burden of user terminal, reduction the user dependence on IT professional knowledge.

2.1.2 Technical characteristics of cloud computing

The definition of cloud computing of National Institute of Standard and Technology has confirmed three service modes, as shown in Figure 1.

a. *Software as a Service (SaaS)*: Consumers use the application, but does not control the operating system, hardware or the operational network infrastructure. It is the basis of service concept. Software service providers offer customer service by the concept of lease, rather than purchase. Common pattern is to provide a set of account password, such as Microsoft CRM and Salesforce.com.

b. *Platform as a Service (PaaS)*: Consumers use host operating application. Consumers control operation application environment (also with the partial control on host), but does not control the operating system, hardware or the operational network infrastructure. Platform is usually the basic infrastructure of application program, such as Google App Engine.

c. *Infrastructure as a Service (IaaS)*: Consumers use basic computing resources, such as *processing* power, storage space and network components or middleware. Consumers can control operating systems, storage space, deployed application program, and network components (e.g., firewalls, load balancers, etc), but does not control the cloud infrastructure, such as Amazon AWS, Rackspace.

2.2 Wireless sensor network

Wireless Sensor Networks (WSN) is a wireless network composed of a large number of sensor

nodes by Adhoc way, whose purpose is to perceive, collect and process detection object information collaboratively in geographical area covered by sensor network and passes the processing results to the observer.

2.2.1 The technical features of wireless sensor network

WSN is composed by a large number of tiny sensor nodes of low price deployed in monitoring area, which form a multiple hops self-organizing network through wireless communication mode. WSN is a kind of brand-new information acquisition platform, which can monitor and gather real-time information of all kinds testing object in network distribution area. And it sends information to the gateway node, so as to realize the target detection and tracking within the complex specified range, which has characteristics of expandability, survivability, autonomy, etc.

2.2.2 Node structure and system composition of wireless sensor network

The structure of WSN is very complex, constrained by space, the node structure and system is mainly involved in this article.

a. *Node Structure of WSN*: e basic composition of node in WSN includes the following three basic units: Sensor Unit (including sensor and D/A conversion function module), Processing Unit (including processor, memory), Communication Unit (including wireless communication module). In addition, the other choosable functional units includes: Node Positioning System, Mobile System, as well as Self-powered System, etc. So the node structure of WSN is shown in Figure 2.

b. *System Composition*: kind of system component of WSN is shown in Figure 3. It is mainly divided into nodes, gateway, and data processing/command center, and so on.

All kinds of data collected will be transmitted to gateway through wireless channel by each node. Then data will be transmitted to data processing/

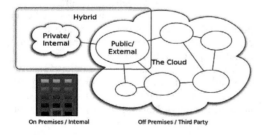

Figure 1. The three types of service models in the definition of cloud computing.

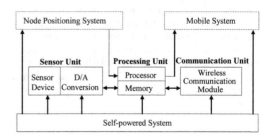

Figure 2. Node structure of WSN.

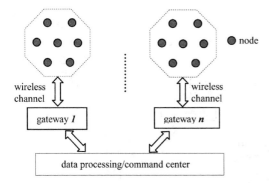

Figure 3. A kind of system component of WSN.

command center by each gateway. Here, the gateway is bidirectional, which linked closely the nodes and data processing/command center through the data serial port. In the general case, the driver of nodes must have the initialization and serial data processing part.

3 WIRELESS SENSOR NETWORK MODEL COMBINED WITH CLOUD COMPUTING

With the rapid development of technology, industry put forward higher requirements for the computing ability of WSN. Therefore, how to apply the huge advantage of cloud computing into WSN is a relatively new subject worth exploring.

3.1 System model

Based on the above techniques, the authors put forward a model of WSN based on cloud computing platform. The system block diagram is shown in Figure 4.

3.2 Working principle

As shown in Figure 4., there are *m* cloud servers. Each cloud server connects to a data processing/command center. Among them, each data processing/command center has *n* gateway. Each gateway connects to each node group respectively through wireless channel.

All kinds of data collected will be transmitted to gateway *1~n* through wireless channel by each node in node group. Then data will be transmitted to data processing/command center. After data fusion and preliminary processing, then the data will be spread to the cloud server. In the end, through the protocol conversion, the data is uploaded to the cloud computing center.

After Distributed Computing, Utility Computing, Parallel Computing, Network Storage Technologies, Virtualization, Load Balance, the data is disposed by expert knowledge system in cloud computing center [2]. Then the required data is sent back to the server, and spread to the data processing/command center, which is used for analysis and decision. And it can made dynamic adjustment by the data processing/command center.

It is not hard to see in Figure 4 the key of cloud computing is protocol conversion, which puts forward very high requirements for the hardware and software.

3.3 The hardware design and working process

We take the temperature data as examples of information awareness. Then the hardware design of the information awareness mainly includes the following several parts in the system.

3.3.1 Terminal node

The hardware principle diagram of terminal node is shown in Figure 5. The main microprocessor is msp430 single chip microcomputer, which connects to CC2430 wireless transceiver module [3]. Temperature sensor connects to the msp430 single chip microcomputer, which is responsible for collecting temperature data. After AD conversion of msp430, the data will be sent to the gateway routing nodes through the wireless transceiver module.

3.3.2 Routing node in the gateway

The routing node hardware principle diagram in the gateway is shown in Figure 6. The main microprocessor is S3C2410, which connects to CC2430 wireless transceiver module at the same time. Meanwhile, the processor also connects to the optical coupling and thermostat through IO port for temperature control.

3.4 The software design and working process

In this system, the software design is very complicated. So it is to design respectively. Software consists of initialization, serial ports data processing and command word of all kinds of units.

3.4.1 Program design of gateway

Gateway is the hub between data processing/command center and terminal node, which receives instructions from the gateway and sends corresponding instructions to the nodes in the next layer. At the same time, it receives data from the nodes in lower layer, and sends data to the processing/command center.

Gateway uses interruption mode, which receives the instructions of host in upper layer and data

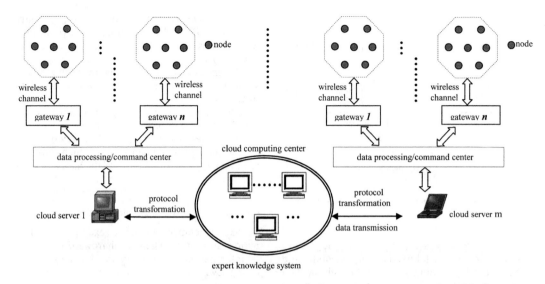

Figure 4. A kind of system block diagram of wireless sensor network model based on cloud computing platform.

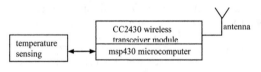

Figure 5. Hardware principle diagram of terminal node.

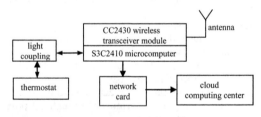

Figure 6. Hardware principle diagram of routing node in the gateway.

from the nodes in lower layer. And it performs the instructions accordingly. Its main program uses the two serial ports to communicate with the host and the nodes in lower layer. The main program workflow chart of routing node in the gateway is shown in Figure 7.

3.4.2 *Program design of terminal node*

Terminal node receives the query command from the gateway. After that, if it needs data of this node, then the required data is packed and sent back to the gateway. If it is lower node data, then it continues forwarding instructions to lower node and receiving the lower nodes data. So it sends data to the upper again. Communication mode adopts

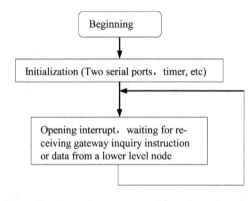

Figure 7. The main program workflow chart of routing node in the gateway.

with interrupt receiving, which is an active sending way. The main program workflow chart of terminal nodes is shown in Figure 8.

3.4.3 *Design of communication protocol*

Data processing/command center communicates with the gateway via serial port, which is completed serial level transformation by MAX3316 [4]. Gateway receives the query command from data processing/command center, and passes back the corresponding data. Inquiry instruction format is shown in Table 1. In the table, ff (expressed by hexadecimal) is the header of instructions. The following bytes mean the required data type. Then it is ID number of query node.

After Gateway receives the inquiry instruction, the route table is found by its inner ID number.

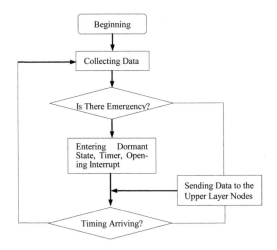

Figure 8. The main program workflow chart of terminal nodes.

Table 1. Inquiry instruction format of data processing/command center.

Inquiry instruction	Function
ff + 01 + ID(X)	Needing temperature data of X node
ff + 02 + ID(Y)	Needing temperature data of Y node

Table 2. Data format of gateway returning data.

Returning data	Function
ff + 01 + ID(X) + data + verification code	Returning temperature data of X node
ff + 02 + ID(Y) + data+ verification code	Returning temperature data of Y node

Then the instruction is forward to the lower nodes, which has been spread to the target node straight through multiple hops routing. Target node packs data, sending the data back by opposite path. The corresponding data format is shown in Table 2.

The first byte of returning data package is the header ff, the second byte is the data type, then is the ID number, the last is the data and the verification code.

4 CONCLUSION

In this paper, on the basis of cloud computing and wireless sensor technology, the authors put forward a model of wireless sensor network based on cloud computing platform, which can be served as a new kind of way for information awareness. The system has the characteristics of low cost, self-organization, low-power consumption, convenient information interaction. So the model can be served as a new kind of way for information awareness.

ACKNOWLEDGMENTS

The research group expresses heartfelt thanks to Professor Qin Tuanfa of Computer and Electronic Information College of Guangxi University, who has made great contributions to WSN. And he has put forward a lot of valuable opinions and suggestions for the work of research group.

The research has also got help from relating staffs of our unit. Hereon, we thank for them all together.

REFERENCES

Chen Lihu, Ye Xiangbin & Hu Gang. The Design of Wireless Sensor Network Node Based on MSP430F149 [J]. Sensor World, 2004, (10): 28–31.

Cui Xunxue, Zhao Zhan & Wang Cheng. Fileds Applications and Design Technologies of Wireless Sensor Networks [M]. National Defense Industry Preess, 2009, 1.

National Institute of Standard and Technology. The Definition of Cloud Computing of National Institute of Standard and Technology [R]. 2011, 9.

Zhang Yaoxiang. Cloud Computing and Virtualization [J]. Network and Computer Security, 2011, (5): 80–82.

Control Engineering and Information Systems – Liu (Ed)
© 2015 Taylor & Francis Group, London, ISBN 978-1-138-02685-8

Research of virtual maintenance laboratory based on wireless network

T. Yuan & B. Shen
Wuhan Mechanical College, Wuhan, Hubei, China

ABSTRACT: With the development of science and technology, more and more advanced equipment is used to the army. The internal structure of new equipment is being more complex with more powerful features of new equipment, which makes the service maintenance and repair of new equipment faced with severe challenges. So in the teaching of new equipment, the integrated use of multimedia tools is necessary. Multimedia teaching essentially enriches the teaching methods of the traditional classroom, and it makes an indelible contribution to the improvement of the standards and effect of teaching. This paper research and discusses how to build the structure of the virtual laboratory system using a wireless network, the key technology and its implementation.

1 INTRODUCTION

Wireless LAN, also known as WLAN (Wireless LAN), is generally used for the broadband home, inside of the building and the park, the typical distance covers from tens of meters to several hundred meters, the current technology uses 802.11a/b/g series. As an alternative or extension of the traditional network cabling, WLAN uses wireless technology to transport data, voice and video signals in the air. The problems encountered in the original wired network are solved immediately with the invention of Wireless LAN, it can make any user expand and extend wired network. As long as wireless communications are achieved on the basis of the wired network through a wireless access point, wireless bridge, wireless card and other wireless devices, it can provide all the features of the wired LAN and achieve Mobile applications with changing and extending the network as the needs of users at random. Wireless LAN liberates individuals from the desk edge, so that they can anytime, anywhere access information, which improves the efficiency of the staff office.

Virtual laboratory technology is a unity of software technology and theory of cognitive simulation in the final analysis. As these two areas are currently in a period of rapid development, any new ideas, new technologies, new methods, new theories proposed will promote the development of intelligent virtual laboratory technology. In recent years, the data trend of network application is growing, pressure of network capacity is increasing, so that the required capacity of the wireless network in the next 10 years will be tens or even

hundreds of times Compared to now. It can override the global science and technology laboratories, science laboratories in the future develop to the digital, networked and intelligent inevitably (Li & Wang 2008).

2 THE KEY TECHNOLOGIES OF BUILDING A VIRTUAL LABORATORY TYPE

The first step of deploying WLAN indoor is to determine the number and location of access points, as well as to expand the area covered by a set interconnect access points. The WLAN can not to be Connectedness if it has gaps in the coverage area. Installation can determine the location and number of access points according to site surveys. Site surveys can weigh the actual environment and user needs, which include the coverage frequency, channel usage and throughput requirements. Direct-sequence technology transmit data used a 22 MHz channel, and the 2.4 GHz band has three non-overlapping 22 MHz channels, ranging from 2.4 GHz to 2.483 GHz. The coverage area can eliminate the channel overlap and gap of coverage area with the use of these three channels. The overlap of these three channels will not interfere with other and can broadcast area overlapping. Through the use of these three channels, direct-sequence WLAN can achieve a certain degree of redundancy.

As well as receiving radio programs, the communication between the mobile users and the access point is becoming more and more difficult with their moving away from the access point, and the WLAN can improve the transmission rate by reducing the reliability, the opposite can also reduce the transmission rate to ensure reliability. Many programs can ensure reliability through the multi-rate technology; this is an important feature of enterprise-class WLAN. As mobile users are moving away from the access point, the transmission throughput will be reduced gradually.

Because of rich virtual teaching content of the new equipment, it requires a higher bandwidth if WLAN want to meet the transmission of data, voice, images and other files, and the number of users more will be intensive. This program will provide maximum bandwidth to all users. It can achieve 11 Mbps wireless coverage, reduce the wireless power output, so that the cell size is substantially reduced; and the use of a large number of access points can meet the needs of users of high-density.

Wireless virtual laboratory structure shown in Figure 1. Teachers and students can implement interactive teaching and learning through a wireless network. Depending on the different of classes opened, the number of students can control. Any spatial arrangement can be achieved between the virtual device (touch screen), wireless projectors, teachers, students according to the needs. The materials of teachers can intercommunicate via a laptop with a wireless network, so that it can ensure the teaching materials are updated. Through wireless extension, wireless virtual network can interact with other Training Room and the Department. As needed, The network can connect through an interface with the campus network, so that it can ensure that students study independently in the dormitory (T & Xu 2008, F & T & KR 2007).

3 VIRTUAL SYSTEM DESIGN

Virtual assembly is the product 3D assembly process which is simulated in truth according to figure and precision characteristic of product design. And user is permitted to control the product 3D assembly process in truth by interaction mode. And it can test whether the product is fit. It stress on the simulation process for product physical assembly process.

All the virtual assembly come turely based on 3D model. Its quality straightly effect on authenticity of the virtual environment. Because the large-scale complex assembly model can not be built on the Virtools5.0 software, the modeling software is need to model three-dimensional entities firstly. The equipment model is a mechanism model, so parts model is built on AutoCAD which is the professional 3D mechanism drawing software. Afterwards, the model built on AutoCAD is imported into the 3dsmax 2009 software on which it is rendered. And it is exported to Virtools 5.0 finally.

Primarily, it must be analyzed elementary before the model being built, and the complicated mechanical equipment must be decomposed to some simple parts which are easy built into 3D model. After the equipment model is built, it is imported into the 3dsmax 2009 software which can endow the model with materials and textures. Accordingly, the model can be made more realistic and decreased some unnecessary polygon. That can improve the display refreshing rat.

Parts collision and distance estimation for assembly are key technologies of the virtual assembly. It can be achieved by the proprietary collision BBS technology and VSL design which are provided with the Virtools 5.0 software. Afterwards, distance estimation and collision estimation for virtual assembly are showed in Figure 2 and Figure 3. Finally, a virtual assembly figure for the automata of a equipment is show in Figure 4, and some virtual tools is put in the left of the Figure 4.

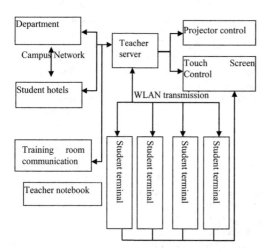

Figure 1.   Struction of the wireless network.

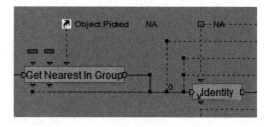

Figure 2.   Distance judgment.

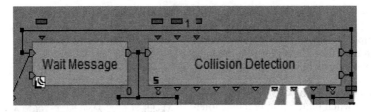

Figure 3. Collision judgment.

Figure 4. Virtual assembly.

4 SUMMARY

With the development of science and technology, more and more advanced equipment is used to the army. The internal structure of new equipment is being more complex with more powerful features of new equipment, which makes the service maintenance and repair of new equipment faced with severe challenges. So in the teaching of new equipment, the integrated use of multimedia tools is necessary. Multimedia teaching essentially enriches the teaching methods of the traditional classroom, and it makes an indelible contribution to the improvement of the standards and effect of teaching. However, the range of modern means of education is extensive, it needs to integrate the variety of resources, starts from a variety of modes such as listening, say, reading, writing, watching etc. The other hand, modern education mode must break the limit of traditional mode fundamentally. It not only enriches the teaching methods, and makes the teaching and learning activities become easy to carry out, so that students learn and teachers answer interaction with students anywhere, the barriers of space and time on the teaching are thoroughly broken. The forms of multimedia teaching is new, but it still have to rely on traditional classroom instruction, subject to the limitations of time, space, teaching conditions and other factors, It only enriches the teaching method, but does not fundamentally change the mode of teaching (Tian & Georganas 2002).

The rapid development of wireless LAN technology is gradually compensating for the functions that the wired network can not do. Ethernet services for desktop PC and mobile users can be achieved, as long as a certain amount of AP (wireless access point) is arranged according to the structure of the building with using wireless technology, plus a small amount of wiring. Wireless users which distributes in the wireless area that covered by the AP access point can connect with the school internal network, and can do online seamless mobility roaming in the inter-AP, as long as the AP is distributed reasonably in accordance with the honeycomb structure, in accordance with the principle of the wireless access point in the same region supporting up to three independent channels (802.11b and 802.11g standard are so). (In China AP wireless access point can support 13 channels, the three channels that are 1Ch, 6Ch; 11Ch is independently without disturbing each other). If the network of the entire Department is unified planned, wireless users of the Department can access into the wireless network in the Department office area. At the same time, the wireless encryption mechanism of 128 WEP can guarantee the security of the network. With the constantly enhance of the needs of Department teaching mission for the new equipment, the mature wireless LAN technology is the best way to achieve this goal. The network architecture of wireless LAN is simple and flexible, low input costs, easy maintenance, and it can also provide all services such as cable network file sharing, office automation, e-commerce and ERP that wire network can.

The software development platform for the virtual system of the laboratory is built on Virtools 5.0 of the 3dvia inc in France and Visual C++6.0. The equipment model is built on Auto CAD 2007 which is the software of three-dimensional design of the AutoDesk inc. 3d animation for the system is designed on 3dsmax 2009 of the AutoDesk inc. The database of the system which deposit falut knowledge base, parts library, assistant tool library, Scanning Tool library and service support samples is built on Access 2003. There are much more parts in an equipment model ant it is always belong to a large-scale complex assembly model. Good effect must be got on designing the large-scale complex assembly model by the technology which provided on Solidworks.

A virtual laboratory with wireless net is made by VR softwares, witch build an interactive, intelligent and wireless Multimedia 3D environment. Each visual 3D object represented an experiment object in the 3D environment. By input device such as mouse, keyboard, touch screen and joy stick, a user can carry through virtual experimentation or long-distance control experimentation. It is the foundation which a virtual laboratory with wireless net can achieve effectively being combined with multimedia, network and virtual instrument technology. The wireless net provides a basic platform for virtual laboratory. It is a basic mode to achieve a virtual laboratory internet. It must be a good study base to new equipment being trained rationally.

REFERENCES

Akyldiz I.F. & Melodia T. 2007. A survey on wireless multimedia sensor network. Computer Networks 51(4):921–960.

Li C. & Wang P. 2008. A cluster based on demand multi-channel MAC protocol for wireless multimedia sensor networks. Proceedings of the IEEE International Conference on Communications: 2371–2376.

Quazi T. & Xu H.J. 2009. Quality of service for multimedia traffic using cross-layer design. IET Communications 3(1): 83–90.

Tian D. & N. Georganas. 2002. A coverage-Preserving Node Scheduling Scheme for Large Wireless Sensor Networks.

A self-repairable and secure data storage scheme in Unattended WSNs

W. Cheng, Y. Li, Y.M. Zhang & D. Li

School of Electronics and Information, Northwestern Polytechnical University, Xi'an, China

ABSTRACT: The mechanisms to maintain redundancy are essential to distributed data storage in Unattended Wireless Sensor Networks (UWSNs). To address this challenge, we propose a self-repairable and secure data distribution scheme in this paper. We take the advantages of both key evolution and Self-Repairing Codes (SRC) to guarantee Forward Secrecy (FSe) and probability Backward Secrecy (BSe) for repairing, during the period of redundancy maintenance. To improve the security level, different strategies are adopted for repairing. Detailed analysis show the superiority of the proposed scheme in comparison with traditional Erasure Codes (EC) based scheme in UWSNs.

1 INTRODUCTION

Recently, Unattended Wireless Sensor Networks (UWSN) has become a subject of attention in the security research community of sensor networks (Pietro & Mancini 2008, Ma & Tsudik 2008, Pietro & Ma 2008). The most important feature of a UWSN is that sink is not continually present in the network. Wang & Ren (2009) take advantage of secret sharing and Reed-Solomon Codes, adding data redundancy to provide resilience to node invalidation and Byzantine failure. Ren & Oleshchuk (2010) have proposed a secure and reliable scheme for distributed data storage in UWSNs. To maximize security level of data and optimize data reliability, a constrained optimization data distribution scheme was proposed which also is based on Reed-Solomon Codes.

However, an important issue is not addressed in all above works, which is *Redundancy maintenance*. Because of sensor node failures, or attacked by adversary, data redundancy usually would be reduced or lost in UWSN. This redundancy can be achieved using either replication (Pietro & Mancini 2008), or erasure coding (Wang & Ren 2009, Ren & Oleshchuk 2010). Erasure coding techniques play a prominent role in providing storage efficient redundancy, and are particularly effective for secure and reliable data storage in UWSN, while redundancy would be lost over time because of various reasons such as data damage, sensor failure or attacked by adversary, so the mechanisms to maintain redundancy are essential.

We firstly describe the network and adversary model, then we highlight the self-repairing codes for distributed storage system, which is followed by the proposal of the novel secure distributed data storage and redundancy maintenance scheme in UWSN.

2 PROPOSAL

2.1 Network model

We consider an UWSN that consists of N sensor nodes. It can be formulated as an undirected graph $G(N, E)$, and each sensor s_i has nb_i neighbors. There is a mobile sink that visits the UWSN periodically to collect data. The time interval between the current visit and the previous visit is denoted as T. The sensor s_i generates data at each round, and the data generated at round r is denoted as D_i^r. Once a data D_i^r is generated, it is stored locally, and waits until an authorized mobile sink offloads them. Each sensor has the ability to perform *one-way* hashing and symmetric key encryption. We assume that the mobile sink is a trusted party which cannot be compromised. Additionally, the mobile sink will re-initialize the secret keys and reset the round counters when the mobile sink visits the network.

2.2 Adversary model

We focus on a Mobile Adversary that prefers roaming in the UWSN while the mobile sink is absent. We refer to it as *ADV* hereafter. *ADV* can compromise a sensor node s_i at round r_1, and release it at round r_2. Between round r_1 and r_2, *ADV* can compromise sensors during a time interval T, while it would not interfere the communication between nodes, and would not rework any data sensed by, or stored on sensors it compromises. *ADV* can monitor and the communications, and can also randomly select some sensor to physically corrupt them. In this occasion, the sensors lose the functionality and data which stored on the node.

2.3 Self-repairing codes

Self-repairing codes (Oggier & Datta 2011) is designed to suit for networked storage systems, which encode k fragments of an object into n encoded fragments to be stored at n sensors; meanwhile encoded fragments can be repaired directly from other subsets of encoded fragments by downloading fewer messages than the size of the complete data. A fragment can be repaired from a fixed number of encoded fragments, the number depending only on how many encoded blocks are missing and independently of which specific blocks are missing.

The data encoding is as follow:

1. Take data D of length M, with k a positive integer that divides M. Decompose D into k fragments of length M/k

$$D = (d_1, ..., d_k), \quad d_i \in F_{2^{M/k}}$$

2. Take a weakly linearized polynomial with coefficients in $F_{2^{M/k}}$ and encode the k fragments as coefficients, namely take $p_i = d_{i+1}, i = 0, ..., k-1$, thus

$$p(X) = \sum_{i=0}^{k-1} p_i X^{2^i} \tag{1}$$

3. Evaluate $p(X)$ in n non-zero values $\alpha_1, ..., \alpha_n$ of $F_{2^{M/k}}$ to get a n-dimensional codeword $(p(\alpha_1), ..., p(\alpha_n))$, and each $p(\alpha_i)$ is given to sensor s_i for storage.

There is a deterministic code construction called Homomorphic Self-Repairing Code (HSRC), which have nice symmetric structure. HSRC self-repair operations are computationally efficient. It is done by XORing encoded blocks, each of them containing information about all fragments of the data.

2.4 Redundancy maintenance

By using erasure code in UWSN, there is a naïve approach on redundancy maintenance. A (k, n) erasure code encodes a block of data into n fragments, so that any k can be used to reconstruct the original block. While the naive approach to repair a single missing fragment will require that k encoded fragments are first fetched in order to create the original data, from which the missing fragment is recreated and replenished.

By using SRC in UWSN, data storage scheme could achieve excellent properties in terms of maintenance of lost redundancy in the storage system, which has low-bandwidth consumption for repairs, with flexible strategy. Consider the following scenario for an example: For any choice of a positive integer k that divides M, we work in the finite field $F_{2^{M/k}}$, and it is convenient to use ω, which is the generator of the multiplicative group $F_{2^{M/k}}^*$ When HSRC(15, 4) was used for distributed data storage in UWSN, first we evaluate $p(X)$ in $\omega^j, 0 = 1, 2, ... 14$, then the 15 encoded fragments $\{p(\omega^0), p(\omega^1), ..., p(\omega^{14})\}$ are distributed to the 15 neighbors. Suppose sensor s_i which stores $p(\omega^5)$ fails. A new neighbor can get $p(\omega^5)$ by asking for $p(\omega^2)$ and $p(\omega)$, Oggier & Datta (2011) show the operation for F_{16} arithmetic, due to the properties of weakly linearized polynomial, we have:

$$p(\omega^2) + p(\omega) = p(\omega^2 + \omega) = p(\omega^5) \tag{2}$$

Thus the lost redundancy of distributed data storage can be retrieved.

2.5 Proposed scheme

To provide FSe for repairing, we adapt key-evolution to update secret key K_i at each round by applying hash function: $K_i^r = h(K_i^{r-1})$. Due to one-way property of hash function, ADV cannot derive the previous rounds key before the sensor was compromised. Thus, FSe is provided. The ADV which holds the secret key $K_i^r, r \in [r_1, r_2]$ still can derive the future key which will be used in the following rounds, since between round r_1 and r_2, the ADV is residing in s_i and it could decrypt the data by mimicking key update. However, BSe can be probabilistically achieved if sensors cooperate with their neighbors.

The new scheme contains the following steps:

Step 1: Initialization.
The mobile sink picks a secure hash function $h(.)$, and a master key s_i. Before deploying each sensor node, the mobile sink preloads to the sensor hash function and initial data encryption keys for each sensor $K_i^0 = K_i$. Here, K_i is computed as $K_i = h(K_m \| i)$. In the end of each round, the encryption key are updated as $K_i^r = h(K_i^{r-1})$. Thus, the mobile sink only needs to store a single master K_m and all round keys K_i can be derived as needed.

Step 2: Distributed data storage.
Each sensor s_i firstly generates a keyed hash value with round key K_i^r by $MAC_i^r = h(D_i^r \| K_i^r)$ and then a plaintext data that consists of D_i^r, MAC_i^r and r and s_i, which is encrypted by using current round key K_i^r. The encryption data is denoted as:

$$EncText_i^r = Enc(K_i^r, \{D_i^r \| MAC_i^r \| r \| s_i\}) \text{ then } D_i^r$$
$$\text{is equipped in } EncM_i^r = \{EncText_i^r, r \| s_i\}$$

Then s_i employs $HSRC(n,k)$ to encode $EncText_i^r$ into n data fragments $\{p_{i,1}^r, ..., p_{i,n}^r$, then s_i selects n neighbors and sends one randomly selected distinct data part $m_{i,j}^r = \{p_{i,j}^r, r \| s_i\}$ to s_i by using pair-wise secret key $K_{i,j}$

Step 3: Redundancy maintenance.

s_i and its neighbors cooperate with each other to monitor the redundancy level of data storage. Once the redundancy is lower, a repairing scheme based on self-repairing codes is carried out in different manner: *direct downloading* or *mobile-agent computing*. In the former way, sensor downloads all corresponding data fragments, such as $\{p(\omega^2), p(\omega)\}$ which are needed for repairing $p(\omega^3)$, directly from other sensors. While in the latter way, a mobile agent is employed for repairing, so that computing of repairing operation is carried out hop by hop: the sensor s_l which has $p(\omega)$ sends a mobile agent to the sensor s_k which has $p(\omega^2)$, and compute the $p(\omega^3)$ on s_k, then s_k sends the result to the sensor which ask for repairing $p(\omega^3)$.

3 PERFORMANCE ANALYSIS

When SRC was used for distributed data storage in UWSN, redundancy maintenance is achieved. However, the repairing operation brings extra risk, that it is easier to compromise BSE during repairing than at other times.

The proposed scheme can guarantee *FSe* of repairing, because The ADV cannot derive the previous key from the current key it holds due to the one-way property of hash function. Hence even the *ADV* compromised the sensor which asks for repairing and other sensors to get enough fragments, it only obtain the whole $EncText_i^r$, but still cannot decrypt the data encrypted and stored in the previous rounds. Therefore, the *FSe* of repairing is guaranteed.

Since the naive approach using erasure code to repair a single missing fragment will require that k encoded fragments are first fetched in order to create the original data, from which the missing fragment is recreated and replenished. This essentially means, once the *ADV* compromise the repairing sensor and the sensor which holds the secret key, the *BSe of* Redundancy repairing can be compromised. Thus the probability that *ADV* compromise the *BSe* of repairing is:

$$P_{RB\_EC} = P_c\{s_i\} \cdot P_c\{s_j\} \tag{3}$$

where is, $P_c\{s_i\}$ $P_c\{s_j\}$ is the probability that *ADV* compromises s_i, *which* holds the secret key, and s_j, which asks for repairing.

In the proposed scheme, the BSe of repairing can be compromised by ADV, only if the following

three conditions are satisfied: 1) sensor s_i which holds the secret key K_i^r is compromised; 2) sensor s_j which asks for repairing is compromised; 3) Except m fragments which are compromised on s_j during the repairing, the *ADV* also compromises $k-m$ neighbors to get enough fragments to recover whole data. Therefore, the *BSe* of repairing can be *probabilistically* achieved through sensors cooperation. Thus the probability that *ADV* compromise the *BSe* of repairing is:

$$P_{RB\_SRC} = P_c\{s_i\} \cdot P_c\{s_j\} \cdot P_c\{nb_i\} \tag{4}$$

where $P_c\{s_i\}$, $P_c\{s_j\}$ is the probability that *ADV* compromises s_i, which holds the secret key, and s_j, which asks for repairing. $P_c\{nb_i\}$ is the probability that *ADV* compromises neighbors to get rest fragments to recover whole data. For different manner, by *direct downloading*, the $P_c\{nb_i\}$ can be expressed as:

$$P_c\{nb_i\} = \rho(k,d,k) \cdot \prod_{j=1}^{k-m} P_c(S_{nb,j}) \tag{5}$$

where $S_{nb,i}$ is the compromised neighbor sensors, while by *mobile-agent computing*, the $P_{c\_MA}\{nb_i\}$ can be expressed as:

$$P_{c\_MA}\{nb_i\} = \rho(k,d,k) \cdot \frac{C_{n-1-m}^{k-m}}{C_{n-2-m}^{k-m}} \cdot \prod_{j=1}^{k-m} P_c(S_{nb,j}) \tag{6}$$

where m is the expected number of fragments that need to be downloaded, $\rho(k,d,k)$ is the fraction of sub-matrices of dimension $k \times d$ with rank k out of all possible sub-matrices of the same dimension, which is special used for reconstruction in SRC (Oggier & Datta 2011). In the scheme using erasure codes, the expected number of fragments for repairing is:

$$m_{EC} = k + n - x - 1 \tag{7}$$

where x is defined as the current number of fragments that are available. While in the scheme using SRC, the repairing traffic is

$$\begin{cases} m_{SRC} = 2 & \text{if } x \geq (n+1)/2 \\ m_{SRC} = 2p_2 + k(1-p_2) & \text{if } x < (n+1)/2 \end{cases} \tag{8}$$

where $p_2 = 1 - (1 - (x/n)^2)^\delta$ is the probability that only two fragments are enough to recreate the missing fragment, and the diversity δ of SRC is defined as the number of mutually exclusive pairs of fragments which can be used to recreate any specific fragment.

In Figure 1, we compare the repair cost per lost fragment, achieved using the proposed SRC with that of traditional erasure codes. We observed that while SRC is effective reducing repair cost to carry out maintenance of lost redundancy in distributed storage for UWSN, when eager repair is used that means the threshold of available fragments is high. However a strategy like lazy repair with low threshold while using traditional EC may outperform SRC in terms of mean cost, but choosing a low threshold of available fragments may lead system vulnerability and have spiky bandwidth usage.

Without loss of generality, we assume all sensors have same probability p to be compromised. As shown in Figure 2, we observe that the proposed scheme can guarantee the better probabi-

listic *BSe* for repairing with respect to the naïve EC scheme, no matter what manner and strategy is used. When mobile agent is adopted with eager repairing, the proposed scheme has the best *BSe*. We also observed that mobile agent and eager repair strategy can help the scheme for UWSN to further improve probabilistic backward secrecy.

4 CONCLUSION

In this paper, we proposed self-repairable and secure scheme for distributed data storage in UWSNs. We take the advantages of both key evolution and Self-repairing Codes to guarantee *FSe* and probability *BSe* for repairing, during the period of redundancy maintenance. To improve the security of repairing, we further adopt different strategies: Direct Downloading (DD) or Mobile-Agent computing (MA), and lazy or eager repair. Through detailed security analysis, we show that compared with Erasure Codes (EC) based approach, the proposed scheme is more robust to support the redundancy maintenance for UWSN.

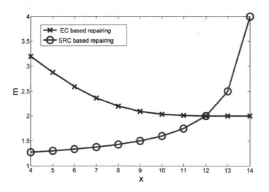

Figure 1. Repair cost of the Erasure Codes (EC) based and the Self-Repairing Codes (SRC) based scheme, where m is repair traffic per lost fragment, x is number of available fragments, when n = 15, k = 4.

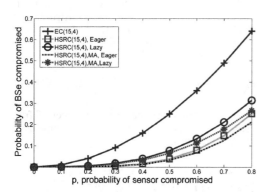

Figure 2. BSe for repairing of the Erasure Codes (EC) based and the Self-Repairing Codes (SRC) based scheme, where different strategies were used: Direct Downloading (DD) or Mobile-Agent computing (MA), and lazy repair or eager repair.

ACKNOWLEDGMENT

This work is supported by China Postdoctoral Science Foundation under Grant No. 2012M512026.

REFERENCES

Ma, D. & Tsudik, G. 2008. DISH: Distributed Self-Healing. In *Proc. 10th Int. Symp. On Stabilization, Safety, and Security of Distributed Systems (SSS '08)*, Detroit, MI, USA, Nov. 2008:47–62.

Oggier, F. & Datta, A. 2011. Self-repairing homomorphic codes for distributed storage systems, in *Proc. IEEE INFOCOM 2011*, Shanghai, China, April, 2011:1215–1223.

Pietro, R.D. & Mancini L.V. 2008. Catch me (if you can): data survival in unattended sensor networks. In *Proc. IEEE PerCom '08*, Hong Kong, Mar. 2008:185–194.

Pietro, R.D. & Ma, D. 2008. POSH: Proactive co-Operative Self-Healing in unattended wireless sensor networks. In *Proc. IEEE Symp. on Reliable Distributed Systems (SRDS '08)*, Napoli, Italy, Oct. 2008:185–194.

Ren, Y. & Oleshchuk V. 2010. A Scheme for Secure and Reliable Distributed Data Storage in Unattended WSNs. In *Proc. IEEE GLOBECOM 2010*, Miami, USA, December 2010: 1–6.

Wang, Q. & Ren K. 2009. Dependable and secure sensor data storage with dynamic integrity assurance. In *Proc. IEEE INFOCOM 2009*, Rio de Janeiro, Brazil, Apr. 2009:954–962.

Control Engineering and Information Systems – Liu (Ed)
© 2015 Taylor & Francis Group, London, ISBN 978-1-138-02685-8

College fixed assets management system based on RFID technology

Z.R. Li, Y. Hou & D.H. Sun
North China Institute of Aerospace Engineering, Langfang, Hebei, China

ABSTRACT: This paper introduces a kind of college fixed assets management system based on RFID technology, the system uses LPC-11C14 chip as the control core of the reader, complete read-write operation with the electronic tag, label and transmitted information to PC for storage and management through the serial communication. The system manage the information in a specific encoding, and put the collision algorithm ensures that the system reliability for reading and writing. Using this system can manage the fixed assets more conveniently and accurate, has a very high practical value.

1 INTRODUCTION

Along with our country higher education scale unceasing expansion, the hardware conditions continue to improve, the amount of fixed assets has sharply increased. In the past, all the fixed assets are accounted for by the school registration, management, asset management and statistical method using manual entry database. Because of this method mainly rely on artificial participation, so that greatly increased the workload of staff, and inevitably the registration error; at the same time, once the fixed assets position changes and did not have time to modify the data information, will produce the asset lookup problem, moreover can cause the loss of state assets (LANDMAC 2003). So, how to complete the state-owned fixed assets use of advanced technology and management is a very meaningful research topic.

This paper introduces a method of using RFID technology to complete the management of fixed assets. RFID technology, radio frequency identification technology, is a new automatic identification technology developed in the nineteen eighties. Radio frequency identification technology without manual intervention, it can work in various environments. In this paper a laboratory equipment management as an example, to verify the feasibility of this method and advanced.

2 SYSTEM COMPOSITION

The system is composed of the electronic tag, the reader and the computer, as shown in Figure 1.

Electronic label is composed of the tag antenna and the tag chip, generally attached to the surface of equipment, store the equipment management information (Rao 1999). Electronic tags commonly

divided into active and passive. The active electronic tags are powered by battery, distance is identified, but due to the consumption of electric energy, so the use of the limited time; and the passive electronic tags are identified in a short distance, but do not need to consume energy, it relying on the accumulation of signal energy accumulation from the reader, so theoretically it can last forever. Therefore, the fixed assets management system in this paper used the passive electronic tags to the equipment information storage.

The reader can read and write on an electronic tag, also can be connected with a computer, put the asset information into the database for record and query. The frequency band of RFID system is determined by the frequency of the reader, the power to decide the effective distance of radio frequency identification. The reader according to the structure and technology use different can be read or read/write device, RFID system information control and processing center.

The system is divided into read or writes operation and information upload operation.

The read and write operation is completed by the reader and the tag (Rekik 2010). When the system works, the reader intermittent signal sent a certain

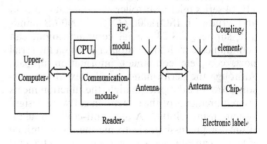

Figure 1. System chart.

frequency, the passive electronic tag with the same resonant frequency absorption and stores the signal energy. When transmitted signal reader after a short time, stop sending and enter the signal receiving mode, when the instrument stop transmitting, passive electronic label the release of stored energy is reflected back to the probe, through RFID technology and electronic label pre-stored information to the detector, or the reader will be stored data to send to the electronic tag.

When the information upload, the reader get the modulation signal sent by the tag, antenna controller is transmitted to the reader the signal processing module, demodulated and decoded by the USB line or the serial communication line and RFID middleware will effectively information sent to the computer.

3 MAIN HARDWARE

3.1 The electronic tag

Usually the frequency of the reader is known as the operating frequency of the RFID system, is usually divided into 3 ranges: 30 kHz~300 kHz, 3 MHz~30 MHz and 300 MHz~3GHz. As everyone knows, under the same transmission power, the low frequency electromagnetic wave spread longer than the high frequency electromagnetic wave. The attenuation when the electromagnetic wave spread in the medium have a direct relationship with the wavelength, the shorter the wavelength attenuation is large. However, in actual use, the choice of operating frequency is restricted by many factors, such as information exchange, power limitation, modulation, radio management regulations, cost etc. Considering various factors, this system chooses ZK-RFID604 passive tag, the tag frequency using 902~928 MHz or 860~870 MHz, communication distance can reach 8 meters, can store 2056 bits data, to meet the requirements of the system.

3.2 The reader

The reader uses the ARM Cortex-M0 LPC-11C14 series microcontroller as a core; it can read and write with ISO18000-6C, SO18000-6B standard, support RS-232 serial communication, can communicate with the computer through the serial port, complete equipment information upload. Although the performance of Cortex-M0 processor is low, but the price is high, the function meets the requirements of the system. In software design, system using slotted Aloha anti-collision algorithm for processing, electronic tags with unique ID code, can effectively avoid each other in the unit area information read-write conflict.

3.3 The computer

The main function of the computer in the college fixed assets management system is read the information and received and stores the information in the database. (Kouvelis and Li 2012) The system management software main interface using VC++ programming language, the database uses SQL Server2000. The main interface calling the communication of 5 main interface function of Interop.vtComRFID. dll dynamic library file by the COM components to achieve logistics control and management system.

4 SOFTWARE DESIGN

Software design includes open/close the serial port, Tag ID read, read and write data, database management etc ... The switch of the serial port can directly call C++ 6.0 function, database management through the completion of the SQL Server2000 operation, therefore no longer tautology, and then focuses on the information code to read and write tag and tag are described.

4.1 Information encoding

When the ZK-RFID604 passive electronic tags leave the factory, all they has a unique 94 bits ID for user identification, so users do not need to code to tag description once again, At the same time, the manufacturers obligate 256 bits memory area for user, allow users to code. The user can be encoded according to the actual demand. This system is convenient; only use 50 bits to describe the device information. Information rules were prepared as shown in Figure 2, the first string of characters written by binary code; denote the device type and unit, the second strings of characters written by BCD code, a detailed description of the current state of equipment (Stavrulaki 2010). Follow the rules not only conveniently identification, classification for the equipment, at the same time, when the information upload, it can also be converted to text information by PC software, convenient user archiving and query.

Device type	Equipment department	
10 bits	8 bits	

Date of acquisition	Warehouse	Device status
24 bits	6 bits	2 bits

Figure 2. Coding rules.

4.2 Information read and write operations

Writes a string with specified length to a specified ID tag, and then read out the string from the ID tags, the function is mainly Write Single Content and Read Tag Content, in addition to call the function about the related peripherals, the main process is shown in Figure 3.

4.3 Anti-collision processing

When the system works, if there are 2 or more electronic tags simultaneously into the operation scope of the reader, information will appear in the process of mutual interference, the interference is called the collision, which is likely to cause the information to read and write failure. The general approach is through software programming, using a certain algorithm to solve this problem. The

improvement of ALOHA algorithm is adopted in this system, the ALOHA algorithm is a method of signal random access, using electronic label control mode, scope is electronic tag into the reader, it automatically to the reader sends its own serial number, then with the reader to communication. In the process of an electronic tag transmits data, if the electronic label other also sending data, and then send the signal overlap caused by collision. Reader once a collision is detected, it will send a command to one tag to suspend data transmission, random wait after a period of time to send data. Because the transmission time for each data frame is only a small part of repetition time, resulting in a considerable interval between two data frame. So there is a certain probability, make the data frames of two tags without collision. The public channel in unit time average sending data in the T frame G and transmission path throughput calculation formula of S were: $G = \sum t *rn/T$ and $S = G * e - 2G$. n is the number of tags, rn is the frames that electronic label n to send data in T time.

According to the relationship between S and G we will know when $G = 0.5$, the maximum S value of 18.4%. This shows that more than 80% of the data path is not used; the method realizes the efficiency high costs prevent collision. But due to the simple ALOHA algorithm, and is suitable for label indefinite number of occasions, can be used as a suitable anti-collision method better to read electronic tag system.

In order to improve the throughput of ALOHA algorithm, the improved ALOHA algorithm. Time slot ALOHA algorithm is divided into a plurality of discrete identical size in ALOHA algorithm based on time, the label can only be to send data for each time point. Such labels or sent successfully or conflict, the original ALOHA algorithm to produce the conflict time interval $T = 2\tau$ reduced to $T = \tau$. According to the formula of $S = G \cdot e (-G)$ can be obtained when $G = 1$ throughput reaches the maximum value is 36.8% S. Slotted ALOHA algorithm than ALOHA algorithm for maximum efficiency doubled, but at the same time requirements of electronic label all must be controlled by the reader, thus can solve the tag collision problem effective.

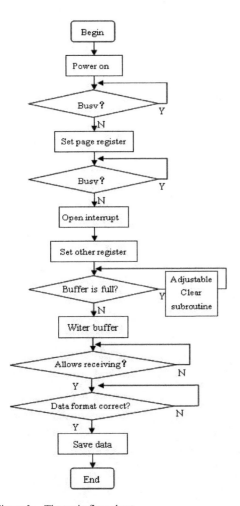

Figure 3. The main flow chart.

ACKNOWLEDGMENTS

This work was based on the research platform in LangFang city of Hebei Province: "Networking and Automation Technology Key Laboratory".

And this work was supported by the key project of the Science and Technology Department of Hebei Province (Project ID: 13214102D), and Doctoral Fund of NCIAE (Project ID: KY201006B).

REFERENCES

Euthemia Stavrulaki. 2010. "Inventory decisions for substitutable products with stock-dependent demand". *International Journal of Production Economics*. 2010 (1).

Lionel M Ni. LANDMAC. 2003. "Indoor Location Sensing using Active RFID". In *IEEE International Conference in Pervasive Computing and Communications 2003*, Dallas, USA.

Panos Kouvelis, Jian Li. 2012. "Contingency Strategies in Managing Supply Systems with Uncertain Lead-Times". *Production and Operations Management*. 2012 (1).

Rao. 1999. "An over view of backscattered radio frequency identificatinn system (RFID)", In *Microwave Conference, 1999* Asia Pacific, Volume: 3, 746–749.

Yacine Rekik. 2010. "Inventory inaccuracies in the wholesale supply chain". *International Journal of Production Economics*. 2010 (1).

Control Engineering and Information Systems – Liu (Ed)
© 2015 Taylor & Francis Group, London, ISBN 978-1-138-02685-8

The AODV routing protocol performance analysis in cognitive Ad Hoc networks

T. Wang & R.H. Qiu

College of Information Sciences and Technology, Donghua University, Shanghai, China

ABSTRACT: In order to simulate cognitive Ad Hoc network routing protocol, build cognitive radio environment based on NS2, the control unit Multi Control Unite (MCU) and the transmission control unit Prop Control Unite (PCU) are proposed. The MCU modulecombined the LL module and NetIF module, collected the usage information of channel, transmitted to RTagent and then decided which channel should be used to transmit data. The PCU module, sets transmission power for different channel, protects the primary user's communication and changes CR users' transmission power dynamically, then run AODV routing protocol on the platform and analysis the latency and throughput of AODV routing protocol. Analytical and simulation results show that the proposed platform can effectively simulate the cognitive network and can greatly improve the network throughput, reduce the average delay of nodes.

1 INTRODUCTION

With the development of wireless networks, the Ad Hoc network technology has got very fast development which does not depend on fixed infrastructure. The cognitive radio technology is mainly proposed for solving the lack of spectrum resources, the unauthorized user, which has the function of cognitive can communicate in the authorized spectrum in a certain "Opportunistic Way" under the condition of no or only a few traffic and achieve to the mechanism of spectrum sharing with the premise of without affecting the main user communication. Cognitive radio technology for free utilization of spectrum resources is very suitable in Ad Hoc network. Cognitive Ad Hoc is based on the wireless Ad Hoc network architecture [1]. When the cognitive radio technology is applied to the low power and multiple hops Ad Hoc networks, which need a new MAC protocol and a new routing protocol to support the realization of the system [2] because of its multiple hops communications, network topology changed dynamically, spectrum status changed over time and physical location features. Cognition Ad Hoc network belongs to the multi-channel Ad Hoc network, and many scholars have been done some researches in spectrum estimating [3,4], also have obtained certain research results. The multi-channel model presented in literature [2], modify the corresponding AODV routing protocol and change channel allocation mechanism at the same time, did not consider power control [5]. Adopted double interface, multi-channel mode of transmission, the transmission power is not controlled. The CM module presented in [6] manages the channel

information, but not introduce the transfer mode and the transmission power, spectrum detection method is proposed in the literature [7], adopt the single interface multi-channel transmission mode.

This paper proposes a network simulation tool of the cognitive radio based on NS2 simulation platform, Using MCU to complete the transmission information about interface and channel. PCU is used for node transmission model controlling and the physical transmission controlling. Finally, imply AODV routing protocol simulation under the simulation platform of network.

2 SYSTEM DESIGN

2.1 Original module

The existing NS2 simulation tools does not support multichannel simulation, many scholars both at home and abroad studied on the multichannel interface, have obtained certain research results. Typically are Hyacinth scheme [8,9] and Ramon scheme [10]. Ramon scheme, widely used in some researches, copy the information from network interface module to the Rtagent, which decide how to manage the usage of channel. In order to complete the function of multichannel, multi-interface, each node needs to add the same interface as the channel, here are the code run on Ramon scheme:

```
if {[info exits channum]} {
for {set i 0} { $ i< $ numifs} {incr i} {
for {set i 0} { $ i< $ numifs} {incr i} {
$node add-interface $chan[expr $i* channum];
  }
}
```

In ns-lib.tcl, channel number "channum"and interface number "numifs" are set to realize the single-channel, single-interface, single-channel, multi-interface, multi-interface, multi-channel switching. If both set "channum" and "numifs" said multi-interface, multi-channel expansion, if only set "numifs" said multi-interface, single-channel expansion, for a single-interface, single-channel, chan (0) is the default channel. The design may also bring bad effects to the effective of the network, for many nodes in the network. Though there may be one node, which has multi-interface can be recognized as many nodes in the network.

The original node structure model is shown in Figure 1. Address Resolution Protocol (ARP), LLmodule, MAC module, NetIF module and channel are included in this model, completes communication between nodes through some network protocols.

1. ARP module: check MAC address of destination node, put packets in Ifq if the ARP knows the MAC address of next node.
2. LL module: run many network protocols; complete the restructuring of data packets.
3. MAC module: monitoring the packets through RTS/CTS/DATA/ACK.
4. NetIF module: receive parameters and record them in packet head.

2.2 Extension module

The Channel Extension Modal is shown in Figure 2. For cognitive Ad Hoc networks, multi-channel, multi-interface model is needed, extent the original module, use MCU and PCU to compact the whole model. The MCU module, combined the LL module and NetIF module collected the usage information of channel, transmitted to

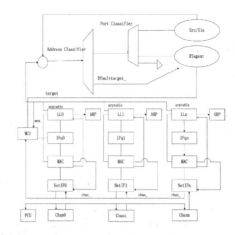

Figure 2. Extension module.

RTagent and then decide which channel should be used to transmit data. The PCU module, sets transmission power for different channel, protects the primary user's communication and change CR users' transmission power dynamically.

2.2.1 MCU module

The MCU module, combined with the LL module and NetIF module, collected the usage information of channel, transmitted to RTagent and then decide which channel should be used to transmit data.

MAC module: MAC module can monitor the packets through RTS/CTS/DATA/ACK. The usage of channel can be monitored by transmitting packets.

The *chan_, defined to point to the current channel and sent to the NetIF module to create the available channel array avail_chan_[]. The available channel can be found in this layer, shown in PCU module. The MCU module received the *chan_ and transmitted to Rtagent, the routing agent choose the proper channel to send packets.

Add cognitive module to the NetIF module: create the available channel array avail_chan_[]. The utlizenum_ is set to record the number of using each channel, the more using channel is used as common channel completing data transmission. If the common channel is busy, the other channel should be taken in avail_chan_[]. The NetIF module will record these parameters and send them to MCU via mcu_.

In ns-lib.tcl, the "MCUmodule" procedure, used to provide the interface information to the MCU module should be added. The mcu_ is defined in node-config:

set args[eval $self init-vars $args]

$self instvar addressType_ routingAgent_ propType_ macTrace_\

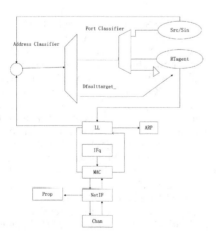

Figure 1. Original module.

routerTrace_agentTrace_movementTrace_
channelType_ channel_ mcu_ \

In ns-mobilenode.tcl, the MCUmodule are added to the procedure of creat-wireless-node.tcl, and then add MCUmodule.h/cc in C++ compiler layer. The MCUClass, inherited from TCLClass and MCU need to be declared in C++ compiler layer. The MCU completed receiving packets from the upper layer, dealing with the inner data and sending packets to lower layer.

The MCU module can handle the packets sent by the lower layer, avoid sending interface information to Rtagent, which must handle such changes immediately, improve the network performance. The module can also help the Rtagent make some decisions about allocating channel to each node.

2.2.2 PCU module

PCU Module PCU module is mainly to set transmission power for different channel, which build on the Prop module, collect the channel information and maintain a list of available channel information. For the method of power control of cognitive model, some do a lot of research at home and abroad, the traditional optimal power control technology mainly include the Lagrangian relaxation method, classical method, quadratic programming, game theory and geometric programming [10]. Power control performance index is mainly the convergence rate of using a power control algorithm and the average transmission power. In [11], the signal-to-noise ratio is the key part of power control, all the messages transferred to MAC layer, completing noise accumulation and power control. Nodes in this design, using multi-channel, multiple interfaces model, you can choose different channel for transmission, choose what kind of channels are determined by the Packet header.

MAC module can monitor the packets through RTS/CTS/DATA/ACK.PVU module receive the *mcu_ from the MCU module and the array prop[] are defined to store the SNR in channel via the *mcu_. The max transmission power maxp_ and the minute power minp_ are changed by the new value of SNR via monitor the packets send from Phy layer to MAC layer. The CR nodes use the minute transmission power when sending data to other nodes in order to protect authorized user normal communication, and use the max power to send data when the channel is available.

3 SIMULATION

In simulation, each user is equipped with two transceivers, one for control channel monitoring control information, CR users transmit through the RTS-CTS selection criteria, choose the data channel with the shortest distance to communicate in the absence of awareness to the main user; When there is awareness to the main user, the CR user nodes can't choose the authorized channel even if there is no channel available in the channel list, and the primary users can use authorization channel at any time. Another transceiver is fixed on the data channel for data transmission on the data channel.

In our simulation, topology environment was set at 1000 m * 1000 m, and the specific parameters are shown in Table 1.

Throughput in multichannel and single-channel represented by the Figure 3, changed with the node numbers. In the simulation model, compared the throughput in multi-channel and signal-channel through switching mechanism along with the change of nodes. Both throughput increases as the number of nodes adds, which means that the times of nodes communicating with each other increases and then the throughput increases. In multi-channel model, nodes can choose different channel to communicate, the value of throughput is higher than the single-channel.

Figure 4 shows the average delay along with the number of nodes increase, comparing the simulation results to the CA-AODV protocol in another multi-channel mechanism presented in [2]. Both are

Table 1. Parameters.

Parameters	Value
The transport layer protocol	UDP
Type of flow	CBR
Packet size	512 bytes
Queue length	50
Simulation time	20s
Sending rate	120 packet/s
Routing protocol	AODV
Moving range	1000 m*1000 m

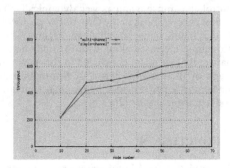

Figure 3. Single-channel and multi-channel.

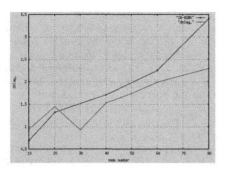

Figure 4. Average delay changed with node number increase.

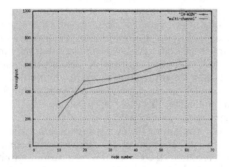

Figure 5. Throughput increased along with the node number change.

generally increased along with the number of nodes added, this is due to increase the number of nodes, corresponding communication times between nodes increases, the average time delay increase. The avail_chan_[] provides the available channel to each node, the time finding channel is reduced. The MCU module can handle the packets sent by the lower layer, avoid sending interface information to Rtagent, which must handle such changes immediately, improve the network performance. The module can also help the Rtagent make some decisions about allocating channel to each node. The PCU control the transmission power dynamically. The average delay is less than the CA-AODV protocol in another multi-channel mechanism. Figure 5 shows throughput increased along with the node number change. By two figure concluded in this paper, the design of the model in terms of throughput and average delay performance better than simulation model in the literature [2].

4 CONCLUSION

The simulation model of cognitive radio presented by this paper adds the MCU module and PCU module. The MCU module, combined with the LL module and NetIF module, collected the usage information of channel, transmitted to RTagent and then decide which channel should be used to transmit data and also can handle the packets sent by the lower layer, avoid sending interface information to Rtagent, which must handle such changes immediately, improve the network performance. The PCU module, in power control, setting different transmission power for each channel, protecting the communication between primary users and guarantying communication between cognitive users more smoothly, and the simulation results show the validity of the cognitive model.

REFERENCES

Chueht, Raniwal A.A. & Akrishnan R. An IEEE 802.11-based multi-channel wireless mesh network.[EB/OL]. [2005-0-12]. http://www.eesl.CS.sunysb.edu/multichannel unican.es/aguerocr/.

Jia Jtmcheng & Zhang Qian HC-MAC:A Hardware-constrained Cognitive MAC for Efficient Spectrum Management[C// Proc. of GLOBECOM'07. Washington D.C., USA: IEEE Press, 2007.

Jia Juncheng & Zhang Qian. HC-MAC:A Hardware-constrained Cognitive MAC for Efficient Spectrum Management[C]// Proc. of GLOBECOM'07. Washington D.C., USA: IEEE Press, 2007.

Li Zongshou. The Study of Thread Hoc Routing Protocol under Cognitive Radio Protocol under Cognitive Radio Circumstances. Nanjing posts and Telecommunications University, 2011.

Liu Yumei, Wu Haowen & Zhao Yijun. Design for cognitive radio Ad Hoc networks. Applied Science and Technology, 2011,38(10):44–49.

Noun C. Maulin P. & Venkatesan S.A. Full Duplex Multi- channel MAC Protocol for Multi-hop Cognitive Radio Networks[C]// Proc. of the 1st Intemmional Conference Oile Cognitive Radio Oriented Wireless Network and Communications. Orlando, Florida, USA:[s.n.], 2006.

Ramon Aguero Calvo & Jesus Perez Campo. Adding Multiple Interface Support in NS-2.[J]. Computer communications. 2007.

Shi Lei & Zhang Zhongzhao. Cooperative detection algorithm of idle spectrum in cognitive radio[J]. Electric Machines and Control. 2009(3):307–311.

Xi Zhihong, Wang Xiaoguan, A cognitive radio power allocation algorithm based on the interface temperature model[J]. Applied Science and Technology, 2010(3), 37(6):13.15.

Yang Chengwei & Len Pengpeng. NS2-based Cognitive Radio Network Model. Computer Technology, 2010, 36(5):111–113.

Zhang Jinyi, Liu Jun & Guo Wei. Cognitive wireless Ad Hoc cognitive architecture research Communication Technology, 2011(2):56–58.

Control Engineering and Information Systems – Liu (Ed)
© *2015 Taylor & Francis Group, London, ISBN 978-1-138-02685-8*

A new generation of Automatic Weather Station network service architecture design and research

G.B. Lei & C.Q. Xiong

School of Computer, Hubei University of Technology, Wuhan, China

ABSTRACT: Automatic Weather Station have been working in China for many years, it has already meets the demands of Meteorological service and environmental requirement. Currently, most of AWS only upload the local monitoring date by GPRS, and have a very single communicate method. In order to satisfied the robust transmission demand for new generation AWS. The authors provides a multi-channel integrated Auto Routing AWS network service framework, first it describe the frame work of the AWS' network service, then it explains the implementation principle of data access layer network control layer and network capacity layer. Finally introduces a model of forming a net of system.

1 OVERVIEWS

Automatic Weather Station have been working in China for many years, it has already meets the demands of Meteorological service and environmental requirement, but there is plenty of room for improvement in it's means of communication [1].

At present, the communicate method for the Automatic Weather Station is very single. Most of them can only upload the local monitoring date by GPRS. So we can't locate at the areas which the base station and wireless base station do not coverage. As a result, AWS can not provide weather information during natural disasters or failures of the device happened [2].

This paper proposes a multi-channel integrated Auto Routing Automatic Weather Station network service framework. Offering a new method to satisfied the new generation AWS's omnibearing and multi-means networked demands.

2 DESIGN FOR NETWORK SERVICE FRAMEWORK

The newly AWS provides multi-channel access for mobile network (GSM, CDMA, 3G), wired network, VHF wireless self-organized network and satellite network. Through monitoring network status, AWS can choose the best Internetwork Routing and make the best decision based on Qos and energy principle. It can also realize autonomic network and ad hoc network protocols has been adopted. According to combination of specified routing logic and dynamic routing, the system complete network agile configuration.

The frame work of the AWS for network service is composed of physical layer, data link layer, network capacity layer, network control layer and data access layer. The physical layer and data access layer are provided by the network communicate model. The network service framework for AMS was shown in Figure 1.

2.1 *Data access layer*

The Date access layer provides date sending and receiving interface for upper applications, it also provide data compression, decompression and fault-tolerant coding for application layer. Figure 2 shows how it works [3].

The data access layer of the system includes functional entities like: data encryption and

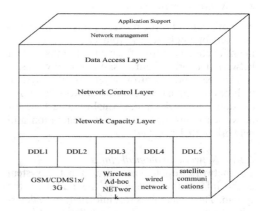

Figure 1. Automatic meteorological station network service architecture.

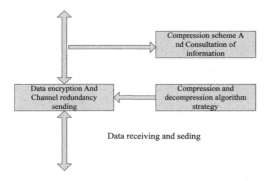

Figure 2.   The schematic of data interface.

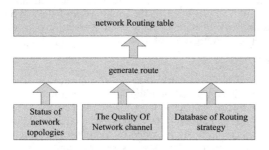

Figure 3.   The schematic of network control function.

channel redundant sending, strategies for compression and decompression, compression scheme and consulting information extraction etc.

- Data encryption and channel redundant sending,
- Strategies for compression and decompression,
- Compression scheme and consulting information extraction.

2.2   Network control layer

Network control layer provide transparent transmission for applications from end to end, their main function is routing the channel, transfer data between networks. Figure 3 shows how it works.

The Network control layer of the system includes functional entities like: network topological status, network channel quality, data base for network routing strategies, route generation and network route list, etc [4].

2.2.1   Network topological status

The network topological status for the system is the information related to the nodes of the system. The system uses this information to determine the next position. These include: vector information, geographical position information, energy information.

2.2.2   Network channel quality

2.2.2.1   Data base for network routing strategies
The routing strategies that the system can choose includes: energy routing, multi-path energy routing, directed diffusion routing, rumor routing, geographic location routing, system-specified routing.

2.2.3   Route generation

Analyze and fuse data through network topological data, network channel quality data and network routing strategies. Create network route list for the system.

2.3   Network capacity layer

Network capacity layer can screening the difference between physic networks' data transmission, it can provide uniform data transmitting capacity, realize the data QoS control and state acquisition and integration, Figure 4 shown how it works.

Network capacity layer includes functional entities like: multi-queue QoS management, QoS strategy management, relay node's state collection and report, adjacent node's state collection and report, link-local state collection and report.

2.3.1   Multi-queue QoS management

Transmit dates for local node and relay node. To different node and local upload data conducting the QoS management for data transmit. This is based on the different priority level of the nodes. Adapt data for bottom multi-data link through traffic shaping.

2.3.2   QoS strategy management

QoS strategy management provides QoS strategies for different bottom channel. Recording local node and relay node's priority level, and bandwidth reservation for system specified date.

2.3.3   Relay node's state collection and report

The delay node's state collected and reported by network capacity layer includes: network address of the node, geographical distance, energy parameter, available channel parameters, running state of the

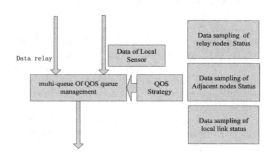

Figure 4.   The schematic of network capacity layer.

functional module of the node. Adjacent node's state collection and report

Network capacity layer collect and report all the state date for the next and previous node. This includes: the physic address of the node, geographical distance, energy parameter, running state of the functional module.

2.3.4 *Link-local state collection and report*

In order to make the right link choose for the system, network capacity layer collect and report link-local state. This includes: on-off state, channel conditions, link date congestion conditions, next node type of the link and physic distance data.

3 NETWORK APPLICATION MODE

The setting circumstance for the AWS is very complex, it often located in the mountain areas where there is a lot of geological disaster or near a river. There are some public networks like: wired access network, PWLAN (GPRS/CDMA1X/3G) in this place, but a lot of sector in this areas do not have a public network to cover. For this reasons, in order to realize a Robust AWS network, make sure the failure of parts of the node do not effect the transmission of the network, we need to use multi-channel deployment method.

Figure 5 is a typical deployment condition for a regional AMS; all the nodes in the picture form a Self-organized Network via VHF channel. At the gateway node 6 and 7, where public network is covered, configured with wired network and public wireless network communication model. In this condition, we access gateway node 6 and 7 into wired network and public wireless network (GPRS/CDMA/3G), and deploy standby gateway nodes 1 and 2 with Satellite communication module as the emergency date communication channel under disasters.

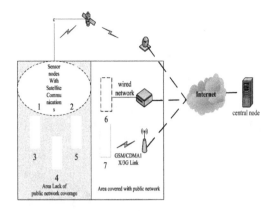

Figure 5. Overall arrangement of automatic weather station.

ACKNOWLEDGEMENTS

This research is supported by National Natural Science Foundation of China under grant number 61075059, and Key Program of Hubei Provincial Department of Education in China under grant number D20101402.

REFERENCES

Benghanem, M, "A low cost wireless data acquisition system for weather station monitoring," Renewable Energy, 2010, Vol. 35 (4), pp. 862–872.

Huang Ling, Wen Chunhua, Zhu Chunqiao, Li Chang, "Availability evaluation for automatic weather station," Meteorological, Hydrological and Marine Instruments, 2012, Vol. 29 (02), pp. 118–122.

Weather Station, Do-It-Yourself Retailing, Dec, 2000, Vol. 179 (6), p. 59, ISSN: 0889-2989.

Yu Hongsen, Liu Dantong, "Implement QoS Routing in Ad Hoc Network," Energy Procedia, 2011, Vol. 13, pp. 9172–9180.

Control Engineering and Information Systems – Liu (Ed)
© 2015 Taylor & Francis Group, London, ISBN 978-1-138-02685-8

Implementation of Linux VPN Gateway based on Netlink communication

L. Zhou

Department of Internet of Things Technology, Suzhou Vocational University, Suzhou, China

ABSTRACT: As a comparatively new IPC, Netlink socket has its own predominance in the communication between user process and kernel space. This paper has analyzed the application of Netlink socket in the IPSec support mechanism, summarized the Netlink socket message format supported by kernel IPSec module, implemented the IPSec VPN gateway with Netlink sockets.

1 INTRODUCTION

IPsec, which provides source authentication, data integrity and data secrecy at the network layer in both the Ipv4 and Ipv6 environments, protects the communication between two hosts, two security gateways and a host and a gateway. It is a comprehensive scheme to construct large-scale VPN.

Linux kernel provides IPSec support mechanism, which uses Netlink socket as the communication interface. User process can configure the Security Association Database (SAD) and Security Policy Database (SPD) with Netlink socket and, consequently, implement the kernel IPSec process of inbound or outbound IP datagrams.

2 VPN GATEWAY OVERALL DESIGN

IPSec Support Mechanism in kernel includes IPSec administer module, the definition of SA and SP, the construction of SAD and SPD, the IPSec inbound and outbound process, and algorithm library. The VPN security gateway consists of kernel space modules and user space modules.

VPN Gateway Console: This module provides a way for Administrator to manually configure gateway with commands, communicates with other gateways to negotiate a new SA in a secure way, sends the encapsulated message to kernel and

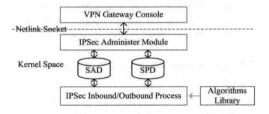

Figure 1. VPN Gateway overall design.

monitoring the feedback of kernel, implements the gateway configuration.

IPSec Administer module: It supports the communication between user process and kernel IPSec module with Netlink socket, handle the user message, implement the user configuration on kernel security component (for example, SA, SP, SAD, SPD), and transfer kernel IPSec prompt to user space.

SA and SAD: Security services are afforded to an SA by the use of AH or ESP [1]. AH is used to provide connectionless integrity and data origin authentication for IP datagram [1]. ESP is used to provide confidentiality, data origin authentication, connectionless integrity, network-layer secrecy and limited traffic flow confidentiality [1]. The kernel builds SAD with two hash lists (xfrm state byspi and xfrm state bydst), one for SPI and the other for IP destination address.

SP and SPD: SP specifies what services are to be offered to IP datagram and in what fashion. The mapping of traffic and security policy depends on sector. The sector that can be defined coarsely or finely is parts of traffic parameters. The kernel builds SPD, xfrm policy list, which is a hash list for the direction of the policy (inbound and outbound).

Inbound AH or ESP process: It includes the judgment on whether the IP datagram can pass the security test, inquiry of SPD, and verify of the security process applied on this datagram. Linux kernel adds security policy inquiry in the process of outbound datagram. If a IP datagram needs security process according to the policy inquiry result, it will be submitted to the security protocol function ah output or esp output before the routing process.

Algorithms Library: IPSec specifies HMAC-MD5-96 and HMAC-SHA-96 as default authentication algorithm and DES-CBC Cipher as default encryption algorithm [2, 3, 4].

3 NETLINK SOCKETS COMMUNICATION

3.1 *Netlink socket APIs in user space*

3.1.1 *Create and close Netlink socket*
Netlink socket supports standard socket APIs. In user space, the creation of Netlink socket is the function socket calling, just like the other common socket, as follows:

int sockfd = socket (AF NETLINK, SOCK
RAW(or SOCK DGRAM), protocol)

The protocols, which Netlink socket supports, are defined in file netlink.h. For IPSec communication, the value of this parameter should be set with NETLINK XFRM.

The function close is used to close the work of Netlink socket and callback the distributed resource.

3.1.2 *Bind address*
Netlink socket needs bind proper Netlink address to define the processes it services, as follows:

struct sockaddr nl
{ sa family t nl family; /* Netlink domain*/
unsigned short nl pad; /* Netlink pad*/
    u32 nl pid; /* Process ID*/
    u32 nl groups; /*Multicast Address Mask*/
}

Netlink socket can unicast (the destination is a process ID) or multicast (the destination is a group ID). The kernel has defined some commonly used multicast group Ids (each with a certain group value) in the file rtnetlink.h. The user process can get the message of a certain event by setting the nl groups with the corresponding group value. If several events messages are needed, the field nl groups should be set with the AND operation result of all the group values.

3.1.3 *Send and receive message*
User process uses the function sendto to send message to Netlink socket and the function recvmsg to receive message from Netlink socket. In order to improve the quality of communication, the size of receive buffer should be set comparatively large.

3.2 *Netlink socket APIs in Kernel space*

3.2.1 *Create and close Kernel Netlink socket*
Kernel space Netlink socket APIs are defined in the file af netlink.c. The other kernel modules use these APIs to implement their communication with user processes.

The kernel module that supports Netlink socket communication uses the function netlink kernel create to create a kernel Netlink socket, as follows:

struct sock * netlink kernel create(int unit,
void (*input)(struct sock *sk, int len));

- The parameter unit has to be set with the value of communication protocol (for example, NETLINK XFRM for IPSec protocol) which Netlink socket supports.
- The parameter input defines which kernel process to handle the message.
- Once the kernel has created a Netlink socket for a certain communication protocol, this Netlink socket will works as an unique entrance of the protocol (all the user message of this protocol will be send to this Netlink socket firstly and then be handled by the function input).

Kernel space Netlink socket is also closed by the system calling close.

3.2.2 *Unicast or multicast message to user process*
The function netlink unicast transfers message from the kernel process to a single user process, the function prototype as follows:

int netlink unicast(struct sock *ssk, struct sk
buff *skb, u32 pid, int nonblock)

- The parameter ssk is set with the socket descriptor return by the function netlink kernel create.
- The data field of the parameter skb stores the Netlink socket message.
- The parameter pid points out which user process to receive the message.
- The parameter nonblock is used to indicate the operation (block the process or return an error value), when the receive buffer is not large enough.

The function netlink broadcast transfers message from the kernel process to a group of user processes, the function prototype as follows:

int netlink broadcast(struct sock *ssk, struct sk
buff *skb, u32 pid, u32 group, int allocation)

- The definition of the parameter ssk and skb is the same as those of the function netlink unicast.
- The parameter pid doesn't have actually purpose in this function.
- The parameter group indicates a group of user process to receive the message. All the user processes registering for the group can receive message from this kernel Netlink socket.

3.3 *Communication between Netlink sockets*

3.3.1 *Message format*
Netlink socket message has two parts, the header and the data. All the message of Netlink socket has the common header, the data structure nlmsghdr.

The message data has different definition according to the communication protocol.

struct nlmsghdr

```
{
    u32 nlmsg len; /* Message Length */
    u16 nlmsg type; /* Message Type */
    u16 nlmsg flags; /* Flag Field*/
    u32 nlmsg seq; /* Sequence Number */
    u32 nlmsg pid; /* Process ID */
};
```

• The field Message Length is the total length of the message (header and data).
• The field Message Type has to do with the communication protocol set by the Netlink socket creation process.
• The field Falg defines additional communication control information.
• The fields Sequence Number and Process ID. These two fields mainly service the application that uses the Netlink socket to send message.

3.3.2 Cooperation of two Netlink sockets

With the cooperation of a pair of Netlink sockets, user process and kernel module can exchange messages, as shown in Figure 2.

The Netlink socket in kernel must be created first. Then just as a server role in the communication, it listens and responds to the Netlink socket in user space.

The Netlink socket of user process is created with the function socket and bound the corresponding address.

The user process uses the function sendto to send message to the kernel. The corresponding kernel module receives message from kernel space Netlink socket, and then call the function input to handle the message.

The kernel process uses the function netlink unicast or netlink broadcast to send message to the user process. The user process gets message from the receive buffer of the Netlink socket with the function recvmsg.

Netlink socket for the IPSec module is created by the function netlink kernel create. The parameter protocol is set with NETLINK XFRM. The parameter input is set with xfrm netlink rcv, which means all the user IPSec message to the kernel will firstly be handled by this function. The IPSec messages initiated by user progress include New SA, Update SP, Update SA, Delete SP, Delete SA, Inquiry SP, Inquiry SA, Export SAD, Get SPI, Export SPD and New SP. The IPSec messages initiated by kernel progress include Acquire SA, SA Expire and SP Expire.

4 CONFIGURATION AND TEST EXAMPLE

The aim of the configuration example is to build an AH tunnel from Gateway 1 to Gateway 2 to

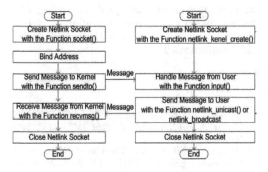

Figure 2.   Cooperation of two Netlink sockets.

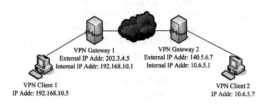

Figure 3.   Configuration topology.

Table 1.   SP fields configuration.

SP fields configuration	VPN Gateway 1		VPN Gateway 2	
	Inbound SP	Outbound SP	Inbound SP	Outbound SP
Source IP	140.5.6.7	202.3.4.5	202.3.4.5	140.5.6.7
Destination IP	202.3.4.5	140.5.6.7	140.5.6.7	202.3.4.5
Upper layer protocol	ICMP	ICMP	ICMP	ICMP
Direction	IN	OUT	IN	OUT
Action	ALLOW	ALLOW	ALLOW	ALLOW

Table 2. Sa fields configuration.

SP fields configuration	VPN Gateway 1		VPN Gateway 2	
	Inbound SP	Outbound SP	Inbound SP	Outbound SP
Destination IP	202.3.4.5	140.5.6.7	140.5.6.7	202.3.4.5
Security protocol	AH	AH	AH	AH
Protocol mode	Tunnel	Tunnel	Tunnel	Tunnel
Algorithm key length (b)	128	128	128	128
Algorithm key	(01234567 01234567 01234567 01234567)16	(76543210 76543210 76543210 76543210)16	(76543210 76543210 76543210 76543210)16	(01234567 01234567 01234567 01234567)16

Table 3. Value of main fields.

IP header fields	Datagram position		
	From Client 1 to Gateway 1	Between Gateway 1 and Gateway 2	From Gateway 2 to Client 2
Header length	60	104	60
Time to live	128	127	126
Upper layer protocol	ICMP	AH	ICMP
Source IP	192.168.10.5	202.3.4.5	192.168.10.5
Destination IP	10.6.5.7	140.5.6.7	10.6.5.7

protect the communication between Client 1 and Client 2. Configuration topology is shown in the Figure 3. The authentication algorithm chosen by the tunnel is HMAC-MD5-96.

On Gateway 1 and Gateway 2, the configuration steps include two "New SP" messages (inbound and outbound) and two "New SA" messages (inbound and outbound). The settings of the main fields of these messages are shown in Table 1 and Table 2.

In the testing of transmission, we catch the IP datagram send from Client 1 to Client 2 and analyze the main fields of the IP header, as shown in the Table 3.

5 CONCLUSION

With the research on the Netlink socket methods, the application of Netlink socket in the IPSec support mechanism is analyzed. The IPSec VPN gateway based on the communication of Netlink socket between user process and kernel IPSec module is implemented.

REFERENCES

Madson, C. and N. Doraswamy, "The ESP DES-CBC Cipher Algorithm With Explicit IV", RFC 2405, November 1998.
Madson, C. and R. Glenn, "The Use of HMAC-SHA-1-96 within ESP and AH", RFC 2404, November 1998.
Madson, C. and R. Glenn, "The Use of HMAC-MD5-96 within ESP and AH", RFC 2403, November 1998.
Stephen Kent and Randall Atkinson, "Security Architecture for the Internet Protocol", RFC 2401, November 1998.

Control Engineering and Information Systems – Liu (Ed)
© 2015 Taylor & Francis Group, London, ISBN 978-1-138-02685-8

Information and communication system based on data processing center for smart grid

Y.H. Han, J.K. Wang, Q. Zhao & P. Han
Northeastern University at Qinhuangdao, Qinhuangdao, China

ABSTRACT: Two-way seamless communication is the key aspect of realizing the vision of smart grid. Reliable and real-time information becomes the key factor for reliable delivering of power from the generating units to the end-users. This paper addresses critical issues on smart grid technologies primarily in terms of information and communication technology and opportunities. The main objective of this paper is to provide a contemporary look at the communication infrastructure in smart grid. Data processing center is considered to include in WAN and HAN to reduce latency and improve real-time data transmitted. And parallel acceleration processing techniques is proposed for data processing center to solve the compute-intensive character of applications of smart grid. It is also expected that this paper will provide a better understanding of the requirement, potential advantages and research challenges.

1 INTRODUCTION

Electricity has been an indispensable resource for living. Power plants, generators, batteries, and large energy storage have been vastly manufactured and built from time to time to ensure that electricity would geographically serve every end point of territories and regions throughout the world.

A traditional electrical grid is a network of technologies that delivers electricity from power plants to end consumers. In most cases, the traditional grid is so dumb that workers still have to walk from house to house to read the electricity meter, and utilities have no clue when the lights go out until customers call to complain. Whereas a Smart Grid is so "smart" that the meters can report in, appliances can control how much energy they use, and end customers can choose cheap electricity to use. In another word, the smart grid has the ability to sense, monitor, and in some cases, automatically control how the system operates or behaves under a given set of conditions [8] (Parikh et al. 2010).

Smart grid will be characterized by two-way flow of power in electrical network, and information in communication network. In the recent report on National Institute of Standard and Technology framework and roadmap for smart grid interoperability standards [10], several wired and wireless communication technologies are identified for smart grid. How to use information and communication technologies is a major challenge in the smart grid to save energy, reduce cost and increase reliability. The Institute of Electrical and Electronics Engineers (IEEE) has recently taken the initia-

tive to define these standards and write guidelines on how the grid should operate using the latest in power engineering, communications, and information technology. Though standard of smart grid are in progress, one thing is certain: a secure and reliable, real-time two-way, high-speed communication infrastructure is an essential part of the smart grid reality [7, 4, 9] (Lo & Ansari 2012, Guggor & Sahin 2011, Roy et al. 2011).

2 CONSIDERED ARCHITECTURE FOR SMART GRID

The smart grid is constructed in four dominant layers as Figure 1. Power layer specifies smart power generation, conversion, transport, storage, and

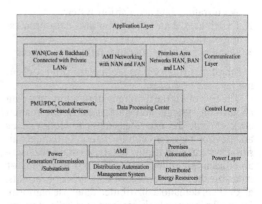

Figure 1. The layered smart grid.

consumption; Control layer involves smart sensors and actuators for data collection and conveyance as well as sensing and control; In the control Layer, it has Data Processing Center, which includes information management systems for key functions such as data management, fusion, mining and control. Communication layer characterized different roles to ensure accurate and effective data transmission; an effective communication infrastructure will operate at multiple levels, connecting end users (customers) to other network elements (distribution transformers, substations, and data processing centers). This infrastructure will need to support moderately high-speed secure data transfers with predictable and controllable latencies specified through the use of quality of service parameters at the message level. And application layer supports decision intelligence with the aid of information technologies for various smart energy applications.

In the smart grid, reliable and real-time information becomes the key factor for reliable delivery of power from the generating units to the end-users. The impact of equipment failures, capacity constraints, and natural accidents and catastrophes, which cause power disturbances and outages, can be largely avoided by online power system condition monitoring, diagnostics and protection. To this end, modern information and communication technologies have become essential to realize the envisioned smart grid. Fast data processing and transmission has evolved over the years into a complex communication, computation and control system. To over the challenges, a frame work of Grid-based future power system control layer has been proposed, using the technologies that are just emerging. The data processing is proposed include in control layer. The data processing center may play an important role in in communication system to reduce latency and improve real-time data transmitted, which is characterized by

1. An ultrafast data acquisition system;
2. Distributed data acquisition and data processing services;
3. Dynamic sharing of computational resources;

Standard Grid services architecture and tools to manage the resources.

The first one is the requirement of additional data acquisition systems and databases. In the market environment, the new measurement requirements and devices, new market information or data acquisition systems, and related databases that will ensure an efficient and fair market. The fast increasing of data size, the large number of data types or formats, all of them demand a distributed database platform. Hence flexible and high-performance collecting and processing system is needed.

In order to solve the compute-intensive character of applications of smart grid, based on advantages of Graphics Processing Unit (GPU) parallel operation, parallel acceleration processing technique is proposed for data processing center, which is a set of multiprocessors. Each multiprocessor is a set of processors with SIMD (Single Instruction Multiple Data) architecture that each processor of the multiprocessor executes the same instruction but operates on different data, at each clock cycle.

The philosophical and architectural underpinning of GPU is to create mass of thread level parallelism that can be dynamically exploited by hardware, which is capable of executing a high number of threads in parallel and operating as a coprocessor to the host CPU. A portion of an application that is executed many times on different data, can be divided into a function that is executed on the device as many different threads. So we can design GPU acceleration algorithms expressing monitoring, diagnostics and protection for smart grid.

3 SMART GRID COMMUNICATION REQUIREMENTS

In a smart grid, sensing and measurement, integrated communications, and advanced control methods are the three fundamental building blocks. Sensing and measurement will have the ability to detect malfunctions or deviations from normal operational ranges that would warrant action. Further, since in a smart grid, a point of electricity consumption can also become a point generation, the sensing and measuring process will be closely linked with the metering process. Integrated communications will allow inputs from sensors to be conveyed to the control centers of the grid which will generate control messages for transmission to various points on the grid resulting in appropriate action. O these three key blocks, the implementation of integrated communications is a fundamental need, required by the others and essential to the modern smart grid. Due to its dependency on data acquisition, protection and control, the smart grid cannot work without an effective integrated communications infrastructure. Thus, establishing the integrated communications infrastructure must be of the highest priority in building the smart grid.

Smart grid communication depends on two important requirements [5] (Hauser et al. 2008), communication latency and large volume of message. The five most important requirements are mentioned here:

1. Two-way communication: The new grid will add thousands of distributed but intermittent generators, turning one-time pure energy

consumers into part-time producers as well. These renewable energy sources are known to be more stochastically time-varying, however. As a consequence, the future grid will need to send a great deal of control information to end nodes (consumers) for the purposes of applications such as demand response. This implies a more symmetrical, two-way flow of information, at least in some segments of the future smart grid, as energy flows from the consumer with spare energy resources toward grid control centers [1] (Aggarwal et al. 2010).

2. Capacity: The communication link should be able to convey the information of load and price to the destinations with negligible error in a realtime manner.

3. Latency: It is one of the most stringent requirements for the grid. If the control center misses any input then it might substitute the missing input with inputs from other sensors which can produce different actions leading to erroneous results. The latency is in the order of a few milliseconds [6] (Kezunovic 2011).

4. Security: It is important to preserve the privacy of the system state in smart grid. If the information is leaked, and eavesdropper could use this information to break the stability of the

power market or steal personal private information [2–3] (Fadlullah et al. 2011, Gharavi & Ghafurian, 2011).

5. Large numbers of messages: As new elements are added to the network with the evolution of the grid added to the network with the evolution of the grid system, the new network should be able to transport more messages simultaneously without any major effect on latency. The numbers of messages will likely increase must faster than the number of elements on the network.

4 A SMART COMMUNICATION INFRASTRUCTURE

Figure 2 shows our considered Smart Grid communication architecture. The lower layer domains related to electric power system are generation domain. Generation domain shares information with regional system operator, power market, and control center. In case of lack of generation or generator failure, immediate actions need to be taken by regional system operator and power market. Transmission domain presented as regional control center and substation automation system. Transmission domain is typically regulated by

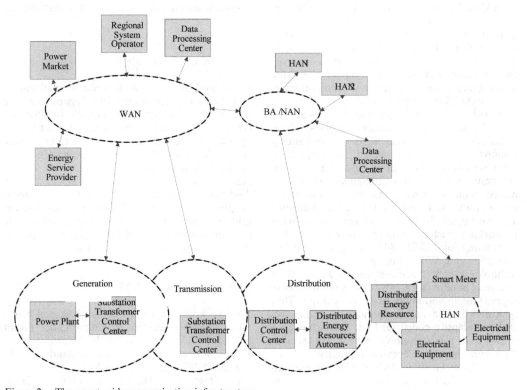

Figure 2. The smart grid communication infrastructure.

regional system operator with the help of information exchange with control center and data processing center. Distribution domain is consists of Distribution Control Center and Distribution Automation System. Customer domain includes distributed energy resources plant automation, and residential or industrial customer automation system.

For communications, however, communications links have different requirements than those of power lines. Transmission domain is connected to one another in a meshed network, which can be built over optical fiber technology. Next, the rest of the considered smart grid communication topology is divided into a number of networks that feature real-life setups of a city or metropolitan area. Broadly speaking, a city has many neighborhoods, each neighborhood has many buildings, and each building may have a number of apartments.

The communication architecture for the lower distribution network is divided into a number of hierarchical networks: the Neighborhood Area Network (NAN), Building Area Network (BAN), and Home Area Network (HAN). For simplicity, each distribution substation is considered to cover only one neighborhood zone. Each NAN can be considered to be composed of a number of BANs. On the other hand, every BAN contains a number of apartments. In Figure 1, the apartments are shown to have their respective LANs referred to as a HAN. In addition, there are advanced meters called smart meters deployed that represent advanced metering infrastructure for enabling automated two-way communication as gateways [6] (Kezunovic 2011).

A wide variety of in premise devices-thermostats, lighting controls, energy displays, appliances, plug-in electric vehicles, distributed energy resources such as premise-based photovoltatic or wind resources, and premise-based energy storage units such as battery appliances-will be connected to the smart meter interface so that energy consumption and generation can be monitored and controlled. Technology choices for in-home networking include broadband-over-power-line communication, IEEE 802.15.4/ZigBee wireless networks and WiFi. Emerging communication technologies seek to achieve peak data rates of up to several hundred megabits per second in order to support audio and multimedia applications. They typically use advanced modulation techniques such as adaptive coded OFDM, which has been shown to be robust in noisy environments such as the power line while achieving scalability.

In order to facilitate BAN-HANs communication, 3G or 4G may be used to cover more areas. It can be also applied as a low-cost solution for substation automation to control and monitor substation performance when small bursts of information are needed. A NAN comprises one or more base stations and a number of BANs.

Utilities have long been operating WANs for a variety of applications such as providing information and access to their plants, offices and supervisory control and data processing center that monitor and control the electricity grid. These legacy networks have incorporated various technologies including power line communications, fiber optics and a variety of licensed and un-licensed wireless devices.

In Figure 2, the Smart Grid WAN consists of two interconnected networks: the core network, and backhaul or distribution network. The core network connects the head offices and substations and commonly uses fiber optics. Where fiber is unavailable or too expensive to deploy, wireless solution using worldwide interoperability for WiMAX technology is a good gift due to ease of deployment and proven reliability. The backhaul or distribution network handles the broadband connectivity to NANs, mobile workforces, and automation and monitoring devices that are located on the distribution or transmission networks. Technologies such as fiber optics, WiMAX, PLC, satellite and cellular communications are widely employed in WAN distribution networks.

5 CONCLUSION

Smart grid relies on a dependable information and communication system. In this paper, we have attempted to highlight the network architectural considerations and communication technology options. New communication infrastructure may be associated with data processing center, which is needed to compute-intensive data. Data processing center is considered to include in WAN and HAN. And parallel acceleration processing techniques is proposed based on GPU for applications in smart grid. It is also expected that this paper will provide a better understanding of the requirement, potential advantages and research challenges.

ACKNOWLEDGMENTS

This work was supported by the National Natural Science Foundation of China under Grant No. 61374097 and 61104005, by Natural Science Foundation of Liaoning Province under Grant No. 201202073, by Research Fund for the Doctoral Program of Higher Education of China under Grant No. 20110042120015.

REFERENCES

Aggarwal A., Kunta S. & Verma P.K. "A Proposed Communications Infrastrucutre for the Smart Grid[C]," Proceedings of Innovative Smart Grid Technologies 2010, Gaithersberg: IEEE, 2010, pp. 1–5.

Fadlullah Z.M., Fouda M.M. & Kato N. et al, "Toward Intelligent Machine-to-Machine Communications in Smart Grid," IEEE Communications Magazien, pp. 60–65, April 2011.

Gharavi H. & Ghafurian R. "Smart Grid: The Electric Energy System of the Future [J]," Proceedings of The IEEE, vol. 99, no. 6, pp. 917–921, 2011.

Gungor V., Sahin D. & Kocak T. et al. "Smart Grid Techinologies: Communication Technologies and Standards [J]," IEEE Trans. on Industrial Informatics, vol.7, no. 4, pp. 529–539, 2011.

Hauser C.H. et al, "Security, Trust, and QoS in Next Generation Control and Communication for Large Power Systems [J]," Int'l. J. Critical Infrastructures, vol. 4. no. 1/2, pp. 3–16, 2008.

Kezunovic M. "Translational Knowledge: From Collecting Data to Making Decisions in a Smart Grid [J]," Proceedings of The IEEE, vol. 99, no. 6, pp. 977–997, 2011.

Lo C. & Ansari N. "The Progressive Smart Grid System from Both Power and Communications Aspects [J]," IEEE Communications Surveys & Tutorials, vol. 14, no. 3, pp. 799–821, 2012.

National Institute of Standard and Technology, Standards Identifed for Inclusion in the Smart Grid Interoperability Standards Framework, Release 1.0, Sept. 2009, [Online]. Available:http://www.nist.gov/smartgrid/standard.html.

Parikh P., Kanabar M. & Sidhu T. "Opportunities and Challenges of Wireless Communication Technologies for Smart Grid Applications[C]," Proceedings of Power and Energy Society General Meeting, Minneapolis: IEEE, 2010.

Roy S., Nordell D. & Venkata S. "Lines of Communication," IEEE Power & Energy Magazine, September/October, pp. 65–73, 2011.

Research on visualization platform for electric mobile access system based on Flex technology

Z.J. Dai, Z.P. Shao, Y.F. Wang & J. Chu
*Research Institute of Information Technology and Communication, China Electric Power Research Institute,
Nanjing, China*

ABSTRACT: Due to the lack of a unified visualization platform, the mobile access system cannot provide an effective method for visualization management. First introduce the background and problem of the mobile access system and analyze why visualization platform should be deployed. Finally, the detailed design of visualization platform for the mobile access system is given.

1 INTRODUCTION

In the recent years, Personal Digital Assistant (PDA) has been widely used in the electricity production, marketing, supplies, emergency command system and other electric internal systems. With the promotion and deployment of electric mobile access system in provincial power company of State Grid, it is difficult to provide users with a clear and intuitive information display scheme due to the lack of effective information management platform, thus the users cannot obtain useful information from the access system. So there is an urgent need to support the electric mobile access system through visualization platform. The visualization platform helps to establish the model of security access system and the unified management of data, graphics, knowledge and information.

Currently, the mobile access system has covered the headquarters of the State Grid Corporation and more than two dozen provincial companies, but there are following problems to be solved.

1. The mobile access system lacks for a unified visualization platform and cannot provide affluent means of visual management: document management of access hardware and software, animated charts of various data indicators.
2. Due to the lack of analysis functions, it is difficult to support the analytical and intelligent management. The analysis of statistical function of mobile access systems is base on respective data and no data can be shared between the various access systems. Thus it's difficult to provide multi-level data analysis and decision support.

2 MOBILE ACCESS SYSTEM AND VISUALIZATION PLATFORM

The mobile access system is the unified entrance of the third-party border which has access to the electric interior-net. It is responsible for unified authentication, regulatory and security of access objects including smart meter, mobile operations, mobile office and staff. The mobile access system is mainly composed of security terminals, mobile access gateway, netgap, authentication system, and centralized monitoring system. It provides an effective solution to protect the confidentiality, integrity and availability of electric data and strengthens the information sharing and integrated applications when mobile terminals visit business systems. Mobile access system is provided with important practical significance and necessity.

3 KEY TECHNOLOGIES OF VISUALIZATION

With the development of computer technology and improvement of network performance, it seems that traditional web applications have rather simple display, poor user experience and have been gradually forsaken by users. A Rich Internet Application (RIA) is a Web application that has many of the characteristics of desktop application software, typically delivered by way of a site-specific browser, a browser plug-in, an independent sandbox, extensive use of JavaScript, or a virtual machine (Gao, 2004). RIAs combine the reach of the internet with compelling rich user interface, the combination of which provides an enhanced level of satisfaction and success in users' web-based interactions.

Apache Flex is a Software Development Kit (SDK) for the development and deployment of cross-platform RIA based on the Adobe Flash platform. The greatest feature of Flex is that it combines the functionality of desktop applications with the flexibility of web application (Jin, 2004). Flex client can store the state information of client, therefore, compared with the traditional HTML page; it reduces the call of server to show more detailed data particularly.

Building visualization platform of the mobile access system using Flex technology has the following advantages: (1) Flex provides users with better experience and supports complex business logic of the client as well as dynamic data update. (2) Visualization platform of the mobile access system can be developed rapidly through Flex technology.

4 ARCHITECTURE OF VISUALIZATION PLATFORM

4.1 Overview of the platform

The realization of visualization platform for the mobile access system in this paper provides a web-oriented RIA application for the network management architecture of State Grid. The management and control of the mobile systems deployed all over the country can be easily implemented through the platform. It also provides a variety of data visualization schemes and shows the operating status of terminals, tunnels, business, systems through the mobile access systems in each provincial company.

According to the functional responsibilities of the platform, the platform is divided into six business modules, as shown in Figure 1.

4.2 Overall architecture of the platform

The overall architecture of the platform is divided into three layers and a library: acquisition source layer, acquisition and processing layer, application presentation layer and unified repository.

Capture source layer: Capture source layer is a list of all collection objects of the mobile access system, whose goal is to show the business information and status of all the related equipment of the mobile access system.

Acquisition and processing layer: Acquisition and processing layer is to collect the status conditions, resource configuration data, asset data, security data of all resource objects of the mobile access system, and its goal is to collect running data of the mobile access system from multiple perspectives and analyze collected data, then transmit and keep in the data. The ultimate target is to find fault and potential problems through the analysis and processing status, and ensure the stable and sustained working of the mobile system.

Application presentation layer: A unified application, management and display interface, built on a unified graphical platform.

Unified repository: The Unified repository is the core of the entire platform data structures and storage, for other applications, display module via a single data bus interface provides a unified, complete and accurate data.

4.3 Design of application presentation layer

Application presentation layer uses flex as the presentation layer implementation, in the foreground between the interface and the business logic implemented by adding a control layer around the station data communications (Linda, 2005).

To build the platform, the following steps must be completed:

– Establish a common model for data services called Eos Model Locator to store the interaction between each module shared data;
– Create a menu tree to the main interface providing the interface for each module entry
– Establish a Front Controller registering all events of listeners. The abbreviation "i.e." means "that is", and the abbreviation "e.g." means "for example".

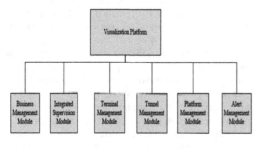

Figure 1. Functional modules of visualization platform.

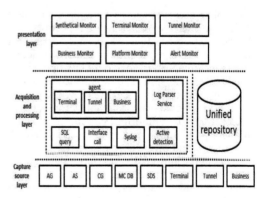

Figure 2. Overall architecture of the platform.

– Establish corresponding ModelLocator, View, Event and Command for the seven functional modules

This design model is not only to ensure the readability of the code, but also allows developers to design from the tedious freed and better focus on the program logic implementation.

4.4 *Design of business logic layer*

Business logic layer is primarily responsible for communicating with the flex client, and the platform uses a service-oriented architecture middleware platform as the business logic of the system technical architecture and development tools. In the service-oriented architecture middleware platform, three tasks are mainly to be completed. First, establish a separate building package and write different workflows for each functional module. Second, achieve a functional interaction between the backgrounds of each module through specific operational procedures for each configuration of the logical flow (Xiao, 2005). Third, establish a connection with Oracle, transfer the XML form into a database SQL statement, store the corresponding data required by the client, and finally return the corresponding data back to the client (Daxi, 2008).

4.5 *Communication between presentation layer and business logic layer*

Flex provides three kinds of data exchanging mode including HTTP service, web service, and remote object. As the data type submitted by the middleware platform of service-oriented architecture is mostly Java type, remote object is adopted as the interactive mode. It helps communicate with Flex and Java, improving the efficiency of data transmission. When the processing of logic flow is completed, encapsulate the output of the logic flow into a large array of data objects and return them to the action script of flex to resolve. Add a remote object component in Flex application and set the remote object destination corresponding to the remoting-config. xml in the configuration.

4.6 *Data layer*

Oracle database is used in the data layer. The connection between the information processing of the Flex presentation layer and the Oracle database is achieved by the middleware platform of the service-oriented architecture. The information stored in the data layer mainly includes detailed information of each item, history information of project, personal information of users, dynamic menus of the platform and records of configuration types, inquiring dictionary information of business, file information of log files and so on.

5 SUMMARY

The platform realized effective management of the mobile system for the headquarters and provincial companies of power grid achieved the desired design goals. With the deepening of Flex applications, enterprises of power grid focus on the development which is highly interactive and of rich user experience and powerful client. To sum up in a word, development of Flex application in the power grid will be more widely applied.

REFERENCES

Jerry Gao, Ph.D., Mansi Modak, Satyavathi Dornadula, and Simon Shim, Ph.D. Rich Internet.
Lai Jin, Takashi Sakairi. An e-Business Framework for Developing Rich and Reliable Client Applications. In Proceedings of IEEE International Conference on Systems, Man and Cybernetics, Volume 5, pp. 4294–4299, 5–8 Oct. 2003.
Linda, Dailey, Paulson. Building Rich web Application with MAXation Computer. 2005, 23(10): 77–78.
Pan Dasi, Research and Implementation of Web Application Support Offline Processing Based on Flex, Computer Science, vol. 37, 2008, pp. 292–301.
Xiao Guozhi, A review on technical characteristics of RIA and its development trend, Journal of ChangJiang University, vol. 18, 2008, pp. 34–36.

A novel authorization scheme for mobile IPv6

C.S. Wan, L. Zhou & J. Huang
School of Information Science and Engineering, Southeast University, Nanjing, China

J. Zhang
Accounting Department, Nanjing University, Nanjing, China

ABSTRACT: To address the authentication and authorization issues in mip6 split scenario, this paper gives a secure authorization solution based on diameter protocol. When the authorization server authorizes MIP6 service to the mobile node, it communicates with the authentication server to verify the mobile node, and then it sets up a trust relationship between the mobile node and the home agent. This paper also proves the security of authorization solution using BAN logic, and compares the performance of this solution with other related solution.

1 INTRODUCTION

The mobile IPv6 Protocol[1] (MIP6) is standardized by IETF for managing the mobility of node in the network layer. With the wide application of MIP6 in mobile networks, the deployment issues of MIP6 have become increasingly prominent. RFC4640[2] proposed three scenarios for the deployment of mobile IP: Static configuration scenario, Integrated scenario and Split scenario. In Static configuration scenario, there is a pre-configured trust relationship between the Mobile Node (MN) and the provider of hometown MIP6 (MSP), MSP completed MIP6 authorizing services of MN by using this trust relationship, and ensure the security of MIP6 services; In Integrated scene, access services and mobile IP services are provided by the same entity, there is a trust relationship between the MN and the accessing server, this trust relationship can be used to authorize the MN, and to ensure the security of mobile IP service; In Split scene, Mobile IP Service Provider (MSP) and Mobile IP Service Administrator (MSA) are provided by different entities, the trust relationship between MN and MSP does not exist. Split scene is conducive to the decoupling of each entity which deploys MIP6 services (MSP, MSA), and can realize multiple MSP share one MSA, form an economic network, therefore adopt Split Scene deploy MIP6 service is the most important scene in three scenarios. In Split scene, MN and MSP the trust relationship between MN and MSP does not exist, so it is more complex than the other two scenarios that how MSP provides MIP6 service to MN safely.

IETF dime Working Group proposed the basic solution based on the diameter against the problems in Split scene[3]. After getting care-of address (COA), MN through the IKEv2 protocol[4] to the home agent (HA) for MIP6 service, when HA receives a service request, by first through the diameter-EAP protocol[5][6] to the authentication server (AAAH-EAP) for sending an authentication request, to identity authentication of the mobile node (MN)[1]; then through the diameter protocol to authorize server (AAAH-the MIP6[3]) for a authorization requisition, after successful authorization, AAAH-MIP6 can build a trust relationship between HA and MN, in order to ensure the security of MIP6 service (Fig. 1).

Literature[3] proposed the basic protocol which assumes AAAH-EAP and AAAH-MIP6 exist in the same device, so the trust relationship exists between the AAAH-MIP6 and MN, when authorized to MN, AAAH-MIP6 can confirm the identity of the MN so as to guarantee the security of the authorization. However the suppose that authentication and authorization server exist on the same device is not conducive to realize a variety of services share one authentication server, and costs higher to form a network, so AAAH-EAP and the AAAH-MIP6 server may be located in two

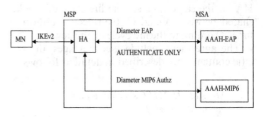

Figure 1. Authentication/authorization architecture in Split scenario.

different device[3]. In the condition of that AAAH-EAP and AAAH-MIP6 are separated, AAAH-EAP does not know the mobile nodes require for MIP6 service when it authenticates the mobile nodes; the AAAH-the MIP6 server neither know whether the mobile nodes have been certified when it authorized MIP6 services to the MN. Thus the authorized program provided by literature[3] has following several security problems to be solved:

1. AAAH-MIP6 does not have identity authentication function and can't prevent malicious nodes impersonate legitimate nodes to steal MIP6 services.
2. There is no trust relationship between AAAH-MIP6 and the MN, after AAAH-MIP6 authorized the MN can't guarantee MIP6 service data transport to the MN correctly.
3. There is no trust relationship between the MN and HA, the HA can't prevent other malicious nodes stealing MIP6 services.

In Split scenes, the authorized security issues caused by separation of AAAH-EAP and AAAH-MIP6 server, which leading MIP6 services can be stolen by malicious nodes easily, and affecting normal service of the legitimate nodes. The IETF DIME Working Group has just pointed out the problem, and the related solutions are very important for saving networking cost, but there is no literature to propose solutions. This paper presents a safety Mobile IP service licensing agreement against licensing issues about the separation of AAAH-EAP and AAAH-MIP6 server.

2 OUR SCHEME

In order to facilitate the description and analysis of the agreement, the following will use (A) represents MN, B represents AAA-EAP, C represents HA, (D) represents AAA-MIP.

When A is authorized by D, using the certification of B to authenticate A, according to certified results complete the authorization process. After the Authorization, generate a shared key for A and C to protect MIP6 signaling between A and C, and to prevent malicious nodes to steal MIP6 services from HA, at the same time set the life cycle of the key shared between A and C, then decide authorized time according to the life cycle of the key (Fig. 2).

The agreement includes six messages, the specific content s are described in detail as follows.

2.1 Message 1: MIP6 service request message (A→C)

After AAA-EAP server (B) completes the authentication of the mobile node (A), A sends request

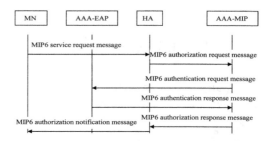

Figure 2. MIP6 authorization procedure.

message of MIP6 services to the home agent (C). Then sends authorization identity Aid, the nonce which generated Aid, and message authentication code to the HA. The service request message is defined as follows:

$$Service\_request = \{Aid, nonce, MN\_ID, \\ Service\_Type, MAC\} \quad (1)$$

In order to ensure that a variety of licensing server can share the same authentication server in case of the separation of authentication and authorization, but also in order to reduce the risk of abuse of EMSK, the ID (Aid) of MIP6 services authorization is generated by MIP6-USRK (k_{ad}). The method of generating Aid is divided into two steps:

1. A uses the method in literature[7] to generate MIP6-USRK[7] (k_{ad}). Generating formula of k_{ad} is defined as follows:

$$k_{ad} = PRF(EMSK, "MIP6USRK" \\ |MSA|MSP|MN|length) \quad (2)$$

EMSK is the key generated after the EAP authentication; optional data MSA, MSP, MN can uniquely identify the entity related to Mobile IP services; PRF is a pseudo-random function[4].

2. The generation formula of Aid as follows:

$$Aid = PRF(k_{ad}, "authorizationID" \\ |nonce|Service\_Type|length) \quad (3)$$

The nonce is used to ensure the freshness of the Aid and prevent our protocol from replay attacks. Service Type is used to identify the MIP6 services.

When service is requesting message, the generated formula of the message authentication code which is used to provide integrity protection of generated as follows:

$$MAC = First(maclength, PRF \\ (k_{ab}, Aid|MN|Service\_Type|nonce)) \quad (4)$$

414

K_{ad} is generated by the MIP6-USRK (k_{ad}), the generated formula as follows:

$$k_{ab} = PRF(k_{ad}, "MIP6TOKEN",$$
$$MN \mid Service\_Type \mid length) \qquad (5)$$

The SERVICE_TYPE is used to identify MIP6 Service.

2.2 Message 2: MIP6 authorization request message (C→D)

After received the service request message from A, C uses the diameter protocol[8] package the service request message for the authorization request message (MIP6_authz_req), then sent to the AAA-MIP server (D). The authorization request message is defined as follows:

$$MIP6\_authz\_req$$
$$= \{Aid, nonce, MN, Service\_Type, MAC\}_{k_{cd}} \quad (6)$$

Aid, nonce, MN, SERVICE_TYPE and MAC are consistent with the service request message. The integrity and confidentiality protection of the authorization request message is provided by the shared key between C and D.

2.3 Message 3: MIP6 authentication request message (D→B)

After received authorization request message from C, D uses the diameter protocol[8] send the authentication request message (MIP6_authe_req) to the AAA-EAP server (B), request B to authenticate the message. The authentication request message is defined as follows:

$$MIP6\_authe\_req$$
$$= \{Aid, nonce, MN, Service\_Type, MAC\}_{k_{bd}} \quad (7)$$

In which Aid, nonce, MN, SERVICE_TYPE and MAC are consistent with the service request message. The integrity and confidentiality protection of the authentication request message is provided by the shared key between D and B.

2.4 Message 4: MIP6 authentication response message (B→D)

After received authentication request message from D, B need to verify the message first:

1. Identity verification: B can calculate the value of the Aid according to the formula (3), and compare with the value of the Aid in the formula (7) to verify the identity of A.
2. Integrity verification: B can calculate the value of k_{ab} according to the formula (5), and calculate the value of MAC according to the formula (4), Compare the calculation results are consistent with the results of the formula (7), and to verify the integrity of the message.
3. Replay attacks Verification: B can determine whether the received message is replay attacks according to the value of the nonce.

After the completion of the verification, B can calculate the value of MIP6-USRK (k_{ad}) according to the formula (2), and use the diameter protocol[8] to return to D. MIP6 authentication response message (MIP6_authe_rep) is defined as follows:

$$MIP6\_authe\_rep = \{k_{ad}\}_{k_{bd}} \qquad (8)$$

2.5 Message 5: MIP6 authorization response message (D→C)

After received authorization response message from B, D parses out k_{ad} from the messages, and uses (k_{ad}) for generating (k_{ac}). The generation formula of k_{ac} is defined as follows:

$$K_{ac} = PRF(k_{ad}, "MNHAKEY" \mid HA \mid MN \mid length)$$
$$\qquad (9)$$

D returned k_{ac} and authorization data result to C by the diameter protocol[8]. MIP6 authorization response message is defined as follows:

$$MIP6\_authz\_rep = \{result, k_{ac}\}_{k_{cd}} \qquad (10)$$

2.6 Message 6: MIP6 authorization notification messages (C→A)

After received authorization response message, C sends the result which is encrypted by k_{ac} to MN. The MIP6 authorization notification message (MIP6_authz_notify) is defined as follows:

$$MIP6\_authz\_notify = \{result\}_{k_{ac}} \qquad (11)$$

After received MIP6 authorization notification message, A calculates (k_{ac}), and parses the authorization result.

3 SECURITY AND EFFICIENCY ANALYSIS OF OUR PROTOCOL

At present, there is a variety of analytical logic which is protocol formalized. Because BAN logic[9] has intuitive logic expression, so it is widely used in the proof of security protocols. BAN logic contains three processing objects: subject, key and formula. According to the general method that BAN logic

415

analyzes protocol, this article proves authorization protocol in three steps: (1) initial assumption; (2) security goals; (3) security analysis. Since the survival time of key can guarantee its freshness, the following proof will not discuss the fresh issue.

3.1 Security analysis of our protocol

3.1.1 Initial assumptions of our protocol

The assumption of this protocol: authorization server stores authorization data of the mobile node, and is responsible for the management of the key related to MIP6 service; authentication server has authentication function; there is a trust relationship between the authentication server and the home agent; after the authentication, there is a trust relationship between the mobile node and the authentication server. The initial assumption of this protocol is formalized as follows:

a. $AB \models A \xleftrightarrow{k_{ab}} B$

b. $CD \models C \xleftrightarrow{k_{cd}} D$

c. $BD \models B \xleftrightarrow{k_{bd}} D$

d. $B \models\Rightarrow k_{ad}$

e. $A \models A \xleftrightarrow{k_{ad}} D$

f. $D \models\Rightarrow k_{ac}$

g. $A \models A \xleftrightarrow{k_{ac}} C$

h. $D \models\Rightarrow result$

3.1.2 Security goals of our protocol

Goal 1: authorization server need to confirm that mobile node has send MIP6 service requests, scilicet

$$D \models A |\sim (Aid, nonce)$$

Goal 2: the authorization server needs ensure that MIP6 service data passed to the mobile node safely, scilicet

$$A \models D |\sim result$$

Goal 3: The home agent has to establish a trust relationship with the mobile node, scilicet

$$C \models A \xleftrightarrow{k_{ac}} C$$

3.1.3 Security analysis of our protocol

By formula (1) to obtain:

$$C \triangleleft \{Aid, nonce\}_{kab} \tag{12}$$

By (12), formula (6) and the assumption (b) to obtain:

$$D \triangleleft \{Aid, nonce\}_{kab} \tag{13}$$

By (13), formula (7) and the assumption (a) to obtain:

$$B \models A |\sim (Aid, nonce) \tag{14}$$

By (14), formula (8) and the assumption (c) and (d) to obtain:

$$D \models A |\sim (Aid, nonce) \tag{15}$$

$$D \models A \xleftrightarrow{k_{ad}} D \tag{16}$$

By (15) and (16), formula (10) and the assumption (b) and (f) to obtain:

$$C \models D |\sim result \tag{17}$$

$$C \models A \xleftrightarrow{k_{ac}} C \tag{18}$$

By (18), formula (11) and the assumption (g) and (d) to obtain:

$$A \models D |\sim result \tag{19}$$

The result is proved by (15), (18) and (19).

3.2 Efficiency analysis of our protocol

The protocol mainly contains two steps: the AAA server authenticates the MN, and the AAAH-MIP6 distributes k_{ac} keys.

The comparison of the computation of certification process between this protocol and the literature[3] is shown in Table 1.

In this protocol, MN only needs to compute the message authentication code once, while in[3], MN needs to calculate IKE_AUTH data twice, the calculation of MN in this protocol is smaller than the authentication method in[3]. The protocol calculates message authentication code in the preparatory stage of the implementation of the protocol, this can greatly improve the speed of implementation of the protocol. In this protocol, the mobile node only needs to send datagram once and receive datagram once; when literature[3] uses the EAP method provided by RFC4306 to authenticate, mobile node needs to send datagram at least four times and receive datagram[4] at least four times. In the case of the access of the mobile node in a wireless manner, the present protocol can save radio resource.

When authentication server and authorization server are separating, the two servers both need to complete the authentication process with the MN, because the authorization server does not have authentication function, therefore, the authentication in the process of authorization provided by this protocol depends on the authentication

Table 1. Comparison of the calculation between new protocol and other similar protocols.

Scheme	Calculation of MN	The amount of interaction of MN and HA	Calculation of HA
[3]	Calculate IKE_AUTH data (twice)	Send datagram (at least four times); Receive datagram (at least four times)	Calculate IKE_AUTH data (twice)
The protocol	Calculate message authentication code (once)	Send datagram (once) Receive datagram (once)	Shared key decryption (once) Shared key encryption (once)

process before the authorization. By comparison, authentication process before the authorization used by this protocol reduces the calculation of the certification in the process of authorization.

4 CONCLUSION

In the Split scenario, when the AAAH-EAP server and the AAAH-MIP6 server are separated, IETF is studying the issue about the MIP6 service authorization. The authorization problems directly affect the deployment of MIP6 in mobile networks. This paper presents a secure MIP6 service authorization protocol to address security issues that malicious nodes impersonate legitimate nodes to steal MIP6 service, and ensure the correctness of transmission of the MIP6 service data. Moreover, our protocol establishes a trust relationship between the HA and the MN for protecting MIP6 service messages, and preventing malicious nodes from stealing MIP6 service. This paper mainly focuses on the authorization scheme of MIP6 service. However, the idea can also be used by other services that want to separate its authentication and authorization process.

REFERENCES

Aboba B. et al, Wxtensible Authentication Protocol (EAP), RFC3748, IETF, June 2004.
Bournelle J. et al, Diameter Mobile IPv6: HA-to-AAAH support, draft-ietf-dime-mip6-split-01, IETF, October 2006.
Burrows M., Abadi M. & Needham R. A logic of authentication. ACM Trans on Computer Systems, 1990.
Calhoun P. et al, Diameter Base Protocol, RFC3588, IETF, September 2003.
Eronen P. et al, Diameter Extensible Authentication Protocol (EAP) Application, RFC4072, IETF, August 2005.
Johnson D. et al, Mobility Support in IPv6, RFC3775, IETF, June 2004.
Kaufman C. "Internet Key Exchange (IKEv2) Protocol", RFC4306, IETF, December 2005.
Patel A. et al, Problem Statement for bootstrapping Mobile IPv6 (MIPv6), RFC4640, IETF, September 2006.
Salowey J. et al, Specification for the Derivation of Usage Specific Root Keys (USRK) from an Extended Master Session Key (EMSK), draft-ietf-hokey-emsk-hierarchy-00, IETF, January 2007.

Control Engineering and Information Systems – Liu (Ed)
© 2015 Taylor & Francis Group, London, ISBN 978-1-138-02685-8

Research and development status of mobile informatization in China power industry

H.F. Yang
NARI IT Project Management and Research Center, China

Q.G. Hu, P. Zhang, X. Zhang & M. Cheng
NARI Accenture Information Technology, China

ABSTRACT: With the development of informatization, mobile informatization is a tendency. This paper presents an in-depth look at the status of mobile informatization in the power industry in China. It studies the mobile application in various fields of electric business, such as power grid operation, electric power marketing, material management, construction, financial management and human resource management. It researches on the mobile application platform, terminals, development technology and security mechanisms. One the one hand, the progress in these areas is analyzed. On the other hand, this paper points out some shortages of the mobile informatization at present. In view of these shortcomings, this paper also proposes the advice for electric enterprises to improve in the future mobile informatization.

1 INTRODUCTION

At present, mobile informatization is developing to drive by technological innovation, industrial application of traction and coordinated development of industry. In the power system, some enterprises have begun to bring mobile informationization into force in some areas, such as grid's operation and maintenance, electric power marketing, electrical material management, construction, financial management and human resources management. Three parts, mobile support platform, communications network and mobile terminal, are all contained in these mobile application system. Some of mobile application systems even consider the security and protective mechanism. Now, mobile informationization is still imperfection, and the range of business is not wide. So with the next development of mobile informationization in power system, we need to further strengthen in terms of business application, development mode, normalization, standardization and safety protection.

2 CURRENT BUSINESS STATUS OF MOBILE INFORMATIZATION IN POWER INDUSTRY

2.1 Operation and maintenance of power grid

By present, the main mobile business application in power grid operation and maintenance should be:

- Patrol for equipment
  Mobile operation terminal is useful to help the staffs complete inspect business in the production field, it covers equipment management, defect management, site boot process, inspected result register, operation data collection and some other functions.
- Equipment maintenance
  Mobile operation terminal is useful to help the staffs complete overhaul business in the production field, it covers equipment management, defect management, site boot process, modified record register, specific data collection and some other functions.
- Equipment test
  Mobile operation terminal is useful to help the staffs complete test business in the production field, it covers equipment management, defect management, site boot process, experiment data register, operation data collection, etc.
- Equipment detection
  Mobile operation terminal is useful to help the staffs complete detection business in the production field; it covers equipment management, defect management, defective site boot process, detection record register, operation data collection, etc.
- Troubleshooting
  Mobile terminals are useful to help the staffs complete detection jobs in the production fields. It covers equipment management, defect

management, site boot process, broken-down rescue register, operation data collection, etc.

The mobile operations application which mentioned before, by the end of 2012, have been use in the company in Beijing, Shanghai, Hunan, Hubei, Anhui, Chongqing, Xinjiang, etc. There are totally 19 companies.

- Scene monitoring
Field monitoring is developed focus on filed remote monitoring technology, which through collecting voice and video frequency, telecommunication and computer information and some other technology to deliver the filed information to the rear to achieve the monitoring. The remote mobile and monitoring platform program in Shenyang Power-Supply Company is trying to investigate field sensing and interaction technology through 3G-VPDN private network, information technology to build the remote mobile and monitoring platform, help the rear management department monitoring the hole filed, and build the communication system between basement layer and management layer, achieve the information sharing. It is helpful to check and give orders, achieve the multi-angle analysis in the operation process, real-time monitoring the normative and security in the operation process (Zhang 2012).

- Emergency command
The mobile command system of electricity emergency repair is covers wireless communication technique, mobile terminal technique, data transmission technique, information processing technique. It is a kind of multifunctional and comprehensive command platform. Shanghai Electricity Company has built its own electricity emergency repair mobile command system (Qian et al. 2009). The emergency repair mobile command system is mainly used in the situation when power grid meets the serious accident, in order to satisfy the need of field commanding and decision-making, use the satellite as a carrier, through the informatization platform, scheduling resources reasonably, organize repairs promptly, help the power grid get right.

2.2 Electricity marketing

At present, the main mobiles business in electricity marketing should be:

- Mobile Field Meter Reading
Meter reading managers distribute the tasks. The staffs receive the information by intelligent terminal, and download the specific data. After arrive the filed, the staff type in the registration. They the print the electricity bill on the scene through build-in mini-printer, offer the better service for the user. And then, deliver the data to the marketing system.

- Mobile Meter Abnormal Registration
Receiving the meter reading abnormal information is mainly cover meter burnout, stop working, idling breaking, backwards, block, seal missing, meter lose, user infringement, stealing electricity, electricity consumption uprush, anticlimax, according to the question classification (measure fault, default and power stealing, electricity quantity and charge error, record error), to visit the marketing operation application, initiate the relevant treatment scheme.

- Dealing with Power Stealing and Default
Against the default, stealing electricity and some other anomalous using situation, use mobile terminal to photograph, pickup, record and some other functions to obtain the evidence on the sense, and record the default equipment, electrical properties, equipment capacitance, time and investigate result on the sense.

- Management of Mobile Operation
Through the mobile terminal, it can achieve to record the situation of stop and stat offering the electricity, record the relevant information of preventability experiment, check the operation archive of ancillary equipment, check the added harmonic detection, voltage detection, reactive compensation detection situation, check the network assess staffs register information.

2.3 Electric power material management

About the material management, current mobile operation application covers site handover, check and accept, warehouse, delivery management. In the repository, receiving, set the disk shelves, pick list, out delivery, check, shipping space regulation.

The performance personal and warehouse staffs establish the letter of advice in ERP, distribute the tasks through middleware, the distributed staff will download the tasks to the PDA terminal to achieve the site handover, check and accept, warehouse and removal.

Warehouse staffs establish the letter of advice in ERP, distribute the tasks through middleware, warehouse manager download the tasks to the PDA terminal the process the production handover, check and accept, warehouse, removal and check. Set the disk shelves, pick list, out delivery, check, shipping space regulation according to the dump application.

2.4 Construction

In the construction business, there is no practically deployed mobile application now. However, there are some modules to be published.

- The function of mobile three-dimensional construction path modules is: equipment research and position, three-dimensional scene interactive, automatic roam, process recorded (model and equipment situation), the contrast between milestone plan and reality, milestone process warning, coverage transparency install, eagle-eye map, and etc.
- The function of digital photograph mobile apply module is: system initialize (contents, schedule subentry, the inventory) browse and check the pictures, browse and check the property of the picture, retrieval the browse mode, increase and compile (property of pictures), the locality picture selected, photo shot, picture management (dispense with uploading, not yet uploaded, already uploaded), picture uploading management, integration with the main system.
- The construction site and the module of conference summary realize the functions of: the list presentation a digital photograph mobile apply module, and editing, changing and checking conference summary, with the main system integration and with the server synchronization.
- The function of capital construction schedule processing module is: schedule process module time specific rate, create schedule with the PDF form, mobile interface display, backlog process information obtain connector, backlog submit connector, tabulate.

2.5 Financial management

In order to solve the assets inventory problems of the constrained condition districts, improve the management level, decrease the mistake in checking process, and increase the efficiency of making inventory, Sichuan, Jilin and some other units begin to experiment, use RFID or bar code in the capital, achieve the fixed assets intelligent checking through mobile terminal application.

2.6 Human resource management

In the human resource, some electricity companies use a sectional mobile informatization. A Guangxi electric construction company (Lin 2012), when developing the staff educating and training, takes full advantages of current network technology, improves the remote education, and builds company training system. It covers managers [project manager, technician], enrich the education methods, develop diversification training method, improve enterprise education and training level, and improve the quality of staffs. According to the case, mobile training can improve the working enthusiasm, safely produce skills and knowledge of the staffs, it is beneficial to develop the training on the construction site. However, there are also having some other problems, such as interface is not concise, friendly enough, courseware deficiency, developed slowly, etc. these problems limited the staffs sustained, embedded using training system, the quantity of the website visited trends prove the opinion. Meanwhile, in the systematic training system, it also requires some other method to support it.

3 CURRENT TECHNOLOGY STATUS OF MOBILE INFORMATIZATION IN POWER INDUSTRY

3.1 The current status of platform and development

The production of mobile operation in power network based on "PI3000 Mobile" mobile application development platform and build runtime environment, the middleware uses the Weblogic, it is a Native development way, by using the mobile communication network and APN to access, and C/S structure as overall architecture. Mobile terminal through reciprocal visits between a agreement which include Web Service and RESTful Service with a server, and adopting the embedded database SQLite3, unstructured data stored in the form of files, using the format of zip for packing data, supporting push and pull these two data synchronization models and providing desktop data synchronization pattern. In terms of GIS application based on the Windows Mobile platform and third-party development (the company of Super Map, eSuperMap) development, supporting online GIS application, using the cache maps to provide GIS application when the signal is poor.

The mobile applications of power marketing have not a unified development and operating environment. Mainly Uses a Native development way, by using the mobile communication network and APN to access, and C/S structure as overall architecture. Mobile terminal through reciprocal visits between a Web Service agreement with a server, and adopting the embedded database SQLite3, and supporting a transform pattern of off-line/on-line, part of their application request data from a server by a way of real-time online, such as query the user profiles data on-line; part of their applications stored data offline to mobile terminal, such as meter application. GIS application are based on Android platform and the HTML 5 technology, providing functions such as GIS path navigation, the trajectory tracking, spatial query, equipment location and information query.

The application of power grid material based on SoTower development platform to build a mobile service, which running on a Weblogic middleware environments, mobile terminal uses the mobile communication network and APN to access, and C/S structure as overall architecture. By reciprocal visiting between a agreement which include Web Service and RESTful Service with a server, and adopting the embedded database SQLite3, create passwords before access to the database, for the key data require encrypted storage, after converted into some relevant SQL commands, by isolation mounting to visit the database of middleware directly, finishing query, download and simultaneously upload.

3.2 *The current status of mobile terminal*

- The industrial-strength PDA mainly based on the system of Windows Mobile, some of them based on Windows CE. In site operations, the PDA mainly used for these following business scenario: equipment ledger management, work and task management, the record of operation, patrol and measurement management, the text management, the record of maintenance management, defect management, hidden risk management, GIS navigation, etc.
- The industrial-strength tablet PC and laptop based on the system of Windows XP and Windows 7. In site operations, the tablet PC and laptop mainly used for these following business scenario: equipment ledger management, work and task management, the record of operation, patrol and measurement management, the text management, the record of maintenance management, defect management, hidden risk management, test report management, operation order management and work ticket management.
- The smart sensing devices, such as RFID, mainly depend on an RFID tag or module that may active or passive. In site operations, it mainly used for these following business scenario: vehicle monitoring management, equipment monitoring management, equipment inventory, etc.

3.3 *The current status of security*

The secure connect platform of State Grid Corporation of China is a representative security protection mechanism. So here we will use it as a typical case for analysis. The secure connect platform of State Grid Corporation of China has been added to the intelligent grid security infrastructure work as an important protective measure of secure access, the secure access platform has accumulated good technical foundation through the technology research of security encryption chip, intelligent security card, the strengthening of mobile terminal, appropriative encrypted communication, etc. With the using of secure access platform, a lot of problems has been solved, transmission of security data in the public Netcom, the security access problem of access objects, identity certify and access authorize, unified supervision and audit, request services and data exchange of mobile access, specification of the access object's self-protection, achieved the access task of intelligent grid mobile business system. Primary system of access include mobile task, state monitoring of transmission and transformation, warehouse management system, financial management and control systems, and other mobile task system, late phase will continue consummating the research and actualization of secure access task system, like unified video platform, integrated payment platform, collection of supply voltage, electricity quality management and so on. Current security access platform has completed the entire network deployment, constitute a preliminary unified and secure and reliable State Grid's access system, and strong support for the intelligent grid business needs.

4 ANALYSIS OF CURRENT MOBILE INFORMATIZATION IN POWER INDUSTRY

From the research on the current status of mobile informatization in China power industry, it can be discovered that there are some problems.

4.1 *There are some mobile applications now, although the scope of them is narrow*

At present, in the power industry, there are some mobile applications in the fields of power grid operation, electric power marketing, material management, construction, financial management and human resource management. However, the current mobile applications are generally driven by some specific business demands; the scope of mobile informatization is still small.

4.2 *The efficiency of enterprise mobile application development is not high*

In electrical enterprises, most of the current mobile applications are individually developed to meet some specific demands in different scenarios. Each time, a suite of individual platform, terminals and network communication mechanism is newly developed. The efficiency of such development model is not high. At present, State Grid Corporation of China is unifying the process of mobile application development. All the mobile applications for different business fields will be developed by one unified platform. The efficiency of the development can become higher.

4.3 The level of standardization of the mobile informatization is not high enough

In the process of current mobile informatization in power industry, because the mobile applications for different business fields are individually developed, the standardization of mobile informatization is relatively low. In the respect of platform, there is no unified technique platform standard. In the respect of network communication, standard communication protocols have not been built up. In the respect of mobile terminals, the terminal techniques, the hardware manufacture, the software systems configuration and the performance requirements are still not standardized.

4.4 Security has been considered, but the current secure mechanisms are not enough

In the process of present mobile informatization in China power industry, for some mobile applications, security has been considered. Some approaches have been proposed to protect the applications from being attacked. However, as the mobile informatization goes far, more and more mobile terminals will connect to the application platform. The platform has to cope with numerous terminals of different types, some of which may be vicious. All of these challenges drive us to build a more secure protection system.

5 RESEARCH ON THE DEVELOPMENT OF MOBILE INFORMATIZATION IN POWER INDUSTRY

5.1 Develop more mobile applications for different electric business

In the future development of mobile informatization, more mobile application with enhanced function should be developed in the fields of power grid operation, electric power marketing, material management, construction, financial management and human resource management. The scope of mobile applications should become larger.

5.2 Build up and enhance the enterprise unified platform for mobile application development

The enterprises in the power industry should first summarize the outcomes and experience of the current enterprise mobile informatization. Then, the enterprises, which have no unified development platform, should build up the enterprise-wide standard development platform for mobile

applications. Based on this platform, various kinds of mobile applications for different business can be efficiently developed. The enterprises, which already have the unified mobile application development platform, should enhance the platform functions, in order to achieve more efficiency. The key points are as follows.

5.2.1 Cross-platform application technique
The enterprises can first research on the advanced and mature framework for cross-platform application development. It can help to cut down the development cost and shorten the overall time.

5.2.2 Mobile work flow
Achieving the mobile work flow can enhance the function of mobile applications. It is very helpful to the business in electric enterprises.

5.2.3 Data presentation on mobile terminals
The electric enterprises can develop a suite of data presentation components, which contains the function of presenting charts, figures and complex data tables on terminals. That will be useful to enhance the mobile applications to support Business Intelligence.

5.2.4 Mobile device management
With the consideration of the business and application platform, the enterprises in the power industry should also pay attention to the mobile device management, in order to realize the safe and convenient management of different kinds of terminals.

5.3 Build up the mobile informatization standards

In order to provide the efficient and unified support for mobile informatization, the enterprises in the power industry should build up a series of technical standards. These standards can guide

Table 1. Table of mobile informatization standards.

Device layer	The technical standards of mobile terminals
Application layer	Application development standards User experience standards Test approach standards Performance index standards
Platform layer	Application operation Standards Platform interface standards Mobile device management standards
Network layer	The network communication standards
Security layer	Secure access standards Secure data transfer standards

and restrain the development and integration of terminals, applications, security components and the platform. At lease, the mobile informatization of an electric enterprise needs the following standards.

5.4 Enhance the security of mobile informatization

The mobile informatization of electric enterprise business may influence the normal power supply, so the security is a very important aspect in the mobile informatization. With the development of mobile informatization, the following points should be emphasized.

1. Research on and develop the software to realize the security enhancement on the smart terminals with Android, Linux and WinCE operating systems.
2. Build the monitoring systems for enterprise internal networks, formulate the indexes for measuring the security on network edge, and develop the modules for collect the monitoring data.
3. Improve the hardware and software interfaces between the safety protection system and the business systems in the electric enterprises.

REFERENCES

Lin T., "The application of mobile training in the electric construction enterprises", Coastal Enterprises and Science & Technology, vol. 11, 2012, pp. 47–49.

Qian W., J. Yang, M. Wang, "Essentials to communication platform construction for mobile emergency response command systems for Shanghai Power Grid", East China Electric Power, vol. 5, 2009, pp. 759–762.

Zhang Y., "Research and application of electric field remote mobile monitoring technology", Power Electronics, vol. 6, 2012, pp. 54–55.

Control Engineering and Information Systems – Liu (Ed)
© 2015 Taylor & Francis Group, London, ISBN 978-1-138-02685-8

The evaluation information system of architectural heritage value in the era of big data

M. Liu
Tourism Institute, Beijing Union University, Beijing, P.R. China

A.L. Liu
College of Resource Environment and Tourism, Capital Normal University, Beijing, P.R. China

ABSTRACT: As an important part of the whole heritage system, the architectural heritage is not only the witness of history and culture carrier, but also a vivid manifestation of the human civilization and spiritual home. The value assessment of architectural heritage is an important step of the value recognition and identification of architectural value and plays a connecting role in protection. Therefore, the value assessment of architectural heritage is the basic work of architectural heritage protection. Due to the deepening of the awareness of the architectural heritage value, the value assessment of architectural heritage also shows different angles and different methods. Based on the analysis of the architectural heritage value, this paper first analyzed value assessment methods, and then built the value evaluation information system of architectural heritage. The findings of the paper would help understand the value recognition and scientific evaluation of architectural heritage.

1 INTRODUCTION

Architectural heritage, being an important part of the cultural heritage, refers to the monomer structures and buildings, building groups, historical section, historical districts, historical cities, towns, villages and the whole context which have survived from history and have outstanding value from the perspective of history, art and science. Architectural heritages is the combination of true history and cultural memory of all the times, a witness to history, the carrier of culture, the coordinates of the ethnic history of the region, and also is a vivid manifestation of human civilization (Zhu Guangya, 2006)[1]. The value assessment of architectural heritage is the basic work of architectural heritage protection. Assessment is a management practice of many international and domestic industries. Decision-making based on assessment could get rid of the impacts brought out by personal preferences and tendentious statement, and make the decisions much more scientific.

Taking architectural heritage into the evaluation system is an indispensable part of the architectural heritage management in society. Therefore, the value assessment of architectural heritage is not only an essential process to make a reasonable judgment, to avoid individual or unilateral factors magnify the building heritage protection and management into quantified, but also a decisive

step in the promotion of scientific and legal track. The basic work of the protection of the architectural heritage is to assess the value of architectural heritage.

It's been many years since the value assessment of architectural heritage was studied. The procedure of value assessment of architectural heritage in Canada is divided into three steps by the Ministry of Environment: Survey-Evaluation-Policy, so as to develop the scientific decision-making and form a more complete architectural heritage protection system. Chinese domestic research on value assessment of architectural heritage began in 1990s, by the means of introducing research cases abroad. In 1995, Chengkan village in Anhui Province conducted the first case of architectural heritage quantitative assessment. From then on, the value assessment of architectural heritage in China has gone through a series of phases, including learning Western experience, understanding China's national conditions, the process of evaluation designing, and into a new stage of practice test, adjustment and improvement (Zhu Guangya, 1996; 1998)[2-3]. Headed by Mr. Zhu Guangya, some other scholars concerned about the issue of architectural heritage value assessment, analyzed the value of architectural heritage and broader cultural heritage value from different perspectives, and proposed appropriate evaluation methods[4-10]. These studies effectively promote the awareness of the architectural heritage value; however, due to the different composition of architectural heritage value, the methods used in the studies also vary and need further study.

With the emergence of new ways of information releasing such as blog, social networking and LBS (location-based services), as well as the rise of cloud computing and things-internet, data accumulation is now growing at an unprecedented rate, which means the era of big data has arrived. Architectural heritage assessment is a huge and complex economic and social activity. Effective evaluation management of different value categories which uses data from all aspects is the necessary means to promote the scientific development of architectural heritage value assessment, test and protection architectural heritage. This is also the only way of architectural heritage value assessment in the era of big data.

2 ANALYSIS OF ARCHITECTURAL HERITAGE VALUE ASSESSMENT

Different regions have a different understanding of different architectural heritage value. Wang Yanan pointed out in the paper that the evaluation criteria of historic buildings in England (Table 1) are different from that in German (Table 2).

Overall, the Western architectural heritage value assessment focus on big events, reflecting the position of the historical subjects; Second, Western architectural heritage value assessment pays more attention to environmental value and the position of architecture in urban space; Third, Western

Table 1. The evaluation criteria of historic buildings in England.

Basic standards	Points
Art	Works of art—building products with unique and creative idea.
Architecture	Should be considered as a part of the history of building development and not be interrupted, instead they should be protected well.
	Some imperfect individuals could be combined with others with the passage of time and have a chance to become an outstanding combination overall.
Technology	Examples in the process of technological development.
Society	Examples that reflecting a certain life style which has disappeared but had sociological meanings.
History	Architectures related to Great men or historical events in history.

Table 2. The evaluation criteria of historic buildings in Germany.

Typologies	Points
Historical value	
Functions	The status of the building's original features, such as public buildings, hotels and residential buildings.
Morphological types	The plan, elevation and other forms of the buildings, as well as the degree of scarcity.
Frames	The structural features of the buildings, such as exposed Akashi structure and mortar stone structure.
Era	The construction periods of buildings, such as the late Middle Ages and the Renaissance.
Artistic value	The level of architectural art, the designer and the accurate time.
Urban space value	The position and influence of architecture or city blocks in the urban space.
Integrity	The integrity of the architectural forms.

architectural heritage shows strong concern on uniqueness and social value.

"Cultural Relics Protection Law of the People's Republic of China" provides that the value of heritage has four aspects: historical value, artistic value, scientific value and historical value. The method of architectural heritage value assessment basically continues the Canadian evaluation system, which conducts value assessment from five perspectives of art, history, environment, use value and integrity. The popular existing value assessment nowadays is based on experts individual data collection and the use of subjective judgments (Table 3).

Based on the comparison of value assessment of architectural heritage between China and Western countries, the roots of the architectural heritage could be traced further. The architectural heritage value not only focuses on architectural heritage itself, but pays more attention to the promotion of human well-being. Development and protection do not exist for the purposes of the architectural heritage, architectural heritage and promotion of human culture, education and well-being is more to be concerned about the ultimate goal. Therefore, the purpose of the value assessment of the architectural heritage should be given more in-depth thinking. Attention should not only be paid statically on the architectural heritage of simple protection, preservation, but also on their contribution to human well-beings. Therefore, the use value and sentimental value of architectural heritage should be put more emphasis (Fig. 1).

Table 3. The typical methods of value assessment of architectural heritage in China.

Researchers	Decomposition perspective of architectural heritage value	Main methods used
Administration of Canadian history and Architectural Information	(1) Artistic value (2) historical value (3) environmental values (4) use value (5) integrity	AHP, weights, expert scoring
Zhu Guangya, etc. (1996)	(1) Historical heritage value, (2) scientific value, (3) Art (4) practical value, (5) uniqueness	AHP, weights, expert scoring
Zhou Shang Yi, etc. (2006)	(1) Historical value, (2) scientific value, (3) artistic value, (4) ecological value, (5) the economic value	AHP, weights, expert scoring, CVM
Lin Yuan, etc. (2007)	(1) Information Value (historical value, artistic value, science and technology value) (2) the emotional and symbolic value, (3) the use of the value	AHP, weights, expert scoring
Yin Zhanqun, etc. (2008)	(1) Intrinsic value (historical value, scientific value), the value of (2) external value (environmental value, use value)	AHP, weights, expert scoring

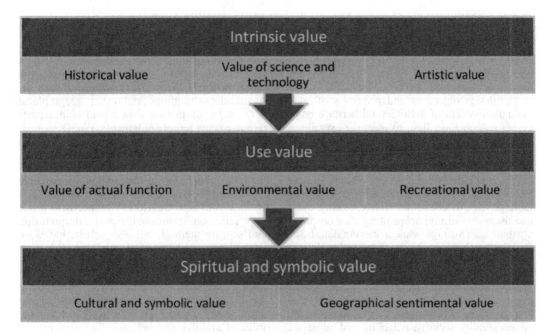

Figure 1. The framework of architectural heritage value system.

3 INFORMATION SYSTEM OF ARCHITECTURAL HERITAGE VALUE ASSESSMENT

The methods used in architectural heritage assessment in 1995's Chengkan case of Anhui Province was the combination of on-site paper questionnaires and evaluation forms for scoring. With the popularization of computer applications, the means of architectural heritage assessment requirements become more advanced and more accurate, as a result, computer-aided architectural heritage assessment emerges. Computer-based quantitative scientific mode will promote the institutionalization of the architectural heritage assessment and scientific management. The Suzhou architectural heritage assessment system software (second edition) won the second prize of scientific and technological innovation of the Ministry of Cultural Heritage, which makes the the scientific computing of original weights and factors. With the perspective changes of the awareness of the architectural heritage value, the simple analytic hierarchy process cannot fully cover the value assessment of the architectural heritage. Therefore, the use of diverse data, complex algorithm becomes the core

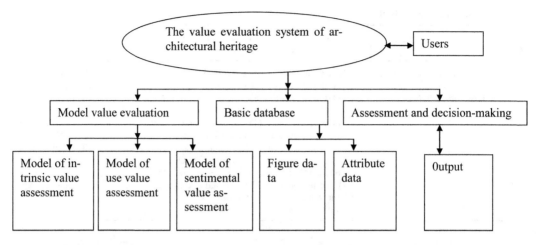

Figure 2.   The structure of architectural heritage value assessment information system.

problems to be solved for the architectural heritage information system with the use of big and diverse data.

The design objectives and principles of value evaluation system of architectural heritage based on big data include: first, to meet the overall system requirements, from channel model running of the basic data processing and analysis to the output of the operation results; Second, to realize the compatibility of a variety of data sources, establishment of a comprehensive architectural heritage value database, the update, modify, query and the assessment report printing, data output of attribute data storage, while achieving data transfer interface between the database and evaluating model; third, to realize the report visualization of assessment, with the function of flexible query, analysis and statistics; fourth, the whole system has a clear modular structure, friendly interface, and should be easy to operate; fifth, to ensure the system's utility, security, reliability and advanced. Through the designing of the value evaluation system of architectural heritage, researchers, managers and other system users could have respective platform to fulfill their aims of research, decision-making and so on.

There are five modules of the value evaluation system of architectural heritage, including the underlying database module, applied analysis model database module, display module, decision support and output modules (Fig. 2).

The underlying database system was built based on the multivariate data in era of big data. The data types include not only the various types of attribute data, but also picture data and GIS spatial data. The data source include not only the historical eras, architectural style, function,

but also other data established by the GIS system, such as location coordinates, distance from the main traffic artery, the relationship with the surrounding buildings and other geographical data. A lot of picture data would help experts to make some remote judgments. What's more, it can not only help determine the integrity of the building through data algorithm, but also correspond to the local social and economic statistics, such as number of visitors, of places where architectural heritage locate. As far as the Valuation Model core algorithm is concerned, it not only relies on traditional expert interpretation and scoring method, but also will be based on the architectural heritage of users—residents and visitors TCM (Travel Cost Method), CVM (CVM method), HPA (the hedonic price method) and market comparison method, the value reduction algorithm for integrated operations, to achieve more systematic and scientific assessment of the value of architectural heritage. By the algorithm can it effectively improve the existing two defects of architectural heritage value assessment, namely: (1) the perspective of experts or government, rather than the general perspective of the user; (2) human intervention exist in whether the weight or experts scoring processing, especially the latter.

4   CONCLUSIONS

The era of big data will bring profound changes in our society and our daily life. The purpose of the paper is to promote the effective application of technology to heritage protection and human well-being. Out of this purpose, this paper built value assessment of the architectural heritage, provided a powerful methodology, and identified the core technical aspects of the architectural heritage assessment, based on the existing architectural heritage value assessment. However, there is no off-the-shelf integrated architectural heritage value assessment information system. The system is still in the trial stage. The ideas, methods, and built process shown in this paper would help explore the subject further by outlining the basic framework, hardware configuration, software development and other aspects of the technology in future.

ACKNOWLEDGEMENTS

The work has been support by Beijing Natural Science Foundation Project of China (grant no.8123042) and Beijing Social Sciences and Natural Sciences Collaborative Innovative Research Base Project (grant no. 2013SZJD005).

REFERENCES

Han Bing, Luo Zhi-Star. Architectural heritage value Evaluation Method [J]. Huazhong Architecture, 2010, (6):116–118.

Li Lili. Guangzhou heritage of historical and cultural heritage and value assessment [J] Journal of Guangzhou University, 2006. Instructor: Ho into.

Lin Yuan Chinese Architectural Heritage Conservation basic theory research [D]. Xi'an University of Architecture and Technology University PhD thesis, 2007.

Liu Min, Chen Tian, Liu Aili the the tourist recreational value assessment study progress [J]. Human Geography, 2008, 23 (1):13–19.

Shang Yi Zhou Jimin, Jiang Miaomiao unmovable heritage value evaluation of the protective effect on the pattern of the ancient capital of cultural space: Xicheng District, Beijing as a case study area [J]. Tourism Tribune, 2006, 21 (8): 81–84.

Wang Yanan. Qingdao Modern Architecture Evaluation and Protection and Utilization [J] Journal of Zhengzhou University, 2005.

Yaojian Yun, Huang Anmin, blue Xiaoyuan Value Assessment the CVM method based cultural heritage: Yungang Grottoes [J] Economic Research Guide, 2011, (5):257–258.

Yin Zhanqun Qian Zhao Yue. Suzhou architectural heritage assessment system issues research [J]. Southeast Culture, 2008, (2):85–90.

Zhang Xiaonan assessment of outstanding universal value and heritage constitutes Analysis Method: Grand Canal as an example [J]. Conservation and Archaeology, 2009, 21 (2):1–8.

Zhu Guangya, a basic work of Jiang Hui development of architectural heritage intensive areas: Architectural Heritage Assessment [J]. Planners, 1996, (1):33–38.

Zhu Guangya, Fang Qiu, Lei Xiaohong architectural heritage assessment of an exploration [J]. New construction, 1998, (2):22–24.

Zhu Guangya, Yang Lixia architectural heritage protection in the Course of Urbanization [J] Architecture and Culture, 2006, (6):15–22.

Control Engineering and Information Systems – Liu (Ed)
© 2015 Taylor & Francis Group, London, ISBN 978-1-138-02685-8

Design and realization of a mineral resources management information system

M.M. Fei
China University of Geosciences, Beijing, China

ABSTRACT: With the rapid development of the industrialization and urbanization and agricultural modernization synchronous, the contradiction between supply and demand of mineral resources is obvious. It is significant to establish an effective management system for mineral resources based on modern information technology. This paper introduces the current situation and basic characteristics of mineral resources in China. The author puts forward a solution to the architecture of mineral resources management information system on the basis of Service-Oriented Architecture principle (SOA). The architecture can be divided into basic services, common services, and business services. The functional modules of the system are designed, which aims to provide visualization dynamic monitoring and management on the mineral resources exploration, development, utilization, protection, examination and approval. Finally, the author elaborates the system realization and the application effect.

1 INTRODUCTION

1.1 The current situation of mineral resources in China

Mineral resources is defined as a concentration or occurrence of natural, solid, inorganic or fossilized organic material in or on the earth's crust in such form, and quantity and of such a grade or quality that it has reasonable prospects for economic extraction. There are a great number of mineral resources have been verified in China, and a fairly complete system for the supply of mineral products has been established, which provide an important guarantee for the sustained, rapid and healthy development of the Chinese economy. According to some statistics, the oil production reached 210 million tons, the gas production is about 105 billion cubic meters, the coal breakthrough 3.5 billion tons, and iron ore production is more than 1.3 billion tons, the ten main non-ferrous metal production is about 34 million tons at the end of 2011[1].

Nonetheless, with the rapid development of the industrialization and urbanization and agricultural modernization synchronous, the contradiction between supply and demand of mineral resource is obvious. In April 2012, The Minister of land and resources stated that the China's mineral resources consumption keep double-digit growth keep double-digit growth over the past 15 years. The major mineral imports increased significantly, among them, more than 50% of oil, iron ore, refined aluminum, refined, sylvite are rely on foreign supplies. The Minister emphasized that the

quantity, ecological trinity of mineral resources management mode should be explore and establish, and the <Outline of prospecting breakthrough strategic action (2011–2020)> has been deliberated and approved by the executive meetings of the State Council, which means the geological prospecting work have a blueprint[2].

1.2 The characteristics of mineral resources in China

The basic characteristics of mineral resources in China are as follow:

1. The total quantity of the resources is rich and there is a fairly complete variety of minerals. China has found a rather complete variety of mineral resources and a fair abundance of mineral resources in total quantity. There are 52 large orefield and 80 medium-sized orefield are proved up in 2011.
2. The per-capita quantity of the resources is small, and there is an imbalance between supply and demand for some of the resources. In February 2012, The Minister of land and resources stated that China's mineral resources reserves per capita world average is only 58% of reserves.
3. Superior mineral ores exist side by side with inferior ones. There are both high-quality ores and those of low grade and complex constituents.
4. The resources with a low degree of geological control account for a greater proportion of the verified reserves of the mineral resources.

5. The conditions for mineralization are good, and there are good prospects for finding more mineral resources. According to the mineral resources statistics bulletin, the proven reserves of national coal, iron ore, copper, lead, zinc mine, bauxite, tungsten molybdenum, tin ore, gold, silver, and pyrite, phosphate rock and potassium mineral resources of such important minerals have risen in varying degrees in 2011.

2 RELATED STUDIES

In order to satisfy the needs of mineral resources management for governments and the general public, it is significant to establish an effective management system for mineral resources based on modem information technology.

From the 2000s until now, research on mineral resource management information system has drawn public attention from many scholars, and has been a new important theme of the study in recent years. Many scholars, domestic and overseas, have achieved some remarkable results.

In this field, foreign researches and developments are earlier, and have already been proved to be a kind of effective mode improving the efficiency and level of mineral resources management. The Canada researched the modeling, quantitative and digital management in mineral resources. Canadian federal government established a "mineral geology database index", "MINTEC", "CANMINDEX", "MINSYS" and other basic database. The United States is a overall information country in mineral resources management and mining management. The computer information database of mineral resources (CRIB) is the largest mineral resources information database in the world. In which, there are more than 40000 deposits and orefield of the USA, and more than 6000 abroad.

In recent years, the domestic scholars have carried on a large amount of research to mineral resource informatization. W. Li et al. (2009)[3] developed a mineral resource management information system by using SuperMap, the system has realized the functions such as management of documents, query and search of mineral resources, and analysis and forecast of mining sections. W. Xun et al. (2010)[4] designed and developed a mineral resources management information system based on C/S and B/S mixed structure. T. Yongjie et al. (2011)[5] put forward to develop mineral resource management system using the data from the mineral rights survey and verification project. It used single mineral rights' data as its basic data source, and functioned in a basic administration unit covering a county.

Judging from the current domestic and foreign researches, we can find that many scholars are faced

with specific application. The architecture based on Service-Oriented Architecture (SOA) principle is rare. This paper proposed the architecture as a variety of services, providing a flexibility mineral resources management information system.

3 THE DESIGN OF MINERAL RESOURCES MANAGEMENT INFORMATION SYSTEM

3.1 *The architecture design*

The paper design the architecture of mineral resources management information system make use of SOA principle. In software engineering, a SOA is a set of principles and methodologies for designing and developing software in the form of interoperable services. These services are well-defined business functionalities that are built as software components that can be reused for different purposes. SOA can help businesses respond more quickly and more cost-effectively to changing market conditions. This style of architecture promotes reuse at the macro (service) level rather than micro (classes) level. SOA realizes its business and IT benefits by utilizing an analysis and design methodology when creating services.

In order to build mineral resources management information system, we should analysis the business process. Because of the mineral resources business model is complicated; the business process should be break down into smaller and more concrete steps. In the top-down approach, we can decompose the

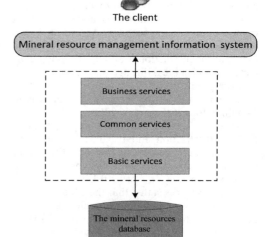

Figure 1. The architecture of mineral resources management information system.

mineral resources business into smaller chunks until reach the level of basic services. In the bottom-up approach, we can build business processes by composing services into more general chunks.

Therefore, the author decompose the business process with a variety of services, and the services can be divided into three types. The architecture of the system is shown as Figure 1.

3.1.1 *The basic services*
At the lowest layer, it provides business data and spatial data, and the data management and data operation is encapsulated as a form of basic service. Basic services read or write data from database. These services typically each represent a fundamental business operation of the backend. Basic services encapsulate platform-specific aspects and implementation details from the outside world. So that developer can request a service without knowing how it is implemented.

Due to the system is closely related to spatial data, the author introducing the basic services with spatial data service. The GIS tasks encapsulate a particular piece of GIS functionality, such as querying or editing, with a user interface in a format that you can easily add to your web application. Tasks make it easy for the end user of the application to perform certain functions, and in many cases, they facilitate the developer's job because they can be added to a web application without writing any code. Basic services must ensure that external systems can access these back ends without being able to corrupt them. That is, they should hide the technical self-contained business functionality.

3.1.2 *The common services*
The common services layer operate at a higher level than basic services. Common services are composed of multiple services, providing some common functions. The purpose of this layer is to ensure the system operation, which have no direct relation with the complex business. It can set up the system parameters and business parameters.

3.1.3 *The business services*
The business service layer are services that contain business logic. According to the demand of the users, the business module can be built as service. Due to the particle size of service is very small and abstract, even if business often changes, the system maintenance is easy. Certainly, from a business point of view, it doesn't matter whether a service is a basic or a common service or business service. The only important point is whether it fulfills the necessary business functionality.

3.1.4 *The application system*
Take advantage of the layers above, the management system is built, and the users can visit to. According to the authority. If business is applied, the system will invoke relevant service and access to the database.

3.2 *The functional modules design*

Through the analysis of the demand, this paper design ten functional modules in mineral resource management information system. It can realize visualization dynamic monitoring and management on the mineral resources exploration, development, utilization, protection, examination and approval. Figure 2 depicts the functional modules.

Firstly, there are five modules directly related to mineral resources, including mineral resources development, mining right, mineral exploration right, planning, and reserves, which have covered the major business for management department.

Secondly, because of mineral is closely related to geology, there are some modules for geology, including geological environment management, geological exploration and geological data management.

Thirdly, it provides file management with a variety of form. It can be divided into the mining right file, mineral exploration right files, the mineral resources exploration files, mining area files, the mining right, annual reports, etc.al.

Finally, in order to assist decision support, it can achieve specialized analysis for system data.

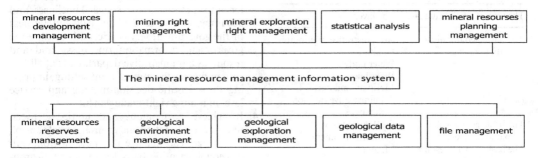

Figure 2.   The functional modules of the system.

On the basis of GIS, the system can provide spatial analysis for the geographical position information of mineral resources, for instance, it achieves auxiliary delimit the boundary of exploitation of mineral resources planning.

4 THE REALIZATION AND APPLICATION OF MINERAL RESOURCES MANAGEMENT INFORMATION SYSTEM

4.1 The construction of mineral resources database

The mineral resources database includes attribute data and spatial data. The author has developed the system within ArcGIS; An ArcSDE data source supports the query processing functions. The spatial data mainly includes basic geographic database, geological information database and mineral resources information database. Table 1 depicts the spatial database.

On the basis of ArcGIS, the system provides this spatial data management capability for mineral resources database, which is a geographic data structure (geodatabase). A geodatabase needs to support the wide variety of spatial data types typically used by today's GIS applications. In addition, the geodatabase also includes directly in the database the rules that govern the behavior of these data types. This eliminates the need for separate custom programming to define even the simplest of data behaviors.

The mineral resources database is viewed as a single database. The actual physical implementation of the database occurs in a variety of deployment approaches. If the system is deployed on a city, the main users are the department of the city, the mineral resources database can managed as a single central database. If the system is deployed on a province, it need to link the database of each city,

it can be distributed to several physical locations but operated "virtually" as though it were a single data resource.

4.2 The application effect

Take advantage of the system, the mineral resources status can be show as visualized model, and it achieve the mineral resources information query, examination and approval, statistics, and other functions. The administrative staff can manage the mineral resources conveniently, and the production and business operator can get effective service to reduce the cost of investment and investment risk, and the public will gain a very good understanding of the business. The system provides the science basis for the further research development and utilization of mineral resources.

In summary, the application effect of the system is as follow:

1. It improves the efficiency of mineral resource development, to shorten the development cycle. In traditional way, due to the backward in technology and information block factors, the foundational work will spend a lot of time, correspondingly longer cycle of each stage. Along with information technology development, The mineral resources information can be acquired quickly, and the further development and utilization can gain strong technical support; meanwhile, it improved the work efficiency, and effectively shorten the cycle.
2. Promote the rationalization of develop mineral resources, reducing the waste of mineral resources. The system put the dynamic monitoring and management effectively unified to a great extent, it enables the development and utilization of mineral resource structure and layout to be optimized. So as to realize the intensive development and utilization, make the resource development and utilization is more reasonable and standardization, reducing the consumption of resources and waste.
3. Reflect the superiority of openness in government affairs. Firstly, the government can transfer information to the social Timeliness and accuracy, to ensure the integrity and fairness of the government services Secondly, realize the government information openness, land and resources administrative departments at all levels can master the mining right setting, improving the scientific decision-making and service level in mining right management.
4. It achieve harmonious and unified between resources exploitation and environmental control. The system can manage the mine environment, providing scientific basis for the mine environment management will the exploitation of mineral

Table 1. The spatial databse of mineral resources.

The type of spatial database	The layer
The basic geographical spatial database	Administrative region unit Habitation Water body Waterline Meridian and parallel Administrative division range
The basic geological spatial database	Geological body Fracture lineament
The basic mineral resources spatial database	Mineral

resources and environmental control effectively to unify, promote the coordinated development on exploitation of mineral resources, the mine ecological construction, environmental protection.

5 CONCLUSION AND FURTHER RESEARCH

It is significant to establish an effective management system for mineral resources based on modem information technology. This paper analyzed the research status in recent years, and proposed to design the mineral resources management information system make use of SOA principle. The author designed and developed the system, and expounded the application effect. Using the system, the management level and social service efficiency of mineral resources can be improved.

Indeed, the researchers only studied the architecture and function of mineral resources management information system, the key technology haven't be analyzed in depth. Future research is needed to study the key technology and to improve the system with the business needs.

REFERENCES

Li W., Fadong W., Mingzhong T., Xuesong W. & Jing L. "Development of Alashan mineral resources management system based on SuperMap", Metal Mine, Mar, 2009(3), pp. 137–139.

The Ministry of Land and Resources, "The mineral resources should be increase income and reduce expenditure", Available from <http://www.mlr.gov.cn>. Apr, 2012.

The Ministry of Land and Resources, "The news conference of oil and gas and the main solid mineral resources reserves of China in 2011", Available from: <http://www.mlr.gov.cn>. Feb, 2012.

Xun W. & Yaohuan Y. "Mineral resource management information system based on C/S & B/S mixed structure", Geospatial Information, vol. 8, Oct, 2010(5), pp. 82–84.

Yongjie T., Jia G., Renyong X. & Wenjian Z. "Development of mineral resource management information system based on national survey and verification data of mineral rights", China Mining Magazine, vol. 20, Jul, 2011(7), pp. 50–54.

Information loss of unintentional disclosure in P2P

L. Xiong
Business School, Jinling Institute of Technology, Nanjing, China

Y. Zeng
Faculty of Computer Science and Technology, Jinling Institute of Technology, Nanjing, China

ABSTRACT: In the information age, the Company faces all kinds information security risk, in which unintentional disclosure information is one of the biggest risks of enterprise. We research on unintentional disclosures through peer to peer file-sharing networks. It demonstrates unintentional disclosures will result in a substantial threat and vulnerability for large financial firms, put forward measure methods for the risk of unintentional disclosure information and how to avoid to the risk of unintentional disclosure information.

1 INTRODUCTION

As firms become ever more dependent on information, it will have the very big economic loss once that information is leaked [1]. So information security has become a new focus of the business news. The firms come under increased pressure to harden their networks and take a more aggressive security posture. However, it is often not clear what security initiatives offer firms the great improvement. Through a close look at the media and network, we find that the reasons of information leak are not hacker attacks on the network, but rather the unintentional disclosures information in the process of work [2]. For examples, laptops at Boeing, were lost and stolen, in each case inadvertently disclosing person and business information. Organization has mistakenly posted on the Web many different types of sensitive information, from legal to medical to financial. Even technology firms such as Google and AOL have suffered the embarrassment of inadvertent Web posting of sensitive information. Still other firms have seen their information. In each case, the result was the same-sensitive information inadvertently leaked creaking embarrassment and financial loss for the firms [3].

This paper discusses a common, but widely misunderstood source of unintentional disclosure peer to peer (P2P) file sharing networks. The P2P network is used to for more clients to transmit the mass data. Many uses pointed out that the evolution of these networks has done little but increase the Risk [4]. Internet service providers, banks and copyright holder for P2P limit P2P both technically and legally and prompt P2P developers to create decentralized, encrypted, anonymous networks that can find their way through corporate and residential firewalls. These networks are almost impossible to track. So risk issues aren't resolved.

2 THE SITUATION OF FILE SHARING IN P2P

P2P networks are a network that is used to exchange data for one client to connect the other client but not browse and download data to connect server [5].

The file sharing in P2P is the service of sharing data between the nodes in the node groups, which can make peers share their data, locate and get date of the other peer, manage the shared data for local peers, and browse and download dates in the other peer [6].

In 1999, Napster used firstly file sharing in P2P networks, and brought the concept of file sharing into the mainstream with its wildly popular music-sharing service. Napster enabled tens of millions of users to share MP3-formatted song files. In its place many other file-sharing systems have emerged, driving an endless debate over the impact of music sharing and a string of legal challenges by the music and video content industry. For thwarting file sharing, Firms, universities, and ISPS block or throttle traffic associated with P2P systems using approaches such as port filtering. Client developers responded by using ports associated with other services to exchange data, blending file-sharing traffic with other data streams. The debate and thwarting don't caused the file sharing to reduce and continues to grow, with usage doubling from less than 4 million in

2003 to nearly 10 million simultaneous users in 2006. The number of types of P2P networks is increasingly added, such as which is one of the most popular applications for very large files such as video. Users of these systems readily adapt and change to new networks based on legal pressure, features, and popularity. These rapid shifts suggest low barriers to entry for new sharing technologies, so there are some legal and Security problem in P2P.

3 UNINTENTIONAL DISCLOSURES ON FILE SHARING

File sharing in P2P networks is one of sources of the unintentional disclosure. There are several routes for confidential data to get on to the network: a user accidentally shares folders containing the information; a user stores music and other data in the same folder that is shared; a user downloads malware that, when executed, exposes files; or the client software has bugs that result in unintentional sharing of file directories. Of course, it is not necessary for a worm or virus to expose personal or sensitive documents because many users will unknowingly expose these documents for many reasons for example, some users mistakenly point to My Documents and end up sharing all of their files. In some cases, the client interface design makes it difficult to see what is being shared. Moreover, P2P file sharing systems often provide incentives for users to share files via faster downloads or broader searches. The clients typically come with wizards that are designed to find all media files and share the directories where media files are located. So a single MP3 file in My Documents can lead to sharing everything in My Documents. Moreover, the clients often share all subdirectories of a shared directory.

Many of these reasons point to the interface design and features of P2P clients that facilitate unintentional sharing. Many persons think that the interface design and features of P2P clients that facilitate unintentional sharing lead to unintentional disclosure. On fact, many information leaks are the result of accidentally shared data rather than the result of malicious outsiders; there are many other trends that are driving more security concerns. They include:

1. Widely usage of network and lack of safety awareness for net work mean more leaks. Many network uses lack of safety awareness and don't know how to avoid to information leak. P2P developer can't think out all factors of the information security. So the above problems lead to information leak when use share files on the P2P networks. P2P clients tend to be "set and

forget" applications that run in the background while the user is not at the computer. This suggests that the user is not carefully tracking the activities of the P2P client increasing the opportunity for abuse.

2. No borders result in global losses. Geography is largely irrelevant in P2P networks, meaning no particular country or region is safer than another. A computer logging on in one place becomes part of the same network as a computer in the other. While the overwhelming majority of traffic on P2P networks is entertainment content (games, movies, music, etc.), also lurking on P2P networks are files that pose severe security risks.

Certain firm mistakenly thinks they are not impacted by unintentional disclosure in the P2P because they make usage of firewall to prevent from the malware being into their network. But as long as your computer connects to the web, even the best perimeter systems fail. More importance, the sensitive information can be leaked on the web because of unintentional disclosure. So the unintentional disclosure is increases the challenge of preventing leaks.

4 RISK ASSESSMENTS AND CIRCUMVENTION FOR UNINTENTIONAL DISCLOSURE

To characterize the risks facing large firm, their partners (suppliers, contractors), and their customers, we examined both the vulnerability and resulting consequences of leaked files and the threat posed by those searching to exploit the vulnerability.

We focused on analysis on the financial chains of the top 30 civil banks and gathered and categorized P2P searches and shared files related to these institutions.

To gather relevant searches and files, we developed a digital footprint for each financial firm. A digital footprint comprises terms that would quickly lead you back to the host firm or important trading partners (suppliers, contractors, vendors). These terms, if googled, would often (but not always) lead you directly back to the host firms. For example, they would include as following: firm names, abbreviations, nicknames, ticker symbol, key brands and subbrands, suppliers, contractors, vendors. Tracking those the term included in the digital footprint, we capture the more sensitive information. Increasing the number of terms included in the digital footprint increases the number of search and file matches found.

Using this approach, we collected over 145933 searches issued by P2P users looking for terms that matched our digital footprints including 13900 unique strings. The resulting searches were then manually analyzed to assess their intent. Our goal was to categorize the searches by a measure of their threat. After studying thousands of searches, we developed a four-point threat scale: high (3), medium (2), low (1), and public (0). Those categorized as high threat (i.e., 3) were searches directed for specific documents or data that could fuel malicious activity. Medium-threat searches were ones targeted generically against the firm. Such searches would uncover sensitive files along with music, video, and so on. Low-threat searches were ones searching for music, picture, or video files related to the bank's footprint. While these searches could be seen as benign, they would also uncover sensitive files and thus expose vulnerabilities that could still represent a threat to the institution and its customers.

Directed searches for databases, account user information, passwords, routing, and personal identification numbers represent clear threats. Medium-threat searches, such as those for bank names, are more generic. Low-threat searches may seem innocent. Each of these low-threat searches would uncover other bank-related files. Of course, for many firms, coincidental association with a popular song or brand represents another problem we call digital wind.

Millions of searches for that song increase the likelihood of exposing a sensitive bank document. Either by mistake or by curiosity, when these documents are exposed, they are sometimes downloaded to other clients, thus spreading the file and making it more likely to fall into the hands of someone, who will try to exploit its information and track those information, and find out the financial information of bank, it is possible that an economic losse is brought to the bank.

It is noted that although the data of low security level ought to be no danger, it is dangerous if the data is from the sensitive document, and so we must be careful to the document of low security level.

5 EXPERIMENTS

Figure 1 shows the result of search for 30 bank institutions.

Bank names are specified in Figure 1 and the number of banks is selected randomly. From Figure 1, bank 2, bank 6, bank 24 and bank 30 experienced a large number of highly threatening searches. Bank 13, bank 17 and bank 20 represents

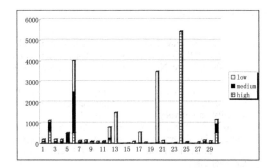

Figure 1. Figure of the classified search for 30 bank institutions.

the case of a bank experiencing significant digital wind. That bank does not have a well-known global brand, but its name and associated products have names that unfortunately share common elements with a popular music group. So the bank in this category experience a low threaten search. The Many of the smaller banks with less recognition still experienced targeted searches and face threat.

6 CONCLUSIONS

Unintentional disclosing insensitive information is a data security issue facing firm. Many firms should take measures to take precautions against data security. (1) The firm should be named according to the digital wind of many entertainment media for successful search in google and yahoo and enhance the brand strength and recognition. Firms could also introduce file-naming conventions and policies to reduce the meta-data footprint of their documents. These types of initiatives reduce the threat of documents being found and spread. (2) employee, contractor, supplier, and customer and educate on the dangers of P2P file sharing. One of the security challenges many organizations face is developing effective strategies to help individuals in the extended enterprise make better information risk decisions. (3) According to the specific the file leak standard established based on the different persons and the different things, the P2P network is monitored. This standard is used to measure the risk of information security.

ACKNOWLEDGEMENT

This work is supported by doctoral research grants jit-b-201113 from Jinling institute of technology.

REFERENCES

Johan Pouwelse, Paweł Garbacki & Dick Epema. The bittorrent P2P file-sharing system: measurements and analysis. Lecture Notes in Computer Science, 2005, 3640: 205–216.

Johnson M.E. & Goetz E. Embedding information security risk management into the extend enterprise. IEEE security and privacy, 2007, 5(3): 16–24.

Mundinger J., Weber R. & Weiss G. Analysis of peer-to-peer file dissemination amongst users of different upload capacities. Poster at IFIP WG7.3 Performance, October 2005.

Oberhozer-Gee F. & Strumpf K. The effect of file-sharing on record sales: An empirical analysis. Journal of political economy, 2007, 115(1): 1–42.

Sun L., Srivastava R.P. & Mock T.J. An information system security risk assessment model under the dempster-shafer theory of belief functions. Journal of management information system, 2006, 22(4): 109–242.

Xiaojun Hei, Chao Liang & Jian Liang. A measurement study of a large-scale P2P IPTV System. IEEE *trans.* on MULTIMEDIA, 2007, 9(8): 1672–1687.

Yue W.T. & Cakanyildirim M. Intrusion prevention in information system: reactive and proactive response. Journal of management information system, 2007, 24(1): 329–353.

Null space based interference cancellation for group femto uplinks

S.Z. Fu & J.H. Ge
State Key Laboratory of Intergrated Service Networks, Xidian University, Xi'an, Shannxi, China

ABSTRACT: In order to cancel the heavy interference experienced by the femtos from the Macro Mobile stations (MSs) in uplink owing to the bandwidth used by the macro MSs is reused by the femto MSs; a novel Cooperative Interference Cancellation scheme based on Null Space theory (CIC-NS) is presented at femtos. CIC-NS is done by selecting Partner femtos to enhance the received signal ranks, a receive vector is then devised to force the interference from the macro MSs to be zero. The signal destined to femto from a special femto MS is separated from the aggregate receive signal by multi-user detection. Each femto maintains a list of UEs that allowed accessing to it and solely operates in the closed-access mode. In line with the current 3rd Generation Partnership Project (3GPP) discussions on the Inter-Cell Interference Coordination (ICIC) topic, the CIC-NS is backward compatible with Rel-8/Rel-9 User Equipments (UEs). Numerical studies illustrate that the capacity of femto and the heterogeneous network (HetNet) exhibit significant improvement with only slight decrease in macro cell capacity.

1 INTRODUCTION

Poor signal reception caused by penetration losses through walls severely hampers the operation of indoor data services[1,4]. Therefore, the femtos attracted considerable interests[2,3]. The femtos, connecting to the core network via broadband link, are overlaid on the existing macro cell network, forms a two-tier HetNet. Femtos share radio frequency resources with macro cells. According to 3GPP discussions, femtos are solely operated in the close access mode, i.e., the femto only grants particular set of MSs to access it.

Unfortunately, the femto deployment of unplanned ad-hoc manners[2–10] and restricted access mode cause destructive interference to macro BS and vice-versa[6,7]. Closed-access femtos cause the most detrimental uplink interference, since a macro MS lying in the coverage area of a femto is not allowed to access to it, but must communicate with the macro BS that lies outdoors, due to wall penetration losses, such macro MS have to transmit in a much higher power than femto MSs in the uplinks.

The problems of interference cancellation at femtos have been addressed in the References. Zubin suggested a dynamic Resource Partitioning (RP) interference avoidance scheme via the X2 interface (Zubin 2009). Accessing the resources that are assigned to the nearby macro MSs are denied by the femtos. The essential of avoiding resource impact in RP algorithm leads to low spectral efficiency in femtos. An Improved Partial Co-Channel deployment (IPCC) was proposed, in which a femto interference pool is maintained (Wu 2007). IPCC algorithm increases the schedule complexity. Moreover, it restricts the resource that macro MS can be assigned, which decreases the macro cell throughput. In order to reduce cross-tier co-channel interference, a scheme which uses Frequency Reuse (FR) with pilot sensing was presented (Kim 2008). The frequency reuse factor in femto and macro cell is 2/3 and 1/3, respectively. The spectral efficiency needs to improve while the interference is alleviated and the aggregation throughput is increased. Similarly, a resource hopping scheme was provided based on the traffic pattern information which reflects channel activities (Lee 2010). However, the macro cell traffic pattern information is difficult to obtain in femtos. Furthermore, resource hopping at femtos always impacted on the macro cell with certain probability.

The issue of uplink interference cancellation at femtos in HetNet is addressed in this paper. A novel Cooperative Interference Cancellation scheme based on Null Space (CIC-NS) theory is presented. Partner femtos are selected to work together and enhance the receive signal ranks. A receiver vectors based on null space is derived to eliminate the uplink interference from macro MS. In order to separate the signal destined to femto from a specific femto MS from the aggregate receive signal, multi-user detection is adopted.

2 HETNET INTERFERENCE

2.1 *System model*

The uplink of an Orthogonal Frequency Division Multiple Access (OFDMA) system is considered, where the system bandwidth B is divided into N Resource Blocks (RBs), i.e. $B = N \times B_{RB}$. A RB represents one basic time-frequency unit with bandwidth B_{RB}. Perfect synchronization in time and frequency is assumed.

The signal received by femto i at RB n is defined by:

$$Y_i^n = H_i^n X_i + I_i^n + n \tag{1}$$

where H_i^n are the channel gains between femto MS and its serving femto. X_i is the signal delivered to the serving femto from its MS. I_i^n is the interference introduced by the MSs working on the same RB in neighbor cells (co-channel MS). Therefore, I_i^n is called co-channel interference. n is the thermal noise per RB.

2.2 *HetNet interference generating mechanism*

Universal frequency reuse is considered so that both macro and femto cells can utilize the entire system bandwidth B. The available RBs are allocated by macro and femto BSs among their associated macro and femto MSs, respectively. The aggregate interference I_i^n is composed of macro and femto MSs interference:

$$I_i^n = \sum_{j \in M_{int}} G_{i,j}^n X_j + \sum_{k \in F_{int}} G_{i,k}^n X_k \tag{2}$$

where $G_{i,\cdot}^n$ are the channel gains from macro MS j or femto MS k to the victim femto i at RB n. All macro MS uplink signals in the system are treated as interference. In addition, femto MS uplink signals other than the serving femto MS are regarded as interference. The sets of interfering macro and femto MSs are denoted by M_{int} and F_{int}, respectively.

3 CIC-NS SCHEME

N femtos which connecting to the local processor form a femto group, as shown in Figure 1. The femto group can be widely used in hotspots such as supermarket, office building, and apartment building to significantly increase the indoor data services.

Each femto MS selects a femto as the primary femto, while the residual femto in the group act as partner femtos. The uplink channel between

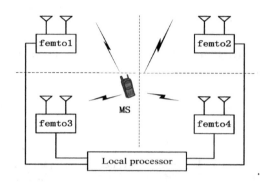

Figure 1.   A block diagram of cooperative femto cell.

femto MSs and femtos are estimated by the femtos. The uplink channel between a femto MS and the femto group can be built in the local processor according to the estimation by all the femtos in the group.

3.1 *Signal analysis*

The uplink interference from the co-channel MS j to the femto group at RB n is a column vector composed of the signal received by all the femtos in the group, formulized as:

$$I_j^n = \begin{bmatrix} I_j^{n,1} \\ \vdots \\ I_j^{n,N} \end{bmatrix} = \begin{bmatrix} G_j^{n,1} \\ \vdots \\ G_j^{n,N} \end{bmatrix} X_j = G_j^n X_j \tag{3}$$

where I_j^n is the uplink interference vector received by the femto group from the co-channel MS j at RB n. $I_j^{n,k} (k = 1, ..., N)$ are the uplink interference received by femto k in the femto group from the co-channel MS j at RB n. $G_j^{n,k} (k = 1, ..., N)$ are the channel gains between the femtos in the femto group and the co-channel MS. X_j is the signal that the co-channel MS j transmits to its serving BS at RB n.

Suppose that the femto group can detect K co-channel MSs. The uplink interference received by the femto group at RB n is given as:

$$I^n = \begin{bmatrix} G_1^{n,1} & \cdots & G_K^{n,1} \\ \vdots & \ddots & \vdots \\ G_1^{n,N} & \cdots & G_K^{n,N} \end{bmatrix} \begin{bmatrix} X_1 \\ \vdots \\ X_K \end{bmatrix} = G^n X^n \tag{4}$$

where G^n is the channel gain matrix between the K co-channel MSs and the femto group. X^n is the K co-channel MSs signal vector transmitted to the serving BSs, respectively.

Suppose that there are M ($M<N$) femto MSs served by the femto group at RB n simultaneously, the desired signal at the femto group is given by:

$$.Y_d = \begin{bmatrix} H_{d,1}^1 & \cdots & H_{d,M}^1 \\ \vdots & \ddots & \vdots \\ H_{d,1}^N & \cdots & H_{d,M}^N \end{bmatrix} \begin{bmatrix} X_{d,1} \\ \vdots \\ X_{d,M} \end{bmatrix} = H_d X_d \quad (5)$$

where H_d is the channel gain matrix between M femto MSs and the femto group, X_d is the signal vector delivered to the primary femto from M femto MSs, respectively.

The signal received by the femto group at RB n is the sum of the desired receiver signal and the aggregation interference:

$$Y = Y_d + I^n + n = \begin{bmatrix} H_{d,1}^1 & \cdots & H_{d,M}^1 \\ \vdots & \ddots & \vdots \\ H_{d,1}^N & \cdots & H_{d,M}^N \end{bmatrix} \begin{bmatrix} X_{d,1} \\ \vdots \\ X_{d,M} \end{bmatrix}$$
$$+ \begin{bmatrix} G_1^{n,1} & \cdots & G_K^{n,1} \\ \vdots & \ddots & \vdots \\ G_1^{n,N} & \cdots & G_K^{n,N} \end{bmatrix} \begin{bmatrix} X_1 \\ \vdots \\ X_K \end{bmatrix} + n \quad (6)$$

The first term at the right hand side of the second equal in Equation (6) is the desired signal, and the second term is the interference which should be eliminated. The local processor filters the received signal with a vector ϖ:

$$\hat{Y} = \varpi Y = \varpi \begin{bmatrix} H_{d,1}^1 & \cdots & H_{d,M}^1 \\ \vdots & \ddots & \vdots \\ H_{d,1}^N & \cdots & H_{d,M}^N \end{bmatrix} \begin{bmatrix} X_{d,1} \\ \vdots \\ X_{d,M} \end{bmatrix}$$
$$+ \varpi \begin{bmatrix} G_1^{n,1} & \cdots & G_K^{n,1} \\ \vdots & \ddots & \vdots \\ G_1^{n,N} & \cdots & G_K^{n,N} \end{bmatrix} \begin{bmatrix} X_1 \\ \vdots \\ X_K \end{bmatrix} + \varpi n \quad (7)$$

If the vector ϖ is selected appropriately, the interference can be cancelled but the desired signal can be recovered by further process. It can be expressed as $\varpi G^n = 0$ but $\varpi H_d \neq 0$.

3.2 Null space-based cooperation

$\varpi G^n = 0$ can rewrite as $(G^n)^H \varpi^H = 0$. Notation $(\cdot)^H$ represents Hermitian transpose. According to the matrix theory, ϖ^H that satisfies $(G^n)^H \varpi^H = 0$ is in the null space of $(G^n)^H$. Employing the Singular Value Decomposition (SVD) of $(G^n)^H$, we arrive at:

$$(G^n)^H = U\Lambda V$$
$$\Lambda = \begin{bmatrix} \Sigma & 0 \\ 0 & 0 \end{bmatrix} \quad (8)$$

where U is a $K \times K$ unitary matrix, V is an $2N \times 2N$ unitary matrix, Λ is a $K \times 2N$ matrix, $\Sigma = diag(\sigma_1, ..., \sigma_\gamma)$ $\gamma = rank((G^n)^H)$. The dimension of the null space of $(G^n)^H$ is $2N - \gamma$. If $2N - \gamma > 0$, there are $2N - \gamma$ normal orthogonal basis. The right singular vectors of the zero singular value $\{v_{\gamma+1}, v_{\gamma+2} ..., v_{2N}\}$ are the normal orthogonal basis vectors.

If $\varpi_1^H \in \{v_{\gamma+1}, v_{\gamma+2} ..., v_{2N}\}$ is selected to filter the receiver signal, the interference is forced to be zero, the filtered signal can be expressed as:

$$\hat{Y} = \varpi_1 Y = \varpi_1 \begin{bmatrix} H_{d,1}^1 & \cdots & H_{d,M}^1 \\ \vdots & \ddots & \vdots \\ H_{d,1}^N & \cdots & H_{d,M}^N \end{bmatrix} \begin{bmatrix} X_{d,1} \\ \vdots \\ X_{d,M} \end{bmatrix} + \varpi_1 n$$
$$= \begin{bmatrix} h_1^1, \cdots, h_1^M \end{bmatrix} \begin{bmatrix} X_{d,1} \\ \vdots \\ X_{d,M} \end{bmatrix} + \varpi_1 n \quad (9)$$

where $[h_1^1, ..., h_1^M]$ is a $1 \times M$ equivalent channel. Equation (9) indicates that the signal $\hat{Y}$ after being filtered is composed of $X_{d,1}, ..., X_{d,M}$ from M femto MSs. In order to detect the signal from a particular femto MS, M different receiver vectors selected from $\{v_{\gamma+1}, v_{\gamma+2} ..., v_{2N}\}$ synchronously filter the received signal as shown in Figure 2 ($M = 4$).

The received signal filtered by M different vectors constitutes a vector as Equation (10):

$$\begin{bmatrix} \hat{Y}_1 \\ \vdots \\ \hat{Y}_M \end{bmatrix} = \begin{bmatrix} \varpi_1 Y \\ \vdots \\ \varpi_M Y \end{bmatrix} = \begin{bmatrix} h_1^1 & \cdots & h_1^M \\ \vdots & \ddots & \vdots \\ h_M^1 & \cdots & h_M^M \end{bmatrix} \begin{bmatrix} X_{d,1} \\ \vdots \\ X_{d,M} \end{bmatrix} + \begin{bmatrix} \varpi_1 \\ \vdots \\ \varpi_M \end{bmatrix} n \quad (10)$$

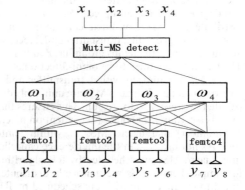

Figure 2.   A block diagram of local processor (M = 4).

Equation (10) is a set of linear equation, which has the unique solution. The signal $X_{d,1}, ..., X_{d,M}$ transmitted by M femto MSs can be recovered by solving Equation (10).

3.3 Selection of femto MSs

A femto in the group can be selected as the primary femto by only one femto MS. The number of femto MSs scheduled simultaneously by femto group at RB n is restricted by:

$$M = \min(N, 2N - \gamma) \tag{11}$$

where γ is the rank of $\left(\mathbf{G}^n\right)^H$. The femto MSs can be scheduled with available algorithms such as max C/I, Round Robin (RR) and Proportion Fairness (PF).

4 NUMERICAL STUDIES

The performance of the proposed CIC-NS scheme is verified by system simulation in a Het-Net composed of LTE (FDD) and femto group. The simulation is run in a full-buffer traffic model. Furthermore, all MSs in the system are assumed to be static during the snapshot, so that the effects due to Doppler spread are neglected. Perfect synchronization in time and frequency is assumed.

The RP scheme from Ref [8] and FR scheme from Ref [10] are simulated for comparison.

4.1 Simulation setup

The simulation area comprises a one-tier, hexagonal grid cell. In order to eliminate edge effects with regards to interference, an additional two tiers are simulated. However, statistics are only taken from the first tiers. The macro BS inter-site distance is 1732 meters. There are 3 sectors per macro BS. The azimuth antenna pattern of the macro BS is 120 degrees. A femto group cell (M = 4) random distributes in each sector. Each femto in the femto group is equipped with 2 transmitter and 2 receiver antennas, so that the femto group has 8 antennas. The femto are assumed to be omni-directional.

The MS is equipped with 1 transmitter and 2 receiver antennas. 50 macro MSs are uniformly distributed over every sector region. This ensures that there is a certain probability (dependent on macro MS number) that macro MSs lie within femto cells. As not involved in the interference cancellation, macro MSs are scheduled by RR algorithm for simple. Femto MSs are uniformly distributed in femto group. Femto MSs are scheduled by RR and PF algorithm.

The indoor hotspot (inH) model is used. The carrier bandwidth is 10 MHz. The threshold of interference is −112 dBm. Other simulation parameters are conformable to R4-092042 (3 gpp 2009).

4.2 Simulation results

Figure 3 depicts the uplink capacity of macro MSs that lie inside the sector where the femto group resides. From this figure, it can be seen that macro MS capacity reduces approximate 11% from about 7.90 Mbps to 6.96 Mbps when proposed CIC-NS scheme is introduced. The capacity reduction comes from the extra interference originating from femto group MSs. In general, the number of the femto group MSs which are scheduled simultaneously is fixed according to Equation (11). Furthermore, the femto MS transmitting power is nearly constant because of small cell size and short transmission distance. As a result, the extra interference to macro MSs introduced by femto MSs is almost invariable. Therefore, The conclusion can be conclude that the macro MS capacity will be approximate constant when the number of the femto MS is keep increasing. The RP algorithm avoids resource impact, the interference only leaked from neighboring RBs, so, the macro MS capacity suffers approximate negligible 4% degradation to 7.61 Mbps. The interference from macro MS to femto MS is the smallest in FR algorithm, which distinguishes between femto and macro in frequency and the interference just appears at the boundary RB. However, only 1/3 frequency band that FR algorithm can fully explored limited the capacity to about 5.00 Mbps.

The impact of CIC-NS scheme on overall uplink performance in HetNet is shown in Figure 4. It is demonstrated that HetNet capacity is up to approximately 19.55 Mbps, which is increased by a factor of approximate 1.24 compared with macro system without femto group. The decrease in macro cell uplink capacity caused by CIC-NS scheme is

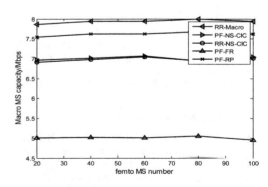

Figure 3. Uplink macro MS capacity comparison for CIC-NS scheme.

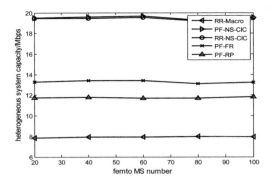

Figure 4. Uplink heterogeneous system capacity comparison for NS-CIC scheme.

negligible regarding the overall system capacity enhancement. The HetNet capacities introducing RP and FR algorithms are about 11.74 Mbps and 13.31 Mbps, respectively. The HetNet capacity of the proposed CIC-NS scheme is improved by 66.5 and 46.9 percentage compared to RP and FR algorithm. Coupled with the concept can be practically realized, highlights the positive impact to be implementation through CIC-NS scheme.

5 CONCLUSIONS

Femto deployment results in a viable complement to cellular networks. The system throughput of cellular network can be significantly improved through the widespread deployment of femto. Femto improve indoor coverage, which brings broadband-like experience directly to the handset. They also offload resources from the macro BS which can be utilized to improve coverage to outdoor users.

It has been seen that in a closed-access system, macro MSs lying in the adjacency of femto cause much interference to femto in the uplink. By introducing CIC-NS scheme, the detrimental uplink interference to femto is cancelled, and the capacity of HetNet can be increased by a factor of approximate 1.24. The incurred throughput degradation of macro cells in doing so is marginal, as macro cell typically loses only a small fraction of their capacity. Therefore, introducing CIC-NS scheme to a closed-access system with femto deployment

is a powerful means to retain ubiquitous coverage by macro-cells, as well as to substantially boost the experience of users located indoors.

REFERENCES

[1] Airvana White Paper, "Femtocells: Transforming the Indoor Experience," Femto Forum, 2007.
[2] Andy Tiller, "The Case for Femtocells," The Basestation Newsletter, June 2007.
[3] H. Claussen, "Performance of Macro- and Co-Channel Femtocells in a Hierarchical Cell Structure," in Proc. of the 18th IEEE international Symposium on Personal, Indoor and Mobile Radio Communications (PIMRC), Athens, 2007, 9: 1–5.
[4] J.D. Hobby and H. Claussen, "Deployment Options for Femtocells and their Impact on Existing Macrocellular Networks," *Bell Labs Technical Journal*, 2009 13(4): 145–160.
[5] Jaeki Lee, Jangho Yoon, "Traffic Pattern based Resource Hopping Schemes of Femto Base Stations" IEEE CCNC 2010 Las Vegas, 2010, 1: 1–5.
[6] L. Ho and H. Claussen, "Effects of User-Deployed, Co-Channel Femtocells on the Call Drop Probability in a Residential Scenario," in Proc. ofthe 18th IEEE International Symposium on Personal, Indoor and MobileRadio Communications (PIMRC), Athens, 2007, 9: 11–15.
[7] M.-S. Alouini and A. Goldsmith, "Area Spectral Efficiency of Cellular Mobile Radio Systems," *IEEE Transactions on Vehicular Technology*, 1999, 48(4): 1047–1066.
[8] R4-092042 Alcatel-Lucent, "Simulation assumptions and parameters for FDD HeNB RF requirements" 3GPP TSG WG4 MEETING 51 San Francisco USA 4–8 May 2009.
[9] Tae-Hwan Kim and Tae-Jin Lee, "Throughput Enhancement of Macro and Femto Networks By Frequency Reuse and Pilot Sensing" IEEE IPCCC08 Atlanta, 2008, 12: 390–394.
[10] V. Chandrasekhar, J. Andrews, and A. Gatherer, "Femtocell Networks: A Survey," *IEEE Communications Magazine*, 2008, 46(9): 59–67.
[11] Yi Wu, Dongmei Zhang, Hai Jiang, Ye Wu, "A Novel Spectrum Arrangement Scheme for Femto Cell Deployment in LTE Macro Cells" in Proc. of the 18th IEEE international Symposium on Personal, Indoor and Mobile Radio Communications (PIMRC), Athens, 2007, 9: 1–5.
[12] Zubin B, Harald H, Gunther A "Femto-Cell Resource Partitioning" IEEE globcom09, Hawaii, 2009, 11: 1–6.

Control Engineering and Information Systems – Liu (Ed)
© 2015 Taylor & Francis Group, London, ISBN 978-1-138-02685-8

Analysis and research of T.37 fax based on email

G. Wang & Y.W. Tang
Zhengzhou Information Science and Technology Institute, Zhengzhou, China

Y.Q. Jin
China National Digital Switching System Engineering and Technological Research Center, Zhengzhou, China

ABSTRACT: With the development of network technology, the traditional way of faxing has gradually developed to the network fax. The network fax technology is discussed in this article, and makes a thorough analysis for the T.37 fax format, encoding rules and TIFF file format based on Email. Meanwhile, we schemed out a new method which could quickly extract the attachment based on non-standard fields. T.37 fax can precisely identify and extract through analyze and filter data step by step.

1 INTRODUCTION

At present, the traditional telephone fax business is gradually transferring to network fax which is based on Internet service. The biggest advantage of network fax is using the remote relay on the Internet. With the continuous development of broadband network, its geographical coverage and bandwidth have made a qualitative breakthrough. It can offer multiple routing protections in the whole network, and make the whole network more reliable and robustness; its business can reliably arrive at the target site. At the same time, because network fax can dynamically allocate bandwidth, its cost is much lower than the 3rd fax which is based on the fixed bandwidth and it became more competitive.

In June 1998, ITU-T discussed and identified three network fax protocols, namely F.185 (ITU-T 1996a), T.37 (ITU-T 1996b) and T.38. F.185 provided an overall description of transmitting fax via the Internet. It supported real-time mode and store-and-forward mode. T.37 described how the network fax is stored and forwarded by Email. T.38 defined the real-time data transmission between fax machines or between a fax machine and a network terminal. Figure 1 shows the T.37 fax system model.

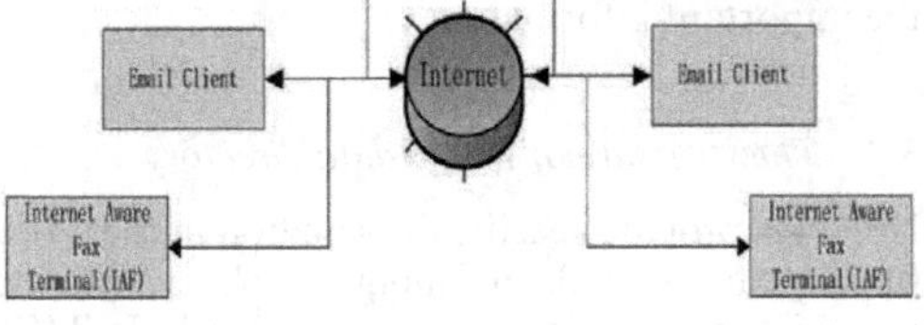

According to T.37 proposal, the 3rd fax machines of the sender and the receiver synchronize and communicate with local fax gateway. Facsimile data is stored in the gateway, and transmitted by the email system. Email system stored and forwarded the data, so it isolated the real-time 3rd fax transmission and non-real-time internet email transmission, which make the facsimile data stored and forwarded via the Internet. At the same time, the email format and TIFF file format are important to extract fax from email. The email format which T.37 proposals required is MIME (Multipurpose Internet Mail Extensions) (Freed 1996a) that is widely used, while the TIFF file format is based on RFC3949.

In this paper, we will make a brief introduction to T.37 and TIFF, then put forward a method of identifying and extracting facsimile data based on the analysis of T. 37 structure.

2 ITU-T.37 ANALYSIS

T.37 proposal is a fax standard based on store-and-forward. It shows how to send and receive fax like emails based on F.185. It allowed facsimile data to form a file which would be sent as an email's attachments. Its implementation would neither change the IETF (Internet Engineering Task Force) standards, nor change the ITU fax protocol; therefore it allowed fax users and email users to communicate with each other (Freed 1996b).

Figure 1. T. 37 fax system model.

T.37 proposal defines detailed technical regulations for sending, forwarding and receiving fax, in view of this research is for the Email data which was already saved, not involving specific process of sending and receiving, we will only discuss part of T.37 proposal.

2.1 The rule of address

T.37 fax allows communication among email client, network fax device and traditional fax machines. The fax terminal uses telephone number as unique identifier, while the email client use Email address as identifier, so the fax gateway requires mutual conversion between email address and fax number. T.37 proposal requires that the formats of fax number and Email address should comply with the RFC2303 (Allocchio 1998a) and RFC2304 (Allocchio 1998b), namely:

- International phone number should be prefixed by a "+";
- International phone number could be supported by a delimiter "–";
- Phone number prefixed by "/ T33S = " means extension;
- The format of Email address using fax number as user name is: fax number + "@" + domain name.

For example, FAX = + 33-1-74983927/T33S = 2587, is a typical of T. 37 fax number, with corresponding email address FAX = + 33-1-74983927/ T33S = 2587 @ faxserv.net. According to this regulation, fax gateway can support the conversion between email address and fax number through parsing user name.

2.2 The rule of image formats

T. 37 suggest using of multi-page TIFF file in sending fax image data. TIFF is a graphic file format which developed for Macintosh machines by Aldus. This image format is supported by Windows and most scanners. TIFF file structure is complicated, and T.37 proposal has provided details according to RFC2301. Overall, RFC 3949 defines six compression modes of TIFF file for network fax, respectively:

- Profile S: this mode is for the black and white pattern, and use MH encoding;
- Profile F: a mode for the extend black and white pattern by using MH, MR and MMR encoding;
- Profile J: a mode for the loss-free JBIG encoding of black and white pattern;
- Profile C: a mode for the lossy JPEG encoding of color and gray mode;

- Profile L: a mode for the loss-free JBIG encoding of color and gray mode;
- Profile M: a mixed compression mode.

2.3 The rule of email encoding

The common Email encoding includes: Base64 encoding, Quote Printable (QP) encoding and Uuencode (UU) encoding, T.37 provides the encoding of image data by using Base64. Currently most Email attachments are encoded by this method.

3 THE ANALYSIS OF TIFF FILE FORMAT IN NETWORK FAX

TIFF has good expansibility and can easily be transplanted; it can either introduce a new compression algorithm at any time, or can be easily exchanged between machines with different processor. However, the structure of TIFF file is complex, it composed of three data structures, respectively for the Image File Header (IFH), one or more Image Files Directory (IFD), and data. IFD is a directory structure with tags which are cores of TIFF files.

3.1 The structure of the image file header

IFH structure is the only part of TIFF file with fixed position. The 8 bytes must be the starting position of file; it correctly explains the important information of the rest parts. The first field containing two byte describes the TIFF file byte order. A field II (0 × 4949 h) indicates Intel format; and a MM (0 × 4d4 dh) means Motorola format. So the TIFF file can exchange information between the IBM and Macintosh PCs. The second field with two bytes is a version field, now is 0 × 002 ah, the change of this field means the fundamentally change of the TIFF file format, so this field can be used to confirm the TIFF file format. The last four bytes of IFH describe the offset from the file header to the first IFD structure, and is measured by byte.

IFH structure is the only fixed position part of TIFF file. The 8 bytes of the starting position of the structure must be in the file, it gives out a correct explanation TIFF files needed for the rest of the important information.

3.2 The structure of image data directory

IFD tags provide specific information in the TIFF file, it can help determining which fields does not exist in decoding program of TIFF. If TIFF images have multi-page, each page have an IFD, but the order of the file are not demanded strictly. IFD structure is as shown in Figure 2, includes:

- Count field: 1 byte, indicate the number of the tag field;
- Tagged field:12 bytes, arranged in accordance with respective order in IFD;
- Ended field: 4 bytes, indicate the next IFD position, all 0 indicate the last IFD.
- Tag: 2 bytes. Currently value of tag is between $254(0 \times FEh)$ and $321(0 \times 141 h)$. The value of a private tag is greater than or equal to 32768 $(0 \times 8000 h)$, it can be assigned to each company for proprietary program rather than an ordinary user.
- Type: indicates the data type of this field, TIFF supports the following data types:
  1 = 1 Byte integer (BYTE)
  2 = 1 byte ASCII (ASCII)
  3 = 2 bytes integer (SHORT)
  4 = 4 bytes integer (LONG)
  5 = 8 bytes scores (RATIONAL).
- Length: It indicates numerical value rather than the length of data, for example, a LONG data with a length of 64 actually has 256 bytes.
- Data and offset: If data is less than or equal to 4 bytes, this part is directly stored, otherwise, this part indicates the offset which shows actual data location related to file head.

3.3 The rules of TIFF format

RFC2301 describes tag field of TIFF file in network fax, as described above. It contains six modes; black and white mode is the basic mode of T.37. All T. 37 terminals support this mode, so we use it as an example to discuss the details of the TIFF format Settings.

In order to smoothly transmit information among terminals, the mode strictly limits the use of TIFF field, only include the necessary field of minimum subset, the value of each field is restricted within a minimum range, and the byte order of

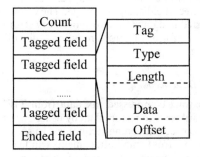

Figure 2. IFD Structure.

TIFF format must be in Intel mode. Below are details of each field requirements.

3.3.1 Basic fields
BitsPerSample (258) = 1: means black-and-white fax;

- G3, Compression will (259) = 3: means G3 Compression, namely one-dimensional or two-dimensional encoding, and should be specified in the T4Option;
- ImageWidth (256) = 1728: this mode is only supported 1728 image width;
- ResolutionUnit (296) = 2: resolution measured in inches;
- SamplesPerPixel (277) = 1: black and white fax;
- Xresolution (282) = 200, 282: horizontal resolution should be 200 or 204 pixels per inch;
- Yresolution (283) = 283, 98, 196, 200: vertical resolution.

3.3.2 Extension fields
T4Options (292) = (Bit 0 = 0, Bit 1 = 0, Bit 2 = 0, 1): this field is required when Compression = 3, "Bit 0 = 0" means MH encode, "Bit 1" must be 0, "Bit 2 = 1" means EOLs (end of line), and 0 vice versa.

The remaining set of other model is extended upon this mode; the method is either enlarging the scope of field, or introducing more tag field to realize the function of the extension. TIFF file structure is relatively complicated; however the format is very strict, so it is easier for emails.

4 HOW TO IDENTIFY AND EXTRACT T.37 FAX

We extract T.37 facsimile data from email based on standards above. Through plentiful analysis of data, we convert standards into feasible solutions. We design three steps to extract T 37 fax: Email-level filtration, File-level Identification and Data Decoding.

4.1 Email-level extract

The object which we deal with is stored Email files, it forms three possible results, through preliminary filtration, include: the facsimile data should/may/cannot be contain in Email file. The Process is shown in Figure 3.

Specific filtrate method is as follows:

1. T. 37 email address format are given in 1.1, It's forbidden for normal Emails to use this format, so we check the user name of senders and receivers, if it comply with format in 1.1, we conclude that this email contains a fax.

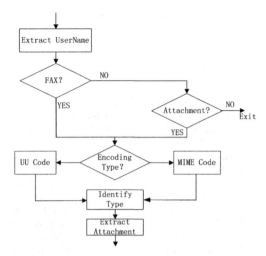

Figure 3.　Email-level extract flowcharts.

2. When T.37 fax is transferred between two computer terminals, the address will not necessarily comply with the provisions in section 1.1. At this time, we should firstly evaluate whether the email contains attachments and the number of attachments. If Email contains only one attachment, we could continue to judge its encoding mode. MIME mail with QP/Base64 encoding is different from the email with UU encoding in file format, which should be handled separately. To Base64 encoding mode, the email which encloses TIFF format attachments may be judged to contain T. 37 facsimile data.

3. If the Email file contains facsimile data, we extract the attachments namely the TIFF file to make further process. As the current mainstream email standards, MIME is widely used and more flexible. Here, in order to improve the process efficiency, we design a fast method based on non-standard field for the extraction of MIME email attachments. The core is to confirm the attachment location in the process; some non-standard field is of higher priority than the standard field. The method is based on the following facts:

Fact 1: The purpose of MIME standard fields is to ensure the realization of multiple components email, rather than considering the realization of the convenience and efficiency.

Fact 2: The purpose of non-standard field is to fill the gaps of standard field. Therefore non-standard field will usually provide us more direct source of information, and can quickly position attachments;

Facts 3: The nonstandard field which this method relies on is widely used. Because not all of the email clients support non-standard field, the use frequency of this method is lower, and the processing efficiency will be affected.

Usually, the email error occurs during the process of data transmission often lead mail client hardly to identify attachments, we found following mail data error type in the research:

Case 1: File finished by accident, attachment data is not complete; these errors take very high proportion;

Case 2: Lack of end logo, similar to case 1, however in fact data integrity is not damaged;

Case 3: File is taken by massive 0; this kind of error can affect the realization of the algorithm;

Case 4: The length of line excesses limit value, or some individual length is not consistent with overall length;

Case 5: Attachment put in the mail directly without encoding;

Case 6: Blank appears in the first line of email file.

For errors above, we usually use software to make fault tolerance analysis and process to improve the ability of data recovery.

4.2　*File-level identify*

This process object is the TIFF file extracted from email. To emails which surely contain facsimile data, we could extract the compressed data directly (Ong, Dawley & Clem 2003) or the emails which may contain facsimile data; we make further confirmation by analysis of the corresponding tag field set and then extracting facsimile data. We have analyzed the simplest mode, in the real process, for taking other model into account; we took the largest subset of the fields as analysis object, and expanded each field scope.

According to T. 37, multi-page TIFF file must be used in transporting multi-page fax. The offset in TIFF file is described in absolute deviation relative to file head, so the influence caused by inversion bits is smaller than caused by missed bits or extra inserted bits.

Meanwhile, the non-data part of TIFF file will affect the display of images. The errors which occurred in data is similar to ordinary error of fax data, the forms of error are different for different coding.

4.3　*Data decoding*

We extract compressed fax data from TIFF file, the purpose is to restore data by using our existing error correction coding technology. In some cases, the errors from email attachments can lead TIFF file error, which are errors of compressed data itself. Now we need to consider using reduction technology which is of higher efficiency, stronger error correction, for users' satisfaction.

5 CONCLUSION

The acquisition of fax data is the precondition of fax classification and selection, we have talked about data accurate and rapid extraction of email in T.37 here, and analyzed relevant suggestion about the store-and-forward type fax, the method of E-mail protocols and encoding and relevant provisions of the TIFF file format in network fax.

Based on the analysis, it has given a nonstandard field based email attachments rapid positioning method, and provided the facsimile data identification and extraction method. Besides, it has specifically discussed the characteristics of the actual data and problems that should be paid attention to in the process.

We believe that through the research and analysis of T.37 fax format, the solution of the T.37 facsimile data extraction and restoration problem in email will provide a certain basis for similar processing system research.

REFERENCES

Allocchio C. RFC2303 Minimal PSTN address format in Internet Mail (1998a).

Allocchio C. RFC2304 Minimal FAX address format in Internet Mail (1998b).

Freed N. RFC2045 Multipurpose Internet Mail Extensions (MIME) Part One: Format of Internet Message Bodies (1996a).

Freed N. RFC2046 Multipurpose Internet Mail Extensions (MIME) Part Two: Media Types (1996b).

ITU-T. Recommendation F.185 Internet facsimile: Guidelines for the support of the communication of facsimile documents (1996a).

ITU-T. Recommendation T.37 Procedures for the transfer of facsimile data via store-and-forward on the Internet (1996b).

Ong R.J., Dawley J.T. & Clem P.G.: submitted to Journal of Materials Research (2003).

Control Engineering and Information Systems – Liu (Ed)
© 2015 Taylor & Francis Group, London, ISBN 978-1-138-02685-8

Statistical analysis of the effective bandwidth for FM HD Radio

F. Wang, G. Yang, W.W. Fang & J. Liu
School of Information Engineering, Communication University of China, Beijing, China

ABSTRACT: The bandwidth of FM signal change according to the variable audio programs, and the dynamic ranges of bandwidth are wide. This paper studies the generation of analog FM signal and its spectrum characteristic, and then analyzes the effective bandwidth statistically of different types of programs. Finally this paper gets the statistical regularities about the effective bandwidth of FM signal. Simulation results show that the bandwidth of FM signal is closely related to the specific audio program. The effective bandwidth of FM signal is much less than the boundary bandwidth of HD Radio Hybrid between the analog and digital signal, and it has a large Mean Square Error (MSE). Under this circumstance, the real-time changes of the FM bandwidth may cause plenty of spare spectra. These analytic results provide the support for the digital signal spectrum dynamic adaptive access in HD Radio.

1 INTRODUCTION

Digital radio broadcasting is now taking off in a big way, and is still in a research stage in domestic. HD Radio system is designed to permit a smooth evolution from current Frequency Modulation (FM) radio to a fully digital In-Band On-Channel (IBOC) system. It uses the same carrier frequency as current FM radio, and uses existing FM radio channel to provide high-definition DAB (Digital Audio Broadcasting) and data service without affecting current FM radio, which suits China's reality.

For Hybrid Waveform, the spectra of analog FM signal and digital signal are combined in a fixed way, which put digital signal on about 130 kHz away from carrier. However, the bandwidth of analog FM signals change as audio signal's frequency and amplitude change. It is much less than 100 kHz and has a large MSE. Under this circumstance, the real-time changes of the bandwidth may cause a lot of spare spectra. The bandwidth, in this paper, means the distance away from the carrier in Single Side Band (SSB).

In this paper, we studies the analog FM signal generation and spectrum characteristic in Section II, then Section III analyzes the effective bandwidth statistically of different types of programs, finally gets the statistical regularities about the effective bandwidth. The conclusions are made in Section IV.

2 FM BROADCASTING SYSTEM FOR HD RADIO HYBRID

RF signals of HD Radio Hybrid consist of two parts, the analog FM signal and the digital signal as shown in Figure 1. This section first introduces the analog

FM signal, and then analyzes the effective bandwidth of the FM signal spectrum. The analog FM signal generation includes stereo encoding and frequency modulation. Stereo encoder codes audio signal into composite stereo signal, and composite stereo signal is modulated by frequency modulation.

2.1 FM signal theory analysis

Stereo encoding uses AM-FM pilot system, and the process of encoding and modulation is shown in Figure 2. $mpx(t)$ is a composite stereo signal, which can be expressed as

$$mpx(t) = 0.9\left(\frac{1}{2}(L+R) + \frac{1}{2}(L-R)\cos(2\pi \times 38k \times t)\right) + 0.1\cos(2\pi \times 19k \times t)$$

(1)

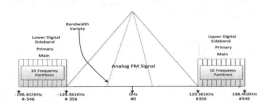

Figure 1. Spectrum of the Hybrid Waveform.

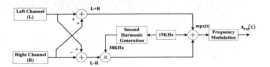

Figure 2. Block diagram of stereo encoding and frequency modulation.

where L is the left channel and R is the right channel.

Frequency modulation conveys information over a carrier wave by varying its frequency. We assume that the initial phase is zero, FM signal $x_{FM}(t)$ is obtained by

$$x_{FM}(t) = A_c \cos[\omega_c t + K_{VCO} \int_0^t m(t)dt] \qquad (2)$$

where $m(t)$ is modulating signal, K_{VCO} is Voltage-Controlled Oscillator (VCO) gain whose unit is Hz/V, and $K_{VCO} \times m(t)$ is instantaneous frequency offset.

When the modulating signal $m(t)$ is

$$m(t) = A_m \cos(2\pi f_m t) = A_m \cos(\omega_m t) \qquad (3)$$

In this way, Equation 2 can be written as

$$\begin{aligned} x_{FM}(t) &= A_c \cos\left[\omega_c t + \frac{K_{VCO} A_m}{f_m} \sin(\omega_m t)\right] \\ &= A_c \cos\left[\omega_c t + \frac{\Delta f}{f_m} \sin(\omega_m t)\right] \\ &= A_c \cos[\omega_c t + \beta \sin(\omega_m t)] \qquad (4) \end{aligned}$$

where $\Delta f = K_{VCO} A_m$ is the frequency deviation, which represents the maximum shift away from the carrier frequency in one direction. It is directly proportional to the amplitude of $m(t)$. $\beta = \Delta f / f_m$ is called modulation index.

As can be seen from Equation 4 that $x_{FM}(t)$ is a periodic function. It can be written in Bessel functions of the first kind form as

$$x_{FM}(t) = A_c \sum_{n=-\infty}^{\infty} J_n(\alpha) \cos[(\omega_c + n\omega_m)t] \qquad (5)$$

where $J_n(\alpha)$ is Bessel functions of the first kind, shown in Figure 3.

$$J_n(\alpha) = \frac{1}{2\pi} \int_{-\pi}^{\pi} e^{j(\alpha \sin x - nx)} dx \qquad (6)$$

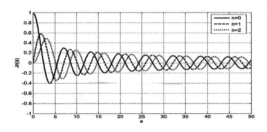

Figure 3.   Bessel functions of the first kind.

2.2   The effective bandwidth of the FM signal

The spectrum of the FM signal is composed of carrier ω_c ($n = 0$) and an infinite number of sidebands $\omega_c + n\omega_m$ ($n \neq 0$) by Equation 5.

The relative amplitude value of the carrier and the sidebands is determined by the corresponding Bessel function. When n is odd, the upper and lower sideband have opposite phase and the same amplitude, but when n is even, the upper and lower sideband have the same phase and amplitude.

The bandwidth of the FM signal should be infinite in theory.

However, when α is a constant, the value of $|J_n(\alpha)|$ reduces with the increase of n, as Figure 3 shows. From the energy point of view, most of the FM signal energy is concentrated on the limited bandwidth near the carrier. The sideband value of $|J_n(\alpha)|$ < 0.1 usually can be negligible, which does not cause significant distortion of the FM signal, and bandwidth power accounts for about 98% of the total power. It has been proved theoretically that when $n > \alpha + 1$, $|J_n(\alpha)| < 0.1$. Carson bandwidth rule is used to estimate the effective bandwidth of the FM signal in engineering as Equation 7.

$$BW \approx 2(\beta+1)f_m = 2(\Delta f + f_m) \qquad (7)$$

where Δf is the peak frequency deviation, and is the highest frequency in the modulating signal.

Generally, the FM broadcasting's Δf is 75 kHz, and f_m of stereo signal is 53 kHz, $BW = \pm(75 + 53) = \pm128$ kHz. So HD Radio system sets ±129.361 kHz as the boundary bandwidth of HD Radio Hybrid between analog and digital signal. Test reports prove that the digital signal, outside the FM effective bandwidth, doesn't interfere the FM signal as long as it satisfies the requirements of transmitting power.

3   THE STATISTICAL ANALYSIS MODEL OF THE FM SIGNAL'S EFFECTIVE BANDWIDTH

3.1   The statistical analysis model

A model is established to analyze the effective bandwidth of the FM signal statistically. The statistical analysis model is shown as Figure 4.

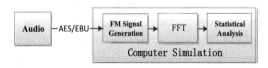

Figure 4.   The statistical analysis model of the FM signal's effective bandwidth.

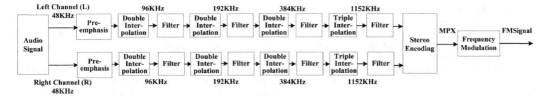

Figure 5.   Block diagram of FM Signal Generation.

Computer simulation contains FM Signal Generation, FFT and Statistical Analysis three parts. In this paper, FM Signal Generation is simulated in MATLAB. Figure 5 shows that the left and right channels of the audio signal pass through pre-emphasis, interpolation and filters separately, and then it codes into a composite stereo signal. Finally the composite stereo signal passes through the frequency modulator.

3.2   Simulation results

FM signal has an infinite number of sidebands and hence an infinite bandwidth, but in practice all significant sideband energy (98% or more) is concentrated within the bandwidth defined by Carson bandwidth rule. Therefore the bandwidth that contains 99% signal energy is defined as the Effective Bandwidth (EBW) of FM signal in this paper.

According to the taxonomy model of music suggested by Tzanetakis, this paper selects the more representative programs to analyze, including music (such as rock, pop, classical and opera) and the talk show (such as storytelling and news). Those program files are both get from the radio station.

Simulation parameters are shown in Table 1.

Figure 6 shows the relationship between the time-varying audio signal and its EBW after modulation. The length of time of each packet is 5 ms. In the upper portion of Figure 6 is the time domain waveform of test music about 1 s. And in the lower portion, abscissa is packet number and the ordinate is EBW. In this way, we can visually denote the time-varying characteristics of EBW.

We measure the statistical parameters of EBW from the simulation results of different audio programs. They are maximum and minimum values of EBW, mean value of EBW (E(EBW)) and MSE of EBW(σ(EBW)). A parameter $\eta = \sigma$(EBW)/E(EBW) is defined to represent the change of EBW in this paper.

3.2.1   Results of spectral analysis with stereo encoding

The program files are encoded in stereo, the other simulation parameters shown in Table 1.

Table 1.   Simulation parameters.

Program time	15 minutes
Sampling frequency	Fs = 48 kHz
Peak frequency deviation	Δf = 75 kHz
Time constant of pre-emphasis	τ = 50 us
Interpolation multiples	N = 24
Points of a packet	NPoint = 240 sample

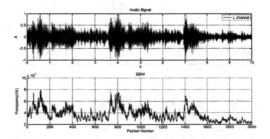

Figure 6.   Audio signal and its EBW after modulation.

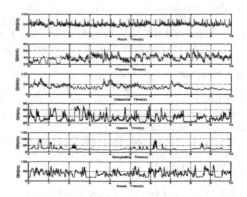

Figure 7.   The time-varying characteristics of EBW about six types of programs.

Figure 6 shows the changes of the effective bandwidth of the FM signal in ten seconds of six types of programs. The time-varying characteristics of EBW are very evident. Furthermore, different types of programs show the different time-varying characteristics from Figure 7.

455

Table 2. Results of spectral analysis with stereo encoding.

Programs	MAX (EBW) [kHz]	MIN (EBW) [kHz]	E (EBW) [kHz]	σ (EBW) [kHz]	Rate of EBW change η	Available spectrum (SSB) [kHz]	Available bandwidth percent
Rock	77.7	28.8	46.076	8.320	0.181	83.924	64.56%
Pop	69.5	21.5	38.839	7.943	0.205	91.161	70.12%
Classical	73.4	20.4	35.034	9.518	0.272	94.966	73.05%
Opera	74.5	20.3	34.845	9.759	0.280	95.155	73.20%
Storytelling	93.0	19.1	30.191	12.395	0.411	99.809	76.78%
News	79.4	19.1	38.571	13.391	0.347	91.429	70.33%

Table 3. Results of spectral analysis with MONO encoding.

Programs	MAX (EBW) [kHz]	MIN (EBW) [kHz]	E (EBW) [kHz]	σ (EBW) [kHz]	Rate of EBW change η	Available spectrum (SSB) [kHz]	Available bandwidth percent
Rock	82.5	15.7	37.654	9.427	0.250	92.346	71.04%
Pop	62.1	6.2	26.946	10.053	0.373	103.054	79.27%
Classical	54.3	3.5	17.068	8.092	0.474	112.932	86.87%
Opera	72.5	4.1	27.418	13.560	0.495	102.582	78.91%
Storytelling	85.9	0.5	22.783	15.657	0.687	107.217	82.47%
News	101.7	0.5	30.750	17.460	0.568	99.250	76.35%

As Table 2 shows, E(EBW) of the difference programs are about 30 to 50 kHz, which are far away from 130 kHz. So it may cause plenty of spare spectra. E(EBW) of storytelling is the least, and its available bandwidth is about 99.809 KHz. E(EBW) of rock is the largest, and its available bandwidth is about 83.924 KHz. For different programs, the difference between maximum and minimum values of EBW has about 55 kHz. EBW has a large MSE which means that the dynamic ranges of EBW are wide. EBW of music is wider than the talk show, but its dynamic range is smaller than the talk show, so η is smaller. News' dynamic range of EBW can be about 27 kHz and η is big; rock music's dynamic range of EBW is about 17 kHz and η is the smallest.

In brief, the largest available spectrum with stereo encoding is about 99.809 kHz compared to HD Radio Hybrid.

3.2.2 Results of spectral analysis with mono encoding

The program files are encoded in mono, the other simulation parameters shown in Table 1 and the results are shown in Table 3.

As Table 3 shows, News' dynamic range of EBW can be about 35 kHz and η is big; rock music's dynamic range of EBW is about 19 kHz and η is the smallest. The simulation data with mono encoding can verify the conclusions which with stereo encoding. While EBW and η with mono encoding are much wider than stereo encoding, and the difference between maximum and minimum values of EBW is wider. It means that the spare spectra with mono encoding are much more than that with stereo encoding.

In brief, the largest available spectrum with mono encoding is about 112.932 kHz compared to HD Radio Hybrid, and 10 kHz wider than that with stereo encoding.

4 CONCLUSIONS

This paper analyzes the effective bandwidth of different audio FM signal by calculating their spectra in real-time, and we can draw the conclusions: EBW of FM signal is much less than the boundary bandwidth of HD Radio Hybrid (130 kHz), and it has a large MSE which means that the dynamic ranges of EBW are wide. The largest available spectrum is about 99.809 kHz with stereo encoding and about 112.932 kHz with mono encoding compared to HD Radio Hybrid. The dynamic range of EBW can be about 27 kHz with stereo encoding and about 35 kHz with mono encoding. Therefore, placing the digital signal by the FM signal bandwidth dynamically can effectively utilize the spare spectrum. These results provide the data for the digital signal spectrum dynamic adaptive access in HD Radio.

REFERENCES

Carson, J.R. 1963. Notes on the theory of modulation, Proc. IEEE, Vol. 51: 893–896.

David P. Maxson. 2007. The IBOC Handbook: Understanding HD Radio Technology, Burlington: Elsevier/Focal Press:National Association of Broadcasters.

Doc. No. SY_IDD_1011s rev. G. 2011. HD Radio™ Air Interface Design Description—Layer 1 FM. iBiquity Digital Corporation.

Hans Dieter Lüke. 1999. The Origins of the Sampling Theorem. IEEE Communications Magazine: 106–108.

He J. & Chen J. 2007. Frequency Demodulation Based on Time Lag Quadratic Transform. Journal of Vibration and Shock.

NRSC-G202 FM IBOC Total Digital Sideband Power for Various Configurations. 2010. iBiquity Digital Corporation.

Ron J. Pieper. 2001. Laboratory and Computer Tests for Carson's FM Bandwidth Rule. System Theory, Proceedings of the 33rd Southeastern Symposium: 145–149.

Wang, A.H. & Huang. Z.H. 2009. Method of Signal Synthesis in Digital Frequency Modulation Exciter. Transactions of Beijing Institute of Technology.

Zhang, Y.B. & Zhou J. 2007. A Review of Content-Based Audio and Music Analysis. China Journal of Computers.

Control Engineering and Information Systems – Liu (Ed)
© 2015 Taylor & Francis Group, London, ISBN 978-1-138-02685-8

Environmental effects on UHF RFID antenna implanted in tire rubber

T.Q. Song
College of Information Science and Technology, Qingdao University of Science and Technology, Qingdao, China

Y. Zhou
College of Electronic Engineering, Qingdao University of Science and Technology, Qingdao, China

ABSTRACT: For the design of an RFID antenna will be directly implant in car tire rubber, the proximity effects due to the rubber material have to be considered. The dielectric properties of different parts of rubber tire are analyzed, and the position of RFID tags implanted into tire is considered. The effect of rubber material dielectric parameters on the dipole antenna was simulated with HFSS. When the dipole antenna planted in rubber material, the antenna resonant frequency was reduced from 915 MHz to 660 MHz, compared with in the air. Several type tags were planted into tire to test the read distance with UHF reader. The performance of RFID tags implanted in the rubber material was tested. The experimental results show the read distance of the RFID tag is decreased from 300 cm to 21 cm, because of the rubber material dielectric properties impacting on the RFID tag.

1 INTRODUCTION

Radio Frequency Identification (RFID) is a contactless data transmission and reception technique between the data carrying device, called a RFID tag, and a RFID reader. The role of the antenna and its impact on the global performance of the tag is very critical, especially when one considers applications in the UHF bands. Indeed, in that case and in order to obtain functional tags, it is necessary to take into account the RFID chip characteristics, environment of the application and some constraints like minimum Read-Range, size of the tag and frequency and power regulations. Passive RFID tags have no on-board battery, but are powered by the energy received by their antenna. The read range of a passive tag is limited by the ability to provide sufficient voltage and power at the antenna to power the tag's integrated circuit, and also if affected by the environment.

Today, the RFID tag has been applied to tires and other rubber products, in order to achieve real-time tire identification and tire pressure monitoring. A tag usually attached to or implant directly in the object it purposes to identify. Many materials maybe metal or other media which has high conductivity and permittivity performance and have strong effects on the performance of UHF tag antenna, so that the antenna is difficult to achieve the best condition in the design, in some cases may cause missing the object. For the tire RFID tag, it implanted directly through vulcanized rubber composites. The rubber materials, which covered the RFID module, have high electrical and dielectric properties. Due to the addition of carbon black and other materials, some of the relative permittivity of the rubber material can reach 11.5. So, it is necessary to study rubber material impact on the RFID tag antenna, find ways to improve the RFID tag performance.

2 TIRE ELECTRONIC TAGS AND IMPLANT POSITION

From the structure, the tire can be divided into bias tires and radial tires, and according to different parts of the functions it can be consist of tread, belt, sidewall, inner liner layer, wire, cord and other materials, as shown in Figure 1. Different parts of tire have different rubber material, and have different characteristics. The tread portion required having abrasion resistance and old drainage properties, tire inner liner layer to ensure good air tightness and

Figure 1. Tire hierarchy structure.

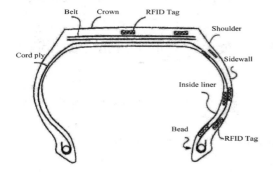

Figure 2.   The location can be implant RFID tags.

maintain the correct tire pressure, and the sidewall should protect the carcass ply of the tire.

When the RFID tag tire implanted in the tire, it need to consider the safety and reliability of the RFID tag. The RFID tag can be implanted in the position shown in Figure 2.

Because there is wire layer in tire crown part, the RFID tag implanted in crown parts will be masked. There are many locations for RFID tag to be implanted in the tire side portion. For example, it can be implanted between the inner liner and the ply or be implanted between the ply and the sidewall rubber.

In the tire manufacturing process, the RFID tag implanted between the ply and the sidewall rubber. This location can be easily lamination produced aided by laser positioning, and can get relatively good signal communication with the RFID tag implanted. Therefore, this is the preferred location when implant RFID tag in tires. In the tire, the location can be implant RFID tags are shown in Figure 2.

3   TIRE RUBBER MATERIAL IMPACT ON THE RFID ANTENNA

The matching between the antennas is particularly important for the implanted RFID tag within the tire. Because RFID antenna is directly implanted into the interior of the tire and is contact with the tire rubber directly, and the tire rubber has high conductivity and dielectric characteristic, the original characteristics of the antenna is changed. When the tire with the RFID tag into the electromagnetic field emitted by the reader, on the one hand a certain amount of energy dissipated due to the antenna and the conductive rubber material contacts, on the other hand the rubber conductive material is formed an electricity conduction path between the RFID two feed point, also consumes a certain amount of radio frequency energy, which

would reduce the RF signal energy, affecting the reading distance. So, it needs to simulate the RFID antenna with the test data of the tire rubber, adjust the antenna parameters to achieve antenna impedance matching purposes.

Polymer dielectric is a property that can make electrostatic energy storage and loss under the electric field. The dielectric permittivity ε is defined as a capacitor ratio between a dielectric media capacitance C and the capacitance of the vacuum. Dielectric permittivity is a measure of the degree of polarization of the macroscopic physical, and can present the electrical energy storage capacity of the dielectric materials. For UHF RFID tag antenna, the operating wavelength of the electromagnetic wave in the medium:

$$\lambda = \frac{\sqrt{2}}{f\sqrt{\mu\varepsilon}}\left(\sqrt{1+\left(\frac{\sigma}{\omega\varepsilon}\right)^2}+1\right)^{\frac{1}{2}} \qquad (1)$$

where, λ is the wavelength of the electromagnetic wave in the medium, f is the frequency of the electromagnetic wave in the medium, ε is the dielectric permittivity, μ is the magnetic permeability and σ is the conductivity.

As can be seen from (1), λ is inversely proportional to $\sqrt{\varepsilon}$, while the antenna size is proportional to λ. Because the dielectric permittivity in the medium is much higher than in the air, so when the tag antenna implanted in dielectric material, it need to reduce the antenna size to meet the requirements of the tag antenna match.

And the electromagnetic wave attenuation coefficient α in the dielectric material can follow (2) to calculate:

$$\alpha = \frac{\omega\sqrt{\mu\varepsilon}}{\sqrt{2}}\left[\sqrt{1+\left(\frac{\sigma}{\omega\varepsilon}\right)^2}-1\right]^{\frac{1}{2}} \qquad (2)$$

As can be seen from (2), only in an ideal antenna substrate conductivity that σ can be 0, and will ensure the attenuation coefficient $\alpha = 0$. And only in this case electromagnetic waves can be achieved without loss in the propagation medium. Therefore, the high dielectric materials must pay attention to the value of parameter σ. The smaller σ is the less electromagnetic losses in the dielectric material. Generally, it uses dielectric loss tangent tan δ to present the dielectric loss of medium. The natural rubber has little dielectric loss and the value is about 10^{-3}–10^{-4}. But when mixed with carbon black in the rubber component, rubber conductive and dielectric properties will change greatly. Figure 3 shows the parameter distribution of dielectric section of a tire. As can be seen from the graph, the dielectric

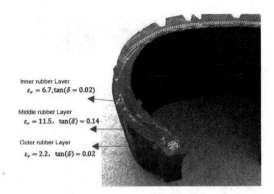

Inner rubber Layer
$\varepsilon_r = 6.7, \tan(\delta = 0.02)$

Middle rubber Layer
$\varepsilon_r = 11.5, \tan(\delta) = 0.14$

Outer rubber Layer
$\varepsilon_r = 2.2, \tan(\delta) = 0.02$

Figure 3.   Tire dielectric material parameters.

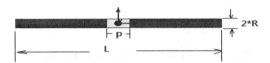

Figure 4.   The dipole RFID antenna.

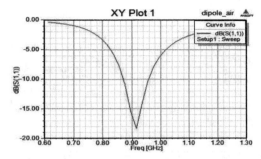

Figure 5.   S11 versus frequencies in air.

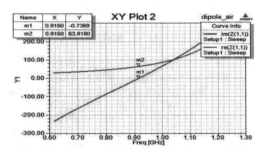

Figure 6.   Complex impedance versus frequencies in air.

parameters of different parts of the tire have great difference. When the RFID tags implanted into different position of the tire, it will cause the antenna performance change obviously, and will lead to the mismatching of the antenna impedance and cause resonant frequency changes. As the result, the reliability and read distance is reduced greatly in contrast with in the air.

In the inner of the tire, there will have high temperature, high pressure, bending and friction phenomena when moving, and it will cause the implanted RFID tag damage. When implant the RFID tag in the tire, people usually chooses deformation is small and not easily damaged parts, as the middle rubber layer shown in Figure 3. The dielectric parameters of this layer is $\varepsilon_r = 11.5$ and tan $\delta = 0.14$. In order to study the effect of rubber material for RFID antenna implanted, dipole antenna simulation model is established with the HFSS simulation software. The normal dipole antenna is chosen for the RFID tag antenna, as shown in Figure 4. The total length of the dipole is L, the radius of the dipole is R and the gap distance of the dipole is P.

Using HFSS to simulate the dipole antenna, and take L = 154 mm, R = 1 mm and P = 3 mm, set the Lumport excitation, we can get the simulation result of S11 and antenna input impendence as Figure 5 and Figure 6. At the goal frequency of 915 MHz, the antenna has gain of −18.5 DB, and has the complex impedance of 63.8-j0.74, and it's −10 DB bandwidth is about 12 MHz.

Implant the dipole antenna into the rubber block of the size 174 mm × 71 mm × 11 mm. The dielectric properties of the rubber are as the same with the outer rubber layer shown in Figure 3. Due to conductive dielectric and electrical performance of the rubber impact on the dipole antenna, the resonant frequency is about 660 MHz, and the deviation from the center frequency is 255 MHz, the frequency decreased significantly. While the impedance changes more obviously, it is obviously beyond 50 ohm impedance matching requirements. Therefore, due to the influence of dielectric parameters of rubber material, it makes the resonant frequency of RFID antenna and matching impedance change significantly, resulting in reading distance is shortened, and in some cases can't read the tag data.

The properties of RFID tag antenna are closely related with the dielectric parameters of the rubber materials covered RFID antenna. Figure 7 shows the S11 parameter changes with RFID coated rubber material relative dielectric various. As see from the Figure, with the increase of relative permittivity of rubber material, the resonant frequency of RFID antenna is reduced, and the gain of the antenna is reduced too. In order to compensate the frequency offset, it can consider reducing the antenna size or adding insulating layer outside of the antenna.

461

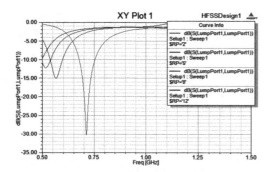

Figure 7.   Simulation result of S11 versus frequencies.

Figure 8.   Vulcanized rubber block.

Table 1.   Read distance test.

RFID type	Read distance (CM)	
	Before implant	After implant
18000-6B	55	20
EPC Class1 G2	300	21
EPC Class1 C1	43	26

4   THE EXPERIMENT AND ANALYSIS

In order to verify the influence of rubber material for RFID tag, the RFID tag is implanted in rubber for test. Three types of tags are encapsulated in the thickness of rubber block of 4 mm. Firstly, making the mixed rubber into thin films, which thickness is about 2.5 mm. Then cut the rubber film into rectangle shape, and the length and width is respectively about 15 CM and 7 CM. Put the RFID tag in the middle of two rubber films. Finally put the film into flat vulcanizing machine. The film after vulcanizing is shown in Figure 8.

This experiment uses 18000-6B, EPC Class1 Gen2 and EPC Class1 C1 three protocol chips, and uses AdvancedID UHF reader for test, recording RFID tag reading distance before vulcanized and recording RFID tag reading distance after vulcanized respectively, as shown in Table 1.

It can be shown from dada in Table 1, that the reading distance of RFID tag which encapsulated with rubber is shortened, especially label G2 is most obvious. The results of experiment show that, rubber has a great impact on the antenna performance of electronic tags. As the above analysis, RFID antenna is easily affected by the electrical conductivity and polymer dielectric of rubber after the UHF RFID is planted into the rubber. On the one hand, a certain amount of energy can be dissipated when the antenna contacts to the rubber which is electrical conductive material; on the other hand, a certain amount of RF energy is also consumed when a loop is formed between the RFID two feed in the electrical conductive rubber material. The energy of RF signal is lower which can affect the reading distance of the reader. Further, because of the effect of the polymer dielectric of rubber, the original impedance matching relationship of RFID antenna changes, causing RF resonance frequency shift. So the antenna simulation and adjusting the antenna parameters are needed according to the specific test data to achieve the purposes of antenna impedance matching.

5   CONCLUSIONS

With the rapid development of radio frequency identification technology, RFID applications are more widely. But in many cases, RFID tags need to be integrated together with identified objects and sometimes may also need to implant in them, such as in tires, or in other organisms. Furthermore, it is needed to carefully analyze and explore solutions to problems because of the UHF RFID antenna being influenced by the media. For RFID tires, the impedance characteristics and the resonant frequency of RFID tags is significant shift, resulting in read performance is greatly reduced. Because of such characteristic of RFID tags, the means that reduce the antenna size or increase the insulation layer can be used to achieve the match between electronic tag and reader.

ACKNOWLEDGEMENTS

This work was financially supported by the Qingdao Municipal Science and Technology Development Plan subject (13-1-4-135-jch), and also supported by the Shandong Natural Science Foundation (ZR2013FL011).

REFERENCES

Basat S., Tentzeris M.M. & Laskar J. (2006). Design and Development of a Miniaturized Embedded UHF RFID Tag for Automotive Tire Applications. IEEE Antennas and Propagation Conference, pp.160–163.

Dobkin D.M. & Weigand S.M. (2005). Environmental effects on RFID tag antennas. in IEEE Microw. Symp. Dig., 135–138.

Foster P. & Burberry R. (1999). Antenna Problems in RFID Systems. IEE Colloquium on RFID Technology, 3/1–3/5.

Grosinger J., Mayer L.W., Mecklenbräuker C. & Scholtz A. (2009). Input impedance measurement of a dipole antenna mounted on a car tire. In Proc. International Symposium on Antennas and Propagation.

Lasser G. & Mecklenbrauker C.F. (2012). Vehicular low-profile dual-band antenna for advanced tyre monitoring systems. IEEE 23rd International Symposium, 1785–1790.

Perret E., Tedjini S. & Fongalland G. (2011). Trends in the design of RFID tags. Microwave & Optoelectronics Conference (IMOC), SBMO/IEEE MTT-S International, 1–3.

Preradovic S., Karmakar N.C. & Balbin I. (2008). RFID Transponders. IEEE Microwave Magazine. vol. 2, 90–103.

Song T.Q., Liu W.B. & Ma L.X. (2013). Experimental Research on RFID Tag Implanted in Tire. Applied Mechanics and Materials, 411, 1637–1640.

Song T.Q., Zhou Y. & Ma L.X. (2014). Study on Relationship between Carbon Black and Dielectric Properties of Tire Rubber in UHF Band. Applied Mechanics and Materials, 536, 1456–1459.

Yanchen Gao, Diancai Yang & Weiwei Ning. (2010). RFID Application in Tire Manufacturing Logistics. IEEE International Conference on Advanced Management Science, Chengdu, 109–112.

Control Engineering and Information Systems – Liu (Ed)
© 2015 Taylor & Francis Group, London, ISBN 978-1-138-02685-8

Allocating channel automatically of IEEE 1394 based on normalized topology

Y.L. Wan
Run Technologies Co., Ltd., Beijing, China

D.G. Duan & Z.M. Han
College of Computer Science and Technology, Beijing Technology and Business University, Beijing, China

ABSTRACT: For practical video surveillance in sensor networks, automatic channel allocation is required. In this paper, we derive a normalize topology from self-ID packets for channel allocation. A superior method is proposed to determine the unique identify of nodes by comparison of the normalize topologies, which doesn't occupy bus bandwidth. Applying the algorithms and using Visual C++ and Windows XP DDK, network architecture visualization is realized. The results show that channels are allocated automatically and in real-time. It is beneficial to monitor network and optimize performance.

1 INTRODUCTION

IEEE 1394 was designed as a high speed (up to 3.2 Gbps) data bus for consumer electronics. It can support powerful isochronous transmission for multimedia streaming, so it is a good candidate for High-bandwidth sensor networks such as video surveillance (Chandramohan et al. 2002). To guarantee quality of service, channel allocation must be correct and real-time (Xiao et al. 2007). But the bus architecture is hot-pluggable, so the bus topology must be reconfigured every time a node is added to or removed from the network (1394B 2002). Topology reconfiguration results in channel reallocation, and makes it more difficult because topology information change occurred. (Chandramohan et al. 2002) propose a hybrid location-centric routing protocol for future WSNs based on IEEE1394 with store and forward nodes, which is more suitable for a large-scale network not a local area network. (Moon et al. 2002) present a Deterministic Multicast Channel Allocation Protocol (DMCAP) that allocates IEEE-1394 logical channels for IP multicast groups, which doesn't guarantee quality of service because of channel collisions.

Channel allocation needs a global topology map. Consulting network routing protocols, the active routing is suitable for maintaining dynamic topology information (Xiao et al. 2007). But with this protocol, node address and link state must be exchange frequently, which brings out additional communication and process overhead (Qin et al. 2007) (Depei et al. 2004). In this paper we propose a topology discovery and maintenance scheme based on self-ID packets. This scheme uses bus manager to maintain the global topology for automatic channel allocation in time. It makes the bus network architecture visible, which is beneficial to monitor network and optimize performance.

2 ANALYSIS OF A BUS NETWORK ARCHITECTURE

In IEEE 1394, two levels of service are available: "asynchronous" packets are sent on a best effort basis while "isochronous" packets are guaranteed to be delivered with bounded latency. The isochronous packets are labeled with 6-bit "channel" numbers. First, the node must acquire the necessary bandwidth and a channel from the Bandwidth Available and Channels Available registers at the IRM by means of asynchronous traffic. Asynchronous transfers use 64-bit fixed addressing, where the upper 16 bits of each address represent the node ID. The higher orders 10 bits specify a bus ID and lower order 6 bits specify a physical ID. The node ID is subject to reassignment each time bus reset (Chandramohan et al. 2002). During bus configuration, a treelike topology is built; each node is assigned a physical node number and also sends self-ID packet(s) that is used by the management layer. A bus manager shall have the capability to monitor self-ID packets. It may set its own force root variable to TRUE and initiate a bus reset in order to become the IRM at the same time, which benefits information sharing.

EUI-64	Physical ID	Port 0	Port 1	Port 2	Channel

Figure 1. Node data structure for normalized topology.

The IEEE 1394 bus network management relies on bus topology building and update. The process of establish bus topology is to determine connection relationship and unique identify of nodes. Connection relationship of nodes can be determined by deriving a normalized topology from self-ID packets. To determine the unique identity of nodes, if there is no previous topology information the only means is to read the configuration ROM bus information block of each node. This flurry of activity occupies bus bandwidth needed for other time critical tasks, such as the reallocation of isochronous resources or net update. Otherwise there is a superior method, which is the comparison of a saved copy of a normalized topology accurate before the bus reset with the newly derived normalized topology.

Part of the information the node requires is contained in any consistent set of self-ID packets collected after bus reset; these completely describe bus topology but they are not in a format useful for topology comparison subsequent to future bus resets. The bus manager shall reorganize this information into a normalized topology viewed from its own perspective.

The data structure that represents each node in the tree is represented graphically by Figure 1. The EUI-64 field represents the nodes unique identifier (ultimately obtained from configuration ROM). The port fields are overloaded in the figures to represent either the port status reported in the self-ID packet (disconnected, child or parent) or a link to another node data structure in the normalized topology. The channel that has been allocated for isochronous talker is recorded to its data structure. So an isochronous listener can use the talker's unique identifier to search the corresponding channel. At the same time the talker's physical ID can be acquired, by which asynchronous transactions can be start to initialize and configure some variable of isochronous operations, such as packet length.

3 TOPOLOGY DISCOVERY ALGORITHM BASED UPON SELF-ID PACKETS

3.1 *Normalization topology*

The self-identify process uses a post-order binary tree traversal algorithm. Based on this algorithm, an intelligent method can deduce the network topology from the self-ID packets. The methods of self-ID packet analysis are discussed with reference to an example topology, illustrated by Figure 2.

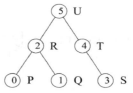

Self-ID packets			
0	P	-	-
1	P	-	-
2	C	C	P
3	P	-	-
4	-	C	P
5	C	C	-

Figure 2. Reference topology (with self-ID packets).

In Figure 2, nodes are represented by circles that contain their physical IDs. The root (physical ID 5 in this figure) is at the top. For the sake of brevity, a single number shown to the right of each node represents its unique identity. The table to the right of Figure 2 contains pertinent details from the set of self-ID packets generated for this topology; child and parent connections on a PHY port are abbreviated as C and P, respectively.

Assuming a node has completed a power reset, there is no previous topology information. The first phase in the process is to analyze the self-ID packets in order to determine which nodes are connected to each other and by which numbered ports. The algorithm starts with the self-ID packets for the node with physical ID zero and concludes with the root. As each node self-ID packets are encountered, transfer the port status into the corresponding node data structure. If the node is childless, also push the nodes physical ID onto a stack; this information is used later in the process when the nodes parent is encountered in the self-ID packets. Whenever self-ID packets for a node indicate connected child ports, the processing is different. As before, copy the port status information from the self-ID packets to the node data structure—but do not push the nodes physical ID onto the stack just yet. For each connected child port, pop a physical ID from the stack and establish a link between this nodes data structure and the node data structure identified by the physical ID obtained from the stack. As the stack values are popped, the connections are made to this node connected child ports in decreasing order. Upon completion, if the node has an unresolved parent connection, push the nodes physical ID onto the stack. Figure 3 shows the results of processing the self-ID packets for node two (shaded); links have been created to its children and its physical ID has been pushed onto the stack.

Eventually the process comes to an end when the root node is encountered. All of its connected child ports are processed and links created in the topology. Since the root, by definition, has no connected parent, its physical ID is not pushed onto the stack. At this time, a complete topology from the perspective of the root has been derived from

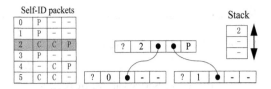

Figure 3. Self-ID packet topology analysis (nodes two).

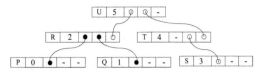

Figure 4. Normalized topology with EUI-64 information.

the self-ID packets. Once configuration ROM has been read for all the unidentified nodes, the normalized topology is complete, as shown by Figure 4. All of the steps described above are programmed with C language excerpt as follow. The normalized topology derived by this code should be saved for comparison with potentially changed topologies after future bus resets.

```
VOID normalize topology() {
INT i, j, m, n;
memset(node, 0, sizeof(node)); /* Clear the
topology at the start */
for (i = 0; i < = rootID; i++) {
for (m = 16; m > = 0; --m) {
node[i].port[m] = selfID[i].port[m]; /* Copy port
connection sttus */
if (selfID[i].port[m] = = CHILD) { /* Found a
child connection? */
j = pop(); /* Yes, set j to child PHY ID */
for (n = 0; n < 16; n++) /* Scan for parent
port */
if (selfID[j].port[n] = = PARENT) {
node[i].link[m] = &node[j]; /* Link from parent
to child ... */
node[j].link[n] = &node[i]; /* ... and from child
to parent */
break; /* Only one parent port per PHY */
}}}
if (i < rootID)
push(i); /* Remember this node for later parent
link resolution */}}
```

3.2 Bus topology reestablishment

If a new node, with a EUI-64 represented by the letter X, is inserted, the topology is altered and a different set of self-ID packets is generated as shown by Figure 5.

First, the new set of self-ID packets is analyzed to produce the normalized topology illustrated by Figure 6. At this point in the process, the only EUI-64 known with certainty is that of the node that observed the self-ID packets—in this case, the same node U used in the first example. Note that the new normalized topology reflects a changed physical ID for node U. The next step is to compare the normalized topologies before and after bus reset

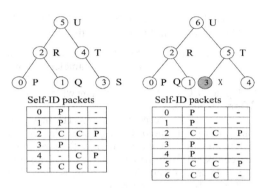

Figure 5. Reference topology (inserted node).

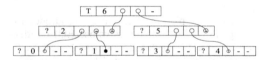

Figure 6. Normalized topology (relative to node six).

in order to transfer as much EUI-64 information as possible from the saved information. The algorithm recursively traverses the two trees, starting from the position of the observing node. If a port is disconnected in the new topology, there is no need for any analysis. If the same port was connected to a child both before and after bus reset, the EUI-64 identity of the child is unchanged. The C procedure shown below describes the algorithm.

```
for (i = 0; i < = 16; i++) {
/* Compare old and new connection status on
all ports */
if (new->port[i] = = DISCONNECTED)
continue; /* Nothing to explore down this
branch ... */
if (new->link[i] = = parent) /* Does connection
point to our parent? */
continue; /* Been there, done that! */
if (prior->port[i] ! = DISCONNECTED) {
/* Was same port connected in the prior topol-
ogy? */
new->link[i]->eui64 = prior->link[i]->eui64; /*
OK! We know it has the same EUI-64 */
```

467

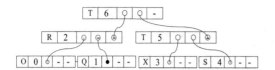

Figure 7. Normalized topology with EUI-64 information (inserted node).

updateEUI64(new->link[i], prior->link[i], new);
/* Recurse down this branch of the tree */}}

The recursive process is started with a call to updateEUI64 () with the parameters shown below:

Update EUI64 (&node [phy IDnew], prior [phy IDprior], NULL);

In the code excerpt above, phy ID new and phy ID prior are the observing nodes physical ID before and after the bus reset. Since the topologies are compared from the perspective of the observing node as if it were the root, there is no parent node and its parameter is null.

Upon completion of the algorithm, the normalized topology is updated with EUI-64 identity information from the prior topology, as shown by Figure 7.

At the completion of the recursive application of update EUI64, the nodes whose EUI-64 identity could not be deduced from topological analysis are left with zero for their EUI-64s. This is an invalid value for a EUI-64; it is necessary to read configuration ROM to obtain the EUI-64s for these nodes. Configuration ROM reads for newly inserted nodes should be deferred until at least one second since the most recent bus reset.

4 AUTOMATIC CHANNEL ALLOCATION

Once the physical ID of the IRM has been determined, a node may attempt the allocation of bandwidth or channels from the BANDWIDTH AVAILABLE and CHANNELS AVAILABLE registers at the IRM, and record the allocated channel number on its date structure. The target may transmit a channel allocation request in order to request IRM to return channel number according to the talker's unique identity. A channel allocation request/response shall conform to the format illustrated by Figure 8. Field usage in a channel allocation request/response is as follows:

Source ID: This fieldshall specify the physical ID of the sender. Tarteg ID: In a channel allocation request, this field shall specify the IP address from which the sender desires a response. In a channel allocation response, it shall be IGNORED. Channel: This field shall specify an allocated channel number.

Source ID	Target ID	Talker EUI-64	Channel

Figure 8. Channel allocation request/response format.

All of the steps described above are programmed with C language excerpt as follow.

VOID normalize topology() {
for (i = 0; i < = rootID; i++) {/* Scan for the talker */
if (node[i].eui64 = = eui64) {
channel = node[i].channel; /* Yes, set channel to allocated channel number */}}}

In order to promote stability of serial bus configurations across a bus reset, the owners of isochronous resources established prior to the bus reset, i.e., channels and bandwidth, are required to reallocate these resources as soon as possible after a bus reset. Whereas any new isochronous operations shall wait until 1000 ms have elapsed before these new isochronous resources could be allocated.

5 ALGORITHM APPLICATION

Applying the algorithm designed in the previous subsection and using Visual C++ and Windows XP DDK, topology discovery and visualization are demonstrated in the case that a node has just completed power reset and a node is inserted.

Figure 9 depicts a normalized topology with five nodes after power reset. Every nodes is labeled its physical ID and unique identify. The root (unique identify 'O' in this figure) is at the top left corner. The color lines show which nodes are connected to each other and by which numbered ports. An application wishes to record video data from the camera (unique identify 'R' in this figure) on the digital DVCR (unique identify 'Q' in this figure). The dashed line on the Figure 9 shows channel '0' is allocated automatically. Figure 10 shows the normalized topology with the addition of a new node. It is compared to the Topology retained from before the node insertion, as shown in figure. After an intelligent topology analysis, the identity of all nodes can be determined except the new node. It is necessary to read configuration ROM to obtain the unique identify for this node. A new node, with a unique identify represented by the number 'S', is inserted; the topology is alerted and a different set of physical IDs and unique identify are generated, as shown in Figure 9. Figure 9 also shows channel '0' is resumed. Meanwhile, a new isochronous operation commences, playback from the camera to the TV monitor (unique identify 'S' in this figure), channel '0' is allocated to node 'S'.

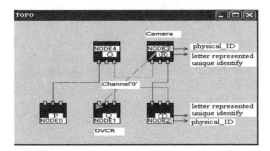

Figure 9. Normalized topology with EUI-64 information (after power reset).

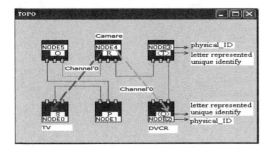

Figure 10. Normalized topology with EUI-64 information (inserted a node).

Table 1. Times consumed by channel allocation.

Node P	Node S
1.5 ms	1001.625 ms

Video data can be transferred though the DMA mechanism. The DMA transfer is important because each time the interface DMA controller fills a DMA buffer, the hardware interrupts the computer. By using the DMA interrupt, the time used by channel allocation will be determined approximately which can verify succeed of the allocation.

Commencing with the time the bus reset is initiated, times consumed by channel allocation have been evaluated, with the results given in Table 1. It is apparent that the channel number allocation is successful, since isochronous traffic on corresponding channel can be realized. The prior isochronous channel for 'record' is reallocated in time and the same isochronous channel for the new operation ('playback') is allocated later.

6 CONCLUSION

With widespread application of IEEE-1394 to High-bandwidth sensor applications, quality of service guarantees is becoming more and more important. So it is necessary to allocate channels automatically in real time. To cope with this, we derive a normalize topology from an intelligent analysis of self-ID Packets, which facilitates channels allocation even if net topology changes after the bus reset. If there is prior topology information, we proposed a superior method to determine the unique identity of nodes through comparing the normalize topologies other than reading the configuration ROM of each node. As such, bus bandwidth is saved. Based algorithm designed in this paper, we develop program using visual C++ and Windows XP DDK to realize channels allocation function, which demonstrates our analyses correctness and makes topology and allocated channel information intuitive. The experiment results show our studies lay a foundation for future routing protocol of hybrid network (IEEE 1394 and Ethernet).

ACKNOWLEDGMENT

This paper is supported by Key Lab of Information Network Security, Ministry of Public Security.

REFERENCES

1394B. 2002.1394bTM IEEE Standard for a High-Permance Serial Bus-Amendment 2[S]. IEEE Computer Society Sponsored by the Microprocessor and Microcomputer Standards Committee.

Chandramohan, V. & Christensen, K. 2002. A First Look at Wired Sensor Networks for Video Surveillance Systems, Local Computer Networks, Proceedings, 728–729, Nov. 2002.

Depei, X.H.Q. & Heng, L.Z.C. 2004. Visualization of network topology with hierarchical display layout. Journal of Beijing University of Aeronautics and Astronautics, Vol. 30, PP. 529–533 (In Chinese).

Moon, G. & Lim, H. 2002. Simple channel allocation method for IP multicast over IEEE-1394 channels. Electronics Letters, Vol. 36, PP. 384–385, Aug. 2002.

Qin, Z.R. Gang, Y.J. & Ji. L.Z. 2007. On Performance of Routing Schemes in Mobile Ad hoc Network. Computer Science, Vol. 34, PP. 55–62 (In Chinese).

Xiao, S. Zhang, L. & Long, C. 2007. An Improvement of TCP Performance on Ad hoc Networks. Microprocessors, Vol. 3, PP. 29–35 (In Chinese).

Control Engineering and Information Systems – Liu (Ed)
© 2015 Taylor & Francis Group, London, ISBN 978-1-138-02685-8

Mining potential determinants of Internet addiction among Chinese cybercafé users

L. Huo, W. Shang & S.Y. Xu
Academy of Mathematics and System Science, Chinese Academy of Sciences, Beijing, China

ABSTRACT: Cybercafés (net bar) in China have been criticized as a bed of Internet addiction. Yet with the maturing of the generation grown with Internet and computers, cybercafés have become an important means of socializing and relaxing in China. There is a call for critical research on the determinants of Internet addiction among Chinese cybercafé users to encourage beneficial use and avoid abuse in the new era. This research surveys cybercafé users' behavioural patterns, their demographics, as well as whether they are addicted to Internet. Besides mining correlations, a decision tree approach is employed to obtain more comprehensive understanding of the addiction phenomena. To deal with the massive surveyed items, and their frequent Internet addicts, variable clustering and Pearson correlation are utilized for dimension reduction, and data resampling is employed for the imbalanced dataset. A set of Internet addiction rules are found, key determinant of which include: years having been using Internet, the sensitivity to the café price, sensitivity to overnight service at cybercafé, distance between the work/study place and the café, and the age of first exposure to Internet. Findings of this research will foster a more critical thinking of the Internet addiction in the new age of Internet.

1 INTRODUCTION

The Internet has become an essential media channel for personal communications, academic research, information exchange and entertainment, such that most people can hardly live without it. China Internet Network Information Centre (CNNIC) reported that 564 million people had gone online in China till late 2012 (CNNIC 2013). As a result of increased Internet penetration, people find themselves become more dependent and sometimes indulged in the virtual online world.

In China, public concerns more about adolescence's obsession in Internet and cybercafés. An NPC (National People's Congress) member proposed to shut down all the cybercafés to prevent adolescence Internet addictions, in the Eleventh National People's Congress, 2010. However, cybercafés are not the only means for people to get addicted to Internet. Furthermore, the function of cybercafés has been evolved to places for young people to get together and socialize. It is essential to explore the distinct behavioural patterns and find predictive determinants of Internet addiction among cybercafé users to have a critical and comprehensive understanding and propose more constructive regulations to reduce the Internet addiction risk without harming the function of socialization and entertainment of cybercafés.

This study researches cybercafé users' behaviours with respect to Internet addiction diagnostics. What we do includes two steps. The first is survey design, and the second is getting knowledge from the swarm of questionnaire results. For the second step, representative variables are selected using variable clustering together with correlation analysis. To compensate the defect of class imbalance, we design a data resampling mechanism to obtain a balanced dataset. The prediction algorithm C4.5 decision tree is utilized to extract predictive variables from the large number of variables, and to generate explicit principles for Internet addiction prediction.

2 LITERATURE REVIEW

Internet addiction disorder refers to the behavior of excessive computer and Internet use, which interferes with people's daily life (Cao & Su 2007). It is considered as a pathological problematic behavior which results in academic, social, and occupational impairment. Students with Internet addiction reported more negative consequences on their studies and daily routines than did those without Internet addiction (Chou & Hsiao 2000). In the workspace, Internet addictive behavior symptoms include a decline in work performance and a withdrawal from co-workers (Young 1999).

To have a more comprehensive and accurate understanding of Internet addiction, Fang et al. (2009) established a decision tree and generated a huge rule set (17 pathways). It is more accurate in prediction, but somehow lack of generalizability.

Two issues need to be conquered to have a more accurate and generalizable Internet addiction prediction rules using a decision tree approach. The first one is the imbalanced dataset, which will hinder the accuracy in the minor group recognition (i.e. the addicted group). The second one is the complexity of the rules. Too many rules (i.e. deepness of the decision tree) tend to over-fit the model to the samples, and lack of overall prediction power. It is necessary to conduct dimension reduction among as many influential factors as possible. This will help to find more generalizable rules, instead of finding the opportunistic rules limited by the candidate factor pool. Therefore, the solution will be to design an all-inclusive survey related to Internet addiction and then use the dimension reduction techniques to find a minimum set of most representative variables.

3 METHODS

3.1 Subjects

Subjects of this study are cybercafé users nationwide in China. One thousand subjects took part in an on-site survey. They were volunteers from 20 cybercafés, which were randomly selected from 10 cities and their nearby counties using a cluster sampling approach. Variety in location and economic status were considered during the sampling. 976 effective questionnaires were finally collected after validation and data cleaning.

3.2 Questionnaire design

An initial questionnaire was designed based on the research questions of TASCHA (2010) considering the China contexts. We applied 20 copies of initial questionnaire on a pilot survey, and revised the initial questionnaire to form the final version.

The questionnaire contains 27 questions and consists of four sections, including Internet use, cybercafé use, personal objectives and Internet/cybercafé use impact, and Demographics. The first Internet use section contains four questions, i.e. 11 items. A question has a title, and items are the sub questions within a question. The second cybercafé use section contains 15 questions, i.e. 41 items. The third section contains 3 questions,

i.e. 58 items. The last section contains 8 questions, i.e. 8 items. The third section contains diagnostic questions that distinguish addicts from common users.

Among the several assessment instruments, YDQ is most commonly used. This paper follows Beard & Wolf (2001) which improved YDQ. The questions are followed:

1. Do you feel preoccupied with the Internet?
2. Do you feel the need to use the Internet with increasing amount of time in order to achieve satisfaction?
3. Have you repeatedly made unsuccessful efforts to control, cut back or stop Internet use?
4. Do you feel restless, moody, depressed, or irritable when attempting to cut down or stop Internet use?
5. Do you stay online longer than originally intended?
6. Have you jeopardized or risked the loss of significant relationship, job, educational or career opportunity because of the Internet?
7. Have you lied to family members, therapist or others to conceal the extent of involvement with the Internet?
8. Do you use the Internet as a way of escaping from problems or of relieving a dysphoric mood (e.g. Feelings of hopelessness, guilt, anxiety or depression)?

Instead of let user answer Yes or No to these questions, we use a five scales: usually, most times, sometimes, seldom and never. Internet addiction index is true when a respondent choose usually or most times in all the first five questions and at least one of the last three question. Consequently, of the valid 976 subjects, 22 could be described to suffer from "Internet addiction", which accounts for 2.3%. The results are accord with existing researches and reports on Internet addiction in China (Cao & Su 2007).

3.3 Analytical instruments

To distinguish the special characteristics and behaviours of Internet addicts in cybercafés, a decision tree approach is employed. There are two primary issues we need to deal with before the classification modeling. The first is that there are 118 items, so we need to focus on a part of them. The second is that the Internet addicts are far less than common users, thus we need to deal with imbalanced class distribution.

When there is a multitude of variables, it probably includes irrelevant and redundant ones. Large attribute dimensionality incurs high computational cost, and, causes over-fitting, resulting in large decision trees that generalize poorly. Variable clustering and correlation analysis can help reduce computational costs and alleviate possible over-fitting (Sebastiani 2002), and result in C4.5 producing smaller trees.

The imbalanced dataset problem (Chawla 2003) is prevalent in many applications, including fraud detection (Fawcett & Provost 1996), risk management, and text classification. Similarly, the cyber-café user survey is typically imbalanced, for there are many more instances of non-addicts than addicts. Machine learning algorithms could be biased towards majority classes due to over-prevalence. Filtering dataset using resampling method is one way to improve learning algorithms' performance (Van et al. 2007).

C4.5 Decision tree (Quinlan 1993) has proven popular in practice. It deals remarkably well with irrelevant and redundant information, and is able to deal with continuous values.

4 RESULTS

To get prepared for decision tree analysis, and extract useful information from the data, we need to pick out representative variables from all the questionnaire variables, and revise the distribution of classes so that the objective class is explicitly represented. To reach the target, variable clustering and data resampling are employed.

4.1 Variable selection

Questionnaire answers were cleaned and transformed before the data analysis and knowledge discover. And after that, the questionnaire is characterized by 252 attributes.

As there is a magnitude of attributes, variable clustering and Pearson correlation were conducted to determine the explained variability in the dataset. As a result, variables in each cluster retain best similarity while correlations between clusters are minimized.

We pick up useful variables from each cluster under three principles: Firstly, select variables with maximum Information Value. Secondly, select variables with minimum 1-R2 ratio. Finally, incorporate any additional business priorities. This logic ensures that the most predictive variables are selected while taking into account the "uniqueness" of each predictor. Finally, we select 25 variables that are potential causative factors. They are as listed in Table 1. The selected variables will become inputs to the following procedures.

4.2 Data resampling

To deal with the imbalanced class distribution, we use a resampling filter to get the instances redistributed. The "Internet addiction" survey dataset is imbalanced, with a majority of non-addicts. The morbid data usually result in inaccurate prediction of the minority class. Resampling is commonly used to compensate for this problem (Sadat et al. 2011). In our study we sample the dataset with a bias toward uniform distribution. It oversamples the minority class and under samples the majority one to create a more balanced distribution for training. Before resampling, the addicts/non-addicts proportion was 22/954. After that, the proportion was 511/465.

4.3 Determinants of Internet addiction

We employ C4.5 decision tree to find the determinative predictive factors of "Internet addiction". According to the effect of classifier, we go back to variable selection and re-select variables.

The resulted decision tree is shown in Figure 1. It provides the combination of significant factors as well as the threshold values that will lead to Internet addiction. Nodes in the tree correspond to features, branches to their associated values, and leaves to classes. The top node Internet Use Year is the best node for classification. The other features in the nodes of decision tree appear in descending order of importance. The target variable is Is Addicted. If a leave is impure (i.e. some records are misclassified), the number of misclassified records will be given after a slash.

The decision rules discovered are as follows:

1. Those that use Internet equal or less than 5 years are not likely to be internet addicts.
2. For those who use Internet more than 5 years, if they are less sensitive to cybercafé price increase, and a cybercafé is less than 300 m from work/study place, they tend to be addicted.
3. For those who use Internet more than 5 years and are more sensitive to cybercafé price increase, if they are sensitive to overnight services at cybercafés, they are likely to exhibit Internet addiction behaviour.
4. For those who use Internet more than 5 years and are more sensitive to cybercafé price increase, if not so sensitive to overnight services at cybercafés, but the first exposure to Internet is less than 10-year-old, they are also predicted addicted.

The information on the classification accuracy of the decision tree is shown in Table 2. A confusion matrix (Table 3) reports the accuracy. It indicates how many data records are correctly

Table 1. Meanings of variables.

Variable	Meaning
Section: Internet use	
LOCATION_Number	Number of places where the subjects go on Internet
WHERE_Calculated	Whether learnt computer and internet at cybercafé
InternetUseYear	Years having been using Internet
InternetStartAge	Age of first exposure to Internet use
LOCATION_Library	Percentage of Internet access at library
LOCATION_Cybercafe	Percentage of Internet access at cybercafé
Section: Cybercafé use	
MoneySpentPerVisit_calculated	CafePrice × DurationOfCafeVisit
RespondToPriceIncrease	Change of purchase predisposition when café charge 1/3 more
FrequencyOfCafeVisit	Frequency to cybercafés in the past 12 months
Activity1	Frequency of using emails at cybercafés
FACTOR2	Sensitivity to Internet speed of cybercafé
DistanceOfCafeToWorkStudy	Distance of workplace/school to cybercafé nearby
CafePrice	Money the café charge per hour
DayTimeOfCafeVisit	Time of the day the user visits a cybercafé
RespondToPriceDecrease	Change of purchase predisposition when café charge 1/3 less
Activity7	Frequency of gaming at cybercafés
Activity8	Frequency of listening to music at cybercafés
FACTOR9	Sensitivity to overnight service at cybercafé
NumberofCafesVisited	Number of cybercafé visited in the past year
Section: Demographics	
Sex	Sex of subject
Age	Age of subject
rural_urban	Living in rural or urban
Education	Education status with 6 levels
Income	Personal income monthly from <500 to >8000
Profession	Profession with 13 options

*All variables are numerical except WHERE_Calculated, Sex, rural_urban and Profession.

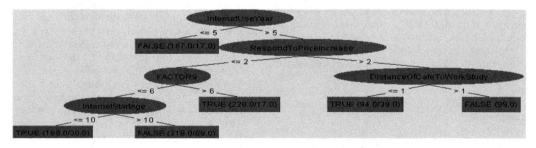

Figure 1. Decision tree predicting the Internet addiction.

Table 2. Detailed accuracy by class.

TP rate	FP rate	Precision	Recall	F Measure	Class
0.811	0.239	0.756	0.811	0.782	FALSE
0.761	0.189	0.816	0.761	0.787	TRUE

Table 3. Confusion matrix.

FALSE	TRUE	← Classified as
377	88	FALSE
122	389	TRUE

classified and misclassified in the class of TRUE and FALSE. Provided m is the different label values, the size of a confusion matrix is M × M.

5 DISCUSSIONS

From this study we found out that people using Internet for more than 5 years had a higher probability of Internet addiction. This indicates that the addiction behavior is gradually formed, which is quite consistent with other "substantial addictions" such as pathological gambling, but contrary to Kraut et al. (1998) which suggested that new users were more inclined to develop problematic behaviors. Subjects in Kraut et al. (1998) were all during their first 1 to 2 years on-line, so he concluded that from a short-term perspective. Furthermore, the Internet is more complex compared to 1998 when Kraut et al. (1998) was done.

We also found out that if there is a cybercafé near the participant's work or study place, he/she tends to be detected as Internet addicts. This is accordant with people's common sense, and also proved it effective of some local governments' policies in China to prevent opening cybercafés around schools.

Most interestingly, addicts attach much importance to a cybercafé's overnight service. This implicates that addicted people are apt to keep staying in cybercafés all-night, while normal Internet usage usually happens in the daytime. This is basically in keeping with Young (1999). According to our study, typical addiction Internet activities are often time consuming, like gaming. A possible measurement to reduce Internet addictions in cybercafés could be imposing more tax on cybercafés' overnight services, instead of shut them down.

Finally, this study suggests that the earlier a person is exposed to Internet, the more likely he/she is to have Internet addiction. This is probably due to the lack of self-control of youth. So the supervision of young people's exposure to Internet is imperative for the youngsters' Internet addiction control.

6 CONCLUSIONS

This study researches the Internet addiction among Chinese cybercafé users to discover comprehensive determinants of Internet addictions. Five predictive factors and four rules are generated. Better understanding of the Internet addiction phenomena and regulation suggestions are discussed. Our results and discussions complement previous studies and may be used to develop intervention measures to prevent Internet addiction.

Further research incorporates psychological factors will motivate deeper insight especially in intervention design for Internet addicts. Moreover, automatic rules discovery schema design will be our future work.

ACKNOWLEDGMENT

This research is partially supported by Amy Mahan Research Fellowship Program to Assess the Impact of Public Access to ICTs and Beijing Municipal Commission of Education (Key Project of Science and Technology Plan, No. KZ201411232036).

REFERENCES

Beard, K.W. & Wolf, E.M. 2001. Modification in the proposed diagnostic criteria for Internet addiction. CyberPsychology and Behavior 4(3): 377–83.

Cao, F. & Su, L.Y. 2007. Internet addiction among Chinese adolescents: prevalence and psychological features. Child Care Health & Development 33(3): 275–81.

Chawla, N.V. 2003. C4.5 and imbalanced datasets: Investigating the effect of sampling method, probabilistic estimate, and decision tree structure. In Proceedings of the ICML'03 Workshop on Class Imbalances Vol3.

Chou, C. & Hsiao, M.C. 2000. Internet addiction, usage, gratification, and pleasure experience: the Taiwan college students' case. Computers and Education 35(1): 65–80.

CNNIC. 2013. The 31st Statistical Report on Internet Development. [WWW document] URL http://www1.cnnic.cn/AU/MediaC/rdxw/2012nrd/201301/t20130116_38529.htm.

Fang, K., Lin, Y.C. & Chuang, T. 2009. Why do internet users play massively multiplayer online role-playing games? A mixed method. Management Decision 47(8): 1245–1260.

Fawcett, T. & Provost, F. 1996. Combining Data Mining and Machine Learning for Effective User Profile. Proceedings of the 2nd International Conference on Knowledge Discovery and Data Mining. Portland, 8–13. OR: AAAI.

Kraut, R., Patterson, M., Lundmark, V., Kiesler, S., Mukopadhyay, T. & Scherlis, W. 1998. Internet paradox. A social technology that reduces social involvement and psychological well-being?. American Psychologist 53(9): 1017–1031.

Quinlan, J.R. 1993. C4.5 Programs for Machine Learning. Morgan Kaufmann, San Mateo.

Sadat, M., Samuel, H., Patel, S. & Zaiane, O. 2011. Fastest association rule mining algorithm predictor. Proceedings of the Fourth International C* Conference on Computer Science and Software Engineering. 43–50.

Sebastiani, F. 2002, Machine learning in automated text categorization. ACM Computing Surveys 34(1): 1–47.

TASCHA (Technology & Social Change Group, University of Washington Information School). 2010. Global impact study. [WWW document] URL http://www. globalimpactstudy. org/surveys.

Van Hulse, J., Khoshgoftaar, T.M. & Napolitano, A. 2007. Experimental Perspectives on Learning from Imbalanced Data. In ICML. 935–942.

Young, K.S. 1999. Internet addiction: symptoms, evaluation and treatment. Innovations in Clinical Practice: A Source Book 17, 19–31.

Control Engineering and Information Systems – Liu (Ed)
© 2015 Taylor & Francis Group, London, ISBN 978-1-138-02685-8

Weighted self-localization algorithm of Networked Munitions

H.W. Liu, C.L. Jiang & M. Li
School of Mechatronic Engineering, Beijing Institute of Technology, Beijing, China

X.Y. Cheng
North Micro-Electro-Mechanical Intelligent Group Corporation Ltd., Beijing, China
School of Mechatronic Engineering, Beijing Institute of Technology, Beijing, China

ABSTRACT: Trilateration is one of the most commonly used methods in the Self-localization of Wireless Sensor Networks and Networked Munitions. In view of the feature that this algorithm is deeply affected by the relative positions of anchors and the ranging errors, which is unable to meet the demand of Networked Munitions, a new algorithm that is based on weight of localization-triangle and residual errors of distances is presented. This algorithm effectively reduces the negative effect on localization caused by relative positions of anchors and the inaccuracy of ranging. Simulation show that this algorithm has the advantages of better accuracy and robustness with lager errors of ranging when compared with the Maximum Likelihood Estimation. This algorithm also reduces communication overhead, meeting the requirement of low power consumption of wireless sensor networks.

1 INTRODUCTION

With the development of information technology, future military reform will promote the mode of combat to transform from Platform Centric Warfare to Network Centric Warfare (Ren et al. 2006). Being a newly arisen information technology which is low-cost and easy maintenance, wireless sensor networks could be used in various kinds of military applications to realize information sharing among dispersed personnel and equipments to increasing the combat capacity. Networked Munitions system is based on wireless sensor network, and it consist of lots of munition nodes that could communicate with each other via radio. Each munition node could realize the sharing of information and resources.

Technology of self-localization is a key technology in networked munition system. The munition nodes will detect target and make attacked decisions. Without the positions the munitions it could not finish the tasks. Networked munitions could work with the GPS system, but it may not receive the satellites signal in some extreme situations and the cost of it is expensive. The only way to locate themselves is using the technology of self-localization of wireless sensor networks.

There are many kinds of self-localization algorithms. In general, they are divided into two types. One is Range-free, such as DV-Hop and Centroid. The other is Range-based, such as AHLoS and the Maximum Likelihood Estimation (MLE) (Jing et al. 2011). But almost all of them bring either huge calculation complexity or large communica-

tion overhead. They could not meet the demand of the network munition system, which has limited hardware resources and energy. There is no self-localization algorithm that could be used in any environment and application at present.

Most self-localization Rang-based algorithms have similar theory. Part of nodes could obtain their position in different ways, such as GPS or artificial arrangement. This kind of node is called anchor nodes. The other parts could not obtain its position directly which are called unknown nodes. The unknown node obtain is position by some algorithms, through communicating with other nodes, measuring the distance between themselves and the anchor nodes, and using the network connectivity.

Trilateration is a basic Range-based localization algorithm. It uses three anchor nodes and the distances between unknown node and each anchor nodes to calculate the unknown node's position. It has many advantages such as simple principle and less calculation. So it meets the requirement of network munitions about the hardware resource and energy. However the ranging error and the relative positions of anchor nodes have great influence on positioning accuracy. It does not have a good robustness.

In networked munition system that deployed densely, there are usually over three neighbor anchor nodes of each unknown node. So the unknown node could choose different groups of anchor nodes to realize its localization. In order to choose anchor nodes with better relative positions and location performance, a new self-localization distributed algorithm based on trilateration and weight is presented.

2 TRILATERATION

Trilateration uses the positions of three non-collinear anchor nodes and distances between the unknown node and the three anchor nodes to calculate the position of the unknown node in two-dimensional space (Li & Zhang, 2009). As shown in Figure 1, making the anchor nodes as centers and the distances as radius, three circles could be described. The intersection point of the circles is the position of the unknown nodes.

The position of the unknown node could be calculated as

$$
\begin{cases}
A = \begin{bmatrix} 2(x_A - x_C) & 2(y_A - y_C) \\ 2(x_B - x_C) & 2(y_B - y_C) \end{bmatrix} \\
B = \begin{bmatrix} x_A^2 - x_C^2 + y_A^2 - y_C^2 + d_C^2 - d_A^2 \\ x_B^2 - x_C^2 + y_B^2 - y_C^2 + d_C^2 - d_B^2 \end{bmatrix} \\
\begin{bmatrix} x \\ y \end{bmatrix} = A^{-1}B
\end{cases}
\tag{1}
$$

where the (x_A, y_A), (x_B, y_B) and (x_C, y_C) are the coordinate of anchor A, B, and C respectively while the (x,y) is the coordinate of the unknown node D. The d_A, d_B and d_C are the distances between the unknown node and the anchor node A, B, and C.

Actually, the distances in (1) should be the actual distances. But in practice, the measuring distances are used. Considering the errors of ranging. (1) will become as follow

$$
\begin{aligned}
\begin{bmatrix} \overline{x} \\ \overline{y} \end{bmatrix} &= A^{-1}B \\
&= A^{-1} \begin{bmatrix} x_A^2 - x_C^2 + y_A^2 - y_C^2 + d'_C^2 - d'_A^2 \\ x_B^2 - x_C^2 + y_B^2 - y_C^2 + d'_C^2 - d'_B^2 \end{bmatrix}
\end{aligned}
\tag{2}
$$

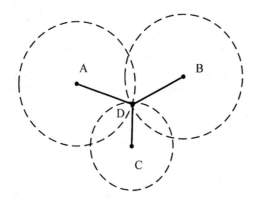

Figure 1. Schematic diagram of trilateration.

where the d'_A, d'_B and d'_C are the measuring distances including ranging errors. The relationship between the actual distance and the measuring distance is

$$
\begin{cases}
d'_i = d_i + \varepsilon_i \\
\varepsilon_i = \gamma_i * d_i
\end{cases}, i = A, B, C.
\tag{3}
$$

where γ_i is a percentage of the actual distance, which distributed normally. Make (2) minus (1), then the follow function could be got.

$$
\begin{bmatrix} \Delta x \\ \Delta y \end{bmatrix} = A^{-1} \begin{bmatrix} d_C^2 \gamma_C (\gamma_C + 2) + d_A^2 \gamma_A (\gamma_A + 2) \\ d_C^2 \gamma_C (\gamma_C + 2) + d_B^2 \gamma_B (\gamma_B + 2) \end{bmatrix}
\tag{4}
$$

(1) and (4) imply that the localization error is relative with the shape of the triangle formed by the three anchor nodes, which is called localization-triangle, and the ranging errors.

3 WEIGHT OF THE TRIANGLE

Inspired by (Vaishali & Bijendra, 2005) and the dilution of precision of GPS system (Xu, 2003), it could be found that the geometry, which formed by the anchor nodes, and the relative position of the anchor nodes and the unknown node deeply affect the error of location. In trilateration, when the anchor nodes form an equilateral triangle, the error of location will be the minimum. In a triangle, the more degree of the inequality of the values of its three angles, the more the triangle closed to an equilateral triangle. So the follow function could be used to describe the degree a triangle closed to an equilateral triangle.

$$
V = \frac{\left(\alpha_A - \dfrac{\pi}{3}\right)^2 + \left(\alpha_B - \dfrac{\pi}{3}\right)^2 + \left(\alpha_C - \dfrac{\pi}{3}\right)^2}{3},
\tag{5}
$$

where the α_A, α_B and α_C are the three angles of the triangle and V is the variance of the angles.

Based on the above analysis, it could be found that when the three anchor nodes form an equilateral triangle, the value of V will be zero. In this case, the weight of the triangle in calculating the position should be set as 1. In another case, if the three anchor nodes collinear, which is a extreme situation, the V will be the maximum as $2\pi^2/9$. As known, in this case the error of localization will be very large, so the weight of the triangle should be set as zero. Use the method presented in (Gao, 2007) and according to the relationship between V and the weight, the follow function could be presented.

$$w_{triangle.m} = \frac{\dfrac{1}{V_m}}{\displaystyle\sum_{i=1}^{n} \dfrac{1}{V_i}}, \quad m = 1, 2, \ldots n. \tag{6}$$

where n is the count of groups of anchor nodes and V_m is the variance of the angles of group m, while $w_{triangle.m}$ is the weight of the group of anchor nodes.

4 WEIGHT OF THE RESIDUAL OF THE DISTANCES

Using three anchor nodes and the corresponding distances, the estimated position of the unknown node could be calculated. Then the follow function could be used to calculate a estimated distance.

$$d_{ei} = \sqrt{(x_E - x_i)^2 + (y_E - y_i)^2}, \quad i = A, B, C. \tag{7}$$

where the d_{ei} is the estimated distance between the anchor nodes and the estimated position of the unknown node, and (x_E, y_E) is the estimated position. The estimated distance d_{ei} could be used to calculate the residual of the distances in the follow function

$$e = \frac{\displaystyle\sum_{i=1}^{3} |d_{ei} - d_i|}{3}, \quad i = A, B, C. \tag{8}$$

where the d_{ei} is the estimated distance calculated in (7), and d_i is the measuring distance between the anchor nodes and the unknown node.

If the errors of ranging do not exist, the three circles in Figure 1 will have one intersection point that will be the actual position of the unknown node without error. In this case the residual of the distances is zero, and the weight should be set as 1. When the errors of ranging exist, the circles may not have intersection point, but (2) still have a solution, which would not be the actual position of the node. It could been seen form (4) that greater errors of ranging would bring a larger error of localization. Thus it will lead to a larger residual of the distances calculated in (8). Under this condition the weight should be set as small as possible. So the residual of the distances could imply the performance of the localization. According to the relationship between the residual and the weight, the follow function could be presented.

$$w_{resdual.m} = \frac{\dfrac{1}{e_m}}{\displaystyle\sum_{i=1}^{n} \dfrac{1}{e_i}}, \quad m = 1, 2, \ldots n \tag{9}$$

where n is the count of groups of anchor nodes and e_m is the residual of the distances of group m, while $w_{resdual.m}$ is the weight of the group of anchor nodes.

5 STEP OF THE ALGORITHM

5.1 *Calculate all of the estimated positions of the unknown nodes, and calculate variance of angles and residual of distances of each group*

Unknown node finds the neighbor anchor nodes by sending broadcast package. All of the found anchor nodes would be divided into groups by three. For example, the count of anchor nodes is n, and count of groups would be Cn3. Then the unknown node uses each group to calculate a estimated position using (1), variance of angles using (5) and residual of distances using (7).

5.2 *Calculate the two weights of all estimated positions*

Using the variance of angles and residual of distances of each group obtained in step A, each unknown node could calculate the two weights described above using (6) and (9).

5.3 *Calculate the position of unknown node*

Unknown nodes get the position of themselves using the follow function

$$\begin{bmatrix} x \\ y \end{bmatrix} = \begin{bmatrix} \dfrac{1}{2} \displaystyle\sum_{i=1}^{n} x_i (w_{triangle.i} + w_{resdual.i}) \\ \dfrac{1}{2} \displaystyle\sum_{i=1}^{n} y_i (w_{triangle.i} + w_{resdual.i}) \end{bmatrix}. \tag{10}$$

In the function above, (x_i, y_i) is one of the estimated positions of the unknown node.

5.4 *Repeat the Step A to Step C to calculate all other unknown nodes.*

6 SIMULATION

Simulations have been organized to verify the performance of the algorithm presented above. The algorithm has been compared with the Maximum Likelihood Estimation (MLE), which is commonly used for localization in wireless sensor networks. For the sake of simple, the process of producing errors of ranging would not be simulated, instead by adding a simulant error to the actual distance. Reference (Langendoen & Reijers, 2003) infers that the errors of ranging distributed normally, whose average is zero and standard deviation is one percentage of the actual distance. In the simulations, the error of localization is calculated as follow

$$E_{Localization} = \frac{\sqrt{(x - x_{actual})^2 + (y - y_{actual})^2}}{R} \quad (11)$$

where (x, y) is the estimated position of each group, and (x_{actual}, y_{actual}) is the actual position of the unknown node. R is the radius of the propagation rang of wireless signal.

The simulations consist of three sections. The first is to compare the localization errors of the two algorithms under the same situation. The second is to compare the influence caused by the increasing of the errors of ranging. The last is to compare the influence caused by the increasing of the number of anchor nodes.

6.1 Accuracy of localization under the same situation

In this simulation, set radius of propagation rang of signal as 100 meters, set the standard deviation of error of ranging as 0.2. The count of neighbor anchor node of each unknown node is 8, and the count of unknown nodes is 50.

The result of simulation could be seen in Figure 2. Under the setting situation, the localization errors of MET is below 0.3, while the algorithm even could make the errors below 0.2. It also shows that the localization errors of different nodes vary irregularly. That is because the ranging error is a random number which distributed normally. However it is obvious that in view of a single node, the localization error of the algorithm presented is always smaller than that of MLE.

6.2 Influence caused by the increasing ranging errors

In this simulation, set the related parameters as the same with the simulation above except the standard deviation of error of ranging. The standard deviation of error of ranging increases form 0.05 to 0.6 by step of 0.025.

Result of simulation could be seen in Figure 3. Under the setting situation, with the ranging errors increasing, the localization errors of the both two algorithms would increase. But the MLE is more sensitive to ranging errors, especially when the ranging errors is greater than 0.5, the increasing rate of localization errors would become much faster. The localization errors of algorithm presented are less influenced by ranging errors and increase gently. The algorithm presented has better robustness than the MLE. Networked munitions usually worked in adverse environment where the ranging errors would be unsatisfactory. The robustness against ranging errors is very suitable for the networked munitions.

6.3 Influence by the number of anchor nodes

Set the related parameters as the same as the first simulation, except the number of anchor nodes, which increases form 3 to 17.

The result could be seen in Figure 4. With the increasing of number of anchor nodes, the localization errors of both algorithms decrease. But the errors of algorithm presented are always smaller than those of MLE. When the number is above 12, the errors would not decrease obviously. Errors of MLE keep below 0.12, while that of algorithm presented keep below 0.08. So it could been concluded that no matter densely or loosely the munitions deployed, the algorithm presented has better performance than the MLE.

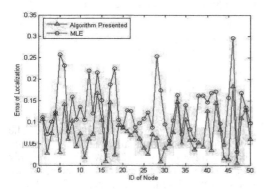

Figure 2.   Comparison of localization performance.

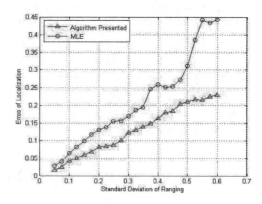

Figure 3.   Influence caused by ranging error.

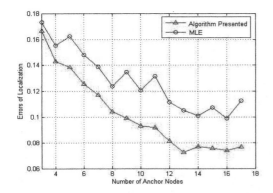

Figure 4. Influence caused by number of anchor nodes.

7 CONCLUSION

In view of the feature of networked munitions, a new distributed self-localization algorithm that is based on trilateration is presented. This algorithm uses weight of the triangle and weight of the residual of distances to obtain a more accurate position and a better robustness against ranging errors. It effectively decreasing the influence caused by some disadvantages. Simulation shows that the algorithm presented has better performance in localization accuracy and robustness compared with MLE, and is more suitable for the networked munitions. Besides, this algorithm has no need to communicate among nodes except finding anchor nodes, which has a better performance on saving energy.

REFERENCES

Gao S. 2007. Three calculating weights methods in analytic hierarchy process, Science Technology and Engineering, 7(20): 5204–5207.

Jing B., Zhang J. & Sun Y. 2011. Smart Networked Sensor and Wireless Sensor Network, Beijing, National Defence Industry.

Langendoen K. & Reijers N. 2003. Distributed localization in wireless sensor networks: a quantitative comparision, Computer Networks, 43: 499–518.

Li J. & Zhang B.H. 2009. Analysis and improvement for Localization Algorithm Based on Trilateration in Wireless Sensor Network, Computer Science. 36(10 A):53–81.

Ren Q., Sheng Y.B. & Mu S. 2006. Study of network centric warfare: Computer Engineering and Design (27):433–436.

Vaishali P.S. & Bijendra J. 2005. Localization accuracy and threshold network density for tracking sensor networks, Proc. IEEE International Conference on Personal Wireless Communications: 408–412.

Xu G.C. 2003. GPS—Theory, Algorithms and Applications, Heidelberg: Springer.

Control Engineering and Information Systems – Liu (Ed)
© 2015 Taylor & Francis Group, London, ISBN 978-1-138-02685-8

A new measurement for the importance of nodes in networks

L.Q. Qiu, Y.Q. Liang, Z.Y. Chen & J.C. Fan
College of Information Science and Technology, Shandong University of Science and Technology, Qingdao, Shandong, China

ABSTRACT: Networks are widely used in a variety of different fields and attract more and more researchers. Centrality analyses, one of the research hotspots, provide answers with measures that define the importance of nodes. However, classical centrality analyses usually have high expensive complexity. Moreover, we propose a new measurement for the importance of nodes in networks to avoid some shortcomings of classical measurements. DC centrality integrates two different measurements-degree measurement and cohesion centrality. We also argue the limitation of relying on single measurement. The experiment results show that DC centrality can get better performance.

1 INTRODUCTION

Networks are useful tools to characterize complex systems. The system components are represented as nodes and their mutual interactions as edges. Finding the important nodes in such networks is therefore of great relevance for understanding the mechanisms that underlie the system evolution. This explains the increasing interest in the topic, peculiarly in the measurement of nodes in networks. There are many measurements for the importance of the nodes [1]. However, a common weakness in these studies, as we will discuss in detail in related work, is that the computation of measurements can be expensive. For large-scale networks, efficient computation of centrality measures is critical and requires further research. In the paper we propose a new method of measuring the importance which proves to be efficient, which will be denoted by the acronym DC centrality from now on.

The rest of the paper is organized as follows: in Section 2 we discuss related work and several measurements for the importance of nodes are introduced. Next, we describe in Section 3 our generalization of new measurement-DC centrality for the importance of nodes. In Section 4, we provide experimental studies. Finally in Section 5, we give the conclusion and future directions.

2 RELATED WORKS

Centrality analysis provides answers with measures that define the importance of nodes. There are many classical and commonly methods used ones: degree centrality, closeness centrality, betweenness centrality. These centrality measures capture the importance of nodes in different perspectives. With large-scale networks, the computation of centrality measures can be expensive except for degree centrality. Given a graph G of n nodes and m edges, then we can get time complexity and space complexity about the centrality measures. Closeness centrality, for instance, involves the computation of all the pairwise shortest paths, with time complexity of $O(n^2)$ and space complexity of $O(n^3)$ with the Floyd-Warshall algorithm [2] or $O(n^2 \log n + nm)$ time complexity with Johnson's algorithm [3]. The betweenness centrality requires $O(nm)$ computational time following [4]. For large-scale networks, efficient computation of centrality measures is critical and requires further research.

Moreover, one measurement only partly indicates the importance while the node is important and its role is multiple and complex, therefore we can not rely on single measurement.

3 DC CENTRALITY

3.1 Degree centrality

For degree centrality, the importance of a node is determined by the number of nodes adjacent to it. The larger the degree of one node, the more important the node is. Node degrees in most networks follow a power law distribution, i.e., a very small number of nodes have an extremely large number of connections. Those high-degree nodes naturally have more impact to reach a large population than the remaining nodes within the same network. Thus, they are considered to be more important. Therefore degree centrality considers the connection model of the nodes.

The degree centrality of node v is defined as:

$$C_D(v) = d/(n-1) \quad (1)$$

where d is defined as the number of nodes adjacent to v, and n is defined as the number of nodes in the network.

3.2 Cohesion centrality

Although degree centrality proves efficient in some networks, but it is not efficient under some scenarios, i.e., some important nodes (i.e., bridge contacts connect with merely two edges) don't have high degree centrality. In other words, degree centrality is one-sided which ignores the role of the nodes. Based on the idea, we argue that the importance of one node is determined by its connection model as well as its role in the networks. Accordingly we consider two factors, namely the connection model of the node and its role in the network. The connection model of one node can be described by its degree centrality, and the role of one node can be described by its cohesion centrality as follows.

Definition 1: The connectivity of node v is defined as the number of the edges between v and the nodes directly connected with it.

The connectivity of a node measures how close it is to the nodes which are directly connected with it, and reflects the local connection property of the node. Obviously, the span of connectivity is between 0 and $C_D(v)(C_D(v)-1)/2$.

Definition 2: The cohesion centrality of node v is defined as follows:

$$C_c(v) = \frac{C_D(v)(C_D(v)-1)}{2c} \quad (2)$$

where $C_D(v)$ is the degree centrality of node v, and c is the connectivity of node v.

According to the relations between the nodes and the edges in the network, the value of $C_c(v)$ satisfies the conditions:

$$C_c(v) \geq 1 \quad (3)$$

We find that the larger the connectivity of one node, the less important the node is. This is because the deletion of the node with larger connectivity will make less affection on the network. Thus according to equation (2), the more importance one node is, the larger the cohesion centrality of the node is. Therefore, the cohesion centrality is the positive evaluation measurement for the node.

3.3 DC centrality

To integrate the two factors (i.e., connection model of one node and its role in networks), DC centrality is introduced to measure the importance of the node, where the importance consists of two parts-a degree centrality and a cohesion centrality:

$$I(v) = \alpha \cdot C_D(v) + (1-\alpha) \cdot C_c(v) \quad (4)$$

where α satisfies $0 \leq \alpha \leq 1$.

In above importance function, the degree centrality $C_D(v)$ measures the connection model of node v, and the cohesion centrality $C_c(v)$ measures the role of node v. The parameter α is set by the user to control the level of emphasis on each part of the total importance function. Thus according to equation (4) we can select important nodes with high value.

How to determine the parameter α in Equation (4) is a challenging issue. The problem of how to automatically find the best α when there is no ground truth is beyond the scope of this paper. To simply the experiments, we will set α as 0.5 in the following example applications.

4 EXPERIMENTS

For our examples of the operation of our method, we apply it to one small network, which is the much-discussed "karate club" network of friendships between 34 members of a karate club at a US university, assembled by Zachary [5] by direct observation of the club's members. This network is of particular interest because the club split in two during the course of Zachary's observations as a result of an internal dispute between the director and the coach. In other words the network can be classified into two communities-one's center is the director, the other's center is the coach.

Firstly, we explore the limitation of single measurement. Secondly, we compare our method-DC centrality with baseline algorithms (i.e., degree centrality and PageRank algorithm).

4.1 First experiment

In the experiment we use there classical measurements (i.e., degree centrality, closeness centrality, betweenness centrality) to analyze "karate club" network. Figure 1 shows the results of the experiment. We use square shape to represent the nodes in the network, but we use different size levels to indicate their values of different measurements. The bigger the size of one node is, the more important the node is.

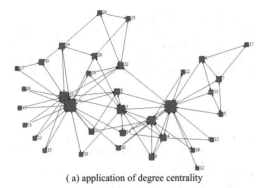

(a) application of degree centrality

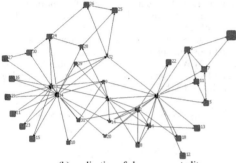

(b) application of closeness centrality

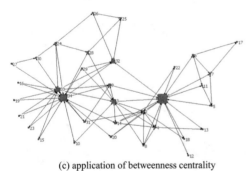

(c) application of betweenness centrality

Figure 1.　The results of the first experiment.

The main observations we would like to emphasize are:

1. Degree centrality and betweenness centrality can detect some important nodes. For example, node 34 and node 1 have the biggest size in Figure 1(a) and 1(c), which respect the director and the coach separately. This is in line with the actual situation obviously. However, closeness centrality can not reflect the importance of nodes, i.e., the more important one node is, the smaller size of the node is.
2. Relying on single measurement to indicate the importance of one node is infeasible. For

example, in Figure 1(a) node 9, node 14 and node 24 have the equal value, but node 9 and node 14 are more important than node 24 actually, because they are on the boundary of two different communities. Closeness centrality and betweenness centrality usually generalize small important values as we can see from Figure 1(b) and 1(c), which result in small differentiation.

4.2　Second experiment

We select degree centrality and PageRank algorithm as baseline methods, which be identified as "degree" and "PageRank" respectively. Table 1 shows the result of the experiment. We select the nodes in the top 10 according to different methods.

The main observations we would like to emphasize are:

1. PageRank algorithm is better than degree centrality.

 Although two methods have the same rank of Top 10, their values are different. As we can see from the table, degree centrality can not distinguish nodes because some nodes have the equal value. For example, node 32 and node 4 have the equal value according to degree centrality, which shows that degree centrality can not distinguish nodes well.
2. DC centrality is better than PageRank algorithm.

 For example, according to PageRank algorithm node 31 is not more important than node 24. However, node 31 is more important than node 24 actually, because node 31 is on the boundary of two communities. According to DC centrality, node 31 is important than node 24 obviously.

Table 1.　The results of the second experiment.

Degree	Node ID	PageRank	Node ID	DC	Node ID
0.515	34	0.101	34	0.694	34
0.485	1	0.097	1	0.569	1
0.364	33	0.072	33	0.485	33
0.303	3	0.057	3	0.414	3
0.273	2	0.053	2	0.365	2
0.182	32	0.037	32	0.297	32
0.182	4	0.036	4	0.248	14
0.152	24	0.032	24	0.209	4
0.152	9	0.030	9	0.194	9
0.152	24	0.030	14	0.187	31

5 CONCLUSIONS

In the paper, we propose a new measurement-DC measurement for the importance of nodes in networks. Rather than relying on only one measurement, DC measurement integrates the two factors (i.e., connection model of one node and its role in networks), which prove efficient than degree centrality and PageRank algorithm. Although DC centrality may avoid some shortcoming of classical measurements, how to achieve better performance to detect importance nodes need to be further studied.

ACKNOWLEDGEMENT

This paper is supported by: The Specialized Research Fund for the Doctoral Program of Higher Education (No. 20133718110014), The Nature Science Foundation of Shandong Province (No. ZR2012FM003 and ZR2013FM023), and The National Science Foundation for Post-doctoral Scientists of China (No. 2013M541938), and Shandong Province Postdoctoral special funds for innovative projects of China (No. 201302036).

REFERENCES

Brandes U. A faster algorithm for betweenness centrality, Journal of Mathematical Sociology, 2001, 25(2):163–177.

Floyd R.W. Algorithm 97: Shortest path, Communications of the ACM, 1962, 5 (6):345.

Johnson D.B. Efficient algorithms for shortest paths in sparse networks, Journal of the ACM, 1977, 24(1):1–13.

Wasserman S. and Faust K. Social network analysis: methods and applications, 1994:3,5,8,13,34,68.

Zachary W.W. An information flow model for conflict and fission in small groups, Journal of anthropological Research, 1977, 33:452–473.

Control Engineering and Information Systems – Liu (Ed)
© 2015 Taylor & Francis Group, London, ISBN 978-1-138-02685-8

Factor analysis for public adoption of Mobile Electronic government*

P. Zhou*
Academic Administration, Jinling Institute of Technology, Nanjing, China

S. Seah*
College of Public Administration, Huazhong University of Science and Technology, Wuhan, China

ABSTRACT: Mobile Electronic government (M-gov) is one of the innovations of government management. The author explores the factors influencing the public adoption of M-gov, integrates the influencing factors as the foundation for the further research, such as perceived usefulness, perceived ease of use, task-technology fit, some eternal variables and forms an integrated model, which is based on the Technology Acceptance Model 3 (TAM3) and Task-Technology Fit model (TTF).

1 INTRODUCTION

With the development of mobile technology, the public are accustomed to using mobile network for their information attainment. Until Dec 2012, the numbers of Chinese mobile internet users have reached to 388 million, accounting for 72.2% of the total number of internet users by statistics from CNNIC.[1] Because there is a rapidly growing trend of the numbers of mobile users, numerous citizens think it is time to call for the public service from government. Presently, the network platform and infrastructure are already provided by Chinese mobile government development (M G). However, M G has not drawn enough attention and the process is still moving slowly at the public adoption work. According to survey, only 2.5% of internet users access government websites.[2] As a result, the huge gap between demand and supply will greatly restrict the development process of M G. Thus, the study of the factors influencing public adoption of M G is beneficial to improve the reasonable public adoption, and ultimately realize service optimization. Based on TAM3 and TTF models, this paper integrates the influencing factors such as perceived usefulness, perceived ease of use, task-technology fit, some eternal variables, and the characteristics of M G. And based on that, it also brings up the public adoption model of M G. In the end, a research framework for the M G public adoption intention is provided.

*This paper supported by the 2012s Technology Innovation Fund "Study on the development of the smart-tourism public service platform from the perspective of smart-city—Taking Haikou as an example" (No. CXY12M014) of Huazhong University of science and technology, and it is one of the achievements.

2 RESEARCH OVERVIEW

2.1 Concept of Mobile Electronic government

(1) Viewpoints from Chinese scholars: M G is development of traditional Electronic Government (E-gov); it is the application of mobile communication technology, such as terminal function, access speed, security and network etc. Based on the technology and the administration innovation, the portability of E-gov has been accomplished. M G is not only an important part of E-gov, but also shows the future trend.[3] From the task function level: Li Mingsheng (2005) pointed out that M G was a series of government activities by mobile communication services. It mainly includes the government management and public service.[4] Yu Lixian and Wang Tingfang (2008) thought that M G would help to solve digitized government affairs by means of wireless information technology, government management & public service, the application of mobile communication network and the internet.[5] From the technology level: Song Gang and Li Mingsheng (2006) considered M G as the functions of public services relying on mobile phone, PDA, wireless network, Bluetooth, RFID etc. Xu Xiaolin and Zhao Xinzi (2007) showed that M G was a new type of E-gov application mode based on traditional political environment. (2) Scholars from other countries also gave the definitions of M G. Kim et al. (2004) argued that the government could improve public service efficiency. On the other side, citizens could also access to government services platform at anytime from anywhere by M G.[6] Antovski (2007) mentioned that M G was the public government affairs mode, which built clear information channel and communication platform between government, organizations and citizens.[7] Ntaliani et al. (2008) pointed out that M G

contained in the E-gov and government could provide more public information and services through the wireless communication technology.[8]

2.2 Technology Acceptance Model 3 (TAM 3)

In the area of information system user acceptance, Davis developed the Technology Acceptance Model (TAM) in 1986s. It is developed by measuring the perception of users in the use of new technologies. Davis tried to explain and predict the reasons why people adopt or disregard information system. Perceived Usefulness and Perceived Ease of Use are two important factors influencing users' attitude.[9][10] TAM2 was founded on TAM. Venkatesh and Davis attempted to figure out the key factors outside of Perceived Usefulness and Perceived Ease of Use, so as to enhance the explanatory power of model.[11] There are two composites: social influence processes and cognitive tool of process, which were used to explain perceived usefulness and use intention in TAM2. TAM3 is a comprehensive model which is found by Venkatesh and Bala. It shows why and how the employees accept and use information technology from the organization level. TAM3 was integrated and improved from TAM2 and the perceived ease of use determinant model.[12]

In order to further explain TAM3, Venkatesh and Bala established TAM3 theoretical framework[13] Perceived usefulness and perceived ease of use are determined by four different types of factors. Individual difference refers to the personality or demographic variables (such as personal characteristics or state, gender and age). They influence the personal perception of perceived usefulness and perceived ease of use. The characteristics of the system are prominent features of the system. It helps training the perceived usefulness or ease of preference (or detest) perception. Social influence gets the user perceived of information from different social processes and mechanism. Convenient conditions refer to the support oriented user use information technology from organization.

In the TAM 3 model (Fig. 1), social norms, image, working relevance, output quality, results demonstration and perceived ease of use determine that the perceived usefulness, experience and voluntary are control factors. Social norms and images belong to the social influence in the Figure 1; others are the parts of characteristics of the system. The effect on perceived usefulness which caused by these factors mainly shows as: mutual influence of working relevance and output quality on perceived usefulness, the higher output quality, the stronger affected from working relevance to perceived usefulness. With more users' experience, the affect from social norms to perceived usefulness gradually weakened. Social pressure (social norm) of information usage

is always influenced by image (which people want to maintain in groups). Anchor means that people are more likely to put on future estimation and use estimates connected, meanwhile, influenced by suggestions from others. In uncertain circumstances, people usually use reference point and anchor to reduce the fuzziness, then, draw the final conclusion through certain adjustment. The effect of perceived usefulness on behavior intention performance as: during the process of information technology personal acceptance and use, perceived usefulness is an important predictor of behavioral intention. Social norms, experience and resources also have important influence on behavior intention. Under a voluntary (Using information technology) context, with increasing experience in usage, the effect from social norms to behavior intention has been weaken, however, it is stronger in the mandatory use context.

In TAM 3, the determinants of perceived ease of use include: computer self-efficacy, computer anxiety, computer playfulness, and perceptions of external controller facilitating conditions, perceived enjoyment and objective usability. The first four variables belong to individual differences in TAM3 theoretical framework. The following two terms are the parts of system characteristics. Venkatesh and Bala described six factors of perceived ease of use with views which were different from other scholars. Computer self-efficacy refers to the individual judgment, such as their own ability of using computer. Computer anxiety refers to the fear and anxiety of individuals from computer using. Computer playfulness means joyful psychological state individuals produced in the process of using computer.[14] The external control (or convenient conditions) means a judgment from the external individuals. Perceived enjoyment refers a judgment about how joyful that specific information technology could bring. Objective possibility refers to the possibility of practical efforts to accomplish certain tasks. Then, the performance of these factors on perceived ease of use as, in the process of personal acceptance and use of information technology, computer self-efficacy, external control, computer anxiety and computer enjoyment are significant predictor of perceived ease of use. With the increased experience, computer anxiety on perceived ease of use is weakened gradually. Just in the initial stage, perceived ease of use could effect on behavior intention. Because of the usage of experience, this effect diminished.

2.3 Technology-Task Fit model (TTF)

TTF was proposed by Goodhue and Thompson in 1995s, which has important influence on the information technology research area (Fig. 3).[15] TTF means the information system (technology) in correspondence with user's task. In other words,

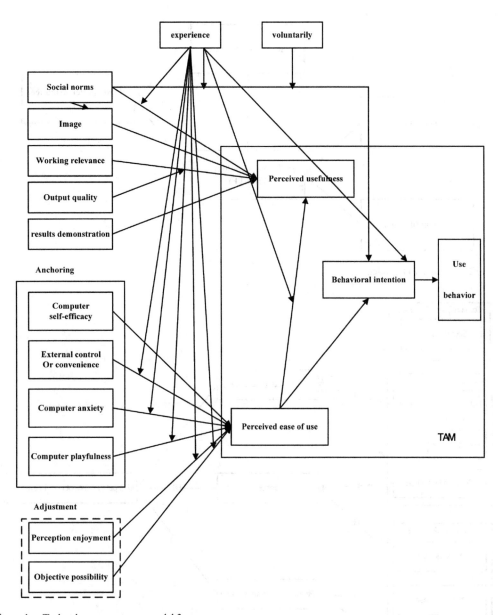

Figure 1. Technology acceptance model 3.

to what extent does information system function support users? Though the TTF model, Goodhue and Thompson pointed out: the practical use of TTF and (user oriented) system jointly affect individual performance (the system using effect). But at least, TTF could influence the practical use of the system under some circumstances.

By considering the function of information system, user's task requirements and the impact on system using, the TTF model made up the main defects

of TAM. Essentially, information system is a tool for users to accomplish a task. Whether the system is useful or easy to use should be aimed at specific user's task, namely, perceived usefulness and perceived ease of use cannot be separated from the user's task and existence. When such technology can support task well, it will soon be adopted. Therefore, it is suitable for interpretation user to use a technology or an information system. M G can be seen as driven by mobile communication technology, then, the user's online

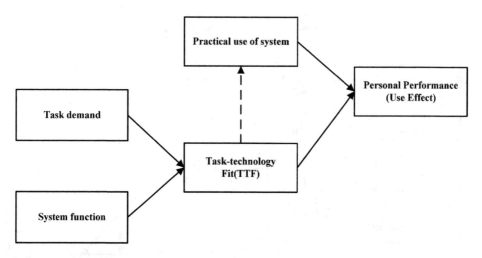

Figure 2.   A basic TTF model.

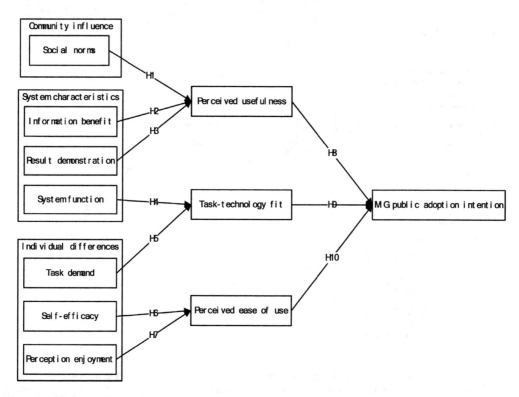

Figure 3.   The integration model of public adopted intentions.

government affairs can also be interpreted as a task. Consequently, applications TTF to analyze the use of M G for granted. Whereas, the application of TTF in M G was not widely than TAM, and a lot of research was still to predict individual's organization information system accepted. This shows that some variable of original TTF model does not adapt to the specific M G environment, and it requires corresponding corrections to follow the demand of M G activity.

According to the literature, the author found that there are few studies of TTF model under M G environment; however, TTF has been widely applied in

Table 1. Influencing factors of M G public adoption.

Factor pattern	Factors
Technical level	System/Service quality, Mobility, Flexibility, Complexity, Connection speed, Screen limit, Regional, Advantage, Convenience, Navigation, Playability, Interactivity;
Organization level	Service provider, Regulation system, Authorized, Authority;
Task level	Entertainment perception, Task type/features, Transaction speed, timeliness, price, Feasibility, Express perception, Reliability, Frequency, Emergency handing ability, performance, efficiency,
Individual level	Self-efficacy, influence, Trust perception, Individual innovation, Behavior control, Satisfaction, Experience, Risk perception, Individual expectations, Knowledge, Individual differences, Financial resources perception, Self-expression;
Environment level	Social influence, Subjective norm, Peer influence, External influence, Use situation influence, Media services, Equipment characteristics.

*Note: this table clears up by author.

Mobile Electronic business (M E-business). Considering the relationship between M E-business and M G, the author will reference the research achievement of M E-business. In the study of electronic business, Chu and Huang also applied the TTF model[16]. Follow the particularity of M E-business, they amended the task, technical characteristics of original model, and considered the effects of personal characteristics.

3 THE INFLUENCING FACTORS OF PUBLIC ADOPTION MOBILE ELECTRONIC GOVERNMENT

Public adoption is the key point to M G development. So far, we found few studies of M G acceptance and influence factors of continuous use, while the study of M E-business and E-gov user adoption is more abundant. Du Zhizhou (2010), He Dehua and Lu Yaobin (2008) analyzed the influence factors on user acceptance of M E-business from individual, organizational, technical, task, environment aspects.[17] Jiang Xiao (2010) partition the public E-gov adoption's factors as four characteristics, such as environment, technology, quality and individual.[18] Since the M G, E-gov and M E-business are interlinked, in consideration of the characteristics of M G, the influence factors of M G public adoption intention are divided into five levels (Table 1).

4 AN INTEGRATION MODEL OF MOBILE ELECTRONIC GOVERNMENT PUBLIC ADOPTION

4.1 *The necessity of the integration of TAM 3 and TTF model*

The task and technology adaptation is an important Prerequisite for M G implementation

effectiveness. Given the context mentioned above, task relevance in TAM 3 only considers its influence on perceived usefulness, through perceived usefulness to affect the user's intention. However whether the task and technology adaptation will directly impact behavior intention has not yet been confirmed. The theory has certain structure defects. The TTF model is explicitly included in the task characteristics, by focusing on the task, task-technology adapted affect information system use to supplement or extend the TAM 3.

On the other hand, the TTF model also has some shortcomings. The main reason is: a lack of Communication Bridge between task-technology fit and information system using, and the internal mechanism can not reflect the task-technology adaptation on individual behavior. This bottleneck was hinder TTF model to deeper develop. To overcome this deficiency, it is necessary to find out the way of task-technology fit influence user information system use. Obviously, TAM 3 can provide perceived usefulness, ease of use and intention of use for reasonable explanation.

4.2 *The model application in M G based on TAM 3/TTF integration*

Factors based on the above research, this study puts forward a research framework for further empirical research. (Fig. 3) (1) In community influence level: The behavior of majority of the member of social group will have an impact on other members of the group. In other words, if the people around potential are using the M G, which will greatly influence their perception of usefulness of M G. (2) In system characteristics level: The improvement of administrative that M G brings is public more concerned about. The results demonstrate refers to the advantage degree that is formed after M G using. Both of them will have an impact on

491

perceived usefulness. For M G, the system function is the most important part, and it is what the public pursuit for. System function is also one of the TTF external variables. (3) In individual differences level: Integrating degree of M G service and personal task demand will directly impact task-technology fit perception. Self-efficacy means a judgment of individual M G using. Perceived enjoyment is a judgment of personal pleasure on the use of M G. Both of them will have an impact on perceived ease of use.

Based on the above model, for the public M G adoption intention, the author puts forward hypotheses to test: (1) Social norms are positively related to M G perceived usefulness. (2) Information benefit is positively related to M G perceived usefulness. (3) Result demonstration is positively related to M G perceived usefulness. (4) System function is positively related to Task-technology fit. (5) Task demand is positively related to Task-technology fit. (6) Self-efficacy is positively related to perceived ease of use. (7) Perception enjoyment is positively related to perceived ease of use. (8) Perceived usefulness is positively related to M G public adoption intention. (9) Task-technology is positively related to M G public adoption intention. (10) Perceived ease of use is positively related to M G public adoption intention.

REFERENCES

Chu Yan, Huang Lihua. Mobile business applications adoption model based on the concepts of task/technology fit [C]. Proceedings of ICSSSM'05, Beijing, China, June 13–15, 2005: 1355–1359.

CNNIC. The report of Chinese mobile phone users online behavior research. [2013-1-1]. http://www.cnnic.cn.

CNNIC. The report of Chinese mobile phone users online behavior research. [2010-1-15]. http://www.cnnic.cn.

Davis F.D., Bagozzi, R.P., Warshaw P.R. User Acceptance of Computer Technology: A Comparison of Two Theoretical Models. Management Science, 1989, 35(08): 982–1003.

Davis F.D., Perceived Usefulness, Perceived Ease of Use, And User Acceptance Of Information Technology. MIS Quarterly, 1989, 13(03): 319–339.

Goodhue D.L., Thompson R.L. "Task technology fit and individual performance [J]." MIS Quarterly, vol.2, pp.213–236,1995.

He Dehua, Lu Yaobin. "Literature Review on Mobile Business Acceptance Research [J]." Journal of UESTC (Social Sciences Edition), vol.5, pp.46–50, 2008.

Jiang Xiao. "An Empirical Study of E-government Adoption: Citizen's Perspective[D]." Shandong, Dalian University of Technology, pp.33–40, 2010.

Li Mingshen, "Mobile electronic government, to expand the government service [J]," E-Government, vol.8, pp.45–46, 2005.

Ljupco Antovski. Improving Service Matching in M-Government with Soft Technologies. [2010-11-21]. http://ieeexplore.ieee.org/stamp/stamp.jsp?arnumber=04394202.

M. Ntaliani, C. Costopoulou, S. Karetsos. "Mobile government: A challenge for agriculture [J]." Government Information Quarterly, vol.25, pp. 699–716, 2008.

Ni Xiang, "The first shade of mobile electronic government [J]," Guanchayusikao, vol.11, pp.62–63, 2004.

Venkatesh V. "Determinants of Perceived Ease of Use: Integrating Perceived Behavioral Control, Computer Anxiety and Enjoyment into the Technology Acceptance Model [J]." Information Systems Research, vol.11, pp.342–365, 2000.

Venkatesh, Hillol Bala. "Technology Acceptance Model3 and a Research Agenda on Interventions [J]." Decision Sciences, vol.2, pp.273–315, 2008.

Viswanath Venkaesh, Fred D Davis. "A Theoretical Extension of the Technology Acceptance Model: Four Longitudinal Field Studies [J]." Management Science 200, vol.2, pp.186–204, 1998.

Webster J., Martocchio J.J. "Microcomputer Playfulness: Development of a Measure with Workplace Implications [J]." MIS Quarterly, vol.16, pp.201–226, 1992.

Yoojung Kim, Jongsoo Yoon, Seungbong Park and Jaemin Han. Architecture for Implementing the Mobile Government Services in Korea. [2010-05-21]. http://www.springerlink.com/content/6 jlg78wv4870wqgj/.

Yu Lixian, Wang Tingfang, "Understanding of Mobile electronic government [J]," Consume Guide, vol.23, pp.61–62, 2008.

Control Engineering and Information Systems – Liu (Ed)

Implementation and analysis of TD-LTE system level simulation in subway system

G.W. Bai & C.G. Cheng
China Railway Engineering Consultants Group Co., Ltd., Beijing, China

S.D. Zhou
State Key Laboratory on Microwave and Digital Communications, National Laboratory for Information Science and Technology, Tsinghua University, Beijing, China

ABSTRACT: The 3GPP LTE has established a worldwide standard of the 4th generation wireless communication system. Aimed at the utilization of TD-LTE in China, network operators tried to introduce TD-LTE in subway systems to provide higher transmission rate and reliable QoS. However, the deployment and performance in subway system are different compared with the traditional usage, both solutions and key techniques are still uncertain. In this work, we studied the scenario of subway transmission and established a simplified TD-LTE system level simulation platform used in subway system under matlab. A performance evaluation based on subway scenario of 3GPP LTE Release 10 has been done using this platform.

1 INTRODUCTION

Advances in personal wireless communication systems lead to the revolution of network architecture and key techniques. As candidates of the 4th generation of wireless communication systems, 3GPP LTE-Advanced [1] and IEEE 802.16 m [2] becomes more and more popular.

In China, TD-LTE becomes popular and be formed proprietary intellectual property rights. In 3GPP release 10 and later, the model of TD-LTE [3] has its unique advantage in resource management and interference cancellation rather than FDD method, which makes TD-LTE [4] more flexible.

Subway system is the important scenario of wireless personal communications. With rapid demand of multimedia content, users even located in subway area will require higher throughput and reliable QoS; however, current subway transmission is poor to maintain reliable QoS for the design of deployment and transmission. The deployment of TD-LTE provides a chance to extend high speed coverage to subway transmission scenario by introducing newly designed antennas and transmission schemes.

Considering the demand on evaluating system performance and transmission techniques, a system level simulation platform is in great need. The system level simulations mainly focus on the total system performance instead of the detail of transmission and link process, which is effective to

reduce complexity. In subway transmission system, the difficult is to build simulation scenarios and design flow chart of the simulation process.

In this work, by analyzing the demand and the real transmission scenario, we established a simplified system level simulation platform using Matlab. The simulation platform can support multi-station and one subway train working at 2 by 2 MIMO transmission modes. Both interference and transmission techniques are quite different from the standard listed in 3GPP TS 36.814 and related models.

The following papers are organized as follows. In part 2, we introduce the design of TD-LTE in subway transmission scenario. In part 3, the modeling and key technique are introduced, which is the fundamental of simulation platform. In part 4, we present the simulation result given by our simulation platform. The last part is the conclusion with acknowledgement followed.

2 SUBWAY TRANSMISSION SCENARIO

The specific content of subway TD-LTE transmission system contains subway base stations and trains respectively. Users located in subway areas have the characteristic of intensive traffic load, so traditional coverage will not satisfy the demand of interference management and transmission power.

The design of TD-LTE subway transmission system is given in Figure 1 provided by 3GPP partner

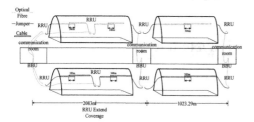

Figure 1. Illustration of subway transmission scenario. The design of transmission accessories such as BBU, RRU and transmission cables are presented in the figure.

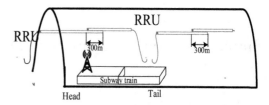

Figure 2. Illustration of subway train antenna. The TD-LTE antenna is deployed at the head of the train, transmit or receive signal from tunnel base station antennas.

specifications [5]. In this figure, the solution that distributed deployment using RRU is introduced. The subway train will equip LTE antennas and forward the signal to all subway cars through access points. The distributed deployment will effectively reduce energy consumption and interference to adjacent cells.

To avoid the interference from adjacent base stations, users will not get access to RRU located in the tunnel of the subway train, but using the remote access point. From system level point of view, the whole subway train is acting as a 'huge user' with high demand on throughput and QoS level. Users in the train car are acting as elements of the huge user, sharing bandwidth and resources. The access point that provides service to users in train car is just like the one located in edifice, with the same base station id as the station located at the head of the train.

The definition of base station is also a little bit different than traditional urban areas [6]. Each train station will establish a BBU working as a real base station, RRU is established in a distributed way connected using optical fiber as is shown in Figure 1.

The antenna is also quite different from traditional sectorized antenna. In Figure 2, the line connected the RRU is the newly designed antenna called leakage cable. The leakage cable is designed to use in tunnel area.

The leakage cable is usually used in tunnel area such as subway train, etc. By connecting the cable directly to the RRH unit, the cable is working as antennas. Along the cable, there exists small radius unit per average distance. The direction of the radiation is directing to the subway track. Interference from adjacent radius unit is controlled by the radius angle θ, respectively. The attenuation of the cable is defined by an attenuation factor γ, which is the function of distance. That means, when the distance is far from the RRH, the more attenuation will be.

In order not to lose coverage in the area of handover, the leakage cable is designed to overlap

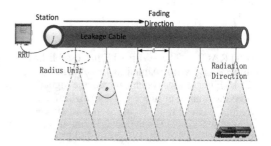

Figure 3. The design of leakage cable used in TD-LTE subway tunnel communication. The cable is directly connected to RRU and working as an antenna.

in a designed distance as is given in Figure 2. In current TD-LTE subway system, the distance is set to 300 m as a result of channel measurement [6]. In the overlapped area, adjacent base stations can work in both same and different carrier frequency, depending on the need of the system. It is quite different that the spectrum efficiency will not decrease obviously by adopting the frequency reuse technique. First, the design of leakage will only affect a short distance for the directional of radius unit. By optimizing the angle θ and the distance between radius unit d, there exists a value that may satisfy the system performance and reduce the interference.

The following part will introduce the principle of subway TD-LTE system level simulation platform.

3 ESTABLISHING SYSTEM LEVEL SIMULATION PLATFORM

Considering the different parameters and scenarios of subway TD-LTE system [7], we introduce our design of system level simulation used in subway scenario. However, we modified the system level simulation platform introduced by 3GPP and 802.16 m, instead of re-design the whole platform. Parameters and scenarios are modified to support TD-LTE subway system.

3.1 System structure

The system level simulation platform used in subway transmission systems are proposed to support multi base stations and 2~4 trains.

The flow chart of the platform is illustrated in Figure 4. Modified from 3GPP TS 36.814 [8], we established the platform by separating the process into large scale and small scale information. The large scale information of the system is mainly about the pathloss, and the small scale information is about Doppler caused by movement.

3.2 Channel model

In 3GPP specifications, the proposed channel model is SCM or SCME according to [8] and [9]. In subway transmission system, the channel model is simple for the use of leakage cable.

Other parameters such as number of antennas, paths, sub paths are simplified to match the scenario. In multi base station single train scenario, we now simplify the channel coefficient matrix according to specifications.

The channel from Tx antenna element s to Rx antenna u for cluster n is expressed as [8][9]:

$$H_{u,s,n}(t;\tau) = \sum_{m=1}^{M} \begin{bmatrix} F_{rx,u,V}(\phi_{n,m}) \\ F_{rx,u,H}(\phi_{n,m}) \end{bmatrix}^{T}$$
$$\begin{bmatrix} \alpha_{n,m,VV} & \alpha_{n,m,VH} \\ \alpha_{n,m,HV} & a_{n,m,HH} \end{bmatrix} \begin{bmatrix} F_{tx,s,V}(\varphi_{n,m}) \\ F_{tx,s,H}(\varphi_{n,m}) \end{bmatrix}$$
$$\times \exp\left(j2\pi\lambda_0^{-1}\left(\bar{\phi}_{n,m}\cdot\bar{r}_{rx,u}\right)\right)\exp\left(j2\pi\lambda_0^{-1}\left(\bar{\varphi}_{n,m}\cdot\bar{r}_{tx,s}\right)\right)$$
$$\times \exp\left(j2\pi\upsilon_{n,m}t\right)\delta\left(\tau-\tau_{n,m}\right) \quad (1)$$

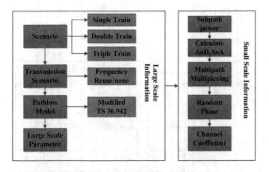

Figure 4. Illustration of subway train antenna. The TD-LTE antenna is deployed at the head of the train, transmit or receive signal from tunnel base station antennas. System structure design. The design is modified from 3GPP TS 36.814 [8], the typical deployment parameter of LTE/LTE-advanced system level simulation.

where $F_{rx,u,V}$ and $F_{rx,u,H}$ are the antenna element u field patterns for vertical and horizontal polarizations respectively [10], n,m,VV and n,m,VH are the gains of vertical-to-vertical and horizontal-to-vertical polarizations of ray n,m respectively, theta is the wave length of the carrier frequency, $\bar{\phi}_{n,m}$ is the AoD unit vector, $\bar{\varphi}_{n,m}$ is the AoA unit vector, $\bar{r}_{tx,s}$ and $\bar{r}_{rx,u}$ are the location vectors of element s and u respectively [11], and n,m is the Doppler frequency component of ray n,m. If the radio channel is modelled as dynamic, all the above mentioned small-scale parameters are time variant, i.e., they are functions of t.

In subway transmission, some parameters could be simplified: the number of paths can be squeezed to 2 due to the transmission environment of tunnels; the polarized antenna can also be canceled by replacing vertical and horizontal gain to $G_{rx,u}$ and $G_{tx,v}$. If the train uses the omnidirectional antenna, the gain of receive antenna can be also replaced by I. The modified channel gain can be written as:

$$H_{u,s,n}(t;\tau) = \sum_{m=1}^{2} G_{rx,u}(\phi_{n,m})^{T} G_{tx,v}(\varphi_{n,m})$$
$$\times \exp\left(j2\pi\lambda_0^{-1}\left(\bar{\phi}_{n,m}\cdot\bar{r}_{rx,u}\right)\right)\exp\left(j2\pi\lambda_0^{-1}\left(\bar{\varphi}_{n,m}\cdot\bar{r}_{tx,s}\right)\right)$$
$$\times \exp\left(j2\pi\upsilon_{n,m}t\right)\delta\left(\tau-\tau_{n,m}\right) \quad (2)$$

The small scale parameter of doppler frequency component caused by movement is calculated from the AOA, train speed v and direction of movement θ_v:

$$\upsilon_{n,m} = \frac{\|v\|\cos\left(\phi_{n,m}-\theta_v\right)}{\lambda_0} \quad (3)$$

After given the channel gain matrix of system level simulation, we hereby present the way to calculate the received SINR. The value is the index of QoS, leading to the decision of AMC and HARQ, etc. The method to get the SINR of target train is MMSE [12], which is widely used in signal detection. In this paper, we only use the conclusion and do not deduce it.

3.3 Adaptive modulation and coding scheme

The adaptive modulation and coding scheme is used to accommodate channel environment to perform different transmission decision. It is commonly the combination of modulation and coding scheme. In system level simulation, the process is commonly using the mapping method called effective SINR mapping, that means the final decision of combination of modulation and coding.

The function of AMC is to decide desired combination of modulation and coding scheme to perform the transmission, e.g. QPSK with 1/2 code rate. In Figure 5, we present the simulation result of AMC module under AWGN channel. We can infer that, with the incensement of system SINR, modulation level and code rate increase to make full utilization of good channel condition. The modulation order is set to BPSK, QPSK and 16QAM, with the code rate listed in Table 1. The combination of modulation order and code rate is provided in [9].

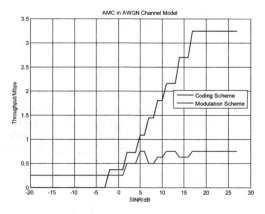

Figure 5. Simulation result of AMC module in AWGN channel. The SINR is the leading factor to decide modulation order and code rate.

Table 1. Simulation assumptions and parameters.

Name	Parameter
Cell layout	10 cell with 2 antenna
Inter-site distance	3000 m
Train configuration	2 antenna, 5 cars, 20 m per car
Train speed	60 Km/h in average
Transmission bandwidth	10 MHz
BS Tx power	38 dBm (6.3 watt)
Interval of radius unit	10 m
Angle of leakage cable	20 degree
Max re-transmission time	4
Carrier frequency	2 GHz
HARQ scheme	IR
Channel model	Simplified SCME
Pathloss	$pl(d) = d^{\alpha}$
Shadowing std	none
Noise power	−104 dBm
Service type	Full buffer
Simulation length	20 min
Train number	2
CQI measurement	Ideal
AMC	QPSK (R = {1/8, 1/7, 1/6, 1/5, 1/4, 1/3, 2/5, 1/2, 3/5, 2/3, 3/4, 4/5}) 16QAM (R = {1/2, 3/5, 2/3, 3/4, 4/5})
Target BLER	0.1

4 SIMULATION AND ANALYSIS

After a full analysis of system concepts and simulation environment, we have established the simplified system level simulation platform to support 20 stations and 4 trains at the most. The 2 by 2 MIMO mode is added to provide higher transmission rate. Train antenna is located at the head of the train, with 2 antennas. To make the simulation simple, we do not consider the stops of different stations, that means the train will run at a given speed unless the simulation end. The other parameters and assumptions not mentioned are configured the same as in 3GPP Urban Micro scenario.

4.1 Simulation scenarios and assumptions

In the simulation part, we simply evaluate the performance of multi station and one train scenario. Parameters and assumptions are given in Table 1. Parameters of subway trains are collected from Beijing subway train with 5 cars and each car is 20 meters long. By using simplified SCME channel model, the pathloss is set to exponential fading and shadowing is set to 0.

4.2 Performance analysis

In Figure 6, we simulated the SINR performance of the train. It is clear that the fading occurred

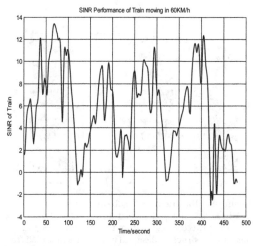

Figure 6. SINR performance of TD-LTE subway transmission system using 2 by 2 antennas.

when meeting the overlap of leakage cable, which reflects the influence of interference. The small fluctuation of the curve reflects the small scale information caused by moving.

We can assume that, if we deploy another antenna located at the end of the train, which may help improve the performance when reaching the point of handover.

5 CONCLUSION

In this paper, we develop and analyze the TD-LTE system level simulation platform used in subway transmission system. The method that provides TD-LTE coverage in subway system will greatly increase system performance. The centralized deployment of train antenna will gather traffic load from users all over the train car, which is effective to reduce interference and save energy. The handover position that the leakage cable overlaps will affect the system performance. So the design to deploy another antenna at the end of the train is sometimes necessary to enhance system performance.

ACKNOWLEDGEMENT

I would like to thank my advisor, Prof. Shidong Zhou, for his kind and wise suggestion to this work, I would also like to thank Mr. Chaogang Cheng, and he also contributes a lot.

REFERENCES

3GPP TR 36.814 V9.0.0 (2010-03), "Further advancements for E-UTRA physical layer aspects (Release 9)."

3GPP TR 36.913, "Requirements for further advancements for Evolved Universal Terrestrial Radio Access (E-UTRA) (LTE-Advanced)."

3GPP TSG RAN WG1 Meeting #56, Sharp, Considerations on precoding scheme for DL joint processing CoMP.

3GPP TSG RAN1#59bis, CoMP Support in Rel-10 with Coordinated Beamforming.

3GPP TSG-RAN WG1 #56, CMCC, Downlink CoMP-MU-MIMO transmission Schemes.

3GPP TSG-RAN-1 Meeting #37,OFDM-HSDPA System level simulator calibration (R1-040500).

3GPP, Available: http://www.3gpp.org/ftp/Specs/Draft IEEE 802.16 m Evaluation Methodology, Available: http://ieee802.org/16.

HTNG, Multi-Mobile Network Operator In-Building LTE Remote Radio Head Technical Requirements, Version 1.00, 3 April 2012.

Motorola, TD-LTE: Enabling New Possibilities and Revenues for Operators Maximizing adaptable DL:UL ratio and lower spectrum cost, Available Online: shttp://www.motorola.com/web/Business/Solutions/Industry%20Solutions/Service%20Providers/Network%20Operators/LTE/_Document/Static%20Files/TD-LTE%20 Apps%20 Solution%20 Paper%20-%20FINAL.pdf.

Motorola, TD-LTE: Exciting Alternative, Global Momentum, Available: http://www.motorola.com/web/Business/Solutions/Industry%20Solutions/Service%20Providers/Network%20Operators/_Documents/_static%20files/TD-LTE%20 White%20 Paper%20-%20FINAL.pdf.

Nokia Siemens, TD-LTE whitepaper, Available: http://www.nokiasiemensnetworks.com/system/files/document/TD-LTE_whitepaper_low-res_Online.pdf.

Overview of 3GPP Release 12 V0.0.2 (2012-01). Available: http://www.3gpp.org/ftp/Information/WORK_PLAN/Description_Releases/.

Tetsushi Abe. "3GPP Self-evaluation Methodology and Results", NTT DoCoMo, 2009.

Control Engineering and Information Systems – Liu (Ed)
© 2015 Taylor & Francis Group, London, ISBN 978-1-138-02685-8

Research on distributed cooperative routing based on network coding

Y. Hui, J.T. Fang & W. Liu
Wuhan Military Commissary Office of the Information Department, GSHQ of PLA, Wuhan, China

R.X. Zhang
Institute of National Defense Information, Wuhan, China

ABSTRACT: By using the network coding, the total throughput, reliability and safety of the network will be improved. In the paper, the advantages and disadvantages were analyzed, the problems of Cooperative routing combined with all sorts of Cooperative technologies and the advantages of distributed Cooperative routing based on network coding were described, and the key technologies of network coding combined with distributed routing were researched with emphasis that had great referenced meaning to accelerate the development of network coding.

1 INTRODUCTION

The open wireless communication method is used in the wireless mobile network. Due to the wireless channel has the following characteristics: the limited network bandwidth, the electromagnetic environment is bad, the channel fading, the collision probability is high, easy to be disturbed and invaded, while transmitting power, communication distance of each node in wireless mobile network is limited, the traditional distributed routing store and forward data according to the routes is very difficult to meet the requirements, in order to improve the throughput, reliability and the security of the entire wireless network, the distributed cooperative routing that have cooperative communication capabilities will be researched widely.

2 DISTRIBUTED ROUTING

In order to support the large-scale wireless mobile network and reduce the centralized computing the amount of computation, the distributed routing technology will be required to use the distributed routing technology can be adaptive to modify the original route when the network topology changes.

Distributed routing technology based on the drive mode is divided into periodic update to maintain the table-driven routing of the node routing table and demand-driven routing. The routing table entries whose timeliness is high and routing overhead is big are maintained by the table-driven

routing, demand-driven routing overhead is smaller but poor timeliness. The table-driven routing includes Destination Sequence Distance Vector (DSDV) routing, Clusters the first Gateway Swap (CGSR) routing, Wireless (WRP) routing, System Throughput Adaptive Routing (STARA) and Optimize Link (OLSR) routing, the demand-driven routing include Dynamic Source Routing (DSR) Algorithm, On-demand Distance Vector (AODV) routing and Time Order (TORA) routing. The hybrid routing combining the table-driven routing and the demand-driven routing uses the table-driven routing in the local scope to maintain the real time efficiency and use the demand-driven routing to reduce the routing overhead when the target node is far.

Distributed routing technology to better support wireless dynamic topology of the network, but with the development of wireless networks, network scale is growing increasingly, Network load is getting heavier and heavier, and the electromagnetic environment is getting worse, while network bandwidth resources and power are limited. The routing technology based on a simple table-driven routing and demand-driven is difficult to adapt. When the network load is heavy, for the table-driven routing technology, the nodes frequently search routing while the timeliness losses, it is difficult to guarantee service quality and reliability of the network. For demand-driven routing technology, due to the expansion of network scale, the network overhead corresponding exponentially increased, thus the actual payload throughput of the network accordingly.

3 COOPERATIVE ROUTING

In cooperative wireless communication, collaborative technology can significantly improve the network quality of service, expand network coverage, save nodes energy, increase network capacity and improve the reliability and security of the information. The cooperative technology include the cooperative multi-antenna technology, the cooperative diversity multiplexing technology, the cooperative space-time encoding technology, the cooperative network coding technology, the cooperative radio resource allocation technology, the cooperative power allocation technology, these cooperative technologies and routing technologies combine to form the cooperative routing technology that is better than the original routing technology.

In [1], new multihop cooperative protocol was proposed using the space-time codes for the purpose of energy savings, subject to a required outage probability at the destination, and two efficient power allocation schemes were derived, which depended only on the statistics of the channels. In [2], a distributed Throughput-Optimized Cooperative Routing (TOCR) algorithm based on the Adaptive Forwarding Cluster Routing (AFCR) algorithm and the cooperative routing technology was proposed to improve network throughput in the mobile Ad Hoc network. Researchers in [3] proposed an approach called PC-CORP (Power Control based Cooperative Opportunistic Routing Protocol) for WSN (Wireless Sensor Networks), providing robustness to the random variations in network connectivity while ensuring better data forwarding efficiency in an energy efficient manner. Based on the realistic radio model, the region-based routing, rendezvous scheme, sleep discipline and cooperative communication were combined together to model data forwarding by cross layer design in WSN. At the same time, a lightweight transmission power control algorithm called PC-AIMD (Power Control Additive Increase Multiplicative Decrease) was introduced to utilize the co-operation of relay nodes to improve the forwarding efficiency performance and increase the robustness of the routing protocol. In [1], a Decentralized Weighted Cooperative Routing Algorithm (DWCRA) through opportunistic relaying was proposed with the weighted metric consisting of remaining energy of the relays and the channel state information between nodes which can obtain a Quality of Service (QoS) tradeoff between MRE and OCS in terms of the delivery ratio and network lifetime.

The cooperative trellis coding process the information accepted at the intermediate nodes, and then sent out, while the receiving nodes restore the information. Through network coding at the intermediate nodes, not only the overall network throughput is improved, but also there are great potential for development in terms of reliability and security. The cooperative trellis coding usually is implemented in the above the network layer protocols, so the network hardware and the corresponding protocol are not greatly modified.

Due to the traditional routing considers the transmission information in the network can not be superimposed, nodes only store and forward, resulting in subsequent node on the path are not the maximum throughput of transmission, the transmission capacity of such networks is far less than the maximum network throughput, and wireless network with broadcast properties, each node between the collision out probability is much larger than the wired network, need to improve the performance of cooperative network coding in wireless mobile networks. The cooperative routing combining routing and cooperative network coding together can make full use of advantages of network coding, make the network throughput reach its maximum throughput of the theory, reduce transmission times in the network, save all kinds of resources, strengthen the security of information hiding, balance network load and improve the robustness of the network.

4 THE COOPERATIVE ROUTING OF NETWORK CODING

4.1 *Analysis of network coding technology*

In the network topology, and not all nodes need network coding, the part of the nodes are dynamically selected to give the network coding function according to the routing information for every network node, and the nodes without network coding store and forward only, so that can reduce the algorithm complexity and hardware requirements. For the wireless mobile network, due to the topology changes frequently, so the nodes need network coding are selected considering various factors of global network topology, the routing congestion, the node computing ability, link service quality and other various factors.

According to the different implementations, network coding can be divided into linear network coding and random network coding. The linear network coding node information transmit linearly mapped to a finite domain, using the linear relationship to the encryption process, the row vector of the k-dimensional information is issued by the source node, the destination node according to the global coding matrix and receiving a plurality of the k-dimensional data decode information of the k-dimensional row vectors included in each dimension. The linear network coding requires all nodes in the global coding vectors must be linearly independent, ie the intermediate nodes are aware of the global encoding vector. The random network coding overcomes the disadvantages that the nodes need to know the entire network topology and the encoding process of other nodes, and proposes the intermediate nodes simply randomly select in the limited domain digital, when Allocating the encoding vector, and update the coding coefficients do not need to verify whether the relevant.

The network coding can greatly improve the throughput of the network, but the network coding for error rate requirements are very harsh, less error rate to ensure the validity and reliability of network coding, so it needs to design a reliable guarantee of network coding with low bit error rate in data transmission, and adopt appropriate techniques to reduce the impact on network coding error rate.

4.2 Analysis of collaborative routing technology

The basic principles of the network coding is to send the information Encoded by a plurality of input information, eliminate the queuing delay of the node and therefore can achieve the shared network resource of the plurality of receiving nodes in the multicast. For the distributed routing in order to reduce the queuing delay to minimize cross paths, and for network coding may not need to avoid cross paths, but it can be used to improve the utilization rate of nodes to achieve using less relay nodes to achieve multicasteffect, and each receiver node must be received to the k-dimensional irrelevant information can be correctly decoded, so the routing of the network coding is different from the traditional routing.

The distributed routing is needed to meet the requirements of rapid convergence, can fast reconstruction available routing control overhead, etc. The distributed routing includes path generation, path selection and path maintenance three core functions. The path generation with table driven routing uninterrupted sends a route request message is not suitable for cooperative routing network coding. The demand-driven routing according to the need to find a route to the destination node, and increase the network coding opportunities field in the sending route request broadcast message, each node cache shortest path set of records may be through the node and its neighbor nodes flow, can immediately send the data in the cache after receiving the routing request. For the path selection, the cross paths are selected according to

the characteristic of network coding, and need to ensure that the K irrelevant path uniformity of the routing path. The route maintenance periodically broadcasting routing request packets between two adjacent nodes exchange their neighbor information for routing maintenance, but also need to maintain the network coding node.

5 CONCLUSION

With the massive growth of the wireless mobile communication business needs, the wireless mobile network is facing a serious shortage of throughput. This paper briefly analyzes the adaptive distributed routing in wireless mobile networks, unable to adapt to the large capacity requirements of wireless mobile network, and the use of cooperative multiple antenna technology, cooperative diversity and multiplexing technology, space-time coding, cooperative radio resource allocation techniques, cooperative power allocation technology, collaborative technology and routing are combined to form a cooperative routing only partially to improve the throughput of network.

The network coding allows the throughput of the entire wireless network to reach its theoretical maximum throughput, but need to solve the impact of the determination of network coding node, the design of random network coding and reduce the error rate for network coding. Need to improve routing application protocol in the cooperative routing, and need to improve network coding efficiency using cross paths in the path selection. This paper discusses the distributed collaborative routing based on network coding, with reference to the development of application and further promote the distributed cooperative routing.

REFERENCES

Hang Wansheng, Liu Kai, Wang LI. 2011. Distributed cooperative routing algorithm for mobile AD Hoc network. Journal of Xidian University, vol 38:34–39.

Hu Haifeng, Zhu Qi. 2009. Power Control Based Cooperative Opportunistic Routing in Wireless Sensor Networks. Journal of Electronics (China). Vol 26:52–63.

Ji Yimu, Zong Ping, Chen Chao. 2009. Weighted Cooperative Routing Algorithm through Opportunistic Relaying. Systems Engineering and Electronics. Vol 31:2728–2746.

Maham B, Narasimhan R, Hjorungnes A. 2009. Energy-efficient Space-time Coded Cooperative Routing in Multihop Wireless Networks, Global Telecommunications Conference: 1–7. Honolulu: IEEE Press.

Control Engineering and Information Systems – Liu (Ed)
© 2015 Taylor & Francis Group, London, ISBN 978-1-138-02685-8

A hybrid approach for performance assessment in mobile commerce

W. Liu
Jiangxi Key Laboratory of Data and Knowledge Engineering, Institute of Information Resources Management, Jiangxi University of Finance and Economics, Nanchang, P.R. China

ABSTRACT: This paper proposes an effective fusion of Analytic Hierarchy Process (AHP) and Grey Relational Analysis (GRA) approach for the risk evaluation in Mobile Commerce (MC) development. The hybrid method employs the complementary strength of these two appealing techniques, in which, the criteria weight obtained by applying AHP approach, and the GRA approach used to get the overall performance measure of MC risk. At last, experimental results are shown in this paper to verify the effectiveness of the hybrid model.

1   INTRODUCTION

With the vigorous global development of the wireless telecommunications and cellular phones sector, mobile commerce (m-commerce) is now attracting significant interest. Today, MC and related strategy concepts are promoted as important components for organizations to survive[1,2]. Although abundant research on the technology side of mobile commerce has been published, managers confront with the difficulty of the decisions of what and how to implement MC for attaining the required performance in a turbulent world. And although MC has many distinct characteristics, and offers various opportunities, we can easily find that some fall into a predicament and face enormous risks, even fail ultimately. Of course, there are many factors responding for this problem. However, a task that is critical to the proper management of MC development is the assessment of risk.

To date, there has not been much research focused on identifying the potential risk factors that threaten successful MC development. The degree of MC risk depends on several factors including technical, organizational, and content risk and so on. Since it is impossible to eliminate the risk, it may be hoped that perceived MC risk can be reduced if advance warning can be obtained through risk evaluation. So, we believe that it is necessary to have a comprehensive risk management for MC development.

There are many risk analysis techniques currently in use that attempt to evaluate and estimate risk. These methods have their own advantages and disadvantages. But MC development is relatively new to most companies, and only limited information is available on the associated risks. These risk factors are complex and usually interdependent on each other. So, this study will apply an Evidential reasoning approach for risk evaluation in MC development. In the process, the hybrid approach of AHP and GRA is used to model the uncertainties inherent in MC risk evaluation and to aggregate the assessments of MC risk elements and components. The advantages of the new model pave a new way for risk assessment in MC development.

The rest of this paper is organized as follows. Section 2 briefly reviews the background of AHP and GRA approach, and presents the proposed hybrid approach for risk assessment. In section 3, we evaluate the performance of the proposed hybrid approach with a substantial number of experiments. Finally, section 4 ends the paper with some concluding remarks.

2   RISK EVALUATION METHODOLOGY

2.1   *Evaluation aspects and criteria*

On a comprehensive literature review, this paper proposes a risk taxonomy showing a hierarchical risk-breakdown structure associated with MC development. This breakdown classifies the risk into three dimensions, namely, technical, organizational, and environmental risk. In our evaluation model, we established the evaluation criteria of MC's risk, which is shown in Figure 1.

2.2   *Analytic Hierarchy Process (AHP)*

The Analytic Hierarchy Process (AHP) was devised by Saaty[6]. It is a useful approach to solve complex decision problems. It prioritizes the relative importance of critical factors and sub-factors through

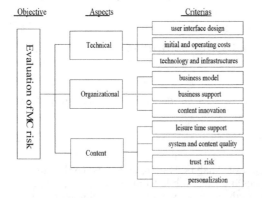

Figure 1. Taxonomy of risk analysis in MC development.

pairwise comparisons amongst the factors by relevant experts. After construction of the hierarchy layer structure, the AHP method is processed using the following stages:

1. Establish the pair-wise comparison matrix P.
   AHP uses pairwise comparison to allocate weights to the elements of each level, measuring their relative importance by using Saaty's 1–9 scale, and finally calculates global weights for assessment at the bottom level.
2. Calculate the eigenvalue and eigenvector of each pair-wise comparison matrix W.

$$W = (W_1, W_2, ..., W_m)^T \qquad (1)$$

where

$$W_i = \frac{W_i'}{\sum_{i=1}^{m} W'}, \quad W_i' = \sqrt[m]{\prod_{j=1}^{m} P_{ij}}, \quad i = 1, 2, ..., m$$

3. Test the consistency of each comparison matrix.
   AHP calculates a Consistency Ratio (CR) to verify the coherence of the judgments, which must be about 0.10 or less. CI denotes the consistency indicator of the level total sequence with RI being the mean random consistency indicator of the level total sequence. The expressions are as follows.

$$CI = \frac{\lambda_{max} - m}{m - 1}, \quad CR = \frac{CI}{RI} \qquad (2)$$

where λ_{max} the largest eigenvalue of judgment matrix P, m is is the rank.

4. Estimate the relative weights and the global weights of the elements of each level.

2.3 Grey Relational Analysis (GRA)

Grey system theory, initiated by Deng[7], can perform grey relational analysis for sequences. The grey relational analysis is used to explore the qualitative and quantitative relationships among abstract and complex sequences and to capture their dynamic characteristics during the development process. It can make use of relatively small data sets and does not demand strict compliance to certain statistical laws, simple or linear relationships among the observable variables. It can analyze a grey system that is of poor, incomplete and with uncertain information. This technique has been successfully applied in many social and technology issues, including design optimization, performance evaluation and factor effect evaluation.

For a given reference sequence and a given set of comparative sequences, grey relational analysis can be used to determine the relational grade between the reference and each element in the given set. Then the best comparative one can be found by further analyzing the resultant relational grades. In other words, grey relational analysis can be viewed as a measure of similarity for finite sequences.

Let x_0 denote the referential series with entities,

$$x_0 = [x_0(1), x_0(2), x_0(3), ..., x_0(n)] \qquad (3)$$

And let represent the compared series,

$$x_i = [x_i(1), x_i(2), x_i(3), ..., x_i(n)]; i = 1, 2, 3, ..., m \qquad (4)$$

The procedures of modeling grey relational analysis are as follows:

Step 1: Dimensionless processing in order to remove anomalies associated with different measurement units and scales. The initial-value processing and the average-value processing are two methods that are widely used in GRA.

The grey relational grade for series to is then given as

Step 2: Calculation of the relational coefficient $L_{0,i}$, expressing the relative distance between two factors:

$$L_{0,i} = \frac{\Delta_{min} + \Delta_{max}}{\Delta_{0,1}(k) + \rho \Delta_{max}} \qquad (5)$$

where $\Delta_{0,1}(k) = |x_0(k) - x_i(k)|$ $\Delta_{max} = \max\max \Delta_{0,1}$ (k), $\Delta_{min} = \min\min \Delta_{0,1}(k)$ and ρ is a predefined constant which is usually set to 0.5. The value of $L_{0,i}$ is always no larger than 1.

Step 3: Calculation of grey relational grade, assuming each point in a sequence of equal weight:

The larger value implies that the series pair $<x_0, x_i>$ has tighter relation. So, the relation factors between any series pair $<x_0, x_i>$, denoted as r_{0i}

$$r_{0i} = \frac{1}{n}\sum_{k=1}^{n} r\big(x_0(k), x_i(k)\big) \qquad (6)$$

Step 4: Data Normalization

Before calculating the grey relational grades, the series data can be normalized by the equation of lower-bound effectiveness of measurement (smaller-the-better).

$$x_i(k) = \frac{\max x_i(k) + x_i(k)}{\max x_i(k) + \min x_i(k)} \qquad (7)$$

Step 5: Ranking the grey relational grade.

The r_{0i} show quantitatively how closely the series x_i is related to the reference series x_0. Generally speaking, $r > 0.9$ indicates a notable influence, $r > 0.8$ a less notable influence, $r > 0.7$ a noticeable influence and $r > 0.6$ a negligible influence.

3 NUMERICAL EXAMPLE

A performance evaluation study has been conducted on three enterprises (A, B, C), which deal with Mobile Commerce in China. At the same time, we chose ten experts including manager and technical advisor as evaluator.

3.1 The application of AHP

The AHP weighting is mainly determined by the decision-makers, who conduct the pairwise comparisons, if there are n evaluation criteria, then while deciding the decision-making the decision-makers have to condusect $C(n,2) = n(n-1)/2$ pairwise comparisons. Figure 1 shows the evaluation hierarchy structure of MC risk factors. There are three evaluation criteria in the objective level of 'technical', including 'user interface design', 'initial and operating costs' and 'technology and infrastructures'. Then the evaluation measurement of ratio scale is employed to conduct pairwise comparison to clarity the relative importance of each attribute. Therefore, the comparison has to make for three times. To have a further explanation of the compassion, the evaluators would make the comparison between that of the importance of 'user interface design' and 'technology and infrastructures', 'initial and operating costs' and 'technology and infrastructures', At last, make the comparison between 'user interface design' and 'initial and operating costs'. By means of the comparative importance derived from the pairwise comparisons allows a certain degree of inconsistency within a domain. We should use the principal eigenvector of the pairwise comparison matrix to find the comparative weight among the criteria. The AHP

method should be an exact measure of the difference of attribute preference for MC developer and results of this approach are better than the others. For the reasons, this study utilizes AHP method to evaluate the criteria weights of MC risk.

After finding the relative weights of the attributes, the relative weights of the sub-attributes are found out in order to obtain the global weights of all sub-criterias. Figure 2 shows the relative weights of the three aspects of MC risk, which are obtained by applying AHP. The weights for each of the aspect are: technical (0.330), organizational (0.336), and environmental (0.334).

3.2 Application of GRA

From the criteria weights obtained from AHP, the performance of alternatives corresponding to each evaluation criterion uses the methodology of smaller-the-better, as in (7), for the normalization purpose. Table 1 lists how each of the three MC systems compares for all criterias.

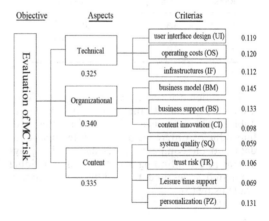

Figure 2. Weights of the criterias.

Table 1. Data for the operational criteria of MC risk.

Evaluation criteria	A	B	C
User interface design	6.78	5.78	4.63
Initial and operating costs	7.93	4.81	8.02
Technology and infrastructures	9.2	9.15	7.15
Business model	9.1	8.6	5.92
Business support	9.5	8.86	7.94
Content innovation	9.63	8.7	7.63
Leisure time support	8.21	7.35	7.13
System and content quality	8.84	7.12	8.6
Trust risk	9.32	7.6	8.78
Personalization	9.51	8.35	7.72

Table 2. Difference series and grey relational grade.

Evaluation criteria	A	B	C
User interface design	0	0.302	0.824
Initial and operating costs	0.635	0.76	0
Technology and infrastructures	0	0	1.14
Business model	0	0.402	0.823
Business support	0.261	0	1
Content innovation	0	0.376	0.534
Leisure time support	1.6	0	1
System and content quality	0	0.298	0.806
Trust risk	0.217	0.508	0.905
Personalization	0	0.493	0.908
$\Delta_{0,1}(k)$	0.2231	0.6014	0.6023
$L_{0,1}(k)$	0.8267	0.5642	0.6413
Normalized $L_{0,1}(k)$	0.4247	0.2624	0.2901

Table 3. Final ranking of three enterprise.

Rank	Enterprise	Risk ranking
1	A	0.3987
2	C	0.2746
3	B	0.2315

The difference series, $\Delta_{0,1}(k)$, grey relational grades $L_{0,i}$, and all the values are shown in Table 2. The difference series $\Delta_{0,1}(k)$ is the absolute value of difference been the reference series data and the compared series data and it plays a vital role for finding the $\Delta_{0,1}(k)$ of individual compared series and to find the grey relation grade using (5).

3.3 Final ranking

The overall comparison of risk ratings using the integrate method of the AHP and GRA are shown in Table 2. We use AHP method in obtaining criteria weight, and apply GRA to assess the final risk in MC development. The MC development risk evaluation results can be seen in Table 3. The highest $L_{0,i}$ value shows the highest risk. So, it can be seen that the company A has the highest MC risk, since it has the highest weight of (0.3987) among the three companies.

4 CONCLUSIONS

MC development takes place in a complex and dynamic environment that includes high levels of risk and uncertainty. This study has outlined an approach to the assessment of the risks associated with MC development using the hybrid approach of AHP and GRA. This hybrid model was proposed to assist MC project managers and decision makers to think and assess the current risk environment of their MC development in a more systematic manner than before.

In this paper, we used the AHP rule and the concept of hierarchical structure to make the pairwise comparison among elements. On other hand, we use the GRA approach to rank the final MC risk of each enterprise. Therefore, according to the result of the empirical study, this hybrid approach is better than the traditional statistic approach.

ACKNOWLEDGMENTS

This research was supported by The National Social Science Foundation of China under Grant 13BTQ059 and Technology Foundation of Jiangxi Provincial Department of Education under Grant GJJ12730 and GJJ12267.

REFERENCES

Deng J.L. Introduction to grey system theory. Syst Control Lett 5(1), (1989), 1–24.

Feng H., Hoegler T. & Stucky W. "Exploring the Critical Success Factors for Mobile Commerce," Proceedings of the International Conference on Mobile Business (ICMB'06), 2006.

Maltz A.C., Shenhar A.J. & Reilly R.R. Beyond the balanced scorecard: refining the search for organizational success measures. Long Range Planning, 36(2), (2003), 187–204.

Martensson M. A critical review of mobile commerce as a management tool. Journal of Mobile commerce, 4(3), (2000), 204–216.

Ngai E.W.T. "review for mobile commerce research and applications," Decision Support Systems, Vol. 43 (3), pp. 3–15, 2007.

Ngai E.W.T. & Chan E.W.C. Evaluation of mobile commerce tools using AHP. Expert Systems with Applications, 29(4), (2005), 889–899.

Saaty T.L. The analytic hierarchy process: planning, priority setting. New York: McGraw Hill International Book Co., 1980.

Session 4: Electrical engineering

Harmonic characteristic of current source inverters based on double Fourier integral theory

J. Bai, J. Liu & S.Q. Lu

Electrical and Information Engineering Department, Beihua University, Jilin, China

ABSTRACT: A mathematical model of Current Source Inverters (CSIs) based on double Fourier integral theory is proposed in this paper because of the applications of CSIs in high-power medium-voltage. Total Harmonic Current Distortion (THD$_i$) and fundamental current are calculated and phase current spectrum is obtained by this model. Simultaneously, harmonic characteristics of current source inverters for different modulation strategies are analyzed in detail as well. This paper finds that current source inverters have the same harmonic characteristic when compared to Voltage Source Inverters (VSIs).

1 INTRODUCTION

The With the development of power semiconductor components and practical application of Superconducting Magnetic Energy Storage (SMES) (Luongo 1996, Karasik et al. 1999) technology, Insulated Gate Bipolar Transistors (IGBT) and Integrated Gate Commutated Thyristors (IGCT) rapidly replace Gate Turn-Off thyristors (GTO) in high-power applications which lead to larger capacity and better stability of Current Source Inverters (CSIs) (Wiechmann et al. 2008). In the meanwhile, SMES system can improve the energy storage inductor efficiency of CSIs. CSIs are found in a number of industrial processes such as high-power adjustable-speed drives, where four-quadrant operation, near-sinusoidal motor terminal voltages, supply voltage variations immunity, and inherent short-circuit protection are required (Lopes & Naguib 2009). CSIs could be a good alternative to Voltage Source Inverters (VSIs) in wind energy power generation and High Voltage Direct Current (HVDC) transmission systems (Shang & Li 2013). Therefore, CSIs are equipped with unique superiority in the motor adjustable-speed system, power system and so on and considered to have a broad development prospects and great market value.

2 MATHEMATICAL MODEL OF CSIS

Double Fourier integral theory is a mathematical tool to analyze harmonic characteristic of all kinds of inverters. Because CSIs have two time variables which are carrier wave and reference wave, a two variable function $f(x,y)$ can be expressed as a summation of harmonic components according to double Fourier integral theory. The formula is given by

$$f(x,y) = \frac{A_{00}}{2} \text{ (DC Offset)}$$

$$+ \sum_{n=1}^{\infty}(A_{0n}\cos ny + B_{0n}\sin ny)$$

(FundamentalComponent and Baseband Harmonics)

$$+ \sum_{m=1}^{\infty}(A_{m0}\cos mx + B_{m0}\sin mx)$$

(Carrier Harmonics)

$$+ \sum_{m=1}^{\infty}\sum_{\substack{n=-\infty \\ (n\neq 0)}}^{\infty}\left[A_{mn}\cos(mx+ny) + B_{mn}\sin(mx+ny)\right]$$

(Sideband Harmonics)

(1)

where

$$A_{mn} = \frac{1}{2\pi^2}\int_{-\pi}^{\pi}\int_{-\pi}^{\pi}f(x,y)\cos(mx+ny)dxdy \quad (2)$$

$$B_{mn} = \frac{1}{2\pi^2}\int_{-\pi}^{\pi}\int_{-\pi}^{\pi}f(x,y)\sin(mx+ny)dxdy \quad (3)$$

or in complex form

$$C_{mn} = A_{mn} + jB_{mn}$$

$$= \frac{1}{2\pi^2}\int_{-\pi}^{\pi}\int_{-\pi}^{\pi}f(x,y)e^{j(mx+ny)}dxdy \quad (4)$$

where $x(t) = \omega_c t + \theta_c$, $y(t) = \omega_0 t + \theta_0$, ω_c is carrier angular frequency, θ_c is phase offset angle for carrier waveform, ω_0 is fundamental angular frequency, θ_0 is phase offset angle for fundamental waveform.

3 ANALYSIS OF SINGLE-PHASE HALF-BRIDGE CICS

Figure 1 shows a single-phase half-bridge CSI. Switches S_1, S_2 control the direction of current; switches S_3, S_4 prevent open circuit of current source during change the direction of current. As shown in Figure 1, M is modulation index ranging from 0 to 1, ω_0 is output current angular frequency, θ_0 is phase offset angle for output current waveform, ω_c is carrier angular frequency, θ_c is phase offset angle for carrier waveform. (These symbols are used throughout this paper.)

Analysis of sine-triangle naturally sampled PWM of the single-phase half-bridge CSI is as follows:

$$A_{mn} + jB_{mn}$$

$$= \frac{1}{2\pi^2} \int_{-\pi}^{\pi} \int_{-\pi}^{-\frac{\pi}{2}(1+M\cos y)} -I_{dc} \cdot e^{j(mx+ny)} dx dy$$

$$+ \frac{1}{2\pi^2} \int_{-\pi}^{\pi} \int_{-\frac{\pi}{2}(1+M\cos y)}^{\frac{\pi}{2}(1+M\cos y)} I_{dc} \cdot e^{j(mx+ny)} dx dy$$

$$+ \frac{1}{2\pi^2} \int_{-\pi}^{\pi} \int_{\frac{\pi}{2}(1+M\cos y)}^{\pi} -I_{dc} \cdot e^{j(mx+ny)} dx dy$$

(5)

harmonic components of phase current is calculated by

$$I(t) = M \cdot I_{dc} \cdot \cos(\omega_0 t + \theta_0)$$

$$+ \frac{4I_{dc}}{\pi} \sum_{m=1}^{\infty} \frac{1}{m} J_0\left(m\frac{\pi}{2}M\right)$$

$$\times \sin\left(\frac{\pi}{2}m\right)\cos m(\omega_c t + \theta_c) + \frac{4I_{dc}}{\pi}$$

$$\times \sum_{m=1}^{\infty} \sum_{\substack{n=-\infty \\ n\neq 0}}^{\infty} \frac{1}{m} J_n\left(m\frac{\pi}{2}M\right) \sin\left[(m+n)\frac{\pi}{2}\right]$$

$$\times \cos\left[m(\omega_c t + \theta_c) + n(\omega_0 t + \theta_0)\right]$$

(6)

Figure 2 shows the phase current spectrum for sine-triangle naturally sampled PWM when $M = 0.9$ and $f_c/f_0 = 21$.

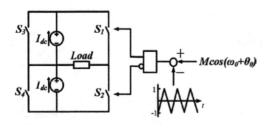

Figure 1. A single phase half-bridge CSI.

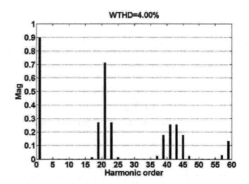

Figure 2. The phase current spectrum for sine-triangle naturally sampled PWM when $M = 0.9$ and $f_c/f_0 = 21$.

All harmonic components have been normalized with the magnitude of output fundamental when $M = 1$ in this paper in order to make the comparison of harmonic components for different topologies easier. All data have been normalized with I_{dc} in Figure 2 and the calculated value for Weighted Total Harmonic Current Distortion (WTHD$_i$) is 4.00%.

Figure 2 shows the fundamental component, carrier harmonics and sideband harmonics by the modulation process under the circumstance of naturally sampled when the carrier ratio is 21 and the modulation index is 0.9. A particular characteristic has been observed that the baseband harmonics don't produced. The even carrier harmonics are eliminated completely due to the existence of $\sin((\pi/2)m)$ in (6). In the meanwhile, because of the existence of $\sin[(m+n)\pi/2]$ in (6), the odd sideband harmonics around odd multiples of the carrier fundamental and even sideband harmonics around even multiples of the carrier fundamental are also eliminated completely.

It is worth mentioning that in order to satisfy the requirement of switch function, the single-phase half-bridge circuit must add switch S_3, S_4. Furthermore, the trigger signal of S_1, S_3 must be the same and the trigger signal of S_2, S_4 should also be the same. Therefore, this circuit doesn't have much value in practical applications. However, this topology is the basic unit of three-phase CSI and harmonic components of three-phase CSIs can be calculated based on this topology. In this regard, this circuit is also significant for the calculation of current harmonic components of the three-phase bridge topology.

4 ANALYSIS OF THREE-PHASE CICS

Figure 3 shows a three-phase CSI, three bridges modulate by the same carrier and three sinusoidal references which displaced in time by 120°. Because

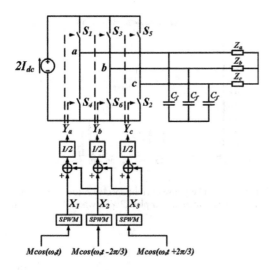

Figure 3.   A three-phase CSI.

Table 1.   Switch function of the CSI.

Bridge			Upper switches			Lower switches		
Y_a	Y_b	Y_c	S_1	S_3	S_5	S_4	S_6	S_2
1	−1	0	1				1	
1	0	−1	1					1
0	1	−1		1				1
0	−1	1			1	1		
−1	1	0		1		1		
−1	0	1			1	1		
0	0	0	1				1	
0	0	0		1				1
0	0	0			1	1		1

θ_c doesn't affect the magnitude of harmonic components but only phase offset angle, θ_c is set to be zero for easier calculation.

Analysis of sine-triangle naturally sampled PWM of the three-phase CSI is as follows.

At any instant of time, only two switches including the upper switch and the lower switch are conducting. The requirement of switch function is listed in Table 1.

Thus

$$\begin{bmatrix} Y_a(t) \\ Y_b(t) \\ Y_c(t) \end{bmatrix} = \frac{1}{2} \begin{bmatrix} 1 & -1 & 0 \\ 0 & 1 & -1 \\ -1 & 0 & 1 \end{bmatrix}$$

$$\times \begin{bmatrix} X_1(t) \\ X_2(t) \\ X_3(t) \end{bmatrix} \quad \text{(Wang \& Ooi 1993)} \quad (7)$$

$X_1(t)$, $X_2(t)$, $X_3(t)$ can be calculated by (6). Bilogic $X(t)$ can be converted into trilogic $Y(t)$ by (7) so as to satisfy the requirement of switch function.

$$Y_a(t) = \frac{1}{2}[X_1(t) - X_2(t)]$$

$$= \frac{\sqrt{3}}{2} M \cdot I_{dc} \cdot \cos\left(\omega_0 t + \frac{\pi}{6}\right)$$

$$+ \frac{4 I_{dc}}{\pi} \sum_{m=1}^{\infty} \sum_{\substack{n=-\infty \\ n \neq 0}}^{\infty} \frac{1}{m} J_n\left(m\frac{\pi}{2}M\right)$$

$$\times \sin\left[(m+n)\frac{\pi}{2}\right] \sin\left(n\frac{\pi}{3}\right)$$

$$\times \cos\left[m(\omega_c t) + n\left(\omega_0 t - \frac{\pi}{3}\right) + \frac{\pi}{2}\right] \quad (8)$$

$$Y_b(t) = \frac{1}{2}[X_2(t) - X_3(t)]$$

$$= \frac{\sqrt{3}}{2} M \cdot I_{dc} \cdot \cos\left(\omega_0 t - \frac{\pi}{2}\right)$$

$$+ \frac{4 I_{dc}}{\pi} \sum_{m=1}^{\infty} \sum_{\substack{n=-\infty \\ n \neq 0}}^{\infty} \frac{1}{m} J_n\left(m\frac{\pi}{2}M\right)$$

$$\times \sin\left[(m+n)\frac{\pi}{2}\right] \sin\left(n\frac{\pi}{3}\right)$$

$$\times \cos\left[m(\omega_c t) + n(\omega_0 t - \pi) + \frac{\pi}{2}\right] \quad (9)$$

$$Y_c(t) = \frac{1}{2}[X_3(t) - X_1(t)]$$

$$= \frac{\sqrt{3}}{2} M \cdot I_{dc} \cdot \cos\left(\omega_0 t + \frac{5\pi}{6}\right)$$

$$+ \frac{4 I_{dc}}{\pi} \sum_{m=1}^{\infty} \sum_{\substack{n=-\infty \\ n \neq 0}}^{\infty} \frac{1}{m} J_n\left(m\frac{\pi}{2}M\right)$$

$$\times \sin\left[(m+n)\frac{\pi}{2}\right] \sin\left(n\frac{\pi}{3}\right)$$

$$\times \cos\left[m(\omega_c t) + n\left(\omega_0 t + \frac{\pi}{3}\right) + \frac{\pi}{2}\right] \quad (10)$$

harmonic components of phase current is calculated by

$$I_a(t) = Y_a(t) - Y_b(t) - Y_c(t)$$

$$= \sqrt{3}M \cdot I_{dc} \cdot \cos\left(\omega_0 t + \frac{\pi}{6}\right)$$

$$+ \frac{8I_{dc}}{\pi} \sum_{m=1}^{\infty} \sum_{\substack{n=-\infty \\ n\neq 0}}^{\infty} \frac{1}{m} J_n\left(m\frac{\pi}{2}M\right)$$

$$\times \sin\left[(m+n)\frac{\pi}{2}\right]\sin\left(n\frac{\pi}{3}\right)$$

$$\times \cos\left[m(\omega_c t) + n\left(\omega_0 t - \frac{\pi}{3}\right) + \frac{\pi}{2}\right] \quad (11)$$

Figure 4 shows the phase current spectrum for sine-triangle naturally sampled PWM when $M = 0.9$ and $f_c/f_0 = 21$. The data in Figure 4 have been normalized with $\sqrt{3}I_{dc}$. WTHD$_i$ = 2.04%, Total Harmonic Current Distortion (THD$_i$) equals 60.16%. Fundamental I$_1$ = 1.559.

As shown in Figure 4, all the carrier harmonics of phase current are eliminated completely because the carrier harmonics of each bridge leg are the same. Since the existence of $\sin[(m+n)\pi/2]$ in (11), the odd sideband harmonics around odd multiples of the carrier fundamental and even sideband harmonics around even multiples of the carrier fundamental are eliminated completely. Moreover, three sinusoidal references which displaced in time by 120° generate $\sin(n(\pi/3))$ in (11), there is no triple sideband harmonics in output. This is the particular characteristic of three-phase CSIs.

Figure 5 shows the phase current spectrum for sine-triangle symmetrical regular sampled PWM when $M = 0.9$ and $f_c/f_0 = 21$. The data in Figure 5 have been normalized with $\sqrt{3}I_{dc}$. WTHD$_i$ = 2.04%, THD$_i$ = 60.75%. Figure 6 shows the phase current spectrum for sine-triangle asymmetrical regular sampled PWM when $M = 0.9$ and $f_c/f_0 = 21$. The data in Figure 6 have been normalized with $\sqrt{3}I_{dc}$. WTHD$_i$ = 2.02%, THD$_i$ = 60.03%.

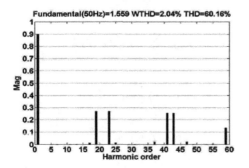

Figure 4. The phase current spectrum for sine-triangle naturally sampled PWM when $M = 0.9$ and $f_c/f_0 = 21$.

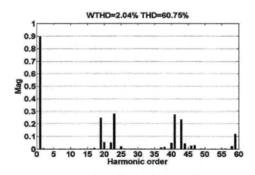

Figure 5. The phase current spectrum for sine-triangle symmetrical regular sampled PWM when $M = 0.9$ and $f_c/f_0 = 21$.

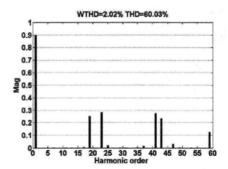

Figure 6. The phase current spectrum for sine-triangle asymmetrical regular sampled PWM when $M = 0.9$ and $f_c/f_0 = 21$.

As shown in Figure 5, the generation of the fundamental component is accompanied by the nontriple sideband harmonics and very small nontriple baseband harmonics when we use the symmetrical regular sampled strategy, but the carrier harmonics don't generated. However, as shown in Figure 6, when the asymmetrical regular sampled strategy is used, it can be observed that the fundamental component and very small odd nontriple baseband harmonics are generated. In the meanwhile, the triple sideband harmonics, odd sideband harmonics around odd multiples of the carrier fundamental and even sideband harmonics around even multiples of the carrier fundamental are eliminated. There is no significant difference among these three modulation strategies after comparison of their performance.

5 COMPARISON BETWEEN THE VSI AND CSI

By means of mathematical calculation of harmonic components of three-phase CSIs based on double Fourier integral theory, it can be found that the harmonic characteristic of current source inverters isthe same to the harmonic characteristic of voltage source inverters (Holmes & Lipo 2003). Figure 7 and Figure 8 show the topologies of the CSI and the VSI.

Although the topologies of the CSI and the VSI are similar, but their switch strategies are different due to the differences between their requirements of switch function. It can be observed that harmonic components of the VSI are equivalent to the voltage by subtracting voltage of one bridge leg form voltage of another bridge leg, for example, $V_a(t) - X_1(t) - X_2(t)$, where $V(t)$ is defined as harmonic components of bridge leg voltage and $X(t)$ is defined as harmonic components of bridge leg. However, harmonic components of the CSI are equivalent to the sum of current of each bridge leg, e.g., $I_a(t) = Y_a(t) - Y_b(t) - Y_c(t) = 2Y_a(t)$ and $Y_a(t) = 0.5[X_1(t) - X_2(t)]$, so that $I_a(t) = X_1(t) - X_2(t)$, where $I(t)$ is defined as harmonic components of phase current, $X(t)$ is defined as harmonic components of bridge leg for the bilogic modulation signal and $Y(t)$ is defined as harmonic components of bridge leg for the trilogic modulation signal after preprocessing of (7). It can be found that result of the CSI is the same to the result of the VSI. In the meanwhile, their vector space is analyzed as well. A static vector space of the CSI consists of six active states and three null states. Simultaneously, six active states of the CSI are in correspondence with

six active states of the VSI and the six space vectors for the CSI will be displaced 30° only compared to the VSI in the complex plane (Zmood & Holmes 1998). The CSI and the VSI are different from the distribution of null states, but this difference doesn't affect their harmonic characteristics. Consequently, CSIs have the same harmonic characteristic to VSIs.

6 SIMULATION RESULTS

The simulation experiment of the CSI and the VSI is conducted under the circumstance of sine-triangle naturally sampled PWM. Simulation parameters are listed in Table 2.

Figure 9 shows the simulation results of the CSI. $THD_i = 59.93\%$, $I_1 = 159.5$ A. The results of current spectrum, total harmonic current distortion and fundamental current are very close to the theory results. Figure 10 shows the simulation results of the

Table 2. Simulation parameters of the CSI and the VSI.

Categories of inverters	CSI	VSI
Modulation index M	0.9	0.9
Output frequency f_0	50 Hz	50 Hz
Carrier frequency f_c	1050 Hz	1050 Hz
DC link current or voltage	200 A	200 V
Single phase load	10Ω, 1 mH	10Ω, 1 mH

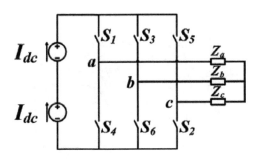

Figure 7. The topology of the CSI.

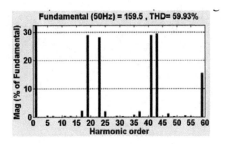

Figure 9. The simulation results of the CSI.

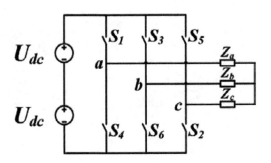

Figure 8. The topology of the VSI.

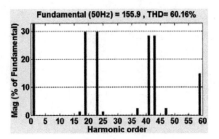

Figure 10. The simulation results of the VSI.

VSI. By comparing Figure 9 with Figure 10, it can be found that the harmonic characteristics of the CSI and the VSI are very similar to each other.

7 CONCLUSION

To sum up, a mathematical model of current source inverters based on double Fourier integral theory is proposed in this paper. Harmonic components of single-phase half-bridge CSIs are calculated and harmonic components of three-phase bridge CSIs have been derived. Total harmonic current distortion and fundamental current are calculated and phase current spectrum is also obtained. In addition, harmonic characteristic of three-phase current source inverters for different modulation strategies is analyzed in detail as well. Finally, this study shows that current source inverters have the same harmonic characteristic when compared to voltage source inverters which can prove the feasibility of the development of current source inverters and provide the theoretical basis for the applications of CSIs in high-power medium-voltage.

REFERENCES

Holmes, D.G. & Lipo, T.A. 2003. *Pulse Width Modulation For Power Converters: Principles and Practice.* NJ: Wiley.

Karasik, V. et al. 1999. SMES For Power Utility Applications: A Review of Technical and Cost Considerations. *IEEE Transactions on Applied Superconductivity* 9(2): 541–546.

Lopes, A.C. & Naguib, M.F. 2009. Space Vector Modulation for Low Switching Frequency Current Source Cnverters With Reduced Low-Order Noncharacteristic Harmonics. *IEEE Transactions on Power Electronics* 24(4): 903–910.

Luongo, C.A. 1996. Superconducting Storage Systems: An Overview. *IEEE Transactions on Magnetics* 32(4): 2214–2223.

Shang, J. & Li, Y.W. 2014. A Space Vector Modulation Method for Common-Mode Voltage Reduction in Current Source Converters. *IEEE Transactions on Power Electronics* 29(1): 374–385.

Wang, X. & Ooi, B.T. 1993. Unity PF Current-Source Rectifier Based on Dynamic Trilogic PWM. *IEEE Transactions on Power Electronics* 8(3): 288–294.

Wiechmann, E.P. et al. 2008. On the Efficiency of Voltage Source and Current Source Inverters for High-Power Drives. *IEEE Transactions on Industrial Electronics* 55(4): 1771–1782.

Zmood, D.N. & Holmes, D.G. 1998. A Generalised Approach to the Modulation of Current Source Inverters. *IEEE Power Electronics Specialist Conference* 1: 739–745.

Control Engineering and Information Systems – Liu (Ed)
© 2015 Taylor & Francis Group, London, ISBN 978-1-138-02685-8

A method of simplifying logic functions

Y.P. Xu & H.W. Li

School of Computer Engineering, Jiangsu University of Technology, Changzhou, Jiangsu, China

ABSTRACT: The paper proposes a method of simplifying logic function based on the minterm combination map that can demonstrate the rules of minterm combination. All of the most simple logic functions are obtained by using the method.

1 INTRODUCTION

There are some methods which are used for the simplification of logic functions such as Karnaugh maps, formula and Q-M. Tao [Tao, Y. M. 2009] proposed a method "3-D simplification" based on the idea of Karnaugh maps to simplify logic function variables increased to six. Xu [Xu, J. P, Cheng, L. X. 2011] proposed an improved Q-M method for simplification of logic functions. This method can reduce the number of iterations, and accelerate the speed of simplification. A method using Ant Colony Algorithm to simplify large scale logical function was put forward [Li, Y. Z, Pan, Q. K, Li, J, Q. 2007]. Experiments showed the evolvable hardware method solve simplification of large scale logic function. Wang [Wang, P, Zeng, S. Y, Yan, J. F. 2006] using thinking of evolvable hardware design had realized a novel method in simplifying the given logic function, which was different from conventional method, such as algebra method and Karnaugh map method. Experiments showed that evolvable hardware method could solve simplification of large scale logic function. Wang [Wang, B. 1987] is given by the adjacent code requires minimum coding method of simplifying logic function algorithm. In this paper, in the given a minimum combination rules that reflect the combined minimum map based on a minimum of the combination map simplification of logic function method. The simplification of logic function method has all of the available logic functions in simplest form, simplification process is simple, intuitive and easy to implement on a computer.

2 MINTERM COMBINATION MAP

The minterm combination map of only one logical variable: The logical variable is A. It has two minterms: A and $\overline{A}$. As shown in Figure 1, Their combined term is $A + \overline{A} = 1$.

The minterm combination map of two logical variables: The logical variables are A and B, They have four minterms: $\overline{A}\overline{B}, \overline{A}B, A\overline{B}$ and AB. As shown in Figure 2, it is the minterm combination map of two logical variables.

$$\overline{A}\overline{B} + \overline{A}B = \overline{A}, \ \overline{A}B + AB = A, \ \overline{A}\overline{B} + A\overline{B} = \overline{B},$$
$$\overline{A}B + AB = B, \ \overline{A} + A = 1, \ \overline{B} + B = 1.$$

The minterm combination map of three logical variables: The logical variables are A, B and C. They have 8 minterms: $\overline{A}\overline{B}\overline{C}$, $\overline{A}\overline{B}C, \overline{A}B\overline{C}, \overline{A}BC, A\overline{B}\overline{C}, A\overline{B}C, AB\overline{C}$, and ABC. As shown in Figure 3, it is the minterm combination map of three logical variables.

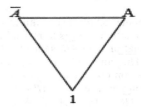

Figure 1. Minterm combination map with one variable.

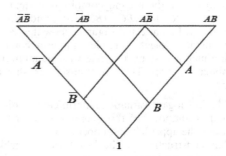

Figure 2. Minterm combination map with two variables.

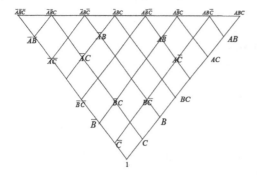

Figure 3. Minterm combination map with two variables.

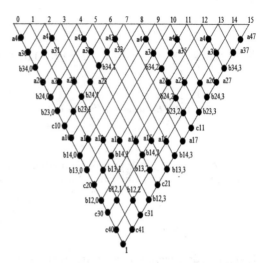

Figure 4. Minterm combination map with four variables.

$$\overline{ABC} + \overline{ABC} = \overline{AB}, \quad \overline{ABC} + \overline{ABC} = \overline{AB},$$

$$A\overline{BC} + AB\overline{C} = A\overline{B}, \quad AB\overline{C} + ABC = AB;$$

$$\overline{ABC} + \overline{ABC} = \overline{AC}, \quad \overline{ABC} + \overline{ABC} = \overline{AC},$$

$$A\overline{BC} + ABC = A\overline{C}, \quad A\overline{BC} + ABC = AC;$$

$$\overline{ABC} + A\overline{BC} = \overline{BC}, \quad \overline{ABC} + A\overline{BC} = \overline{BC},$$

$$\overline{ABC} + AB\overline{C} = B\overline{C}, \quad \overline{ABC} + ABC = BC;$$

$$\overline{AB} + \overline{AB} = \overline{AC} + \overline{AC} = \overline{A};$$

$$A\overline{B} + AB = A\overline{C} + AC = A;$$

$$\overline{AB} + A\overline{B} = \overline{BC} + \overline{BC} = \overline{B},$$

$$\overline{AB} + AB = B\overline{C} + BC = B;$$

$$\overline{BC} + B\overline{C} = \overline{AC} + A\overline{C} = \overline{C}.$$

The combination of the two minterms is called a first combination item, as shown in Figure 3, the first combination item $\overline{AB}$ is the result of combination of $\overline{ABC}$ and $\overline{ABC}$. The second combination item is composed by two first combination items, four minterms, e.g. $A = A\overline{B} + AB = A\overline{C} + AC$, $A = A\overline{BC} + \overline{ABC} + ABC + AB\overline{C}$. And so on, we can define the nth combination item. On the contrary, minterms and combination items are called coverage of combination item, e.g. The covers of the second combination item A are $A\overline{B}, AB, A\overline{C}, AC, \overline{ABC}, A\overline{BC}, ABC$, and $AB\overline{C}$.

Figure 1, Figure 2 and Figure 3 show that constructing k+1 logical variables minterm combination map is easy by k logical variable minterm combination map. Their relationship is shown below.

1. The k+1 logical variable minterm combination map is composed of three compositions of the lower, the upper left and upper right.
2. The construction of the k+1 logical variable minterm combination map is the same as the k logical variables minterm combination map, and is shown as follows:

- The lower is the k logical variable minimum combination map.
- Add NOT of the k+1th logical variable before the k logical variables minterm and its combination items. It is the upper left of the k+1 logical variables minterm combination map.
- Add the k+1th logical variable before the k logical variables minterm and its combination items. It is the upper right of the k + 1 logical variables minterm combination map.

The minimum combination map of the four logical variables A, B, C, and D is shown in Figure 4.

In Figure 4, the minterm is represented by its decimal number. The symbol "•" marks the combination item position. Let 1, 2, 3, and 4 represent variable respectively for A, B, C, and D. The first combination term b_{ij} represents the ith variable is absent and its value is j. The second combination term $b_{ij,k}$ represents the ith variable and the jth variables are absent and its value is k. The third combination term c_{ij} represents only the ith variable is present and its value is j.

3 A SIMPLIFICATION METHOD OF LOGIC FUNCTION BASED ON MINTERM COMBINATION MAP

The steps of logic function simplification are as follows:

1. A logic function is expressed as a sum of minterms.
2. To draw the minterm combination map according to the logic function contained in the number of logical variables.

3. To mark all of the minterms in the figure according to the logic function.
4. To mark all of the combination terms of minterms and the combination terms of combination t terms in the figure.
5. To remove all marked combinations are covered by marked items in the figure.
6. Minterms marked by all cover the same marked item become product of terms.
7. Constitute the simplest expresses of the logic function.

Example: Simplify the following logic function using minterm combination map.

$$F(A,B,C,D) = \overline{AB}\overline{D} + \overline{BC}\overline{D}$$
$$+ A\overline{B}C + \overline{A}BD + AB + \overline{A}CD$$

1. The logic function $F(A,B,C,D)$ is expressed as a sum of minterms.

$$F(A,B,C,D) = \overline{A}\overline{B}\overline{C}\overline{D} + \overline{A}\overline{B}C\overline{D} + \overline{A}BCD$$
$$+ \overline{A}B\overline{C}D + \overline{A}BCD + \overline{A}B\overline{C}\overline{D}$$
$$+ A\overline{B}C\overline{D} + A\overline{B}CD + AB\overline{C}\overline{D}$$
$$+ AB\overline{C}D + ABC\overline{D} + ABCD$$

$$= \sum m \binom{0,2,3,5,7,8,10,}{11,12,13,14,15}$$

2. There are four variables in the logic function $F(A,B,C,D)$, to draw the minterm combination map of the four variables, shown as Figure 5.
3. Using the symbol "O" marks minterms of $F(A,B,C,D)$.
4. Using the symbol "O" marks all of combination terms.
5. Using the symbol "⊗" marks all of combination terms which are covered by other combination terms.

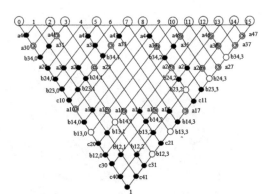

Figure 5. Simplifying logic function using minterm combination map with four variables.

6. The remainder combination terms marked as "O" are stored to a set S.

$$S = \{[b34,3], [b24,3], [b23,2], [b14,1],$$
$$b13,0], [b13,3], [b12,3]\}$$
$$= \{AB, AC, \overline{A}\overline{D}, \overline{B}C, \overline{B}\overline{D}, BD, CD\}$$

7. To process the set S to produce the most simple expressions. The steps are as follows:
   – To calculate the minterms that the elements of S imply, minterm coverage of combination item, as shown in Table 1.
   – To simplify the Table 1. Find combination items by that a minterm is only implied. e.g. the minterm 0 is only implied by the $\overline{B}\overline{D}$, and the minterm 5 is only implied by the BD. They are the members of the most simple expressions. To delete the minterms that have the $\overline{B}\overline{D}$, and BD in their combination item. And sort the table according to the number of combination items, the simplified table as shown Table 2.
   – To traverse the tree in Table 2, each selected a combination item, and eliminated the minimum entry contains the item, if the table is not empty then continue processing by the method until the table empty.
   – To find a function simplified expressions that included the minimum number of items. Table 3 shown the most simple expression, they are:

$$F(A,B,C,D) = \overline{B}\overline{D} + BD + \overline{B}C + AB$$
$$F(A,B,C,D) = \overline{B}\overline{D} + BD + \overline{B}C + A\overline{D}$$
$$F(A,B,C,D) = \overline{B}\overline{D} + BD + CD + AB$$
$$F(A,B,C,D) = \overline{B}\overline{D} + BD + CD + A\overline{D}$$

Table 1. Minterm coverage of the combination terms.

Minterm	Combination item	Minterm	Combination item
0	$\overline{B}\overline{D}$	10	$AC\ \overline{B}C\ A\overline{D}\ \overline{B}\overline{D}$
2	$\overline{B}\overline{D}\ \overline{B}C$	11	$AC\ \overline{B}C\ CD$
3	$\overline{B}C\ CD$	12	$AB\ A\overline{D}$
5	BD	13	$AB\ BD$
7	$BD\ CD$	14	$AB\ AC\ A\overline{D}$
8	$\overline{A}\overline{D}\ \overline{B}\overline{D}$	15	$AB\ AC\ BD\ CD$

Table 2. Simplification of Table 1.

Minterm	Combination item	Minterm	Combination item
3	$BC\ CD$	11	$AC\ BC\ CD$
12	$AB\ AD$	14	$AB\ AC\ AD$

Table 3. Simplification of logic function.

Minterm 0,5	Minterm 3	Minterm 12	Is it the most simple expression?
$\overline{BD} + BD +$	$\overline{B}C +$	AB	Yes
		$A\overline{D}$	Yes
	$CD +$	AB	Yes
		$A\overline{D}$	Yes

4 CONCLUSION

This paper proposes a minterm combination map that does not limit the number of logical variables, and intuitively reflects the law of minterm combination. When the number of variables is not too much, we can quickly simplify logic function by using it. However, when the number of variables increases, the corresponding minterm combination map become very large. So, we can use the recursive nature of minterm combination to design program to achieve rapid simplification of logic functions. Next, we will optimize algorithm.

REFERENCES

Li, Y.Z., Pan, Q.K., Li, J.Q. 2007. Implementation of Logic Function Simplification Using Ant Algorithm (in Chinese). Computer Engineering and Design, Vol.28-12, 2788–2789.

Tao, Y.M. 2009. 3-D Method: A New Method to Simplify Boolean Logic FormulaStack (in Chinese). Computer and Information Technology, Vol.17-1, 4–7.

Wang, B. 1987. An Algorithm of Encoding Operation for Simplifying Boolean Functions (in Chinese). Chinese Journal of Computers, Vol. -7, p.434–437.

Wang, P., Zeng, S.Y., Yan., J.F. 2006. Implementation of Logic Function Simplification Using Genetic Algorithm (in Chinese). Computer Engineering and Design, Vol.27-3, 365–375.

Xu, J.P., Cheng, L.X. 2011. Improved Q-M Method for Simplification of Logic Functions. Computer Engineering (in Chinese), Vol. 37-20. 30–32.

Control Engineering and Information Systems – Liu (Ed)
© 2015 Taylor & Francis Group, London, ISBN 978-1-138-02685-8

Thermal conductivity determinator for potato starch wastewater

W.J. Ding & H.T. Hao
Mechanical Engineering College of Ningxia University, Yinchuan, China

J. Xu
Chemical Engineering College of Beifang University of Nationality, Yinchuan, China

ABSTRACT: In order to quickly measure the thermal conductivity of potato starch processing wastewater, using transient and comparison method, with the use of the design concept of the virtual instrument, a PC-based liquid thermal conductivity detector is designed. Specific content, including hardware parts of virtual instrument, the power supply, constant current source, the V-F converter design and calculations, procedures of the software platform for voice collecting, data analysis program. On the conditions of 25 °C water bath, a series of temperature of measured liquid while it is heated by the control system, were acquired and analysis by software in PC, and then, get the liquid thermal conductivity quickly and accurately, especially the potato starch wastewater.

1 INTRODUCTION

Potato starch processing wastewater is the main by-product in potato processing, which has high chemical oxygen demand and is one of the most polluted wastewater in the food industry. Potato starch processing wastewater thermal conductivity is an important design parameter in the design of processing wastewater treatment device, calculating the parameter by experiment helps to verify the calculation model of the thermal conductivity, which conforms to the need of wastewater treatment device design. The portable measurement device can be realized quickly measure the thermal conductivity in waste water treatment field, and it is the basic tool to mass collect different samples' wastewater thermal conductivity.

2 DESIGN PRINCIPLE AND METHOD OF OPERATION

2.1 Design principle

Hot wire using in transient state method is the most widely used method in laboratory, and its structure is simple, but the determination process is difficult to control and the solution process is complex. The thermal conductivity λ's measurement is based on the schematic formula of liquid thermal conductivity.

$$q = -\lambda \frac{\partial t}{\partial n} \tag{1}$$

In order to get conductivity, the temperature gradient must be measured, and at the same time the heat transferred q need to be known.

Therefore, the design used a thermal resistance as a point heat source, put it into heat conduction trough which had liquid to be measured, and put the conduction trough into thermostat water bath. When the thermal resistance was energized, it can formed temperature change and a certain temperature gradient field in measured liquid.

Affected by self-heating effect, the thermal resistance had changed. Thermal resistance is powered by constant current source of Wheatstone bridge; the change in resistance can lead to the bridge imbalance voltage.

Document (Li, Zhang & Guo 2010) suggested that the relationship between the voltage change rate with time and the liquid thermal conductivity, which can be expressed as:

$$\lambda = A \cdot \frac{\mathrm{d}V}{\mathrm{d}t} + B \tag{2}$$

The voltage change was transformed into a frequency signal though voltage-frequency converter module, and realized timing acquisition by the computer sound card. Computer software is responsible to control heating, measuring, recording, data processing, stopping the heating process and analysis of the calculated results.

Therefore, selecting two standard liquid which thermal conductivity known to determine their $\mathrm{d}V/\mathrm{d}t$ value, and A, B can be obtained by the method of undetermined coefficients. Thereafter,

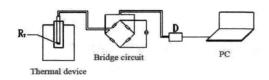

Figure 1. Schematic diagram of measurement.

according to the known function relationship which had been obtained, the voltage change rate with time of the unknown liquid thermal conductivity dV/dt can be measured (measuring a voltage value at a certain time point and another voltage value in the next time point, dV/dt is the ratio of the difference between the two voltage and time interval). Thereby, the liquid thermal conductivity λ can be measured, accuracy of measurement is related to the standard liquid, and it should be higher than the accuracy of engineering applications.

2.2 *Principle of measurement*

The experiments carried out at 25 °C thermostat water bath, after the sample in constant temperature for 30 min, started computer controlled circuit-changing switch and sent 0.04 mA current to bridge circuit, adjusted the bridge to make it balanced (after balanced, bridge resistance shouldn't make any adjustment). Then, turned the current to be 1.5 mA, for the heat, the thermal resistance in circuit occurred continuously changes, and caused the output voltage change in a bridge circuit. The change in voltage used the V-F conversion module to make the frequency signal inputted to the computer for continuous acquisition, and converted to a voltage. After a certain length of time (5–10 s), cut off the heating current, and completed the acquisition. Analysis and calculated dV/dt, and automatically calculated the thermal conductivity λ. Figure 1 shows the principle of the measurement.

3 DESIGN AND SELECTION OF KEY ELECTRONIC COMPONENTS

3.1 *The constant current source*

The constant current source is the power supply of bridge circuit; it supplies constant current to bridge circuit. Using the high-precision constant current source LM334, the supply current was about 1.5–2.0 mA and the voltage was about 12 V.

3.2 *The design and calculations of thermistor*

Platinum resistance's physical and chemical properties are very stable at high temperature and oxidation conditions, and be used as industrial temperature measurement device and the temperature standard. International temperature standard IPTS-68 requirements, at −259.34–630.74 °C, make the platinum resistance thermometer as a reference.

The relationship between Platinum resistance temperatures, at 0–630.74 °C:

$$R_t = R_0(1 + At + Bt^2) \tag{3}$$

R_t is resistance at t °C, R_0 is resistance at 0 °C, t is one of any temperatures, A is temperature coefficient, it is $3.940 \times 10^{-2}/°C$. In the design, the initial temperature is 25 °C, the resistance value is 198.4635 Ω.

3.3 *The design calculations of bridge circuit*

When listing facts use either the style tag List signs or the style tag List numbers.

The maximum voltage of the constant current source is 12 V; the allowed maximum current is 2 mA. Total resistance is 6000 Ω; therefore, the resistance of the circuit should be combined into 6000 Ω at least.

In the case of the circuit worked properly, bridge circuit diagram (Qin 2007) (Fig. 2) could be regarded as the series $R_1 + R_T$ and R_2, the series R_3 and R_4, and both series. Therefore, $R_{\text{Total}} = (R_1 + R_T + R_2)//(R_3 + R_4) \geq 6000\ \Omega$. And then, it must meeted the conditions: $(R_1 + R_T + R_2) \geq 6000\ \Omega$ and $(R_3 + R_4) \geq 6000\ \Omega$.

At the beginning of experiment, the bridge should be transferred to the equilibrium position, and valued $R_4 = 20$ kΩ, $R_1 = R_2 = 8200\ \Omega$, $R_3 = 5100\ \Omega$. According to the above-launched bridge balance conditions, the value of the resistance box:

$$(R_1 + R_T)R_4 = R_2 R_3^{\sim}.$$

$$R_4 = 4979.5\ \Omega. \tag{4}$$

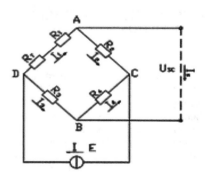

Figure 2. Bridge circuit diagram of the constant current source.

When the bridge is balanced, $R_{\text{Total}} = 6271.25\ \Omega$.

After 1.5 mA constant current passes through the thermistor and other resistor in the circuit resistance never change. Therefore, the total resistance in the circuit is 6271.25 Ω, at this time, the voltage of the constant current source:

$$E = IR_{\text{Total}}$$
$$= 9.406\text{V}$$

$$U_{\text{sc1}} = \frac{R_2R_3 - (R_1 + R_T)R_4}{(R_1 + R_2 + R_T)(R_3 + R_4)}\tilde{E}$$
$$= 8.4249122 \times 10^{-6}\text{V} \qquad (5)$$

When the thermistor resistance changed, the voltage output also changed. According to the formula:

$$U_{\text{sc}} = \frac{R_2(R_1 + R_T)\Delta R_T}{(R_1 + R_2 + R_T)^2(R_1 + R_T)}\tilde{E} \qquad (6)$$

The change value in voltage could be calculated. Within the specified time period, dV/dt could be calculated.

3.4 The design and calculations of V-F converter module

The output voltage of bridge circuit could not be directly used for signal acquisition; voltage signal (analog signal) should be turned into a frequency signal (digital pulse signal), and sent it to computer through the sound card acquisition system and analyzed it. VFC32 module is a conversion unit of voltage to frequency, it is used in the design to convert voltage signal into frequency signal.

3.4.1 The design and calculations of resistors and capacitors

In the VFC32 application circuit (Fig. 3), the resistance of R_1 determined the voltage range, a 10 V

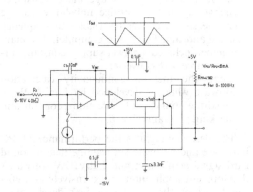

Figure 3. Bridge circuit diagram of the constant current source.

full-scale input voltage need a 40 kΩ resistance; the maximum voltage of the input voltage frequency converter module can not exceed 10 V, make V_{FS} as 10 V, therefore, $R_1 = V_{\text{FS}}/0.25 = 40$ kΩ.

The input value of the maximum voltage is 0–10 V, the corresponding frequency output is 0–1000 Hz. When the frequency is 1000 Hz, the value of the corresponding capacitance is about 0.0033 pF, the existing error can be corrected in the software.

3.4.2 Frequency output terminal

Frequency signal output terminal is open-collector output; an adjustable rheostat usually associated with a 5 V regulated supplied power to connect the standard logic pulse. However, the voltage of the both ends in adjustable rheostat shouldn't be higher than the value of V_{cc}. There would be a long time stable and positive output pulse during the delay time. The current flows through the open-collector output and conducts feedback through the transistor which has the same end. This terminal is connected with the ground through the load.

After the voltage is converted to the frequency, the design and calculations base on the empirical formula:

$$F_{\text{out}} = \frac{V_{\text{IN}}/R_{\text{IN}}}{I(C_1 + 44) \cdot 6.7} \qquad (7)$$
$$k = \frac{1}{R_{\text{IN}}I(C_1 + 44) \cdot 6.7} \qquad (8)$$

For R_{IN}, I, C_1 in the formula are constants, so k is also constant. The input voltage and the output frequency are converted by the proportional coefficient k.

4 PROGRAM DESIGN

Computer uses sound card to collect frequency signal, program for data acquisition and processing was developed on LabVIEW, it converted frequency signal into voltage-time ratio to calculate thermal conductivity. Program includes module of frequency-time signal acquisition, module of calculation for A and B, and correction module.

4.1 Sound collection program

Frequency acquisition: Sending the frequency—time signal which is converted by V-F conversion module to the computer's sound card from MIC channel, and using sound card collection Program (Fig. 4) which is developed by LabVIEW to realize pulse signal acquisition (Chen & Zhang 2007).

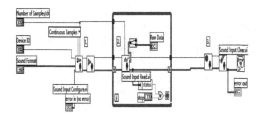

Figure 4. Program diagram of sound card signal acquisition.

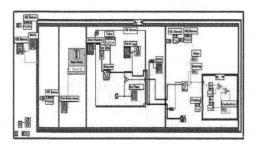

Figure 5. Determination program for thermal conductivity.

4.2 Determination program of thermal conductivity

Figure 5 is the determination program of thermal conductivity.

4.3 Determine A and B, and check the program of A, B

Collecting the frequency-time signal of two standard liquid (Benzene and Glycerol), and turning them to voltage/time parameters. And then, two linear equations which include A, B as variables are built. Therefore, calculating the value of A, B could obtain the slope and intercept of the thermal conductivity equation (Xu, Zheng & Xiao 2009). In the subsequent measuring process, as long as the process of change of the voltage with time obtained, the thermal conductivity could be calculated though the acquisition of the voltage/time relationship (Fig. 5 Frame 4).

5 DEVICE STRUCTURE

The device mainly consists of measuring bath cup, measuring circuit and PC (Fig. 6).

6 DETERMINATION

Take benzene and glycerin as reference, and measure 4% concentration waste water (Table 1).

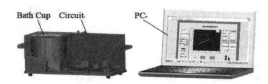

Figure 6. Device structure.

Table 1. Experimental determination.

Substance	Standard W/(m·K)	dV/dt V/s	Measured W/(m·K)	A B
Benzene	0.1442	0.0190	1.997	
Glycerin	0.2900	0.0092		0.1063
4% waste water		0.1922	0.4901	

Problems: The selected standard substance is able to obtain the instrument constant, but waste water is high thermal conductivity liquid, and has the problem of thermal conductivity extrapolation; the content substances have greater impact in it. Therefore, improving the measurement accuracy still has to select other criteria which had higher thermal conductivity.

7 CONCLUSIONS

The design adopt transient thermal conductivity measurement principle, and used comparison method of two standard liquid to determine the thermal conductivity calculated parameters, therefore, the equation of the liquid thermal conductivity could be obtained.

- The device design, adopts resistive elements of the thermal step signal bridge, makes resistance change signal form bridge imbalance voltage change signal, converts variables to frequency signal and transfers it into computer. The main member included: a power supply, constant current source, V-F converter.
- Using the thermal conductivity calculation software which is developed by LabVIEW to constitute sound card collection program and checking program of A, B.

The design integrates the performance of PC hardware and software sufficiently, it uses sound card signal acquisition and LabVIEW software's development applications, inexpensively and efficiently settled the portable device's development of potato starch processing wastewater thermal conductivity detector. From the view of practical application, the next step will need to explore the standards and the compact structure.

ACKNOWLEDGEMENT

Thanks funds support from:

– Ningxia Natural Science Foundation of China (NZ1131).
– Chemical Technology Key Laboratory of State Ethnic Affairs Commission (2011SY03).

REFERENCES

Chen X.H. & Zhang Y.H. 2007. *Labview 8.20, program design from entry to master*. Beijing: Tsinghua University Press.
Li K.N., Zhang J.G. & Guo N.N. 2010. The transient comparison method to measure the liquid thermal conductivity. *Low Temperature Physics*: 32 (3) pp.231–234.
Qin Z.H. 2007. *Electrical engineering*. Beijing: Higher Education Press.
Xu X.D., Zheng D.Y. & Xiao W. 2009. *LabVIEW8.5, commonly used functions and programming examples to explain*. Beijing: Electronic Industry Press.

Control Engineering and Information Systems – Liu (Ed)
© 2015 Taylor & Francis Group, London, ISBN 978-1-138-02685-8

Missile avoidance trajectory optimization based on Gauss Pseudospectral Method

J.F. Cheng, X.M. Dong, J.P. Xue, H.T. Wang & J.H. Zhi
Aeronautics and Astronautics Engineering College, Air Force Engineering University, Xi'an, Shaanxi, China

ABSTRACT: Aim to realize aircraft maneuver evasive in the situation of missile pursuit effectively, a trajectory optimization method based on Gauss Pseudospectral Method (GPM) is proposed. The aircraft and missile are modeled as mass point, and aerodynamic angle differential equations of the aircraft are taken into consideration to satisfy the agility and multi-constraint requirements. The missile avoidance optimal performance index is designed, then, the pursuit-evasion control is described as a Two Point Boundary Value Problem (TPBVP). The GPM is used to equally convert the original continue TPBVP to a discrete nonlinear programming problem, whose initial solution is preprocessed through intelligent optimization algorithm, and then the sequential quadratic programming algorithm is applied to solve it. Finally, a numerical simulation example is given and the result shows that the optimization trajectory is in accordance with the real combat manipulation and it can provide an effective guiding strategy for pilot to avoid missile's attack.

1 INTRODUCTION

In modern air combat, the avoidance of guided missiles is extremely important for the aircraft's survival. Now, it mainly includes three ways that pilot can use to evade missile's attack, they are maneuver avoidance [1–2], releasing decoy 3 and launching defensive missile [4–5]. Meanwhile, within its tactics executes directly and unnecessary of any other assistant equipment, maneuver avoidance becomes the first, basic, and mostly applied methods for missile avoidance.

Basically, missile avoidance is a differential game problem, while, owing to proportional navigation law adopted by majority missile, it can be described as a single aspect trajectory optimization problem. In reference [1], direct multiple shooting method is applied to solve endgame evasion problem based on the performance index of maximized miss distance in three dimensions reference frame. Reference [2] introduces the receding horizon control scheme to obtain near-optimal controls in a feedback form for the aircraft trying to avoid a closing air-to-air missile. Considering aircraft's agility limitation, angular accelerations of attack angle and bank angle are defined as parameter optimization decision variables aiming to decrease constrained conditions and enhance optimization convergence, and then Particle Swarm Optimization (PSO) arithmetic is applied to gain the optimal evasive trajectory. [6]

The common essence of these methods is to equally convert the original continue optimal control problem to a discrete parameterized optimization problem, so the accuracy of transcription and convergence rate of optimization procedure are main key elements determining the arithmetic good or bad. Gauss Pseudospectral Method (GPM)7 is one of the parameterized method through discretizing control variables and state variables simultaneity, which is the only one that provides equivalence between the costates of the optimal control problem and the KKT (Karush-Kuhn-Tucker) multipliers of the transcribed Non-Linear Programming (NLP) problems 8, so it can acquire higher accuracy than any other methods.

Therefore, the paper transforms the trajectory missile avoidance optimization to a NLP problem by GPM, further more, to enhance the convergence rate, PSO algorithm is used to gain the initial solution of transformed NLP. The paper is structured as follows: In the following section, the aircraft and the missile models are presented, and the pursuit–evasion problem along with the performance measures are formulated. The GPM transcription scheme for continue optimal control problem is introduced in Sec. III. The initial solution and final NLP solving arithmetic is presented briefly in Sec. IV, followed by a numerical example in Sec. V. Finally, concluding remarks appear in Sec. VI.

2 VEHICLE MODELS

2.1 *Aircraft model*

It is assumed that aircraft's thrust vector is in accordance with pitch axis and through the mass center, so the motion of the aircraft is described in wind-axes reference frame by the following differential equations 1–2

$$\frac{dx_a}{dt} = V_a \cos \gamma_a \cos \chi_a$$

$$\frac{dy_a}{dt} = V_a \cos \gamma_a \sin \chi_a$$

$$\frac{dh_a}{dt} = V_a \sin \gamma_a$$

$$\frac{dV_a}{dt} = \frac{1}{m_a}\left(\eta_a T_{a,\max} \cos \alpha - D_a\right) - g \sin \gamma_a$$

$$\frac{d\gamma_a}{dt} = \frac{1}{m_a V_a}\left(\eta_a T_{a,\max} \sin \alpha + L_a\right)\cos \mu_a - \frac{g}{V_a}\cos \gamma_a$$

$$\frac{d\chi_a}{dt} = \frac{1}{m_a V_a \cos \gamma_a}\left(\eta_a T_{a,\max} \sin \alpha + L_a\right)\sin \mu_a$$

$$\tag{1}$$

where the state variables are aircraft forward range x_a, cross range y_a, altitude h_a, velocity V_a, flight path angle γ_a and heading angle χ_a. The attack angle α, bank angle μ_a and throttle setting η_a are control variables. $T_{a,\max}(h_a, M_a)$ represents the maximum available thrust force in current altitude h_a and mach M_a. $L_a = 1/2\rho V_a^2 S C_{La}$ denotes the aircraft lift force and $D_a = 1/2\rho V_a^2 S C_D(C_{La})$ is drag force.

In the situation of aircraft missile duel, aircraft needs to maneuver fast and greatly to avoid missile's attack, so the mass point model can't satisfy this property completely. This paper introduces the aircraft rotation kinematics in the form of aerodynamic angle differential equations.

$$\frac{d\alpha}{dt} = q_a - \frac{\eta_a T_{a,\max} \sin \alpha + L_a - m_a g \cos \gamma_a \cos \mu_a}{m_a V_a}$$

$$\frac{d\mu_a}{dt} = p_a \cos \alpha + r_a \sin \alpha$$
$$+ \frac{L_a + \eta_a T_{a,\max} \sin \alpha}{m_a V_a}\sin \mu_a \tan \gamma_a$$

$$\frac{d\beta_a}{dt} = p_a \sin \alpha - r_a \cos \alpha + \frac{g}{V_a}\sin \mu_a \cos \gamma_a$$

$$\tag{2}$$

It assumes that aircraft sideslip angle is zero, then plugging (2.3) into (2.2)

$$\frac{d\mu_a}{dt} = \frac{p_a}{\cos \alpha} + \frac{g \sin \mu_a \cos \gamma_a \tan \alpha}{V_a}$$
$$+ \frac{L_a + \eta_a T_{a,\max} \sin \alpha}{m_a V_a}\sin \mu_a \tan \gamma_a$$

$$\tag{3}$$

Therefore, (1), (2.1) and (3) constitute new augmented aircraft dynamic differential equations, and the control variables become pitch angle rate q_a, roll angle rate p_a and throttle setting η_a.

In addition, aircraft dynamic response must satisfy such constrained conditions.

1. Control variable limited

$$|q_a| - Q_{a,\max} \leq 0; \quad |p_a| - P_{a,\max} \leq 0 \quad 0 \leq \eta_a \leq 1 \quad \tag{4}$$

where $Q_{a,\max}$ and $P_{a,\max}$ denote the maximum pitch angle rate and roll angle rate, which reflect agility requirement.

2. Considering the structure stress limited on the aircraft and the load a pilot can tolerate, the normal load $n_a(\alpha, h_a, V_a)$ restriction is introduced as a path constraint

$$n_a(\alpha, h_a, V_a) - n_{a,\max} \leq 0 \tag{5}$$

3. In addition, the altitude and the dynamic pressure $q(h_a, V_a)$ are constrained by

$$h_{a,\min} - h_a \leq 0; \quad q(h_a, V_a) - q_{\max} \leq 0 \tag{6}$$

where $n_{a,\max}$, $h_{a,\min}$ and $q_{\max}$ denote the maximum load factor, minimum altitude, maximum dynamic pressure.

2.2 Missile model

The motion property of the missile is described by three degrees-of-freedom mass point model, and introducing the dynamic lag between the commanded accelerations and the actual accelerations due to the missile seeker, guidance system and actuator dynamics, then the state equations of the missile system are 1–2

$$\frac{dx_m}{dt} = V_m \cos \gamma_m \cos \chi_m$$

$$\frac{dy_m}{dt} = V_m \cos \gamma_m \sin \chi_m$$

$$\frac{dh_m}{dt} = V_m \sin \gamma_m$$

$$\frac{dV_m}{dt} = \frac{1}{m_m}\left(T_m(t) - D_m\right) - g \sin \gamma_m$$

$$\frac{d\gamma_m}{dt} = \frac{1}{V_m}\left(a_p - g \cos \gamma_m\right) \tag{7}$$

$$\frac{d\chi_m}{dt} = \frac{a_y}{V_m \cos \gamma_m}$$

$$\dot{a}_p = \frac{a_{pc} - a_p}{\tau}$$

$$\dot{a}_y = \frac{a_{yc} - a_y}{\tau}$$

The missile state variables are defined as same as aircraft. The control variables are commanded pitch acceleration a_{pc} and commanded yaw acceleration a_{yc}, furthermore a_p and a_y are corresponding to the actual value respectively. τ denotes the dynamic lag.

The Ideal Proportional Navigation (IPN) is adopted by missile, and the expressions are induced as follows.

Define the distance vector between the missile and target aircraft as:

$$\mathbf{r} = \begin{bmatrix} x_a - x_m & y_a - y_m & h_a - h_m \end{bmatrix}^T \qquad (8)$$

The closing velocity is the negative of the time derivative of distance vector $\mathbf{r}$, that is

$$\mathbf{v}_c = -\dot{\mathbf{r}} = \begin{bmatrix} \dot{x}_m - \dot{x}_a & \dot{y}_m - \dot{y}_a & \dot{h}_m - \dot{h}_a \end{bmatrix}^T \qquad (9)$$

Then the angular rate of the Line-Of-Sight (LOS) vector from the missile to the target aircraft is

$$\omega = \frac{\mathbf{r} \times (-\mathbf{v}_c)}{\mathbf{r} \cdot \mathbf{r}} \qquad (10)$$

According to the definition of IPN, the commanded accelerations are given by

$$\mathbf{a}_c = N\boldsymbol{\omega} \times \mathbf{v}_c \qquad (11)$$

where N is the IPN constant, usually is 3 to 5.

Considering the influence of gravity acceleration, the real missile command acceleration is

$$\mathbf{a} = \mathbf{a}_c + \mathbf{g} = \begin{bmatrix} a_{mx} & a_{my} & a_{mh} \end{bmatrix}^T + \begin{bmatrix} 0 & 0 & g \end{bmatrix}^T \qquad (12)$$

The commanded acceleration $\mathbf{a}$ is defined in earth-axes reference frame, which must be transformed to wind-axes reference frame that real autopilot can use, then, the vertical and horizontal accelerations perpendicular to the velocity vector of the missile are

$$a_{pPN} = \mathbf{a}_c \cdot \mathbf{e}_{a_p} + g\cos\gamma_m; \quad a_{yPN} = \mathbf{a}_c \cdot \mathbf{e}_{a_y} \qquad (13)$$

where

$$\mathbf{e}_{a_p} = \begin{bmatrix} -\sin\gamma_m \cos\chi_m & -\sin\gamma_m \sin\chi_m & \cos\gamma_m \end{bmatrix}^T \qquad (14)$$

$$\mathbf{e}_{a_y} = \begin{bmatrix} -\sin\gamma_m & \cos\gamma_m & 0 \end{bmatrix}^T \qquad (15)$$

Meanwhile, the commanded accelerations are limited to values not imposing structural damage, that is

$$a_{PN} \le g \cdot n_{m,\max} \qquad (16)$$

where $a_{PN} = \sqrt{(a_{pPN})^2 + (a_{yPN})^2}$, $n_{m,\max}$ denotes maximum load.

2.3 Optimization cost index

In modern air combat, the strategy of aircraft avoiding missile attack can be summarized as follows: according to missile's guidance style, using its energy and seeker's constraints, adopting the performance index such as maximum miss distance, maximum capture time, minimum closing velocity and so on, through implementing barrel roll maneuver, diving maneuver, split-s maneuver etc, aiming to the object of evading missile's attack.

The optimization avoidance trajectory by different index is generally identical, and this paper adopts the minimum closing velocity as the cost measure.

Redefine the scale closing velocity

$$\vartheta_c = -\dot{r} = \mathbf{r} \cdot \mathbf{v}_c / r \qquad (17)$$

in which

$$r = \sqrt{(x_a - x_m)^2 + (y_a - y_m)^2 + (h_a - h_m)^2} \qquad (18)$$

In the terminal time, the cost measure can be described as

$$J = \vartheta_c(t_f) \qquad (19)$$

where t_f denotes the time of terminal distance between aircraft and missile arriving to h_f.

$$r(t_f) = h_f \qquad (20)$$

and $r(t_f)$ can be calculated from (18).

2.4 Missile avoidance optimization problem model

The missile avoidance problem can be formulated as a nonlinear optimal control problem, which possesses the characteristic of terminal time free, terminal state fixed, control variables, state variables and path limited.

Define the performance index

$$J = \varphi(\mathbf{x}(t_0), t_0, x(t_f), t_f) + \int_{t_0}^{t_f} L(\mathbf{x}(t), \mathbf{u}(t), t) dt \qquad (21)$$

determine the control $\mathbf{u}(t)$, subject to differential equation constraint

$$\dot{x} = f\left(\mathbf{x}(t),\mathbf{u}(t),t\right) t \in [t_0,t_f] \qquad (22)$$

path constraint

$$\mathbf{g}\left(\mathbf{x}(t),\mathbf{u}(t),t\right) \leq \mathbf{0} \qquad (23)$$

boundary condition

$$\mathbf{h}\left(\mathbf{x}(t_0),t_0,\mathbf{x}(t_f),t_f\right) = \mathbf{0} \qquad (24)$$

where $\mathbf{x}(t) \in \mathbb{R}^n$ is state vector of aircraft and missile, $\mathbf{u}(t) \in \mathbb{R}^m$ denotes aircraft control, t_0 and t_f represent initial and final evasion time respectively. The corresponding relation is as follows: performance index, (19); dynamic differential function limited, (1), (2.1), (3) and (8); Path constraint $g \in \mathbb{R}^q$, (5), (6) and (16); boundary condition $h \in \mathbb{R}^c$, (20); control variable restricted, (4).

3 OPTIMAL PROBLEM TRANSFORMED TO NLP VIA GPM

The principle of GPM is using Lagrange interpolation polynomials approximating state and control variables, then converting the continue TPBVP to a NLP. This paper applies GPM to transform the missile avoidance trajectory. [7–8]

Step 1: Time scale transformation.

The GPM requires to map the general time interval $t \in [t_0 \quad t_f]$ to the fixed one $\tau \in [-1 \quad 1]$ via the affine transformation

$$t = \frac{t_f - t_0}{2}\tau + \frac{t_f + t_0}{2} \qquad (25)$$

Step 2: Approximation state and control variables using global interpolation polynomials.

The GPM selects the initial point $\tau_0 = -1$, Legendre-Gauss (LG) points $\tau_i(i=1,...,N)$ and terminal point $\tau_{N+1} = 1$ as collection points, and LG points are defined as the roots of the Nth-degree Legendre polynomial $P_N(\tau)$, where

$$P_N(\tau) = \frac{1}{2^N N!}\frac{d^N}{d\tau^N}\left(\tau^2 - 1\right)^N \qquad (26)$$

Based on the initial point $\tau_0 = -1$ and LG point $\tau_i(i=1,...,N)$, a N+1th degree Lagrange interpolation polynomials $L_i(\tau)$ is construct, and the state variables approximation are formed as follows

$$\mathbf{x}(\tau) \approx \mathbf{X}(\tau) = \sum_{i=0}^N L_i(\tau)\mathbf{x}(\tau_i) \qquad (27)$$

On the same principle, the control variables approximation are formed with a basis of Nth Lagrange interpolation polynomials

$$\mathbf{u}(\tau) \approx \mathbf{U}(\tau) = \sum_{i=1}^N L_i^*(\tau)\mathbf{u}(\tau_i) \qquad (28)$$

where

$$L_i(\tau) = \prod_{j=0,j\neq i}^N \frac{\tau - \tau_j}{\tau_i - \tau_j}, \; L_i^*(\tau) = \prod_{j=1,j\neq i}^N \frac{\tau - \tau_j}{\tau_i - \tau_j}.$$

Step 3: Transform differential function limited to algebraic constraints

Differentiate (27), the derivative of interpolation function

$\mathbf{X}(\tau)$ at LG points τ_k are

$$\dot{x}(\tau_k) \approx \dot{\mathbf{X}}(\tau_k) = \sum_{i=0}^N \dot{L}_i(\tau_k)\mathbf{X}(\tau_i) = \sum_{i=0}^N \mathbf{D}_{ki}\mathbf{X}(\tau_i) \quad (29)$$

in which the differential matrix $\mathbf{D}_{ki}$ can be determined offline from (30).

$$D_{ki} = \dot{L}_i(\tau_k)$$
$$= \begin{cases} \dfrac{(1+\tau_k)\dot{P}_N(\tau_k)+P_N(\tau_k)}{(\tau_k - \tau_i)\left[(1+\tau_i)\dot{P}_N(\tau_i)+P_N(\tau_i)\right]} & i \neq k \\ \dfrac{(1+\tau_k)\ddot{P}_N(\tau_k)+2\dot{P}_N(\tau_k)}{2\left[(1+\tau_i)\dot{P}_N(\tau_i)+P_N(\tau_i)\right]} & i = k \end{cases} \qquad (30)$$

Combining the (22), then the differential function constraints can be transformed to algebraic constraints

$$\mathbf{R}_k \equiv \sum_{i=0}^N \mathbf{D}_{ki}\mathbf{X}(\tau_i)$$
$$- \frac{t_f - t_0}{2}f(\mathbf{X}(\tau_k),\mathbf{U}(\tau_k),\tau_k;t_0,t_f) = \mathbf{0} \qquad (31)$$

Equation (31) only defines the state constraints on LG points, while ignoring the terminal point $\mathbf{X}(\tau_{N+1})$. According the system state function (22), combining the time transformation (25), it can be derived:

$$x(\tau_{N+1}) = x(t_f)$$
$$= x(t_0) + \frac{t_f - t_0}{2}\int_{-1}^1 f(\mathbf{x}(\tau),\mathbf{u}(\tau),\tau)d\tau \quad (32)$$

Discrete and Gauss quadrature (32), then the terminal variable constraints are

$$\mathbf{R}_f \equiv \mathbf{X}_f - \mathbf{X}_0$$

$$-\frac{t_f - t_0}{2} \sum_{k=1}^{N} w_k f(\mathbf{X}(\tau_k), \mathbf{U}(\tau_k), \tau_k; t_0, t_f) = \mathbf{0}$$

$$(33)$$

where $\mathbf{X}_0 = \mathbf{X}(-1) = x(t_0)$, w_k is Gauss quadrature weights.

$$\omega_k = \int_{-1}^{1} L_i(\tau) d\tau$$

$$= \frac{2}{(1 - \tau_k^2)\left[\dot{P}_N'(\tau_k)\right]^2} (k = 1, ..., N) \qquad (34)$$

Step 4: Approximation performance index.

The integral term in the cost function of (21) can be approximated with a Gauss quadrature.

$$J = \varphi(\mathbf{X}_0, t_0, \mathbf{X}_f, t_f)$$

$$+ \frac{t_f - t_0}{2} \sum_{k=1}^{N} w_k f(\mathbf{X}(\tau_k), \mathbf{U}(\tau_k), \tau_k; t_0, t_f) \qquad (35)$$

Discrete the path and boundary constraints, and it can gain

$$g(\mathbf{X}_k, \mathbf{U}_k, \tau_k; t_0, t_f) \leq 0 (k = 1, ... N) \qquad (36)$$

$$h(\mathbf{X}_0, t_0, \mathbf{X}_f, t_f) = 0 \qquad (37)$$

To sum up, the object function (35) and algebraic constraints (31), (33), (36) and (37) together define a NLP, whose solution is an approximate result of the continuous optimal problem (21)–(24). In the GPM, the variables that used for discretization are as follows: system state $(\mathbf{X}_0, \mathbf{X}(\tau_1), ..., \mathbf{X}(\tau_N), \mathbf{X}_f)$, control variables $(\mathbf{U}(\tau_1), \mathbf{X}(\tau_2), ..., \mathbf{U}(\tau_N))$ and time (t_0, t_f). Similarly, the constraints are as follows: dynamic differential equation constraints$(\mathbf{R}_1, ..., \mathbf{R}_N, \mathbf{R}_f)$, path constraints $(\mathbf{g}_1, \mathbf{g}_2, ..., \mathbf{g}_N)$ and boundary conditions $\mathbf{h}$.

4 SOLVING THE NLP PROBLEM

4.1 *The initial solution setting*

For the missile avoidance NLP problem, within aircraft dynamic response changing fast and system state forecasting difficultly, so the number of collection points and the feasibility of initial solution become main key factors influencing the optimization efficiency. When more collection points be selected, the NLP solving precision becomes higher, whereas, it is based on the assumption of reasonable initial solution, otherwise, not only the

solving efficiency will decreasing, even resulting in infeasibility solution.

Intelligent optimization algorithm possesses unique superiority in solving parameter optimization problem, which doesn't require an initial guess of the discrete optimal trajectory, but only needs ranges of the parameters as a search area. Also, it didn't require the gradient of the objective function and constraint condition, and this information existing in the form of implicit function isn't easy to be obtained in this paper. Furthermore, it tolerates the discontinuities encountering which usually causes slow convergence or divergence. Most importantly, it possesses a higher global searching ability. Based on these advantages, this paper applied PSO to acquire NLP initial solution and the solving procedure is as follows. [6, 9]

Step 1: Set a terminal time t_f and discrete the whole time span $[t_0, t_f]$ on N nodes $t_0 < t_1 < ... < t_N = t_f$. Then, acquire the N+1 collected points and corresponding decisive variables are $u_i (i = 0, 1, ..., N)$.

Step 2: Given control vector $\mathbf{u}_i (i = 0, 1, \cdots, N)$ and initial solution $\mathbf{x}_0$, integrate the state differential function (22) using Runge-Kutta method to obtain the each collected point and terminal states.

Step 3: Based on the cost index of closing velocity (21), boundary conditions (24) and path constraints (23), using PSO arithmetic to solve the parameter optimization problem so as to obtaining the control sequence $\mathbf{u}_i$.

Step 4: According to the solved $\mathbf{u}_i$, reintegrate the (22) to acquire state value $\mathbf{x}_i$ for each discrete point.

Step 5: Apply cubic spline interpolation approximation variable $\mathbf{u}_i$, $\mathbf{x}_i$ and t_i from discrete points to LG collected points, and finally constitute the initial solution of NLP.

4.2 *Solving the optimiation problem*

The numerical arithmetic solving NLP mainly includes penalty function, feasible direction methods and so on. Han and Powell extend the quasi-Newton method to the constrained optimization problem, then, develop the SQP arithmetic, which outperforms many other well-known conventional algorithms and has been considered as one of the most efficient methods for NLP problems. The general ideal is carried out by iteratively solving a sequence of quadratic programming, and this paper applies SQP arithmetic to solve the obtained NLP. [10]

5 SIMULATION AND ANALYSIS

This section presents a numerical example to verify the efficiency of the method mentioned above.

The aircraft parameter is referring F161,2,11, and the maximum pitch rate, maximum bank angle rate, maximum dynamic pressure, minimum altitude, maximum load factor are set to $Q_{a,\max} = 30^o/s$, $P_{a,\max} = 80^o/s$, $q_{\max} = 80\,kPa$, $h_{a,\min} = 500\,m$, $n_a \leq 9$.

The missile's engine is off being endgame evasion, so the thrust is 0, and mass is constant. The IPN constant, time delay, and load factor are set to $N = 4$, $\tau = 0.3s$, $n_m \leq 40$. The drag force satisfies $D_m = k_1 V_m^2 + k_2(a_p^2 + a_y^2)/V_m^2$, where $k_1 = 1/2\rho(h_m)S_m C_{Dm0}$ and $k_2 = 2k_m m_m^2/\rho(h_m)S_m$. The initial combat condition is listed in Table 1.

While selecting the terminal reference distance $h(\mathbf{x}_f) = 150\,m$, using PSO arithmetic to get the initial solution, and the red line represents the simulation curve. Based on the initial value assumptions above, the GPM and the SQP are integrated, and the blue line shows the simulation result. The aircraft control, relative distance, closing velocity, missile load factor, and velocity results are plotted in Figures 1 and 2.

Comparing the simulation graph, the initial solution from PSO and the final optimization result from GPM are consistent basically, so the GPM initial solution solved by PSO is feasible and efficient. Through 10.38 s simulation, GPM opti-

Table 1. The initial combat state of aircraft and missile.

	$x\,(m)$	$y\,(m)$	$h\,(m)$	$v\,(m/s)$	$\gamma\,(°)$	$\chi\,(°)$
Aircraft	0	0	10000	250	0	180
Missile	−6000	0	10000	600	0	0

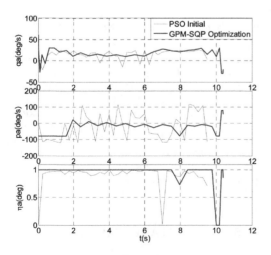

Figure 1. The graph of pitch rate, roll rate and thrust angle caption.

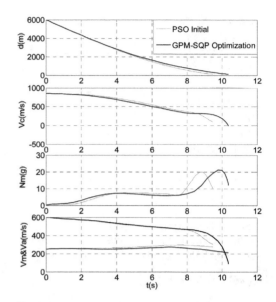

Figure 2. The graph of relative distance, closing velocity, missile load factor and velocity.

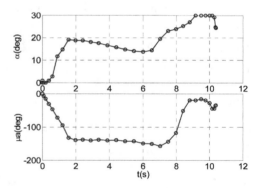

Figure 3. The graph of attack angle and bank angle.

mization relative distance is up to 150 m, and the closing velocity is about to −14.18 m/s. However, the corresponding PSO initial simulation time is 9.50 s when the relevant distance is 150 m, and the closing velocity is 94.88 m/s at this moment. This result shows that GPM-SQP arithmetic can acquire higher accuracy in solving pursuit-evasion NLP problem.

The attack angle, bank angle and 3D flight trajectory base on GPM-SQP arithmetic are displayed in Figures 3 and 4.

The simulation can also show that: in the situation of head-on shot, when the performance index is defined as minimum closing velocity, the throttle setting is constantly at $\eta_a = 1$ in the whole flight stage to acquire a faster evasion speed. In the initial

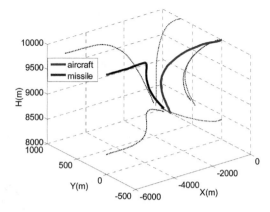

Figure 4.   3D flight trajectory of aircraft and missile.

evasion stage, the mainly strategy is rapidly turning downwards, and the bank angle is almost constant value of 160°. At the same time, increase the attack angle to decrease the flight altitude, because the lower altitude has denser atmosphere density which can increase the missile's drag force, and weaken the energy and speed advantage quickly. Furthermore, aircraft engine can obtain bigger thrust force because of dense atmosphere that can further avoid missile reach to the target.

In the combat final stage, when the relative distance is small, aircraft should turn quickly and execute high attack angle and maneuver greatly so as to acquire bigger LOS angle rate, at this time, the missile load factor is up to 20 g, and the velocity advantage is weakened more quickly. When the relevant distance is about 150 m, the closing velocity is up to −14.18 m/s, and the aircraft evades the missile's attack successfully. In sum, the simulation result is in consistent with real combat pilot maneuver that is turning, diving, and high attack angle.

6   CONCLUSIONS

This paper mainly researches about aircraft maneuver evasive problem in the situation of missile attack. Firstly, construct the extended mass point aircraft model considering agility, in the assumption of missile proportional navigation, the aircraft missile duel is modeled as a single trajectory optimization problem. Then, the GPM is used to convert the continue TPBVP to a discrete NLP problem, whose initial solution is gained by PSO algorithm, and the SQP is applied to solve it. Finally, numerical simulation example verifies the

adopted algorithm validity and the result shows that GPM-SQP is an efficient tool for pursuit-evasion trajectory optimization.

REFERENCES

Andrey Perelman, and Tal Shimaz, "Cooperative differential games strategies for active aircraft protection from a homing missile," Proceedings of AIAA Guidance Navigation and Control Conference, Toronto, Ontario, Aug. 2010, pp. 1–22; also AIAA Paper 2010-7878.

Arthur Vermeulen, "Missile avoidance maneuvers with simultaneous decoy deployment," Proceedings of AIAA Guidance Navigation and Control Conference, Chicago, illinois, Aug. 2009, pp. 1–17; also AIAA Paper 2009–6277.

David A. Benson, Geoffrey T. Huntington, Tom P. Thorvaldsen, and Anil V. Rao, "Direct trajectory optimization and costate estimation via an orthogonal collection method," Proceedings of AIAA Guidance Navigation and Control Conference and Exhibit, Keystone, Colorado, Aug. 2006, pp. 1–14; also AIAA Paper 2006-6358.

David Benson, "A gauss pseudospectral transcription for optimal control," Aeronautics and Astronautics Massachusetts Institute of Technology, Massachusetts, 2005, pp. 49–55.

Henri Hytönen, "Utilization of air-to-air missile seeker constraints in the missile evasion," Helsinki University of Technology Department of Engineering Physics and Mathematics Systems Analysis Laboratory, Helsinki, 2004, pp. 1–37.

Janne Karelahti, Kai Virtanen, and Tuomas Raivio, "Near optimal missile avoidance trajectories via receding horizon control," Journal of Guidance Control and Dynamics, Vol. 30, No. 5, 2007, pp. 1287–1298.

Kazuhiro Horie, "Collocation with nonlinear programming for two-side flight path optimizaion". University of Illinois, Illinois, America, 2002, pp. 81–96.

Philip E.G, "SNOPT: An SQP algorithm for large scale constrained optimization," SIAM Review, Vol. 47, No. 1, 2005, pp. 99–131.

Tuomas Raivio, and Jukka Ranta, "Optimal missile avoidance trajectory synthesis in the endgame," Proceedings of AIAA Guidance Navigation and Control Conference and Exhibit, Monterey, California, Aug. 2002, pp. 1–11; also AIAA Paper 2002-4947.

Vitaly Shaferman, and Tal Shima, "Cooperative multiple model adaptive guidance for an aircraft defending missiles," Proceedings of AIAA Guidance Navigation and Control Conference, Toronto, Ontario, Aug. 2010, pp. 1–25; also AIAA Paper 2010-8320.

Wang, X.P., Lin, Q.Y., Dong X.M., "Aircraft evasive maneuver trajectory optimizaion based on QPSO," 2010 IEEE International Conference on Ultra Modern Telecommunication and Control System and Workshops, 2010, pp. 416–420.

Control Engineering and Information Systems – Liu (Ed)
© 2015 Taylor & Francis Group, London, ISBN 978-1-138-02685-8

A self-biased wideband amplifier with high linearity

C.P. Liu & Z.R. He
Sichuan Institute of Solid State Circuits, China Electronics Technology Group Corp., Chongqing, China

ABSTRACT: A novel configuration for self-biased wideband amplifier with high linearity is presented in this paper. This amplifier has been realized by 2 μm InGaP HBT process: This amplifier exhibits high performance including: over DC ~6 GHz, input return loss is well blew 14 dB, out return loss is well blew 9.8 dB. At 2 GHz, the typical power gain is 19.9 dB, 1 dB compression point is 16.12 dBm. with a insensitivity to temperature. while consuming 56.2 mA from 5 V supply at 25°C.

1 INTRODUCTION

With the needs for wireless and portable communications, there is steady increase in the demand for low-cost, miniaturized RF and microwave devices and components. which have been leading us to the development of the wideband devices (Aemijo et al. 1989).

Wideband amplifiers are important functional blocks in RF applications. Such as microwave wireless receiver front-end. As a key component in communication systems, the amplifiers need to amplify the received small signals with sufficient gain and as high Linearity as possible for Wideband communications products. therefore, there is all urgent need to develop the RF equipments for a high-gain level of the analogue part. Among various technologies, the excellent performance of Darlington amplifiers as a wide-band stage Call satisfy the requirement (Liu et al. 2013).

Integrated Darlington amplifiers are often employed in transistor-like integrated circuit package. such amplifiers packages are attractive because they typically have a small outline, a low cost, and a user-friendly implementation.

In this paper, detailed analysis and general design procedures about novel self-biased wideband amplifier are given. Using this reconfigurable architecture, the amplifiers have a lower sensitivity to temperature, higher output voltage headroom, output power and so on (Li et al. 2010).

2 GETTING STARTED

Darlington configuration is a compound structure consisting of two transistors connected in such a way that the Darlington amplifier has been shown to have a wideband higher gain characteristic when

compared to the single transistor (Ding et al. 2010, Armijo et al. 1989).

2.1 Conventional Darlington amplifiers

Conventional Darlington amplifiers have one or more of the following disadvantages:

a. Across a wide range of temperature, the total current change significantly;
b. A limited output voltage swing and the maximum output power capability;
c. A unsafe voltage pulse on the output is formed during the start-up process (Nisbet et al. 2003).

Referring to Figure 1, a conventional Darlington amplifier is shown. Normally the Darlington amplifiers are designed to operate with an ideal current source. In practice the current source is often substituted with a voltage source and

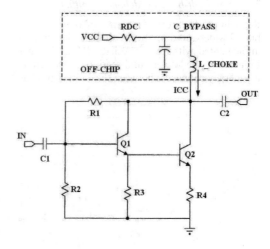

Figure 1. Conventional Darlington amplifiers.

an off-chip bias set series resistor RDC, which converts it to a reasonably good current source. RDC value might be few hundred ohms, typically, the more voltage drop across the resistor RDC, the better the stability of the bias current Icc over temperature and supply variations; therefore, the DC voltage required to operate the amplifier will be higher than the device voltage. An optional L_ CHOCK may be connected in series with the RDC to increase the total shunting impedance in order to minimize gain and power loss, provide a RF chock (Kobayshi et al. 2005).

If a low frequency response is desired, the de-coupling capacitors C1 and C2 can have a significant value. When the voltage is turned on, capacitors C1 and C2 will charge at a differing rate due to the difference in time constants because series resistors source impedance+R1 is different from load impedance. The total current will be split into two unequal parts, initially in a proportion to approximately (source impedance+R1)/load impedance. Figure 2 presents the Conventional Darlington amplifiers device voltage versus time. The voltage bump can overload and possibly damage the circuitry connected (Mordkovich et al. 2005).

With increasing temperature, the voltage across Vbe of each of the transistors Q1 and Q2 decreases, and vice versa. So, assuming that the resistor values do not change with temperature, the voltage across resistor R3 and resistor R4 is higher at high temperature, due to the lower Vbe of transistors Q1 and Q2. Hence more current flows through the collectors of transistors Q1 and Q2 at high temperature. At lower temperature, the opposite holds true.

2.2 Novel self-biased wideband amplifier

Referring to Figure 3, a diagram of the novel self-biased wideband amplifier is shown. The circuit is implemented without the resistor RDC shown in Figure 1 of the background section. A bias circuit is generally connected between the emitter and base

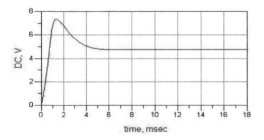

Figure 2.   Device voltage versus time.

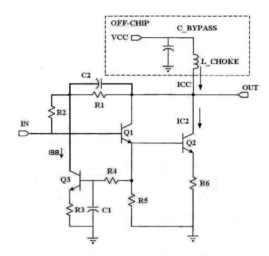

Figure 3.   Novel self-biased wideband amplifier.

of the transistor Q1. The bias circuit may be used to stabilize the bias of the wideband amplifiers without relying on an external resistor RDC. Without the voltage drop across the resistor RDC, the device voltage can be the same with the DC voltage VCC, and without the unsafe voltage pulse.

The bias circuit generally comprises three resistors R2, R3, R4, a transistor Q3 and a capacitor C1. The resistor R2 may be implemented as an RF isolation. The resistor R3 may be coupled between the emitter of the transistor Q3 and ground. The resistor R4 may be implemented as an RF blocking resistor. The transistor Q3 may be implemented as a biasing transistor. The capacitor C1 may be implemented as an AC bypass. The transistor Q3 generally operates as a pseudo mirror bias transistor of the transistor Q2. The bias circuit generally works in conjunction with the parallel feedback resistor R1 to set up a reference current IBB. The current IBB is approximately mirrored to the output transistor Q2. The relationship between IBB and IC2 is only a approximate, but generally mirror each other in current over temperature, supply voltage, and input drive level variations. The ratio of the areas of the transistor Q3 and the transistor Q2, and the emitter resistors R3 and R6, are generally scaled in proportion to the bias currents IBB and IC2, respectively. However, other ratios may be implemented to meet the design criteria of a particular implementation.

The bypass capacitor C1 and the RF blocking resistor R4 set the lower frequency limit of operation. The lower frequency may be extended by increasing values of the capacitor C1 better, because increasing the values of the resistor R4 will generally degrade the bias mirroring relationship between the transistor Q3 and the transistor Q2.

3 RESULTS

Using the novel configuration, a amplifier has been realized. The circuit with package measurement was completed by using a computer controlled test system, Which consists of computer platform, GPIB interface, Agilent PNA-X network analyzer and data acquisition card. Input and output

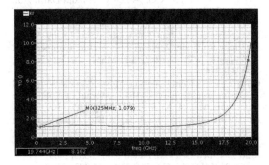

Figure 4.   Simulated stable factor K.

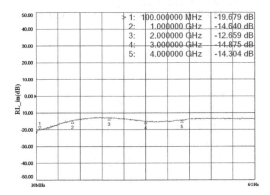

Figure 5.   Input return loss.

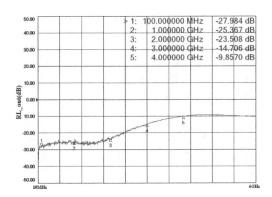

Figure 6.   Output return loss.

inductance of 0.5 nH included. The simulation of our amplifier have been presented based on the Cadence spectre. Using a inductor the proposed amplifier keeps unconditional stable over DC-20 GHz (Fig. 4), over DC ~6 GHz, input return loss is well blew 14 dB (Fig. 5); out return loss is well blew 9.8 dB (Fig. 6); At 2 GHz, the typical power gain is 19.9 dB (Fig. 7); output 1 dB compression point is 16.12 dBm (Fig. 8); current variation of 3 mA at 5 V over 0°C to 85°C (Fig. 9); Figure.10 presents the layout of the amplifiers with a chip size of 0.47*0.72 mm².

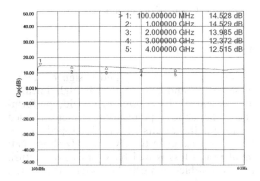

Figure 7.   Power gain.

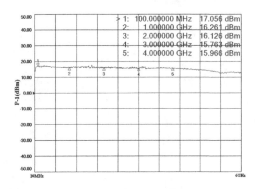

Figure 8.   Output 1 dB compression point.

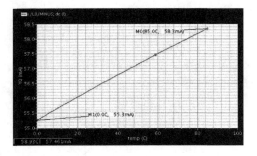

Figure 9.   Current variation over temperature.

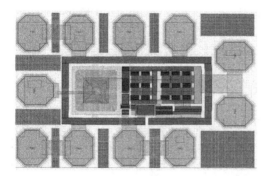

Figure 10.   Layout of the amplifiers.

4   CONCLUSION

In this paper, the theory and design of a self-biased wideband amplifier with high linearity are presented. To ensure high yield, performance redundancy optimization strategy is used in design. the results of the developed amplifier with package in the DC ~6 GHz show that: input return loss is well blew 14 dB; out return loss is well blew 9.8 dB; At 2 GHz, the typical power gain is 19.9 dB; output 1 dB compression point is 16.12 dBm; current variation of 3 mA at 5 V over 0°C to 85°C. the proposed one is good candidate for Darlington wideband amplifiers applications.

REFERENCES

Armijo, C.T., Meyer, R.G. "A new wide-band Darlington amplifier," Solid-State Circuits, IEEE Journal of vol. 24, Issue 4, pp. 1105–1109, Aug 1989.

Cheng Peng Liu; Jian Gang Shi; Guo Qiang Wang; Zheng Rong He,"A Novel Temperature Compensated Amplifier" Proceedings of MSIT, 2013: 756–759, p781–784.

Chris T. Aemijo and Robert G. Meyer, "A new WideBand Darlington Amplifier". IEEE Journal of Solid-State Circuits. Vol 24, N0 4. Aug 1989, p1105–1109.

Ding Hua-feng. "A 0.1–4 GHz Darlington Cascode Broadband Gain Block". Telecommunication Engineering, Nov. 2010: p80–84.

Joho J. Nisbet, CA, U.S. Patent 6,665,353. (2003).

Kevin W. Kobayshi, CA, U.S. Patent 6,972,630. (2005).

Li Xiao qian. "High Linearity-wideband PHEMT Darlington Amplifier". RESEARCH & PROGRESS OF SSE, jun,. 2010:p218–221.

Mikhail Mordkovich, NY, U.S. Patent 6,943,629. (2005).

Control Engineering and Information Systems – Liu (Ed)
© 2015 Taylor & Francis Group, London, ISBN 978-1-138-02685-8

The application design of transfer lines in automatic workshop based on FCS

J.H. Wang, Z.R. Li, D.H. Sun & G. Chen
North China Institute of Aerospace Engineering, Langfang, P.R. China

ABSTRACT: This paper introduces the design of automatic workshop transfer lines based on FCS, points out the meaning and function of transfer lines in the factory. In order to ensure the safe and reliable operation of transfer lines, to achieve efficient and safe production, the automatic transfer system is designed, which based on Field Bus and PLC, in which the detection and hydraulic technology have been applied. Transfer lines motor starting method Includes manual and automatic startup started. Automatic transmission tasks can be done automatically by a specific plan; manual single motor can be run separately, deliver, repair purpose.

1 INTRODUCTION

With the development of digital level instrument and equipment in the industrial production process continuously improve; many factories are in the construction of the automatic production line based on the digital bus. This is an automatic transfer lines based on field bus, the system uses the FCS structure, namely the bus control system, mainly composed of PLC control system and field bus.

Comprehensive workshop of modern production workshop for processing, transmission, processing, and delivery of one, lifting the traditional production mode using the crane hoisting of semi-finished products, crane need hand operation, but also has special hoisting, workshop environment with high requirement of space, but also the existence of security risks. Using the FCS system to realize the workshop automatic transfer lines is feasible, each area of the whole transfer lines covering the whole workshop, transfer lines can be the whole operation, can also operate independently of each area, so as to improve the work efficiency, the overall transfer lines is controlled by PLC, are connected by a field bus car each area control unit, design and implementation individual independent, integrated modernized production workshop.

2 SYSTEM REQUIREMENTS

The whole transfer lines consists of 6 separate transfer lines, each line by connecting each conveying vehicle transportation, transfer lines and contains 7 work stations, stations and transfer lines through the transport of materials transport vehicle.

Each transfer lines using AC motor and transmission chain connection form, divided into motor drive transfer lines independently at each station, the station is designed to achieve the independent operation better. Transport by rail vehicles run way, driven by AC motor, hydraulic lifting implementation of goods transport. The sensor at the same time in each independent transfer lines began operation, intermediate and end are equipped with detection items, determine whether the goods in line, at the same time in the transportation car rail ends, hydraulic lifting the upper and lower as the position sensor, accurate, fast, safe production and processing of the objective. Workshop requirements shown in Figure 1 (Hu, Duan and Ding 2005).

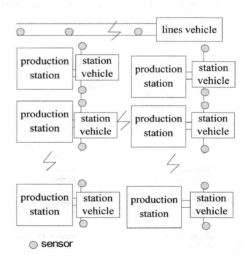

Figure 1. Workshop requirements.

3 SYSTEM DESIGN

The design requirements for workshop, design the application of fieldbus, PLC, sensor, hydraulic technology, motor drive technology and configuration monitoring technology. Used in the design of network structure of three layers, namely the application layer, management layer and field bus. The field application layer belongs to the sensor detection layer application of AS-I network to collect data on the sensor, and the site for inverter motor drive control, including start, stop, frequency conversion speed regulation. Fieldbus layer realization is based on data communication layer of PLC and remote module, complete control of the area, including data sharing, remote control etc.. The management of using Ethernet as the media, to achieve on-site data monitoring, finishing, trend analysis, instruction etc.

Which based on the part of the system transfer lines with Siemens S7-300 programmable controller as the core, build together with the industrial Ethernet and Profibus bus as the main communication forms. System structure diagram is shown in Figure 2 (He, Huang and Shi 2007).

Only by adopting information acquisition technology, fully, real-time, accurate state information of the manufacturing process, and on this basis to realize the effective control and management of the manufacturing process, in order to improve the flexibility, robustness and fault handling capacity of the whole system.

The system mainly includes industrial control computer, PLC, remote communication module, power supply, input and output module, analog input and output module, sensor etc.

Industrial control computer as the monitoring equipment system, application of Siemens configuration software WINCC to design the production flow chart, the computer can realize the monitoring of the area sensor, operation of equipment, has the function of alarm information, remote statements, production, production trend summary series function.

PLC can be programmed logic controller, is a specially designed for applications in the industrial environment and the design of digital system, integrated computer technology, automatic control technology and communication technology, used for the control process, the user oriented "natural language" programming, meet industrial environment, simple and easy to understand, easy operation, high reliability, is a new generation of general industrial control device. It uses programmable memory, used to implement logic operations in the internal memory, sequence control, timing, counting and arithmetic operation instruction, and through digital, analog input and output, machinery or production process control of various types, programmable controller and related equipment, should according to with the industrial control system as a whole, easy to expand functions of the principle of design. The system uses S7-300 PLC as the control core, complete the whole transfer lines state collection, judgment, operation control and communication.

The remote communication module based on SIEMENS ET200M. Based on Profibus field bus network technology of SIEMENS company to realize the solution of data communication system. ET20OM distributed I/O system is a part of the SIEMENS automation system, it open based on PROFIBUS bus technology, can realize the scene from the signal to a data communication control cabinet, because the data transmission rate is very high, to ensure the reliable communication between controller and I/O equipment. Through the field bus system has realized remote digital and analog input module, output module, signal intelligence apparatus and process adjustment device with Programmable Logic Controller (PLC) data transmission between and computer, the input and output channels in the field of the actual needs of the equipment, reduces the installation and wiring costs, reduce the cost.

The power supply adopts PS307 power module, the 220V AC voltage into DC voltage required by 24V. Provide power for the controller and sensors and actuators.

Automatic transfer workshop will be a large numbers of I/O equipments in workshop, so the system by use of the remote I/O ET200M module, connection with the master through field bus.

The sensor adopts a non-contact sensor, according to the need to choose the electromagnetic or photoelectric sensor, and can be directly

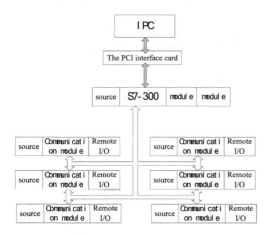

Figure 2.   System structure diagram.

transmitted to the PLC production line state, a state judge, and based on the results of the production line for motion control.

The working process of the system, in the production process, real-time call workers, can be manually, automatic selection, in the automatic mode, the position in accordance with the technological advance design and requirements, automatically conveying articles for each station by PLC (Heng 2007). In manual mode, in accordance with the station needs manual control conveying vehicle implementation of station and transfer lines items between the transfer, can also manually control transfer lines running, the station or other work items or shipped.

4 SOFTWARE DESIGN

4.1 *System functions*

The workshop production information acquisition and monitoring system mainly realize the following functions (Zhu 2005):

1. The production of information analysis and coding, including the monitoring (current position and assembly state), the post condition monitoring (the running state of the equipment and emergency requests), production process monitoring (logistics, production system of each production unit complete variety and quantity of the workpiece etc.), fault monitoring;
2. Information acquisition based on field bus;
3. The use of high-level language programming realization of supervisory control and data acquisition.

4.2 *Software structure*

The software structure of the system can be divided into three parts (Zhu 2005):

1. The data acquisition layer, which is the programmable controller by bus to realize the field data collection. The programmable controller used language, the main flow chart as shown in Figure 2;
2. Data transmission layer, is mainly the programmable controller and the host computer for data exchange between. The programmable controller part of the programming language using programmable controller, the central computer using C++ advanced language.
3. The data management layer, mainly through the Ethernet to achieve the upper center computer and remote management computer terminal data interaction, using the advanced C++ language. This part of the software, custom data communication protocol, in order to achieve

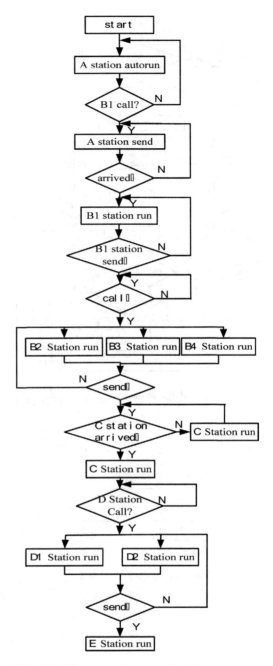

Figure 3. The computer program flow 1.

the distinction between data, instructions; make more timely, reliable data interaction.

Software design includes the monitoring software of PC and PLC software design and PC design. According to the actual operation process in response to the position requirements, require-

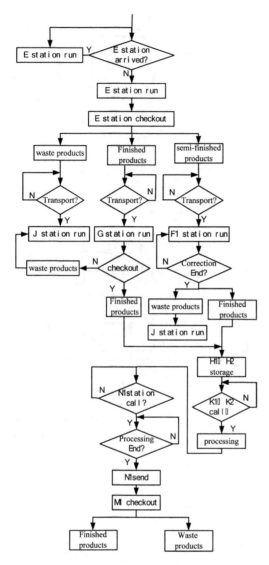

Figure 4. The computer program flow 2.

ments of real-time response field, at the same time display operation status and response operation content in the upper computer, the dynamic form of the response to the operation. The upper computer is mainly based on WINCC configuration software, real-time communication with client, will be the production of real-time dynamic simulation on a computer screen. WinCC developed by SIEMENS, is the first to use the 32 latest technologies in process monitoring system, and has good openness and flexibility. The upper monitoring software and data trend chart, data sharing and other workstation. The computer program flow diagram of Figure 3, shown in Figure 4.

The machine adopts PLC as the main control unit, the completion of the main production site acquisition, transfer lines condition, and according to the control strategy to control the production line motor to carry out actions, production.

ACKNOWLEDGEMENT

This paper is supported by Hebei provincial key disciplines detection technology and College Fund of North China Institute of Aerospace Engineering Nciae.

REFERENCES

Hengyue. "The PLC automation control system of modern manufacturing" Modern manufacturing 2005 11;

Li Jingfang; facing Fieldbus Industrial Age [J]; electronic technology; in 2005 10.

Liang Hu, Fajie Duan, Keqin Ding. "Development of on-line computer vision detection system of steel strip surface defects" Iron and steel, 2005, 40(2): 59–61.

Yonghui He, Shengbiao Huang, Guifen Shi. "Research on Application of cold steel strip surface defect detection system online". China Metal Institute. 2007 China steel annual meeting proceedings. Beijing: 2007.

Zhu Qiubo. "Communication network of distributed control system" control engineering; 2005 02.

Research on adaptive gateway math model of Coalmine Monitoring System

Y.G. Xu, G. Hua & Y.J. Zhao
School of Information and Electronic Engineering (SIEE), China University of Mining and Technology (CUMT), Xuzhou, Jiangsu, China

ABSTRACT: This paper chiefly researches how to build a math model of adaptive gateway for Coalmine Monitoring System (CMS), following analysis and studying the communication protocol between gateway and PC and existed systems, we propose the conceptions of data frame, unified data model, characteristic pick-up matrix, and characteristic vector, and lucubrating the property of them. Basing on the results, the general math model of CMS is build up. The model sets up theory foundation for adaptive gateway's development, and it is applied in project of Coalmine personnel positioning system to Huaibei' mines successfully.

1 INTRODUCTION

With meticulous management of production safety of coalmining enterprises, more and more types of control systems into the monitoring, however, these systems are all the independent agents, most of them are charged with different controlling and monitoring missions, when integrating these systems, it is necessary to understand the communication protocols of each system, concretely, grasping the framing. All of these are not difficult to approach with systemic methods. But there are not uniform protocols and date analyzing methods, when a new system is accessed in, the relative accessed gateway is demanded naturally, so, the workload is heavy. If the system has some changes, adapting the former program to the changes, it makes the usage and maintenance more difficult. The purpose of this paper is to find a kind of generally procession method for data frame of monitoring system on algebra transform. Building up foundation of theory and providing viable model for developing accessed gateway.

2 MONITORING SYSTEM ACCESSED MODEL

Most of monitoring systems are similar to data transmission, processing, storage and similar hardware framework, which makes that adaptive gateway can adopt unified protocol structure with deferent hardware interface and core software. Based on it, a adaptive gateway has been proposed, which is used to interconnection different monitoring systems in coalmine, which has 2-CAN, 2-232, 2-485, 1-RJ45 ports, the Ethernet port links ring network underground, the others link monitoring systems. The device plays a important role showing as Figure 1.

The gateway connects other monitoring systems with 232/485/CAN to gather data. The advantage

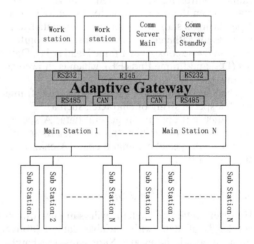

Figure 1. The adaptive gateway of monitoring system.

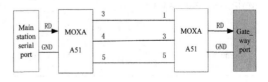

Figure 2. The application model of adaptive gateway.

of this model does not influence original systems separated by the main station. To prevent physical damage in interconnection, Opt electronic Isolator is adopted. Showing as Figure 2.

3 CONCEPTS AND DEFINITIONS

3.1 Adaptive gateway

Adaptive gateway is a kind of intelligent device, which connects two or more different systems with standard hardware interface and adaptive protocol. During the system integration, generally, the IP network acts as a unified integration platform, so various monitoring systems must exchange information on IP network with a unified information structure by a special adaptive gateway, which the integrated system can recognize correctly.

3.2 Monitoring system data frame

A monitoring system is generally composed by a main station, substations, and/or wireless nodes, communication bus and etc.; the substation goes out a bunch of data per cycle, which is called a data frame. According the usages, it is generally divided into two major classes: controlling frame and data frame. Controlling frame is used to distribute commands from the main station to substations or to reply the main station's commands. The data frame is the result gathered by substations; one monitoring data frame consists of Frame Head (FH), Data Set (DS) and Frame End (FE), which is showed as Figure 3.

The frames generally have a same fixed frame head, a special string or character, such as ".", "^" etc, which not appears in general data. A special frame end character like "return" etc. Its length is fixed with special command char. (A channel is supposed a Byte).

3.3 The data model of data frame

However, different system will samples different kinds of data with variable data, adaptive frame model is necessary. Supposing the length of per sample data is S bytes, then, every frame $X = \{x_i \in R^S \mid x_1, x_2, ..., x_s\}$ can be shown as following column vector:

Frame Head (FH)	Data Set(DS)	Frame End (FE)

Figure 3. Frame structure of fixed length.

$$X = \left(x_1, x_2, .., x_S \right)^T. \tag{1}$$

3.4 Unicode data model

Without loss of generality in system analyzing, supposing there are M analogy, N digital inputs, and K digital outputs in a substation, $A_{M \times 1}, D_{N \times 1}, C_{K \times 1}$ respectively, and the expression of the sample value is made of row vector F:

$$F_{(M+N+K) \times 1} = \begin{pmatrix} A_{M \times 1} \\ D_{N \times 1} \\ C_{K \times 1} \end{pmatrix}. \tag{2}$$

4 OFFSET SHIFTING OPERATOR

1. *Left shifting operators $LS^n(x)$*: The function of left shifting operator $LS^n(x)$ is to carry out the left shifting n one bit of the byte variable x.
2. *Right shifting operators $RS^n(x)$*: Similliar to the left shefting operator with n bits right-shift by $RS^n(x)$.
3. *Obtaining string*: Supposing getting bit clusters of 5th and 6th bits of x, can be showed as following formula: $Y = RS^6(x)LS^2(x)$, it means data is left shifted two bits first, then right shifted six bits, the result is the bit clusters we need. If getting the 3th bit value of variable with x, showing as the following formula: $Y = RS^2(x)LS^5(x)$, with simple expression $Y = R^6 L^2$ and $Y = R^2 L^5$ corresponding.

5 IDENTIFICATION OF DATA FRAME

Supposing the data can be sampled continuously, then, a continuous data stream can be got in gateway. The purpose of frame identification is that recognizing data frame correctly, which comes from different substations, special identified format in the data stream coming from different substation, after adaptive frame analyzer, then gets sampling value or control value of per in-out channel. Before identification, assuming the character and composing of data frames are absolutely clears; the problem now is how to draw out them from data stream.

5.1 Method

Supposing the length of data frame is S, then, getting length of S vector X from data stream:

$$X = (x_1, x_2, ..., x_S)^T \quad x_i \in (0,255). \tag{3}$$

Supposing characteristic pick-up matrix of some data frame is H_i, if X is the kind of data frame, then it must be

$$H_i X - X_{iT} = 0. \qquad (4)$$

Here, H_i is Characteristic pick-up matrix, which is of a kind data frame character; X_{iT} is vector, which is of a kind data frame character. For data frames of different types, H_i and X_{iT} are different, thus distinguishing problem of data frames transform into solving H_i and X_{iT}.

5.2 Characteristic pick-up matrix H_i of data frame

Defining Characteristic pick-up matrix of data frames is $S \times S$ phalanx. It picks up the character of data frame which would be judged, then compare the result, which is multiplied H_i and X, with character vector X_{iT}, if the result is accordant, it show this data frame is from substation which is of the type. If not, it need be picked up again; for the right data frames, it will be sent to frame analyzer.

Supposing the structure of data frames shown as Table 1, the head of the frame can be found 126 from it, the sum of the second and third bytes are $0 \times ff$, the fourth byte is the type of substation, the fifth byte is the type of order 0×02 (data collecting order), the end of the frame is 0×51. Based on the characters, the character picked up matrix is

$$H_i = \begin{pmatrix} 1 & & & & & & & & \\ & 1 & 1 & & & & & & \\ & & 0 & & & & & & \\ & & & 1 & & & & & \\ & & & & 1 & & & & \\ & & & & & 0 & & & \\ & & & & & & 0 & & \\ & & & & & & & \ddots & \\ & & & & & & & & 0 \\ & & & & & & & & & 1 \end{pmatrix}_{23 \times 23} . \qquad (5)$$

Put H_i into the equation (4) upward, then: if X satisfies:

$$x_1 = 126$$
$$x_2 + x_3 = 255$$
$$x_4 = 2$$
$$x_5 = 2$$
$$x_{23} = 85.$$

Then it can be confirmed that is data frame of the ith substation. H_i corresponds a set of equations; it actually indicates the restricted relation between data of the frame. As long as the structure of data frame and H_i are found, it's not difficult to judge its type.

5.3 Character vector X_{iT} of data frame

Definition: the character vector of data frame is a nonzero vector; the dimension is L. It

Table 1. The data frame structure of some sub station.

FH: AE, Thereunto: 32 is the style of substation; 53 correspond responding of order	Analogy (low precision), low byte of Analogy (high precision) A01~A08	High 4 bits (d01~d08) state; low 4 bits, high precision Analogy high 4 bits (A01~A08)	Controlling quantity	End of frame
1~5	6~13	14~21	22	23
7E	A01	D01	C01~C08	51 h
sNo	A02	A01 H4		
rsNo	A03	D02		
32	A04	A02 H4		
53	A05	D03		
	A06	A03 H4		
	A07	D04		
	A08	A04 H4		
		D05		
		A05 H4		
		D06		
		A06 H4		
		D07		
		A07 H7		
		D08		
		A08 H4		

expresses essential character of data frame. From formula (4)

$$X_{iT} = H_i X. \tag{6}$$

The condition is that $\mathbf{X}$ must be a standard data frame. The key of extraction of data frames is to find H_i, put a standard data frame in the formula (6), then, for the frame of Table 1:

$$X_{iT} = (126, 255, 0, 2, 2, 0, ..., 85)^T. \tag{7}$$

Obviously, component of the character vector indicates the structure of frame and characteristic value, such as the sum of the second and third bytes are 255, the head of the frame is 126, the end of the frame is 85 and so on.

6 ANALYSIS OF DATA FRAMES

6.1 *The definitions of data frames analysis*

The so-called data analysis is to find out the $(M + N + K) \times L$ rank transformation matrix T, and satisfy formula followed:

$$F = TX. \tag{8}$$

In the formula: X is legal data frame, for different substation types, there are different T.

6.2 *Examples*

Supposing a frame structure of some substation is shown in Table 1 (sNo: substation code rsNo: substation contrary code).

There are eight analogy imports, eight digital imports, and eight digital outputs. Supposing $M = 8, N = 8, K = 8$ then T is 24×24 matrix. It is not difficult to write out from the frame structure:

$$a_i = x_{6+i-1} + 256 * R^4 L^4 x_{14+i-1}, \, i = (1,2,3,...,8). \tag{9}$$

$$d_i = R^4 x_{14+i-1}, \quad i = (1,2,3,...,8). \tag{10}$$

$$c_i = R^7 L^{8-i} x_{14+i-1}, \quad i = (1,2,3,...,8). \tag{11}$$

Changing the equation group upwards into the style of matrix, then T is

$$T = \begin{pmatrix}
0\,0\,0\,0\,0\,1\,0\,0\,0\,0\,0\,0\,0 & 256R^4L^4 & 0 & 0 & 0 & 0 & 0 & 0 & 0 & 0\,0 \\
1 & & 256R^4L^4 & & & & & & & \\
& 1 & & 256R^4L^4 & & & & & & \\
& & 1 & & 256R^4L^4 & & & & & \\
& & & 1 & & 256R^4L^4 & & & & \\
& & & & 1 & & 256R^4L^4 & & & \\
& & & & & 1 & & 256R^4L^4 & & \\
& & & & & & 1 & & 256R^4L^4 & \\
& & & & & & & R^4 & & \\
& & & & & & & & R^4 & \\
& & & & & & & & & R^4 \\
& & & & & & & & & R^4 \\
& & & & & & & & & R^4 \\
& & & & & & & & & R^4 \\
& & & & & & & & & R^4 \\
& & & & & & & & & R^4 \\
& & & & & & & & & R^7L^7 \\
& & & & & & & & & R^7L^6 \\
& & & & & & & & & R^7L^5 \\
& & & & & & & & & R^7L^4 \\
& & & & & & & & & R^7L^3 \\
& & & & & & & & & R^7L^2 \\
& & & & & & & & & R^7L \\
& & & & & & & & & R^7
\end{pmatrix}$$

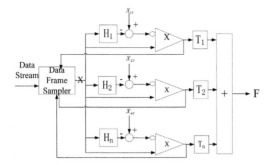

Figure 4. Calculation model of system accessed gateway.

The above transformation matrix is the algebra transformation matrix that aims at the data frame upwards. While carrying on design of gateway driving programs, as long as setting right matrix **T** upwards, then can analyze this type of data frame. Therefore, uniform system accessed program can realize multi system access, and not need alter the driver.

7 THE MATH MODEL OF SYSTEM ACCESSED GATEWAY

Through the discussion of the 3rd, 4th two sections, the identification and analysis problems of data frame have been solved correctly, organically combining with the two conclusions together, the math model is shown as Figure 4.

The data stream comes across the data frame sampler, which is separated into sampling frame of different length, then sent to character analyzing matrix to verify. If the result vector is consistent with character vector of some frame, then it indicates that this frame is standard data frame. Putting it into corresponding analysis implement to get the channel data F, and feeding back the judged result to data frame sampler to adjust the data queue. Thus the monitoring system recognizes one frame, deals with one frame, and corresponds data queue moving along the length of corresponding one frame to analyze data continuously.

8 CONCLUSION

The model upwards has been used by security monitoring system and mine personnel positioning system of Huaibei mineral bureau in China successfully; the effect is proved very well.

ACKNOWLEDGMENT

This project supported by:

1. National Natural Science Foundation of China No. 51204186.
2. The Project Supported by National Science and Technology Support Program No. 2012BAH12B02.
3. The Fundamental Research Funds for the Central Universities No. 2010QNB30.

REFERENCES

Gang Hua, Key Technology on Synthetically Dispatching of Safety and Production in Coalmine [D], Phd Dissertation of CUMT, 2002.6, 15~18 1955.

Gang Hua, Research and Development on mine monitoring and control system connecting with gateway of internet [J], Coal Science and Technology, 2005.11. Vol 33, 19–21.

Jing Guan; Xianjun Wang, International Workshop on Intelligent Systems and Applications [C], IEEE Trans, 2009.

Mohammed, O.A., Nayeem, M.A.; Kaviani, A.K. A laboratory based microgrid and distributed generation infrastructure for studying connectivity issues to operational power systems [C];IEEE PES General Meeting, PES 2010.

Nakandala, Dilupa; Samaranayake, Premaratne; Lau, H.C.W. Network reliability analysis of complex systems using a non-simulation-based method [J],. European Journal of Operational Research 2013 Vol. 225(3):507–517.

Piccioni, M., Oriol, M.; Meyer, B. Class Schema Evolution for Persistent Object-Oriented Software: Model, Empirical Study, and Automated Support [J], IEEE Transactions on Software Engineering, 2013 v39(2), 184–96.

Tu, Xuyong.; Du, Shuxin; Tang, Lie; Xin, Hongwei; Wood, Ben. A real-time automated system for monitoring individual feed intake and body weight of group housed turkeys [J]: Computers and Electronics in Agriculture, 2011 Vol. 75(2):313–320.

Yao, Qing; Sun, Yuqing. Applying case-based reasoning technique for customized management [C]: Proceedings2012 International Conference on Intelligent Systems Design and Engineering Applications, ISDEA 2012:783–786.

Yazhou Jiao; Zhigang Jin; Zhuoqun Ma; A cross-layer method to improve mobile database synchronization performance [C]; Proceedings of the 2009 5th International Conference on Wireless Communications, Networking and Mobile Computing (WiCOM), 2009.

Control Engineering and Information Systems – Liu (Ed)
© *2015 Taylor & Francis Group, London, ISBN 978-1-138-02685-8*

Impact of FSIG-based wind turbine on overcurrent protection

H. Yuan, W.M. Wu & L.Y. Huang
Hainan Power Grid Corporation, Haikou, China

D.Y. Guo
Xi'an Jiaotong University, Xi'an, China

ABSTRACT: The wind farm's integration into distribution network may cause overcurrent protections' incorrect operations. Based on analysis and simulation, our paper studies the fault feature of the Fixed Speed Induction Generator (FSIG)-based wind turbine and its impact on distribution network protection. Because of the in feed current or shunting effect brought by the wind turbine, the overcurrent protections may mal-operate or mis-operate. Further, the range of incorrect operation is related to the fault location, the fault condition and the wind turbine capacity. By changing these influencing factors during the simulation, we obtain the maximum short-circuit capacity ratio at the wind farm integration point in case of a typical distribution network simulation example.

1 INTRODUCTION

With the development of wind power technology, the rated capacity of single wind turbine is augmenting dramatically. Under this background, the wind farm's integration into distribution network may change the network's original structure and result in relay protection's incorrect operations (Chi, 2006, Liu et al. 2007). Because of the in feed current or shunting effect, the fault current flowing through the protection may increase or decrease. All these impacts depend on the fault location and the fault conditions (Sun et al. 2009, Feng et al. 2010, Tao et al. 2012, Zhang et al. 2012, Brand et al. 2007, Sun et al. 2007). Undeniably, the study on the protection's behavior in the wind farm connected distribution network is of great significance.

Based on the electro-magnetic transient model of the FSIG-type wind turbine established on PSCAD/EMTDC, we analyze its fault feature as well as its probable impact on overcurrent protections; then, we discuss in detail the influencing factors on the overcurrent protections' incorrect operations when the wind turbine is connected in front of or behind the fault. In addition, we derive the relationship between the fault current flowing through each protection and the wind turbine's capacity. Finally, we obtain the limitation of the short-circuit capacity ratio at the wind turbine integration point.

2 FSIG-BASED WIND TURBINE FAULT FEATURE

Under normal operating condition, FSIG's rotation speed almost remains the same as the synchronous speed, performing as a Constant Speed Constant Frequency (CSCF) power generation system. In this paper, we establish the FSIG electro-magnetic transient model on PSCAD/EMTDC in order to study its three-phase and two-phase short-circuit fault features. See the FSIG parameters in (Zhang et al. 2012, Zhang et al. 2013). Rated power = 1600 kW; rated voltage = 0.39837 kV; $R_s + jX_s = 0.00488 + j0.01386$; rotor impedance: $R_r + jX_r = 0.0549 + j0.105$; excitation reactance: $X_m = 3.9527$; $T_J = 2$ s.

2.1 *Three-phase short-circuit fault*

At the moment t = 0 s, three phase short-circuit fault occurred at the exit of FSIG based wind turbine. The simulation waveform of three-phase fault current at the generator terminal is shown in (Fig. 1). Without independent excitation windings, the FSIG stator windings have high short-circuit current which will decrease rapidly. If the fault remains, the short-circuit current will gradually reduce to 0.

During 1–2 cycles after the fault, the FSIG short-circuit current augments promptly and then decreases, which could be influential on instantaneous overcurrent protections.

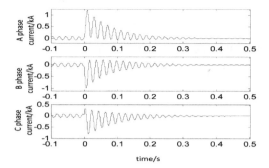

Figure 1. FSIG three phase fault current.

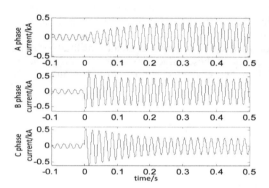

Figure 2. FSIG two phase fault current.

2.2 Two-phase short-circuit fault

At the moment t = 0 s, BC phase short-circuit fault occurred at the exit of FSIG based wind turbine. The simulation waveform of two-phase fault current at the generator terminal is shown in (Fig. 2). Unlike three-phase fault, during the two-phase fault the generator terminal can remain a constant short-circuit voltage and current. Especially, for the fault phase, high short-circuit current appears at the beginning of the fault and then decreases rapidly; for the non-fault phase, the short circuit current increases gradually. Both the fault and non-fault phase possess constant short-circuit current (Ma, 1998).

At the beginning of the fault, the short-circuit current of fault phase increases instantaneously, which may result in instantaneous overcurrent protection's mal-operations; the non-fault phase short-circuit current increases gradually to a constant value which is higher than that of normal operating condition, leading to mis-operations of time-delay instantaneous overcurrent protection.

It can be concluded that because of the fault characteristic of the FSIG-based wind turbine, its integration into distribution network may cause the overcurrent protection's mal-operation of zone I and mis-operation of zone II. Under different operating conditions, we analyze in detail the influencing factors of these problems.

3 IMPACT OF FAULT LOCATION ON OVERCURRENT PROTECTION

In this paper, we apply the two-zone overcurrent protection scheme which composes of instantaneous overcurrent protection and time delay instantaneous overcurrent protection, where the instantaneous overcurrent protection fails to protect the whole length of the line where the protection itself locates, and the time delay instantaneous overcurrent protection is able to protect the whole length of the line within a delayed operating time (Zhang, 2005).

Assume that the wind farm is connected to a 10.5 kV distribution network as shown in (Fig. 3). The system impedance under maximum operation mode $Z_{smin} = 0.91\ \Omega$, and under minimum operation mode $Z_{smax} = 1.16\ \Omega$; the transmission line's impedance is $(0.169 + j0.394)\ \Omega/km$. Taking the FSIG based wind turbine's integration into the distribution network as an example, we analyze the protection's behavior when the wind turbine is connected to bus B (Zhang, 2013).

According to the setting principles of overcurrent protections, the operating current's setting value of instantaneous overcurrent protection R_2 $I_{set.2}^I$ should be superior to the fault current when three-phase short circuit fault occurs at bus C under system's maximum operating mode, and the operating current is

$$I_{set.2}^I = K_{rel}^I I_{k.C.max} = \frac{K_{rel}^I E_\phi}{Z_{s.min} + Z_{B-C}} \quad (1)$$

where the reliability coefficient $K_{rel}^I = 1.2 \sim 1.3$; E_ϕ is the phase electromotive force of the system equivalent source; Z_{B-C} is impedance of line BC.

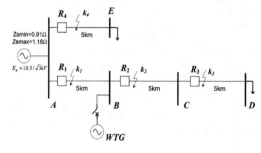

Figure 3. Wind farm connected distribution network simulation model.

The operating current's setting value of time delay instantaneous overcurrent protection R_1 is

$$I_{set.1}^{II} = K_{rel}^{II} I_{set.2}^{I} \tag{2}$$

where the reliability coefficient $K_{rel}^{II} = 1.1 \sim 1.2$; $I_{set.2}^{I}$ is the zone I setting value of protection R_2.

The operating time limit of time delay instantaneous overcurrent protection R_1 t_1^{II} should be longer than that of instantaneous overcurrent protection R_2 by Δt, where $\Delta t \approx 0.5$ s.

$$t_1^{II} = t_2^{I} + \Delta t \tag{3}$$

Without regard to the wind turbine integration, the protections' setting value is shown in (Table 1) ($K_{rel}^{I} = 1.2$, $K_{rel}^{II} = 1.1$).

3.1 Impact of wind turbine integration on overcurrent protection when the integration point is in front of the fault

As shown in (Fig. 3), three-phase short-circuit fault k_3 occurs at the beginning of line CD under the system's maximum operating mode. When the wind turbine's capacity changes, the value of fault current flowing through each protection is shown in (Table 2).

Influenced by the shunting effect, the fault current flowing through R_1 decreases with the increase of the wind turbine's capacity, thus the protection's sensitivity diminishes. Because of the in feed current supplied by the wind turbine, the fault current flowing through R_2 increases while the wind turbine's capacity increases. When the capacity reaches to 5 MW, the fault current flowing through R_2 is already higher than its zone I setting value,

Table 1. Protection setting values without connected wind turbines.

	R_1	R_2	R_3
I_{set}^{I}/kA	2.424	1.416	1.000
I_{set}^{II}/kA	1.558	1.100	–

Table 2. Fault current according to different wind turbine capacity during k_3 fault.

Fault current/kA Wind turbine capacity	R_1	R_2	R_3
0 MW	1.162	1.162	1.162
1 MW	1.133	1.214	1.214
3 MW	1.068	1.336	1.336
5 MW	0.994	1.471	1.471

Table 3. Fault current flowing through R_1 according to different wind turbine capacity during k_1 fault.

Wind turbine capacity	Fault phase short-circuit current/kA	Non-fault phase short-circuit current/kA	$I_{set.1}^{II}$/kA
0 MW	1.645	0.143	1.558
1 MW	1.597	0.229	1.558
3 MW	1.563	0.471	1.558
5 MW	1.549	0.673	1.558

which can result in protection's mal-operation. Furthermore, the range of mal-operation augments as the wind turbine's capacity augments. Affected by the in feed current, the fault current flowing through R_3 increases while the wind turbine's capacity increases. In addition, protection R_4 will not be influenced because there is no wind turbine current flowing through.

3.2 Impact of wind turbine integration on overcurrent protection when the integration point is behind the fault

As shown in (Fig. 3), two-phase short-circuit fault k_1 occurs at the end of line AB under the system's minimum operating mode. When the wind turbine's capacity changes, the value of fault current flowing through R_1 0.4 s after the fault is shown in (Table 3).

Protection R_2, R_3 and R_4 will not be influenced by the wind turbine because there is no wind turbine fault current flowing through. Influenced by the shunting effect, the fault current flowing through R_1 decreases leading to the protection's mis-operation of zone II. When the wind turbine's capacity is 5 MW, the fault current flowing through R_1 is already lower than its zone II setting value.

It can be concluded that when the FSIG-cased wind turbine is connected to the distribution network in front of the fault, the zone I of the protection on the line next to the one where the fault occurs may mal-operate because of the in feed current supplied by the wind turbine; when the wind turbine is connected behind the fault, the zone II of the protection on the same line where the fault occurs may mis-operate because of the shunting effect caused by the wind turbine. Furthermore, the degree of mal-operation or mis-operation depends on the fault condition and the wind turbine's capacity.

4 IMPACT OF SHORT-CIRCUIT RATIO AT THE INTEGRATION POINT ON OVERCURRENT PROTECION

The short-circuit capacity ratio refers to the ratio of wind turbine's capacity on the short-circuit capacity at the integration point. The short circuit capacity of bus B is 36.7 MVA. In (Fig. 3), three-phase short-circuit fault occurs at the beginning of line CD under system's maximum operating mode. When the wind turbine's capacity changes, we can obtain the relation curve between the fault current flowing through R_2 and the short-circuit capacity ratio at the integration point (bus B). When the short-circuit capacity ratio at bus B is over 11.72%, the fault current flowing through R_2 is higher than its zone I setting value, leading to the mal-operation of protection.

As shown in (Fig. 3), two-phase short-circuit fault occurs at the beginning of line BC under system's minimum operating mode. When the wind turbine's capacity changes, we can obtain the relation curve between the fault current flowing through R_1 at 0.4 s after the fault and the short-circuit capacity ratio at the integration point. When the short-circuit capacity ratio at bus B is over 21.8%, the fault current flowing through R_1 will be lower than its zone II setting value, leading to the mis-operation of protection.

It can be concluded that the change of wind turbine's capacity has impacts on overcurrent protections. Synthesizing each kind of situation, when the short-circuit capacity ratio at the wind turbine's integration point is over 11.72%, the integration of wind power will cause the protections' incorrect operations.

5 CONCLUSION

In conclusion, the integration of FSIG based wind turbine may cause protections' mal-operation or mis-operation according to different system operating mode and different fault location. Besides, the range of mal-operation or mis-operation depends on the fault conditions and the turbine's capacity. To prevent the incorrect operation of overcurrent protection, it is necessary to limit the short-circuit capacity ratio at the wind turbine integration point to below 10%.

ACKNOWLEDGMENT

This project is supported by the National High Technology Research and Development Program of China (863 Program) (Grant No. 2012 AA-050201), and the support is gratefully acknowledged.

REFERENCES

Brand, A.J. 2007. Impacts of wind power on thermal generateon unit commitment and dispatch. In IEEE transactions on Energy Conversion 22(1): 44–51.

Chi, Y.N. 2006. Studies on the stability issues about large scale wind farm grid integration. In China Electric Power Research Institute. Beijing.

Feng, X.K. & Tai, N.L. 2010. Research on the impact of DG capacity on the distribution network current protection and countermeasure. Power System Protection and Control 38(22): 156–165.

Liu, W.K. & Zhang, Z.Y. 2007. Wind energy and wind power technology: 205–230. Beijing: Chemical Industry Press.

Ma, Z.Y. 1998. The transient characteristics of electric motor. CEPP: 88–94.

Sun, J.L. & Li, Y.L. 2009. A Protection Scheme for Distribution System with Distributed Generators. In Automation of Electrical Power System 33(1): 81–89.

Sun, Y.Z. & Wu, J. 2007. Influence research of wind power generation on power systems. In Power system technology 31(20): 55–62.

Tao, S. & Guo, J. 2012. Analysis on Allowed Penetration Level of Distributed Generation and its Grid-Connected Position Based on Principles of Current Protection. In Power System Technology 36(1): 265–270.

Zhang, B.H. & Guo, D.Y. 2013. Impact of wind farm on relay protection(5): Impact of wind farm on distribution network protection. In Electric Power Automation Equipment 33(5): 1–6.

Zhang, B.H. & Li, G.H. 2012. Affecting factors of grid-connected wind power on fault current and impact on protection relay. In Electric Power Automation Equipment 32(2): 1–8.

Zhang, B.H. & Li, G.H. 2013. Impact of wind farm on relay protection (1): electromagnetic transient equivalent model of FSIG-based wind farm. In Electric Power Automation Equipment 33(1): 1–6.

Zhang, B.H. & Wang, J. 2012. Cooperation of relay protection for grid-connected wind power with low-voltage ride-through capability. In Electric Power Automation Equipment 32(3): 1–6.

Zhang, B.H. & Yin, X.G. 2005. Power System Protective Relaying. CEPP.

Control Engineering and Information Systems – Liu (Ed)
© 2015 Taylor & Francis Group, London, ISBN 978-1-138-02685-8

The response analysis of the transmission lines in the cavity with apertures based on the network BLT equation

Y. Li
The State Key Laboratory of Complex Electromagnetic Environmental Effects on Electronics and Information System, Luoyang, Henan, China
College of Science, National University of Defense Technology, Changsha, Hunan, China

L. Hong
The State Key Laboratory of Complex Electromagnetic Environmental Effects on Electronics and Information System, Luoyang, Henan, China

M. Zhou & X.D. Chen
College of Science, National University of Defense Technology, Changsha, Hunan, China

ABSTRACT: Based on the network BLT equation, combining aperture equivalence, electromagnetic field penetration and field-to-line coupling, this paper presents a new method to analyze the induced load responses arising from illumination by an incident field that penetrates through the aperture into the electrical system interior and the influence of radiation from the line currents on the shielding effectiveness of the cavity.

1 INTRODUCTION

There are many transmission lines located in an electronic system. If the system is well shielded, the internal transmission lines may not be affected by the external excitation; however, all systems must have imperfections in the shielding. The apertures are needed for such various reasons as input and output connection, ventilation panels, etc. An external electromagnetic (EM) source produces an incident field that penetrates through the aperture into the system interior, where it couples to the Transmission Line (TL). Their effect is to induce currents and voltages on the line and in the load impedances at the ends. Generally, the behavior of the induced responses can be evaluated using EM scattering theory and coupling theory. However, in many cases of practical interest, fast network BLT equation that provides approximation is sufficient. To provide some advice for EMC design of equipments and systems, it is necessary to compute the induced load responses of transmission line inside a rectangular cavity with a rectangular aperture by an incident EM field and analyze the influence of the transmission line currents on the shielding effectiveness.

In the recent work, the aperture coupling and transmission line responses are hot issues (Tesche & Butler 2004, Tesche et al. 1997, Robinson et al. 1998, Li et al. 2009, Li et al. 2010). For dealing with aperture coupling problems, the aperture is considered as a length of coplanar strip transmission line, shorted at each end (Robinson et al. 1998). The coupling problems of an exterior EM field to line are analyzed by transmission line theory (Tesche et al. 1997). Using the dyadic Green's function and the equivalent magnetic current, the author derived expressions for the coupling electromagnetic field in metallic enclosures with apertures excited by an external source (Li et al. 2009, Li et al. 2010).

Based on the network BLT equation, this paper combines aperture equivalence, EM field penetration and field-to-line coupling to analyze the induced load responses arising from the illumination of the line by an interior coupling field and the influence of radiation from the line on the shielding effectiveness.

2 A CALCULATION METHOD BASED ON THE NETWORK BLT EQUATION

2.1 The network BLT equation

The network BLT equation can be used to describe the voltage and current responses at junctions of the network (Tesche & Butler 2004).

Consider a closed cavity with an aperture illuminated by a plane wave, shown in Figure 1. The aperture, waveguide and transmission line are considered as a transmission line network, denoting the ends of each of the transmission line segments

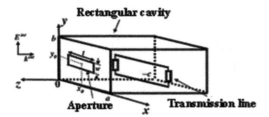

Figure 1. A cavity with an aperture under illumination.

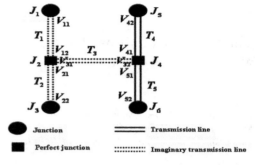

● Junction	═══ Transmission line
■ Perfect junction	⋯⋯ Imaginary transmission line

Figure 2. The TL topology network for Figure 1.

in by J_i. The TL network contains 5 TL segments, in which T_1, T_2, T_3 are imaginary lines. All TL induce currents and voltages at ends for the entire network are represented by two supervectors.

$$\bar{V} = [V_{11}, V_{12}, V_{21}, V_{31}, V_{22}, V_{41}, V_{51}, V_{32}, V_{42}, V_{52}]^T,$$
$$\bar{I} = [I_{11}, I_{12}, I_{21}, I_{31}, I_{22}, I_{41}, I_{51}, I_{32}, I_{42}, I_{52}]^T.$$

The network BLT equation for the currents and voltages on the network (Tesche et al. 1997) can be written as.

$$\bar{V} = [\bar{\bar{U}} + \bar{\bar{\rho}}][\bar{\bar{\Gamma}} - \bar{\bar{\rho}}]^{-1}\bar{S} \tag{1}$$

$$\bar{I} = \bar{\bar{Y}}[\bar{\bar{U}} - \bar{\bar{\rho}}][\bar{\bar{\Gamma}} - \bar{\bar{\rho}}]^{-1}\bar{S} \tag{2}$$

where $\bar{\bar{\rho}}$ is a scattering supermatrix, $\bar{\bar{\Gamma}}$ is a admittance supermatrix, $\bar{\bar{Y}}$ is a characteristic admittance supermatrix, $\bar{\bar{U}}$ is the 10×10 identity matrix, $\bar{S}$ is the excitation supervector. The locations of the various coefficients in these supermatrix and supervector depend on how the junctions in the network are numbered and interconnected. For an incident field excitation of this topology network, the current and voltage responses at the junctions of the network are computed by solving the network BLT equation.

2.2 The determination of parameter matrix

As shown in Figure 2, junctions J_1 and J_3 are short circuits ends of aperture and their reflected coefficients are $\rho_1 = \rho_3 = -1$. Junction J_2 represents a middle point of aperture and is a perfect point and its scattering matrix (Parmantier 1998) is given by

$$\rho_2 = \begin{pmatrix} \dfrac{-Y_g}{2Y_c + Y_g} & \dfrac{-Y_g}{2Y_c + Y_g}+1 & \dfrac{Y_g - 2Y_c}{2Y_c + Y_g}+1 \\ \dfrac{-Y_g}{2Y_c + Y_g}+1 & \dfrac{-Y_g}{2Y_c + Y_g} & \dfrac{Y_g - 2Y_c}{2Y_c + Y_g}+1 \\ \dfrac{-Y_g}{2Y_c + Y_g}+1 & \dfrac{-Y_g}{2Y_c + Y_g}+1 & \dfrac{Y_g - 2Y_c}{2Y_c + Y_g} \end{pmatrix},$$

where Y_c is the aperture characteristic admittance, Y_g is the waveguide characteristic admittance. Junction J_4 represents the connection between transmission line and waveguide, a perfect point, and its scattering matrix can be used as

$$\rho_4 = \begin{pmatrix} \dfrac{-Y_g}{2Y_t + Y_g} & \dfrac{-Y_g}{2Y_t + Y_g}+1 & \dfrac{Y_g - 2Y_t}{2Y_t + Y_g}+1 \\ \dfrac{-Y_g}{2Y_t + Y_g}+1 & \dfrac{-Y_g}{2Y_t + Y_g} & \dfrac{Y_g - 2Y_t}{2Y_t + Y_g}+1 \\ \dfrac{-Y_g}{2Y_t + Y_g}+1 & \dfrac{-Y_g}{2Y_t + Y_g}+1 & \dfrac{Y_g - 2Y_t}{2Y_t + Y_g} \end{pmatrix},$$

where Y_t is the transmission line characteristic admittance. Junction J_5 and J_6 represent the ends of the transmission line inside the cavity and their reflected coefficients are

$$\rho_5 = \frac{Z_{L_1} - Z_t}{Z_{L_1} + Z_t}, \quad \rho_6 = \frac{Z_{L_2} - Z_t}{Z_{L_2} + Z_t},$$

In which Z_{L_1} and Z_{L_2} are the load impedance of the line. For the entire network, these scattering parameters can be put into a 10×10 scattering supermatrix

$$\bar{\bar{\rho}} = diag(\rho_1, \rho_2, \rho_3, \rho_4, \rho_5, \rho_6).$$

The propagation matrix of each TL segment is given by

$$\Gamma_1 = \Gamma_2 = \begin{pmatrix} 0 & e^{jk_0 l/2} \\ e^{jk_0 l/2} & 0 \end{pmatrix}, \quad \Gamma_3 = \begin{pmatrix} 0 & e^{jk_g p} \\ e^{jk_g p} & 0 \end{pmatrix},$$

$$\Gamma_4 = \Gamma_5 = \begin{pmatrix} 0 & e^{jk_0 L/2} \\ e^{jk_0 L/2} & 0 \end{pmatrix},$$

where p refers to the distance between the center point in the aperture and the middle point of the line. For the entire network, these propagation parameters can be put into a 10×10 propagation supermatrix $\bar{\bar{\Gamma}}$ according to the order of these junctions. The characteristic admittance supermatrix of the entire network is written as

$$\bar{\bar{Y}} = diag(Y_c, Y_c, Y_c, Y_g, Y_c, Y_t, Y_t, Y_g, Y_t, Y_t).$$

The excitation supervector of this network is given by

$$\bar{S} = [S_{11}, S_{12}, S_{21}, S_{31}, S_{22}, S_{41}, S_{51}, S_{32}, S_{42}, S_{52}]^T.$$

For each TL segment, it is desired to compute the source vectors by using the Agrawal method (Tesche et al. 1997, Agrawal et al. 1980)

$$\begin{pmatrix} S_{11} \\ S_{12} \end{pmatrix} = \begin{pmatrix} \frac{1}{2}\int_{x_l}^{x_0} e^{jk_0 x} V_s(x)dx - \frac{V_1}{2} + \frac{V_2}{2}e^{jk_0 l/2} \\ -\frac{1}{2}\int_{x_l}^{x_0} e^{jk_0(l/2-x)} V_s(x)dx + \frac{V_1}{2}e^{jk_0 l/2} - \frac{V_2}{2} \end{pmatrix},$$

$$\begin{pmatrix} S_{21} \\ S_{22} \end{pmatrix} = \begin{pmatrix} \frac{1}{2}\int_{x_0}^{x_h} e^{jk_0 x} V_s(x)dx - \frac{V_2}{2} + \frac{V_3}{2}e^{jk_0 l/2} \\ -\frac{1}{2}\int_{x_0}^{x_h} e^{jk_0(l/2-x)} V_s(x)dx + \frac{V_2}{2}e^{jk_0 l/2} - \frac{V_3}{2} \end{pmatrix},$$

$$\begin{pmatrix} S_{31} \\ S_{32} \end{pmatrix} = \begin{pmatrix} 0 \\ 0 \end{pmatrix},$$

$$\begin{pmatrix} S_{41} \\ S_{42} \end{pmatrix} = \begin{pmatrix} \frac{1}{2}\int_{x_{tl}}^{x_{r0}} e^{jk_0 x} V_{ts}(x)dx - \frac{V_4}{2} + \frac{V_5}{2}e^{jk_0 L/2} \\ -\frac{1}{2}\int_{x_{tl}}^{x_{r0}} e^{jk_0(L/2-x)} V_{ts}(x)dx + \frac{V_4}{2}e^{jk_0 L/2} - \frac{V_5}{2} \end{pmatrix},$$

$$\begin{pmatrix} S_{51} \\ S_{52} \end{pmatrix} = \begin{pmatrix} \frac{1}{2}\int_{x_{r0}}^{x_{th}} e^{jk_0 x} V_{ts}(x)dx - \frac{V_5}{2} + \frac{V_6}{2}e^{jk_0 L/2} \\ -\frac{1}{2}\int_{x_{r0}}^{x_{th}} e^{jk_0(L/2-x)} V_{ts}(x)dx + \frac{V_5}{2}e^{jk_0 L/2} - \frac{V_6}{2} \end{pmatrix},$$

where the distributed voltage source can be given in terms of the incident and interior coupling E-fields. The interior coupling E-fields can be computed by expressions from (Li et al. 2009, Li et al. 2010).

Solving the network BLT equation (1) and (2), the responses at each junction can be obtained. Especially, by combing expression (Azaro et al. 2002) for voltage and E-field and uncontinuity between waveguide and aperture gives an equivalent E-field $E_{J_4} = V_{32}\sqrt{2/wl}$ at J_4. Therefore, the shielding effectiveness can be obtained at this junction (imply that we consider the radiation from the transmission line currents).

3 NUMERICAL SIMULATION AND ANALYZE

Calculate the load responses of the transmission line inside a cavity with an aperture under a plane wave illumination. The parameters are as follows: the cavity's size $0.3\,\text{m} \times 0.12\,\text{m} \times 0.3\,\text{m}$, the aperture's size $0.1\,\text{m} \times 0.005\,\text{m}$, the line length $L = 0.1\,\text{m}$, the line separation distance $d = 0.01\,\text{m}$, the line radius $a_1 = a_2 = 3\,\text{mm}$, matched load. The incident field is normally incident onto the side containing the aperture for the frequency range up to 2 GHz. Figure 3 shows the results of the terminal response at junction J_6.

At center of the cavity with the aperture, Figure 4 compares the shielding effectiveness of cavity in which the transmission line resides with reference data of empty cavity those from (Robinson et al. 1998).

Figure 3 shows that the induced current peaks occur at 0.7 GHz and 1.6 GHz over the 0–2 GHz band, respectively. It is associated with the resonant

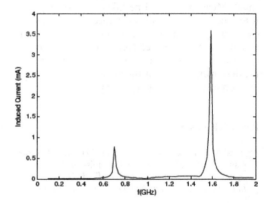

Figure 3. Induced current at the load.

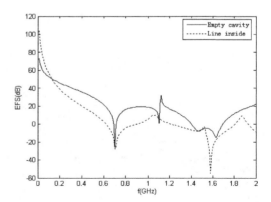

Figure 4. Comparison of EFS between the cavity with the transmission line inside and empty cavity.

frequency of the cavity 707 MHz, while another correspond to the resonance of the aperture 1.5 GHz. Near the aperture resonance and the cavity resonance frequency, Figure 4 also shows that the strongest coupling at the center of the cavity. Hence the external EM source produces interference on the transmission line inside cavity. Comparison of EFS shows that the influence of the radiation from induced currents on the shielding effectiveness is not small.

4 CONCLUSION

For these problems of the terminal response of the transmission lines inside the rectangular cavity with apertures illuminated by an exterior excitation, this paper presents a calculation method based on the network BLT equation. The results show that the EMI from the external excitation on the transmission line reaches its maximum value near the aperture resonance and the cavity resonance. Moreover, the influence of the radiation from the transmission line on the shielding effectiveness of the cavity is not small. Measures should be taken to restrain cavity resonance and aperture resonance. In dealing with electromagnetic protection problems of electronic systems in complex electromagnetic environment, consideration should be given to the influence of the design and layout of the aperture and the transmission line on electromagnetic protection.

REFERENCES

Agrawal A.K., Price H.J. & Gurbaxani S.H. 1980. Transient response of multiconductor transmission lines excited by a nonuniform electromagnetic field. *IEEE Transaction on Electromagnetic Compatibility*, 22: 119–129.

Azaro R., Caorsi S. & Donelli M. et al. 2002. A circuital approach to evaluating the electromagnetic field on rectangular apertures backed by rectangular cavities. *IEEE Transaction on Microwave Theory and Techniques*, 50(10): 2259–2266.

Li Y. Luo J., Ni G. & Shi J. 2010. Electromagnetic topology anaysis to coupling wires enclosed in cavities with apertures. *Mathematical Problems in Engineering,* Beijing, China, 2010: 1–12.

Li Y., Ni G., Luo J., Shi J. & Zhang X. 2009. Coupling onto the Two-wire Transmission Line Enclosed in Cavities with Apertures. *PIERS2009, Moscow*, Russia, 2009: 628–633.

Parmantier J.P. 1998. An efficient technique to calculate ideal junction scattering parameters in multiconductor transmission line networks. *Interaction Note* 536:1–13.

Robinson M.P., Benson T.M. et al. 1998. Analytical formulation for the Shielding effectiveness of enclosures with apertures. *IEEE Transaction on Electromagnetic Compatibility*, 40(8): 240–248.

Tesche F.M. & Butler C.M. 2004. On the addition of EM field propagation and coupling effects in the BLT equation. *Interaction Note* 588: 1–43.

Tesche F.M., Ianoz M.V. & Karlsson T. 1997. *EMC Analysis Methods and Computational Models*. New York: John Wiley and Sons.

Xu L. & Cao W. 2006. *The Theory of Electromagneitc Field and Electromagneitc Wave*. Beijing: Science Press. (in Chinese).

Control Engineering and Information Systems – Liu (Ed)
© 2015 Taylor & Francis Group, London, ISBN 978-1-138-02685-8

Study on application of optimizing resonance-demodulation technology in mechanical fault diagnosis

C. Wang, D.P. Jiang & Z.C. He

College of Mechanical and Power Engineering, Chongqing University of Science and Technology, Chongqing, China

ABSTRACT: In mechanical fault diagnosis especially of with gear case and antifriction bearing, demodulation is wildly used to analyze and diagnose the problems. Nowadays, resonance demodulation technology can extract the feature of early micro shock. But in terms of other methods, we find that it will rise the speed and accuracy if we combine the bandpass filter with Hilbert transform and refining resampling.

In mechanical equipment fault especially of gear case, such as gear broken, corrosive pitting, fatigue flaking of rolling bearings, shaft bending, will all cause periodic pulse impact and phenomena of vibration signal modulation. It will appear evenly spaced modulation sideband in the meshing frequency or natural frequency on both sides. It's the most used method of extracting modulated information extraction to analyze the frequency and intensity in mechanical fault diagnosis. Above all, the high frequency resonance method is still the best.

1 DEMODULATION PRINCIPLE AND ARITHMETIC OPERATORS OF HILBERT TRANSFORM

If signal AM-FM $x(t) = a(t)\cos[\varphi(t)]$, it's Hilbert transform is as following:

$$\hat{x}(t) = \frac{1}{\pi} \int_{-\infty}^{\infty} \frac{x(\tau)}{t - \tau} dt = x(t) * \frac{1}{\pi t} \tag{1}$$

Due to the Fourier transform of $\frac{1}{\pi t}$ is:

$$F\left(\frac{1}{\pi t}\right) = -j sign(\omega) \tag{2}$$

$sign(f)$ is a sign function, to make Fourier transform of $x(t)$ is:

$$F[x(t)] = X(\omega) \tag{3}$$

So the Fourier transform of $\hat{x}(t)$ is:

$$\hat{X}(\omega) = X(\omega) \cdot [-j sign(\omega)] \tag{4}$$

So, $\hat{x}(t)$ is the result of a phase shift in the frequency domain of $x(t)$, $\pi/2$ delay in the positive frequency, $\pi/2$ advance in a negative frequency domain.

From above we can know the concrete steps of Hilbert are as follows:

Make $x(t)$ to $X(\omega)$ via FFT;
Phase shift $X(\omega)$ to $\hat{X}(\omega)$;
Make $\hat{X}(\omega)$ to $\hat{x}(t)$ via contrary FFT

We make use of the signal from $x(t)$ and $\hat{x}(t)$ to get that:

$$z(t) = x(t) + j\hat{x}(t) = r(t)\exp[j\theta(t)] \tag{5}$$

and $\dot{\theta}(t)$ are used to estimate the signal amplitude envelope and instantaneous frequency. In this case, Hilbert Transform Separation Algorithm (HTSA) can be stated as:

$$r(t) = \sqrt{x^2(t) + \hat{x}^2(t)} \approx |a(t)| \tag{6}$$

$$\dot{\theta}(t) = \frac{d}{dt}\left\{\arctan\left[\frac{\hat{x}(t)}{x(t)}\right]\right\} \approx \omega_i(t) \tag{7}$$

At the end of the envelope signal spectrum analysis, envelope spectrum of the signal can be obtained.

Schematic diagram of Hilbert transform:

A modulated signal $x(t) = a(t)\cos[\varphi(t)]$, $a(t)$ is modulating signal. $\cos[\varphi(t)]$ is carrier signal, their frequency spectrums are Figure 1(c), 1(a), 1(b). The modulating signal and the multiplication operator of $a(t)\cos[\varphi(t)]$ with Hilbert, according to the time-domain multiplication correspond to the frequency domain convolution theorem. We can get the frequency spectrum as Figure 1(e). If we make $a(t)$ as modulating signal, $\sin[\varphi(t)]$ in Figure 1(d) is carrier signal. We can get the frequency spectrum of $a(t)\sin[\varphi(t)]$ the same. Obviously, 1(e) is the same as 1(f). That means $a(t)\cos[\varphi(t)]$ via Hilbert is $a(t)\sin[\varphi(t)]$.

We can know from above, Hilbert transform is just folding the original signal with $1/\pi t$, and makes a phase shift in the frequency domain. In the discrete time domain, $\hat{x}(n) = x(n) * h(n)$, this IIR filter is called arithmetic operators of Hilbert transform, also Ideal Hilbert converter or 90° phase shifter.

The description of Hilbert transform operator: (Fig. 2(a))

$$H_d(j\omega) = \begin{cases} -j & 0 \le \omega \le \pi \\ +j & -\pi \le \omega \le 0 \end{cases} \tag{8}$$

The Impulse response of $H_d(j\omega)$ is:

$$h_d(n) = \frac{1}{2\pi}\int_{-\pi}^{0} je^{j\omega n}d\omega - \frac{1}{2\pi}\int_{0}^{\pi} je^{j\omega n}d\omega \tag{9}$$

or

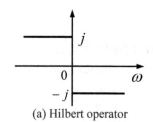

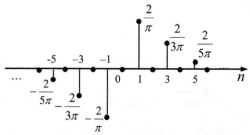

(a) Hilbert operator

(b) Hilbert impulse response operator

Figure 2. The Hilbert operator and its pulse response.

$$h_d(n)\begin{cases} \dfrac{2}{\pi}\dfrac{\sin^2(\pi n/2)}{n} & n \ne 0 \\ 0 \end{cases}$$

$$h_d(n)\begin{cases} \dfrac{2}{\pi n} & (n\ is\ an\ even\ number) \\ 0 & (n\ is\ an\ odd\ number) \end{cases}$$

The actual implementation of the Hilbert transform is by using a FIR approximation of IIR $h(n)$, such FIR filters can be designed by window function or the ripple method.

2 CLASSICAL HILBERT TRANSFORM AND ALGORITHM

$$\hat{x}(j) = \frac{2}{\pi} \cdot \sum_{k=1}^{M/2} \frac{\overline{x}(j+2k-1) - \overline{x}(j-2k+1)}{2k-1}$$
$$(j = M, M+1, ..., N-M) \tag{10}$$

$$z(j) = \sqrt{\overline{x}^2(j+M) + \hat{x}^2(j)}$$
$$(j = M, M+1, ..., N-M) \tag{11}$$

Obviously, Hilbert transform raises the operation speed very much. Broadband Hilbert transform demodulation is not used for band-pass digital filtering, but if in frequency modulation signal as the difference of solution without two temporal summation signal of the breakdown, it will cause errors. So band-pass filter can overcome

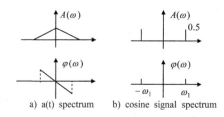

a) a(t) spectrum   b) cosine signal spectrum

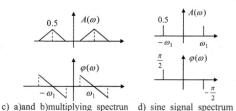

c) a)and b)multiplying spectrun   d) sine signal spectrum

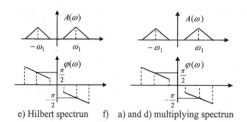

e) Hilbert spectrun   f) a) and d) multiplying spectrum

Figure 1. Principium diagram of Hilbert transformation.

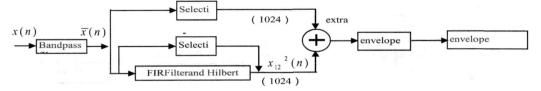

Figure 3.   The principle chart of classical Hilbert transform method.

this drawback. But it can not be refined and need data segment of the convolution, which costs more time.

3   OPTIMIZE HILBERT TRANSFORM AND THE ALGORITHM

For conventional method's boundedness, here is a way of combining the band-pass filtering and Hilbert transform. Here it is.

$$[h_1(t) * x(t)] * h_2(t) = [h_1(t) * h_2(t)] * x(t) \quad (12)$$

For the folding of symmetric odd point of Bandpass filter factor and odd symmetric odd point Hilbert transform factor, the last impulse response function factor is an odd symmetric odd point. And even value is zero. Because of the only original time series of the selected snapshot, it greatly reduces the number of operations.

4   SPECTRUM ZOOM AND ALGORITHM

The envelope spectrum from envelope analysis can not intuitively judge the fault type bearing. The fault characteristic frequency is not shown obviously in envelope spectrum. In the practical application, we often use a higher sampling frequency. In the case of the spectral lines, the fault characteristic frequency gathers in the low-frequency region which is hard to recognize. So it's necessary to thin the envelope spectrum.

Basic principle of the C-Zoom algorithm:

An "N" sequence, x (n),
$0 \leq n \leq N - 1$ (Amplitude Co.)

Then

$$X(Z) = \sum_{n=0}^{N-1} x(n) Z^{-n} \quad (13)$$

The transformation point as Z_k

$$Z_k = A W^{-k}, k = 0, 1, 2, ..., M - 1$$
$$(W = e^{-j\phi_0} \quad A = e^{j\theta_0}) \quad (14)$$

$$Z_k = e^{j\theta_0} e^{jk\phi_0} = e^{j(\theta_0 + k\phi_0)} \quad (15)$$

Then

$$X(Z_k) = \sum x(n) \cdot A^{-n} \cdot W^{nk} \quad (16)$$

Using identities:

$$nk = \frac{1}{2} \left[n^2 + k^2 - (k - n)^2 \right] \quad (17)$$

Substitution type and then:

$$X(Z_k) = \sum_{n=0}^{N-1} x(n) \cdot A^{-n} \cdot W^{n^2/2} \cdot W^{k^2/2} \cdot W^{-(k-n)^2/2}$$
$$= W^{k^2/2} \sum_{n=0}^{N-1} x(n) \cdot A^{-n} \cdot W^{n^2/2} \cdot W^{-(k-n)^2/2} \quad (18)$$

To make $g(n) = x(n) \cdot A^{-n} W^{n^2/2}$, n = 1, 2, ..., N − 1, and

$$X(Z_k) = W^{k^2/2} \sum_{n=0}^{N-1} g(n) \cdot W^{-(k-n)^2/2}$$
$$= W^{k^2/2} \cdot g(n) * W^{-n^2/2}$$

(k = 0, 1, 2, ..., M − 1) (* is convolution symbol) (10).

From above we can see, C-Zoom algorithm is the discrete convolution of $g(n)$ and $W^{n^2/2}$. C-Zoom algorithm increases the sampling Z transform flexibility, which is not like Fourier transform that needs N = M. In selecting sampling ϕ_0 interval, it doesn't need $\phi_0 = 2\pi/n$, ϕ_0 can be random. Refining envelope spectrum. Is using C-Zoom algorithm to select part of original statistics in order to get a series of new sampling interval. Since the initial phase angle θ_0 and interval ϕ_0 can be arbitrarily set, thus we can get a certain band refinement purposes.

557

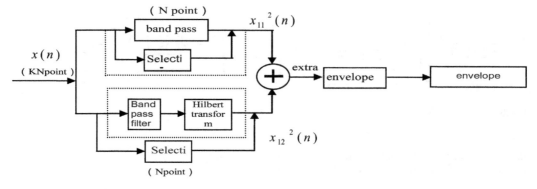

Figure 4. The optimize chart of Hilbert transform arithmetic.

5 THE SIMULATION CALCULATION AND EXAMPLES OF ENGINEERING APPLICATION

5.1 The simulation calculation

Use the composite signal of addition and multiplication from the computer.

$$x(t) = \sin(2\pi \cdot 80t) + \sin(2\pi \cdot 100t)$$
$$+ \sin(2\pi \cdot 4000t) + \sin(2\pi \cdot 4020t) + n(t)$$

($n(t)$ is random noise signal)

Using 20 Hz sampling, the sampling point number is 2048. The time domain chart is as shown in Figure 5(a), The fast algorithm of Fourier, then Figure 5(b).

5.2 Examples of engineering application

In fault diagnosis of rolling bearings of a test bench with the demodulated resonance technique, Bandpass filter using Chebyshev filter, the optimization of Hilbert operator demodulating method to envelope demodulation, we'd use the C-ZOOM algorithm in the frequency spectrum refinement this link.

Figure 6(a) is the time domain of 6406 fault bearing vibration signals of a test. We can see that the vibration signal of rolling bearing has obvious impact signal. We can conclude that the bearing is defective. But unable to judge what kind of fault. Figure 6(b) is the Energy Operator Demodulation after optimizing and the spectrum zoom envelope spectrum. There are peaks when $f = 88\ Hz$, $176 (= 2 \times 88)\ Hz$, $264 (= 3 \times 88)\ Hz$. According to the calculation formula of the fault characteristic frequency, the available roller fault characteristic frequency is 81.9 Hz, which is approximately equal to 88 Hz. And the rolling body 10 Hz and

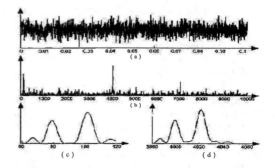

Figure 5. Result comparison of C-ZOOM arithmetic and FFT arithmetic.

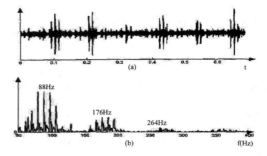

Figure 6. Time-field wave-form and envelope spectrum of vibration signal of fault bearing 6406.

calculated the frequency of 12.3 Hz is close to. Therefore, the rolling body we can determine the bearing fault. Figure 7(a) is a domain vibration signals of rolling bearings, Figure 7(b) is the envelope spectrum. There exists obvious peak amplitude fading when $f = 85\ Hz$, $171 (\approx 2 \times 85)\ Hz$, $257 (\approx 3 \times 85)\ Hz$, $343 (\approx 4 \times 85)\ Hz$. According to

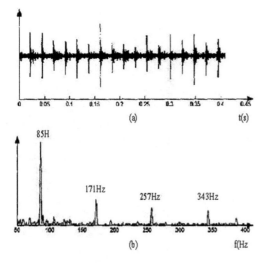

(a)   t(s)

(b)   f(Hz)

Figure 7. Time-field wave-form and envelope spectrum of vibration signal of fault bearing 6406.

the calculation formula of 406 bearing type, the outer ring fault characteristic spectrum of 86.4 Hz 5 Hz with the calculated is very close. Comparing (a) and (b), Frequency of each fault characteristic frequency on both sides basically doesn't appear modulation sideband. Therefore, we can determine the bearing outer ring of a fault.

6 CONCLUSION

If we make use of the law of convolution to enhance the choice of the optimum pumping, filtering and Hilbert transform, we can achieve refinement purposes and greatly improve the speed of computation. Because of the odd point filter solution envelope, we can guarantee the filtering and time sequence signal without phase shift and the high precision. The optimization algorithm is simple in principle. As long as the choice of band-pass filter parameters suitable, Filter out the time domain signal can be added without fault. Used in the simulation calculation and engineering practice the situation is satisfactory.

REFERENCES

Chen Jin, Vibration monitoring and fault diagnosis of mechanical equipment, 1999.P68–75.

Ding kang, Xie Ming, Zoom complex analytical band-pass filter spectrum principle and method of analysis, 2001.14(1).P29–35.

Jiang Ping, Mechanical equipment fault diagnosis technology and its application, Northwestern Polytechnical University press, 2000.

Li Guang, Research and implementation of fault diagnosis of rolling bearings based on the resonance demodulation, 2006, 129–131.

Li Li, Spectrum zoom method based on Z transform and its application, 1997(2):P44–46.

Wang Qingsong, Peng Donglin, Automatic diagnosis system of rolling bearing based on demodulated resonance technique, 2003(2):P45–47.

Yang Dingxin, Yang Yingang, Application of stochastic resonance in gearbox fault detection, 2004, 17(2); 201–204.

Yang Yu, Yu Dejie, Cheng Junsheng, Characteristics of energy transformation of Hilbert-Huang and application of the fault diagnosis of rolling bearing, 2004, 6–8.

Influences of analog quadrature modulator on DPD technology

M.D. Zhao & Z.B. Zeng

Communication University of China, Beijing, China

ABSTRACT: Due to hardware design flaws, there is always IQ imbalance of quadrature modulator, which will always introduce interference, worsen the communication system bit error rate and the performance of digital predistortion. This paper focuses on the influences derived from IQ mismatch of quadrature modulator on performance of digital predistortion using Matlab simulation.

1 INTRODUCTION

Power amplifier linearization technology in broadband wireless communication has become a key technology, in which digital predistortion (DPD) linearization technology is currently a research hotspot. In the digital predistortion system, in addition to the RF power amplifier, quadrature modulator IQ imbalance distortion will have a greater impact on the transmitter performance, resulting in predistortion linearization performance degradation. In this paper, influences of modulator on DPD performance are analyzed and verified through Matlab simulation tool.

2 DPD OVERVIEW

The basic principle of predistortion technique is: adding a Predistortion Module (PD) between the input signal and the power amplifier which can produce the reverse characteristics of the power amplifier. The predistortion module preprocesses the input signal to compensate for the two distortions caused by nonlinear power amplifier: AM/AM distortion and AM/PM distortion, and finally, making a linear relationship between the output and input of power amplifier. The principle of predistortion linearization is shown in Figure 1. In Figure 1, $F(\bullet)$ is the transfer function of predistortion module (PD), $G(\bullet)$ is the transfer function of power amplifier, so as to satisfy:

$$F(|V_i|) * G(|V_d|) = K \qquad (1)$$

where K represents the system gain. When the input signals pass through the predistorter and the amplifier, the final output can be linearly amplified if K is a constant.

3 THE BASIC PRINCIPLES OF ANALOG QUADRATURE MODULATOR

The basic principle of the Analog Quadrature Modulator (AQM) can be represented by Figure 2.

As shown in Figure 2, assuming that the input signal I and Q are respectively expressed as:

$$I = V_i \cos(\omega_{bb}t),\ Q = V_q \sin(\omega_{bb}t) \qquad (2)$$

where ω_{bb} is the angular frequency of baseband signal, V_i and V_q are the amplitudes of I and Q.

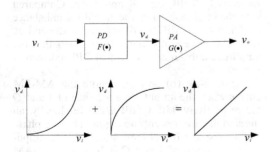

Figure 1. The predistortion principle block diagram.

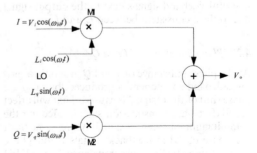

Figure 2. The basic principles of AQM.

The oscillation signals are $L_i \cos(\omega_l t)$, $L_q \sin(\omega_l t)$, then the signals after quadrature modulation are:

$$I_m = \frac{V_i V_{Li}}{2} \left[\cos(\omega_l - \omega_{bb})t + \cos(\omega_l + \omega_{bb})t \right] \quad (3)$$

$$Q_m = \frac{V_q V_{Lq}}{2} \left[\cos(\omega_l - \omega_{bb})t - \cos(\omega_l + \omega_{bb})t \right] \quad (4)$$

The total output is: $V_o = I_m + Q_m$.

Ideally, $V_i = V_q, V_{Li} = V_{Lq}$, in this case, $V_o = V_i V_{Li} \cos(\omega_l - \omega_{bb})t$ does not exist the nonideal characteristics. But in practice, the nonideal characteristics are common.

4 NONIDEAL CHARACTERISTICS OF THE QUADRATURE MODULATOR

4.1 Amplitude imbalance of I and Q

Because of the amplitude imbalance of analog signal I and analog signal Q, the sideband signals cannot be completely offset. Within the emission spectrum, there will be sideband signals next to the useful signal spectrum, and the quality of the transmission signal will be decline. This effect would be more serious through the power amplifier, affecting the EVM of transmission signal severely. At the same time, in the DPD system, the amplitude imbalance of signal I and Q is reflected in the amplitude inconsistent between the feedback signal Q and the baseband signal Q, and therefore, will affect the result of the correction.

For the inconsistent between L_i and L_q, the inconsistent between I and Q, the modulator output signal can be derived from Equation (4) and (5):

$$V_o = I_m + Q_m = \left(\frac{V_i V_{Li}}{2} + \frac{V_q V_{Lq}}{2} \right) \cos(\omega_l - \omega_{bb})t$$
$$+ \left(\frac{V_i V_{Li}}{2} - \frac{V_q V_{Lq}}{2} \right) \cos(\omega_l + \omega_{bb})t \quad (5)$$

We can seen from Equation (5) that there are unuseful sideband signals next to the output signal due to the inconsistent between L_i and L_q.

4.2 Phase imbalance of I and Q

As the phase imbalance of I and Q, not only a serious sideband component is produced, but also the phase shift of the Q signal is caused. That will affect the SNR of the transmission signal, affecting the demodulation of received signal. At the same time, the phase of DPD feedback signals I and Q are shifted caused by the phase imbalance. When DPD adaptive algorithm does the correlation operator,

the baseband signals and the feedback signals cannot be alignment accurately in the time domain that will worsen the result of the DPD correction.

Assuming that baseband signal I and Q:

$$I = V_i \cos(\omega_{bb}t) , \quad Q = V_q \sin(\omega_{bb}t) \quad (6)$$

And the local oscillator signals:

$$V_{Li} \sin(\omega_l t), V_{Lq} \cos(\omega_l t) \quad (7)$$

Then the results of multiplied are:

$$I_m = \frac{V_i V_{Li}}{2} \left[\cos(\omega_l - \omega_{bb})t + \cos(\omega_l + \omega_{bb})t \right] \quad (8)$$

$$Q_m = \frac{V_q V_{Lq}}{2} \{ \cos[(\omega_l - \omega_{bb})t - \theta_1] - \cos[(\omega_l + \omega_{bb})t + \theta_1] \} \quad (9)$$

Assuming that $V_i = V_q = V$, $V_{Li} = V_{Lq} = V_L$, the output signal after modulation is:

$$V_o = I_m + Q_m$$
$$= \frac{V V_L}{2} \{ \cos(\omega_l - \omega_{bb})t + \cos[(\omega_l - \omega_{bb})t - \theta_1]$$
$$+ \cos(\omega_l + \omega_{bb})t - \cos[(\omega_l + \omega_{bb})t + \theta_1] \} \quad (10)$$

Seen form Equation (10), there are sideband signals next to the modulator output signal brought by phase imbalance.

5 SIMULATION AND COMPARISON

OFDM modulation signal with bandwidth of 8 MHz is the input. The model of PA is the normalization memory polynomials with series order 7 and memory depth 2.

5.1 Amplitude imbalance of I and Q

Figure 3(a), (b) and (c) denote the AM-AM curves when the amplitude imbalance of I and Q are respectively 0 dB, 0.5 dB and 1 dB. Compared with the ideal case, when the amplitude imbalance of I and Q is 1 dB, the divergence at the end of AM-AM curve is obvious, which shows that the amplitude imbalance has affected DPD correction performance.

Figure 4(a), (b) and (c) denote the AM-PM curves when the amplitude imbalance of I and Q are respectively 0dB, 0.5dB and 1dB. Ideally, all the samples are concentrated around a fixed phase shift can be seen from Figure 4(a). When the amplitude imbalance of I and Q is 1dB, the AM-AM curve diverges obviously, which shows that DPD

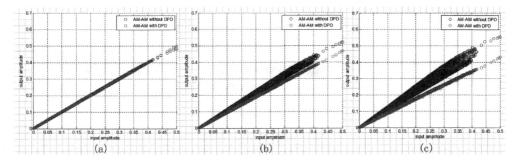

Figure 3.　Effects of amplitude imbalance on the AM-AM characteristics.

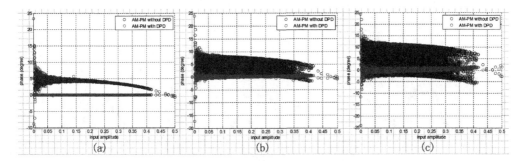

Figure 4.　Effects of amplitude imbalance on the AM-PM characteristics.

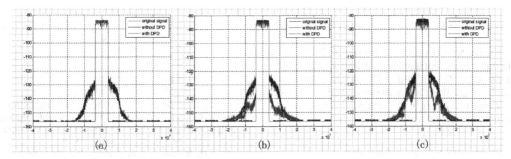

Figure 5.　Effects of amplitude imbalance on DPD results.

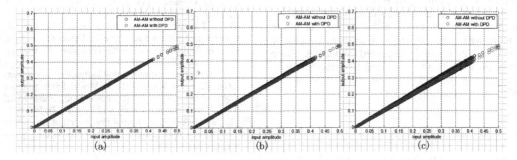

Figure 6.　Effects of phase imbalance on the AM-AM characteristics.

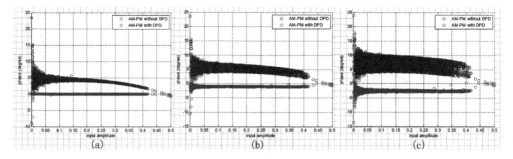

Figure 7.　Effects of phase imbalance on the AM-PM characteristics.

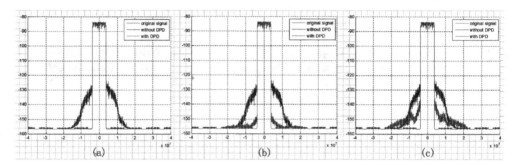

Figure 8.　Effects of phase imbalance on DPD results.

correction performance is worsened by the amplitude imbalance.

Figure 5(a), (b) and (c) denote the PSDs when the amplitude imbalance of *I* and *Q* are respectively 0dB, 0.5dB and 1dB. Above simulation results show that, when the amplitude imbalance of *I* and *Q* reaches 0.5dB, the performance of the system has seriously declined.

5.2 *Phase imbalance of I and Q*

When the phase imbalance of *I* and *Q* are respectively 0 degree, 1 degree, 5 degrees, the simulation results are given in Figure 6, 7, 8.

Through the comparison of the simulation results above, we can see that when the phase imbalance reaches to 5 degrees, the performance of DPD will be seriously impacted. When the phase imbalance exists, that will lead to a serious deterioration of the DPD correction effects if coupled with the nonlinearity of PA.

6 CONCLUSION

Simulation results show that compared with the ideal situation, when the imbalance of *I* and *Q* is

less than a certain range, the impacts on the system are acceptable.

REFERENCES

Angrisani, L. & Ghidini, I. & Vadursi, M. 2006. A New Method for I/Q Impairment Detection and Evalution in OFDM Transmitters. *IEEE Trans. on Instrumentation and Measurement*. 55(5): 1480–1486.

Chen, H.H. & Chen, J.T. & Huang, P.C. 2004. Adaptive I/Q imbalance compensation for RF transceivers. Global Telecommunications Conference. 2: 818–822.

Ding, L. 2004. Digital Predistortion of Power Amplifiers for Wireless Applications. Georgia Institute of Technology.

De Witt, J.J. & van Rooyen, G.J. 2009. A Blind I/Q Imbalance Compensation Technique for Direct-Conversion Digital Radio Transceviers. *Vehicular Technology, IEEE Trans*. 58: 2077–2082.

Lei, G. 2012. Optimmized Low-Complexity Implementation of Least Squares Based Model Extraction for Digital Predistortion of RF Power Amplifiers. *IEEE Trans. On Microwave Theory and Tech*. 60(3): 594–603.

Patel, J. 2004. Adaptive Digitial Predistortion Linearizer for Power Amplifiers in Military UHF Satelite. [doctor thesis], 10–34.

Control Engineering and Information Systems – Liu (Ed)
© 2015 Taylor & Francis Group, London, ISBN 978-1-138-02685-8

Foxjit ERP system web service-based electronic Kanban's research and implementation

L. Zhang
Computer Science, Sichuan Top IT Vocational Institute, Chengdu, Sichuan, China

Y. Liu, F.Y. Liu & J.Y. Li
Computer Engineering, Chengdu Technological University, Chengdu, Sichuan, China

ABSTRACT: JIT in time for the complement (Just In Time) after the process forward by pulling the material process, how much "pull" How much, in order to reflect the Kanban production control materials issued to the kanban quantity. Traditional JIT kanban calculation of the number of sheets is mainly based on experience, the lack of scientific method, and the control function is poor. This paper presents a theoretical and Web Service 3C JIT Electronic Kanban way then consider the future demand for materials sharing, material values and tables and other materials, with the scientific method to calculate the number of sheets Kanban and be adjusted according to the actual needs supplementary materials. Our newly developed Web Service based electronic kanban FoxERP enterprise resource planning JIT—FoxJIT system features proved that the theory and the method are feasible.

1 FOXERP JIT SYSTEM COMPONENTS, OPERATING ENVIRONMENT AND FEATURES

JIT is the time for the complement (Just In Time) for short, is a kind of Kanban as the core management system, which requires the production according to market demand determines what, when produced, how much to produce, this JIT "pull" production system is In Kanban way (to billboards and sheets) in the control material flow. After the spirit is what the pre-production process need only give it what, do not give back, not much to give. JIT production is a kind of Oriental mode of production, Toyota Motor Corporation was first proposed, it overcomes the West based on MRP (Material Requirements Planning) is not accurate market forecasts for production deficiencies. Currently at home and abroad is in the ascendant.

FoxERP enterprise resource planning system is the author of the research group for the little fox Chengdu software development company after several years of the launch of new products. It is about small and medium manufacturing enterprise resource planning system, subsystem by the master production schedule, inventory management subsystem, management subsystem in products, manufacturing order management subsystem, ledger management subsystem, personnel management subsystem, quality management subsystem other 25 sub-systems. From the orders, manufacturing

to shipping to the overall business process of a comprehensive digital management, which FoxJIT timely management subsystem for the complement of the "pull" mode of production has been managed. FoxJIT in time for complementary management subsystem is the enterprise resource planning FoxERP (Fox Enterprise Resources Planning) is an important part.

FoxJIT in time for complementary management system consists of user management, project management, system management, Kanban management, product warehousing and storage of raw materials and other modules. Our FoxJIT kanban system is an electronic kanban, and the signage is based on Web Service that allows vendors in the world to use in their systems, real-time visual tracking (factory every hang out a kanban supply Manufacturers will see the growth of Kanban rectangles, after rectangles be filled, can delivery) manufacturers need material situation is very convenient, its timeliness, the number of sheets to adjust than traditional Kanban good, there is rare.

FoxJIT system uses ASP.NET4.5, C # mainstream language and SQL Sever database technology, good support for the B/S mode structure, and its interactive capability, user-friendly, fast, and also help with the system's total pages and each page help to facilitate the majority of users. Set up a multi-level user permissions, to ensure system security.

This system not only for the actual business management, but also for classroom teaching and experiment can also be used in the Internet environment in different regions as distance education software for students to use.

2 FOXJIT 3C AND ELECTRONIC KANBAN SYSTEM TECHNOLOGY

JIT is a process after the process pull the material forward, this "pull" action exists in each two adjacent processes (process definitions vary) between. In order to reflect the Kanban production control materials issued to the kanban quantity. Traditional JIT kanban calculation of the number of sheets is mainly based on experience; the lack of scientific methods to control the function is poor. The fusion of "3C theory" JIT way then consider the future demand for materials sharing, tables and other material values and materials, so use the scientific method to calculate the number of sheets Kanban and adjusted at any time, according to the actual demand for supplementary materials. FoxJIT technology is based on this theory and Web Service 3C electronic Kanban JIT mode.

3C theoretical goal is to find the material supplied to the supply chain, the point of use of the material in the best way. In the repetitive production environment, 3C can be used to plan material requirements, instead of the traditional MRP. 3C refers capacity (capacity), sharing (commonality) and consumption (consumption). And the entire 3C core of the theory is to consider the capacity (MSR), consumption (TOP multiply BOM) and sharing (check each product consumable materials m maximum) "speed material table."

2.1 FoxJIT electronic Kanban system technology

Kanban management is the core of our FoxJIT subsystem, JIT production management model revolves around the whole Kanban expanded. Rather Kanban can manage the entire production of the central nervous system. The enterprise Kanban management is an important component. We hang on billboards, remove, view to coordinate operations between the various processes within the enterprise production (Figure 1).

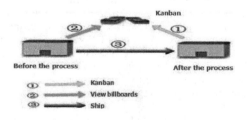

Figure 1. Kanban processes.

When this process (assuming work number is 002) production of A material needed 1000 parts, so hang billboards (① operation performed). Before the process (assuming work numbers 001) View billboards (for ② operation), according to information provided to the kanban process 002 shipped (Material Name A number 1000) at the same time remove the billboards.

FoxJIT electronic Kanban system technology is the real-life Kanban technology uses a computer to simulate. So that it can efficiently serve the production. In between suppliers and secondary positions using electronic Kanban benefit is that suppliers can readily understand the factories of various materials consumption behavior, we can pre-prepare and punctual delivery. Traditional Kanban approach can only be recovered when the next delivery Kanban material has been consumed, timeliness poor.

2.2 Electronic Kanban sheets and calculation rate bill of material

a. The number of sheets Kanban

Kanban sheets should expect sales pace with changes in TOPm and RBOMm change. RBOM (rate bill of material) is expected from the sale of each product multiplied by a unit of the speed of the product use of the material m, and the sales rate exactly equal to the production rate, and then take the rate of consumption of each product m maximum. It is used to calculate the number of sheets Kanban.

Find the number of sheets Kanban formula: Kanban sheets = [REPm * RBOMm + LSm]/ tankage

- REP: from supply room (or former process) to the factory (or post-process) the time between the supplement (replenishment time)
- LS: standard batch that suppliers have to wait until a batch when shipped.

Find the number of sheets Kanban
int a = 0; int b = 0; int c = 0; int d = 0;
a = Convert.ToInt32 (TextBox4.Text); get LS (standard lot)
b = Convert.ToInt32 (TextBox5.Text); get REP
c = Convert.ToInt32 (TextBox6.Text); get the container capacity
d = (b * ss + a)/c; calculating Kanban sheets (where SS is the rate bill of material)

b. Computational Rate bill of material

Seeking rate bill of material formula: RBOMm = max {TOPp * BOMpm}

- expected sales pace or demand-pull table (TOP, table of pull): It was agreed that the product p peak sales rate (peak sales rate) is TOPp.

- Summary of Materials (BOM, summarized bill of material): APICS is defined as the aggregate structure of a product and the quantity of all materials used in material form. It does not consider the material class; any material listed only once, the unit dosage amount of the total cases.

FoxJIT system rate bill of material obtained using the following algorithm:

① Let RBOMm = 0, the bottleneck capacity (the maximum speed of sales MSRp are based processes, assuming that products have a common bottleneck process) utilization CU = 0;

② Calculate the material of each product P m, peak consumption TOPp * BOMpm, and follow the results in descending order according to p.

③ Select the first p;

④ RBOMm ← RBOMm + TOPp * BOMpm.

⑤ CU ← CU + TOPp/MSRp, if CU <1 and there is p option, then select the next p, back ④; otherwise ends. The spirit of the above algorithm is to use the product p output speed limit, that capacity constraints, to calculate the various materials m maximum consumption rate.

3 FOXJIT SYSTEM WEBSERVICE TECHNOLOGY

3.1 Providers, enterprises between

In JIT production mode FoxJIT system supplier, the relationship between enterprises in Figure 2.

When times barn stock is lower than reorder point (in the order analysis calculated), they hang in the library at the same time a kanban (① operation performed). Suppliers through business to suppliers "Web Service" Interface view kanban (for ② operation), when the number reaches a certain volume Kanban when supplier shipment to the second position (for ③ operation).

Materials ordering point is only recently reorder point (because as material changes in the impact of rate bill of material, the material Order-to-point and order point will change), the largest in-progress

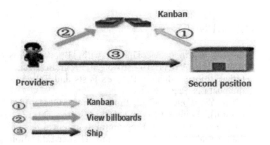

Figure 2. FoxJIT system supplier, the relationship between enterprises.

inventory REPm * RBOMm + LSm and REPm * RBOMm is actually very small, the great law's Order-to-point and order point. Inventory below reorder point (a material is consumed, resulting in an empty kanban), the Kanban process flows back to front, the equivalent of issuing an order to set back the inventory point.

3.2 Programming deploy web service Kanban

Its programming instructions:

This program uses a programming language is C, externally callable class name is DataService. Importing the WebService requires the use of the name space, System and System. WebService, as well as access to the SQL database used namespaces System.Data and System.Data.SqlClient. Create a Web Service class, Kanban business logic layer, this new class must be declared as public, but also need to inherit from the Web Service class. Such methods or properties defined, if coupled with [WebMethod] mark indicates that the Web service outside the program can access the method or property.

This Web Service has been completed, save it as DataService.asmx.

3.3 The client to use web service Kanban

Using a service part of the work requires the following two: Create a Service Agent and programming client users to use the program.

a. Create a service agent

Just start writing on a server to call a Web Service (assuming the server's IP address is 192.168.80.111). In the browser address bar http://192.168.80.111/jit/Service1.asmx?wsdl.Found on the server Web Service, directly in the browser's. Asmx followed by "? Wsdl". This is automatically generated wsdl file in XML format to save the contents to a local project, named DataService. Wsdl.

Add Web Reference, a reference to the service just saved find, and given a name, such as "WebService", added successfully, the service will automatically have a local class namespace, which will have a "Reference.cs" files.

In the "Reference.cs" command "csc t: library out: bin Reference.dll reference.cs" Reference.dll compiled into a dynamic link library. This has been completed the proxy service programming.

b. The client interface and procedures

In FoxJIT system home input materials name and press the OK button, you can view the latest status of the material, including progress bar and signage details, shown in Figure 3.

"Kanban situation" is a form of progress bar image shows that the current kanban posted situation. Which is described as percentages and time schedule when the corresponding percent "real time."

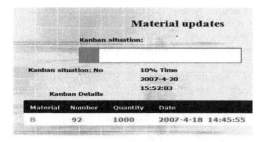

Figure 3. The latest status of the material Kanban including progress bar and signage details.

"Kanban Details" is the name of the material in the form of tables, kanban number, quantity, time and other details send messages displayed on the client.

When the progress bar reaches full grid Kanban, the left of the "last time the full grid" will prompt "progress bar is full time."

In Kanban progress bar reaches full grid explained that it had reached a batch, this time should be shipped to the sub-suppliers warehouse storage. When the second warehouse storaged, progress bar will go blank, and a rectangular table will be empty.

On the client using Web services, key code:

```
Using System.Data.SqlClient;
Using System.Web.Services;
Namespace JIT
{
Private void Page_Load (object sender, System.
EventArgs e)
{
Label5.Text = "last full-time:" + Session ["time"].
ToString ();
localhost1.DataService datasets2 = new JIT.
localhost1.DataService invoke the Web Serv-
ice compiled dynamic link library object is
instantiated.
int ss = datasets2.Accept1 (Session ["cailiao"].
ToString inside Accept1 invoke the Web Service
method.
if (! Page.IsPostBack)
{
str4 = 10; maximum number of kanbans
str5 = Convert.ToInt32 (ss); actual number of
kanbans
str5 = str5% 10; get the number of the actual
kanbans modulo operating
if (Convert.ToBoolean (str5) = = false)
{Str5 = 0;}
Else
{If (Convert.ToBoolean (str5% 10))
{}
Else
{Str5 = 10;
```

Session ["time"] = DateTime.Now.ToString use Session ["time"] to the time at full frame.

```
str6 = Convert.ToInt32 ((str5 * 100)/str4);
str1 = "jitbar1/jitjd";
str2 = ". jpg";
str3 = "%";
Image1.ImageUrl = str1 + str6 + str2; In the
Image inside a progress bar shows the current
situation.
Label1.Text = str6 + str3 + "time" + DateTime.
Now; Label1 progress bar in each moment in time.
}
DataGrid1.Visible = true;
localhost1.DataService datasets1 = new JIT.
localhost1.DataService invoke the Web Service
compiled dynamic link library to instantiate an
object.
DataSet ds = datasets1.GetDataSet (Session
["cailiao"]. ToString ());inside GetDataSet invoke
the Web Service method.
DataGrid1.DataSource = ds;
DataGrid1.DataBind ();
}
}
```

4 CONCLUSION

We develop FoxJIT system has been on Chengdu little fox software company successfully test run (the next step to prepare the bar code input and output for signage), which applies to small and medium manufacturing enterprises, will be further promotion for our enterprise information over-taken contributions.

ACKNOWLEDGEMENTS

This work was supported by Sichuan applied basic research project "based on multi-source informa-tion fusion and dynamic cloud computing intel-ligence service applied research" (project number 2013JY0059).

REFERENCES

[America] Matthao MacDonald with, Booz studios translated. "ASP.NET4 Advanced Program Design (4th Edition)." Beijing: People's Posts and Telecom-munications Press, 2011.

Liu Fu Ying, etc. "C # Programming Guide (3rd edition)." Beijing: Electronic Industry Press, 2012.

Liu Fu Ying, etc. "Web Programming practical techni-cal tutorial (2nd Edition)." Beijing: Higher Education Press, 2008.

Ye die. "Enterprise resource planning ERP". Beijing: Electronic Industry Press, 2002.

Control Engineering and Information Systems – Liu (Ed)
© 2015 Taylor & Francis Group, London, ISBN 978-1-138-02685-8

Improved self-adaptive differential evolution based on exponent crossover

Y.L. Xu
Institute of Information and Technology, Henan University of Traditional Chinese Medicine, Zhengzhou, China
College of Information and Technology, DongHua University, Shanghai, China

J.A. Fang, Z.Y. Liu & W.X. Cui
College of Information and Technology, DongHua University, Shanghai, China

ABSTRACT: Differential Evolution (DE) is well known as a simple and one of the most powerful intelligence algorithms in current. Based on exponent crossover, this paper address to enhance the performance of *j*DE, which is a famous variant of original DE with self-adaptive control parameter. Firstly, we introduce a new method to measure population diversity by the value of object function. Then we provide a crossover scheme according to aforementioned diversity degree, when above degree is smaller than a threshold value our scheme deploys part of superior target vector and inferior mutant vector for crossing, otherwise, the scheme controls part of superior target and superior mutant vector for crossing. Furthermore, combining this scheme with exponent crossover we can distinctly improve performance of original *j*DE algorithm. Finally, the comparative study indicates that our new scheme can enhance the performance of conventional *j*DE on a suite of 6 numerical optimization problems.

1 INTRODUCTION

In recent years, Differential Evolution (DE) has been demonstrated to be one of the most excellent evolutionary algorithms [1–3], which is simple to implement and convergent speed is quite fast. However, the faster convergence may also lead to a higher probability of trapping in local minimum because the diversity of population descends quickly during evolution. Therefore, many variants of DE were proposed to enhance its performance by self-adaptively controlling the strategies and parameters.

Literature [4] proposed the variants FADE, which changes the values of F and Cr according to the behavior of the algorithm. The adaptive variant SaDE [5] adjusts parameters during the evolution by using normal distribution. Inspires by the idea of SaDE, an ensemble of mutation strategies and control parameters (EPSDE) was employed in [6]. A pool of mutation strategies competes to produce successful offspring population in this algorithm. Unlike the SaDE, the target vector of the EPSDE should be randomly reinitialized with a new mutation strategy from the respective pools or from the successful combinations stored with equal probability when the trial vector performs poorer.

Recently, J. Brest and S. Greiner et al. [7] proposed self-adapting control parameters in differential evolution (*j*DE). They encoded control parameters F and Cr into the individual and adjusted them by introducing two new parameters τ_1 and τ_2. In the *j*DE, a set of F and Cr values was assigned to each individual in the population, augmenting the dimensions of each vector. The better values of these encoded control parameters lead to better individuals that in turn, are more likely to survive and produce offspring and, thus, propagate these better parameter values. In literature [8] a new self-adaptive variants, namely JADE, is proposed to improve optimization performance, the algorithm applies a new mutation strategy, referred by the authors as DE/current-to-*p*best and utilizes an optional external archive to provide information of progress direction. In addition, a simple strategy adaptation mechanism named (SaM) [9] was implemented for a family of DE variants to select an appropriate strategy adaptively, and the proposed SaM was then combined with JADE to verify the efficiency of their approach. Moreover, similarly with above DE variant such as CoDE [10] was proposed. In CoDE, three well-studied offspring generation strategies were combined in a random way to generate trial vectors to enhance performance.

Above works can help to enhance the search ability and performance of DE, but there still remains research room to improve it and we find that all of these researches are performed by using binomial crossover. Hence, this paper tries to use exponent crossover to improve the performance of original DE and *j*DE. In addition, we introduce a new cross scheme based exponent crossover and a population diversity measure method.

2 OUR METHOD AND ANALYSIS

In this section, we introduce the canonical *j*DE algorithm briefly, and then propose a new method to measure population diversity. Thirdly, we give the crossover scheme according this degree of diversity, when diversity degree is smaller than a threshold value our scheme deploys part of superior target vector and inferior mutant vector for crossing, otherwise, the scheme controls part of superior target and superior mutant vector for crossing. Finally, we use our scheme to improve original *j*DE based on exponent crossover.

2.1 The canonical jDE algorithm

Brest et al. [7] proposed a self-adaptation scheme for the DE control parameters. They encoded control parameters F and Cr into the individual and adjusted them by introducing two new parameters τ_1 and τ_2. In their algorithm (called "*j*DE"), a set of F and Cr values was assigned to each individual in the population, augmenting the dimensions of each vector. The better values of these encoded control parameters lead to better individuals that in turn, are more likely to survive and produce offspring and, thus, propagate these better parameter values. The new control parameters for the next generation are computed as follows [3]:

$$F_{i,G+1} = \begin{cases} F_l + rand_1 * F_u, & \text{if } rand_2 < \tau_1 \\ F_{i,G}, & \text{otherwise} \end{cases} \quad (1)$$

$$CR_{i,G+1} = \begin{cases} rand_3, & \text{if } rand_4 < \tau_2 \\ CR_{i,G}, & \text{otherwise} \end{cases} \quad (2)$$

where F_l and F_u are the lower and upper limits of F and both lie in [0, 1]. In [7], Brest et al. used $\tau_1 = \tau_2 = 0.1$. As $F_l = 0.1$ and $F_u = 0.9$, the new F takes a value from [0.1, 0.9] while the new Cr takes a value from [0, 1]. As $F_{i,G+1}$ and $CR_{i,G+1}$ values are obtained before the mutation is performed, they influence the mutation, crossover, and selection operations for the new vector $X_{i,G+1}$.

2.2 Measuring population diversity by fitness value

In recent year, many researchers have studied to how to measure the diversity of population. Literature [11] claimed that the diversity of population can be considered from two aspects, one is

on the fitness landscape (e.g., distance between each pairs of individuals), and the other way is through its fitness values.

Obviously, considering object fitness values to compute population diversity is simple and direct.

In Ref. [12], a measurement of the fitness diversity as the difference in performance between the fittest individual and other members of the population has been introduced. Literature [13] and [14] present ξ to measure the diversity of population, which can indicate how close the average fitness to the best one. Similarly, in Ref. [15], a parameter ψ is given to measure the population diversity, which means that the position of average fitness when all fitness values sorted a line. In order to consider the degree of fitness values sparse in population, V. Tirronen et al [16] proposed parameter v to measure the diversity of population. Parameter χ in [12] is defined to answer to the question "How much better is the best individual than the average fitness of the population with respect to the history of the optimization process?" Combining parameter v and ψ, V. Tirronen and F. Neri [11] proposed φ to measure the diversity of population, the φ implies the sparse are the fitness values with respect to the range of fitness variability at the current generation. It can be perceived that all of aforementioned parameters consider the fitness values to measure the diversity. Beyond that, in [17] Ali W.M and Hegazy Z.S inspired from nature and human behavior proposed use the best and worst individuals to measure diversity based on fitness landscape.

$$\varphi = \frac{\sigma_f}{\left| f_{worst} - f_{best} \right|} \quad (3)$$

In this paper, inspired from article [11] we apply to indicate the diversity in Eq. (3). Where σ_f is the standard deviation of fitness values over individuals of the populations, f_{worst} and f_{best} are the worst and best fitness value of individual in population, respectively. Obviously, the value of φ is between 0 and 1.

2.3 Our superior-inferior crossover scheme based on exponent

Aim to maintain the population diversity during evolution. We present a new crossover scheme based on exponent crossover. In our scheme, when the degree of population diversity is smaller than a threshold value, means individuals excessive gather and population diversity lower, thus we employ superior and inferior individual in *p*% part of population to cross for improving diversity. Otherwise, we use superior and superior individuals in *p*%

part of population for crossing to enhance exploit ability.

Note that our scheme is not a new cross operation but adjust individual's index before exponent crossover operation. In addition, the value of p is in the interval [0, 1], based on simulation experiment in next section we use $p = 0.1$ as initial value.

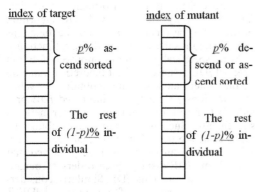

index of target   index of mutant

$p\%$ ascend sorted $p\%$ descend or ascend sorted

The rest of $(1-p)\%$ individual   The rest of $(1-p)\%$ individual

Figure 1. Superior-inferior crossover.

Algorithm 1. Improved *j*DE algorithm based on our scheme using exponent crossover.

/* improved *j*DE algorithm based on our scheme */
1. Initialization parameters F, Cr, τ_1, τ_2, popsize, D.
2. $FES = 0$; % the generation of evolution
3. **While** FES $D*100$
   % D is number of problem dimension
4. *Update F and Cr* by Eq. (1) and Eq. (2);
5. Mutation operation using "DE/current-to-best/1" strategy;
6. Compute diversity of population by Eq. (3);
7. Superior-inferior scheme function ()
   {
8. If $\varphi < \varepsilon$ then
9.   Implement superior-inferior scheme;
10. $p\%$ front of target vector ascending sorted and
11. $p\%$ front of mutant vector descending sorted;
12. else
13.   Implement superior-superior scheme;
10. $p\%$ front of target vector ascending sorted and
11. $p\%$ front of mutant vector ascending sorted;
12. end if
   }
13. Crossover operation using **exponent** crossover;
14. Select operation
15. Update *popold, F, Cr*;
16. $FES = FES + popsize$;% next generation
   % Judgment of terminal condition
17. **if** $FES \geq m*100$
18.   **break**;
19. **end**
20. **end**

After initialization the value of p is alterable in every generation. Inspired by the literature [18], we apply following Eq. 4 to vary the value of p_i in every generation i.

$$p_i = \max\left(2,\ ceil\left[NP \cdot p\% \cdot \left(\frac{G_{current}}{G_{max}}\right)\right]\right) \qquad (4)$$

Moreover, we named aforementioned threshold value as tolerance degree of diversity, which denotes ε. Based on the simulation experiment in next section and the literature [19], we set $\varepsilon = 0.05$ as empirical value.

Specifically, the proposed crossover idea is shown in Figure 1, the adjusting operation on mutant vector and target vector is executed before crossover operation. After that the crossover operation will enhance the diversity of population and increase the exploration of the search space, the diversified population will not only become a set of newly promising solutions but also easily escape the local minimum trap.

More specifically, to express our idea of improving self-adaptive differential evolution based on exponent crossover clearly. The whole improved algorithm is provided in Algorithm 1.

3 SIMULATION EXPERIMENT AND ANALYSIS

3.1 *Test functions setup*

As discussed in literature [20], many benchmark numerical functions commonly used to evaluate and compare optimization algorithms may suffer from two problems. First, global optimum lies at the center of the search range. Second, local optima lie along the coordinate axes or no linkage among the variables exists. To overcome above drawbacks, we use shift or rotate the conventional benchmark functions. The detail feature of shift or rotate functions are presented in [20–21], which is not repeated here. In this paper, we just use functions 1–10 for testing, and Table 1 provide these detail information.

All the functions are tested on 30 dimensions. For each algorithm and each test function, 30 independent runs are conducted with 300 000 function evaluations (FES) as the termination criterion.

3.2 *Comparison with original jDE and DE variants*

In this subsection, we apply the proposed scheme on original DE and *j*DE for comparison. In our scheme we use $p = 0.1$ and $\varepsilon = 0.2$ as empirical value, the others algorithm's parameters are shown in Table 2. In addition, we shall provide the detail analyses about p and ε in next subsection.

We evaluate the performance of above algorithms over 30-dimensional version of the 6 benchmark functions. Table 3 reports the experimental results of the mean error and standard deviation of the solutions in 30 independent runs. The best result among those DEs in indicated by Boldface in the table. Rows "+ (better)," "≈(Same)," "−(Worse)" give the number of functions that the SI scheme performs significantly better than, almost the same

Table 1. Benchmark functions configurations.

Functions	Name	Search space	f_{bias}
f_1	Shifted sphere	$[-100,100]^D$	−450
f_2	Shifted schwefel's problem 1.2	$[-100,100]^D$	−450
f_3	Shifted rotated high conditioned elliptic	$[-100,100]^D$	−450
f_4	Shifted schwefel's problem 1.2 with noise in fitness	$[-100,100]^D$	−450
f_5	Schwefel's problem 2.6 with global optimum on bounds	$[-100,100]^D$	−310
f_6	Shifted rosenbrock	$[-100,100]^D$	390

Table 2. EA algorithms for comparison.

Algorithm	Parameters	Reference
DE/rand/bin	$F = 0.5, CR = 0.9$	[1–3]
DE/best/1	$F = 0.5, CR = 0.9$	[1–3]
DE/current-to-best/1	$F = 0.5, CR = 0.9$	[1–3]
jDE	$\tau_1 = \tau_2 = 0.1$	[7]
Improved jDE	$p = 0.1, \varepsilon = 0.2$	This paper

as, and significantly worse than the compared algorithm on fitness values in 30 run, respectively. The last three rows is shows the difference between the number of +'s, −'s and ≈'s, which is used to give an overall comparison between the two algorithms. The best result among those data is indicated by **Boldface** in the table.

The results in Table 3 show that our scheme influences distinctly the performance of original jDE. Specifically, our improved jDE algorithm provides the best performance on the functions f_1–f_4, then ranks the second on the f_5, f_6 function.

More specifically, both our improved jDE and original jDE achieve the best theoretical value on f_1, but others method cannot obtain this value. On functions f_2 and f_4, our improved jDE provides better performance than the other methods significantly. Such as on f_2 our scheme obtain the mean value is 8.04e−13, but original jDE only get 1.24e−06 as mean value, it can be perceived that our scheme jDE display seven orders of magnitude better than original jDE. Similarly, considering f_4, the performance of our improved jDE is 4.41e−04, which two orders of magnitude better than original jDE (2.81e−02). With respect to function f_3, f_5 and f_6, our scheme does not yield marked performance enhance them, but which can attain an equal performance compare with original jDE.

As conclusion according to above analyzed for Table 3, it is observed that our scheme improved jDE can enhance the performance of original algorithm distinctly. Therefore, the proposed crossover scheme is effective and efficient.

3.3 Parameters setup in our crossover scheme

In this subsection, we shall study the parameters in the proposed scheme improved jDE. There are two parameters in the proposed scheme: the diversity tolerance degree of population ε. Parameters

Table 3. Experimental results of 30-dimensional problems F1-F6, averaged over 30 independent runs with 300000 FES.

	DE/rand/1 Mean ± Std	DE/best/1 Mean ± Std	DE/current-to-best/1 Mean ± Std	jDE Mean ± Std	Our improved jDE Mean ± Std
f_1	1.78e−30 ± 9.22e−30+	6.14e+03 ± 3.39e+03+	3.41e+03 ± 1.92e+03+	0.00e+00 ± 0.00e+00≈	**0.00e+00±0.00e+00**
f_2	4.94e−05 ± 3.54e−05+	1.51e+04 ± 5.51e+02+	6.47e+04 ± 2.89e+04+	1.24e−06 ± 1.64e−06+	**8.04e−13±1.27e−12**
f_3	4.94e+05 ± 1.34e+06+	3.42e+07 ± 3.86e+07+	8.11e+06 ± 4.41e+06+	1.44e+05 ± 1.64e+06≈	**1.32e+05±8.34e+04**
f_4	1.51e−02 ± 1.27e−02+	3.31e+03 ± 6.16e+03+	3.77e+02 ± 2.87e+02+	2.81e−02 ± 6.97e−02+	**4.51e−04±5.28e−04**
f_5	**2.07e−01±1.43e−01−**	1.29e+04 ± 2.52e+03+	7.22e+03 ± 1.53e+02+	1.17e+03 ± 4.33e+02≈	1.15e+03 ± 5.47e+02
f_6	**1.98e+00±1.18e+00−**	7.25e+08 ± 5.98e+08+	1.81e+09 ± 7.18e+08+	3.08e+01 ± 2.88e+01≈	3.04E+01 ± 2.55E+01
+	4	6	6	2	*
−	2	0	0	0	*
≈	0	0	0	4	*

Table 4. Effects of paramaters on performance of our improved JDE, averaged over 30 independent runs with 300 000 FES.

	Our scheme improved jDE					Mean(Std)	
	p=0.1, ε=0.05	p=0.1, ε=0.1	p=0.1, ε=0.2	p=0.1, ε=0.3	p=0.1, ε=0.5	p=0.2, ε=0.05	p=0.2, ε=0.1
f_1	0 (0)	0 (0)	0 (0)	0 (0)	0 (0)	0 (0)	0 (0)
f_2	1.91e–12	1.53e–12	**8.04e–13**	1.47e–12	2.49e–12	5.79e+00	7.22e+00
	(2.27e–12)	(2.67e–12)	**(1.28e–12)**	(4.46e–12)	(7.36e–12)	(2.46e+00)	(3.56e+00)
f_3	1.42e+05	**1.53e+03**	1.32e+05	1.68e+05	1.59e+05	1.22e+07	1.17e+07
	(7.46e+04)	**(6.53e+04)**	(8.34e+04)	(8.48e+04)	(9.21e+04)	(3.78e+06)	(4.08e+06)
f_4	6.47e–04	2.33e–03	**4.51e–04**	6.81e–04	6.19e–04	7.91e+02	7.97e+02
	(5.48e–04)	(9.21e–03)	**(5.28e–04)**	(8.55e–04)	(9.49e–04)	(2.91e+02)	(3.78e+02)
f_5	8.25e+02	1.25e+03	1.14e+03	1.19e+03	**1.04e+03**	3.55e+03	3.58e+03
	(6.92e+02)	(6.33e+02)	(5.47e+02)	(6.23e+02)	**(5.76e+02)**	(5.22e+02)	(6.07e+02)
f_6	**2.96E+01**	3.27e+01	3.04e+01	3.18e+01	3.85e+01	3.69e+01	3.02e+01
	(2.47E+01)	(2.62e+01)	(2.55e+01)	(2.52e+01)	(2.93e+01)	(1.85e+01)	(2.15e+01)

p is the number of individuals which participate in adjusting before crossover.

First of all, with respect to parameter p, whose value is between [0, 1]. However, in the "current-to-best" mutation strategy, there are have corresponding relationship in the same index between target and mutate vector, thus, when after adjusting operator these relationship will be discard. Hence, for keeping that relationship a small p is good choice, that is the reason why we set $p = 0.1$.

Secondly, the diversity tolerance degree of population ε, which is an important factor for judge the population excessive gather or not. Frankly speaking, how to judge the gather degree for population is a vast research field. In [15] the author utilizes 0.05 as empirical value to experiment and obtain good results. However, in order to find reasonable ε in this paper, we test many combination of these parameter and provide the result in Table 4.

Table 4 demonstrates that when $p = 0.1$ and $\varepsilon < 0.5$ the performance of our scheme improved jDE is good. Specially, if $p = 0.1$ and $\varepsilon = 0.2$, the result is a litter better than others parameter applied, thus as a conclusion, $p = 0.1$ and $\varepsilon = 0.2$ is the empirical value we obtained.

4 CONCLUSION

In this paper, we proposed a new crossover scheme to improve jDE algorithm and obtained good result.

Firstly a method to compute diversity degree of population is introduced. And then, according this degree we compare it with diversity tolerance to judge implement which crossover operation to enhance original jDE performance, the operation above mention is between target vector and mutation

vector, and the individual participated in operation is part of population. Generally speaking, when above diversity degree is smaller than a threshold value our scheme deploys part of superior target vector and inferior mutant vector for crossing, otherwise, the scheme controls part of superior target and superior mutant vector for crossing.

Furthermore, we combine proposed scheme with exponent crossover to improve original jDE algorithm.

To verify our scheme's effectiveness, an extensive performance comparison has been conducted over 6 commonly used CEC2005 contest test instances. The experimental results suggested that its overall performance was better than the original jDE algorithm significant.

ACKNOWLEDGEMENT

This paper was supported by the Key Creative Project of Shanghai Education Community (13ZZ050), the Key Foundation Project of Shanghai (12 JC1400400), and the Nursery Research Project of Henan University of Traditional Chinese Medicine under Grant MP2013-36. The authors are grateful to the Editor-in-Chief, Associate Editor and anonymous reviewers for their constructive suggestions that helped to improve the content as well as the quality of the manuscript.

REFERENCES

Brest J., Boškovič B., Greiner S., Žurner V. & Maučec M.S. Performance comparison of self-adaptive and adaptive Differential Evolution Algorithms, Soft Comput. 11 (7), 617–629, 2007.

Caponio A., Cascella G.L., Neri F., Salvatore N. & Sumner M. A fast adaptive memetic algorithm for on-line and on-line control design of PMSM drives. IEEE Transactions on System Man and Cybernetics-part B, special issue on Memetic Algorithms 37(1), pp:28–41 (2007).

Caponio A., Neri F. & Tirronen V. Super-fit control adaptation in memetic differential evolution frameworks. Soft Computing-A Fusion of Foundations, Methodologies and Applications 13(8), pp: 811–831 (2009).

Gong W., Cai Z., Ling C.X. & Li H. Enhanced Differential Evolution with Adaptive Strategies for Numerical Optimization, IEEE Transactions on Systems, Man, and Cybernetics, Part B: Cybernetics, 41(2): 397–413, 2011.

Jingqiao Zhang, Arthur C. & Sanderson, JADE: Adaptive Differential Evolution with Optional External Archive, IEEE Transactions on Evolutionary Computation, Vol. 13, No. 5, 945–958, 2009.

Liang J.J., Suganthan P.N. & Deb K. "Novel composition test functions for numerical global optimization" in Proc. IEEE Swarm Intell. Symp, Pasadena, CA, Jun, pp:68–75, 2005.

Lin Y.C. & Hwang K.S. A Mixed-Coding Scheme of Evolutionary Al-gorithms to Solve Mixed-Integer Nonlinear Programming Problems, Computers and Mathematics with Applications 47, 1295–1307 (2004).

Liu J. & Lampinen J. A fuzzy adaptive Differential Evolution, Soft Comput. 9, 448–462, 2005.

Mallipeddi R., Suganthana P.N., Pan Q.K. & Tasgetiren M.F. Differential evolution algorithm with ensemble of parameters and mutation strategies, Applied Soft Computing, 11: 1679–1696, 2011.2

Minhazul SK., Das S. & Ghosh S. A Adaptive Differential Evolution Algorithm With Novel Mutation and Crossover Strategies for Global Numerical Optimization, IEEE Transactions on systems. Man. And Cybernetics-Part B: Cybernetics. Vol. 42, No. 2:482–500, 2012.

Mohamed A.W., Sabry H.Z. & Khorshid M. An alternative differential evo-lution algorithm for global optimization. Cairo University, Journal of Avanced Research. 3, pp:149–165 (2012).

Neri F., Toivanen J., Cascella G.L. & Ong Y.S. An adaptive multimeme algorithm for designing HIV multidurg therapies. IEEE/ACM Transactions on Computational biology and bioinformatics, Special issue on computational intelligence approaches in computational biology and bioinformatics, 4(2), pp:264–278 (2007).

Neri F., Toivanen J. & Makinen R.A.E. An adaptive evolutionary algorithm with intelligent mutation local searcher for designing multidrug therapies for HIV. Applied Intelligence, Special issue on computational intelligence in medicine and biology, 27(3), pp:219–235 (2007).

Qin A. & Suganthan P. Self-adaptive differential evolution for numerical optimization, in: Proceedings of CEC 2005, vol. 1, IEEE Computer Press, pp: 630–636, 2005.

Storn R. & Price K. Minimizing the real functions of the ICEC'96 contest by differential evolution, Proc. IEEE Int. Conf. Evolutionary Computation, 842–844, 1996.

Storn R. & Price, K. Differential evolution: A simple and efficient heuristic for global optimization over continuous spaces, Y. Global Optimization 11, 341–369, 1997.

Suganthan P.N., Hansen N., Liang J.J., Deb K., Chen Y.P., Auger A. & Tiwari S. "Problem definitions and evaluation criteria for the CEC 2005 special session on real-parameter optimization," Nanyang Technol Univ., Singapore, May 2005.

Swagatam Das & Suganthan P.N. Differential Evolution: A survey of the Satate-of-the-Art, IEEE Transactions on evolutionary computation, 15(1): 4–31, 2011.

Tirronen V. & Neri F. A memetic differential evolution in filter design for defect detection in paper production. In: Giacobina, M.(ed.) Evo Work-shops 2007. LNCS, vol. 4448, pp:320–329. Springer, Heidelberg (2007).

Tirronen V. & Neri F. Differnetial evolution with fitness diversity self-adaptation. Nature-Inspired Algorithms for Optimisation, SCI 193, pp: 199–234 (2009).

Wang C.Y., Cai Z. & Zhang Q. Differential Evolution with Composite Trial Vector Generation Strategies and Control Parameters, IEEE Transactions on Evolutionary Computation, 15(1): 55–66, 2011.

Control Engineering and Information Systems – Liu (Ed)
© 2015 Taylor & Francis Group, London, ISBN 978-1-138-02685-8

Development and application of static compensator filter for distribution

H.C. Ma
Fushun City Power Supply Corporation, Fushun, Liaoning Province, China

F.Q. Quan
Liaoning Electric Power Development Stock Co., Ltd., Shenyang, Liaoning Province, China

D.L. Zhang
Fushun City Power Supply Corporation, Fushun, Liaoning Province, China

Z.Y. Zhang & T.W. Bai
Liaoning Electric Power Development Stock Co., Ltd., Shenyang, Liaoning Province, China

ABSTRACT: The SCF consists of three single-phase VCS and numbers of MTSC. The VCS is able to dynamically analyze the harmonic current component, fundamental reactive current component as well as the current three-phase unbalance degree of the load current. Through tracking compensation the target current in real time, to compensate for harmonic current, dynamic reactive power compensation and the compensation of each phase unbalance current purpose. The MTSC includes capacitors, a composite switch including thyristors and reactance composition. By dynamic controlling of the composite switch, the MTSC can achieve dynamical and grading reactive power compensation to the system. The SCF combines the technical characteristics of existing equipment like the APF, the STATCOM. It can achieve three-phase separation harmonic filtering; split phase dynamic reactive compensating and three-phase unbalanced improve better. Actual operating results proved the feasibility and practicality of the device.

1 INTRODUCTION

Accompanied by power system continuous development, serious problem like the increase of harmonic sources, the lack of reactive power, serious three-phase unbalanced in the system endanger the safe and stable operation of the system. Harmonics cause the waveform distortion, electrical installations overheating, burning, misoperation and communication interference. The power factor drops will cause the increase of loss and voltage fluctuations in system. Power quality problems have drawn increasing attention of the majority of electricity practitioners.

At present, there are types of device for governance in power quality as follow: The Static Var Compensator (SVC) and its subclasses have been used. Mechanically switched capacitors, reactors and saturable reactor are earlier applied; Thyristor switched capacitors, reactors, etc. As research further, static synchronous compensator (STATCOM) using full-controlled power electronic devices comes out, and it achieved better application of results in the field of reactive power compensation (Yuan et al. 2008).

Currently, the class of harmonic treatment equipment at power quality product mainly bases on the principle of the LC resonant passive filter. In recent years, there have been Active Power Filter (APF) based on power electronic devices. The application practice has proved that the APF has a good prospect for development in the field of harmonic treatment.

The class of power quality equipment for governing three-phase unbalanced is on the basis of TCR and TSC (MSC). It adopts a special algorithm, change wiring of TSC, to achieve three-phase unbalance compensation.

Although the products above can govern a certain problem in the grid, but there is no one can solve most of the problems affecting the power quality in the grid. In this paper, a device named SCF (Static Compensate Filter) will be proposed that focus on governing power quality problems in the grid (Wang et al. 2011) (Sun 2012). It sets a fixed capacitor grading reactive power compensation (Zhang 2009), static synchronous compensation; harmonic compensation (Yang 2006) and three-phase unbalance compensation in one. It can dynamically continuous regulate reactive power output, improve power factor, eliminate the current harmonics, balance three-phase load and mitigate voltage flicker. It can significantly improve the power quality and the stability of the low-voltage power distribution system.

2 THE STRUCTURE AND WORKING PRINCIPLE OF SCF

The wiring of SCF in the three-phase four-wire system shown in Figure 1 (Tang et al. 2011) (Liu et al. 2009). It consists three single-phase Voltage Source Converters (VSC) and two groups (the number depending on the compensation capacity) three-phase Multiple Thyristor Switched Capacitor (MTSC) unit in parallel. SCF can be regarded as a fast controllable reactive power source in its adjustable range. MTSC be used to meet the steady-state reactive power demand. VSC is used for dynamic reactive power compensation, harmonic, three-phase current imbalance in the system. At the same time, it retains enough controllable and dynamic reactive power reserves in the steady-state case. When the reactive power, harmonics, phase imbalance in system needs dramatic changes instantly, SCF change its output quickly by VSC at first. It compensated for reactive power; harmonics, phases unbalance fast, dynamically, continuously to provide dynamic support and steady-state regulation for the access point voltage, thereby improving the power quality of the system. When the disturbance causes the system to enter a new stable operating point, SCF re-enter or resection MTSC unit to bear part of the reactive load removed by VSC. This makes the SCF not only can meet the demand for different reactive power compensation, improve operational flexibility and speed of response, but also to meet the real-time compensation of harmonics and unbalanced three-phase in system. All above make the SCF as a comprehensive management device has a high performance. Under the unified deployment of SCF, the VSC and the MTSC unit not only coordinate with each other but also work relatively independently.

2.1 *The structure and working principle of MTSC*

The internal structure of MTSC is shown in Figure 2. A basic MTSC unit contains a capacitor, a composite switch and a reactor in series. The capacitor linked together by a triangle. The composite switch has played the role of the capacitor connected to the grid or disconnected from the grid. The impedance value of the reactor in series with the capacitor of is small. The value is generally chosen as the natural resonant frequency of the LC branch in order to be more than 4 times of the rated system frequency. This can avoid parallel resonance between the capacitor and AC system impedance in certain frequency, but also can not affect the MTSC control. The working principle of MTSC which is similar to the conventional TSC will not repeat them in this article.

2.2 *The structure and working principle of VSC*

The VSC unit is based on a full-controlled power electronic voltage source converter (shown as Fig. 3). The storage capacitor provides DC voltage. The unit was in parallel with the grid through the reactor in its AC side. The working principle

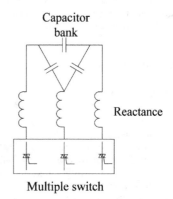

Figure 2.   The structure of MTSC.

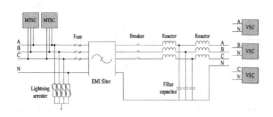

Figure 1.   The main circuit wiring diagram of the SCF.

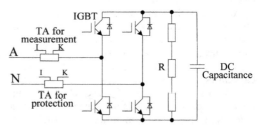

Figure 3.   The structure of single-phase VSC.

of VSC is similar to APF. Its rectifier circuits are both loads of consuming fundamental active current and generating sources of fundamental reactive current and harmonic current. In the output, the fundamental reactive current can be considered as the source of dynamic compensation reactive power in system, but the harmonic current is one of the target components which need to be compensated.

The VSC compensates system current by generating a target current. First, according to the demand of the reactive, harmonics and phase imbalance in system, the required current vector I1(t), I2(t), I3(t) for compensation of the reactive power, harmonic, phase unbalance can be calculated out. The target current vector I(t) has the following relationships:

$$I4(t) = I1(t) + I2(t) + I3(t) \qquad (1)$$

I4(t) is used to support a DC voltage component. Its adjustment is based on the target of the DC voltage which has been set. Through PI regulator, the value of I4(t) is always in dynamic adjustment.

Because of the expression of I(t) contains the components of I1~I4, the target current can compensate reactive power, harmonic and phase unbalance in system at the same time.

3 THE DETECTING METHOD OF VSC TARGET COMPENSATION CURRENT

3.1 *Reactive current and harmonic compensation current detection*

More commonly used in reactive current and harmonic current detecting method includes Fourier expansion method, complex power method and so on. When the three-phase voltage is distortion because of harmonic, the compensation current compensation results calculated by the method of complex power have deviation. That is because the distortion of the supply voltage is directly involved in the operation. When the compensating current calculated by Fourier expansion method, the compensation current is independent of voltage waveform distortion. That is because just sinωt and cosωt used instead of the distortion of the supply voltage during the calculation process. Author selects the Fourier expansion method for harmonic compensation current in this paper (Zhang & Zhang 2012).

Figure 4 is the block diagram of use of Fourier expansion for the compensation current component. In Figure 4, i_L is load current, ω1 is system fundamental angular frequency, I_p is the amplitude of the a single cycle average active current i_p, I_q is

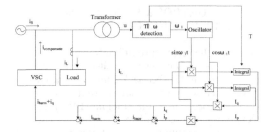

Figure 4. The block diagram of using Fourier expansion for the compensation current component.

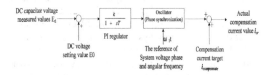

Figure 5. The control block diagram of VCS loss compensation.

the average reactive current iq amplitude for a single cycle.

In the block diagram we can see that after transform, VSC can detect the fundamental reactive current component iq, harmonic current components iharm and fundamental current component ibase.

3.2 *Calculation on the loss compensation of VSC and the unbalanced current in system*

The calculation of the three-phase current imbalance is to provide a three-phase symmetrical voltage for the capacitor bank △ connected in MTSC switching. This can reduce the damage caused by three-phase voltage imbalance to the capacitor. Above we have obtained the fundamental component of the load current by Fourier expansion method. So it is easy to calculate the required compensation current of each phase relative to the equilibrium (Liu et al. 2003). We will not repeat them here.

It is noteworthy that, due to using active power filter technology, VSC do generate circuit losses in the actual operation. So the compensation current will be affected by the changes of the DC voltage. Therefore, in order to ensure the stability of the output of the compensation current, it is necessary to set the control loop to ensure that the DC-side voltage is kept constant.

As shown in Figure 5, the voltage of the DC capacitor compared with set values. Their deviation transforms to the deviation of current though hysteresis control, resulting in a fundamental current signal i_c.

4 THE CONTROL METHOD OF SCF COMPENSATION CURRENT

SCF uses the direct current control mode for current compensation. The so-called direct current control, that is, the feedback control to the current waveform instantaneous value output by VSC. The control process uses the tracking PWM control technology. The control method in this paper is difference to commonly method used in the direct current control such as the hysteresis comparator method (Gu et al. 2006), the triangular wave comparison method. Based on the principle of average current be equal in each sampling period, SCF set DC voltage, AC voltage, the value of connected inductance and the expected value of current compensation in current system as input conditions. Calculated by the software, the PWM value for controlling IGBT obtained directly. By the control of the IGBT working, the expected compensation current value obtained directly.

SCF uses direct current control mode quote the d-q transform and the instantaneous reactive power theory (shown in Fig. 6). The reference value I_Q, I_P (as mentioned before, the reactive and active current amplitude) and the feedback value (i.e., I_Q, I_D) in the steady state are DC signals. Therefore the PI regulator can achieve current tracking control without steady-state deviation. Compared to the tracking PWM technology based on the triangle wave comparison method, the direct current control mode based on the d-q transform does not require the introduction of a sinusoidal signal. This makes the final current control to avoid the steady-state deviation.

5 THE EFFECT OF DEVICE

In order to verify the effect of the SCF device, we selected a 380V transformer belong to Fushun Power Supply Company as an experimental point. The line is long-term in the three-phase imbalance. The loads carried by are a number of the small steel plants using medium frequency induction furnace (non-linear loads and harmonic source). The voltage, current waveform measured before treatment in this line shown in Figure 7 and Figure 8.

Governed by the SCF device, voltage and current measurement waveform show in Figure 9 and Figure 10. As we can see from the figure, the three-phase load imbalance problem in the line has been better solved. Comparing the data by

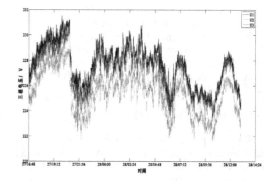

Figure 7. The real-time three-phase voltage waveform before treatment.

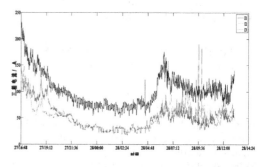

Figure 8. The real-time three-phase current waveform before treatment.

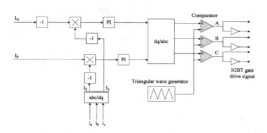

Figure 6. The VCS current control mode.

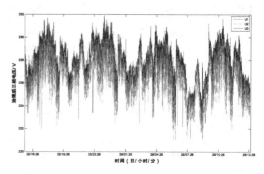

Figure 9. The real-time three-phase voltage waveform after treatment.

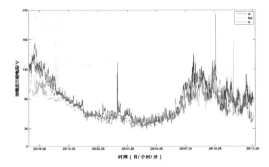

Figure 10. The real-time three-phase current waveform after treatment.

Table 1. The statistic table of each phase voltage and current distortion rate before and after being governed.

	THD$_{max}$		THD$_{min}$		THD$_{agv}$	
	Before	After	Before	After	Before	After
U$_A$	18.46%	12.34%	1.91%	1.00%	4.80%	3.31%
U$_B$	19.18%	13.22%	2.41%	1.86%	5.11%	3.97%
U$_C$	21.61%	15.57%	1.86%	1.38%	5.10%	3.76%
I$_A$	35.07%	20.11%	2.07%	2.05%	11.14%	6.74%
I$_B$	121.6%	52.28%	7.99%	3.73%	31.84%	8.89%
I$_C$	126.5%	49.96%	7.16%	2.99%	34.82%	8.03%

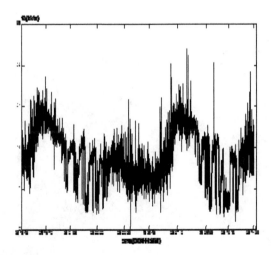

Figure 11. The reactive power of the SCF output.

harmonic analysis between governed and not, we can discover the SCF device plays a good compensation effect to harmonic (see Table 1). Figure 11 is a reactive power of the SCF output. Can be seen from Figure 11, the reactive power of the SCF output have a high dynamic adjustment capability and it is able to meet the reactive power demand of the line better.

6 CONCLUSIONS

1. The VCS unit uses Fourier expansion method can detect the fundamental reactive current component, harmonic current component and the fundamental current component. It can dynamic compensate system harmonic current better to achieve the purpose of filtering.
2. As a result of using single-phase compensation, the VCS can generate compensation current for each phase according to three-phase current imbalance in the system. It solve the three-phase current imbalance to create a better environment for the MTSC unit capacitor switching.
3. The VCS unit can output real-time small-capacity reactive power compensation to the system. The MTSC can provide the basic reactive power compensation to the system. With the two units cooperating, SCF can realize real-time dynamic reactive compensation continuously.
4. Experimental results show that SCF be able to complete harmonic compensation, reactive power compensation and three-phase imbalance adjustment in the low voltage distribution network.

REFERENCES

Gu H.R. & Yang Z.L. 2006. Research on Hysteresis-band Current tracking Control of Grid-connected InVerter, Proceedings of the Chinese Society for Electrial Engineering. China, Vol. 26. No. 9 May 2006, pp. 108–112.

Liu S.M., Guo C.L., Xu Y.H. & Xiao X.N. 2009. Research of Shunt Power Quality Conditioner Start-up Charge Strategy, Power System Protection and Control. China, Vol. 37. No. 22 Mov. 16, 2009, pp. 87–90.

Liu W.H., Liu W.H., Lu J.F. & Liu B. 2003. Balancing Compensation of Three-phase Unbalanced Load with Single Phase STATCOM, Power Electronics. China, Vol. 37. No. 4 May 2003, pp. 10–12.

Sun J. 2012. Application of Power System Compensating Device (SVG) and Its Simulation, Taiyuan University of Technology. China. 2012.

Tang Z., Jiang Y.D., Yao Y.S. & Guo W. 2011. Direct-side Voltage Control of Mid-point Capacitor Tree-phase Four-wire Shunt Active Power Filter, Electrical Measurement & Insturmentation. China, Vol. 48. No. 545 May 2011, pp. 41–44.

Wang Z.H., Gu Y., Shi J.J. & Li G.Q. 2011. Research on Simulation Experiment of 10 kV Static Var Generator, Journal of Northeast Dianli University. China, Vol. 31. No. 5 Oct 2011, pp. 107–111.

Yang F. 2006. Research on the Device of Power System Active Power Filter and Reactive Compensation Based on DSP Control System, Tianjin University. China. 2006.

Yuan Y.C., Wang J.Y. & Liu X.C. 2008. Development and Application of Static Synchronous Compensator for Distribution, Journal of Xihua University: Natural Science. China, Vol. 27. No. 2 Match 2008, pp. 52–54.

Zhang M. 2009. Research on the Hybrid Dynamic Var Compensation Technology for Power Distribution Systems, Dalian University of Technolog. China. 2009.

Zhang X. & Zhang C.W. 2012. PWM Rectifier and Control, Mechanical Industry Press, Mar. 2012.

Control Engineering and Information Systems – Liu (Ed)
© 2015 Taylor & Francis Group, London, ISBN 978-1-138-02685-8

Study on the evaluation model of the interval type electronic commerce system

H. Shen
School of Economy and Management, Hunan Institute of Technology, Hengyang, Hunan Province, China

A.M. Liu
School of Mechanical Engineering, Hunan Institute of Technology, Hengyang, Hunan Province, China

ABSTRACT: In this paper, with the analysis of the electronic commerce rapid development trend, the electronic commerce evaluation index system is discussed and the index weight is determined by using combination weighing method. It is presented that the interval constituted with group expert assessing information can be used as evaluation scale. And finally the evaluation model is given based on the interval type e-commerce system. As a result, theoretical knowledge of the e-commerce evaluation system is richened.

1 INTRODUCTION

E-commerce usually means that under the environment of the opening up of internet, based on browser and server application method, in widely global commercial and trade activities, the buyers and sellers realize a new-type commercial buy and sale pattern, which includes customers and agent's internet purchasing and trading, online electronic paying, and other financial and related synthetic service activities. In recent years, under the background of stably increasing global economy and the rapid use of the internet and broadband technology, the e-commercial market with its high-tech and high human resources and high added values, has kept a high-speed increasing trend. The internet market in our country has boomed and the transaction quota of e-commercial market has increased stably since 2005. After several years' development in e-commerce, it has already gradually matured and stabilized. In 2011, the e-commerce in our country took a fast development and more and more commerce's entered e-commercial business. Many giant sellers, for example GoMe, Suling and the like, have transferred into internet shop sales from the real shop sales, followed by Dongjing, taobao, PaiPai and the like. At the beginning of 2012, GoMe, Suling and Dongjing started a marketing war, and make the dealing quota go up directly and the internet business bills spurt. It showed that there was a big market in internet purchases. As the internet purchasing environment is getting better year by year and the middle and big companies do their business relying

on e-commerce more and more with the wide use of computers and costing down in use of internet. E-commerce has been developing as a necessary part in modern human life. As the development of e-commerce, network users have a higher and higher expectation. They not only acquire e-commerce suppliers to offer a justice and safe services and easy express services, but also need the network platform simple to search for related information. As a result, e-commerce business dealers must evaluate their e-commerce systems timely and understand their weakness.

The study of the e-commerce system evaluation in our country now is mainly focused on the establishment of its evaluation system, namely, on the theory of how and where to do the evaluation. However, other countries emphasize their studies on the analyses of cases of e-commerce system from e-commerce. In view of the related research works in and out of our country in this field, whether or not in theoretical or applicable studies they are all based on real number information with little consideration of uncertainties. In fact there are many uncertain factors in the process of evaluation of e-commerce system, especially in lack of complete information and in complication for the acquisition of evaluation information[5][6]. It is unable to acquire accurate evaluation information. Therefore it is more reasonable and scientific to take uncertainty evaluation scale in the process of evaluation of e-commerce system. With the consideration of limited work in uncertainty evaluation methods, the establishment of the e-commerce evaluation system based on uncertain environment is therefore given in this paper to make contribution to e-commerce evaluation theoretical system.

2 E-COMMERCE EVALUATION INDEX SYSTEM AND EVALUATION SCALE

E-Commerce Evaluation Index System mainly consists of evaluation index system, index weight determination, and the choice of evaluation methods. The choice of evaluation index is mainly based on the composition of e-commerce system to decide which aspects should be considered for the evaluation. Some achievements have been made in the research of e-commerce evaluation index. Yao yuan [1] introduced e-commerce evaluation system in the procedure based on the hierarchy analysis method and established matching evaluation index system and evaluation method. Fu Li Fang, Feng Yu Qan, and Wu Qiu Fen [2] have analyzed e-commerce evaluation index system and presented DEC-based e-commerce system evaluation method and quality diagnosing method. This method avoided subjectively determining index weight and finally verified it effective in case studies. Chen Wen Li [3] analyzed those influencing factors in e-commerce system and proposed that the e-commerce operating differentiations could be analyzed by using both DEA and ratio analysis methods. In view of the research achievements obtained [4] [8], e-commerce evaluation index system is generally influenced by the following factors:

(C_1) Client Satisfaction: User's satisfactions on e-commerce network and services.

(C_2) Paying Complete Extents: Convenience and safety for customer's purchasing in paying ways as internet bank and or other paying means.

(C_3) Materials Circulation Allocation Complete Extents: Needs of meeting clients in e-commerce materials circulating construction and management.

(C_4) After-Sale Services: Product purchasing in e-commerce is first a virtual product, i.e. first paying and then seeing the real product. Then after-sale service is a necessary part of e-commerce system.

(C_5) E-commerce System's Harmonization: E-Commerce System is a responsible one that deals on internet and allocates in reality. Therefore its operating harmony is a relatively important factor in an e-commerce system.

(C_6) Financial Situation in E-Commerce System: Guarantee of the policy performance and capital continuity of e-commerce system.

From the analysis of evaluation index, it can be seen that these evaluation indexes are all not quantitatively analyzed data. It is impossible to obtain accurate data to describe them although it is necessary to use quantitative analysis for evaluation information. In fact it should be done by selecting an interval number to scale it, namely to select percentage in process of evaluating index. Many experts should be employed and their opinions are used for evaluation. These experts' evaluation data are scattered in order to constitute final evaluation information. Normally not all experts have the same evaluation results. Therefore an interval is formed by scattering and this interval number can use as evaluation value of evaluation index. Such an evaluation value includes all evaluation information of experts and is scientific and reasonable.

Summarizing the above mentioned, the main factors that influence e-commerce system are (C_1) Client Satisfaction, (C_2) Paying Complete Extents, (C_3) Materials Circulation Allocation Complete Extents, (C_4) After-Sale Services, (C_5) E-commerce System's Harmonization, and (C_6): Financial Situations in E-Commerce System. In the selection of evaluation value, the interval number is adopted as final evaluation scale after concentrating group expert's evaluation information.

For the evaluation of n e-commerce system, the interval number evaluation scale is used for evaluation and the following matrix is therefore obtained:

$$A = \begin{bmatrix} & C_1 & C_2 & \cdots & C_6 \\ X_1 & [a_{11}^-, a_{11}^+] & [a_{12}^-, a_{12}^+] & \cdots & [a_{16}^-, a_{16}^+] \\ X_2 & [a_{21}^-, a_{21}^+] & [a_{22}^-, a_{22}^+] & \cdots & [a_{26}^-, a_{26}^+] \\ \vdots & \vdots & \vdots & \vdots & \vdots \\ X_n & [a_{n1}^-, a_{n1}^+] & [a_{n2}^-, a_{n2}^+] & \cdots & [a_{n6}^-, a_{n6}^+] \end{bmatrix}$$

Considering the interval number is not suitable for operation and comparison, generally the following function is then used for deleting the uncertainty of the interval number:

$$b_{ij} = f\left([a_{ij}^-, a_{ij}^+]\right) = \int_0^1 \frac{d\rho(x)}{dx}\left(a_{ij}^+ - x(a_{ij}^+ - a_{ij}^-)\right)dx$$

Normally choosing $\rho(y) = y^r (r \geq 0)$, then we have

$$f\left([a_{ij}^-, a_{ij}^+]\right) = \frac{a_{ij}^+ + ra_{ij}^-}{r+1}$$

After deleting the uncertainty the judgment matrix is then transferred into the following matrix

$$B = \begin{bmatrix} & C_1 & C_2 & \cdots & C_6 \\ X_1 & b_{11} & b_{12} & \cdots & b_{16} \\ X_2 & b_{21} & b_{22} & \cdots & b_{26} \\ \vdots & \vdots & \vdots & \vdots & \vdots \\ X_n & b_{n1} & b_{n2} & \cdots & b_{n6} \end{bmatrix}$$

3 THE DETERMINATION OF EVALUATION INDEX WEIGHT

After the determination of the e-commerce system's evaluation index, the second thing to do is to determine index weight. Some achievements have been made in this research field [7]. Generally speaking it can be divided into subjective weighing method, objective weighing method, and synthetic weighing method. In general, synthetic weighing method can effectively overcome the weakness of subjective and objective weighing methods, and make full use of their merits to synthesize the information of them. Subjective weighing method is usually that the experts, based on their own experiences and knowledge, give the importance of each index in e-commerce system evaluation. But this way has certain subjection. Objective weighing method is however to obtain the importance of each index by refining the evaluation information... Therefore it is should be done by synthesize considering subjective and objective weighing methods for combination weighing. The most used in the objective weighing method is the entropy weighing method that is based on the information entropy.

The calculation formula of information entropy is $e_i = -1/\ln n \sum_{j=1}^{n} p_{ij} \ln p_{ij}$, Among it, $p_{ij} = b_{ij} / \sum_{k=1}^{n} b_{ik}$, which presents the normalized value for the jth evaluation index in the ith evaluated e-commerce system.

After calculating out the information entropy we can thus calculate out the entropy weight by using the following formula:

$$\lambda_i = \frac{1 - e_i}{\sum_{k=1}^{6} (1 - e_k)}$$

The differential degree of evaluation information in evaluation index of each e-commerce evaluation system can be determined by the use of information entropy. It makes full use of what the experts give in the e-commerce system evaluation information.

After determining the objective weight λ and making the subjective and objective weight, the final synthesized weight of each evaluation index can therefore be obtained as follows

$w = \alpha\omega + (1 - \alpha)\lambda$, in which $\alpha \in (0,1)$.

4 INTERVAL TYPE E-COMMERCE SYSTEM EVALUATION MODEL

After determining e-commerce evaluation index system and index weight, what the final thing remaining is how to concentrate the expert's evaluation information. There are certain research achievements in the determination of the evaluation information. Generally it can be divided into the linear and non-linear synthesizing methods. Considering that in e-commerce system evaluation not only the position and the present situation as well of each evaluated e-commerce system in the total evaluation process need to be known, therefore the use of linear weighing method is the most suitable for it. As a result in the process of evaluating e-commerce system the simple weighing should be adopted as modeled as:

$$Effect = BW = \begin{bmatrix} b_{11} & b_{12} & \cdots & b_{16} \\ b_{21} & b_{22} & \cdots & b_{26} \\ \vdots & \vdots & \vdots & \vdots \\ b_{n1} & b_{n2} & \cdots & b_{n6} \end{bmatrix} \times \begin{bmatrix} w_1 \\ w_2 \\ \vdots \\ w_6 \end{bmatrix}$$

$$= \begin{bmatrix} Ef_1 & Ef_2 & \cdots & Ef_n \end{bmatrix}$$

In which,

Effect—Company e-commerce system synthetically evaluation value.

b_{ij} ($j = 1, ..., 6$, $i = 1, 2, ...,$ —The evaluation value in the jth evaluation index of the ith evaluated e-commerce system.

w_i ($i = 1, 2, ..., 7$)—The combination weight values of the index.

After the synthetic evaluation value of e-commerce system is calculated by using the above linear weighing model, the values of Ef_1 Ef_2 $\cdots$ Ef_n can be compared to realize the ordering of e-commerce system and more importantly to make grading. In general, the grading of the e-commerce system can be made according to its synthetic evaluation value and in a way of Table 1.

Table 1. E-Commerce system grades.

Points	A Class	B Class	C Class	D Class
Grades	Over 90	80–90	60–80	Under 60

The grading of e-commerce system can be realized by the above grading table. The feedback should be done timely after the grading of e-commerce system to encourage the good ones and qualify the bad ones.

5 CONCLUSIONS

With marketing economic system's reform and completion further, whole social e-commerce consciousness's strengthening, and gradual stipulation and completion of state's policy and regulations in e-commerce, the developing environment and motivation of e-commerce will be getting better and better. Our country should enhance the development and application of the e-commerce industry, and make the combination between e-commerce and conventional companies, promote with each other and develop harmonically, and let e-commerce become a new increasing engine of national economy. Managers make evaluation for the e-commerce system of local regional companies. This can effectively enhance the development of e-commerce and promote the regional economical development with more profits and economic increasing of companies.

ACKNOWLEDGMENT

Special thanks are given for the financially supported project: Study on Bachelor Application-Oriented Student Education Majored in Economy and Management in Local Colleges (Xian Jiao Tong [2012] 401).

REFERENCES

Chen wen Lin, E-commerce evaluation and e-commerce achievement industry comparison [J], Shen Zhen University Journal (Human Social Science Version), 2009(2).

Chian-Son Yu, Chien-Kuo Li, A group decision making fuzzy AHP model and its application to a plant location selection problem[C], Joint 9th IFSA World Congress and 20th NAFIPS International Conference, Vancouver, Canada, 2001: 76–80.

Chu Xiao Yan, Zhan Xin Zhen, Calculation of weight and concentration of interval number judgment matrix in group decision [J], System Engineering, 2005,9 (23).

Fu Li Fang, Feng Yu Qan and Wu Qiu Fen, DEC-based e-commerce system evaluation method and quality diagnosing method [J], Harbin Normal University Natural Science Journal, 2007(3).

Li Bin Ju, Liu Shi Fen, The concentration method based on interval number judgment matrix group information [J], Jan Nan University Journal (Natural Science Version), 2004,3(5):23–26.

Wu Jian, A concentration method of the favorable information of group interval number complementary judgment matrix Application of Theoretical Methods in System Engineering, 2004, 13(6).

Yao Yua, Study on e-commerce evaluation system based on AHP [J]. Scientific progress and strategies, 2009, 26(10):129–133.

Yin Chun Wu, Study on fuzzy multiple criteria group decision method [D], Xi An University of Science and Technology, 2007.

Control Engineering and Information Systems – Liu (Ed)
© 2015 Taylor & Francis Group, London, ISBN 978-1-138-02685-8

Intelligent and function-oriented testing system based on characteristic keyword

W.Q. Hu, P. Wang & Y.P. Wu
College of Photonic and Electronic Engineering, Fujian Normal University, Fuzhou, China

ABSTRACT: As the traditional testing system was time-consuming, manual, expensive and difficult, this paper designs an intelligent and function-oriented testing system based on characteristic keyword for embedded system as a promising approach to generate test-data intelligently, which design can precisely locate the fault and shorten the detection time of malfunction. According to the functional relationship between the inputting and outputting, there are two kinds of generation schemes of the testing case, one is manually inputting and the other is intelligently generating, and in order to improve the coverage of testing case, the design makes them permutated and comminute randomly. Because the design uses the stable procedure to process the unstable data streams, it can avoid writing special parse and conversion program for different communication protocol, and make the program more flexible and more adaptable. What's more, the design uses the multithreading test-data management technology to realize a PC controlling many lower machines. This design has been used for the ongoing projects in the laboratory, and the effectiveness of the running proved the feasibility of the design.

1 INTRODUCTION

Along with the arrival of the digital age, a large number of complex, increasingly powerful embedded system are constantly entering the market, while its applications are becoming increasingly complex, which put forward higher requirements for developing technology and testing capabilities of testing system of embedded system [1],[2]. As embedded system is becoming more intelligent and integrated, the software part of the system plays an increasingly important role, and often can achieve the function of the hardware [3]. The embedded system is an extremely dedicated computer system, which is provided with high reliability [4], and the reliability of software is one of the bottlenecks restricting its reliability, which is also the key factor of system-dependent [5][6].

Since the advent of the computer, experts and scholars have never stopped the pursuit of program correctness and reliability, and testing system is a proven method that can effectively helps people identify and ensure the software quality of embedded system. As early as 1979, Myers proposed that testing is a process of running a program or system in order to find the error, and broadly speaking the purpose of the test is that software errors [7]. At the same time, software testing is a very difficult and complicated work, and regardless of calculating from either time-consuming or resource-consuming it has reached more than 50% of the entire software

project [8][9]. Therefore, the testing system has been active in the area of research in software engineering, which is widely used to improve the efficiency and reliability of the software execution.

However, in the embedded system applications, an upper machine often need to test many lower machines simultaneously, whose functions are not same, fox example, some of them are responsible for data acquisition, some other are used to convert protocol. What's more, Different application of the various protocols has different functions. Therefore, in this case, it has some practical application significance to achieve an upper machine testing multi-audience-bit machine online, and to improve the accuracy of fault location and the coverage rate of test case.

2 SYSTEM FRAMEWORK

Testing is a process, whose central task is to find the flaws in the system. The design regard the functional requirements as a starting point, and regard the implemented as the end, which can precisely locate the fault and performance statements, and its framework of the system is shown in Figure 1. Each module of upper machine designed based on the test requirements of the lower machines and testing capabilities, whose purpose is to effectively solve the contradiction between demand and supply, and enhance the versatility of the system.

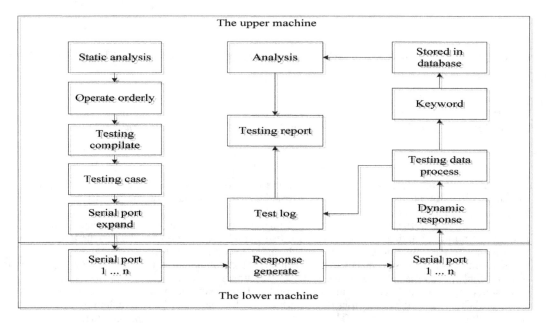

Figure 1. The system framework.

The Communication between the upper and lower machine by the serial port and serial communication has many advantages, such as reliable data transmission, simple line, flexible, easy maintenance, has been widely applied in the field of industrial control, data acquisition, real-time monitoring [10]. Due to the increasing scale of embedded systems, more and more integrated module, a host computer often requires simultaneously with multiple lower position machine communication message. When a computer as upper machine, it generally only comes with two serial ports, and that can not meet the demand for testing multiple lower machines. Therefore, the design uses MSComm control serial communication serial port expansion combined with C#.

3 OVERVIEW OF DESIGN

The design considers the embedded system as a whole, which do not regard its internal structure and characteristic in the operation of generating at all, sending, and analytical test packets, and consider whether it can receives and responses the message correctly only from the view of the actual functional requirements. At the same time, the design is closely related with the functional requirements, and the quality of the requirements document directly affects the results of the test. During the functional test of embedded system, the desired functionality of the system of

the system is an important basis, and according to the actual needs of the behavior, performance, and acceptance criteria we can determine whether the embedded software meet the requirements and specification. In addition, in order to guarantee high-quality test, we need to check the tools of burning process as well as the interface between hardware and software.

3.1 Test-data generation

The process of software testing is a process of running program and the test case is the data of testing. In order to improve the coverage of test cases, we provide two schemes about how to produce test cases: 1) manually inputting test cases; 2) intelligently generating test cases. The purpose of manual inputting is checking the function of a special module, which is according to the relationship between input and output in the embedded system; While the automatic testing not only can test all modules in the system, but also test any one function module in the system, and it generate test cases according to the combination between the functional relationship of the input and output in the system with a random function, which can continuously test without human intervention.

3.2 Multi-serial communication

Serial communication is classic, which has many advantages, such as high reliability, operability, simple lines, and it has been playing a very important role in communication. Nowadays, on the background that the requirement of decentralized control and centralized management becomes increasingly high, a computer often has to monitor and manage many lower machines. However, the serial number of the PC is limited, so extending the number of serial has important practical significance. This design uses the serial port expander to become one serial to eight, you can achieve a serial port of the computer serial features expanded into eight, which is able to meet the demand for communication between the computer and more serial peripherals.

3.3 *Multi-threading management*

An application is equivalent to a process, and in order to improve the efficiency of the program, the process often needs to be divided into different priority thread. The management of thread includes the creation, activation, hang of thread. The design open reading thread and writing thread after finishing the configuration of the serial port, and the reading thread is mainly responsible for reading data from the eight peripheral serial, and the writing thread is responsible for writing data to the peripherals, while the main thread is responsible for processing all data receiving from the peripherals. If the operation that peripheral devices write data to the serial port is not so frequent, the writing thread can only open when the design need sending data to the serial, and the writing operation is suspended after the completion. The purpose of doing so is to use system resources rationally and improve the efficiency of the software execution.

3.4 *Intelligent keyword extraction*

Regardless of the data flow transmission between the different application layer protocols, separating the description of data frame format of the communication protocol with the conversion code of the data stream can make the resolution conversion process has better flexibility and universality, which do not care whether the data frame format is explicitly defined. The paper put forward to sending and receiving mechanism based on characteristics keyword, which use the characteristics keyword to describe the protocol data frame format, and that has nothing to do with the specific data frame format.

4 CASE STUDY

The testing system has been used in the laboratory's ongoing project, and it is realized on the platform of Visual studio C# without manipulating to manage heap directly, which realized a host computer testing on the number of lower machines through RS232 serial port. The packet structure of the testing system is shown in Figure 2, and it consists of Start flag, Packet header and Packet body, in which Start flag and Packet header are required, and Packet body is optional. Start flag is fixed, and ther are AA and BB. The length in Packet header is the legth of the whole packet. CRC is XOR checksum. Packet body contains content and abstract, whose length is not fixed.

The case convert data stream between the different protocols based on characteristics keyword, in which the data-flame format is described by characteristic keywords, and the data stream parse and conversion program are independent of the communication protocol by separating the description of communication protocol data flame format from the parse and conversion program. If the agreement is changed, on the one hand, we need to add a new protocol in the protocol description file description or moderately change the description of the existing agreement, on the other hand, we need to

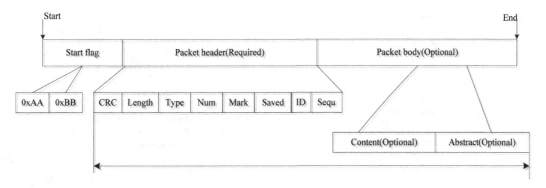

Figure 2. Packet structure of the test system.

add a new configuration in the configuration file or change the existing configuration. If the data type has been added, we can simply add a new conversion module of data type in the data-type conversion process.

4.1 *The flow of case*

The testing system can simultaneously send and receive multiple serial data, and the serial number can intelligently retrieve and save, and the baud rate can intelligently match and preserve. The sending process is shown in Figure 3. The packets of function test generate in two ways, manually sending and intelligently sending. Checking intelligently the control of the sending checkbox is activated, and the system will generate characteristics keyword if it has been activated. Test system package packets according to the time interval. CRC check code can intelligently generate and test packets are randomly generating by the Random function, what is more the sending option can also be selected at random and multiple messages can be sent simultaneously, which can add different types of test cases and increase coverage rate. System will check whether the serial port is empty, and will send the message directly if it is empty, otherwise packets will be stored in the database, which includes sending database, receiving database and configuration database of serial port, and the system will orderly send the packets lately.

The protocol of the case includes same items of packet header. The receiving process of packet header of the testing system is shown in Figure 4. When the system analyzes intelligently the packet header, it would find start flag firstly, that is "AABB". If the start flag is right, the system judge

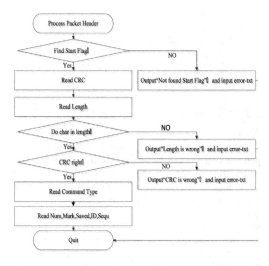

Figure 4. Receiving process of packet header.

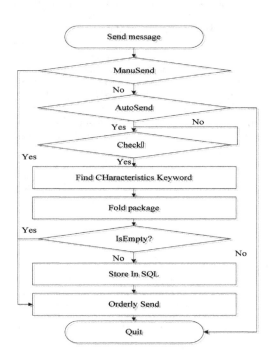

Figure 3. Sending process.

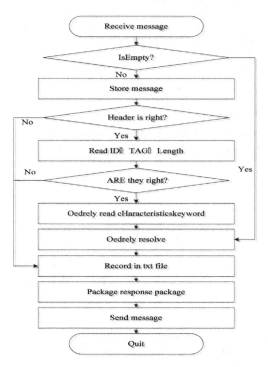

Figure 5. Receiving process of packet body.

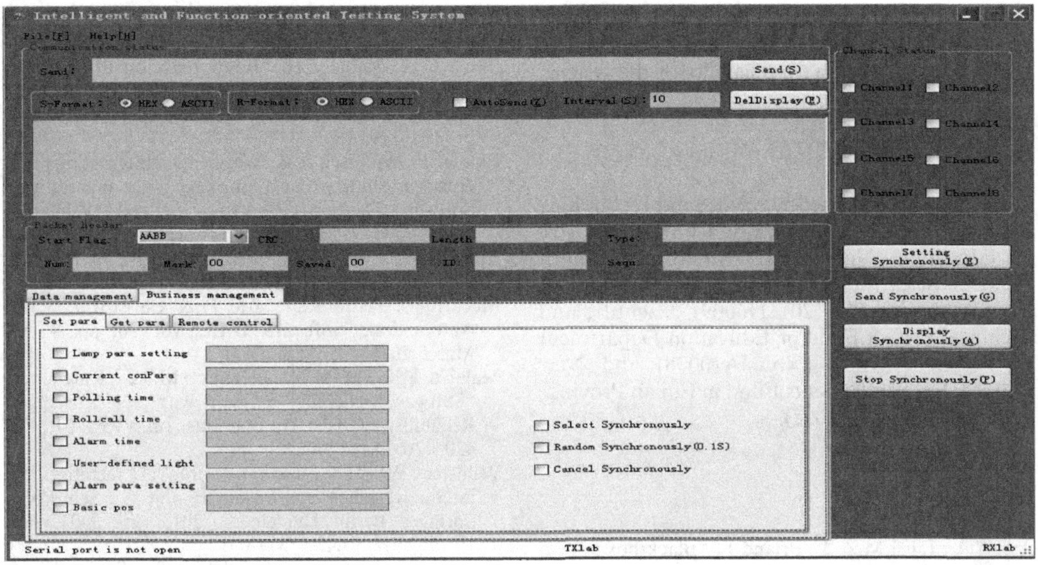

Figure 6. Execution interface.

that they are communicating, and it uses XOR checksum to confirm whether the CRC is right. If CRC is right, the system checks the length, and judges all chars. If the above description is correct, the system orderly check whether command type, serial number, mark, reserved words, product number, terminal serial number are right. After judging packet header is right, the system will analyze the body of the packet.

The receiving flow of the testing system is shown in Figure 5 and the message system received stored in the database in chronological order. The design sets two receiving database table, whose purpose is to prevent conflicting as multiple serial ports simultaneously receive data. Then the design judges whether it is the response message, and finds a corresponding send flag, if there is no flag, the packet will be stored in a txt file to record the wrong, and intelligently interlace to specify the cause of the error. On the contrary, the system starts parsing packets, and determines the length, the CRC check code and packet flag is correct. If it is correct, the system extracts the contents of the packet according to the demand for the corresponding function. Otherwise, the packet will be stored in the txt file.

4.2 Interface of case

The execution interface of the testing system is shown in Figure 6. The toolbar of File includes serial operation, the exit option, and the serial operations include serial port configuration and close the serial port operation. The receiving message and sending message have two message format, one is Hex and the other is ASCII. The case has a fixed packet header, and in order to facilitate the operation we design a specialized module to display it. All channels can be operate synchronously, and they also can be operate separately.

5 CONCLUSION AND FUTURE WORK

In this paper we have presented the intelligent and function-oriented testing system for embedded system. The testing system has been successfully used for the ongoing projects in the laboratory. Regarding to the quality of the final out-come, two key contributions are presented in this paper. Firstly, considering the embedded system as a whole, we just need to focus on functional requirements, rather than regard its internal structure and characteristic, which can precisely locate the fault and performance statements. Secondly, the generation of random test cases has two methods, one is manually inputting test cases and the other is intelligently generating test cases, which can greatly improve the covering of the test cases. In addition, using the stable procedure to process the unstable data streams, the design can avoid writing special parse and conversion program for different communication protocol, and make the program more flexible and more adaptable.

Immediate future work is to expand the application filed of the testing system and shorten the time of test accomplishment and discovering defect.

We also like to add watchdog mechanism to monitor the status of thread in accordance with the thread ID, which can improve the reliability of the system.

ACKNOWLEDGMENTS

This work was supported by Science and Technology Research Key Project Fund of Chinese Ministry of Education (No. 212087), Important Project Fund of Scientific and Technical Department of Fujian Province (No. 2011H6009), Scientific and Technical Project Fund of Education Department of Fujian Province (No. JA10078) and New Century Excellent Talents Fund in Fujian Province University (No. JA11037).

REFERENCES

Arcuri A., Iqbal M.Z. & Briand L. "Black-box system testing of real-time embedded systems using random and search-based testing," Testing Software and Systems, Springer Berlin Heidelberg, March. Nov. 2010, pp. 95–110, doi: 10.1007/978-3-642-16573-3_8.

Blanco R., Tuya J. & Adenso-Díaz B. "Automated test data generation using a scatter search approach," Information and Software Technology, vol. 51, pp. 708–720, April. 2009.

Ebert C. & Jones C. "Embedded software: Facts, figures, and future,". Computer, vol. 42, pp. 42–52, April. 2009.

Liang Q. & Rubin S.H. "Randomisation in designing software tests for systems of systems," International Journal of Information and Decision Sciences, vol. 4, pp. 108–129, May. 2012.

Lindlar F., Windisch A. & Wegener J. "Integrating model-based testing with evolutionary functional testing," Software Testing, Verification, and Validation Workshops (ICSTW), 2010 Third International Conference on. IEEE, Paris, April. 2010, pp. 163–172, doi: 10.1109/ICSTW.2010.10.

Macario P., Pedro R., Mario P. & Christof E. "Test Automation," Software, IEEE, vol. 30, pp. 84–89, March.2013.

Malaiya Y.K., Li M.N., Bieman J.M. & Karcich R. "Software reliability growth with test coverage," Reliability, IEEE Transactions on, Vol. 51, pp. 420–426, Dept. 2002.

Wahlster W. "The semantic product memory: an interactive black box for smart objects," SemProM. Springer Berlin Heidelberg, 2013: pp. 3–21, doi: 10.1007/978-3-642-37377-0_1.

Yunping Wu, Shengzhen Cai, Weida Su, Jinying Wu & Wangbiao Li. "A Design and Application of Multi-RS232 of System Based on MCS-51," Journal of Fujian Normal University (Natural Science Edition), vol. 22, pp. 29–33, April. 2006.

Zhian Sun, Xiaoli Pei & Xin Song, "Software Reliability Engineering," Beijing University of Aeronautics and Astronautics Press, March. 2009, pp. 285.

Exploratory search oriented concepts latent relations discovering

L. Gao, Y. Zhang & B. Zhang
College of Information Science and Engineering, Northeastern University, Shenyang, Liaoning, China

K.N. Gao
Computer Center, Northeastern University, Shenyang, Liaoning, China

ABSTRACT: Exploratory search, in which a user solves complex information problems, is cumbersome with today's search engines. We propose collecting hidden concepts and latent relations (i.e. effective information) from Knowledge-based Question Answering System and system log to study ontology content extension methods, so as to better support exploratory search. We present a model to discover whether two concepts have relevance or not. Methods based on the statistic analysis are introduced to give explanations about the latent relations between concepts. We conduct experiments to evaluate the contribution of the effective information through simulating the exploratory search process. Direct at two search tasks, two different evaluation approaches are introduced.

1 INTRODUCTION

Exploratory search is a specialization of information exploration which represents the activities carried out by searchers who are either: a) unfamiliar with the domain of their goal; b) unsure about the ways to achieve their goals; c) or even unsure about their goals in the first place [1]. As the above definition illustrates, people engaged in exploratory search require support that goes beyond the known-item retrieval well-handled by many modern search systems. They may need to help discovering new associations and kinds of knowledge, resolving complex information problems, or developing an understanding of the terminology and the information space structure [2]. For example, what if we want to find something from a domain where we have a general interest but no special knowledge? The answer may be that we usually submit a tentative query and take things from there, exploring the retrieved information, selectively seeking and passively obtaining cues about where the next steps lie [10], even the retrieved results deviate from what we have expected in the first place.

Therefore, in order to support the user's exploratory search process, we need to find the concept corresponding to the query, discover other concepts related to it, so as to conduct query expansion, help users to find the expected results as soon as possible. Ontology is a formal explicit representation of concepts in a domain [4]. Features of domain ontology determine that it can effectively support the process of exploratory search.

But stem from respective problems and concrete engineering consideration; concepts in ontology may not agree on the users' exploring needs, and also ontology refresh cycle is longer relatively. People, engaged in the process of exploratory search, may input various key words and certain relations that aren't reflected in the ontology may emerge. Hence, concepts and the relations in the ontology appear a little weak in the help of exploratory search process.

For purpose of supporting users' exploratory search better, helping them to filter the search results, recommending the information users may be interested in, in this paper, we base on an ontology that has already existed (called base ontology), study ontology content extension methods, which are suitable for the exploratory search system through updating the knowledge reflected in the base ontology. Specifically, this paper extracts the effective information from the Knowledge-based Question Answering System (KQAS) and system log, i.e. discovers entity concepts and relations between concepts as well as the types, which are not represented in base ontology, discovers directed relations that formed through cascading indirect relevance; and then generates human-readable explanations (such as natural language description) for the concepts and latent relations discovered.

To achieve the goals, in this paper, we base on two mild assumptions (more details in section 4.1), utilize simple but effective Co-Occurrence analysis method to extract effective information. But simultaneously considering the problems of data sparseness etc, we present an algorithm referred as Conceptual Characteristic Vocabulary (CCV) to discover the latent concepts and relations, and also it can make corrections and supplement the relations computed through Co-Occurrence analysis. Based on the idea of the two algorithms, we present a model to discover the relations between concepts. And then in order to add the explanatory description about the relations obtained and help users understand better in an intuitional way, we put forward methods based on frequency statistics and knowledge of base ontology. Finally, we apply our methods to extract effective information from Baidu Zhidao and system log to conduct evaluation experiment.

The rest of this paper is organized as follows. Section 2 reviews the related work. Section 3 describes the overall framework about the role of effective information. Section 4 presents the algorithms, and Section 5 gives experimental evaluation. Finally, Section 6 concludes with future directions discussed.

2 RELATED WORK

In this section, we cover research related to this paper at a high level, with additional related work specific two aspects of the paper discussed in context.

Our research falls into the area of exploratory search. White et al. write that exploratory search systems aid users with information seeking problems that are open-ended, persistent and multi-faceted [1, 6]. This stands in contrast to traditional Web search, which is primarily concerned with navigational queries and closed information request [6]. Despite the prevalence of exploratory queries, exploratory search is a relatively new research area with many open questions [1]. In the field of discovering the relations between concepts, simply speaking according to the methods, prior work pursed three main directions: comparing text fragments as bags of words in vector space [11], using lexical resources, and using Latent Semantic Analysis [12]. Evgeniy Gabrilovich et al. [13] proposed Explicit Semantic Analysis (ESA), a novel method that represents the meaning of texts in a high-dimensional space of concepts derived from Wikipedia, and then compared the corresponding the vectors using conventional metrics. There are many semantic relatedness measures in the literature, even limiting our attention to Wikipedia-specific semantic relatedness measures, which have been shown to be better [14] or as good as Word-Net-based measures [6]. Three such measures—WikiRelate [16], MilneWitten [17], and Explicit Semantic Analysis [13] are among the best known semantic relatedness measures and each uses a

different Wikipedia lexical semantic resource [15], thereby capturing different types of relationships between concepts [6]. According to the features of our information source and the process of the exploratory search, we utilize lexical Co-Occurrence analysis and present Conceptual Characteristic Vocabulary algorithm.

Another set of studies related to our work is the description of the relations between concepts. In order to help users to understand better, after obtaining the relations, it is necessary to give the corresponding description. WikiRelate [16] used the Wikipedia Category Graph (WCG) structure as its lexical resource to display WikiRelate explanations; MilneWitten [17] gave the explanations using the links, and descriptions derived from ESA. While [6] generated interactive visualizations of query concepts using thematic cartography and gave the descriptions with geographic reference systems. All these studies focused on descriptions of the relations using the human-readable fashions such as natural language descriptions and visualization. Direct at the process of exploratory search, we adopt the entirely different information source and give the descriptions based on the statistical analysis and the base ontology we have mentioned earlier.

3 OVERALL FRAMEWORK

Query expansion places a critical role in the modern information retrieval system. This paper extracts effective information from information sources, updates them to the base ontology, and studies ontology content extension methods, so as to conduct query expansion. As shown in Figure 1.

In Figure 1, we can see that on the condition of existing base ontology, a critical point is the extraction of the effective information, which is the main work of this paper. Note that the key to the extraction is to discover the hidden concepts and latent relations between concepts as well as types, and meanwhile we give the description of the relations obtained, considering the reason that direct at the given concept, system should explain to users the

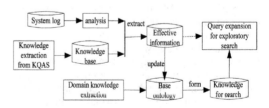

Figure 1. Structure of query expansion in exploratory search system. Note that the effective information plays a certain role.

reason to expand new concepts and give users an intuitive understanding. While the explanations give the reference to users and the process reflects the exploration.

Specific to the above two main aspects, we utilize the Co-Occurrence analysis method and algorithm CCV to extract effective information, and then put forward methods based on the frequency statistics and base ontology to give explanations (more details in Section 4).

3.1 Model of relations between concepts

Based on the idea of algorithms above, in this subsection, we present a novel model based on the probabilistic analysis to discover the concepts that have the relevance, i.e. to judge whether two concepts have relations or not. The concrete form of this model is as follows:

$$p(c_k|c) \rightarrow \sum_j \sum_i^{l_j} p(sen_{ij}|sor_j, c)p(c_k|sen_{ij}) \qquad (1)$$

Here, c denotes an already given concept, c_k denotes a concept collected from the information sources, sen_{ij} denotes the sentence containing the concept c in the corpus sor_j of the information sources, and l_j denotes the sum of sentence sen_{ij} in the corpus sor_j. The value of formula in the right is the approximate value of $p(c_k|c)$, which reflects the strength of the relations.

4 ALGORITHMS

In this section, we introduce the computational algorithm referred as lexical Co-Occurrence analysis and present Conceptual Characteristic Vocabulary (CCV) to discover the hidden concepts and the latent relations between concepts, and then put forward methods to describe the above relations obtained with human-readable natural language description. In the following, we first introduce the basic assumptions on which the algorithms are based.

4.1 Basic assumptions

Like any useful models, we need to make appropriate assumption in order to perform meaningful estimation and discover useful patterns. Specifically for our problem of discovering the hidden relations between concepts, it suffices to make the following two mild assumptions:

1. In order to make problem simplicity and also illustrate substance of the problem, we concentrate to the information about only one target domain.

2. If people, engaged in the process of exploratory search, are faces with confusion and difficulties, they tend to ask for help or search useful information in KQAS.

We argue that the above assumptions are reasonable for discovering the hidden relations between concepts. Without loss of generality, condition of the first assumption can easily expand to other informative domain, even multi-domain. The second assumption can guarantee that our sources of information are rich and can be enough to contain the relations that useful to the users.

4.2 Co-occurrence analysis method

In this sub-section, we utilize simple but effective Co-Occurrence analysis method to discover the hidden concepts and the latent relevance between concepts, and then give the concrete processing steps of this method. The reason is that due to the features of KQAS and system log [3], it is more likely that the concepts that have stronger relevance co-occur in the same window.

On the basis of above, the processing steps are as follows:

1. After attaining the concepts, it computes the co-occurrence frequency of every two concepts, and then makes use of Jaccard coefficient to compute the relevancy to get the correlation matrix, which used to analyze the degree of relevance between two concepts. Formula computing Jaccard coefficient is as follows:

$$Corr_{Jaccard} = \frac{c_{ij}}{c_i + c_j - c_{ij}} \qquad (2)$$

Here, c_{ij} denotes the co-occurrence frequency of concept i and concept j, c_i and c_j separately denotes the frequency of concept i and concept j in the information sources.

2. For the expression of Chinese language, concrete meaning of the vocabulary can be explained with context that it has. Thereby, we base on the correlation matrix already computed in step 1 to get every vector representing concept meaning. This so-called vector is composed of every row of matrix, and the dimensionality is the sum of concepts. Here, cosine similarity is used to compute the degree of relevance. Finally we select the concept pair whose relevance strength is larger than threshold.

4.3 Conceptual characteristic vocabulary

Similar to the idea of the second step in the above method, in this sub-section, we present an algorithm CCV to discover the hidden concepts and the latent relevance between concepts and the algorithm can also make corrections and supplement the relations computed through Co-Occurrence analysis method.

This algorithm represents concepts as characteristic vocabulary vectors through mutual information method, and then computes the similarity of the vectors to determine the relevance.

1. The conceptual model module

Direct at every concept, this module represents them as conceptual model reflected by characteristic vocabularies on the basis of context they have in the information sources. Conceptual model can be regarded as a vector, and the component of the vector is the weight value of a characteristic vocabulary, describes the degree of representing the assigned concept. The module utilizes the mutual information between concept and characteristic vocabulary as the component of the conceptual vector. The bigger the mutual information is, the more consistent the concept and characteristic vocabularies convey.

To be precise, for a given concept c_i, its conceptual model can be expressed as:

$$\vec{C}_i = (MI_{i,1}, MI_{i,2}, \ldots, MI_{i,k}, \ldots, MI_{i,n})^T \quad (3)$$

Here n denotes the base number of characteristic vocabularies dictionary. The symbol $MI_{i,k}$ denotes the degree of characteristic vocabulary t_k describing the concept C_i. Among this, the average mutual information is utilized, whose Formula is as follows:

$$MI_{i,k} = p(C_i, t_k) \log_2 \frac{p(C_i, t_k)}{p(C_i) p(t_k)}$$

$$E(p(C_i, t_k)) = \frac{\left|\{s \mid C_i \in s\} \cap \{s \mid t_k \in s\}\right|}{l},$$

$$E(p(C_i)) = \frac{\left|\{s \mid C_i \in s\}\right|}{l}, \quad E(p(t_k)) = \frac{\left|\{s \mid t_k \in s\}\right|}{l} \quad (4)$$

Here, $p(C_i, t_k)$ denotes the probability of C_i and t_k occurring together in the same sentence of the information sources (we here regard a sentence as a window), $p(C_i)$ and $p(t_k)$ respectively denotes the probability of C_i and t_k occurring in the information sources, $\{s \mid C_i \in s\}$ and $\{s \mid t_k \in s\}$ respectively denotes the set of sentences in which C_i and t_k occurs. The symbol l denotes the total number of the sentences in the information sources.

2. The concept relevance calculation module

In this module, the input is the conceptual model of every concept, and the output is the set of concept pair with the value of relevance strength,

which have the strong relevance. According to the conceptual vector model, this module utilizes the cosine similarity to compute the degree of relevance for every concept pair, and then finds out the concept pair whose strength of relevance is larger than the threshold we have already set.

4.4 Relations describing

In order to add explanations to the direct concept pair that has latent relations and express them more explicit, in this sub-section, we utilize method based on statistics to realize this function. We count the frequency of sentence, which contains the concept pair, and then regard the sentence that has the largest frequency as the explanations. For example, concept pair <Ireland, Beer> is obtained; the concrete explanation process is summarized in Table 1 and the first explanation is accepted.

In order to add explanations to the indirect concept pair which has latent relations but has indirect relations in base ontology, apart from the statistics frequency analysis, we utilize method based on the ontology to realize this function. For concept pair <a, b>, its relations in the ontology can be expressed as $a \rightarrow c \rightarrow b$ ($a \rightarrow c \leftarrow b$ etc. can be discussed analogously), in which $c(c = \{\{c_1, c_2\}, \{c_3, \ldots, cm\}, \ldots\})$ denotes a set that contains multi-group concepts set. Thus, c is selected as the auxiliary concept of explanation (c contains the least number of concepts in c). In Figure 2, the solid line with arrow represents

Table 1. The explanations for the concept pair <Ireland, Beer>. For this case, we select the first explanation.

Explanations	Frequency
Today, Ireland produces approximately 8.5 m hectoliters of beer annum. Over 42% is exported, mainly to the UK.	3
Though Ireland is better known for stout, 63% of the beer sold in the country is lager.	1
By 1814 Ireland was exporting more beer to England than it imported.	1

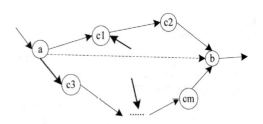

Figure 2. The relations of concept pair <a, b> in base ontology.

Table 2. The explanations for the concept pair <ipad, iTunes>. For this case, $c' = \{Apple\}$.

Explanations (ipad and iTunes link to Apple)

IPAD:
The ipad is a line of tablet computers designed and marketed by Apple Inc., which runs Apple's iOS.
ITUNES:
iTunes is a media player and media library application developed by Apple Inc.

---

already existing relations, while the dashed line with arrow represents latent relations discovered. For example, the explanation of concept pair <ipad, iTunes> is summarized in Table 2.

5 EVALUATION EXPERIMENTS

Evaluating exploratory search system is a notoriously difficult problem [1, 18]. In this paper, our evaluation strategy is to evaluate the contribution of the effective information obtained from our information sources. Specifically, we focus on difference before and after adding the effective information within a search using two evaluation approaches.

Actually, an ideal way would be to develop a real exploratory search system, which direct at a specific domain, so as to accumulate information and directly engage users. With such a system, aiming at the specific query users input, we can depend on the effective information conducting the query expansion, so as to help users to filter the search results, recommend the search results users may be interested in. However, it needs substantial engineering efforts and resources to build [19].

The alternative is that we utilize the local search engine to simulate the process of exploratory search, so as to achieve our evaluation strategy.

5.1 Human evaluators

There were 30 human evaluators, actually graduate students, between the ages of 22 to 30 who are recruited in this study. Fifteen were male and fifteen were female. They were from various departments and academic majors. Before being accepted, they were asked to insure they are unfamiliar with the topics we used or unsure about how to achieve those, so as to guarantee they would conduct exploratory search [1].

5.2 Data collection

As mentioned above, our information sources contain Knowledge-based Question Answering

System (KQAS) and system log. Here, for KQAS, we collected data from Baidu Zhidao (*http://zhidao. baidu.com/*); for system log, we used a web search log from the local search engine that spanned five months. Each query was associated with a user id, so we could recover the complete search history of a particular user, and then we could collect data direct at our goals. In addition, we also utilized characteristic vocabularies dictionary mentioned in the algorithm CCV, which contained more than 190,000 manually selected vocabularies [5], including the concepts we needed.

Note that, although our experimental data was based in Chinese, the algorithms in context are independent of language and it does not require any language or cultural-specific processing.

5.3 Tasks

The search tasks of our experiments were to collect information for writing a report and to get correct information about what they expected. Specifically, there were two corresponding tasks:

1. To gather information on the Web as preparation for writing a report on Political topic, namely about the effect of the change in government (just the same topic in [20]).
2. Imagine you are planning a trip to Beijing. Please write a summary of the following: what cultural events are there (for example in May 2013)? What sightseeing to do? Cover also hotel, flights, traveling to Beijing, weather.

These two tasks were chosen because they are not fact-finding tasks and there were diverse discussions and opinions on these topics, especially for the first tasks [20]. Actually, direct at the two tasks, we adopted different evaluation approaches.

5.4 Evaluation approaches

In the experiment, we focused on evaluating the contribution of the effective information. Due to lacking the real exploratory search system, and the local search engine conducts query expansion without using the base ontology, and yet we utilized the local search engine to simulate the process of exploratory search.

In order to solve the situation, before the search, we provided the base ontology, more accurately, the relations and the concepts related to the tasks (called *pre-search concepts sets*), to the evaluators. Evaluators could refer to *pre-search concepts sets* to achieve the tasks using the local search engine. From another point of view, this reflected the effects of the base ontology, which are supposed to be used for the query expansion. So before the search, this processing mode was applied to both the two tasks.

In the following, we discussed the evaluation approaches. Direct at the first task, after the search, evaluators were required to draw similar pre-search concepts sets (called *Spost-search concepts sets*. Of course, we would teach the relative knowledge to them). After adding the effective information to the base ontology, we called it *post-search concepts sets*. We analyzed the difference between *Spost-search concepts sets* and *post-search concepts sets*. The factors we considered were concepts (including the count, the same and new concepts) and the relations between concepts (including the count of latent relations).

Direct at the second task, different from the above, we utilized a new evaluation approach. Indeed, subject-matter learning has been proposed as a way to evaluate exploratory search systems as a function of exploration time and effort expended [18]. Evaluators needed to conduct two search processes: in the first, they referred to the *pre-search concepts sets*, while in the second they referred to the *post-search concepts sets*. After they found the expected results, we compared the count of web pages clicked, the number of queries submitted to search engine and the cost time.

5.5 Procedure

To sum up, the procedure of our experiments could be summarized in the Figure 3. All evaluators were required to use Windows XP and the IE browser. At last, twenty-five evaluators used Baidu search engine, and the rest used Google.

5.6 Results and analysis

5.6.1 Task 1

Table 3 lists the latent concepts and relations for task 1 in the effective information collected from our information sources. We set threshold 0.6 in the algorithms (Section 4) to insure that the

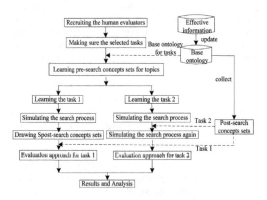

Figure 3. The procedure of our experiments.

Table 3. Latent concepts and relations for task 1.

Concept pair	Degree of relevancy	Type
<effect of change in government, attitude>	0.72	Direct
<democratic party, change in government>	0.91	Indirect → direct
<effect of change in government, politicians rights>	0.89	Indirect → direct
<effect of change in government, wasteful budget>	0.81	Direct
<effect of change in government, government>	0.65	Indirect → direct
<change in government, corruption>	0.89	Direct
<effect of change in government, construction project>	0.82	Direct
<effect of change in government, planning>	0.80	Direct
<change in government, friendship with other country>	0.72	Direct

latent concepts have stronger relations. Note that, italics in Table 3 are the latent concepts. From Table 3, we can see that some latent concept pairs were extracted, such as <effect of change in government, construction project> (the degree of relevancy is 0.82), the reason can be that it has time-validity.

We updated the contents in Table 3 to the base ontology, and then obtained the *post-search concepts sets* for task 1, summarized in Figure 4. From Figure 4, we can see that there are 25 concepts (the count of nodes) and 27 relations (the links between nodes). Here we no longer gave the *pre-search concepts sets* and *Spost-search concepts sets*.

Table 4 lists the summary of evaluation factors, specifically, includes minimum, maximum, mean, and median for every factor for task 1. From Table 4, we can see that the mean of the count of concepts in *Spost-search concepts sets* and new concepts is respectively 23.13 and 6.23, while it is 25 and 6 in *post-search concepts sets*. The effective information plays a certain role in the process of exploratory search, in spite of the rate of same concepts and discovered latent relations is respectively about 61.32% (15.33/25) and 58.56% (5.27/9). But note that, the local search engine conducts query expansion without ontology and

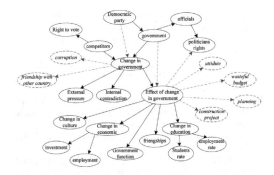

Figure 4. *Post-search concepts sets* for task 1. Note that we translated the original concepts in Chinese into English. Concepts encircled by a dashed line represent latent concepts, and the red dashed lines with arrow represent the latent relations.

Table 4. Summary of evaluation factors for task 1.

Factors	Min	Max	Mean	Median
The count of concept in Spost-	12	38	23.13	22.50
The count of same concept (vs. post-)	10	20	15.33	12.50
The count of new concept (vs. pre-)	0	20	6.23	5
The count of latent relations (vs. pre-)	0	16	5.27	4

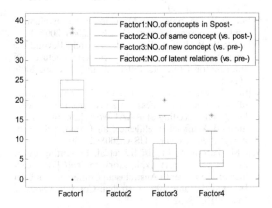

Figure 5. Concrete numbers of evaluation factors for task 1.

actually evaluators could select to whether refer to the base ontology or not. The time of exploratory search process is relatively shorter and evaluators easily were restricted to some certain corpus, while we based on the information sources and the width and depth of web information is larger.

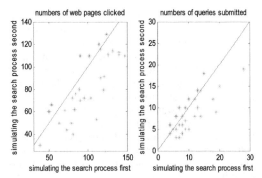

Figure 6. Numbers of web pages clicked and queries submitted for task 2.

Table 5. Summary of web pages clicked and queries submitted for task 2.

Factors	Min	Max	Mean	Median
The count of web pages clicked first	30	147	95.37	92.5
The count of web pages clicked second	24	129	79.53	74
The count of queries submitted first	4	28	10.47	9
The count of queries submitted second	3	19	8.5	8

Figure 5 gave the concrete numbers of evaluation factors in Table 4.

5.6.2. *Task 2*

Direct at task 2, we considered the cost time and the numbers of web pages clicked and queries submitted in two search process. Figure 6 and Table 5 respectively shows numbers of the latter two factors and the summary. 80% (24/30) and 76.67% (23/30) of evaluators respectively clicked numbers of web pages and submitted queries more in the first search process than in the second. Of course Table 5 also shows the same tendency. In the first search, the longest cost time was 2.4 hours, while it was 2.0 hours in the second. We can see that the effective information does play a certain role, although the factors may be affected by other factors of influence.

6 CONCLUSION AND FUTURE WORK

In this paper, we updated the hidden concepts and latent relations extracted from KQAS and system log to the base ontology. We utilized Co-Occurrence analysis method and an algorithm

CCV to discover the effective information. Based on the idea of the above methods, we presented a probability model to judge the strength of relevance between concepts. We gave explanations in order to help users to understand better the new concepts expanded from query, while the process just reflected the information-seeking process in an exploratory way. Through some quantitative analysis about the experiments, we could see that the effective information did play a certain role in helping users to filter the search results and recommending the information users may be interested in.

In future work, we plan to conduct further analysis on the methods discovering latent relations and hidden concepts to support exploratory search better.

ACKNOWLEDGMENTS

This work was supported by National Natural Science Foundation of China (No. 61073062), Science and Technology Funds of Liaoning Province (No. 20102060), the Key Technologies R&D Program of Shenyang City (F11-264-1-33) and the Fundamental Research Funds for the Central Universities (No. N120304002).

REFERENCES

Baeza-Yates R, Ribeiro-Neto B. Modern information retrieval[M]. New York: ACM press, 1999.

Carpineto C, Romano G. A survey of automatic query expansion in information retrieval[J]. ACM Computing Surveys (CSUR), 2012, 44(1): 1.

Deerwester S, Dumais S.T, Furnas G.W, et al. Indexing by latent semantic analysis[J]. Journal of the American society for information science, 1990, 41(6): 391–407.

Egusa Y, Saito H, Takaku M, et al. Using a concept map to evaluate exploratory search[C]//Proceedings of the third symposium on Information interaction in context. ACM, 2010: 175–184.

Fensel D, McGuiness D.L, Schulten E, et al. Ontologies and electronic commerce[J]. Intelligent Systems, IEEE, 2001, 16(1): 8–14.

Gabrilovich E, Markovitch S. Computing semantic relatedness using wikipedia-based explicit semantic analysis[C]//Proceedings of the 20th international joint conference on artificial intelligence. 2007, 6: 12.

Hecht B, Carton S.H, Quaderi M, et al. Explanatory semantic relatedness and explicit spatialization for exploratory search[C]//Proceedings of the 35th international ACM SIGIR conference on Research and development in information retrieval. ACM, 2012: 415–424.

Juan Yu. Learning Domain Ontologies from Chinese Text Corpora[D]. Dalian University of Technology, 2010, pp. 60–63.

Radinsky K, Agichtein E, Gabrilovich E, et al. A word at a time: Computing word relatedness using temporal semantic analysis[C]//Proceedings of the 20th international conference on World wide web. ACM, 2011: 337–346.

Rohde D.L.T, Gonnerman L.M, Plaut D.C. An improved model of semantic similarity based on lexical co-occurrence[J]. Unpublished manuscript, 2005.

Singer G, Norbisrath U, Vainikko E, Kikkas H. Search-logger analyzing exploratory search tasks[C]//Proceedings of the 2011 ACM Symposium on Applied Computing. ACM, 2011: 751–756.

Strube M, Ponzetto S P. WikiRelate! Computing semantic relatedness using Wikipedia[C]//Proceedings of the National Conference on Artificial Intelligence. Menlo Park, CA; Cambridge, MA; London; AAAI Press; MIT Press; 1999, 2006, 21(2): 1419.

Tan B, Lv Y, Zhai C.X. Mining long-lasting exploratory user interests from search history[C]//Proceedings of the 21st ACM international conference on Information and knowledge management. ACM, 2012: 1477–1481.

White R.W, Kules B, Drucker S.M. Supporting exploratory search, introduction, special issue, communications of the ACM[J]. Communications of the ACM, 2006, 49(4): 36–39.

White R.W, Kules B, Drucker S.M. Supporting exploratory search, introduction, special issue, communications of the ACM[J]. Communications of the ACM, 2006, 49(4): 36–39.

White R.W, Muresan G, Marchionini G. Evaluating exploratory search systems[J]. EESS 2006, 2006: 1.

White R.W, Roth R.A. Exploratory search: Beyond the query-response paradigm[J]. Synthesis Lectures on Information Concepts, Retrieval, and Services, 2009, 1(1): 1–98.

Witten I, Milne D. An effective, low-cost measure of semantic relatedness obtained from Wikipedia links[C]//Proceeding of AAAI Workshop on Wikipedia and Artificial Intelligence: an Evolving Synergy, AAAI Press, Chicago, USA. 2008: 25–30.

Wu M, Turpin A, Puglisi S.J, et al. Presenting query aspects to support exploratory search[C]//Proceedings of the Eleventh Australasian Conference on User Interface-Volume 106. Australian Computer Society, Inc., 2010: 23–32.

Zesch T, Gurevych I, Mühlhäuser M. Analyzing and accessing Wikipedia as a lexical semantic resource[J]. Data Structures for Linguistic Resources and Applications, 2007: 197–205.

Control Engineering and Information Systems – Liu (Ed)
© 2015 Taylor & Francis Group, London, ISBN 978-1-138-02685-8

Fault diagnosis of mine hoist system based on fault tree analysis method

Z.S. Yang & X.M. Ma
Xi'an University of Science and Technology, Institute of Electrical and Control Engineering,
Xi'an City, Shanxi Province, China

ABSTRACT: In order to realize the fault diagnosis of mine hoist, a braking system fault tree of mine hoist is established. First of all, working principle of the fault tree analysis method is discussed in this paper, it is based on the mechanism of system working of an analysis in the fault diagnosis method, secondly, brake motor stop to work when happened the failure, the fault tree is established. Finally, the influential factors of the failure has carried on the qualitative analysis by minimum cut sets, and the various factors cause the probability is calculated when the failure occurs, and then according to the size of the probability, for breakdown maintenance of the equipment. It has proved that the method is feasible in Practice.

1 INTRODUCTION

Hoist is a "throat" of the mine, it is a major key equipment in mining production. Its' task are lifting materials, lifting personnel and so on, it is the only way to contact with above mine and underground [1]. Promotion chance the safety and reliability directly affect not only the economic benefits of the mine production, and the most important is related to the miner's life safety. So far, mining production due to the lifting equipment failure and many of the major accident was happened in China, it has caused great economic losses and casualties, as a result, for mine hoist fault diagnosis has important significance.

According to the statistics data analysis, 75% of the failures are caused by the braking system failure [2]. Therefore this article mainly aims at the malfunctions of mine mechanism dynamic system of overwinding accident caused by insufficient braking force to establish the fault tree, and the qualitative and quantitative analysis of fault mechanism for artificial intelligence diagnosis fault lay a certain foundation.

2 CLASSIFICATION OF MINE HOIST SYSTEM FAILURE

Hoist system fault is refers to the hoist system itself because of the abnormal running status, parameter or failure of external conditions change. Such failure including travel monitoring class failure, speed monitoring failure, position monitoring and synchronous fault correction [3].

A. Trip monitor such failure. Trip fault is mainly refers to class to improve the electric control system to improve the container (such as skip) for the entire monitoring process of fault, such as including overwind fault, the overwind fault is a control system from two aspects of schedule arithmetic and shaft actual position monitoring promotion in order to judge the container overwind or not. From the travel operations on the roll failure is the soft overwind failure; Overwind from shaft actual position (shaft switch) to produce the roll failure is the shaft hoist fault (hard roll failure).

B. Speed monitoring fault. In all stages of operation, the speed is keep watch on of the whole operation process of elevator in hoist system, the fault is called speed monitoring, the method is by measuring the actual speed is more than 15% of the given speed, according to the different phases of the operation is divided into constant velocity section speeding of failure, deceleration period of rapid failure and crawling period of rapid failure [4].

C. Reduction monitoring fault is on when the elevator running to the deceleration section is slow or reduce the speed value is declined to set data for monitoring, include slowdown point failure of monitoring and deceleration period of rapid.

D. Position monitoring and synchronous fault correction, there are encoder failure and terminal switch failure in position monitoring protection; PLC in the elevator control system, improve container location arithmetic values consistent with the actual value is extremely important, in the actual operation, caused by fault parking slide wire rope on the drum, and power outages, cause of wire rope elongation operation value and the actual value in the position of hoist vessel produces deviation, the deviation needs to be in the running to the finish location prior to correct, proofreading point location is called synchronization proofreading. If the system is not correct or proofreading switch does not work, it produces synchronization failure. Synchronous fault include shaft synchronous switch and wire rope magnetic synchronous switch failure.

3 BASIC THEORY AND METHOD OF FAULT TREE ANALYSIS

Fault tree analysis is a kind of used for large complex system's reliability, security and risk evaluation methods, it is a kind of graphic deductive method, it is the reasoning method of fault event in certain conditions [5]. Fault Tree Analysis (FTA) is often called the most don't want to happened the parent node fault (or events), events that will not get as leaf node failure (or events), and between the parent node failure (or top) and leaf node failure all event (or events) is called the child node failure events (or middle), and step by step, until find out the basic reason of the fault occurred, the fault tree of the leaf node failure (or events). On behalf of these events with a corresponding symbol (or failure), then use the appropriate logic regarding the parent node failure, the failure of child nodes and leaf nodes connected into inverted tree, the fault tree is FT. To FT as tools to analyze equipment system failure (fail) of a variety of reasons, ways, puts forward effective prevention measures of system reliability research methods for fault tree.

Fault tree of the cut set is a set of events, cut set to reflect the system state of a failure or a failure mode, all cut set to reflect the system failure or failure mode.

The minimum cut set is the most important basic concepts in the FTA and failure analysis and the basic method. When cutting focus each leaf node failure exist at the same time, the parent node fault occurred, then the cut set is called minimum cut sets. In other words, the minimum cut set is without any one leaf node failure is no longer a collection of cut set of leaf node failure [6].

4 BRAKING SYSTEM AND ESTABLISHMENT OF FAULT TREE ANALYSIS

Hoist overwind failure is mainly caused by two aspects: one is the brake system failure caused by insufficient braking force, the second period of over speed protection failure is slowing down the speeding. Below for the braking system are analyzed.

4.1 Qualitative analysis

Overwind fault mechanism, through the analysis of its fault tree of nodes, listed in Table 1.

In the brake failure, the reason there is brake motor to stop working, first analysis affecting the failure event, listed in Table 2.

According to Table 2, the establishment of the fault tree as shown in Figure 1.

4.2 Quantitative calculation

In order to get promotion mechanism of the Figure 1 move when the motor stops leaf node status (parent) distribution table, the fault tree of the minimum cut set must be find out.

Its method has two kinds: ascending method (Semanderes method, or Boolean algebraic reduction method) and descending method (Fussell vesely method). Downward method the result of minimum cut sets and the ascending-method results are the same, here no longer to research.

Table 1. Mine mechanism dynamic system fault tree node.

Coding	Event
F	Mechanism of promoting mine dynamic system failure
A	Brake failure
B1	Brake is not opened
B2	Misfunctioning of brake
B3	Brake shoe wear and tear
B4	Loose brake failure
C1	Low oil pressure
C2	Low friction coefficient
C3	Spring fatigue
C4	residual pressure is high
C5	Great shoe clearance
C6	Deflection of brake disc
C7	There is air in the hydraulic system
C8	Oil pressure instability
C9	Pressure slowly
C10	Overflow valve failure
C11	Pressure reducing valve failure
D	Hydraulic station fault
E1	Oil pump failure
E2	Electric hydraulic pressure regulating device failure

Table 2. Break motor stops working fault events.

Coding	Event
F	Brake motor to stop working
G1	Positive and negative switching circuit contactor contactor does not work properly
G2	Main circuit contactor contactor does not work properly
G3	Direction of contacts of contactor does not work properly
G4	Direction of contacts of contactor contact undesirable
G5	ZC contactor contactor does not work properly
J1	XLC contactor does not work properly
J2	XC self-preservation contactor does not work properly
J3	Reverse contacts of contactor FC poor contact
J4	SDZJ relay does not work properly
J5	ZC contactor contacts of ZC poor contact

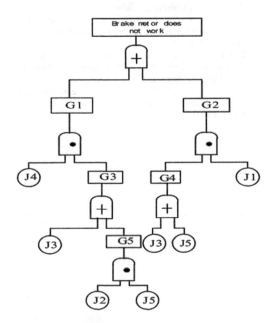

Figure 1. The fault tree of brake motor to stop working.

1. Ascending method for minimum cut sets

$$G5 = J2J5$$
$$G3 = J3 + G5$$
$$G1 = J4G3$$
$$G4 = J3 + J5$$
$$G2 = G4J1$$
$$F = G1 + G2$$

Substitution and simplified step by step, and finally get it,

$$F = G1 + G2 = J1J3 + J1J5 + J3J4 + J2J4J5 \qquad (1)$$

Therefore, there are four minimum cut sets: $\{J1,J3\}, \{J1,J5\}, \{J3,J4\}, \{J2,J4,J5\}$.

2. Take the leaf node status (parent) distribution table:

On Formula (1), ask for leaves (parent) node status distribution table in Table 3.

Leaf node failure (or events) is based on the fault tree structure important degree, by calculating method to determine the location of the point of failure for different, leaf node failure cause the size of the parent node failure probability. In view of the leaf node failure Ji takes only two values (0, 1), so

$$Ji = \begin{cases} 1 & \text{The failure has not occurred} \\ 0 & \text{The failure has not occurred} \end{cases}$$

By probability theory, the leaf nodes of crucial importance (harm) probability:

$$P(i) = \frac{1}{2^{m-1}} \left[\sum_{x/xi=1} F(1_i, X) - \sum_{x/xi=0} F(1_i, X) \right]$$

m is the number of leaf nodes, $\sum_{x/xi=1} F(1_i, X)$ is set number, when the leaf node failure i is happened, the parent node failure, $\sum_{x/xi=0} F(1_i, X)$ is set number, when the leaf node failure i is not happened, the parent node failure.

Table 3. Leaf node state distribution table (parent).

J1	J2	J3	J4	J5	F	J1	J2	J3	J4	J5	F
0	0	0	0	0	0	1	0	0	0	0	0
0	0	0	0	1	0	1	0	0	0	1	0
0	0	0	1	0	0	1	0	0	1	0	0
0	0	0	1	1	0	1	0	0	1	1	1
0	0	1	0	0	0	1	0	1	0	0	1
0	0	1	0	1	0	1	0	1	0	1	1
0	0	1	1	0	1	1	0	1	1	0	1
0	0	1	1	1	1	1	0	1	1	1	1
0	1	0	0	0	0	1	1	0	0	0	0
0	1	0	0	1	0	1	1	0	0	1	1
0	1	0	1	0	0	1	1	0	1	0	0
0	1	0	1	1	1	1	1	0	1	1	1
0	1	1	0	0	0	1	1	1	0	0	1
0	1	1	0	1	0	1	1	1	0	1	1
0	1	1	1	0	1	1	1	1	1	0	1
0	1	1	1	1	1	1	1	1	1	1	1

According to type and Table 3 leaf node may be calculated on the critical importance of (harm), respectively:

$$P(J1) = \frac{1}{2^{5-1}}(12-5) = \frac{7}{16}$$

$$P(J2) = \frac{1}{2^{5-1}}(9-8) = \frac{1}{16}$$

$$P(J3) = \frac{1}{2^{5-1}}(12-5) = \frac{7}{16}$$

$$P(J4) = \frac{1}{2^{5-1}}(11-6) = \frac{5}{16}$$

$$P(J5) = \frac{1}{2^{5-1}}(11-5) = \frac{3}{8}$$

From the calculation result, $J1$ and $J3$ are most importantly, $J5$ and $J4$ take second place, in the leaf node failure (events), $J2$ is Minimal factor to effect the parent node failure is happen. According to the size of the probability of failure of successive order of fault diagnosis, namely first diagnosis fault probability of big leaf incident, successive recursion. At the same time, in the monitoring system, can be ignored for the state of small failure probability of detection, in this way, the monitoring can be reduced in the monitoring system, shorten the time of fault diagnosis.

5 CONCLUSION

This article focuses on the braking system failure which is caused by a fault, the braking system is established in the brake motor does not work, established the fault tree, the qualitative and quantitative analysis of fault occurs at the same time, the probability of the factors influencing it, and through the Wang village in shanxi and be applied in the mine, it has obtained the good effect, the diagnostic accuracy was 94.7%, and provide a theoretical basis for subsequent intelligent fault diagnosis.

REFERENCES

Guo-Jun Tan, The mine DC hoist computer control technology [M] China university of mining press 2003.6 Version 1.

Qiang-Niu, The fault diagnosis of mine hoist based on semantic study [D] China University of Mining and Technology: 2011.

Xiao-Huan Jiang, Research of vertical shaft hoisting system safety and reliability [D] China University of Mining and Technology: 2006.

Xiao-Lin Xu, A mine hoist fault diagnosis method of support vector machine (SVM) [J] Coal mine machinery 28 volumes (8): 202–204.

Ya-Wei Liu, Based on the fault tree analysis method analysis of the elevator fault diagnosis [J] Industry and Mine Automation 77–79.

Yu-Jin Li, Hoisting system of the mine safety accident analysis and prevention [J] Coal Engineering 2012.1:100–102.

Control Engineering and Information Systems – Liu (Ed)
© 2015 Taylor & Francis Group, London, ISBN 978-1-138-02685-8

Reliability statistical analysis on k/n (G) system of objective test questions

Y. Wang
Department of Mathematics, Qinghai Normal University, Xining, China

L.F. Ji
Foreign Language Department, Qinghai Normal University, Xining, China

X.Q. Zhang
Financial and Economic College, Qinghai University, Xining, China

ABSTRACT: Shock model is an important model in reliability theory; it is mainly about system reliability problem in shock environment. In this paper, we studies reliability statistical analysis on k/n (G) system of objective test questions. The system is composed of n objective test questions, when the number of right answers is less than k; it means not passing the examination (system failure). That is the new definition of k/n (G) system problem. When n test question are under Poisson shock at the same time, we get the conclusion that the availability of k/n (G) learning system reduce with the increase of learning force or review time.

1 INTRODUCTION

Shock model is an important model in reliability theory; it is mainly about system reliability problem in shock environment. Early research mainly discusses the life problem of single component systems under the Possion shock; Barlow and Proschan [1] have studied this kind of problem. They point out that the losses on the system caused by shock can be overlaid, when losses exceeds to a certain limit, the system fails, But this model does not contain all of the possible problems, so Shamthikumar and Sumita [2][3] introduced a special kind of shock model: the system fails when single shock value exceeds the critical value. Z.H. Li [4], G.J. Wang and Y.L. Zhang [5–6] studied another special shock model: when continuous shock interval is less than the critical value δ the system fails. Wang and Zhang [7] studied the shock model with mixed failure modes.

K/n (G) System is composed of N objective test questions. The reliability system is acceptable, if and only if, having at least k test question answers. The basic theory of this kind of system can refer to K. Cheng [8]. The further study into the system, such as continuous study by Kontoleon [9], repairable continuous study k/n (G) System by Zhang and Lam [10], k/n (F) system studied in the incomplete information condition by Utkin [11], This paper studies reliability statistical analysis on k/n (G) system of objective test questions.

2 MODEL HYPOTHESIS AND MODEL ANALYSIS

First let's do the following assumption of the reliability model:

Assumption 1. Reliability system is composed of N objective test questions, when the numbers of correct test questions are less than k and the system fail, namely it is k/n (G) system.

Assumption 2. k/n (G) system is impacted by a homogeneous Poisson stream, the impact of flow intensity is λ, each of the impact quantity $\hat{x}$ obeys F distribution, the amount of impact is actually the loss amount of answering capability.

Assumption 3. n objective test questions of the system are subjected to impact at the same time, the fault threshold of each question under impact is a random variable τ, it obeys Φ distribution.

Assumption 4. System can malfunction only when the impact occurs.

If the value of the impact is $\hat{x}$, failure probability of each test question is

$$P_{\hat{x}} = P(\tau \leq \hat{x}) = \Phi(\hat{x}) \tag{1}$$

From the hypothesis 2, hypothesis 3 we know that the failure probability of a single test question from one impact is

$$P = P(\hat{x} > \tau) = \int_0^\infty \Phi(\hat{x}) dF(\hat{x}) \tag{2}$$

Arriving impact influences all the questions independently because the system consists of N objective test questions. The size of effect is inter-relating with impact value.

For impact value $\hat{x}$, it makes the probability of test failure be $\Phi(\hat{x})$. Denoted as $\Phi(\hat{x}) = x$, where

$$P(X \le x) = P(\Phi(\hat{x}) \le x) = F(\Phi^{-1}(x)) = F\Phi^{-1}(x)$$
(3)

Let $G = F\Phi^{-1}(x)$, so X obeys distribution. $X = \Phi(\hat{x})$ is monotonically decreasing function of $\hat{x}$, its value represents the size of the failure proba-bilities of each test question caused by one impact. In order to discuss the problems we might as well let X be the value of impact, i.e. lack of quantity of the answering capability.

Let T as the operating time of system, $N(t)$ is the value of impact during $(0,t]$ time interval, the impact model for the reliability function is

$$\bar{H}(t) = P(T > t) = \sum_{l=0}^{\infty} P(T > t \mid N(t) = l) \cdot P(N(t) = l)$$
(4)

Use B_j, $j = 1, 2, ..., n$ to represent answering all the specified j questions correctly, P_j is the probability of occurrence of B_j, X represents the amount of shock, then we have (5)

Use A_j, $i = 1, 2, ..., n$ to represent answering i questions correctly after 1 impacts, $m = 1, 2, ..., n$; $S = 1, 2, ...$ represents answering m questions cor-rectly after s impacts, then the corresponding prob-ability events can be expressed as follows

$$P(A_i) = P\left(\bigcap_{S=1}^{l} C_S^{(1)}\right) = P_1^l, i = 1, 2, ..., n.$$
(5)

$$P(A_1 A_2 \cdots A_n) = P\left(\bigcap_{S=1}^{l} C_S^{(n)}\right) = P_n^l$$
(6)

Lemma 2.1 $\xi_1, \xi_2, ..., \xi_n$ is any n event, its occur-rence probability of r event is just

$$P = \sum_{i=0}^{n-r} (-1)^i \binom{r+i}{r} \sum_{j_1 < j_2 <} \cdots \sum_{\cdots < j_{r+i}} P(\xi_{j_1} \xi_{j_2} \cdots \xi_{j_{r+i}})$$
(7)

Theorem 2.1 System is influenced by the param-eters for homogeneous Poisson flow impact, in the model of 1–4, reliability function for a system is

$$\bar{H}(t) = P(T > t) = \sum_{r=k}^{n} \binom{n}{r} \sum_{i=0}^{n-r} (-1)^i \binom{n-r}{r} \cdot e^{-\lambda t(1 - P_{r+i})},$$

which is given by (5).

Proof: By lemma we can get conditional probability

$P(\text{At } t \text{ just answering } r \text{ questions correctly} \mid N(t) = l)$

$$= \sum_{i=0}^{n-r} (-1)^i \binom{r+i}{r} \sum_{j_1 < j_2 <} \cdots \sum_{\cdots < j_{r+i}} P(A_{j_1} A_{j_2} \cdots A_{j_{r+i}})$$

$P(\text{At } t \text{ just answering } r \text{ questions correctly} \mid N(t) = l)$

$$= \sum_{i=0}^{n-r} (-1)^i \binom{r+i}{r} \binom{n}{r+i} P_{r+i}^l = \binom{n}{r} \sum_{i=0}^{n-r} (-1)^i \binom{n-r}{i} P_{r+i}^l$$
(9)

Denote $D_r(t)$, $r = 1, 2, ..., n$, n represents just answering r questions correctly at t, then we get:

$P(T > t \mid N(t) = l)$

$$= \sum_{r=k}^{n} P(D_r(t) \mid N(t) = l)$$

$$= \sum_{r=k}^{n} \binom{n}{r} \sum_{i=0}^{n-r} (-1)^i \binom{n-r}{i} P_{r+i}^l$$
(10)

where $l = 1, 2, 3, ...$. When $l = 0$, $P(T > t \mid N(t) = 0) = 1$.

Due to the impact flow is the Poisson process with parameter λ, reliability function of system at t is:

$$\bar{H}(t) = \sum_{l=0}^{\infty} P(T > t \mid N(t) = l) \cdot P(N(t) = l)$$

$$= \sum_{r=k}^{n} \binom{n}{r} \sum_{i=0}^{n-r} (-1)^i \binom{n-r}{r} \cdot e^{-\lambda t(1 - P_{r+i})}.$$

Then from reliability function of system $\bar{H}(t)$ we can get:

Corollary 2.1 Average working time of the sys-tem is:

$$ET = \int_0^{\infty} \bar{H}(t) dt = \sum_{r=k}^{n} \binom{n}{r} \sum_{i=0}^{n-r} (-1)^i \binom{n-r}{i} \cdot \frac{1}{\lambda(1 - P_{r+i})}$$
(11)

Corollary 2.2 For series system of n question, when $k = n$, reliability function is

$$P(T > t) = e^{-\lambda t(1 - P_n)}$$
(12)

Proof: Series system of N question fault as long as there is a question that fault system

$$P(T > t \mid N(t) = l) = \left[\int_0^l (1 - x)^n dG(x)\right]^l$$

604

Then the reliability function for a system is:

$$P(T > t) = \sum_{l=0}^{\infty} P(T > t \mid N(t) = l) \cdot P(N(t) = l)$$

$$= \sum_{l=0}^{\infty} P_n^l \cdot \frac{(\lambda t)^l}{l!} e^{-\lambda t} = e^{-\lambda t(1 - P_n)} \qquad (13)$$

It is obvious that system life obeys exponential distribution with parameter $\lambda(1 - P_n)$, further we can get its average response time:

$$E(T) = \int_0^{\infty} \bar{H}(t) dt = \frac{1}{\lambda(1 - P_n)} \qquad (14)$$

Corollary 2.3 For the parallel system of n test questions when $k = 1$, the reliability function for a system is:

$$P(T > t) = \sum_{i=1}^{n} (-1)^{i-1} \binom{n}{i} e^{-\lambda t(1 - P_i)} \qquad (15)$$

Proof: In parallel system of n test questions, as long as there is one question to be answered, it is able to work, and so:

$$P(T > t \mid N(t) = l) = P\left(\bigcup_{i=1}^{n} A_i \right)$$

$$= n P_1^l - \binom{n}{2} P_2^l + \binom{n}{3} P_3^l + \cdots + (-1)^{n-1} \binom{n}{2} P_n^l$$

$$= \sum_{i=1}^{n} (-1)^{i-1} \binom{n}{i} P_i^l,$$

so the reliability function for a system is:

$$P(T > t) = \sum_{l=0}^{\infty} P(T > t \mid N(t) = l) \cdot P(N(t) = l)$$

$$= \sum_{l=0}^{\infty} \sum_{i=1}^{n} (-1)^{i-1} \binom{n}{i} P_i^l \cdot \frac{(\lambda t)^l}{l!} e^{-\lambda t}$$

$$= \sum_{i=1}^{n} (-1)^{i-1} \binom{n}{i} e^{-\lambda t(1 - P_i)} \qquad (16)$$

The average response time for the corresponding system:

$$E(T) = \int_0^{\infty} \bar{H}(t) dt = \sum_{i=1}^{n} (-1)^{i-1} \binom{n}{i} \frac{1}{\lambda(1 - P_i)} \qquad (17)$$

If the threshold of 3 is a constant, when the coming impact value is greater than τ, all questions

fault the system faults; when the coming impact value is less than τ, all questions don't fault the system doesn't fault. Conditional probability is: $P(T > t \mid N(t) = l) = [F(\tau)]^l$.

We can get:

Corollary 2.4 If the threshold of 3 is a constant, so the reliability function for a system is:

$$\bar{H}(t) = P(T > t) = e^{-\lambda t(1 - F(\tau))}$$

Proof:

$$P(T > t) = \sum_{l=0}^{\infty} P(T > t \mid N(t) = l) \cdot P(N(t) = l)$$

$$= \sum_{l=0}^{\infty} \left(F(\tau)^l \right) \cdot \frac{(\lambda t)^l}{l!} e^{-\lambda t} = e^{-\lambda t(1 - F(\tau))} \qquad (18)$$

The average response time is:

$$E(T) = \int_0^{\infty} \bar{H}(t) dt = \frac{1}{\lambda(1 - F(\tau))} \qquad (19)$$

Corollary 2.5 In hypothesis 2, if its impact value $\hat{x} = x_0$ is a constant, so the reliability function for a system is:

$$\bar{H}(t) = P(T > t)$$

$$= \sum_{r=k}^{n} \binom{n}{r} \sum_{i=0}^{n-r} (-1)^i \binom{n-r}{i} \cdot e^{-\lambda t \left(1 - (1 - \Phi(x_0))^{r+i} \right)} \qquad (20)$$

Proof: If impact value $\hat{x} = x_0$ is a constant, the probability of a single objective test under one impact is: $P(\hat{X} < \tau) = P(\tau > x_0) = 1 - \Phi(x_0)$, P_j can be transformed into $P_j = (1 - \Phi(x_0))^j$ in form (5), the substitution of theorem conclusion form (8) will be confirmed.

3 THE LEARNERS LEAK CHECKING K/N (G) SELF TESTING REPAIRABLE SYSTEM

Availability is an important index of reliability in repairable system. In this section, we will first narrate the concept of availability. For the self testing repairable system having only answer and fault two possible states, let

$$X(t) = \begin{cases} 1, & \text{the system is in answer state at } t \\ 0, & \text{the system is in fault state} \end{cases}.$$

The definition of instantaneous availability of system is: $A(t) = P\{X(t) = 1\}$, it refers to the system probability of working state in at t. If the limit

605

$A = \lim_{t \to \infty} A(t)$ exits, it is called the limit of steady state availability.

Hypothesis 5. If the learners repair the old fault module as new in accordance with the leak checking principle.

The state of definition system is:

0: answering all n test questions; having i faults in n test questions, $i = 1, 2, \ldots, n$

Because the arrival time of the impact obeys Poisson process, and time for learning repairing obeys the exponential distribution, so if we use $M(t)$, $t \geq 0$ to form a Markov process, its state space is $F = \{n - k + 1, \ldots, n\}$.

Using $P_{ij}(\Delta t)$ to express the probability system transforming from state I to state j, so we have:

$$P_{ij}(\Delta t) = p(N(t + \Delta t) = j | N(t) = i), \forall i, j \in \Omega \quad (21)$$

Using α_{ij} to express the probability of fault number having i test question j in the testing system, then we have:

$$\alpha_{ij} = \int_0^1 \binom{n-i}{j-i}(1-y)^{n-j} y^{j-i} dG(y) \quad (22)$$

By definition of the state transition probability we have:

$$P_{0j}(\Delta t) = \lambda \alpha_{0j} \Delta t + 0(\Delta t),$$
$$j = 1, 2, \ldots, n,$$

$$P_{ij}(\Delta t) = \lambda \alpha_{ij} \Delta t + 0(\Delta t), i = 1, 2, \ldots, n - k; i < j \leq n,$$

$$P_{ii}(\Delta t) = 1 - \left(\lambda \sum_{j=i+1}^{n} \alpha_{ij} + \mu\right)\Delta t + 0(\Delta t), i = 1, 2, \ldots, n - k$$

$$P_{ii-1}(\Delta t) = \mu \Delta t + 0(\Delta t), i = 1, 2, \ldots, n.$$

Among which $0(\Delta t)$ represents the higher order infinitesimal of Δt.

From the above expressions we can write the transition probability matrix Q, denote $P = (P_0, P_1, \ldots, P_n)$ as the probability of system stability, from $PQ = 0$, $PI' = 1$, $I = (1, 1, \ldots, 1)_n$ we can write out equations and get:

$$\mu P_1 = \lambda \sum_{j=1}^{n} \alpha_{0j} \cdot P_0,$$

$$\left(\lambda \sum_{l=j+1}^{n} \alpha_{jl} + \mu\right)P_j = \lambda \sum_{i=0}^{j-1} \alpha_{ij} P_i + \mu P_{j+1},$$
$$j = 1, 2, \ldots, n - k,$$

$$\mu P_j = \lambda \sum_{i=0}^{n-k} \alpha_{ij} P_i + \mu P_{j+1},$$
$$j = n - k + 1, \ldots, n - 1,$$

$$\mu P_n = \lambda \sum_{i=0}^{n-k} \alpha_{in} P_i,$$

$$P_0 + P_1 + \cdots + P_n = 1 \quad (23)$$

From the form above we can get recursive relation as follows:

$$P_1 = \frac{\lambda}{\mu} \sum_{j=1}^{n} \alpha_{0j} P_0,$$

$$P_{j+1} = \left(\frac{\lambda}{\mu} \sum_{l=j+1}^{n} \alpha_{jl} + 1\right)P_j - \frac{\lambda}{\mu} \sum_{i=0}^{j-1} \alpha_{ij} P_i,$$

$$j = 1, 2, \ldots, n - k$$

$$P_{j+1} = P_j - \frac{\lambda}{\mu} \sum_{i=0}^{n-k} \alpha_{ij} P_i, j = n - k + 1, \ldots, n - 1 \quad (24)$$

If we have (24), all P_j can be expressed as a function of P_0, then all can be gained by (23).

Steady state availability of the system is

$$A = P_0 + P_1 + \cdots + P_{n-k} \quad (25)$$

From Markov process repairing theory (see Chengkan [8]), average working time can be obtained under the steady state system, average stoppage time and the average cycle length respectively are:

$$ET_w = \frac{P_w I}{P_w Q_{Bw} I} = \frac{A}{\left(\lambda \sum_{i=0}^{n-k} \sum_{j=n-k+1}^{n} P_i \alpha_{ij}\right)},$$

$$ET_B = \frac{P_B I}{P_w Q_{Bw} I} = \frac{(1-A)}{\left(\lambda \sum_{i=0}^{n-k} \sum_{j=n-k+1}^{n} P_i \alpha_{ij}\right)}$$

$$ET = ET_w + ET_B = \left(\lambda \sum_{i=0}^{n-k} \sum_{j=n-k+1}^{n} P_i \alpha_{ij}\right)^{-1} \quad (26)$$

The steady-state control frequency is:
$$\gamma = \lambda \sum_{i=0}^{n-k} \sum_{j=n-k+1}^{n} P_i \alpha_{ij}.$$

4 NUMERICAL EXAMPLES: 2/3(G) SYSTEM ANALYSIS

System state space is $\Omega = \{0, 1, 2, 3\}$, working set $W = \{0, 1\}$, fault set $F = \{2, 3\}$, suppose the impact value $\hat{x}$ of F distribution and distribution Φ of threshold τ are the same, the distribution of X $G = F\Phi^{-1}$ is uniformly distributed on the interval $(0, 1)$, so we get $\alpha_{01} = \int_0^1 \binom{3}{2}(1-y)^2 dy = \frac{1}{4}$.

606

Similarly we can get: $\alpha_{02} = \alpha_{03} = 1/4$,

$$\alpha_{12} = \int_0^1 \binom{2}{1}(1-y)y\,dy = \frac{1}{3},$$

$$\alpha_{13} = \int_0^1 \binom{2}{2}y^2\,dy = \frac{1}{3}$$

The system state transition matrix is:

$$Q = \begin{bmatrix} -\frac{3}{4}\lambda & \frac{1}{4}\lambda & \frac{1}{4}\lambda & \frac{1}{4}\lambda \\ \mu & -\left(\frac{2}{3}\lambda + \mu\right) & \frac{1}{3}\lambda & \frac{1}{3}\lambda \\ 0 & \mu & -\mu & 0 \\ 0 & 0 & \mu & -\mu \end{bmatrix} \quad (27)$$

Denote $P = (P_0, P_1, P_2, P_3)$ as stability probability of system, we can get by $PQ = 0$, $PI' = 1$:

$$\frac{3}{4}\lambda P_0 = \mu P_1, \quad \left(\frac{2}{3}\lambda + \mu\right)P_1 = \frac{1}{4}\mu P_0 + \mu P_2,$$

$$\mu P_2 = \frac{1}{4}\lambda P_0 + \frac{1}{3}\lambda P_1 + \mu P_3,$$

$$\mu P_3 = \frac{1}{4}\lambda P_0 + \frac{1}{3}\lambda P_1, \; P_0 + P_1 + P_2 + P_3 = 1 \quad (28)$$

By writing out equations we can get:

$$P_1 = \frac{3}{4} \cdot \frac{\lambda}{\mu} P_0,$$

$$P_2 = \frac{1}{2} \cdot \frac{\lambda}{\mu} P_0 + \frac{1}{2} \cdot \left(\frac{\lambda}{\mu}\right)^2 \cdot P_0, \quad (29)$$

$$P_3 = \frac{1}{4} \cdot \frac{\lambda}{\mu} P_0 + \frac{1}{4} \cdot \left(\frac{\lambda}{\mu}\right)^2 \cdot P_0$$

Then by (27), we have:

$$P_0 = \left[1 + \frac{3}{2} \cdot \frac{\lambda}{\mu} + \frac{3}{4} \cdot \left(\frac{\lambda}{\mu}\right)^2\right]^{-1} \quad (30)$$

And then back to the equations, we get the corresponding state probability is P_1, P_2, P_3
System availability is:

$$A = P_0 + P_1 = \left(1 + \frac{3}{4} \cdot \frac{\lambda}{\mu}\right)\left[1 + \frac{3}{2} \cdot \frac{\lambda}{\mu} + \frac{3}{4} \cdot \left(\frac{\lambda}{\mu}\right)^2\right]^{-1} \quad (31)$$

The average smooth working time and average stoppage time respectively are

$$ET_W = A \bigg/ \left(\frac{1}{2}\lambda P_0 + \frac{2}{3}\lambda P_1\right) = \frac{3\lambda + 4\mu}{2\lambda(\lambda + \mu)},$$

$$ET_B = (1 - A) \bigg/ \left(\frac{1}{2}\lambda P_0 + \frac{2}{3}\lambda P_1\right) = \frac{3}{2\mu} \quad (32)$$

System average answer cycle length is:

$$ET = ET_W + ET_B = \frac{3\lambda^2 + 6\lambda\mu + 4\mu^2}{2\lambda\mu(\lambda + \mu)} \quad (33)$$

System failure frequency is:

$$\gamma = \frac{2\lambda\mu(\lambda + \mu)}{3\lambda^2 + 6\lambda\mu + 4\mu^2} \quad (34)$$

Let $Z = \lambda/\mu$, then (30) can be expressed as $\Psi(Z) = \left(1 + \frac{3}{4}Z\right)\bigg/\left(1 + \frac{3}{2}Z + \frac{3}{4}Z^2\right)$, derivation is available and we get $\Psi'(Z) < 0$, $Z = \lambda/\mu$, increase means increasing the impact strength or average repair time, along with the increase of Z, system availability decreases, that is with the increase of study strength or review time, the availability of learning system reduce.

REFERENCES

Barlow, R.E. and Proschan, F. Statistical Theory of Reliability and Life Testing, Hot, Rinehart and Winston, Inc., New Yor, 1975.

Chen, K. Life distribution type and reliability mathematical theory, science press, 1999.

Kontoleon, J.M. Reliability determination of a r-successive-out-of-n: F system, IEEE Transactions on Reliability, 29(1980), 437–440.

Li, Z.H., Huang, B.S. and Wang, G.J. Life Distribution and Its Statistical Properties of Shock Model with Shock Source, Journal of Lanzhou University, 35(4) (1999), 1–7.

Shanthilumar, J.G. and Sumita, U. Distribution properties of the system failure time in a general shock model, Adv. Appl. Prob., 16(1984), 363–377.

Shanthilumar, J.G. and Sumita, U. General shock models associated with correlated renewal sequences, J. of Appl. Prob., 20(1983), 600–614.

Utkin, L.V. Reliability models of m-out-of-n systems under incomplete information, Computers and Operations Research, 31(2004), 1681–1702.

Wang, G.J. and Zhang, Y.L. A shock model with two-type failures and optimal replacement policy, International Journal of systems Science, 36(4)(2005), 209–214.

Wang, G.J. and Zhang, Y.L. General δ-Shock Model and Its Optimal Replacement Policy, Or Transactions, 2003,7(3):75–82.

Wang, G.J. and Zhang, Y.L. δ-Shock Model and the Optimal Replacement Policy, Journal of southeast University (Natural science Edition), 31(5)(2001), 121–124.

Zhang, Y.L. and Lam, Y. Reliability of consecutive-k-out-of-n: G repairable system, International Journal of systems science, 29(12)(1998), 1375–1379.

Control Engineering and Information Systems – Liu (Ed)
© 2015 Taylor & Francis Group, London, ISBN 978-1-138-02685-8

Real-time system for overload and overspeed detection of large trucks based on ARM

B. Jiang

Computer School, Panzhihua University, Panzhihua City, Sichuan Province, China

ABSTRACT: With the continuous development of modern society, transportation plays a vital role. While highway transportation group composed of large transport vehicles become the most fast, the most convenient means of transport, it can reach areas inaccessible to other modes of transport. The more and more big trucks on the road caused a lot of problems of highway transportation. Overload and overspeed not only bring serious damage to the pavement, but also the greatest source of traffic accident. To solve this problem, a design based on ARM9 embedded system which can test weight and speed automatically in the car and sent the data via GPRS was proposed in this paper. It can greatly reduce the cost of traffic management, significantly improve management efficiency and effectiveness, and reduce traffic accidents to supervise the running of large trucks using modern network.

1 INTRODUCTION

The core of embedded system is EMPU (Embedded Microprocessor Unit), which have been widely used in the car, all kinds of electronic products and even the field of household electrical appliances. The system used ARM9 series microprocessor which is mainly used in wireless devices, instrumentation, security systems, set-top box, high-end printer, digital camera and digital camcorders etc.

GPRS (General Packet Radio Service) as a transitional technology from the second generation mobile communication technology (GSM) to the third one (3G), is a kind of mobile packet data service based on GSM, provides user with mobile packet IP or X.25 connection. GPRS supports calls, text messages and internet access business, which has the advantage of always-on, with greater bandwidth and higher speeds.

The real-time system for overload and overspeed detection of large trucks presented in this paper was based on ARM9 embedded system. A design based on ARM9 embedded system which can test weight and speed automatically in the car and sent the data via GPRS was proposed. It can greatly reduce the cost of traffic management, significantly improve management efficiency and effectiveness, and reduce traffic accidents to supervise the running of large trucks using modern network.

2 DESCRIPTION OF SYSTEM FUNCTIONALITY

With embedded microprocessor ARM9 as core hardware, applying wireless communication network technology and Ethernet technology, through the transplant of embedded Linux operating system, a portable real time overload and overspeed monitoring system for large truck was realized. The system collected the weight and speed of trucks in real time, connected the center of the system with CAN bus, obtained the result through processing and analysis of system. When trucks are overweight or exceeding the speed limit, warned the cab, at the same time, the analysis results were transmitted to traffic control center through GPRS, as a punishment and management basis. The overall system structure is shown in Figure 1.

3 SYSTEM STRUCTURE

This system was based on UP-NETARM2410-S development platform of Beijing Universal

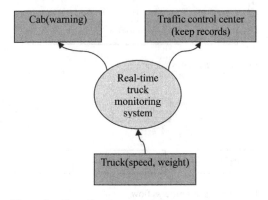

Figure 1. Overall system structure.

Pioneering Technology Co., Ltd as the center, added external speed test module, Weight measurement module, GPRS module and storage module, etc. constituted a system as a whole. The running of the whole system was controlled by S3C2410 ARM CPU. The overall structure is shown in Figure 2.

3.1 System hardware structure

3.1.1 Hardware structure

Hardware mainly contains two parts, one is the hardware system of development platform, the second is expanded hardware system.

The structure is shown in Figure 3, the core-board, LCD, audio, power and various interfaces were used in development platform; expanded hardware system mainly included speed test module, weight measurement module and GPRS module.

3.1.2 Introduction of main hardware

3.1.2.1 Weight measurement module

Gravity signals were collected with pressure sensors. The collected signals were processed with Application Specific Integrated Circuit for HX711 electronic scales and resulted in digital signals. The four-way signals were transferred to SPACE061 A, after conversion processing, the accurate quality of trucks was obtained and then transferred to the S3C2410 ARM.

Weighing module had hardware and software components, hardware part was responsible

for collecting the vehicle body weight data and bodywork tilt Angle data, the software part was responsible for data preprocessing and processing, instead of introducing hardware part, the software part was mainly introduced. The process flow of preprocessing data is shown in Figure 3.

3.1.2.2 Speed test module

Placed the four magnets in the turntable connected with the transmission shaft, hall switch generated a negative pulse when the magnet through the Hall switch. The output of the AH44E was connected to the IOB3 (external interrupt 2 input port) of SPACE061 A. The SPACE061 A counted the pulses, and at the same time, the accurate velocity value was obtained by the processing of SPACE061 A and was transferred to S3C2410 ARM CPU through bus.

This method measured the speed of a truck indirectly by measuring a turntable speed, therefore, the turntable was connected with the rotating shaft of the truck, and the speed of the turntable and that of the shaft formed a certain relationship. The turntable was divided into four equal parts, installing a magnet in the boundary of each equal part, as shown in Figure 4.

The VCC terminating of hall switch connected to 5~12 V DC power supply, OUT terminating connected to IOB3 of SPCE061 A, the other end connected to public ground. Every time the magnet steel through the below of hall switch the OUT end will generate a negative pulse with a pulse width.

3.1.2.3 Display module

Acquisition S3C2410 with LCD display, can display the speed and weight to tip the driver.

3.1.2.4 Sound module

The module also used built-in strong sound processing features of S3C2410. The sound module was also used to alert the driver in order to achieve the purpose of reducing traffic accidents.

3.1.2.5 Bus module

This module used CAN (Controller Area Network) bus which was widely used in automobile.

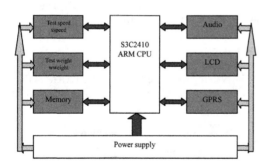

Figure 2.   System structure.

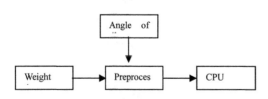

Figure 3.   Data process flow.

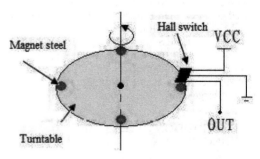

Figure 4.   The principle of the speed test hardware.

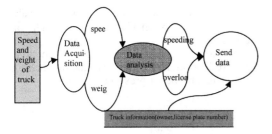

Figure 5. Program flow.

CAN bus is a kind of field bus, with a stability and reliability, simple connection (only two lines), low network consumption, faster than serial port, long distance transmission characteristics.

3.1.2.6 GPRS module

GPRS module was developed by self with SIM300_ v7.03 of Beijing Universal Pioneering Technology Co., Ltd. It can realize the function of the phone, SMS, Internet, through the UART interface communicated with ARM9.

3.2 *System software architecture*

Real-time detection systems for large trucks which installed in the large trucks, constantly tested the weight and speed of the trucks. When there were overweight or speeding, such information will be recorded, and sent to the Supervision Center (such as Traffic Management Bureau). This can greatly reduce the truck management processes, reduce costs, and can greatly improve the efficiency of supervision, reduce the occurrence of traffic accidents.

This system mainly used C language, and a small amount of assembly language. The software program was divided into three parts, respectively was: data acquisition, data analysis and data transmission. The transmitting data included some information of truck itself to determine the identity of the truck. The program flow diagram is shown in Figure 5.

4 OVERALL FLOW OF SYSTEM

The system, mainly using embedded technology and GPRS network technology, through the wireless network, realized real-time tracking for large trucks to prevent the damage for public traffic road and potential traffic safety hazard caused by overloading and speeding. To complete the system functions, adopting modular program design, the system was divided into five modules: weight acquisition, speed acquisition, data analysis, data

transmission and warning sending. The program flow chart is shown in Figure 6.

4.1 *Speed acquisition*

After starting the system, read a truck speed at set intervals. The main function of this module was to obtain velocity and digitalized it for judgment and further processing.

4.2 *Weight acquisition*

Weight acquisition must meet certain conditions. During the high-speed traveling, because the truck body constantly shaken, then the weight deviation measured will be relatively large; while in low running speed, the accuracy is greatly improved. Weight acquisition module was to gaining the speed back into the system, and digitalized it to get the accurate velocity values for further processing.

4.3 *Warning sending*

The purpose of warning sending was to timely remind drivers to reduce the accidents. Sending a warning was based on speed and weight, displaying the current weight and speed can be the form, outputting the audio was also allowed.

4.4 *Data transmission*

Data transmission provided the basis for management and punishment. When the traffic violation reached a certain level, the illegal information will be sent to supervision center (e.g., traffic center) and archived in it.

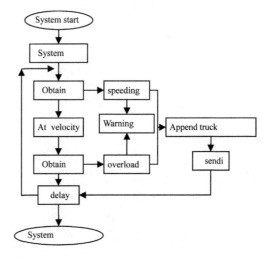

Figure 6. Work principle of truck monitoring system.

4.5 *Data analysis*

Data analysis was the system's main decision module, by judging weight, speed values etc. to made a certain rule to decide when to start speed, when weighing, when transmitting data.

5 CONCLUSION

Overweight speeding real-time monitoring system provided great convenience of supervision and monitoring for Transportation Bureau. The system had the following advantages.

The continuous data acquisition and data analysis guaranteed the accurate and reliable information.

Anytime, anywhere monitoring of system avoided many opportunistic behaviors.

Reducing staff workload, a staff can easily know where and when the truck overweight or speeding.

With the constant improvement of its functionality, overload and speeding real-time monitoring system will have better prospects.

REFERENCES

ARM7TDM 1 Technical Reference Manual 2000. http://www. alTl1.com/arm/documentation open documentatlon.

S3 C44B0X RISC MICROPROCESSOR SUMSUNGC0.

Ligong Zhou et al. Foundation and Practice of ARM microcontrollers [M]. Beihang university press, 2003.

Control Engineering and Information Systems – Liu (Ed)
© 2015 Taylor & Francis Group, London, ISBN 978-1-138-02685-8

Power management system for microbial fuel cells

R.H. Huang, Q. Zheng & B. Mo
Department of Electrical Engineering, Huaqiao University, Xiamen, China

F. Zhao
Institute of Urban and Environment, Chinese Academy of Sciences, Xiamen, China

ABSTRACT: Microbial fuel cells produce electrical energy from biomass energy, which is considered as a promising and sustainable energy source, especially for wastewater treatment and remote monitoring sensors. This paper designs a power management system for microbial fuel cells. The electrical energy is stored in super-capacitors and the voltage is boosted to power the load. The system realizes the application of microbial fuel cells.

1 INTRODUCTION

Microbial Fuel Cells (MFCs) are regarded as a sustainable and promising approach for wastewater treatment and bio-production. MFCs transform biomass energy in organic matters into electricity[1]. The cells can also be used to produce bio-production or bioremediation at the same time, such as hydrogen production[2], desulfuration[3], denitrification[5]. The waste water in the sewage plant or benthal deposit is able to be the fuel for MFCs, which is abundant in daily life. On account of the low power output, the electricity generated by MFCs was not utilized to drive loads several years ago. Multiple factors will influence the operation of MFCs, including microorganisms, electrodes, fuels and proton exchange membrane[1]. MFCs are attractive for powering remote wireless monitoring system. Coupled with the development of MFCs and electronic technique, some electricity energy harvesting systems have been proposed in recent years. Yangming Gong et al. used MFC and backup battery to charge a 200 F ultra-capacitor and the voltage was boosted to 7 V to power an acoustic modem and temperature sensor[6]. Lewandowski et al. put forward an energy harvesting system consisting of charge pump and capacitor[7]. Peter K. Wu et al. proposed a booster circuit to increase the voltage of a cell to a maximum of 3 V, yet the oscillator controller needs applied voltage to activate[8]. Jae-Do Park et al. presented a synchronous boost converter based on inductance and transistor, the power conversion efficiency was 75.9%, not including the energy consumption of the hysteresis controller[9].

The systems above are all aimed at harvesting energy from a single MFC, and the energy conversion efficiency is around 5%, some microcontroller of the circuits need external power source. Due to the low power output of a single MFC, connecting MFCs in serials or in parallel will boost the voltage and current theoretically, however, voltage reversal will appear in experiments. B.E. Logan et al. investigated the causes of the voltage reversal, the result shows that fuel starvation and current densities are primary factors[10]. Chen xi et al. investigated the mechanism of voltage reversal by means of connecting diodes in stacked cells, the results showed that the imbalance of electrons and protons was the essence of voltage reversal[11].

This paper proposes a power management system to harvest electrical energy from MFCs. The system realizes the application of microbial fuel cells.

2 THE WORKING PRINCIPLE OF MFCS

Microbial fuel cells consist of anode, cathode and Proton Exchange Membrane (PEM), as shown in Figure 1. Organic matter is decomposed by

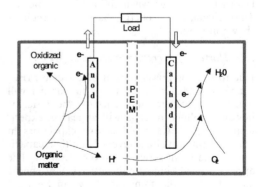

Figure 1. The working principle of microbial fuel cell.

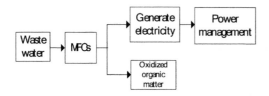

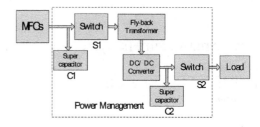

Figure 2. Sewage disposal system of the microbial fuel cell.

Figure 3. The diagram of power management system.

microorganism at anode, generating electrons and protons. The electrons transfer to cathode by means of external circuit, protons transmit to cathode through the proton exchange membrane. The electrons, protons and oxygen appear combination reaction at cathode. The fuels for MFCs are widespread in daily life, reaction conditions are temperate and the products are environmental friendly[1].

Sewage disposal system of microbial fuel cell is shown in Figure 2. MFCs are able to decompose organic matter in wastewater on the one hand, electrical energy is generated at the same time. When MFC is in benthal deposit, the generated electricity could be the power source for remote monitoring system.

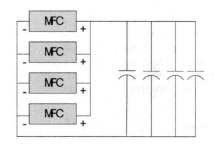

Figure 4. MFCs charge ultra-capacitors in parallel.

3 POWER MANAGEMENT SYSTEM FOR MFCS

3.1 Harvesting energy from a single MFC

The diagram of power management system is shown in Figure 3. The open circuit voltage of a single MFC is around 0.6 V to 0.8 V. MFCs are connected to super capacitor C1 and switches are all open at first. When the voltage of C1 reaches a set threshold, switch S1 closed, C1 starts to discharge. The voltage is boosted by means of a fly back transformer coupled with a DC/DC converter. The boosted voltage is 5 V and stored in capacitor C2. Switch S2 closed and C2 begins to discharge when the load need power. The schematic of energy harvester circuit is shown in Figure 4.

3.2 Harvesting energy from arrays of MFCs

Due to the low power output of a single MFC, connecting MFCs in serials or in parallel will boost the voltage and current theoretically, however, voltage reversal will appear in experiments. The MFCs could charge capacitors in parallel without voltage reversal and then capacitors discharge in series to power the load. The approach would improve the efficiency of energy harvester. The diagrams are shown in Figure 4 and Figure 5.

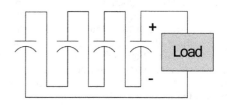

Figure 5. Capacitors discharge in series.

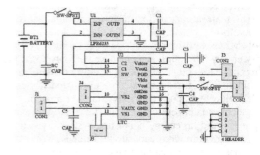

Figure 6. The schematic of energy harvester circuit.

3.3 Experiments and results

The schematic of energy harvester circuit is shown in Figure 6. LPR6235 can be used as fly-back transformer in DC DC converters, which is perfect for low voltage step-up in energy harvesting applications. The turn ratio of the LPR6235-752SML is 1:100. The schematic of transformer is shown in Figure 7.

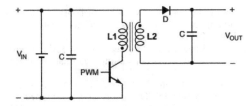

Figure 7. Schematic of fly back transformer.

Figure 8. The photograph of the microbial fuel cell.

Figure 9. The photograph of the energy harvester circuit.

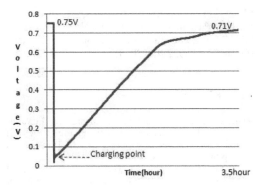

Figure 10. Charging curve of capacitor C1 (5 F).

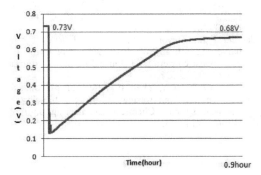

Figure 11. Charging curve of capacitor C1 (1 F).

shown in Figure 11. 5 F capacitor cost 3.5 hour and 1 F capacitor cost 0.9 hour to fully charge. The charging speed is inverse proportion to the capacitance of the capacitor. Finally, the voltage is boosted to 5 V and the load works intermittently.

4 CONCLUSION

Microbial fuel cells are regarded as a promising and environmental friendly method for wastewater treatment and bio-production. A power management circuit is designed to accumulate energy from MFCs. The electrical energy is stored in super capacitor and the voltage is boosted by DC/DC converter. The system demonstrates the feasibility of microbial fuel cells.

ACKNOWLEDGMENT

The authors would like to show their grateful to Huaqiao University and Chinese Academy of Sciences, the research is support by "the National Natural Science Foundation of China" (Grant No.

MFCs use 27 mL two chamber reactors, carbon felt electrode and proton exchange membrame are employed. The photograph of micobial fuel cell is shown in Figure 8. The fuel for MFCs are wastewater from sewage treatment plant, electron acceptor in the cathode are K3[Fe(CN)6], Na2HPO4 • 12H2O and NaH2PO4 • 2H2O. MFCs are operated at room temperture and fed batch mode. The generated electrical energy is input to the circuit as shown in Figure 9.

The open circuit voltage is 0.75 V. MFC charges capacitor C1 (5 F) at charging point, the overall voltage drops to 0.2 V immediately, then the voltage of C1 ascend to 0.71 V gradually as shown in Figure 10. When C1 is 1 F, the charging curve is

61204122), "the Fundamental Research Funds for the Central Universities" (Grant No. JB-ZR1206) "the Science and Technology Project of Xiamen" (Grant No. 3502Z20123034).

REFERENCES

Andrew Meehan, Hongwei Gao, Zbigniew Lewandowski., "Energy harvesting with microbial fuel cell and power management system". IEEE Transactions on power electronics, 2011(26):176–181.

Chen Xi, Zhu Nengwu, Li Xiaohu, "Voltage reversal behavior during stacking microbial fuel cells in series". Environmental science & technology. 2011.34(8):139–142.

Feng Zhao, Robert C.T. Slade, John R. Varcoe., "Techniques for the study and development of microbial fuel cells: an electrochemical perspective". Chem. Soc. Rev. 2009(38):1926–1939.

Jae-Do Park, Zhiyong Ren, "High efficiency energy harvesting from microbial fuel cells using a synchronous boost converter", J. Journal of power sources, 2012(201):322–327.

Liang F.Y, Xiao Y, Zhao F., Effect of pH on sulfate removal from wastewater using a bio-electrochemical system. Chemical Engineering Journal, 2013(218):147–153.

Logan B.E, Regan J.M., "Microbial fuel cells—challenges and applications," Environ Sci Technol, 2006, vol. 40 (17):5172–5180.

Oh, S.-E., Logan, B.E., "Voltage reversal during microbial fuel cell stack operation", Journal of Power Sources, 2007(167):11–17.

Peter K. Wu, Justin C. Biffinger, Lisa A. Fitzgerald. "A low power DC/DC boost circuit designed for microbial fuel cells", J. Process Biochemistry. 2012. 47(11):1620–1626.

Tartakovsky, B., Mehta, P., Santoyo, G. and Guiot, S.R., "Maximizing hydrogen production in a microbial electrolysis cell by real-time optimization of applied voltage," International journal of hydrogen energy. 2011(36): 10557–10564.

Yanming Gong, Sage E. Radachowsky, "Benthic Microbial Fuel Cell as Direct Power Source for an Acoustic Modem and Seawater Oxygen Temperature Sensor System", Environmental Science & Technology, 2011. 45(11):5047–5053.

Zhao, F., Rahunen, N., Varcoe, J., Roberts, A., Avignone-Rossa, C., Thumser, A. and Slade, R., "Factors affecting the performance of microbial fuel cells for sulfur pollutants removal, Biosens. Bioelectron", 2009(24):19–31.

Control Engineering and Information Systems – Liu (Ed)
© 2015 Taylor & Francis Group, London, ISBN 978-1-138-02685-8

The research on airfoils between Hex and NACA

N. Zhang

Changchun Institute of Optics, Fine Mechanics and Physics, Chinese Academy of Sciences, Changchun, China

ABSTRACT: The paper introduces structured 2D-grid division, use fluent software to conduct the simulation calculation for flow field of subsonic wing section, which research the lift and drag characteristics of it. To better illustrate the problem, we use the naca and hex wing sections. This paper calculate lift-drag characteristics of each wing section that in different Mach Numbers, different attack angles, then compare the lift-drag characteristics in different conditions and analyze from the perspective of flow field.

1 INTRODUCTION

Computational Fluid Dynamics (CFD) is a rapidly developing discipline, new theory; algorithms and software related to this subject are endless. As an independent discipline, computational fluid dynamics appeared in the 1960s, after decades of development, it has become an important subject of separate categories of fluid sciences and theoretical fluid dynamics and experimental fluid mechanics [1]. Currently, CFD numerical simulation technology for aircraft, which is gradually developing. The constant improvement of the turbulence model, solver and high-performance computer can achieve a greater number of mesh model appear, making the calculation results are compared with the early development of the CFD technology, its accuracy has been greatly improved, the results are more accurate and trustworthy.

For subsonic aircraft, the design requirements of the missiles' lift-drag characteristic is high, so the missile main wing section design is very important, but in some cases, high lift to drag ratio of the airfoil is often difficult to process, it always have to adopt expensive CNC machine tools for processing, increases the cost.

In this paper, the two airfoil contained 0.3 Ma and 0.5 Ma, under the attack angle from 0 to 12 degrees, 2-dimensional CFD numerical simulation analysis and compare with the different wing section of lift-drag characteristics, afford a reference for selecting aircraft wing section to the future.

2 CALCULATION MODEL

2.1 N-S equations

The conservation of momentum is another common law of fluid movement followed, described as follows: in a given fluid system, the time rate of change of momentum is equal to the sum of the external force acting thereon, the mathematical expression is the equation of conservation of momentum, known as the motion equations, or N-S equations.

For compressible viscous fluid motion equations:

$$\begin{cases} \rho\dfrac{du}{dt} = \rho F_{bx} + \dfrac{\partial p_{xx}}{\partial x} + \dfrac{\partial p_{yx}}{\partial y} + \dfrac{\partial p_{zx}}{\partial z} \\[2mm] \rho\dfrac{dv}{dt} = \rho F_{by} + \dfrac{\partial p_{xy}}{\partial x} + \dfrac{\partial p_{yy}}{\partial y} + \dfrac{\partial p_{zy}}{\partial z} \\[2mm] \rho\dfrac{dw}{dt} = \rho F_{bz} + \dfrac{\partial p_{xz}}{\partial x} + \dfrac{\partial p_{yz}}{\partial y} + \dfrac{\partial p_{zz}}{\partial z} \end{cases}$$

F_{bx}, F_{by}, F_{bz} are on the unit mass of fluid mass force component in three directions; p_{yx} is flow of the body stress tensor components [1].

2.2 Boundary conditions

Boundary conditions is to control the boundary of the fluid motion equations should satisfy the conditions, generally interested in numerical calculations have an important impact. In this paper, we use the pressure far-field boundary conditions, and the wall use a standard wall function.

The pressure far-field boundary conditions used in the Fluent used to simulate the free-flow conditions at infinity, wherein the free stream Mach number and the static conditions are specified. Boundary condition that applies only to the density law and the ideal gas [4].

2.3 Turbulence model

Turbulence model uses Spalart-Allmaras (S-A) model equations, the SA model equations comes from experience and dimensional analysis, the single-equation model applicable to solid wall turbulent flow with laminar flow for simple flow and then gradually added development, amount of eddy viscosity transport equation, the variable characterization of a turbulent kinematic viscosity near the wall outside the region. Transport equation is

$$\rho\frac{d\tilde{v}}{dt}=G_v+\frac{1}{\sigma_{\tilde{v}}}\left\{\frac{\partial}{\partial x_j}\left[\mu+\rho\tilde{v}\right]\frac{\partial\tilde{v}}{\partial x_j}+C_{b2}\left(\frac{\partial\tilde{v}}{\partial x_j}\right)\right\}-Y_v$$

G_v is turbulent viscosity; Y_v is wall blocking and viscous damping caused by the turbulent viscosity reduction; $\sigma_{\tilde{v}}$ and C_{b2} is a constant; v is molecular kinematic viscosity [1].

Relative to the algebraic turbulence model, S-A model at each time step for the entire flow field is more than solving a set of partial differential equations, it takes more time, but the S-A model of separated flow in a range of simulation capabilities better than B-L algebraic model. Relative to the two equations turbulence model, the calculation of the amount of the S-A model is smaller, better stability, and does not need to be fine in the object plane at the computational grid, with the grid order of magnitude of the algebraic model can, it takes more and more attention in practice engineering.

2.4 Mesh and topology

Using fluent software for CFD numerical simulation, ICEM meshing tools is for the pre-processing, meshing the external fluid with structured mesh. Accurate calculation, high accuracy is the

Figure 1.   Mesh of NACA0003 airfoil.

Figure 2.   Topology of NACA airfoil.

Figure 3.   Mesh of Hex airfoil.

Figure 4.   Topology of Hex airfoil.

advantages, and the disadvantage is the relatively poor adaptability, meshing the complex shape of the external flow field is very difficult. In this paper, two-dimensional airfoil calculated computational model mesh is shown in Figure 1 and 3, the topology shown in Figure 2 and 4.

3 RESULTS

3.1 Text and indenting

For the two different airfoils, we calculated the lift and drag coefficients under 0.3 and 0.5 Ma, 0–12 degrees attack angles, and comparative analysis of the results as shown in Figures 5 to 12.

The picture above shows, in a feature (in 0.5 Ma, 4 degrees attack angle), the NACA airfoil static pressure distribution curve of the upper and lower surface along to the X-axis, it can be seen from the figure the pressure distribution of NACA airfoil was better than Hex airfoil, the NACA0003 airfoil lift coefficient is higher than the Hex airfoil, the most important thing of the loss lift of Hex airfoil is that there is a turning point on each the upper and lower surfaces, leading the flow tube cross-sectional area due to changes. The following contrast two airfoil lift-drag characteristics.

By calculating the results, the lift-to-drag ratio of NACA0003 airfoil in the small attack angle is relatively large, the zero-lift drag is relatively small, but the value of C_D^α is relatively large, resulting in the attack angle of more than 4 degree, the lift-drag ratio of Hex airfoil is as well as NACA0003 airfoil, even smaller, the drag coefficient of NACA0003 airfoil were significantly higher than the Hex airfoil. Overall, the lift-to-drag ratio of NACA0003

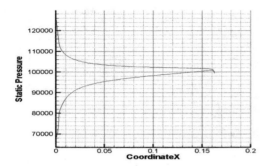

Figure 5. The NACA airfoil static pressure of the upper and lower surface.

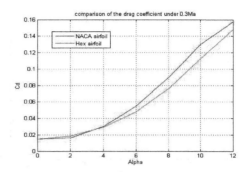

Figure 8. Comparison of the drag coefficient under 0.3 Ma.

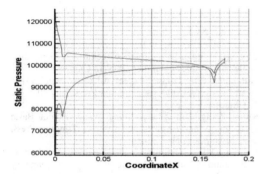

Figure 6. The Hex airfoil static pressure of the upper and lower surface.

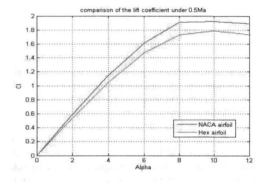

Figure 9. Comparison of the lift coefficient under 0.5 Ma.

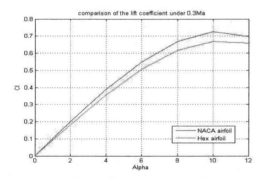

Figure 7. Comparison of the lift coefficient under 0.3 Ma.

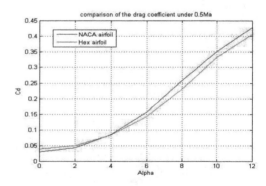

Figure 10. Comparison of the drag coefficient under 0.5 Ma.

airfoil is better than the Hex airfoil, but because of its drag value, to a certain extent also limits its application in missile, especially in the missile propulsion system has been established, a large drag coefficient means that the overall speed of missile is reduced, the dynamic pressure reduced. Although the lift coefficient is higher, but the maximum

overload of the missile may not be superior to Hex airfoil, through multiple aerodynamic design with approximation of the trajectory simulation, give full consideration to the advantages and disadvantages of various airfoil in order to ultimately determine the main wing airfoil required in different situations.

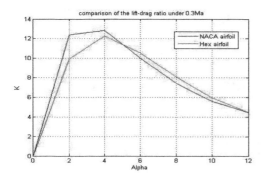

Figure 11. Comparison of the lift-drag ratio under 0.3 Ma.

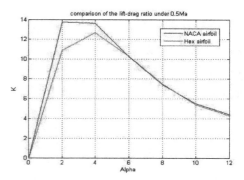

Figure 12. Comparison of the lift-drag ratio under 0.5 Ma.

4 CONCLUSION

CFD numerical simulation shows that the NACA airfoil is not better than the Hex airfoil in various conditions, the missile require large overload, takes a larger attack angle flight conditions, the NACA airfoil may not be appropriate, the lift coefficient of Hex airfoil is poor, but low drag coefficient in the high angle of attack, and ease of processing should be given full consideration in the design of the aircraft design. In future works, we should calculate the missile body—wing combination, compare with the NACA airfoil and traditional Hex airfoil, to give a foundation in the future works.

REFERENCES

Chu Xie, Hongyu Jia, Aerospace Technology Generality. Beihang Press, September 2005 (In Chinese).

Fan Jiang, Peng Huang, Fluent Advanced Application and Example Analysis. Tsinghua Press, July 2008 (In Chinese).

Ruijin Wang, Kai Zhang, Gang Wang, Fluent Basic Technique and Application Examples. Tsinghua Press, February 2007 (In Chinese).

Ruisheng Miao, Xianming Ju, Jiasheng Wu, Missile Aerodynamics. National Defense Industry Press, January 2006 (In Chinese).

S.S Qin, Missile Externality Design. National Defense Industry Press, July 1965 (In Chinese).

Control Engineering and Information Systems – Liu (Ed)
© 2015 Taylor & Francis Group, London, ISBN 978-1-138-02685-8

Algorithms and reducts for a double-quantitative model regarding difference integration

X.Y. Zhang

College of Mathematics and Software Science, Sichuan Normal University, Chengdu, China

ABSTRACT: Double quantification plays an important role to completely describe the approximate space in rough set theory, and GDP-Model (a model regarding difference integration of grade and precision) serves as a fundamental double-quantitative model. According to this model, this paper aims to construct algorithms and further explore reducts. Based on the classified structure and calculation formulas of model regions, the approximation algorithm and microscopic algorithm are proposed, analyzed and compared, where the microscopic algorithm exhibits more advantages in calculation complexity. Furthermore, reducts regarding region preservation are proposed and investigated in the two-category case. This study provides some fundamental thoughts for the optimal calculation and reduction application of double-quantitative models.

1 INTRODUCTION

Rough Set Theory (RST) is a data analysis theory and a new mathematical tool for dealing with vague, inconsistent, incomplete information. The classical model (Pawlak-Model) (Pawlak 1982) is a qualitative one so has some limitations, such as no fault-tolerant capabilities. Thus, the construction and development of quantitative models hold significance. The Variable Precision Rough Set (VPRS) (Ziarko 1993) and graded rough set (GRS) (Yao & Lin 1996) serve as two fundamental quantitative models. VPRS and GRS have the relative and absolute fault-tolerant capabilities, respectively, and they exhibit both in-depth studies and wide applications. VPRS-Reduction were studied (Inuiguchi et al. 2009, Wang & Zhou 2009, Mi et al. 2004) and VPRS was applied to practical geoscience and psychology fields (Yanto et al. 2012, Xie et al. 2011), while the GRS model construction was also discussed (Liu et al. 2012). VPRS and GRS exhibit a close relationship due to basic quantitative expansion; thus, a systematic comparative study of both models was made to provide their relationship, transformation, and similar properties (Zhang et al. 2012).

VPRS and GRS utilize precision and grade to make the relative and absolute quantification, respectively. The RST approximate space is essentially two dimensional (Zhang & Miao 2013). Thus, the two kinds of quantification regarding precision and grade exhibit completeness and complementarity for the approximate space, and the double quantification usually has the same linear computation level when compared to the single quantification system (Zhang & Miao 2013) so holds great significance. In particular, the double-quantitative models inherit accuracy so can highly improve upon the performance of both VPRS and GRS. Based on the Cartesian product combination, two basic double-quantitative models were constructed to make granular computing investigation (Zhang & Miao 2013); based on difference integration of grade and precision, a basic double-quantitative model—GDP-Model—was proposed (Zhang et al. 2010). According to GDP-Model (Zhang et al. 2010), this paper aims to make a further study on algorithms and reducts.

2 GDP-MODEL

GDP-Model, a model regarding difference integration of grade and precision, mainly utilizes the logical difference for integration construction, and it is a basic double-quantitative model. This section makes some review on this model (Zhang et al. 2010).

Suppose U is the finite and nonempty universe, and R is an equivalence relation; thus, (U, R) constitutes the approximate space, and equivalence class $[x]_R$ is also called the information granule. Now, $X \subseteq U$. In VPRS, the misclassification ratio $c([x]_R, X) = 1 - |[x]_R \cap X|/|[x]_R|$ is related to precision: $p([x]_R, X) = |[x]_R \cap X|/|[x]_R|$, and threshold β is usually in the range $[0, 0.5)$; in GRS, measures $|[x]_R \cap X|$ and $|[x]_R| - |[x]_R \cap X|$ correspond to grade in two different directions, and threshold k is a nonnegative integer.

Definition 1. In GDP-Model, there are only three regions (i.e., the upper and lower approximations, the negative region), and they are defined as follows:

$$\overline{R}_{k-\beta}X = \{x : |[x]_R \cap X| > k, c([x]_R, X) \geq 1 - \beta\}$$

$$\underline{R}_{k-\beta}X = \{x : |[x]_R| - |[x]_R \cap X| \leq k, c([x]_R, X) > \beta\}$$

$$negR_{k-\beta}X = \sim (\overline{R}_{k-\beta}X \cup \underline{R}_{k-\beta}X).$$

$\overline{R}_{k-\beta}X$ is collection of equivalence classes, whose number of elements inside X is greater than k but whose misclassification ratio with respect to X is not smaller than $1 - \beta$; $\underline{R}_{k-\beta}X$ is collection of equivalence classes, whose number of elements outside X is not greater than k but whose misclassification ratio with respect to X is greater than β; moreover, all the surplus parts belong to the negative region, which also has the relevant logical explanation regarding double-quantitative semantics. Thus, GDP-Model's construction utilizes the logical difference integration of grade and precision (or the misclassification ratio), thus exhibiting the double-quantitative semantics of grade and precision and double-tolerance mechanism of the relative and absolute errors. Therefore, GDP-Model becomes a basic double-quantitative model and provides an initial model example for double-quantification research.

Proposition 1. In GDP-Model, the upper and lower approximations, the negative region constitute a complete classification of the universe.

Proposition 1 exhibits the special region structure regarding the three-region classification, and this result in fact originates from a particular viewpoint regarding information difference integration of grade and precision.

Proposition 2.

$$\overline{R}_{k-\beta}X = \overline{R}_kX - \overline{R}_\beta X, \quad \underline{R}_{k-\beta}X = \underline{R}_kX - \underline{R}_\beta X.$$

Based on Proposition 2, GDP-Model's approximations correspond to the set difference operation of the GRS and VPRS approximations.

Proposition 3. When $\beta \in (0, 0.5)$ and $k > 0$,

$$\overline{R}_{k-\beta}X = \{x : |[x]_R| > k/\beta, k < |[x]_R \cap X| \leq \beta |[x]_R|\},$$

$$\underline{R}_{k-\beta}X = \{x : |[x]_R| < k/\beta,$$

$$|[x]_R| - k \leq |[x]_R \cap X| < (1 - \beta) |[x]_R|\}$$

$$negR_{k-\beta}X = \{x : |[x]_R| = k/\beta\}$$

$$\cup \{x : |[x]_R| > k/\beta,$$

$$|[x]_R \cap X| \leq k, \text{or}, |[x]_R \cap X| > \beta |[x]_R|\}$$

$$\cup \{x : |[x]_R| < k/\beta,$$

$$|[x]_R \cap X| < |[x]_R| - k, \text{or},$$

$$|[x]_R \cap X| \geq (1 - \beta) |[x]_R|\}.$$

By measures $|[x]_R|$ and $|[x]_R \cap X|$, Proposition 3 provides the microscopic structural formulas of the three regions. Note that this paper mainly focuses on a general case $-\beta \in (0, 0.5)$ and $k > 0$.

3 ALGORITHMS FOR GDP-MODEL

Based on the macroscopic formulas (Proposition 2) and the microscopic formulas (Proposition 3), this section will naturally propose two algorithms for region calculation and emphatically make their analyses and comparison.

Algorithm 1 (Approximation algorithm).

Input: $|[x]_R|, |[x]_R \cap X|, \beta, k$.
Output: GDP-Model's upper and lower approximations, negative region.
Step 1. Calculate VPRS approximations $\overline{R}_\beta X, \underline{R}_\beta X$, and GRS approximations $\overline{R}_k X, \underline{R}_k X$;
Step 2. Calculate $\overline{R}_{k-\beta}X, \underline{R}_{k-\beta}X$ by the set difference operation (Proposition 2);
Step 3. Yield $negR_{k-\beta}X$.

Algorithm 2 (Microscopic algorithm).

Input: $|[x]_R|, |[x]_R \cap X|, \beta, k$.
Output: GDP-Model's upper and lower approximations, negative region.
Step. Directly calculate the three regions: $\overline{R}_{k-\beta}X, \underline{R}_{k-\beta}X, negR_{k-\beta}X$ by the microscopic formulas (Proposition 3).

The approximation algorithm utilizes both the VPRS and GRS approximations to construct GDP-Model's approximations, and the negative region can be simply obtained. In contrast, the microscopic algorithm utilizes the microscopic structural formulas to make the direct calculation. Thus, the former resorts to the VPRS and GRS bases so is easy to be understood, particularly in the macroscopic aspect; the latter resorts to the basic measures description, thus exhibiting profundity and fundamentally. In particular, the following algorithm analysis will illustrate the calculation advantage of the microscopic algorithm.

The calculation process of the approximation algorithm is clear. According to Proposition 3, the microscopic algorithm concerns two comparison process: (1) $|[x]_R|$ should be compared with k/β; (2) $|[x]_R \cap X|$ should be compared with several parameters. For determination, we will precisely describe the comparison order of parameters in the second comparison process. Here, (1) if $|[x]_R| > k/\beta$, then $|[x]_R \cap X|$ should be compared with k first and $\beta|[x]_R|$ second; (2) if $|[x]_R| < k/\beta$, then $|[x]_R \cap X|$ should be compared with $(1-\beta)|[x]_R|$ first and $|[x]_R| - k$; (3) moreover, the case $|[x]_R| = k/\beta$ is clear.

The core task of both algorithms is to study whether each information granule belongs to a specific set. For each information granule, there are two input data: $|[x]_R|$ and $|[x]_R \cap X|$. Suppose there are n information granules, then $2n$ input data are needed. To analyze and compare both algorithms, we will choose the relevant operations (related to $|[x]_R|, |[x]_R \cap X|, n$) as the basic operation, including the comparison, multiply, division, and subtraction. In particular, for simplification, the VPRS approximations mainly utilize the precision form:

$$\overline{R}_\beta X = \{x : p([x]_R, X) > \beta\},$$
$$\underline{R}_\beta X = \{x : p([x]_R, X) \geq 1 - \beta\}.$$

In the approximation algorithm, each information granule first needs calculating to show its belonging feature for the GRS and VPRS approximations. This process needs four comparison times, as well as two auxiliary variables $p([x]_R, X), |[x]_R| - |[x]_R \cap X|$ and two relevant operations. Thus, time and space complexity of approximation algorithm are $T(n) = 6n, S(n) = 2n$, and the result is invariant in the best, worst or average cases.

In the microscopic algorithm, $|[x]_R|$ needs comparing once with k first. (1) If $|[x]_R| > k$, then $|[x]_R \cap X|$ usually needs comparing with k and $\beta|[x]_R|$, and an auxiliary variable $\beta|[x]_R|$ and a relevant operation are required. (2) If $|[x]_R| < k$, then $|[x]_R \cap X|$ usually needs comparing with $(1-\beta)|[x]_R|$ and $|[x]_R| - k$, and two auxiliary variables and two relevant operations are required. (3) If $|[x]_R| = k$, then no comparison and no auxiliary variables are required, and $[x]_R \subseteq negR_{k-\beta}X$. Obviously, the three cases need at most four basic operations and an auxiliary variable, five basic operations and two auxiliary variables, a basic operation and no auxiliary variables, respectively. Thus, time and space complexity of microscopic algorithm are $T(n) = 5n, S(n) = 2n$ in the worst case, which is in case (2).

In the worst case, the asymptotic analyses of time and space complexity become the same in the approximation algorithm and microscopic algorithm, i.e., $T(n) = \Theta(n), S(n) = \Theta(n)$. Thus, both algorithms exhibit the linear feasibility, thus becoming effective. However, the microscopic algorithm exhibits more advantages on the computation complexity, and its calculation time and space are not worse than those of the approximation algorithm even in the worst case, let alone other cases. In particular, we provide two points. (1) Compared to the approximation algorithm, the microscopic algorithm has the same calculation space but the less calculation time in the worst case. (2) For the approximation algorithm, the calculation time and space do not change, and they are the upper bounder of relevant items of the microscopic algorithm. For the microscopic algorithm, there are many cases where the calculation time and space can be further reduced, such as the cases $|[x]_R| > k/\beta$ and $|[x]_R| = k/\beta$. Obviously, that all information granules satisfy the condition $|[x]_R| = k/\beta$ becomes the best case for the microscopic algorithm, where $T(n) = n, S(n) = c$.

The microscopic algorithm has more advantages on algorithm analyses. This is mainly attributed to the accurate microscopic description of the three regions, i.e., Proposition 3, where the microscopic structural formulas directly utilize the initial input data: $|[x]_R|, |[x]_R \cap X|$. Thus, based on the range constructed by the thresholds or parameters, the microscopic algorithm can directly determine the belonging with respect to the three classified regions, and the basic operations and auxiliary variables can be lesser in most cases. However, the approximation algorithm must check the belonging with respect to the four basic regions—the VPRS approximations and GRS approximations, so calculation redundancy inevitably exists; thus, the calculation time and space do not change and become the upper bound of those of the microscopic algorithm.

In practice, information granules usually appear much more in the universe. Hence, using the microscopic algorithm would decrease computation complexity in applications, especially when dealing with massive data. Moreover, the program is easy to be implemented for the microscopic algorithm.

4 REDUCTS FOR GDP-MODEL

Based on algorithm construction and analyses, the above section has provided a basis for region calculation. Herein, we discuss attribute reduction for GDP-Model based on the three classified regions, and the symbol "GDP-Reduction" is given for description. For simplification, only one basic set X is concerned in the approximate space (U, R); thus, the decision table is in the two-category case. Suppose decision table $(U, C \cup D)$, where C, D represent the condition and decision attribute sets respectively, and $U / IND(D) = \{X, \sim X\}$. Moreover, $C \rightarrow B$ denotes the attribute deletion from C to B. Note that the attribute subset corresponds to the equivalence relation.

Regions underlie practical applications of RST, and the relevant region criterion becomes the primary factor for attribute reduction. Pawlak-Reduction mainly utilizes the region-preservation criterion. Thus, region-preservation becomes a fundament reduction principle, and in GDP-Reduction, this strategy is also adopted. In GDP-Model, there are only three regions, so the reduction rule means the three-region preservation; in other words, in $C \rightarrow B$:

$$\overline{B}_{k-\beta}X = \overline{C}_{k-\beta}X, \ \underline{B}_{k-\beta}X = \underline{C}_{k-\beta}X,$$

$$negB_{k-\beta}X = negC_{k-\beta}X.$$

Definition 2 In GDP-Model B is a reduct of C, if the following two conditions are satisfied: (1) $C \rightarrow B$ preserves the three regions; (2) $\forall B' \subset B$, $C \rightarrow B'$ does not preserve the three regions. Let $\text{Re}d(C)$ be all reducts of C.

By the reduction criterion on the three-region preservation, Definition 2 naturally proposes the reduction notion in GDP-Model. Item (1) means the region preservation feature of GDP-Reduct, while Item (2) shows the GDP-Reduct maximality with respect to the three-region preservation. Moreover, GDP-Reduct can be also defined by the independence. Next, the core notion is proposed, and its significance for reduct will be further analyzed.

Definition 3. $\forall c \in C$, if $C \rightarrow C - \{c\}$ does not preserve the three regions, then c is called a necessary attribute in C. Furthermore, the set composed of all necessary attributes in C is called the core of C, denoted by $Core(C)$.

Theorem 1. $Core(C) = \cap \text{Re}d(C)$.

According to Theorem 1, the core is in the intersection set of all reducts. Therefore, the core underlies the reduct calculation, which is similar to the relevant result in classical Pawlak-Reduction. Thus, based on the fundamental role of core for reducts, we next construct the attribute addition algorithm based on the core.

Algorithm 3 (Attribute addition algorithm based on the core).

Input: Decision table, $(U, C \cup D)$, thresholds β, k.
Output: GDP-Reduct B.
Step 1. Calculate $Core(C)$;
Step 2. $B = Core(C)$;
Step 3. *While* $C \rightarrow B$ does not preserve the three regions *do*
Step 4. c is randomly chosen in $C - B$, and let $B = B \cup \{c\}$;
Step 5. *end while*
Step 6. return B.

This algorithm adopts the similar thought of the relevant algorithm in Pawlak-Reduction. Step 1 is to obtain the core, while Steps 3–5 seek the three-region preservation subset including the core, where the added attribute is random. Thus, this algorithm is convergent and usually can obtain one GDP-Reduct when neglecting the maximality requirement.

5 CONCLUSION

GDP-Model is a fundamental double-quantitative model. According to this model, we investigate its algorithms and reducts in this paper. The obtained results, such as the conclusion "the microscopic algorithm has more advantages in calculation complexity", have deepened the basic GDP-Model's results (Zhang et al. 2010); in particular, attribute reduction proposed in this paper underlies GDP-Model applications. Furthermore, this paper's study provides some basic thoughts for the optimal calculation and reduction application of double-quantitative models.

ACKNOWLEDGEMENTS

This work was supported by the National Science Foundation of China (61203285 & 61273304), China Postdoctoral Science Foundation Funded Project (2013T60464 & 2012M520930), Shanghai Postdoctoral Scientific Program (13R21416300), and Key Project of the Sichuan Provincial Education Department of China (12ZA138).

REFERENCES

Inuiguchi, M., Yoshioka, Y. & Kusunoki, Y. 2009. Variable-precision dominance-based rough set approach and attribute reduction. *International Journal of Approximate Reasoning* 50: 1199–1214.

Liu, C.H., Miao, D.Q., Zhang, N. & Gao, C. 2012. Graded rough set model based on two universes and its properties. *Knowledge-Based Systems* 33: 65–72.

Mi, J.S., Wu, W.Z. & Zhang, W.X. 2004. Approaches to knowledge reduction based on variable precision rough set model. *Information Sciences* 159(3–4): 255–272.

Pawlak, Z. 1982. Rough sets. International Journal of Computer and Information Sciences 11: 341–356.

Wang, J.Y. & Zhou, J. 2009. Research of reduct features in the variable precision rough set model. *Neurocomputing* 72: 2643–2648.

Xie, F., Lin, Y. & Ren, W.W. 2011. Optimizing model for land use/land cover retrieval from remote sensing imagery based on variable precision rough sets. *Ecological Modelling* 222(2): 232–240.

Yanto, I.T.R., Vitasari, P., Herawan, T. & Deris, M.M. 2012. Applying variable precision rough set model for clustering student suffering study's anxiety. *Expert Systems with Applications* 39(1): 452–459.

Yao, Y.Y. & Lin, T.Y. 1996. Generalization of rough sets using modal logics. *Intelligent Automation and Soft Computing: an International Journal* 2(2): 103–120.

Zhang, X.Y. & Miao, D.Q. 2013. Two basic double-quantitative rough set models of precision and grade and their investigation using granular computing. *International Journal of Approximate Reasoning* 54(8): 1130–1148.

Zhang, X.Y., Mo, Z.W., Xiong, F. & Cheng, W. 2012. Comparative study of variable precision rough set model and graded rough set model. *International Journal of Approximate Reasoning* 53(1): 104–116.

Zhang, X.Y., Xiong, F., Mo, Z.W. & Cheng, W. 2010. Rough set model of logical difference operation of grade and precision. *Journal of University of Electronic Science and Technology of China* 39(5): 783–787. (in Chinese).

Ziarko, W. 1993. Variable precision rough set model. *Journal of Computer and System Sciences* 46(1): 39–59.

Control Engineering and Information Systems – Liu (Ed)
© 2015 Taylor & Francis Group, London, ISBN 978-1-138-02685-8

Oil enrichment regularity of Putaohua formation in Daqingzijing area

S. Yan
Earth Sciences Institute, Northeast Petroleum University, Daqing, China

ABSTRACT: Due to Putaohua formation is non-essential reservoir in Daqingzijing area, less attention is paid to the basic geology research. In order to boost the exploration development of the oil layer; this paper adopts a comprehensive analysis of the seismologic means, geology, logging and well testing. Meanwhile, the pointed recognition methods and forecasting techniques of different types of hydrocarbon reservoir are proposed. Finally, the paper concludes the horizontal and vertical distribution, and then the oil enrichment regularity of Putaohua formation is build up according to priority, which is directly used in the practical exploration of lithologic deposit in Daqingzijing area.

1 INTRODUCTION

The Daqingzijing area is located Songnan the best SOURCE depression in—Changling north. The northern part in Changling is Qian'an secondary depression and the southern part is Heidimiao secondary depression. Daqingzijing well structure is in the saddle between the two depressions. The overall pattern is the long axis of the north-trending syncline. The east and west wings of the syncline are asymmetrical with steep west wing and moderate east wing. In the Mesozoic and Cenozoic, the district long-term development is in the basin sediments, sedimentation axis, with a more complete strata development, local magmatic intrusions and volcanic eruptions. The Putaohua oil layer is in the first section of Yaojia group. Songliao basin is in the water withdrawal period between the two largest qn1, n2 transgressive periods which are the period of low water lever with the main branch channel deposits. Late period with transgressive water, the depression center moving towards the south, the river retreating, the early delta sandstone was water influx destruction, forming residual mouth bar, coastal dam. Reservoir sand is longitudinal and misfolding with low lateral connectivity, formation of the longitudinal alternating layers of sandstone and mudstone, and the horizontal sandstone and mudstone at the pattern of connection, thus Putaohua reservoir plane distribution reflects faults—lithologic reservoir at the background of local uplift or nose structure.

As Putaohua reservoir area isn't the main oil-bearing formation, we put less emphasis on it and have a serious shortage of basic geological research. In order to promote the exploration and development of the layers, it is necessary to go further study of the Hydrocarbon Distribution for the intervals. Practice has proved that the subtle reservoir characteristics determine that its exploration is much more difficult than anticline. With relatively low success rate, it requires the use of a variety of methods, a variety of techniques, multidisciplinary, multi-means of joint research. As in the past, it is not enough to only rely on sequence stratigraphy analysis techniques to prove low abundance and complex mechanism of accumulation of non-structural reservoirs. In order to clarify the favorable rich oil and gas region, through a comprehensive analysis of microfacies, the top surface of the structure, production test data, we can determine the distribution characteristics of different types of oil and gas reservoirs and summarize the plane and vertical distribution pattern of oil and gas, then establish the gathering mode of Putaohua oil and gas reservoir, while predict favorable oil and gas area.

2 OIL AND GAS RESERVOIRS MODE

2.1 Structure reservoir mode

The majority of Putaohua layers structural reservoirs in Daqingzijing oilfield is broken nose and block reservoir of faulted reservoirs, of which the lowest common denominator in the formation updip is fault trap, including the following three types. (1) Broken nose structural reservoirs. Due to the regional squeeze and faulted reverse drag, it formed nose uplift. Nose uplift matched with different direction faults formed broken nose structure. Broken nose structure is a large number of types of traps in this area. The hydrocarbon accumulation in the nose structure traps formed fault nose oil and gas reservoirs. Although there are a lot of fault nose reservoirs in this area, the height and overall scope of it are small, scattering at parts of the syncline graben. (2) Arc fault block reservoir. In the updip direction of the tilt reservoirs, surrounded by a convex curved fault plane, it behaves as the flat structure contour intersecting to bending fault lines on the structural map. Forming a trap condition, the horizontal distribution law is similar to the broken nose structural reservoir. (3) cross-fault block reservoir. In the updip direction of the inclined reservoir, surrounded by two intersecting faults, it behaves as the flat structure contour intersecting to cross fault lines on the structural map. Forming a trap condition, it mainly spread over the more developed parts of central and southern fault in the study area.

2.2 Lithologic reservoir

Lithologic reservoir is formed because of the gathering of oil and gas in reservoir related with lithology traps. This area with main delta front is the favorable position for lithological trap development with large thin layers interbedded sandstone and mudstone. Having favorable conditions for the formation of lithologic traps: delta front with vertical interoperability between sandstone and mudstone, horizontal interaction of sand-shale, fore-end confined by lacustrine mudstone, the formation of sedimentary background which is favorable for the lithologic trap development. Shallow water delta front gives priority to a large area of distribution of distributary channel sandstone between which is packed by bay mudstone between distributary, constituting a lateral sealing; the trending direction along the distributary channel sand bodies is packed or blocked by faults or local structures, forming traps. Practice has proved that the development of lithologic reservoir in this area is very extensive, including reservoir of sandstone updip pinch out, lenticular lithologic reservoir, fault lithologic reservoir and rare structural—lithologic reservoir. Sandstone updip pinch-out and lenticular reservoirs are mainly distributed in the north and northeast of the study area, relatively developed areas of the southern fault forming a large area of misfolding distributing fault—lithologic reservoir.

3 RESERVOIR DISTRIBUTION REGULARITY

3.1 Horizontal distribution regularity of oil and gas regularity

It can be seen from the 29 tested well for oil gas and water layer maps (Fig. 1) which shows the oil

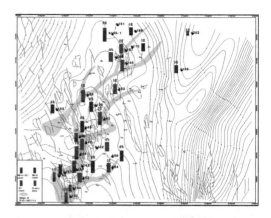

Figure 1. Test wells maps of Putaohua reservoir in Daqingzijing oilfield.

and water distribution regularity. In Daqingzijing oilfield, the oil and gas plane distribution of PuTaohua reservoir is obviously influenced by tectonics, which mainly shows that in parts of the syncline graben, oil and gas enrichment near the fault zone is nearly north-south distribution. Fault mainly control the distribution of oil and gas, forming tectonic, lithology and other types of oil and gas reservoirs. Not only fault near the fault zone was the channel for oil and gas migration, playing an excellent role in communications between the source rock and reservoir sands, but also deactive faults play a shielding effect. Faults cutting sand forms traps, and it aggregating oil and gas forms oil and gas reservoirs.

There is good chance of capturing oil near the synclinal graben fault zone, so it is favorable and potential area. But the south and north reservoir forming rules of the fault zone are quite different. Southern oil and gas mainly gathered at local uplift of the graben fault zone, belonging to the depression in the uplift traps accumulation. Oil and gas was dredged through the fracture to gather in the local uplift traps forming oil and gas reservoirs; Northern oil and gas are mainly distributed on the eastern slopes near the graben fault zone, belonging to traps accumulation where the faults cut and block the updip reservoir sand body.

3.2 Vertical distribution law of oil and gas

Reservoir cross-sectional results of Putaohua reservoir in Daqingzijing oilfield show that the most oil and gas on the vertical distribution is the fourth layer, 3, 5 taking second and the oil and gas distribution of the remaining small layers are the least. This oil and gas vertical distribution law and the sand bodies' developing law are corresponding, while the fourth reservoir sand bodies is the most developed. Through the fine reservoir anatomy of the key blocks' production wells combined with the reservoirs' cross-sectional view in the whole region, we conclude the oil-water vertical distribution law of the Putaohua reservoir: The whole region's oil-water vertical distribution is complex. Pure reservoir is rare. Oil and water are commonly at the same layer. The portfolios of oil under the stem or oil under water are common. The vertical distribution of oil and water is strongly affected by local tectonic and fault. The region doesn't have the same oil-water interface; Vertical distribution of oil and gas is strongly influenced by the reservoir properties. Generally, if the physical properties are better, the oil bearing will be better. The region's oil layer, water layer and the reservoir of the same layer's physical properties are different. The oil layer's physical properties are generally better than the same layer and the water layer. The difference of physical properties affects the small layer's oil bearing. Distribution of oil and gas is obviously influenced by faults. Under the condition of fault development and better communication with the sand body, the distribution of oil and gas is better. Without the sandstone communicated with fault, sandstone lenses are usually water layer or dry layer.

4 THE ESTABLISHMENT OF THE OIL AND GAS-RICH MODE

The process of the establishment of oil and gas accumulation mode is just like the process of looking for hydrocarbon accumulation in favorable locations, which would look for favorable traps and favorable reservoir microfacies with parts of better combination. According to the oil test and production test results, we can verify whether there is the oil and gas enrichment. Then optimize the oil and gas-rich region again. Now illustrate the preferred embodiment of the oil and gas-rich region.

Figure 1 the yellow area is the favorable oil and gas area forecast range of the fourth layer. In the range, 4 small layer of sand bodies are developed, the average thickness of the sand body is about 3 m, individual up to over 6 m, high thickness, giving priority to distributary channel microfacies. Distributary channel in part of the regional was cut into several segments discontinuous stripped sand body. Poor horizontal connectivity creates conditions for lateral sealing, so the reservoir conditions are better. From the trap condition, we can see that northeast fault development is less. Lithology updip pinch out is the main types of traps in the context of the monoclinic structure. We can occasionally find the arc fault block reservoirs and cross-fault block reservoirs in

the central. The relatively developed fault areas in the south are mainly fault—lithologic reservoirs, scattering a large number of broken nose reservoir and cross-fault block reservoirs. The production test data verification shows that the conclusions of the majority of oil test is reservoir or oil-water layer. Some wells' oil production is more than 10t/d, and only the Qian139 and the Hei 45-well are water layers. Analysis shows that the two wells located downthrown. Fault sealing is not conducive to the formation of faults—lithologic reservoir.

5 CONCLUSION

Through anatomical analysis of block research data in the Qian157 oil test, we can determine that reservoir types in Putaohua reservoir have two categories named tectonic, lithology reservoir, and five kinds: broken nose, arc fault block, cross-fault fault block, lithologic updip pinch out and faults—lithologic reservoirs. (2) The region's oil-water vertical distribution is complex. Pure reservoir is rare. Distribution of oil and gas is strongly influenced by the reservoir properties and fault. There is no uniform oil-water interface; Plane distribution is significantly affected by tectonics, mainly reflecting in that hydrocarbon enrichment is in syncline depression fault zone and the surrounding area, the highest probability of capturing oil, a slight difference between the southern part of the fault zone and north accumulation. Oil and gas mainly gathered in the local uplift of the graben, belonging to the bulge accumulation in the depression. The northern oil and gas are mainly located on the eastern slope near the depression, belong to fault trap accumulation.

REFERENCES

Chen Guiju, Zhang Weimin, Song Xinmin, analysis of enrichment patterns in Jilin Daqingzijing Oilfield [J]. Fault-Block Oil & Gas Field, 2005, 12 (6):1–3.

Guo Fangda, Chen Cheng. Daqingzijing the nature of the oilfield fluid identification method [J]. Fault-Block Oil & Gas Field, 2008, 15 (3):1–3.

Jia Chengzao, Zhao Wenzhi, Zou Caineng, the lithologic stratigraphic reservoir, geological theory and exploration technology [J]. Petroleum Exploration and Development, 2007, 34 (3):257–271.

Meng Qi-an, Huang Wei, LinTie feng, et al. Songliao Basin lithologic reservoirs in northern formation conditions and the distribution [J]. Petroleum Geology, 2004, 4:6 11.

Tian Zaiyi, Chang Cheng yong, Qian Shao xin exploration of subtle reservoirs [J]. DaQing Petroleum Geology and Development, 1984, 3 (1):39–50.

Wang Yongchun southern Songliao Basin lithologic reservoir formation and distribution [C] Beijing: Petroleum Industry Press, 2001.

YangMingda Yang Ming Hui, Tang Zhenxing, et al. The controlling factor analysis of oil and gas distribution in southern Songliao Basin Daqingzijing region [J]. Petroleum Geology, 2003, 25 (3):252–256.

Zhao Wenzhi, Dou Lirong remaining onshore oil and gas resource potential and its distribution and exploration strategy [J]. Petroleum Exploration and Development, 2001, 28 (1):1–5.

Zou Caineng, Jia Chengzao, ZhaoWenzhi, et al. Possession of power and distribution pattern of the southern Songliao Basin lithology—stratigraphic reservoirs [J]. Petroleum Exploration and Development, 2005, 32 (4):125–130.

Control Engineering and Information Systems – Liu (Ed)
© 2015 Taylor & Francis Group, London, ISBN 978-1-138-02685-8

Reasoning based on the E (T)-ALC time event and rules

C. Huang & X.L. Zeng
Department of Computer, Gannan Teacher's College, Ganzhou, Jiangxi, China

J.H. Yang
Information Engineering School, Jiangxi University of Science and Technology, Ganzhou, Jiangxi, China

ABSTRACT: This paper will analyze the expression of time and propose a formal representation of time events, starting from the time element of event ontology; it analyses different time period (point) inference relations between events, taking the time of occurrence of events as the main line; analysis of different time period (point) inference relations between events, combining the rules detailed reasoning the relation between time events, mining more implicit information from the time events.

1 INTRODUCTION

Time is one of the important elements of the event ontology. It plays a very important role in describing the event status and reasoning of operation. Time has a linear characteristic, while influencing factors of event action are diversified. But information of time in the event reports is often inaccurate. The inaccurate time information contains much implicit information associated with the event. The information can also have great influence on the reasoning of event operation. Starting from the time element event ontology, this paper will analyze expression of time, propose the formal representation of time events, establish relations between rules and different time periods (point) events, and mine more implicit information from the time events which are going to be verified by an example.

2 STUDY ON TEMPORAL REASONING

Study on time is not a recent phenomenon. Allen [1983] studied the relationship between any two time intervals, putting forward thirteen kinds of possible relationships. Schockaert [2006] proposed a relationship based on a linear function of the fuzzy time interval and listed the fuzzy time relationship. Jerry [2004] proposed the representation of time within Semantic Web ontology. Artale [1989] proposed TL-ALCF temporal description logic decidability by reducing the knowledge presentation ability of this logic; C. Zhang [2011] analyzed the representation of Chinese time ontology from the perspective of Chinese culture under the Semantic Web. C. Lutz [2001] suggested that the reasoning should combine the description of logic

TBox with time interval. Villafane [2000] proposed to express the inclusion relation between time events by algorithm. Mo Sunzhi et al. [2005] proposed a model to express inaccurate time and tense relations based on the fuzzy sets. Lin Chuang et al. [2001] proposed a method to represent imprecise time interval by increasing extended period of time on the right of a certain time interval, however, the model does not explain how to determine the length of extended periods. Liu Wei [2012] proposed a time-dimensional description logic T-ALC from the perspective of event action and put forward reasoning relations between actions with regard to time. However, no scholars use the main line of the time of occurrence of events and according to the relationship of the time point and time interval and combined with the rules of the relationship to research the relationship of events.

3 REPRESENTATION AND REASONING OF E (T)—ALC TIME EVENT

Definition 1 The event occurs in a specific time and circumstances participated by several roles, showing some action characteristics [2009]. Formally, it is represented by e and defined as the structure of six tuples:

$$e = def(A, O, T, V, P, L)$$

Elements in the six tuples of event are called the event elements, which respectively denote action, object, time, environment, assertions and language performance. As one of the six elements of the event, time is represented by the unit of time and continuous digits.

Definition 2 The relationship between the two time periods is mainly based on Aratle[4]. It is defined and analyzed by time point and time period, time segments time segments. Its semantics are as follows:

- The relationship between time point and time point

  Follow(t1, t2), {t1 Follow t2 | $t^2 < t^1$}
  Before(t1, t2), {t1 Before t2 | $t^1 < t^2$}
  Match (t1, t2), {t1 Equals t2 | $t^1 = t^2$}

- The relationship between time point and time interval

  Follow(t, T), {t Follow T) | $t^1 > T_{end}$}
  Before(t, T), {t Before T | $t^1 < T_{start}$}
  Match(t, T), {t Equals T| $t^1 = T_{start} \wedge t^1 = T_{end}$}

- The relationship between time segment and time interval

  Follow(T1, T2), {T1 Follow T2 |$T_{end}^1 < T_{start}^2$}
  Before(T1, T2), {T1 Before T2 |$T_{end}^1 < T_{start}^2$}

  During (T1, T2), {T1 During T2 | $T_{start}^1 > T_{start}^2 \wedge T_{end}^1 < T_{end}^2$}
  Match (T1, T2), {T1 Equals T2 | $T_{start}^1 = T_{start}^2 \wedge T_{end}^1 = T_{end}^2$}
  Meets (T1, T2), {T1 Meets T2 | $T_{end}^1 = T_{start}^2$}
  Overlaps(T1, T2), {T1 Overlaps T2 | $T_{start}^1 < T_{start}^2 \wedge T_{end}^1 < T_{end}^2$}
  Contain (T1, T2), {Contain (T1, T2) | $T_{start}^1 < T_{start}^2$ and $T_{end}^1 > T_{end}^2$}
  Left-Contain (T1, T2), {Left-Contain (T1, T2) |$T_{start}^1 = T_{start}^2, T_{end}^1 > T_{end}^2$}
  Right-Contain (T1, T2), {Right-Contain (T1, T2) | $T_{start}^1 < T_{start}^2, T_{end}^1 = T_{end}^2$}
  Where $[T_{start}^1, T_{end}^1] = T, T_{start}^1$ is the start time of T1, T_{end}^1 is the end time of T1, and $T_{start}^1 \leq T_{end}^1$.

When $T_{start}^1 = T_{end}^1$, T is the time point, denoted as t. Otherwise, for the time period T, it is referred to as T.

Definition 3 Event collection represents a collection of events with common features. Event collection is often made up of several different events. Different events do not necessarily occur at the same time. The collection of events with time can be expressed as:

$E(T) = \{E_1; E_2; :::; E_i, T_i\}, i \in N$

The E_i is a triple (ei; T_{start}^i, T_{end}^i), $T_{start}^i, T_{end}^i \in T_i$. ei is the event label, T_{start} is a start time of an event, T_{end} is end time of the event. For each Ei in the event ε, its start time is ordered. If an event occurs at the time point, then $T_{start}^i = T_{end}^i$. If a set of events contain m events, then it is called m layer events.

Definition 4 Sequential event relations, for event groups (sets) events are defined as s = {$E_{(T)}$, R_T }, where S is the time period (point) event $E(T)$, |$E_{(T)}$| = n, R_T = {R_T (E1;E2);R_T (E1;E3);::::; R_T (E1;En); R_T (E2;E3);::::; R_T (E2;En);::::; R_T (En–1;En)}, R_T

is the relationship of (Ei;Ej) in time, i = 1;::::; n, j = i+1;::::; n–1.

Definition 5 Time interval events: In one form of action system based on E (T)-ALC, the time interval event refers to an status of event in a certain interval of time continued. Its Formalization could be understood as: The event is one of axiom ABox E (T) and one of explain I (E (T)) in T, where C (a) is concept assertions and R (a, b) is role assertions.

Definition 6 Time point event: In one form of action system based on E (T)-ALC, Time point event refers to a status of event change in a time point. Its Formalization could be understood as: The event is one of axiom ABox E (t) and one of explain I(e(t)) in t, where C (a) is concept assertions and R(a,b) is role assertions.

Definition 7 Event knowledge base of $K_{E(T)\text{-}ALC} = <\mathcal{T}, \mathcal{A}_{E(T)}>$ consists of two parts: The TBox $\mathcal{T}$ is sets of event consisting by finite general concept implication type GCIs, it describes event application domain of set. Therefore does not change with time. ABox $\mathcal{A}_{E(T)}$ is sets of consisting by finite event instance assertions with time information. The event instance assertion with time includes concept assertions shape such as $a^{e(t)}$:C or $a^{e(T)}$: C and role assertion such as $(a,b)^{e(t)}$: R or $(a,b)^{e(T)}$: R, where a, b $\in N_O$, C $\in N_C$, R $\in N_R$. E(T) is time interval of events, e(t) is time point events.

Definition 8 ABox $\mathcal{A}_{E(T)}$ at from can explained two tuple such as $\mathcal{I}(e(t)) = (\triangle^{\mathcal{I}(e(t))}, \cdot^{\mathcal{I}(e(t))}).\triangle^{\mathcal{I}(e(t))}$ is domain of events asserting, it is a nonempty set of asserting event at time interval T or point time t. $\cdot^{\mathcal{I}(e(t))}$ is a mapping function of event asserting at time interval T or point time t.

Definition 9 TBox $\mathcal{T}$ at form can explain two tuple such as $\mathcal{I}(e(t)) = (\triangle^{\mathcal{I}(e(t))}, \cdot^{\mathcal{I}(e(t))})$. The $\triangle^{\mathcal{I}(e(t))}$ is domain of axiom, point to the axiom domain of event is nonempty at time interval $\mathcal{T}$ or point time t. T(t) $\in$ [0, V], indicate knowledge base existent at scope of all valid time, $\triangle^{\mathcal{I}(e(t))}$ not subject to time constraints, can be simplified as $\triangle^{\mathcal{I}}$. $\cdot^{\mathcal{I}(e(t))}$ is a mapping function at time interval $\mathcal{T}$ or point time t, similarly $\cdot^{\mathcal{I}(e(t))}$ also is not limited by time, can be simplified $\cdot^{\mathcal{I}}$, the concept of C $\in N_C$ can mapping each subset of C $^{\mathcal{I}(e(t))}$($C^{\mathcal{I}}$), each character of R $\in N_R$ mapping a binary between R $^{\mathcal{I}(e(t))}$ (R $^{\mathcal{I}}$) of $\triangle^{\mathcal{I}(e(t))} \times \triangle^{\mathcal{I}(e(t))}(\triangle^{\mathcal{I}} \times \triangle^{\mathcal{I}})$.

In the knowledge base of $K_{E(T)\text{-}ALC}$, although defined the explanation of ABox $\mathcal{A}_{E(T)}$ and TBox $\mathcal{T}$, but the event and time information representation in ABox is a restricted range, relative ABox the TBox no time limit, that is to say the time limit for TBox can be see established at scope of all valid time in the knowledge base, this characteristic is decided by the invariance of event knowledge structure. Therefore, it can will both unified

632

explanation, common representation event knowledge base $K_{E(T)\text{-}ALC}$ interpretation.

4 REPRESENTATION AND REASONING OF TIME EVENTS BASED ON THE T (E)—ALC RULES

4.1 Representation of time events based on the T (E)—ALC rules

Rules can describe various types of knowledge, such as some factual knowledge, procedural knowledge and rules knowledge etc. The specific representation from of rules is logic programming; it is composed of a set of rules set. Each rule in the logic program can be regarded as a sequence $H \leftarrow B$ established by the premise B and conclusion H, if B was established, then the conclusion H is also established.

Definition 10 Program Datalog [1996] is a collection which is composed of finite rules like $H(X) \leftarrow B1(X1), ..., Bn(Xn)$. Where in:

- $H(X)$ is rule head
- $B1(X1), ..., Bn(Xn)$ is rule body
- $H, B1, ..., Bn$ is predicate symbols
- $X, X1, ..., Xk$ is event items.

Definition 11. Assume there is a knowledge base $K = <K_{E(T)\text{-}ALC}, K_R>$, where $K_{E(T)\text{-}ALC}$ denoted knowledge based on E(T)-ALC; K_R denoted knowledge based on rules, in which K_R knowledge also contains time information, a set of form as $H(a_1, a_2, ..., a_n)^t$ or $H(a_1, a_2, ..., a_m)^T$, denoted as $F_{T(t)}$.

Definition 12. Assume $K = <K_{E(T)\text{-}ALC}, K_R>$ is an event formalism Knowledge Base. Giver an acyclic TBox T, fact F_T and interpretation $\mathcal{I}(t)$ of ABox $\mathcal{A}_{E(T)}$, K_R in time T (t), and $\mathcal{I}(t)| = T, \mathcal{A}_{E(T)}, K_R$. Event E occurs at time T(t). After the event occurs, some new events in T'(t') will produce, that is, $\mathcal{A}_{E(T)}$ is to be converted into $\mathcal{A}_{E'(T')}$, F_T will be changed to F_T', interpretation $\mathcal{I}(t)$ will also be converted into interpretation $\mathcal{I}'(t')$ (t'>t) of another time point, and $\mathcal{I}'(t')| = T, \mathcal{A}_{E'(T')}, K_R'$.

In this definition, semantic of event time conversion can be expressed as: the conversion process from one interpretation to another in time, with ABox and F_T changes, that is to say event led to the change of ABox and F_T.

4.2 Reasoning algorithm of integrated E(T)—ALC and rules

Definition 13. Assume existence a knowledge base of $K = <K_{E(T)\text{-}ALC}, K_R>$ representation by based on the E(T)-ALC and the rules of formal, which E(T)-ALC representation the knowledge of E(T)-ALC and K_R represent regulation knowledge, it not only contains the fact F_T of form for $H(a_1, a_2, ... a_n)^t$ or $H(a_1, a_2, ... a_n)^T$, but also contains the regulation of form for $H(Y) \leftarrow B_1(X_1), ..., B_n(X_n)$. if the priority of R_1 greater R_2, the denote $R_{T1} > R_{T2}$, corresponding $B_1(X_1)$ priority $B_2(X_2)$, denote $B_1(X_1) > B_2(X_2)$ and $E_1(T_1) > E_2(T_2)$ in the execution time.

According to the time variation of the time events, combining the process of reasoning events of rules timing relationships, the algorithm can be described as follows:

Input: E (T)-ALC knowledge base $K_{E(T)\text{-}ALC} = < T, \mathcal{A}_{E(T)}>$ and rule knowledge base K_R.

Output: For H(Y), find the operable rules to obtain satisfiable H (E) (E is a corresponding event of the Y, t the representation form of case rule-based reasoning are events in this paper.)

Begin:

Step 1: Traverse all rules in R, then find out the rules which head of H (Y);

Step 2: If there is only one rule (m = 1), then determine the Satisfiability under the $K_{E(T)\text{-}ALC}$ rules of R_1. If the predicate is satisfiable in the rule body, then apply the rule and return $H(\bar{E})$; If there exists an unsatisfiable or undecidable predicate in the rule body, then delete the rule and turn to step 4.

Step 3: If there are multiple rules (m > 1), determination satisfiability of all predicates exists time priority rules of $R_1, ..., R_t$ under the $K_{E(T)\text{-}ALC}$ according to the time priority order. If all the predicates are satisfiable of a rule which contains the time priority in a rule body, then apply the rules and return $H(\bar{E})$; If there is an unsatisfiable or undecidable predicate in the rule body which contains the time priority, then delete the rule. To determine the subsequent rules, until all rules which contain time priority are deleted;

Step 4: All rules are determined, no rules available, H (Y) is undecidable.

End.

5 EXPERIMENTS AND DISCUSSIONS

Take a traffic accident report as an example to construct a time event form system based on E (T)–ALC and its rules.

Multi-car rear-end event happened on highway in Shandong has caused 9 deaths and 69 injuries:

At 8:45 on the morning of 14th, Traffic accidents occurred at the border of KongCun-XiaoZhi because of rain, snow and heavy fog in resulting more than twenty cars rear-end crash. Half an hour later, police and rescue workers rushed on the spot to start rescue, by 17 PM, the accident caused 9 people dead and 69 people injured. At present, the road traffic has been restored; other wounded is still in the active treatment. According to the Xinhua News Agency (December 14, 2012 20:11).

First of all, knowledge base of the event description logic is to be given $K_{E(T)\text{-}ALC} = <T, A_{E(T)}>$:

The term set = {traffic accident = death ⊔ injury ⊔ doctor ⊔ police}

E_T (t1) = {Occurred (accident, place),

E_T (t2) = {rush (policeman and doctor, accident), treated (doctor, injure),

E_T (t3) = {die (people, 9), injure (people, 69),

E_T (t4) = {road (pass through), treated (doctor, injure)

Secondly, the rule knowledge base is applied by KR = <(R, Q), FT > when the knowledge is difficult to be represented by description logics in system:

The set of rules = {

R1: Formed (accident group, 13) ← Occurred (accident, t1) ∧ Reported (accident, t, t2)

R2: HurryTo (doctor and traffic policeman, accident scene, t4) ← Formed (accident group, t3)

R3: Traffic control (traffic policeman, car, t5), shoot (traffic policeman, scene, t5), inquiry (traffic policeman, witness, t5), treated (doctor, injure, t5) ← At (doctor and traffic policeman, accident, t4)

R4: Recover pass (road, t7) ∧ leave (injure, t7) ← Clean (scene, t6) ∧ dredge traffic(car, t6) }

The finite order is Q = {R1 > R2 > R3 > R4}.

Finally, according to the describing process of events, the different time points (segment) information is to be given to can infer the specific events contained, that is E(T(t)). Three time phrases given in the traffic accident: "t_1:At 8:45 on the 14th" "t_2:half an hour later", "t_3:up to 17 PM", "t_4:At present (reported time)", Combining rules, it can infer traffic accidents occurrence at time $R_{t1} = t_1$;[R_{t2}, R_{t4}] ⊆ (t1,t2), that is to say "report the specific information event to the traffic police department" among t1 and t2; "The traffic police department set up group and discuss the solutions of events", "The hospital sent staffs to rush to the scene of the accident event and the traffic police department sent police to the scene"; (R_{t4}, R_{t6}]⊆(t2, t3), events such as "the police rushed to the scene, traffic control, shooting, questioning the witnesses, clean up the site, guide the traffic; Medical staff statistics, treat the injured" occur among t2 and t3. $R_{t7} = t_4$, events such as vehicular traffic, wounded removal took place at t4.

6 CONCLUSION

Based on representation method of the E (T)-ALC and the rules of time event, it can described process of events through the time and combine rules for reasoning about time events to extract more event information, thereby enhance the knowledge conveying capacity. But the events described are frequently encountered with the representation of fuzzy time, such as "about during the Olympic Games". The further research will focus on how to use the formal representation and corresponding reasoning relations.

ACKNOWLEDGEMENT

The work of this paper is fund by the project of The Education Department of Jiangxi province science and technology project (No. GJJ09246). The authors are also grateful to the anonymous reviewers for their helpful comments.

REFERENCES

Allen, J.F. 1983. Maintaining Knowledge about Temporal Intervals. Communications of the ACM, Vol. 26 No. 11, pp. 832–843,

Artale A, Franconi E. 1989. A temporal description logic for reasoning about actions and plans [J]. Journal of Artificial Intelligence Research (JAIR), 9:463–506.

Jerry R. Hobbs. 2004. An ontology of time for the semantic web. Communications of the ACM, Vol 3 No.1, pp 66–85.

Lifschitz, V. 1996. Foundations of Logic Programming [G], //Principles of Knowledge Representation. CSLI Publications, Stanford, CA, U.S.:69–127.

Lin Chuang, et al. 2001. Extended Interval Temporal Logic for Undetermined Interval: Modeling and Linear Inference Using Time Petri Nets. Chinese J. Computers 24(12):1299–1309.

Liu, Z., Huang, M., et al. 2009. Research on event-oriented ontology. Computer Science 36(11), 126–130.

Lutz, C. 2001. Interval-based temporal reasoning with general TBoxes, in: B. Nebel (Ed.), Proceedings of IJCAI-01, Seattle, WA, Morgan Kaufmann, San Mateo, CA, pp. 89–94.

Mo Sunye, Lin Jiayi, et al. 2005. Expression of Uncertain Time and Tense Relation Based on Fuzzy Set. Computer Engineering 31(18):197–199.

Schockaert, S., De Cock, M., Kerrre, E.E., 2006. Imprecise temporal interval relations, Lecture Notes in Artificial Intelligence, vol. 3849, pp. 108–113.

Villafane, R., Hua, K., Tran, D., Maulik, B. 2000. Knowledge discovery from series of interval events, Journal of Intelligent Information Systems, vol. 15, no.1, pp. 71–89.

Wei Liu, Wenjie Xu, Dong Wang, Xujie Zhang, Zongtian Liu. 2012. An extending description logic for action formalism in event ontology. International Journal of Computational Science and Engineering.

Zhang, C., Cao, C., Sui, Y., Wu, X., 2011. A Chinese Time Ontology for Semantic Web. Knowledge-Based Systems vol. 24, no. 7, pp. 1057–1074, October.

Based on Support Vector Machine (SVM) of fatigue life prediction research

Y.F. Yang
Southwest Forestry University, Kunming, China

R.P. Xu
Kunming University of Science and Technology, Kunming, China

S. Chen & X.L. Zhao
Southwest Forestry University, Kunming, China

ABSTRACT: The Support Vector Machine (SVM) is applied to fatigue life's prediction, and with RBF neural network and the traditional linear regression methods are compared. Through contrasting and analyzing examples, the result shows that this method has high accuracy, good generalization, and the characteristics which are easy realized by engineering. It provides a new method for fatigue life's prediction.

1 INTRODUCTION

Under the action of alternating load, most materials of engineering structures, such as metal, plastic, wood, mixes clay and composite materials can produce fatigue damage. Fatigue damage is a main form which belongs to structural failure of damage, according to the U.S. relevant statistics, more than 80% of the structural damage associates with fatigue[1]. So the predictable method of fatigue life which has been researched is very meaningful.

The fatigue life of structure is effected by many factors, the traditional method is based on a lot of fatigue experiments, in order to find out the relationship between various influencing factors with fatigue life or fatigue strength. The error is much bigger in practical application, and counted complex. Support vector machine is a new learning method of machine[3] which was put forward by Vapnik etc. in the early 1990s. Support Vector Machine (SVM) is built on the statistical theory basis. It is designed to learn machine which uses principle of Minimize risk, empirical risk and complexity of model are comprehensive considered, over-learning, nonlinear, high dimension etc. can better be solved. Finally, it is attributed to solve a convex quadratic programming problem, it can obtain the global optimal solution, it avoids local problems of extremum which are about neural network and others methods, therefore, it has generalization ability with better, it has become one of the research focus in Machine learning field.

2 PRINCIPLE OF SUPPORT VECTOR MACHINE AND ITS ALGORITHM

2.1 *The core idea of support vector machine*

Support vector machine is based on the minimal principle of VC dimension theory and structure risk in the statistics, the model is established through the finite sample, and between its complexity and learning ability to find the best balance, to get the best ability of generalization. The characteristic vector of input space is mapped to high dimension though nonlinear mapping, make nonlinear into a linear problem in the high dimension space, complexity of point set computing can be avoided in the high space.

2.2 *Regression algorithm of Support Vector Machine (SVM) and its realization*

Give a data set $G= \{x_i, d_i\}$, $i = 1, 2, 3, ..., n$, x_i is input vector, d_i is expectations, n is the total number of data points. Fitting function of data is:

$$y = (\omega \bullet \phi(x)) + b \ \phi: R^n \to F, \ \phi \in F \qquad (1)$$

Among them, $\phi(x)$ is from the input space to high dimension, is a function column vector; ω is parameters listed vector, ω and b are estimated by next type (2):

$$R_{svm}(c) = c\frac{1}{n}\sum_{i=1}^{n} L(d_i, y_i) + \frac{1}{2}\|\omega\| \qquad (2)$$

In type (2), L(d_i, y_i) is insensitive loss function about ε, it is defined as

$$L(d,y)=\begin{cases} d-y|-\varepsilon, & |d-y|\geq\varepsilon, \\ 0, & |d-y|<\varepsilon. \end{cases} \quad (3)$$

It needs to introduce relaxation factorial, solving its minimum value:

$$R_{svm}(\omega,\xi)=\frac{1}{2}(\omega\bullet\omega)+C\sum_{i=1}^{n}\left|\xi_i+\xi_i^*\right|_{\varepsilon} \quad (4)$$

Constraint conditions are

$$\begin{aligned} &\omega\bullet\phi(x)+b-d_i\leq\varepsilon+\xi_i; \\ &d_i-\omega\bullet\phi(x)-b\leq\varepsilon-\xi_i; \\ &\xi_i\geq0, \xi_i^*\geq0 \end{aligned} \quad (5)$$

Finally, Lagrange equation is established:

$$\begin{aligned} L=&\frac{1}{2}(\omega\bullet\omega)+C\sum_{i=1}^{n}\left|\xi_i+\xi_i^*\right|_{\varepsilon} \\ &-\sum_{i=1}^{n}\alpha_i(\varepsilon+\xi_i-d_i+\omega\bullet x+b) \\ &-\sum_{i=1}^{n}\alpha_i^*(\varepsilon+\xi_i^*+d_i-\omega\bullet x-b) \\ &-\sum_{i=1}^{n}(\gamma_i\xi_i+\gamma_i^*\xi_i^*) \end{aligned} \quad (6)$$

Among them, α_i and α_i^* are Lagrange factors. The dual problem can be got by adopting the optimal methods

$$\begin{aligned} Q(\alpha_i,\alpha_i^*)=&-\varepsilon\sum_{i=1}^{n}(\alpha_i^*+\alpha_i)+\sum_{i=1}^{n}d_i(\alpha_i^*-\alpha_i) \\ &-\frac{1}{2}\sum_{i,j=1}^{n}(\alpha_i^*-\alpha_i)(\alpha_j^*-\alpha_j)K(x_i\bullet x_j) \end{aligned} \quad (7)$$

$$\sum_{i=1}^{n}(\alpha_i^*-\alpha_i)=0$$

Constraint conditions are

$$\begin{aligned} &0\leq\alpha_i^*\leq C, \quad i=1,2,...,n \\ &0\leq\alpha_i\leq C \end{aligned} \quad (8)$$

Through solving the maximum of (8), and the type parameter α_i and α_i^* are obtained, therefore, linear regression function is obtained:

$$f(x,\alpha_i,\alpha_i^*)=\sum_{i=1}^{n}(\alpha_i-\alpha_i^*)K(x_i,x_j)+b \quad (9)$$

Among them, α_i and α_i^* only a few are not 0, the corresponding sample is support vector, $K(x_i,x_i)=\varnothing(x_i)*\varnothing(x_j)$ is kernel function. Different kernel functions can create different support vector machines. Common kernel functions include: RBF kernel function, polynomial kernel function, sigmoid kernel function, linear function etc. RBF kernel function (10) is adopted to build a regression model of support vector machine.

$$K(x,x_i)=\exp\left(-\frac{|x-x_i|^2}{2\sigma^2}\right) \quad (10)$$

3 EXAMPLE

To introduce the validity of the method, for example Table 1, Test data of fatigue crack growth length of aluminum alloy is from reference [3] 2024-T42.

The data of Table 1 are analyzed, the cycle times (N) are as inputted variables, the fatigue crack length (a) is as output variable, a regression forecast model of support vector machine is established about single input and output. Former 16 groups are trained, surplus 5 groups are tested data, the model is tested.

3.1 Pretreatment of data

In order to avoid differences among the magnitude in various factors, eliminating effects which are caused by different dimensions and units in various factors, eliminating effects which are these factors, therefore, before model is established, each factor need normalize, numerical range is [−1,1].

$$X_i=\frac{2(x_i-x_{min})}{x_{max}-x_{min}}-1 \quad (11)$$

$$Y_i=\frac{2(y_i-y_{min})}{y_{max}-y_{min}}-1 \quad (12)$$

Among them, x_i stands for data of cycle load times (N) in Table 2, x_{min} and x_{max} separately stand

Table 1. Test data of fatigue crack growth length of 2024-T42 aluminum alloy [crack length/mm].

N	a	N	a	N	a
0	5.30	35000	9.35	60000	18.05
5000	5.60	40000	10.35	61000	19.45
10000	6.05	45000	11.50	62000	20.45
15000	6.55	50000	13.00	63000	21.50
20000	7.25	52500	13.85	64000	22.55
25000	7.85	55000	15.00	65000	24.00
30000	8.60	57500	16.75	66000	26.25

Table 2. Prediction results and the error of the fatigue crack length.

N	a	SVM	Error	Linear	Error	Bp	Error
62000	20.45	20.49	−0.002	21.34	−0.044	21.40	−0.047
63000	21.50	21.64	−0.007	21.88	−0.018	22.62	−0.052
64000	22.55	22.90	−0.018	22.42	0.006	24.19	−0.075
65000	24.00	24.26	−0.011	22.99	0.042	25.56	−0.065
66000	26.25	25.72	0.020	23.56	0.102	27.63	−0.053

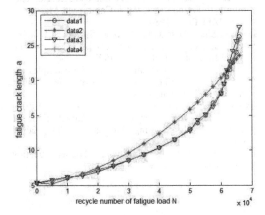

Figure 1. Forecast of the fatigue crack length compared with the measured ones.

for the maximum and minimum value of cycle load times; y_i stands for actual measurements (a) of fatigue crack length. y_{min} and y_{max} stand for the maximum and minimum value of actual measurements in the fatigue crack length (a) respectively; X_i and Y_i stand for the data of normalized. After prediction is finished, samples need reverse normalization, and it can be compared and analyzed with the actual samples.

3.2 Parametric selection of Support Vector Machine

Values of Punish coefficient c affects fitting accuracy and predictive ability of regression function, normally, when c increases, Fitting error and forecasting error will be drab decline, but when c increases to specific value, errors will decline very few, if c value is too big, it also can rise the training time. Insensitive coefficient reflects boundary of noise amplitude in data which is influenced by SVM, when data's variance is smaller, the smaller ε should be used. ε still controls the ability of generalization about model, when ε is too big, it will probably happen owe fitting, reduces the precision of result,

when ε is too small, it will reduce the ability of generalization. The article uses MATLAB, on the foundation of repeated experiments, chooses $c = 1000$, and $\varepsilon = 0.01$.

3.3 Forecasting model of linear regression

In order to compare the ability of prediction about the SVM regression and the traditional linear regression, the cycle load times are used as forecasting factor, fatigue crack length is used as forecasting quantity, confidence is 95%, the model of prediction about fatigue crack length is built. The equation of prediction is:

$$a = 4.59e^{1.64N} \tag{13}$$

a is fatigue crack length, N is recycle number of fatigue load.

In the equation (13), the parameters are calculated by least-square method, the related coefficient is R = 0.991, using significant test F = 6.64 > F0.05 = 6.39, therefore, the equation is right. Table 2 and Figure 1 separately give forecasting values and fittings effect of three models about fatigue crack length.

The inline equations (equations within a sentence) in the text will automatically be converted to the AMS notation standard.

3.4 Forecasting results are analyzed

The errors of SVM and BP neural network are smaller on the Figure 1, about fitting accuracy, methods of linear regression can only satisfy one part, the best way of sample complexity and degree of error is difficult to be found. Forecasting results aren't well.

SVM and BP neural network methods are compared, they are used to predict the fatigue crack length, maximum error is 2.0%, average error has only 1.16%, but maximum error of BP neural network is 7.5%, average error is 5.84%. Thus it can be seen, After degree of error in the sample is calculated, fitting error of SVM regression forecast is much smaller than BP.

4 CONCLUSIONS

Sample space is mapped to higher-dimensional space by SVM regression, using nonlinear mapping of kernel function, the nonlinear relation of original sample space is turned into the linear relation of higher-dimensional space, therefore, process of solution which is about explicit expression of nonlinear mapping can be avoided, the form of inner products sum are turned into the form of matrix multiplication by MATLAB. Numerical calculation is greatly simplified. In addition, according to sample information, the best way can be sought between complexity of model and learning ability. Adaptability is very good, the shortcomings of BP neural network, for example, the over-learning and convergence rate is slower and the local minimum and others are overcame, the fatigue crack length is respectively predicted by models of SVM regression and linear regression, the nonlinear relation of the fatigue crack length and cycle load times can be better simulated. The results of calculation show that SVM regression is better in the forecasting precision and adaptability than the traditional linear regression and the BP neural network forecasting methods.

REFERENCES

Gao Zhentong. Fatigue statistics [M]. Beijing: National defence industry press, 1986.

Kolmogorov, A.N Doklady Akademii Nauk [M]. SSR, 114 (1957), 953–9561.

Rumelhart, D.E and McClelland, J.L Parallel distributed processing: exploration in the microstructure of cognition [J]. Foundations, Vol.1, Cambridge, Massachusetts, MIT Press, 1986.

Shaowen Shao, Yoshisada Murotsu. Structural Reliability Analysis using a neural network [J]. JSME International Journal. Vol. 40, No. 3, 1997.

Shi Feng, Wang Xiaochuan, Yu Lei. 30 examples are given in the MATLAB neural network [M] Beijing: Aerospace university press, 2010.

Tian Xiuyun, Du Hongzeng, Sun Zhiqiang. Curve Fitting of The Fatigue Crack Growth Characterristic of Metals [J]. Engineering Mechanics: 2003, 20(4): 136~140.

Xu Hao. Fatigue strength [M]. Beijing: Higher education press, 1998.

Session 5: Computer security

Control Engineering and Information Systems – Liu (Ed)
© 2015 Taylor & Francis Group, London, ISBN 978-1-138-02685-8

Design and analysis of the key stream generator based on the KASUMI

Y.Q. Deng, J.Y. Li, H. Shi, J. Gong, Z. Sun & W. Ding
College of Communications Engineering, PLAUST, Nanjing, Jiangsu, China

ABSTRACT: The keystream generator is the most important unit in the stream cipher algorithm, because its properties will determine the security of the cipher. This paper designs a keystream generator based on the KASUMI algorithm, and analyzes the randomness of its keystream. The result proves that the keystream generator has very good security.

1 INTRODUCTION

The keystream generator is the most important unit in the stream cipher algorithm, because its properties will determine the security of the cipher algorithm.

At present, there are four methods to design the keystream generator. They are as follows:

- Design based on the Nonlinear Feedback Shift Register (NLFSR). It uses the shift register as the core, with nonlinear feedback such as combined generator, filtering generator and clock generator, etc.
- Design based on the state table. It uses the state table to construct the stream cipher, such as the PY, the HC-256 and the PolarBear algorithms, etc (Biryukov 2005).
- Design based on the block cipher. This method is not only very convenient, but also very secure, such as the Mir-1, the Phelix and the LEX algorithms (Biryukov 2005, Hastad & Naslund 2005).
- Design based on the other new theories. It uses artificial intelligence, neural network and chaos theory to design the keystream generator, such as the TRBDK3 YAEA algorithm.

This thesis uses the KASUMI algorithm (3rd Generation Partnership Project 2011) as a core algorithm to design the keystream generator, the block diagram of the KASUMI algorithm is shown in Figure 1.

2 DESIGN OF THE KEYSTREAM GENERATOR

In the block cipher, the CTR, OFB and CFB operation modes can be only used to generate the keystream (Stallings 2011, Deng 2011). Because of the

error diffusion in the CFB mode, it will deteriorate the quality of communications.

This paper comprehensively uses the CTR, OFB modes of the KASUMI algorithm, in order to improve the security of the keystream generator as soon as possible. The keystream generator based on the hybrid modes of the KASUMI algorithm is shown in Figure 2, the IV is the initial vector of the 64 bits counter CTR and the 64 bits shift register SREG, and R is a 64 bits random number used for randomization of the CTR number, and K is 128 bits seed key.

In Figure 2, the feedback data to the SREG can choose different length according to different applications, such as 8 bits or 16 bits, even 64 bits if necessary.

3 RANDOMNESS TEST OF THE KEYSTREAM

3.1 The methods of randomness tests

The security of the stream cipher depends largely on the randomness of the keystream. In general, the better of the randomness of the keystream, the stronger of the security of the stream cipher. Therefore, the randomness of the keystream must be tested in the stream cipher.

There are many methods to test the local randomness of the keystream (Murphy 2000, Juan, Andrew 2010). Here only gives the most common five kinds of test methods, and α represents the confidence level.

3.1.1 Frequency test

The purpose of the frequency test is to detect the balance of 0's and 1's numbers of the keystream sequence.

Assuming the size of the tested sequence is n, and the 0's number and 1's number of the tested

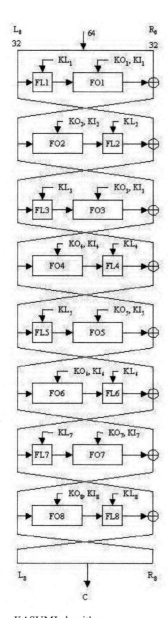

Figure 1.   KASUMI algorithm.

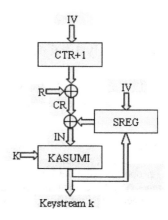

Figure 2.   Keystream generator algorithm.

3.1.2   *Transition probability test*

The transition probability test is used to examine whether a keystream sequence has the appropriate transition probability.

Suppose the number of occurrences of combination 00, 01, 10 and 11 of a tested sequence of length n are n_{00}, n_{01}, n_{10} and n_{11} respectively, and the 0's number and 1's number of the tested sequence are n_0 and n_1 respectively. According to the formula

$$x^2 = \frac{4}{n-1}(n_{00}^2 + n_{01}^2 + n_{10}^2 + n_{11}^2) - \frac{2}{n}(n_0^2 + n_1^2) \qquad (2)$$

We compute the value of the x^2 and compare it with $x_\alpha^2(2)$. When $x^2 < x_\alpha^2(2)$, it means this sequence passes the transition probability test.

3.1.3   *Poker test*

The poker test is used to examine whether the frequency of various combinations of the tested n-bit sequence is approximately equal when divided in m-bit size.

Assuming that the number of occurrences of i in each group is n_i. According to the formula

$$x^2 = \frac{2^m}{F} \sum_{i=0}^{2^m-1} n_i^2 - F \qquad (3)$$

where $F = \sum_{i=0}^{2^m-1} n_i = \lfloor m \mid n \rfloor$, we compute the value of the x^2 and compare it with $x_\alpha^2(255)$. When $x^2 < x_\alpha^2(255)$, it means this sequence passes the Poker test.

3.1.4   *Autorelevantion test*

The autocorrelation test is used to examine the correlation property between a n-bit sequence and its shifted sequence.

sequence are n_0 and n_1 respectively. According to the formula

$$x^2 = \frac{(n_1 - n_0)^2}{n} \qquad (1)$$

We compute the value of the x^2 and compare it with $x_\alpha^2(1)$. When $x^2 < x_\alpha^2(1)$, it means this sequence passes the frequency test.

Assumed that the sequence S is shifted by 1 bit, and the number of bit different from original sequence is $c(1)$. According to the formula

$$C = 2(c(1) - \frac{n-1}{2})/\sqrt{n-1} \qquad (4)$$

We compute the value of the C and compare it with the bilateral quintile x_α in $\phi(x)$-table. When $|C| < x_\alpha$, it means this sequence passes the auto-correlation test.

3.1.5 Runs test

The runs test is used to examine whether the 0's and 1's distribution of a n-bit sequence is reasonable.

Assuming the numbers of i-bit 0-run and i-bit 1-run are Z_i and L_i, and their mathematical expectation is e_i. According to the formula

$$x^2 = \sum_{i=0}^{k} \frac{(Z_i - e_i)^2}{e_i} + \sum_{i=0}^{k} \frac{(L_i - e_i)^2}{e_i} \qquad (5)$$

Here k is the largest integer of i that meets $e_i \geq 5$, we compute the value of the x^2 and compare it with $x_\alpha^2(\nu)$ ($\nu = 2$ if 128 bits, 6 if 640 bits). When $x^2 < x_\alpha^2(\nu)$, it means this sequence passes the runs test.

3.2 Data of randomness tests

Selecting

IV = 0
R = 34222bc8f7c39416
K = 2bd6459f82c5b300952c49104881ff48

Based on Figure 2, the corresponding 640 bits keystream sequence are as follows:

af24 cc029ac39d08
23dd1041aeecae7b
d95cdad24bc7162f
3f9faa1c80d1db1b
87782a2c1dc93006
e49bac44f71b868c
a5398989e10adfb3
e07fea9c2c20914a
0f437466f0c8a81d
1bf4536e2d9900c4

When testing the randomness of the keystream, we select confidence level $\alpha = 0.01$, and divide the above 640 bits keystream into groups by 128-bit block. The test data is shown in Table 1.

The abbreviations in Table 1 are as follows:

DF: Degrees of freedom
THD: Threshold
Freq: Frequency test

Table 1. Test data.

Test	DF	THD	B1	B2	B3	B4	B5
Freq	1	6.635	0.5	0.781	1.125	0.281	1.125
Trpr	2	9.210	0.846	0.502	1.796	0.498	1.166
Pok	255	310.5	272	240	240	272	240
Autc	1	2.575	0.177	0.177	0.710	0.355	0.532
Runs	2	9.210	0.188	1.438	1.563	2.813	2.438
Con			P	P	P	P	P

Test	DF	THD	640b		B6	B7	256b
Freq	1	6.635	1.225		0.281	0.281	0
Trpr	2	9.210	2.210		0.498	0.372	0.012
Pok	255	310.5	220.8		240	240	224
Autc	1	2.575	0.875		0.355	0.355	0
Runs	6	16.81	3.001		2.375	7.75	3.313
Con			P		P	P	P

Trpr: Transition probability test
Pok: Poker test
Autc: Autocorrelation test
Runs: Runs test
Con: Conclusion
P: Pass
B1~B5: 5 groups, 128-bit every group (above 640 bits)
B6~B7: 2 groups, 128-bit every group (other 256 bits), shown as follows:
B6: 86067e967716d478 f505df4423a90714
B7: d9b372769f8dde65 a16de0d38a2d6310

The B6 and B7 are continuous 256 bits key-stream. They are used to further confirm the key-stream's randomness.

From Table 1, we can get the following conclusions:

- Both 128-bit group and 640-bit as a whole, they all pass the keystream randomness tests.
- Another continuous 256 bits keystream, which is an arbitrary choice, also passes all randomness tests.
- From what has been discussed above, the key-stream generator shown in Figure 2 has very good security.

4 CONCLUSION

The keystream generator is the most important unit in the stream cipher algorithm, because its properties will determine the security of the cipher algorithm.

This paper comprehensively uses the CTR, OFB modes of the KASUMI algorithm and designs a hybrid Keystream generator, and analyzes the randomness of its keystream. The analysis results prove that the keystream generator shown in Figure 2 has very good security, because its keystream passes all randomness tests.

ACKNOWLEDGMENT

This paper is supported by the science and technology innovation fund of PLAUST.

REFERENCES

3rd Generation Partnership Project. Specification of the 3GPP Confidentiality and Integrity Algorithms, Document 2: KASUMI Specification, 3GPP TS 35.202, Release 10.0.0, 2011.

Andrew, R. A Statistical Test Suite for Random and Pseudorandom Number Generators for Cryptographic Applications [DB/OL]. (2010-11-06). http://www.nist.gov.

Biryukov, A. A New 128 bit Key Stream Cipher LEX. ECRYPT Stream Cipher Project Report 2005. Available at http://www.ecrypt.eu.org/stream.

Hastad, J. and M. Naslund. The Stream Cipher Polar Bear. ESTREAM, ECRYPT Stream Cipher Project, Report 2005/02 (2005). http://www.ecrypt.eu.org/strearm/.

Juan, S. Statistical Test of Random Number Generators [DB/OL]. http://www.nist.gov.

Murphy, S. The Power of NIST's Statistical Testing of AES Candidates. (2000-03-07). http://citeseery,ist,psu.edu/viewdoc/ summary?doi=10.1.1. 42.8668.

William Stallings. "Cryptography and Network Security Principles and Practice" (5th Edition), Publishing House of Electronics Industry, 2011.

Yuanqing Deng, Jing Gong, Hui Shi, "Concise Course on Cryptography", Publishing House of QingHua University, 2011 (In Chinese).

Research and analysis of the cloud era various network authentication and accounting technology

J.J. Zhou, K. Yu & J.F. Liao
Huazhong University of Science and Technology Wenhua College, Wuhan, China

ABSTRACT: In this paper, through research and analysis of the cloud era various network authentication and accounting technology such as PPPoE + Radius, DHCP + WEB + Radius, 802.1X + Radius, Kerberos and SSO, we deems that Kerberos as a new mode is suitable for the application of the cloud network system identity Authentication; in many cases SSO can access all the mutually trusted application system by a single login; in modern broadband Ethernet access, optimized IEEE 802.1x protocol + Radius server authentication mode is recommended, which can solve the bottleneck problem of the traditional authentication methods such as PPPoE and Web/Portal.

1 PPPoE + RADIUS MODE

1.1 Briefing and principles

Came out in late 1998, Point to Point Protocol over Ethernet (PPP over Ethernet, PPPoE) technology was developed jointly based on IETF RFC by Redback Networks company, client software developer RouterWare company and the subsidiary of Worldcom—UUNET Technologies company. Its main purpose is combining the most economical LAN technology, extendibility and management control functions of Ethernet and Point to Point Protocol together. It allows service providers to provide more user-friendly support in multi-user broadband access services through digital subscriber line, cable modem or wireless connections, etc.

The establishment of PPPoE requires two stages, namely the Discovery stage and PPP Session stage. When a host wants to initiate a PPPoE session, it must first complete the Discovery stage to determine the terminal Ethernet MAC address, and establish a PPPoE session number (Session-ID). At the time of PPP protocol defining a relationship between two terminal devices, the Discovery stage is a client—server relationship. In the process of the Discovery Stage, the host (client) searches and finds a network device (server). In the network topology, the host can communicate with more than one network devices. In the Discovery stage, the host can find all the network devices but can only choose one. At the successful completion of the Discovery stage, the host and network equipment will have all the information to establish the PPPoE. Discovery stage will exist until the PPPoE session is built. Once the PPP session is established, host and network devices must provide resources for the virtual interface of the Session stage.

1.2 The advantages and disadvantages of PPPoE

Advantages:

- An extension of traditional PSTN narrowband access technique on the Ethernet access technology.
- Consistent with the original narrowband network access authentication system.
- End users are relatively easy to accept.

Disadvantages:

- PPP protocol has essential difference with Ethernet technique. PPP protocol needs to be encapsulated into Ethernet frame again so the efficiency is very low.
- PPPoE generates a lot of broadcast traffic in Discovery stage which have great impact on network performance.
- The multicast business has many difficulties, while video business is mostly based on multicast.
- Require operators to provide the client terminal software; the maintenance workload is excessive.

Fund: Hubei Provincial Colleges and Universities outstanding young scientific and technological team plans funded project (No. T201431); Hubei Province Natural Science Foundation General project (No. 2013CFC113); Huazhong University of Science and Technology Wenhua College Youth Foundation project (No. J0200540119); Hubei provincial Colleges and Universities 2013 Innovation and Entrepreneurship Training project (No. 201313262005).

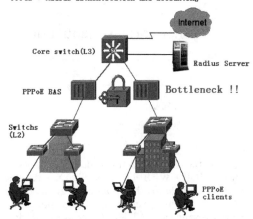

PPPoE + Radius authentication and accounting

Figure 1. The PPPoE + Radius authentication and accounting mode.

PPPoE authentication generally needs external BAS (Broadband Access Server, the core equipment certification system). After the authentication the business data flow must also get through the BAS equipment, which is easy to cause the single point bottleneck and fault, and usually the equipment is very expensive.

2 DHCP + WEB + RADIUS MODE

2.1 Briefing and principles

Web authentication, which has already been the authentication mode of operator network platform, implements authentication through the Web page by judging whether users are granted with the right to access.

The main process of Web authentication is listed below.

When the user computer first starts, system program makes the DHCP-relay by BAS according to the configuration, and then asks the DHCP Server for IP address;

BAS structures table information (based on the port number and IP) for the corresponding user and adds user ACL (Access Control List) service strategy (so that users can only access the portal server and some internal servers, or individual external servers such as DNS);

The portal server provides the authentication page for users, in this page the user inputs account and password and clicks the button of login, or directly clicks the button without inputting. The login button starts a program on the portal server, which sends user information (IP address, account and password) to network center equipment BAS;

BAS uses IP address to get user layer 2 address and physical ports (such as VLAN ID, ADSL, PVC ID, or PPP session). By such information BAS checks the legitimacy of the user. If the user inputs the account, it is considered as card user and authenticated by radius server with the account and password. If the user does not input the account, it is regarded as fixed user, and then the network equipment looks up the user table for user's account and password by VLAN ID or PVC ID before it is authenticated by Radius Server;

Radius Server returns the authentication result to the BAS;

After authentication, BAS modifies the user's ACL; Then users can access the Internet or external specific network services;

Before a user leaves the network, it connects to the portal server and clicks the button of logout to stop charging system, which deletes a user's ACL and forwarding information and restricts user's access to the external network.

In this process, attention must be paid to cases of users leaving network abnormally, such as user host crash, network dropping, or direct power off.

2.2 The advantages and disadvantages of DHCP + Web

Advantages:

• Do not need special client software. Reduce maintenance workload.
• Provide business authentications such as portal.

Disadvantages:

• Web hosts in the layer 7 protocol, which needs higher devices and more cost of network construction.

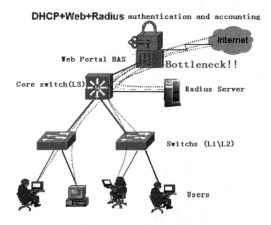

Figure 2. DHCP + Web + Radius authentication and accounting mode.

- Worse standardization and openness. Use private communication protocol between the authentication equipment and Web.
- Poor user connectivity. Not easy to detect whether the user is offline. Difficult to achieve time-based accounting.
- Ease of use is not enough. When the user accesses the network, not matter by Telnet, FTP, or other business, it must use a browser for Web authentication.
- The allocation of IP address is previous to user authentication. If the user is not the dial-up user, it causes address waste. Also, it is not convenient for multi-ISP support.
- Business flow and control flow can not be distinguished before and after authentication, so a lot of CPU resources are consumed.

3 802.1X + RADIUS MODE

3.1 What is IEEE 802.1X protocol

IEEE 802.1X is a port-based network access control protocol.

The architecture of IEEE 802.1X protocol includes three important parts: Supplicant System, Authenticator System and Authentication Server System.

3.2 The IEEE 802.1X protocol technical characteristics

3.2.1 The implementation simpleness

IEEE 802.1X protocol is a layer 2 protocol and does not concern layer 3, so it is less demanding on the overall performance of the device, resulting in effectively decrease of network construction cost.

3.2.2 The authentication and business separation

IEEE 802.1X authentication architecture adopts "controlled port" and "uncontrollable port" logic

functions, which achieves the separation of business and authentication. After the user is authenticated, business flow and authentication flow are detached. There are no special requirements on the following packet processing. The business can be very flexible. Especially, the method has great advantage in carrying out the business of broadband multicast, and no business is limited by the authentication mode.

3.2.3 The comparison with other modes

IEEE 802.1X protocol is derived from IEEE 802.11 wireless Ethernet (EAPOW). However, its introduction in the Ethernet solved the problems brought by traditional PPPoE and Web/Portal authentication, eliminated network bottlenecks, reduced encapsulation overhead of the network, and abated the cost of network construction.

3.3 Advantages of IEEE 802.1X

Concision and efficiency: pure Ethernet technology kernel; remain connectionless characteristic of IP network; remove redundancy expensive multiservice gateway device; eliminate network authentication accounting bottleneck and single point of failure; easy to support multi-service.

Ease of implementation: it can be implemented on ordinary L2, L3, IP, DSLAM; the network comprehensive cost is low.

Safety and reliability: achieve user authentication on layer 2 network; combined with MAC, port, account and password, etc., the binding technology is highly secure.

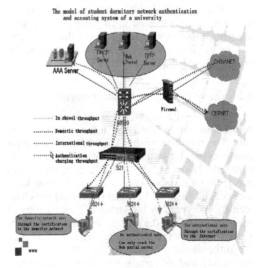

Figure 4. 802.1X + Radius security authentication and accounting system mode.

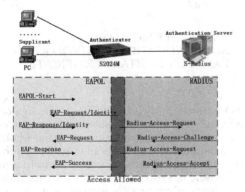

Figure 3. 802.1X + Radius authentication mode.

Industry standards: IEEE standard, built-in support by Microsoft operating system.

Ease of operation: complete separation of control flow and business flow; easy to implement multi-service operation; can be upgraded into a carrier-grade network by small amount of transformation from the single fee system such as the traditional monthly subscription network.

3.4 802.1X + Radius security authentication and accounting system solutions

According to user needs, the SAM system of RG company can be used to build a student dormitory network security authentication and accounting system. The core technology uses 802.1x + Radius constitution; the whole authentication and accounting system consists of several parts of the safety switches (such as S2126G), S-Radius server, DHCP server and portal server. The specific implementation mode is shown below.

4 KERBEROS NETWORK AUTHENTICATION

4.1 What is kerberos protocol

Kerberos is a network authentication protocol. Its design objective is to provide strong authentication for client/server applications through the key system. Implementation of the certification process does not rely on the authentication of the host operating system, needs not trust based on host address, does not require the physical security of all the hosts on the network, and assumes that the data packets transmitted on the network can be freely read, modified and inserted with data. In the above cases, as a trusted third party the Kerberos authentication service performs certification services by traditional cryptographic techniques, for example, shared key.

4.2 The kerberos authentication process

First of all, the client sends a request to the Authentication Server (AS) for a server certificate. The response of the AS contains the certificate encrypted by the client key. The certificate contains: 1, the server "ticket"; 2, a temporary encryption key (also known as the session key). The client transmits the ticket (including the client identity encrypted by the server key, and a copy of the session key) to the server. Session key (now be shared between the client and the server) is used for authentication of the client or the authentication server, and also can be used to provide encryption services for the subsequent communication, or by exchanging the independent sub-session key for

the two communicating parties to provide further communication encryption services.

4.3 Two parts of the kerberos protocol

1. The client sends its own identity information to key distribution center KDC. KDC gets TGT (Ticket-Granting Ticket) from the Ticket Granting Service, and replies the TGT encrypted by key between Client and KDC before start of the protocol.

   At this time only true client can use the encryption key between itself and KDC to decrypt the encrypted TGT. This process avoids unsafe way: client sending the password directly to the KDC.
2. The client uses the TGT obtained before to request the ticket of other service from KDC.

4.4 The defects of kerberos

1. Single point of failure: it requires a sustained response to the central server. Before the end of the Kerberos service, no one can connect to the server. This defect can be compensated by using composite Kerberos server and the defect certification mechanism.
2. Kerberos request hosts participating in communication to be clock-synchronized. The ticket is of timeliness; therefore, if the host is not synchronical with the Kerberos server, authentication will fail. The default configuration requires that clock time do not differ by more than 10 minutes. In practice, usually the network time protocol daemon is used to keep the hosts synchronization.
3. Management protocol is not standardized; there are some differences in the server implementation tools.
4. All user keys are stored on a central server; acts that endanger the security of the server will endanger the entire keys.
5. A dangerous client will hazard the user password.

5 THE SINGLE SIGN-ON AUTHENTICATION TECHNOLOGY BASED ON DIGITAL CERTIFICATE

5.1 SSO profile

SSO is based on digital certificate single sign-on technique, assembles the information resources and the protection system station as organic whole, communicates with authentication server of protection system by installing access control agent middleware in various information resource ends, and shares advantages by security bulwark and information services provided by the system.

5.2 SSO principle

Its principles are as follows:

1. Configure an access agent for each information resource, and allocate different digital certificate to different agent, which used to guarantee the secure communication between itself and system service.
2. When the user login the center, confirm the identity of the user according to the digital certificate provided by the user.
3. When accesses a specific information resource, system service sends encrypted user identity

information by the agent-corresponding digital certificate in the form of digital envelope to the related information resource server.

4. After the information resource server received the digital envelope, it performs decryption and validation via access agent to get the user identity, on which it authenticates the internal permission.

5.3 SSO application advantages

1. Single sign-on: By a simply log in once, the use can access multiple applications of the background through a single sign on system, and the second landing needs not re-enter the user name and the password.
2. C/S single sign on solution: does not need to modify any existing application system service and client to implement the C/S single sign on system.
3. Self installation: by simple configuration, it can be utilized without modifying any of the existing B/S or C/S application systems.
4. Application flexibility: Embedded dynamic domain name system (for example: gnHost) can be implemented independently, or combined with other products.
5. Role-based access control: access control functions are based on the user's role and URL.
6. A comprehensive audit of the log: accurately record a log of the user query; statistics and analysis on the log are according to the dates, addresses, users, resources and other information
7. Cluster: Via clustering capabilities, fulfill dynamic load balancing between multiple servers.

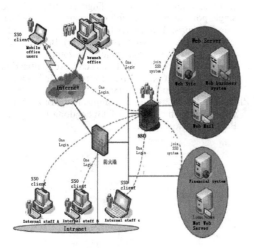

Figure 5. The SSO authentication technology system interaction.

Table 1. Technical comparison of the various authentication modes.

Modes	PPPoE	WEB Portal	802.1X	Kerberos	SSO
Standard level	RFC 2516	Private	IEEE standard	RFC1510/4120	No
Encapsulation overhead	Relatively hign	Low	No	High	Relatively hign
Access control	User	VLAN-MAC-IP	User	Centralized user/ system default	User/system control/agent
IP address allocation	After authentication	Before authentication	After authentication	May not allocate	Not need to allocate
Multicast support	Poor	Good	Good	Support	Support
VLAN demand	No	Much	No	No	No
Client software	Need	Browser	Need	OS build-in	Need/browser
Device support	Public protocol	Manufacturer private	Public protocol	Public protocol	Public and compatible
User connectivity	Good	Poor	Good	Relatively good	Good/with DB ticket user can shares resources
Device performance requirements	Relatively high (BAS)	High (whole course VLAN)	Low	High	Relatively high (multi-system)

8. Transmission encryption: Support for a variety of symmetric and asymmetric encryption algorithms; prevent user information from theft and tampering during transmission.
9. Scalability: good compatibility with subsequent expansion and extension of the business system.

6 CONCLUSION

We know that in the modern network system, usually a variety of components such as user terminals, NAS, Radius Server and database are used to build authentication and accounting system, which aims to achieve network security access control, billing, operations management, and provide a AAA (Authentication, Authorization, and Accounting) network security services framework. In many cases, the AAA needs to manage the security protocols, for instance, RADIUS, TACACS, and Kerberos, etc. This paper researches the secure authentication and accounting technologies in modern network, analyzes protocols comparatively, including PPPoE, WEB Portal, Radius, 802.1X, Kerberos and SSO, and summarizes their features as followed in Table 1, for better service applied to network system of the cloud era.

REFERENCES

Bian, Y. "QinQ solve PPPoE broadcast before authentication," Computer and Information Technology, 2008.

Bi, Q. "PPPOE Broadband Access Technonogy and Ordinary Failure Analyze, " China Cable Television, No. 8, 2004.

Mishra, A. and W.A. Arbaugh, "An initial security analysis of the IEEE 802.1X Standard," http://citeseer.nj.nec.com/, 2002.

Xiao, G., Z. Zhang and F. Wang, Digital certificate based single-point sign-on technology research and application Enterprise Technology Development, 2009.

Xie, X. Computer Network (fifth edition). Beijing: publishing house of electronics industry, 2008.

Zhang, L. "Discussion about Kerberos Authentication and its Application in Middleware Tuxedo," Science Mosaic, No. 3, pp. 2, 99–100, 2008.

Zhang, L. The 802.1X-based campus network authentication Academic Journal of Guangdong College of Pharmacy, No. 6, 2004.

Control Engineering and Information Systems – Liu (Ed)
© 2015 Taylor & Francis Group, London, ISBN 978-1-138-02685-8

The randomness test of the key stream based on the counter mode of AES

H. Shi, X.C. Zhang, J. Gong, Y.Q. Deng & Y. Guan
College of Communications Engineering, PLAUST, Nanjing, China

ABSTRACT: The encryption of the stream cipher is simple, so the safety of which mainly depends on the key stream generator. It is an important direction in current cryptography field to construct the stream cipher and improve its safety with the complicated block cipher algorithm. The paper studies the key stream generator based on the counter mode of AES algorithm. The randomness of the key stream generated is tested using the general test software. The paper puts forward a new key stream generator thought which is a kind of both CTR (Counter) + OFB (Output Feedback) model.

1 INTRODUCTION

The safety of the stream cipher mainly depends on the randomness of the key stream sequence or the characteristics of the key stream generator. For the key stream sequence, the more close to the true random sequence, the algorithm is more secure, so the key to design the stream cipher is to design a high performance key stream generator. AES is the most famous block cipher algorithm in the 21st century (Stalling 2006, Hu & Wei 2005, Courtois & Pieprzyk 2002, Zhang & Jiang & Xiao & Huang 2009). Its biggest advantages are high security performance, strong ability of anti-attacking, high encryption and decryption speed, easy to implement by software and hardware. The counter mode of AES algorithm can be used to generate the key stream of the stream cipher (Good & Benaissa 2006).

2 THE KEY STREAM GENERATOR BASED ON THE COUNTER MODE OF AES

2.1 The AES algorithm

AES is a symmetric block cipher with a block length of 128 bits and supports for key lengths of 128, 192, or 256 bits (Stalling 2006). The Figure 1 shows the overall structure of AES (Hu & Wei 2005). The input to the encryption and decryption algorithms is a single 128-bit block. AES does not use a Feistel structure but process the entire data block in parallel during each round using substitutions and permutation.

2.2 The local randomness test of the key stream

For the security consideration of the stream cipher (Chen & Hnricksen & Millan 2005/2006, Good

& Benaissa 2006, Andrew 2010), it requires that the local of the key stream should be good at randomness. Therefore, it is need to test the local randomness of key stream. Although there are many methods can be used to test the local randomness of key stream, just five test methods most commonly used are given, respectively, as follows.

2.2.1 Frequency test
To examine the equilibrium of the 0 s and 1 s of the key stream Sequence.

$$x^2 = \frac{(n_1 - n_0)^2}{n} \qquad (1)$$

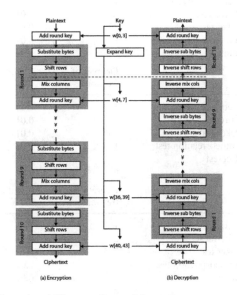

Figure 1. AES encryption and decryption.

2.2.2 Paired test

To examine whether a key stream sequence with the appropriate transition probability or not, namely to examine whether the numbers of occurrence of the different combinations between two adjacent elements are equal.

$$x^2 = \frac{4}{n-1}(n_{00}^2 + n_{01}^2 + n_{10}^2 + n_{11}^2) - \frac{2}{n}(n_0^2 + n_1^2). \quad (2)$$

2.2.3 Poker test

To examine the number of occurrence of every combination, which is an n-bit key stream grouped by m-bit unit, is equal.

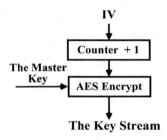

Figure 2. The key stream generator based on the CTR mode of AES.

$$x^2 = \frac{2^m}{F} \sum_{i=0}^{2^m-1} n_i^2 - F \quad (3)$$

2.2.4 Autocorrelation test

To examine the correlation between the n-bit of the key stream sequence after shift change and the original sequence.

$$C = 2\left(c(d) - \frac{n-d}{2}\right)\Big/\sqrt{n-d} \quad (4)$$

2.2.5 Run test

To examine the distribution of 0 s and 1 s of the n-bit of the key stream sequence is reasonable.

$$x^2 = \sum_{i=0}^{k} \frac{(B_i - e_i)^2}{e_i} + \sum_{i=0}^{k} \frac{(G_i - e_i)^2}{e_i} \quad (5)$$

2.3 Test method and data

Figure 2 depicts the key stream generator based on the Counter Mode of AES. The Counter is initially set to some Initialization Vector (IV) and then incremented by 1, which as input to AES for encryption with the Master Key (K) to generate the key Stream.

In the Figure 2, the IV of the counter and the master key are as follows:

Table 1. The randomness test data of the key stream sequence.

	Degrees of freedom	Significance level	1st 128-bit	2nd 128-bit	The threshold value
Frequency test	1	0.01	**7.03125**	1.53125	6.6349
Paired test	2	0.01	7.54355	2.01993	9.2103
Poker test	255	0.01	272	304	310.4574
Autocorrelation test	1	0.01	1.2423	0.177471	2.575
Run test	2	0.01	**12.375**	4.125	9.2103
	Degrees of freedom	Significance level	3rd 128-bit	4th 128-bit	The threshold value
Frequency test	1	0.01	0.03125	0.125	6.6349
Paired test	2	0.01	0.496309	1.47373	9.2103
Poker test	255	0.01	272	240	310.4574
Autocorrelation test	1	0.01	0.532314	1.06483	2.575
Run test	2	0.01	5.4375	5.6875	9.2103
	6	0.01			
	Degrees of freedom	Significance level	5th 128-bit	Total 640-bit	The threshold value
Frequency test	1	0.01	0.5	1.40625	6.6349
Paired test	2	0.01	0.468504	1.42787	9.2103
Poker test	255	0.01	272	259.2	310.4574
Autocorrelation test	1	0.01	0.532414	0.158238	2.575
Run test	2	0.01	3.4375		9.2103
	6	0.01			16.8119

The IV of CTR:
c143f6a8885a328d313198a2e03707d4
The master key:
6cef1516 28aed2a6abf7158809cf4f7e
The test data in the Table 1 illustrates that 1st data does not pass the Frequency Test and Run Test.

The Key stream generator, shown in the Figure 2, generates 5 groups of 128-bit the key stream sequence, respectively, as follows (the counter incremented by 1 every group):

2f 28 b0 85 01 12 46 54 c9 9e f0 22 42 a3 49 54
d7 68 98 f7 83 50 6b 11 68 55 26 36 31 30 1f 11
0f c6 33 39 c8 35 3b 6e b9 1a e2 f8 65 b6 e2 30
eb 64 39 92 9e 72 1d d3 ae 7a f9 87 c9 80 44 3e
b9 9a 98 5b 98 f1 b2 78 95 a6 27 ad f5 12 63 9f

The randomness test data of the above 5 groups and the total 640-bit key stream sequence are shown in the Table 1.

3 THE ALTERED KEY STREAM GENERATOR BASED ON THE COUNTER MODE OF AES

In the Figure 3, 128-bit random number is XORed with the key stream sequence generated by the key stream generator shown in the Figure 2, and the results are used to be the key stream.

Here, the 128-bit random number is:
30 a2 5d 6a 5c 95 dd e2 39 07 58 b1 50 ff 70 38
In the Figure 3, the IV of the counter and the master key are as follows:
The IV of counter:
c143f6a8885a328d313198a2e03707d4

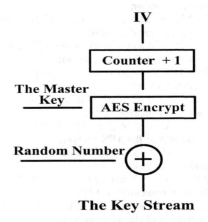

Figure 3. The altered key stream generator based on the CTR Mode of AES.

Table 2. The randomness test data of the key stream sequence.

	Degrees of freedom	Significance level	1st 128-bit	2nd 128-bit	The threshold value
Frequency test	1	0.01	0.78124	3.78125	6.6349
Paired test	2	0.01	1.13214	3.17151	9.2103
Poker test	255	0.01	240	272	310.4574
Autocorrelation test	1	0.01	0.177471	0.177471	2.575
Run test	2	0.01	1.6875	0.625	9.2103
	Degrees of freedom	Significance level	3rd 128-bit	4th 128-bit	The threshold value
Frequency test	1	0.01	2.53125	0.125	6.6349
Paired test	2	0.01	2.21678	2.41831	9.2103
Poker test	255	0.01	272	240	310.4574
Autocorrelation test	1	0.01	0.532414	1.41977	2.575
Run test	2	0.01	**10.125**	8.265	9.2103
	6	0.01			16.8119
	Degrees of freedom	Significance level	5th 128-bit	Total 640-bit	The threshold value
Frequency test	1	0.01	0	0.50625	6.6349
Paired test	2	0.01	0.212598	0.512529	9.2103
Poker test	255	0.01	240	278.4	310.4574
Autocorrelation test	1	0.01	0.532414	0	2.575
Run test	2	0.01	1.75		9.2103
	6	0.01		6.77361	16.8119

The master key:
6cef1516 28aed2a6abf7158809cf4f7e

The altered key stream generator, shown in the Figure 3, generates 5 groups of 128-bit the key stream sequence, respectively, as follows (the counter incremented by 1 every group):

The outputs of the AES encrypt:
2f 28 b0 85 01 12 46 54 c9 9e f0 22 42 a3 49 54
Random Number:
30 a2 5d 6a 5c 95 dd e2 39 07 58 b1 50 ff 70 38
The key stream after XOR:
1f 8a ed ef 5d 87 9b b6 f0 99 a8 93 12 5c 39 6c
The outputs of the AES encrypt:
d7 68 98 f7 83 50 6b 11 68 55 26 36 31 30 1f 11
Random Number:
30 a2 5d 6a 5c 95 dd e2 39 07 58 b1 50 ff 70 38
The key stream after XOR:
e7 ca c5 9d df c5 b6 f3 51 52 7e 87 61 cf 6f 29
The outputs of the AES encrypt:
0f c6 33 39 c8 35 3b 6e b9 1a e2 f8 65 b6 e2 30
Random Number:
30 a2 5d 6a 5c 95 dd e2 39 07 58 b1 50 ff 70 38
The key stream after XOR:
3f 64 6e 53 94 a0 e6 8c 80 1d ba 49 35 49 92 08
The outputs of the AES encrypt:
eb 64 39 92 9e 72 1d d3 ae 7a f9 87 c9 80 44 3e
Random Number:
30 a2 5d 6a 5c 95 dd e2 39 07 58 b1 50 ff 70 38
The key stream after XOR:
db c6 64 f8 c2 e7 c0 31 97 7d a1 36 99 7f 34 06
The outputs of the AES encrypt:
b9 9a 98 5b 98 f1 b2 78 95 a6 27 ad f5 12 63 9f
Random Number:
30 a2 5d 6a 5c 95 dd e2 39 07 58 b1 50 ff 70 38
The key stream after XOR:
89 38 c5 31 c4 64 6f 9a ac a1 7f 1c a5 ed 13 a7

The randomness test data of the above 5 groups and the total 640-bit key stream sequence after XOR operation are shown in the Table 2.

The test data in the Table 2 illustrates that 3rd data does not pass the Run Test.

4 CONCLUSIONS

Based on the randomness test of the key stream generated by the two key stream generator, we can get the following conclusions:

As can be seen in Table 1, the randomness of the key stream with direct use of AES algorithm, is not particularly good, and sometimes some randomness tests can't pass.

As can be seen in Table 2, although the randomness of the key stream with the AES algorithm is XORed with random number can be improved, but some randomness test also can't pass.

The method only using the CTR mode of AES algorithm or supplemented by the random number XORed with AES algorithm, can't generate random particularly good key stream. It is need to take more complicated methods to generate the key stream sequence of the stream cipher, such as using the CTR + OFB mixed model XORed with random number, in order to improve the performance of the key stream.

REFERENCES

Andrew, R. A Statistical Test Suite for Random and Pseudorandom Number Generators for Cryptographic Applications [DB/OL]. (2010-11-06). http://www.nist.gov.

Chen, K., M. Hnricksen, W. Millan, J. Fuller, L. Simpson, E. Dawson, H. Lee and S. Moon. Dragon: A Fast word based stream cipher. ECRYPT Stream Cipher Project Report 2005/2006.

Courtois, N.T., J. Pieprzyk, "Cryptanalysis of Block Cipher with Overdefined Systems of Equation", Advances in Cryptology-CRYPTO 2002, LNCS2501, pp:267–287, Springer-Verlag, 2002.

Good, T., M. Benaissa. AES as Stream Cipher on a Small FPGA, Circuits and Systems, 2006. ISCAS 2006. Proceedings. 2006 IEEE International Symposium 29–532.

William Stalling (US). Cryptography and Network Security Principles and Practices, Fourth Edition [M]. Beijing: Publishing House of Electronics Industry, 2006.

Xiangdong Hu, Qinfang Wei. Application Cryptology Course, [M]. Beijing: Publishing House of Electronics Industry. 2005.

Xuewang Zhang, Hong Jiang, Changjun Xiao, Liangyou Huang. "Study on Analysis of AES's S-Box and its Improvement", Information Safety [C], pp:51–53, 2009.

Control Engineering and Information Systems – Liu (Ed)
© 2015 Taylor & Francis Group, London, ISBN 978-1-138-02685-8

The application of Information Classified Security Protection on the information systems of Urban Mass Transit

T. Zhao & Y.Y. Yang

Quality Supervision and Testing Center of Security Products for Computer Information, System of the Ministry of Public Security, The Third Research Institute of the Ministry of Public Security, Shanghai, China

ABSTRACT: As the information systems of urban mass transit are getting much bigger and more complex, the application of information classified security protection on the systems is extremely important and necessary. The authors introduce the basic concepts of Information Classified Security Protection at first. Then, based on the characters of the urban mass transit industry, the security status and the security problems of the urban mass transit information systems are discussed. After that, integrating the Information Classified Security Protection and the security status of the urban mass transit, the key problems are thoroughly analyzed and corresponding countermeasures are proposed. Finally, the authors make a conclusion and emphasize the core points of the systems.

Keywords: Urban Mass Transit; information system security; Information Classified Security Protection

1   INTRODUCTION

Along with the rapid development and the wide application of information technology, the development of the national Urban Mass Transit (hereafter abbreviated as UMT)[1–3] steps into a new phase. In this phase, the information system of UMT has become an important carrier to transfer the UMT business operation data, i.e., the UMT depends more and more on its information system. Correspondingly, the security of the UMT information system is getting more and more important.

In order to reduce the security risks of public infrastructure networks and critical information systems like UMT, Chinese government has issued a series of laws, regulations, and policy papers. In these papers, the Information Classified Security Protection (hereafter abbreviated as ISCP)[4–8] system, which is used to protect the security of information systems, is the most important and basic in China.

In this paper, we will introduce the general concepts of ISCP in section 2. In section 3, the characters of the UMT are stated and the application of ISCP on the UMT is explained. After that, in section 4, the core security problems of the UMT are analyzed and the countermeasures are proposed. Finally, a brief conclusion is made in section 5.

2   ISCP REQUIREMENTS

2.1   *The definition of ISCP*

ISCP is the basic policy and method for Chinese national information security system. It is composed of a series of management specifications and technical standards. Generally speaking, it includes three aspects.

- The systems that store, transfer and deal with the national secret information or civic public information must be protected classifically according to the importance of the information.
- The information security products used in the information system must be managed according to the class of the information system.
- The information security events occurred in the information systems must be hierarchically responded and disposed.

According to the importance of the information systems for the national security, economic construction and social life, ISCP defines five different classes, from the 1st class to the 5th class. Each information system is given a certain class after classification evaluation, and the system must be constructed and managed according to the requirements of its class.

Among the information systems of UMT industry, office network systems are usually defined as the 2nd-class system, while signal control systems (such as the automatic train control system) and operational support systems (such as the passenger information system) are mainly regarded as the 3rd-class system. These information systems should be protected according to national management specifications and technical standards of the ISCP. Moreover, the national supervision departments will guide, supervise and check the ISCP work of these systems.

2.2 The implementation of ISCP

There are five key nodes in the implementation of the ISCP.

The first node is the classification evaluation of information systems. According to "the Management Approaches of the ISCP" (MPS Communication [2007] No.43, hereafter abbreviated as "the Management Approach") and "Information security technology—Classification guide for classified protection of information system" (GB/T 22240-2008), there are two factors that influence the classification of an information system. The first one is the importance of bearer services of the information system. The second one is the business dependency on the system. These two factors further influence the business information security class and system services security protection class of the information system. The two factors together determine the class of the information system eventually.

The second node is to file the class of information systems to the government. More precisely, the 2nd-class information system or above must be filed to the city-level or above public security organs by its operation and use organizations.

The third node is the security construction and rectification of information systems. After the class of an information system is determined, its operation and use organization should purchase and deploy information security products to meet the requirements of the system-class, and then construct the information security facilities comforting to the requirements of system-class according to the "Management Approach" and related management specifications and technical standards.

The fourth node is the evaluation of the information systems. After the third node is completed, the operation and use organizations of the information systems should conduct the class evaluation from the management and technical status. They should verify the results of construction the rectification by GB/T 22239-2008 and related technical standards. The 3rd-class information systems or above must be evaluated by the qualified evaluation agencies.

The fifth node is the supervision and inspection. The public security organs will supervise and inspect the ISCP work according to "the Management Approach". Forth more, the 3rd-class information systems or above must be evaluated annually by the public security organs.

3 THE ACHIEVEMENT OF ISCP WITH THE CHARACTERISTICS OF UMT INDUSTRY

3.1 The characteristics and security status of UMT information systems

UMT is a complex giant system. Since the passengers must be transported safely and fast to the destination, it need all subsystems such as traffic scheduling, passenger organizations, rolling stock, line, power supply and communication signals to coordinate smoothly and efficiently. Any components failure may lead to accidents[9]. Besides conventional vulnerability protection for computer operating system, the UMT information systems security also should take the other inspection into account, such as the physical security of the hardware facilities and components, anti-electromagnetic interference, personnel operating procedures, security management systems etc.

After an intensive research of the information systems (such as the ATC system, PIS system and office network systems, etc.) of the UMT companies, we found that there were many high-risk vulnerabilities, including weak passwords in servers' operating systems and applications, buffer overflow vulnerabilities, worms, etc. But these vulnerabilities were neglected by management personnel. Once these vulnerabilities are exploited, it may cause serious problems and finally endanger the normal operations of the UMT.

Therefore, the security goal of UMT information systems is to strengthen the security of networks and information systems. It aims not only to protect the infrastructure of networks to ensure the security and reliability of networks, but also to protect the services and applications. It emphasizes on real-time control of the security situations.

3.2 Application of ISCP on the UMT information systems

ISCP requires different security capabilities for information systems with different security classes. According to the implementation method, the basic security requirements can be divided into two types: technical requirements and management requirements[10].

Technical requirements relate to the technical security mechanisms of information systems, including physical security, network security, host security, application security and data security. They are implemented mainly by the deployment of hardware and software, and by correct configuration of the security features of the system.

Management requirements relate to the activities of different roles of the system, including security management system, safety management agencies, personnel security management, system construction and management, and system operation and maintenance management. They are implemented by the policies, systems, norms, processes and records provisions.

In the following section, we will take a 3rd-class information system as an example to study the application of ISCP on the UMT information systems.

4 THE APPLICATION OF ISCP ON A 3RD-CLASS INFORMATION SYSTEM

4.1 Physical security

Physical security aims to protect the security operation of information system with operating environment, security facilities, and other aspects. Physical security requires the security of the machine room where the information system equipments are placed. It includes server room security and office space security. Chinese national standard GB/T 22239-2008 mainly concerns the server room security, which includes the selection of physical location, environmental security controls and protection. As many influence factors are involved, it commonly chooses the combination of management and technical means as the security protection means in the ISCP work. For the evaluation of the information systems, we determine the security status of each evaluation points by means of interviews, on-site inspections and through professional instruments to obtain data[11].

Followings are the most of the nonconformities of physical security, for example, the server room has no safety zoning, server room or important regions have no strict access control, accessing to registration information is missing, no waterproof detection and alarm facilities are applied, no electromagnetic shielding of key equipment and magnetic media, server room and their integrated wiring have no professional third party acceptance testing, etc. These problems are often caused by the historical issues of construction, the high cost of the transformation, management not in place, and so on.

With the rapid development of UMT network, the scale of the information systems is growing rapidly, which makes the area of server rooms and the amount of equipments greatly increase. So it puts forward higher requirements on the server room space, temperature and humidity, power supply, etc. But part of server rooms is transformed from the office rooms. Therefore, their physical security is difficult to fully meet the ISCP requirements, which can be fundamentally resolved only through new construction or relocation.

4.2 Network security

Network is the most fundamental media and is the foundation that carries the UMT applications. Network security directly determines the overall security level of UMT network, and is the kernel of the construction and evaluation of ISCP. Construction and evaluation of ISCP mainly concerns about the network structure, network boundaries and network equipment security, etc.

Most of the network security problems focus on the following aspects: the granularity of network segment division is too coarse, a number of different application servers are placed in the same network segment, no boundary integrity check protection, no network malicious code prevention products, and inadequate security measures for wireless network access, etc.

According to the structure security requirements of the 3rd-class system, the network is divided into different zones by the business class to determine the network boundary. Combined with the characteristics of UMT applications, the separation of the office network, business bearing network, and data center network is necessary. Access control equipments (e.g. firewalls and gaps, etc.) are deployed at the network boundary, which apply the port-level access control policy settings and other ways to achieve the isolation between the internal network and external networks in order to improve the security of critical networks, information systems and business data.

Network architecture of the 3rd-class information systems should not only meet the basic protection requirements of network security running, but also consider setting the priority in network processing capabilities to ensure that important business service hosts can operate normally when the network jams. Network boundary access control is not only able to "defense", but also able to respond proactively. Therefore, it is necessary to deploy the Intrusion Detection System (IDS) on the core switch of UMT companies, or to deploy the Intrusion Prevention System (IPS) between core switches and firewalls, and timely updates its feature library to prevent important hosts and servers from malicious code attacks.

The followings are the common problems on network equipment security, e.g. remote management session is not encrypted, local authentication passwords is stored in clear text, password length and strength is not enough, SNMP service uses default password, authentication does not use a combination of two or more authentication technologies.

Some of above problems (e.g. two-factor authentication, etc.) can be solved through technical means, and the others can be solved by the combination of management and technical means.

4.3 Host security

As main carriers of information systems and critical business data, hosts and databases carry a variety of applications. The host system security is the operating system and database system security of computer equipments including servers, management terminals and workstations. Therefore the host system security is the focus of the information systems security protection.

However, according to the third-class ISCP requirements, host security control points emphasized on server-monitoring, monitoring and alarming of the minimum service level.

Most of the nonconformities on host security include: security policy on operating system is the default configuration; default operating system users are not renamed or limited privileges; unnecessary services and ports are not turned off; no sensitive mark on important information resources; no malicious code prevention software installed on hosts, etc. There are common problems of database system, e.g. no division of authority, no using two-factor authentication technology to manage users, no authentication fail processing, remote management identification information transmitted in the clear text. Part of the FAQs above can be solved by system configuration and third-party software deployment and other technical means.

4.4 Application security

Application security is the final defense line of business information systems. In accordance with the 3rd-class ISCP requirements, the authentication of the application system security requires a combination of two or more authentication mechanism to administrators.

The main nonconformities on application security include, no password complexity check, using weak authentication mechanism in part of the application systems, lack of communication integrity and confidentiality measures, etc.

Above problems mainly because that the application software in the UMT systems are custom developed. In the early design and construction process of the application systems, the developers focused more on business functions rather than security requirements. In recent years, with the highlight in the importance of information security, the security of the UMT applications is gaining increasing attentions.

To solve above problems, we recommend using passwords, USB-Key, biometrics-based method, digital certificates and other multiple authentication mechanisms to strength the security. Besides, using technologies such as encryption, digital certificates and access control to ensure the integrity and confidentiality and anti-denial of the data transmission process are also important. Moreover, if conditions permit, application software is needed to enhance the security features and to carry out the third-party security testing.

4.5 Data security and backup

During operation, information systems need to deal with large amounts of data. Various types of data are transmitted, stored and processed on the network, host, and application. Once these data are destroyed, it will affect or endanger the normal operation of the UMT information systems. Therefore, we need to consider the factors of the physical environment, network, operating systems, database and applications together to ensure the data security. For the situation of clients' login backend servers in the 3rd-class systems, it is recommended that the IPSec protocols and cryptographic algorithms are adopted to protect data confidentiality and integrity during the data-transmission.

Data backup is an important measure to prevent system management data, authentication information and important business data from being destroyed. The offsite timely backups can be used in the third-class information systems to effectively prevent and control the system harm once disaster occurs. The hardware redundancy of the major network equipment, communication lines and data processing systems is an important measure to ensure systems availability.

4.6 Security management

The management requirements of the 3rd-class ISCP emphasize to set up information safety management departments, and to establish the comprehensive information security management system involving the security policies, management system, and the operation procedures etc. The system must arrange full-time security administrators, clearly stipulate the periodic safety education and training specified in writing, and develop different training programs for different posts.

The most prominent problem is that part of the staff has weak awareness of information security and prevention. The management departments must carry out regularly information security awareness education to the key staff and the important sector employees. Correspondingly, a working mechanism for prevention and treatment is gradually formed.

More importantly, the UMT companies must establish a set of management system from the top management to the implementation of management, which will constraint and guarantee all security management measures to implement effectively. At the same time, it is necessary to strengthen communication and coordination between the inner related business departments and the security management department. The system should actively seek opportunities to cooperate with the various organizations of the security industry, invite professional evaluation agency and experts to give the guidance on information security construction, participate in safety planning and safety assessment; and carry out safety checks on a regular basis.

5 CONCLUSION

Information security is a comprehensive, dynamic security system. The 3rd and higher class information systems require to be evaluated once a year. While the ISCP construction is a sustaining dynamic, long-term and unremitting work.

The policy documents and technical standards have specifically described the security requirements of information systems. When these policy documents and technical standards apply to the UMT information system, not only must the requirements of the specific security items have to be implemented, but also the characteristics of the UMT systems and security scheme needs to be taken into account. At the same time, the security, operability and economy have to be balanced in order to properly carry out and achieve the ISCP work of the UMT information systems.

REFERENCES

GB/T 22239-2008, Information security technology—Baseline for classified protection of information system.

GB/T 28448-2012. Information security technology—Testing and evaluation requirement for classified protection of information system.

Hu C. & Lv C. "Method of risk assessment based on classified security protection and fuzzy neural network", 2010 Asia-Pacific Conference on Wearble Computing Systems, Shenzhen:IEEE, pp. 379–382.

Jin J., Teo K. & Sun L. "Disruption Response Planning for an Urban Mass Rapid Transit Network", The Transportation Research Board (TRB) 92nd Annual Meeting, Washington D.C.: TRB, 2013, pp. 1–18.

Liu H., Mao B., Ding Y., Jia W. & Lai S. "Train energy-saving scheme with evaluation in urban Mass transit systems", Journal of Transportation Systems Engineering and Information Technology, Vol. 7, pp. 68–73, Oct. 2007.

Liu J., Xu G., Yang Y. & Gao Y. "The analysis of classified protection compliance detection based on dempster-shafer theory", 2010 IEEE International Conference on Progress in Informatics and Computing, Shanghai:IEEE, pp. 537–541.

MPS Information Classified Security Protection Evaluation Center, Information Security Classified Evaluators Training Tutorials (primary). Beijing Publishing House of Electronics Industry, 2011.

Tian Z., Wang B., Ye Z. & Zhang H. "The survey of information system security classified protection", Electrical Engineering and Control, Vol. 98, pp. 975–980, Dec. 2011.

Tian Z., Ye J., Wang B. & Liu F. "A security BLP model used in clssified protection system", 2011 6th IEEE Joint International Information Technology and Artificail Intelligence Conference, Chongqing: IEEE, pp. 211–215.

Zhang L., Li Y. & Suo T. "Study on the classified protection evaluation methods of rail traffic signal system", Netinfo Security, pp. 122–124, Dec. 2012.

Zhang Y. "Coherent network optimizing of rail-based urban mass transit", Discrete Dynamics in Nature and Society, Vol. 2012, pp. 1–8, Dec. 2012.

Control Engineering and Information Systems – Liu (Ed)
© *2015 Taylor & Francis Group, London, ISBN 978-1-138-02685-8*

The key frame interpolation algorithm by conditional choosing quaternion interpolation equations

J. Yu & M. Zhao
Department of Computer Science and Technology, Yanshan University, Qinhuangdao, China

ABSTRACT: With the continuous development of the Internet and virtual reality technology, three-dimensional network is becoming more and more obvious, the urgent demand in these studies can have greater progress, so the 3D internet emerge as the times require. Virtual human motion control on 3D internet becomes increasingly the focus of attention. In this paper, the key frame interpolation algorithm by conditional choosing quaternion interpolation equations is proposed to achieve an efficient generation of in-between frames, obtain coherent, realistic animation motion effect. And by using a new method of JNI combined with Java and OpenGL, to realize real-time control of virtual human motion in 3D Internet.

1 INTRODUCTION

Since entering the new century, with unknown constant exploration and the rapid development of computer technology, in the highly dangerous environment or complex scientific experiment, application of virtual simulation experiment was carried out in order to reduce the risk of the increasingly strong demand. Virtual human technology in manned spaceflight, virtual reality, 3D animation, military, medical and other fields has been widely applied. At the same time, with the rapid development of Internet, WWW has become linked to a global bond; the existing Text and pictures interface allows visitors to easily and freely roaming virtual world. Although a variety of multimedia technology and Java extension technology on Hypertext Technology for expansion can be used, but the growing needs of people are still difficult to satisfy. With the development of virtual reality technology, 3D network is becoming more and more obvious. There is an urgent demand in these areas, so the concept of the 3D Internet emerges as the times requiring.

3D Internet can be understood as Internet content of the three-dimensional. Internet content are developed from the original text, pictures to intuitive frames of video and audio, and introducing virtual reality technology to realize.

At present, there are several 3D internet schemes:

1. Cult3D[1]: Cult3D technology introduced by Cycore Company and now has been widely applied. Cult3D is based on Java kernel, so it can be embedded in Java, use Java to enhance the interaction and function expansion.

2. Java3D[2]: Java3D is the Java language expansion program introduced by adapting to the 3D graphics, is essentially a set of application programming interface, offers interface which can be prepared to display virtual scenes based on the webpage.

Hebei Province Natural Science Foundation Project of China, Project Number: 07213530.

3. WebGL[3]: WebGL is a new 3D drawing standards, this drawing techniques standard combine JavaScript with OpenGL ES 2.0. Through to increase a binding JavaScript in the OpenGL ES 2.0 to achieve providing 3D hardware accelerated rendering to HTML5 Canvas.

In this paper, the following sections are elaborated through the following aspects of the virtual human motion control on 3D internet: first, linear modeling, and followed by the description of key frame extraction, then The key frame interpolation algorithm by conditional choosing quaternion interpolation equations is proposed in the paper, through experimental analysis, by applying a new method of JNI combined with Java and OpenGL, to control virtual human motion for real-time on 3D internet, and finally conclude.

2 VIRTUAL HUMAN MODELING

With the development of the 3D internet, the traditional virtual human's walking control method can not meet the needs of people. Completely real-time calculation type control method can not satisfy the demand of computing speed; while the preformed key frame approach because of its rigid motion can not meet the requirements for "user be personally on the scene". According to the quaternion linear interpolation and spherical interpolation method is the core algorithm key frame interpolation to realize virtual human movement control.

Common virtual human geometrical modeling method is mainly divided into: wire frame modeling, solid modeling and surface modeling. According to the high real-time requirement of the network virtual environment on the data transmission speed and model rendering, in order to simplify the description, this paper adopts the wire frame modeling method. Wire frame model is characterized by small data memory capacity, fast speed on the edit and modify. Of course there are shortcomings of two meanings and realistic deviations, but in the network environment this drawback can be ignored.

For a concise description of the human body motion, not considering the complex bone, muscle and cartilage system, the human leg skeletal system is simplified as shown in Figure 1.

One hip is hip joint, expressed with H; left (right) the knee joint for leftknee (rightknee), short for LK (RK). Distinguish between hip point to allow the left (right) leg L1 (R1) in the vertical plane of the swing forward and backward, and LK (RK) only allows the calf to thigh swinging backwards. Assuming the left (right) leg length L1 (R1), left (right) leg length L2(R2). LF (RF) left (right) foot ankle joint, Figure 1(b) from Z axis positive look model.

The human walking on the ground full cycle is divided into two step cycle; each step can be divided into five steps:

Step 0: initial state, still standing, as shown in Figure 1(b);
Step 1: assumes that the first step right foot, then the center of gravity does not move, keep left leg upright, as shown in Figure 2(a).
Step 2: the right foot not to land, but the centre of gravity of the body begins to move forward, to

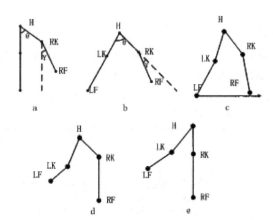

Figure 2. The decomposition of the human walking on the ground.

maintain the center of gravity moment balance, as shown in Figure 2(b).
Step 3: the right foot landing, the left foot has not yet been off the ground, as shown in Figure 2(c).
Step 4: left foot off the floor, right leg not straight. Then the centre of gravity of the body began to move ahead, ready to transfer to Upper right of the support leg (right leg), as shown in Figure 2(d).
Step 5: right leg straight, left foot off the floor, the centre of gravity of the body shifted to Just above the support leg, the legs have not yet shut, as shown in Figure 2(e).

Overall, the human walking on the ground is the five steps; the rest of the action can be obtained by interpolating the five steps.

3 KEY FRAME ACQUISITION

First of all, the BVH file is analyzed to obtain key frame group. BVH (Biovision Hierarchy) hierarchical model developed by Biovision Company is a description of the motion capture data file format. Because it is captured by the human body models wearing special costumes with sensor, which performance human motion lifelike and vivid. The BVH file can be exported by 3DMax, POSER tools, and stored as text, so it is very easy to be analyzed. A BVH file is usually composed of two parts: frame information and data block. After the BVH document analysis, then the corresponding virtual human skeletal data and the action data can be obtained, then these data are processed to build corresponding skeletal relationship structure and motion files, the time and motion corresponding to each frame can be got by the two file.

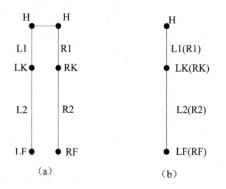

Figure 1. The wire frame modeling of the human leg skeletal system.

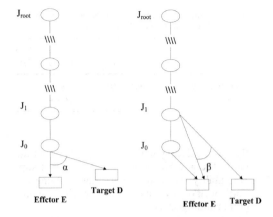

Figure 3. The CCD algorithm diagram.

Then, the reverse dynamics calculation of the key frame is applied, which uses an iterative search algorithm (cyclic-coordinate and descent, CCD[4]) for key frame inverse dynamic load. It is an approximation algorithm, which uses changing direction line to guide the lower node changes, so that the end effector would approach the target, the CCD Algorithm diagram as shown in Figure 3.

The specific processes are as follows: first move the joint J0, from the J0 to the end effector is hand position E vector VE, then from J0 to the target point of F to make vector VF, obtained 2 vector included angle α and axis of rotation VR (using the right-hand rule).Let J0 child bone rotates about the axis of rotation angle α, then the end effector E to a new location. If E reached the target point F, then stop, otherwise the J1 which is the father of J0 do the same operation, iterative calculating. If iteration to the root node, Jn still do not reach the target point, then again starting a new round of iteration from the J0, until it reaches the position of the target point. If still not get the result, the target point is out of the accessible range, the root node need to be moved, then the iteration.

4 KEY FRAME INTERPOLATION ALGORITHM

After the key frames have been got, the intermediate frames between the key frames should be interpolated to achieve higher fidelity. Key frame interpolation problem can be attributed to the parameter interpolation problem; all parameters affecting the picture image, such as position, rotation angle and so on are the key frame parameters. Generally, the traditional interpolation methods can be applied to the key frame method, but the key frame interpolation has its particularity, so

the interpolation calculation is divided into two main methods: linear interpolation method and spherical interpolation method. Linear interpolation method is a fast direct method, but have their limitations, in describing the rotation, although could obtain the intermediate frames, but will make intermediate accelerated motion, the rotation angular velocity is not uniform, the reduction of skeletal animation is very adverse. So in relation to describe rotating motion, the spherical interpolation method should be used to calculate intermediate frames, and spherical interpolation method use Euler angle[5] to describe the joint rotation, transforming the Euler angle into quaternion, and using quaternion spherical interpolation equation to calculate interpolation. Therefore, this paper combined with linear interpolation method and the spherical interpolation method characteristic, the key frame interpolation algorithm by conditional choosing quaternion interpolation equations is proposed, by judging the conditions to select between the linear interpolation equation and spherical interpolation equation, thereby generating high efficiency, consistency, lifelike continuous frames, and the algorithm is described as follows:

KeyFrameInterpolationbyQuaternionInterpolation(K)
// Import key frame array K; Export continuous pictures;
{For(i=0;i<=num(K);i++)
{M[i]=RotationMatrix(α, β, φ);
//Euler angle (α, β, φ) is converted to the corresponding rotation matrix M, because with the direction of rotation axis is related to the order, so set the rotation order: first rotating around the Y axis β, and rotating around the X axis α, finally, rotating around the Z axis φ;
Q[i]=Quaternions(M[i]);
//Transform rotation matrix M to quaternion Q;
If(i!=0)
{For(j=0;j<=G;j++)
//According to the motion acceleration between the key frames, get the number G of the interpolated intermediate frames;
{if(Judge(Q[i-1],Q[i],W)
//The function Judge calculates q1·q2, and compared with constant W to choose from linear interpolation equation and spherical interpolation equation;
q[j]=Q[i-1]·sin(1-t)θ+Q[i]·sinθ;
q[j]=(Q[i-1]·sin(1-t)θ+Q[i]·sinθ)/sinθ;
}
P=Insert(Q[i-1],q,Q[i]);
// Insert interpolation quaternion q into Q;
}
}
}

663

N=QuaternionstoRotationMatrix(P);
//Transform interpolated quaternion array P to rotation matrix array N;
E=RotationMatrixtoEuler(N);
// Transform rotation matrix array N to Euler angle array E;
Regeneration(E);
//generate continuous pictures;
}

5 EXPERIMENTAL ANALYSIS

Through the application of the key frame interpolation algorithm by conditional choosing quaternion interpolation equations, this paper presents a new method by using JNI combined with Java and OpenGL, to realize the real-time motion control of the virtual human in a 3D internet by experiment analysis. The method is based on the OpenGL method similar to the Java3D, different is all OpenGL call no package in Java language, but through the Java native interface to transfer data from Java code to the OpenGL display program delivery, and accept the data feedback, in Java interface to achieve the operation effect of OpenGL. First, in the Java program package an JavaBean to call the Java3D OpenGL, and the biggest difference lies in that the JavaBean is open, can be freely combined according to need by the programmer edit, in the Java program to call the JavaBean to finish OpenGL operation. In this mode, the programming mode and traditional writing OpenGL mode is basically the same. Simply means that the Java and OpenGL compilation process are separated, and JNI is used to coordinate them.

The experimental results show that the key frame interpolation algorithm can effectively improve the fidelity of virtual human movement; make the virtual human motion effect is more vivid, flexible. Adding judges function and redraw process, to effect the virtual human motion much more obvious.

Using JNI with Java and OpenGL method in computing efficiency is better than the direct use of Java3D approach, contrasting results as shown in Table 1.

In addition, compared with Java3D, JNI combined with Java and OpenGL, also has the following advantages:

1. The independence of the platform: the standard OpenGL and JNI is independent of platform, which can guarantee on the basis of the Java application truly multi-platform support;

Table 1. Comparison of the results between Java3D and JNI with Java and OpenGL method.

	Java3D	JNI with Java and OpenGL
CPU occupancy rate	65%–90%	5%
Language	Java	VC++

2. By 3D calculating on the local OpenGL runtime bases, the graphics hardware acceleration support ability also can be got, thus avoiding the bottleneck of Java calculation capacity, the program running speed is improved.
3. The process of programming can be separated, reducing the programmer's requirements and increasing the writing speed and the reuse rate of the code.

6 CONCLUSION

In the paper, the key frame interpolation algorithm by conditional choosing quaternion interpolation equations is proposed, achieving an efficient generation of intermediate frame, and improving the continuity and realism of the animation. Finally, by applying JNI combined with Java and OpenGL to real-time control virtual human movement in a 3D network. The operational efficiency, independence of platform, hardware acceleration capacity and program dissociated have been effectively improved.

REFERENCES

Grochow K., Martin S.L., Hertzmann A., et al. Style-based inverse kinematics[J]. ACM Transactions on Graphics, 2004, 23(3): 522–531.

Huang You-Qun, Bing Yu-Lei. Design and implement of virtual sculpture system based on Java3D. Journal of Shenyang University of Technology. 2009, 31(1): 73–76+120.

Kwak Hwy-Kuen, Lyou, Joon. Euler angle-based global motion estimation model for digital image stabilization. Journal of Institute of Control, Robotics and Systems. 2010, 16(11): 1053–1059.

Robles-Ortega María Dolores, Ortega Lidia, Feito Francisco R, González Manuel J. Navigation and interaction in urban environments using WebGL. IVAPP 2012: 493–496.

Yun Ruwei, Zhang Baoyun, Pan Zhigeng. Research on using Cult3D and java to realize virtual assembly. LNCS, 2009: 363–370.

Control Engineering and Information Systems – Liu (Ed)
© 2015 Taylor & Francis Group, London, ISBN 978-1-138-02685-8

Complementary S-Box design for resistance to power analysis attack

N.H. Zhu, Y.J. Zhou & H.M. Liu
Department of Electronic Engineering, Shanghai Jiao Tong University, Shanghai, China

ABSTRACT: Power analysis attacks aims at recovering the secret key of a cryptographic core from measurements of its consumed power when the cryptographic core is in encryption or decryption operation. Dynamic power analysis attack exploits the dependence of the dynamic power and the processed intermediate data to recover the secret key, while leakage power analysis attack focuses on the dependence of leakage power and intermediate data. In this paper, we proposed a complementary S-Box design for block cipher to resist not only dynamic power analysis attack but also leakage power analysis attack. The implementation of the critical S-Box of the Advanced Encryption Standard (AES) algorithm shows that using countermeasure of complementary S-box can thwart power analysis attack, not only dynamic but also leakage power analysis attack. Moreover the countermeasure circuit can be mounted into different symmetric algorithm which has S-Box architecture. Simulation results show that complementary S-Box proposed in this paper is a promising approach to implement a power analysis resistant crypto processor.

1 INTRODUCTION

With the massive spreading out of inexpensive integrated circuits which are able to store and process confidential data, the phenomena that more and more research on information security issues has been sprung up [1]. Side-channel attacks exploit the leaked physical information from chips to analyze the cryptographic devices and recovery the secret key stored in cryptographic devices. Research in [2] proposed attacks that utilize the timing or power information with controlled data from the attacked devices. Since the power information can easily be obtained by existing equipment, power analysis attack has become the most common attacking method [3]. Recently, power analysis attacks have been extensively shown to be a major threat to the security of data that processed and stored in cryptographic devices, such as smart cards, because power analysis attacks can generally be performed using relatively cheap equipment [4]. Block cipher cryptosystems embedded in cryptographic device are susceptible to attacks which show that security cannot be an afterthought. Like as performance and testability are important issues which the designer takes care in the design cycle, security also has to be taken into consideration early in the design cycle [5].

Dynamic power analysis attack which is a type of power analysis attack, is well known because there are massive amount of papers are researched in this attack [6]. Literature shows that asymmetry in the power consumption of transitions from 0 to 1, and 1 to 0 in the Complementary Metal Oxide Semiconductors (CMOS) library are the fundamental cause for dynamic power attacks [7].

As CMOS technology is scaled down, leakage power is predicted to become dominant than dynamic power. Especially, in sub-100-nm technologies, the dynamic power is no longer the dominant contribution to the chip power consumption, because of the much faster increase of the leakage power at each technology generation. For example, at the 65-nm technology node, the leakage power is in the order of half the chip power consumption and is planned to be an even greater fraction in successive technologies [8]. Hence, the leakage power can be easily measured in the traditional power analysis attacks. Nowadays, leakage power analysis attacks which are another type of power analysis attack are reported and analyzed. Due to the strong leakage dependence on the input digital circuits [9], leakage power can also provide a significant amount of information on the secret key. Literature in [10] [11] proposed leakage power analysis attack which is a novel class of attacks to nanometer cryptographic circuits. Leakage power analysis attacks are recently shown to be a new serious threat to information security of smart cards.

This paper is structured as follows. In Section 2, theoretic power model and procedures of power analysis attack are introduced. In Section 3, we simulate the dynamic and leakage power of DES and AES S-Box, and power characteristics of S-Box are shown in this section. We use different ideal power models to implement power analysis attacks on DES and AES S-Box, the proper ideal power model will be proposed in Section 4. And Section 5 concludes all of our work.

2 REVIEW OF POWER ANALYSIS ATTACK

2.1 *Theoretical dynamic power model*

Literature shows that asymmetry in the power consumption of transitions from 0 to 1, and 1 to 0 in the Complementary Metal Oxide Semiconductors (CMOS) library are the fundamental cause for dynamic power attacks. Dynamic power current will be consumed when the output capacitance is charged or discharged.

The overall dynamic current is equal to the sum of the charged or discharged current. Because the number of charged or discharged output is dependent on the hamming distance of output data, so the overall dynamic current $I_{dynamic}$ results to

$$I_{dynamic} = HD.I_{charge} \qquad (1)$$

In equation (1), HD stands for hamming weight of CMOS output, and I_{charge} means current is consumed when output is charged or discharged. From equation (1), it is clearly apparent that the overall dynamic current linearly depends on the hamming distance HD of the output patterns.

2.2 *Theoretical leakage power model*

The leakage current of static CMOS logic gates strongly depends on their inputs [9]. The related work [11] expresses that the overall leakage power of a real circuit linearly depends on the hamming weight HW of the input word, rather than the specific value of each bit.

In bit-sliced structures, the overall leakage current is equal to the sum of the leakage current of the m-bit slices, each of which is assumed to be equal to the high (low) level $I_H(I_L)$ when the corresponding input bit is high (low). Since the number of bit slices having a high leakage current I_H is equal to the number of input bits equal to 1, or equivalently, the hamming weight w of the input word, the overall leakage $I_{leakage}$ results to

$$I_{leakage} = HW \cdot (I_H - I_L) + m \cdot I_L \qquad (2)$$

From equation (2), the overall leakage power linearly depends on the hamming weight HW of the input patterns, rather than the specific value of each bit.

2.3 *Steps of power analysis attack*

As was discussed in above, the dynamic power and leakage power reveals the hamming distance and hamming weight of the m-bit data X that is processed within a given circuit block. Hence, power provides useful information to recover the secret key K of a cryptographic device if the processed

data X under attack are a function of K. We will briefly introduce the procedures in the following.

In the first step, the adversary chooses an internal m-bit signal X that is physically generated within the cryptographic circuit under attack. In general, signal X depends on both the input I and the secret key K of the cryptographic algorithm according to a well-defined function (3), which is known by the adversary.

$$X = f(I,k) \qquad (3)$$

In the second step, the adversary applies 2^m different input values I_i (with $I = 1 \ldots 2^m$) and measures the corresponding power P_i of the cryptographic chip at the point of time in which X is physically evaluated. The way of measuring the power is different between these two attacks, one is measuring the dynamic power, and the other is measuring the leakage power.

In the third step, the physical value of X within the chip is estimated for each input I_i. According to equation (3), since the generic input I_i is applied by the adversary, the only unknown variable in equation (3) is the secret key k; hence, it must be guessed. For each possible guess k_j of the secret key, the resulting value of $X_{ij} = f(I_i,k_j)$ under of the generic input I_i is found according to equation (3).

In the fourth step, the dynamic or leakage power of the block generating X is estimated. In particular, owing to the linear relationship between the leakage current within the block generating X and the hamming distance $HD(X)$ in equation (1) or hamming weight $HW(X)$ in equation (2), the current leakage is estimated by $HD(X)$ or $HW(X)$.

In the fifth step, the measured power P_i and the estimated $H_{i,j}$ are compared. For a given key guess k_j, the sequences P_i and $H_{i,j}$ associated with the random (but known) sequence of input I_i can thought of as random variables. When the key guess is correct, the estimated and measured power is maximally correlated. According to function (1) and (2), the correct guess of k is that the leading to the highest value among all possible guesses.

3 COMPLEMENTARY S-BOX DESIGN

In symmetric cryptography, S-Box is a very important part which implements non-linear transformation. And S-Box also consumes most power of the whole cryptographic engine [12]. In traditional dynamic power attacks, S-Box is the often chosen as attack point by the adversary.

The fundamental ideal of a power resistance circuit is to break the dependency between intermediate values and power traces. There are two basic ways to break the dependency between intermediate values and power traces. One is masking the intermediate values. However, modifications to the S-Box are necessary in these proposals. The other way is to balance the dynamic or leakage power. Figure 1 shows the block diagram of our proposed complementary S-Box. The countermeasure circuit is designed to work in parallel along with the S-Box module without any modification to the S-Box. This is the big advantage than the first way. To balance the power consumption of the S-Box, a complementary S-Box is designed to provide with the compensate current.

3.1 Proposed countermeasure circuit

Figure 1 shows the architecture of our proposed countermeasure circuit, not only for dynamic power analysis but also for leakage power analysis. Figure 1 shows the S-Box which can balance dynamic power and leakage power.

The countermeasure circuit consists of a basic S-Box and a complementary S-Box. The basic S-Box is a standard S-Box which is defined in cryptography, such as DES and AES. The complementary S-Box achieve bitwise negative function to basic S-Box. That is to say, the outputs of basic S-Box and complementary S-Box are always bitwise negative. So, it is easy to get the function in equation (4).

$$BAC\_SBox(I) = \overline{CMP\_SBox(I)} \qquad (4)$$

In equation (4), BAC_SBox stands for basic S-Box, CMP_SBox is the complementary S-Box, and I is the input data to basic S-Box and complementary S-Box.

There are two design considerations for our proposed countermeasure circuit: 1) the hamming weight of countermeasure circuit are independent on the output of countermeasure circuit and 2) the hamming distance of countermeasure circuit are independent on the output of our proposed countermeasure circuit.

To thwart dynamic power analysis attack, dependency between hamming distance and power must be removed. From equation (4), we can easily assert that the hamming distance of countermeasure circuit is always zero. That is because the outputs of basic S-Box and complementary S-Box are always bitwise negative.

To resist leakage power analysis attack, relationship between hamming weight and power must be removed. From equation (4), it is apparent that the hamming weight of countermeasure circuit always equals to a fixed number. That is also because the outputs of basic S-Box and complementary S-Box are always bitwise negative. Take AES for example, the hamming weight of countermeasure circuit is always equal to 8.

3.2 Attack resistance analysis

For the sake of illustration, we simulated the AES S-Box without countermeasure circuit and AES S-Box with our proposed countermeasure circuit to evaluate our proposed countermeasure. The circuit is simulated in SPICE to obtain exact simulated power traces. Two circuits all achieve the AES S-Box function.

The crypto core in Figure 2, based on the AES S-Box transformation without countermeasure was considered. This cryptographic core was synthesized using a 65-nm CMOS cell library. For the sake of simplicity, we fixed the secret key as $0 \times 2b$, and simulated on this crypto core in Figure 2.

We applied all possible plain words to the AES S-Box crypto core without countermeasure, simulated and stored all power traces. There are 256 power traces in total. The simulation time is from 0 ns to 1 us by the step of 10 ps. So every power trace has 100000 power points. All plain words are zero in the time range from 0 ns to 500 ns. At the time 500 ns, plain word has a transition to the different possible values. And in the

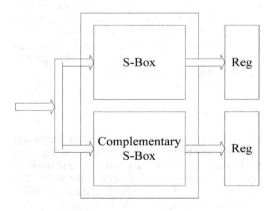

Figure 1.   Architecture of complementary S-Box countermeasure circuit.

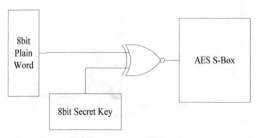

Figure 2.   Crypto core based on AES S-Box.

time range from 500 ns to 1000 ns, plain word remains unchanged.

The power of the circuit is estimated for each input plain word according to theoretical power model. Since the plain words are known, we guess all possible secret key from 0×00 to $0 \times ff$. And the corresponding hamming weight of plain word exclusive OR with the secret key is calculated for each guessed possible secret key. For each guessed possible secret key, the correlation coefficient between hamming weight with measured real power is calculated at each simulation time. Figure 3 is the correlation coefficients for all possible key guesses. There are all 256 correlation coefficient cures corresponding to 256 possible guessed key respectively. We plotted correlation coefficient curve as red curve that the guessed key is $0 \times 2b$. Figure 3 shows the dynamic power analysis attack. From the Figure 3, it is very apparent that the guessed key $0 \times 2b$ leads to the highest correlation coefficient among the other possible keys. And key $0 \times 2b$ is precisely the secret key.

Figure 4 shows the leakage power analysis attack. From the Figure 4, it is very apparent

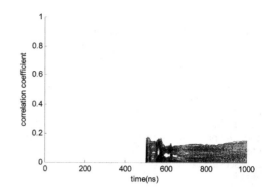

Figure 5. Power analysis attack to crypto core with countermeasure.

that the guessed key $0 \times 2b$ leads to the highest correlation coefficient among the other possible keys. And key $0 \times 2b$ is precisely the secret key.

In the same way, we also simulated the AES crypto core with our proposed countermeasure circuit in Figure 1. The secret key is also fixed at $0 \times 2b$, and all possible plain words are applied to the circuit, the power traces are recorded and stored.

Figure 5 shows the analysis result of AES crypto core with our proposed countermeasure. The guessed key $0 \times 2b$ is plotted as read line. It is apparent that the $0 \times 2b$ cannot lead to the highest correlation coefficient. That is to say, the correct key now does not result in the highest correlation at this condition. The correct key is now hidden in the analysis result. Therefore, even if attacks can find a peak in the analysis result, the still cannot find the correct key.

From the above analysis results, our proposed countermeasure circuit can resist power analysis attacks by balancing the power characteristic of the S-Box. Thus, the correlation between power traces and hamming weights can be effectively broken to hide the correct key. In addition, since the countermeasure circuit of our approach is independent of the operation frequency, the proposed countermeasure circuit can provide better protection over decoupling capacitors.

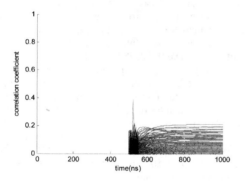

Figure 3. Dynamic power analysis attack to crypto core without countermeasure.

4 CONCLUSION

Power analysis attacks have become an important threat against cryptographic chips. In this paper, we proposed a complementary S-Box countermeasure circuit working in parallel along with the S-Box. The countermeasure circuit can be easily mounted on different implementations of the S-Box to resist power analysis attack. No throughput degradation and low area overhead. The analysis results of our proposed

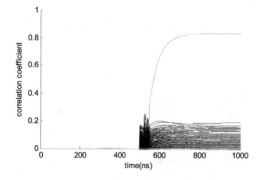

Figure 4. Leakage power analysis attack to crypto core without countermeasure.

countermeasure circuit showed that countermeasure using the complementary S-Box can resist the dynamic power analysis attack and leakage power analysis attack. Our proposed countermeasure is a promising way to thwart power analysis attack.

REFERENCES

Abdollahi A., Fallah F. & Pedram M. "Leakage current reduction in CMOS VLSI circuits by input vector control," Very Large Scale Integration (VLSI) Systems, IEEE Transactions on, vol. 12, pp. 140–154, 2004.

Alioto M., Giancane L., Scotti G. & Trifiletti A. "Leakage Power Analysis Attacks: A Novel Class of Attacks to Nanometer Cryptographic Circuits," Circuits and Systems I: Regular Papers, IEEE Transactions on, vol. 57, pp. 355–367, 2010.

Alioto M., Giancane L., Scotti G. & Trifiletti A. "Leakage Power Analysis attacks: Well-defined procedure and first experimental results," in Microelectronics (ICM), 2009 International Conference on, 2009, pp. 46–49.

Alioto M., Poli M. & Rocchi S. "A General Power Model of Differential Power Analysis Attacks to Static Logic Circuits," Very Large Scale Integration (VLSI) Systems, IEEE Transactions on, vol. 18, pp. 711–724, 2010.

Benini L. et al., "Energy-aware design techniques for differential power analysis protection", in Design Automation Conference, 2003. Proceedings, 2003, pp. 36–41.

Bodhisatwa M., Debdeep M. & Indranil S. "Design for Security of Block Cipher S-Boxes to Resist Differential Power Attacks", in VLSI Design (VLSID), 2012 25th International Conference on, 2012, pp. 113–118.

Djukanovic M., Giancane L., Scotti G., Trifiletti A. & Alioto M. "Leakage Power Analysis attacks: Effectiveness on DPA resistant logic styles under process variations," in Circuits and Systems (ISCAS), 2011 IEEE International Symposium on, 2011, pp. 2043–2046.

Kocher P., Jaffe J. & Jun B. "Differential power analysis", in Advances in Cryptology—CRYPTO'99, 1999, pp. 789–789.

Po-Chun Liu, Hsie-Chia Chang & Chen-Yi Lee. "A Low Overhead DPA Countermeasure Circuit Based on Ring Oscillators", Circuits and Systems II: Express Briefs, IEEE Transactions on, vol. 57, pp. 546–550, 2010.

Tillich S., Feldhofer M., Popp T. & Großschädl J. "Area, Delay, and Power Characteristics of Standard-Cell Implementations of the AES S-Box," Journal of Signal Processing Systems, vol. 50, pp. 251–261, 2008.

Tiri K. & Verbauwhecle I. "Simulation models for side-channel information leaks", in DAC'05, Proceedings of the 42nd annual Design Automation Conference, 2005, pp. 228–233.

Wagner M. "700+ Attacks Published on Smart Cards: The Need for a Systematic Counter Strategy", Constructive Side-Channel Analysis and Secure Design, Lecture Notes in Computer Science, vol. 7275, 2012, pp. 33–38.

Control Engineering and Information Systems – Liu (Ed)
© 2015 Taylor & Francis Group, London, ISBN 978-1-138-02685-8

Research on resource consumption attack detection and defend

X.X. Hu

Harbin Finance University, Harbin, China

ABSTRACT: Based on the pattern of resource consumption, the attack process is modeled with queuing theory and the technology of detection and defend are discussed. The resource consumption attack detection method based on CUSUM algorithm is presented, which detects network traffic mutation using statistics matrix. With the queuing model optimization, a defend mechanism against resource consumption attack is presented. The experiment results show that the scheme can detect and defend the resource consumption attack, achieves high detection rate with low computation cost, and hence has good practical value.

1 INTRODUCTION

Internet has gradually become an indispensable tool in people's daily life. However, with the deepening of Internet applications, network paralysis will result in huge economic losses. The Internet is a huge open system, there are many defects in the design and implementation of the agreement by self, together with the vulnerabilities in upper layer application software, the security issues of the Internet has increasingly become the focus. Compared with other network security threats such as network intrusion, viruses, etc, the network resource consumption attacks are reaching a wider, faster attack speed, more destructive. Although this type of network attacks could not control the information system, the service capabilities of information system reduced or lost completely due to the consumption of large quantities of resources, such as memory, computing, bandwidth and so on.

The resource consumption attack principle is very simple, represented by network worms and DDoS[1][2], but so far no one technology can prevent such attacks better. The main means of suppression to resource consumption attacks has been based intrusion detection technology. Network-based intrusion detection such as IPv6 or IPSec protocol[3] needs simultaneous implementation on the Internet, and router-based filtering needs to modify the TCP/IP protocol and implementation, so it is more difficult to implement[4]. Host-based detection techniques, such as the firewall deployed with strict ACL rules, not only affect the real-time of complex communication, and is not able to identify the type of resource consumption attacks[5] based on traffic.

In order to solve the above problem, based on the modeling and analysis of resource consumption attack to host, combined with the law, analysis of the differences between normal Internet stream and attack flow, we first proposed an attack detection method DMBC based on the CUSUM algorithm. DMBC analyzes network traffic mutation using matrix multi-statistic method, better alleviates the interference in which background traffic of legitimate users extracts attack feature, and detects resource consumption attacks accurately. Next, by optimizing the queuing theory model, we further proposed defense mechanism of resource consumption attack.

2 ANALYSIS OF RESOURCE CONSUMPTION ATTACK QUEUING THEORY

In order to research the detection and defense mechanisms of resource consumption attack better, we first model the processing of receiving packet on the host when attacked, aiming at analysis of the system behavior before implementation to determine the method of optimization. Due to resource consumption attacks sending large amounts of data packets to the attacker by using spoofed source IP address, both the server bandwidth capacity of communications link and router packet forwarding capabilities reduce ultimately, and thus could not provide normal service. Obviously, the queuing and congestion arising in the process of dealing with server packets, is all the characteristics of a random system, so queuing theory is as the theoretical tools for modeling[6]. Simulating the process of host dealing with packets to system model shown in Figure 1, this model is composed of S1 and S2 services, which indicate NIC hardware interrupt and user layer processing respectively. It is assumed that arrival of the data is the Poisson process with

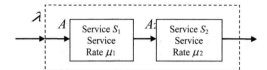

Figure 1. The random service system model of host receiving packages.

parameter λ: the service time of each subsystem is independent, exponentially distributed, service rate is μ_1 and μ_2 respectively, and the customer can only enter the internal network from S1, in addition, leave from S2. So the random service system is an open Jackson queuing network[7].

Set A_1 and A_2 as the average arrival rate of customer for each subsystem, so

$$A_1 = A_2 = \lambda \tag{1}$$

Set the random variable Ni as the length of the i-th node in steady state, the probability of joint distribution in steady-state is

$$P\{N_1 = m, N_2 = n_2, L, N_k = n_k\}$$
$$\equiv P_{m, n_2, L, nk} = \prod_{i=1}^{k} (1 - \rho_i)\rho_i^{n_i} \tag{2}$$

Seen from the formula (2), the probability of the joint distribution has the product form, So S1 and S2 can be regarded as independent M/M/1 queuing system.

S1 is used to simulate processing section of NIC hardware interrupt, it can accommodate the number of customers l_1 is generally small, but μ_1 is large, so it does not generally turn out that customer leaves because the queue is full in S1, pursuant to which you can get the probability of S1 busy named resource utilization[8], it is

$$\rho_1 = \lambda / \mu_1 \tag{3}$$

The customer reach of service system S2 is still the Poisson process with strength λ. It is assumed that the maximum number of packets accommodated by S2 is L, S2 is a M/M/1/L random service system, and the probability of S2 busy named resource utilization, it is

$$\rho_2 = \lambda / \mu_2 \tag{4}$$

3 THE DETECTION METHOD OF RESOURCE CONSUMPTION ATTACK BASED ON CUSUM ALGORITHM

As can be seen from the foregoing queuing theory, the key to the success of resource consumption

attack is whether or not data traffic is very considerable proportion in attack. So the core idea of DMBC is that cluster the traffic divided by controlled domain at the IP layer, and timing count and update the ratio of attack traffic in each controlled domain to legitimate traffic using CUSUM algorithm[9][10]. In specific implementation, we sample the number of the connection SYN and FIN (RST) every t0 time. In order to count FIN together with SYN in the same connection, the sampling time of FIN (RST) lag td interval than SYN. Selection of td depends on the interval between most of SYN and FIN (RST) connected. The recent traffic measurement of Internet network shows that: most of the TCP connection for 12–19 seconds, so we will set td as 10 seconds. In order to balance detection resolution and the stability and accuracy of the algorithm, we will set t0 as 20 seconds.

We define $\{\Delta n, n = 1, 2, \ldots\}$ as the difference between the number of SYN and FIN (RST) in a sampling period, regard the number of FIN (RST) packets in the n-th sampling period as $F(n)$, next standardize $\{\Delta n\}$ with $F(n)$, define $X_n = \Delta_n / F(n)$, regard the average value of X_n as c, $c \ll 1$ and $c \to 0$. So $\{X_n\}$ is independent of time and network traffic, and its dynamic characteristics depend entirely on the specification of TCP protocol. Therefore, we regard $\{X_n\}$ as a stationary random process, i.e. $\{X_n\}$ is considered to satisfy the following two conditions:

Condition 1: $\{X_n\}$ is ψ—mixing, meaning that parameter $\psi(s)$ defined below closes to 0 when $s \to \infty$:

$$\psi(s) = \sup_{t \geq 1} \sup_{\substack{A \in F_1^t, B \in F_{t+s}^{\infty} \\ P(A)P(B) \neq 0}} \left| \frac{P(AB)}{P(A)P(B)} - 1 \right|, \tag{5}$$

In above formula, F_1^t is algebra σ produced by $\{X_1, X_2, \cdots X_t\}$, F_{t+s}^{∞} is algebra σ produced by $\{X_{t+s}, X_{t+s+1}, \cdots\}$. $\psi(s)$ is affected by the correlation of samples $\{X_n\}$. Due to the highly correlation of $\{X_n\}$, $\psi(s)$ is decaying slowly when $s \to 0$.

Condition 2: The marginal distribution of $\{X_n\}$ satisfies the following law: $\exists t > 0$, $E(e^{tX_n}) < \infty$.

Generally, $E(X_n) = c \ll 1$. We choose parameter a as the upper limit of c, that a > c, and define $\tilde{X}_n = X_n - a$, make that $\tilde{X}_n$ is negative normally. When the attack takes place, $\tilde{X}_n$ will suddenly become a large positive value. It is assumed that in an attack, the lower limit incremental of $\tilde{X}_n$ is h. Our detection method to resource consumption attack is based on observable facts: $h \gg c$. Make

$$y_n = (y_{n-1} + \tilde{X}_n)^+, \tag{6}$$

In above formula, $y_0 = 0$, when $x > 0$, $x^+ = x$, when $x \leq 0$, $x^+ = 0$. Significance of y_n can also be understood in the following way: define $S_k = \sum_{i=1}^{k} \tilde{X}_i$, $S_0 = 0$, it is easy to get:

$$y_n = S_n - \min_{1 \le k \le n} S_k \qquad (7)$$

It is the maximum continuous incremental before time n. Great $\{y_n\}$ is the strong signal that attack occurred. That is, y_n is able to clearly reflect the changes of $\{X_n\}$, and y_n showes a rising trend at time m. Accordingly, set N as the threshold value, when y_n is greater than N, it is considered that resource consumption attack happens. In addition, DMBC detects the cumulative effect of the attack. Therefore, it can detect the attack with flooding rate less than h by a longer period. In fact, if $c \approx 0$, the method is able to detect that the lower limit of resource consumption rate is a. Additionally, DMBC method is not sensitive to the mode of flooding: It detects not only the persistent flood attacks but also sudden ones[11].

4 THE DEFENSE MECHANISMS AGAINST RESOURCE CONSUMPTION ATTACK BASED ON QUEUING THEORY ANALYSIS

When the outbreak of network packet arrival or the system is very heavy loads, network card interrupt, it is that S1 will produce packet loss, S1 can be seen as M/M/1/l queuing system, and its packet loss rate[12] is

$$P_{drop1} = \rho_1^l \frac{1 - \rho_1}{1 - \rho_1^{l+1}}, \qquad \rho_1 = \lambda / \mu_1 \qquad (8)$$

When the service system is attacked by SYN Flood[13], it meet $0 \le \rho_2 < 1$, the packet loss rate of S2 is

$$P_{drop2} = \rho_2^L \frac{1 - \rho_2}{1 - \rho_2^{L+1}} \qquad (9)$$

When the service system overloads by attacked, it is that $\rho_2 > 1$, and the throughput of S2 is:

$$T = \lambda \left(1 - \frac{\rho_2^L (1 - \rho_2)}{1 - \rho_2^{L+1}} \right) \qquad (10)$$

The formula (10) is simplified as follows, in the case of heavy load ρ_2 is much greater than 1, it can be considered:

$$1 - \rho_2^{L+1} \approx -\rho_2^{L+1}$$

Thus:

$$T \approx T' = \lambda / \rho_2 = \mu_2 \qquad (11)$$

The formula (11) can be understood as this: when S2 is under a heavy load, the probability is close to zero that there is no customer in S2, and the waiter always works almost, so the throughput is equal to the actual service rate approximately.

According to the above model optimization, we propose the following defense mechanisms against resource consumption attack:

A. Increase the capacity named L of queue
   Seen from the formula (9) and (10), when the load of S2 is normal, increasing the capacity named L of queue and improving service speed named μ_2 can reduce the packet loss rate and increase efficiency. When the server handles network connection, the length named L of model is the one of memory half-connection queue[14]. When the system is subject to resource consumption attacks and $\rho < 1$, the packet loss rate of server can be reduced by increasing the length of half-connection queue L. But to prevent the half-connection queue is too large, memory-intensive and lead to the decline in the overall performance of the system, the length of half-connection is restricted by the operating system. However, if you have a large enough memory space, you can use half-connection queue technology through memory expansion. As shown in Figure 2, open up storage area to store half-connection information within the range of memory allowed. This queue is connected to the original half-connection queue, if the original queue is full; the new semi-connection information is stored in the extended one.

B. Increase the processing rate
   Seen from formula (11), improving service speed named μ_2 is able to raise system throughput, and reduce packet loss rate when S2 is overload. From the analysis of the three-way handshake process of TCP protocol established by connection[15], it shows that server waits the user to respond to the ACK, if it is timeout and server did not receive ACK packet, server will release half-connected information

Client Syn packets

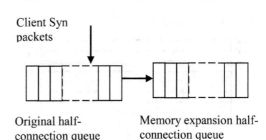

Original half-connection queue

Memory expansion half-connection queue

Figure 2. The extension methods of half-connection queue.

and corresponding memory space. It can be seen that retransmission timeout of ACK is an important factor to μ_2'. The case of Redhat 9.0 shows that, if ACK is not received from client during three-way handshake, server will retransmit SYN/ACK five times interval of 3 s, 6S, 12 s, 24 s, 48 s and total timeout is about 90 s. In the resource consumption attack, shorten the timeout of waiting for ACK packet and change the times of retransmissions to be 3. So the time is $3\,s + 6\,s + 12\,s$ of $= 21\,s$, and it is apparent that μ_2 is improved greatly.

5 EXPERIMENTAL DATA AND ANALYSIS OF RESULTS

To further analyze the functionality and performance of the proposed method, performance testing has been completed at the Sugon server (CPU-PIV2.0GHz × 2, memory-4G) with Gigabit Fast Ethernet card, associating the hair fiber with dedicated packet-sending hardware SmartBits-6000 (B) and Cisco switches.

We use four Sugon servers to build test environment with the role of attacking, be attacked, detection (DMBC) as well as sending background traffic, and use SmartBits (dedicated packet-sending hardware) to send interference data. In order to reflect the actual results, we use the real data admitted from campus network in advance for background traffic, moreover, HGod is used for attack tools to detect and analyze the detection results of the decision engine, finally, the test results are shown in Figure 3. When the attack starts in FIG, y_n appears an obvious concave (threshold N is 0.98) because the distribution of network traffic generates mutations. Accordingly, we can easily find the network anomalies, and locate the specific abnormal IP to obtain more detailed test results by further analysis.

The above test results show that: the flow is greater and the demographic characteristics are more obvious. Thus, the characteristics of y_n are the ones of the TCP protocol itself. There is no correlation between the network traffic and period. Obviously, when resource consumption attack occurs, these characteristics will inevitably change significantly.

Figure 4 shows the changes of HTTP request success rate affected by the time-out of half-connection queue, when the length of half-connection is 128, the rate of HTTP request is 1600/s and the power of resource consumption attack is 400/s. As can be seen, the Http request rate will be increased with the decrease of the timeout period.

Figure 5 shows the changes of HTTP request success rate affected by the length of half-connection queue, when the power of resource consumption attack is 1000/s. In order to get more significant influence by queue length, we will set the queue timeout to 1 s. As can be seen from Figure 5, with the length of memory half-connection queue increasing, the connection rate is improved greatly, so the length of memory queue is less impact on overall performance.

Based on the modeling and analysis of attack to the host, combined with its law, we researched on the detection and defense technology against resource consumption attack, firstly, we proposed an attack detection method based CUSUM algorithm, then, by optimizing the queuing theory model, we further proposed the defense mechanism. Enhance the feasibility of detection and prevention against

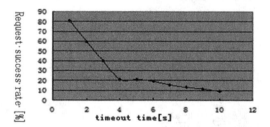

Figure 4. Request success rate affected by timeout.

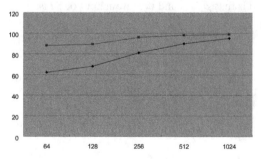

Figure 5. Request success rate affected by queue length.

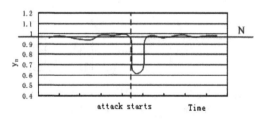

Figure 3. The detection results of resource consumption attack.

resource consumption attack in large traffic network environment overall.

Project Name: (The Construction and Optimization of Public Opinion Monitoring System about Heilongjiang Higher Education Information), the Project of Science and Technology Department of Heilongjiang Provincial Department of Education. Project NO: 12531091.

REFERENCES

Bencsáth B. & Vajda I. Protection Against DDoS Attacks Based On Traffic Level Measurements. SIMULATION SERIES Bibliographic Details. 2004, 36(1):22–28.

Bo Chen. Detection and Mitigation of Abuse Attack for Large Scale Network. Harbin Institute of Technology. 2007, 10:2–20.

Chuan Lai Lu queuing theory. Beijing University of Posts and Telecommunications Press, 1994:54–58.

Chuang Lin. Performance evaluation of computer networks and computer systems. First Edition. Tsinghua University Press, 2001:47–104.

Dejun Wen, Daiyuan Zhang. Research of Computer Network Security and Defence Technology. Computer technology and development. 2012, 12:171–174.

Fen Lin, Hong Zhang. Prevention technology research of terminal server network resource consumption attack. Computer engineering and design. 2011, 12:3997–4001.

Hui Sun, Long Fan. Principle, Detection and Defense of SYN Flood Attack. Modern computer (Professional), 2012,

Jian Pan. Research on SYN Flood attack detection and defense method based on campus network. Maritime Affairs University of Dalian, 2012, 6:29–32.

Jianqiang Li, YuShun Fan. A Method of Workflow Model Performance Analysis, Chinese Journal of computers, 2003, 5:513–522.

Liu ChunFeng, Shu YanTai, etc. Detection of Selfish Behavior in Wireless Ad Hoc Networks Based on CUSUM Algorithm. Tianjin University and Springer-Verlag Berlin Heidelberg 2010, 16:104–108.

Nianmin Yao, Jiubin Ju. The performance study of overload server, Journal of Software, 2003, 14 (10): 1781–1786.

Popstojanoval K.G. & Wang F. Characterizing intrusion tolerance systems using a state transition model. DARPA Information Survivability Conference and Exposition, 2001, 2:211–221.

Sasha, Beetle. A Strict Anomaly Detection Model for IDS. Phrack Magazine, 2000, http://artofhacking.com/files/phrack/phrack56/index1.html.

Shuo Wang, RongCai Zhao. Distributed Denial of Service Attack Research Based on Proxy Servers. Journal of Information Engineering University, 2012, 6:365–369.

Zhixin Sun, YiWei Tang. Router Anomaly Traffic Detection Based on Modified-CUSUM Algorithms. Journal of software. 2005, 12:2117–2123.

Session 6: Algorithm and simulation

Control Engineering and Information Systems – Liu (Ed)
© 2015 Taylor & Francis Group, London, ISBN 978-1-138-02685-8

A hybrid filter-wrapper feature selection algorithm for SVM classification and its application to credit risk analysis

W.Z. Zhang
School of Software and Microelectronics, Peking University, Beijing, China

J.Y. Yang & J. Chen
Department of Computer Science, University of California, Los Angeles, USA

ABSTRACT: Support vector machine has been used in credit risk assessment as a powerful classifier. However, high dimensional training data not only results in time-consuming computation but also affects the performance of the classifier. Feature selection technique can help reduce the size of the feature set of the training data before the classifier is trained. In this paper, we propose a SU-GA-wrapper feature selection method by combining the advantage of filter and wrapper. In our proposed method, the original features are evaluated by Symmetrical Uncertainty, and the resulting estimation is embedded into the genetic algorithm applied to search optimal feature subset with train accuracy of SVM. We test our method in a real-world credit risk prediction task, and our empirical results demonstrate the advantage of our method over other competing ones.

1 INTRODUCTION

Credit risk management has played a key role in financial and banking industry. An accurate estimation of risk, and its use in corporate or global financial risk models, could be translated into a more efficient use of resources. One important ingredient to accomplish this goal is to find useful predictors of individual risk in the credit portfolios of institutions. However, accessing credit risk is very challenging because many factors may contribute to the risk and their relationship is very complicated to capture and measure. Recent years have witnessed a growing trend in applying machine learning methods for credit risk analysis. These methods can automatically learn from historical data, and yield highly accurate predictions in many practical tasks. One powerful machine learning method is Support Vector Machine (SVM) [1].

SVM is a kind of machine learning technique [2] to train a powerful classifier which is a good way to address credit assessment problem. However, high dimensional data from credit assessment problem may bring difficulty to the classification training as well as influence the classification accuracy. Feature selection, as a preprocessing step to train a classifier, is effective in reducing dimensionality, removing irrelevant data, and increasing classification accuracy. It plays a key role in the data analysis process since irrelevant features often degrade the performance of algorithms devoted to construction of predictive models, both in speed and in predictive accuracy. Irrelevant and redundant features interfere with useful ones, so that most supervised learning algorithms fail to properly identify those features that are necessary to describe the target concept.

Feature selection algorithms [3, 4] fall into two broad categories, the filter model or the wrapper model [5, 6, 7]. Filter feature selection can be accomplished independently from the performance of a specific learning algorithm. Optimal feature selection is achieved by maximizing or minimizing a criterion function. Conversely, the effectiveness of the performance-dependent wrapper feature selection model is directly related to the performance of the learning algorithm, usually in terms of its predictive accuracy. Generally, a wrapper approach should give better results than a filter method; however, it also has larger computation costs which may make it prohibitive.

2 COMBINATION OF SU AND GA-WRAPPER

2.1 *Information gain and symmetrical uncertainty*

Information gain is based on well-known information theory measure entropy, which characterizes the purity of an arbitrary collection of items and is considered as a metric of system's unpredictability. Information gain measuring the expected reduction of entropy caused by partitioning the examples according to feature A is given by:

$$IG(S,A) = H(S) - \sum_{v \in V(A)} \frac{|S_v|}{S} \cdot H(S_v),$$

$$H(S) = -\sum_{c \in C} \frac{|S_c|}{|S|} \log_2 \frac{|S_c|}{|S|},$$

where S is the item collection, $|S|$ its cardinality; $V(A)$ is the set of all possible values for feature A; S_v is the subset of S for which A has value v; C is the class collection; S_c is the subset of S containing items belonging to class c. In the process of IG-based feature selection [8, 9], the features are ranked with their information gains, and then filter out the nonsignificant features by setting an appropriate threshold in the ranking.

Symmetric uncertainty is a method of eliminating redundant features as well as irrelevant ones is to select a subset of features that individually correlate well with the class but have little intercorrelation. The correlation between two nominal features X and Y can be measured using the Symmetric Uncertainty (SU) criterion, which also compensates for the inherent bias of Information Gain by dividing it by the sum of the entropies of X and Y:

$$SU(X,Y) = 2 \frac{H(Y) + H(X) - H(X,Y)}{H(X) + H(Y)}$$
$$= \frac{2 \times IG}{H(Y) + H(X)},$$

where H is the entropy function. The entropies are based on the probability associated with each feature value; $H(A,B)$, the joint entropy of A and B, is calculated from the joint probabilities of all combinations of values of A and B. Owing to the correction factor 2, SU gets values, which are normalized to the range [0, 1]. A value of $SU = 0$ indicates that X and Y are uncorrelated, and $SU = 1$ means that the knowledge of one feature completely predicts the other. Similarly to GR, the SU is biased toward features with fewer values.

2.2 Information gain and symmetrical uncertainty

Genetic Algorithms (GAs) [10], a form of inductive learning strategy, are adaptive search techniques based on the analogy with biology, in which a set of possible solutions evolves via natural selection. There are three fundamental operators in GA: selection, crossover and mutation within chromosomes. As in nature, each operator occurs with a certain probability. There must be a fitness function to evaluate individuals' fitness. The evaluation function is a very important component of the selection process since offspring for the next generation are determined by the fitness values of the present population. Crossover and mutation are used to generate new individuals (offspring) for the next

generation. Crossover operates by randomly selecting a point in the two selected parents and exchanging the remaining segments of the parents to create new individuals. Mutation operates by randomly changing one or more components of a selected individual. Figure 1 provides a simple diagram of the iterative nature of genetic algorithms.

Feature selection problem can be addressed by using genetic algorithms in a wrapper approach. In these works, genetic algorithms were used to explore the space of all possible subsets of feature in order to obtain a set of features which can maximize the accuracy of a specific learning algorithm. In this method, the time required for reaching a subset of relevant features depends on the complexity of the learning algorithm used for fitness evaluation. Experiments evidences that, using such an approach, genetic feature selection can require some hours of CPU time to obtain a good feature subset on large datasets.

2.3 Combination of filter and wrapper

As we known, a wrapper approach give us better results than a filter method, while the filter approach gives a fast solution to address the feature selection problem. This paper focus on hybrid feature selection framework by combining these two methods. We propose a two-phase feature selection method, the idea of which is to use the feature estimation from the filter phase as the heuristic information for the wrapper phase. In the first phase we adopt filter approach based on Symmetrical Uncertainty (SU) to get feature estimation; in the second phase, Genetic Algorithm (GA) is adopted as search algorithm for wrapper selector, which makes use of learning algorithm as part of its evaluation. The feature estimation obtained from SU-based filter approach provides a fast feature ranking, which can not only eliminate the redundant and irrelevant features as preprocessing, but also guide the initialization of the population for genetic algorithm.

Figure 2 illustrate the SU-GA-wrapper algorithm, namely symmetrical uncertainty filter adheres to GA-Wrapper process. In first phase, we compute Symmetrical Uncertainty (SU) between features and the target concept. Second phase is the application of genetic algorithms in optimal feature subset searching. The fitness function of GA consists of the classifier accuracy and the size of the features.

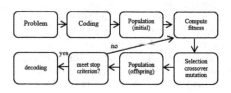

Figure 1.   Flow diagram of genetic algorithm.

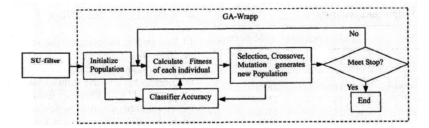

Figure 2. SU-GA-wrapper.

Coding: A feature subset is represented by a binary string with length of n (n is the number of features), with a zero or one denoting corresponding feature whether to be selected. The structure of the chromosome in GA for feature selection is shown in Figure 3.

Initialization of GA population: The population of GA is initialized based on the rank of the features according to the evaluation results gotten by SU filter. The rank in the front means the feature has stronger distinguishing quality. The features in the top rank should have bigger probability to be selected, which means the corresponding bit the chromosome should have more chance to be one. The individuals are initializes according to the probabilities based on the rank of corresponding features. We rank the features with SU value greater than threshold according to their weight. Then we set the selection probabilities of each feature: set the probability to be p1 for the feature ranking first and p2 for the feature ranking last, and then generate probabilities for the other features according to arithmetic sequence.

Design of fitness function: The fitness of solutions is mainly evaluated by training classifier on the training data using only the features corresponding to 1 s in the chromosome, and returning the classification accuracy as the fitness. Besides, the size of feature subset is also a factor affecting the fitness of solution [4]. A single objective fitness function that combines the two goals into one was designed to solve the multiple criteria problem. The formula is as below.

$$fitness = w_a \times accuracy + w_f \bigg/ \sum_{i=1}^{n_f} f_i$$

where w_a represents the weight value for classification accuracy, w_f for the number of features; f_i is the mask value of the i-th feature, '1' represents that feature i is selected; '0' represents that feature i is not selected. It can be inferred that high fitness value is determined by high classification and small feature number. Figure 1 illustrates the principle of GA-based feature selection.

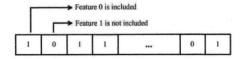

Figure 3. GA chromosome.

Selection, crossover and mutation: We adopt tournament selection, uniform crossover, and standard mutation. Tournament selection operates by randomly selecting a set number of candidates, from which the two fittest chromosomes survive. The survivors, called the parent chromosomes, are then subjected to crossover and mutation [10].

3 EXPERIMENTS AND ANALYSIS

3.1 Dataset

The dataset used in this paper comes from the private label credit card operation of a major Brazilian retail chain. There are 50,000 instances in the original dataset, each being labeled as positive (good) and negative (bad). In this experiment, we use a subset of the data which contains 5,000 instances with balanced positive and negative labels. Each instance has 32 features, including client ID, sex, age, education, shopping history, monthly income, etc. In our experiments, we use 22 features which we believe are most relevant to credit risk.

3.2 Experimental setup

We randomly split the data into a training set (60% of points), a validation set (20% of points) and a testing set (20% of points). We use the training set to do feature selection and record the optimum feature subset, and finally use this small subset of feature to train the classifier for credit prediction. The decision tree classification accuracy with 10-folds cross validation on the whole dataset is used to measure the effect of the feature subset we select. The parameter setting in genetic algorithm is the following: the Population size: 20; Number of generations: 20; Probability of crossover: 0.6; Probability of mutation: 0.033; and Random number seed: 1.

Table 1. Performance comparison.

Method	# of features	SVM accuracy
Raw	22	71.150
ReliefF	10	58.214
Information gain	11	78.416
Gain ratio	15	75.808
Chi-squared	15	75.808
Symmetrical uncertainty	13	79.440
GA-wrapper + SVM	12	80.183
SU-GA-wrapper + SVM	10	81.091

We compare several feature selection methods as follows:

- **ReliefF** is an instance-based learning methods that sample instances randomly from the training set and check neighboring records of the same and different classes—"near hits" and "near misses." If a near hit has a different value for a certain attribute, that feature seems to be irrelevant and its weight should be decreased [8].
- **Information Gain (IG)** is based on well-known information theory measure entropy, which characterizes the purity of an arbitrary collection of items and is considered as a metric of system's unpredictability [8].
- **Gain Ratio (GR)** evaluates the worth of an attribute by measuring the gain ratio with respect to the class by the following formula: GainR(Class, Attribute) = (H(Class) − H(Class | Attribute))/H(Attribute).
- **Chi-square** (χ^2) evaluates the worth of an attribute by computing the value of the chi-squared statistic with respect to the class.
- **PCA** performs a principal components analysis [11, 12] and transformation of the data. Use in conjunction with a Ranker search. Dimensionality reduction is accomplished by choosing enough eigenvectors to account for some percentage of the variance in the original data. Attribute noise can be filtered by transforming to the PC space, eliminating some of the worst eigenvectors, and then transforming back to the original space.
- **GA wrapper + SVM search** is a wrapper method for feature selection, which uses SVM classifier as evaluator and searches the space of attribute subsets by GA.

3.3 *Result*

The result of experiments indicates that SU-GA-wrapper feature selection algorithm can be used to deal with feature selection problem, and it is able to reduce the number of selected features significantly and produces obvious improvement in the classification accuracy. It gains better than other simple filter methods or wrapper models.

4 CONCLUSIONS AND FUTURE WORK

In this paper, we address the credit risk analysis problem which is a crucial task in finance and management. Our work is based on a machine learning method SVM. We have shown that how efficient our proposed SU-GA-Wrapper feature selection algorithm generate a subset of features for SVM classifier. SU-GA-Wrapper is a two-phase algorithm combining the filter with the wrapper, in which the wrapper process act as the amender for filter and the result of filter provides heuristic information for the search of the wrapper procedure. Our empirical study shows that the selected group of feature can significantly improve SVM classification accuracy, compared to several competing methods.

REFERENCES

Abdi H. & Williams L.J. (2010). "Principal component analysis." Wiley Interdisciplinary Reviews: Computational Statistics, 2: 433–459.

Cortes, Corinna, Vapnik & Vladimir N. "Support-Vector Networks", Machine Learning, 20, 1995.

Dash M. & Liu H. Feature Selection for Classification. Intelligent Data Analysis, 1997, Vol. 1, No. 3, pp. 131–156.

De Jong K. "Learning with Genetic Algorithms: An overview". Machine Learning Vol. 3, Kluwer Academic publishers, 1988.

Fukunaga K. Introduction to Statistical Pattern Recognition. Academic Press, San Deigo, California, 1990.

Jolliffe I.T. "Principal Component Analysis, Series: Springer Series in Statistics", 2nd ed., Springer, NY, 2002, XXIX, 487 p. 28.

Kittler J. Feature set search algorithms. Pattern Recognition and Signal Processing. 1978, 41–60.

Lewis P.M. The characteristic selection problem in recognition system. IRETransaction on Information Theory, 1962, 8, pp. 171–178.

Pang-Ning Tan, Michael Steinbach & Vipin Kumar. Introduction to data mining. Addison Wesley Longman, 2006.

Wei Han, Jian Pei & Yiwen Yin. Mining frequent patterns without candidate generation, Proceedings of the 2000 ACM SIGMOD international conference on Management of data, p. 1–12, May 15–18, 2000, Dallas, Texas, United States.

Yu L. & Liu H. Efficiently handling feature redundancy in high-dimensional data, in: Proceedings of The Ninth ACM SIGKDD International Conference on Knowledge Discovery and Data Mining (KDD-03), Washington, DC, August, 2003, pp. 685–690.

Zexuan Zhu, Yew-Soon Ong & Manoranjan Dash. Wrapper-filter feature selection algorithm using a memetic framework. IEEE transactions on systems, man, and cybernetics. Part B, Cybernetics: a publication of the IEEE Systems, Man, and Cybernetics Society, 2007;37(1):70–6.

Study on the effectiveness equivalent of army weapons based on the sos simulation

Z.Q. Zhao
Science and Technology on Composite Land Systems Simulation Laboratory, Beijing, China
College of Mechaeronics Engineering and Automation National University of Defense Technology,
Changsha, Hunan, China

X.Z. Zhang
Science and Technology on Composite Land Systems Simulation Laboratory, Beijing, China

ABSTRACT: Under the background of the system of systems (Sos) confrontation of army weapon, owing to being enclosed in the sealed fighting subsystem, and the effectiveness matrix is degrading, so the effectiveness equivalent on the varieties of the weapon in the system of systems cannot be conducted for many years as a difficult problem in the field of assessment on the effectiveness of weapon systems. The principle of multiple effectiveness equivalent aiming at expanding the traditional effectiveness equivalent theory was presented, filling in gap existing in the original theory, and satisfactorily realized the effectiveness equivalent of varieties of main weapon systems in army's weapon system of systems, which is an important breakthrough and innovation in the study on assessment of effectiveness of weapon systems, enjoying broad utilization prospect.

1 INTRODUCTION

The effectiveness of weapon system is an important attribute of weapon system, as well as a potential capacity of itself. Under certain conditions, this capacity can be measured by the extent to which the regulated tasks could be completed. Presently, the most difficult problem existing in the assessment on the effectiveness of weapon systems is how to determine the weighting coefficient of the effectiveness among the different weapon systems. The cause lies in the differences in the design concept of all kinds of weapon systems, the differences in the indicators of tactical and technical performance indicators, which causes poorer comparability among the different weapon systems, which is presently also the research subject in the technical field of assessment of effectiveness of weapon systems.

The assessment of effectiveness equivalent of weapon systems mainly include weapon index method, test statistic method, ADC method, economic value method, etc. (Guo & Yuan & Zhi, 2004).

1.1 *Weapons index method*

This method is to measure combat capability of the weapon with a dimensionless index. The index method contains scoring method, firepower index method, weapon index method, etc. To determine weapon systems capability index, it should take one kind of weapon as standard, setting the capacity index as 1, and then a capacity index is able to be given to each kind of weapon by combining the qualitative comparison and quantitative calculation. The drawback is that it cannot answer the issue of effectiveness equivalent, especially cannot reply the effectiveness equivalent question among different kinds of weapon systems.

1.2 *Test statistical method*

The test statistical method is such a method that will, on the regulated site or under the test conditions, collects sufficient data reflecting the performance and attribute of weapon systems, and then analyzes these data and conducts assessment on the effectiveness of system. The commonly-used methods are sampling survey, parameter estimation, hypothesis testing, regression analysis, correlation analysis, etc. The disadvantage of this method is that a great deal of statistical information is needed, therefore, needing lots of massive manpower and material resources. The test statistical method is suitable to be applied in the operation or exercise. But it is very difficult to obtain reliable and overall information.

1.3 *ADC method*

The ADC method—Model of Weapon System Effectiveness Assessment, was put forward by the Weapon System Effectiveness Industry Advisory

Committee (WSEIA) in January 1965, which aims at integrating effectiveness, reliability and capacity into a singular effectiveness measurement which can demonstrate the comprehensive performances of weapon system. The shortcoming of this method lies in the compositeness for determining the process of ADC matrix. Some factors are drawn from experiences or predictions, as well as the poor objectivity. Because the ADC method is based on building model by dividing according to system's states and by the transition probability of its conditions, so when the quantity of weapon systems increases, the matrix dimension will expand rapidly, not suitable for the assessment on army weapon system confronting multiple states.

1.4 *Economic value method*

Set the weighting coefficient of each weapon being directly proportional to its price. The disadvantage of this method is that the price index of weapon systems is difficult to be accurately determined, and the effectiveness of weapon is not exactly directly proportional to its price.

Because the modeling and simulation method can realistically simulate large-scale, composite military operations under the confronting conditions of weapons, can fully take the impact levied by manpower, weapon equipment and combating environment on effectiveness of weapon into sufficient consideration, can faithfully reflect the effectiveness of all kinds of weapon systems in the weapon systems, so it is presently the most effective method to obtain the basic data using for assessing the effectiveness of weapon systems, in addition to the real-facility combat or drill, as well as a reliable effectiveness assessment method. The effectiveness weighting coefficient method for assessing weapon systems put forward in this paper is a more scientific and more rational method for obtaining weapon effectiveness coefficient, by conducting effectiveness equivalent and multiple equivalent based on the test result of confrontation simulation of weapon system of system. It can overcome the shortages of the above-mentioned methods to a great extent.

2 MODEL OF EFFECTIVENESS EQUIVALENT OF WEAPON SYSTEMS AND ITS SOLUTION

2.1 *Matrix representation model of effectiveness equivalent of weapon systems*

Set up sets of weapon systems which effectiveness equivalent should be assessed according to the research target, establish or utilize the existing simulation system of army weapon system of

systems counterwork, carry out simulation system and make statistical on simulation results.

Introduce matrix notation

$$A = \left(A_{ij} \right)_{m \times n} \quad B = \left(B_{ji} \right)_{n \times m} \tag{1}$$

where, A is $m \times n$ matrix, whose elements are the number of each kind of weapons of Blue Army destroyed by the various weapons of Red Army, which is called the effectiveness matrix of Red Army; while B is $n \times m$ matrix, whose elements are the number of each kind of weapons of Red Army destroyed by the various weapons of Blue Army, which is called the effectiveness matrix of Blue Army. The weighting coefficient and comprehensive effectiveness of Red Army's and Blue Army's weapons can be expressed as a single row matrix

$$d = \begin{pmatrix} d_1 \\ d_2 \\ \dots \\ d_m \end{pmatrix} \quad H = \begin{pmatrix} H_1 \\ H_2 \\ \dots \\ H_m \end{pmatrix} \quad c = \begin{pmatrix} c_1 \\ c_2 \\ \dots \\ c_n \end{pmatrix} \quad L = \begin{pmatrix} L_1 \\ L_2 \\ \dots \\ L_n \end{pmatrix}$$

$$\tag{2}$$

In (2), H and C are comprehensive effectiveness vectors, c and d are the weighting coefficient vectors, and satisfy

$$c'c = 1 \quad d'd = 1 \tag{3}$$

Here, c' and d' which are row vector, respectively refer to the transposed vectors of column vector c and d.

In the above notations, we have

$$H = Ac \quad L = Bd \tag{4}$$

where, α and β are proportionality constant.

The above equations have given the mathematical model for determining the weighting vector c and d.

If weighting vector c and d exist, then

$$\alpha d = H = Ac \tag{5}$$

$$\alpha \beta d = A(\beta c) = AL = ABd \tag{6}$$

This indicates that the weighting vector d is eigenvector of m-order square matrix AB corresponding to the eigenvalue $\alpha \beta$. Similarly, we have

$$\alpha \beta c = BAc \tag{7}$$

i.e. the weighting vector c is the eigenvector of n-order square matrix BA corresponding to the eigenvalue $\alpha \beta$.

2.2 Solution to effectiveness weighting coefficient when effectiveness matrix is regular

In order to study the existence of the solution of effectiveness equivalent model, some of natures of positive matrix and nonnegative matrix should be used.

If the elements in the square matrix D are all positive ones, it is called positive matrix, note $D > 0$; if elements in the square matrix D are all nonnegative ones, it is called nonnegative matrix, note $D \geq 0$; if each component of vector is positive, then it is called positive vector; while if each component of vector is nonnegative, then it is called nonnegative vector. In addition, for a square matrix, the eigenvalue, which has largest modulus, is called the first eigenvalue or the largest eigenvalue.

Lemma 1: The largest eigenvalue of positive matrix is unique and should be a positive number, each component of the corresponding eigenvector is positive number, and should be a unique one if regardless of length. Corresponding to the nonnegative matrix D, if natural number m exists and $D_m > 0$, then the largest eigenvalue of matrix D is a positive one and the solely one. Each component of corresponding eigenvector is positive and is a unique one if regardless of length.

Lemma 2: The largest nonnegative eigenvalue exists in the nonnegative matrix, corresponding to this eigenvalue; the eigenvector with all its components being nonnegative must exist.

Lemma 3: For the given effectiveness matrix A and B of Red Army and Blue Army, the product matrix AB and product matrix BA have the same largest eigenvalue.

If effectiveness matrix A and B of Red Army and Blue Army was given, then the product matrix AB and BA are all square matrices. If their largest eigenvalue has positive eigenvector corresponding to it, then the matrix A and B are called regular, otherwise, they will be called non-regular or degraded. In case of regular, varieties of weapon systems used in the combat should be included in the scope of effectiveness equivalent, while in case of non-regular, the weapon systems corresponding to the components with their value being zero in the weighting vector are not put in the scope of effectiveness equivalent. In case of non-regular, effectiveness equivalent can only be realized in party of the weapon systems instead of realizing in overall weapon systems which are used in the combat. That is what we dislike to see. Obviously, when AB and BA satisfy $AB > 0$ and $BA > 0$, then the effectiveness matrix A and B are regular, particularly, if the elements of effectiveness matrix A and B are all positive, then they are certainly regular.

If the effectiveness matrix A and B of Red Army and Blue Army are given, for obtaining the weighting coefficient of effectiveness equivalent of varieties of weapon systems, it needs to seek the eigenvalue of product matrix AB and BA. If the matrix is regular matrix, there are usually two methods to seek weighting coefficient vector: One is to obtain by using the iterative algorithm of eigenvector corresponding to the first eigenvalue of matrix, the other one is to seek by using approximate equation. (Zhao & Huang & Ni, 2008).

By using the iterative algorithm for seeking eigenvector corresponding to the first eigenvalue of matrix, not only the weighting vector of effectiveness equivalent of various weapons of Red Army and Blue Army, and a proof was given to the above-mentioned Lemma 2. A description on this algorithm is given bellow by taking the solution of seeking the largest eigenvalue in Matrix AB and its corresponding eigenvector as example

The iterative algorithm is as follows

$$x_0 = \frac{1}{\sqrt{m}} \begin{pmatrix} 1 \\ 1 \\ \cdots \\ 1 \end{pmatrix}, \quad \lambda_0 = x_0' AB x_0$$

$$x_1 = AB x_0 / |AB x_0|, \quad \lambda_1 = x_1' AB x_1$$

$$\cdots$$

$$x_{k+1} = AB x_k / |AB x_k|, \quad \lambda_{k+1} = x_{k+1}' AB x_{k+1} \qquad (8)$$

If accuracy ε was given, carry out the Iteration calculation till the following equation established.

$$|\lambda_{k+1} - \lambda_k| \leq \varepsilon \qquad (9)$$

x_{k+1} is the effectiveness equivalent weighting vector of weapons. Obviously, when matrices A and B, are positive matrix, the approximation of eigenvalue to seek for is positive, each component of the corresponding eigenvectors is also positive.

3 SOLUTION TO THE EFFECTIVENESS WEIGHTING COEFFICIENT BASED ON THE MULTIPLE EFFECTIVENESS EQUIVALENT METHOD

Seen from the studies on the effectiveness equivalent of various belligerent relations, (Zhao & Ni & Huang, 2009) we can know, when effectiveness matrix presents a form of diagonal block or downward diagonal block, then it will be a degraded one, equivalent cannot be conducted in the whole war weapon systems involved in combat. Therefore, the existing effectiveness equivalent theory cannot be applied in the effectiveness equivalent of various weapon systems in the confronting simulation of weapon system of systems. In order to get rid of this dilemma, we propose the multiple effectiveness equivalent method.

3.1 Solution to the effectiveness weighting coefficient when the effectiveness matrix is diagonal block effectiveness matrix

In the confrontation of Red Army and Blues Army, the Red Army has m kinds of weapon systems, and the Blues Army has n kinds of weapon systems. Evenly divide the weapon systems of Red Army and Blue Army into k groups. The numbers of weapon system kinds in each group are marked as $m_1, m_2 \dots m_k$ and $n_1, n_2, \dots n_k$. The Red Army just attacks the No.i group of weapons of Blue Army with its own No.i weapons; on the contrary, the No.i group weapons of Blue Army just attacks the No.i group of weapons of Red Army. Thus, the effectiveness matrix can be expressed as

$$A = \begin{pmatrix} A_1 & 0 & \dots & 0 \\ 0 & A_2 & \dots & 0 \\ \dots & \dots & \dots & \dots \\ 0 & 0 & \dots & A_k \end{pmatrix} \quad B = \begin{pmatrix} B_1 & 0 & \dots & 0 \\ 0 & B_2 & \dots & 0 \\ \dots & \dots & \dots & \dots \\ 0 & 0 & \dots & B_k \end{pmatrix}$$

(10)

At this time, the effectiveness matrix of Red Army and Blue Army is a diagonal block matrix, but the product $A_i B_i, B_i A_i (i=1,2,\dots k)$ of each diagonal block matrix is regular, therefore, the effectiveness equivalent on various weapon systems in each weapon group can be conducted. Because the system effectiveness matrix A and B of Red Army and Blue Army are a degraded one, so the corresponding equivalent relation cannot be established in the overall m kinds of weapons of Red Army and overall n kinds of weapons of Blue Army. To overcome this obstacle, we first create an effectiveness equivalent relationship in each group of weapons, then establish corresponding equivalent relationship between each group of weapons, and finally realize the effectiveness equivalent among overall weapons by combining the equivalent relationships at two layers together.

The effectiveness matrices of the confrontation between Group i weapons of Red Army and the Group i weapons of Blue Army are respectively A_{ii} and B_{ii}, a kind of matrix with its elements being all positive. The weighting coefficient vector of various weapons of the group are respectively d_i and c_i, which satisfy.

$$A_i B_i d_i - \lambda_i d_i \quad B_i A_i c_i = \lambda_i c_i \quad (11)$$

The comprehensive effectiveness of the group of weapons of Red Army and Blue Army are respectively as follows

$$\alpha_i = d_i' A_i c_i, \; \beta_i = c_i' B, \; \alpha_i \beta_i = \lambda_i \quad (12)$$

We note that comprehensive effectiveness of each group of weapon system was obtained in the same operational context. It has comparability. Thus, α_i and β_i can be taken as the weigh coefficient for the Group i weapon system of Red Army and Blue Army. Thus, the ratio of weighting coefficient of each group of weapon systems between Red Army and Blue Army are as follows

$$\alpha_1 : \alpha_2 : \dots : \alpha_k$$
$$\beta_1 : \beta_2 : \dots : \beta_k \quad (13)$$

Note

$$\alpha = (\alpha_1, \alpha_2, \dots, \alpha_k)^{-1}, \; \beta = (\beta_1, \beta_2, \dots, \beta_k)^{-1} \quad (14)$$

Thus, by normalization processing, the length of the vector α and β can be written as $|\alpha|$ and $|\beta|$, the composite weighting coefficient vector of No.i group of weapons of Red Army can be obtained

$$\bar{d}_i = \alpha_i d_i / |\alpha| \quad (i = 1, 2, \dots k) \quad (15)$$

The composite weighting coefficient vector of No.i group of weapons of Blue Army can be obtained

$$\bar{c}_i = \beta_i c_i / |\beta| \quad (i = 1, 2, \dots k) \quad (16)$$

3.2 Solution to the effectiveness weighting coefficient when the effectiveness matrix is downward diagonal block effectiveness matrix

In the confrontation of Red Army and Blue Army, the Red Army has m kinds of weapon systems, and the Blues Army has n kinds of weapon systems. Evenly divide the weapon systems of Red Army and Blue Army into k groups. The numbers of weapon systems kinds of each group are marked as $m_1, m_2 \dots m_k$ and $n_1, n_2, \dots n_k$. The Red Army just attacks the No.i group of weapons of Blue Army with its own No.i weapons; on the contrary, the No.i group of weapons of Blue Army just attacks the No.i group of weapons of Red Army. Thus, the effectiveness matrix can be expressed as

$$A = \begin{pmatrix} A_{11} & 0 & \dots & 0 \\ A_{21} & A_{22} & \dots & 0 \\ \dots & \dots & \dots & \dots \\ A_{k1} & A_{k2} & \dots & A_{kk} \end{pmatrix} \quad B = \begin{pmatrix} B_{11} & 0 & \dots & 0 \\ B_{21} & B_{22} & \dots & 0 \\ \dots & \dots & \dots & \dots \\ B_{k1} & B_{k2} & \dots & B_{kk} \end{pmatrix}$$

(17)

At this time, the effectiveness matrix of Red Army and Blue Army is a downward diagonal block matrix, but the product $A_{ii}B_{ii}, B_{ii}A_{ii} (i = 1, 2, \cdots k)$ of each diagonal block matrix is regular, therefore, the effectiveness equivalent on various weapon systems in each weapon group can be conducted. Because the system effectiveness matrix of Red Army and Blue Army is a degraded one, so the corresponding equivalent relations cannot be established in the overall m kinds of weapons of Red Army and overall n kinds of weapons of Blue Army. To overcome this obstacle, we first, follow the above example, create an effectiveness equivalent relationship among various weapons in each group of weapons following the multiple equivalent principles, then establish corresponding equivalent relationship between each group of weapons, and finally realize the effectiveness equivalent among overall weapons by combining the equivalent relationships at two layers.

The oppositional effectiveness matrix of No.i group of weapon of Red Army and the No.i group of weapon of Blue Army are respectively A_{ii} and B_{ii}. The weighting coefficients vectors of the said group of various weapons are respectively d_i and c_i, satisfy

$$\alpha_i = d_i'A_ic_i, \quad \beta_i = c_i'B, \quad \alpha_i\beta_i = \lambda_i \tag{18}$$

The comprehensive effectiveness of the said group of weapons are respectively as follows

$$\alpha_i = d_i''(A_{i1} \ A_{i2} \ \cdots \ A_{ii})*(c_1 \ c_2 \ \cdots \ c_i)^{-1}$$
$$= d_i'\sum_{j=1}^{i} A_{ij}c_j$$
$$\beta_i = c_i'(B_{i1} \ B_{i2} \ \cdots \ B_{ii})*(d_1 \ d_2 \ \cdots \ d_i)^{-1}$$
$$= c_i'\sum_{j=1}^{i} B_{ij}d_j \tag{19}$$

For the diagonal block matrix and downward diagonal block matrix, the effectiveness of equivalent of each group of weapon systems can be obtained by the corresponding block matrix on the diagonal line. Its model and algorithm are the same. When calculating the comprehensive effectiveness of each group of weapons, only the combating results within each group of weapons need to be considered when using diagonal matrix. As for the downward diagonal block matrix, the confronting effectiveness of each group of weapons with other group of weapons should be considered. Thus, the equations used in calculating comprehensive effectiveness under the above two circumstances are different. Doing so will better reflect the combating effectiveness of each group of weapons, therefore more rational.

Similarly, the comprehensive effectiveness of each group of weapon systems was gained in the same operational context, so which is comparable. Thus, α_i and β_i can be taken as weighting coefficient of No.i group of weapons of both Red Army and Blue Army. Thus, the ratios of weighting coefficient of each group of weapon systems of both Red Army and Blue Army are

$$\alpha_1 : \alpha_2 : \cdots : \alpha_k$$
$$\beta_1 : \beta_2 : \cdots : \beta_k \tag{20}$$

Note

$$\alpha = (\alpha_1, \alpha_2, ..., \alpha_k)^{-1}, \ \beta = (\beta_1, \beta_2, ..., \beta_k)^{-1} \tag{21}$$

Thus, by normalization processing, the lengths of the vector α and β, are $|\alpha|$ and $|\beta|$, and the composite weighting coefficient vector of the No.i group of weapons of Red Army is

$$\bar{d_i} = \alpha_i d_i / |\alpha| \quad (i = 1, 2, ... k) \tag{22}$$

The composite weighting coefficient vector of the No.i group of weapons of Blue Army is

$$\bar{c_i} = \beta_i c_i / |\beta| \quad (i = 1, 2, ... k) \tag{21}$$

4 THE MULTIPLE EFFECTIVENESS EQUIVALENT IN THE CONFRONTATION OF ARMY WEAPON SYSTEM OF SYSTEMS

The confrontation and belligerent relations of army weapon system of systems may be divided into four basic types: direct-aim weapons, suppression weapons, air defense weapons and air strike weapons. The direct-aim weapons include light weapons, anti-armor weapons and armor weapons. Each kind of direct-aim weapon of both Red Arm and Blue Army can and can only confront with the direct-aim weapon of the other side. The suppression weapons of both armies can combat each other and can suppress the direct-aim weapons, air defense weapons, information equipment and support equipment. The air defense weapons of one side can only combat with the air strike weapons of the other side. The air strike weapons of the two sides cannot only raid the air defense weapons of the other side, but can also raid the direct-aim weapons, suppression weapons, information equipment and support equipment.

With the confronting simulation of army weapon system of systems as background, and by the consideration of the mutilates caused by the main weapons of the two sides, in accordance with the belligerent relations, the main weapons can be divided into four groups, to express with the effectiveness matrix of Red Army and Blue Army as

$$
A = \begin{pmatrix} A_{11} & 0 & 0 & 0 \\ A_{21} & A_{22} & A_{23} & 0 \\ 0 & 0 & 0 & A_{34} \\ A_{41} & A_{42} & A_{43} & 0 \end{pmatrix}
$$

$$
B = \begin{pmatrix} B_{11} & 0 & 0 & 0 \\ B_{21} & B_{22} & B_{23} & 0 \\ 0 & 0 & 0 & B_{34} \\ B_{41} & B_{42} & B_{43} & 0 \end{pmatrix} \tag{24}
$$

According to the structure of effectiveness matrix, the main weapons can be adjusted into two groups: a group is direct-aim weapons, the other group is suppression weapons, air defense weapons and air strike weapons. Then effectiveness matrix of Red Army and Blue Army can be adjusted to

$$
A = \begin{pmatrix} A_1 & 0 \\ A_3 & A_2 \end{pmatrix} \quad B = \begin{pmatrix} B_1 & 0 \\ B_3 & B_2 \end{pmatrix} \tag{25}
$$

Where,

$$
A_1 = A_{11} \qquad\qquad B_1 = B_{11}
$$

$$
A_2 = \begin{pmatrix} A_{22} & A_{23} & 0 \\ 0 & 0 & A_{34} \\ A_{42} & A_{43} & 0 \end{pmatrix} \quad B_2 = \begin{pmatrix} B_{22} & B_{23} & 0 \\ 0 & 0 & B_{34} \\ B_{42} & B_{43} & 0 \end{pmatrix}
$$

$$
A_3 = (A_{21}\ 0\ A_{41})^{-1} \quad B_3 = (B_{21}\ 0\ B_{41})^{-1} \tag{26}
$$

They are the downward diagonal block effectiveness matrix, and the weighting coefficients of the weapon system can be determined by the model of multiple comprehensive effectiveness. First, the matrix product $A_1 B_1$ and $B_1 A_1$ are regular, the largest eigenvalue λ_1 and eigenvector d_1 and eigenvector c_1 can be obtained, namely

$$
A_1 B_1 d_1 = \lambda_1 d_1, \quad B_1 A_1 c_1 = \lambda_1 c_1 \tag{27}
$$

Thus, the weighting coefficient c_1 and d_1 of effectiveness equivalent of direct-aim weapons system of both Red Army and Blue Army can be obtained. On the other side,

$$
A_2 B_2 = \begin{pmatrix} A_{22}B_{22} & A_{22}B_{23} & A_{23}B_{34} \\ A_{34}B_{42} & A_{34}B_{43} & 0 \\ A_{42}B_{22} & A_{42}B_{23} & A_{43}B_{34} \end{pmatrix} \tag{28}
$$

$$
B_2 A_2 = \begin{pmatrix} B_{22}A_{22} & B_{22}A_{23} & B_{23}A_{34} \\ B_{34}A_{42} & B_{34}A_{43} & 0 \\ B_{42}A_{22} & B_{42}A_{23} & B_{43}A_{34} \end{pmatrix} \tag{29}
$$

Because there are zero matrix blocks in the two product matrices, so they cannot be judged as regular. But, it is easy to know that the elements of the square of these two product matrix $(A_2 B_2)^2$ and $(B_2 A_2)^2$ are all positive, thus, by extending the Perron Theorem, we can see that the two effectiveness matrices are regular, thus the maximum eigenvalue λ_2 and its corresponding eigenvector c_2 and d_2 can be obtained

$$
A_2 B_2 d_2 = \lambda_2 d_2, \quad B_2 A_2 c_2 = \lambda_2 c_2 \tag{30}
$$

These two eigenvectors yield the weighting coefficients of effectiveness equivalent in the weapon systems for the suppression weapons, air defense weapons and air strike weapons of both Red Army and Blue Army.

In accordance with the principle of the multiple equivalent, then

$$
\alpha_1 = d_1' A_1 c_1, \quad \beta_1 = c_1' B d_1
$$
$$
\alpha_2 = d_2'(A_3 c_1 + A_2 c_2), \quad \beta_2 = c_2'(B_3 d_1 + B_2 d_2) \tag{31}
$$

Thus, multiple effectiveness equivalent coefficients

$$
d_i^* = \frac{\alpha_i}{\sqrt{\alpha_1^2 + \alpha_2^2}} d_i, \quad c_i^* = \frac{\beta_i}{\sqrt{\beta_1^2 + \beta_2^2}} c_i \tag{32}
$$

where, $i \in [1,2]$, from which we can yield effectiveness equivalent weighting coefficients of various weapon systems of direct-aim weapons, suppression weapons, air defense weapons and air strike weapons in army's weapon and facilities systems under the background of confrontation of army's weapon and facility systems.

5 CONCLUSION

Since the seventies of last century, the effectiveness equivalent principle of weapon systems could only be applied to equivalent of direct-aim weapons. Under the background of confrontation of army weapon system of systems, and owing to the existence of sealed combating subsystem, the effectiveness matrix is degrading, so the effectiveness equivalent could not be implemented in various weapon systems in the system. This is a difficult problem in the field of effectiveness assessment on weapon systems for many years, which resulted in the stagnant of the equivalent research on weapon systems. The multiple effectiveness equivalent principle contributed by us has successfully gotten rid of this barrier, effectively expanded the scope of traditional effectiveness equivalent theory, creatively filled the gap in the original theory and finally realized the effectiveness equivalent in each main weapon system in the army weapon system of systems. This is an important breakthrough and innovation in the study of effectiveness equivalent assessment on weapon systems, enjoying a broad application prospects.

The weighting coefficient of the weapon systems obtained on the basis of the effectiveness equivalent of confrontation simulation of weapon system of systems, like weapon index, has universal useful value and important significance in the demonstration of weapon systems development, military training and operations analysis. In the analysis of weapon system of systems, weighting coefficient supplies effective tool for the static assessment of weapon system of systems, and the optimization of the structure of the weapon system of systems, as well as provides support for the scientification of decision-making on weapon systems development. In the operation analysis, weighting coefficient can supply scientific means for unfolding troop planning and optimizing force structure.

REFERENCES

Guo Q.S., Yuan Y.M. & Zhai Z.G. 2004. Study on Effectiveness of Military Equipment and its Assessment Methods. *Journal of The Academy of Armored Forces Engineering* 18(1): 1–5.

Zhao Z.Q., Huang K.D. & Ni Z.R. 2008. Effectiveness Equivalent Model and Algorithm of Weapon Systems. *Journal of System Simulation* 20(5): 1103–1106.

Zhao Z.Q., Ni Z.R. & Huang K.D. 2009. Study on the Degradation of Effectiveness Equivalent Model of Weapon Systems. *Journal of System Simulation* 21(21): 6716–6720.

Control Engineering and Information Systems – Liu (Ed)
© 2015 Taylor & Francis Group, London, ISBN 978-1-138-02685-8

An improved variable step-size LMS adaptive filtering algorithm and its application simulation

Y.F. Li
Zhejiang Business Technology Institute, Ningbo, China
College of Information Science and Engineering, Ningbo University, Ningbo, Zhejiang, China

Z.W. Zheng
College of Information Science and Engineering, Ningbo University, Ningbo, Zhejiang, China

ABSTRACT: This paper presents an improved normalized variable step size LMS adaptive filtering algorithm, a normalized LMS algorithm with variable step size iterative formula is deduced. The simulation results prove that the new algorithm has good convergence speed and smaller steady-state error and good performance on real time tracking. At the same time also analyzed and applied the algorithm in system identification, noise cancellation and adaptive notch and other aspects of the implemented through the MATLAB simulation, results show that, the proposed algorithm has been applied. The LMS adaptive filtering algorithm has been widely used in many applications such as system identification, noise cancellation and the adaptive notch filter, the paper analyses the application and implement the simulation by matlab. The result shows the proposed algorithm has been applied well.

Keywords: adaptive filtering; variable step size; Least Mean Square algorithm

1 INTRODUCTION

Adaptive filter theory is an important research topic in the area of adaptive signal processing. It call adapt the filter coefficients according to some criterions and approach optimal filtering. The design of adaptive filter algorithm is an important part within the design of adaptive filter. The performance of the adaptive algorithm decides the performance of the adaptive filter. Least Mean Square (LMS) algorithm is a classical adaptive algorithm which has simple structure, good stable property, and low computational complexity, thus is easy to be implemented. It has been widely used in many applications such as system identification, noise cancellation, speech signal prediction, adaptive channel equalization, adaptive antenna array and so on. The main drawback of the LMS algorithm is its slow convergence rate. Which deteriorates its performance in many applications.

In order to solve this problem, there are many improved adaptive algorithm is proposed, such as variable step size algorithm in time domain, the transform domain LMS algorithm and so on. An improved variable step size LMS adaptive algorithm is proposed in this thesis, it has better convergence speed and relatively smaller offset steady-state in the steady state Simulation results show that the proposed algorithms perform better in some applications such as System identification, noise canceller and adaptive notch filter.

2 ADAPTIVE FILTER

The adaptive filter is a kind of special Wiener filter and it can adjust the parameters automatically. When designed it need not to know the statistical characteristics of the input signal and noise in advance, it can gradually "understanding" in the work process or to estimate the desired statistical properties, and then automatically adjust its parameters, in order to achieve the best filtering effect.

There are two input: $X(n)$ and $D(n)$, two output: $Y(n)$ and $E(n)$, both are time series in the Figure 1. The $X(n)$ can be not only a single input signal but also multiple input signals. These signals represented different content In different application background.

The input vector is

$$X(n) = \left[x(n)x(n-1) \ldots x(n-M+1) \right]^T \qquad (1)$$

As shown in Figure 2.

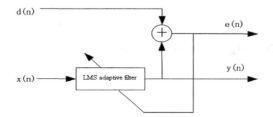

Figure 1. The schematic diagram of adaptive filter.

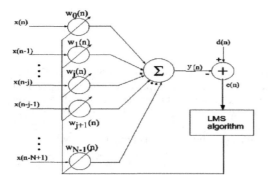

Figure 2. Adaptive linear combiner.

The weighted vector (i.e. filter parameter vector) is

$$W(n) = \left[w_1(n)w_2(n) \dots w_M(n) \right]^T \qquad (2)$$

The output of the filter is

$$y(n) = \sum_{i=1}^{M} w_i(n)x(n-i+1) = W^T(n)X(n)$$
$$= X^T(n)W(n) \qquad (3)$$

$Y(n)$ with respect to filter the desired output $d(n)$ error is

$$e(n) = d(n) - y(n) = d(n) - W^T(n)X(n) \qquad (4)$$

According to the minimum mean square error criterion, the optimal filter parameters should keep the mean square error is the least, in $W(n)$ is a constant vector case, the mean square error expression of n is at the moment

$$\xi(n) = E\left[e^2(n) \right]$$
$$= E\left[d^2(n) \right] - 2P^T W(n) + W^T(n)R_x W(n) \qquad (5)$$

Among them, $E[d^2(n)]$ is the variance of expected response $D(n)$, $P = E[d(n)X(n)]$ is the

cross-correlation vector of input vector and the expected response $D(n)$, $R_x = E[X(n)X^T(n)]$ is the autocorrelation matrix of input vector $X(n)$. On $W(n)$ derivative, and take the derivative equal to zero, then the canonical equation will be obtained.

$$\frac{\partial \xi}{\partial W} = -2P + 2R_x W_{opt} = 0 \qquad (6)$$

When R_x is the full rank, the canonical equation has a unique solution

$$W_{opt} = R_x^{-1}P \qquad (7)$$

This solution is called Wiener solution. When $W = W_{opt}$ the minimum of mean square error function (i.e., the minimum mean square error) is equal to

$$\xi_{min} = E\left[e^2(n) \right]_{min} = E\left[d^2(n) \right] - P^T W_{opt} \qquad (8)$$

The computational method for inverse of Rx will be get the big, Gradient method is often used in practice.

3 GRADIENT METHOD AND IMPROVED LMS ADAPTIVE ALGORITHM

Gradient method is a kind of method of no matrix inversion for solving canonical equation, it seeks the best value for weighting vector by recursive method. It is the basis of LMS algorithm which is the most widely used.

As the name suggests, the gradient method is the negative gradient direction along the surface of the performance, that is searched the lowest point downward along the steepest direction of performance surface. This is an iterative search process. Expression to compute the weight vector for the gradient method

$$W(n+1) = W(n) + \mu[-\nabla(n)] \qquad (9)$$

Among them, μ is a positive number, $\nabla(n)$ is the mean square error, $\xi = E[e^2(n)]$ relative to the weight vector $W(n)$ gradient, the expression is

$$\nabla = \frac{\partial \xi}{\partial W} = -2P + 2R_x W \qquad (10)$$

Type (9) is a recursive expression. As long as the value μ is chosen appropriately, regardless of the initial value $W(0)$, the mean square error ξ will be tend to its minimum value. In fact it is impossible to rigorously obtained gradient vector, then the gradient vector can be made the estimation of

$$\hat{\nabla}(n) = -2X(n)d(n) + 2X(n)X^T(n)W(n)$$
$$= -2X(n)e(n) \tag{11}$$

Then the weight vector recursion formula can be obtained

$$W(n+1) = W(n) + 2\mu X(n)e(n) \tag{12}$$

Equation (12) update type which is expressed by convergence and tracking ability better, at the same time, small amount of calculation.

The step parameter μ is the key factor that influence the convergence of LMS algorithm, which updates the tap weight vector at each iteration of the volume. The mean weight vector converges to W_{opt}, that is

$$\lim_{n \to \infty} E\left[\hat{W}(n)\right] = W_{opt} \tag{13}$$

Parameters μ must satisfy the following formula

$$0 < \mu < \frac{1}{\lambda_{max}} \tag{14}$$

λ_{max} is One of the biggest eigen value of R_x

The disadvantages of the classical LMS algorithm is slow of convergence speed, so the emergence of new LMS algorithm continuously. In this paper, in order to ensure the stable convergence of adaptive filter, the convergence factor is normalized, the normalized convergence factor of the algorithm is expressed as

$$\mu' = \frac{\mu}{\sigma_x^2} \tag{15}$$

In the formula, σ_x^2 is the variance of the input signal $x(n)$. It is hard to get the results from direct calculation, we often use the time average to replace the statistical variance, that is

$$\sigma_x^2 = \sum_{i=0}^{M} x^2(n-i) = x^T(n)x(n) \tag{16}$$

the normalized convergence factor is substituted into LMS algorithm, it will be got

$$W(n+1) = W(n) + 2\frac{\mu}{x^T(n)x(n)}e(n)x(n) \tag{17}$$

In order to avoid the emergence of the 0 value, we add the small positive constant ψ in the formula in the denominator. So the normalized LMS algorithm for iterative formula will be get

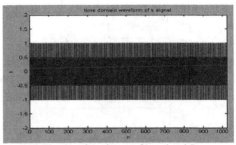

The time domain waveform signal S

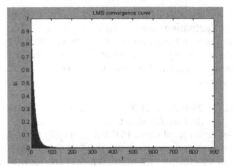

LMS algorithm convergence curve

Figure 3. The time domain waveform signal S. LMS algorithm convergence curve.

$$W(n+1) = W(n) + 2\frac{\mu}{\psi + x^T(n)x(n)}$$
$$\times e(n)x(n) \tag{18}$$

The following is the use of statistical methods, LMS algorithm on this variable step size convergence curve to the simulation.

4 SIMULATION OF AN IMPROVED LMS FILTERING

4.1 System identification based on improved LMS algorithm

4.1.1 System identification

The unknown system identification is shown as the Figure 4. It is found that the adaptive filter and another unknown transfer function of the filter were inputted at the same time $x(n)$. The output $d(n)$ of unknown controlled object is the output of the all system.

In the convergence, the output of adaptive filter $\hat{d}(n)$ is in an optimal manner similar to the $d(n)$. It is matched that the order of adaptive filter which provided and unknown controlled object, and the input signal $x(n)$ is the generalized steady-state, coefficient of the adaptive filter will be convergent

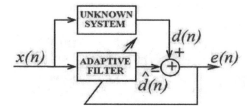

Figure 4.   System identification principle diagram.

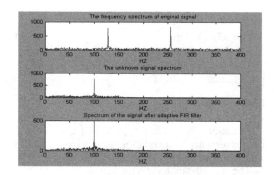

Figure 5.   The spectrum of signal processing system.

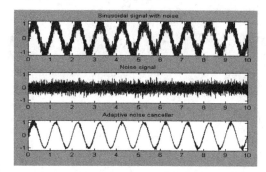

Figure 6.   Noise canceller.

to the same value of unknown. controlled object. The actuating signal $x(n)$ is usually random sequence in Figure 3 because of its wide bandwidth and traversing the entire band, it can encourage all modal system and obtain system parameters more accurate.

4.1.2   *System identification modeling and matlab simulation*

Through the adaptive FIR filter, modify the system function constantly, make the full approximation with parameters of unknown system, reduce the error, and achieve the purpose of system identification. In view of unknown system of Butterworh type, using LMS algorithm, simulation results using Matlab as shown in Figure 5.

It can be seen from the Figure 5 that the adaptive FIR filter can simulation the unknown system well, and Close to processing effects of the original signal. Thus It can be got the system function of the unknown system through the parameters of the adaptive FIR filter and It can be carried out the function of the same hardware reconfiguration of the unknown system.

4.2   *Adaptive noise canceller based on improved LMS algorithm*

Adaptive Noise Cancellation (ANC) system is a kind of special adaptive optimal filter, It is the first successful research at the Stanford University in the United states in 1965. The basic principle of ANC is a cancellation algorithm between pollution signal by noise and reference signal, so as to eliminate the noise in noise signal. The key problem is it must have a certain correlation with reference signal and the noise to be eliminated, and it is irrelevant with signal to check.

The improved LMS algorithm is proposed in this paper, it is the design of a 2 order weighted adaptive noise canceller, filtering the sinusoidal signal with white noise channel of Gauss. The results are shown as follows.

In the Figure 6, the signal source generates a sine signal, and overlay with Gauss white noise produced by noise source into main channel of

ANC. The input of the adaptive filter is a single noise signal generated by noise source, It can adjust weight coefficient of linear combiner adaptively by the improved LMS algorithm, cancel the noise signal in the main channel and the reference channel. And the Output error signal is the expected sinusoidal signal generated by the signal source.

4.3   *Adaptive notch filter based on improved LMS algorithm*

In the communication system and electronic systems, it is often interfered by 50 Hz, such as single frequency or narrow band. This interference serious impacts on the reliability and validity of signal receiving or detecting. So it need to adopt the adaptive notch filter to eliminate this interference. When reference input of the adaptive noise cancellation system is a single frequency sine signal, the system can be adaptive notch filter.

The adaptive notch filter has two advantages, one is the frequency characteristic can have a very narrow stop band, which is closed to the ideal characteristics, and the width of stop band is easy to control; the another advantage is when there are

694

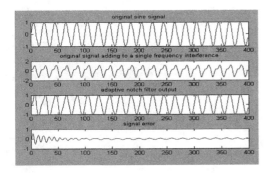

Figure 7. Implement by the improved LMS algorithm.

changes in interference frequency, the position of stop band can track this change.

The following design is to use the common ideal low-pass filter based on the 50 Hz frequency interference notch filter, and implement by the improved LMS algorithm, as shown in Figure 7.

In Figure 7, the original input signal is a sine signal. The input of the adaptive filter is the same to the interference of single frequency signal in frequency but is different in phase.

It can adjust weight coefficient of linear combiner adaptively by the improved LMS algorithm, and the adaptive notch filter can be used to output the original input sinusoidal signal. Changes in the signal error, when the number of iterations increases, the error tends to zero, so as to meet the design requirements.

5 CONCLUSION

This paper presents an improved normalized variable step size LMS adaptive filtering algorithm, analyzes the application of the algorithm in system identification, noise cancellation and adaptive notch etc. and using MATLAB software modeling and simulation, the simulation results show that the algorithm has better feasibility and applicability in different systems.

ACKNOWLEDGMENTS

This work was supported in part by the National Science Foundation of China (No. 60972063), the National Science and Technology Major Project (No. 2011ZX03002-004-02), the Program for New Century Excellent Talents in University (No.NCET-08-0706), the Science Foundation of Zhejiang Province (No. R1110416), the Specialized Research Fund for the Doctoral Program of Higher Education (No. 20113305110002), the Program for Technology Innovation Team of Ningbo Government (No. 2011B81002), Zhejiang Scientific and Technical Key Innovation Team of New Generation Mobile Internet Client Software (2010R50009) The Provincial Department of education scientific research project of Zhejiang (Y201330021).

REFERENCES

Abry P., R. Baraniuk, P. Flandrin, R. Riedi and D. Veitch, "Multiscale nature of network traffic," IEEE Signal Process. Mag., vol. 19, no. 3, pp. 28–46, May 2002.

Chang Y.C. and S. Chang, "A fast estimation algorithm on the Hurst parameter of discrete-time fractional Brownian motion," IEEE Trans. Signal Process. vol. 50, no. 3, pp.554–559, Mar. 2002.

Flandrin P., "On the spectrum of fractional Brownian motions," IEEE Trans. Inf. Theory, vol. 35, pp. 197–199, Feb. 1989.

Liu S.C. and S. Chang, "Dimension estimation of discrete-time fractional Brownian motion with applications to image texture classification," IEEE Trans. Image Process., vol. 6, pp. 1176–1184, Aug. 1997.

Loussot T., R. Harba, G. Jacquet, C.L. Benhamou, E. Lespesailles and A. Julien, "An oriented fractal analysis for the characterization of texture: Application to bone radiographs," EUSIPCO Signal Process., vol. I, pp. 371–374, Sep. 1996.

Mandelbrot B. and J.W. Van Ness, "Fractional Brownian motions, fractional noises and applications," SIAM Rev., vol. 10, pp. 422–437, Oct. 1968.

Pentland P., "Fractal-based description of natural scene," IEEE Trans. Pattern Anal. Mach. Intell., vol. PAMI-6, no. 6, pp. 661–674, Nov. 1984.

Perrin E., R. Harba, C.B. Joseph, I. Iribarren and A. Bonami, "nthorder fractional Brownian motion and fractional Gaussian noises," IEEE Trans. Signal Process., vol. 49, no. 5, pp.1049–1059, May 2001.

Wornell G.W. and A.V. Oppenheim, "Estimation of fractal signals from noisy measurements using wavelets," IEEE Trans. Signal Process. vol. 40, no. 3, pp. 611–623, Mar. 1992.

Wornell G., Signal Processing with Fractals: A Wavelet-Based Approach. Englewood Cliffs, NJ: Prentice-Hall PTR, 1996.

Control Engineering and Information Systems – Liu (Ed)
© 2015 Taylor & Francis Group, London, ISBN 978-1-138-02685-8

Coordinate transformation of 3D laser scanning based on nonlinear least squares algorithm in MATLAB

P.F. Yan
Jiaozuo Coal Industrial Company, Henan Coal and Chemical Industry Group Co., Ltd., China

J.X. Zhang, G.Q. Zhao & Z.Y. Zhang
School of Surveying and Land Information Engineering, Henan Polytechnic University, China

ABSTRACT: As a high-technology in surveying and mapping field, 3D laser scanning has the advantages of high efficiency and high precision, however, it acquires data still relatived to the instrument coordinate system. In this paper, the aim of experiment is to get multi-station scanning data of one high-slope, simultaneously survey the space retangular coordinates of targets with none cooperation target total station. Through nonlinear least squares algorithm in MATLAB, the paper complete Big Euler Angle coordinate transformation from scanner coordinate system to space coordinate system. According to the difference between the two kinds of coordinates, it gets the point error approximately. It is learned that this method is applicable to coordinate transformation, accurately extracting the target center from the point cloud and accurately surveying target center with total station is the key to achieve the coordinate transformation accurately.

1 INTRODUCTION

Now 3D laser scanning is a non-contact, fast getting high density and high precision point cloud of high-technology. It could proceed quick scan to space objects in complex scenes, directly get great point cloud. Therefore, it is also called "Imaging duplication technology" [1]. The advantages make it being used widely in cultural relic protection, deformation monitoring, digital city modeling. etc.

In many cases, one station can not acquire the complete object 3D information under the influence of scanning view angle, the object shape size, trees blocked. So the scanner is located in different stations from different directions, then match the point cloud of different stations and get the complete object point cloud model. Additionally matching different parts of point cloud is putting different point cloud data to the same coordinate system so that the point coordinate is converted. Currently three-dimensional coordinate is basically converted with the seven parameters method, which has three translational parameters, three rotation parameters and one telescopic parameter. With the emergence of various measurement new technology, the main problem we faced is three-dimensional space coordinate conversion of Big Euler Angle and linearization simplified model is used for coordinate conversion of Small Euler Angle [2]. Today relatively mature algorithm has adjustment of iterative method based on

improved Gauss-Newton method raised by Luo Changlin etc.

This paper choose one bare highslope as research object lie in the Fengshanzhen garden in Jiaozuo, set four scanning stations to acquire scan data, simultaneously, survey coordinates of 4 control points and all the targets with high precision and none cooperation target of total station. In order to achieve the conversion from scanning coordinates to space coordinates, the paper get the optimal solution iteratively through nonlinear least squares algorithm in MATLAB, then put all the target scanning coordinates into MATLAB, calculate corresponding space retangular coordinates in Bursa model. Acquired coordinates are used to contrast to the total station coordinates ultimately, better results are acquired.

2 3D POINT CLOUD COORDINATE TRANSFORMATION MODEL

Assuming 3D laser scanner coordinate system is O-XYZ, and space retangular coordinate system is O-XYZ. Between the process of coordinate transformation, firstly, move scanner coordinate origin to space retangular coordinate origin in X, Y, Z directions and rotate it three times will achieve conversion. Then asumming three translational parameters are g1, g2, g3, the telescopic parameter is g4, three rotation parameters are g5, g6, g7 in the paper.

According to Bursa model:

$$\begin{bmatrix} X \\ Y \\ Z \end{bmatrix} = \begin{bmatrix} g1 \\ g2 \\ g3 \end{bmatrix} + (1+g4)R(g)\begin{bmatrix} x \\ y \\ z \end{bmatrix} \qquad (1)$$

In the formula (1), $[x\ y\ z]^T$ is point cloud coordinates, $[g1\ g2\ g3]^T$ is translation matrix, $R(g)$ is rotation matrix, $R(g) = R(g5)R(g6)R(g7)$;

$$R(g5) = \begin{bmatrix} 1 & 0 & 0 \\ 0 & \cos(g5) & \sin(g5) \\ 0 & -\sin(g5) & \cos(g5) \end{bmatrix}$$

$$R(g6) = \begin{bmatrix} \cos(g5) & 0 & -\sin(g5) \\ 0 & 1 & 0 \\ \sin(g5) & 0 & \cos(g5) \end{bmatrix}$$

$$R(g7) = \begin{bmatrix} \cos(g7) & \sin(g7) & 0 \\ -\sin(g7) & \cos(g7) & 0 \\ 0 & 0 & 1 \end{bmatrix} \qquad (2)$$

Calculating seven parameters need three double corresponding points at least in least squares algorithm [3]. According to three double corresponding points, one over-determined system of $f(x) = 0$ is comprised of nine equations. Convert it into square functional form and is:

$$\varphi(x) = \frac{1}{2}f(x)^T f(x) = \frac{1}{2}\sum_{i=1}^{m} f_i^2(x) \qquad (3)$$

Obviously, the problem of calculating over-determined system of $f(x) = 0$ is transformed into nonlinear least squares problem of calculating the minimum point of $\varphi(x)$ [4].

To calculate seven parameters and transformation coordinates, program is compiled based on nonlinear least squares algorithm in MATLAB. According to the measured real situation can choose the initial value of seven parameters. Program can calculate the approximation and do it again with this approximation as initial value in a recycling program. When the norm of the difference between the latest two seven parameters is smaller than a certain threshold, seven parameters can get the optimal values.

3 DATA ACQUISITION

3.1 Total station data acquisition

The no prism total station is GPT3002 LN in experiment, which has 1200 m of maximum measuring distance, (3 + 2ppm*D) mm of precision and 2″ of angle measurement precision. Operators set A, B, C and D of 4 scanning stations in measurement area, then do the control measurement among the stations. While arranging 25 blue targets, then survey their spatial coordinates with total station. The coordinate data is shown in Table 1.

3.2 3D laser scanning data acquisition

The terrestrial laser scanning is Scanstation 2 of Leica. Its maximum scanning distance is 300 m and can acquire 50000 points per second. Erect scanner on A, B, C, D, then set the scanning distance for 150~200 m, the scanning density is 2 cm × 2 cm. Eventually gain point cloud which is shown in Figure 1, retalively true image is shown in Figure 2.

At the same time, the target should be scanned alone with high resolution, then extract the system coordinate of each target in different station

Table 1. Part of target coordinates.

Point name	X(m)	Y(m)	H(m)
X2	1058.809	1048.992	421.241
X3	1061.383	1045.886	423.276
X10	1043.908	896.998	432.518
X11	1025.018	879.346	429.814
X13	1095.433	1007.691	432.973
S1	1097.572	1024.250	461.811
S2	1109.078	993.502	462.126
S5	1086.448	914.504	453.326
S7	1068.420	892.342	453.164
S9	1062.020	886.579	452.265

Figure 1. Point cloud data.

Figure 2. True image.

Table 2. System coordinates of target in station A.

Point name	X(m)	Y(m)	Z(m)
S1	27.7313	−96.6199	60.3253
S2	60.1089	−91.2265	60.6398
S5	117.2433	−32.1556	51.8399
S7	127.4257	−5.4867	51.6797
X2	−13.0331	−75.4146	19.7600
X3	−9.0885	−76.0866	21.7847
X13	41.0079	−86.5095	31.4737

Table 3. System coordinates of target in station B.

Point name	X(m)	Y(m)	Z(m)
S1	46.6986	117.0088	56.7725
S2	13.8635	116.7676	57.0765
X5	52.1769	100.9500	23.2641
X11	−62.7119	−2.5651	24.7745
X13	31.9280	109.1370	27.8759

in cyclone 6.0. System coordinates are shown in Table 2 and 3.

4 BASED ON MATLAB FOR PARAMETER CALCULATION AND COORDINATE TRANSFORMATION

The steps for conversion parameters calculation is as follows:

1. According to the measurement range of A, three targets are choose uniformly in measurement area, and these system coordinates are constituted of three double corresponding points with total station coordinates.

2. The three double corresponding points are put into 'change', which is compiled in MATLAB. The original value of g0 for seven parameters is given with: 1000,1000,400,0,0.79,0.79,0.79.
3. The function of 'lsqnonlin' is used in the model as below. Finally the approximate matrix of 'g' for seven parameters is calculated.
   [g, resnorm] = lsqnonlin('change', g0)
4. The cycling function was running until meet the threshold.
5. Finally, the iterations and the optimal solution of seven parameters are output in MATLAB.

The transformation parameters of station B can be calculated with the same method as station A. The calculated transformation parameters are as shown in Table 4. In the first step, because scanner coordinate system is the right hand coordinate system, and space rectangular coordinates is the left hand coordinate system, the scanner axes should be converted in calculating [5].

As the coordinate transformation parameters are obtained, according to formula 1, program of 'calculate' can be compiled in MATLAB. After that all the system coordinates can be converted into space rectangular coordinates. The converted coordinates and the difference between the two kinds of coordinates in station A,B are shown in Table 5 and 6.

As shown in Table 5 and 6, the difference between the two kinds of coordinates has different size. The coordinate difference of some points is particular. To obtain the approximate global assessment of the point position after converting, the point error can be calculated with formula 4.

$$m_p = \sqrt{\frac{\sum_1^n (\Delta x^2 + \Delta y^2 + \Delta z^2)}{n-1}} \quad (4)$$

By calculating result, the point error of station A is 0.0414 m after converting, and the point error of station B is 0.0317 m. In the whole process of data acquisition and processing, data accuracy is influenced not only by the point positioning error of instrument, but also by the conversion error and point distinguishing error of the target [6]. Assuming each error sources has the same weight, error produced by converting can be calculated with the formula of $\sqrt{3}/3 * m_p$, then the conversion error of station A is 0.0239 m, the conversion error of station B is 0.0183 m. In the actual experiment, the scene of measurement area is a little bigger, vegetation blocking is much more, the used targets are not very standard, and the targets scanning are not sufficient and accurate. All these problems are the main reason which produced the larger data error.

Table 4. Coordinate transformation parameters.

	g1(m)	g2(m)	g3(m)	g4	g5(d)	g6(d)	g7(d)
A	999.9901758	999.9842025	401.5786	−0.000136269	3.141788661	3.140800863	−0.52315894
B	1004.914858	938.8514664	405.171134	−9.79e-05	−0.002128613	−0.001255414	−0.366139233

Table 5. Coordinate conversion and the coordinate difference station A.

	X(m)	Y(m)	Z(m)	ᐃX(m)	ᐃY(m)	ᐃZ(m)
S1	1097.5757	1024.2455	461.8137	−0.0037	0.0044	−0.0027
S2	1109.0788	993.5082	462.1261	−0.0008	−0.0062	−0.0001
S5	1086.4506	914.5116	453.3630	−0.0026	−0.0076	−0.0369
S7	1068.4385	892.3699	453.2219	−0.0184	−0.0278	−0.0579
X2	1058.8136	1048.9504	421.2788	−0.0045	0.0415	−0.0377
X3	1061.3677	1045.8700	423.3019	0.0152	0.0159	−0.0258
X13	1095.4286	1007.6891	432.9715	0.0044	0.0018	0.0015

Table 6. Coordinate conversion and the coordinate difference station B.

	X(m)	Y(m)	Z(m)	ᐃX(m)	ᐃY(m)	ᐃZ(m)
S1	1097.55	1024.25	461.8404	0.022	0.00034	−0.02942
S2	1109.0795	993.5072	462.1248	−0.0015	−0.0052	0.001196
X5	1080.5312	1023.667	428.3172	0.04179	−0.0182	−0.02322
X11	1025.0172	879.3458	429.8129	0.00078	0.00022	0.001122
X13	1095.4323	1007.686	432.9752	0.00068	0.00502	−0.00218

5 CONCLUSIONS

According to the experiment, control points which are arranged in scanning scene are measured by high-precision of no prism total station, moreover the target coordinates can be converted into rectangular space coordinates quickly. Coordinates accurate transformation is based on accurate acquisition for the center coordinate of target from point cloud and accurate measurement for the center coordinate of target by total station. So the following problems should be mainly pay attention to:

1. For flat-screen target, the obliquity in target flat and scanning direction should be as small as possible, else center coordinates of target can not be extracted accurately in software.
2. The further scanning distance is, the weaker scanning signals will be. So the position of scanner should be as close as possible to targets.
3. The point cloud density should be improved in target scanning.
4. Each target is scanned twice. Weather scanning is disturbed by external condition or not can be determined by the target center coordinate difference of two scans [7].

This paper adopts nonlinear least squares algorithm of coordinate transformation to avoid excessive Euler Angle. Coordinate transformation parameters can be acquired accurately through MATLAB programming computation, and all scanner coordinates can be converted accurately according to these parameters in MATLAB.

In paper [8], coordinates transformation parameters are calculated according to adjacent last station, and all the data in every station is converted into the same project coordinate system, so multi stations scanning point cloud data can be merged seamless. In this paper, the multi stations scanning data is converted into space rectangular coordinates and can be merged too. Of course, the error distribution and stitching precision after transformation is the main facing problem.

Terrestrial laser scanning is a relatively new spatial data access method. Through this experiment research, point cloud data conversion and registration requires further research. In the same time, to improve data registration precision, point cloud data noise removal and empty repair will be the main research direction.

REFERENCES

Chen Yu and Bai Zhengdong. An nonlinear least squares algorithm for spatial coordinate transformation [J]. Journal of Geodesy and Geodynamics, 2010, 30(2):129–130.

Dong Xiujun. Research on application of 3D laser scanning technology in acquiring DTM with high accuracy and resolution [J]. Journal of Engineering Geology, 2007,15(3):431.

Liu Dajie, Shi Yimin and Guo Jingjun. GPS principles and data processing [M]. Shanghai: Tongji University Press, 2006.

Sheng Yehua, Zhang Ka and Zhang Kai, et al. Seamless multi station merging of terrestrial laser scanned 3D point clouds [J]. Journal of China University of Mining and Technology, 2010, 39(2):236–237.

Shi Guigang. Research of data processing technology and operation method about terrestrial laser scanning [D]. Tongji University, 2009.

Shi Yimin. Modern geodetic control survey [M]. Beijing: Survey and Mapping Press, 2003.

Su Xiaopei and Hao Gang. Research of the target center identifying method about terrestrial 3D laser scanning [J]. Urban Geotechnical Investigation and Surveying, 2010(3):70.

Xie Rui, Cheng Xiaojun and Zhang Hongfei. Reasearch of point clouds registration based on reflectorless total station [J]. Geotechnical Investigation and Surveying, 2010(9):60.

Control Engineering and Information Systems – Liu (Ed)
© 2015 Taylor & Francis Group, London, ISBN 978-1-138-02685-8

Tomography imaging in near-field based confocal filtered back-projection algorithm

B.L. Tian, H. Yang, H.P. Wang, Y.J. Luo & H.Q. Liang
Xi'an Communications Institute, Xi'an, China

K.Z. Li
Henan Polytechnic University, Jiaozuo, China

Y. Chen
Research Institute of China Ordnance Industries, Xi'an, China

ABSTRACT: Tomography imaging of synthetic aperture radar can realize 3D tomography although its algorithm is simple, computation quantity is small, but is mainly suitable for the far field conditions. In addition, by increasing the number of baseline to improve the resolution will greatly increase the system complexity. According to the above problems, a tomography imaging technique in near-field using confocal circular SAR which on a partial-circle or curved orbit is proposed. Two techniques of synthetic aperture and confocal imaging are combined to achieve space-time. Namely, confocal filtered back-projection algorithm which usually used in two-dimensional imaging was extensively applied in confocal imaging. The results of digital simulation and experiment show that the proposed algorithm overcomes altitude ambiguity and has the capability to extract quasi-3D imaging information from a target scene.

1 INTRODUCTION

The Synthetic Aperture thought has been widely used in microwave Imaging. With regard to the regular Linear Synthetic Aperture Radar (LSAR), moving along the line, having an equivalent focal linear array, the azimuth resolution is determined from the effective aperture of the antenna, and the range resolution is determined by launching pulse width. Confocal tomography imaging which has been widely applied in the medicine is according to the measured scattering data to reconstruct the image of material dielectric constant, and for capturing target chip image by focusing on different layer tissues. Its basic principle is using the collected radiation energy data to obtain coherent superposition according to the scheduled tracks. In the central scattering region, signal stacks in same phase. The energy can reach the effect of resonance, conversely, it may be counteracted each other due to the random character of the phase. In conventional radar imaging system, through calculation, radar can obtain distribution of scattering center by gathering backscattering data of single inclined plane. However, in the real scene, this type of imaging means that the target is not placed in the focal plane. For example, the height of target is lost when target projected on the inclined plane on vertical direction. This is the so-called Radar Layover.

Aiming at limitations of being hard to meet the far-field condition in the microwave darkroom, this article puts forward idea of the construction of confocal tomography imaging. Combining synthetic aperture with focal tomography methods, the paper presents a diagnosis scheme which is confocal tomography imaging in near-field based on circular scanning, the radar antenna on the arc path to complete the space-time onfocal imaging. Unlike the traditional 2D and 3D imaging, the radar platform is moving over the target area, the trace and imaging district center formed a confocal cone shape. Collect the near-field downward-looking CSAR (Circular Synthetic Aperture Radar) data along the circumferential scan path. Due to the spectral components of the echo in the height direction, it has the capability of extracting 3d information from 3d image. Moreover, different from LSAR, it will not generate phase ambiguity in the height direction.

2 CONFOCAL IMAGING BY SCANNING ALONG ARC

2.1 *Geometric relation of confocal imaging by scanning along arc*

In the scene if the near-field downward-looking confocal tomography imaging, the radar platform is moving over the target area, the trace and imaging district center formed a confocal cone shape. The geometric relation of confocal imaging by scanning along arc as shown in Figure 1, mapping relation and the projection integral lines are occurring change.

2.2 The algorithm description

SAR flight along the circular grinding path, height relative to the ground plane H, and the horizontal distance between center of the rotating table and SAR is defined as R_g. Then, the slant range is:

$$R_0 = \sqrt{H^2 + R_g^2} \tag{1}$$

Slant angle of depression is defined as this form:

$$\alpha = \arctan\left(\frac{H}{R_g}\right) \tag{2}$$

Mapping relationship between the two coordinates is:

$$u = -x\cos\alpha\sin\theta + y\cos\alpha\cos\theta - z\sin\alpha$$
$$v = x\sin\alpha\sin\theta - y\sin\alpha\cos\theta + z\cos\alpha \tag{3}$$

In polar coordinate system, the distance from SAR to any point of the target point S_θ is:

$$R = \sqrt{R_0^2 + \rho^2 - 2R_0\rho\cos(\angle AOS_\theta)} \tag{4}$$

According to the spatial relation described in Figure 1, we can get the follow mathematical relationship:

$$\cos(\angle AOS_\theta) = \cos(\angle AOS)\cdot\cos(\angle SOS_\theta)$$
$$= \cos\alpha\cdot\cos\theta \tag{5}$$

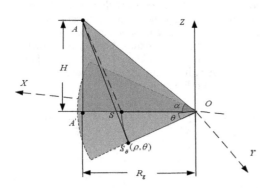

Figure 1. Geometric relation of confocal imaging by scanning along arc.

By the substitution of Equation 5 and Equation 4 is turned into:

$$R = \sqrt{R_0^2 + \rho^2 - 2R_0\rho\cos\alpha\cos\theta} \tag{6}$$

The projection line of target scattering function in the sight line direction of SAR is:

$$l = \sqrt{R_0^2 + x^2 + y^2 - 2R_0(y\cos\theta - x\sin\theta)} - R_0 \tag{7}$$

Mapping to 2-dimensional plane:

$$l_\alpha = \sqrt{R_0^2 + x^2 + y^2 - 2R_0\cos\alpha(y\cos\theta - x\sin\theta)} - R_0 \tag{8}$$

The target model located in the plane of $Z = 0$ as an example, we elaborated model and reconstruction of tomography imaging in near-field based confocal filtered Back-Projection algorithm. One-dimensional range profile can be completed the transverse direction focusing by using integral algorithms along the characteristic direction. Namely, we can obtained two-dimensional digital image focused on the plane. However, at the other height of Z, image is defocused by using objective function $f(x, y, z)$.

The reconstructed image may be regarded as three-dimensional objective function $f(x, y, z)$, which has a small oscillatory component in the plane of $Z = 0$.

Through transforming, we can get the object functions reconstruction at any height of $Z = h$:

$$R_{c;z=h} = \sqrt{R_g^2 + (H - h)^2} \tag{9}$$

$$\alpha_{z=h} = \arctan\left(\frac{H - h}{R_g}\right) \tag{10}$$

$$R_{z=h} = \sqrt{R_g^2 + \rho^2 - 2R_c\rho\cos\alpha\cos\theta} \tag{11}$$

$$l_h = \sqrt{R_{c;z=h}^2 + x^2 + y^2 - 2R_{c;z=h}\cos\alpha_{z=h}(y\cos\theta - x\sin\theta)} - R_{c;z=h} \tag{12}$$

Then, the modified Filtered Back-Projection algorithm is:

$$P_\theta(l_h)$$
$$= \int_0^B (k_0 + k_{0\min})G(k_0 + k_{0\min}, \theta)\exp(j2\pi k_0 l_h)\,dk_0 \tag{13}$$

$$\hat{g}(x, y, z = h) = \int_{\theta_{\min}}^{\theta_{\max}} P_\theta(l_h)\exp(j2\pi k_{0\min}l_h)\,d\theta \tag{14}$$

$$l_h = \sqrt{R_{c;z=h}^2 + x^2 + y^2 - 2R_{c;z=h}\cos\alpha_{z=h}(y\cos\theta - x\sin\theta)} - R_{c;z=h}$$

(15)

In summary, when we have known the data near-field measurement, in order to obtain the far-field image, mainly concludes the following steps:

Step 1, high resolution one-dimensional range profile in the sight line direction of Radar can be obtained through FFT's filtering, and the conditions of FFT can be satisfied by spacial frequency coversion and integral interval transform.

Step 2, to determine height of focusing imaging plane, to calculate the slant range and slant angle of depression.

Step 3, to modify the projection line, the projection integral line, to map the sight line direction of Radar to focusing imaging plane.

Step 4, to obtain two-dimensional digital image focused on the plane through one-dimensional range profile being completed the transverse direction focusing by using integral algorithms along the characteristic direction.

3 DIGITAL SIMULATION

3.1 *Text and indenting*

The data collection is based on the "stop and go" manner, and the setting of simulation parameters were as follow:

Radar frequency range: 8 GHz~12 GHz
Tracking point number: 801
Rotation angle range: Π rad

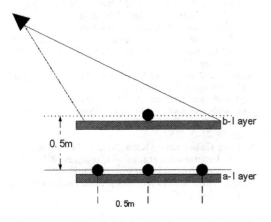

Figure 2. Positional relationship between the simulation objectives.

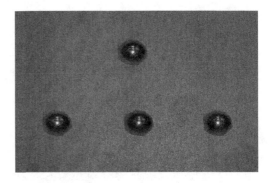

Figure 3. Positional relationships between the experimental objectives.

Gap of rotation angle: 2Π/360 rad
Distance between the radar and simulation objective: 10 m, and the height is 4 m.

The positional relationship between the simulation objectives as shown in Figure 3. Some additional remarks.

Measurement units are meter in the direction of the x or y axis for all the results of the simulation, yet, the direction of the z axis is relative value.

4 EXPERIMENTAL TEST

Two coaxial waveguides which are close to each other are used as a receiving or radiation antenna in the test system. Radar frequency range is from 9 GHz to 11 GHz and frequency spacing is 25 MHz. The target is placed on the low scattering foam scaffold of rotary table. The distance between antenna and the above center of rotary table is 4 meters. In specific implementation, the replacement of the circular-scanning antenna with the movement of the rotating platform in practical applications. The parameters were as follow: angle ranges is Π rad, the angle interval is Π/180 rad. The positional relationships between the experimental objectives as shown in Figure 4.

Figure 2 shows the positional relationship between the simulation objectives. Figure 3 shows the experimental objectives' positional relationship which is the same to the simulation objectives'. Figure 4 showed the different layer effect of digital simulation of focusing imaging using algorithm proposed in this paper. Figure 5 showed the different layer effect of experimental test. In Figure 5, we can see that serious defocusing in scattering center of the other layer, yet entirely focusing only in scattering center of the target layer. Consequently, it can realize target focus at any altitude and can reconstruct quasi-three dimension images.

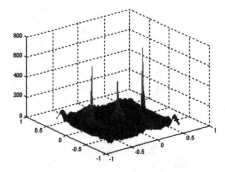

(a) The simulation effect of a-layer

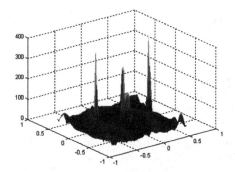

(a) Experimental test effect of a-layer

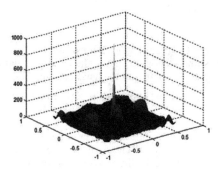

(b) The simulation effect of b-layer

Figure 4.   Digital simulation of focusing imaging.

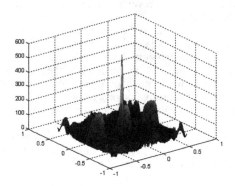

(b) Experimental test effect of b-layer

Figure 5.   Experimental test of focusing imaging.

5   CONCLUSIONS

The results of simulation and experiment prove the algorithm of tomography imaging in near-field based confocal filtered Back-Projection is of application value in practice. By using the modified mapping relation, scanning once by a single-baseline, it can realize near-field quasi-three dimensional imaging and overcome altitude fuzzy.

Experimental test still correspond well with and layout of target. Because of scanning along arc, and the accumulation of observing angle is relatively small, focusing effect with the proposed algorithm in this paper is little worse than that with spherical wave focusing convolution algorithm which the radar antenna on the circular path.

REFERENCES

Akira Ishimaru, F Tsz-King Chan, Yasuo Kuga. 1998. An Imaging Technique Using Confocal Circular Synthetic Aperture Radar. IEEE Trans. on geoscience and remote sensing, 36(5): 1524–1530.

Berardino P., Fornaro G. & Lanari R. 2002. Multi-pass synthetic aperture radar for 3-D focusing. Toronto: Int. Proc. Geoscience and Remote Sensing Symp.

Broquetas A. & De Porrata R. 1997. Circular synthetic aperture radar (C-SAR) system for ground-based applications. Electronics letters, 33(11): 988–989.

Huang Ping-ping Deng Yun-kai, Qi Wei-kong. 2010. Based on Intra-pulse Beam Steering to Solve Layover Lead by Digital Beam Forming in SAR. Journal of system Simulation, vol. 22 (1): 254–255.

Liu Xiang-le. 2007. Study on Multibaseline Tomography Synthetic Aperture Radar. Beijing: Institute of Electronics, Chinese Academy of Sciences.

Mehrdad Soumekh. 1996. Reconnaissance with Slant Plane Circular SAR Imaging. IEEE Trans. on image processing, 5(8): 1252–1265.

Reigber A. 2001. Airborne polarimetric SAR tomography. Stuttgart: University of Stuttgart.

Reigber, A. and Moreira, A. 2000. First Demonstration of Airborne SAR Tomography Using Multibaseline L-Band Data, IEEE Trans. Geosci. Remote Sensing, vol. 38: 2142–2152.

Tsz-King Chan, Yasuo Kuga, Senior, Akira Ishimaru. 1999. Experimental Studies on Circular SAR Imaging in Clutter Using Angular Correlation Function Technique. IEEE Trans. on geoscience and remote sensing, 37(5): 2192–2197.

Control Engineering and Information Systems – Liu (Ed)
© 2015 Taylor & Francis Group, London, ISBN 978-1-138-02685-8

Simulation environment for a missile based on Matlab/Simulink development

Z.Y. Luo & M. Li

Changchun Institute of Optics, Fine Mechanics and Physics, Chinese Academy of Sciences, China

ABSTRACT: The paper takes some ballistic missile as the object of study; a mothed which deals with the trajectory modeling and simulation is presented by adopting rapid control prototype. First, a six-degree-of-freedom nonlinear mathematical model for the ballistic missile is built; and then, the modular-ized modeling and simulation is set up by using Matlab/Simulink, additionally, a research on interface for simulation model is conducted; finally, realtime simulation environment is set up, and the process of auto code generation based on RTW is introduced. The trajectory model is verified by realtime simulation. The verification results show that the rationality and effectiveness of the method.

1 INTRODUCTION

Ballistic simulation technology plays an important role in the process of missile weapon development. How to efficiently build simulation platform to promote the development of trajectory simulation test has become a key problem. The develop-ing mode of traditional simulation platform has separated the design, implementation and testing process, once the problems appeared in the proc-ess of implementation, the design needs to start again, and this will result in time and cost over-head. The introduction of rapid control prototyp-ing technology can achieve rapid transformation from model design to goal realization [1,2], and provides a guarantee to accelerate product devel-opment. The technology has been applied more and more widely in the system with high perform-ance. Ballistic modeling and realtime Simulation Based on Matlab/Simulink, uses the design idea of rapid control prototype. Through establishing the model of simulation system in Simulink, and using RTW automatically generates C code which runs in real-time simulation platform. The accuracy of the model is validated by the realtime code at last.

In the paper, ballistic modular design is carried out based on Simulink, automatic code transfor-mation of model is completed by RTW, and the simulation platform of missile is set up efficiently. This method can realize the simulation software module division and management at the top level design, at the same time; the problem of traditional simulation process due to the upper model changes lead to a large number of changes is solved. In this way, errors that introduced by the manual pro-gramming is greatly reduced.

2 MATHEMATICAL MODEL

The mathematical model is an abstract descrip-tion of motion law and essence of the objective things using mathematical language, it is the base of simulation [3,4]. In the description of missile mathematical model of 6DOF, the paper ignores the curvature of the earth, assumes that the mis-sile is rigid body and the ground system is the inertial reference system. The 6DOF kinematic and dynamic equations of missile is established by analysising the force and moment of the missile in flight and using Newton's second law and moment of momentum theorem.

2.1 Dynamic equations of missile centroid

$$\begin{cases} m\dot{V}_{x1} = P - qS(C_{x1} + C_{x1\delta_x}) - mg \cdot 2(q_0q_3 + q_1q_2) \\ \qquad - m\omega_{y1}V_{z1} + m\omega_{z1}V_{y1} \\ m\dot{V}_{y1} = qSC_{y1} - mg\left(q_0^2 + q_2^2 - q_1^2 - q_3^2\right) \\ \qquad - m\omega_{z1}V_{x1} + m\omega_{x1}V_{z1} \\ m\dot{V}_{z1} = qSC_{z1} - mg\left(2(q_2q_3 - q_0q_1)\right) \\ \qquad - m\omega_{x1}V_{y1} + m\omega_{y1}V_{x1} \end{cases} \quad (1)$$

where, V_{x1}, V_{y1}, V_{z1} are missile velocity compo-nents, $\omega_{x1}, \omega_{y1}, \omega_{z1}$ are angular velocity compo-nent, $C_{x1}, C_{x1\delta_x}, C_{y1}, C_{z1}$ denote respectively axial force coefficient, elevator angle axial force coeffi-cient of aileron, normal force coefficient and lat-eral force coefficient, q is dynamic pressure, S is reference area, P is thrust.

2.2 Dynamic equations of missile body around centroid movement

$$
\begin{cases}
J_{x1}\dot{\omega}_{x1} = qSL(m_{x1} + m_{x1}^{\bar{\omega}_{x1}}\bar{\omega}_{x1} + m_{x1}^{\beta}\beta + m_{x1}^{\bar{\omega}_{y1}}\bar{\omega}_{y1} \\
\qquad + m_{x1}^{\bar{\omega}_{z1}}\bar{\omega}_{z1}) + (J_{y1} - J_{z1})\omega_{y1}\omega_{z1} \\
J_{y1}\dot{\omega}_{y1} = qSL(m_{y1} + m_{y1}^{\bar{\omega}_{y1}}\bar{\omega}_{y1} + m_{y1}^{\beta}\bar{\beta} + m_{y1}^{\bar{\omega}_{x1}}\bar{\omega}_{x1}) \\
\qquad + (J_{z1} - J_{x1})\omega_{z1}\omega_{x1} \\
J_{z1}\dot{\omega}_{z1} = qSL(m_{z1} + m_{z1}^{\bar{\omega}_{z1}}\bar{\omega}_{z1} + m_{z1}^{\bar{\alpha}}\bar{\alpha} + m_{z1}^{\bar{\omega}_{x1}}\bar{\omega}_{x1}) \\
\qquad + (J_{x1} - J_{y1})\omega_{y1}\omega_{x1}
\end{cases}
\quad (2)
$$

where, J_{x1}, J_{y1}, J_{z1} is moment of inertial in body coordinates, m_{x1}, m_{y1}, m_{z1} denote respectively rolling moment coefficient, yawing moment coefficient and pitching moment coefficient, $m_{x1}^{\bar{\omega}_{x1}}, m_{x1}^{\beta}$ denote respectively rolling damping moment coefficient, the derivative of β by rolling moment coefficient, $m_{x1}^{\bar{\omega}_{y1}}, m_{x1}^{\bar{\omega}_{z1}}$ is spiral derivative coefficient, $m_{y1}^{\bar{\omega}_{y1}}, m_{y1}^{\beta}, m_{y1}^{\bar{\omega}_{x1}}$ denote respectively yawing damping moment coefficient, the derivative of β by yawing moment coefficient and the derivative of $\bar{\omega}_{x1}$ by yawing moment coefficient, $m_{z1}^{\bar{\omega}_{z1}}, m_{z1}^{\bar{\alpha}}, m_{z1}^{\bar{\omega}_{x1}}$ denote respectively pitching damping moment coefficient, the derivative of $\bar{\alpha}$ by pitching moment coefficient and the derivative of $\bar{\omega}_{x1}$ by pitching moment coefficient, L is the length of missile, α is angle of attack, β is sideslip angle.

2.3 Kinematic equation of missile centroid movement

$$
\begin{cases}
\dot{x} = V_{x1}(q_0^2 + q_1^2 - q_2^2 - q_3^2) + V_{y1}[2(q_1q_2 - q_0q_3)] \\
\qquad + V_{z1}[2(q_1q_3 + q_0q_2)] \\
\dot{y} = V_{x1}[2(q_0q_3 + q_1q_2)] + V_{y1}(q_0^2 + q_2^2 - q_1^2 - q_3^2) \\
\qquad + V_{z1}[2(q_2q_3 - q_0q_1)] \\
\dot{z} = V_{x1}[2(q_1q_3 - q_0q_2)] + V_{y1}[2(q_0q_1 + q_2q_3)] \\
\qquad + V_{z1}(q_0^2 + q_3^2 - q_1^2 - q_2^2)
\end{cases}
\quad (3)
$$

where, x, y, z are position component of the earth coordinate system.

2.4 Kinematic equation of missile body around centroid movement

$$
\begin{cases}
\dot{q}_0^* = -0.5(q_1\omega_{x1} + q_2\omega_{y1} + q_3\omega_{z1}) \\
\dot{q}_1^* = 0.5(q_0\omega_{x1} - q_3\omega_{y1} + q_2\omega_{z1}) \\
\dot{q}_2^* = 0.5(q_3\omega_{x1} + q_0\omega_{y1} - q_1\omega_{z1}) \\
\dot{q}_3^* = 0.5(-q_2\omega_{x1} + q_1\omega_{y1} + q_0\omega_{z1})
\end{cases}
\quad (4)
$$

where, q_0, q_1, q_2, q_3 are normalized component of quaternion.

3 THE MODULAR DESIGN

3.1 The design of ballistic model on Simulink

An important feature of the system design which using rapid control prototyping technology is that algorithm is not completed by writing C code, instead of using the diagram method [5]. Therefore the ballistic model is established by Simulink. Graphical user interfaces provided by Simulink can quickly realize modeling, simulation and analysis of dynamic systems [6].

Ballistic model is divided into aerodynamic, engine, structure, atmospheric environment and motion equations module by Simulink diagram. Where aerodynamic module calculates aerodynamic force and moment according to the missile attitude, force condition and control command. In the process of calculating the aerodynamic force and moment lift coefficient, drag coefficient, moment coefficient is supplied by wind tunnel test. The engine module calculates the thrust according to pressure and temperature which provided by environment module. The structure module calculates missile mass, center of gravity, moment of inertia. The atmospheric environment module calculates atmospheric density, pressure, velocity and acceleration of gravity accoding to the parameters of missile flight height, flight speed, pressure and temperature. Motion equations module implements the mathematical model of the missile by S function. S function uses non-graphic way to describe a function module. Figure 1 provides the ballistic simulation model block diagram based on Simulink.

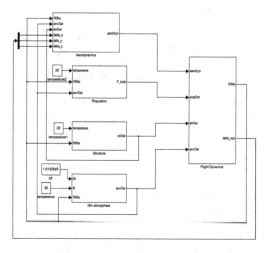

Figure 1. The ballistic simulation model based on Simulink.

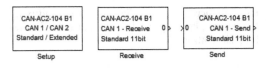

Figure 2. The communication interface model based on Simulink.

3.2 Interface module design

In the real time simulation process, ballistic model will communicate with external avionics devices. The accuracy of simulation is affected by the interface module. The interface module is designed by S function in Simulink. S function contains non-embedded S function and embedded S function [7]. Non-embedded S function which is suitable for digital simulation completes the corresponding interface operation through the MdlStart, MdlOutputs, MdlUpdate function. However, in the realtime simulation, if code conversion is completed by non-embedded S function, not only need to apply for a lot of memory space and computing resources, but also some unnecessary function calls is imported, this will result in the decrease of code execution efficiency. On the other hand, embedded S function change the RTW automatic code generation mode by introducing the TLC object file, so the code efficiency is impoved greatly. Sampling rate consistency of data communication between the interface module and the ballistic model should be considered carefully. If the sampling rate is inconsistent, the entire model should be set to multiple rate mode, and the both ends are connected through the rate conversion module. CAN bus interface module as shown in Figure 2 is established according to the needs of real-time simulation platform. The setup module is used to initialize the CAN card, the send module and receive module is used for sending and receiving the CAN message respectively. CAN bus interface module is established by embedded S function. The key question is to establish a TLC file which can directly invoke the actual functions to access the hardware interface. In the TLC file, the executive function should be declared in Block-TypeSetup funtion, and should be called in Ouputs function.

4 HARDWARE-IN-THE-LOOP SIMULATION

Hardware-in-the-loop simulation is a technique that use actual sensors replace mathematical model to ensure high reliability, it has the advantages of low cost, high efficiency and reproducibility [8].

Table 1 provides the physical environment of simulation. High-performance industrial computer of X86 architecture is chosen as simulation computer, not only ensure the ability of the simulation calculation, but also can satisfy the stability of the operation. With the strong real-time request, VxWorks is chosen as operating system. VxWorks operating system has advantages of high real time capability, stable and reliable, scalable, widely used in the field of missile, aircraft, etc [9]. The Hardware-in-the-loop simulation system contains missile-borne computer and actuator. The whole system connect with each other through CAN bus.

The structure of the hardware-in-the-loop has been assigned as shown in Figure 3. In the host computer, ballistic model and interface model are transformed into executable code, and are loaded into the target computer. After received the sync signal of integrated control platform, target computer calculated the flight state parameters according to the initial state of the dynamics and kinematics model of missile body, then the information is transmitted to the missile-borne computer through CAN bus. Missile-borne computer set the position information of target, and calculated the missile-target relative motion information according the missile flight parameter send by target computer, then calculated the control information combining with inertial navigation model. The actuator received the control signal and send rudder feedback to the target computer, thus a closed-loop

Table 1. The physical conditions of Simulation.

Host computer	Target computer	Others
Windows XP Matlab 2011a Tornado 2.2	Industrial computer VxWorks Interface card of CAN bus	Missile-borne computer Actuator

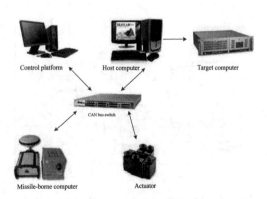

Figure 3. The hardware-in-the-loop platform.

709

system was constituted. After the simulation, the target computer sends the simulation data to the host computer for analysis through FTP.

If your text starts without a heading you should place the cursor on the I of INTRODUCTION, change the tag to First paragraph and type your text after deleting the word INTRODUCTION.

5 SIMULATION VERIFICATION

5.1 *Test conditions*

Simulink ballistic model and interface model base on Simulink are loaded into VxWorks simulator. The initial conditions are set without interference, pitching angle is set to 50 degrees and the simulation step size is set to 5 ms. Missile-borne computer controls the roll, pitch and yaw channels. The entire flight simulation is carried out in target computer. The simulation data is stored in the target computer.

5.2 *Simulation analysis*

The simulation test shows the interface of model is stable and the simulation step size meets the requirements. Figure 4 provides the simulation results curve of the hardware-in-the-loop simulation. The flight velocity, flight distance and flight altitude has the same result with the digital simulation, so the correctness and validity of Ballistic model is proved. Where the pitch angle is the main object of the guidance and control, it is kept at −20 degree from the results, complied with the design requirements. The roll and yaw angle changes also meet

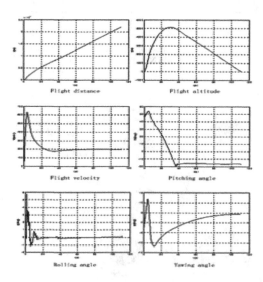

Figure 4.   The flight state curve of the harware-in-the-loop simulation.

the guidance control requiements. The validity of the control law is proved in the hardware-in-the-loop simulation.

6 CONCLUSION

The paper proposes using rapid prototype method to build realtime simulation platform aiming at the requirement of the ballistic simulation. First, according to the mathematical model of missile, the modular design of the ballistic model and communication interface model based on Simulink is established. Then the accuracy of the model and the reliability of the simulation platform is validated by hardware-in-the-loop simulation. This shows that with the rapid control prototyping technology introduced, system development will be more quilckly and the system verifiability will be improved, in another word, it provides effective guarantee for the research on guidance and control of missile. All in all, rapid control prototyping technology is effective for hardware-in-the-loop simulation of missile.

REFERENCES

Chen X.M., Gong X.L. & Zhou H.X. An Economical Rapid Control Prototyping System Design with Matlab/Simulink and TMS320F2812 DSP. IMECS 2010, 3:951.

Darko Hercog, Karel Jezernik. Rapid Control Prototyping using MATLAB/Simulink and a DSP-based Motor Controller. Int. J. Eng. Ed., 2005(21):96.

Dingyu Xue, Yangquan Chen. System simulation techonlogy based on MATLAB/Simulink [M]. Tsinghua University Press, 2002.

Gu F.M., Harrison W.S. & Tilbury D.M. Hardware-in-the-loop for manufacturing automat ion control: Current status and identified needs [C]. Proceedings of the 3rd Annual IEEE Conference on Automation Science and Engineering. Scottsdale, AZ, USA: IEEE Press, 2007, 5(1):1105.

Jang J., Ahn C.K., Han S. & Kwon W.H. Rapid Control Prototyping for Robot Soccer System using SIMTool[C]. SICE-ICASE International Joint Conference 2006, Busan, Korea, 2006, 10(2):3035.

Ru Zhang, SongLin Sun, Xiaogang Yu. Base of the embedded system Technology. Beijing Universigy of Posts and Telecommunication Press, 2005.

Tadeusz Michalowski. Applications of MATLAB in Science and Engineering, 2011(9):318.

The MathWorks Inc. Simulink Code User's Guide [M]. Version 8. USA: The MathWorks Inc, 2011.

Xiaoyuan Peng. System simulation technology. Beihang University Press, 2006:16.

Xingren Wang, Chuanyuan Wen, Bohu Li, etc. The development of the techology of system modeling and simulation in China [J]. Journal of system simulation, 2009, 21(21):6683.

Control Engineering and Information Systems – Liu (Ed)
© 2015 Taylor & Francis Group, London, ISBN 978-1-138-02685-8

Improved DCT digital watermarking algorithm

Z.H. Xiao

School of Computer Science and Engineering, Chongqing University of Technology, Chongqing, China

ABSTRACT: The paper mainly research is based on the DCT (Discrete Cosine Transform) color image digital watermarking blind detection, the research object is the color image of the robust watermark, adding two random key, Amold transformation and color image transformation to improve common watermarking discrete cosine transform technology, and select the Y component for the embedded watermark signal, in the watermark embedding after adding noise, cropping JPEG compression, filtering, attack, and blind extraction of digital watermarking. From the supply to the test results, to ensure the system in the embedded watermark imperceptibility but also maintained a good robustness, the accept various supply after the test, can also maintain a good visual effect.

1   INTRODUCTION

In the process of transformation from traditional business to electronic business affairs, a large number of excessive electronic document will appear, such as a variety of paper bill scanning image. Even the network security technology is mature, a variety of electronic bill also need some no password authentication method. Digital watermarking technology can provide invisible certification mark for various bill, so greatly increase the forge difficulty[1–5].

Discrete cosine transform (Discrete Cosine Transform) DCT[6–8]. It is the mathematical operation closely related to Fourier transform, and is the important method simplified o Fourier transform. In the Fourier series expansion, if the function was expanded real even function, then its Fourier series only contains cosine, then its discretization can export cosine transform, because it is based on the real orthogonal transformation.

With a length of N sequence $s(x)$ one-dimensional discrete cosine transform $S(u)$ is defined as:

$$S(0) = \frac{1}{\sqrt{N}} \sum_{x=0}^{N-1} s(x) \qquad (1)$$

$$S(u) = \sqrt{\frac{2}{N}} \sum_{x=0}^{N-1} s(x) \cos \frac{(2x+1)u\pi}{2N} \qquad (2)$$

The Inverse Discrete Cosine Transformation (IDCT) is expressed as:

$$s(x) = \sqrt{\frac{1}{N}} S(0) + \sqrt{\frac{2}{N}} \sum_{x=0}^{N-1} s(u) \cos \frac{(2x+1)u\pi}{2N} \qquad (3)$$

Two-dimensional DCT is used in the digital image processing, with N rows and N columns of the image s (x,y) matrix, its DCT transform is:

$$S(u,v) = \frac{2}{N} c(u)c(v) \sum_{x=0}^{N-1} \sum_{y=0}^{N-1} s(x,y) \cos \left(\frac{\pi u(2x+1)}{2N} \right)$$
$$\times \cos \left(\frac{\pi v(2y+1)}{2N} \right)$$
$$(4)$$

Its Inverse Discrete Cosine Transform (IDCT) is expressed as:

$$s(x,y) = \frac{2}{N} \sum_{u=0}^{N-1} \sum_{v=0}^{N-1} c(u)c(v) S(u,v) \cos \left(\frac{\pi u(2x+1)}{2N} \right)$$
$$\times \cos \left(\frac{\pi v(2y+1)}{2N} \right)$$
$$(5)$$

2   BASED ON THE DCT WATERMARK EMBEDDING ALGORITHM

The typical DCT digital watermark embedding algorithm limit the embedded watermark signal volumes because the embedded watermark signal frequency position is fixed. On the other hand it limits the watermark embedding volumes, namely the watermark capacity. Therefore, the improved digital watermarking embedded system, the embedded position is not the particular frequency location, such as high low frequency watermarking embedding point location for encryption, and instead of using the high frequency of compromises, so as to reach the middle frequency coefficients encryption, so that not only guarantees the robust digital watermarking system, but also meet the transparency of the digital watermarking system.

In addition to the above two problems, further improvement is needed, such as the algorithm is based on the two values grayscale vector image, the watermark image is also based on the two values gray image to achieve. In this study, it is based on color vector image works, and is not the traditional double gray images.

In this study, we embedded two value gray watermark in the color carrier image, embedding algorithm is shown in Figure 1.

The K is key, W is watermark image, G is watermark signal generation algorithm, The watermark W and key K generates the watermark signal W' by watermarking algorithm G, the watermark signal W' get W'' by Arnold scrambling; I is the original carrier image, the original carrier image get YCBCR images by YCBCR transform, The Y' is obtained Y component image after block and DCT transformation, The Y' and the watermark signal W' generates the watermark image I' by watermarking algorithm Em.

Its steps are follow.

2.1 Key

Key K, is two groups of uniform distribution and small correlation random number sequence, is convenient for watermark embedding and extraction. Especially in the realization of blind detection, is achieved by the correlation between two groups random numbers.

2.2 The watermark signal generation algorithm

The watermark signal generation algorithm turns the original watermarking image into 2D gray image.

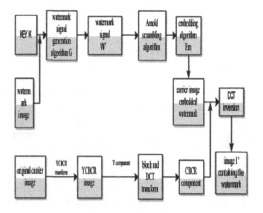

Figure 1. The DCT watermark embedding algorithm.

2.3 Arnold scrambling algorithm

The watermark signal pretreatment, the safety has been further enhanced.

2.4 YCBCR transform

Because the carrier image is color, does not directly embedded watermark, first the color carrier image is transformed by YCBCR, in which the Y is the brightness component, Cb refers to the blue chrominance components, and Cr refers to red chrominance components.

2.5 Block and DCT transform

Block, is that the carrier image is divided into 8*8 blocks, in order to resist JPEG compression attack, the block bigger will get better effect, but it requires the sacrifice of information capacity. DCT transform will transform the 8*8 blocks carrier image by two-dimensional discrete cosine transform function.

2.6 The embedding location selection

Because the human eye is more sensitive to the Y component video, so in the chrominance components are sub-sampled to reduce the chroma components, and the eye will not aware of changes in image quality, so the watermark signal is embedded in Y component of the carrier image. Because DCT transform in low and middle frequency coefficients are able to fight off Gauss noise attack and JPEG attack performance, so in this study, the watermark embedding positions, no longer is embedded into low frequency or intermediate frequency or high frequency coefficient of the carrier image, but choose a compromise embedded point.

2.7 Embedding algorithm

In the embedded watermark signal, selected embedding rule is a wig criterion, $v'_i = v_i + aw_i$. When the watermark image element is 0, it is encrypted with the generated one-dimensional random number matrix RAND1 of little correlation, the a value will be RAND1; When the watermark image element is 1, it is encrypted with the generated one-dimensional random number matrix RAND2 of little correlation, the a value will be RAND2.

3 DCT WATERMARK EXTRACTION ALGORITHM

A typical DCT digital watermark extraction algorithm is directed to take non blind algorithm to achieve, because of the need for the original

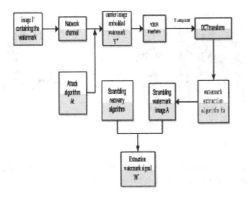

Figure 2. DCT watermarking algorithm.

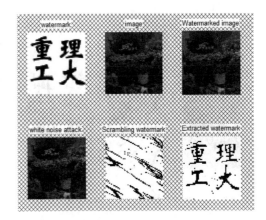

Figure 3. Attacks white noise test chart.

watermark extraction image works, which brings about great inconvenient and security problems, so here mainly studies a blind algorithm to achieve for extracting the watermark algorithm.

In this experimental study, mainly uses small correlation two groups of random one-dimensional array to achieve. Through the calculated respectively containing watermark signal carrier works and correlation between two groups of random numbers, then compare the correlation between them to determine the size of the watermark signal is present or not and the specific element of the image, then the image elements are combined, thus we get the Arnold scrambled watermark image. If you need to get a clear image, only the Arnold decryption can be extracted from the watermarked image.

Blind detection of watermark extraction is adopted, the watermark extraction does not need the original carrier image. So, for those simply reverse the watermark extraction algorithm, it ensure the watermark security, no longer worry that the third party can obtain secret information of hidden watermarking signal because we obtained the original carrier image, so the watermark signal get effective protection in a certain extent. The watermark extraction algorithm is shown below.

In the network communication the containing watermark signal carrier image I′ will become the watermark carrier image I″ after attacking by watermark attacking algorithm At, I ″ will be transform into YCBCR, it will get Y component, and gets the watermark scrambling image A through watermark extraction algorithm Ex, then gets clear watermark image after Arnold scrambling recovery algorithm.

4 ALL KINDS OF ATTACK TEST

4.1 *White noise attack*

Noise attacking is the most common channel transmission of inadvertent attack, in this experiment,

the noise is the white noise attacking. Figure 3 is shown as white noise attacking test chart, the image clarity is high after the experiment of added white noise, and the embedding strength factor is 10. And can clearly extract watermarks, although the extracted watermark image and the original watermarking image exists interference information, but we can clearly see the "重理工大", does not affect the watermark information identification.

4.2 *Gauss low-pass filter attack*

Filtering attack is one of the common attack in digital watermark. In this experiment, uses the 4*4 median filter to experiment, and verify its robustness. Gauss low-pass filter attack test Figure 4, we see that the Gauss low-pass filter attack have good robustness for digital watermarking system, the extracted watermark can be clearly identified and has nothing to do with the watermark information confirmation.

4.3 *JPEG compression attack*

Experiment is the simulation of the JPEG compression attack, the embedded watermark image carrier works use JPEG compression, in order to verify the robustness of digital watermarking system. JPEG compression attack test shown in Figure 5, In the image of the 85% JPEG compression attack testing, experimental image was no great change, the watermark image after JPEG compression is clearly visible, and the information interference is small, it proves that digital watermarking system has good robustness.

Common DCT algorithm JPEG compression attack test shown in Figure 6, in the image of the 85% JPEG compression attack testing, experimental image was not clearly visible than the image after the improved algorithm, and the informa-

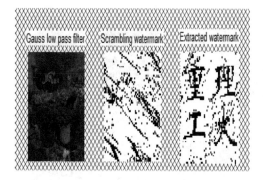

Figure 4. Gauss low pass filtering attack test chart.

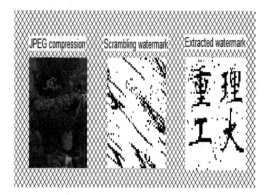

Figure 5. JPEG compression attack test chart.

Figure 6. Common DCT algorithm.

tion interference is more, although does not affect the digital watermark image identification, but it proves that the common DCT algorithm robustness is not good than the improved algorithm.

4.4 Clipping attack

Clipping attack is a common geometric attacks, and many of the digital watermarking geometric attack system to keep a good robustness. Experiment is the carrier of the image with the watermark works of cutting, the extracted after cutting watermark

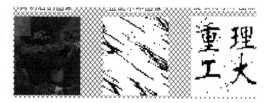

Figure 7. Clipping attack test chart.

image shows as in Figure 7, thus the robustness of a digital watermarking system is very good.

5 SUMMARY

To sum up, this article proposed watermarking algorithm has good encryption type, the algorithm to resist various attacks has obvious advantages, its robustness is good.

Study of digital watermarking technology in recent years is rapidly developing field, with the in-depth study, the digital watermarking technology is constantly developing an important role in the subject in the copyright security and authentication, as a new subject, the digital watermark technology development will be faster and better.

REFERENCES

Computer Engineering and Applications; 2004–11:34–35.
Digital Image Watermarking Based on Streak Block in the Discrete Wavelet Domain[J]; Microelectronics & Computer; 2004–07;23–24.
Ge Xiuhui, Tian Hao, Guo Lifu et al., <the principle of information hiding and its application>, Tsinghua University press, 2008.
Image Watermarking Based on Wavelet Transform[J].
Jin Cong, <the theory and technology of digital watermarking>, Tsinghua University press, 2008.
Nuo Li. A New Algorithm of 2-Dimension Color Image Digital Watermarking[J]. Computer Engineering and Applications, 2007, 43(2): 43–44.
Wang Bingxi, Chen Qi, Deng Fengsen Bian Zhu, <digital watermarking technology>, Xi'an Electronic and Science University press 2003.
Wang Bingxi, Peng Tianqiang, <information hiding technology>, National University of Defense Technology press, 2006.
Yang Hong-ying, Wang Xiang-yang, Zhao Yan, Zhang Wei-dong, Chne Li-ke, Zhao Hong.
Yang Yu editor, <information hiding and digital watermarking experiment tutorial>, National University of Defense Technology press, 2009.
Zhang Xiaofeng Duan Huilong (College of Biomedical Engineering. Zhejiang University, Hangzhou 310027);

A novel text clustering algorithm treated attributes differently

Y.J. Tan
Department of Information Engineering, Henan Polytechnic University, Zhengzhou, China

C.X. Li
School of Software and Technology, Zhengzhou University, Zhengzhou, China

ABSTRACT: The traditional algorithms based on partition treat all the attributes equally in clustering process. They all suppose that the importance of each attribute is equal, which contradict with the real world. So, These algorithms will have a lower accuracy. In order to handle this problem, this paper provides a new text clustering algorithm. During the iteration of this algorithm, it can find the important attributes. Moreover, this algorithm also can find the cluster structure hiding by the unimportant attributes. The simulation of this algorithm on test documents set prove that the algorithm in this paper can gain a good computation speed, found the latent structure and can remark the different importance of each attribute.

1 INTRODUCTION

With the appearance of computer and expanding of information on Internet, the demand of how to organize and manage texts becomes more and more exigent. The technology of text clustering is different from the one used by text mining. Without the known subjects, the text clustering's purpose is to divide the text sets into several clusters. The similarity of the texts in one cluster is greater than the texts in different clusters. The clustering algorithms being used widely can be divided into two categories. The one is hierarchical clustering algorithm represented by G-HAC and the other is partitional clustering algorithm represented by C-means. Though hierarchical clustering algorithms can obtain higher precision, they are not suitable to cluster the sets with a large number of documents as for their slow speed. The partitional clustering algorithms can obtain a rapid speed. So the partitional clustering algorithms are more suitable to cluster the documents set[1]. The outstanding representations of partitional clustering algorithms are C-means and fuzzy C-means. The advantages of C-means algorithm are its simplicity, speed, reliability in theory and easiness of realization[2]. But it also has some drawbacks, such as hard partition, being sensitive to noise. Hard partition makes the objects in the dataset only belong to one cluster. It is inconsistent with the real world. According to Zedeh's fuzzy theory, membership is used to describe the relationship between objects and cluster. Fuzzy C-means algorithm that first proposed by Dunn and expanded by Bezdek is one of the most efficient ones among fuzzy clustering algorithms.

But the traditional partitional clustering algorithms represented by C-means and fuzzy C-means take the same assumption that the attributes of objects play the same role in clustering. This is not desirable in documents clustering. Sometimes, the part attributes contribute more than others in deciding the cluster structure. But how to distinguish the importance of these attributes? This paper proposes a new fuzzy c-means text clustering algorithm. This algorithm can remark the importance of each attribute at the same time to realize the soft partition. It can find the real cluster structure hiding by the noises data.

2 C-MEANS AND FUZZY C-MEANS TEXT CLUSTERING ALGORITHM

C-means text clustering algorithms cluster a group of texts into a predefined number of clusters. It starts with randomly initial cluster centroids and keeps reassigning the data objects in the dataset to cluster centroids based on the similarity between the data objects and the cluster centroid. The reassignment procedure will not stop until a convergence criterion is met (e.g., the fixed iteration number, or the cluster result does not change after a certain number of iterations).

Suppose that $D = (d_1, d_2, ..., d_n)$ is a document sets with n objects, and is divided into c clusters $V = \{v_1, v_2, ... v_c\}$, each cluster can be represented by its cluster center v_i. In general, we always hope that the sum of square error between the interior clusters reaches its minimum. In mathematical language, we can retell the objective function of C-means as follows[3]:

$$J(u,v) = \sum_{k=1}^{n}\sum_{i=1}^{c} u_{ik}\left\|x_k - v_i\right\|^2,$$

where $u_{ik} \in \{0,1\}, \sum_{i=1}^{c} u_{ik} = 1.$ (1)

The iterative process of C-means algorithm can be described as follows:

Step 1: Choose c cluster centers to coincide with c randomly chosen patterns or c randomly defined points inside the hyper volume containing the pattern set.

Step 2: Assign each pattern to the closest cluster center.

Step 3: Recomputed the partition matrix using (2a), and then recalculated cluster centers using (2b)

$$u_{iik}^{(l+1)} = \begin{cases} 1 \text{ if } i=\text{argmin}\left\{\left\|x_k - v_i\right\|^2\right\} \\ 0 \text{ otherwise} \end{cases}$$ (2a)

$$v_i = \sum_{k=1}^{n} u_{ik}^{(l+1)} x_k \Big/ \sum_{k=1}^{n} u_{ik}^{(l+1)}$$ (2b)

Step 4: If a convergence criterion is not met, go to step 2. Typical convergence criteria are: no (or minimal) reassignment of patterns to new cluster centers, or minimal decrease in squared error.

With the theories of Zedeh, we use membership to denote the relationship between clusters and objects. Then $u_{ik} \in (0,1)$ subject to $\sum_{i=1}^{c} u_{ik} = 1$. Let U be a partition matrix, $1 < m < +\infty$ and $2 \leq c < n$, then the objective function of the FCM is:

$$J(u,v) = \sum_{k=1}^{n}\sum_{i=1}^{c} u_{ik}^m \left\|x_k - v_i\right\|^2$$ (3)

The iterative process of FCM is similar to that of C-means. The only difference is the computing of partition matrix and cluster center. The iterative formulas of u_{ik} and v_i in FCM are:

$$u_{ik} = \left(\left\|x_k - v_i\right\|^2\right)^{1/(1-m)} \Big/ \sum_{j=1}^{c}\left(\left\|x_k - v_j\right\|^2\right)^{1/(1-m)}$$ (4a)

$$v_i = \sum_{k=1}^{n} u_{ik}^m x_k \Big/ \sum_{k=1}^{n} u_{ik}^m$$ (4b)

3 ATTRIBUTE-WEIGHTED FUZZY C-MEANS TEXT CLUSTERING ALGORITHM

Considering that in some applications, finding out the structure of some attributes is valuable, we propose a novel algorithm based on FCM. In that algorithm, we can find different weight of each attribute. Moreover, this algorithm can find the latent structure of some attributes. In this paper, we call the new algorithm as AWFCM algorithm.

The objective function of AWFCM can be defined:

$$J(u,v,w) = \sum_{i=1}^{c}\sum_{k=1}^{n}\sum_{j=1}^{s} u_{ik}^m w_j^\beta \left(x_{kj} - v_{ij}\right)^2$$ (5)

subject to $\sum_{i=1}^{c} u_{ik} = 1, \sum_{j=1}^{s} w_j = 1, 1 \leq i \leq c, 1 \leq j \leq m$ where c is the number of clusters, m called weighting exponent, u_{ik} denotes the degree of object k belongs to ith cluster, w_j is the weight of jth attribute.

By Lagrange multiplier's approach, we can obtain the necessary conditions for the minimum of $J(u,v,w)$ as follows:

$$v_{ij} = \sum_{t=1}^{n} u_{ik}^m x_{kj} \Big/ \sum_{t=1}^{n} u_{ik}^m$$ (6a)

$$u_{ik} = \left(\sum_{j=1}^{s} w_j^\beta \left\|x_{kj} - v_{ij}\right\|^2\right)^{1/(1-m)}$$
$$\times \left(\sum_{t=1}^{c}\left(\sum_{j=1}^{s} w_j^\beta \left\|x_{kj} - v_{ij}\right\|^2\right)^{1/(1-m)}\right)^{-1}$$ (6b)

$$w_j = D_j^{1/(1-\beta)}\left(\sum_{t=1}^{m} D_t^{1/(1-\beta)}\right)^{-1},$$
where $D_j = \sum_{i=1}^{c}\sum_{k=1}^{n} u_{ik}^m \left(x_{kj} - v_{ij}\right)^2$ (6c)

Consequently, the procedure of the AWFCM can be described as follows:

Step 1: Fix the number of clusters c, the weighting exponent m, the iteration limit $Tcount$, the attribute's weight β and the tolerance ε, initialize the partition matrix U and attribute weight W;

Step 2: Update the cluster center v_{ij} by (6a), if convergence criterion is met, stop;

Step 3: Update the membership function u_{ik} by (6b), if convergence criterion is met, stop;

Step 4: Update the attribute weight W by (6c), if convergence criterion is met, stop;

Typical convergence criteria are: the value of J between two successive iterations is less than ε, or arrive to maximum iterations $Tcount$. By the same way in[4], we can say that given m and β, the AWFCM algorithm converges to a local minimal solution or a saddle point in a finite number of iterations.

The computational complexity of the algorithm is O($cns \times Tcount$), where c is the number of clusters, $Tcount$ is the number of iterations, n is the number of objects, s is the number of attributes.

Compared with traditional C-means and FCM text clustering algorithm, our algorithm treats attributes differently. It is clear that an attribute that has a smaller sum of the within cluster distance should be assigned a larger weight and be assigned a smaller one to an attribute that has a larger sum of the within cluster distances.

4 EXPERIMENTS

Based on the AWFCM proposed in section III, we use experiments to demonstrate the performance of this algorithm in discovering clusters and identifying insignificant (or noisy) attributes from given text sets. The experiment system includes: (1) word segmentation; (2) preprocessing; (3) characters reduction; (4) clustering; (5) results analysis.

1. **word segmentation** After segmenting the document into words with the aid of dictionary, the stop words and ordinary words are deleted. There will be a segmented document corresponding to the original document. Each line in the segmented document represents a word[6].
2. **preprocessing** The purpose of the selection of index terms is to determine which words will be used as an indexing elements. Usually, the decision on whether a particular word will be used as an index term is related to the syntactic nature of the word. In fact, noun words frequently carry more semantics than adjectives, adverbs, and verbs. After preprocessing, the words set of the documents in set will come into being. We can statistic the frequencies of each word occurring in each document. The frequencies just is the weight in document's Vector Space Model (VSM).
3. **characters reduction** In order to gain a simpler vector space, the words in word set will be deleted according to a predefined threshold. The frequencies of each word in the set will be statistic secondly. The vector space model will also be represented once more[6].
4. **clustering** Given the original cluster centroids, the number of clusters, the fuzzy exponent m and parameter β, the AWFCM is implemented many times. The C-means and FCM text clustering algorithms are all start with random cluster centroids, the results are provided by section V.

5 EXPERIMENTS' RESULT AND ANALYSIS

The D is a Chinese document set including 2400 documents. The categories' subject and the number of each category are known with the aid of some websites. The information is supplied by Table 1.

Table 1. Summary of testing documents set.

Category	Number
Computer	400
Art	400
Education	400
Transportation	400
Medication	400
Economic	400

Table 2. The results of different conditions.

	Parameters	Min-number	Max-number	Avg-number
AWFCM	$m = 2, \beta = 2$	0	199	23.82
	$m = 2, \beta = -2$	0	191	21.9
	$m = 2, \beta = -1.5$	0	191	34.63
	$m = 3, \beta = 2$	0	197	25.78
	$m = 3, \beta = -2$	0	171	15.52
	$m = 3, \beta = -1.5$	0	213	31.78
C-means	$\times$	46	226	124.56
FCM	$m = 2$	35	273	29.82

After deleting stop words, the word set of the test documents includes 10526 words. The number of words its frequencies more than 3 is 3805. The number of words its frequencies more than 8 is 1602. The word occurs most frequently is "educator". The frequency is 568.

After being represented by the vector space model, the text set is clustered by traditional C-means, FCM algorithm and AWFCM algorithm. In this experiment, we can use the wrong number of clustering to evaluate.

In order to compare the average performance, repeated experiments are implemented. These algorithms are all start with random centroids. After implementing the C-means algorithms 100 times, we can have the maximum wrong cluster, the minimum wrong cluster and the average wrong cluster. These numbers is given in Table 2.

Although FCM algorithm is influenced by parameters, many researchers have studied the parameters and given that 2 is the suitable value of parameter m. We implements FCM algorithms 100 times with $m = 2$.

AS for the performance of AWFCM is influenced by m and β, we implement AWFCM with six different group of m and β. They also start with the different cluster centroids, are implemented 100 times.

Table 2 gives the maximum wrong cluster number, minimum wrong cluster number and

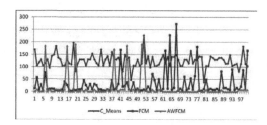

Figure 1. Wrong cluster numbers of C-means, FCM ($m = 2$) and AWFCM ($m = 3$, $\beta = -2$).

average wrong cluster number with different parameters. The "×" in Table 2 means that the C-means algorithm doesn't need parameters. As shown in Table 2, the average performance of AWFCM is better when m = 3 and $\beta = -2$. In order to compare the three algorithm, Figure 1 gives the change curve of C-means, FCM with m = 2 and AWFCM with m = 3 and $\beta = -2$.

As shown in Figure 1, the results of C-means and FCM text clustering algorithms can't be satisfied by people. At the same time, the results of AWFCM are better. Although the wrong cluster number will change greatly during implementing the AWFCM sometimes, the value of maximum wrong cluster number is lower than the one of C-means and FCM. Thus it can be seen that the AWFCM text clustering algorithm is better than other traditional partitional text clustering algorithms.

6 CONCLUSIONS

This paper studies the partitional text clustering algorithms in the literatures. Based on the analysis of these algorithms' drawbacks, we propose a new attribute-weighted text clustering algorithm-AWFCM. After taking different experiments, the result shows the validity of this algorithm. This algorithm also can remark the importance of each attribute which can help the researchers to reduce the dimensions of the Vector Space Model of documents.

REFERENCES

Han I, Kamber M. 2000. Data Mining: Concepts and Techniques [H]. Berlin: Morgan Kaufmann Publishers, 335–389.

Huang Joshua Zhexue et al. 2005. Automated Variable Weighting in k-Means Type Clustering [J]. IEEE Transactions on Pattern Analysis and Machine Intelligence, vol. 27, NO. 5, 657–668.

Jia Wei Han et al. 2001. Research on Web Mining: A Survey [J]. Journal of Computer Research & Development, 38(4)405–411.

Michael Steinbach. 2000. A Comparison of Document Clustering techniques [C]. KDD'2000. Technical report of University of Minnesota.

Runkler T.A., J.C. Bezdek. 2003. Web mining with relational clustering [J]. International Journal of Approximate Reasoning 32, 217–236.

Shen Huang et al. 2006. Multitype Features Coselection for Web Document clustering [J]. IEEE Transactions on Knowledge and Data Engineering, vol 18, NO. 4, 448–459.

Control Engineering and Information Systems – Liu (Ed)
© 2015 Taylor & Francis Group, London, ISBN 978-1-138-02685-8

Entropy-based global K-means algorithm

H.F. Feng, J. Zheng & Z.H. Wu
Computer Center of East China Normal University, Shanghai, China

ABSTRACT: K-means algorithm is one of the most commonly used in the clustering analysis. As the clusters center and the number of clusters are uncertain, which lead the results of K-means algorithm clustering uncertain. To solve this problem, this paper proposed an Entropy-based Global K-means clustering algorithm, which adopted incremental algorithm and given a certain weight to each attribute based on Entropy. The proposed algorithm not only has a good performance in determining the starting points, but also unnecessary to specify the clusters number, which make the Entropy-based Global K-means algorithm has a good stability in clustering. Theory analysis and experimental data results show that the algorithm can improve the clustering results stability and acquire a good performance.

1 INTRODUCTION

Cluster analysis is one of the primary methods for exploring the underlying structure and identifying hidden knowledge of a given dataset, which widely used in data mining, image processing, pattern recognition, machine learning and statistics. Clustering aims at to seek a partition of the given dataset into several clusters in which data objects in the same clusters are homogenous as soon as possible while data objects in different groups are well separated. Such algorithms can be categorized into several classes: hierarchical clustering, partitional clustering, neural network-based clustering and kernel-based clustering. With the increasing emphasis on the clustering results, the stability of the clustering results directly affects decision-making. Among the existing clustering algorithms, K-means algorithm[1],[2] is a typical partitional clustering algorithm and widely used since its simplicity and general run fast, which seeks an optimal partition of the dataset into k clusters by minimizing the sum-of-squared-error criterion in Eq. 1, which each cluster is represented by cluster center namely centroid[3].

In this paper, Suppose a dataset consists N objects where each object $d_i \in R^n$, $i = 1, .., N$. By using K-means algorithm clustering the dataset into k disjoint clusters C_i, $i = 1, ..., k$, each containing n_i data points, where $1 \leq n_i \leq N - 1$, The sum-of-squared-error criterion defined as:

$$E = \sqrt{\sum\nolimits_{i=1}^{k} \sum\nolimits_{d \in C_i} (d - c_i)^2} \qquad (1)$$

Supported by the National High-Technology Research and Development Program of China under Grant NO. 2013AA01A211.

where d is a data point in the dataset with n dimensional; $c_i = 1/N_i \sum\nolimits_{d_j \in C_i} d_j$ is mean for the ith cluster with N_i objects.

The basic clustering procedure of K-means algorithm is summarized as follows:

Step 1: Randomly initialize the k cluster center. Calculate the cluster center list $C = [c_1, ..., c_k]$.
Step 2: Partition each data object into the nearest cluster C_i, yields:

If $d_j \in C_i$, subject to for any m, where $m = 1, ..., k$ and $m \neq i$ and

$$\left\| d_j - c_i \right\|^2 < \left\| d_j - c_m \right\|^2 \qquad (2)$$

Step 3: Recalculate the cluster center list C based on the current partition,

$$c_i = \frac{1}{N_i} \sum\nolimits_{d_j \in C_i} d_j \qquad (3)$$

Repeat Step 2 and Step 3 until there is no change for each cluster or the sum-of-squared-error less than the threshold.

While the empirical speed and simplicity of the K-means algorithm at the cost of the results accuracy. Its final cluster structure relies on the choice of initial seeds [4],[5]. Besides K-means general converge to a local optimum instead of converging to a global optimum as different initial points lead to different convergence centroids, which makes it important to select a high quality set of starting points.

In order to solve the problem, A number of methods have been proposed. R-MEAN initializes the starting points by adding a Gaussian noise to the mean vector [6].CCIA adds an extra parameter

k' (where in general $k' > k$), which selects k' starting candidates, and then, these candidates are down to k starting candidates by Density-based Multi Scale Data Condensation[7]. In order to extract the density information, Redmond and Heneghan used kd-tree to decide which points will be selected based on the density information[8]. Although some progresses have been made by using above algorithms, these methods still have some limitations. Such as R-MEAN still exists randomness problem and CCIA have some difficulties to decide the number of k' as we do not know how many cluster in the dataset in practical.

In addition, the number of clusters k also plays an important role in results accuracy, as different k value often lead to different results. Therefore, identifying k becomes a very important topic in clustering.

Aristidis Likas et al. proposed an incremental approach namely Global K-means algorithm[9], which gradually increases the number of cluster from 1 to k instead of identifying k in advance. Furthermore Adil M. Bagirov proposed algorithm, which reduce a new cluster center starting point candidates and computational complexity over the Global K-means algorithm[10].

Another typical algorithm RPCL is proposed by L. Xu et al. that for each input, not only the winner seed is updated to adapt to the input, but also its rival is penalized by a learning rate[11].

The rest of paper is organized as follows. Global K-means algorithm and Modified Global K-means algorithm (MGKM) described in Section II. Section III presents the Entropy-based Global K-means. Section IV constructs the experimental analysis. Two real datasets are used to demonstrate the effectiveness of the new algorithm. Finally, Section V concludes the paper.

2 GLOBAL K-MEANS AND MGKM

In this section, we give a brief introduction of the Global K-means algorithm and the MGKM.

2.1 Global K-means algorithm

For the Global K-means clustering algorithm, the new cluster starting point decided by each step of cluster split.

Suppose the dataset S be partitioned into $(k-1)$ clusters, and $C = \{c_1, c_2, ..., c_{k-1}\}$ denote the set of cluster centers. In order to determine the starting point of k-th cluster, a value of b_i is calculated by the following equation.

$$b_j = \sum_{i=1}^{m} \max\left\{0, d_{k-1}^i - \left\|d_j - d_i\right\|^2\right\} \quad j = 1, 2, ..., m$$

$$(4)$$

where d_{k-1}^i is squared distance between d_i and the closest center among $(k-1)$ cluster center obtained so far:

$$d_{k-1}^i = \min\left\{\left\|d_i - c_1\right\|^2, \left\|d_i - c_2\right\|^2, ..., \left\|d_i - c_{k-1}\right\|^2\right\} \quad (5)$$

A data point $d_i \in S$ with the largest value selected as starting point for the k-th cluster center. Once the starting point d_i is decided, the algorithm takes $\{c_1, c_2, ..., c_{k-1}, d_j\}$ as the initial set of cluster center to partition the dataset S. The Global K-means algorithm proceeds as follows:

Step 1: Computing the centroid c_1 from the dataset S:

$$c_1 = \frac{1}{m}\sum_{i=1}^{m} d_i \quad (6)$$

where m is the number of data points. Set $j = 1$.
Step 2: Set $j = j + 1$.if $j > k$, then stop; otherwise go to Step 3.
Step 3: Find the data point d_j with largest value of b_j, thus obtaining the set of initial cluster centers $\{c_1, c_2, ..., c_{k-1}, d_j\}$ and apply K-means algorithm to find $\{c_1, c_2, ..., c_{k-1}, d_k\}$, which partitions S into k clusters. then go to Step 2.

2.2 MGKM algorithm

MGKM algorithm takes different methods to determine the starting point for a new cluster center. The rest parts are as similar as the Global K-means algorithm. Here is the initial point selection algorithm:

Step 1: For any data point d_j, compute the set $S(d_j) = \{d_i \| d_j - d_i \|^2 < d_{k-1}^i, \forall d_i \in S\}$, then compute the set S center c_j.
Step 2: For the set $S(d_j)$ and the center c_j compute the following measure:

$$F(c_j) = \sum_{i=1}^{m} \min\left\{d_{k-1}^i, \left\|d_j - d_i\right\|^2\right\} \quad (7)$$

Step 3: Determine c_j' such that $F(c_j) \geq F(c_j')$, where $j' \neq j$ and $1 \leq j', j \leq m$
Step 4: Compute the set $S(d_j')$, and the Set $S(d_j')$ center c_j'
Step 5: Repeat Step 4 until no more data points return to the set $S(d_j')$

3 ENTROPY-BASED GLOBAL K-MEANS

Suppose given a dataset $S = \{d_i \| d_i \in R^n, i = 1, 2 ..., m\}$. The i-th cluster denoted by C_i, and the centroid of i-th cluster represented by c_i.

720

In this paper, we take each attribute weight into account. A weighted Euclidean distance adopted in this paper, which defined as

$$WD_{ij} = \sqrt{\sum_{k=1}^{n} W_k (d_{ik} - d_{jk})^2} \tag{8}$$

where W_k is the k-th dimensional weight, which can be calculated by following steps:

Step 1: Normalizations of each attribute in order to avoid the effect of the different scale. The normalization scheme can take the form as:

$$N_{ij} = \frac{d_{ij}}{\sum_{k=1}^{m} d_{kj}} \tag{9}$$

where N_{ij} is the normalized value, and $\sum_{k=1}^{m} d_{kj}$ represents the sum of column j

Step 2: Compute the Entropy of j-th column, which can defined as

$$Entropy_j = -p \sum_{k=1}^{m} N_{ij} \ln N_{ij} \tag{10}$$

where $p = 1/\ln m$

Step 3: In order to describe the Entropy affect the weight. Diff was introduced to solve this problem, which defined as

$$Diff_j = 1 - Entropy_j \tag{11}$$

The bigger Diff is, the more contribution to classify the data points

Step 4: Calculate each column weight, which can illustrated as:

$$W_j = \frac{Diff_j}{\max_{k=1,2,\dots,n} Diff_k} \tag{12}$$

where $0 \leq W_j \leq 1$

Furthermore, in order to reduce the algorithm complexity, an adjusted sum-of-squared-error measure proposed as follows

$$fc(k) = \sum_{i=1}^{m} \min_{j=1,\dots,k} \|d_i - c_j\|^2 \tag{13}$$

Here k is the number of clusters

Whether the function of Eq. 1 or $fc(k)$, Both are decreases monotonically with the increase of cluster number. In this paper, we adopted a new method to compute the new cluster center, which maximize the value of Δ:

$$\Delta = fc(k) - fc(k,d) = \sum_{j \in S_2} d_{k-1}^j - \|d - d_j\|^2 = b_d \tag{14}$$

data point d selected as the new cluster center when Δ get the max value.

At the same time, we proposed a method to divide the dataset into two sets, such that to reduce the number of candidates data point when decide the new cluster center. Two datasets constructed as

$$S_1 = \left\{ i \in (1,2,\dots,m) : \|d_i - c_j\|^2 < d_k^i \right\} \tag{15}$$

$$S_2 = \left\{ i \in (1,2,\dots,m) : \|d_i - c_j\|^2 \geq d_k^i \right\} \tag{16}$$

where only the data points in S_1 have the chance to become the new cluster starting point

In order to reduce the algorithm computing time, we proposed a function:

$$f_i(b) = \sum_{i=1}^{m} \min \left\{ d_k^i, \|b - c_i\|^2 \right\} \tag{17}$$

where c_i is the i-th cluster center, Furthermore b_d can be simplified according to S_1, S_2 as follows:

$$b_d = \sum_{j \in S_2} d_{k-1}^j - \|d - d_j\|^2 \tag{18}$$

The Eq. 14 shows that if we maximize the b_d so that the Δ can get max value. So the way to decide the new cluster centroid defined as follows:

Step 1: For any data point $d \in S_1$, compute the set S_0 and the cluster center c

Step 2: Compute the min value of $fi(d)$

Step 3: For the data point $d_{\min}$, compute the set S_1

The Entropy-based Global K-means algorithm is presented as follows:

Step 1: Compute the starting point c_1 of the dataset S, then set $k = 1$ and compute the value of $fc(1)$. Initialize the max running time $t_{\max}$ and threshold value ε, the max number of clusters $c_{\max}$

Step 2: Set $k = k + 1$, if $k \geq c_{\max}$ or running time $t \geq t_{\max}$ then stop

Step 3: Compute the new cluster center by using the above method

Step 4: If $\Delta \div fc(1) < \varepsilon$ then stop, otherwise go to Step 2.

4 EXPERIMENTAL ANALYSIS

In order to evaluate the Entry-based Global K-means algorithm in practical, the algorithm has been implemented and tested it in Python+Fortran and was run on a PC, which operating system is Windows Seven, CPU consists Intel Core 2 Duo CPU E7400 2.8 GHZ and the capacity of the RAM is 4 GB. We compared the Entropy-based Global K-means algorithm with the MGKM.

4.1 Datasets

We evaluated the performance of Entropy-based Global Algorithm and MGKM on two UCI datasets.

The first dataset, Iris consists of 150 points. Each point has four attributes. The dataset contain three categories Iris-setosa, Iris-versicolor and Iris-virginica, which each category consists of 50 instances.

The second dataset, Wine consists of 178 points, which are the results of a chemical analysis of three different cultivars' wines grown in the same district in Italy. Each data point in the dataset has 13 attribute.

4.2 Results

The results of accuracy for the MGKM and Entropy-based Global K-means are listed in Table 1 and Table 2. As the experimental results acquired by both MGKM and Entropy-based Global K-means are unchanged with the variation of starting points and the number of clusters k, for each algorithm needs only trial.

Table 1. Experimental results of accuracy on the Iris dataset.

| Category | MGKM | | Entropy-based Global K-means | |
	Instances	Experimental data	Instances	Experimental data
Iris-setosa	50	50	50	50
Iris-versicolor	50	61	50	57
Iris-virginica	50	39	50	43
Acc	92.67%		95.33%	

Table 2. Experimental results of accuracy on the Wine dataset.

| Category | MGKM | | Entropy-based global K-means | |
	Instances	Experimental data	Instances	Experimental data
Class 1	59	62	59	62
Class 2	71	69	71	69
Class 3	48	47	48	47
Acc	98.31%		98.31%	

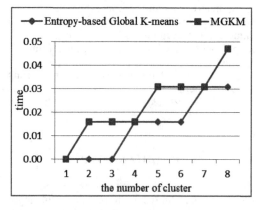

Figure 1. Experimental results of running time on Iris dataset.

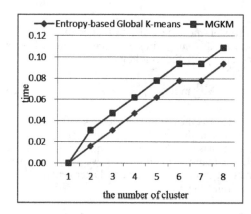

Figure 2. Experimental results of running time on Wine dataset.

In order to measure the accuracy, Acc was used in the metric. Acc was defined as follows:

$$Acc = 1 - \left| \frac{\Delta}{\#IN} \right| \qquad (19)$$

where Δ is the number of error data points, $\#IN$ is number of instances contain in a certain category.

The difference between MGKM and Entropy-based Global K-means on the Iris dataset was significantly. Entropy-based Global K-means achieved about 2.7% accuracy improvement over MGKM. Although on the Wine dataset MGKM acquired very high accuracy. Entropy-based Global K-means also achieved the same accuracy.

In order to compare the running time between MGKM and Entropy-based Global K-means, the number of k is fixed. The results of running time for MGKM and Entropy-based Global K-means are displayed in Figure 1 and Figure 2. As seen in Figure 1 and Figure 2, The experimental results were encouraging; for example, for the Iris dataset and Wine dataset the solution were obtained with the Entropy-based Global K-means and MGKM. The results of Entropy-based Global K-means outperformed MGKM all the time.

5 CONCLUSION

A new clustering algorithm named Entropy-based Global K-means is proposed which performs better clustering results than MGKM without pre-determining the exact number of clusters and the starting points. On the other hand, MGKM has a significant improved performance over K-means, which has already proved by Adril M. Bagirov[10].

Future research directions will take large scale dataset and data streams into account. We also plan to further enhance the accuracy of the Entropy-based Global K-means algorithm through make a deep research on Entropy.

ACKNOWLEDGMENT

This work is supported by the National High-Technology Research and Development Program of China under Grant NO. 2013AA01A211.

REFERENCES

[1] J. MacQueen. Some methods for classification and analysis of multivariate observations. In L.M.L. Cam and J. Neyman, editors, Proc. of the 5th Berkeley Symposium on Mathematical Statistics and Probability, Volume I, Statistics. University of California Press, September 1967.

[2] A.K. Jain, "Data clustering: 50 years beyond k-means," Pattern Recognition Letters, vol. 31, no. 8, pp. 651–666, 2010.

[3] X. Wu, V. Kumar, J. Ross Quinlan, J. Ghosh, Q. Yang, H. Motoda, G. McLachlan, A. Ng, B. Liu, P. Yu, Z.-H. Zhou, M. Steinbach, D. Hand, and D. Steinberg, "Top 10 algorithms in data mining," Knowledge and Information Systems, vol. 14, pp. 1–37, 2008.

[4] Xu, R., Wunsch, D.: Survey of clustering algorithms. IEEE Transactions on Neural Networks 16(3), 645–678 (2005).

[5] Han, J., Kamber, M.: Data Mining: Concepts and Techniques. Morgan Kaufmann, San Francisco (2006).

[6] Khan, S.S., Ahmad, A.: Cluster center initialization algorithm for K-means clustering. Pattern Recognition Letters 25(11), 1293–1302 (2004)

[7] Khan, S.S., Ahmad, A.: Cluster center initialization algorithm for K-means clustering. Pattern Recognition Letters 25(11), 1293–1302 (2004).

[8] Redmond, S.J., Heneghan, C.: A method for initialising the K-means clustering algorithm using kd-trees. Pattern Recognition Letters 28(8), 965–973 (2007).

[9] Aristidis Likas, Nikos Vlassis, Jakob J. Verbeek.The global k-means clustering algorithm[J]. Pattern Recognition, 36(2003):451–461.

[10] Adril M.Bagirov*, Modified global k-means algorithm for minimum sum-of-square clustering problems[J]. Pattern Recognition, 41(2008):3192–3199.

[11] L.Xu, A. Krzyzak, E. Oja. Rival penalized competitive learning for clustering analysis, RBF net, and curve detection. IEEE Transactions on Neural Networks[J], 1993,4(4):636–648.

Control Engineering and Information Systems – Liu (Ed)
© 2015 Taylor & Francis Group, London, ISBN 978-1-138-02685-8

Object extraction by superpixel grouping

H. Cheng & J.H. Shang
Department of Mathematics, Institute of Natural Sciences, Shanghai Jiao Tong University, Shanghai, China

C. Zhang
School of Electronic, Information and Electrical Engineering, Shanghai Jiao Tong University, Shanghai, China

ABSTRACT: Detecting a number of unknown objects in cluttered scenes is an important and difficult problem in computer vision filed. Several previous approaches have been proposed to manage object detection by adopting powerful perceptual grouping cues, such as the gestalt laws of proximity and continuity etc. In this paper, we propose a fragments grouping method to extract object contour via superpixel, which has strong edge support in the image. Moreover, a new cost function has been introduced to promote spatially coherent sets of superpixel with object boundary. We evaluate the proposed method against two leading contour closure approaches in the literature on the BSDS500. The results demonstrate that the proposed object detect method performs both good accuracy and time efficiency against other state-of-the-art methods.

1 INTRODUCTION

Detecting a number of unknown objects in cluttered scenes is an important and difficult problem in computer vision research and finding a specified object or area is a pre-process to the following steps in image processing. If multiple objects are obtained from the image, we can significantly improve image segmentation, surveillance, and semantic analysis etc. In particular, it also can help doctors make correct diagnosis from medical images.

In Human Visual System (HSV), an object boundary has a set of non-accidental contour relations, namely Gestalt laws, which were articulated by the Wertheimer (Wertheimer 1938), such as visual cues: proximity, good continuation, and symmetry, etc. Therefore, the main idea of extracting objects is to use the perceptual grouping method does, such as virtual link to complement a set of fragmented contours into a cycle and then separating the salient object from its background. What makes the problem particularly hard is the intractable number of cycles, which may exist in the contour extracted from the image of a real scene.

In this paper, we introduce a framework that builds a regularization function for efficiently searching an optimal closure contour from another perspective. An overview of the proposed approach is illustrated in Figure 1. We restrict object boundaries close to the boundaries of superpixel, and the ideal case is that missing boundaries can follow the boundary of superpixel. In frame-

work, our first is to reduce the problem of finding cycles to the problem of finding a closure boundary that align well with some subset of superpixels boundary which has strong contour support in the P-map image. Furthermore, the accuracy of P-map image is very critical in searching, because many details will interfere with the real contours and cause inaccurate result. However, there does not have a method where all details will vanish and only true object edges remain. This is why we use point information of all scaled images in a probability map, where all relevant information exists for further (sub-segment) analysis. An algorithm proposed in (Levinshtein, Sminchisescu, & Dickinson 2010), called SC, which is very time-consuming to compute probabilities map and superpixel, especially in large data set. Therefore, we use a new method called CDRF contour detection which via Random Forest (Zhang, Ruan, Zhao, & Yang 2012) and is quite fast. On average, CDRF method takes 10 seconds but gPb method (Arbelaez, Maire, Fowlkes, & Malik 2011) (Maire, Arbelaez, Fowlkes, & Malik 2008) of SC algorithm takes 180 seconds in one image. In addition, CDRF algorithm achieves the highest average precision among all methods. After the formation of sub-segments, the proposed object extraction method can rely on the scale information in searching. In order to improve the accuracy of the searching for object boundary over sub-segments, some preprocessing steps are introduced in (Kiranyaz, Ferreira, & Gabbouj 2006, Wang, Kubota, Siskind, & Wang 2005, Stahl & Wang

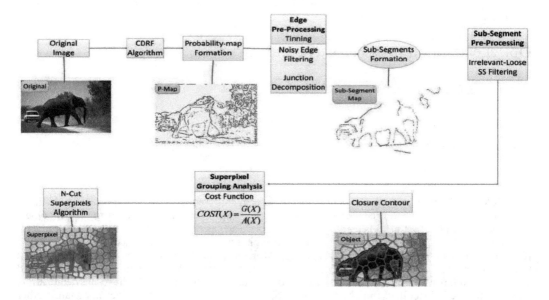

Figure 1. Overview of the proposed approach: (1) contour image: it is the Probability-map obtained by M-mPb and Compass Edge Detector as the first step; (2) Sub-Segment: many details edge in P-map that un-useful to object extraction, therefore use preprocessing step to remove some details, noisy edge, and irrelevant SS that remain accuracy contour fragments. (3) Superpixel Segmentation: Superpixel ensures that target boundaries are reasonably well approximated by superpixel boundaries. (4) Superpixel Grouping Analysis: a cost function construct reflects the extent to which the superpixel boundary is supported by evidence of a real image contour. (5) Object: find a global optimal cycle base on cost function that make the largest set of superpixels bounded by contours that have the least gaps and complement.

2007). Then, we use least relevant Sub-Segments filter in sub-segments preprocessing step.

The reformulation needs a mechanism to obtain superpixel subsets which are spatially coherent (Levinshtein, Sminchisescu, & Dickinson 2010). It is a property of cost function that computes the ratio of perimeter to area. We build a ratio cost function based on Stahl and Wang (Wang, Kubota, Siskind, & Wang 2005) and Levinshtein (Levinshtein, Sminchisescu, & Dickinson 2010) to operate on superpixels rather than contours. In addition, we consider a similar pixel which is not considered in (Levinshtein, Sminchisescu, & Dickinson 2010). The function represents that boundaries of superpixel and contour will have more prominently spatial coherence. Next, we use five features to illustrate the "gap function" which is the distance between image contour and superpixel boundary, the strength of nearby image contour, the orientation and curvature of those pixels on the two boundaries and the similar pixel factor. Those features play an important role in our reformulation.

The final is that we use parametric maxflow (Kolmogorov, Boykov, & Rother 2007) to get the global optimum based on our cost function, which contains area and gap factor. Those solutions are the complete object boundaries which have largest set of Superpixel area and least gap. Moreover, the parametric maxflow not only generates the minimum cost solution, but also generates serial cost solutions (Kolmogorov, Boykov, & Rother 2007).

Therefore, SC algorithm in (Levinshtein, Sminchisescu, & Dickinson 2010) may be more reasonable if they had considered least relevant filter in preprocessing and Similar Pixel Feature in gap function.

2 RELATED WORKS

Several approaches have been proposed by researchers to extract object contour from images. In (Wang, Kubota, Siskind, & Wang 2005), Wang had summarized some methods detailed and we will introduce them briefly. The earliest attempt at finding salient boundaries was based on edge detection (Canny 1986, Heath, Sarkar, Sanocki, & Bowyer 1997, Konishi, Yuille, Coughlan, & Zhu 2003) and edge-linking methods (Hart, Nilsson, & Raphael 1968) which use edge or local search techniques to link clustered edges into closed boundaries. However, it is not certain whether those boundaries are accurate.

In recent years, Kiranyaz (Kiranyaz, Ferreira, & Gabbouj 2006) use uniform cost search algorithm to make automatic object extraction based on linking clustered boundary fragments.

A closed contour is represented by a parameterized curve, which is a classical model, called 'snake' (Kass, Witkin, & Terzopoulos 1988). However, this kind of method has a larger challenge from the change of topology and the presence of corners. To resolve these problems, level set approach has been proposed by Osher. This approach can handle topological changes by the curve in an implicit form. Another salient measure was proposed based on Bayesian variational problem, including the Theater-Wing model (Nitzberg, Mumford, & Shiota 1993) and the Region Competition model (Zhu & Yuille 1996), but it is usually difficult to find optimal solutions.

Due to the drawbacks of aforementioned methods, graph theoretic methods were introduced to solve these problems. Graphic method is to find a boundary for partitioning the graph, which makes the cost function optimal. These methods include Minimum Cut (Wu & Leahy 1993), Normalized Cut (Shi & Malik 2000), Average Cut (Sarkar & Soundararajan 2000), Ratio Cut (Wang & Siskind 2003), and JI algorithm (Jermyn & Ishikawa 2001). But graph constructed by these methods always use pixels or small regions as vertices, which makes it difficult to consider many Gestalt cues.

Our aim is to obtain closure boundary. Therefore, in the context of graph-based optimization algorithms, the constraint corresponds to finding cycles in a graph. Elder and Zucker (Elder & Zucker 1996) define boundary saliency that connect between two adjacent contour fragments, and find optimal closure boundary by using the shortest path algorithm. Williams and Thornber method (Williams & Thornber 1999) (Mahamud, Williams, Thornber, & Xu 2003) has the similar model and use spectral analysis techniques and a strongly-connected-component algorithm to obtain closure boundary. Wang et al. (Wang, Kubota, Siskind, & Wang 2005) use the Minimum Weight Perfect Matching to identify the alternate cycle with minimum cycle ratio that operate on contour fragments, however, Jermyn (Jermyn & Ishikawa 2001) works directly with pixels in a 4-connected image grid. Nevertheless, if a measure only depends on the total boundary gap, it is insufficient for perceptual closure, and the distribution of gaps along the contour is also important to analysis perceptual closure, which was argued by Elder and Zucker. However, those methods suffer from the high complexity of choosing the right closure from a sea of contour fragments. In addition, some elegant combinatorial optimization measures are available.

In recent years, superpixel is becoming increasingly popular for use in computer vision applications. The idea of superpixel was originally developed by Ren and Malik (Ren & Malik 2003). Some research makes contour grouping based on superpixel, such as Levinshtein, (Levinshtein, Dickinson, & Sminchisescu 2009) constrain the symmetric parts to be collections of superpixels, and (Levinshtein, Sminchisescu, & Dickinson 2010) obtain optimal contour closure base on superpixel grouping, We will draw on this idea of superpixel grouping, and we will improve its performance by changing gap function.

In this paper, our goal is to find closed object boundary in an efficient manner. Drawn on (Wang, Kubota, Siskind, & Wang 2005) (Levinshtein, Sminchisescu, & Dickinson 2010) (Levinshtein, Dickinson, & Sminchisescu 2009), we use sub-segments and superpixel to constrain the search space of the resulting closure. Moreover, the gap computation is also easy by superpixel boundary. On the optimization side, we will use parametric maxflow problem method as used in (Levinshtein, Sminchisescu, & Dickinson 2010) to obtain a global optimum of closure cycle.

3 PROBLEM FORMULATIONS

We refer to the process of identifying a subset of fragments, which is produced by preprocessing, and building a cost function to form a closed boundary. The framework is mentioned in Sec. 1.

3.1 Contour image and sub-segments preprocessing

In the framework, we should obtain P-map at first. Complications and degradations in the segmentation accuracy begin to occur when the image gets more and more "detailed". Therefore, we use the probability of a pixel belonging to a contour map as the scale information. An algorithm for estimating the probability of a pixel on contour boundary called gPb was proposed, which combines multiple local features in a probabilistic framework that contains two main components: the mPb detector based on local image analysis at multiple scales and the sPb detector based on mPb and the normalized cut segmentation results, which is the spectral of affinity matrix. The gPb used in (Levinshtein, Sminchisescu, & Dickinson 2010) is time-consuming, although gPb has high accuracy with a higher F-measure. So instead we use a new efficient and effective algorithm for contour detection called CDRF. After we obtain P-map, the next step is sub-segments formation and post-processing. In this problem, we filter the least relevant sub-segments by an experimental formula to reduce noisy disturbance and then use maxflow operation to obtain the finally results.

3.1.1 *Contour detection via random forest*

In order to efficiently and effectively grouping, a learning algorithm for contour detection based on random forest was proposed. We use six features of a training image, which is formed by the magnitude and the direction of image gradients (MG, DG), Inhibition Terms (IT) (Papari & Petkov 2011), Brightness Gradients (BG), color gradient (CG) (Arbelaez, Maire, Fowlkes, & Malik 2011) and Compass Operator (CO) (Ruzon & Tomasi 1999), to construct a classifier within the random forest (Breiman 2001) learning framework. With the learning classifier, we accumulate and normalize the voting results of all the trees for probabilistic output.

Figure 2 show the precision-recall curves and the F-measures with different methods. A detailed description of the algorithm can be found in Ref (Zhang, Ruan, Zhao, & Yang 2012).

The *gPb* method performs a little well in terms of accuracy, but the speed is much slower than that of the proposed CDRF algorithm. On average, it takes 180 seconds for the *gPb* method to detect contours in one image but the proposed CDRF algorithm only requires 10 seconds.

In addition, the proposed CDRF algorithm achieves the highest average precision among all methods. We note that both the *gPb* and the CDRF methods are able to extract object contour with fewer spurious edges. Therefore, in order to efficiently and effectively grouping, we choose the proposed CDRF method instead of *gPb* to achieve contour image.

3.1.2 *Least relevant sub-segments filter*

Because the P-map has some incorrect results by noisy disturbance, that the sub-segments based on

inaccurate P-map maybe result in a wrong closure boundary. So, we hope to find a method that can alleviate its destruction.

Once all the sub-segments are formed from the edge pixels of the P-map, the most relevant sub-segments that bear the major object boundaries are usually longer with higher probability. Therefore, the relevance, R, of a sub-segment SS, can then be expressed as Eq. (1) (Kiranyaz, Ferreira, & Gabbouj 2006):

$$R(SS) = \sum_{e \in SS} p(e) \qquad (1)$$

where $p(e)$ is the probability factor of an edge pixel e in SS. Sorting all the sub-segments formed over the P-map and removing the least relevant ones, the threshold is chosen by an empirical Eq. (2).

$$N_T = \begin{cases} x_1 \| \, |x_1 - 100| < |x_2 - 100|, \\ x_1, x_2 \in (N\_mean, N\_std) \end{cases} \qquad (2)$$

where, N_mean is the mean of all sub-segments, N_std is the standard deviation. Consequently, a more important sub-segment map can obtain by this processing.

3.2 *Superpixel grouping analysis*

Our framework reduces grouping complexity by restricting closure to lie along superpixel boundaries. According to Figure 1, we will make superpixel grouping analysis to construct cost function after obtain sub-segments and superpixel.

We define closure cost function as Equation (3), which draw on Stahl and Wang (Wang, Kubota, Siskind, & Wang 2005), Levinshtein (Levinshtein, Sminchisescu, & Dickinson 2010).

$$COST(\mathbf{X}) = \frac{G(\mathbf{X})}{A(\mathbf{X})} \qquad (3)$$

where $\mathbf{X}$ is a vector indicator that labels all superpixels of image I as Figure (1) or ground (0). $G(\mathbf{X})$ is the boundary gap along the perimeter of $\mathbf{X}$, and $A(\mathbf{X})$ is its area. The boundary gap is defined to be $G(\mathbf{X}) = P(\mathbf{X}) - E(\mathbf{X})$, it is a measure of the difference between boundary of $\mathbf{X}$ and contour fragments. $P(\mathbf{X})$ is the length of $\mathbf{X}$, and $E(\mathbf{X})$ is the number of boundary $\mathbf{X}$, which satisfies the rules in Section 4.

In order to solve cost function conveniently, we use a method in (Kolmogorov, Boykov, & Rother 2007). Let X_i be a binary variable indicator for the i-th superpixel, P_i be the perimeter length of superpixel i and P_{ij} be the length of the shared edge between superpixel ι and φ. Similarly, let E_i

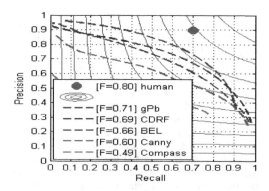

Figure 2. Precision-recall curves of contour detectors.

728

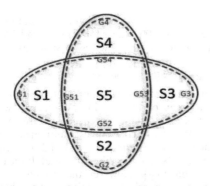

Figure 3. Boundary gap computation over superpixel graph. S_1, S_2, S_3, S_4, and S_5 correspond to superpixels that were selected. G_i and G_{ij} are the boundary gap of superpixel ι and the gap on the edge between superpixels ι and φ respectively. The gap along the outermost ring is then $G_{1234} = \sum_{i=1}^{4} G_i - 2(G_{51} + G_{54} + G_{52} + G_{53})$.

be the conditional edges of Superpixel i's boundary, and E_{ij} be the conditional edge for the shared boundary between superpixel ι and φ. Then let $G_i = P_i - E_i$ and $G_{ij} = P_{ij} - E_{ij}$ as the boundary gaps between superpixel and contour fragment. Above all, the final formula is Eq. (4):

$$Cost(\mathbf{X}) = \frac{\sum_i G_i X_i - 2\sum_{i<j} G_{ij} X_i X_j}{\sum_i A_i X_i} \quad (4)$$

Eq. (4) illustrate that an optimal closure boundary should have a small gap between boundary of superpixel and sub-segments. We wish the true closure boundary is along superpixel boundary. Most of other grouping approaches are to complement the missing contour fragments and find the missing fragments between two solid fragment endpoints. However, those ideas are complex because they try to find missing fragments in a sea of possible superpixel boundary. On the contrary, we inverse those ideas and try to use superpixel as basis to find a closure contour along the boundary. The gaps measure will be discussed detailed in the next section. It is the reason that the smaller the gaps, the better it is. Because it's may exist some shared boundary of superpixels that can be computed in gaps if we only use $\Gamma\iota$ for every internal boundary. Finally, the gaps will be wiped out. We subtract the gaps for all internal boundaries. In general, we are always sensitive for an object with larger area, and then we add the individual areas of all the superpixels as the denominator. By this formation, the areas not only promote spatial coherence but also promote compactness.

4 GAP MEASURE AND OPTIMIZATION FRAMEWORK

The critical part is the gap measure and the method to find optimal closure contour. Levinshtein et al. proposed a gap measure (Levinshtein, Sminchisescu, & Dickinson 2010) that can obtain a good results at most time. However, it maybe fails when pixel information is lost in contour image and preprocessing step. Therefore we introduce a feature that is in terms of the similar point feature. An example as Figure 4 shows.

4.1 Gap measure

The gap $G(\mathbf{X})$ computation by incorporate multiple contour features. For a pair of superpixels i and j, the gap on the edge between them is $G_{ij} = P_{ij} - E_{ij}$, Where $P_{ij} = |\text{LB}_{ij}|$, the $|\text{LB}_{ij}|$ is the length of Superpixel boundary (i, j) and $E_{ij} = \sum_{p \in LB_{ij}} E_{ij}^p$, where $E_{ij} = [\text{Logistic}(f^p) > T_e]$ is an edge indicator for pixel p, in which f^p is a feature vector for the pixel p, Logistic is a logistic regression, and T_e is a threshold that can determine whether current pixel belongs to closure contour. This feature vector consists of five features (See Fig. 4).

In all five features mentioned above, the first fourth features are defined as (Levinshtein, Sminchisescu, & Dickinson 2010) and the fifth feature is Eq. (5):

$$f_a(p,q) = \frac{2}{3\pi}\{\arccos[d_{e1}(p,q)] + \arccos[d_{e2}(p,q)]\} \quad (5)$$

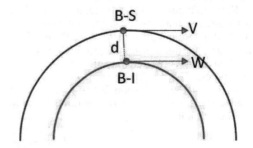

Figure 4. Five features illustrate. $B - S$ is boundary of superpixel, $B - I$ is boundary of image object. Black curves correspond to Superpixel boundaries, while the red curve corresponds to detected image edges. All features are computed at superpixel boundary pixel $B - S$: Distance between $B - S$ and $B - I$, this is mean if the pixel at superpixel boundary is closed to the detected image edges, it is more likely the edges along superpixel boundary; Strength use image edge strength at $B - I$; Alignment, computed as the absolute value of the cosine of the angle between v and w; the fourth feature is curvature that computed as the squared curvature at $B - S$; Final feature is similar pixel feature, we use p and q instead of $B - S$ and $B - I$ in Eq. (5).

where:

$$d_{e1}(p,q) = D'(p) \cdot L(p,q)$$
$$d_{e2}(p,q) = D'(q) \cdot L(p,q) \tag{6}$$

$$L(p,q) = \begin{cases} \dfrac{1}{\|p-q\|}(q-p), & \text{if } D'(p) \cdot (q-p) \geq 0 \\ \dfrac{1}{\|p-q\|}(p-q), & \text{if } D'(p) \cdot (q-p) < 0 \end{cases} \tag{7}$$

$$D'(p) = (-I_y(p), I_x(p))$$
$$D'(q) = (-I_y(q), I_x(q)) \tag{8}$$

Then we use $f_d(p,q)$ as the similar feature, which indicate two pixels are similar when $f_d(p,q)$ is smaller. All the pixels are from original image and the fifth feature reflect contour edge or missing edge information. Next, we use logistic classifier over a feature vector to obtain E_{ij}.

4.2 A parametric maxflow problem

In this paper (Wang, Kubota, Siskind, & Wang 2005), Wang constructs a graph to find a perfect matching that make the solution be the optimal cycle constraints on ration minimum. However, instead of minimizing the ratio in Eq. (3), we reduce it to a parametric energy function $E(\mathbf{X}, \lambda) = G(\mathbf{X}) - \lambda A(\mathbf{X})$. Therefore, it can obtain optimal solution according to optimal λ. The constraints on the ratio guarantee that the resulting difference is the global minimization.

Ratio minimization can be reduced to solving a parametric maxflow problem and making the method in (Kolmogorov, Boykov, & Rother 2007) applicable for minimizing the ration $Cost(\mathbf{X})$. The method can not only find one solution, but also can find others. The details can refer to (Levinshtein, Sminchisescu, & Dickinson 2010). In our experiments, we choose 150 superpixels to compute the cost function in each image.

5 EVALUATIONS

We evaluate the proposed method (MSC) against two other contour grouping methods: one version of ratio contours (RRC) (Wang, Kubota, Siskind, & Wang 2005) and the other method of superpixel grouping (Levinshtein, Sminchisescu, & Dickinson 2010), called SC. We provide a qualitative evaluation on various images (see Fig. 6), as well as a quantitative evaluation on the Berkeley Segmentation Data Set (BSDS500) which includes 500 images with human labeled segmentation

results, but the human segmentation is not binary image. We use 300 images for training and 200 for test (Arbelaez, Maire, Fowlkes, & Malik 2011).

5.1 Quantitative evaluation

For a quantitative evaluation of the results, we use Variation of Information (VI) metric, which measures the distance between two segmentations in terms of their average conditional entropy, Segmentation Covering (Cover), RI (Arbelaez, Maire, Fowlkes, & Malik 2011) and F-measure as the benchmark. However, the ground truth in BSDS500 is several segmentations but not binary segmentation. So the quantitative evaluation is not like binary segmentation benchmark. We average the entire image F-measures in BSDS500 and show some results in the following.

The whole benchmark is based on (Arbelaez, Maire, Fowlkes, & Malik 2011). We chose the best parameters for all three algorithms and fixed them for the entire experiments. For RRC, we used $\lambda = 0$ and $\lambda = 1$, the all sets as (Wang, Kubota, Siskind, & Wang 2005). For SC, we used 150 superpixels and other sets as (Levinshtein, Sminchisescu, & Dickinson 2010). For our method, we fixed the number of Superpixel to 150 and set $T_\varepsilon = 0.05$ as SC.

Figure 5 shows the superpixel number influence the MSC results. As above mentioned, Figure 5.1 shows the cover is larger than other number of superpixels along with the increasing of solutions under 150 superpixels. Figure 5.2 shows that different number of superpixel could not largely influence on the results of RI.

As (Arbelaez, Maire, Fowlkes, & Malik 2011) shows, when the measures of the cover and RI are much larger and the VI is much smaller, the final result much well. In Figure 5.4, we find the change is not critical for F-measure under 150 or 200 superpixels.

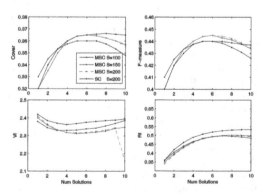

Figure 5. Evaluate MSC results use cover, RI, VI and F-measure measures.

All these evaluation results are based on the number of solutions, we set the number of the solutions to 10 in maxflow computation, but the optimal result is not automatic selection. Therefore, we manually choose best solutions from the whole results, and evaluate those results to compare with *SC* results at 100, 150, 200 superpixels respectively. The evaluation results are shown in Table 1.

As shown in Table 1, we can find the MSC has a well performance than *SC*, because the *MSC* F-measure is 0.49 under 150 superpixels, yet *SC* F-measure is 0.48. The other index of measures is secondary. Therefore *MSC* has a slighter performance than other two algorithms, such as *RRC* and *SC*. In addition, we use CDRF instead of *gPb* to improve time cost and hardly change any accuracy.

5.2 Qualitative evaluation

In order to compare those three algorithms, we also provide a qualitative evaluation. Figure 6 illustrates the performance of our method comparing with the other two competing approaches. We manually select best solution for each method in BSDS500. We can see our method is well than *RRC*. Because we use CDRF algorithm as contour detector, which is very fast and make the detected contours be closer to the true object contour that make gap computation more accuracy. We observe that our framework is more effective, and it can accurately extract object from background quicker than *SC*. This is clearly visible in the image of goose and horse, where unbroken goose (Right) was obtained by *MSC*, but it couldn't find an unbroken object in *SC* and *RRC*. However, this is not the usual case. When there is more compact contour which is not lost on gap, it will be preferred. This is the reason why the filled gap is between the horse's legs in second image. In contrast, *SC* obtains a better solution than *RRC*. However, the *SC* couldn't find the whole horse in second image and fifth image, because it finds many uncorrelated areas that make incorrect object boundary. *RRC* is much worse. In the third image,

Table 1. The comparison between the best MSC results on different scale of superpixels and best SC results used in region benchmarks on the BSDS500.

BSDS500

	Cover	RI	VI	F-measure
MSC S = 100	0.40	0.58	2.22	0.48
MSC S = 150	0.41	0.58	2.2	0.49
MSC S = 200	0.40	0.58	2.21	0.48
SC	0.40	0.59	2.28	0.48

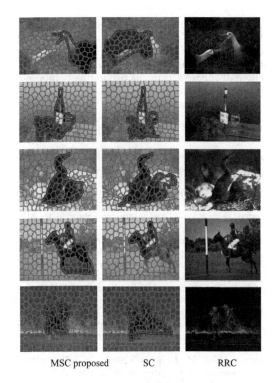

| MSC proposed | SC | RRC |

Figure 6. Sample experimental results. We compare our results MSC to two other algorithms: SC and RRC.

two people couldn't perfectly be extracted by *SC*; But *MSC* can obtain a compact result.

6 CONCLUSIONS

In this paper, we propose a method of contour fragment grouping via superpixel, in which boundary has strong edge support in the image. While we use superpixel properties with an ideal scope and a convenient mechanism for incorporating appearance information, this method yields an optimal framework for closure detection that compares favorably with two leading prior approaches. In this work, we have some contributions as following. Firstly, although *gPb* have high F-measure, our method, CDRF, can also obtain much well results with quick time, and so we use a competitively effective and efficient CDRF contour detector instead of *gPb* to obtain P-map; Secondly, we use a new irrelevant loose SS filtering formula to reduce some noisy edges jamming in Sub-Segment map as post-process; Finally, we introduce a similar pixel formula to improve gap measure. Based on the above, a correct object can be obtained by *MSC* algorithm, and all these results are demonstrations as quantitative and qualitative evaluation. In the future, we will plan to pursue a more elegant coarse-to-fine framework for object extraction by using multiple superpixel scales to improve the precision of superpixel, and incorporating homogeneous property in object appearance.

REFERENCES

Arbelaez, P., M. Maire, C. Fowlkes & J. Malik (2011). Contour detection and hierarchical image segmentation. PAMI 33, 898–916.

Breiman, L. (2001). Random forests. Machine Learning 45, 5–32.

Canny, J. (1986). A computational approach to edge detection.

Elder, J. & S. Zucker (1996). Computing contour closure. Volume 1064, pp. 399–412.

Hart, P., N. Nilsson & B. Raphael (1968). A formal basis for the heuristic determination of minimum cost paths. Systems Science and Cybernetics, IEEE Transactions on 4, 100–107.

Heath, M., S. Sarkar, T. Sanocki & K. Bowyer (1997). A robust visual method for assessing the relative performance of edgedetection algorithms. PAMI 19, 1338–1359.

Jermyn, I. & H. Ishikawa (2001). Globally optimal regions and boundaries as minimum ratio weight cycles. PAMI 23, 1075–1088.

Kass, M., A. Witkin & D. Terzopoulos (1988). Snakes: Active contour models. IJCV 1, 321–331.

Kiranyaz, S., M. Ferreira & M. Gabbouj (2006). Automatic object extraction over multiscale edge field for multimedia retrieval. TIP 15, 3759–3772.

Kolmogorov, V., Y. Boykov & C. Rother (2007). Applications.

Konishi, S., A. Yuille, J. Coughlan & S.C. Zhu (2003). Statistical edge detection: learning and evaluating edge cues. PAMI 25, 57–74.

Levinshtein, A., C. Sminchisescu & S. Dickinson (2010). Optimal contour closure by superpixel grouping.

Levinshtein, A., S. Dickinson & C. Sminchisescu (2009). Multiscale symmetric part detection and grouping. pp. 2162–2169.

Mahamud, S., L. Williams, K. Thornber & K. Xu (2003). Segmentation of multiple salient closed contours from real images. PAMI 25, 433–444.

Maire, M., P. Arbelaez, C. Fowlkes & J. Malik (2008). Using contours to detect and localize junctions in natural images. pp. 1–8.

Nitzberg, M., D. Mumford & T. Shiota (1993). Filtering, Segmentation, and Depth, Volume 662 of Lecture Notes in Computer Science. New York: Springer-Verlag. of parametric maxflow in computer vision. PAMI 8, 679–698.

Papari, G. & N. Petkov (2011). An improved model for surround suppression by steerable filters and multilevel inhibition with application to contour detection. Pattern Recognition 44, 1999–2007.

Ren, X. & J. Malik (2003). Learning a classification model for segmentation. ICCV 1, 10–17.

Ruzon, M.A. & C. Tomasi (1999). Color edge detection with the compass operator. CVPR, 2160–2166.

Sarkar, S. & P. Soundararajan (2000). Supervised learning of large perceptual organization: graph spectral partitioning and learning automata. PAMI 22, 504–525.

Shi, J. & J. Malik (2000). Normalized cuts and image segmentation. PAMI 22, 888–905.

Stahl, J. & S. Wang (2007). Edge grouping combining boundary and region information. TIP 16, 2590–2606.

Wang, S. & J. Siskind (2003, june). Image segmentation with ratio cut. PAMI 25, 675–690.

Wang, S., T. Kubota, J. Siskind & J.Wang (2005). Salient closed boundary extraction with ratio contour. PAMI 27, 546–561.

Wertheimer, M. (1938). Laws of organization in perceptual forms. A source book of Gestalt psychology 4, 71–88.

Williams, L.R. & K.K. Thornber (1999). A comparison of measures for detecting natural shapes in cluttered backgrounds. IJCV 34, 81–96.

Wu, Z. & R. Leahy (1993). An optimal graph theoretic approach to data clustering: theory and its application to image segmentation. PAMI 15, 1101–1113.

Zhang, C., X. Ruan, Y. Zhao & M.-H. Yang (2012). Contour detection via random forest. In Pattern Recognition (ICPR), 2012 21st International Conference on, pp. 2772–2775.

Zhu, S.C. & A. Yuille (1996). Region competition: unifying snakes, region growing, and bayes/mdl for multiband image segmentation. PAMI 18, 884–900.

Control Engineering and Information Systems – Liu (Ed)
© 2015 Taylor & Francis Group, London, ISBN 978-1-138-02685-8

A novel eigenvalue source enumeration algorithm based on eigenvectors

W.Z. Huang
Department of Engineering Technology, Xijing University, Xi'an, China

G. Wang
Institution of Information and Navigation, Air Force Engineering University, Xi'an, China

ABSTRACT: The performance of classical eigenvector source estimation algorithm drops quickly in case of low Signal-to-Noise Ratio (SNR). For more, when the estimated source number is distorted or even deviated, the orthogonality between different subspaces will be damaged, which will lead to Difference Of Arrival (DOA) estimation performance degradation directly. To solve the problem, a novel eigenvalue algorithm is presented based on the theory of source enumeration. In the algorithm, the eigenvectors of sample covariance matrix are employed as the decision value, which is insensitive to SNR. In detail, the criterion of improved Predictive Description Length (PDL) is adopted to enumerate source number. Theoretical analysis and simulation results demonstrate the validity of the proposed algorithm. When input SNR is low, estimation success probability of proposed algorithm is higher than those of Minimum Description Length (MDL) and PDL.

1 INTRODUCTION

Among classic subspace decomposition estimation algorithms for Difference of Arrival (DOA), source number is critical in dividing signal eigenvector from noise eigenvector, which is applied in spanning the independent noise subspace, and signal subspace as well [1, 2]. When estimated source number is distorted or even deviated, the orthogonality between the two subspaces will be damaged, which will lead to DOA estimation performance degradation directly [3, 4]. For source number estimation, the algorithm based on information theory criterions is now declared the most common and effective, such as the sequence hypothesis criterion, Akaike Information Criterion (*AIC*) and Minimum Description Length (*MDL*) criterion [5, 6, 7]. However, for information theory criterion, the estimation performance reduction is more serious under the condition of low *SNR* or missing snapshot. The algorithm based on *MDL* criterion is used to multiple coherent source number's estimation and positioning [3], and an improved algorithm applying random geometry-array and covariance matrix noise is proposed [7], but a large number of arrays are required and difficult to realize effective estimation at low *SNR*. As for the performance and calculation speed, the recursion method of PDL criterion provided lately makes have great improvement than *MDL* and G&T criterion [8, 9].

All the above algorithms are based on covariance matrix's eigenvalue, and the performance is decided by the eigenvalue capability in dividing signal eigenvalue and noise eigenvalue. However, in case of limited snapshot data, the division process is quite difficult, especially at low *SNR*. The solution is to change eigenvalue, in which a reconstructed eigenvalue cluster at low *SNR* is used to extract coherent and non-coherent signal in *BEM* criterion [7, 8]. The performance of eigenvalue is depressed seriously by various noises, and the eigenvector is also used to replace eigenvalue to estimate source number [10].

In the paper, a source number estimation algorithm is presented based on sampling covariance matrix eigenvector. Eigenvector of sampling covariance matrix is used to construct judgment variable in the algorithm, and the number of source is estimated according to improved *PDL* criterion. The paper is organized as follows. The second part gives the preliminary of signal model, and the eigenvector algorithm and source enumeration is provided in third part. The fourth part demonstrates the simulation and the results, and the last gives the conclusion.

2 SIGNAL MODEL

Denote N-Array spaced array has $M(M < N)$ signals with narrow-band signal and fixed center frequency from far-field source shooting with angles θ_k

$(k = 1, 2, 3, ..., M)$. Denote $X(t)$ the complex envelope receiving vector of antenna arrays, and can be expressed as follows at snapshot time t [10].

$$X(t) = A(\theta)S(t) + N(t)$$
$$= \sum_{i=1}^{M} \alpha(\theta_i) \cdot s_i(t) + N(t) \quad t = 1, 2, 3 \ldots \quad (1)$$

In which, $S(t) = [s_1(t), s_2(t), ..., ss_N(t)]^T$ is the complex envelope vector of M receiving source signals, $N(t) = [n_1(t), s_2(t), ..., n_N(t)]^T$ complex Gauss noise vector of antenna arrays, and $A(t) = [\alpha_1(\theta_1), \alpha_2(\theta_2), ..., \alpha_N(\theta_M)]$ manifold array matrix with $N \times M$ dimensions. Element $\alpha(\theta_k)(k = 1, 2, ..., M)$ is a steering vector with $N \times 1$ dimensions aiming at wave direction.

Covariance matrix of original receiving data vector $X(t)$ can be expressed as

$$R = E[X(t)]X^H(t) \quad (2)$$

and can be decomposed as

$$R = U\sum{}^2 U^H = U_s \sum{}^2_N U_S^H + U_N \sum{}^2_N U_N^H \quad (3)$$

In which, U_S is the matrix consisting of eigenvectors corresponding with top M large eigenvalues,

$$\sum{}^2 = diag(\lambda_1, \lambda_2, ..., \lambda_N) = diag(\sigma_1^2 \sigma_2^2 ... \sigma_N^2)$$

The diagonal matrix consisting of eigenvalues, U_N the matrix consisting of eigenvectors corresponding with last $N - M$ small eigenvalues. Elements of diagonal matrix

$$\sum{}^2_S = diag(\lambda_1, \lambda_2, ..., \lambda_M)$$

Consist of the top M large eigenvalues corresponds with U_S's eigenvalues, elements of diagonal matrices

$$\sum{}^2_S = diag(\lambda_{M+1}, \lambda_{M+2}, ..., \lambda_N)$$

The last $N - M$ small eigenvalues corresponds with U_N's eigenvalues. Eigenvalues meet the condition

$$\lambda_1 > \lambda_2 > L > \lambda_M > \lambda_{M+1} = \lambda_{M+2} = L = \lambda_N = \sigma^2$$

3 SOURCE ENUMERATION VIA EIGENVECTORS

3.1 Transform of eigenvector

Define signal subspace as the subspace spanned by U_S, and noise subspace spanned by U_N, where

the two subspaces are orthogonal. For the subspace spanned by U_S and A belongs to the same signal subspace, and U_S's column vectors are basic vectors for signal subspace, the steering vector $\alpha(\theta_i)(i = 1, 2, ..., M)$ can be expressed as

$$\alpha(\theta_i) = \sum_{j=1}^{M} c_{ij} \cdot s_j \quad (4)$$

In which, S_j is the basic vector for signal subspace and column vectors of U_S, $c_{ij}(i, j = 1, 2, ..., M)$ the linear combination index for U_S.

As the noise subspace and steering vector subspace spanned from A are mutually orthogonal

$$n_K^H \alpha(\theta_i) = 0, (i = 1, 2, ..., N - M) \quad (5)$$

Define vector $\overline{y_k}(k = 1, 2, ..., N - M)$ the function of column vectors n_K^H and $X(t)$

$$y_k(t) = n_K^H X(t)$$
$$= n_K^H \cdot \left(\sum_{i=1}^{M} \alpha(\theta_i)s_i(t) + N(t) \right)$$
$$= n_K^H N(t) = \omega_{NK}(t) \quad (6)$$

Similarly, eigenvectors of U_S's column vectors can be defined as standard orthogonal basis, and a new vector $\overline{z_i}(t)(i = 1, 2, ..., M)$ can be gained

$$\overline{z_i}(t) = S_i^H X(t)$$
$$= S_i^H \left(\sum_{j=1}^{M} \alpha(\theta_j)s_j(t) \right) + S_i^H N(t)$$
$$= \sum_{j=1}^{M} \left[\sum_{\rho=1}^{M} c_{j\rho} s_i^H s_\rho s_j(t) \right] + S_i^H N(t)$$
$$= \sum_{j=1}^{M} c_{jt} s_j(t) + \omega_{Ni}(t) \quad (7)$$

In which, $\omega_{Nk}(t)$ and $\omega_{Ni}(t)$ are independent and identically distributed complex Gauss random vector, with zero mean and σ^2 variance.

From (6) and (7), it can be concluded that if the steering vector is weight disposed by noise subspace eigenvectors, the output variable will not comprise any signal component. Otherwise, if the steering vector is weighted by signal subspace eigenvectors, the output variable will comprise both signal component and noise component. Therefore, column vector of snapshot time t can be expressed as

$$d(t) = [z_1(t), ..., z_M(t), y_1(t), ..., y_{N-M}(t)]^t$$

And observation matrix $D = [d_1, d_2, ..., d_T]$ is constructed by T snapshots. Each column's power spectrum and peak value P_n of matrix D can be obtained via Fast Fourier Transform (*FFT*). Judgment variable $q_n(n = 1, 2, ..., N)$ are defined as

$$q_n = N \cdot \frac{P_n}{\sum_{i=1}^{N} P_i} \tag{8}$$

3.2 Improved judgment criterion

According to (2), the covariance matrix of original data $X(t)$ can also be expressed as

$$R = AR_s A^H + \sigma^2 I \tag{9}$$

In which, R_s is the covariance matrix, σ^2 noise variance, and matrix A the function of incident azimuth vectors $\theta_k = [\theta_1, \theta_2, ..., \theta_M]^T$. Covariance matrix R can also be viewed as the function of parameter set $\mathbf{\Phi} = [M, \theta_k, \sigma^2, R_s]$, and the conditional probability density function of receiving data vector is as [5]

$$f(X \mid \mathbf{\Phi}) = \frac{1}{\pi^n |R|} \exp\{-X^H (R)^{-1} X\} \tag{10}$$

where symbol $|R|$ stands for matrix's determinant. The aim of source number estimation is to estimate M and then the incidence angle vector $\theta_k = [\theta_1, \theta_2, ..., \theta_M]^T$.

Define *PDL* of L data column vector as

$$P(L) = -\sum_{i=1}^{L} \log f(X(t) \mid \mathbf{\Phi}_{t-1}) \tag{11}$$

In which, $\mathbf{\Phi}_{t-1}$ is *ML* estimator of all receiving data vector before time t. For any given moment, parameter vectors' estimators are based on previous observation, thus it is called prediction of description length estimation.

To estimate source number, the following cost function is designed [5]

$$PDL(m) = \arg\min_{m \in K} P(L), K = \{0, 1, ..., N-1\} \tag{12}$$

And source number m is estimated only when $PDL(m)$ is minimized.

From the classic PDL judgment criterion provided in [8], $P(L)$ can be redefined as

$$P(L) = -\sum_{i=1}^{L} \log f(x_i(t) \mid \lambda_1) \tag{13}$$

Among which, λ_i is the eigenvalue of $\mathbf{\Phi}_{t-1}$, $x_i(t)$ the ith receiving data of antenna arrays. Replace

λ_i with judgment variable q_n. The source number estimation criterion can be expressed as

$$P(L) = -\sum_{i=1}^{L} \log f(x_i(t) \mid q_1) \tag{14}$$

Therefore, the process of source number estimation algorithm can be concluded in following steps.

Step 1: Construct covariance matrix R;
Step 2: Calculate corresponding eigenvectors by the eigenvalues of R;
Step 3: Weight $X(t)$ with eigenvectors and obtain observation matrix D;
Step 4: Calculate each column's power spectrum of D;
Step 5: Calculate the peak value and q_n in (8);
Step 6: Confirm source number according to (14).

4 SIMULATION RESULTS

Monte-Carlo numerical simulation method is adopted to estimate source number estimation algorithm's performance. Simulation parameters are set as follows.

- Antenna arrays is an eight uniform array, and array element spacing is 0.5λ.
- At every Monte-Carlo simulation, snapshot number is set as 8192.
- Input signals are three incoherent narrow-band signal (BPSK signal with 32 Kbps), and the incidence angles are 45°, 80° and 89°.
- The channel is AWGN channel.

For given SNR, average value of 200 times Monte-Carlo simulation results is employed to obtain the testing success probability. When $Eb/N0 = 0dB$, the distribution result and comparison of judgment variable q_n and eigenvalue in proposed algorithm are shown in Figure 1. From Figure 1, it can be seen that after the arrays output vector weighted, the difference between q is quite obvious. For the value of q array, the top three ones are larger than the last five ones, and the last five ones are almost equal. All the eigenvalue are almost equal expect the first one, which provides guarantee for next source number estimation.

Simulation provides the change between provided algorithm's, *MDL*'s and *PDL*'s estimation success probability in Figure 2. In Figure 2, mean value is used to balance the probability at each $Eb/N0$.

In detail, Figure 2 shows that when $Eb/N0 \geq 6$ db (signal input SNR is high), the proposed algorithm has similar estimation success probability with that of MDL and PDL. When input SNR is low, estimation success probability of proposed algorithm

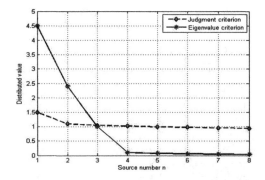

Figure 1. Distribution of judgment variance and eigenvalue.

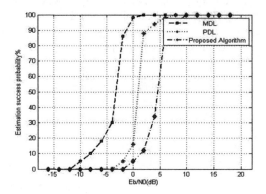

Figure 2. Estimation success probability.

is higher than those of MDL and PDL. Under the condition that estimation success probability is 90%, proposed algorithm has almost 4 dB improvement than PDL and 6 dB improvement than MDL.

5 CONCLUSION

A new source number estimation algorithm is provided to solve the problem in case of low SNR. In the proposed algorithm, eigenvectors of covariance matrix is adopted to replace eigenvalue to obtain a new judgment variable, and then the characteristics of source number are more effectively displayed. Therefore, the improved PDL judgment criterion is used to implement the estimation.

The simulation results shows that the algorithm provided has obvious improvement in SNR compared with classic MDL and PDL at the same estimation success probability.

ACKNOWLEDGMENT

The results discussed in this paper are part of the theses works of UAV data link. This work is supported by National Natural Science Foundation of China (61174162), and Shaanxi Provincial Science & Technology Department (Program No. 2011K06-36).

REFERENCES

Cozzens, J.H. & Sousa, M.J. 1994. Source enumeration in a correlated signal environment. *Signal Processing, IEEE Transactions on* 42(2): 304–317.
Green, P.J. & Taylor, D.P. 2002. Dynamic signal enumeration algorithm for smart antennas. *Signal Processing, IEEE Transactions on* 50(6): 1307–1314.
Gu, J.F., Wei P. & Tai, H.M. 2007. Detection of the number of sources at low signal-to-noise ratio. *Signal Processing, IET* 1(1): 2–8.
Huang, L., Wu S. & Li, X. 2007. Reduced-rank MDL method for source enumeration in high-resolution array processing. *Signal Processing, IEEE Transactions on* 55(12): 5658–5667.
Krim, H. & Cozzens, J.H. 1994. A data-based enumeration technique for fully correlated signals. *Signal Processing, IEEE Transactions on* 42(7): 1662–1668.
Lei, J., Ping C. & Juan, Y. 2007. The Source Numer Estimation Based on the Beam Eigenvalue Method. In Harbin (CN): *2007 Second IEEE Conference on Industrial Electronics and Applications. IEEE Conference on Industrial Electronics and Applications, 20070523-25*: 2727–2731.
Liavas, A.P., Regalia, P.A. & Delma, J.P. 1999. Blind channel approximation: Effective channel order determination. *Signal Processing, IEEE Transactions on* 47(12): 3336–3344.
Valaee, S. & Kabal, P. 2004. An information theoretic approach to source enumeration in array signal processing. *Signal Processing, IEEE Transactions on* 52(5): 1171–1178.
Yazdian, E., Gazor, S. & Bastani H. 2012. Source enumeration in large arrays using moments of eigenvalues and relatively few samples. *IET Signal Processing* 6(7): 689–696.
Zhou, Q.C., Gao H. & Wang, F. 2012. Modified DOA estimation methods with unknown source number based on projection pretransformation. *Progress In Electromagnetics Research B* 38: 387–403.

Control Engineering and Information Systems – Liu (Ed)
© 2015 Taylor & Francis Group, London, ISBN 978-1-138-02685-8

A Class of generalized dependence measures based on inequality theory

Z.H. Cheng
College of Textile and Clothing Engineering, Soochow University, Suzhou, China
School of Information Science and Technology, Donghua University, Shanghai, China

J.A. Fang
School of Information Science and Technology, Donghua University, Shanghai, China

ABSTRACT: Measures of dependence are fundamental in areas of signal and image processing. In this paper, the definition of generalized dependence measure, which is based on inequality theory, is given. Several classic dependence measures which meet the definition of generalized dependence measure have been integrated in a unifying framework. With the comparison of some classic dependence measures, it has been proved that adopting the generalized dependence measures can expand the range to construct objective function in modeling independence component analysis. Although the work is preliminary, the method is rather general and will have many applications.

1 INTRODUCTION

The dependence measure is a fundamental and widely used concept referred to any statistical relationship between two random variables or two sets of data (Vapnik, 1995). For the solution of blind source separation problem with the technique of Independent Component Analysis (ICA), it is a crucial step to estimate, that is, to minimize the statistical dependence between different components with an objective function (Hyvärinen et al., 2001). There are many dependence measures so far that have been introduced and extensively exploited to construct the objective function in ICA model (Bell and Sejnowski, 1995, Cardoso, 1997). The application of these measures can also be found in many fields such as signal processing, pattern recognition and machine learning etc.

Among classic statistical dependence measures, a class of information-theoretic dependence measures, which has a sound mathematic foundation due to their connection to Information theory, for example, Shannon mutual information and negentropy, is often the nature choice in ICA model (Hyvärinen et al., 2001, Xu et al., 1998). Meanwhile, other novel dependence measures underlying in information theory have also been exploited in literatures (Jenssen et al., 2003, Hild et al., 2001).

However, these dependence measures usually lead to some drawbacks when they are adopted to deal with real-life application. In practice, some dependence measures, especially measures of non-gaussianity commonly rely on simple assumptions on the source statistics. As a direct consequence,

when the assumed statistical model is inaccurate, the ICA algorithms can only achieve just suboptimal results or even fails to produce desired source separation. Unfortunately, such situation seems very likely to occur when recover the original source signals without the benefit of any prior knowledge about the mixing operation or the sources themselves. Another importance problem that has to be faced is that some dependence measures, such as Shannon mutual information, have a high computational complexity.

In this paper, we investigate the intriguing relationship between the Shannon mutual information and Shannon inequality using Kullback-Leibler divergence as a medium. Inspired by its connection, we propose and define a concept of general-propose dependence measure, namely generalized dependence measure. The motivation of this concept is to construct novel independence measures based on inequality theory (Rassias, 2000). According to definitions of generalized dependence measures, several classical dependence measures are unified into a framework of inequality theory. Although the concept of generalized dependence measure does not rigorously satisfy the definition of conventional distance, it expands the possibility of measuring the independence of random variables.

2 SHANNON MUTUAL INFORMATION AND SHANNON INEQUALITY

We first clarify the relationship between Shannon mutual information and Shannon inequality and then extend this relationship to more general concept.

Let X be a discrete random variable and let $p(x), q(x), x \in X$ be two probability density functions (pdf) of X, that is, estimated distribution and assuming distribution respectively. The Kullback-Leibler divergence, namely KL distance or relative entropy, is a measure of the distance between two distributions and it is defined as follows (Cover and Thomas, 2006):

Definition 1: Kullback-Leibler divergence between probability density functions $p(x)$ and $q(x)$ is:

$$D(p \| q) = \sum_{x \in X} p(x) \log \frac{p(x)}{q(x)} \tag{1}$$

Note that the logarithmic base 2 is used throughout this paper unless otherwise specified. It is well known that KL divergence is always non-negative, additive but not symmetric and equal to 0 if and only if $p(x) = q(x)$ for all $x \in X$. The KL divergence can be interpreted as a measure of the distance between two distributions. In practice, it illustrates the inefficiency in statistics between the assuming distribution $p(x)$ and the true distribution $q(x)$. However, the KL divergence is not a true distance metric between distributions since it is not symmetric and does not satisfy the triangle inequality.

Shannon mutual information is a measure of the amount of information that one random variable contains about another random variable. Consider two discrete random variables X and Y, let $p(x)$ and $p(y)$ be the two marginal probability density distribution of X and Y, respectively, and $p(x, y)$ the joint probability density distribution of random variables X and Y.

Definition 2: The Shannon mutual information $I_s(X;Y)$ is the Kullback-Leibler divergence between the joint distribution $p(x, y)$ and the product of two distributions $p(x)$ and $p(y)$:

$$I_s(X;Y) = \sum_{x \in X} \sum_{y \in Y} p(x,y) \log \frac{p(x,y)}{p(x)p(y)} \tag{2}$$

According to the above definition, we could conclude that Shannon mutual information and KL divergence are intimately connected and the Shannon mutual information between two discrete random variables X and Y can be characterized with the KL divergence measure. The Shannon mutual information has following properties:

Property 1: $I_s(X;Y) = D(p(x,y) \| p(x)p(y))$ i.e. Shannon mutual information is nonnegative. Equal-

ity holds if and only if $p(x,y) = p(x)p(y)$, i.e. the random variables X and Y are independent.

Property 2: Shannon mutual information is additive.

Property 3: $I_s(X;Y) = I_s(Y;X)$ i.e. Shannon mutual information is symmetric.

In information theory (Cover and Thomas, 2006), for two discrete probability distributions $P = (p_1, ..., p_n)$ $Q = (q_1, ..., q_n)$ $n = 2, 3, ...$, a fundamental result related to the notion of the Shannon entropy is the inequality:

$$\sum_{i=1}^{n} p_i \log p_i \geq \sum_{i=1}^{n} p_i \log q_i \tag{3}$$

which is valid for all positive real numbers and with:

$$\sum_{i=1}^{n} p_i = \sum_{i=1}^{n} q_i, i = 1, 2, ..., n.$$

Equality holds in (3) if and only if $p_i = q_i$ for all i. This result, sometimes called the fundamental lemma of information theory.

From Shannon inequality in (3), it is obvious to derive that the KL divergence can be expressed with following inequality relationship due to its nonnegative property:

$$\begin{aligned} D(P \| Q) &= \sum_{i=1}^{n} p_i \log \frac{p_i}{q_i} \\ &= \sum_{i=1}^{n} p_i \log p_i - \sum_{i=1}^{n} p_i \log q_i \geq 0 \end{aligned} \tag{4}$$

Thus, it could be concluded that the Kullback-Leibler divergence meet the relationship derived from Shannon inequality (3), that is, the Shannon mutual information implies the relationship of Shannon inequality.

3 GENERALIZED DEPENDENCE MEASURES

After analyzing the relationship between Shannon inequality and Shannon mutual information above, the fact that Shannon inequality could be derived from the definition of Shannon mutual information, whiles Shannon mutual information could be interpreted as a distance metric, can be found. This implies that Shannon mutual information measure has a tightly connection with Shannon inequality. In essence, variables which are not statistically independent between each other suggest the existence of some functional relation between them. Those functional relations in this context could be evaluated with inequalities. We are engaged in an inverse problem that whether or not measures of dependence could be constructed from inequality and hope to unify some classical dependence measures within the framework of inequality.

3.1 Definition of generalized dependence measures

By generalizing the relationship between Shannon mutual information and Shannon inequality to other inequalities, we have a motivation of constructing the dependence measures, namely generalized dependence measure, with those classical inequalities.

Lemma 1: Let $F(X,Y), G(X,Y)$ be two functions with respect to two random variables X and Y, where $X, Y \in R^n$. An inequality involved X and Y can be expressed as:

$$F(X,Y) \geq G(X,Y) \tag{5}$$

Definition 3: The generalized dependence measure between two random variables X and Y can be quantitatively defined as:

$$D(X;Y) = F(X,Y) - G(X,Y) \tag{6}$$

It is obvious to note that the generalized dependence measure is nonnegative since it satisfies the inequality given by (5). Usually the definition of generalized distance does not rigorously satisfy the metric of distance since it does not satisfy the triangle inequality. However, it has no significant impact to use measures of kind in the theoretic analysis and application in practical engineering problem.

3.2 Classic generalized dependence measures

It is easily to see from (2)(4) that Shannon mutual information measure satisfy the definition of generalized dependence measure such that Shannon mutual information measure is a special case of generalized dependence measure. Many other dependence measures which also satisfy the definition of generalized mutual information, such as Euclidean distance mutual information, Cauchy-Schwartz mutual information, Jensen mutual information, Minkowski mutual information, can be derived from corresponding inequalities.

3.2.1 Cauchy-Schwartz dependence measure and mutual information

Lemma 2: In Euclidean space R^n with the standard inner product, the Cauchy-Schwartz inequality is defined as:

$$\left(\sum_{i=1}^{n} x_i^2 \right) \left(\sum_{i=1}^{n} y_i \right) \geq \left(\sum_{i=1}^{n} x_i y_i \right)^2 \tag{7}$$

where equality holds if and only if X and Y are linearly dependent.

Definition 4: According to (5) (6) (7), Cauchy-Schwartz dependence measure can be defined as:

$$D_{cs}(X;Y) = \left(\sum_{i=1}^{n} x_i^2 \right) \left(\sum_{i=1}^{n} y_i^2 \right) - \left(\sum_{i=1}^{n} x_i y_i \right)^2 \tag{8}$$

where $D_{cs}(X;Y) = 0$ if and only if X and Y are linearly dependent.

Furthermore, providing to impose a strong constrain on inequality (7), for example, based on the Cauchy-Schwartz inequality $\|X\|^2 \|Y\|^2 \geq \left(X^T Y \right)^2$, take logarithm on this inequality so that inequality (9) holds,

$$\log \frac{\|X\|^2 \|Y\|^2}{\left(X^T Y \right)^2} \geq 0 \tag{9}$$

then replace the inner product between vectors by inner product between pdfs, the Cauchy-Schwartz mutual information can be obtained as deriving in literature (Xu et al., 1998):

$$I_{cs}(X;Y) = \log \frac{\sum_{x \in X} \sum_{y \in Y} P(x,y)^2 \sum_{x \in X} \sum_{y \in Y} p(x)^2 p(y)^2}{\left(\sum_{x \in X} \sum_{y \in Y} p(x,y) p(x) p(y) \right)^2} \tag{10}$$

As is apparent in (9), Cauchy-Schwartz mutual information $I_{cs}(X;Y) \geq 0$, and the equality holds if and only if $p(x,y) = p(x)p(y)$, that is, the two random variables are statistically independent.

3.2.2 Jensen mutual information

Lemma 3: For a convex function $f(\cdot)$, the following Jensen's inequality is valid for any x_1 and x_2:

$$\alpha_1 f(x_1) + \alpha_2 f(x_2) \geq f(\alpha_1 x_1 + \alpha_2 x_2) \tag{11}$$

where parameters α_1 and α_2 is called convex coefficient, which meet constrains $\alpha = (\alpha_1, \alpha_2) \geq 0$ and $\alpha_1 + \alpha_2 = 1$.

As we know, the function $-\log(\cdot)$ is a popularly adopted convex function. Thus replacing the notation of $f(\cdot)$ with $-\log(\cdot)$ and considering the joint distribution and the product of marginal distributions $p(x_1, x_x)$ as inputs of convex function $f(\cdot)$, we could gain following definition.

Definition 5: Based on (5) (6) and (11), the Jensen mutual information proposed by Liu (Lin, 1991) is defined as:

$$I_J(X;Y) = \sum_{x \in X} \sum_{y \in Y} \begin{pmatrix} \log(\alpha_1 p(x,y) + \alpha_2 p(x)p(y)) \\ -\alpha_1 \log p(x,y) \\ -\alpha_2 \log(p(x)p(y)) \end{pmatrix}$$

(12)

It is obvious that $I_J(X,Y) \geq 0$ with equality if and only if $p(x,y) = p(x)p(y)$, that means two random variables X and Y are independent.

3.2.3 Minkowski mutual information

Lemma 4: For two set of real number sequence, the Minkowski inequality is valid:

$$\left(\sum_{i=1}^n (f(x_i))^\alpha \right)^{\frac{1}{\alpha}} + \left(\sum_{i=1}^n (g(x_i))^\alpha \right)^{\frac{1}{\alpha}}$$

$$\geq \left(\sum_{i=1}^n (f(x_i) + g(x_i))^\alpha \right)^{\frac{1}{\alpha}}$$

(13)

Definition 6: Similarly to the definition 4, the Minkowski dependence measure can be defined as:

$$D_M(X;Y) = \left(\sum_{x \in X} \sum_{y \in Y} (p(x,y))^\alpha \right)^{\frac{1}{\alpha}}$$

$$+ \left(\sum_{x \in X} \sum_{y \in Y} (p(x)p(y))^\alpha \right)^{\frac{1}{\alpha}}$$

$$- \left(\sum_{x \in X} \sum_{y \in Y} (p(x,y) + p(x)p(y))^\alpha \right)^{\frac{1}{\alpha}}$$

(14)

where coefficient $\alpha \geq 1$. Minkovski dependence measure $I_{cs}(X;Y) \geq 0$, and the equality holds if and only if $p(x,y) = p(x)p(y)$, that is, the two random variables are statistically independent.

3.2.4 Euclidean distance mutual information

Lemma 5: for any nonnegative real number x_i and y_i $t = 0, 1, ..., n$, the following inequality is valid

$$\sum_{i=1}^n \left(\frac{x_i + y_i}{2} \right) \geq \sum_{i=1}^n \sqrt{x_i y_i}$$

(15)

where equality holds if and only if $x_i = y_i$.

Definition 7: According to (5) (6) (15), Euclidean dependence measure can be defined as:

$$D_{ED}(X;Y) = \sum_{x \in X} \sum_{y \in Y} \left(\frac{\frac{p(x,y) + p(x)p(y)}{2}}{-\sqrt{p(x,y)p(x)p(y)}} \right)$$

(16)

where Euclidean dependence measure $D_{ED}(X;Y) \geq 0$, and the equality holds if and only if $p(x,y) = p(x)p(y)$, that is, the two random variables are statistically independent.

Furthermore, the quadratic mutual information based on the Euclidean distance, namely Euclidean distance mutual information was proposed in (Xu et al., 1998) as follows:

$$I_{ED}(X;Y) = \sum_{x \in X} \sum_{y \in Y} (p(x,y) - p(x)p(y))^2$$

(17)

which has a similar property with Euclidean dependence measure.

4 COMPARISON OF GENERALIZED DEPENDENCE MEASURES

For the propose of illustrating the relationship among generalized dependence measures above, i.e. Euclidean distance measure I_{ED}, Shannon mutual information I_S, Cauchy-Schwartz mutual information I_{CS}, Jensen mutual information I_J and Minkovski dependence measure D_M, we use a simple case with two discrete random variables X and Y to verify the property of convergence for five generalized dependence measures. Provided the joint probability of two events A and B, the case of a marginal probability of X with $P_X(A) = 0.7$, $P_Y(B) = 0.3$, a marginal probability of Y with $P_Y(A) = 0.5, P_Y(B) = 0.5$, the joint probabilities $P_{XY}(A,A)$ and $P_{XY}(B,A)$ in interval [0, 0.7] and [0, 0.3] respectively, take convex coefficient $\alpha_1 = \alpha_2 = 0.5$ for Jensen mutual information, coefficient $\alpha = 2$ for Minkovski dependence measure D_M, the five generalized dependence measures is illustrated in Figure 1.

It is significant to note that different mutual information measures reach the same minimum point where the condition of independence between X and Y happens. Also, among these five measures, the flattest curve and the steepest curve are attained by the Minkovski mutual information and the Jensen mutual information, respectively. This evaluation implies that the implementation of ICA based on the Jensen mutual information achieves the minimum value of mutual information effi-

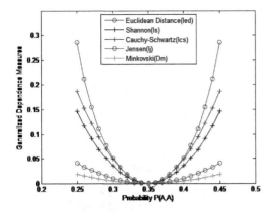

Figure 1. Comparison of five generalized dependence measures.

ciently. Comparably, a small number of iterations are needed in estimating Jensen ICA demixing matrix owing to the close relation between probability model and demixing matrix.

5 CONCLUSION

In this paper, starting with the analysis of the relationship between Shannon mutual information and Shannon inequality, we propose a universal definition of generalized dependence measures in terms of inequality theory. Several classic dependence measures have been integrated into a unifying framework of inequality. A comparison among five classic dependence measures revealed that the Jensen mutual information achieves the minimum

point of mutual information efficiently. The universality of the definition is demonstrated by its applications in different domains.

ACKNOWLEDGMENT

This research has been partially supported by CNTAC Research Grant 2011091.

REFERENCES

Bell, A.J. & Sejnowski, T.J. (1995) An Information-Maximization Approach to Blind Separation and Blind Deconvolution. *Neural Computation,* 7, 1129–1159.

Cardoso, J.F. (1997) Infomax and maximum likelihood for blind source separation. *Signal Processing Letters, IEEE,* 4, 112–114.

Cover, T.M. & Thomas, J.A. (2006) *Elements of Information Theory,* Wiley.

Hild, K.E., Erdogmus, D. & Principe, J.C. (2001) Blind Source Separation using Renyi's Mutal Information. *Ieee Signal Processing Letters,* 8, 174–176.

Hyvärinen, A., Karkunen, J. & Oja, E. (2001) *Independent Component Analysis,* John Wiley & Sons, Inc.

Jenssen, R., Principe, J.C. & Eltoft, T. (2003) Cauchy-Schwartz pdf Divergence Measure for non-Parametric Clusting. *Proc. NORSIG'03.* Bergen, Norway.

Lin, J. (1991) Divergence measures based on the Shannon entropy. *Information Theory, IEEE Transactions on,* 37, 145–151.

Rassias, T.M. (2000) Survey on classical inequalities, Springer.

Vapnik, V.N. (1995) *The Nature of Statistical Learning Theory,* NY, Springer-Verlag.

XU, D., Principe, J.C., Fisher, J.I. & Wu, H. (1998) A novel measure for independent component analysis (ICA). *Proc. of ICASSP.*

Control Engineering and Information Systems – Liu (Ed)
© 2015 Taylor & Francis Group, London, ISBN 978-1-138-02685-8

Missile guidance law parameter identification using fixed memory least squares

X.P. Wang
Electronic and Information Engineering College, Xi'an Jiaotong University, Xi'an, China
Aeronautics and Astronautics Engineering College, Air Force Engineering University, Xi'an, China

Y.L. Cai
Electronic and Information Engineering College, Xi'an Jiaotong University, Xi'an, China

Q.Y. Lin & F.W. Wang
Aeronautics and Astronautics Engineering College, Air Force Engineering University, Xi'an, China

ABSTRACT: Firstly, this thesis researched the difference among missile characteristic parameter with IPN, PP and CLOS guidance law, the method of Bayes theory to is used to identify the guidance law accurately. Secondly, missile guidance law parameters are identified accurately with method of fixed memory least squares and the identification for time constant in navigation loop is achieved. Lastly, the simulation results show that this identification method is fast and effective for guidance law with obvious characteristic. The parameter identification arithmetic is simple and fast to execute in computer, it could identify real-time parameter accurately and the problem of "data saturation" is solved, so this identification arithmetic is a real-time and effective guidance law parameter method.

1 INTRODUCTION

The rapid development of aeronautic technology lead air fight attack mode to transform from using traditional aero automatic gun to using long-range, intermediate-range and short-range air-to-air missile[1] (Shaw R.L. 1985). In the actual combat, it's so important for air fight decision-making and maneuverability escape to identify the state of missile and grip missile's information.

As for the identification of guidance law and parameters, there is little literature openly at home and abroad. In literature (Lin L., Kirubarajan T.& Barshalom Y. 2005), maximum likelihood estimation is used to estimate the PPN guidance law parameters in the Gapa. Based on guidance law model, the method of Bayes theory to create likelihood function is used in literature (Janne K. & Kai V. 2007) to identify guidance law and optimize track. In literature (Shaferman V. & Shima T. 2010.), combining Multi-Model Adapt Estimate (MMAE) with Bayes theory, identification for fixed missile guidance law is achieved. While above methods put emphasize on identification of guidance law, consideration for time constant in navigation loop is deficiency and precision for guidance law parameter is not precise enough.

Based on missile characteristic parameter, accurate identification for guidance law is realized using Bayes theory. The ARMA model of Missile's navigation loop is deducted, fixed memory least squares is used to avoid the phenomenon of "data saturation" in general identification arithmetic. The simulation results show that real-time parameters in guidance law are identified exact and fast.

2 IDENTIFICATION OF MISSILE GUIDANCE LAW

2.1 Analysis of guidance law characteristic

Before to identify the real-time parameters, we need to identify missile guidance law, several typical guidance law equation is given as:

Inverse Proportion (IPN) guidance law (Yuan P.J. & Chern J.S. 1992) (Li X.M & Su Jin. 2007)

$$\mathbf{a}_c = N\boldsymbol{\omega} \times \mathbf{V}_c \tag{1}$$

$\mathbf{a}_c$ is guidance law control input, N is Inverse proportion guidance coefficient, V_c is relative velocity between missile and target, $\boldsymbol{\omega}$ is missile rotation angle rate around the line of sight.

Pure Pursuit (PP) guidance law (Shneydor N.A. 1998)

$$\mathbf{a}_c = N\boldsymbol{\omega} \times \mathbf{V}_c \qquad (2)$$

$e_{(V_m \times r) \times V_m}$ is a unitage vector along $(\mathbf{V}_m \times r) \times \mathbf{V}_m$, it's a vector plumbing to $\mathbf{V}_m$ and pointing to r. V_m is missile velocity, δ is a angle between missile velocity vector and line of sight.

CLOS guidance law (Zarchan P. 1997)

$$\mathbf{a}_c = k_1 \mathbf{d} + k_2 \dot{\mathbf{d}} \qquad (3)$$

where:

$$\mathbf{d} = \mathbf{r}_A - \mathbf{r}_{ML}, \dot{\mathbf{d}} = \mathbf{V}_{A\perp} - \mathbf{V}_{ML\perp}, \mathbf{r}_A = (\mathbf{e}_{AL} \cdot \mathbf{r}_{ML})\mathbf{e}_{AL},$$

$$\mathbf{e}_{AL} = \frac{\mathbf{r}_{AL}}{|\mathbf{r}_{AL}|}, \mathbf{r}_{ML} = [x_m - x_l \quad y_m - y_l \quad h_m - h_l],$$

$$\mathbf{V}_{ML\perp} = (\mathbf{e}_{AL} \times \mathbf{V}_m) \times \mathbf{e}_{AL}$$

$$\mathbf{r}_{AL} = [x_a - x_l \quad y_a - y_l \quad h_a - h_l], \mathbf{V}_{A\perp} = \boldsymbol{\omega}_{AL} \times \mathbf{r}_A$$

$\boldsymbol{\omega}_{AL}$ is angle velocity of guidance aircraft and target.

The angle δ is between missile velocity vector $\mathbf{V}_m$ and target LOS, the distance d is from missile to the beeline between aircraft and target. They have different characteristic and trend with the different missile guidance law. On some extent, they also have different probability distributing as Figure 1. And under the initial state of Table 1, method is quantum PSO optimal escape tracking.

Result from optimization arithmetic:

At the same time, through Figures 1–3, we could know that when the missile is in the end navigation stage, if target plane escape with great maneuverability, no matter what kind of guidance law we use, the parameter especially $\delta(k)$ would be nearly to 90 degree. So, in the end navigation stage, the identification of missile guidance law should be shut down and use former result to do the latter work.

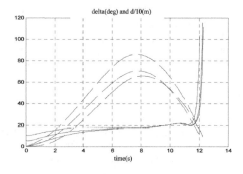

Figure 1. The curve of δ and d with IPN.

Table 1. Arget, missile and aircraft conditions.

	X (m)	y (m)	h (m)	V (m/s)	γ (0)	χ (0)
T	0	0	10000	250	0	45
M	−10000	0	10000	250	0	0/5/10
A	−10000	0	10000	250	0	0

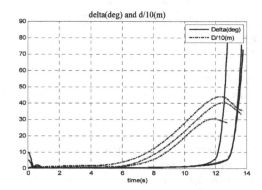

Figure 2. The curve of δ and d with PP.

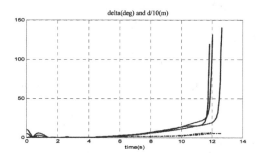

Figure 3. The curve of δ and d with CLOS.

Choose $\zeta_k = [\delta(k) \quad d(k)]$ as guidance law correlative characteristic parameter, for IPN guidance law, the $\delta(k)$ is a constant nearly to zero, and independent of $d(k)$, the probability function is defined as:

$$f^\delta(\delta|IPN) = \frac{\delta}{\delta_0^2} e^{-\frac{\delta}{\delta_0}}, f^d(d|IPN) = C \qquad (4)$$

For IPN guidance law, $\delta(k)$ is nearly to zero, and independent of $d(k)$, the probability function is defined as:

$$f^\delta(\delta|PP) = \lambda_1 e^{-\lambda_1 \delta}, f^d(d|PP) = C \qquad (5)$$

For IPN guidance law, the guidance law could lead missile to fly along the flight path, so the $\delta(k)$

nearly to be zero, while the $d(k)$ nearly to be zero, he probability function is defined as:

$$f^\delta(\delta|CLOS) = \lambda_2 e^{-\lambda_2\delta}, \quad f^d(d|CLOS) = \lambda_3 e^{-\lambda_3 d} \quad (6)$$

Comparatively, the probability distributing of $f^\delta(\delta|CLOS)$ is more gently than $f^\delta(\delta|PP)$, generally, $\lambda_2 < \lambda_1$.

2.2 Identification of missile guidance law

Assume at sampling moment k, the probability of guidance law is $P_k(\theta|\zeta_k)$, $\zeta_k = \begin{bmatrix} \delta(k) & d(k) \end{bmatrix}$ means guidance law's characteristic parameter at moment k, $\theta \in G = \{IPN \quad PP \quad CLOS\}$, while the probability of guidance law at moment $k+1$:

$$P_{k+1}(\theta|\zeta_{k+1}) = \frac{P_k(\theta|\zeta_k)f(\zeta_{k+1}|\theta)}{\sum_{\xi \in G} P_{k+1}(\xi|\zeta_k)f(\zeta_{k+1}|\xi)} \quad (7)$$

where:

$$f(\zeta_k|\theta) = f^\delta(\delta(k)|\theta)f^d(d(k)|\theta) \quad (8)$$

At initial state, we assume that every probability of guidance law is equal to 1/3. Formula 11 reflect that every probability of guidance law depend on the historical probability and current parameter which matches the variety of actual guide law parameters.

3 PARAMETER IDENTIFICATION BASED ON FIXED MEMORY LEAST SQUARES

Through above method, we could complete the identification to three kind of missile guidance law (IPN, PP, CLOS). While in order to optimize the plane's escape track, we need to ensure the type of guidance law and the parameter of missile and its guidance law. The mainly parameters are:

$$\begin{cases} \theta_1 = \begin{bmatrix} \gamma_m & \chi_m & V_m & \delta_0 \end{bmatrix}, \\ \theta_2 = \begin{bmatrix} m_m(t) & T_m(t) & D_m \end{bmatrix} \\ \theta_3 = \begin{cases} \begin{bmatrix} \tau_1 & \tau_2 & N \end{bmatrix} & IPN \\ \begin{bmatrix} \tau_1 & \tau_2 & k \end{bmatrix} & PP \\ \begin{bmatrix} \tau_1 & \tau_2 & k_1 & k_2 \end{bmatrix} & CLOS \end{cases} \\ \theta_4 = \begin{bmatrix} \omega & \vec{r} & V_c \end{bmatrix} \end{cases} \quad (9)$$

θ_1 is missile's locomotion parameter, δ_0 is projection angle when missile fire. θ_2 is Missile's noumenal parameter and pneumatic parameter: θ_3 is Missile's navigation system parameter: θ_4 is Target line of sight parameter LOS. θ_1 and θ_4 can be directly estimated according to correlative method such as IMM-UKF, θ_2 and θ_3 can be calculated through parameter identification.

3.1 Estimation of missile's locomotion parameter

Missile is described by 3-DOF particle model, they are:

$$\begin{cases} \dot{x}_m = V_m \cos\gamma_m \cos\chi_m \\ \dot{y}_m = V_m \cos\gamma_m \sin\chi_m \\ \dot{h}_m = V_m \sin\gamma_m \\ \dot{V}_m = \begin{bmatrix} T_m(t) - D_m \end{bmatrix}/m_m - g\sin\gamma_m \\ \dot{\gamma}_m = \begin{bmatrix} a_p - g\cos\gamma_m \end{bmatrix}/V_m \\ \dot{\chi}_m = a_y/V_m \cos\gamma_m \end{cases} \quad (10)$$

In above state equations, $[x_m, y_m, h_m, V_m, \gamma_m, \chi_m]$ is missile's three-dimensional coordinate, velocity, track lean angle, rack yaw angle, T_m is engine thrust, D_m is air resistance.

The formula of missile's locomotion parameter estimation:

$$\hat{\gamma}_m(k) = arctg\frac{\hat{h}_m}{\sqrt{\hat{x}_m(k)^2 + \hat{y}_m(k)^2}},$$

$$\hat{\chi}_m(k) = arctg\frac{\hat{y}_m(k)}{\hat{x}_m(k)} \quad (11)$$

Missile target line of sight angle motion parameter identification formula:

$$\begin{cases} \hat{\mathbf{r}}(k) = \begin{bmatrix} x_a(k) - \hat{x}_m(k) & \begin{matrix} y_a(k) \\ -\hat{y}_m(k) \end{matrix} & h_a(k) - \hat{h}_m(k) \end{bmatrix} \\ V_c(k) = \begin{bmatrix} \dot{x}_a(k) - \hat{x}_m(k) & \begin{matrix} \dot{y}_a(k) \\ -\hat{y}_m(k) \end{matrix} & \dot{h}_a(k) - \hat{h}_m(k) \end{bmatrix} \end{cases} \quad (12)$$

$$\hat{\omega}(k) = \frac{\hat{\mathbf{r}}(k) \times (-\hat{\mathbf{v}}_c(k))}{\hat{\mathbf{r}}(k) \cdot \hat{\mathbf{r}}(k)} \quad (13)$$

From the missile's three DOF locomotion (10), we know that missile's mass, missile's engine thrust and the resistance, all have directly influence on missile's velocity V_m, assume:

$$a_m(k) = \sqrt{\hat{\ddot{x}}_m(k) + \hat{\ddot{y}}_m(k) + \hat{\ddot{h}}_m(k)} + g\cos[\hat{\gamma}_m(k)] \quad (14)$$

3.2 Missile and guidance law parameter ARMA model

Considering the inertia when missile search the target, generally we predigest missile's pitching channels and yaw channels as a one order

745

subsystem to describe, τ means inertial time constant, describe as:

$$\dot{a}_p = \frac{a_{pc} - a_p}{\tau} \quad \dot{a}_y = \frac{a_{yc} - a_y}{\tau} \tag{15}$$

where, a_{pc} means acceleration order in missile's pitching channel, a_{yc} means acceleration order in missile's yaw channel.

Discrete (15), and put it into guidance law:
IPN guidance law:

$$
\begin{bmatrix} a_p(k+1) \\ a_y(k+1) \end{bmatrix} =
\begin{bmatrix} 1 - \dfrac{T}{\tau_1} & 0 \\ 0 & 1 - \dfrac{T}{\tau_2} \end{bmatrix}
\begin{bmatrix} a_p(k) \\ a_y(k) \end{bmatrix}
$$

$$
+ \begin{bmatrix} \dfrac{TN_1}{\tau_1} & 0 \\ 0 & \dfrac{TN_2}{\tau_2} \end{bmatrix}
\begin{bmatrix} u_{pc}(k) \\ u_{yc}(k) \end{bmatrix}
$$

$$
+ \begin{bmatrix} \dfrac{T}{\tau_1} g \cos(\gamma_m(k)) \\ 0 \end{bmatrix} \tag{16}
$$

where: N_1, N_2 are inverse proportion coefficient in the navigation system pitching channels and yaw channels. τ_1, τ_2 are inertial constant in the navigation system pitching channels and yaw channels. Generally, the two channels are the same, so we could identify the parameters through ARMA (1,1) model in yaw channels.

PP guidance law:

$$
\begin{bmatrix} a_p(k+1) \\ a_y(k+1) \end{bmatrix} =
\begin{bmatrix} 1 - \dfrac{T}{\tau_1} & 0 \\ 0 & 1 - \dfrac{T}{\tau_2} \end{bmatrix}
\begin{bmatrix} a_p(k) \\ a_y(k) \end{bmatrix}
$$

$$
+ \begin{bmatrix} \dfrac{TK_1}{\tau_1} & 0 \\ 0 & \dfrac{TK_2}{\tau_2} \end{bmatrix}
\begin{bmatrix} u_{pc}(k) \\ u_{yc}(k) \end{bmatrix}
$$

$$
+ \begin{bmatrix} \dfrac{T}{\tau_1} g \cos(\gamma_m(k)) \\ 0 \end{bmatrix} \tag{17}
$$

where: K_1, K_2 are inverse proportion coefficient in the navigation system pitching channels and yaw channels. τ_1, τ_2 are inertial constant in the navigation system pitching channels and yaw channels. Generally, the two channels are the same, so we

could identify the parameters through ARMA (1,1) model in yaw channels.

CLOS guidance law:

$$a_p(k) + \left(\frac{T}{\tau_1} - 1 \right) a_p(k-1)$$

$$= \frac{k_1 T + k_2}{\tau_1} u_{pc}(k-1) - \frac{k_2}{\tau_1} u_{pc}(k-2)$$

$$+ \frac{T}{\tau_1} g \cos(\gamma_m(k)) a_y(k) + \left(\frac{T}{\tau_1} - 1 \right) a_y(k-1)$$

$$= \frac{k_1 T + k_2}{\tau_1} u_{yc}(k-1) - \frac{k_2}{\tau_1} u_{yc}(k-2) \tag{18}$$

These are two controlled ARMA (1,2) model, since to the coherence between pitching channels and yaw channels, the yaw channel can be identified.

Through above analyze, for different guidance law (IPN, PP, CLOS), when we predigest navigation loop as one order subsystem, the parameters model waiting for identification could be expressed as controlled ARMA model.

3.3 Fixed memory least squares parameter estimation

In navigation loop, the controlled ARMA's equation:

$$A\left(q^{-1}\right) y(t) = B\left(q^{-1}\right) u(t) + \xi(t) \tag{19}$$

where: $y(t)$ is system output, $a_p(k)$ and $a_y(k)$ is acceleration in pitching channels and yaw channels: $u(t)$ is system input, $u_{pc}(k)$ and $u_{yc}(k)$ is control-input in pitching channels and yaw channels, $u(t)$ is different when guidance law changes. $\xi(t)$ is zero average white noise.

Since the inertial constant in missile navigation loop $\tau(h_m, M_m(h_m, V_m))$ is relevant to missile's flight parameter, so it could be regarded as time-vary parameter.

As for the identification of time-vary parameters, generally there are two ways: The one are is gradually fading memory, the other is fixed memory. This thesis use the second one.

3.3.1 Basical idea

"Fixed Memory" is a method that beforehand we regulate the quantity of observation data group. Every time when estimate parameter, it would add a new group of data and abandon the most ancient data, and we get the new data describe as:

$$\hat{\theta}_{LS}^N(N) \rightarrow \hat{\theta}_{LS}^{N+1}(N+1) \rightarrow \hat{\theta}_{LS}^N(N+1)$$

3.3.2 Arithmetic

Step 1: Given initial condition $\hat{\theta}_{LS}(0) = e$ (e is a quite small real vector), $\mathbf{P}(0) = \alpha^2 \mathbf{I}$ (α is a quite small data), according to recursive least squares, from $u(0), u(1), \ldots, u(N-1), y(0), y(1), \ldots, y(N)$ we can get $\hat{\theta}_{LS}^N(N)$ and $P(N)$ to be the initial value for fixed memory recursive least squares arithmetic.

Step 2: When get a new group of data $u(N)$, $y(N+1)$, with recursive least squares, from $\hat{\theta}_{LS}^N(N)$ and $P^N(N)$ we get $\hat{\theta}_{LS}^{N+1}(N+1)$ and $P^{N+1}(N+1)$.

$$P^{N+1}(N+1) = P^N(N) + P^N(N+1)\varphi(N)$$
$$\times\left[1 + \varphi^T(N+1)P^N(N)\varphi(N+1)\right]^{-1}$$
$$\times \varphi^T(N)P^N(N) \tag{20}$$

$$\begin{cases}\hat{\theta}_{LS}^{N+1}(N+1) = \hat{\theta}_{LS}^N(N) + K(N+1) \\ \left[y(N+1) - \varphi^T(N+1)\hat{\theta}_{LS}^N(N)\right] \\ K(N+1) = P^N(N)\varphi(N+1) \\ \left[\mathbf{I} + \varphi^T(N+1)P^N(N)\varphi(N+1)\right]^{-1}\end{cases} \tag{21}$$

Step 3: Since

$$\hat{\theta}_{LS}^N(N+1) = \hat{\theta}_{LS}^{N+1}(N+1)$$
$$- L^*(N)\left[y(n) - \varphi^T(n)\hat{\theta}_{LS}^{N+1}(N+1)\right] \tag{22}$$

$$L^*(N) = P(N+1)\varphi(n)\left[\mathbf{I} - \varphi^T(n)P(N+1)\varphi(n)\right]^{-1} \tag{23}$$

$$P^N(N+1) = P^{N+1}(N+1) + P^{N+1}(N+1)\varphi(n)$$
$$\times\left[1 - \varphi^T(n)P^{N+1}(N+1)\varphi(n)\right]$$
$$\times \varphi^T(n)P^{N+1}(N+1) \tag{24}$$

Get fixed memory least squares estimate value $\hat{\theta}_{LS}^N(N+1)$ and $P^N(N+1)$.

Step 4: Back to step 2, we get the identification parameter.

3.3.3 Parameter calculation

$$\theta_{IPN} = \begin{cases} \tau_1 = \dfrac{T}{a_{p1}+1} \\ \tau_2 = \dfrac{T}{a_{y1}+1} \\ N_1 = \dfrac{\tau_1 b_{p1}}{T} \\ N_2 = \dfrac{\tau_2 b_{p2}}{T} \end{cases} \quad \theta_{PP} = \begin{cases} \tau_1 = \dfrac{T}{a_{p1}+1} \\ \tau_2 = \dfrac{T}{a_{y1}+1} \\ K_1 = \dfrac{\tau_1 b_{p1}}{T} \\ K_2 = \dfrac{\tau_2 b_{p2}}{T} \end{cases}$$

$$\theta_{CLOS} = \begin{cases} \tau_1 = \dfrac{T}{a_{p1}+1} \\ \tau_2 = \dfrac{T}{a_{y1}+1} \\ k_1 = \dfrac{\tau_1\left(b_{p1}+b_{p2}\right)}{T} \\ K_2 = -b_{p2}\tau_1 \end{cases}$$

4 ANALYSIS OF SIMULATION

4.1 Identification and simulation of missile guidance law

Simulate conditions: the same to Table 1, (1) IPN guidance law, (2) PP guidance law, (3) CLOS guidance law, (4) At fifth second, CLOS transform IPN guidance law, (5) At fifth second, PP transform IPN guidance law, guidance law identification arithmetic, where $\lambda_1 = 1.5$, $\lambda_2 = 0.4$, $\lambda_3 = 0.0075$. The results of guidance law identification are shown as:

Through above simulation results, the identify method is fast and effective for guidance law with obvious character, and IPN is 0.5 s, PP is 1 s, CLOS is 0.7 s. Moreover, when the missile transfers its guidance law, it also could be identified. And the time CLOS transfer to IPN is about 0.6 s.

4.2 Simulation of missile parameter identification

Simulate conditions: the same to Table 1, (1) IPN guidance law with guidance parameter $N_m = 4.0$, $\tau = 0.3$, (2) PP guidance law with guidance parameter $K_{pp} = 5$, $\tau = 0.3$. The results of guidance law identification are shown as Figures 8–11.

From figures, we know that as for IPN guidance law, the identify result N_m Nm is between 3 to 3.8, $\tau \approx 0.36$, it's basically the same to real value. As

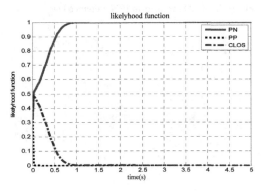

Figure 4. IPN guidance law.

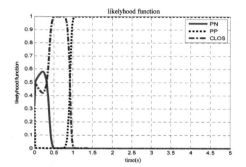

Figure 5.　PP guidance law.

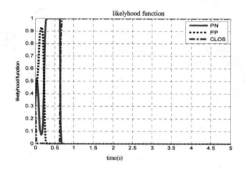

Figure 6.　CLOS guidance law.

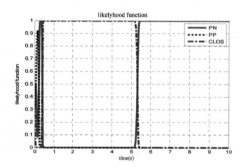

Figure 7.　CLOS transform IPN guidance law.

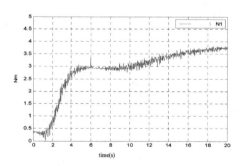

Figure 8.　IPN guidance law.

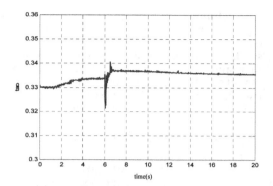

Figure 9.　IPN guidance law τ.

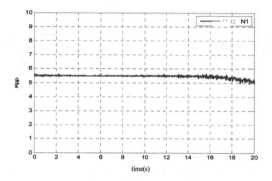

Figure 10.　PP guidance law K_{pp}.

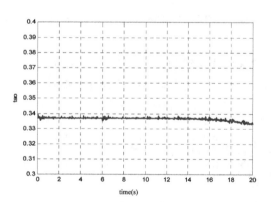

Figure 11.　PP guidance law τ.

for PP guidance law, the identify result $K_{pp} = 5.4$, $\tau \approx 0.37$, it's very close to the real result.

When identify missile state with IMM-UKF method, through matching missile's trajectory characteristic, the identification for missile guidance law is achieved. Combining with fixed memory least squares or other identify methods, the estimation for missile's guidance law parameter and the time constant in could be achieved. Above simulation results show the method is effective.

5 CONCLUSION

This thesis researched the difference among missile characteristic parameter with IPN, PP and CLOS guidance law, through the method of Bayes theory the guidance law is identified accurately. The ARMA model of Missile's navigation loop is deducted, missile guidance law parameters are identified accurately with method of fixed memory least squares and the identification for time constant in navigation loop is achieved. The simulation results show that: This identification method based on difference between missile characteristic parameter the is fast and effective for guidance law with obvious characteristic. The parameter identification arithmetic is simple and fast to execute in computer, it could identify real-time parameter accurately and the problem of "data saturation" is solved, so this identification arithmetic is a real-time and effective guidance law parameter method.

REFERENCES

Janne Karelahti & Kai Virtanen. 2007. Adaptive Controller for the Avoidance of an Unknownly Guided Air Combat Missile. *The 46th IEEE Conference on Decision and Conrol.*

Li, Xiangmin & Sun, Jin. 2007. Firepower Control Theory. *Defence Machine Press.*

Lin, L., Kirubarajan, T. & Barshalom, Y. 2005. Pursuer identification and time-to-go estimation using passive measurements from an evader. *IEEE Transactions on Aerospace and Electronic Systems,* vol 41 no 1, pp 10–204.

Shaferman Vitaly & Shima Tal. 2010. Cooperative Multiple-Model Adaptive Guidancefor an Aircraft Defending Missile. *Journal of Guidance, Control, and Dynamics.* Vol 33, No 6, November–December.

Shaw, R.L. 1985. Fighter Combat: Tactics and Maneuvering [M]. *Naval Institute Press, Annapolis, MD.*

Shneydor, N.A. 1998. Missile Guidance and Pursuit: Kinematics, *Dynamicsand Control. Chichester,* England: Horwood Publishing.

Yuan, P.J. & Chern, J.S. 1992. Ideal proportional navigation, *Journal of Guidance, Control, and Dynamics.* vol 15, no 5, pp 1161–1165.

Zarchan, P. 1997. Tactical and Strategic Missile Guidance, *3rd ed., ser. Progress in Astronautics and Aeronautics.* Reston, VA: AmericanInstitute of Aeronautics and Astronautics, Inc. vol. 176.

Control Engineering and Information Systems – Liu (Ed)
© 2015 Taylor & Francis Group, London, ISBN 978-1-138-02685-8

The research of emergency rescue optimal path model based on analytic hierarchy process and genetic algorithm

H.Y. Yang
College of Information Engineering, Inner Mongolia University of Technology, Hohhot, China

H.H. Li
College of Electric Power, Inner Mongolia University of Technology, Hohhot, China

W.Y. Lv
College of Information Engineering, Inner Mongolia University of Technology, Hohhot, China

J.Y. Li
College of Electric Power, Inner Mongolia University of Technology, Hohhot, China

ABSTRACT: Emergency rescue evacuation as an important part of The City's Emergency Command System, Evacuation model and algorithm selection directly affect the implementation of the population emergency evacuation plan. Evacuation, that is, how to transfer the people to a place of refuge fastest, that requires us to establish a rational evacuation strategy. In this paper, first of all, use Analytic Hierarchy Process to determine the weight of all paths in the road network, and introduce it into the objective function, then establish a mathematical model, and finally solve the emergency rescue path optimization problem with genetic algorithm, that is to find the optimal path. Upon the study of above all, we use MATLAB simulation software to simulate the evacuation path planning program to verify the reasonableness and effectiveness of this plan.

1 AN OPTIMAL PATH PROBLEM

To solve the optimal path problem, using graph theory terminology is described as follows: In Figure G (V, R), where V represents the set of all nodes in the figure, R represents the set of path weights between any two nodes, If the path does not exist between the two nodes, expressed with infinity. The weight of a path is defined as consisting of all the path of the weights of the Sections.

Depending on the optimization goal, the path weights can represent different variables; it can be distance, time or cost. Consider the characteristics [1] of this emergency evacuation, the most important factor is the time [2], without having to consider the comfort of the journey, fees and other issues. The so-called optimal path problem is to seek out the path using the least evacuation time in the path set that consisting of a given start point to the end point.

2 DETERMINING THE WEIGHTS OF THE PATH—ANALYTIC HIERARCHY PROCESS

Analytic Hierarchy Process is a decision-making method on the basis of quantitative and qualitative analysis and a kind of co-ordinating the theory [3] [4], which decompose the element related to strategic decision into the goal, guidelines, programs and other levels.

Road attribute data weighting factor lies in the choice of indicators [5]. This article After a lot of literature to read, as well as to verify the actual situation after the accident to determine following four main factors having impact on the emergency evacuation time: (1) Path distance; (2) Road grade; (3) Extent of damage to the road (4) Weather conditions. Weighting factors affecting road have been identified. The hierarchical structure is divided into two layers (Fig. 1), the target layer Z and guidelines layer A. The path weights of the target layer is W; criterion layer for each indicators target layer weights are w1, w2, w3, w4.

2.1 Judgment matrix

Judgment Matrix elements are pairwise comparison result of the weights that four factors. According to the literature data and experience of decision makers to construct judgment matrix A as shown in Table 1.

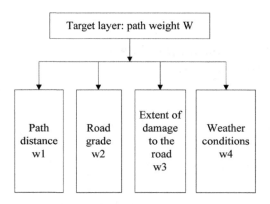

Figure 1. Hierarchical model.

Table 1. Comparison matrix.

Z	A1	A2	A3	A4
A1	1	1	3	6
A2	1	1	3	6
A3	1/3	1/3	1	2
A4	1/6	1/6	1/2	1

2.2 *Consistency checker*

1. The first step: Calculate the consistency index

$$CI = \frac{\lambda_{max} - n}{n - 1} \qquad (1)$$

2. Step Two: Find the corresponding average random consistency index, which is determined by the order of the decision matrix. As shown in Table 2:

3. The third step: Calculate the consistency ratio

$$CR = \frac{CI}{RI} \qquad (2)$$

When $CR < 0.1$, the matrix A is considered acceptable, otherwise, modified contrast matrix A.

2.3 *Calculation criteria layer weights*

Here use Eigen value method to Calculate criterion level weights, as follows:

$$AW = \lambda_{max} W \qquad (3)$$

where, λ_{max} is the maximum Eigen value of A, W is the corresponding eigenvector, it can be taken as a criterion layer weight vector by normalizing.

Table 2. Average random consistency index.

n	1	2	3	4	5	6	7	8
RI	0	0	0.58	0.90	1.12	1.24	1.32	1.41

It can calculate largest Eigen value and the corresponding eigenvectors of judgment matrix based on MATLAB programming.

2.4 *Determining the weights of the path*

Get criterion layer weight of each factor value, then we can calculate every path weight values for the entire network, that is the target layer weights: method is as following:

$$W_{ij} = s/v + s/v \times w_1(w_2 \times r + w_3 \times f + w_4 \times w) \qquad (4)$$

where, w_1, w_2, w_3, w_4 are the criteria layer for the target level of each factor weight coefficient; s is the path distance; v is the average speed of road design, according to the design data obtained; r as road grade, according to road user tasks function and adaptation of traffic, divide in four grade which were assignment (1,2,3,4); f for road damage, will be set at several situations: no damage, minor damage, moderate damage, serious damage and in turn its assigned (0,0.5,2,∞); f is weather conditions, according to the weather conditions, the suitability of vehicles assigned to low (0-1) interval of real numbers.

3 GENETIC ALGORITHM FOR OPTIMAL PATH

When the evacuation zone have been chosen, that mean the beginning and end of the evacuation has been determined, and the weight of each path has also calculated use the AHP, which requires us to determine an mathematical model for solving the optimal path problem, and finally use a kind of algorithm to obtain the optimal solution. After a lot of literature review, and the relevant comparison to determine the genetic algorithm to solve the optimal solution.

Genetic algorithms as a new global optimization search algorithm [6] [7], has incomparable superiority compare with the classic path optimization algorithm. Because it use the way of population to search, so the algorithm is very suitable for large-scale parallel operation, and the search process exhibit greater flexibility. Therefore, this article was selected Genetic Algorithm as the best evacuation path algorithm.

3.1 Objective functions

Emergency evacuation route selection should use the weight of the path as its index parameters, so we can get the evacuation objective function is:

$$\min Z = \min \sum W_{ij} \qquad (5)$$

Equation 5 represents the sum of the weights of all sections is minimum from the evacuation point from the point to be evacuated to the evacuation zone, also known as the optimal evacuation paths.

3.2 Determine the encoding method

For optimal path problem, we use the most common and simplest binary coding, for a given graph model, Do natural ordering of the graph vertices, then Make each vertex to be selected in this order as a chromosomal gene, when the values of genes is 1, which indicates that the corresponding vertex has been elected to the path, Otherwise, it means have not passed through the vertex, Chromosome length is equal to the number of all the vertexes in the figure, the sequence of the Chromosomal genes value is 1 represent the order of the vertexes when walking through the path.

3.3 Individual evaluation methods to determine

The fitness function in this paper is directly called the objective function, show as the following equation.

$$f(x) = \min Z = \min \sum W_{ij} \qquad (6)$$

It can be seen from the above equation, the smaller min Z is, and the smaller the value of fitness function is, the closer the corresponding solution close to the optimal solution.

3.4 Design of genetic operators

3.4.1 Selection
The purpose of selection is to select excellent individuals from the current group as a parent Involved in multiplication. This paper chooses roulette wheel selection method to finish individuals' selection. If produce the same individual, retain only one, delete the remaining, repeat the above process and complete selection operating.

3.4.2 Crossover
Crossover operation plays and a key role in the genetic algorithm and it's the main method to generate new individual. On the basis of the correlation of the serial number of path node, at first, crossover operation generates a random number and confirm the cross location, then exchange the gene on this location of chromosome.

3.4.3 Mutation
After selection and crossover operation, variation can avoided the loss of some information; it's an essential step in genetic algorithm to ensure the ability of local search.

The problem of the optimal path is to add or remove a point on some certain path, but at the same time, it must ensure that the offspring chromosomes formed after vitiation only has a path, the variation point is neither source nor end point. When doing arithmetic crossover and mutation operators, it maybe in the form of chromosome loop, if it happened, the loop must be deleted.

4 GENETIC ALGORITHM STEPS

The general steps of genetic algorithm [5] shown in Figure 2.

1. algebra t = 0;
2. to generate the initial population P(t), the optimal solution x'
3. Calculating P (t) in each individual fitness value of x, if f (x) > f (x'), then x' = x
4. If you meet the stopping criterion, the algorithm terminates, x is the output, otherwise go to (5)
5. On p (t) to achieve selection, crossover and mutation operations to obtain p (t + 1), so that, t = t + 1, go to (3).

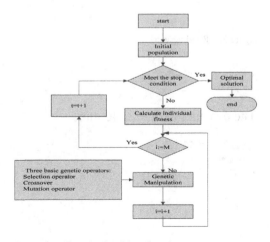

Figure 2. Genetic algorithm flowcharts.

753

5 OPTIMAL ROUTE SELECTION SIMULATION

In this paper, we use MATLAB software as a simulation; realize the process to find the optimal path in AHP and genetic algorithm. First, we use the AHP to determine the weight of the path. Classifying the matrix can be determined by adjusting the weight coefficients w_1, w_2, w_3, w_4; Determined according to the formula 4 in road network weights of each path, where we use min as unit. Road network schematic identified as Figure 3 shows. v1 indicates accident place of occurrence, v8 indicates evacuation sites namely endpoint.

Figure 4 is the result of running, find out the optimal path. And calculates all the weights of sub-paths and the time spent. In addition, the speed of running is very fast. Over evolved, we find out the sum of the weights from the entire optimum path finally. The resulting optimal path is:

Figure 5 and Figure 6 are the changing map of the chromosomal location which show the before and after of implementation of genetic operators respectively, Its horizontal and vertical coordinates, represent the individual and the fitness value of the corresponding. The comparison of before and after of the evolution about chromosomal

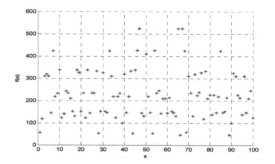

Figure 5. Chromosome initial positions.

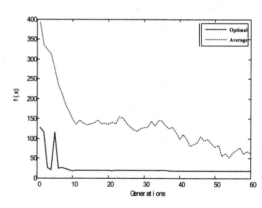

Figure 6. Chromosomes final positions.

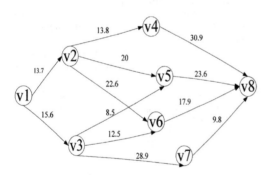

Figure 3. Road network.

The evacuation time is 46 min.

Figure 4. The optimal path.

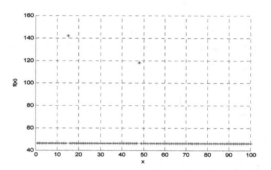

Figure 7. Evolutionary curve.

location shows that, after evolution, the fitness value of new individuals produced keeps a straight line at 46 basically, just like Figure 4 which shows the genetic algorithm for solving the optimal path. Figure 7 shows the optimal individual fitness function and the average fitness function curve, the evolution of the group curve, where the abscissa represents the number of iterations, the vertical axis represents the adaptation value. As can be seen from the figure, there are eight nodes of road network, the fitness function and the average fitness function have a greater change at the beginning, however, after about 50 generations of the function, the optimal is obtained. Basically become a stable linear, that is, the path and the path of least weight.

6 CONCLUSIONS

According to emergency rescue incidents, this paper makes a selection of optimal path under the actual situation. First, we base on AHP and solve the path weight, then use the genetic algorithm to find the optimal path, putting the analytic hierarchy process and genetic algorithm together effectively, and finally, we find out the optimal corresponding path and travel time corresponding to our experiment. In this paper, not only the path distance but also the actual road conditions under the state of emergency contribute to the final result. Therefore, the model has a certain value.

REFERENCES

Huang Feng, Li Xiaoli. Optimization model of urban population emergency evacuation plan l[J]., *Education and culture*, 2011 HEAD, 205–206.

Huang Meng, Zhang Zhen. The Research of Digital Emergency Management Platform Based On GIS, He Bei Yan Jiao as a Case Study[J]. IEEE, 2010, 15:490–493.

LiuYong. Research based ant colony optimization algorithm for emergency rescue Path[D]. *China University of Geosciences*, 2010.

Tian Junqi Problem solving based on genetic algorithm to achieve the shortest path[J]. *Science and Technology Forum*, 2005, 10:33–34.

Tu Danna. Key technology research based on available disaster contingency sparse road network[D], *Beijing University of Posts and Telecommunications*, 2012.

Wang Hong, Lu Yushi and Wang Shasha. MATLAB-based emergency rescue optimal route selection[J]. *Industrial Safety and Environment*, 2009, 35(2): 48–53.

Zhang Kan. Analysis based on Genetic Algorithm network path [D]. Sun Yat-sen, 2006.

Probability iterative algorithm for ultrasonic computerized tomography on concrete

X.Z. Qi & X.M. Zhao
School of Information Engineering, Chang'an University, Shan'xi, Xi'an, China

ABSTRACT: In order to enhance the speed and to improve the accuracy of the ultrasonic computerized tomography of concrete, an Improved Algebraic Reconstruction Technique (PART) from a view of probability for concrete image reconstruction is presented. In the PART algorithm, the defect probability of each unit is accounted based on the detected data. Then, the images of concrete structures are initialized, and the error of detected data is distributed depending on the probability. The results show that the PART algorithm is effective to control wave velocity dispersion and to improve the accuracy of computation and the quality of image reconstruction on concrete; the PART algorithm is more accurate for reflecting the internal structure of concrete.

1   INTRODUCTION

CT (Computerized Tomography) is an inversion technique, which can rebuild the internal image of the "object" by use of a particular ray projection data with certain mathematical model (Papoulis, 1975). Acoustic tomography is generated by acoustic testing techniques and CT technology, it can be used for the internal quality testing of the mass concrete bridge engineering and bored piles; and it provides information for the acceptance of project (Cao, 1998). Tomography can get the degree of decline in the quality of the defect, which can provide accurate data for engineering reinforcement (Kwon, 2005).

In concrete CT, complete projection data cannot be obtained because of the limit of the objective conditions of the concrete structure and environment. The quantity and structure of the obtained projection data can't meet the requirements of the analytical imaging techniques conditions, so, iterative reconstruction methods (Jing, 2002, Rosalie, 2005) were used to get a clearer image on ray theory. The commonly used inversion algorithms of concrete ultrasonic CT include the ART (Algebraic Reconstruction Technique), SIRT (Simultaneous Iterative Reconstruction Techniques) and LSQR (Liu, 1997).

Although the simulation results and the test results are valid, they require a large amount of the iterative and a long time to calculate, and unable to meet the requirements of real-time, accuracy and rapidity of the algorithms are not solved (Yang, 1994). In this paper, by using the probability methods, the conventional ART algorithm is improved;

the speed and precision of image reconstruction are further improved.

2   ART ALGORITHM

ART (Wu, 1997) is developed based on physical considerations. The basic idea is to give reconstruction area an initial value, and the projection residuals are projected back evenly one by one along the ray direction, the reconstructed image is corrected constantly, until it meets the required requirements.

The reconstruction area is meshed, the model of the ultrasonic computerized tomography on concrete is

$$\tau_i = \sum_{j=1}^{m} a_{ij} f_j \qquad (1)$$

where T_i = the walking time (sound travel time) of the ray i from the excitation point to the receiving point; f_j = the wave slowness (reciprocal velocity) of the imaging unit j; a_{ij} = the length of the ray i in the imaging unit j.

By equation (1), ignore the error, modify increments of wave slowness is Δf, so,

$$\sum_{j} \Delta f_j a_{ij} = \Delta \tau_i \qquad (2)$$

In ART iterative algorithm, wave slowness's modify increments (Δf_j) of all grid that the ray i passes through, are calculated by the walking time difference (ΔT_i) of the ray i. Since equation (2) is highly under determined, the L^{2p} module minimum value ofΔf_j are obtained by this constraints. By Lagrange multiplier method, objective function can be written as follows,

$$\min(Q) = \min\left[\left(\sum_{j=1}^{m}|\Delta f_j|^{2p}\right)^{1/2p} + \lambda(\Delta \tau_i - \sum_{j=1}^{m}a_{ij}\Delta f_j)\right] \quad (3)$$

where λ = Lagrange multiplier; p = The number of order of the mold.

By $\partial Q/\partial(\Delta x_j) = 0$ and $\partial Q/\partial \lambda = 0$, the equation (4) can be obtained.

$$\Delta f_j = \frac{a_{ij}^{1/(2p-1)}}{\sum_{j=1}^{m}a_{ij}^{2p/(2p-1)}}\Delta \tau_i \quad (4)$$

$p = 1$ or $p \rightarrow \infty$, two modify incremental formulas of the wave slowness can be obtained.

$$\Delta f_j = a_{ij}\Delta \tau_i \Big/ \sum_{j=1}^{m}a_{ij}^2 \quad (5)$$

$$\Delta f_j = \Delta \tau_i \Big/ \sum_{j=1}^{m}a_{ij} \quad (6)$$

The difference of the formula (5) and (6) is that travel time error is weighting assigned to each unit in formula (5), but it is evenly distributed to each unit in formula (6). The former is widely applied because of the higher inversion accuracy, but the calculation speed is slower.

In order to increase the stability of the calculation, in formula (4), the modify incremental Δf_j multiplied by the relaxation factor μ ($0 < \mu \le 1$) because of the ill-posed problem of image reconstruction.

3 PROBABILITY ART ALGORITHM (PART)

In ART algorithm, the initial values of the wave slowness are selected just by prior knowledge for the concrete imaging units; then, in iterative process, the influences on travel time error of all grids that the ray passing through, are seen as the same; the travel time error is equally assigned to all grids. However, in the actual process, the projection (travel time) error is caused by defective unit. If we can predetermine the probability of each grid that is defective unit, treated differently to each grid in the iterative process and initial value selection,

the speed and accuracy of the algorithm will be improved.

In PART: by the projection (travel time), the probability of each ray that pass through the defective unit is calculated preliminary, and then the probability of the each unit that is defective unit is determined. By this probability, the iterative initial value is selected, and the projection (travel time) error is rational assigned. More travel time error is assigned to the unit that has high defect probability. The image reconstruction speed and accuracy are improved.

3.1 The probability of ray passing through the defective unit

$$v_i = \sum_{j}a_{ij}\Big/\tau_i \quad (7)$$

$$\bar{v} = \sum_{i=1}^{n}v_i/n \quad (8)$$

$$s_v = \sqrt{\sum(v_i - \bar{v})^2/(n-1)} \quad (9)$$

where v_i = the velocity of ray i; v = the mean of velocity; s_v = standard deviation of velocity; n = the number of ray.

In the same testing area, the velocity of various ray obey normal distribution. The equation (10) is the normal distribution function.

$$\phi(\lambda) = \int_{\lambda}^{\infty}\frac{1}{\sqrt{2\pi}}e^{-u^2/2}du = P(Z \ge \lambda) \quad (10)$$

The values of λ_t ($t = 1,2, ..., 10$) are respectively obtained when confidence probability P is 100%, 90%, ... 10%, and then the velocity lower limit value V_t in normal regional are respectively calculated by confidence probability P_t ($t = 1,2, ..., 10$) as follows:

$$V_t = \bar{v} + \lambda_t \cdot s_v/\sqrt{n}(t = 1,2, ..., 10) \quad (11)$$

If the velocity is less than the lower limit value of confidence probability P_t, the probability of ray pass through the defective unit is P_t, the probability of each ray can be obtained. As follows:

$$p_i = P_t|_{v_i < V_t} \ (i = 1,2, ..., n) \quad (12)$$

3.2 The probability of image unit that is the defective unit

If the probability of a ray that pass through the defective unit is p_i, the probability of each unit that the ray pass through is p_i. As be passed through

758

by different rays, each unit will get a set of probability data. The problem is transformed into how to determine the probability of the unit that is defective unit according to a set of different probability.

If p_i is small, indicating that the probability q_j ($j = 1, 2, ..., m$) of all the units that the ray pass through are small, and conversely, the larger p_i only shows that the probability q_j of individual unit is larger. So, we should select a relatively small data in $(p_{(1)}, p_{(2)}, ..., p_{(l)})$ for q_j, simultaneously, because a unit has different effects on the different rays, the selection of q_j should consider the ratio of a_{ij} and the total length of the ray,

$$\varpi_i = a_{ij} \Big/ \sum_j a_{ij} \tag{13}$$

We select a smaller probability data for q_j by the filtering method, ω_i is the weight coefficient.

For each unit, p_i and ω_i are assigned one-to-one correspondence to it. We get a set of data $\{(p_{(1)}, \omega_{(1)}), (p_{(2)}, \omega_{(2)}), ... (p_{(lj)}, \omega_{(lj)})\}$ in order by p_i, in which, lj is the number of the ray that pass through the image unit j. According to formula (15), we can obtain the q_j for each image unit.

$$\bar{\varpi} = \sum \varpi_{(i)} / lj; \, p_{(1)} \le p_{(2)} \le \cdots \le p_{(lj)} \tag{14}$$

$$q_j = p_{(r)} \Big|_{\varpi_{(r-1)} < \bar{\varpi} \times \alpha \le \varpi_{(r)}} \quad 0 < \alpha \le 1 \tag{15}$$

where α = the parameter factor.

3.3 The selection of iterative initial value

When the probability of the image unit that is defective unit is determined, the initial value of wave velocity is selected for each image unit by the probability.

$$v_j^{(0)} = V_1 \cdot q_j + V_{10} \cdot (1 - q_j) \, j = 1, 2, ..., m \tag{16}$$

Then, the initial value of wave slowness is determined as formula (17).

$$\hat{f}^{(0)} = 1 / v^{(0)} \tag{17}$$

3.4 The modify incremental of wave slowness

The travel time error is weighted assigned, the weighted coefficient is the product of a_{ij} and q_j. The effects of normal unit and defective unit on travel time are fully considered. We can obtained the modify increments of wave slowness by formula (18).

$$\Delta f_j = \mu (b_{ij} \cdot q_j \Big/ \sum_{j=1}^m b_{ij}^2) \Delta \tau_i \tag{18}$$

where

$b_{ij} = a_{ij} \cdot q_j$ ($i = 1, 2, ..., n; j = 1, 2, ..., m$).

3.5 The PART algorithm steps

1. Using the projection (travel time), we determine the probability $p_i (i = 1, 2, ..., n)$ of each ray that pass through the defective unit;
2. By p_i, we can obtained the probability q_j ($i = 1, 2, ..., m$) of each unit that is defective unit;
3. The initial values $f^{(0)}$ of wave slowness are determined according with q_j;
4. In q round, $\hat{f}_j^{q,i}$ is the estimated value of unit j with ray i, by formula (19), wave slowness is modified one by one ray $i(i = 1.2 ..... n)$;

$$\hat{f}_j^{q,i+1} = \hat{f}_j^{q,i} + \mu \left(b_{ij} \cdot q_j \Big/ \sum_{j=1}^m b_{ij}^2 \right) (\tau_i - \hat{\tau}_i^q) \tag{19}$$

5. If the convergence criteria are met, the program is terminated, otherwise, we carry on the $q+1$ round iteration.

4 PERFORMANCE EXPERIMENT

4.1 Computer simulation experiment

Model A:
The Model A is the measurement area of 100 (cm) × 60 (cm), and it is divided into 10 × 6 = 60 grids. The No. 22, 23, 28, 29, 34, 35 grids are defective units. The defective velocity is 4050 m/s, and the normal wave velocity is 4500 m/s.

In Table 1, the iterative results of ART and PART are compared.

In the Figure 1 (a–c), the three-dimensional wave velocity of the model, ART (100 iterations) and PART (50 iterations) is shown. Wherein, (x, y) are the horizontal and vertical coordinates of the

Table 1. Comparison of iterative results.

Algorithm	Average value normal area	Average value defective area	Velocity dispersion	Worst distance
ART (100)	4489.47	4135.77	69.56	122.46
PART (50)	4499.54	4053.55	7.69	17.11

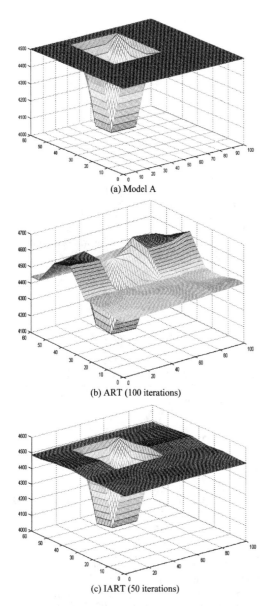

(a) Model A

(b) ART (100 iterations)

(c) IART (50 iterations)

Figure 1. Three-dimensional wave velocity in computer simulation.

survey area, z values are the wave velocity of the corresponding points.

Model B:
The Model B is shown in Figure 2, the measured area is 120 cm × 120 cm, the size of image unit is 10 cm × 10 cm, the velocity of the normal area is 4000 m/s. It has a high-velocity area and three low-velocity areas, the high velocity is 4600 m/s, and the low velocity is 3400 m/s, the differences is 15%. The results are shown in Table 2 and Figure 3.

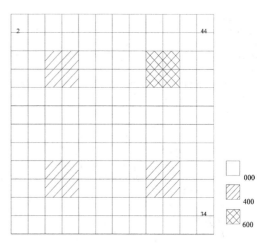

Figure 2. Model B.

Table 2. Comparison of iterative results.

Algorithm	Average value normal area	Average value defective area	Velocity dispersion	Worst distance
ART	4351.3	4023.6	492.59	1200.0
PART	4373.7	3788.7	295.01	695.3

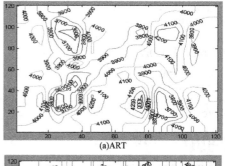

(a)ART

(b) PART

Figure 3. Contour map of wave velocity.

The tables and figures show that the speed and accuracy of iteration are improved in PART algorithm.

4.2 Concrete specimen test

The concrete specimen is shown in Figure 4, the cross-section is 36 cm × 40 cm, and it is divided into 9 × 10 grids. The central part is a circular defect that making with the foam. The diameter is 15 cm. We arrange the emission and receiving transducer on both sides of the test block in longitudinal direction. The spacing is 4 cm. We can get 9 × 9 = 81 sound times. The wave velocity of cross-sectional is obtained by ART and PART. The distributions are shown as Figure 5(a) and Figure 5(b).

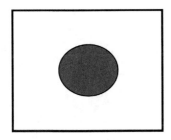

Figure 4. Concrete specimens.

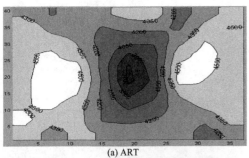

(a) ART

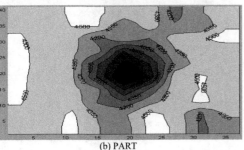

(b) PART

Figure 5. Wave velocity chromatogram of cross-section.

The Figure 5 further illustrates that the PART is more accurately than ART on internal structure of concrete, and the results of reconstruction are better.

5 CONCLUSIONS

1. The computer simulation and the test results of concrete specimen show that the iterative speed and accuracy (especially accuracy) have been improved in the PART.
2. The improved tomography algorithm PART achieved good results in the concrete tomography detection system, and the real-time nature of concrete tomography is implemented.

ACKNOWLEDGMENTS

This research is supported by the National Natural Science Found in China (No. 51278058 and No. 50978030) and Changjiang Scholars and Innovative Research Team Development Program (No. IRT0951). It is also supported by the Special Fund for Basic Scientific Research of Central Colleges, Chang'an University in China with number: CHD2011 JC045 and 0009-2014G1241044.

REFERENCES

Cao J.X. & Nie Z.P. 1998, Model Based Evaluation Method for the Imaging Inversion Algorithms and SASART Algorithm[J]. Journal of Chengdu University of Technology, No 25(4):473~479.
Jing X.L., et al. 2002, Study of Tomographic Inversion in the Construction of Velocity Model. Geophysical Prospecting for Petrole, 41(1):72~75.
Kwon J, Choi, S.J. Song, S.M.H. 2005, Implementation and evaluation of the ultrasonic TOF tomography for the NDT of concrete structures[J]. Computational ImagingIII. 5674:47~54.
Liu Z.Q. & Sun J.P. 1997. Structured image retrieval[J]. Journal of visual languages and computing. 8(3):333~357.
Papoulis A. 1975, New algorithm in spectra analysis and band-limited signal extrapolation[J]. IEEE Transaction on Circuits and systems. 22(9):735~742.
Rosalie C. et al. 2005, Structural health monitoring of composite structures using stress wave methods[J]. COMPOSITE STRUCTURES. 67(2):157~166.
Wu L.1997, The foundation and application of tomography in cross-well seismic[M], Petroleum Industry Press.
Yang W.C. & Du J.Y. 1994, New Algorithms of Tomographic Inversion and application in project examination. Chinese Journal of Geophsics. No 37(2): 239~244.

Control Engineering and Information Systems – Liu (Ed)
© 2015 Taylor & Francis Group, London, ISBN 978-1-138-02685-8

Observability research and simulation experiment about appliances identification in NILM by dispersed data

L. Chen

Electrical and Instrument Control Department, Henan Electric Power Survey and Design Institute, Zhengzhou, China

X. Yi

Technology and Quality Control Department, Henan Electric Power Survey and Design Institute, Zhengzhou, China

ABSTRACT: This paper is focused on the observability of appliance identification problem in Non-Intrusive Load Monitoring (NILM) system. According to the existing situation, set the original data is from dispersed energy-metering data, the observability is analysis by the permutation, combination and probability method, and the simulation experiment is designed for testing and verification of the observability analysis at the end of this paper.

1 INTRODUCTION

Demand-Side Management (DSM) is viewed as the least-cost energy resource plan when both environmental costs and welfare needs are taken into account. This conception is proposed for about twenty years and many policies have been studied under this conception [1,2,3]. Non-Intrusive Load Monitoring (NILM) system is one of the important studies under DSM, and how to do the load identification by a financially viable and easily applicable way is one of core problem in NILM.

The methodology of NILM and its load identification model problem are simply introduced in second section. Then observability analysis in appliances identification problem in NILM and basic mathematics model used in this paper is described. In fourth section, simulation experiment is designed and analysis for testing and verification. The final section is the conclusion.

2 IDENTIFICATION PROBLEM IN NILM

NILM is monitoring system for electrical circuit that contain several different appliances which can switch on and off independently, by a non-intrusive method. One of the earliest approaches was developed in the 1980s at MIT by George Hart which had its origins in load monitoring for residential buildings. These findings and its process are described in [4]. Both USA and France have developed their NILM system at 90's of last century. Since, several methods have been developed to improve this system [5].

For load monitoring systems, most methods of analyze and identification faced steady state or transient state of appliances. Steady state analysis, each individual load or group of loads were determined by identifying times at which electrical power measurement changes from a steady state value to another. The variations related to the current, the active and reactive power, the admittance related to either turning on or off of loads [4,6,7]. Steady state analysis can not identify nonlinear appliances which limit its universal. For transient analysis, loads are identified by their spectral analysis [8,9], high frequency responses [10] or wavelet transform, but they usually needs high sampling frequency which limit its universal as well.

There are two problems in all these methods above. One is the contradiction between the precision/cost of monitoring and the accuracy of identification. The other one is how to get the key parameters, especially for steady state identification, the crucial data of events always occurs in the end of events, but the last time of events are uncertain which leave the problem to parameter collectors. The collector needs face the mass information from sensors, so the thinking model of data mining for data stream is helpful in this situation.

3 OBSERVABILITY ANALYSIS

As the analysis above, the monitoring and identification problem in NILM is caused by the original data obtaining problem under the existing situation, therefore, the original data of observability analysis are from dispersed data collected by energy-metering of the electric system, and the problem is summarized as the following questions.

How many appliances in monitoring system boundary?
What is the power rating of each appliance?
What is the working status of the appliances at each interval?

Each appliance has its unique power rating and working status, and Each NILM system has its sampling interval. The presupposition that the only data we got is dispersed energy-metering data of the whole electric system. Because the lack of system details and some odd situations of mathematics, we can only get the most reliable guess, not the exact answer. But if the sampling interval is small enough and the observability research of discrete system can be seen as the real system.

3.1 Basic math model

Set $P_i(S_i^m)$ The power function of i-th appliance
S_i^m The state of i-th appliance at time interval m, usually 0 for off and 1 for on.
m The sequence of measurement interval.
Y The total work measured by watt-metering.
v The error in the measurement.

So Y is system output, the status T of the appliances is the input U. Then we can tell that the object of the research is find the input and some parameters of the system, that is we called observability.

If the number of the appliances is n, then we can formulate

$$\Delta Y^m = \sum_{i=1}^{n} P_i(S_i^m) * T_i^m(S_i^m) + v^m \qquad (1)$$

Easily we know that Y is an monotone increase function.

Set, $P = [P1\ P2\ ...\ Pn]$, $V = [v1\ v2\ ...\ vm]$ and $U = [T11, T21, ..., Tm1; T12, T22, ..., Tm2; ...; T1n, T2n, ..., Tmn]$. The function (1) can be rewrite as follow.

$$\begin{cases} \dot{X} = P * U + v \\ Y = X \end{cases} \qquad (2)$$

3.2 Details discussion

1. Suppose precision of the energy-metering is 1.0 (1%).
2. The number of appliance in the system and the difference of the power of each appliance are not large, for the smaller power appliance will not be detected.

3. The change frequency of the appliance status would be better not larger than the data collect interval. If the sampling interval is small enough, this hypothesis can be ignoring.
4. The maximum power of the system would be better given by the user for it is an important data for the model. We will use the maximum data the energy-metering collected instead if there is no exact maximum power, but it will influence the precision of the model.

When the data collect interval is almost equal and the system is not too far from we supposed, we got the data $z(k)$ collected by energy-metering and maximum power max of the system from user. We can go next step to use the model to find out what we concerned.

3.3 Precision analysis

Before dealing the data, error bring by existing situation (metering) should be account as follows:

1. The precision of the energy-metering own, take the energy-metering we used for example. The precision is 1:0 and that means may be there is 1% error of the data collect, either bigger or smaller.
2. Just as the energy-metering collect data can not be perfectly precise, the sampling interval can not be exact equal. If we collect data every 30 second, actually the point of data collect may be 1 second earlier or later, that means the data we got may have 1/180 bias, either bigger or smaller.

The two errors above are brought by metering. Huge data collecting may decrease their influence of the last result, but it can not be ignored here. We still have to integrate them as the measure error by using the following calculations.

$$(x/181 - 1.01 * x/180)/x/181 = -1.56\%$$

$$(x/179 - 0.99 * x/180)/x/179 = 1.65\%$$

The final measure error is between -1.56% and 1.65%. The error of the result is not only brought by measure, but also by mathematics model.

Actually, the interval between we collect data is much longer than the status change of appliances, but we have to simple the model for calculation, that is why the error will bring by mathematics model. In my following work, the error of model will be a same level, 3% *max in the following experiments.

3.4 Basic math model

After decided the precision of the model, we can begin to process the data. The main flow has 9 steps as follows.

S1. Calculate the power of every interval and keep it in v (k).

S2. Trim data. Take 3% of max as a limit of a section and find the number of interval power in the section. Calculate the arithmetic mean of every section and replace the original data.

S3. Half of this step can be done with last step in the programme for the mission is note the arithmetic mean of every section and the number of power appearance in every section; we named the two-dimension array powerdata first. Then we can note n as the number of section.

S4. Set the kind of combination of different appliances that appear in the collection is between 25% and 100% of the whole combination. The parameter should be a integer, if note m as the int (log2N) + 1, and the number of appliances should be m or m + 1.

S5. Create a new array tempdata to note the minimum power in powerdata and the difference between max and the maximum power in powerdata, the kind of appliance is M.

S6. Note the change power of every neighbor measure point in another array diffdata, the number of these power appearances should be noted together in diffdata.

S7. Then tempdata and diffdata will be database for power guessing of single appliance, every element in the two arrays may be the power of single appliance in the system. The main mission in this step is listing the possible combination of these elements. The detail of find the combination will be discussed in the appendix.

S8. Test the combination by certain ways and calculate the reliability for each combination.

S9. Find the most reliable combination as the final result.

3.5 Reliability calculation

The formulation of Reliability Calculation is

$$\text{Reliability } (i) = (qa*a(i) + qb*b(i) + qc*c(i)) / (qa + qb + qc) \quad (3)$$

The weight parameter can be given by user depending on the circumstances. If there is no weight information, it can be set as the same weight in the calculation.

Take the i-th combination for example:

1. Calculate the relative error $a(i)$ between the sum of combination and total powermax, the weight of this parameter is qa. The influence of the parameter will be grown with the increase of weight qa.

The weight qa should be rise when the precision of energy-metering is high or the total power max is certain.

2. List the possible power of system and contrast against the powerdata we measured, the degree of matching will be note as $b(i)$.

3. The emerge times of combination elements in diffdata will be note in $c(i)$ in the form of percent. The weight qc should be rise when the status of appliance is not change frequent.

4 SIMULATION EXPERIMENT DESIGN AND ANALYSIS

In this section, a simulation experiment is for proving the processing flow designed above.

First, we have to make a system to test the model first for we do not have an actual data from electric system. But we just use the output and the data we can actually collect by energy-metering. Then we will follow the steps above and get the result by the model.

4.1 System output simluate

Suppose there are four appliance in the system and their power is d = [40; 60; 100; 120] respectively. We make them work like the following states, randomly. Actually, the working states may not change so frequently if the sampling interval is properly, which makes the simulation more persuasion.

$$State = \begin{bmatrix} 000111000100111111110100 \\ 011110000011010111011101 \\ 111100000001111001011111 \\ 100000111000111100001010 \end{bmatrix} \quad (4)$$

The main energy cost of the system is added random error, and the final data are as shown in the Figure 1.

4.2 Data process

The system output above is seen as the original data which trimmed as the flow steps. We will see the untrimmed data and trimmed data in contrast in Figure 2. Blue line for untrimmed data and red for trimmed data.

Statistic about the data in two-dimension array powerdata is shown as the Table 1.

Next step is create the guess power of single appliance. Two different elements will be shown in Table 2. Things we should notice here is we will get 0 as a power in Diffdata for the status may not change between some interval, but we would be

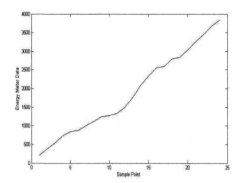

Figure 1. System output simulation.

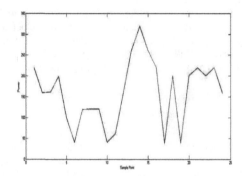

Figure 2. Data trimmed contrast.

Table 1. Power data.

POWER	40	60	100	121	161	199	220	259	281	324
TIMES	3	1	2	3	4	4	3	2	1	1

Table 2. Guess power.

		POWER	40		61		100		121	
Tempdata		TIMES	2		2		2		2	
Diffdata		POWER	0	21	39	60	82	99	120	159
		TIMES	3	2	2	5	4	4	1	2

better kick it out of the guess power array for simple calculation.

4.3 Combination generate and reliability calculate

All the possible combinations are noted in result, 0 in the array stand for null instead of actual power.

Table 3. Guess power.

Result					Reliability
21	39	60	82	99	0.8024
39	60	82	120	0	0.7469
21	39	60	82	120	0.8201
21	60	99	120	0	0.7125
39	60	99	120	0	0.8312
21	39	60	99	120	0.8024
21	82	99	120	0	0.6507
39	82	99	120	0	0.6986
21	39	82	159	0	0.7524
21	60	82	159	0	0.7118
39	60	82	159	0	0.7931
21	39	99	159	0	0.8035
21	60	99	159	0	0.7608
21	39	120	159	0	0.6774
40	61	100	121	0	0.8312

Reliability of these combinations is also in the Table 3. The two most reliable combinations in the table are [39 60 99 120] and [40 61 100 121]. We can easily find that the two combination is nearly the same with the power we supposed in simulation system, and the error is in accepted range.

5 CONCLUSION AND IMPROVEMENTS

This paper proposed methods for load identification and parameter recognition by the permutation, combination and probability, and design a simulation experiment to verify the observability in the extreme case. The proposed method can be used in the NILM system and still needs improvements in the following aspects.

1. The presumption of model is that appliances in the system have different power. Actually, appliances have same power rating is possible; it will extremely limit the using of the model if the problem is not solved. The reliability record of single power elements may be a solution.
2. The model is built on the supposition of unchanged status of appliance for the whole measure interval. The actual situation will not be such coincidence. How to find the data has duality of status is another important improvement.

REFERENCES

A Kalman-Filter Spectral Envelope Preprocessor Shaw, S.R.; Laughman, C.R.; Instrumentation and Measurement, IEEE Transactions on Volume 56, Issue 5, Oct. 2007 Page(s): 2010–2017.

Hart G.W.; Nonintrusive appliance load monitoring, Proceedings of the IEEE Volume 80, Issue 12, Dec. 1992 Page(s):1870–1891.

Ibrahim, M.; Jaafar, M.Z.; Ghani, M.R.A.; Demand-side management. TENCON '93. Proceedings. Computer, Communication, Control and Power Engineering.1993 IEEE Region 10 Conference on: 572–576 vol. 4.

Leeb, S. James L. Kirtley, Jr., Michael S. LeVan, Joseph P. Sweeney, Development and Validation of a Transient Event Detector, AMP Journal of Technology Vol. 3 November, 1993.

Majumdar, S.; Chattopadhyay, D.; Parikh, J.; Interruptible load management using optimal power flow analysis, IEEE Transactions on Power Systems, May 1996, PP. 715–720.

Najmeddine, H.; El Khamlichi Drissi, K.; State of art on load monitoring methods, Power and Energy Conference, 2008. PECon 2008. IEEE 2nd International, 2008, Page(s): 1256–1258.

Pihala H. Non-intrusive appliance load monitoring system based on a modern KWh-meter, technical research center of Finland ESPOO 1998.

Sultanem F. Using appliance signatures for monitoring residential loads at meter panel level, IEEE Transactions on Power Delivery, Vol. 6, pp. 1380–1385, 1991.

Turvey R.; Peak-load pricing, The Journal of Political Economy, 1968, pp. 101–113.

Wang zhenyu; Zheng guilin; Residential appliances identification and monitoring by a nonintrusive method, IEEE Transactions on Smart Grid, Volume 3, Issue 1, March 2012, Page(s): 80–92.

Wenders; Experiments in seasonal-time-of-day pricing of electricity to residential users; The Bell Journal of Economics, 1976, pp. 531–552.

Session 7: Computer science and its application

Design of mine safety management information sharing system

S.W. Qin

School of Environment and Civil Engineering, Wuhan Institute of Technology, Wuhan, China

ABSTRACT: Aiming at information sharing of heterogeneous mine safety management system, Author analyzed current multi-source data sharing solution method; On the basis of the analysis, proposed a new schema of heterogeneous system integration; Using CAS single sign-on technology, realized data access control; Using the model of Third Information Flow, designed the framework of heterogeneous system data exchange. This schema proposed a method to solve data exchange problem of heterogeneous mine safety management system.

1 INTRODUCTION

Affected by natural geography conditions, working environment and security conditions are extremely complicated, which threatened safety production. Most mine safety accidents were caused by weak safety basis, backward technology, loose safety management and lack of professional skill (Hou et al. 2009). Mine safety management includes 14 items, they are production safety policy and objectives, safety production rules, guarantee of safety production, risk management, safety training, process safety management of production, equipment safety management, job site safety management, safety technology and investment, occupational health management, daily safety inspection, accidents handling, performance evaluation and emergency management. All information of the above mentioned originated from office system, production command system, equipment management system, human resources system, major hazard monitoring system, etc. First of all the problems of mine safety management system is how to implement these multi-source data visit and exchange.

Now major solving method of heterogeneous integration is to adopt middleware such as OGSA-DAI, provide SQL query in relational database or XPATH query in XML database. The middleware is only used for data extraction. It does not achieve the writeback data (Tian et al. 2010).

Depending on Single Sign-On (SSO), this paper provided a multi-source data visit module based on third information flow, solved the problem of data update from heterogeneous data source, enhanced interactive, implemented cooperative work between systems.

2 SYSTEM DESIGN

Third information flow means that all information originated from different application systems was entrusted to a separate platform that could manage and control these consolidated data (Qin 2010). Key point is taking separate application system as data file, third platform played a role of data base management. All heterogeneous data accesses are done through the third platform. The third platform as a database provides data service for all systems.

Figure 1 shows the system framework, it has 4 levels.

2.1 Data Access Layer (DAL)

This layer is formed by business systems or data base. Each of these business systems or data base as a data file provides a data access interface such as web service, ODBC, html, txt and so on. Data exchange center implements data management

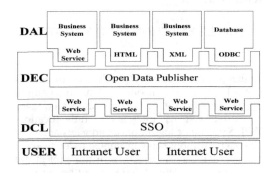

Figure 1. System framework.

through these interfaces. The business system controls data security via it.

2.2 Data Exchange Center (DEC)

In DEC, all of data access interfaces are packaged as uniform format web service which is used to implement the conversion between the user request and the business system request. When DEC receives a user request, DEC transforms it into a serial of separate business system request and sends them to DAL by data access interface. After DEC receives responses from DAL, DEC converts them to one uniform format response and returns to the user. With twice transformations, DEC realizes data exchange in heterogeneous systems.

2.3 Data Control Layer (DCL)

DCL contains user authentication and authorization. These two parts ensure Enterprise Data security. Authentication is the act of confirming the truth of the identity of a user or software program. Single sign-on technology supports one identity accessing different application systems. Authorization is the function of specifying data access right which follows the business system access policy.

2.4 User

A user is an agent, either a human or a software, who read data from Data Exchange Center (DEC) by Internet or Intranet.

2.5 The workflow of system

Data request process is as follow: Firstly, DCL verify the identity of the user. If the user had permission to request data, DCL sends data requests to DEC. According to user request, DEC transforms and distributes it to DAL. Then DEC merges the responses returned by DAL. Finally, it returns the result to the user.

3 IMPLEMENTATION OF KEY MODULES

3.1 DCL

Authentication is the key problem of the access control of heterogeneous system integration. Single Sign-On (SSO) is a common technique for enterprise systems integration authentication. With SSO, a user logs in once and gains access to all systems without being prompted to log in again at each of them (Wang 2013). In this project, we use CAS to implement SSO for authentication. CAS is an authentication system originally created by Yale University. It provides a trusted way

to authenticate a user (Jing 2009). CAS became a Jasig project in December 2004.

The workflow of DCL is shown in Figure 2.

1. *Request*: The user sends a request to DCL. This request includes a Ticket-Granting ticket (TGC) and the application.
2. *Identification*: DCL sends TGC to CAS server for verification. If TGC is valid, CAS generates a new random Service Ticket (ST) and returns it.
3. *Authentication*: DCL sends ST and the requested service name to CAS server for authentication. If it is authenticated, CAS returns the authenticated User Identity (User ID).
4. *Authorization*: After DCL received user identity, it has two methods to finish the user authorization. One is reading user's roles from business systems, then authorizing according to the policy of DCL's data access control. The other is authorized directly by business systems.

In this workflow, DCL visits CAS server twice. But it is transparent to the user. CAS server only finishes identification and authentication. Authorization can be finished by DCL or business system. This method can reduce the system coupling.

3.2 DEC

DEC specifies a number of different request types. Each type provides a Web service, three of which are required by DEC server:

5. *GetCapabilities*: It returns information on DEC (such as service list) and the available data source (data source name, table list of data source, field list of table and so on).
6. *GetInfo*: If a data source is marked as 'queryable' then you can request data according to fields and filters.
7. *UpdateInfo*: If a data source is marked as 'updatable' then you can update (add, delete) data according to value and filters.

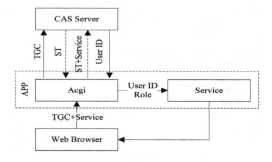

Figure 2. Workflow of Single Sign-On authentication.

If needed, user can expand services for particular business system in DEC.

The data exchange involves performing six steps: request, distribute, access, collect, arrange and response. It is shown in Figure 3.

1. *Request*: The user submits the request for data exchange. The request parameter, based on XML, contains two major sections, that is, serivce and data. For example, the following XML document defines a query for GetInfo service:

```
<?xml version="1.0"?>
<DECReqeust>
  <Service name="GetInfo">
    <Datasource > ems</Datasource>
    <Fields>
      <Field name="equipment.name">
      <Field name="department.name">
    </Fields>
    <Filter>
      <And>
      <FieldIsLike
        name="equipment.name"
        value="%abc%" />
      <FieldIsLike
        name="department.name "
        value="%abc%" />
      </And>
    </Filter>
  </Service >
  <Data />
</ DECRequest >
```

The request parameters are XML document that can be verified by the request DTD (Document Type Definition).

2. *Distribute*: Determine how to distribute according to the request. Firstly, Analyze the request to get the returned data object list. Then scan the list of data relationships in DEC to find their relation, determine data source and access sequence. Based on data source requirements, split user request into data resource requests at last. For complex query, create view in DEC to save mapping and query path.

3. *Access*: Send request to DAL and wait for response. There are two ways to submit a request. One is via HTTP (such as SOAP). The other is through database connection (such as ODBC, JDBC).

4. *Collect*: Format response data from DAL into XML. If the response data format is XML, this step can be skipped. If it is HTML or TXT, extract data from strings by regular expressions, then transform data into XML. For returned data table, take row as main element and column as child element, transform table into XML. For all response data, final converted format is XML as below, children elements of root element are data row, children elements of data row element are data field, children elements of data field element are value, as in the following example:

```
<?xml version="1.0"?>
< Equipment >
  <Row>
    <Name > equipment name</Name>
    <Department > department</Department>
  </Row>
</ Equipment >
```

5. *Arrange*: According to data relation, merged multiple XML documents into one XML document. For only one document, this step can be skipped. For multiple documents,

a. One-to-one relation: merge these two tables into one, then follow steps of single table.

b. One-to-many relation: take data row element of child document as a field element of parent document, insert it into data row element of parent document.

c. Many-to-many relation: convert to one-to-many relation tables and then proceed.

The result of this step is as in the following example:

```
<?xml version="1.0"?>
< Department >
  <Row>
    <Name> department </Name>
    < Equipment >
      <Name > a</Name>
    </Equipment >
    < Equipment >
      <Name > b</Name>
    </Equipment >
  </Row>
</Department >
```

6. *Response*: Return the result to the user. The result includes two parts, exception and result data. If the data exchange is success, exception is null, otherwise it includes failure reason.

4 CONCLUSION

Author analyzed demand of mine enterprises for safety management system; Aiming to multi-source data exchange in mine safety management, designed system framework by using the model of

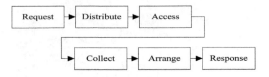

Figure 3. Workflow of data exchange.

Third Information Flow; realized integration and implementation of multi-source data. The system structure is clear, crossing platform, and is easy to manage, has good extensibility and maintainability. It solves the information island problem of mine safety management.

ACKNOWLEDGMENT

This work is supported by the Natural Science Foundation Committee of China under Grant no. 50874080.

REFERENCES

Minchang Jing, Diguan Tang 2009. Research on CAS Single Sign-On Login System of Open Source. Journal of Modern Information 29(3):125–127.

Ming-yan Tian, Shui-ping Zhang, Lu-ping Liu 2010. Multi-resource Data Access Based on Grid. Computer Engineering 36(7):84–86.

Qian Hou, Yunhai Wang, Wuyi Cheng, Chunhua Cheng, Luming Wang 2009. Construction and Evaluation of the Safety Standardization in Metal and Non-metal Mines. METALMINE 398(8):140–148.

Sun-wei Qin, Jing-yue Sun, Xian-fu Li 2010. Enterprise Information Management Mode Based on Third Information Flow. Journal of Wuhan Institute of Technology 32(1):6–9.

Xianhua Wang, Chaoyan Li 2013. The Application of Single Sign-On Based on Filter Technology. Computer Applications and Software 30(3):327–329.

Control Engineering and Information Systems – Liu (Ed)
© 2015 Taylor & Francis Group, London, ISBN 978-1-138-02685-8

Analysis and research of parallel programming techniques based on multicore platform

Y.Y. Teng & Z.G. He

Department of Computer Science and Technology, Dalian Neusoft University of Information, Dalian, Liaoning Province, China

ABSTRACT: Multi-core processor has become the main body of the processor architecture, which demands more for operating system and application software. The original serial programming techniques on multi-core platform is no longer applicable, replaced by a parallel programming technology. This paper focuses on two kinds of parallel programming techniques: Windows API and OpenMP. Compare the characteristics and differences of the two technologies in use. Both techniques were used to complete parallel processing of the numerical integration algorithm for calculating the value of π. Through the analysis of parallel program performance, put forward reasonable proposals using two techniques.

1 INTRODUCTION

Currently, multi-core processor has become the main body of the processor architecture. Multi-core processor will be two or more independent execution of core embedded to the same processor chip. Each execution core has its own execution units and architecture resources. Therefore, the sequence of instructions that run on each core having an independent and complete hardware execution environment, you do not need to hang waiting for some resources. Through division of tasks among multiple execution cores, multi-core processor can perform more tasks in a particular clock cycle to improve the performance of the program.

Multi-threading can achieve a true parallel on multi-core platform. After the increase in the quantity of core, number of threads to use hardware gets more. On the other hand, multi-core processor now uses shared memory architecture. Thus, Multi-threaded program design based on multi-core is a fast method to improve program performance.

2 MULTI-CORE PARALLEL TECHNIQUES TYPES

Multi-core parallel programming model is divided into explicit model and implicit model.

The explicit model programming model provides an API to explicitly manage threads. Thread creation, synchronization, mutual exclusion and other operations can be achieved by calling an API.

Developers in the application of this programming model need to be very clear when the program should create a thread, how to create the thread, the thread when to synchronize, when mutually exclusive. In short, application developers can be done through the API parameter settings on the details of thread management, more conducive to the development of high-performance programs. Explicit model includes Win32 API, threads and so on.

Implicit model is simple to use. Many technical details such as thread creation, thread destruction, synchronization can be ignored in the application process. Application developers can focus on to consider what code should be multi-threaded manner, as well as the reconstruction algorithm in order to get better performance on multicore platform. In this model some templates are also provided for developers to quickly develop parallel programs. Implicit model includes OpenMP, TBB and so on.

3 PERFORMANCE OF MULTI-CORE PARALLEL PROGRAM

Parallel program in multi-core platform is reflected in the thread-level parallelism. Performance indicators include speedup, efficiency, granularity, and load balance.

And traditional application development process, the parallel programs follow the same software development life cycle. Program optimization is the process of iterative process. Each iteration cycle is divided into the analysis, design, implementation, correctness checking and optimization.

3.1 *Speedup*

To quantitatively determine the performance benefit of parallel computing, you can compare the elapsed run time of the best serial algorithm with the elapsed run time of the parallel program. This ratio is known as the speedup and it measures the time required for a parallel program execute versus the time the best serial code requires to accomplish the same task.

Therefore,

Speedup = Serial Time/Parallel Time.

To determine the theoretical limit on the performance of increasing the number of processor cores and threads in an application, Gene Amdahl in 1976 examined the maximum theoretical performance benefit of a parallel solution relative to the best case performance of a serial solution. According to Amdahl's Law, speedup is a function of the fraction of a program that is parallel and by how much that fraction is accelerated. Therefore,

Speedup = $1/[S + (1 − S)/n + H(n)]$

where, S is the percentage of time spent on executing the serial portion of the parallelized version, n is the number of cores, and $H(n)$ is the parallel overhead. The numerator in the equation assumes that the program takes one unit of time to execute the best sequential algorithm.

3.2 *Parallel efficiency*

Parallel efficiency is a measure of how efficiently core resources are used during parallel computations. It is expressed as a percentage:

Efficiency = (Speedup/Number of Threads) * 100%

Low efficiency may prompt the user to run the application on fewer cores and free up resources to run something else, maybe another threaded process or other user's codes.

3.3 *Granularity*

Granularity is more difficult to quantify than parallel speedup or efficiency. It has been defined as the ratio of computation to synchronization. Concurrent calculations that have a large amount of computation between synchronization operations are known as coarse grained. Cases where there is very little computation between synchronization events are known as fine-grained. Synchronization, by definition, serializes execution. Therefore, programmers should strive for more coarse-grained activity in threads.

3.4 *Load balance*

Another challenge in writing efficient multithreaded applications is balancing the workload among multiple threads. The most effective distribution is to have equal amounts of work per thread. If more work is assigned to some threads than to other threads, the threads with less computation will sit idle waiting for the threads that have more work to finish. Load balance refers to the distribution of work across multiple threads so that they all perform roughly the same amount of work.

4 WINDOWS API PARALLEL PROGRAMMING TECHNIQUE

The Windows API is interface between Microsoft's Windows operating system kernel and the upper application. Application gains system ability by calling the appropriate interface. Directly use the Win32 API to write code requires developers to have some knowledge of the operating system, but a small amount of program code, high operating efficiency.

With the advent of Windows NT, Windows application developers can take advantage of the Windows thread library to develop multithreaded programs. With the rapid development of processor technology, Windows multi-threaded API has also become one of the ways to develop multithreaded programs on the Windows platform. Simultaneously critical region, event, semaphore and other mechanisms are provided to effectively solve conflicts between threads.

4.1 *Algorithm description*

A. Algorithm description

Numerical integration method of calculating π, is also known as trapezoidal calculation method. In the algorithm, calculate the definite integral of the formula from 0 to 1 to estimate the value of π, ie. Typically, the calculation of definite integral dx, is calculated curve $y = f(x)$ with the line $y = 0$, $x = 0$, $x = 1$ surrounded by the curved edge of the area of trapezoid T. To this end, a set of lines parallel to the y-axis will be curved trapezoid T is divided into several small curved trapezoidal Ti. While the width of Ti becoming smaller and smaller, the rectangle can be calculated as Ti approximate. Of course, the smaller the width of Ti, the more accurate the calculated π. Algorithm description chart, as shown in Figure 1.

In numerical integration algorithm, the process of integration is completed by the loop iteration. The number of iterations is determined by the

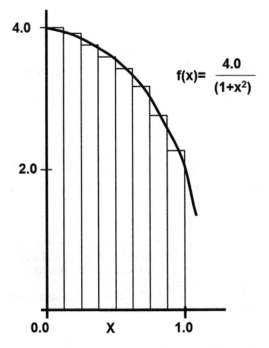

$$f(x)= \frac{4.0}{(1+x^2)}$$

Figure 1.   Algorithm description char.

number of Ti. The loop area calculates the area of a rectangle and sums of all rectangles. Core code is shown below.

Code 1:
```
static long num_steps = 1000000000;
        step = 1.0/(double) num_steps;
        for (i = 0; i< num_steps; i++){
        x = (i+0.5)*step;
        sum = sum + 4.0/(1.0 + x*x);
}
        pi = step * sum;
```

4.2   Parallel algorithm

First, analyse the parallelism of the program. Because the program's most calculations done by the loop iteration, the loop taken in parallel loop is divided into several parts, each by a thread, which is a data decomposition method, the iterative decomposition process should pay attention to the data relevance. After determined parallelism, design the parallel algorithm. Pay attention to protect the share data while designing algorithm. This example uses critical region to resolve data conflicts. Parallel code is shown below.

Code 2:
```
static long num_ steps = 1000000;
const int g Num Threads = 4;
```

```
double step = 0.0;
double pi = 0.0,sum = 0.0;
CRITICAL_SECTION gCS;
DWORD WINAPI thread Function(LPVOID p Arg)
{
    double partial Sum = 0.0, x;
    for (int i = *((int *)p Arg); i < num_ steps; i+ = gNum Threads )
    {
x = (i+0.5)*step;
partial Sum = partial Sum + 4.0/(1.0 + x*x);
    }
    Enter Critical Section (&gCS);
    sum + = partial Sum;
    Leave Critical Section (&gCS);
    return 0;
}
int main()
{
    HANDLE thread Handles[g Num Threads];
    int tNum [g Num Threads];
    Initialize Critical Section(&gCS);
    step = 1.0/(double) num_steps;
    for ( int i = 0; i < gNum Threads; ++i )
    {
    tNum[i] = i;
    thread Handles[i] = Create Thread( NULL, 0,
Thread Function,
&tNum[i],
```

Table 1.   Test data of numerical integration.

Iterations (million)	Serial run time	Parallel run time	Speed up	Parallel efficiency
10	0.066	0.048	1.375	68.75%
100	0.551	0.366	1.5	75%
1000	5.296	3.507	1.51	75.5%
10000	7.558	4.797	1.576	78.8%

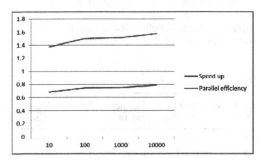

Figure 2.   Windows API speedup and parallel efficiency comparison chart.

```
0,
NULL);
    }

    Wait For Multiple Objects (gNum Threads,
threadHandles, TRUE, INFINITE);
    Delete Critical Section (&gCS);
    pi = step * sum;
}
```

4.3 *Analysis*

On dual-core platform, we tested the results, and
the results shown in Table 1.

Speedup and parallel efficiency comparison
chart, as shown in Figure 2.

5 OpenMP PARALLEL PROGRAMMING TECHNIQUE

5.1 *Notes parallel algorithm*

First determine the parallel region, and then create
several threads in parallel. The number of threads
is decided by the number of cores. In parallel
threads, pay attention to private and shared data.
Parallel code is shown below.

```
Code 3:
long long num_steps = 1000000000;
omp_set_num_threads(4);
#pragma  omp   parallel   for   private(x)
reduction(+:sum)
    for (i = 0; i < num_steps; i++)
    {
        x = (i + .5)*step;
        sum = sum + 4.0/(1.+ x*x);
    }
    pi = sum*step;
```

5.2 *Analysis*

On dual-core platform, we tested the results, and
the results shown in Table 2.

Through the analysis of experiment results we
know, when calculating the scale of hours, due to
thread creation, synchronization overhead is relatively large, so the speedup and parallel efficiency

Table 2. Test data of numerical integration.

Iterations (million)	Serial run time	Parallel run time	Speed up	Parallel efficiency
10	0.066	0.046	1.435	72%
100	0.551	0.309	1.78	89%
1000	5.296	2.879	1.84	92%
10000	7.558	3.929	1.924	96.2%

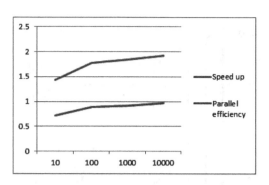

Figure 3. OpenMP speedup and parallel efficiency
comparison chart.

is not high, but when calculating the size increases
and then close to the ideal speedup 2, parallel efficiency close to 100%, indicating that resource utilization is an ideal.

Speedup and parallel efficiency comparison
chart, as shown in Figure 3.

6 CONCLUSION

Multi-threaded programming on multi-core and
multi-threaded programming on a single core are
very different. Parallel programming based on
multi-core platform, not only need to consider the
lock contention problem, but also consider CPU
idle caused by lock contention, load balance on
each CPU core, and task scheduling and other
issues. With the Windows API supports multi-
threaded programming interface perfection, it has
become a powerful tool on the Windows operating system to develop multi-threaded applications.
OpenMP is standard appears as a shared storage
and it is designed as an application programming
interface in order to develop parallel programs on
shared storage environment.

It can be seen from algorithm of numerical
integration calculating Π values that OpenMP
provides a simple, portable way, by using the compiler supports OpenMP to parallel serial code.
Because it hides the technical details related to
the thread, it is so much easier than the Windows
API. If the performance of the application in a
multi-core has already reached saturation, then
make it multi-threaded execution using OpenMP
will almost enhance its multi-core system performance. OpenMP can uses the appropriate number of
threads based on the target system automatically.
Developers should give priority to consider covert
serial code into parallel code conversion by using
OpenMP, and makes the code more portable and
easier to maintain.

REFERENCES

Gonzalez M, Oliver, Martorell X, et al. "OpenMP Extensions for Thread Groups and Their Run-time Support", 13th Int'l Workshop on Languages and Compilers for Parallel Computing (LCPC'2000), New York (USA), pp. 317–331, 2000.

Leslie L. "Arbitration-free synchronization," Distributed Computing. vol. 16(2–3), September 2003, pp. 219–237.

Liu Zhangqiao, WANG Cheng-liang, Jiao Xiao-jun, "Research of Parallel Algorithm for Image Segmentation Under Multi-core Environment", Computer Engineering, Aug 2011, pp: 197–200.

Shameem Akhter, Jason Roberts. "Multi-core programing: Increasing performance through software multithreading", Intel Presss Business Unit, 2004, pp. 72–175.

Tan Hai, "Research of object-based implicit parallel program many-core architecture", Computer Engineering and Design, Vol. 34 No. 2:623–626, Feb 2013.

Wang Jue, Hu ChangJun, Zhang JiLin, LI JianJiang, "OpenMP compiler for distributed memory architectures", Science China(Information Sciences), Vol. 53, No. 5: pp. 932–944, May 2010, doi: 10.1007/s11432-010-0074-0.

Wang Jue, Hu ChangJun, Zhang JiLin, LI JianJiang, "OpenMP compiler for distributed memory architectures", Science China (Information Sciences), Vol. 53, No. 5: pp. 932–944, May 2010, doi: 10.1007/s11432-010-0074-0.

Yang XueJun, Tang Tao, Wang GuiBin, Jia Jia, Xu Xin-Hai, "MPtostream: an OpenMP compiler for CPU-GPU heterogeneous parallel systems", Science China (Information Sciences), Vol. 55 No. 9: 1961–1971, September 2012, doi: 10.1007/s11432-011-4342-4.

Control Engineering and Information Systems – Liu (Ed)
© 2015 Taylor & Francis Group, London, ISBN 978-1-138-02685-8

Analysis of fault tolerance in MPI programs

Y. Zhang & Z.J. Hu
Information Department, Hospital of Beijing PAPF, Beijing, China

H.B. Niu
China Defense Science and Technology Information Center, Beijing, China

ABSTRACT: With the growing scale of high performance computing platforms and the increasing scientific computing demands, fault tolerance in MPI programs has become a major issue. This paper proposed an analytical performance model to help checkpointing mechanisms to get better performance. In this paper, after the introduction of different fault tolerant mechanisms, we focus on the segment-level and variable-level checkpointing approaches and analyze how the fault tolerant mechanisms affect the performance of current MPI systems, then we introduced the analytical performance model we proposed. Last we evaluated and analyzed the performance of two classic fault tolerant approaches and propose what should be done to advance variable-level checkpointing mechanisms.

1 INTRODUCTION

Over the past few years fault tolerance has become a major issue for HPC systems, in particular in the perspective of large Petascale systems and future Exascale ones. These systems will typically scale to half a million or even several millions of cores, for example, the Tianhe-1 A supercomputer created by the National University of Defense Technology (NUDT) beats out of its rivals in the Top 500 rankings last year, which comprises 14,336 Xeon processors and 7,168 of NVidia's Tesla M2050 fanless GPU coprocessors, and the following champion Japanese "K computer" has even 548,352 processors (TOP500,2011).

As the rapid expansion scale of these systems, the inherent reliability of these systems will be rapidly decline, and resulting in the decline of their performance. In addition, with the increase in size of a single chip, the reduction of the size of integrated circuits and the influence of noise in the chip, the failure rate of a single device such as processor core and memory will also increase. According to the actual operation of the Teraflop system, the IBM BlueGene/L with 131,072 cores has MTBF (Mean Time Between Failures) of 53–158 h, and the CRAY XT3/XT4 with 10,880 cores has MTBF of 7–72 h, for the existing Petascale systems, their MTBFs are in the tens of minutes to several hours. According to the formula MTBF $\approx 1/(1 - R^N)$, the actual MTBF of the Japanese "K computer" with 548,352 cores should be in several minutes to tens of minutes.

Nowadays, scientific computing jobs usually last several days to several months, for example, the protein folding program running on the Blue-Gene needs several months (Du, 2008), these existing MTBFs of these systems obviously cannot meet the application needs. Taking into account the above problems, as an important method to improve the system reliability, fault tolerant technology deserves more in-depth and valuable investigation.

There are many ways to achieve various degrees of fault tolerant ability, such as proactive task migration (Smith, 1988; Chakravorty et al. 2006), message logging (Bouteiller et al. 2009), replication (Genaud & Rattanapoka, 2007; Walters & Chaudhary, 2009), checkpointing (Litzkow et al. 1997; Rodrıguez et al. 2010), etc. According to the abstract level at which the fault tolerant mechanisms take effect, these approaches can be divided into two categories: system-level (Elnozahy et al. 2002) and application-level (Bronevetsky et al. 2003).

The widely used approaches in system-level are hardware fault tolerance, rollback-recovery mechanism, I/O fault tolerance. System-level fault tolerance makes failures transparent to users, which makes the system easy to use and also makes the system more complex and less scalable. These ways also bring a great influence to performance and power consumption.

In application-level, the fault-aware error handling gets failure information through fault tolerant interfaces provided by system, and handles the failure according to its types in the application program. Current programming methods provide little support for fault tolerant interfaces. Comparing with system-level's approaches, application-level mechanisms have the advantages of high performance and low overhead, but these ways are less transparent to users and often required to modify the programs, and these methods are always associated with specific applications, which limits its application scope.

Among all these methods, supporting fault tolerance at the programming level is an import direction of solving the reliability problem of HPC systems. MPI is currently the de-facto standard system to build high performance applications, which integrated the most effective methods in many message passing libraries, but its current standard is designed without fault tolerant interfaces.

So we focus on the MPI systems, first we introduce current widely used fault tolerant approaches, especially the approaches MPICH checkpointing and CPPC that we interested, then we propose an analytical model for checkpointing to guide the checkpointing systems to get better performance, then we compare MPICH and CPPC performances by using NPB applications, give our conclusions and present our next work.

2 RELATED WORK

There are many approaches based on MPI standard providing the fault tolerant ability, such as message logging, checkpointing, modifying MPI semantics, extending the MPI specification (Lusk, 2002), etc. In this paper we focus on the checkpointing mechanism, from the point of view of the data stored in state files, which can be classified into segment-level and variable-level. Segment-level checkpointing stores the whole application states while variable-level stores user variables only. Comparing with variable-level techniques, segment-level checkpointing tools usually have complex, non-portable state.

MPICH-V (Bosilca et al. 2002; Bouteiller et al. 2003; Bouteiller et al. 2006) provides complete checkpointing and message logging to enable replacement of aborted processes; the checkpoints avoid reconstructing computations from the beginning through the message logs.

Recent MPICH release also provides checkpoint/restart capability (Duell & Hargrove, 2003; Sankaran et al. 2003). Its method is achieved completely by integrating BLCR routine, which means that it supports fault tolerance by storing and recovering every process's memory image.

FT-MPI (Dewolfs et al. 2006; Fagg & Dongarra, 2000; Fagg & Dongarra, 2004) explores the approach of modifying some of the standard MPI

semantics. However, modifying too much semantics may sacrifices too much in the area of time-tested semantics of MPI objects and functions to be realistic for writing production applications.

CPPC (Rodrıguez et al. 2006; Rodrıguez et al. 2010) supports certain degrees of fault tolerance for widely used SPMD programs, which uses compiler technology auto-detect "safe point" (where guaranteed that there are no in-transit, nor ghost messages) and stores certain process's data used for recovery. Other non-portable states created in the original execution, such as MPI communicators, are recovered by code re-execution, the same means used originally to create them.

Above all, what we want is the implementation of variable-level fault tolerance with better characteristics, for simplicity we focus on the widely used SPMD programs and limit our problem on the stopping model. So we specially analyze these two typical methods, naming MPICH BLCR approach and CPPC, which is respectively in segment-level and variable-level, and find out what should be done in future variable-level fault tolerant systems.

3 PROBLEMS AND SOLUTIONS

It is common belief that segment-level checkpointing mechanisms can't be the widely used methods for providing fault tolerant ability in future HPC systems. So we focus on the two common checkpointing implementations: MPICH BLCR and CPPC, find their different problems encountered and corresponding solving ways in these two types of approaches.

For checkpointing MPI systems, fault tolerance is achieved by storing necessary data and rollback-recovery method. When failure occurs, after the recovery of the failed application, the accurate execution of the application should be ensured. To achieve this goal, these two approaches are different in implementation strategy, providing global consistency, and execution recovery.

3.1 Implementation strategy

Current MPICH release is implemented according to the MPI-2 standard, which has no considerations on fault tolerance. However, widely used BLCR routine was added to the current MPICH release, which could provide segment-level fault tolerance. When MPI program runs, the system can checkpoint the program and then recover the original program using the checkpointed files if failure occurs.

However, CPPC has two different phases: compile time and runtime. CPPC approach precompiles the parallel MPI program and adds user-inserted directives into a fault-tolerant program at compile time. Then at runtime, the states of running application will be dumped into checkpoint file.

MPICH BLCR (BLCR, 2011) approach stores the whole memory image of the program, which conducts the dumped state file large and non-portable. Meanwhile, CPPC approach works at variable-level storing only portable file formats, which makes it possible to use portable file formats, like HDF-5. Of course, there is still storage problem of non-portable states, which will be discussed later.

3.2 *Global consistency*

Storing the state of a parallel application and restarting it later is more complicated than in the sequential case. Considering the passing messages storage, there are two situations that require actions to be performed in order to achieve a correct restart: existence of in-transit messages (sent but not received), and existence of ghost messages (received but not sent) in the set of processes states stored. Figure 1 illustrates these message types.

In MPICH BLCR checkpointing, every process's memory image of the whole program is saved, which provides the exact re-execution ability of the failed program, so this approach didn't need to consider the above situation. When the failure occurs, just reloading the dumped file into memory and then the whole program would correctly execute as original.

However, the CPPC approach doesn't store the program's whole memory image which leads to this mechanism must take the above two situations into consideration. A universal but complex algorithm is proposed in (Lynch, 1996), which classifies passing messages according to their epochs into three categories: late (in-transit) message; early (ghost) message; intra-epoch message. Then design special message protocol according to these three message types and provide global consistency.

The approach in CPPC uses a different way. For its application limit to SPMD programs, so it can insert checkpoints when compiling time in the "safe point", where has neither in-transit nor ghost messages, through communication analyze of well-organized MPI programs. By this way, CPPC saved a lot of efforts to ensure global consistency, which also limit its application scope.

3.3 *Execution recovery*

If certain failure of normal execution program occurs, the checkpointed program should be able to recover the normal execution of the failed program, which includes: data recovery and computing recovery. Data recovery includes the data structure in applications and the data structure in MPI library. Computing recovery is the execution logic how to handle these data structure.

For this problem, the segment-level checkpointing mechanisms are rather simple, as these methods store the whole memory image of the application and don't care which part is the data or execution logic, all they know is after loading all the memory image, the application will restart just as it didn't occur failure.

However, the variable-level checkpointing methods save only restart-relevant and portable states to stable storage. So these methods must provide certain way to rebuild the execution logic how to handle these data structure stored and the data structure in MPI library. For example, CPPC rebuilds the execution logic and data structure in MPI library by code re-execution, which is determined at the compile time.

So, the differences between the above two types of execution recovery are obvious. The segment-level mechanisms simplify the implement of recovery but bring in larger space and time overhead. And the variable-level mechanisms like CPPC are complex and agile can achieve better performance.

4 ANALYTICAL PERFORMANCE MODEL

In this section, we will introduce our analytical performance model for the traditional periodic checkpointing and rollback recovery strategy, to illustrate how to achieve the best performance after bringing in the checkpointing mechanism.

4.1 *Model*

Let T be the checkpointing period, i.e., the time between two checkpoints, and there is C seconds of checkpointing every T seconds. From the work in (Wingstrom, 2009), the expected percentage W of time lost, or "wasted", is

$$W = \frac{C}{T} + \frac{T}{2\mu}. \tag{1}$$

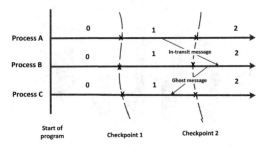

Figure 1. In-transit message and ghost message.

The second term accounts for the loss due to faults and is explained as follows: every μ seconds, a fault occurs and we lose an average of $T/2$ time-steps.

However, this expression for W doesn't account for the recovery time R, so we use a more accurate expression $T_C + T_W$ to replace C, T_C is defined as the cost checkpointing brings except for the time T_W that used to store the checkpointing files. Therefore, a more accurate expression is

$$W = \frac{T_C + T_W}{T} + \frac{T/2 + R}{\mu}. \qquad (2)$$

According to the work in (Young, 1974), W is minimized for

$$T_{opt} = \sqrt{2(T_C + T_W)\mu} \qquad (3)$$

Therefore, we have the corresponding minimum waste

$$W_{min} = \sqrt{\frac{2(T_C + T_W)}{\mu}} + \frac{R}{\mu}. \qquad (4)$$

It turns out that W_{min} may become larger than 1 when μ becomes very small, a situation which is more likely to happen in HPC systems with too many processors. As μ indicating the MTBF of certain HPC systems, is usually a constant, so if we want to reduce the time lost percentage W, the system should have smaller T_C, T_W and R.

5 EXPERIMENTS AND RESULTS

To test the performance between these two mechanisms according to our analytical model, the time and space overheads introduced by checkpointing are evaluated, which includes: the file sizes generated by the checkpointing, the extra time used to perform checkpointing operations. For this test we don't take R into account for this variable takes small part comparing with the periodic checkpointing cost ($(T_C + T_W)$ in expression (4)).

Tests were performed on a cluster of Intel Pentium 1.80 GHz E2160 nodes, 2 GB of RAM, running with Ubuntu 10.04 operating system and connected through Ethernet.

To compare these two types of mechanisms, we select five applications contained in the NAS Parallel Benchmarks (NPB): BT, CG, LU, MG and IS. In order to simplify our tests, the class size of each application is selected by their execution time, which keeps them in a proper range, neither too short nor long.

If we want to compare the dumped sizes of these checkpointing files, we should compare the checkpointing files generated with these two mechanisms when the program executes to the same point. Since the BLCR mechanism uses time interval to guide the creation of checkpoints while CPPC inserts checkpoint directives in the loop of applications to checkpoint the necessary data, so in our experiments, we try to achieve this through carefully control the time interval of BLCR checkpointing.

5.1 State file sizes

In each NPB application, we generated two checkpoint files, their test results are showed in Figure 2.

We can see that using variable-level mechanism (CPPC) generates much smaller checkpointing files in most tests, especially in BT, CG and IS, these applications have less variables to store. Smaller generated state files will save the I/O time consumed (T_W) as well as disk space, and less file I/O operations would reduce large consumed time especially in large scale supercomputers.

We can also see the sizes of checkpointing files are relative with the tested applications, and the advantage of CPPC is more obvious when the application is less compute-intensive. Contrast with CPPC, the sizes of BLCR checkpointing files only depend on the problem size of the application, that is, bigger problem sizes generate bigger checkpointing files. And we can find that when using with applications like BT, the saved space to store state files is very considerable.

5.2 Checkpoint overhead

The time overheads introduced by these two fault tolerant mechanisms are showed in Figure 3.

We can easily know that T_W of CPPC is less than BLCR approach, however, the result shows CPPC didn't win all the tests and its time advantages are

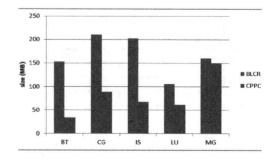

Figure 2. State file sizes.

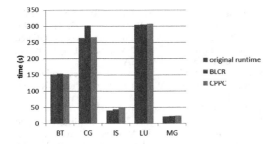

Figure 3. Checkpoint overhead.

not so obvious comparing with space performance. From expression (4), we can see that T_C of CPPC is larger than BLCR's. T_C of CPPC includes: the time used to determine what portable states should be stored; the time used to store these chose states; and the time used to wait for "safe point". Meanwhile, T_C of BLCR is approximate to zero. So when the program generates lots of data when execution (less execution time, larger state file size), the result of CPPC is less advantage.

So if current mechanisms of variable-level checkpointing want to get better performance, T_C used to determine the necessary states must be reduced and the global consistency should be ensured using better strategy other than CPPC. And when the applications have fewer states to store, the variable-level checkpointing mechanisms could get better performance.

6 CONCLUSION

With the development of HPC systems, fault tolerance in MPI programs must find a proper and efficient mechanism to ensure the correctness of long time execution. In the early times, these segment-level approaches like BLCR are widely used for their implement simplicity and mature technique. However, with the higher demand of fault tolerant ability, the approaches using other mechanisms like CPPC in variable-level are received more and more attention for its low overhead and flexible implements.

In this paper, we illustrated different critical issues that must be solved to achieve fault tolerance in two types of different current MPI systems. Then after proposing an analytical performance model to evaluate the performance of checkpointing mechanisms, we evaluated and analyzed the performance of two classic fault tolerant approaches. And we found better global consistency mechanisms and better ways to determine necessary states in variable-level fault tolerant mechanisms are needed. In our future work,

we want to implement a better a better variable-level fault tolerant MPI system, which is easily used and provide enough fault tolerant support for MPI programs.

REFERENCES

BLCR http://ftg.lbl.gov/CheckpointRestart/Checkpoint Restart.shtml.
Bosilca, G., Bouteiller, A., Cappello, F., Djilali, S., Fedak, G., Germain, C., Herault, T., Lemarinier, P., Lodygensky, O., Magniette, F., Neri, V., Selikhov, A. 2002. MPICH-V: Toward a scalable fault tolerant mpi for volatile nodes. In Proceedings of the 2002 ACM/ IEEE conference on Supercomputing, pp. 1–18, Los Alamitos, CA, USA.
Bouteiller, A., Cappello, F., Hérault, T., Krawezik, G., Lemarinier, Pierre., Magniette, F. 2003. MPICH-V2: a Fault Tolerant MPI for Volatile Nodes based on the Pessimistic Sender Based Message Logging. In proceedings of The IEEE/ACM SC2003 Conference, Phoenix USA.
Bouteiller, A., Herault, T., Krawezik, G., Lemarinier, P., Cappello, F. 2006. MPICH-V: a Multiprotocol Fault Tolerant MPI. In International Journal of High Performance Computing and Applications. 20(3):319–333.
Bouteiller, A., Ropars, T., Bosilca, G., Morin, C., Dongarra, J. 2009. "Reasons for a pessimistic or optimistic message logging protocol in mpi uncoordinated failure recovery," in CLUSTER, pp. 1–9.
Bronevetsky, G., Marques, D., Pingali, K., Stodghill, P. 2003. Automated application-level checkpointing of mpi programs. In Principles and Practice of Parallel Programming.
Chakravorty, S., Mendes, C. & Kale, L. 2006. Proactive fault tolerance in mpi applications via task migration. In International Conference on High Performance Computing.
Dewolfs, D., Broeckhove, J., Sunderam, V., Fagg, G. 2006. FT-MPI, Fault-Tolerant Metacomputing and Generic Name Services: A Case Study. Lecture Notes in Computer Science, Springer Berlin/Heidelberg, ICL-UT-06–14, Vol. 4192, pp. 133–140.
Duell, E.R.J., Hargrove, P. 2003. The design and implementation of berkeley lab's linux checkpoint/restart. Technical Report publication LBNL-54941, Berkeley Lab.
Du, Y.F. 2008. 容错并行算法的研究与分析, 博士学位论文, 国防科技大学.
Elnozahy, E.N., Alvisi, L., Wang, Y., Johnson, D. 2002. A survey of rollback-recovery protocols in message-passing systems. ACM Computing Surveys., 34(3):375–408.
Fagg, G. & Dongarra, J. 2000. "FT-MPI: Fault tolerant MPI, supporting dynamic applicaions in a dynamic world," in 7th Euro PVM/MPI User's Group Meeting 2000, vol. 1908/2000. Balatonfred, Hungary: Springer-Verlag Heidelberg.
Fagg, G.E. & Dongarra, J.J. 2004. Building and using a fault tolerant MPI implementation, International Journal of High Performance Computing Applications 18, 3, 353–361.

Genaud, S. & Rattanapoka, C. 2007. Fault management in P2P-MPI. In In proceedings of International Conference on Grid and Pervasive Computing, GPC'07, LNCS.

Litzkow, M., Tannenbaum, T., Basney, J., Livny, M. 1997. Checkpoint and migration of UNIX processes in the condor distributed processing system. Technical Report 1346, University of Wisconsin-Madison.

Lusk, E. 2002. Fault Tolerance in MPI Programs. Special issue of the Journal High Performance Computing Applications, 18:363–372.

Lynch, N. 1996. Distributed Algorithms. Morgan kaufmann, San Francisco, California, first edition.

Rodriguez, G., Martin, M.J., Gonzàlez, P., Touriño, J. 2006. Controller/Precompiler for Portable Checkpointing IEICE Transactions on Information and Systems, E89-D(2):408–417.

Rodrıguez, G., Martın, M.J., González, P., Touri˜no, J., Doallo, R. 2010. "CPPC: A Compiler-Assisted Tool for Portable Checkpointing of Message-Passing Applications"; Concurr. Comp.: Pract. Exp.

Sankaran, S., Squyres, J.M., Barrett, B., Lumsdaine, A., Duell, J., Hargrove, P., Roman, E. 2003. The LAM/MPI checkpoint/restart framework: System-initiated checkpointing. In Proceedings, LACSI Symposium, Sante Fe, New Mexico, USA.

Smith, J.M. 1988. A survey of process migration mechanisms. SIGOPS Operating Systems Review, 22(3):28–40.

TOP500. 2011. http://www.top500.org.

Walters, J. & Chaudhary, V. 2009. "Replication-Based Fault Tolerance for MPI Applications," Ieee Transactions On Parallel And Distributed Systems, Vol. 20, No. 7.

Wingstrom, J. 2009. "Overcoming The Difficulties Created By The Volatile Nature Of Desktop Grids Through Understanding, Prediction And Redundancy," Ph.D. dissertation, University of Hawai'i at Manoa.

Young, J.W. 1974. "A first order approximation to the optimum checkpoint interval," Communications of the ACM, vol. 17, no. 9, pp. 530–531.

Control Engineering and Information Systems – Liu (Ed)
© 2015 Taylor & Francis Group, London, ISBN 978-1-138-02685-8

Preparing a two column paper with MS Word for Windows

B.A. Chen, L.L. Liu & S.Y. Deng
Zhong Nan University of Economics and Law, Wuhan, China

ABSTRACT: Considering the non-coal mines' serious safety problems and the defects in traditional assessment method, this paper develops an on-line assessment system using information technologies. It sets up a sensible assessment index system, and uses Analytic Hierarchy Process (AHP) to calculate each index's weight. Then by using distributed computing, heterogeneous data integration and other technologies, this paper develops an information system, which can accomplish the on-line assessment, statistics and analysis of non-coal mines' safety production. Experiments show that, this system makes the assessment easier and more standard. It provides a good solution for the standardization construction of non-coal mines' safety production.

1 INTRODUCTION

Non-coal mining industry is an important and basic industry in our country. Up to 2005, there are more than 100 thousands non-coal mines in China (Zhou, 2007). Along with economy's rapid development, the demand of mineral resources is increasing. While, non-coal mines in China are diverse and widely distributed. Mining technical equipment and technology is backward, employees' safety consciousness is poor, and basic management is weak. Therefore, there were many mining production accidents in China, some of which are serious accidents (Wang, 2005). Production safety concerns people's life and property safety, which influence reform and society's stability. Thus, setting up a good assessment and management system, prompting the standardization construction of safety production is a task which needs to be done urgently.

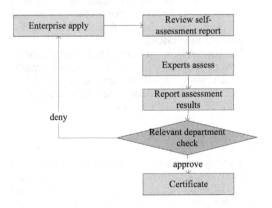

Figure 1. General process of traditional assessment.

The traditional assessment process of safety production mainly contains experts' on-site inspection, discussing, and assessing. The general process is shown in Figure 1.

In the process of traditional assessment, experts' on-site inspection is the most important step. It needs a group of experts to inspect the enterprise's safe management, emergency management, and main production systems seriously and deeply. And it requires the experts to give a totally objective result after that. Besides the large amount of manpower, material and financial resources consumption, it's inefficient and nonstandard. And in the traditional method, assessment standard, process, and results are mostly recorded in paper, which is difficult to preserve and find out in future.

Because of these defects, traditional assessment cannot effectively complete the large amount of assessment work. In order to prompt the standardization construction of non-coal mines' safety production, this paper applies information technology to assessment process, develops an automatic and distributed system. It can simplify assessment process, reduce resource consumption, and improve work efficiency. It provides enterprise and supervision departments with a better assessment and management tool.

In the rest of this paper, it first describes the system's function modules in section 2, then sets up an index system in section 3, and introduces the key technologies in section 4. Section 5 is a conclusion.

2 FUNCTION MODULE FRAME

Safety production assessment system contains experts' management, assessments' management,

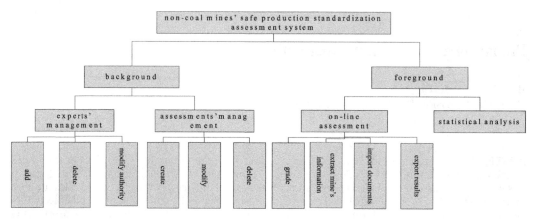

Figure 2. Function module frame.

on-line assessment, and statistical analysis of assessment results. These modules are shown in Figure 2.

Assessment system can be divided into two parts, background and foreground. The background is used by manager to manage information of experts and assessments. And the foreground realizes experts' on-line assessing and assessment results' statistical analysis. Before experts execute their assessments, system will extract some useful information from the related existing information systems and provide references.

3 ASSESSMENT INDEX SYSTEM

According to some related legal norms and technical standards, this paper formulates a two-level assessment index system for non-coal mines' safety production assessment. It contains 14 top level indexes and 53 second level indexes (Hou, 2009). They are shown in Table 1.

4 KEY TECHNOLOGIES

4.1 Distributed processing

Because of non-coal mines' large quantity and wide distribution in China, assessment of them involves the storage, process, and analysis of mass data. Traditional integrated data processing cannot complete this. Here, this system use distributed processing technology to realize the cooperation and information sharing of computers in different locations. This technology divides a complex problem into small parts, assigns them to many computers, and combines results from these computers to get the final result (Wu, 1999).

Compared with the traditional processing method, distributed processing has the following advantages:

1. Scare resources can be shared.
2. Computing load can be balanced on multiple computers.
3. Program will be run on the most suitable computer.

The standardization assessment system of non-coal mines' safety production use Jini[8] to realize the distributed processing of information. Jini provides facilities for dealing with some of the fallacies of distributed computing, problems of system evolution, resilience, security and the dynamic assembly of service components. Code mobility is a core concept of the platform and provides many benefits including non-protocol dependence. A running Jini system consists of three components:

1. The Jini Client—anything that would like to make use of the Jini service
2. The Service Locator—anything that acts as a locator/trader/broker between the service and the client, and is used to find services in a distributed Jini system.
3. The Jini Service—any entity that can be used by a client program or another service (for example, a printer, a DVR (this used to say "VCR"— that's how old this software is), or a software entity like an EJB service).

4.2 Web service

Assessment system of non-coal mines' safety production needs to integrate these mines' existing information resources. These resources are stored in application systems independent of each other. Each non-coal mine is a data source. And these

Table 1. Assessment indicators table.

Top level indexes	Second level indexes
Safety production's policies and goals	Policies, goals
Safety production's laws, regulations, and other requirements	Consciousness of laws and regulations, recognize and obtain demands, integration, assessment and update
Safety production's organization guarantee	Safety production responsibility system, safe organization settings and staff appointment, staff participation, document control, outside relationship and inside communication, assessment of system management, supplier and contractor management, safety approval and reward, safety management after work
Risk management	Requirements of risk sources recognition and risk evaluation, risk evaluation, recognition and analysis of mission critical
Safety education and training	Staff safety consensus, training
Safety management of production technology system	Design requirement, mining technology, production guarantee system, change management
Equipment management	Basic requirements, equipment maintaining
Safety management of working site	Working conditions, working process, labor protection articles
Occupational health management	Health guardianship, occupational hurts control, occupational hygiene monitoring, work efficiency of person and machine
Safety investment, safety technology and work-related injury insurance	Safety investment, safety technology, work-related injury insurance
Inspection	Common requirements, patrol inspection, routine inspection, professional inspection, comprehensive inspection, corrective and preventive actions
Emergency management	Emergency preparedness, emergency plan, emergency reaction, emergency guarantee, emergency assessment and improvement
Accident, accident report, investigate and analysis	Report, investigate, statistical analysis, accident review
Performance measurement and assessment	Performance measurement, system assessment

data sources are heterogeneous. So, it's a key problem to use these resources and build a integrate assessment system. To avoid rebuilding non-coal mines' existing information systems, this paper use Web Service to process theses heterogeneous information.

Web Service is an important technology used for integrating applications. It's a new platform for building interoperable and distributed application. It's generally considered to be a new application providing an API which can be called through web. It's self-contained, self-described, and modularized. And it can be published, searched, and invoked through web (Dignum, 2009).

4.3 Ontology

Besides the different storage schema, data in different information systems may have semantic conflict. For example, two different words may have the same meaning. So, the integration result may contain many redundancies, which will interfere with the data processing. This paper uses Ontology technology to resolve this problem. By building and using non-coal mines' domain ontology, the system can effectively avoid semantic conflicts during integration (Li, 2011).

5 CONCLUSION

According to related laws, regulations, and technique specifications, this paper formulates an index system. It contains 14 top level indexes and 53 second level indexes. Using AHP, they are given weights and used to calculate standardization degree of non-coal mines' safety production. Besides, instead of on-site assessing, this paper develops an assessment system. It allows experts to conduct assessment anywhere and anytime. And it realizes statistics, analysis, and management function. It provides an effective and practicable solution for standardization construction of non-coal mines' safety production.

Because the research of this paper is still in its infancy, the assessment system is not perfect. In future studies, we will focus on distributed data processing and ontology technologies. And find out how to process data more quickly and how to create domain ontology automatically. It will be a meaningful research direction.

REFERENCES

Andrew S Tenanbaum, Maarten van Sten, Distributed Systems: Principles and Paradigms, Prentice Hall, London, 2006.

AnHai Doan, Jayant Madhavan, et al. "Learning to match ontology on the Semantic Web", The VLDB Journal, pp. 303–319, 2003.

Chu, W.W. "Task Allocation in Distributed Data Processing", Computer, pp. 57–69, 1980.

Frank Dignum, Virginia Dignum, et al. "Organizing web services to develop dynamic, flexible, distributed systems", Proceedings of the 11th International Conference on Information Integration and Web-based Applications & Services, pp. 225–234, 2009.

Hou Qian, Wang Yunhai, et al. Construction and Evaluation of the Safety Standardization in Metal and Nonmetal Mines [J]. Metal Mine, pp. 140–148, 2009.

Jie Wu, Distributed system design, CRC press, United States, 1999.

Li Sheng, Technique of Semantic Desktop Search, Wuhan University of Technology Press, China, 2011.

Mei Yuan, Guiyi Wu, Jianxi Li, "The application of AHP in coal mines", Mining machinery, pp. 43–45, 2008.

State Production Safety Supervision and Management Administration. Evaluation of the Safety Standardization in Metal and Nonmetal Mines [S], 2009.4.

Wang Qiming, "Situation of Safety Production in Non-coal Mines and Its Problems and Countermeasures", Metal Mines, pp. 1–4, 2005.

Zhou Jianxin, Zhang Xingka, et al. "Status and Role of Safety Standardization in Safety Production of Non-Coal Mine", Metal Mines, pp. 1–5, 2007.

Research on Prognostic and Health Management technology of Unmanned Aerial Vehicle

G.S. Chen & S.S. Ma
Institute of Ordnance Technology, Shijiazhuang, P.R. China

ABSTRACT: Prognostic and Health Management (PHM) is an advanced technology of test and maintenance for Unmanned Aerial Vehicle (UAV). PHM is a new type maintenance mode; PHM system is based on open system architecture for Condition Based Maintenance (CBM), and is critical for improving mission perform capability and reducing maintenance costs of UAV. The paper studies the effectiveness and practicability of PHM technology applied to UAV maintenance support and designed PHM system structure in maintenance support for UAV. This paper innovatively presents a methodology using trend analysis with regression to build the reference indicator, and produces a degradation model to health monitoring UAV system. This method is concerned with trend analysis for estimation of the future condition of the system and prediction of the time-to-failure. The proposed approach is effective and can be easily applied to UAV systems, for which health state indicator are monitored.

1 INTRODUCTION

With the increasing complexity and powerful function of UAV, more research has become increasingly concerned about reducing the operational and maintenance costs. To meet this demand, UAV and systems are always needed for new technologies and procedures to incorporate into their maintenance.

In recent years, Prognostics and Health Monitoring (PHM) has been focus of research with the challenge of monitoring the life of equipment and systems. This technology can lead to potentially considerable savings in equipment maintenance, reducing the number of delays and augmenting safety. The knowledge of the current and the prediction of the future health state of components in equipment may guide the maintenance activities and spare parts logistics [1].

This task is currently and gradually being accomplished, as more and more field data become available, correlated with the failure reports and the maintenance actions history. PHM is based on the concept of CBM and is about making equipment more powerful by making them more maintainable, and will probably represent an important competitive advantage in the near future [2].

In order to determine the health status of a UAV, two types of information are mandatory: the indication of the current state of UAV and a reference indicator. The indication of the current state of UAV is found through the parameters associated with failure modes related to the degradation of the UAV. The reference indicator can be obtained from a model that represents the physical limits or a known tolerance threshold of the UAV can be used. Often, such reference value does not exist or it is not available to estimate the UAV life consumption. This fact leads to arbitrary definitions of degradation [3].

In situations where sophisticated prognostic models are not warranted due to the lower level of criticality or low rates of failure occurrence and/or there is an insufficient sensor network to assess health condition, a statistical reliability or usage-based prognostic approach may be the only alternative. This form of prognostic algorithm is the least complex and requires component/LRU failure historical data of the fleet and operational usage profile data. Although simplistic, a statistical reliability-based prognostic distribution can be used to drive interval-based maintenance practices that can then be updated on regular intervals [4]. Typically, fault and inspections data are compiled, and employed to fit a probability distribution of time-to-failure. Weibull distributions are typically employed for this purpose [5].

The paper aims to present a probability-based methodology using regression to build the reference health value and produce a degradation model to be used for health monitoring on UAV systems.

2 SYSTEM FUNCTION OF UAV

A key difference between a UAV and a conventional aircraft is that the UAV is part of a total system comprising aerospace vehicle sub-system, control and navigation sub-system, integrated radio sub-system and mission facility sub-system, and every sub-system themselves is consisted of some devices and parts, each with specific maintenance requirements [6]. The topological structure map of UAV system must be build, so as to confirm the fault propagation path of devices. And on the foundation of this map, the structure model of UAV system is established, which can be seen in Figure 1.

In which, the health management of system level mainly acquires the health state and its development trend of the whole UAV system, and carries out maintenance decision based on the health condition of UAV, and then forms maintenance advice for the maintainers.

The health management of sub-system level performs relativity analysis with all classes of condition data, and solves the inconsistency for those data, then confirms and isolates fault, and receives more reliable health information of sub-system. As the real time actuator, the health management of device level acquires the sensor information of key parts, and carries out fusion treatment to realize abnormality detection, enhanced fault diagnosis, health assessment, and life-time prediction. The health management of key parts level actualizes security monitoring and fault early-warning through reading sensor data real-time of key parts and performing feature extraction for those data.

This paper does not enumerate all sub-system failure modes and health assessment method, only to fuel sub-systems of engine which is the key part of aerospace vehicle sub-system, has a certain reference; the remaining sub-systems can refer to the analysis and evaluation methods for analysis.

From the perspective of function, the modern UAV fuel system was composed of fuel supply sub-system, fuel delivering sub-system, refueling sub-system, and the host computer controller sub-system, the lower fuel delivering controller, lower fuel supply controller and so on.

From the common faults of fuel system of UAV occurring in practical application, its fault is generally manifested in two aspects: First, the fault of the fuel components, namely the fault occurred in UAV fuel delivery process, caused by a mechanical failure due to the operational parts (mainly for pumps, valves, regulators, and sensors). Second, the sub-system failure (such as controllers), and that the sub-systems multiple attachments failure, caused the sub-systems decline or loss.

3 PHM WITH TREND ANALYSIS METHOD

Regression is a generic term for all methods attempting to fit a model to observed data in order to quantify the relationship between two groups of variables. In statistics, regression analysis refers to techniques for the modeling and analysis of numerical data consisting of values of a dependent variable and of one or more independent variables [7]. The fitted model may then be used either to merely describe the relationship between the two groups of variables, or to predict new values.

Empirically, the relationship between the time t and a degradation index y can be approximated as

$$y = \beta_0 + \beta_1 t + \varepsilon \tag{1}$$

where β_0 is the linear coefficient, β_1 is the angular coefficient and ε is the model residue. The coefficients β_0 and β_1 are unknown but constant. The residue is also unknown, but it is different for each y observation. Estimates of β_0 and β_1 can be obtained by least-squares. It may be argued that the long-term evolution of degradation with time will seldom be linear, as assumed in Eq. 1. However, this simplified model may be adequate for medium and short-time predictions, which may be enough for CBM purposes.

For estimation purposes, the following notation will be employed:

$$b_0 = \widehat{\beta_0}, b_1 = \widehat{\beta_1}, \hat{y} = b_0 + b_1 t \tag{2}$$

where $\hat{y}$ denotes the predicted value of y for a given t. Thus, given a failure threshold L for the degradation index, the predicted time of failure $\hat{t}_f$ can be calculated as

$$\hat{t}_f = \frac{L - b_0}{b_1} \tag{3}$$

Once the coefficients b_0 and b_1 have been obtained, it is interesting to analyze whether the

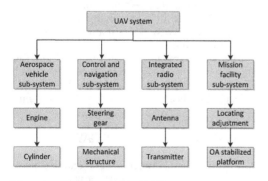

Figure 1. Structure model of UAV.

linear trend observed in the data is statistically significant, as described below.

The significance of regression can be assessed by an Analysis of Variance (ANOVA). Such an analysis is based on the following identity:

$$\sum_{i=1}^{n}(y_i - \overline{y})^2 = \sum_{i=1}^{n}(\hat{y} - \overline{y})^2 + \sum_{i=1}^{n}(y_i - \hat{y}_i)^2 \quad (4)$$

where:

$$\overline{y} = \frac{1}{n}\sum_{i=1}^{n}y_i \quad (5)$$

and n denotes the number of available observations. The terms in Eq. 4 can be disposed in standard ANOVA format as shown in Table 1.

The significance of the regression can be evaluated by comparing MSReg with s2. Under usual assumptions for linear regression [8], it can be shown that the ratio

$$F = \frac{MSReg}{s2} \quad (6)$$

follows an F-distribution with 1 and $(n-2)$ degrees of freedom if $\beta_1 = 0$. This fact can thus be used as a test for the null hypothesis H0: $\beta_1 = 0$ (no linear trend), versus H1: $\beta_1 \neq 0$. Comparing the ratio (6) with the $100(1-\alpha)$% point of the tabulated F $(1, n-2)$ distribution, it is possible to determine whether $\beta 1$ can be considered nonzero. In addition, a p-value can be obtained as the area under the tail of the distribution $F(1, n-2)$ for $F > F_{calculated}$. The p-value can be interpreted as the chance that the observed trend is fortuitous.

Under the usual assumptions for linear regression, the estimate b_0 follows a normal distribution with mean β_0 and variance given by

$$\sigma^2\left[\frac{1}{n} + \frac{\overline{t}^2}{S_{TT}}\right] \quad (7)$$

σ^2 is the variance of the residue ε, $\overline{t}$ is the mean value of t. The mean and variance of estimated b_0 and b_1 are given by:

$$b_0 \sim N\left(b_0, \sigma^2\left[\frac{1}{n} + \frac{\overline{t}^2}{S_{TT}}\right]\right) \quad (8)$$

$$b_1 \sim N\left(\beta_1, \frac{\sigma^2}{S_{TT}}\right) \quad (9)$$

S_{TT} is:

$$S_{TT} = \sum_{i=1}^{n}(t_i - \overline{t})^2 \quad (10)$$

Eq. 7 shows that the accuracy of the estimates may be improved by reducing the noise level, increasing the number of observations and increasing the timespan of the historical record. A covariance value for b_0 and b_1 can also be calculated. As in practice the value of σ is not known, it can be replaced with the standard error (s) obtained as

$$s = \sqrt{\frac{RSS}{n-2}} \quad (11)$$

In this case, a Student-t distribution for b_0 and b_1 would have to be used. However, for large values of n, the Student-t distribution may be approximated by a normal distribution.

In order to obtain a probability distribution for the time of failure t_f, it is possible to use a Monte Carlo simulation method, a class of algorithms based on repeated random sampling. Using a multivariate random number generator, it is possible to obtain random values of b_0 and b_1 that are distributed as described above. The random number generator can also be employed to obtain different realizations of the residue ε over a sequence of future time instants. For each pair (b_0, b_1) and each realization of the residue, a sequence of future degradation values is obtained. The failure time is then defined as the earliest instant for which the degradation is larger than the failure threshold L.

By repeating this kind of simulation, it is possible to obtain a probability distribution around the predicted failure time, and assuming that this distribution is Normal or Weibull, for instance, the confidence interval can be obtained.

Table 1. Terms involved in the analysis of variance for the regression.

Source of variation	Degrees of freedom	Sum of Squares (SS)	Mean Square (MS)
Due to regression	1	$SSReg = \sum_{i=1}^{n}(y_i - \overline{y})^2$	MS_{Reg}
About regression (residual)	$n-2$	$RSS = \sum_{i=1}^{n}(\hat{y}_l - \overline{y})^2$	S^2
Total	$n-1$	$S_{YY} = \sum_{i=1}^{n}(y_i - \hat{y}_l)^2$	S_y^2

4 SIMULATION STUDY

In order to illustrate the technique described above, a case study involving actual data from a UAV fuel system will be presented. The degradation index was obtained from features related to measurements of fuel content in equipment. In this case, the degradation represents the fuel consumption. Appropriate prognostics of such consumption would be of value to schedule a maintenance intervention (replenishment of the reservoir) for a convenient time. The data consists of a historical record of 250 values for degradation index, acquired over time.

An algorithm was used in order to provide the results. Briefly, it first calculates values for b_0 and b_1, by least square linear regression method, refer to Eq. 2. Secondly, values for MS_{Reg} and s^2 are obtained, refer to Table 1. An F-test was initially carried out to assess the significance of the linear regression. The critical value of the F distribution, for $\alpha = 0.05$, was calculated as $F_{crit} = 3.9$. The F-value obtained by using Eq. 5 was 1727, which is considerably larger than F_{crit}. Therefore, the null hypothesis (there is no linear tendency) can be rejected with $100(1 - \alpha)\% = 95\%$ of confidence. The associated p-value was very close to zero, which means that one can be quite sure that the trend observed in the data record is not fortuitous.

Latter, to obtain the probability distribution of the estimated coefficients, the algorithm uses Eq. 8, 9 and 10. In addition, the failure time forecasts are simulated by Monte Carlo technique. For each interaction, a failure time is taken from a linear extrapolation with noise. A vector with the simulated failure times is provided.

The probability densities obtained by the proposed technique may be used, for instance, to calculate the probability of failure within a given time window in the future. Such a result could be used to drive a cost-benefit analysis in order to establish maintenance recommendations.

5 CONCLUSION

This paper has briefly reviewed PHM and the UAV structures, describing in more detail the theoretical and notional foundations associated with regression-based predictions. Moreover, a regression-based technique was presented and illustrated by a simulation study simulated with real data from a UAV system. The failure instant was calculated and the confidence associated with this estimation was evaluated by using a Monte Carlo technique.

The ultimate target is to enhance the armament operational readiness and maintenance support efficiency, to decrease the maintenance cost, and to realize the information-based autonomous logistics support of UAV system.

REFERENCES

Draper, N.R. Smith, H. 1998. Applied regression analysis. 3rd edition, Wiley-Interscience.

Gregory J.K, Michael J.R. 2002. Health Management System Design: Development Simulation and Cost/Benefit Optimization [J]. IEEE, 3065–3072.

Groer, P.G. 2000. Analyses of time-to-failure with a weibull model. In: Maintenance and Reliability Conference (MARCON), 2000, Knoxville. Proceedings.

Keith M.J., Raymond R.B. 2007. Diagnostics to Prognostics—A product availability technology evolution. The 53rd Annual Reliability and Maintainability Symposium (ARMS 2007), Orlando, FL, USA:113–118.

Nishad P., Diganta D., Goebel K., 2008. Identification of Failure Precursor Parameters for Insulated Gate Bipolar Transistors (IGBTs). 2008 International Conference on Prognostics and Health Management (PHM 2008), Denver, CO, USA, 2008: 1–5.

Pendse R., Thanthry N., 2009. Aircraft health management network: A user interface[C]. IEEE Aerospace and Electronic Systems Magazine, New York.

Taylor, R.M., Abdi, S., Drury, R. & Bonner, M.C. 2001. Cognitive Cockpit Systems: Information Requirements Analysis for Pilot Control of Automation. Engineering psychology and cognitive ergonomics, 5, 81–88.

Venkatasubramanian, V. 2003. A review of process fault detection and diagnosis: part 1: quantitative model based-methods, Computers & Chemical Engineering, v. 27, n. 3, p. 293–211.

Control Engineering and Information Systems – Liu (Ed)
© 2015 Taylor & Francis Group, London, ISBN 978-1-138-02685-8

Designing software architecture to support knowledge integration under heterogeneous sources environment

B. Ma & Q. Liu
Glorious Sun School of Business and Management, Donghua University, Shanghai, China

ABSTRACT: With the gradual development of enterprise information systems, the enterprises' knowledge is in the environment of heterogeneous sources. This paper analyzes the barriers to knowledge integration under such condition. It is pointed that software systems, as support of the knowledge integration, is the key enabler of knowledge integration, and that the software usability is the major issue of the system design. In addition, the relationship between the software usability (focused on knowledge aggregation, knowledge reuse and personalization) and the software architecture is discussed. Furthermore, recommendations for the software architecture design are provided to support effective knowledge integration.

1 INTRODUCTION

Knowledge has become the source of enterprise competitiveness. Technology cannot be sustainable sources of competitive advantages. But the advantage of knowledge can be sustained, because knowledge can create profit and advantages (Davenport & Prusak 1998). The scholars at the Harvard University believe that current enterprise management has entered the stage of globalization and become more knowledgeable, and in such stage the sustainable growth turns out to be the management objective. In knowledge management, the topic of management, the ability of knowledge integration and application becomes the sticking point of enterprise in creating and sustaining its competitiveness (Grant 1996, Brown & Duguid 1998). Pisano (1994) considered that enterprise can promote its competitiveness through effective use of internal knowledge sources. If knowledge is not shared or integrated, its value for the organizational objectives can be very limited. Therefore, knowledge integration and application pose a great challenge to knowledge management. Effective knowledge integration and sharing plays a key role in the long-term development of the enterprises. Knowledge integration is strategically important for the enterprises.

Thus, there is an important research question: how to implement the knowledge integration initiatives successfully? The answers are various and involve many influence factors, such as tradition, culture, learning capabilities and information technology etc. (Shankar & Gupta 2005, Chang Lee et al. 2005). Many organizations are becoming increasingly dependent on IT in order to satisfy their business aims and meet their needs. In this paper, we consider the software system should play a very important role and the software usability is the key issue.

The willingness and ability of sharing and using knowledge are generally accepted as main barriers of knowledge integration. So the software systems should have a capability of helping people to overcome those barriers in the satisfied way. It means your systems must be usable because people will refuse to use the systems which are not user friendly, cannot access and use high quality knowledge easily and effectively. So we discuss the most relevant elements of software usability, which are knowledge aggregation, knowledge reuse, and personalization, and try to give some useful recommendations for the software architecture design of knowledge integration support system.

2 HETEROGENEOUS KNOWLEDGE SOURCES AND KNOWLEDGE INTEGRATION

2.1 Heterogeneous knowledge sources

Concerning the definition of knowledge, many literatures have discussed it from different points of view, such as philosophy, economy, and information technology, and strategic management, financial or organizational theory. Among them, there are two main definitions: 1) from information technology aspect, knowledge is defined according to the hierarchical structure of data, information and knowledge; 2) Knowledge is defined from strategic point of view, such as thinking, process and capability. Most scholars believe that knowledge is a personal or organizational understanding of an object, carriers of knowledge is various, and those knowledge and application scenarios are closely linked. In this paper, data, information and knowledge are considered as knowledge in broad sense, each has different carriers, with different structures (structured or non-structured knowledge) and types (explicit or tacit), and all of them belong to heterogeneous knowledge.

Managing knowledge sources is one of the key functions in modern organizations (So & Bolloju 2005). Effective knowledge management ensures that every employee has access to appropriate and the highest quality of information available at the time when a decision needs to be made (Walczak 2005). The heterogeneous knowledge sources environment poses a great challenge to Decision Makers (DMs), because, timely and correct decision depends on the acquisition of the right knowledge under the sharply changing market conditions. Decision failure is usually ascribed to the lack of information, but actually it is due to the lack of knowledge, which can provide insight of overall situation and foresee the knowledge to come. In research of knowledge integration, the major issue is to support DMs to retrieve integrated and appropriate knowledge embedded in the heterogeneous knowledge sources.

2.2 *Knowledge integration*

Until now, there is no generally accepted definition of knowledge integration. It was first used by DEMSETZ in 1991 and first formally proposed by Grant. He believed that knowledge integration, as a vital function of enterprises, should create advantages for the enterprises (Huang & Newell 2003). Knowledge integration implies that timely insights can be made available to be drawn at the right juncture for sense making by the transistors, i.e. knowledge can be exchanged, shared, evolved, refined and be made readily available at the point of need (Badii & Sharif 2003). Some scholars thought that knowledge integration is an ongoing collective process of constructing, articulating and redefining shared beliefs through the social interaction of organizational members (Huang & Newell 2003).

Generally, knowledge integration means identification, acquisition and application of inter- and extra-organizational knowledge related to daily routine or special tasks. The objective of knowledge integration is to overcome the knowledge barriers (or boundaries), solve the knowledge silo problem in organizations, improve knowledge sharing, generate new architectural knowledge and store the knowledge as memory of organization. Knowledge integration provides the support for

the effective (and efficient) knowledge sharing and knowledge application, to improve executive power and competitiveness of the organization.

For an organization or an individual, knowledge integration is not just the simple aggregation of available knowledge, but a process of knowledge innovation. Successful knowledge integration requires various kinds of managerial techniques, including identification and discrimination of learning, clearness of knowledge, etc. Only after real knowledge integration, can the enterprise accomplish knowledge management.

One of the main tasks of knowledge management is to identify the barriers between information technology and knowledge management, and to overcome the barriers (Lang 2001). Knowledge integration is affected by many factors, such as cultural, organizational factors, etc. The key Enabler of knowledge integration in enterprises remains software system, providing support for knowledge integration. From the software requirement of knowledge integration, how to ensure the software usability is the most crucial task. It is decisive whether the knowledge integration support system operates properly, and whether it guarantees efficient, effective and correct access to right knowledge.

2.3 *Software architecture and software usability*

In many literatures, quite a lot concepts and definitions of software architecture have been proposed (Sun et al. 2002). In essence, the core of software architecture describes the organizational structures of a software system, including the description of components, interactions among components and constraints. Interactions among components are the abstract interaction of higher level of the system. The choice of the components should be subject to certain constrains and comply with certain design principles, and evolve in certain environment (Sun et al. 2002). In this paper, software architecture mainly considers the functions of components, interaction among components and architecture, which can meet the requirement of key functions. Concerning the system architecture, much research has been done with respect to hardware, software and networks. This paper discusses the software architecture of the presentation layer and intermediate layer, which facilitates the access and share of knowledge.

Usability means that a product can be used with effectiveness, efficiency and satisfaction (ISO 1998), software usability includes easiness and speed of learning of system use, efficiency to use, reduced number of user errors and subjective satisfaction of users (Nielsen 1993). In recent years, there has been some research about the relationship between usability and architecture. Bosch (2000) described the direct relationship between the architecture decision and capacity, with quality requirement met. Juristo et al. (2004) studied the implementation of usability of system structure. In their view, it is dangerous to consider that the usability only affects the presentation components of software systems, even though the functions are correct and the user interface has been detached by functional components. Architectural decision can limit the capability of usability required for implementation (Hoffman et al. 2005). Bass et al. (2001) have given a collection of architecture patterns, which help software architect to foresee and meet usability requirements.

Most enterprises undergo gradual process of building their information systems, which inevitably leads to a heterogeneous source environment for enterprise knowledge. Not only the knowledge sources are very complex (such as database, document, homepages, even brain of people), but also the way to retrieve knowledge, data or information becomes various. For example, employees can communicate with project team by E-mail, or by using OLAP (On-Line Analytical Processing) system to retrieve data from DW (Data Warehouse), or IRS (Information Retrieval Systems) to search document in enterprise document libraries. Thus, it is more important to provide DMs satisfactory information or knowledge effectively and efficiently, which presents higher requirement for the system design and development.

3 DESIGNING SOFTWARE ARCHITECTURE TO SUPPORT KNOWLEDGE INTEGRATION

3.1 Barriers to knowledge integration

A survey conducted by Gartner Group in 2000 shows, each knowledge employee spends eight hours on dealing with non-value added documents every week. The research by Forrester Research shows that the enterprise information volume increases by 200% each year. On one hand, it is due to the low efficiency in dealing with the documents, on the other hand, it is because of the rapidly growing volume. Knowledge sources exist everywhere, i.e. human, books, articles, homepage and database, which leads to the difficulty in knowledge sources management (Gordan & Holden 2004). Furthermore, in heterogeneous environment, there are different ways to acquire knowledge and information. Thus, the software system should support efficient aggregation and personalization of knowledge, to reduce user access barriers. This can further ensure the users to acquire the right knowledge conveniently. To reflect its value, enterprise knowledge should be shared and reused. Besides the reuse of dominant knowledge, recessive knowledge in brains of employees should also be exchanged in a proper way. Therefore, knowledge integration should incorporate interaction, cooperation and support team-work.

Knowledge integration support system act as facilitator between knowledge sources (tangible and intangible) throughout the organization, and those who require knowledge exist in these sources. Usability of support system for knowledge integration is of great importance. It means that the users can complete their tasks more conveniently and more effectively. During this process, failure should be minimized and the fear of using the systems should be avoided, and do not make user frustrated. For software systems, the challenge is to provide the right knowledge at right time. Accordingly, three core elements of software usability of knowledge integration support system: knowledge aggregation, knowledge reuse and personalization are discussed in the following part. Their influence on the design of software architectures is presented and some practical recommendations are made. The feature models (Kang et al. 1998) and patterns oriented approach (Bass et al. 2001, Adams et al. 2001) are adopted, with design of software architectures recommended to enhance the knowledge access and satisfaction.

3.2 Knowledge aggregation

Problem solving usually requires knowledge from various aspects. The sources of knowledge needed by different problems may differ from case to case, and the knowledge requirement is difficult to predict. Its means need of knowledge is dynamic. The system allows the user to search the contents and presents the content in an organized way. It may also provide the user with information in a proactive manner, e.g. by alerting, notification, etc. (Hasse et al. 2005).

Therefore, software systems, supporting knowledge integration, should show the users how to aggregate knowledge from various sources and how to operate based on that, i.e. generating summary or exporting knowledge. The feature model of knowledge aggregation is given in Figure 1. In order to meet the needs, the search engine and aggregation components are the core of the system. Those two components gathering knowledge from organization knowledge sources according to metadata/ontology, then present them to user with consistent interface. Figure 2 shows the recommended software architecture for knowledge aggregation.

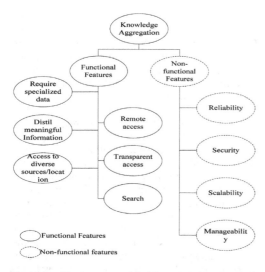

Figure 1.   The feature model of knowledge aggregation.

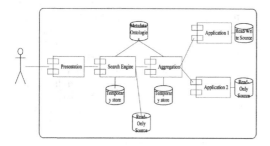

Figure 2.   The recommended software architecture for aggregating knowledge.

3.3   Knowledge reuse

It is believed that employees' knowledge would not be successfully exploited if either knowledge sharing or knowledge reuse is overlooked (So & Bolloju 2005). Enterprise knowledge is distributed in different systems. Users of one system may need to import data from other systems, but users are not granted the access to all systems. Despite the growing awareness of the benefits of knowledge sharing, the accessibility of knowledge is still limited because most knowledge resides in the head of people or in documents or repositories not readily accessible to other (Riege 2005). Consequently, the support for the knowledge sharing and reusability should be firstly strengthened. Secondly, how to acquire and use the knowledge efficiently should be taken into account in the software system design, while new knowledge about how to solve problem by using software components emerges during the use of software. Thirdly, efficient knowledge

integration requires information about who possesses the knowledge, where the expert is located and where the knowledge is called for.

Accordingly, rich communication and collaboration are considered necessary (Alavi & Tiwana 2002). Management initiatives can involve the utilization of expert directories or maps which provide a contact mechanism of seekers of knowledge with the experts who possess the tacit sources (Kudyba 2005). The feature model of knowledge reuse is shown in Figure 3. In this situation, shared applications including email, instant messaging, online conference, and team workplace etc., the server provides Single Sign-On (SSO) and experts contact information through the whole organization. While the workflow components act as organizer of knowledge during the work process, and generate new workflow rules. Figure 4 illustrates the recommended software architecture for knowledge reuse.

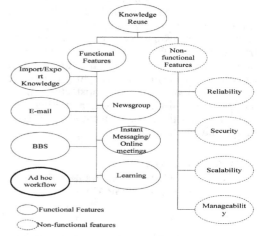

Figure 3.   The feature model of knowledge reuse.

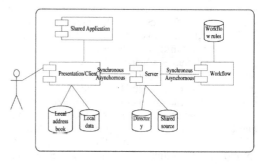

Figure 4.   The recommend software architecture for knowledge reuse.

798

3.4 *Personalization*

With more than 8 billion documents on the web and billions more in corporate and government intranets, personalized information delivery is an important step in providing relevant and timely knowledge to the right people, a key concern of knowledge management (Davies et al. 2005). With a lot of knowledge available, how to acquire the business related knowledge in an effective way is the most concern of the users. In the personalization strategy, knowledge is closely tied to the person who created it and disseminated through person-to-person knowledge sharing networks (Shankar & Gupta 2005).

Therefore, the major issue for the software architect is to provide personalized knowledge for the users. Personalized interface enables the users to construct knowledge application of different levels. Further improved interface can learn and analyze users' habits and fondness to help them make better use of knowledge (Tang & Jin 2000).

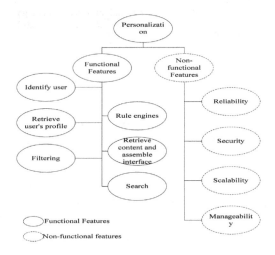

Figure 5. The feature model of personalization.

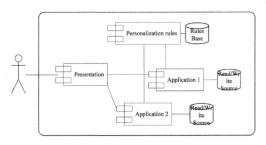

Figure 6. The recommended software architecture for personalization.

The feature model of personalization is shown in Figure 5. The corresponding software architecture for personalization is presented in Figure 6.

4 CONCLUSION

Despite the complexities and difficulties of achieving a successful knowledge integration implementation, seeking a usable software system to support the task is still very important job. As a research hotspot in the field of software engineering, software architecture can provide architectural guidance for the design of software systems, and the framework of software development is given. This paper presents software requirements for knowledge integration, discusses the relationship between software usability and software architecture in the development of knowledge integration support system. The feature models and the recommended software architectures of the three aspects of usability (knowledge aggregation, knowledge reuse and personalization) are provided. Based on the integration of the architectures, future research can be done in usability assessment and design of knowledge integration support system.

ACKNOWLEDGMENT

This work was financially supported by the Fundamental Research Funds for the Central Universities (No. 2232013H-19).

REFERENCES

Adams, J. et al. 2001. *Patterns for e-business: A Strategy for Reuse*. MC Press.

Alavi, M. & Tiwana, A. 2002. Knowledge integration in virtual teams: the potential role of KMS, *Journal of the American Society for Information Science and Technology*, 53(12): 1029–1037.

Badii, A. & Sharif, A. 2003. Information management and knowledge integration for enterprise innovation, *Logistics Information Management*, 16(2): 145–155.

Bass, L. et al. 2001. Achieving Usability through Software Architecture, Technical Report, CMU/SEI.

Bosch, J. 2000. *Design and Use of Software Architecture*. Harlow, England: Addison Wesley.

Brown, J.S. & Duguid, P. 1998. Organizing knowledge, *California Management Review*, 40(3): 90–11.

Chang Lee, et al. 2005. KMPI: measuring knowledge performance, *Information & Management*, 42(3): 469–482.

Davenport, T.H. & Prusak, L. 1998. *Working Knowledge: How Organizations Manage What They Know*. Boston: Harvard Business School Press.

Davies, J. et al. 2005. Next generation knowledge access, *Journal of Knowledge Management*, 9(5): 64–84.

Gordan, J.L. & Holden, M. 2004. Understanding the Human Knowledge Resource using Knowledge Structure Mapping, Technical Report, Applied Knowledge Research Institute: Merseyside England.

Grant, R. 1996. Toward a Knowledge-Based View of the Firm: Implications for Management Practice, *Strategic Management Journal*, 17(winter): 109–122.

Hasse, P. et al. 2005. Management of dynamic knowledge, *Journal of Knowledge Management*, 9(5): 97–105.

Hoffman, D. et al. 2005. Design software architectures to facilitate accessible Web applications, *IBM Systems Journal*, 44(3): 467–482.

Huang, J.C. & Newell, S. 2003. Knowledge Integration Processes and Dynamics within the Context of Cross-functional Projects, *International Journal of Project Management*, 21(3): 167–176.

ISO. 1998. ISO 9241-11 Ergonomic requirements for office work with visual display terminals (VDTs)— Part 11: Guidance on usability.

Juristo, N. et al. 2004. Clarifying the Relationship between Software Architecture and Usability, *Proc. 16th International Conf. on Software Engineering & Knowledge Engineering*, Banff Alberta, Canada, 378–383.

Kang, K.C. et al. 1998. FORM: A feature-oriented reuse method with domain-specific reference architectures, *Annals of Software Engineering*, 5: 143–168.

Kudyba, S. 2005. Enhancing the Transfer of Knowledge Resources through Effective Utilization of Labor and Technology in a Global Organization: A Case Study of Bovis Lend Lease Inc.'s Global Knowledge Transfer System, *Knowledge and Process Management*, 12(2): 132–139.

Lang, J.C. 2001. Managerial concerns in knowledge management, *Journal of Knowledge Management*, 5(1): 43–59.

Nielsen, J. 1993. *Usability Engineering*. London: Academic Press.

Pisano, G. 1994. Knowledge, integration, and the locus of learning: an empirical analysis of process development, *Strategic Management Journal*, 15(Special Issue): 85–100.

Riege, A. 2005. Three-dozen knowledge-sharing barriers managers must consider, *Journal of Knowledge Management*, 9(3): 18–35.

Shankar, R. & Gupta, A. 2005. Towards Frameworks for Knowledge Management Implementation, *Knowledge and Process Management*, 12(4): 259–277.

So, J.C.F. & Bolloju, N. 2005. Explaining the intentions to share and reuse knowledge in the context of IT service operations, *Journal of Knowledge Management*, 9(6): 30–41.

Sun, C.A. et al. 2002. Overviews on Software Architecture Research, *Journal of Software*, 13(7): 1228–1237.

Tang, H.S. & Jin, Y.H. 2000. The Research on Knowledge Management and Knowledge Portal Framework based on Agent Mechanism, *Proc. 3rd World Congress on Intelligent Control and Automation*, Hefei, P.R. China, 263–267.

Walczak, S. 2005. Organizational knowledge management structure, *The Learning Organization*, 12(4): 330–339.

Control Engineering and Information Systems – Liu (Ed)
© 2015 Taylor & Francis Group, London, ISBN 978-1-138-02685-8

Research on Digital House design and application integration mode

L.L. Zhao, P. Yao, T.X. Wei & S. Liu
School of Engineering, Honghe University, Mengzi, P.R. China

H.F. Xing
School of Info-Physics and Geomatics Engineering, Central South University, Changsha, P.R. China

ABSTRACT: "Digital House (or Digital Real Estate)" is real estate data for the digital, networked, intelligent, integrated management information system of the managed object, which is a general term for real estate information systems and related data infrastructure. The paper is focused on the integration of digital house construction research, which is based on the above facts; attempting to resolve the integration issues of technology, data and applications in the Digital house construction process.

Keywords: Digital House; integration mode; spatio-temporal database; the three-tier structure

1 INTRODUCTION

With the further opening up of China's real estate market, real estate has been the rapid development and has become one of the important pillar industries of national economy [1]. "Digital House (or Digital Real Estate)" is real estate data for the digital, networked, intelligent, integrated management information system of the managed object, which is a general term for real estate information systems and related data infrastructure. It takes spatial information as the core, using GIS (geographic information system), MIS (management information system), OA (office automation), WFS (work flow) the integration of advanced technology in order to achieve the various types of real estate information management and services optimization [2].

With strong investment in the urban infrastructure, the speed of urban economic construction continues to accelerate. As the development of urban housing construction is rapid, the type and complexity of the real estate management business is growing, business scope is expanding, such a background to carry out the "Digital house" construction projects not only can interconnect the original dispersed in urban real estate departments under the Bureau of heterogeneous systems, implementing data exchange and information sharing. Combing through real estate-related business, the use of technologies such as GIS, MIS, OA, WFS achieve policy of housing management, real estate transactions, real estate property registration, property management and housing management and so on to achieve inter-departmental business systems

integration transformation, and build cooperative office system to solve the poor system interaction, inconsistent data standards.

Graphics attributes which can not exchange. Give full play to the network, and enhance the city's real estate business management level to achieve the integration objectives of property management and real estate services [3] [4]. The paper is focused on the integration of Digital house construction research, which is based on the above facts; attempting to resolve the integration issues of technology, data and applications in the Digital house construction process.

2 THE INTEGRATION ARCHITECTURE FOR DIGITAL HOUSE PROJECT DATABASE

Data and data sharing are the basis of the Digital house projects. Urban real estate is the distribution of urban land on the ground material, which is located in a specific space of the city buildings; real estate data is typical of spatial data. Traditional database management system is emphasis on the management of non-spatial attribute data, while the rapid development of GIS applications and researches are to promote the emergence of a spatial database technology for spatial data storage, management, analysis and application. The application of GIS and spatial database technology has become the main support of the current real estate information technology.

Digital house data-sharing infrastructure is including two goals: one hand, property management related to housing space objects reasonably classified, unified storage, unified management, the basis of spatial and temporal database. On the other hand, it requires the establishment of spatio-temporal database integration of the related attribute database management and access mode, and thus lays a solid foundation for the Digital house construction projects.

2.1 A spatio-temporal database for housing

Spatio-temporal database for housing is one of the underlying database of digital house projects, it is used to store the record housing where the parcel, housing, distribution of rooms and basic description.

Housing is a general term for a range of housing graphical elements, which is composed of parcel, real estate, layer, households. Parcel refers to a closed block of the property line, is a spatial entity in the Digital house project, it is a closed polygon in the GIS. Real estate is located in the parcel within an independent, the same structure, including different levels of housing. Real estate number is the key identifying information in the same manner, it is preparation of unique. At least one contains one layer; the layer contains at least one.

The household is the minimum property of the real estate management unit, the relationship between ownership is based on households to be embodied. Besides real estate number, real estate also has contained a ground total number of layers, the attribute information of the total underground layers, a total construction area of housing location and the housing structure. Real estate is a spatial entity, the performance of a closed polygon in GIS. The name of the layer is a layer property in the same storied building layer unique. Layer performance is a spatial entity in the database design. Each layer corresponds to a hierarchical splitting of graphical information.

The user information is the most important and most abundant among property management storied building, floors and indoor room. The household is a spatial entity, the performance of a closed polygon in GIS. Parcel, real estate, layer and households are space entity of the Digital house; they correspond to the appropriate real estate graphic data in the map space. The parcel information is represented by information topographic maps patch measured building distribution data or building census data, layer and the user information is represented by real estate layered floor plan and property households plan. Maps of these four categories of basic graphical data are in the real estate management; their relationship is shown in Figure 1.

Time element is very important to the changeable geographic features, the introduced time

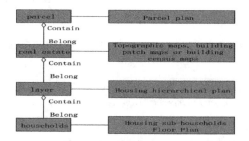

Figure 1. A real estate management graphical data.

element in the Digital house project helps to record every moment of the ins and outs, thus can be a true record of housing, combined with real estate prices, such as spatial distribution of effective analysis and forecasting. Housing is the core of the Digital house project management objects, digital property management objects of housing is due to the merger, division, or the loss causes changes in the spatial features and attribute features, this paper is based on snapshot sequence model and the base state with amendments model proposed by Langran [5] to establish a spatio-temporal database for housing, the specific ideas are as follows:

1. Abstract time for the current and historical moment, which corresponds to set the current layer and the historical layers. Store all graphic objects in the current layer, the layers of history is to store all the history of graphic information, that is, to change the store multiple historical moment T1, T2, T3, ... Tn snapshot, and store an abstract historical snapshot of the moment;

2. Record a graphic object in the layers of history in order to improve the time resolution of the model;

3. Take the entire real estate graphics as the ground state. Points, line, surface are a whole object to correct units for the ground state amendment, the amendment process is actually a new object to replace the old object, the new object is stored in the layers of the current moment, while old object storage layer in the historical moment, by recording the time of the old object and destruction time, their layers are to achieve the effect of the base state with amendments. Chronological backtracking, correction object in the query time period is to replace the overlay to the ground state of the layer in order to achieve the purpose of the spatio-temporal data management.

2.2 The establishment of the basic properties table database

In accordance with the principle of centralized management of the data, the Digital house is including the main database, which is composed of management database of the real estate industry, real estate registration database, mortgage seizure database, pre-sale contracts for the record database, database leasing, property management database, housing the basis of spatial, temporal database and so on.

These databases can be divided into two kinds of spatial databases and operational databases, data representation is independent of their physical, and Digital house construction needs a unified data base in order to achieve a true sense of the data sharing.

This requires a Digital house construction projects need to build housing the basis of spatial and temporal database and the business database logically interrelated, spatial databases and operational databases are into a unified, organic whole, which is a Digital house projects.

Establishment of various business systems are in different periods without considering the correlation between the systems; In addition, due to historical reasons many cities do not have a very mature encoding to be used to identify the buildings, which lead to business systems and the spatial and temporal database, or business systems are difficult to establish the association in the logical sense. Basic real estate database in this context for index database is put forward, the database role of basic real estate table in the form shown is in Figure 2.

2.3 Digital three-tier real estate database schema

On the traditional concept of the network its structure is usually divided into three layers: data layer, logic layer and presentation layer. This article designs a kind of three-tier database architecture for digital property database considering of a network "three-tier structure", the underlying housing is the basis of spatio-temporal database, middle-tier database is real estate table, the top is the business system database. Basic housing spatio-temporal database is on the bottom of the housing, because spatio-temporal database is a database of record housing status quo and housing history, it is

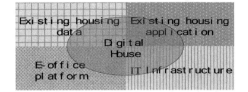

Figure 3. Three-tier database architecture diagram.

the basis of the property management business systems and business data source. The basic properties used to connect the housing basic spatio-temporal database and business database table database is in the middle layer, its role is equivalent to the business logic layer in the network hierarchy, it is from the logic of housing the basis of spatio-temporal database and business database, they associate to achieve the unity of the database, and provide a basis for data sharing in the true sense of the Digital house; business database is directly as real estate management, management of data, real-time more, and designed for the upper.

The three-tier structure of the Digital house database is shown in Figure 3, the application service platform is to access the top business system database, if both the information of the underlying housing and the basis of spatio-temporal database are needed, it is through the correlation structure of the real estate table, and the basic spatio-temporal database is to access.

3 DIGITAL HOUSE PROJECT "INTEGRATION" ARCHITECTURE

The Digital house to be an organic part of the one city's Municipal Bureau of Land Real Estate Board e-office platform, we must consider four aspects of the network, platform, data, applications integration. To achieve the integration, we must consider the following four aspects:

1. How applies the deployment of Digital house on the existing network;
2. Consider the use of the existing platform security system to implement digital property rights management;
3. In accordance with the standards of real estate information technology, management of existing real estate data should beto migrate to a unified database platform of the National Housing Board, and through the existing data engine to spatial data access, data dissemination, data mining and other functions;
4. In the realization of real estate management applications, similar to other applications of existing e-government platform architecture and operating style.

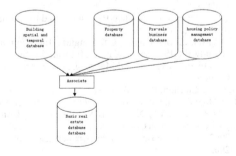

Figure 2. The basic properties database role.

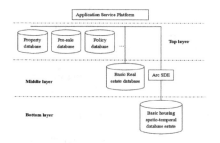

Figure 4. Digital house integrated schematic.

To achieve Digital house "integration" building, also consider the land and resources management, planning and management, property management and other business information sharing and seamless integration. Seamless integration of land, planning, real estate and other business.

Take full account of the four aspects of the network, platform, data, applications integration and land planning, real estate and other business integration on the basis of the overall design of the Digital house project, it can be summarized as follows:

1. To establish the local standards of one city real estate information technology, and migrate existing real estate data in accordance with the standard clean-up, set up the database of housing space, to achieve unified management of digital house data set;
2. In the framework of the existing e-government office platform, design digital house system framework, enhance digital property maintenance and scalability, so that the system can adapt to the changing real estate business needs;
3. The three-tier architecture is used in the new digital house system to implement restructuring and optimization of existing real estate management business systems;
4. Make full use of existing information fruits of construction to implement the digital house security system and data dissemination;
5. Inside and outside the existing e-government platform data exchange interface for third-party data services. Such as: bonding companies, banking, taxation and other departments to provide relevant data query.

4 CONCLUSIONS

The digital real estate project is a huge project, compared to traditional property management systems, its technology, data and application integration are the difficulty lies. The paper is to establish the basic properties table database, so that the spatial building foundation, temporal databases and operational databases is made into a unified whole, thus provides convenient access for housing business systems. The paper proposed to establish three-tier integrated data architecture, and applied to the construction of one City Digital House, effectively to achieve the integrated management of the spatial database and attribute database, and real estate data integration shared. In addition, through the use of technologies such as GIS, MIS, OA and WFS, we proposed the integration of digital real estate project architecture system, integrating real estate business systems and related systems as a whole, the establishment of a unified real estate industry a unified platform for the construction of digital cities and lay a solid based on a model for future industry integration to provide.

ACKNOWLEDGMENTS

This research is supported by the National Natural Science Foundation of China (41201418, 41301442), by Yunnan University Innovation Fund and by the Department of Education Research Fund of Yunnan Province (2011Y298, 2012C198).

REFERENCES

Anonymous, Bluetooth Wireless Technology Enhances Applications for Mobil Telephony, Business Wire, Feb. 21, 2002.

Chasin M.S. 2001. Computers. How a palm-top computer can help you at the point of care. Family Practice Management 8(6), 50–51. G. Henkelman, G. Johannesson and H. Jónsson, in: Theoretical Methods in Condensed Phase Chemistry, edited by S.D. Schwartz, volume 5 of Progress in Theoretical Chemistry and Physics, chapter, 10, Kluwer Academic Publishers (2000).

Fischer S., Stewart T.E., Mehta S., Wax R. & Lapinsky S.E. 2003. Handheld computing in medicine. Journal of the American Medical Informatics Association 10 (2), 139–149.

Hochschuler S.H. 2001. Handheld computers can give practitionersan edge. Orthopedics Today 21 (6), 56. Information on http://en.wikipedia.org/wiki/NMEA_0183.

Ong R.J., Dawley J.T. & Clem P.G.: submitted to Journal of Materials Research (2003).

Shi Weiwei, Zhang Tingyu, Cai Yangjun, Tan Xiujuan. A Study of GIS-based Real Estate Information System, Bulletin of Surveying and Mapping, 2006(8):56–60 (in Chinese).

Xuexiang Li, Fulin Bian, Yongge Shi. System Integration of Digital Real Estate-Management Based on Service, ICAPIE, 2012 T.G. Xydis, S.B. Wilson, Security Comparison: Bluetooth Communications vs. 802.11, Feb. 1, 2002, http://bluetooth.com/dev/wpapers.asp, Feb. 17, 2003.

Control Engineering and Information Systems – Liu (Ed)
© 2015 Taylor & Francis Group, London, ISBN 978-1-138-02685-8

Technologies for detecting data conflicts in distributed situation

Z.P. Zhong

College of Physics and Electrical Information, Anhui Normal University, Wuhu, China

ABSTRACT: Data conflicts are one of the key problems in a distributed situation. When the data of the database is divided horizontally or vertically and are distributed across different nodes, the test data will be faced with greater challenges of conflict which often need to move data from one site to another site. In this paper, an algorithm for detecting conditional functional dependencies conflicts in distributed database is presented which not only can detect the conflicts of conditional functional dependencies in horizontally divided data, but also reduce effectively the data transmission. Experimental results demonstrated that the algorithms are effective.

1 INTRODUCTION

Data quality is one of the primary tasks of data management and it involves the data conflict detection, that means find tuple of Certain Rules in database D which violates CFDs (Conditional Functional Dependencies). For instance, a given relational schema R (*id, CC, AC, city, area, zip, Tel, title, salary*). Where in, *id* (employee identifier, primary key), contact information (*CC, AC, Tel*) and home address (*city, area, zip*). Table 1 and Figure 1 given a relationship example D_0 and definition of CFDs in R respectively. CFD_1 declared functional dependencies: when

Table 1. Relational instance D_0 of schema.

Id	CC	AC	City	Area	Zip	Tel	Title	Sal.
1	86	05	WH	NL	411	345654	Inst.	2k
2	86	05	WH	JH	410	123432	Prof.	4k
3	86	05	WH	JH	410	234543	Prof.	4k
4	86	05	WT	JJ	410	789876	Asso.	3k
5	31	21	AM	Kru.	012	543234	Inst.	4k
6	31	21	AM	Spu.	012	456765	Prof.	5k

CFD_1: ([CC=86, *zip*]→[*area*])
CFD_2: ([CC=31, *zip*]→[*area*])
CFD_3: ([CC, *title*]→[*salary*])
CFD_4: ([CC=86, AC=05]→[*city*='*WH*'])

Figure 1. The set of conditional functional dependency.

$CC = 86$, *area* depends on *zip*, the tuple set $\{t_i \in D_0 | i \in {}^{[1,4]}\}$ in the relationship instance D_0 meet CFD_1; CFD_3 is a traditional functional dependencies, explains that employees of the same country, *title* determine *salary*; CFD_4 declares any *CN* ($CC = 86$ represented countries) employee, if the *area* code is 05, the city must is *WH*. The example requires seeking tuple in D_0 which violates of CFD_1-CFD_4. Assuming t_i represent *id* = *i* corresponding tuple in D_0, conflict set is composed of t_2-t_6. The D_0 meet CFD_3, but t_2-t_4 violate of CFD_1: Because, t_2-t_4 three triples *zip* values are equal (i.e., *zip* = 410), but the *area* have different value *JH* and *JJ*; Similarly t_5-t_6 violates of CFD_2; the t_4 violate of CFD_4: $CC = 86$, $AC = 05$, but, *city*≠*WH*. When D_0 is a centralized database, SQL-based technology to find the conflict is very effective [1,2]. However, in a distributed system, a relationship often be splited and assigned on different nodes, when detect the conditions functional dependency conflict set CFDs of the entire distributed database, often need to move data from one node to another node. How to reduce the number of mobile data and reduce the cost of network transmission, faster system responsiveness has a very important practical value. Previously, research on distributed database data conflict detection technology: Local effectiveness studies (i.e. no data the mobile collision detection), the inconsistent due to update by using trigger processing[3]. The CFDs conflict found[4]; using the CFDs correct data[5]; CFD spread through the view, and to minimize the communication cost as the goal, detecting constraint conflict[7] in the distributed database system etc. But all of these can't directly apply to the distributed database CFD conflict detection.

2 CONDITIONAL FUNCTIONAL DEPENDENCY AND DIVISION OF RELATION

2.1 Conditional functional dependency

Definition 1. Conditional functional dependencies CFDs on the mode R, Supposing the set of attributes of mode R are $attr(R)$, the domain of property A ($A \in attr(R)$) is $dom(A)$, relationship instance tuple t, the value of t on attribute A is $t[A]$, X represents a subset of $attr(R)$, t on the X projection for $t[X]$, define a conditional functional dependencies on R as follows: φ: $R(X \to Y, Tp)$. Wherein, 1) $X, Y \subseteq attr(R)$; 2) $X \to Y$ is a functional dependency; 3) T_p represent mode list of tuples containing the X and Y properties. For each attribute $A \in (X \cup Y)$, mode tuple $t_p \in T_p$ about the value of A $t_p[A]$ is a constant 'a' in $dom(A)$, or an unmarked variable '_'.

In the mode tuple, Use the symbol '||' split attribute in the X and Y. For example, in the above example 4 dependence can be expressed into three conditional functional dependencies CFDs:

φ_1: ($[CC, zip] \to [area]$, T_1),
φ_2: ($[CC, title] \to [salary]$, T_2),
φ_3: ($[CC, AC] \to [city]$, T_3).

where in, T_1 consist of two mode tuples $(86, \_||\_)$ and $(31, \_||\_)$; $T_2 = \{(\_, \_||\_)\}$, $T_3 = (86, 05||WH)$.

Definition 2. Operator$\asymp$. If $\eta_1 = \eta_2$, one of η_1, η_2 is variable '_', then $\eta_1 \asymp \eta_2$. Extended the operator to the tuple, For example, $(JJ, WH) \asymp (\_, WH)$, but $(JJ, WH) \not\asymp (\_, WT)$. For any mode tuple $t_p \in T_p$, instance tuple t_1, $t2 \in D$, if $t_1[X] = t_2[X] \asymp t_p[X]$, then $t_1[Y] = t_2[Y] \asymp t_p[Y]$, that examples of mode R satisfy the condition φ, expressed $D| = \varphi$. In fact, the tuple in t_p is constraints defined in the tuple sub-set $D_{tp} = \{t | t \in D, t[X] \asymp t_p[X]\}$, rather than on the entire D.

2.2 Division of relation

1. Horizontal division: The relationship D are divided into n disjoint subset $(D_1, ..., Dn)$, Wherein, $D_i = \sigma_{Fi}(D)$, $D = \cup_{i \in [1,n]} D_i$, F_i is conditional predicate. Meet, $i \neq j$, $D_i \cap D_j = \varphi$, and all fragments D_i share the same pattern R. For example, by selecting the operating, $\sigma_{title='Inst.'}(D_0)$, $\sigma_{title='Asso.'}(D_0)$ and $\sigma_{title='Prof.'}(D_0)$, split D_0 into three fragments DH_1, DH_2 and DH_3 (due to space reasons, DH_1, DH_2 and DH_3 not listed here) settled in different nodes. Then detect CFD_1-CFD_4, in detection CFD_1, the fragment DH_1, DH_2 and DH_3 information must gathered.

2. Vertical division: The relationship D are divided into n disjoint subset $(D_1, ..., D_n)$, Wherein, $D_i = \pi_{Xi}(D)$, $D = |\times|_{i \in [1,n]} D_i$, $X_i \subseteq attr(R)$, and

Table 2. Vertical division DV_1.

Id	City	Area	Zip	Title
1	WH	NL	411	Inst.
2	WH	JH	410	Prof.
3	WH	JH	410	Prof.
4	WT	JJ	410	Asso.
5	AM	Kru.	012	Inst.
6	AM	Spu.	012	Prof.

contains the $key(R)$, key attribute of R. Each vertical fragment D_i has its own mode R_i, meet, $attr(R_i) = X_i$, $attr(R) = \cup_{i \in [1,n]} attr(R_i)$. For example, through the projection operation $\pi_{id,city,area,zip,title}(D_0)$, $\pi_{id,CC,AC,Tel}(D_0)$ and $\pi_{id,salary}(D_0)$ vertical divided D_0 into DV_1, DV_2 and DV_3 three fragments (see Table 2–4), Respectively allocated to three different nodes S_1, S_2 and S_3. In order to detect CFD_3 conflict, the column *title* and CC data must be moved from node S_1 and S_2 to node S_3 respectively; or CC column and *salary* column data must be moved from node S_2 and S_3 to S_1; or *title* column and *salary* column data must be moved from node S_1 and S_3 to S_2.

2.3 CFDs conflicts

Definition 3. CFDs conflicts. Given a CFD φ, R $(X \to Y, Tp)$ and an instance D of R, tuples violateed φ in D expressed as $Conf(\varphi, D)$. Meet, $t \in Conf(\varphi, D)$, If and only if there is a tuple $t' \in D$ and a pattern tuple t_p, making that $t[X] = t'[X] \asymp t_p[X]$, but, $t[Y] \neq t'[Y]$ or $t[Y] = t'[Y] \not\asymp t_p[Y]$. All tuples violated conditional functional dependencies set Σ in D expressed as $Conf(\Sigma, D)$. In studying the CFD conflict, the tuple which is violate to the CFD schema is genuinely concerned, rather than the entire tuple. Defined $Conf^*(\varphi, D)$ as $\pi_X Conf(\varphi, D)$, i.e. the $Conf(\varphi, D)$ is the projection on the X property, in addition to the X, all other attributes in the $attr(R)$ filled by null. In fact, $Conf^*(\varphi, D)$ is instance of mode R, but significantly less than $Conf(\varphi, D)$. For example, condition function dependencies φ_2 and an instance D_2 of R in the example above, assume that, D_2 contains a tuple $t[CC, title] = (86, Prof.) \wedge t[salary] = 4k$, If there are k different tuples $t'[CC, title] = (86, Pr of.) \wedge t'[salary] = 5k$, then $Conf(\varphi_2, D_2)$ consists of at least $k+1$ tuples, and $Conf^*(\varphi_2, D_2)$ contains only a tuple t, represent exists tuple $t[CC, title] = (86, Prof.)$ violate to φ_2 in D_2.

3 LOCAL TEST CONDITIONS AND RESPONSE TIME

From the above analysis, in order to detect a the CFDs conflict in a distributed database, data need to be moved from a node to another node. A simple conflict detection algorithm is to move the fragment allocated to each node D to the coordinator node (i.e. the node of data fragments centralize); Reconstructed D by fragment, then detected the CFDs conflict $Conf^\pi(\Sigma, D)$ in the coordinator node. However, this method often causes expensive network transmission. In order to describe the cost of communication, use $cost\ (i, j, t)$ to express the overhead costs of moving the tuple t from the node S_j to S_i. In a distributed database to detect conflicts often need to move a tuple sets M. Reduce the number of moved tuple M, the cost of network transmission will be reduced. For example, 1) Assuming the instance D of mode R is horizontally divided into $(D_1, ..., D_n)$, the moved tuple set is M, $M(i)(i\in[1, n])$ expressed all the tuples which is moved to node S_i, $D'_i = D_i \cup M(i$. Moving data set M can local test the response time of function dependencies Σ Use (1):

$$cost(D,\Sigma,M) = \frac{1}{c_t}\max_{j\in[1,n]}\{sum_{i\in[1,n]}\mid M_{(i,j)}\mid/p\}$$
$$+ \max_{i\in[1,n]}\{check(D'_i, \Sigma)\} \qquad (1)$$

where c_t (the actual network traffic c_t) indicates that the data transfer speed, p represents the size of each data packet, $check(D'_i, \Sigma)$ expresse the time to find Σ conflict in the fragment D'_i. Due to data movement and conflict detection is carried out in parallel, so the response time is equal to the sum of the maximum time which all nodes send data to other nodes taked and local conflict detection taked. 2) If D is vertically divided into$(D_1, ..., D_n)$, $M(i, j)$ expressed all the tuples in M which is moved from node S_j to S_i, i.e. a set of tuples in the form of $cost\ (i,j,t)$ Use (2):

$$D'_i = D_i \mid \times \mid\ j \in [1,n] \wedge M(i,j)$$
$$\neq \Phi \cdot M(i,j)(i \in [,1,n]) \qquad (2)$$

Whether it is horizontal division or vertical divided, after moving data M, the local detected φ conditions Use (3):

$$conf^\pi(\phi,D) = \bigcup_{i\in[1,n]}conf^\pi(\phi,D'_i) \qquad (3)$$

4 CONFLICTS DETECTION AND OPTIMIZATION IN HORIZONTALLY DIVIDED DATA

4.1 Determined the data without moving that is local detection CFD

An example, in the Detect the CFDs conflict in horizontal division data, first need to figure out which

tuples need to move, and which one will be moved to. According to the law of a set of decomposition, the conditional functional dependencies $(X\rightarrow Y,T_p)$ can be converted to the CFDs set in form of equivalent $(X\rightarrow A,t_p)$. Wherein, $A\in Y$, $t_p = \pi_{X,A}(T_p)$. If $t_p[A]$ is constant, said $(X\rightarrow A,t_p)$ constant CFD; if $t_p[A]$ is '_', called $(X\rightarrow A,t_p)$ the variable CFD. In the above example, the schema of conditional functional dependencies φ_4, $t_p = (86,05\|WH)$, so φ_4 is constant CFD, φ_1, φ_2 and φ_3 are variable CFDs. If φ is a constant CFD, without moving the data the CFD conflict can be detected at each node. For example, Table 1 tuples $T_4 = (86,05\|WT)$ contrary to φ_4. If in the variable CFD $\varphi:(\ X\rightarrow A,t_p)$, $t_p[A] = $ 'a', $A\in X$, the $F\varphi$ represent set of all that $t_p[A] = $ 'a', by each fragment the $D_i = \sigma_{Fi}(D)$ contains only the tuples satisfy F_i. If $F_i \wedge F_\phi = \varphi$, then there do not exist tuple in D match $t_p[X]$, Therefore, when the check φ tuple do not need to be moved to the node S_i or removed from the node S_i.

4.2 The algorithm on detecting conditional functional dependency conflicts

The algorithm on detecting conditional functional dependency conflicts: *ImproveCentreDetect* is an improved algorithm for the previous algorithm *CentreDetect*. The main distinction between *ImproveCentreDetect* and *CentreDetect* is the choice of the node in the conflict detection. Choose the node which has matched maximum tuple of *LHS* mode in T_p as coordinator S_j of φ, tuple in the other nodes will be send to the node S_j to get conflict detection, this thereby reducing the network transmission and improveing the response time. The execution of *ImproveCentreDetect* discribed as Algorithm 1: example above, $\varphi_1=$ $([CC, zip] \rightarrow [area], T_p= \{(86, \_\|\_), (31, \_\|\_)\})$. The coordinator of φ_1 is the node S_3, as DH_3 have three tuple of which attributes CC is 86 and 31, while the DH_1 and DH_2 have two and one matching tuple. So

Algorithm 1. Improve Centre Detect (φ,D)

Input: conditional functional dependency $\varphi:(X\rightarrow Y,Tp)$ and the relation horizontally divided D = (D1, ..., Dn).
Output: $conf^\pi(\phi,D'_j)$.
1: $lstat_i = cnt(\pi_{X\cup A}(D_i[T_p[X]]))$;
2: Each node S_i send the local statistics lstati to the other nodes;
3: The node S_i received statistical data from other nodes, specified the node which has maximum lstatj value as coordinator;
4: All node S_i $(S_i\neq Sj)$ send $M(j,i)$ $\pi_{X\cup A}(D_i[T_p[X]])$ to coordinator S_j;
5: Coordinator S_j using SQL to detect conflict sets $conf^\pi(\phi,D'_j)$.

807

S_1 move tuple $\{t_1,t_5\}$ to node S_3, S_2 send a tuple t_4 to S_3, 3 data tuples have be moved. If other nodes selected as coordinator, the number of moved tuples are higher than this value.

The inadequacies of the of the Algorithm 1 lies in that it may caused large group moved, resulting in a system bottleneck. This proposed multi coordinator, Conflict detection work assigned to multiple nodes, increased parallelism, and reducing the number of data movement to reduce the response time. For example, DH_1 and DH_2 contain only a single tuple $CC = 86$ and DH_3 contain three such tuples. Similarly, DH_1 and DH_3 contain tuple $CC = 31$, DH_2 don't. When Processing two modes tuple of φ_1, S_3 and S_1 is specified as a mode tuple $(86, \_\|\_)$, $(31, \_\|\_)$ coordinator, this can reduce the mobile data amount and the response time.

Algorithm $PartialDetectS$ and $PartialDetectRT$ have been proposed based on this idea. In order to reduce the number of data movement and system response time respectively, the difference between them is how to specify coordinator for the T_p mode tuple. Algorithm $PartialDetectS$ based on the cost function specify the coordinator node S_l for mode t_p^j. First specified coordinator S_i for each mode tuple arbitrarily using function $\lambda: T_p \rightarrow \{1, ..., n\}$; then other node $S_j(j \neq i)$ send $M(i,j) = \cup_{t_p^l \in T_p, \lambda(t_p^l)=i} H_{jl}$ tuple to S_i. Therefore, the set of tuples which be sended to S_i represent $M(i) = \cup_{j \in [1,n]} M(i,j)$, the move cost Use (4):

$$\cos_s(\lambda) = sum_{i=1}^n |M(i)| = sum_{i=1}^n sum_{j=1}^n |M(i,j)|$$

(4)

Since $M(i,j) = sum_{l=1}^k latat[j,l]$, cost optimization function can be set as $\lambda(t_p^l) = m$. Where, S_m is the node which requires the biggest number of data movement (i.e. the value of $latat[m,l]$ is biggest) to detect t_p^l. Then, send all matching tuple (X,A) to the mode t_p^l corresponding to the coordinator, all modes tuples t_p^l of φ can work in parallel on different nodes. In coordination node of t_p^l, detected $(X \rightarrow Y, t_p^l)$ by using SQL technology. $PartialDetectRT$ assign a coordinator node for each mode to optimize $cost_{RS}(\lambda)$ by using Greedy algorithm, the calculation methods of $cost_{RS}(\lambda)$ Use (5):

$$cost_{RS}(\lambda) = \frac{1}{c_t} \max_{j \in [1,n]} \{sum_{i \in [1,n]} |M_{(i,j)}| / p\}$$
$$+ \max_{j \in [1,n]} \{check(D_j \cup M(j),\phi)\}$$

(5)

wherein,
$M(i,j) = \cup_{t_p^l \in T_p, \lambda(t_p^l)=i} H(j,l)$, $M(i) = \cup j \in [1,n]$ $M(i,j)$. $|M(i,j)|$ and $|M(i)|$ can be calculated partially by $lstat [j,l]$. Assuming λ_{l-1} expressed the

coordinator specified by the $(l-1)$-th mode tuple of T_p, t_p^l is l-th mode tuple. If λ_l^i match the first $(l-1)$ mode tuple λ_{l-1}, then $\lambda_l(t_p^i)$ is set to coordinator node which makes $cost_{RS}$ least, until λ_k is assigned.

4.3 Algorithm on condition functional dependencies set CFDs conflict detection

The CFDs conflict detection algorithm can use the *SequenceDetect* or *ClustDetect*. *SequenceDetect* use the algorithm *PartialDetectS* or *PartialDetectRT* deal with the individual CFD conflict in turn in CFDs. Once the current CFD (tuple division or conflict detection) dealt, enter the next CFD processing immediately. Before all CFDs is finished, not the node is idle, the disadvantage is that may lead to the same tuple moved frequently, adds network transmission. The ClustDetect algorithm reduced the unnecessary data movement by aggregating the CFDs public properties. If $X \subseteq X'$ or $X' \subseteq X$, "merge" $\phi = (X \rightarrow A, T_p)$ and $\phi' = (X' \rightarrow B, T_P')$ into one. If some overlap condition above is satisfied, divided D base on the projection mode list $T_p[X \cap X'] \cup T_p[X \cap X']$, Then assign a coordinator for each pattern tuple in the projection list. Finally, local detecting at each node of the CFD conflict and implement CFDs conflict detection.

5 ANALYSIS OF EXPERIMENTAL RESULTS

To test the effectiveness of the CFD conflict detection in horizontal division data, we have built Local Area Network (LAN) connection from 6 machine, each machine running SQL Server 2000 as a local DBMS, Algorithm implementation on the VC++6.0. Test data sets using synthetic data sets "company sales", among them, contains 1200 k tuples. Verify the relationship between the

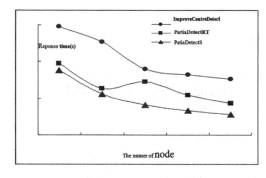

Figure 2. Response time and the number of node.

response time and the number of nodes of the algorithm *ImproveCentreDetect*, *PartialDetectS* and *PartialDetectRT*, The experimental results are shown in Figure 2. The algorithm response time decreases with the increase in the number of nodes.

6 CONCLUSION

This paper studies the CFD conflict detection method in distributed database. For horizontal partition data, first, identify the CFD which can be conducted local conflict detection without data movement. For those CFD who need to move data to detect conflicts, the single CFD and CFDs detection algorithm is proposed, using fragment statistics, CFD model and allocated detection to process multiple nodes to reduce the number of data movement and response time. Based rely maintain, described the CFD which can be partially detected in vertical division relations. For vertically partitioned data studied the conflict refining method of local detection in vertical segments CFDs. Here, our study focus on the CFD detection algorithm in pure horizontal or vertical partition data, how to detect the CFD conflict which has both horizontal and vertical division of data will serve as the next research work.

ACKNOWLEDGEMENTS

We would like to thank the anonymous reviewers. This work was supported by the Foster Foundation of Anhui Normal University under grant No. 2011xmpy010, and by National Natural Science Foundation of China under grant No. 1107403.

REFERENCES

Agrawal S., Deb S., Naidu K.V.M. et al. 2007. Efficient detection of distributed constraint violations. *in ICDE*.

Cong G., Fan W., Geerts F., Jia X. et al. 2007. Improving data quality:Consistency and accuracy. *in VLDB*.

Dahav B., Etzion O. 2003. Distributed enforcement of integrity constraints. *Distributed and Parallel Databases*. 3(3): 227–249.

Fan W., Geerts F., Jia X. et al. 2008. Conditional functional dependencies for capturing data inconsistencies. *TODS, vol.* 33(2): 1–44.

Fan W., Ma S., Hu Y., Liu J. et al. 2008. Propagating functional dependencies with conditions[C]. *in VLDB*.

Golab L., Karloff H., Korn F. et al. 2008. On generating near optimal tableaux for conditional functional dependecies. *in VLDB*.

Mingyu Feng, Zheng Zhao, Gang Zhang. 2002. Data Collsionand It's Solution in Distributed Situation. *Application Research of Computers*. 02: 72–74.

The use of biodata analysis to reconstruct the personnel file indicator system

L.S. Pan

Chinese Academy of Personnel Sciences, Beijing, China

ABSTRACT: Compare to the rapid development of personnel file digitalization, however, the personnel file indicator system which remains unchanged over decades would hardly satisfy the demands of in-depth utilization of the information resources. By using the biodata analysis, this study reconstructed the personnel file indicator system which would be used in developing electronica personnel file management system. In this study, 97 items of biographical information was generated from personnel files. Personnel files were targeted to aspects of five life experiences (personal information, background, education, training and work experiences). 106 Personnel files of financial professionals and 89 personnel files of automotive engineers were coded. Degree of item validity was associated with salary growth and promotion rate which were also gathered from the records of personnel files. Results indicated that the factor structure of dimensions for valid items was different with that of non-valid items. Implications for the construction of biodata forms and the limitations of the study are discussed.

I INTRODUCTION

According to the Twelfth Five-Year Guideline for national archive development, the works of personnel file digitalization which are parts of national personnel management system in china will be conducted in next 10 years to fully exploit the information resources of personnel files (SAAC, 2011). Much more comprehensive sub-files will be developed, such as the file of integrity and the file of performance, to meet the demands of public sectors and corporations. Compare to the rapid development and popularization of personnel file digitalization, however, the personnel file indicator system which remains unchanged over decades would hardly satisfy the demands of exploit the information resources of personnel files. A lot of key items which have high correlations with job performance and career success, such as the number of job/position changed (Oviedo-Garcia, 2007), work-related training (Lefkowitz, Gebbia, balsam & Dunn, 1999) are not directly recorded in the files, so the file of integrity or the file of career span performance could not be developed based on the existing indicator system.

The purpose of this study is to reconstruct the personnel file indicator system to satisfy the demands of in-depth utilization of the information resources from personnel files by using the biodata technology.

2 BIODATA

Biodata research has evolved over 90 years. The first published study, Goldsmith (1922) examined the ability of nine "personal history" items to predict the first-year sales of insurance agents. She found that using a person's data score would improve the hiring decisions made. Specifically, 58 of the 259 individuals (22%) receiving a score 4 or above were considered successful (first-year sales was dichotomized in order to allow for expectancy charts to be constructed). In contrast, 11 of 243 individuals (4%) receiving a score less than 4 were considered successful.

Over the years, a number of strategies have been used for biodata analysis (Hough & Paullin, 1994). There are four main strategies (i.e., empirical, behavioral consistency, rational, and factorial). Researchers sometimes use a combination of them.

An empirical strategy for biodata analysis is based upon establishing a statistical relationship between a biodata item and the criterion of interest. Frequently, a large pool of items is used, and those items that were predictive are chosen for use. This subset of items is used to arrive at an overall biodata score.

A second strategy, behavioral consistency, is based on the adage that past behavior is the best predictor of future behavior. With this strategy, a researcher selects of develops items that are consistent with the criterion of interest. For example, given they were interested in predicting turnover, Barrick and Zimmerman (2005) utilized a biodata item that asked how many months a subject had been in his/her most recent job.

A third strategy for biodata analysis has been called the rational or deductive approach. This approach generally involves conducting a job analysis to determine the knowledge, skills, abilities, and other characteristics that are important for the criterion of interest. Having determined these KSAOs, a researcher uses items that are thought to reflect them. For example, in attempting to reduce voluntary turnover, Barrick and Zimmerman (2005) thought that having a realistic view of the job being applied for was important, they hypothesized that subjects who knew people who worked for the organization would be more likely to have realistic job expectation. This biodata item did, in fact, predict voluntary turnover.

A commonly used strategy (e.g., Schmitt et al., 1999) for biodata analysis is the factorial approach. Typically, this strategy involves the use of principal axis factor analysis or principal components analysis to extract the dimensions underlying the biodata items used. The hope is that the factors extracted will shed light on the constructs that explain why biodata scales predict the criterion of interest (Hough & Paullin, 1994).

According to the purpose of this study, an approach combined with empirical strategy and factorial strategy was used.

3 METHOD

3.1 Sample

195 personnel files were analyzed in this study (131 men, 64 women), including 106 Personnel files of financial professionals and 89 personnel files of automotive engineers. 10.3% had ten years and below of working, 75.9% had 11 to 20 years of working, 13.8% had more than 20 years of working.

3.2 Procedure

In the first step, this study coded the personnel files by using content analysis method to transform the qualitative information into quantitative data. This contains twice code steps. The first coding collected the contents related with biodata items including personal information, background, education, training and work experiences from personnel files, and the second coding transformed the biodata items into a pool of quantitative indicators. The second step collected the criterions including salary growth and promotion rate from the personnel files. The third step, this study used the factorial

approach to examine the construct validity of the pool of indicators. The fourth, this study used the empirical strategy to examine the criterion-related validity of the indicators.

4 RESULTS

After carefully coded the 197 personnel files, this study conducted a combination of factorial approach and empirical strategy to examine the construct validity and criterion-related validity of the indicators.

4.1 Construct validity

To examine the construct validity of the personnel file indicator system, this study used the cluster analysis method. Table 1 shows the results of cluster analysis.

After compared the cluster analysis results of 6 to 11 clusters, the data showed that the 9 clusters solution could be more stable and comprehensive than others. As Table 1 shows, items in the cluster1 involved the learning and education experiences, thus this cluster was labeled *School Achievement*. Items in cluster2 involved the interpersonal and social activities, thus this cluster was labeled *Social skills*. Cluster3 was labeled *Achievement Motivation* because the achievement motivation has a significant correlation with the economic situation of the family (Liu & Wu, 2009) and Entrepreneurial experience (Littunen, 2000). Cluter4 was labeled *Professional integrity* because this cluster involved the behavioral records related to honest. Cluster5 involved the level of professionalism in workplace, thus this cluster was labeled *Specialization*. Cluter6 was labeled *Work Effectiveness* because this cluster involved the performance in the workplace. Cluster7 involved the breadth of the industries, organizations, position experiences, thus this cluster was labeled *Breadth of experience*. Cluster 8 was labeled *background information*. Cluster9 was the Outlier.

4.2 Criterion-related validity

This study addressed the major research purpose: reconstruct the personnel file indicator system, so that more comprehensive sub-files could be developed, such as the file of integrity and the file of performance. To do so, the indicator system should has the ability to predict career span performance (Breaugh, 2009). This study use the salary growth and promotion rate as the criterions.

Table 2 shows the significant predictor variables by using linear regression. Cluster3 achievement motivation has no significant effect on both salary growth and promotion rate because we had not get

Table 1. The results of cluster analysis.

Cluster	Indicators		
C1	University level	Education background	Type of university education
	Duties in high school	Academic reward	
C2	Social activities in school	Reward of social activities	Political status
	Marital status	Duties in university	Family structure
C3	The economic situation of the family	Type of graduate education	Entrepreneurial experience
C4	Drop-out experience	Record of bad behavior	Frequece of job hopping
	Punishment in works		
C5	Duration in one position	Person-vocation fit	Person-job fit
	Frequecy of job training		
C6	Performance awards	The number of projects	The number of papers
C7	Experience of different industries	Experience of different organizations	Experience of differents positions
	Non-work related training	Overseas experience	
C8	Sex	Birth date	Non-academic award
	Interests		
Outlier	Place of birth	Parents' occupations	Foreign language proficiency
	Physical health		

Table 2. The results of linear regression.

Criterion	
Salary growth	Promotion rate
Performance awards**	Performance awards**
Academic reward**	Experience of different industries (R)**
Social activities in school**	Experience of different organizations (R)*
Reward of social activities**	Reward of social activities*
Experience of differents positions**	The number of projects*
Education background*	The number of papers*
The number of projects*	University level*
Frequecy of job training*	Education background*
Duration in one position (R)*	
Political status*	

*, P < 0.05. **, P < 0.01. R indicates a negative correlation.

enough subjects who provided the data about situation of the family and Entrepreneurial experience. We also had not get enough data about cluster4 professional integrity. Cluster8 background information has no significant effect on both criterions. In cluster1 school achievement, academic reward has a significant effect on salary growth (p < 0.01), education level has a significant effect on both salary growth (p < 0.05) and promotion rate (p < 0.05). University level has a significant effect on promotion rate (p < 0.05). In cluster2 Social skills, social activities in school has a significant effect on salary growth (p < 0.01). Reward of Social activities has a significant effect on both salary growth (p < 0.01) and promotion rate (p < 0.05). Political Status has a significant effect on salary growth (p < 0.05). In

cluster5 Specialization, Duration of one Position has a significant negative effect on salary growth (p < 0.05), Frequence of job Training has a significant effect on salary growth (p < 0.05). In cluster6 Work Effectiveness, performance awards has a significant effect on both salary growth (p < 0.01) and promotion rate (p < 0.01), the number of projects has a significant effect on both salary growth (p < 0.05) and promotion rate (p < 0.05), the number of papers has a significant effect on promotion rate (p < 0.05). At last, cluster7, both Experience of different industries (p < 0.01) and Experience of different organizations (P < 0.05) has a significant negative effect on promotion rate, Experience of different positions has a significant effect on both salary growth (p < 0.01).

5 DISCUSSION

The purpose of this study is to reconstruct the personnel file indicator system, so that more comprehensive sub-files could be developed, such as the file of integrity and the file of performance. The first goal is to examine the construct validity of the personnel file indicator system. Suffice it to say that the indicator system has a stable and comprehensive eight-factor structure which was different with non-valid items including personal information, background, education, training and work experiences.

The second goal is to examine the criterion-related validity of the indicators. Academic reward, education level and university level in cluster1 have a significant effect on salary growth or promotion rate. The achievement in university tend to be more predictable than that in high school. Social activities in school and political Status in cluster2 social skill have a significant effect on salary growth, and reward of social activities have strongly effects on both salary growth and promotion rate. This results show that the social skill could be very useful in predictor career span performance. Cluster 3 and 4 have no significant effect on both salary growth and promotion rate because we had not get enough subjects who provided the data. But we speculate that Entrepreneurial experience has a significant effect on performance based on the previous research results (Littunen, 2000). And the items in cluster4 Professional integrity would be a veto. Frequency of job Training has a significant effect on salary growth, however duration in one position has a significant negative effect on salary growth. This results indicated that such as job rotation, work enrichment instead of just stay in one position would be significantly raise one's career span performance. The person-vocation fit (manipulate as the similarity between the major learned in university and the specialty in vocation) and person-job fit (manipulate as the similarity between the major learned in university and the specialty in current position) have no significant effect on both criterion. This results indicated that from the view of entire career, the effect of the similarity between what you learn in the university and what you do in your job on career span performance would not obvious. Experience of different positions has a significant effect on salary growth, however, the experience of different organizations and industries have significant negative effects on promotion rate. Changing organizations and industries often leads to reconstruction of person-environment fit, and the work experience might not match the new job requirement, thus subjects who experienced too many organizations or industries would have lower promotion rate. The cluster8 background information has no significant effect on both criterions.

There are, of course, some limitations in this study. One limitation is the sample that were used. In this study, we only chose the financial professionals and automotive engineers as research samples. It is clear that before definite judgments on the personnel file indicator system, these findings need to be replicated in other setting with other samples. Another practical concern is the criterion. At this time, only salary growth and promotion rate had brought in consideration, the indicator system could lend substantial support to develop the file of career span performance instead of the file of integrity.

REFERENCES

Barrick, M.R. & Zimmerman, R.D. "Reducing voluntary, avoidable turnover through selection." Journal of Applied Psychology, vol. 90, 159–166, 2005.

Breaugh, J.A. "The use of biodata for employee selection: Past research and future directions." Human Resource Management Review, vol. 19, 219–231, 2009.

Goldsmith, D.B. (1922). "The use of the personal history blank as a salesmanship test." Journal of Applied Psychology, vol. 6, 149–155, 1922.

Hough, L. & Paullin, C. "Construct-oriented scale construction". In G.A. Stokes, M.D. Mumford, & W.A. Owens (Eds.), Biodata handbook. 1994, (pp. 109–145). Palo.

Lefkowitz, J., Gebbia, M.I., Balsam, T. & Dunn, L. "Dimensions of biodata items and their relationship to item validity." Journal of Occupational and Organizational Psychology, vol. 72, 331–350, 1999.

Littunen H. "Entrepreneurship and the characteristics of the entrepreneurial personality." Journal of Entrepreneurial Behaviour & Research, vol. 6, 295–310, 2000.

Liu, P., Wu Y. "On the psychological characteristics of Needy College Students and Countermeasures of educational management in Chinese and English." Education Research Monthly, vol. 9, 112–118, 2009.

Oviedo-Garcia, M.A. "Internal validation of a biodata extraversion scale." Social behavior and personality, vol. 35, 675–692, 2007.

Schmitt, N., Jennings, D. & Toney, R. "Can we develop measures of hypothetical constructs?" Human Resource Management Review, vol. 9, 169–183, 1999.

The state archives administration of the people's republic of china. "the Twelfth Five-Year Guideline for national archive development." http://www.saac.gov.cn/.

Fault-tolerant model of web service composition based on Aspect Oriented Programming

L.L. Gu

Computer Engineering Department, Guangdong Industry Technical College, Guangzhou, China

ABSTRACT: Fault-tolerant technique is an effective means to recover web service composition runtime error. In the service composition process, functional business logic code is entangled with fault-tolerant logic code, it causes the code maintenance difficulties and poor system scalability. Aiming at the problem, fault types that cause service failure in web service composition runtime are analyzed, appropriate fault-tolerant strategy are presented, and fault-tolerant model of web service composition runtime are built based on separation of concerns. While ensuring the service function and quality of service, function concerns and fault-tolerant concerns are separated. Finally, the analysis for case study of setup planning is presented to verify the feasibility of the method. By the example analysis the feasibility and efficiency of the method are verified.

1 INTRODUCTION

With the development of the network, web service has become the best implementation technique of service-oriented architecture. It can eliminate difference between systems that have different component models, operating systems and programming languages, is really in a loosely coupled way to achieve interoperability between distributed applications [1]. Web service is an object component, which is deployed on the network, in order to build large and complex distributed application system over the Internet. In order to meet the business needs, usually requires multiple basic web services in accordance with specific granularity and certain logic rules are combined to form more complex web services. The reliability of web service is the key that web service is applied successful. Because the web service is open, distributed, heterogeneous, autonomous and changeable, network environment instability, service update or upgrade, the web service description error and server failure be to cause web service is not available; or UDDI's web service update failure, web service program and process design error will to cause mistake of running. Resulting in quality cannot be guaranteed. Therefore, research on fault-tolerance mechanism for when the service running is very important.

Software fault-tolerance is the effective way to ensure the system that can still work normally in some crashes by redundant. Zhong Hangqi et al. [2] used N version programming fault-tolerant mode for each active candidate services in order to optimize the reliability of web services composition. If the number of same function services is more, inevitably lead to much additional runtime overhead. Zhang Fuzhi [3] builds web services fault-tolerant architecture based on redundancy server idea. According to information of services group, client would access to member of services group in tune until the available services be found. But this needs the client component to act in the failure detector role on server. The method will destroy the loose coupling characteristics of web service. Based on service component interface model, Wang Sheng et al. [4] proposed fault-tolerant strategy that candidate component with same behavior are selected to dynamically replaces fault component, this requires that each service must be inserted into an error capture and exception handling logic. In addition, some researchers started from the extension of the standard protocol [5–6], but because the standard system itself is constantly changing, and compatibility problems. It's not easy to implement.

Analysis of existing studies, the service replacing method has better usability and feasibility of implementation [4], but the fault-tolerant processing will crosscut each function module, leading to code scattering and confusion code, and engender a detrimental effect on code reusability and system scalability. Aspect-Oriented Programming (AOP) provides a mechanism of modularizes crosscutting concerns (such as security, authorization, log, capture and performance issues etc.), effectively solves code scattered or entangled problems by crosscutting concern. Mi Yang et al. constructed software development framework based on AOP. Exception capture codes and crosscutting concerns codes are separated in object-oriented programs, then software fault tolerance is realized, and code-scattering and code-tangling problem is resolved [5]. This paper presents a web service composition model for fault tolerant based on AOP. Put the AOP into BPEL process, failure judgment crosscutting concerns and fault-tolerant processing crosscutting concerns are separated from BPEL process, fault tolerant aspect is built. The model effectively ensures the reliability of web service, improves the code reusability and scalability of the system.

2 AOP TECHNOLOGY

2.1 Basis elements of AOP

Aspect-Oriented Programming (AOP) is a technology based on separating crosscutting concerns. A complex software system can be regarded as implementation for many core concerns and crosscutting concerns, the core concerns is the core business of the system function, and crosscutting concerns are the system peripheral requirements, such as authorization, log, security and exception capture. AOP can separate the crosscutting concern from the core concern implementing code to a separate model, and automatically wove crosscutting concerns into the system, thereby effectively separate the crosscutting concerns from the core concerns solve the problem of code chaotic and dispersion in traditional software development process, enhance the reusability of code, and the maintainability and scalability of system. AOP implementation principle as shown in Figure 1.

2.2 Basis elements of AOP

The main constitution elements of AOP have:

1. Aspect: Module unit of implementing crosscutting concerns, it encapsulates pointcut, advice and inter-type declarations of crosscutting concerns.

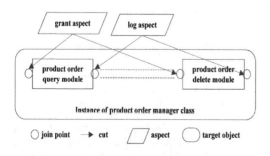

Figure 1.   AOP implementation principle.

2. Join point: It can be woven into aspect during the execution of the program. For example, method invocation, access to members of the class and the exception is thrown.
3. Advice: Action in some join point of the aspect, including before, after and around three types, namely in the join point before, after or around to call advice.
4. Pointcut: Join point matching assertion. The advice is associated with a pointcut expression, and run on the join point. It is the key to cut crosscutting logic, decided to the weaving location.
5. Weaving: It is process that aspect is applied to the target application section type or object, and create a target object. The process can be done at compile or run time, class loading.

3 FAULT-TOLERANT MODEL OF WEB SERVICE COMPOSITION

3.1 Fault-tolerant strategy

The web service structure includes three roles: service provider, service broker and service requester. Service provider registers and deploys service on service broker, and function of service is described with Web Services Description Language (WSDL); By the API the service requester searches for the service on service broker, and get services they need, to realize binding service requester to service provider; then the service requester executes services. The corresponding actions are publishing, searching, binding and executing, errors of any link can result in the service failure. Based on service failure reason analysis of literature [8], this paper summarizes the faults likely occur in web service composition runtime.

Fault Type I: Network environment and hardware failure, such as network delay, network congestion, server etc.

Fault Type II: Service unavailable, such as service description, service does not exist, wrong authentication failure, etc.

Fault Type III: Service execution failure, such as the parameter mismatch, the invalid results returning, the problems of service development and combination process design, etc.

This paper assumes that the service itself and the combination process design are correct, designs the corresponding fault-tolerant design strategy for the three fault types:

1. Retry Service: For failure of response, this paper adopts retry policy that it is the minimum overhead, and sets the threshold value of retry times to avoid affecting the execution of the whole combination process.

2. Failure Replacement: It is a local optimization strategy. During the execution of service composition, the highest score in redundancy service set is selected to perform when the currently executing service fails. To retry operation failed or fault type II or III, the failure service is replaced by the service which is optimal QoS in redundancy services set.

3. Local Business Process Reengineering: Due to web service composition dynamics, atomic service failure may affect on other parts of the process, so in the cases of replacement invalid, the related processes need to be reconstructed. This paper adapts the local reengineering strategy that control structure is regarded as a unit, so as to ensure normal execution of the combination process.

3.2 Fault-tolerant model

In writing the BPEL process, the process designer needs to consider the business, the error finding and the recovery processing as a whole.

But if it includes the exception handling codes in the business process, namely the crosscutting concern, will lead to the business process design is very complicated, and the implement codes become cumbersome, besides limiting portability. Crosscutting concerns is encapsulated exploiting AOP, to constitute a separate section, in order to separate the exception handling and the business process. This paper chooses AO4BPEL [9] to realize that crosscutting concerns is added in the service composition processes. AO4BPEL is an extension of the BPWS4 J, it extends the BPEL in aspect oriented structure, and BPEL aspect runtime which supports aspect. Each activity in the BPEL process can be used as a connection point, and the pointcut is an activity set crossing process which uses an XPath expression to associated pointcut and advice, to determine the location weaving advice is before, after and around, finally the aspect runtime finishes weaving aspect. Aspect is allowed loading or unloading in process running. The fault-tolerant model of web service composition based on AOP (as shown in Fig. 2).

The model includes UDDI, process pretreatment manager, AO4BPEL engine, the fault-tolerant manager aspect, the log aspect, the component for replacing service, the component for local process reengineering, and redundant service storage.

UDDI provides publishing function for service providers, and accepts service request by the service composition process (service consumer), and provides query function. The AO4BPEL engine is responsible for the deployment of BPEL services composition process, aspect, and weaving aspect into process running, to constitute a final

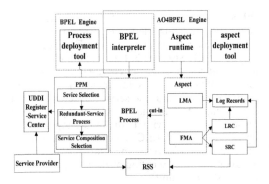

Figure 2. Fault-tolerant model of web service composition based on AOP.

system. The Pretreatment of Process (PPM) is responsible for selecting and finding web service by UDDI, generating redundancy services set. The Redundancy Service Storage (RSS) is responsible for saving redundancy services. The Log Manager Aspect (LMA) is responsible for recording execution status of activity and results data of activity, in order to provide basis for fault processing of the fault-tolerance manager aspect. As a service broker, Fault-tolerant Manager Aspect (FMA) is responsible for invoking web service and judging fault type of runtime error. According to the log data, FMA can achieve error repairing by Service Replacement Component (SRC) and local process reengineering component (LRC).

Workflow of model: While web service composition process are designed, business requirements are mapped to abstract BPEL service composition process, and WSDL is extended that QoS constraints is added to select and find service. At the same time a set of backup process are designed to reengineer local process. While processes are combined, the service selecting module of PPM searches and obtains constituent web service by UDDI on the basis of WSDL document. Redundant service processing module generates redundant services with constraints, and saves services to the service warehouse sort by QoS. Service composition selecting module chooses the best QoS to Process Deployment Tool (PDT) of BPEL engine for deployment, saves IOPE of the constituent service and control structure type of the constituent service to the log records. While the process is executed, logging crosscutting concerns is added in each activity to record execution results and status information of every activity. Fault-tolerant process crosscutting concerns is added in service composition process to calling service and judging fault type by response results when exception occurred and turning over BPEL engine if the exception is logic error that

817

WDSL document describes, then retry strategy is adapted to recalling the service when the fault is type I. If the retry operation is invalid or the fault is type II, III, SRC reads IOPE of failure service and result information of previous activity from log records, then looks for match service from redundant services in the service warehouse, then the optimal QoS alternative service is chosen for replacing failure service. If the failure replacement operation is invalid, LRC determines control structure type for re-factoring the control structure, so as to continue the execution of the process from the control structure unit.

4 KEY IMPLEMENTATION OF WEB SERVICE COMPOSITION FAULT-TOLERANT MODEL

In order to validate the presented fault-tolerant model, there is an application example for verifying the feasibility and effectiveness of the model.

4.1 Example analysis

Example is a combination of meeting schedule service composition, including four separate services: booking ticket of a airline company service, booking ticket of B airline company service, booking hotel service, booking room service. The business process is shown in Figure 3, while user plans to travel, these services are used in the following procedure, and failure of any one can not satisfy the needs of the user. Booking air tickets, booking hotels and booking a meeting room in the process is sequential executed. Obtaining ticket information from A company and B company are parallel, then booking ticket of the most preferential price by comparing A with B. Booking conference hall service is used to reserving time until booking ticket and booking hotel are done.

4.2 RSS

RSS is responsible for saving the alternative services and related information of each constituent service, include: service ID, service IOPE, service URL, service QoS, call state *in_status*. Where

in_staus can be 1 or 0 to indicate if the service is called.

4.3 PPM

PPM include: service selection module, redundant service process module and service composition selection module. Before the process executes, PPM discoveries and selects service, then generates service composition plan, working sequence is shown in Figure 4. Service selection module is responsible for reading function and quality requirements of every service from WSDL, making the service request to UDDI. UDDI is responsible for finding the service which meets the functional and non functional requirements, returns a set of consistent functional entity service description. Redundant service process module saves the candidate services and the related information to RSS, and sorted the services according to QoS. Service composition selection module selects a set service of the optimal QoS to combine, and in_status = 1, saves IOPE and control structure type of every constituent web service to log. Next step is to deploy BPEL process by Process deployment tool of BPEL engine, and then the process is executed by the engine.

4.4 FMA

While the process is executed, FMA is responsible for invoking web service and catching exception, and achieving fault-tolerant of service composition through interaction with the SRC, LRC (as shown in Fig. 5). In executing *invoke* activity by BPEL engine, FMA intercepts *invoke* activity, invokes web service as proxy of engine, and processes the response results. If the response results are consistent with *output* that WSDL document declared, the results are returned to BPEL engine, and this call is completed; If the response results are exception

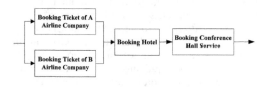

Figure 3. Meeting schedule service composition.

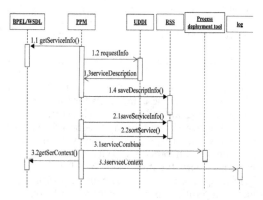

Figure 4. Process pretreatment manager sequence.

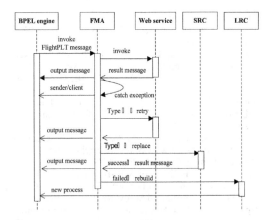

Figure 5. Fault-tolerant manager sequence.

that WSDL document describes, the exception is referred to the BPEL engine; If the response result is empty and not timeout, it is fault type I and retry strategy is adapted that the service is called again; If retry operation is invalid, or the response results is that the host cannot be reached, uncertainty and parameter mismatch, for type II, call the service replacement components to achieve service replacement; If replacement operation failed, then invokes LRC to return the re-factoring process to a BPEL engine.

1. Service replacement: During the replacement of web service, SRC finds appropriate set of redundant services from RSS in accordance with IOPE of failure service, supplied by FMA, obtains the alternative service of optimal QoS and in_status = 0, then sets in_status to 1, gets the result data of the previous service from the log, invokes new service, return results if the new service is invoked successful, or reelection. If no optional services (in_status = 1 or the set is empty), returns a failed message.
2. Local process reengineering: When reengineering local process, LRC obtains the control structure type of failure service by the log, reads the backup process. If the process is a sequence then the process is updated from the failure service, and sends the result data of the previous service to the new process by the log. For other control structures, LRC re-factors entire control structure unit, and sends the result data of the previous service to the new process.

...
```
<!-- exception process -->
<faultHandlers>
<catch faultName="ca:NoReply" faultVariable=
"replyResult">
<switch>
```

```
<case condition="bpws:getVariableData('reply
Result','flightInfo')==null and
bpws:getVariableData('replyResult','flightInfo')!=
'TimeOut'">
<proceed/> <!-- re-invoke service -->
...
<catch faultName="ca:UnknownServer" fault-
Variable= "replyResult">
<switch>
<case condition="bpws:getVariableData('reply
Result', 'flightInfo') =='UnknowSrv' or
bpws:getVariableData('replyResult','flightInfo')==
'ParamMismatch'">
<!-- invoke service replacement component -- >
<invoke partnerLink="FaultWS" portType=
"ReplacePT" .../>
<case condition="bpws:getVariableData('reply
Result', 'replaceInfo') != 'failed'>
<reply.../> <!-- successful replacement -->
<!-- when replacing failure, local process reengi-
neering is invoked -- >
<otherwise>
<invoke partnerLink="FaultWS" portType=
"RebuildPT" .../>
...
</faultHandlers>
<pointcutandadvice>
<pointcut name="fault process" context
Collection="true">
    //invoke[@operation="findAFlight"]||
    //invoke[@operation="findARoom"]
</pointcut>
<advice type="around"> <!-- around advice -- >
<!-- invoke join point service -- >
<sequence><proceed/><reply.../></sequence>
...
```

To achieve interception for each *invoke* activity, it needs that FMA to adapt *around* crosscutting association rules for each *invoke* activity, By executing crosscutting behavior to replace the join point, so as to dynamically change the behavior of the system. Below is FMA fragment of flight booking and hotel reservation sequence logic.

4.5 *LMA*

In the execution of process, LMA records the service execution state of each service and the results data of each service, in order to provide service for fault-tolerant processing. LMA adapts *after* crosscutting association rules that every *invoke* join point is cut *after* advice which records the execution of service. LMA code fragment is as follows.

```
...
<pointcut name="logging Service"
contextCollection="true">
```

```
//invoke[@operation="findAFlight"]|
//invoke[@operation=" findARoom"]|
//invoke[@operation="bookForMeetingroom"]...
</pointcut>
<advice type="after"> <!-- after type advice -- >
<invoke    partnerLink="loggingWS"    portType=
"loggingPT"
operation="loggingProcess"
outputVariable="loggingresponse"/>
...
```

5 CONCLUSION

Web services running environment and features
determine that the service in web service composition running is incredible, and this makes the quality
of the service composition can not be guaranteed.
Fault-tolerant technology can effectively improve
the reliability of the service, but the implementation of fault-tolerant processing as crosscutting
concerns has also troubled the service combination
process design. In this paper, service failure reason
in service composition running is analyzed, classified, the corresponding fault-tolerant strategy is
given, fault tolerance function is stripped from the
business functions by AOP, the web service composition fault-tolerant model based on AOP is built.
Meanwhile the paper considers service failure factor in web service composition running, then the
problem is solved that business logic is entangled
with fault-tolerance code in programming.

REFERENCES

Andrews, T. et al. 2007. BPEL4ws.http://www.ibm.com,/
developerworks/library/specification/ws-bpel/.
Bruning, S. et al. 2007. A fault taxonomy for service-
oriented architecture. Proceedings of 10th IEEE
International Symposium on High Assurance Systems
Engineering.Dallas, TX, USA:367–368.
Charfi, A. & Mezini, M. 2007. AO4BPEL: An Aspect-
oriented Extension to BPEL. World Wide Web,
10(3):309–344.
Diego, Z. & Maria, B. 2007. A Fault Tolerant Web
Service Architecture. Proc. of Latin American Web
Conference. Washington D.C., USA: IEEE Computer
Society: 42–49.
Huang, Y.H. & Li, G.Y. 2008. WSMF Model: Elements
Decomposed and Expressed in Code. Computer
Applications. 28(8):2156–2169.
Liang, D. 2003. Fault-tolerant Web Service. Proceedings
of the 10th Asia Pacific Software Engineering Confer-
ence. Chiangmai, Thailand:310–319.
Mi, Y.H. & Lin, P. 2012. An AOP-Based Software Fault-
tolerance Approach in Pervasive Environment. Com-
puter Applications and Software, 29(9):87–91.
Papazoglou, M.P. & Heuvel W.J. 2007. Service oriented
architectures: approaches, technologies and research
issues. International Journal on Very Large Data
Bases. 16(3):389–415.
Shen, G.F. & Li, X.K. 2008. Research on a Web service composition model with AO4BPEL and context-
awareness. Journal of Hefei University of Technology,
31(6):866–870.
Tang, J.J. et al. 2013. Approach for Web Service Composition Trustworthiness Evaluation. Computer Science,
40(2):163–171.
Wang, S. et al. 2009. A Strategy for Service Fault Tolerance in SOA. Journal of SUZHOU University (Natural Science Edition). 25(3):36–41.
Zhang, F.Z. et al. 2010. Dependable Web Services Composition Method to Guarantee Response Time. Journal
of Chinese Computer Systems. 31(10):1959–1964.
Zhong, D.H. et al. 2008. Approach to Reliability Optimization of Web Service Composition with Redundancy.
Computer Engineering. 34(4):31–33.
Zou, F. & Gao, C.M. 2008. Fault-tolerant Framework
at Runtime of Web Service Composition. Computer
Engineering, 34(18):89–92.

Control Engineering and Information Systems – Liu (Ed)
© 2015 Taylor & Francis Group, London, ISBN 978-1-138-02685-8

The auto data acquiring method for un-sole field name of same data item in dataset with many items

Y. Huang
Network Centre, Beijing City University, Hangtiancheng Region, Beijing, China

L.Q. Hu
Education Management Office, Beijing City University, Beijing, China

ABSTRACT: In actual water resource informational management, there are usually multi names for the same primary key or multi names for the same detected data item in different original batch data. This problem increases the complexity of data processing and the burden of data processing. In this paper, an auto data acquiring method for un-sole field name of same data item in the dataset with many data items is published. It works according to data acquisition method based on primary key mapping policy and attribute mapping policy as well as professional knowledge accumulating policy. The formula description about this auto data acquiring method is demonstrated. And the data structure which can realize the data acquiring method is discussed. The application effects prove that it is feasible.

1 PRE-PROCESSING OF WATER QUALITY DATA WITH MANY ITEMS OCCUPIES MUCH MORE MANPOWER

Water resource is an important factor for the living of people and development of society. And Water quality information data is also an essential element for water resource management, for the deployment and modulation of water production device as well as for the dispatching of water collection. In water quality information data, there are much more data items of water quality need to be obtained so as to evaluate and describe the overall condition of water resource. According to the water quality detecting criterion of different countries, the data quality detecting items include inorganic substance items, organic substance items, and microorganism items, etc. For example, in the surface water environment quality standard of China, there are 24 items of basic quality item, 9 items of supplied drinking water source items and 80 special items of concentrated living drinking surface water source[1]. In the drinking water standard of the Unite States, there are 79 items for the first level drinking water and 19 items for the second level drinking water[2]. In drinking water quality standard of Canada, the physical and chemical detected norms include 80 items[3]. As many data items are demanded and the detecting task for all the water quality items may be executed by different branches, it is rather difficult to collect values of all the water quality items

in one short time period, especially for collecting the historic water quality data.

For so many data items, the typical way is organizing them to be structured schema and importing them in database. Importing batch data into database by computer is a high efficient meaning, but in fact, those original water quality data obtained by various data sources may be scattered. To change them into ordered, integrated and formal structured forms usually need the professional members to pour much working by hands, such as original data normalizing, data extracting and relational searching among batches data, etc. Hence, the high efficient running mode does not remain in the whole data processing loop actually. It is statistics that thus workings occupy 85 percent of all the data processing tasks[4]. Pre-processing of water quality data with many items occupies much more time and manpower.

2 THE REASON FOR THE DECREASING OF WATER QUALITY DATA PROCESSING EFFICIENT

2.1 *The different original batch data contain different data items*

Different property data items of the same data record may be saved in different data collections which are obtained from various data sources in different time. For each original data batch, the data items may be not integral. Some data items may be vacant. So data gathering is necessary working in data processing. These would occupy some time period and delay the data utilizing.

2.2 The ID forms of different original batch data are various

The ID forms of original data record, which act as the primary key, are not identical between different batches usually or sometimes. This condition requests data demander to make the id item out among all the items and to normalize various original data ID forms of record into the standard form. The diversity of ID forms leads to the complexity of recognizing.

2.3 Diversity original name for one target attribute field

The data item name of the same attribute, which are not primary key field, are not identical between different data batches usually or sometimes. This condition requests data demander to recognize the different data item name among so many data items and normalize them to be normal name. The lager amount of data item name is, the more items name normalizing working is need.

2.4 Various data formats of different original batch data

For some data items, the data format in the different batches are not completely identical, such as the format of date, or the format of 'year. month. day', or the format of 'mm/dd/yyyy', or the other date formats.

Thus troublesome tasks raise the complexity of the data processing. Accordingly, the contents of data processing vary with different original data batches. The repeatable processing mode is only available in limited extend. The processing efficiency of water quality data decreases correspondingly.

3 THE CURRENT EXISTING ETL TECHNICAL AND THEIR INSUFFICIENCE FOR WATER QUALITY DATA

Under the condition of diverse data content and variety data format, in order to collect the data from different data sources, seeking the target data items and merging those incomplete dataset which comes from different batches always need people to do data processing. People need to juxtapose the similar items, to pick out the different data items, to make data join or union, and to supply the

essential information if necessary. To finish upper task, some acquiring technical measure should be adopted.

For warehouse, the classified method is utilizing the ETL technique. Current ETL technique includes the DTS technical of Microsoft Corp., Oracle Warehouse Builder of Oracle Corp., and Visual Warehouse of IBM Corp. Some data transform regular are provided by them. Though these common data ETL tool can provide powerful supports for general data extracting, importing and outputting and integration, for the special-purpose of the management of water source basement data, especially the water quality data, the roles that these ETL tools play are limited. The reason includes that configuration in the utilization is too complex, lower repeatable data transform due to the mutation of the data forms between different data resource batches, such as came from historical accumulation or gathered from the wild working, too much data items needed to be extract and the utilizing cost of these common software tool is expensive. Hence, for the water resource professional region, the adaptable of upper ETL software tool is not sufficient. To develop one artificial means which can take automatic policies to recognize the effective data items from original and mess dataset, to pick them out and to organize them into the formal structured formed is a necessity.

4 THE TASK OF WATER SOURCE BASEMENT DATA ACQUIRING

In order to improve the data obtaining ability, to speed up the data collection in sea amount diverse data environment and to reduce the workload of data processing, the special data acquiring technique which can extract and collect data to fit the actual water source and water quality data management business shall not be neglected.

The task of extracting and collecting of water source basement data, such as attribute and property date, the detected water quality data as well as analysis data, is to find an effective method and to develop a corresponding software tool to recognize and to extract the demanded water quality data items from the original scattered data pile and to collect them together according to the relationships of them, to normalize them, so as to provide the fore support for importing them into the water source basement data warehouse, to accumulate basis for the water data management, data analysis and decision-making of water production and water providing[5].

The specific and extended task includes drawing up the data extracting regular, recognizing and extracting demanded data items, verifying the extracted data items, merging the different data items of the same primary data record, classifying the extracted data for actual demand.

- Ascertaining the primary key mapping solution correspond with the actual original dataset.
- Ascertaining and accumulating the attribute name mapping solution correspond with the actual original dataset.
- Searching and extracting the available data items from original dataset according to the primary key mapping solution and attribute name mapping solution in the searching data pool on a felicitous moment.
- Making data auditing according to the extracting clue reserved in the process of data searching and extracting.
- Merging and organizing the diverse data items of same primary data record and importing them into the water resource informational database.
- Establishing the technical hierarchical framework for water source basement data extracting and collection.

5 THE DESIGN LOGIC OF DATA ACQUIRING

According to upper tasks, the data acquiring means should follow specific water business logic and rules, includes establishing the mapping of data identify, establishing the mapping relationship of attribute data items, auditable for the data extracted and data collecting, and cost saving for data acquiring.

5.1 Mapping method of primary key

Means of data identifying in different original data collections is not sole in actual. For example, in water resource region, the means of water well identifying can be the code of well, or the well name, or the simple spelling of Pinyin Code, etc. The fact that multi identifying means of same targeted object are adopted determines that the design of the data acquiring method must consider the various data identify forms. Of course, all the present forms of data identity should be normalized for sole standard primary key form in the end. And for the normalization of data identity, data dictionary for primary key mapping relationship, which can reserve the correspondence relationship between original data identity name and the standard primary key name, should be established.

5.2 Mapping of attribute data items

For one attribute data item of water quality data table, the original items names of different data batch may be not identical. There may be more than one name forms can be adopted in actually. For example, to the iron element, we can recognize it with chemical element code name of 'Fe', or the name of 'Fe ions', or the name in Chinese character. The same as the other water quality item name. So the data dictionary for attribute data items, which establishes the mapping relationship between original attribute data item name and standard field name of relationship table, should be constructed.

5.3 Auditable for the data extracting

The data which is picked out by means of data extracting and collecting should be correct and can be verified. In order to ensure the quality of extracted data, the clue of every data extracting task should be recorded so as to provide the auditing basis. And only passing the collection auditing, the extracted data can be imported into the formulate water source basement database. So the database structure should meet the auditing-able demand. The extracting verifying data table should be appended in the database.

5.4 Cost saving for data extracting

On one hand, if the data extracting is started, the seeking program will search the data items from every data tables with the paying of extracting cost. With the original data batch is accumulated gradually, the bore of data extracting will increase correspondingly. To some extend, if there is no new items sought, the data table which has experienced data extracting is valueless and their re-extracting will take redundant excess time and cost. So they should be separated from the later original batch data which haven't traversed data extracting.

On the other hand, data extracting should be limited in the specialty and professional fields, such as underground water, surface water, water resource management, prevention and curing of water pollution, in which some professional regulars and rules can be utilized to direct the data collection. If exceed these fields, the cost will be too expensive.

6 FORMAL DESCRIPTION OF DATA EXTRACTING AND COLLECTION OF WATER QUALITY DATA

For data object entity E, 'id' is the identity item of E. There exists the structured table which consists of a series of attribute fields of E. $A = \{a_1, a_2, a_3, ..., a_n\}$ $(n \in N)$ is the attribute set of E.

For $\forall a_u \in A, a_v \in A,$

$1 \le u \le n, 1 \le v \le n \cdot u \ne v, a_u \ne a_v.$

If $u \ne v$, then $a_u \ne a_v$. That is to say, any two attributes of E are different with each other.

The corresponding time set

$T_{set} = \{t_1, t_2, t_3, \ldots t_z\}, (z \in N).$

DS is the sample dataset of E, which varies with time.

$DS = \{(id_i, a_1(t_i), a_2(t_i), a_3(t_i), \ldots, a_n(t_i))$
$\quad | n, i \in N, t_i \in T_{set}\}.$

$a_1(t_i), a_2(t_i), a_3(t_i), \ldots, a_n(t_i)$ are attributes items value at the time of t_i.

If exits a mapping Object OE of E in reality, OE owns its identity item 'oid' and attribute set OA.

$OA = \{oa_1, oa_2, oa_3, \ldots, oa_m\} \ (m \in N)$

$\forall oa_u \in OA, oa_v \in OA,$

$1 \le u \le m, 1 \le v \le m, u \ne v, oa_u \ne oa_v.$

That is to say, any two attributes of OE are different with each other.

DR is the OE's sample dataset which varies with time.

$DR = \{(oid_i, oa_1(t_i), oa_2(t_i), oa_3(t_i) \ldots, oa_m(t_i))$
$\quad | m, i \in N, m \le n\}$

6.1 Primary key mapping relationship

For $oid_j \in OA \ (j \in N)$ and $id_i \in A \ (i \in N)$, if there is one to one mapping relationship between Oid_j and id_i, $id_i \leftrightarrow oid_j$, then

$(oid_j, oa_1(t_i), oa_2(t_i), oa_3(t_i) \ldots, oa_m(t_i))$
$= (id_i, oa_1(t_i), oa_2(t_i), oa_3(t_i), \ldots, oa_m(t_i))$ \hfill (1)

6.2 Attribute fields name mapping relationship

For $oa_p \in OA$ and $a_q \in \{a_1, a_2, a_3, \ldots, a_n\}$,
If $oid_i \leftrightarrow id_i$ and $oa_p \leftrightarrow a_q$, then

$(oid_i, oa_p(t_i)) = (id_i, a_q(t_i)).$

If $oid_i \leftrightarrow id_i$, and each 'oa' item in DR can establish one to one mapping relationship with one sole attribute item a in A, that is to say,

$(oa_1, oa_2, oa_3, \ldots, oa_m) = (a_1', a_2', a_3', \ldots, a_m')$

in upper formula, $(a_1', a_2', a_3', \ldots, a_m') \in A$, then

$(oid_i, oa_1(t_i), oa_2(t_i), oa_3(t_i), \ldots, oa_m(t_i))$
$= (id_i, a_1'(t_i), a_2'(t_i), a_3'(t_i), \ldots, a_m'(t_i))$ \hfill (2)

6.3 Attribute fields name mapping relationship among different original batch data

For $\forall a_k \in A$, there are data group (id_i, uod) which comes from a original dataset and data group (id_i, vod) which comes from another original dataset, if $uod \leftrightarrow a_k$ and $vod \leftrightarrow a_k$, then $uod = vod$.

That is to say, $(id_i, uod) = (id_i, vod)$. So $(id_i, uod(t_k))$ and $(id_i, vod(t_l))$ $t_k \ne t_l$ can be merged into one same relationship data table.

6.4 Data extracting

For dataset DS, which saves all the existing data tuple, $DS = \{(id_i, a_1(t_i), a_2(t_i), \ldots, a_n(t_i)) | n, i \in N, n \le n\}.$

If the time t_i, in which the data group is obtained, does not belong to the time set Tset, then data group acquired at moment t_i should be collected in the DS. We can expend this operation as below:

$\forall ds = (oid_k, oa_1(t_p), oa_2(t_p), \ldots, oa_n(t_p))n, k \in N,$

if ds satisfies equation (2) and $t_p \notin T_{set}$, then ds should be absorbed in the DS, and new dataset DS' is produced.

$DS' = ds \cup DS$

6.5 Data union

6.5.1 The case of single attribute union

For two data group $(id, oa_i(t_k))$ which was picked out in different extracting processing, if another data group $(id, oa_j(t_k))$ which is synchronized to oa_i at the moment of t_k, $i \ne j$, they can make union on the same id value. So the group $(id, oa_i(t_k), oa_j(t_k))$ is generated.

6.5.2 The case of multi attributes union

For one data tuple s1 in Dataset DR1,

$s_1 = (oid_i, oa_k(t_i), oa_{k+1}(t_i), \ldots, oa_l(t_i))$

$DR_1 = \{(oid_i, oa_k(t_i), oa_{k+1}(t_i), \ldots, oa_l(t_i))$
$\quad | i, k, l \in N, 1 \le k < n, 1 \le l < n, k < l\}$

there is another data tuple s2,

$s2 = (oid_j, oa_u(t_j), oa_{u+1}(t_j), \ldots, oa_v(t_j))$

in another dataset DR2,

$$DR_2 = \{(oid_j, oa_u(t_j), oa_{u+1}(t_j), ..., oa_v(t_j))$$
$$| j, u, v \in N, 1 \le u < n, 1 \le v < n, u < v\}$$
$$\{k, k+1, ..., l\} \cap \{u, u+1, ..., v\} = \varnothing.$$

When DR1 and DR2 both satisfy the equation (2), if $oid_i \leftrightarrow id_k$ and $oid_j \leftrightarrow id_k$, then $oid_i \leftrightarrow oid_j$. If $t_i = t_j$, then s1 and s2 can combine each other, their union is

$$s_m = (id_k, oa_k(t_i), oa_{k+1}(t_i), ..., oa_l(t_i),$$
$$..., oa_u(t_i), oa_{u+1}(t_i), ..., oa_v(t_i)) \tag{3}$$

Furthermore, equation (3) can combine with equation (2) below.

If each upper oa item in equation (3) can establish one to one mapping relationship with one sole attribute item in A, that is to say, if

$$(oa_k, oa_{k+1}, ..., oa_l, ..., oa_u, oa_{u+1}, ..., oa_v)$$
$$= (a_1'', a_2'', a_3'', ..., a_w'')$$

in upper formula $(a_1'', a_2'', a_3'', ..., a_w'') \in A$, then

$$s_m = (id_k, oa_k(t_i), oa_{k+1}(t_i), ..., oa_l(t_i),$$
$$..., oa_u(t_i), oa_{u+1}(t_i), ..., oa_v(t_i))$$
$$= (id_k, a_1''(t_i), a_2''(t_i), a_3''(t_i), ..., a_w''(t_i))$$

6.6 Generalization about upper demonstration

If the value of primary key of one tuple in one original dataset is equal to the value of primary key of another tuple in another original dataset, the two data tuples can be merged into one new data group on their common primary key value, even if the primary key name is different. If the mapping item name of one attribute field in one original dataset agrees with the mapping item name of one attribute field in another original dataset, then two attribute names which came from different original dataset can be normalized to be one common attribute field name. After upper data processing, the data value would be normalized into standard data value and increase the identical element of different data value.

7 THE COMPONENTS STEP OF DATA ACQUIRING

If regulars of primary key mapping of original and attribute data items mapping as well as data extracting and collecting have been finished, the data obtaining process could be implemented after original dataset has been provided. The components step of data obtaining includes locating the original dataset object, designating of data item which should be sought, data extracting in original dataset, auditing of data which has been picked out, data merging and collection, archiving the dataset which have been extracting.

7.1 To locate the original dataset object

When original batch data are receiving form various channel, they should be changed into the forms of relationship type table and be pushed into one seeking data pool. The seeking data pool, which is a structured relationship type database, plays the role of original data container. Some states of seeking and extracting flag should be set accompanied by the original data batch importing. After data extracting, the state should be set or cleared corresponding the result of data extracting.

7.2 To definite the search data items

To designate the reference standard tables which need to be filled data, so as to definite which target data items should be sought and extracted in the seeking data pool. Of course, if for the flexibility of data extracting, the data items which need to be extracted can be configuration to adjustable. If so, before starting the data extracting, an extracting data items list should be drawn up so as to provide direction for the data extracting.

7.3 Data extracting

According the seek data item list and the region of data searching, the data extracting is be implemented accord with the mapping rules of primary key and the mapping rules of attributing items. Every field name of every data table which belongs to the seeking region in the seek data pool should be participated and do matching with the mapping rules.

To make a traversal in one data table in the original data pool, and to check if there is the individual primary key name or the names combination of primary key matched with the data item name in primary key mapping dictionary. If the match condition is held, the original primary key mapping relationship condition is established. The mapping relationship condition becomes the next step basis of data extracting. Based on the primary key matching relationship, a traversal, which checks the field name of the other fields to judge the matching result between the seeking field name and the attribute field mapping dictionary, is executed. If the matching condition is fitted, the value of the field matched to one data item of the attribute field mapping tale can be picked out and be saved in the candidate table in the seeking data pool. Of course, the value should satisfy logical relation of the water source expertise area. At the same time, the mapping relationship extracted should be recorded in one clue information table, include the name of table which owns the attribute field matching, the matching primary key name and the matching attribute field name. The later data auditing can be carried out based on the relationship storage information.

When extracting of all original data tables is finished, one data extracting task is over.

7.4 Data auditing

To ensure the quality of data which has been picked out in the data extracting, the data auditing according to the extracting relationship information reserved in the clue information table should be implemented to verify the data value. Data verifying can be carried out by software or by hands. The data which can satisfy the verifying conditions should be set auditing flag. The others data should be retreated from the candidate table with the clearing the crediting of extracting clue. Only passing the data verifying, the data extracted can be processed in next step.

7.5 Data item gathering

Merging different attribute values of one data record is demanded, so as to assemble the data value into one row of the targeted standard table according to the relationship of the primary key as well as reference standard table. When the amount of data items which approaches data value increase step by step, the outline and overview of the data record gradually complete. And it will gradually meet the demands of normal quality data record format.

7.6 Importing in the normal database

After experiencing the auditing and merging process, the batch data which are saved in the candidate table can be imported into the corresponding normal data table of water quality database. Then, the data demander could browse the data and utilizing them in normally business, such as data management, data re-process, data analyzing, data evaluating and data predict analysis.

8 THE APPLICATION OF DATA ACQUIRING METHOD

For the basement data of riverside water source in some place, the water quality data items came from different batch data between an interval period were not identical. For example, to the water well whose name is 'Liguan1#', sometimes the primary key value of it in one batch data is 'Liguan 1#', in another batch data the value may be 'LG1#', or the name of 'liguan01'. For data water quality items which need to be collected, different original data batches maybe contain different amount of data items. These bring troublesome for the data accumulating. Every data importing demands the data manager to do the different process working. This leads to the fussy data technical process and low efficient. The actual example is shown in Table 1 and Table 2.

In upper two tables, their label row is not identical obviously. And amount of their items are different, too. For the same ID of records in the two tables, their identity name is not the same, such as the value of 'liguan2-1' in Table 1 is 'LG02-1' in Table 2. Their attribute names of the same item are different, such as the item name of 'Redox potential' in Table 1, is 'Eh_mv' in Table 2. If importing them in same table in relationship type database, the data processing contents like these two tables will be obviously different.

By means of the software with auto data acquiring method for un-sole field name of same data item, after the implementing of data seeking and acquiring in the original data pool, not only the value of data item which are in different original data tables are picked out and are gathered together according to the reference standard table, but also the names of data item are normalized to consistent names. The application with analog water

Table 1. Original table sample 1.

Date	Sample name	CO_3^{2-}	Ammonia	Fe	Hexavalent	Redox potertia	Nitrite	...
2011-3-30	Liguan2-1	0	0.45	2.58	<0.004	180	0.011	...
2011-3-30	Liguan3-1	0	0.09	3.42	<0.004	166	0.023	...
2011-3-30	Liguan4-1	0	<0.02	0.19	<0.004	195	<0.001	...
2011-3-30	Liguan5-2	0	0.02	0.25	<0.004	250	<0.001	...
2011-3-30	Liguan6-2	0	0.07	131	<0.004	140	0.013	...
2011-3-30	Liguan38#	0	<0.02	0.29	<0.004	163	0.012	...

Table 2. Original table sample 2.

Label	Time	Eh_mv	pH	Na	K	Ca^2	NH$_4$	Fe
LG01	2012-9-11	306.0000	6.9700	39.5600	1.8060	35.7700	0.138001	<0.04
LG02-1	2012-9-11	348.0000	6.5300	61.4800	3.1470	71.6100	0.268122	0.0424
LG03	2012-9-11	361.0000	6.6600	64.1300	3.1380	71.7200	0.119988	<0.04
LG04-1	2012-9-11	273.0000	6.7200	58.4600	3.1080	64.6300	0.140470	<0.04
LG05-2	2012-9-11	356.0000	6.5500	138.1000	3.2560	134.7000	0.336153	<0.04
LG10	2012-9-11	392.0000	6.6000	55.2300	2.6850	85.5600	0.103959	<0.04
LG12-1	2012-9-10	484.0000	6.6100	51.1900	2.6540	80.8200	0.035549	<0.04
LG12-2	2012-9-10	387.0000	6.7600	51.8200	2.7870	80.3100	0.051426	<0.04
LG24	2012-9-10	286.0000	6.8000	51.3200	2.7560	78.6500	0.137611	0.1085

quality data proves that the auto data acquiring method for un-sole field name of same data item is feasible.

9 THE DATA STRUCTURE FOR DATA ACQUIRING

The structure of database need to be special designed for the aid of auto extracting and collecting usage.

9.1 For the viewpoint of per data processing loop

There are original batch data, extracted batch data which are audited or not audited in the database. If they are mixed with each other without taking any measure, the data management, table indentifying even data extracting is not convenient. So the database should be split into several parts in logical. They are original data pool, extracted data pool, audited data pool, normal data pool and other assist parts.

9.1.1 Original data pool
The original batch data is firstly imported into this data pool. When the original data, include attributes data of detected object, historic reserved data, real detected data are received from different data provider, they should be changed into structure schema and be stored in the original data pool where the data waits for later extracting. In order to provide convenient for the later data extracting, the data record id of original forms should be remain.

9.1.2 Candidate data pool
Extracted data pool can be view as candidate data pool. After each traversal data seeking and extracting in the original data pool, the attribute data items which were picked out are saved in the extracted data pool. These extracted data will undertake data verifying in the later data auditing according to the extracting clue reserved while data extracting. The data in candidate data pool are in standard pattern after processing of extracting.

9.1.3 Audit data pool
After data verifying is finished, the extracted data are split into two parts, the part of passing the verifying and the part of no passing the verifying. The former should be waiting for data items merging and importing to the water quality standard database. The latter should be feedback to the original data pool and the corresponding extracting clue should be clear.

9.1.4 Normal data pool
After auditing, the data items are credible and normal. Multi attribute items can be merged on the same standard record id. These data items can be imported in the normal database in individual mode or in batch mode.

9.2 The division of work of data table in the extracting database

According to the describing above, the data tables in the database play multi roles in the data extracting and collecting. The core roles include the reference standard table, primary key mapping table, attributes mapping table, clues data table, extracted data item table, merging data table.

10 CONCLUSIONS

With the aid of auto data acquiring method for un-sole field name of same data item in the dataset with many data items, the data pre-processing for different batch data with diverse original field names of actual same data item can decrease the workload of un-repeating data arranging evidently and speed up original batch data collecting business. It provides effective support for water resource basement data management business. This auto data acquiring method improves the actual data collecting ability, increases the available of some original batch data as well as the flexibility of data processing.

REFERENCES

America Environment Protection Agency. 2001. American drinking water standard.

China National Environment Protection Avenue. 2002. Chinese surface water environment quality standard (GB3838-2002): 5–8.

Environnement Canada. 1996. Canada drinking water quality standard.

Weiming Zhang. 2002. The Principle and Application of Data Warehouse: 90–93. Beijing: Electronic Publish.

Yong Huang & Liqin Hu. 2011. Based information management of water source district near the river combined with GIS technology. Computer Engineering and Design 32 (6): 2007–2010.

The access-management mechanism of web-based titanium metallurgy database

T. He, Y.Q. Hou & X.H. Yu
Faculty of Metallurgical and Energy Engineering, Kunming University of Science and Technology, Kunming, China

G. Xie
Faculty of Metallurgical and Energy Engineering, Kunming University of Science and Technology, Kunming, China
The Technique Center of Yunnan Metallurgy Co., Ltd., Kunming, China

ABSTRACT: The every kind of application system based on Web must deal with large of data transport and data interchange between the present databases with the development of Internet/intranet/extranet. Therefore, the efficiency and safety of data transport should be improved. The access-management mechanism of Web-based titanium metallurgy database has been investigated in the paper. The interesting design that a Database access Connection Manager (DCM) is added in the middle-ware has been proposed. Based on the design, the connection of database can be managed individually and the connection mechanism of backstage thread can be applied. Thus, the same database can be shared by different visitors and the optimum access of database can be reached. Furthermore, the high-efficient data linker has been investigated. The structure of safe four arrays based linked list structure has been proposed.

1 INTRODUCTION

The every kind of application system based on Web must deal with large of data transport and data interchange between the present databases with the development of Internet/intranet/extranet. Therefore, the efficiency and safety of data transport should be improved [1]. It is noted that the response rate can be decreased and the system may not work normally because of the increasing transport data and visitors. Therefore, it is very important to develop a flexible database system which can share the data [2]. The following three problems, which is present in the development of application systems based on Web, should be solved in time: (1) the efficiency of data interchange, (2) safety of date save and interchange, (3) the share of data in the system [3, 4]. The access techniques of database have been developed quickly in the past years. The requirement of the third layer, middle layer, has been obvious with the improvement of web interchange and the extensive application of web.

The middle layer is logic layer, on which the data can be visited usually. Lots of problems have been proposed with the development of middle layer, for example how to visit database, how to open the connection, whether to maintain an out-line record and so no. And the problems are the "hot

problems" which have been investigated. Therefore, the access-management mechanism of Web-based titanium metallurgy database has been studied in the paper.

2 THE ACCESS-MANAGEMENT MECHANISM OF WEB-BASED TITANIUM METALLURGY DATABASE

The access of Web based titanium metallurgy database is achieved by web in the modified multylagers Web based titanium metallurgy database. Therefore, access-management mechanism has large effect on the whole system, even is the bottleneck of the system performance. The main compose of Web based titanium metallurgy database is the permanent data saved on the disc. The data must been connected with the backstage when the data are dealled with. Therefore, how to manage the access of those databases has been the main assignment in the Web based titanium metallurgy database system.

The main cost is to establish the access of the Web based titanium metallurgy database in the operation process and the waiting time is even long for users. It is noted that the waiting time is long and even the system maybe stopped because of large data resource consumption if every assess

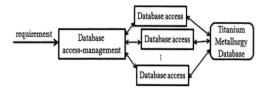

Figure 1. Access-management mechanism of Web-based titanium metallurgy DB.

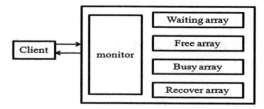

Figure 2. Improved access-management mechanism of Web-based titanium metallurgy DB.

must be established when one wants to access the database. The interesting method of re-use the present data access can solve the problems mentioned above. The access technique can improve the characters of the system because it can reduce the data burden.

Although there are also data buffering in the present techniques, for example ADO, EJB etc, the flexibility and efficiency can be improved through the design of data buffering controlled by application system in own. Therefore, the Database access Connection Manager (DCM) can be added in the system and the access mechanism can adopted backstage line link, as shown in Figure 1. Then the access of database can be share by some users and the access of database can be up to the best.

From figure 5.9, it is obvious that the connection of access-management machine and database is physical connection, the connection of users and database is logical connection. And the access-management machine can obtain the following interesting functions: (1) the protection of database resource, (2) the efficient utilize the database resource, (3) simply the data port of applications, (4) manage and monitor in efficient the access of database resource in the backstage and improve the re-use ratio of the connection database resource.

3   HIGH-EFFICIENT DATA LINKER OF WEB BASED TITANIUM METALLURGY DATABASE

Based on the principle of first access and first served, the safe four arrays based linked list structure have been adopted in the access-management machine, namely waiting array, occupation array of database access, free array of database access and recover array, as shown in Figure 2. The waiting array can be used to store the requirement of database temporarily. The free array of database access can be used to store the present database connection, which is not used by users. The occupation array of database access can be used to store the data access used by users. And recover array can be used to recover the no responding data access which is not closed. Furthermore, there

is monitor procedure in the access-management machine. The administrator can monitor the data-source by monitor procedure, for example the valid access number, the access number in use, the state of access, etc. Meanwhile, the administrator can close the database in hand and maintain the all applications.

Some database connections are established when the system is initialized. These database connections are stored to free array of database access in order to be used by users. Then the cost of the system is reduced because it is not necessary to open a new connection when the database is used. As a result, the source occupied by Web applications is reduced. The source occupied when data connects or disconnects is larger than that when data interchanges. In the common Web database, a database connection is established in the system when the requirement is emitted by user, then close/disconnect database when the operation is over. However, the present system uses the data-base access-management machine. The requirements are inserted in the waiting array, then check there are database connections occupied or not in the free array of database access. The connection can be used if there are free database connections. If not, the new database connections will be established and the new connections will be inserted to occupation array of database access. Then the new database connections can be used by users.

For recover array, the system can discover the errors by monitor and recover these errors when there are some errors in the recover array. These errors can be dealt with in the recover array and then added to free array of database access. The used connections which are closed/disconnect normally can be inserted in free array of database access. Therefore the performance of the system can be improved and the resource can be reduced.

4   CONCLUSIONS

1. The access-management mechanism of Web-based titanium metallurgy database has been investigated in the paper. The interesting design

that a Database access Connection Manager (DCM) is added in the middle-ware has been proposed. Based on the design, the connection of database can be managed individually and the connection mechanism of backstage thread can be applied. Thus, the same database can be shared by different visitors and the optimum access of database can be reached.

2. The high-efficient data linker has been investigated. The structure of safe four array based linked list structure have been proposed.

ACKNOWLEDGMENT

This paper is sponsored by National Natural Science Foundation of China (50574045).

REFERENCES

Alexis D. GutZman. Selecting the appropriate middleware for your web-to-database application, Acm Sigu-CCS XXV, 1997.

Fong W., L. Syyicl, J.S. Liu. Providing for an open real-time CORBA. In Proceedings of the Workshop on Middleware for Real-Timc Systems and Services, (San Francisco, CA), IEEE December 1997.

Li Mian, Zhang Zuo and Wu Qiufeng. The Application of Database Middleware in the Normal Entertainments Network and MIS. Computer Engineering and Applications, 2001, 14: 100–102.

Meng Xianfei, Li Hao and Sun Tongfeng. Model Research of Database Accessing on ADO.NET. Microcomputer Development, 2003, 13(6), 94–97.

Control Engineering and Information Systems – Liu (Ed)
© 2015 Taylor & Francis Group, London, ISBN 978-1-138-02685-8

The application of the slight structure in the southern eight districts of Sanan oilfield

Z.D. Li, P. Liang, S. Liu & M. Li
Petroleum Engineering Institute, Northeast Petroleum University, Daqing, China

ABSTRACT: After years of exploration and development of the one district of Sanan Oilfield South Middle East. The more substantial structural reservoir has been fully developed into the extra high water cut stage. To achieve the efficient development purposes, it must have the finer prediction of the remaining oil. Analysis the causes of the area Sa I reflective layer micro-amplitude structural, the classification and characterization, and use the seismic, logging and other data, to have the trend surface analysis, time slices, coherent body and well point data grid mapping and other technical means to identify and explore the micro-amplitude structural traps. This forms a set of targeted technical methods, guiding the role of injection and production development of the remaining oil, which play a significant part in the scrolling exploration and development of oil and gas fields in the old quarter.

1 INTRODUCTION

So far with the development of Daqing Changyuan, the structure has been recognized gradually in-depth. Reservoir geological studies are now unable to meet the requirements of oil field development. Their main contradictions are: relying solely on the well logging data to predict the well sand will have the low accuracy, inter-well space remaining oil is large, and need well earthquake combined figure out the slight structural characteristics. Since the completion of the 3D seismic acquisition work of Daqing Changyuan Sanan oil field and the Saertu field, the development of seismic reservoir prelude. With the deepening of the degree of exploration, micro-amplitude structure exploration and development is increasingly broad, the required accuracy is also getting higher and higher, and the local structures of micro-amplitude discovery and implementation of oil and gas exploration and deployment is of a non-negligible problem, to study the micro-amplitude structure mapping method has become the key to the exploration and development.

2 GEOLOGY

Sanan Oilfield: The center of the Songliao basin depression Daqing placanticline, on a three-tier structure. In the northernmost of the Daqing placanticline, south with Saertu oilfield phase. Sanan oil field is a structurally controlled short axis of the anticline gas cap reservoir. Exploration and development wells drilled show that La South zone in the Middle East is located in the oilfield south eastern part of the block, the Micro-amplitude structure is not prominent in the work area well, micro-amplitude structural identification is difficult to determine.

3 SLIGHTLY TECTONIC ORIGIN AND CLASSIFICATION

3.1 Slightly tectonic origin

There are two main types of Sa I emission layer of micro-amplitude structure: One class is due to downcutting before the deposition of the sand body, differential compaction and sedimentary terrain formed slightly structure has nothing to do with the tectonism. The other is due to the faulting often associated small micro nose along both sides of the fault or micro-off recess, its possible causes for renting in different parts of the descending speed is decreased resulting in unequal. Decreased slowly part of convex, faster part of the concave; up thrown due to uneven drag force, drag and strong at the concave weak at relatively convex. The above factors often interact, the combined effects of. In addition, due to the impact of macro-tectonic setting, constructed flat, small ups and downs to the formation of small high or low, steep structure of these fluctuations often form small nose structure and small groove.

3.2 Slightly structured classification

Different micro-structure type has different flooded law, the remaining oil distribution are quite different. Based on sand body top, bottom micro fluctuation form the micro-structure type is divided into positive, negative to two slightly constructed types, specific categories as shown in Figure 1. Forward micro structure due to the higher structural position, the low degree of flooding, water cut rising slower development of the late distribution of remaining oil and tapping the potential favorable position, while the negative to the opposite type.

Usually developed in the reservoir top and bottom undulating form with the surrounding terrain is relatively high compared to the region, according to its form can be divided into micro anticline, micro nose, slightly broken nose these three types.

Micro-nose: The nose structure refers to the undulating shape of the top of the reservoir bottom is relatively high compared with the surrounding terrain contour is not closed micro-geomorphic units.

Slightly broken nose: A combination of microgeomorphic units by faults and structural nose, closed height. Generate in the downthrown settlement slower part on the convex formed micro-off nose structure; disk on the rise due to the uneven drag force, the drag force is weak at relatively convex form the structure of the micro-broken nose. The nose structure constructed micro-off amplitude and relatively large scale; explore the potential of the remaining oil.

Micro-syncline: The small syncline often developed in the bottom surface of the sand, to river down cutting often formed in the bottom of the thick sand amplitude larger syncline. Sometimes the small syncline and small anticline inter-phase appear. The amplitude difference is generally 2 to 4 m, a closed area of 0.2 km^2.

Microgroove: Micro-geomorphic units corresponding to the nose structure, morphology and nose echoes, but in the opposite direction, it is not closed at the low-lying. The common micro grooves of the study and micro-structural nose interphase appear, is the local low areas.

Micro-off ditch: Similar to the micro-off nose of tectonic origin, caused mainly due to uneven force on both sides of the fault. Generated concave in the downthrown settlement faster part of the formation of micro-off ditch; disk on the rise due to the uneven drag force, the drag force strong at relatively concave form micro-off ditch.

4 THE IDENTIFICATION OF THE SLIGHTLY CONSTRUCTOR

Due to the reflective layer TS1-phase axis continuity and wave group is preferably characterized, and the trap area is small, low amplitude structure constructed for slightly structure. Manually identify is the main method. Trend surface method, time slicing and coherent identification means supplemented by progressively recognized. Finally, through a small mesh contour validation to determine the distribution of the final micro-amplitude structure.

4.1 Tectonic trend surface analysis

Application of trend surface analysis methods highlight the Bureau edged construct local highs and lows, and further help user the identification of the slightly constructor, its method is the application multiple regression principle, calculate a mathematical curved surface to fit the trend of regional changes in the data, this mathematical curved surface is called "trend surface". For a group of geological data, calculate its trend surface. As a basis, break out will rest on these data (Fig. 2a).

4.2 Data volume time slice

Horizontal slice has a unique role in the identification of small faults and slight structure; it shows a time all seismic phase axes, lines of intersection of each phase axis is inclined reflecting interface with respect to the horizontal plane to display, which indicates the direction of the reflecting interface. The width of the phase axis can indicate the inclination of the strata and the frequency variation of the reflected wave in-phase axis (Fig. 2b).

4.3 Coherence technology

Coherence technology, quantify dealing with earthquake coherent properties, generating interpretable faults and hidden stratum structure (slightly constructor) image, to highlight irrelevant earthquake data. Basically, this process does not explain the

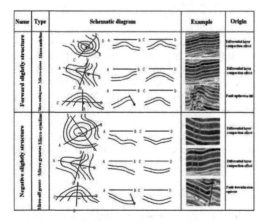

Figure 1.   Slight structure classification model.

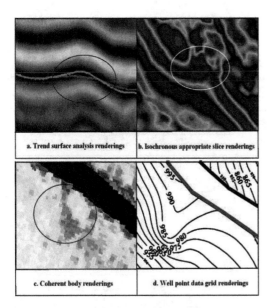

a. Trend surface analysis renderings   b. Isochronous appropriate slice renderings

c. Coherent body renderings   d. Well point data grid renderings

Figure 2. The slight structural identification renderings.

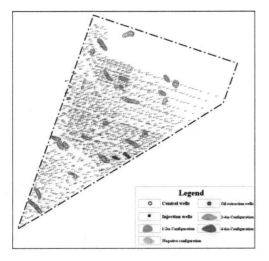

Legend

○	Control wells	●	Oil extraction wells
•	Injection wells		2-4m Configuration
	1-2m Configuration		4-6m Configuration
	Negative configuration		

Figure 3. Sa I pay sets the top interface slightly structure maps.

experience and subjective judgment of the personnel participation, is entirely dependent coherence degree of difference in the data body, so that more reliable interpretation of the results and geological knowledge (Fig. 2c).

4.4 Well point data grid mapping

Slightly trap area is small, the structure amplitude and low. Using by well altitude and make up the heart height and hole deviation correction, in the dense well under the conditions of the use of small mesh 1 m contours depict the top surface of the structure of reservoir group distribution pattern, according to the contours of jitter, offset and circle (Fig. 3).

5 SLIGHTLY TECTONIC CONTROL OF THE REMAINING OIL

There is a close relationship between micro-amplitude structure and remaining oil distribution, looking slightly structure aimed at the analysis of micro-structures on the remaining oil distribution, to find out the distribution of residual oil. For low-amplitude structural reservoir, micro-structure of the original oil and gas aggregation plays an important role in controlling, and has some control over the oil-water movement in the development process. Craig said: "at a certain flow rate, with the increase of the formation dip, oil is driving up the water-flood will be improved, but if the oil drives down its drive efficiency is reduced". Water-flood development changed the original oil-water equilibrium conditions, the injected water under the action of gravity, first in the structural low (negative micro-amplitude structure) wells onrush, first formed in the structural low flooded area, and the first to reach the higher the degree of flooding, then the remaining oil is mainly distributed in the structural highs (positive micro amplitude structure).

Therefore Sa I pay sets of multiple broken nose the positive micro amplitude structures (Fig. 3). Well L10-X3122 slightly high point, for example, the geological prediction that the structural highs in Sa I top sandstone pinch-out type was well-based, well slightly high point sandstone, in accordance with the oil-water transport principle, easy to form immobile remaining oil.

REFERENCES

Chen Zhi Hong. High set high oil field area the Funing group of reservoir heterogeneity and remaining oil distribution [J]. Offshore Oil, 2003; 23 (2):51–54.

Hesth Amm Er J. Evaluation of the timedip, correlation and coherence reaps for structural interpretation of seismic data [J]. First Break, 1998; 16 (5):151–167.

Knipe R.J. Fault prediction: methodologies, applications and successes [C]. Advances in Reservoir Technology, 1997.

Li Xingguo. Micro-structure of the reservoir New Exploration [J]. Petroleum Exploration and Development, 1996; 23 (3):80–86.

Li Xingguo. Supplement micro-structure of the reservoir [J]. Petroleum Exploration and Development, 1993; 20 (1):83–89.

Liu Duo, Chen Bucco, Dongying. 61 fault block of Shahejie Formation Characteristics of sedimentary microfacies and reservoir formation control action [J]. Mineralogy and Petrology, 1998; 18 (01): 33–39.

Ma Haizhen, Yong Hok Good, Yang Wu-Yang. Seismic velocity field to establish with variable velocity ways [J]. Geophysics, 2002; 37 (1):53–59.

Ma Tao. Establishment and application of the velocity field of the Tarim Basin [J]. Geophysics, 1996; 31 (3): 382–393.

Xie Cong-Jiao, Yang Jun-Hong, Liu Ming-Sheng. Microstructure and oil and gas accumulation in China [J]. Fault-Block Oil & Gas Field, 2001; 8 (04):4–7.

Zhao Yanchao, Wang Lijun, Peng Dongling. Reservoir micro-structure of the remaining oil distribution [J]. Xinjiang Petroleum Institute, 2002; 14 (3):41–44.

Zhu Hongtao, Hu Xiaojiang, Shinco. Reservoir microstructure research and its application [J]. Offshore Oil, 2002; (01):30–37.

Control Engineering and Information Systems – Liu (Ed)
© 2015 Taylor & Francis Group, London, ISBN 978-1-138-02685-8

The Geoprobe software 3D visualization application of coherence technology to identify small faults Fang 231 block in Songfangtun oilfield

Z.D. Li, S. Liu, P. Liang & M. Li
Petroleum Engineering Institute, Northeast Petroleum University, Daqing, China

ABSTRACT: In this paper, it mainly elaborates that the seismic exploration has nearly made a new breakthrough in the seismic data interpretation with the 3D visual coherence cube technique and Geoprobe software system based on the study of coherence cube technique carries out in fang231 block of Songfangtun oilfield, with the probe curved along horizon, we can explore small fault and analyze the geological structure by analyzing the similarity of seismic signal between adjacent trace on the seismic cub. By analyzing and the actual effect, we may find coherence cube technology is an available fault's interpret method. It can improve the interpretable precision, and it also plays a guided role in the oil-gas reservoir.

1 INTRODUCTION

Since the seismic coherence cube technique was officially launched at the 65th annual meeting of SEG, the technique has been widely used and it has more obvious strengths in fault identification and special lithologic interpretation than the conventional three-dimensional data. Coherence cube technique uses 3D data to compare the similarity of the local seismic waveform. The lower coherent value point is related to the discontinuities of reflection waveform. It mainly experienced the following ages; the AMOCO Petroleum Company is the first company to propose the coherent technology. Its first-generation algorithm C1-three normalized mutual coherence processing-has a very good detection for high-quality information and has the highest resolution. The C1 algorithm is characterized by calculating the correlation coefficient of the inline and cross-line direction and composing the correlation coefficient of the primary direction. The advantage is that it has a small amount of calculation and is easy to implement. The disadvantage is limited by the material and has a large time window. The second-generation algorithm C2 any multi-channel coherent—is characterized by calculation any multi-channel seismic data coherent. It is a calculation method which based on horizontal slices or a certain time window within layers. It has the advantage of stability, noise immunity and variable window. The disadvantage is that cannot reflect the formation dip. Visualization technology is not only a means of interpretation, but also display means. Third-generation algorithm C3 which is called feature construction uses multi-channel

seismic data to compose the covariance matrix, applying multi-channel characteristics decomposition technique to obtain the correlation between the multi-channel data. The characteristics of calculation are three-dimensional seismic data coherent calculation and do not need high resolution of layers constraints, considering the dip and azimuth. It is a coherent algorithm of inclination and azimuth. Making the horizontal slices for coherence cube, it can reveal geological phenomena such as faults and lithology edge and it can provide a favorable basis to solve special problems in the oil and gas exploration.

2 TECHNICAL METHOD

Hou Bogang and Zhang Junhua has described in detail about coherence techniques principle. Domestic and foreign software company has made great achievements in the application of the theory implemented on a computer. For example the postack module of openwork software made by Landmark Company, Semblance Volum of Geoprobe software and Coherence module of Paradigm Company. Zhang Junhua in our country also improved a lot about research and methods. However, the domestic and foreign software companies which can made the full three-dimensional visualization of the coherence technology excellently is the Semblance Volum module of Geoprobe software made by Landmark company. It is based on holographic and a sense of perspective seismic data. For any seismic trace in the three-dimensional data, when underground fault and strata change, seismic reflection characteristics of the channel will be different from reflection characteristics of the seismic trace hereabout. So it can lead to the discontinuity of the local seismic trace. By detecting the degree of difference between the seismic trace, the seismic coherence cube technique can detect the information of fault or discontinuous change.

Fault detection of Semblance Volum module is based on the joint data (Combo Volume) Coherent technology. Its coherence value can be 3 or 9 modes calculation method. The 3 modes method calculates the fastest. This approach helps brows data body and check the work area roughly. The 9 mode method is more accurate which is suitable for accurate fault interpretation, but the calculation speed is very slow, so we usually select the mode method according to seismic interpretation in practical applications.

3 EXAMPLES OF APPLICATION

Study area is located in the center depression of the northern Songliao Basin; it is the interchange of the oilfield Songfangtun, Shengping oilfields and Xujiaweizi oilfields. The purpose of the study area is Putaohua reservoir. The reservoir is thin which is belonging to low permeability reservoir. Overall it has the trend that the middle is low and the both sides are high. From west to east it can be divided into the west fault zone, central fault zone and the eastern fault zone (Fig. 1). The west side is east plunging structure of the nose structure Songfangtun and most of the fractures are NNW-trending. The east side is nose structure and west plunging part of Xujiaweizi and the fractures are NE-trending. But it is controlled by the NW-trending faults on the whole. The faults of the east and west sides which cut formation into fault blocks of different sizes are more intensive and strip trend is more obvious. It forms the pattern

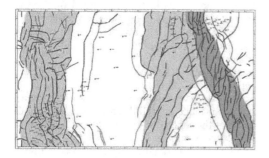

Figure 1. The breaking congruency map of Fang 231 block Putaohua reservoir t2, t1-reflective layer in Songfangtun oilfield.

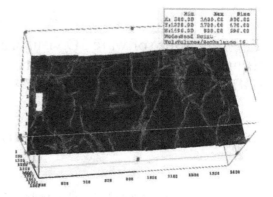

Figure 2. The three-dimensional coherent technical perspective view of Fang 231 block in Songfangtun oilfield.

of horst alternated with graben. Most of the fractures are syngenetic forward faults. The trends of faults are SN, NW, NE, EW. These faults control the direction of the strata.

By researching fault characteristics and fault tectonic evolution history, we found that the disconnecting T1-1 reflecting surface faults disconnected T2 surface and the throw ratio on T2 surface is larger than the above interface. The data indicates that the left-twist tectonic movement in the period of Qingshankou is a regional and the entire northern Songliao Basin Develops. The final tectonic stress of Yaojia group and Nenjiang group is given priority to with twist, but tectonic stress of Yaojia group in the later period is significantly weaker than the end of the Nenjiang.

4 EFFECT ANALYSIS

Tectonic fault is the permanent deformation when rock in the ground stresses reaches broken condition. The difference between small faults and the conventional interpretation of fault are throw size and plane extending shorter. We ofen called the faults whose throw distance is less than 1.5 m and extension length is less than 300 m small faults. Small faults are difficult to explain on the seismic section and have poor lateral continuity.

Fault detection and recognition is very important in the oil and gas exploration. Faults can form oil, gas, water sealing and blocking, it may be the migration pathway of oil, gas, and water as well. Ascertaining the size and distribution of the fault is one of the main objectives of the seismic exploration.

The coherent data is used to explain the fault structure, not subject to human impact. Its Interpreting results are objective. The data in a spatially

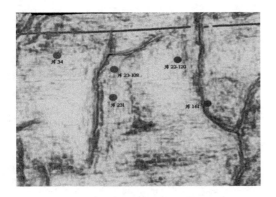

Figure 3. The local coherent perspective view of Fang 231 block in Songfangtun oilfield.

coherent is continuity so that the smaller faults are not easy to miss to play monitoring surefire interpretation results of the original interpretation of fault. But on the small faults, the application of three-dimensional coherent technology can not only study the spatial distribution of the fault, but also can study the details of the characteristics of the fault. By doing experiments again and again, we realized small time window can produce a clear exception, but it also reduces the signal-to-noise ratio and is easy to interfere with the understanding of the geology. Large time window can increase the signal-to-noise ratio, but it is easy to e blur earthquake discontinuities. Finally, the seismic trace 9, time window of 11 ms derived coherent better maps (Fig. 3). We can be seen from the figure, On the basis of the overall gray-black background, the fault is clearly visible and there are some scattered white or red light areas, for example A, B and C judged as the tiny fault or fracture development areas. After confirming the position of the small faults or cracks the tiny fault or fracture development areas, we use hour window to make local analysis processing in the region again (Fig. 3). The results are clearer and the minor faults we identified in the study area are more than 10.

The Fang 231 well block is located in fault-lithologic hydrocarbon reservoirs. This area has good physical properties and the porosity is about 20%. At the place of A, small faults have the trending of SN and the same trend with the sand body. It cannot play a blocking role.

As the compensation of low porosity and low permeability reservoir oil storage, small faults and fractures, it is important to obtain good the effect of the injection oilfield recovery. Limited by information awareness, we judged C as fracture zone. This achievement can provide a strong basis for studying the future development of injection and production and the structural study between wells.

5 AWARENESS AND DEFECTS

The coherence technology is a mature technology, playing a key role in resolving the geological problems of the faults and fractures, changes in lithology. Now we usually use a coherence technology to distinguish small faults and cracks. By analyzing coherent data we can develop breaking space grid and select the seed point automatic track of the destination layer. So the interpreters can be free from heavy work. The results will be more efficient and accurate. However, the tomography interpretation of seismic data is basically still in the stage of qualitative and semi-quantitative. At the same time due to the complexity of the changes of subsurface reservoir space and the immaturity of the geophysical techniques, manual working in the identification of small faults still has many difficulties and we seldom do quantitative study of faults work.

REFERENCES

Bogang, H., Wudabala & Zaiyan, Y. 1999. Seismic coherence technology introduction and applications, Geoscience, 121–124 March 1999.

Guisheng, X. & Yumei, S. 2000. The application of double coherent correlation method to fault detection and gap estimates, petroleum geophysical exploration, 719–723 December 2000.

Junhua, Z. & Yonggang, W. 2002. Improvement and application of coherence technology algorithm, Geophysical and Geochemical Exploration, 50–52 February 2002.

Zuchuan, C. 1997. Latest technological developments of geophysical and future trends [Chinese], geophysical equipment, 1–8 June 1997.

Control Engineering and Information Systems – Liu (Ed)
© 2015 Taylor & Francis Group, London, ISBN 978-1-138-02685-8

Visual analysis on the research of big data

C.J. Ran

Center for Studies of Information Resources, Wuhan University, Wuhan, China
Civil and Economic Law College, China University of Political Science and Law, Bejing, China

Y. Chen

School of Information Management, Wuhan, China

ABSTRACT: We choose ISI Web of Knowledge as the data source of literatures about big data and use the information visualization software CiteSpace II as the tool to analyze. The paper maps the co-citation network map to reveal the representative people and literatures of big data. Then draws the countries and institutions co-citation network map and find out the United States and the People's Republic of China are in the leading positions of the big data research. At last, the research focuses and frontier of big data are discussed.

1 INTRODUCTION

With the emergence of new ways of information dissemination, such as blog and SNS, and the rise of cloud computing, data is being accumulated at an unprecedented rate. Academia, industrial world and even government agencies are beginning to focus on big data. In 2008, Nature launched a special issue of "Big Data" [1], Computing Community Consortium published a report about "Big data computing: Creating revolutionary breakthroughs in commerce, science, and society" [2]. Today, big data has drawn increasing attention and the documents of big data research experiencing a dramatic growth, because of which, it is hard for people to grasp the core knowledge, posing difficulties for further study. This paper uses the method of information visualization to figure out the research focuses and frontiers of big data.

2 RESEARCH METHODS AND DATA

2.1 *Research methods*

CiteSpace II is an information visualization software developed by Professor Chaomei Chen from the College of Information Science and Technology, Drexel University. In recent years, it is the most influential software in the area of information visualization, which combines information visualization methods, bibliometrics, and data mining algorithms in an interactive visualization tool for extraction of the patterns in citation data. We can use this software to identify the fast-growth topical areas, find the hotspots and detect the research front [3].

2.2 *Data source*

We use the "Science Citation Index", published by the institute of Scientific Information, as the data source. In order to ensure the accuracy of the data, the search item is "TS = ((big) near/8 (data))", the retrieve time period starts form 2008 to 2013 and the retrieve time is May 10th, 2013. Totally 1067 articles are targeted. Every data unit includes author name, title, summary, data, document type, address, and reference.

3 RESULT

3.1 *Literatures co-citation network of big data*

Drawing co-citation network map is an important part of citation analysis, which can help us to analyze the literatures of high cited frequency and development process of the research. In co-citation network map, the line between two nodes means the co-citation frequency of the two literatures. The co-citation intensity is an indicator of the co-citation frequency between literatures [4]. We run the download data through the software CiteSpace II, choose "Cited Reference" as node types. Then we can get the co-citation network map as Figure 1.

From Figure 1, we can clearly see that the vast majority of the research focuses are on the clustering of "data". The circles have a growing trend

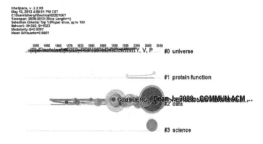

Figure 1. Literature co-citation network of big data.

over the years, which show the big data research arouses more and more attention in recent years.

Figure 1 shows the time line of leading people and literatures. The first cited literature is "The principle of relativity: a collection of original memoirs on the special and general theory of relativity" written by Einstein A in 1952. Maybe it is not the first real research paper about big data, but Einstein's theory of relativity has reference value on many research fields, so the researchers use the relevant principles to study big data.

According to the literature co-citation network, the biggest node is "MapReduce: simplified data processing on large clusters", written by Dean J in 2008, and the citation frequency is up to 14. The article introduces that mapreduce is a programming model and an associated implementation for processing and generating large datasets amenable to a broad variety of real-world tasks, and more than ten thousand distinct MapReduce programs have been implemented internally at Google over the past four years, and an average of one hundred thousand MapReduce jobs are executed on Google's clusters every day, processing a total of more than twenty petabytes of data per day [5].

The second one is "Big data: The next frontier for innovation, competition, and productivity". This is a famous report, written by McKinsey Global Institute in 2011. The report points out that the amount of data in our world has been exploding. Companies capture trillions of bytes of information about their customers, suppliers, and operations, and millions of networked sensors are being embedded in the physical world in devices such as mobile phones and automobiles, sensing, creating, and communicating data. Multimedia and individuals with smartphones and on social network sites will continue to fuel exponential growth. Big data is now part of every sector and function of the global economy. This study examines the potential value big data can create for organization and sectors of economy and seeks to illustrate and quantize that value, and also explore what leaders of organization and policy makers need to do to capture it [6].

Another important literature is "Big data: The future of biocuration", written by Howe D, Costanzo M, Fey P and other ten authors in 2008 on Nature. This is one literature of the special issue "Big Data" which we have talked above. The article is mainly in the perspective of biology to discuss big data. The exponential growth in the amount of biological data means that revolutionary measures are needed for data management, analysis and accessibility. Online databases have become important avenues for publishing biological data, and the authors also said that the field that links biologists and their data urgently needs structure, recognition and support [7].

3.2 Countries and institutions co-citation network of big data

Big Data has aroused worldwide attention, and we can use CiteSpace II to make countries and institutions co-citation networks map to explore the differences between different countries and institutions. We run the download data through the software CiteSpace II, choose "Country" and "Institution" as node types. Then we can get the network map on countries and institutions (Fig. 2).

From the countries and institutions network map, we can see that USA locates the core position and leads the research fronts of big data. In the map, the United States has more co citations with other countries and institutions, and the node is far bigger than others', so it shows that the United States is in the higher frequency of citation in the field of big data research. The centrality of the United States in the network is also the biggest one, which shows that the United States is on the core position in this field.

From the exported data from CiteSpace II, we can get detailed information about frequency and centrality of the countries and institutions (Fig. 3).

It is worth mentioning that the centrality of Chinese Academy of Science is in the second place. The largest single research institution in China,

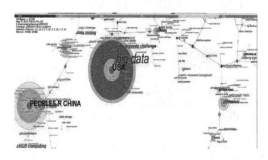

Figure 2. Countries and institutions co-citation network of big data.

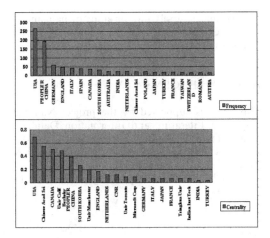

Figure 3. Citation frequency and centrality of countries and institutions.

Chinese Academy of Science makes great contribution in many research fields of natural science. This also shows that China attaches much importance to the research of big data, meeting the new challenges of science and technology activity.

3.3 The international research hotspots and research frontier of big data

Key words and terms are the core of articles, highly condensed summary of the articles. The high frequency words are often used to identify a hot area of a research field [8]. We import all the data into CiteSpace II and determine the network nodes as "Keywords" and "Term". Then we can get the network map on terms and keywords (Fig. 4).

From the exported data from CiteSpace II, we can get detailed information about hotspots of big data (Table 1).

Each round node represents a keyword and each triangular node a noun phrase. Different colors represent different years, and the thickness of the circle is corresponding with the frequency in that year. The bigger the node, the more it appears.

There is no doubt that "big data" is the most important word in this field. But beyond that, "big challenge" is the second important word which appears 44 times. Big data is very important to the development of today's society, which brings us opportunities as well as big challenges, such as lack of data scientists around the world, putting forward newer and higher requirements for data management, posing new threats to the network security and so on. If we cannot solve these properly, the terrible consequences will certainly result in "big data is a big risk". We can also conclude Figure 4 and Table 1 that the research hotspots of

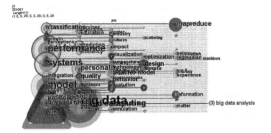

Figure 4. Hot words co-appearance network of big data.

Table 1. List of hot words with high frequency of big data.

Hot words	Frequency	Hot words	Frequency
Big data	183	Personality	17
Big challenge	44	Models	17
Model	33	Mapreduce	15
Data mining	31	Personality traits	15
Performance	30	Data analysis	15
Experimental data	27	Quality	14
Systems	25	Management	14
Cloud computing	22	Networks	14
Biggest challenge	21	Design	13

big data are the techniques, such as model, data mining, cloud computing, design and so on.

Research frontier was first introduced by Price in 1965, and it used to describe the natural transition of a research area. On the basis of the hot words network, we select "Burst" to do cluster analysis, and the timezone map of frontier can be generated. From this map, we can more clearly see the evolution trend about big data research from 2008 to the present (Fig. 5).

Kleinberg's burst detection algorithm is adapted to identify emergent research front concept. From the Figure 5, "big data" has the highest burst rate which is 47.93. So we can conclude that the basic theory of big data can also be hot in the future. The word "mapreduce" has the second high burst rate which is 5.41. Mapreduce is a functional programming model that splits the traditional group-by-aggregation computation into two steps: map and reduce [9]. In the era of big data, the traditional SQL has been unable to satisfy all the requirement, and mapreduce becomes the new trend of the data analysis.

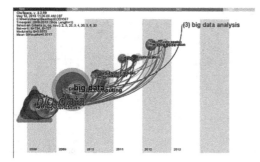

Figure 5.   Research frontier network of big data.

In addition, we can see "big data analysis" in the upper right hand corner of the Figure 5. We can say that the development of data analysis tools and the study of data analysis techniques may be the focus of future research, and it also explains why "mapreduce" has a high burst rate.

4   CONCLUSION

In this paper, the authors reveal the representative literatures, countries, institutions, research hotspots and research frontier of big data using the software CiteSpace II. In the end, we can get the following conclusions:

1. Einsein's theory of relativity has effects on big data research. In today's society, the important leading people are Jeffrey Dean, Doug Howe and so on.
2. The big data research has aroused the worldwide attention, and the United States and the People's Republic of China are in the forefront of the big data research. Chinese Academy of Science has a good performance in the study of big data.
3. Big data brings us opportunities as well as big challenges, and how to solve these challenges may the biggest problem in today's research. We can also conclude that the development of data analysis tools and the study of data analysis techniques may be the focuses of future research.

ACKNOWLEDGEMENT

This work is supported by the project National Copyright Literature and Information Database Construction of National Social Science Foundation of China Major Program (Grant No.10 & ZD133).

Chen Yi is the corresponding author of this paper.

REFERENCES

"Big data: The next frontier for innovation, competition, and productivity," McKinsey Global Institute, 2011.

Bryant R.E., Katz R.H. & Lazowska E.D. "Big-Data computing: Creating revolutionary breakthroughs in commerce, science, and society," [2013-05-10].http://www.cra.org/ccc/docs/init/Big_Data.pdf.

Chen C. "CiteSpace II: Detecting and visualizing emerging trends and transient patterns in scientific literature," Journal of the American Society for Information Science and Technology, vol. 57, 2005, pp. 359–377.

Dean J. "MapReduce: simplified data processing on large clusters," Communications of the ACM-50th anniversary issue: 1958–2008, 2008, vol. 51, pp. 107–113.

Dean J. & Ghemawat S. "MapReduce: simplified data processing on large clusters," In Proceedings of the 6th conference on Symposium on Opearting Systems Design & Implementation, 2004, vol. 6, pp. 10.

Hou Jianhua. "Visual Analysis of the Evolution and Research Fronts of Business Management," Dalian University of Technology, 2009.

Howe D., Costanzo M., Fey P., Gojobori T., Hannick L. & Hide W. et al. "Big data: The future of biocuration," Nature, 2008, vol. 455, pp. 47–50.

Nature Big Data [EB/OL]. [2013–05–10]. http://www.nature.com/news/specials/bigdata/index.html.

Zhao Rongying & Wang Ju. "Visualization Analysis on the International Information Retrieval Models," Library and Information Service, 2010, vol. 18, pp. 61–65.

Control Engineering and Information Systems – Liu (Ed)
© 2015 Taylor & Francis Group, London, ISBN 978-1-138-02685-8

An improved information content measure for semantic similarity based on biomedical ontology

Y.Y. Xing, T.L. Sun, F.Q. Yang & H.G. Sun
School of Computer Science and Information Technology, Northeast Normal University, Changchun, China

ABSTRACT: Many semantic similarity measures have been proposed to determine how similar one concept is to another within the context of ontology. Recently, researchers use intrinsic information content to compute semantic similarity only based on ontology structure. And these measures show their promising results. Intuitively, the height of a concept has an effect on the IC of the concept. Concepts with smaller height are usually more specialized and have more semantic information. In the paper, we propose a new IC computation measure combining the hyponyms of a concept, the height of each hyponym and the depth of the concept. The proposed measure is evaluated against human semantic similarity scores and compared with the existing measures using a standard biomedical ontology SNOMED CT as the input ontology. Results obtained for two benchmarks show that our measure makes semantic similarity measures better correlated with human judgments.

1 INTRODUCTION

Semantic similarity is understood as the degree of taxonomic proximity between terms. The quantification of semantic similarity between concepts is a very important problem in many research areas such as Biomedicine, Cognitive Science, and Artificial Intelligence. Due to the advantages of ontologies, many similarity measures based on ontology have been proposed these yeas such as path-based similarity measures, information-content-based measures. And they are used in sense disambiguation [1, 2], information extraction and retrieval [3, 4], classification and ranking, ontology merging [5] and so on. Because of the appearance of the biomedical ontologies such as SNOMED CT and MeSH in the Unified Medical Language System (UMLS) [6], semantic similarity shows its effect in the biomedical field.

Due to limitations of path-based semantic similarity, researchers use Information Content (IC) to compute semantic similarity between concepts. IC is a very important dimension in measuring semantic similarity between two concepts. It expresses the amount of information provided by the concept when appearing in a context, which contributes to better understanding of concepts' semantic information. Many authors [7–13] used extrinsic or intrinsic information content for semantic similarity measures. Resnik, Lin and Jiang et al. used the external source to measure the information content [7–9]. Seco et al. [10] measured the information content within ontology and they used hyponyms of concepts only. Zhou et al. [11] extended Seco

et al. measure using depth of concepts. David et al. [12] used integrated subsumers and Leaves of concepts to compute IC.

In this paper, we improve the IC measure proposed by Meng et al. [13] and propose a new IC measure using only ontology. They considered the depth of a concept, the number of its hyponyms, and the depth of every hyponym for a given concept in their measure. Intuitively, the height of a concept has an effect on the IC of the concept. We extend their measure by taking the height (height from the bottom concept) of a concept into account. Our method combines the depth of a concept, its hyponyms and the height of the concept. The experimental results show that our approach is able to provide more accurate similarity assessments with regards to human judgments and has significant performance in SNOMED CT. The proposed IC measure makes obvious improvement to Meng et al. measure.

2 RELATED WORK

2.1 IC-based similarity measures

According to the information theory, IC quantifies the amount of information that a given concept expresses when appearing in a taxonomy. Resnik [7] stated that concept similarity depends on the amount of shared information between them. The IC of their Least Common Subsumer (LCS) (i.e., the most specific common ancestor that subsumes both concepts) is used to represent the common information of concepts. If the LCS of the two concepts exists, the semantic similarity between them is defined as:

$$sim_{res}(c_1, c_2) = IC(LCS(c_1, c_2)) \qquad (1)$$

According to Resnik's measure, we find that the similarity value will be the same if any two concepts have the same LCS. To tackle this problem, Lin [8] and Jiang et al. [9] improved Resnik's measure by considering the IC of each of the two concepts.

Lin measured the similarity as the ratio between the information content of the LCS of the two concepts and the summation of information content of the two concepts.

$$dis_{\&c}(c_1, c_2) = \frac{2 \times IC(LCS(c_1, c_2))}{(IC(c_1) + IC(c_2))} \qquad (2)$$

Jiang et al. calculated the dissimilarity between concepts illustrating the similarity of concepts as follows:

$$dis_{\&c}(c_1, c_2) = (IC(c_1) + IC(c_2)) \\ - 2 \times IC(LCS(c_1, c_2)) \qquad (3)$$

2.2 IC computation

IC can help to compute the similarity between concepts and IC computation plays a vital part in IC-based similarity measures. According to Resnik's seminal work [7], IC is computed according to $p(c)$ representing the probability of occurrence of a concept c in a corpus.

$$IC(c) = -\log p(c) \qquad (4)$$

To guarantee the consistency of similarity computation, coherence of $p(c)$ computation based on taxonomical structure should be took into account. Meanwhile, both all the explicit appearances of concept c and its specializations must be considered. Thus, Resnik [7] proposed the measure for calculating $p(c)$ showed as follows:

$$p(c) = \frac{\sum\limits_{n \in w(c)} count(n)}{N} \qquad (5)$$

where $w(c)$ is the set of words subsumed by concept c and N is the total number of observed corpus terms that are contained in the taxonomy.

The classical way of measuring IC of concepts combines knowledge of their hierarchical structure from an ontology with the statistics on their actual usage in text as derived from a large corpus. However, Meng et al. [13] proposed an intrinsic measure of IC by using only the hierarchical structure

of ontology. The measure considers not only the hyponyms of the evaluated concepts but also the depth of concepts. It satisfies the explanation that concepts with more hyponyms express less information and concepts at deeper location convey more information. The IC measure is defined as:

$$IC(c) = \frac{\log(depth(c))}{\log(depth\_\max)} * \\ \left(1 - \frac{\log\left(\sum\limits_{a \in hypo(c)} \frac{1}{depth(a)} + 1 \right)}{\log(node\_\max)} \right) \qquad (6)$$

where $depth(c)$ is the depth of concept c (by node counting) in the taxonomy, $depth\_$max is the max depth of the taxonomy, a is a concept of the taxonomy, which satisfies $a \in hypo(c)$ and $node\_$max represents the maximum number of concepts in the taxonomy.

3 A NEW PROPOSED MEASURE OF INFORMATION CONTENT

Our IC measure is based on the IC measure proposed by Meng et al. [13] and the theory of Valerie Cross et al. [14]. The original method proposed by Meng et al. considers the depth and hyponyms of a concept. However the height of a concept from the artificial bottom concept might be relevant to a concept's IC. We redefine the specificity of a concept in the ontology as [15]:

$$spec(c) = \frac{height(c)}{heightC} \qquad (7)$$

where $heightC$ is the height of root concept node (the depth of the ontology), $height(c)$ is the height of concept c (by node counting), and $spec(c) \in (0,1]$. When the concept c is a root node, $spec(c) = 1$. The smaller the $spec(c)$, the more specialized the concept is. For example, in Figure 1, we can compute specificity of concept c_3 and c_4 as follows:

$$spec(c_3) = \frac{2}{4} = 0.50, spec(c_4) = \frac{1}{4} = 0.25$$

The specificity of c_3 is more than that of c_4, although they have the same depth, and c_4 with a smaller height is more specialized than c_3. The observations show that the more specialized the concept is, the smaller the height of the concept is. In other words, if a concept can not be specialized more, it has a small height from the bottom concept. So it may have more information than another concept at the same depth but at a much greater height from

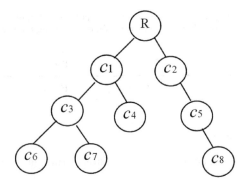

Figure 1. Ontology example.

the bottom concept. Thus, the height of a concept may have an effect on the IC of concepts.

In order to get more accurate IC, we combine the depth of a concept, the number of its hyponyms, and the height of each hyponyms according to the theory that concepts with more hyponyms express less information and concepts at deeper location convey more information. We define our IC measure as follows:

$$IC(c) = \frac{\log(depth(c))}{\log(depth\_\max)} *$$

$$\left(1 - \frac{\log\left(\sum_{a \in hypo(c)} height(a) + 1\right)}{\log(node\_\max)}\right) \quad (8)$$

Note that, if the given concept c is a root $depth(c) = 1$, and the IC of the root will be 0. If the given concept c is a leaf, it has no hyponym, and $hypo(c)$ is 0, then its IC is decided by its depth. When the depth of a leaf is the same as the depth of the ontology, its IC is 1. Thus, $IC(c) \in [0,1]$. For any other concept, its IC will be decided by not only its depth but also the height of each hyponym.

4 EVALUATION AND DISCUSSION

In order to evaluate the performance of our IC computation method, we compare our IC similarity measure with Meng et al. IC measure using classical IC-based semantic similarity measures introduced in section 2 by replacing Resnik IC with Meng et al. IC measure and ours. We calculate correlation values obtained for each of three measures against human judgments.

In our experiments, we use two biomedical datasets. Dataset 1 contains the set of 30 concept pairs and its first 10 medical term pairs are showed in

Table 1 which was annotated by three physicians and nine medical coders. The average of the similarity values of 30 concept pairs were normalized in a scale between 1 and 4. The average correlation between physicians is 0.68 while the average correlation between coders is 0.78. It shows that the correlation between coders is higher than that between physicians and we will assume that the coders rating scores are more reliable than the physician rating scores [15].

Dataset 2 contains 36 biomedical term pairs and its first 10 medical term pairs are showed in Table 2. And the human scores are the average evaluated scores of reliable doctors.

The absolute correlations with human scores using the two datasets are shown in Table 3.

Table 1. Dataset 1: first 10 medical term pairs with averaged experts' similarity scores (extracted from [16]).

Term 1	Term 2	Physician ratings (averaged)	Coder ratings (averaged)
Renal failure	Kidney failure	4.0	4.0
Heart	Myocardium	3.3	3.0
Stroke	Infarct	3.0	2.8
Abortion	Miscarriage	3.0	3.3
Delusion	Schizophrenia	3.0	2.2
Congestive heart failure	Pulmonary edema	3.0	1.4
Metastasis	Adenocarcinoma	2.7	1.8
Calcification	Stenosis	2.7	2.0
Diarrhea	Stomach cramps	2.3	1.3
Mitral stenosis	Atrial fibrillation	2.3	1.3

Table 2. Dataset 2: first 10 medical term pairs with averaged human' similarity scores (extracted from [17]).

Term 1	Term 2	Human
Anemia	Appendicitis	0.031
Dementia	Atopic dermatitis	0.062
Bacterial pneumonia	Malaria	0.156
Osteoporosis	Patent ductus arteriosus	0.156
Amino acid sequence	Anti bacterial agents	0.156
Acquired immunodeficiency syndrome	Congenital heart defects	0.062
Otitis media	Infantile colic	0.156
Meningitis	Tricuspid atresia	0.031
Sinusitis	Mental retardation	0.031
Hypertension	Kidney failure	0.5

Table 3. Correlations values obtained for measures against ratings of physicians, coders and both on 29 pairs of dataset 1 and correlations values obtained for measures against ratings of human on 34 pairs of dataset 2 using snomed ct as ontology (*p-values* are presented in parentheses).

| Measure | Dataset 1 | | | | Dataset 2 |
	Evaluated in	Physicians	Coders	Both	Human
Resnik (Corpora-based IC)	[16]	0.45	0.62	0.55	N/A
Lin (Corpora-based IC)	[16]	0.60	0.75	0.69	N/A
Jiang and Conrath (Corpora-based IC)	[16]	0.45	0.62	0.55	N/A
Resnik (IC computed as Meng et al.)	This work	0.4902 (0.0069)	0.5714 (0.0012)	0.5516 (0.0019)	0.6131 (0.0001)
Lin (IC computed as Meng et al.)	This work	0.5097 (0.0047)	0.6049 (0.0005)	0.5790 (0.0010)	0.6744 (0.0000)
Jiang and Conrath (IC computed as Meng et al.)	This work	0.4948 (0.0064)	0.5771 (0.0010)	0.5569 (0.0017)	0.7190 (0.0000)
Resnik (IC computed as our measure)	This work	0.5545 (0.0018)	0.6705 (0.0001)	0.6362 (0.0002)	0.6482 (0.0000)
Lin (IC computed as our measure)	This work	0.5835 (0.0009)	0.7123 (0.0000)	0.6729 (0.0001)	0.7121 (0.0000)
Jiang and Conrath (IC computed as our measure)	This work	0.5321 (0.0030)	0.6344 (0.0002)	0.6059 (0.0005)	0.7028 (0.0000)

There are 29 out of 30 the biomedical concept pairs in dataset 1 and 34 out of 36 the biomedical concept pairs in dataset 2 in SNOMED CT. We apply Meng et al. IC measure and our new IC measure to biomedical ontology SNOMED CT. Table 3 shows the correlation between each method against human judgments. The experimental results show that the correlations with coders are higher than that with physicians and it seems that the coders rating scores are more reliable than the physician rating scores. Our approach is able to improve Meng et al. IC similarity measure, achieving higher correlations for all three measures in the experiments. The *p-value* is used to measure the significance of the correlations, and *p-value* < 0.001 indicates that the results obtained by our measure are significant.

5 CONCLUSIONS

In the paper, we propose a new IC computation measure combining the depth of a concept, the number of its hyponyms, and the height of each hyponym. The experimental results show that our IC measure achieves significantly better results than the original method proposed by Meng et al.

As future work, we will extend our measure to multiple ontologies and try to find other semantic evidence that has effect on the IC of concepts in ontologies.

ACKNOWLEDGMENTS

This paper is sponsored by the Doctoral Program of Higher Education of China (No. 20110043110011), Jilin Provincial Science and Technology Department of China (No. 20120302, No. 20090503), respectively.

REFERENCES

Al-Mubaid H. & Nguyen H.A. "A cluster-based approach for semantic similarity in the biomedical domain", In: Conference proceedings of the IEEE engineering in medicine and biology society. New York, USA, pp. 2713–2717, 2006.

Budanitsky A. & Hirst G. "Evaluating WordNet-based measures of semantic distance," Comput. Linguistics, vol. 32, no. 1, pp. 13–47, 2006.

Cross V. & Anurekha C.-T. "Measuring Information Content for an Ontological Concept", Fuzzy Information Processing Society (NAFIPS), 2012 Annual Meeting of the North American, pp. 1–6, 2012.

Gaeta M., Orciuoli F. & Ritrovato P. "Advanced ontology management system for personalised e-Learning", Knowl.-Based Syst. Vol.22, no.4, pp. 292–301, 2009.

Hliaoutakis A. "Semantic similarity measures in MeSH ontology and their application to information retrieval on Medline," Master's thesis, Tech. Univ. Crete, Chani'a, Crete, 2005.

Hliaoutakis A. "Semantic Similarity Measures in the MESH Ontology and their Application to Information Retrieval on Medline", Technical Report, Technical Univ. of Crete (TUC), Dept. of Electronic and Computer Engineering, 2005.

Jiang J. & Conrath D. "Semantic similarity based on corpus statistics and lexical taxonomy", In: Proceedings of the international conference on research in computational linguistics (ROCLING X). Taiwan;. pp.19–33, 1997.

Kleinsorge R., Tilley C. & Willis J. (2000) Unified Medical Language System (UMLS) Basics [Online]. Available: http://www.nlm.nih.gov/research/umls/pdf/UMLS_Basics.pdf.

Lin D. "An information-theoretic definition of similarity", In: Shavlik J, editor. Fifteenth International Conference on Machine Learning, ICML 1998. Madison (Wisconsin, USA): Morgan Kaufmann, pp. 296–304, 1998.

Meng L., Gu J. & Zhou Z. "A New Model of Information Content Based on Concept's Topology for Measuring Semantic Similarity in WordNet", International Journal of Grid and Distributed Computing Vol. 5, No. 3, September, 2012.

Patwardhan S. "Incorporating dictionary and corpus information into a context vector measure of semantic relatedness," Master's thesis, Univ. Minnesota, Minneapolis, 2003.

Pedersen T., Pakhomov S., Patwardhan S. & Chute C. "Measures of semantic similarity and relatedness in the biomedical domain", J Biomed Inform, vol. 40, pp. 288–99, 2007.

Resnik P. "Using information content to evaluate semantic similarity in taxonomy", In: Proceedings of the 14th international joint conference on artificial intelligence (IJCAI 95). Montreal, Canada, pp. 448–53, 1995.

Sánchez D., Batet M. & Isern D. "Ontology-based information content computation", Knowl-based Syst, vol. 24, 297–303, 2011.

Seco N., Veale T. & Hayes J. "An intrinsic information content metric for semantic similarity in WordNet", in: Proc. of 16th European Conference on Artificial Intelligence, ECAI 2004, including Prestigious Applicants of Intelligent Systems, PAIS 2004, IOS Press, Valencia, Spain, pp. 1089–1090, 2004.

Sim K.M. & Wong P.T. "Toward agency and ontology for web-based information retrieval," IEEE Trans. Syst., Man, Cybern. C, Appl. Rev., vol. 34, no. 3, pp. 257–269, Aug. 2004.

Zhou Z., Wang Y. & Gu J. "A new model of information content for semantic similarity in WordNet", in: Proc. of Second International Conference on Future Generation Communication and Networking Symposia, FGCNS 2008, IEEE Computer Society, Sanya, Hainan Island, China, pp. 85–89, 2008.

Control Engineering and Information Systems – Liu (Ed)
© 2015 Taylor & Francis Group, London, ISBN 978-1-138-02685-8

Scheduling space for advertising on web pages

Z.H. Zhang & L. Wang
School of Electronic and Information Engineering, Liaoning University of Science and Technology, Anshan, China

Y.K. Wu
School of Architecture and Environmental Engineering, Shenzhen Polytechnic, Shenzhen, China

ABSTRACT: In order to effectively use the limited advertising space on the web page, scheduling web advertising space was studied in this paper based on the sizes and the display numbers of the advertisings. On the basis of previous studies, the planning of a single Web page was extended to the planning of multiple pages in the paper. A decision-making model of multiple pages was established and two heuristic algorithms were proposed to solve the model. By introducing the heuristic algorithm into genetic algorithm, the hybrid genetic algorithm was proposed. Simulation results verify that the hybrid genetic algorithm is more effective than the heuristic algorithms.

1 INTRODUCTION

At present, Internet users in China have reached more than 500 million. So many Internet users will lead to a huge potential advertising market. Because of being the characteristics which are traditional advertising can not match with, Internet advertising has shown tremendous commercial potential and widely development prospects. Increasing web page space utilization is an important way to improve the site effectiveness and can bring considerable economic benefits. Therefore, the problem of scheduling advertising space on web page will be concerned by the website administrator and has very important practical significance.

Web space configuration is that several different ads can be displayed on the same position of the advertising space. It can be realized by periodically updating advertisings on the same advertising space. Such space configurations that make the users see a variety of advertisings in the same position on the advertising space and improve the overall effectiveness of the ads. The issues for minimizing and maximizing advertising space scheduling on a web page were proposed (Adler et al. 2002). Minimization problem is to find a scheme for minimizing the height of the ads placement in the scheduling cycle when all advertisings can all be placed on the web page. Two problems are similar to the packing problems. LSLF and SUBSET–LSLF algorithms were respectively proposed based on the existing algorithms of the packing problem. Dawande et al. (2001a, b 2003 2005) & Kumer et al. (2000 2001 2006) proposed respectively the

variety of the improved algorithms. LFLF algorithm was improved algorithm of minimization problem. MIN algorithm was proposed for solving minimization advertising space scheduling (Zhao et al. 2010). In the paper, the problem of minimizing advertising space scheduling was further researched and extended to multiple pages.

2 PROBLEM DESCRIPTIONS

Scheduling ads space on web pages was the period scheduling. This scheduling time is called the scheduling cycle. In order to display a variety of ads to the user, it requires that advertisings be updated periodically. So the scheduling cycle is further divided into equal slots. It is known to place a group of ads in the scheduling cycle. Advertising A_i is specified by its geometry s_i and display numbers ω_i during this period.

Generally, there are two kinds of relationships between the geometry of the ad and the ad space. A case is that the width of the advertisement is equal to the width of the scheduling spatial position, such as advertisements on the side of the web page. Another case is that the height of the advertising is equal to the height of the scheduling spatial position, such as advertisements on the top and bottom of the web page. The first case was only studied in the paper.

Since the width of each ad is equal to the width of the ad spatial location, it only considers the height of ads for ad sizes. The height of the advertisement represents the size of ads in the web page.

At the same time, advertisers do not wish that the ads display more than once within the same slot. Therefore, advertisement A_i is only displayed at most once within each slot and displayed in ω_i different slots. So the scheduling height of the web page is the maximum height of all slots scheduling. Therefore, the minimization problem of the web advertising space scheduling is to find the project which makes the placement height of the web advertisings smallest.

3 MODEL PROPOSAL

A group of ads $A = \{ A_1, A_2, \ldots, A_n \}$ are known to put in certain type of pages. This type of pages is m pages. Supposing that ad A_i is only allocated to one of the pages, the frequency of ad A_i is ω_i in one page. In practice, the same ad might be put on multiple pages. We can think of the same ad on different pages as different ads. According to the above description, the mathematical model is as follows:

$$\min(\max F_j)$$

$$s.t. \ F_j = \max_k \left(\sum_{i=1}^{n} s_i x_{ij} y_{ik} \right) j = 1, 2, \ldots, m, \ k = 1, 2, \ldots, N \tag{1}$$

$$F_j \leq H_j \ j = 1, 2, \ldots, m \tag{2}$$

$$\sum_{j=1}^{m} x_{ij} = 1 \ i = 1, 2, \ldots, n \tag{3}$$

$$\sum_{j=1}^{m} \sum_{k=1}^{N} x_{ij} y_{ik} = \omega_i \ i = 1, 2, \ldots, n \tag{4}$$

$$x_{ij} = 0, 1 \ i = 1, 2, \ldots, n, \ j = 1, 2, \ldots, m \tag{5}$$

$$y_{ik} = 0, 1 \ i = 1, 2, \ldots, n, \ k = 1, 2, \ldots, N \tag{6}$$

where F_j = the scheduling height of page j in a certain scheduling period; H_j = the fullness height of page j; s_i = the size of ad A_i, which also refers to the height of ad A_i; ω_i = the display numbers of ad A_i in a certain scheduling period, or also known as the frequency of ad A_i; N = the numbers of slots divide a certain scheduling period for each page; $x_{ij} = 1$, if ad A_i is assigned on page j; otherwise, $x_{ij} = 0$; $y_{ik} = 1$, if ad A_i is assigned on slot k; otherwise, $y_{ik} = 0$.

Objective function of the problem is to find minimum height of ads placement on a certain type of web pages in certain scheduling period. Equation (1) shows that the scheduling height of ads should be the maximum height of the ads scheduling height in each slot on page j. Equation (2) ensures that the fullness of any slot should be less than the

fullness height of page j. Equation (3) guarantees that ad A_i only display on one of the pages. Equation (4) guarantees that ad A_i is assigned to exactly ω_i slots on page j if it is selected.

4 RESEARCH METHOD

Minimizing space scheduling is equivalent to the parallel machine scheduling. Based on LPT and Multifit algorithms about the parallel machine scheduling problem (Zhao 2001), two algorithms were proposed in this paper. First, LHD (Largest Height Decreasing) heuristic algorithm was proposed, and then FFHD (First Fit Height Decreasing) algorithm with more effective was proposed based on LHD algorithm, and finally hybrid genetic algorithm GA-FFHD was proposed by introducing FFHD algorithm into genetic algorithm.

4.1 LHD algorithm

The specific steps of LHD algorithm are as follows:

Step 1 For $A = \{A_1, A_2, \ldots, A_n\}$, ad A_i placement size is $v_i = s_i \times \omega_i$.

Step 2 According to v_i, the group of ads are sorted from largest to smallest. If $v_i = v_i + 1$, and $s_i < s_{i+1}$, A_{i+1} rows in front of A_i.

Step 3 According to F_j, all pages are sorted from smallest to largest, $P = \{P_1, P_2, \ldots, P_j, \ldots, P_m\}$, let $j = 1$.

Step 4 The slots on page P_j are sorted from smallest to largest according to the scheduling height.

Step 5 According to the arranged order, ads are placed. If ad A_i can be scheduled on page P_j, it is arranged in the top ω_i slots on page P_j, go to Step 7.

Step 6 Let $j = j + 1$, go to Step 4.

Step 7 The set of ads $A = A - \{A_i\}$, if $A \neq \Phi$, go to Step 3; otherwise, scheduling height SH = max F_j.

SH which is a type of pages space planning height minimum in a certain planning cycle is obtained.

4.2 FFHD algorithm

Based on LHD algorithm, FFHD algorithm improves the quality of the solution. The specific steps of FFHD algorithm are as follows:

Step 1 Let $p = 1$, iteration numbers K is set, let

$$CL = \max \left\{ \frac{1}{mN} \sum_{i=1}^{n} s_i, \max_{1 \leq i \leq n} \{s_i\} \right\}$$

let $CU = OPT_{LHD}$,

Which, CL denotes the initial lower bound of web page space scheduling. CU denotes the initial upper bound of web page space scheduling. OPT_{LHD} is the scheduling height which is solved by LHD algorithm. m denotes the numbers of the scheduling page.

Step 2 For $A = \{A_1, A_2, ..., A_n\}$, ad A_i placement size is $v_i = s_i \times \omega_i$.

Step 3 According to v_i, the group of ads are sorted from largest to smallest. If $v_i = v_i + 1$, and $s_i < s_{i+1}$, A_{i+1} rows in front of A_i.

Step 4 Let C = (CL+CU)/2 is the maximum scheduling height on current web page.

Step 5 According to F_j, all pages are sorted from smallest to largest, $P = \{P_1, P_2, ..., P_j, ..., P_m\}$, let $j = 1$.

Step 6 The slots on page P_j are sorted from smallest to largest according to the scheduling height.

Step 7 According to the arranged order, ads are placed. If ad A_i can be scheduled on page P_j, it is arranged in the top ω_i slots on page P_j, go to Step 9.

Step 8 If $j < m$, let $j = j+1$, turn to Step 7; else let CL = C, turn to Step 11.

Step 9 The set of ads $A = A - \{A_i\}$, if $A \neq \Phi$, go to Step 5.

Step 10 If $A = \Phi$, let CU = C.

Step 11 If $p = K$, terminate, the scheduling height SH = max CU; else $p = p+1$ go to Step 4.

4.3 GA-FFHD algorithm

The solution of Heuristic algorithm has often a lot of limitations. GA is an effective the global parallel optimization search tool. But it has lack of local search ability, and heuristic algorithm based on knowledge has strong local search ability (Li et al. 2002). In the paper, a heuristic hybrid genetic algorithm was proposed by introducing FFHD algorithm into GA.

For the space scheduling problem, the symbol coding method was adopted in this paper. Each gene of the chromosome represents the page on which the ad is placed. The sorting selection method and the optimal preservation strategies were used in this paper (Zhou & Sun 1999).

Web advertising minimization space problem is extended from single page to the type of pages in this paper. For solving this problem, it firstly finds out each set of ads which will be scheduled on denoted scheduling page, and then each set of the selected ads are scheduled according to single page scheduling.

The concrete steps of GA-FFHD algorithm are as follows:

Step 1 Initialize the crossover probability p_c, the variation probability p_m, the maximum of generation max_gen, the size of population pop_size, set the initial number of generation $gen = 1$.

Step 2 Generate the pop_size number of chromosomes.

Step 3 Schedule ads on the pages for each chromosome.

Step 3.1 Set $j = 1$.

Step 3.2 Find out the set of ads $A_j = \{A_{j1}, A_{j2}, ..., A_{jr}, ..., A_{jR}\}$, which will be scheduled on page j. R is the number of the placement ads on page j.

Step 3.3 Adopt FFHD algorithm to schedule the set of ads A_j on page j.

Step 3.4 If the set of ads A_j can be all placed on page j, go to Step 3.11.

Step 3.5 Find ad A_{jr} from ads, if the set of ads can not be all placed on page j. Ad A_{jr} denotes the first ad of A_j which can be placed on page j.

Step 3.6 Sort all pages except page j in ascending sequence according to the scheduling height F_j. Obtain the sequence $T = \{T(1), T(2), ..., T(t), ..., T(m-1)\}$, $t \neq j$.

Step 3.7 Set $t = 1$.

Step 3.8 Take out page P correspond to $T(t)$, arrange ad A_{jr} in the ω_{jr} slots which are sorted in ascending sequence on page P. ω_{jr} denotes the display frequency of ad A_{jr}.

Step 3.9 If page P has not enough space to hold ad A_{jr}, set $t = t + 1$, go to Step 3.8.

Step 3.10 If page P has enough space to hold ad A_{jr}, set $A_j = A_j - A_{jr}$, $A_p = A_p + A_{jr}$. If all ads are scheduled in the set of ads A_j, go to Step 3.11; otherwise go to Step 3.3.

Step 3.11 Calculate the scheduling height F_j of page j.

Step 3.12 If $j < m$, set $j = j+1$, go to Step 3.2.

Step 4 Calculate the fitness function Fitness = −max F_j.

Step 5 Adopt the sorting selection method to select the chromosomes in the population.

Step 6 Adopt two-point crossover according to the crossover probability p_c.

Step 7 Adopt single-point variation according to the variation probability p_m.

Step 8 Adopt the optimal preservation strategies.

Step 9 If $gen = max\_gen$, obtain current optimal individual and the height scheduling SH, terminate; else, $gen = gen+1$, go to Step 3.

5 RESULTS AND DISCUSSION

5.1 Parameters setting

The above algorithm was simulated by MATLAB7.0. The three cases of $m = 5$, $n = 40$; $m = 10$, $n = 80$ and $m = 20$, $n = 160$ were respectively calculated. The scheduling period is divided into $N = 5$ slots. The maximum height of space scheduling on all pages, $H_j = 15$. In FFHD algorithm, set

Table 1. GA parameters setting.

Setting	max_gen	pop_size	p_c	p_m
$m = 5, n = 40$	100	100	0.8	0.05
$m = 10, n = 80$	100	100	0.8	0.05
$m = 20, n = 160$	100	120	0.8	0.1

Table 2. Algorithms comparison.

Setting	LSLF	LFLF	LHD	FFHD	GA-FFHD
$m = 5$, $n = 40$	9	10	8	8	7
$m = 10$, $n = 80$	10	11	8	8	7
$m = 20$, $n = 160$	9	11	9	8	7

$K = 10$. The size and the display frequency of ads are set as follows:

$s_i = 4, w_i = 1, i = 8(j - 1)+1, s_i = 4, w_i = 2,$
$i = 8(j - 1)+2,$

$s_i = 3, w_i = 2, i = 8(j - 1)+3, s_i = 3, w_i = 3,$
$i = 8(j - 1)+4,$

$s_i = 2, w_i = 2, i = 8(j - 1)+5, s_i = 2, w_i = 1,$
$i = 8(j - 1)+6,$

$s_i = 1, w_i = 1, i = 8(j - 1)+7, s_i = 1, w_i = 1,$
$i = 8j,$

$j = 1,2, ..., m.$

GA parameter settings are shown in the Table 1.

5.2 Simulation results

According to the settings of the above parameters, the algorithms of LSLF, LFLF, LHD, FFHD and GA-FFHD were compared. The results were shown in Table 2.

From Table 2, it can see that LSLF algorithm is better than LFLF algorithm, and the solution is closer to the optimal solution. On the whole, LHD algorithm is better than LSLF algorithm. For the cases of $m = 5$, $n = 40$ and $m = 10$, $n = 80$, LHD algorithm have found better solution than LSLF algorithm. But for the large scale ($m = 20$, $n = 160$), LHD algorithm doesn't find better solution than LSLF algorithm. For solving the case of the large scale, FFHD algorithm is better than the LHD algorithm. GA-FFHD algorithm is best in several algorithms and finds the optimal solutions for three cases.

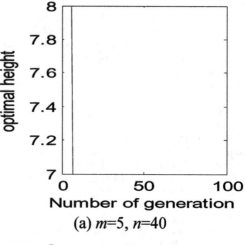

(a) $m=5, n=40$

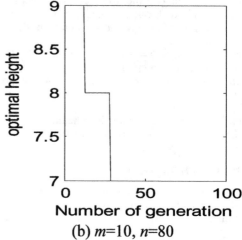

(b) $m=10, n=80$

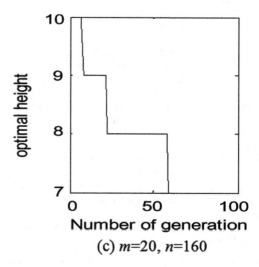

(c) $m=20, n=160$

Figure 1. The optimization performance curve of GA-FFHD.

The optimization performance curve of GA-FFHD algorithm is shown in Figure 1.

The abscissa is the evolution of GA and the ordinate is the optimal solution of the problem. With the expansion of the problem scale, the generation of GA-FFHD algorithm is also increasing. But it ultimately finds the optimal solutions in the 100 generations.

6 CONCLUSIONS

The space scheduling for multiple pages was studies in the paper. Through modeling and solving the problem, it shows that GA-FFHD algorithm is better than previous heuristic algorithms.

Owing to the complexity and randomness of internet, the above study is just preliminary theoretical research. In order to making better use of internet and developing of e-commerce, many researches need to be further improved. The problems to be solved in the future are mainly reflected in the following aspects:

- Web advertising space scheduling will associate with the pricing models of ads. By comparing with the pricing models, it studies the effects of different pricing models on the website income.
- The theories of web advertising decisions are applied to the actual website and combine with the actual situation, which can better solve the problem.

REFERENCES

Adler, M., Gibbons, P.B. & Matias, Y. 2002. Scheduling space-sharing for Internet advertising. *Scheduling* 5(2): 103–119.

Dawande, M., Kumar, S. & Sriskandarajah, C. 2001a. Scheduling advertisements on a web page: new and improved approximation algorithms. *Electronic notes in discrete mathematics* 7: 59–62.

Dawande, M., Kumar, S. & Sriskandarajah, C. 2001b. Improved performance bounds for web advertisement scheduling. *The 11th annual workshop on information technologies and systems; Proc. conf., New Orleans, December 2001.*

Dawande, M., Kumar, S. & Sriskandarajah, C. 2003. Performance bounds of algorithms for scheduling advertisements on a web page. *Scheduling* 6(4): 373–393.

Dawande, M., Kumar, S. & Sriskandarajah, C. 2005. Scheduling web advertisements: a note on the MINSPACE problem. *Scheduling* 8(1)1: 97–106.

Kumar, S., Jacob, V.S. & Sriskandarajah, C. 2000. Scheduling advertising at a web site. *The 10th annual workshop on information technologies and systems; Proc. conf., Brisbane, December 2000.*

Kumar, S., Jacob, V.S. & Sriskandarajah, C. 2001. Hybrid genetic algorithms for scheduling advertising on a web page. *The 22th international conference on information systems; Proc. Intern. conf., New Orleans, 31 December 2001.* Berkeley: AIs elecronic library.

Kumar, S., Jacob, V.S. & Sriskandarajah, C. 2006. Scheduling advertisements on a web page to maximize revenue. *Operational research* 173(3): 1067–1089.

Li, M.Q., Kou, J.S., Lin, D. & Li, S.Q. 2002. *The basic theory and application of genetic algorithms.* Beijing: Science.

Zhao, M.Y. 2001. *The introduction of combination optimization.* Hangzhou: Zhejiang science and technology.

Zhao, W.D., Wang, D.W., Guo, X.P. & Wang, G.C. 2010. Modeling and algorithm of web advertising resources optimization. *The 29th Chinese control conference; Proc. conf., Beijing, 29–31 July, 2010.* Beijing: The electronic magazine.

Zhou, M. & Sun, S.D. 1999. *The principles and applications of genetic algorithm.* Beijng: National defense industry.

Control Engineering and Information Systems – Liu (Ed)
© 2015 Taylor & Francis Group, London, ISBN 978-1-138-02685-8

Contactless smart card Mifare DESFire EV1—multi-application platform

R. Holý & M. Kalika
Faculty of Transportation Sciences, Czech Technical University in Prague, Prague, Czech Republic

ABSTRACT: Contactless chip technology has its supporters and opponents. Supporters favor the clearance rate, longer life and other advantages, opponents refer to "chip totality" and loss of privacy. To further extension, it is necessary to define an adequate level of safety in unique user identification and rules that are defined in the security implemented in all systems where the identifier is used. For this purposes, specific infrastructure information systems are used.

1 INTRODUCTION

Currently, when development of applications and technologies is definitely aimed at global use, such trend must be supported by unequivocal identification of users. The basic principles support safety elements (integrity, indisputableness, resistance to known and supposed forms of attacks) and do not link application to a physical carrier (plastic). The application can be hosted on a bank payment card or NFC, etc.

To reduce safety threats the concept of unequivocal identification using ID cards and conditions, which they technologically produce, are used. Individual technologies used can be defined in ISO/IEC 14443 contactless chip and infrastructure based in HSM (Hardware Security Module) and individual SAM (Secure Application Module) modules can be defined by key management of the chip content.

2 CONTACTLESS CHIP

When designing the systems we worked on the assumption that all functions should be authorized using the keys. So the keys must be present in the chip card, reading module and back-end system. Due to susceptibility to attack, the keys in chip cards should be derived from safely stored main keys (e.g. in HSM).

To maintain safety of cryptographic mechanisms, cryptographic keys must be sufficiently protected. Protection at software level is ensured by firmware and card settings in compliance with FIPS 140-2 Level 3 certificate.

There is a point in talking about administration of keys only if all parts of the system, including the chip card, provide safety storage for stated keys, and especially support adequate cryptographic operations. Chip cards equipped with MCU meet these requirements; they support symmetrical encryption using 3DES, AES or asymmetrical encryption using RSA, elliptic curves. For implementation we compared two possible forms of authentication scheme:

- Application of more expensive chip cards and cheaper control panels using RSA encryption,
- Application of cheaper cards and more expensive control panels (with SAM modules) and symmetrical encryption using 3DES or AES.

In case of application of asymmetrical cryptography, there is no need to keep master keys in the reading modules, which significantly eases design of the system and subsequent maintenance, as there is no need to take care of SAM modules.

2.1 Technology

As for technology, we used contactless chips with Mifare DESFire 8 K EV1 MOA4 modules. Structures of the applications are based on the record as per MAD3 (MIFARE Application Directory ver. 3).

The card contains $0 \times FFFFFF$ application with files

- Of 0 type Value and the value of 0×03
- Of 1 type "Standard Data File", size = 48B (CHS)
- Of 2 type "Standard Data File", size = 24B (CPS)
- Of 3 type "Standard Data File" size = 56B

Keys:

MAD V.3 AMK	APL_FFFFFF_00
MAD V.3 (R CPS)	APL_FFFFFF_01
MAD V.3 (W CPS + CHS)	APL_FFFFFF_02
MAD V.3 (R CHS)	APL_FFFFFF_03
MAD V.3 (R version)	APL_FFFFFF_04

In MAD application, there is a file of Standard Data type, which contains electronic signature of UID chip serial number. The file has set the access rights for "free access" ($0 \times 0E$) and "plain mode" and can be read without necessary authentication to MAD 3.0 application. Information about the cardholder is stored in CHS (Card Holder Sector) and information about the card publisher and card validity is stored in CPS (Card Publisher Sector). In case of identification of the publisher, we draw from the specification of the norm of ČSN ISO/IEC 7812.

2.2 Intelligent SAM

Access of various subjects (service providers) to applications in the card is ensured especially by the logic of SAM not by assigning reading/recording keys and card structures. This means that various subjects will be assigned SAMs with the same file of keys (set of keys) but with different logic (configuration) for work with data in the card. The service provider that accepts the conditions of the system assumes responsibility for SAM.

Acceptance Devices (AD) working with the card do not know its structure. By this reason the risk of errors incurred during work with the card in different systems due to possible different interpretation of the application structure elements is reduced. Simultaneously, unique interface to SAM, which applies for all card types (DESFire, other clever cards), is ensured.

When changing content of the data contained in the memory of SAM chip in virtue of enhancing safety, it simultaneously writes its identification, identification (signature) of the provider that carried out relevant record, into the card, administers its own counters and counters in the card, validates integrity and other safety procedures so that AZ cannot evade them.

Thanks to the intelligent structures in the card and involvement of the logic of SAM, we can ensure that data of various organizations (universities) are parallel stored in one application. Concurrently, various levels of sharing or limitation of access to data between different organizations (universities) can be defined.

Behaviour of SAM is driven by its configuration; hence the possibilities of usage and behaviour of SAM can be made more precise even for individual operations. So SAM intended for devices with various technical possibilities and ways of usage can be differentiated just by configuration. For example SAM can be configured for devices regarding whether:

- It allows entering PIN (allows to carry out operations requiring user's confirmation),
- It is permanently connected online to HSM (if not, time can be saved and no attempts to connect are carried out),
- The device allows reading of other providers' records (—allows displaying complete information about the card too).

As for the system of multi-application cards with more card publishers, building central administration/ distribution of SAM is necessary.

2.3 Safety risks of DESFire EV1 card

When implementing the system using DESFire EV1 card, it is required to consider several safety restrictions, by which the card is limited. If the design of structures in the card does not take them into account, potential attacker could relatively easily use them to gain personal benefits or to damage other people's cards. Especially the following situations are considered:

- Moving the card from the field of the reader/ turning off the field of the reader (e.g. in combination with wiretap it allows the attacker to prevent from recording some data into the card.
- Counterfeiting the commands of Commit Transaction, Abort Transaction by the attacker anytime during communication of SAM with the card. As no operation with DESFire EV1 card is atomic and these commands are not protected by CMAC, the attacker can set the card to inconsistent state or can have only specific data confirmed to the card.
- Other way to discontinue communication (unauthorized incorporation of a command not requiring authentication).
- Alienation of SAM and its misuse for attacks to the system. By moving the logic of work with card to SAM the radius of possible attacks to the card and system is reduced because, in general, no (potentially problem) data cannot be written to the card. In addition, SAM is equipped with a counter that allows only certain number of operations of given type. Then it is blocked and its counter must be reset using the cryptogram from HSM.

2.4 Process safety principles

- Blacklists of cards and applications. The option to make individual applications inaccessible is the basic rule when blocking in the card. Blacklist of entire card takes precedence over blacklist of application (blocking of entire card blocks other applications).
- Blacklist of SAMs—When loosing SAM, it is blocked and a record to the blacklist of SAM is carried out. Acceptance Device (AD) will control the blacklist of SAM during each transaction:
  - Identity of SAM inserted in relevant AZ
  - Identity of SAM in the data stored in the card during card acceptance.

  AZ will reject any logic application recorded by SAM, which was in the blacklist in the moment of record creation.
- Transactions—SAM signs the means of asymmetrical cryptography as transactional (signature key to SAM enters subject connected to the system during SAM authorization) with no chance of influencing by the acceptance device. Thanks to the possibilities, which are offered by PKI we can verify indisputableness of the transaction sentence anytime, without previously knowing the key of SAM used for signature. Reliable infrastructure in the currency of transactions among any number of participants to the system can be built on such basis. (Knowing the weak point that the acceptance device can intentionally provide SAM with wrong information concerning the time of transaction.)

2.5 Architecture of a card with dynamic structures

The lowest possible number of (physical) applications is stored at CC; it means minimum of required keys and so minimum have required authentications (with the aim to reach minimizing of the time needed for work with the card). Assignment of keys to files is not carried out ad-hoc but it is based on logic range of files (static information about the card, studies…) and access roles of the device (e.g. of the card manufacturer, card publisher, HSM, seller, etc.).

The files that are logically related and worked with simultaneously are associated under the sole ID of the application (AID) and common keys (where possible). This saves time (for authentication) and size of the memory (stored keys). Different files that are in various applications (different AID) cannot use common authentication because it is the application, not the file, that is the authentication unit in DESFire/EV1 (the process of authentication always proceeds against the application and then it is possible, within this application, to work with all the files, to which the given key grants a right with no need to authenticate again).

Structures in the card are created especially with regard to universality of their use (trivial usability of the same space in a file for various data structures) and simultaneously for the fastest possible speed of work with the card. That is why so-called dynamic structures were used when designing. Drafts of structures had to be optimized with respect to analysis of behaviour of DESFire EV1 card where the speed of reading and recording of differently extensive blocks of data located in different sized files of any type supported by the card were compared.

2.6 Basic logic applications in the card

Individual logic applications must ensure maximum possible level of interoperability among individual providers across the wide spectrum of the types of services. On the other hand, it can be expected that the cardholder will, in most cases, use a small subset from potentially big offer of services. That is why the structure of the card must be designed so that the space that remains unutilized in the card is minimised.

2.7 Application

Applicability of the chip memory is bound to its size, or the number of addressable applications. Architecture is designed so that the method of progressive building of infrastructure for individual applications could be used. Each application has one AID assigned as per specification of NXP for Mifare DESFire—in total 3 bytes.

Communication is solved in compliance with ISO 14443 A. Operating system of the designed Contactless Chip (CC) separates data spaces in its memory so that the card allows work with independent applications. Access to separated data spaces is controlled according to the type of operations. Implemented solution of CC allows multifunctional use, i.e. parallel location, use and administration of various subjects' applications. In addition to standard safety of Mifare DESFire cards, it offers own native safety elements too—encryption of content, content description using symmetrical and asymmetrical cryptographic mechanisms.

We designed the applications on the basis of such defined conditions.

Service application—connects information about the card and its holder across all other applications. Among others, it contains:

- Logic number of the card
- Elements for securing the card against counterfeiting

- Holder's data (with different degree of making data on the level of the card as well as on the level of data provided to the service provider by SAM, anonymous, card holders' profiles are necessary as minimum).

STUDIUM application, which contains the structure of file (see below) with the following hierarchy:

Studium AS application

File name: ParametersStatus, File number: 0, File type: SDF

Description	Name	Format	Size [Byte]	Value default
Version of AS	VersionAS	Binary	2	1
Owner of AS	OwnerAS	Binary	2	1
Publisher of AS	PublisherAS	Binary	2	1
State of studies (Byte mask file 1 to 8)	StudyStatus	Binary	1	0
Presence form (Byte mask 1 to 8)	PresentStatus	Binary	1	0
Signature of Parameters Status With UID of the card	ValueSign	Binary	56	0
Blocking of study application	FreezeAS	Binary	1	0
Reserve 1	ReserveOne	Binary	10	0
Reserve 2	Reseve-Two	Binary	4	0

File number 0 is defined so that it is considered as available for third parties (reading key). The content is filled from subsequent files number 1..N (structure—see below), which that are used for primary data of schools. The logic of filling in is stored in SAM modules, which are available at individual workplaces responsible for primary data. The structure of keys is designed so that the data stored into files number 1..N, is available only to the author of such data.

File name: Studium1..n, File number:1..N, File type: SDF

Description	Name	Format	Size [byte]	Value default
University ID [numeral]	CollegeID	Binary	8	0
Name of the University [text]	ColegeName	Binary	20	0
State ID as per EAN [numeral]	CountryID	Binary	4	859
Name of the state as per MRZ (MPZ) [text]	CountryName	Binary	4	CZ
Study programme ID [numeral]	StudyID	Binary	8	0
Short description of the study programme [text]	StudyName	Binary	20	_
Form of study (Combined, Attendance, Short-time scholarship, Long-time scholarship, etc.)	StudyForm	Binary	1	0
Status (Active, Suspended, Terminated)	StudyStatus	Binary	1	0
Date of valid study from...	StudyFrom	Date	4	0
Date of valid study till...	StudyTill	Date	4	0

During implementation, the speed of response of application in the chip was emphasised, this was experimentally measured as the time for execution of SAM operation by Secure JCOP 31 module by means of APDU commands with the following results:

- Time for function—Read Card Info File: 452 ms
- Time for function—recording data into file number 0: 987 ms.

3 CONCLUSION

This paper presents safety elements usable for identification of persons and operations by ID card. The scope of implementation of given elements is bound to the requested level of securing and, in case of contactless technology, also by the time of access to data. The contactless chip in the card is designed as multi-application platform applicable for more institutions where registration of application ID in NXP is assumed for uniqueness of the application. The whole concept is applicable as standardized environment in the area of safety and its further development and creation of more detailed implementing documentation for further possible extensions are planned.

REFERENCES

Carter S., Kilvington S., Lockhart H.W., Woollard S. & Nicolls W.; "Ask The Experts—When and why should I chose hardware encryption rather than software encryption?", ITsecurity.com Security Clinic: May 2001, The Encyclopedia of Computer Security.

Holy R., Scherks J. & Kalika M. "Application of ID cards—security components" In: 6th International Conference on Signal Processing and Communication Systems [CD-ROM]. New Jersey: IEEE, 2012, p. 1–250. ISBN 9781467323918.

Holy R., Scherks J. & Kalika M. "Biometric ID cards at CTU in Prague" Proceedings of DICTAP 2012 (Bangkok, Thailand), May 2012. ISBN: 9781467307321 (IEEE Thailand Section, NECTEC, SDIWC).

Holy R., Scherks J. & Kalikova J. "Security concept of individual identification in academic environment of CTU in Prague" Proceedings of WorldCIS 2012 (Guelph, Canada), June 2012. ISBN: 9781908320049 (IEEE UK/RI Section).

Kalika M. & Holy R. "Present and futute of ID cards at CTU", In Cipove karty a elektronicky podpis na vysokych skolach, Pilsen: Zapadoceska univerzita v Plzni, 2009. ISBN 9788070437995.

Schneier B.; "Security in the Real World: How to Evaluate Security Technology". Computer Security Journal (Volume XV, Number 4, 1999), June 1999.

Effect of structure and cooling strategy on the temperature field of Ni/MH battery pack

S.Y. Chen, M.X. Zheng, B.J. Qi & B. Li
School of Mechanical Engineering and Automation, Beijing, China

Y.W. Lou
Shanghai Institute of Micro-system and Information Technology, Shanghai, China

ABSTRACT: A simplified battery thermal model is firstly built in this work. And based on the model of battery pack, the finite volume method is used to simulate temperature distributions in the battery pack at different charging rates; it is found that the simulation results and experimental measurements have a good consistency. Considering concentrated heat and uneven thermal distribution in the battery pack, optimization solutions have been proposed, which are active convection cooling and improving batteries arrangement. An effective control strategy is determined according to simulation results and analysis of the temperature distribution. The strategy is verified by the experiment that the temperature of the battery pack can be controlled within a certain range in the operating conditions.

1   INTRODUCTION

As environment and energy issues deteriorate, the development of electric vehicles attracts widespread concern from all countries and industries for its saving-energy feature. Power batteries act as the main energy source of electric vehicles, and Ni/MH batteries and lithium-ion batteries have been widely researched and applied due to their excellent performances. The characteristics of the Ni/MH battery include high specific energy and power, long cycle life, stable performance and the ability to endure over-charge and over-discharge. Therefore, the Ni/MH batteries have already got extensive application in hybrid electric vehicles.

Since the voltage of a single Ni/MH cell is quite low, in order to provide sufficient power for electric vehicles, it is necessary to place hundreds of batteries both in serial and in parallel connection to form a battery pack. Besides, the security, stability and cycle life of the battery pack are major considerations during operations. Considering that the temperature has an extremely adverse impact on the battery charge and discharge efficiency, as well as the battery capacity and its security, thus analyzing the temperature distribution in battery pack is of great importance.

It is found that for the Ni/MH battery, the appropriate temperature range for charging is 0~40°C and for discharging is −20~65°C (Fu et al. 2005). However, during the charging and discharging processes, Ni/MH battery pack may be highly exothermic, and the scale-up batteries may be closely arranged which would result in difficulty in heat dissipation. Furthermore, the accumulation of heat probably lead to a local sharp temperature rise in the battery pack, which would definitely affects adversely overall battery working performance or even lead to accidents. Therefore, in order to ensure batteries operating in the optimal state, the following conditions are required: (1) The battery pack should operate within an appropriate temperature range of 0~40°C; (2) the battery temperature difference should be lower than 5°C (Pesaran 2002). In this work, the thermal behavior of the Ni/MH batteries and battery pack thermal control strategies will be discussed.

2   BATTERY PACK MODEL AND EXPERIMENTAL SETUP

The research target in this study is a Ni/MH battery pack already used in hybrid vehicles, there is 4×6 which is 24 battery modules set in one battery pack, each module comprises a number of single cells connected in series, the model of the battery pack shows in Figure 1. The single Ni/MH cell is cylindrical and its nominal capacity is 15 Ah. Since a battery box contains two packs, a single battery pack is investigated and a 1/2 battery box model is built in this study. As Figure 2 shows, the air inlet is on the side, the symmetry surface is the intermediate

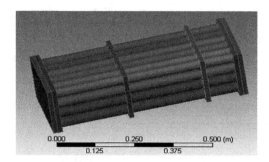

Figure 1. Model of the battery pack.

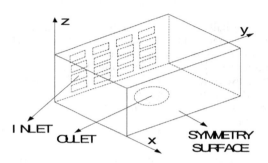

Figure 2. 1/2 battery box schematic diagram.

section of two packs, and in the center of the box bottom an exhaust fan can be installed.

To set up the experiment, electronic load devices is used to charge and discharge the battery, and thermistors (MF58-10k) are used to measure the batteries surface temperature. The batteries voltages, current and temperature are monitored and controlled through battery management system developed by our team (Zheng et al. 2008, Zhang et al. 2010). The accuracy of the temperature data collected is up to 0.125°C. Besides, data real-time recording is realized via the upper computer of the battery management system.

The chemical reactions occurring in the Ni/MH battery during the discharge process is endothermic, while the reaction in the charging process is exothermic. Consequently, there is more heat generated in the charging process than discharging under the same current, so our work will focus on the temperature distribution in the battery pack during charging process. On the other hand, to ensure a good performance and long-lasted lives of the batteries, the SOC (State of Charge) of the battery pack during operation is controlled within the range of 25%~85% by battery management system (Li et al. 2013), so the analysis of the battery pack temperature distribution will be concentrated on the period that the battery pack is charging from its cut-off voltage to 80% SOC. Before a charging

measurement, the batteries are discharged to the cut-off voltage with a trickle current.

3 THERMAL MODEL OF NI/MH BATTERY PACK

3.1 *Thermal model of NI/MH battery*

The heat transfer process in the battery is unstable; heat balance equation of the cylindrical battery can be expressed in cylindrical coordinates as following:

$$\rho c_p \frac{\partial T}{\partial t} = k_r \left[\frac{\partial^2 T}{\partial r^2} + \frac{1}{r} \frac{\partial T}{\partial r} \right] + k_z \frac{\partial^2 T}{\partial z^2} + Q \quad (1)$$

where ρ is the battery density, T the temperature, C_p the heat capacity, z, r the radial and axial length of battery, and k_r, k_z the radial and axial thermal conductivity, respectively, and Q the heat generation rate per unit volume.

According to the Equation (1), it is essential to estimate the exothermic mechanism and thermophysical parameters, however, the internal multi-slice spiral structure in the cell, as well as complex electrochemical reactions and thermodynamic processes during charging and discharging, increases the difficulty in accurately and quantitatively analyzing both thermophysical parameters and the heat generation.

In addition, the accurate prediction of temperature distribution in the battery pack depends greatly on the prediction of the cell thermal behavior. According to a comparative study by Y.Y. Wang on various cell models, it is found that appropriate model simplification is possible and accelerates the calculation (Chen et al. 2005). So in this work the battery model is simplified for application, and the assumptions and simplifications adopted are: (1) Considering the heat convection and heat conduction, while neglecting the radiative heat transfer; (2) Taking the overall cell structure as the homogeneous materials; (3) Battery resistance, density, thermal conductivity and specific heat capacity are constant values within a certain temperature range and does not vary with other characteristics such as SOC.

Many methods have been developed to calculate the heat generation Q, and two theories are most widely developed: the Bernadi's heat generation model emphasizes the thermodynamics effect of the entropy change and enthalpy change in the electrochemical reaction (Bernardi et al. 1985), while the Noboru Sato model stresses the battery electrochemical reaction process, details heat sources and analyses each of them (Sato & Yagi 2000, Sato 2001). In this work, Noboru Sato model is used to calculate the heat generation, its calculation method descripts in Equation (2)~Equation (6):

$$Q = Q_r + Q_p + Q_s + Q_j \qquad (2)$$

$$Q_r = \frac{Q_1}{nFV} = 0.547I \ (kJ/h) \qquad (3)$$

$$Q_p = I^2 R_t(W) = 3.6I^2 R_p \ (kJ/h) \qquad (4)$$

$$Q_j = I^2 R_e(W) = 3.6I^2 R_e \ (kJ/h) \qquad (5)$$

$$Q_s = \frac{Q_2}{nFV} = 5.334I \ (kJ/h) \qquad (6)$$

where the I represents the current for both charging and discharging process, Q_r the reaction heat value, Q_p the polarization heat value, Q_j the joule heat value, and Q_s is heat value of side reaction during overcharge; Q_1, Q_2 are heat generation of chemical reaction and side reaction, kJ/mol, F the Faraday constant. The internal resistance of the battery is divided into two parts: the electrical resistance component R_e and polarization resistance component R_t, which affects the value of Q_j and Q_p, respectively.

The thermophysical parameters of the battery are main reason of the temperature gradient inside the battery, and the battery specific heat capacity and thermal conductivity are two main parameters. The specific heat capacity of the battery can be measured by experiment, and its value is a linear function of the temperature: $Cp = \alpha + \beta T$ [10]. The battery thermal conductivity is anisotropic, which can be obtained by analyzing the thermal conductivity of each material in the battery and estimating from axial and radial directions, respectively.

3.2 Battery pack simulation set up

Based on the model of battery pack, the GAMBIT software is used to generate grids, and the FLUENT software is used to calculate the temperature distribution in a battery pack with the finite volume method. To set up the simulation, various material properties as well as the battery heat generation rate are set, the air inside the pack is taken as the ideal incompressible fluid, and the k-epsilon turbulence model is adopt to explore the case of fluid flow. In addition, for the set up of boundary conditions, the inlet pressure is set to be one atmospheric pressure, the walls are set to be adiabatic and the temperature of the inlet and walls are ambient temperature. Furthermore, the implicit equation is used for time discretion and the first-order upwind equation is utilized to discretize the pressure, momentum and energy. And the pressure

and velocity in the momentum equation can obtain coupled solutions with SIMPLE algorithm.

4 THE BATTERY PACK MODEL VALIDATION

To validate the battery pack model, four groups of charge and discharge experiments are carried out. The temperature is measured experimentally at different charging rates which are 0.5C, 1C, 1.5C, and 2C, respectively. Since the internal temperature of the battery is hard to obtain through experimental methods at present, the batteries surface temperature are measured and recorded. Simultaneously, based on the structure model of the battery pack and the simulation, the temperature distribution in the battery pack is predicted. The initial temperature of the experiments and simulations is 22°C, and room temperature fluctuates within ±1°C.

Figure 3 shows comparison of simulation predictions with experimental results at different charging rates, the temperature shown is the highest temperature detected in the pack. Since the highest temperature measured may occurs in different cell surfaces during experiment, the curves of solid line show slight fluctuations. At 1C and 1.5C rates, the simulation predictions correlate well with measurements, the deviation is less than 0.5°C. However, at 0.5C and 2C rates, the deviation reaches 1.34°C. This phenomenon occurs probably because the Joule heat has a greater effect than the prediction of the model, thus at a larger charge rate the prediction goes lower, while smaller charge rate goes higher. Considering the prediction deviation is less than 1.34°C, the battery pack simulation model can be applied to predict the temperature distribution.

According to the measurements, the highest temperature in the battery pack at 2C charge rate can reach as much as 40.8°C, and the maximum

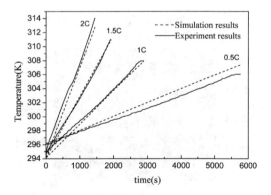

Figure 3. Simulation results against temperature predicts at different charging rates.

temperature rise is 19.5°C. When the battery pack is working under stressful condition such as higher ambient temperature and high power draw, the highest temperature may exceed 50°C. As appropriate working temperature for the battery pack is below 40°C, it is critical to make proposals to control the temperature in the pack.

5 RESEARCH ON THE TEMPERATURE CONTROL STRATEGIES

In an air-cooled battery pack, the temperature distribution in the pack is influenced by many factors such as the battery size, batteries arrangement, the position of the vents, air-duct size, etc. In order to control the temperature in the battery pack, two simple and feasible strategies are implemented, which are active convection cooling and improving the batteries arrangement.

5.1 Effect of active convection cooling on temperature distribution

Considering the heat generated is difficult to dissipate at present, a fan of 40 mm diameter is installed to control the airflow in the battery pack. The centrifugal fan at the battery pack bottom will be turned on when the highest temperature in the pack reached 38°C. The air speed through fan is 5.98 m/s.

A stressful condition is set up during simulation, in which 28°C is taken as the ambient temperature and the 2C charging process is implemented. The simulation results verified that exhausting fan can enhance the convection of air and alleviates the high temperature phenomenon that exists in the battery pack. However, although the temperature near the outlet is suppressed, the maximum temperature is still high, which results in a temperature discrepancy among batteries. Figure 4 shows the

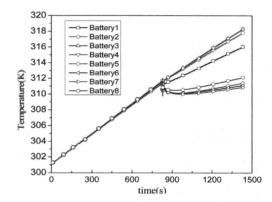

Figure 5. Predicted batteries surface temperature changes with time under strategy I.

temperature distribution of the simulation result in the midsection of longitudinal direction of the battery pack.

Eight batteries of both sides are numbered as Figure 4 shows. During the charging process, the variations of batteries surface temperature with time are shown in Figure 5. The temperature of battery 5~8 of inlet side are suppressed significantly after the fan ventilates, while the temperature of battery 1~4 of symmetry side is little affected. Based on the analysis, the maximum temperature difference between battery surfaces reaches 7.5°C, which should be lower than 5°C. Therefore, there is still room for improvement.

5.2 Effect of improving batteries arrangement on temperature distribution

Considering the gap between the batteries is small, which results in poor ventilation, so improving batteries arrangement may has a great impact on thermal distribution in the pack. Therefore, in this work batteries are cross-arranged and the fan will be turned on when the highest temperature detected reaches 38°C. The temperature distribution results are predicted by simulation.

The predicted temperature distribution in the midsection of the battery pack shows in Figure 6. Analysis reveals that the cross-arrangement enhances the fan cooling effect, accelerates the heat dissipation, and leads to a markedly air temperature decrease. Simultaneously, the variations of 8 batteries surface temperature with time shown in Figure 7 are analyzed. The temperature distribution within the battery pack exhibits good uniformity, the temperature difference is controlled within 5°C. But batteries on the symmetry side such as the battery 3 represents a temperature rise of 14.2°C, which illustrates that the batteries

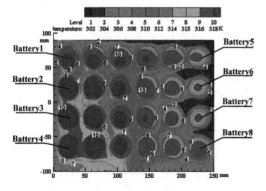

Figure 4. Predicted temperature distribution in the midsection of the battery pack under strategy I.

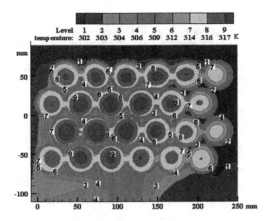

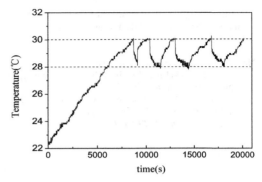

Figure 8. The highest temperature changes with time under operating condition.

Figure 6. Predicted temperature distribution in the midsection of the battery pack under strategy II.

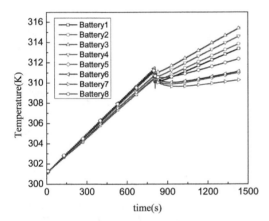

Figure 7. Predicted batteries surface temperature changes with time under strategy II.

surface temperature probably exceed 40°C when both the ambient temperature is above 26°C and there is high power draws out of the batteries.

To verify whether the strategy II can meet the temperature requirement of the Ni/MH batteries under operating condition, an experiment is carried out. This operating condition derives from data recorded in a real operation of an electric bus, including vehicle startup, running, acceleration and braking process. In this experiment, the batteries surface temperature are measured, the fan turns on when the highest temperature measured exceed 30°C and turns off when the temperature is lower than 28°C. The results shown in Figure 8 demonstrate that the temperature in the battery pack can be effectively controlled within the range of 28~30°C when the operating condition is favorable.

6 CONCLUSION

In order to explore the thermal distribution in the Ni/MH battery pack, the thermal model of the battery pack is built. The predictions of temperature distribution based on simulation results correlate well with the measurements at different charging rates, which supported the feasibility of theoretical models. To solve the problem that heat is concentrated and unevenly distributed in the battery pack, two strategies are proposed. Analysis on the simulation results reveals that the strategy I of active convection cooling can accelerate the flow of air within the battery pack, and temperature near inlet are suppressed significantly, but high temperature zone still exists. Furthermore, the strategy II of improving batteries arrangement enhances the fan cooling effect, high temperature is effectively suppressed and the temperature rise is controlled within 15°C. The operating experiment under 22°C ambient temperature further proved that the temperature in the pack can be controlled within the range of 28~30°C.

REFERENCES

Ahmad A. Pesaran. 2002. Battery thermal models for hybrid vehicle simulations, *Power Sources*: 377–382.

Bernardi D., E. Pawlikowski, J. Newman. 1985. A General Energy Balance for Battery System *J. Electrochem Soc*: 132.

Chen S.C., C.C. Wan, Y.Y. Wang. 2005. Thermal analysis of lithium-ion batteries, Journal of Power Sources: 111–124.

Fu Zhengyang, Lin Chengtao, Chen Shiquan. 2005. Key technologies of thermal management system for EV battery pack. *Journal of Highway and Transportation Research and development*: 119–123.

Li B., Zheng M.X., Qi B.J. 2013. The Battery Management System Applied in Smart Grid Energy Storage System, Advanced Materials Research.

Lou Yingying, Wang Wen, Lou Yuwan. 2007. The Thermal Simulation of Ni/MH Battery in the Process of Charge and Discharge, *Journal of Shanghai Jiaotong University*.

Noboru Sato & Kazuhiko Yagi. 2000. Thermal behavior analysis of nickel metal hydride batteries for electric vehicles, *JSAE Review*: 208–209.

Noboru Sato. 2001. Thermal behavior analysis of lithium-ion batteries for electric and hybrid vehicles, *Journal of Power Sources*: 70–71.

Zhang Huahui, Qi Bojin, Zheng Minxin. 2010. Development and application of a management system based on DSP processor for lithium-ion battery series, *Journal of Liaoning Technical University*: 132–135.

Zheng Minxin, Qi Bojin, Wu Hongjie. 2008. The lithium-ion battery management system for hybrid buses, *High Technology Letters*.

Control Engineering and Information Systems – Liu (Ed)
© 2015 Taylor & Francis Group, London, ISBN 978-1-138-02685-8

Gene-gene interaction computing model on BOINC, a volunteer computing system

X.K. Zhang, R. Xu & X.Y. Zhao
Department of Computer Science and Technology, Beijing Electronic Science and Technology Institute, Beijing, P.R. China

ABSTRACT: Volunteer computing can utilize idle computational resources on the Internet to provide massive power to solve the computational problems which can apply to grid and clusters. Gene-gene interaction in bioinformatics is one such problem; solving it would be an important step to search disease associated genes and predict the risk of affection in populations. This paper proposes a computing model for gene-gene interaction analysis on BOINC, a volunteer computing platform, and tests its speedup performance on 40 computing nodes. The results show that it can serve the present need of massive computation brought by gene-disease association analysis with good speedup ratio and running efficiency.

1 INTRODUCTION

In recent years, bioinformatics, as a fast developing discipline, has attracted the interest of many researchers from the fields of computer science, biology, chemistry and physics. Many bioinformatics problems are complicated and computation-intensive, therefore requiring massive application performance. Gene-gene interaction [1] is one such problem which often serves in bioinformatics as an important step of identifying and locating disease associated genes as well as predicting affection risk in populations.

Volunteer computing [2], or public computing, is playing an increasingly important role in engineering or scientific computation as a type of grid computation. This technique solves a big computational problem by dividing it into large amounts of distributable sub-problems, assigning them to the participant nodes (also called 'volunteers') on the Internet, and at last merging the results of the sub-problems on the server side. BOINC (Berkeley Open Infrastructure for Network Computing) [3] is one such system which allows research institutes or individuals to create grid computing projects, and attracts volunteers to share their resources to assist the projects.

In this paper, we propose a BOINC-based computing model to identify and locate disease associated gene-gene interactions, and evaluate the

This paper is supported by Information Security Key Lab Foundation, Beijing Electronic Science and Technology Institute with Project No. YZDJ1104.

model with multifactor dimensionality reduction [4] as a statistical model on a cluster of 40 nodes. The result shows that our computing model is efficient, and that the BOINC system can be effectively deployed for bioinformatics research such as gene-gene interaction studies.

2 PROBLEM DESCRIPTION

Usually when bioinformatics researchers search for disease associated genes, they would use two types of data: a genotype data which codes the genetic information of each subject in the sample, and a phenotype data which codes the affection status of each subject. A genotype data is a $m \times n$ two-dimensional array, in which every element is 0, 1 or 2 (coding for three kinds of genotypes). A phenotype data is a $m \times 1$ vector, in which every element is either 0 (unaffected) or 1 (affected). Here m is the number of individuals in the sample, while n is the number of loci (gene position in the chromosomes) collected.

If we denote the loci (columns in the genotype data) as $X_1, X_2, ..., X_n$, and the phenotype data as Y, then a gene-gene interaction problem can be described as follows: given an integer d where $1 \leq d \leq m$, find the d-dimensional locus combinations which are significantly associated with Y, where a d-dimensional locus combination is in the form of $X'_1, X'_2, ..., X'_d$. The definition of being 'significantly associated' depends on the detailed statistical model and measurement chosen to evaluate each combination. Common approaches include regression-based tests, chi-square tests, multifactor dimensionality reduction, and so on [1]. Here we use multifactor dimensionality reduction (or MDR) [4] as an example, so the definition of 'significantly associated' combinations is the k combinations which have the highest prediction rate, where the value of k is specified by users. Whichever method is used, the scale of the problem is usually $O(C_m^d n)$, but because some methods as MDR require additional sorting procedure to find the top k combinations, the scale in those approaches becomes $O(k C_m^d n)$.

3 BOINC ARCHITECTURE

BOINC is one of the mainstream computing platforms distributed throughout the world. It was initially developed by University of California, Berkeley for SETI@home [5] to detect intelligent life outside Earth, and to prove the viability and feasibility of the idea of volunteer computing. Nowadays, it is widely used in other fields such as mathematics, Medicine, astronomy and meteorology to reap the massive computing power embedded in the Internet to support researchers.

3.1 *Basic concepts*

Some important related concepts on BOINC includes: 1) projects, which are entities doing distributed computing on BOINC. Projects are independent from each other. Every project contains its own applications, database, website, and a URL identity. 2) Applications, which are a set of executables, workunits and results. 3) workunits, or jobs, which are computations to be performed. A workunit may include any number of input arguments or files. 4) results, which are computing statuses of workunits. A result can be either unstated, in progress, or completed.

3.2 *Communications between clients and servers*

A BOINC computing platform is organized as clients and servers, in which clients are volunteers. The server end is identified by its master URL, and includes a scheduling server and a data server. The clients send computing requests to the project's scheduling server. The scheduling server sends computing tasks to clients based on their available hardware and software computing resource, which are described in the computing requests sent to the server. The clients download executables and input data related to their assigned tasks from the data server, then starts computation. When a computing task is completed, the client uploads the output files to the data server. Later the clients reported the completed tasks to the scheduling server, and

request for new tasks. This cycle is repeated until the whole project is completed. In other words, a BOINC project works in 'pull' way to get computing tasks, in which clients 'pull' tasks from the server end; and in 'push' way to get computing result, in which clients 'push' results to the server end.

4 BOINC MODEL OF GENE-GENE INTERACTION ANALYSIS

Generally speaking, in gene-gene interaction analysis, the computation of every gene-combination is independent from each other. Also, the number of gene-combinations is usually huge. Given n gene-locus, the number of d-dimensional gene-combinations is $C\_n^d = n\,!/d!(n—d)!$. In a typical genome-wide association study (GWAS), there're millions of SNPs (a type of gene-locus). Therefore, we can parallelize a gene-gene interaction analysis based on gene-combinations, and distribute the computations of different gene-combinations to different computing units, which in our case are volunteer computers.

The scheduler server must guarantee that every gene-combination will be traversed and dispatched to volunteers for computation. Therefore, we need logic to traverse all the combinations. We can arrange the combinations so that they become a serially ordered set based on some rule. For two combinations $[X_1', X_2', ..., X_d',]$ and $[X_1'', X_2'', ... X_d'']$ from n genes where $X_1', X_2', ..., X_d'$, and $X_1'', X_2'', ... X_d''$ are gene indices and $0 \le X_1', X_2' < \cdots < X_d' < n$, and $0 \le X_1'' X_2'' < \cdots < X_d'' < n$, we define that $[X_1', X_2', ..., X_d'] < [X_1'', X_2'', ... X_d'']$ if and only if there is some integer i where $1 \le i \le d$ so that $X_j' = X_j''$ for $1 \le j \le i-1$ and $X_k' < X_k''$ for $i \le k \le d$, so that all the C_n^d combinations become a serially ordered set and can be arranged from the 'smallest' to the 'largest'. Suppose that the indices of genes start from 0, then the 'smallest' combination would be $[0, 1, ..., d-1]$, and the 'largest' combination would be $[n-d, n-d-1, ..., n-1]$. This rule is easy to implement in code in the traversing logic of the scheduler.

It is impractical to first traverse all the combinations and then stores them in an array in memory or even on disk storage. In a typical GWAS, when there're a million of SNPs, if we want to search for 2-dimensional gene-gene interactions, there would be $C_{1M}^2 \approx 500\,G$ combinations. If every index is stored as a 4-byte integer, it would take $4 \times 2 \times 500\,\text{GB} = 4\,\text{TB}$ space to store all the combinations. Therefore, the scheduler must traverse all the combinations while at the same time dispatch those combinations which have been traversed to the volunteers. For this purpose, the scheduler can maintain a state machine as a combination generator. There're $C_n^d + 2$ states in total of this state machine, in which there is a begin state and an end state. Each of the remaining C_n^d states represents a SNP-combination. When the combination generator is created, it is initialized as the begin state. Whenever it steps to the next state, it generates the next 'larger' combination. The 'largest' combination steps to the end state, then the combination generator won't change its state.

The computing resource of volunteers would not be efficiently utilized if every workunit represents just one combination, because the computing time of each workunit would be too short and most of the project time would be spent in communication. According to our experiment, in 2D scans, it just takes about 2 ms to traverse the genotypes of all the samples of a given combination on an Intel 2.7 GHz CPU.

Therefore, each workunit represents achunk of continuous combinations. In fact, we can still use a combination $[X_1', X_2', ..., X_d',]$ to identify a workunit. Also, we can set that each workunit shall contain Q combinations except for the last workunit. The last workunit may contain less than Q combinations.

Each volunteer also maintains a state machine as a combination generator. When a volunteer gets a workunit, it traverses the Q combinations from $[X_1', X_2', ..., X_d',]$ with the state machine. For threshold-based statistical models such as chi-square test, the volunteer calculates a statistic (typically p value) of each of the Q combinations, then sends those combinations whose statistics pass the threshold (typically $p < 0.05$) back to the assimilator. For sorting-based models such as MDR, the volunteer calculates a statistic (e.g. accuracies in MDR) of each combination of the Q combinations, and sorts the combinations based on their statistics, and sends the top K combinations and their statistics back to the assimilator, where K is a user-specified value. The server shall maintain a sorted list of top K combinations. Whenever a volunteer sends back its result unit to the assimilator, the assimilator shall merge the K combinations in the result unit with the maintained sorted list.

5 PERFORMANCE RESULT

We run our performance experiment on our local cluster with 40 volunteers to evaluate our BOINC model of gene-gene interaction analysis. Each volunteer consists of an Intel 2.7 GHz CPU with

Table 1. Speed-up ratio test of the BOINC model of 2D scan with MDR.

# of SNPs	500	5,000	50,000	500,000
Sequential	6 sec	10 min	12 hr	50 d (est.)
BOINC model with 40 nodes	15 sec	1 min	25 min	1.7 d

12 GB memory. We simulate a data which consists of 500,000 SNPs and 5,000 subjects. We perform 2D interaction analysis with MDR as the statistical model, and select the top 100 SNP-combinations. Our BOINC model only takes 1.5 days to complete the processing. We also simulate three other data which consist of 500, 5,000 and 50,000 SNPs, respectively, to evaluate the scalability of our BOINC model. And we also write a sequential version of the same algorithm to evaluate speed-up ratio of our model. The result is shown in Table 1.

It can be seen from Table 1 that the gene-gene interaction analysis starts to benefit from the performance of the BOINC model from 5,000 SNPs. The speed-up ratio becomes stable and close to ideal (about 30:1) when the problem scale reach to 50,000 SNPs. This shows that our BOINC model is effective.

REFERENCES

[1] Cordell & H.J., Detecting gene-gene interactions that underlie human diseases. Nat Rev Genet, (2009). 10(6): p. 392–404.
[2] Sarmenta & L.F.G., Volunteer Computing, (2002), Massachusetts Institute of Technology.
[3] Anderson & D.P. BOINC: A System for Public-Resource Computing and Storage. in Proceedings of the 5th IEEE/ACM International Workshop on Grid Computing. (2004). IEEE Computer Society.
[4] Ritchie et al., Multifactor-dimensionality reduction reveals high-order interactions among estrogen-metabolism genes in sporadic breast cancer. Am J Hum Genet, (2001). 69(1): p. 138–147.
[5] Anderson et al., SETI@home: an experiment in public-resource computing. Commun. ACM, (2002). 45(11): p. 56–61.

Control Engineering and Information Systems – Liu (Ed)
© 2015 Taylor & Francis Group, London, ISBN 978-1-138-02685-8

Radon measurement technique based on embedded method

X.B. Wang, J.H. Wang, G. Chen & D.H. Sun
North China Institute of Aerospace Engineering, Langfang, Hebei, China

ABSTRACT: Based on the nature of the decay characteristics and radon radioactive material, the application of embedded method to construct measurement system using activated carbon to radon, radon adsorption on characteristics of strong adsorption force collection of radon, the gamma ray decay measuring active carbon radon emitted intensity, so as to obtain the method of radon concentration as the measurement basis, real-time detection of temperature, humidity, correction of radon concentration measurement results, overcome the effect of environmental and climatic factors of radon measurement. Relying on the system to realize continuous measurement of radon online, to reduce the radiation caused by radon personal health problems.

1 INTRODUCTION

There are three natural radioactive Series in the earth's crust: uranium, thorium, uranium actinium series, each series has a radioactive gas radon isotopes, these natural radioactive series will release radon atoms can transfer in attenuation of fission, Radon atoms may enter the connected micro crack can transfer free radon formed gap the migration of radon atoms, in porous media, micro-fracture and fracture by diffusion, convection mechanism of migration to the soil surface, eventually leaving soil surface into the surrounding air. So the radon distribution in the nature is very wide, each corner exists in human activities. Radon is a recognized carcinogen, radon and its progeny is human long-term inhalation, the body will increase the incidence of cancer (Liu 2009). Important means of efficient and accurate measurement of radon concentration in the environment become the environmental assessment, disease control.

Development of high detection sensitivity, high accuracy, stable and reliable work, low power consumption, simple operation, equipment of radon has communication function is an important direction of the present. The radon measurement method of four kinds of commonly used, ionization chamber method and scintillation chamber method, double filter membrane method, active carbon absorption method. This paper adopts the embedded method, the activated carbon adsorption for the measurement basis; realize the measurement of radon and its progeny, advantage and disadvantage of radon measuring technology is analyzed, providing the basis for the selection of appropriate measurement methods of radon concentration.

2 THE PRINCIPLE AND METHOD OF MEASUREMENT RADON

2.1 Ionization chamber method

With the method is reliable, fast measuring speed, can directly collect air samples were measured, also can make the constant air flow through the measuring device for continuous measurement, can quickly give the radon concentration and its dynamic change etc.. But it has low sensitivity, not suitable for low level measurement, defect site and inconvenient to use.

The detection principle for containing radon gas into the chamber, ionizing alpha particles emitted by radon and its progeny in the air, electron-hole pair (Liu 2009). The positive charge accumulation of quartz ribbon central electrode ionization chamber of the electrometer, in under the action of external electric field, quartz fiber deflection, deflection speed and amount of charge accumulation is proportional to, is proportional to the concentration of radon, so the measured deflection velocity can be calculated indirectly radon concentration.

2.2 Scintillation chamber method

With the advantages of simple operation, high accuracy, low limit of detection. But there are also long measuring time, auxiliary equipment, is not convenient for on-site detection of defects, and the scintillation chamber of the bottom increases with time; the detection efficiency decreases with time, its application in the long-term continuous measurement, need regular online calibration; radon daughters deposited on the interior wall body is difficult to remove, use should always use radon or aging of fresh air.

Test method for radon into scintillation chamber, alpha particles of radon and its decay produces the scintillation chamber wall ZnS (Ag) to produce optoelectronic, through the back of the photomultiplier tube optoelectronic into electric pulse signal, test the pulse frequency indirectly measure radon concentration.

2.3 Double filter membrane method

The detection principle is the air containing radon into the double membrane tube through the entrance filter; filter progeny produced new pure radon progeny in the process through the double membrane tube, which was part of the export filter collection (Lu 1994). Based on the accumulation, the decay law of radon progeny inherent can determine the concentration of measured gas radon.

The advantage of this method is that it can be used to measure the concentration of radon progeny, also can measure radon concentration, the detection limit is low, convenient and quick. Weaknesses are: to ensure the export filter is not radon pollution two membranes outside; influenced by humidity is bigger, the relative humidity is higher, the membrane measured alpha radioactivity greater; device needs power, inconvenience for field use.

2.4 Activated carbon adsorption method

The radon measurement is a static, cumulative radon measurement method; the principle is the use of activated carbon adsorption of radon has the characteristics of strong adsorption force collection of radon, the ray strength decay instruments measuring active carbon radon emitted, so as to obtain the radon concentration.

3 EMBEDDED RADON MEASUREMENT SYSTEM

3.1 The radon measurement principle

This system uses the active carbon adsorption method, mainly by the embedded processor as the core control unit, combined with other peripheral circuits, the application of activated carbon has a strong adsorption capacity of radon as a sampler, the detector using a NaI crystal, complete automatic measurement, display and store the data of radon concentration, temperature and humidity of the environment at the same time acquisition the twice amendment, implementation of the measurement data, ensure the accuracy of measurement, and can store the data through the serial interface to the computer for further data analysis and processing.

The measurement principle is mainly by gamma ray detection of radon daughters in soil body strength, radon concentration measurement of size, presence of gamma rays, detector in NaI scintillation of the optical signal in gamma rays into the role, through the photomultiplier tube will flicker signals produced by the body into electrical signals, because the electrical signals produced by the photomultiplier is weak, therefore need to increase the amplifier amplifies the signal, the pulse shaping, the intensity information into gamma rays into electric pulse, the pulse can be directly embedded processor is obtained by counting way, in a unit of time, gamma ray radiation is strong, strong, the number of pulses per unit time; gamma ray radiation is weak, weak pulse, reduce the number of units of time. Therefore, by measuring the pulse number per unit of time, they can see gamma ray intensity. Through the measurement, recorded at different locations throughout the detection area of the sample, can be in different locations of gamma ray strength, in the twice amendment of the temperature and humidity of the environment, we can obtain accurate measurement results.

3.2 System hardware

Radon measurement method of embedded hardware block diagram is shown in Figure 1 based on.

The core processing unit adopts the embedded processor ARM8400 as control core, ARM8400 has very low power consumption, rich interface, fast speed, independent memory characteristics, industrial design based on embedded core board, able to adapt to the harsh environmental

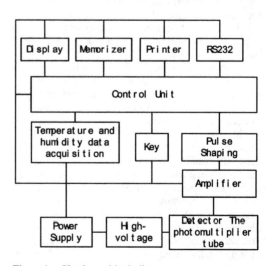

Figure 1. Hardware block diagram.

requirements, with speed and stability is very high (Zhao and Xiao 2007). The use of embedded processor ARM8400 as the control core with other peripheral circuits can be accomplished for the gamma ray activated carbon sampler measurements, correction, display, storage and communication. The circuit comprises a power switching circuit, scintillation counting circuit, temperature and humidity detection circuit, LCD display, keyboard input, storage unit, communication interface and control circuit.

High voltage circuit provides power supply for 1500 V photomultiplier circuit, a low-voltage power supply adopts an integrated DC-DC converter converts the DC voltage is transformed into DC low voltage battery voltage stable, preamplifier, main amplifier, embedded system and display circuit provides power.

Detector scintillator counters by NaI scintillator and photomultiplier tube, when ray sampler into the scintillator, flashing light, light to the photomultiplier tube time bombarding electron pole, electrons in the multistage multiplication, formation of electrical pulse, after the amplifier and pulse shaping circuit, meet the requirements of the pulse signal measurement.

Storage unit mainly to achieve preservation at data of measurement, in order to do simulation in computer (Jia 1998). This system using flash memory as storage unit, flash memory has a power-down data is not lost, repeated reading and writing.

Display part adopts LCD display, the main variety of input parameters and measurement data, correction parameters can be input by a key, completes the function shift, adjustment and confirmation.

Data communication interface to RS-232 and host computer, with automatic handshake function, to realize the automatic connection detection system and the computer, will be stored in the data storage unit are uploaded to the host computer, for further analysis and processing of data.

Printing equipment using micro printer, can directly print measurement data, convenient and intuitive.

3.3 The software design

Software design adopts module design method, the software is divided into nine modules, including pulse data acquisition module, temperature and humidity data acquisition module, interrupt processing module, initialization module, keyboard module, serial communication module, data storage module, data processing module, print module etc. (Ji and Zhang 2010). Is mainly done on the temperature and humidity real-time acquisition, detection output radon concentration pulse, data

correction and data storage, the correction coefficient selection, a variety of interface display, data printing function. The main program flow shown in Figure 2, the interrupt handler flow as shown in Figure 3.

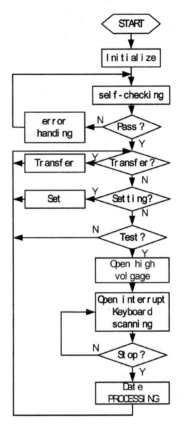

Figure 2.   Main program flow.

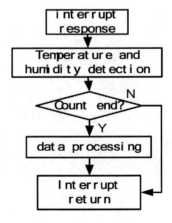

Figure 3.   Interrupt handler flow.

4 CONCLUSION

In application of embedded processor ARM8400 as control core, method using activated carbon adsorption method for measuring radon concentration, with detection is simple and convenient, functional expansion, but also increase the real-time temperature and humidity correction, accurate measurement, stable operation, good repeatability.

ACKNOWLEDGEMENT

This work was supported by the key project of the Science and Technology Department of Hebei Province (Project ID: 13214102D).

REFERENCES

Ji Changsong, Zhang Shuheng. 2010. "Radon measurement instrument of the direct method". *In The fifteenth session of the National Nuclear Electronics and nuclear detection technology academic annual meeting 2010.*

Jia Wenyi. "Cup radon measurement principle and application". "Geophysical and geochemical exploration". 1998, (6): 191–198.

Liu Fangping. 2009. "An activated carbon box method measurement of indoor radon and radon progeny". *Editorial Department of Jilin University Journal.* Jilin University, 2009, 23 (2):20–24.

Quan Lu. "Radon measurement method of communication". *Radiation protection*, 1994, 14 (6): 34–40.

Zhao Guizhi, Xiao Detao. 2007. "Method for measuring radon concentration in soil". *Nuclear electronics and detection technology*, 2007, 27 (3): 583–587.

Research of real-time data warehouse storage strategy based on multi-level caches

X.J. Yao
College of New Energy Engineering, Shenyang University of Technology, China

Y.C. Shao, X.D. Wang & L.W. Tian
School of Information Engineering, Shenyang University, Shenyang, China

H. Chen
Huachuang Wind Energy Company, Qingdao, China

ABSTRACT: Real-time data warehouse extend the application of traditional data warehouse. It can not only support tactical queries for enterprise but also provide much variable tactical decision support effectively. For these reasons, it is very meaningful to research on the structure of real-time data warehouses. This paper introduced the background of real-time data warehouse and proposed the strategy of real-time data warehouse which is based on double mirror replication mechanism. The strategy is composed of two steps. First we used double mirror replication mechanism to enable continuous loading data in the real-time data warehouse with minimum impact in query execution time. Second we proposed incorporating multi-level caches into the data warehouse structure which is based on real-time partition and gave the process of design and implementation with details. We differentiated between queries with various data freshness requirements, and used multi-level caches to satisfy these different requirements.

Keywords: real-time data warehouse; double mirror replication mechanism; multi-level caches

1 INTRODUCTION

Recent years, as computer science and information technology developed rapidly, the electronic information data has become more and more important in the enterprise daily management. Effective, accurate and timely data analysis is required imminently by enterprises. Traditional data warehouse commonly just support some analysis and queries on history data while the variety of commerce information can't be displayed in real-time. Real-time data warehouse extend the application of traditional data warehouse and can support tactical queries for enterprise. Real-time data warehouse effectively cut the time-lapse and can provide much variable tactical decision support. For these reasons, it is very meaningful to study on the structure of real-time data warehouses.

This research is supported by the International S&T Cooperation Program of China (ISTCP) under Grant 2011DFA91810-5 and Program for New Century Excellent Talents in University of Ministry of Education of China under Grant NCET-12-1012 and is also supported by Program for Liaoning Excellent Talents in University.

The conflict between real-time data import and real-time data query will arise in the real-time data warehouse [1]. One of the most difficult parts in the construction of any data warehouse is to import data from different business systems to ECCL. More additional difficulties will be increased if the process becomes real-time. Almost all systems are in operation in batch mode. In these systems, the available data is assumed to exist in a place in a kind of extracted file within a certain definite time schedule. Then, the systems transform and clean the data, and then import it to the data warehouse [2]. The processing exerts a significant influence on the nonresponse period of the data warehouse, for no users can access the data warehouse in the process of import. But with the continuous and real-time import of data, the system could not have any nonresponse periods. The most cumbersome retrieval time period for the data warehouse may be consistent with the time when the data is imported most frequently. Therefore, a fundamental contradiction has been produced between the system and this requirement for continuous update without nonresponse periods [3].

In this article, a data warehouse architecture design based on the real-time storage area has been proposed. Different from traditional data warehouses, the subsystem of this architecture is separated into real-time ETL (Extract/Transform/Loading) and periodic ETL [4]. Periodic ETL refers to periodic batch import of data in data sources, which is finally loaded into the data warehouse. But for real-time ETL, CDC (Change Data Capture) tools directly and automatically capture change data from the data source, which is then loaded into the real-time storage area. Through query, update, deletion and other operations in the real-time storage area, when the system's trigger conditions are satisfied, the data is then loaded into the data warehouse in batches. The storage part can be divided into real-time data storage area and static data storage area. Real-time data query and storage are implemented based on the real-time data storage area. Static data is just the data warehouse. Static data or historical data query is implemented based on the static data storage area. Update query scheduling components can determine the operation order of query and update through the users' demand for them. Data quality components could generate a report when showing query results, so that users can know those data the query contents refer to and the freshness of these data. Through query control components, real-time data query can be implemented in the real-time storage area. Compared with the traditional data warehouses, the response time for query is significantly improved. For historical data query, seamless links [5] are needed between the real-time storage area and data warehouse, so as to ensure the accuracy of query. In this article, a replication mechanism based on double mirror partition alternation has been put forward to realize the real-time storage area and satisfy users' demand for data of different freshness, in order to reduce the response time for query.

2 DOUBLE MIRROR REPLICATION MECHANISM

Real-time storage area is made up of double mirror partitions based on data warehouse replication mechanism, which are operated alternately [6]. This plays a vital role in solving query competitions.

With the continuous and real-time import of data, the query operation from users will be much affected, including its accuracy and query efficiency. Therefore, a conflict between real-time data import and real-time data query will arise in the real-time data warehouse. When data is loaded to the real-time storage area in ETL, it is likely for users to be making many inquires about the real-time storage area. These inquires will be brought into the same statistical result, and then we shall have to consider whether the new data will be brought into the result.

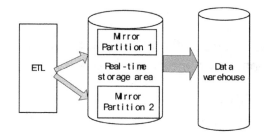

Figure 1.   Examples of real-time storage area structure.

If we do not make any improvement, the accuracy of query results with OLAP is usually barely satisfactory, which thus requires us to design a new structure to solve this problem.

So in this paper, double mirror partitions are constructed in the real-time storage area, whose logic structures and physical structures are all the same. Then, when the records in Mirror Partition 1 are loaded to the data warehouse in batches owing to trigger conditions or operations from the database administrator, the new data in OLTP server will be loaded into Mirror Partition 2. Namely, Mirror Partition 1 is used for query, and Mirror Partition 2 is used for receiving new data, so as to prevent conflicts. Until next time Mirror Partition 2 meets conditions and the records need to be loaded to the data warehouse, the contents in Mirror Partition 1 are depleted, and new data needs to be reloaded into Mirror Partition 1. Query is made in Mirror Partition 2. The two parts of the real-time storage area are working alternately in this way. One partition is used for query, another for receiving ETL update data.

So, the real-time storage area jointly constructed by double mirror partitions can achieve real-time loading of continuous data. This alternate use of Mirror Partition 1 and 2 can not only effectively solve the conflict between update and query, and also meet the query requirement from the real-time storage area when data is loaded in batches.

However, this structure also has certain disadvantages. Although we have reduced operations on the data warehouse by batch loading, due to the mirror partition alternate update mechanism, we cannot get a very good guarantee for the freshness of query results. Thus, based on the double mirror partition alternate replication mechanism, multi-level caching mechanism is proposed in this paper to solve the freshness problem of data.

3 MULTI-LEVEL CACHES MECHANISM

The core of multi-level caching mechanism is multi-level caching with a queue and freshness scheduling module. Data will be loaded into the data storage area by queues and finally stored in the real-time storage area. It is the fresh data that can be stored there, so today's data should be kept in the real-time storage area. Then data there is loaded periodically into the data warehouse in batches. When we position query through the freshness scheduling module, we shall also consider that the recent data should be inquired in the real-time storage area, and historical data should be inquired in the data warehouse. If necessary, effective seamless links are needed at the same time for the contents of multi-level caching, the real-time storage area and the data warehouse. The architecture is shown in Figure 2.

Through the query caching mechanism, we can effectively control the influence of a lot of inquires on the system performance, and improve the query speed. The key to increasing query caching is to apply a useful query matching algorithm, with which we can quickly and accurately identify similar inquires, and guarantee the settlement of query conflicts as well as the accuracy of the results.

Since real-time data changes and data query will influence the performance of the data warehouse, in case the real-time property is not reduced, the most simple solution is to try to reduce the complexity of inquires, restrict users to perform very complex inquires, or lower the refresh frequency of real-time data and support complex inquires. For the modeling method of separated real-time data, real-time data and historical data query can be handled separately. That is, when real-time data is inquired, historical data will not be inquired, and vise versa.

Historical data is deposited in a cache pool in result sets [7], which can improve the query and analysis efficiency with OLAP, so as to provide timely analysis results for users. However, if we ignore the operating characteristic of data cache technology, the performance of query and analysis with OLAP will be declined. For example, due to fast changes of the data items in the buffer,

too long or short cache time, or other factors, the response speed from the system to user query and analysis will be reduced. The major indexes for the evaluation of the cache are: buffer capacity, hit rate, access delay, cache replacement algorithm, cache synchronous algorithm, etc. There exists a certain contradiction between hit rate and buffer capacity. The increased buffer capacity will cause a reduced cache hit rate, declined cache efficiency and increased system access delay. So buffer capacity is generally fixed in a certain size. When the buffer capacity is inadequate, a certain cache replacement algorithm is needed to remove those cache objects of less "value" for the system from the cache, but put those objects of more "value" in the cache. A good replacement strategy, namely cache updating algorithm, plays a decisive role in the efficiency of the cache.

4 THE UPDATING ALGORITHM OF THE MULTI-LEVEL CACHE

The updating algorithm of the first-level cache has been described in Algorithm 1. Q_1 contains the continuous data from the source system, which are those that will be loaded into the first-level cache. Meanwhile, these data will be inserted into the queue Q_2, convenient for data updating in the second-level cache. In this way, data has backups in Q_2, and then the data in C_1 can be directly removed when it is refreshed, which can reduce operations on the cache.

Algorithm 2 is the updating algorithm of the ith-level cache, where i belongs to [2,n], and its difference from the first-level updating algorithm is: in the first-level cache, data in Q1 is all directly loaded into the cache; for the data in the ith-level cache, it can be loaded into the ith-level cache until necessary preprocessing is made. The preprocessing mainly includes deleting the old version of the data, which aims to guarantee the freshest data storage in the cache, and can reduce operations on the cache.

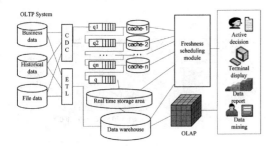

Figure 2.   The architecture of real-time data warehouse with multi-level cache.

Algorithm 1.   Update method of level-1 cache.

INPUT: Input data queue Q1
Input data queue Q2
The dataset C1 in the level-1 Cache
OUTPUT: the updated C1 and Q2
    1. **while** (Q1 in not enpty)
    2.   d = Q1, pop();
    3.   integated d into C1;
    4.   Q2, put(d);
    5.   return C1,Q2

Algorithm 2.   Update method of level-i cache.

INPUT: input data queue Qi
          Input data queue Qi+1
          Updating cucle Ti
          The dataset Ci in the level-i Cache
OUTPUT: the updated Ci and Qi+1
          5. **while** (Ti begins)
          6.   **while** (qi is not enpty)
          7.   d = qi.pop();
          8.   put d into set M;
          9.   **if**(d_{old} M)
/* d_{old} is a version of d,and $F(d_{old})$>$F(d)$*/
          10.   Delete d_{old} from M;
          11.   **for**(each mM)
          12.   integrate m into Ci;
          13.   Qi+1.put(m);
          10.   **return** Ci, Qi+1

For a query Q, if it requires at most one hour for the fresh degree, the fresh levels for the caches C_1, C_2, C_3, C_4 and C_5 are 0 minutes, 10 minutes, 30 minutes, 60 minutes and 240 minutes. We can position this query directly into the fourth-level cache C_4, which then responds to this query. In this way, C_4 just meets the needs of Q.

Here, Ti is the updating cycle of each queue. Each level of queues update the cache based on different cycles, so as to avoid frequent operations on the cache.

5   COMBINATION OF MULTI-LEVEL CACHE AND REAL-TIME STORAGE

The real-time data warehouse has a three-tier storage structure. In order to simplify the operation and improve performance, multi-level cache has the same data structure as the real-time storage area.

In a table with relatively small data volume, it is unnecessary for us to establish such a data structure like indexing or aggregation that is used to optimize the performance of the data warehouse, which will instead increase the operation complexity of multi-level cache. So the multi-level cache has the same data structure as the real-time storage area, which facilitates the cache to be regularly imported to the real-time data storage area. The three-tier storage structure has effectively shared the load, and improved the query response time.

6   EXPERIMENTS

In the case of a lot of queries at the same time, compare the differences in query performance

between the multi-level cache and single cache. Here, we set two query sets of S1 and S2, which are both based on the multi-level cache. They are both composed of the four queries of QS1, QS2, QS3 and QS4, whose proportions are seen in Table 1. In this experiment, 50 records of updating take place per second, and a lot of queries also compete with them in the meantime. So we set 5~100 queries per second in the cache. In Figure 3, the query execution time in different conditions has been shown. SC-S1 indicates that S1 is running in the single cache, and MC-Si indicates that query is running in the multi-level cache. From the result, we can obviously find that the processing time of the multi-level cache is lower than that of the traditional single cache structure. Meanwhile, the reason why S1 is less efficiency than S2 is that this multi-level caching mechanism can position a query to a cache it is fit for. Also, the proportions of the four queries in S1 are not balanced, of which QS1 has accounted for more than 80%. This 80% of queries are allotted to one cache, namely that the same cache has stored most queries in S1. In contrast, for S2, its four queries are equally distributed in four levels of caches, which have effectively shared the query load. So S2's query execution time is less than S1's, namely that S2's efficiency is higher than S1's. It has also reflected where the multi-level cache has advantages.

Figure 4 shows the query execution time under different updating frequency. In the case of a lot of updates at the same time, compare the differences in query performance between the multi-level cache and single cache. The number of queries has all been changed into 100, but the frequencies of updates differ from one another, updating from

Table 1.   Query sets composition.

	QS1	QS2	QS3	QS4
S1	80%	10%	10%	0
S2	10%	30%	30%	30%

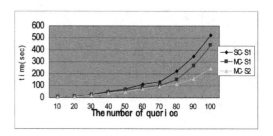

Figure 3.   Query execution time under different number of queries.

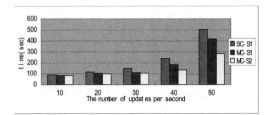

Figure 4. Query execution time under different updating frequency.

10 records per second to 50 records. In different update frequencies, when the multi-level cache and single cache are solving query conflicts, it is still the multi-level cache that has an obvious advantage over the query execution time, which is particularly outstanding when the query frequency is high.

From the experiments on the general real-time storage area and the real-time storage area based on the multi-level cache, we are informed that the real-time storage area solution in this article has a very reliable advantage over the speed in dealing with affairs, and the performance will be better if memory has a larger capacity. We found that at least 8% of the response time would be spent with this method, which was independent of memory and had nothing to do with the memory size. The experimental results also told us that, in the same condition, the larger memory would have less response time, namely rapid response. Meanwhile, for this architecture, with the increasingly frequent occurrence of affairs, the response time extended. With a multiple increase in affairs, the response time growth can still be controlled within 25%, which can be totally accepted by people and satisfy users.

7 CONCLUSION

In this article, the multi-level cache method has been used to solve query competitions in the real-time data warehouse. In the experiment of the real-time storage area based on the multi-level caching mechanism, the multi-level cache is apparently much higher efficiency than the single caching mechanism, and its processing time is less than that of the traditional single cache structure. For a large number of queries produced simultaneously, if there are more kinds of queries, the multi-level cache has more obvious advantages. It can locate queries which require different freshness to different levels of caches, so as to avoid too many queries in the same cache, which may save system resources effectively and reduce those unnecessary query operations on the data warehouse.

REFERENCES

Abadi D.J., Carney D. et al. "Aurora: A New Model and Architecture for Data Stream Management" [J], the VLDB Journal, 2003, 12(2):120–139.

Babu S. and Widom J. "Continuous Queries Over Data Streams" [J], SIGMOD Record, 2001, 30(3): 109–120.

Bruckner R.M., List B. and Schiefer J. "Striving Towards Near Real-Time Data Integration for Data Warehouses" [J], International Conference on Data Warehousing and Knowledge Discovery (DAWAK), 2002, 11(5):136–141.

Kuhn E. "The Zero-Delay Data Warehouse: Mobilizing Heterogeneous Databases" [J], International Conference on Very Large Data Bases (VLDB), 2003, 4(05): 155–162.

Yichuan Shao, Xingjia Yao, Liwei Tian. Peano Space Filling Curve Applied in Managing P2P Service Resources. [J] Applied Mechanics and Materials. Volume 223, pp, 2508–2511, 2012.

Yichuan Shao, Xingjia Yao, Liwei Tian. A kind of Intelligent Client Flow Engine Based on Ajax. Applied Mechanics and Materials, Volume 223, pp, 2504–2507, 2012.

Yichuan Shao, Cao Hui. Rotate Coordinates Fold Line Applied in Multidimensional Data Visualization. Future Control and Automation, Volume 172, pp, 105–114, 2012.

Hazard research on reservoir flood limit water level adjustment by flood prediction model

R.J. Zhang
Fujian College of Water Conservancy and Electric Power, Yong'an, China
Hohai University, Nanjing, China

Z.S. Li & F.C. Zhang
Fujian College of Water Conservancy and Electric Power, Yong'an, China

ABSTRACT: After introducing the influence of flood forecast error, a hazard rate calculation method of flood control is proposed. By determining the hazard error margin and their probabilities of over-designed and designed flood control hazard, and then comparing the probabilities of flooding and the relating probability of over-designed flood hazard, the over-designed hazard rate of flood control level adjustment by flood control forecast and dispatching mode can be determined. As an example, the over-designed hazard rate of Saishan Reservoir flood is 0, 0, 0.0014% and 0 for designed 5%, 2%, 1% and 0.05% flood respectively. The result is between the maximum and the expected over -designed hazard rate, and shows that the method is feasible and reasonable.

1 INTRODUCTION

Flood control forecast dispatch is a scheduling mode of flood resource utilization based on the pre-flood forecast information, Improved flood Criterion of original plan and Advance pre-discharge realization, thereby raising the flood limited level of the reservoir [1–4]. Since Flood forecast information exists forecast error, whether will the reservoir and its upstream and downstream flood risk increase after using Flood control forecast? Whether can the increased risk be acceptable? This is the key to the implementation of flood control forecast, but also the focus of attention of scheduling decision makers.

At present, Risk researches of Reservoir flood control operation by domestic scholars are mainly discriminating flood prevention risk according to the size of flood control capacity corresponding planning eigenvalues (Characteristic level or downstream safety discharge amount) of different flood control standards. But I believe that flood control risk at various levels of flood control standard can also be determined through eigenvalues of flood regulation Correspond relevant order of magnitude flood. If the highest water level (or discharged volume) of flood regulation of a design flood exceeds the corresponding planning eigenvalue by the use of flood forecast operation adjustment. Then flood prevention risk can be considered to be ultra-design level; On the contrary, that the flood risk is not beyond design

levels. Therefore Risk analyses for flood control operation are to examine that whether the highest water level (or discharged volume) of flood regulation is over the Characteristic level (or downstream safety discharge amount), when Suffered a magnitude of flood? How much is the probability of exceeds? Furthermore, because of to the error in the forecast information, flood forecast operation increased a risk source than the conventional scheduling, which also needs to be considered in the risk analysis.

2 HYPER DESIGN RISK RATE

2.1 Type

There are generally two cases about the impact of the flood forecast error to flood control operation:

a) The risk of reservoir flood control Increase, due to the lag of the change under Vent FLOW may make the required flood control capacity increase, when the flood forecasting value is smaller than the actual value
b) Increase the discharged flow may advance more than the amount of the downstream safety vent, this adverse downstream flood control. When the flood forecasting value is larger than the actual value; In this case, the super design risk is divided into super-reservoir design risk and super downstream design risk.

2.2 *The calculation*

a) The division of Flood forecasting error domain super-design risk, which adjust the flood water level by the way of Flood control forecast is a conditional probability; At this time The reservoir suffered flood control standard middleweight flood, and The forecast error Discriminating flood middleweight indicators is within a certain range. Based on this, Prediction error range is divided into design risk does not exceed the error domain E1 and risks Hyper Design error domain E_2, E_3; Where, E_1 is The scope of the flood forecasting errors when the highest water level of the flood (or discharged flow) does not exceed the value of the original design features; E_2 and E_3 are The scopes of the flood forecasting errors when the highest water level of the flood (or discharged flow) does exceed the value of the original design features;

b) The calculation

According to the division of flood forecasting errors domain, the calculation formula for the P^Z_{mF} is:

$$P^Z_{mF}\left(Z > Z_m \mid P_m, Z_{1d}, Z_m, R_F\right)$$
$$= P_m \Phi\left\{\Delta E \in \left(E^Z_2 \cup E^Z_3\right)\right\} \tag{1}$$

In which, Z: the highest water level of Regulating flood; Z_m: water level for the planning characteristics; P_m: designed flood frequency; Z_{Ld}: the domain of adjusted Flood water level; R_F: ways of scheduling Using Flood forecast; E^z_2, E^z_3: The risk error domain of beyond design Corresponding to the P_m flood, When considering the risks of reservoir flood control; $\Phi\left\{\Delta E \in \left(E^Z_2 \cup E^Z_3\right)\right\}$: The probability of Flood forecasting error ΔE is Within The risk error domain of Hyper Design Corresponding to the Pm flood, When considering the risks of reservoir flood control;

Similarly, The calculation formula for the P^Q_{mF} is:

$$P^Q_{mF}\left(Q > Q_m \mid P_m, Z_{1d}, Q_m, R_F\right)$$
$$= P_m \Phi\left\{\Delta E \in \left(E^q_2 \cup E^q_3\right)\right\} \tag{2}$$

In which, Q: the discharged volume; Q_m: Corresponding Downstream safety discharge amount; E^q_2, E^q_3: The risk error domain of beyond design Corresponding to the P_m flood, When considering the risks of downstream flood control; $\Phi\left\{\Delta E \in \left(F^q_2 \cup F^q_3\right)\right\}$: The probability of Flood forecasting error ΔE is Within The risk error domain of Hyper Design Corresponding to the P_m flood, When considering the risks of downstream flood control;

c) The solution step

i. Analyze the distribution of flood forecasting errors. Generally divided into two cases to consider: (1) When insufficient historical data, you can use statistical analysis to study the error series, find its distribution. Recommended methods of statistical analysis are the maximum entropy method. (2) When the lack of historical data, distribution can be directly using normal or lognormal distribution to describe.

ii. Solve the Hyper Design risk error domain. E_2, E_3 May be discrete interval directly Deriving more complicated. E_2, E_3 and E_1 Are complementary and do not intersect, the domain of the entire flood forecasting errors. Therefore, it is possible to determine E_1, thrust reversers E_2 and E_3.

To determine the flood forecasting error limit domain E_2, E_3, calculated using the risk-free rate of increase in flood control forecast. Flood forecasting error limit domain E_2, E_3 intersect range, super design risk was 0; flood forecasting error limit domain E_2, E_3 disjoint intervals, according to the flood forecasting error distribution of flood forecasting errors occur the probability of the risk of Hyper Design error the domain, and then according to equation (1), (2) calculate the rate of design risk.

3 THE EXAMPLE ANALYSIS

The reservoir Saishan annual regulating reservoir of the Huaihe River Basin Shahe River on a large (two) type. Drainage area of 2740 km² and a total capacity of 922 million m³. The main flood season, flood control water level of 102.00 m. Reservoir flood control standard designed of a 100-year flood in 2000, a flood check. The specific scheduling planning results in Table 1.

3.1 *Distribution law of the forecast error*

Flood control operation of Saishan select the forecast cumulative net rainfall as indicators to determine the flood magnitude to determine the flood water level dynamic control the domain [102.00 m,

Table 1. The feature control value of Saishan reservoir scheduling planning.

Flood control standard/%	The characteristic level/m	Downstream safety discharge amount/ $m^3 \cdot s^{-1}$
5	105.38	600
2	105.90	3 000
1	106.19	*
0.05	109.56	*

Note: * indicates do not use this as a control value.

102.60 m], the corresponding control parameters are shown in Table 2 [10]. The results of this study, the reservoir flood control operation caused by the adverse effects of forecast information runoff forecast error. Reservoir runoff forecast error of Saishan x limit value of 35 mm, the corresponding probability density function f(x) [8]

$$f(x) = e^{-2.9589 - 0.012x - 0.0085x^2} \qquad (3)$$

White Turtle reservoir 15 games (second) rainfall greater than 100 mm of historical flood the forecast error analysis, that the results of the study to meet the designed magnitude flood the runoff forecast error requirements.

3.2 Super design risk error domain of flood forecast operation

Reservoir upstream of Saishan Zhaopingtai Reservoir form a cascade reservoirs, flood area can be divided into Zhaopingtai—Saishan Reservoir with frequency and Chao White interval—Saishan Reservoir the same frequency. IN WITNESS white range—Saishan Reservoir same frequency region, for example calculated super the reservoir design risk, which Saishan Reservoir flood level dynamic control domain up to 102.60 m from the transfer, Zhaopingtai Reservoir maintain flood control, flood control forecastwater level 167.00 m, the conventional scheduling.

The flood forecast system simulation scheduler, inquire Saishan flood control operation risk error does not exceed designed domain and the design risk the error domains (see Table 3).

3.3 Super reservoir design risk rate of flood forecast operation

According to Table 3, E_2, E_3 and (3), (1) calculate the risk-free rate flood forecast operation Hyper Design (see Table 4).

Tables 3 and 4 shows that: (1) the frequency of 1% flood, when underreporting of runoff forecast error is greater than 23 mm design risk events occur, design risk, rate of 0.001%; frequency of 0.05%, 2% When the runoff forecast underreport-

ing error is greater than 35 mm, and 5% of the flood will occur beyond design risk events, the the corresponding ultra reservoir design risk rate is 0. It can be seen, the flood control operation of adjusting the frequency of design flood flood reservoir water level over the planning characteristic level of risk events occur within a particular range of error. Solving E_2, E_3 on the basis of the design risk rate calculated for flood control operation is feasible. (2) use of flood control forecast Saishan reservoir flood limit water level elevation from 102.00 m to 102.60 m, suffered Zhao White interval—the reservoir with Saishan frequency flood, the Reservoir defense frequency design flood risk does not increase the basic, only defense frequency of 0.001 4% increased risk of 1% of the design flood. From the safety point of view, the flood water level elevation value is slightly lower some of the more reasonable.

3.4 Comparison of different calculation methods

To further demonstrate the feasibility of this method [5] referred to flood control operation of a risk (expected value) and the maximum risk-free rate (see Table 5). Among them, from the transfer water level still take 102.60 m, the probability of occurrence of flood forecasting errors calculated using equation (3).

As can be seen from Table 5, as the water level corresponding to the different characteristics of design flood control capacity beyond design criteria to judge the risk-free rate, the three methods to calculate the frequency of 5% and 0.05% of the risk of floods rate is less than or equal to the design flood risk, The frequency is calculated according to the proposed method the super design risk rate of 2% and 1% flood exceeding the design value is less than the maximum rate of risk beyond design value is greater than the expected risk. The analysis of the main reason is, expected risk rate calculation method, the error does not cause a risk event occurs, the risk-free rate than the design of the risk-free rate is too small. Although the large error rate risk, but a small error probability of occurrence is much larger than the probability of occurrence of large errors, the result of expectations calculate the risk-free rate

Table 2. Saishan flood control operation of the control parameters.

Flood control standard/%	The highest water level of regulating flood/m	Forecast cumulative net rain/mm	Discharged flow/m³·s⁻¹
≥5	≤105.38	340	600
5~2	(105.38, 105.90]	(340, 420]	3 000
2~1	(105.90, 106.19]	>420	Sluice fully open
1	>106.19	*	Sluice fully open, at the same time to open the north channel. Increase vent 200 m³/s

Table 3. The different prediction error domain considering the risk of reservoir flood control.

The characteristic level/m	Flood control standard/%	E₁/mm	E₂, E₃/mm
105.38	5	[−35, 35]	−
105.90	2	[−35, 35]	−
106.19	1	[−23, 35]	[−35, 23]
109.56	0.05	[−35, 35]	−

Table 4. The risk-free rate flood forecast operation Hyper Design.

The characteristic level/m	Flood control standard/%	Risk-free rate/%
105.38	5	0
105.90	2	0
106.19	1	0.0014
109.56	0.05	0

Table 5. Results of the calculation method for different risk.

The characteristic level/m	105.38	105.90	106.19	109.56
Flood control standard/%	5	2	1	0.05
Expected risk/%	1.732	1.171	0.995	0.026
Maximum risk rate/%	2.738	2.279	1.702	0.029
Value of the expected risk Hyper Design/%	0	0	0	0
Value of greatest risk Hyper Design/%	0	0.279	0.702	0
Super design risk/%	0	0	0.0014	0

tends to be significantly small error, calculated the increase in risk-free rate beyond design value is too small, conducive to flood safety. Maximum risk rate is calculated the most unfavorable consider the design flood is always a small probability of error at the same time suffered enlarge the probability of occurrence of risk events, the calculation of the increase in risk-free rate beyond design value is too large. Therefore, this method results in between, more reasonable, the use of this method to analyze the risk of flood control operation is feasible. As for the results of the risk-free rate calculated by different methods the difference between how much flood water level more reasonable, should be based on the specific circumstances of each reservoir, considering the risks and benefits evaluation, decision-making.

4 CONCLUSION

In This article, The flood forecasting error domain is divided into the super design risk error domain and the unexceed design risk error domain, and Forecast error level allowed to withstand reservoir Is Given, When not exceed the design risk. On this basis, Calculation method of super design risk, When the reservoir adjust the flood water level with Way of flood forecast operation Was established. For the design flood Whose frequency is 5%, 2% and 0.05%, The runoff forecast error will not lead to the design risk event, the Super design risk was zero. But the design flood, Whose frequency is 1%, May lead to super Design risk event, When underreporting errors of the runoff yield forecast range (23, 35], The probability of occurrence is 0.0014%. The Design risk rate Corresponding to the different flood control standards is All Greater than Hyper Design values of the corresponding Expected risk rate, and Less than Hyper Design value of the maximum risk rate, This result is reasonable. Therefore, computation methods of Hyper Design risk rate is feasible. This provides a new idea For risk researches of flood control operation.

REFERENCES

Directors four-Hui, Zhou Rong, Hui-Cheng Zhou. The way of the flood control operation of reservoirs downstream flood risk analysis-to Shenwo, for example [J]. Safety and Environment, 2006, 37 (6):128–130.

High Bo, Wang Yintang, Hu. The reservoir flood water level adjustment and application [J]. Advances in Water Science, 2005, 16 (3):326–333.

Hou Zhao-Di Yifeng, Yin Jun Siem. Flood control operation risk analysis [J]. China Water Conservancy and Hydropower Research and Technology, 2005, 3 (1):16–21.

Hui-Cheng Zhou. Change the red king Bend. Adjusted based on the flood control operation of reservoir flood control level [J]. Hydroelectric, 2006, 32 (5):14–17.

Hui-Cheng Zhou. Directors four-Hui, Deng Chenglin, et al. Risk analysis based on stochastic hydrological process for flood control [J]. Journal of Hydraulic Engineering, 2006, 37 (2):227–232.

Li Liqin. Entropy and fuzzy set theory in flood forecasting and reservoir regulation [D]. Dalian: Dalian University of Technology, 2006.

Li Wei, Guo Sheng-Lian, Liu Pan. The flood water level of the reservoir to determine the method of analysis and forecast [J]. Hydroelectric, 2012, 31 (1): 66–70.

Liu Pan, Guo Sheng-Lian, Wang Cai Jun, et al. The flood water level of the reservoir and real-time dynamic control model [J]. Hydroelectric, 2005, 31 (1): 8–11.

Wang Bend, Hui-Cheng Zhou, Wang Guoli, et al. Henan Province Saishan reservoir flood control level design and the use of research reports [R]. Liaoning: Dalian University of Technology, 2004.

Wang Bend, Hui-Cheng Zhou, Wang Guoli, et al. The flood water level of the reservoir and dynamic control theory and its application [M]. Beijing: China Water Power Press, 2006.

Control Engineering and Information Systems – Liu (Ed)
© 2015 Taylor & Francis Group, London, ISBN 978-1-138-02685-8

On counter-measure to flight colleges modern long-distance education existing state and developing

S. Ye & Q.G. Zhang
Training Department, Aviation University of Air Force, Changchun, China

ABSTRACT: Modern long-distance education is a new, long-range, mutual education way based on modern education technology as computer-tech, net-communication-tech, multi-media-tech, etc. The headquarters of the General Staff has drawn up medium and long-term plan for army modern long-distance education development, and take it as a focal point. This article, beginning from the existing state of flight college modern long-distance education, analyses its insufficiency existed, puts forward a contour-measure of 4 aspects as long-distance-edu manage institution, construction of teacher ranks, construction of education resources, and building cultural atmosphere to long-distance education.

Keywords: flight colleges; modern long-distance education; network

1 INTRODUCTION

The 21st century is the century of space. Future air force operations will not only be implemented in airspace, but will also be extended into near and outer space. The integration of air and space will inevitably represent the future trend in air force development and bring new challenges for flight crew training. Moreover, modern information technology, together with Multimedia Technology, Telecommunications and Network Technology as the core parts, is constantly penetrating into various aspects of flight personnel training, leading to profound changes in the traditional education, and it is also the background where the army modern long-distance education emerges and progresses. Through introducing new teaching method and approach, This new education model greatly improves the education quality and efficiency by the transformation from model of 'Teacher Centered' to 'Student Centered'. However, the development process of any innovation is tortuous, in the process of promoting long-distance education, we must pay continuous attention to sum up experience, identify problems and find out the corresponding countermeasures, and only by this way can we push forward the application of this new teaching model, bringing a rapid and healthy development of army long-distance education [1].

2 CURRENT DEVELOPMENT OF FLIGHT COLLEGE LONG-DISTANCE EDUCATION

The project of Flight College modern long-distance education officially launched in 2001, has now opened undergraduate and postgraduate level professions, and initially realized various forms of non-degree continuing education. Flight College long-distance education provides a new perspective to the development of flight education, and at the same time generates great impact on the traditional education. The development of long-distance education brings wonderful opportunities as well as great challenges to the Flight College education.

2.1 *Though developing rapidly, hardware construction still needs to be further strengthened and the software construction is relatively lagging behind*

In order to adapt to the development trends of modern education, and to build a long-distance education system with distinct Chinese Air Force characteristics and sustainable development, the Air Force Flight institutions have been concentrated on construction according to the requirements of army long-distance education development plan. Various supporting platforms of networks including interconnected campus network, multi-media classrooms, and electronic reading networks have been roughly completed. Network infrastructure is further improved in order to meet the requirements of online teaching, all of which has laid a solid material foundation for long-distance education. But with the rapid update of the network technology, the hardware facility construction needs continually to be strengthened so as to achieve the long-term development of the army long-distance education. On the other hand, compared with the hardware construction, it appears that software construction, though as the key part of long-distance education, is lagging behind. Many educators still do not have enough in-depth study on long-distance education theory, and there is still lack of reflections on pilot projects of long-distance education bas on practice and argumentation; network multimedia courseware are comparatively underdeveloped and network teaching methods are monotonous; there are limited inter-disciplinary personnel who are not only proficient in computer network technology, but also understand the modern educational theory and technology. Thus, the incompatibility of hardware and software in a large extent restricts the healthy development of army long-distance education.

2.2 Teaching subject is not yet fully adapted to the new education model

The subject of Flight College modern long-distance education mainly refers to the teachers and flight students. Currently long-distance education teaching methods are based on sub-point pedagogy, in which teaching evolves into the process organized by using multimedia tools and network technology. In this process, the teaching and learning are carried out at different locations and time, and the teaching can be realized by multimedia technology. Students are able to download study software, search related information online, or communicate with classmates and teachers through email and BBs, and in such way, they can control their study progress and complete learning by themselves. This education model provides study opportunities for pilots and flying cadets who cannot leave their jobs but desire for further educations, which is the key motivation for promoting army long-distance education. However, the current faculties and students are not yet able to fully accommodate to this new online learning environment. From the objective view, this is because in traditional teaching mode, the teachers' job is mainly to preach on a professional field by using textbooks, blackboard, chalks, and they are just not accustomed to using computers.

While from the Subjective viewpoint, the faculties have long been teaching in the traditional model and have accumulated a lot of experience, and they hold deep affection for traditional teaching model and are less willing to make changes. For the above reasons, the long-distance education is widely advocated however not yet popularized. In addition, long-distance education bring obstacles between students and teachers, who are from different locations, and both of them might be used to face-to-face communication and be discouraged by on-line education. Besides, since the teachers and students are separated, the students are expected to finish their work in dormitory or library, leading to the lack of effective teaching supervising mechanism, so that it is difficult to control the teaching process, and the education assessment is also unavailable. Thus teaching quality will not be effectively guaranteed [2].

2.3 The educational resource is not complete and needs updates and improvements

Educational resources generally include curriculum resources, literature resources and long-distance education software supporting platforms. As for curriculum resources, different institutions have built a variety of online teaching information database according to their professional teaching needs. The teachers can select and organize teaching contents conveniently and the students can also freely choose the courses they prefer and have an easy access to the related information through the internet, which breaks the limitations of traditional education that only can provide teaching information through textbooks. However, because of the short time after the long-distance education initiated as well as the special nature of Flight College education, the educational resources are relatively scarce and not systematic. There is no essential improvement and difference in teaching plans, contents, forms and textbooks between current long-distance education and conventional fulltime face-to-face education. The courseware and course contents are obsolete, and the online information resources are limited, which do not highlight the characteristics of long-distance education. As for literature resource, there is not a system for providing comprehensive, complete army public literature service.

Despite the existence of the military training network, it still has deficiencies such as: though with much general information, the professional information is scare, and inquiry waits long time for responds and so on. In addition, we must also take great efforts to further develop long-distance educational software supporting platforms.

3 MEASURES TO STRENGTHEN FLIGHT COLLEGE LONG-DISTANCE EDUCATION

3.1 Building and continuously improving flight college long-distance education management system

To give full play for the advantages of long-distance education, the managers of Flight College should first establish the modern conception of educational management to adapt to network technology, absorb and learn successful experience from foreign and local universities, and through integrating characteristics of Flight College, try to find a management system that accommodates to our country's Air Force long-distance education. Second, we must strengthen the microscopic study of online education, including in-depth and in-breath research of long-distance education, the study of network educational functions and its teaching efficiency, and the study of network courseware's design, dissemination, and degree of acceptance etc. According to the openness, long-period and multilevel teaching objects features in long-distance education, online teaching management should aim at properly handling the relationship between degree education and non-degree education in aspects of curriculum setting, teaching contents, teaching method and so on; developing genuine teaching quality assessing standards, strict assessment criteria and exam system; strengthening network information management by technical, administrative and legislative means, to prevent intrusion of bad information and maintain a clean, efficient and orderly network education environment.

3.2 A key part of developing long-distance education: establishing teachers group adapted to modern long-distance education

3.2.1 Changing the teachers' education philosophy

Teaching faculties have been long under the influence of the traditional education model, and to accommodate to the modern long-distance education, they must first make two changes in their conception of education. First, they must realize the transformation from teacher as the main body of education to student as the center of the education. The teacher should be switched from the role of main information provider in traditional teaching to the role of guiders to help students find the correct method of obtaining information and skills. In Modern long-distance education, the network system can provide better information both in quantity and quality than that from the instructors. The teacher's main task is to guide students to properly and effectively use the information and to acquire knowledge and solve problems by the Internet and multimedia technologies, so as to convert the traditional faculty-centered, face-to-face educational model to students-centered, interactive-approach-oriented new teaching model. Meanwhile, in the process of receiving modern long-distance education, There are limitations in students' recognition of the teaching objectives and the teaching process and they cannot systematically consider and organize the whole learning process as well as the related details, thus inevitably they are sometimes incline to misunderstand the focus and difficulties of the course. So, the teachers must pay more attention than that in the traditional classroom education to organize the whole learning process, and actively guide the students not to get lost in the network knowledge ocean. Second, they must realize the transformation from the main body of authority to the main body of equity. In the traditional teaching mode, the instructors are the authority and they choose most of the study materials for the students, so that students can only learn depending on the preset framework of the teaching program. In the army modern long-distance education, teachers should be student-oriented so as to create an open, democratic learning environment. This is conducive to help faculties examine their teaching performance from the perspective of students, and to help student change from passive learning to independent learning. Meanwhile, teachers must also position themselves as both educators and learners, by communicating with different types of students, learn and gain new knowledge, and continuously improve their intellectual level and teaching ability.

3.2.2 Improving teachers' teaching ability of using modern educational technology

The application of a large number of high-tech methods in modern long-distance education ask for instructors to have teaching ability of 'professions-based and network-equipped' and to have a solid professional knowledge foundation, which are also the minimum requirements of modern long-distance education. The faculty should not only master the theoretical framework of the discipline, track the developing frontier, but also have to get sufficient understanding of the military educational technology. Only by better combining solid professional knowledge with educational technology, can the teachers really meet the needs of the modern long-distance education of military academy to improve online teaching ability is a higher demand for teachers, which is brought by The development of the military academy modern long-distance education. With the widespread application of network and multimedia technology in teaching activities, the time and space structure of teaching has been re-located, and the classroom is not the only place of learning. The Flight College instructors should master modern educational technology, especially network technology, and be skilled at using e-mail, bidirectional video, online discussions, and BBS, which are essential capabilities for performing remote teaching. Moreover, it should be faculties' working focus to develop and produce courseware to promote the teaching and learning under the modern long-distance education environment [4].

3.3 Strengthening the construction of modern long-distance education resource in Flight College

3.3.1 The construction of curriculum resources

Curriculum teaching is the main form of teaching in Flight College and curriculum resources is the basis of Flight College's modern long-distance education, which must be emphasized at any time as the core of modern long-distance education construction. The main contents of modern long-distance education resource includes: the construction of text, graphics, images, animation, audio, video and other media material libraries and the development of related management system; the built of practice library, case base, scenario library, past war library, teaching strategies template Library, courseware library, test database, and other 'micro-teaching unit' and the development of related management system; the construction of network curriculum library and the development of its management system. We should establish and improve the multifunctional platforms of teaching system basing on the repository, so that the instructor can design courseware of different disciplines on the platform through modular filling. This not only improves the production efficiency, but also achieves teaching resources sharing, and thus makes them reusable, reducing production costs. The construction level of curriculum resources is essential to the overall advancement of military education informalization. In order to build high level of curriculum resources in a short time, we can take the following three steps: First, concentrating on building curriculum resources of both key and fundamental courses. By promoting the 'military network teaching system', combining with the construction of '211 Project', and by the policy, equipment, funds and manpower support, giving full play of the advantages of different institutions, to construct a number of military characteristic courses in fields of military basis, command and management, and high-tech equipment. Second, strengthening the co-building and sharing of curriculum resources. Military system can take full advantage of the existing big unite systems, the military academy collaborating centers and connections with the same type institutions, by systematical, professional, and geographical means to promote 'comprehensive' and 'multidimensional' curriculum resources sharing and co-building. Third, fully utilizing the national education resources. The military academy can select from the national education resources a number of military and local sharing fundamental or professional courses, so as to fast build army long-distance education curriculum resources with high quality [3].

3.3.2 Literature resource construction

Whether from the perspective of its supporting role in long-distance education or from its features and ways of construction, literature resources are independent from media materials. In the construction system of long-distance education resources, we must separately formulate construction plans of literature resources that aim at supporting students' 'self-learning' process, and build supporting system for Flight College literature resource of modern long-distance education. First, using the existing literature resource structures. 'Digital access and network transmission' is the basic feature of literature resources organization and utilization in long-distance education. Literature resources can provide support for the army modern long-distance education in three main categories: ① Introduction or download of network digital resources including electronic books, electronic journals, periodicals and papers database, conference papers databases, dissertation databases, patent literature databases, audio-visual resources library, discipline or professional literature data, Internet resources, and other forms of digital document resources. ② The literature information resource of military academies, including publications of the military academies, teaching reference materials, collections of digital information resources, military academies and libraries bibliographic data. ③ Military characteristic digital training information resources, including information databases of military training, military discipline digital information. These three literature resources provide military long-distance education a 3-D interchangeable literature resource supporting system by different types of information organization and information exchanging channels. Second, using digital library resources of various institutions. Literature Resource Construction is a systematical project with high cost and long cycle length that cannot just rely on one or two funding, neither can be completed in a short-term. Therefore, in the process of establishing literature resources supporting system for the army long-distance education, it is better to rely on the already completed libraries as much as possible, fully use their digital library resources, bringing libraries and long-distance education technology platforms together to form long-distance learning support and services system. Third, relying on the military public literature information service system.

The construction of long-distance education software supporting platform.

First, unify software platform structures. Long-distance education software supporting platform is composed of a series of teaching software, tools and management software that can support a variety of teaching modes, which includes: real-time and non-real-time lecturing systems, discussion and communication tools, teaching and educational administration, long-distance education assessment, homework review, tutoring, remote examination, virtual experiments, as well as educational resources searching engine systems. From the development of domestic long-distance education pilot, we can see that most of them use self-developed software platforms. Compared with the foreign long-distance education platforms, the construction of domestic software support platform should continue to be enhanced in systematicality, creativity, practicality. Second, improve the functional structure of the software platforms. Army long-distance education software platforms in general should have the following functional structures: First, the various types of media, including Audio, which supports real-time audio teaching, audio teaching recording and online player, download and playback; Video, which supports video teaching and program design, on-line player, download; Shared screen, which supports screen teaching and learning. Second, various teaching aids, which includes personal planning tools, which provide individual user login and authentication, individual plan according to teaching requirements, and planned assessment; personal practice tools, which provide classified exams, automatic test paper generator, automatic assessment and feedback and personal learning portfolios; remote test tools, which provide test papers distribution, candidates identity authentication and cheat preventing measures, marking function, and the grades publishing. Third, teaching managements, including student administration, which consists of registration enrollment, student files classification, and student personal information reservation; courseware management, which consists of courseware classification, query, download and delivery; educational administration, which includes faculty allocation, curriculum configuration, and teaching assessment. At last, further improve software platforms with interactive structures.

Interactive method is the main feature of modern long-distance education. Army long-distance education software platforms generally have the following interactive features: First, the planning and notification system for the implementation of teaching management and process control; Second, text forum, where the discussion can be organized, downloaded, and forwarded according to course teaching and research; Third, the whiteboard, which provides text, graphics, images services and supports simultaneous access operation and real-time refreshment; Fourth, the E-mail, which is used for information exchange. Fifth, the chat rooms enable students to discuss in long text, and send links and messages point to point.

3.4 Creating a good long-distance learning campus culture

The first is to incorporate the real and the virtual world and bring about interactivity between the learning and practicing. As in the long-distance education, students spend long time staying in the virtual internet world, so the campus culture construction in modern long-distance education should be more practice-oriented.

We should create a real campus atmosphere by technical tools. For example, publicizing campus photographs or floor plans, using 3D technology to build virtual campus, shooting campus movies that records students and teachers' communication in the teaching process, community activities, and various campus daily life, and such movies can be released online after carefully edited in multimedia forms; arranging students who are receiving remote education to make short visit to campus so as to learn and enhance understanding of the college, which would help them develop the sense of honor and belonging; in the aspect of curriculum designing and testing, we should encourage and guide students to transform their knowledge into actual working ability, basing on the characteristics of military unit the students belong to and through the working feedback ultimately achieve bidirectional improvement.

The second is to construct inspiring-teaching platforms in order to achieve self-learning supervision. How to ensure the teaching quality of long-distance education is the key whether long-distance education development is sustainable and can be widely accepted. Therefore, we should build inspiring-teaching platforms as soon as possible which can monitor the entire process of teaching and learning and make the students more motivated in study. We should at the same time implement entertaining promotional strategy, properly reflect all the rules and regulations in the teaching platform in a timely manner so that students can see everywhere, by such way further strengthen the publicity effect and increase the transparency of the education management and administration [5].

The third is to strengthen security. Due to its special nature of teaching content, military academy requires high security and thus we must pay attention to guarantee the confidentiality of course content. In army long-distance education, free access to the Internet should be denied, and we must seek break-through in the application methods to strengthen the firewall construction and safety education, prevent any disclosure of confidential information and any intentional damages.

REFERENCES

Chongming Bai, Dewei Liu, The Theory and Practice Research of Flight Talented Person in Air-sky Times, LanTian Publishing House pp1–2, 2010(6) (In Chinese).

Jie Jian, The Influence to the Teaching and learning by Army School Net Teaching Application [J]. China Military Education 2004 (2) (In Chinese).

Lipeng Wan, Hai Liu, The Conception of Army Modern Long-distance Education Resources Construction [J]. China Military Education 2003(5) (In Chinese).

Wang Zhuzhu, Injury on Modern Long-distance Education Development and Problems Faced [J]. E-Education Research 2001(11) (In Chinese).

Zhu Ruke, Army Colleges Should Put Forth Effort to Develop Net Teaching Based on the Campus Net [J]. High Education Research Journal, 2005(1) (In Chinese).

Control Engineering and Information Systems – Liu (Ed)
© 2015 Taylor & Francis Group, London, ISBN 978-1-138-02685-8

Research on events recognition from data stream in non-intrusive load monitoring

L. Chen

Electrical and Instrument Control Department, Henan Electric Power Survey and Design Institute, Zhengzhou, China

X. Yi

Technology and Quality Control Department, Henan Electric Power Survey and Design Institute, Zhengzhou, China

ABSTRACT: This paper is focused on the data mining from data stream, events recognition identification problem under the non-intrusive measuring environment. In order to expanding serviceable range of NILM system, both computed strength and memory requirement are take into account in the algorithm design, and the experiments are also shown in this paper.

1 INTRODUCTION

Demand-Side Management (DSM) is viewed as the least-cost energy resource plan when both environmental costs and welfare needs are taken into account. This conception is proposed for about twenty years and many policies have been studied under this conception [1,2,3]. Non-Intrusive Load Monitoring (NILM) system is one of the important studies under DSM, and how to do the load identification by a financially viable and easily applicable way is one of core problem in NILM.

This paper proposes an algorithm for parameter acquisition from data stream in real time mode. Usually, identification needs complete feature parameters for higher accuracy, but the complete feature extraction time always uncertain, which bring about practical problems such as mass data process and storage. These problem lead to hardware of the monitoring system and limit the serviceable range. At the present stage, most research are processed in the laboratory, make use of high performance equipments, which can not be reappeared in the real scene. Under the situation above, the algorithm proposed in this paper refines parameters from data stream, both computed strength and memory requirement are take into account.

The methodology of NILM and its load identification model problem are simply introduced in second section. Then basic mathematics model used in this paper is described. In fourth section, the details of algorithm and corresponding parameters setting are presented. The final section is the conclusion.

2 NILM AND LOAD IDENTIFICAITON PROBLEM

NILM is monitoring system for electrical circuit that contain several different appliances which can switch on and off independently, by a non-intrusive method. One of the earliest approaches was developed in the 1980s at MIT by George Hart which had its origins in load monitoring for residential buildings. These findings and its process are described in [4]. Both USA and France have developed their NILM system at 90's of last century. Since, several methods have been developed to improve this system [5].

For load monitoring systems, most methods of analyze and identification faced steady state or transient state of appliances. Steady state analysis, each individual load or group of loads were determined by identifying times at which electrical power measurement changes from a steady state value to another. The variations related to the current, the active and reactive power, the admittance related to either turning on or off of loads [4,6,7]. Steady state analysis can not identify nonlinear appliances which limit its universal. For transient analysis, loads are identified by their spectral analysis [8,9], high frequency responses [10] or wavelet transform, but they usually needs high sampling frequency which limit its universal as well.

All the methods above have their gradual achievements, but the limitation for the widely application is still not solved for many reasons. The research in this paper is based one of latest NILM method which set financially viable and easily applicable as the ultimate goal of the research [11]. It establishes on a new classification of residential appliances, and a new power consumption model to analyze power curve. The processed description data will be used in appliances identification by mean-shift clustering and linear discriminant with a priori knowledge base. This new method, belongs to auto-setup, is non-intrusive in every processes and its sampling instrument is a multi-function meter which ensure financially viable and easily applicable.

There are two problems in all these methods above. One is the contradiction between the precision/cost of monitoring and the accuracy of identification. The other one is how to get the key parameters, especially for steady state identification, the crucial data of events always occurs in the end of events, but the last time of events are uncertain which leave the problem to parameter collectors. The collector needs face the mass information from sensors, so the thinking model of data mining for data stream is helpful in this situation, which is also the core idea of algorithm proposed in this paper.

3 MATHEMATICS MODEL

The basic math model used in this paper is from [11], which set target parameters as two kind events, the triangles and the rectangles. The classification of loads and feasibility of events is also discussed in [11], the details are not repeated in this paper for the length.

The record of triangle event need 4 data items, they are *starttime, peaktime, peakvalue* and the *endtime*. Each rectangles stand for a state change (not necessarily caused by a user's behavior) of certain appliances, and it usually come with instability, which means a rectangle always begins with a half triangle, the key parameters including five data *starttime, peaktime, peakvalue, steadtime, steadpower* can express this basic unit perfectly. The schematic diagram of the two unit graphics is shown in Figure 1.

Loads are classed into three categories. In the first category, there are appliances such as washing

machine and the like, whose main power unit will power on and off at a high frequencies (contrast human reflection speed). It will add many peaks to the power map during its working cycle. The second category are appliances those do not have certain power, such as AC (not include Frequency-alterable AC) and many electronic products which have high power consumption like a PC, TV or large stereo units. The third category is typical for most types of appliances, their main power unit is relatively pure, whose power consumption is stable with an almost constant voltage.

The three categories loads are expressed by different combinations of triangle and rectangle events, which decided the accuracy of identification.

4 ALGORITHM DESIGN

The algorithm has four different states to complete decomposition of power map. Figure 2 is the flow chart of main programme. Bold state in the following figures stand for system state mark.

Initial system state value is 1, power is in the steady state in the former sampling point, the task of this state is determining whether the start event is triangles or rectangles, changes system state to 2.

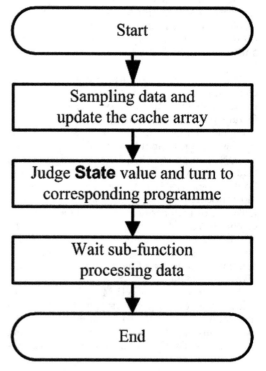

Figure 2. Main process.

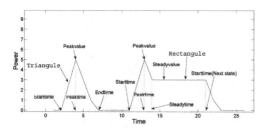

Figure 1. Parameters of two kind events.

894

The system will calculate the fundamental feature of electricity consumption. State 2 means state of appliances is in a change, finding local power extremum and change system state to 3 if the local extremum is confirmed. Extremum here is the collectively called for the biggest or smallest power of a time section. State 3 means the vertex of the change has been found. System state will return to 1 if the power drops to the former average power (triangle), or change to 4 which means a new steady state will appear (rectangle). State 4 is a transient state feature recorder, the task is finding the appearing time of new steady state and record the starting feature of the new steady state, then set system state to 1.

4.1 Peak finder and fundamental feature (state 1)

Figure 3 shows a programme flow chat of this state.

The core of this state is the key point calculation of the change. A constant (20) is set as the boundary of judgement, and the object of judgment is absolute value of difference between current power and predict average power. The reason of parameter selection is stemming from experience's consideration. Other methods are under consideration too, such as the percent of predict average (not selected for increasing the computations and not good for rise precision obviously).

Fundamental feature calculation stop after change begin, because fluctuation in a change is bigger than steady state.

4.2 Local peak point detector (state 2)

Figure 4 shows a programme flow chat of this state.

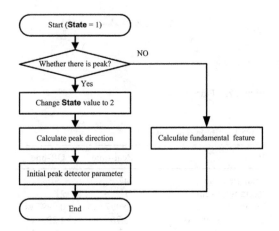

Figure 3. State one.

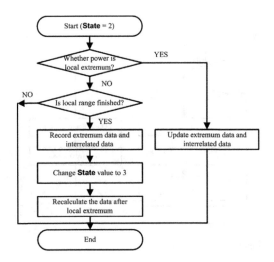

Figure 4. State two.

The core of this state is criterion of the local extremum confirming. If we define local in "local extremum" stand for certain range τ in time domain, the system will have to wait τ to judge whether the current power is local extremum, because the future data is unpredictable. Parameter selection about τ is 5 s.

Data after local extremum will be extracted from cache array and recalculated as new data in other state. The result of this state algorithm can guarantee that local extremum in range τ will be found and no sampling data is discarded without calculation.

4.3 Distinction of triangle or rectangle (state 3)

Figure 5 shows a programme flow chat of this state.

As the title of this subsection, the most important task of this state is distinction of triangle or rectangle. From description of math model, the difference between triangle and rectangle is whether the active power will return to initial predict average power in a certain period of time.

The selection of parameters (power fluctuation tolerance in amplitude and time domain range τ) in this state is consistent.

4.4 Peak finder and fundamental feature (state 4)

Figure 6 shows a programme flow chat of this state.

SAP stands for slide average power and its mathematical expression is as follow.

$$SAP = (Pn + Pn-1 + \ldots + Pn-N+1)/N \qquad (1)$$

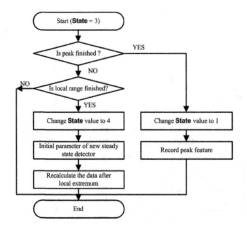

Figure 5. State three.

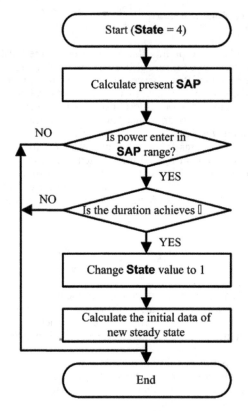

Figure 6. State four.

The core of this state is criterion of the steady state judging. Steady state here refers to state that the momentary power fluctuation is small, which has several kinds of mathematical expressions, each expression has its own advantage and disadvantage. In this paper we use a inequality group as the criterion.

$$\begin{cases} |P - SAP| < 20 \\ \quad T > 1S \end{cases} \quad (2)$$

The word expression of (2) is that the new steady state starts when both the absolute value of difference between new measure data and SAP is less than 20 and the duration time is longer than 1S are satisfied. Advantage of this mathematical expression is simple computation and zero cumulation of error and the disadvantage is that system could not find the steady state when large power fluctuation happen frequently, this may caused by extreme case or the steady state is not apparent.

5 EXAMPLE

Set LG Plasma TV as typical appliance and its rated active power is 220 W. The experiment data are shown as Figure 7, which is more complicate than most situations, even multiple loads.

The parameter extract by real time mode and off-line model are shown in Table 1. The off-line mode use the same pre-setting parameter of events (Triangle and Rectangle), the raw data are one-time given, so the algorithm do not need any cache for events parameter judgment, which can reduce the computational complexity and improve algorithm efficiency.

Similarity in Table 1 is a relatively abstract concept, so another parameter is used here instead, which is easily computed by its power consumption. The similarity is given by percentage, setting the initial power consumption value as the standard 1, and the power consumption of rebuild data

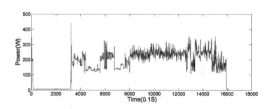

Figure 7. Plasma TV.

Table 1. Table type styles.

Item	Real-time	Off-line
Triangle events	347	361
Rectangle events	21	21
Similarity (vs. raw data)	0.9997 vs 1	0.9997 vs 1
Computing time (second)	13.76	4.22
Memory cost	453	$>3.6 \times 10^4$

is calculated in proportion. The computing time is Matlab runtime in the same notebook. Memory cost stands for the max number of data need in the memory.

In the table, the real-time mode records 347 Triangle events and 21 Rectangle evens, the off-line mode records 361 Triangle events and 21 Rectangle evens. The similarity with raw data is the same. The computing time of real-time mode is about three time than off-line mode, but considering the computation lasts about 1800 seconds, it is acceptable. Real-time mode excelled off-line mode at memory cost, this experiment only runs half an hour, and this advantage will be more obvious when the running time expands.

6 CONCLUSION AND PERSPECTIVE

This paper proposes an algorithm for parameter acquisition from data stream in real time mode. Usually, identification needs complete feature parameters for higher accuracy, but the complete feature extraction time always uncertain, which bring about practical problems such as mass data process and storage. These problem lead to hardware of the monitoring system and limit the serviceable range. At the present stage, most research are processed in the laboratory, make use of high performance equipments, which can not be reappeared in the real scene. Under the situation above, the algorithm proposed in this paper refines parameters from data stream, both computed strength and memory requirement are take into account.

This research is focused on parameter acquisition from data stream in real time mode, which keep the monitoring system deep the economys. The other problem such as how to avoid the parameter space overlapping or missing and building appliance knowledge base, are our future work.

REFERENCES

A Kalman-Filter Spectral Envelope Preprocessor Shaw, S.R., Laughman, C.R. Instrumentation and Measurement, IEEE Transactions on Volume 56, Issue 5, Oct. 2007 Page(s):2010–2017.

Hart G.W. Nonintrusive appliance load monitoring, Proceedings of the IEEE Volume 80, Issue 12, Dec. 1992 Page(s):1870–1891.

Ibrahim, M., Jaafar, M.Z., Ghani, M.R.A. Demand-side management. TENCON '93. Proceedings. Computer, Communication, Control and Power Engineering, 1993 IEEE Region 10 Conference on:572–576 vol.4.

Leeb, S., James L. Kirtley, Jr., Michael S. LeVan, Joseph P. Sweeney. Development and Validation of a Transient Event Detector, AMP Journal of Technology Vol. 3 November, 1993.

Majumdar, S., Chattopadhyay, D., Parikh, J. Interruptible load management using optimal power flow analysis, IEEE Transactions on Power Systems, May 1996, pp.715–720.

Najmeddine, H., El Khamlichi Drissi, K. State of art on load monitoring methods, Power and Energy Conference, 2008. PECon 2008. IEEE 2nd International, 2008, Page(s):1256–1258.

Pihala, H. Non-intrusive appliance load monitoring system based on a modern KWh-meter, technical research center of Finland ESPOO 1998.

Sultanem, F. Using appliance signatures for monitoring residential loads at meter panel level, IEEE Transactions on Power Delivery, Vol. 6, pp.1380–1385, 1991.

Turvey, R. Peak-load pricing, The Journal of Political Economy, 1968, pp.101–113.

Wang Zhenyu, Zheng Guilin. Residential appliances identification and monitoring by a nonintrusive method, IEEE Transactions on Smart Grid, Volume 3, Issue 1, March 2012, Page(s):80–92.

Wenders. Experiments in seasonal-time-of-day pricing of electricity to residential users; The Bell Journal of Economics, 1976, pp.531–552.

Control Engineering and Information Systems – Liu (Ed)
© 2015 Taylor & Francis Group, London, ISBN 978-1-138-02685-8

The temperature prediction in blast furnace based on fuzzy clustering analysis

Q.X. Zhang, D.H. Wang & C. Zhou
Shenyang Aerospace University, Shenyang, China

ABSTRACT: For the blast furnace temperature forecast in iron and steel enterprises, this paper divides the train samples into several categories basing on Fuzzy C-means clustering (FCM). Then we train the samples by Least Squares Support Vector Machine (LSSVM). According the result of cluster analysis, we classify the predict samples. This improved method can avoid the shortcomings of training the samples blindly and improve prediction accuracy of the LSSVM model effectively. According the result of simulation, this method has a significant improvement in the trends and the accuracy of temperature forecasting. It also can play a guiding role in the blast furnace management operations.

1 INTRODUCTION

The temperature of blast furnace as an important indicator of blast furnace production directly affects the quality of molten iron. We should adjust the furnace temperature in advance so that can ensure the temperature change within the normal range. So only predicting the furnace temperature can finish the adjustment. At present, there are many methods of blast furnace temperature prediction research, for example, Bayesian networks, neural networks, support vector machines and wavelet analysis and so on. However, the predicted methods above usually appear deviation when furnace temperature tends to fluctuate.

In this paper, the sample data are collected from a blast furnace of Baosteel. Firstly, we divide the train samples basing on the theory of fuzzy C-means clustering (FCM). Then we make using of the multi-LSSVM to train the samples divided. Finally, we combine with the model after trained and the predicted samples to forecast the change of furnace temperature. The simulation result validates the feasibility and effectiveness of the model.

2 FUZZY C-MEANS CLUSTERING

Training samples are $X = \{x_1, x_2, ..., x_n\}$ and each object has n characteristic indicators. According the requirement, we divide the X into numbers of C. u_{ik} describes the membership of X_k belonged to class i. The classification results can be expressed as a n order Boolean matrix. Corresponding fuzzy C-divided space need to meet the following requirement:

$$\sum_{i=1}^{N} u_{ik} = 1 \quad \forall k$$

$$0 < \sum_{k=1}^{N} u_{ik} \quad \forall i \tag{1}$$

$$u_{ik} \in [0,1] \quad \forall i,k$$

The nature of Fuzzy C-Means transforms the cluster analysis into nonlinear optimization problems. According to iterations, we solve the minimum of objective function which is about membership U and cluster centers. The objective function as following:

$$J(U, c_1, ..., c_c) = \sum_{i=1}^{c} J_i = \sum_{i=1}^{c} \sum_{j}^{n} u_{ij}^m d_{ij}^2 \tag{2}$$

where U is the membership matrix; C_i is the cluster center; u_{ij} is the membership; m is the weighted index. When m is equal to 1, it is equivalent to the hard clustering. Usually we set m being equal to 2. $d_{ij}^2 (d_{ij}^2 = \|x_k - c_i\|^2)$ represents the distance between sample and the cluster center C_i.

Lagrange multiplier method for solving the objective function, we can obtain the necessary conditions to meet the minimum value of the objective function, as following:

$$c_i = \frac{\sum_{j=1}^{n} u_{ij}^m x_j}{\sum_{j=1}^{n} u_{ij}^m} \tag{3}$$

$$u_{ij} = \frac{1}{\sum_{k=1}^{c}\left(\dfrac{d_{ij}}{d_{kj}}\right)^{2/(m-1)}} \qquad (4)$$

FCM clustering analysis of the specific steps is as follows:

1. The initialization part of the matrix U, required to satisfy the formula (1).
2. According to the formula (2), we calculate the cluster center.
3. Calculating the value of the objective function $J(U, c_1, ..., c_c)$. If the calculated value is less than a determined threshold or the change relative to the last value of the objective function is less than a threshold, then the algorithm stops.
4. According to formula (3) to calculate the new membership matrix U. Return to formula (2).

3 LEAST SQUARES SUPPORT VECTOR MACHINE

Assume the samples are $\{(x_1, y_1), ..., (x_l, y_l)\}$ and $x \in R^n, y \in R$. (l is the total number of samples). Since the samples distribute in the high dimensional space, the sample should be mapped into higher dimensional space to realize the nonlinear fitting. According to the principle of structural risk minimization, we can change to optimization problem.

$$\min \quad \frac{1}{2}w^T w + \frac{1}{2}\gamma\sum_{i=1}^{l}\xi_i^2,$$
$$s.t. \quad y_i = w \cdot \varphi(x_i) + b + \xi_i, \quad i = 1, ... l. \qquad (5)$$

where ξ_i is Slack variable, γ is Penalty factor. Then according the formula (5), we can get the Lagrange, as following:

$$L(w, b, \xi, a) = \frac{1}{2}w^T w + \frac{1}{2}C\sum_{i=1}^{l}\xi_i^2$$
$$- \sum_{i=1}^{l}\alpha_i(w \cdot \varphi(x_i) + b + \xi_i - y) \qquad (6)$$

According to the partial derivative of formula (6) for w, b, ξ and α are equal to zero, we can conclude that, as following:

$$w = \sum_{i=1}^{l}\alpha_i\varphi(x_i), \sum_{i=1}^{l}\alpha_i = 0, \alpha_i = C\xi_i,$$
$$w \cdot \varphi(x_i) + b + \xi_i - y_i = 0 \qquad (7)$$

Then we change to solve linear equation:

$$\begin{bmatrix} 0 & 1 & \cdots & 1 \\ 1 & K(x_i, x_j) + 1/c & \cdots & K(x_i, x_j) \\ \vdots & \vdots & \ddots & \vdots \\ 1 & K(x_i, x_j) & \cdots & K(x_i, x_j) + 1/C \end{bmatrix}\begin{bmatrix} b \\ \alpha_1 \\ \vdots \\ \alpha_l \end{bmatrix} = \begin{bmatrix} 0 \\ y_1 \\ \vdots \\ y_l \end{bmatrix}$$

$$(8)$$

where $K(x_i, x_j)$ ($K(x_i, x_j) = \varphi(x_i) \cdot \varphi(x_j)$) is a symmetric function which satisfies the Mercer condition. Finally, the output of the classifier can be expressed:

$$f(x) = \sum_{i=1}^{l}\alpha_i k(x_i, x) + b. \qquad (9)$$

The number of support vector in support vector machine SVM is limited. Because it removed a lot of non-support vector samples, it has sparsity. As a research hotspot of the support vector machine SVM, the least squares support vector machine LS-SVM sees all vectors as support vector. It don't excluded any information of samples and use the method of solving linear equations replacing the solving the convex programming. Comparing the standard SVM, it reduces the computational complexity greatly.

Kernel function $K(x_i, x_j)$ is one of the most important factors of LS-SVM model. According the kernel function, the original sample can be mapped to the high-dimensional space and classified easily. Kernel functions commonly use linear kernel, polynomial kernel function and RBF kernel function, and sigmoid kernel function. The choice of kernel function is different in different situation. In many practical engineering problems, we usually choose the Radial Basis function as the kernel function. Choice of kernel width and punishment factor is important to the establishment of the LS-SVM model. The two parameters relate directly to the accuracy and precision of the model. Usually the initial value of the penalty factor should be selected appropriately. If it is selected too large, the model becomes the problem of smallest error optimization. On the contrary, make the model ignore the error impact. Similarly, nuclear width value has to be appropriate. If the value is very small or far less than the sample minimum distance, all the training samples are correctly classified, that is, there will be over fitting. If the nuclear width value is significant or much larger than the sample the maximum spacing, it will make the performance of model poor, misclassify samples easily or lose the ability to classify. Usually nuclear parameters and penalty factor can be adjusted according to the simulation results. Some optimization algorithms can also be used to select the two parameters, such as Cross validation, "leave-one".

4 FCM-LSSVM

4.1 Sample pretreatment

Basing on the blast furnace production, we select several factors as inputs to predict the furnace temperature which greatly relate to the temperature of blast furnace. Such as, Heat load, pulverized coal injection, humidity, air temperature, the first three furnace hot metal silicon content. Since the silicon content and the oven temperature are directly related to a parameter, we select the silicon content of the iron[S_i] times as the output of the model. Considering the hysteresis characteristics of the blast furnace, the input factors all have some time lag to the temperature of blast furnace. For example, heat load's is 60 min, pulverized coal's is 210 min; humidity is 15 min and so on.

For the samples which deviate from the normal should be removed from the model trained by some methods. Here, we introduce the five-number sum method to remove outlier samples exceed the upper and lower extreme.

Five numbers include: Median F_{mi}; four points up and down F_U, F_L; extreme value up and down f_L, f_U. As following:

$$F_{mi} = \frac{\sum_{i=1}^{n} x_i}{n} \tag{10}$$

$$F_L = \frac{[(x_i)_{\min} + F_{mi}]}{2} \tag{11}$$

$$f_L = F_L - (F_U - F_L) * 1.5 \tag{12}$$

$$f_U = F_U + (F_U - F_L) * 1.5 \tag{13}$$

where x_i represents the input factor i and $i = 1, 2, ..., l$. l represents the total numbers of input factors.

Because each input data exist larger difference in the magnitude, the normalization process for the samples is expressed as following:

$$x_i' = \frac{x_i - x_{\min}}{x_{\max} - x_{\min}} \tag{14}$$

4.2 Modeling

Select 500 samples after deleted as train sample. Set relative parameters of FCM: The number of categories C is equal to 3; Weighted index M is equal to 2; The threshold of iterative algorithm stopping is 1.0e-6. Corresponding LSSVM models are trained basing on clustering results. To kernel function of LSSVM, we choose the Gaussian radial basis function.

Important parameters of LSSVM the width of the nuclear σ and penalty factor γ, we set (σ, γ) = (50,10) by Cross validation. Lastly, we classify 100 predicted samples basing on the clustering results and use corresponding LSSVM to forecast the output. The model structure is shown in Figure 1.

4.3 Simulation results

Compare prediction of LSSVM model and FCM-LSSVM model prediction in Figure 2. Figure 3 shows the deviation statistics of the two prediction methods. As the simulation result, the prediction accuracy of FCM-LSSVM is 88%, better than LSSVM (82%). The silicon content

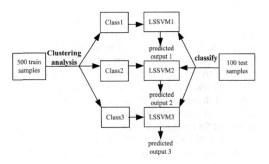

Figure 1.   Structure of FCM-LSSVM.

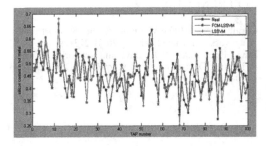

Figure 2.   Predicting silicon content.

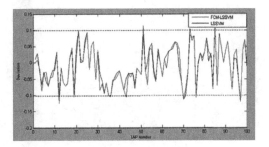

Figure 3.   Predicted deviation.

is subject to significant volatility, FCM-LSSVM forecast value deviation is smaller and the furnace temperature trend forecast more desirable, specifically when the silicon content changes obviously.

5 CONCLUSION

This paper introduces a furnace temperature forecasting method based on FCM and LSSVM. According to the simulation result, the forecast effect of FCM-LSSVM is better than the signal LSSVM. It has guidance significance in management of furnace temperature. In this paper, there is more room for improvement. The selection and pretreatment of input factors remains to be studied.

Only more and accurate related factors are considered, can we predict the output truly. Because of train sample's size limited, the degree of application and generalization ability of the model need to be improved; If the new data can be got in the site database and used to train the model, the precision of model must be promoted. The numbers of categories and the basis of classification need to be furtherly researched. As the important parameters of LSSVM, the width of the nuclear σ and penalty factor γ should be selected by the ideal algorithm.

REFERENCES

Corinna Cortes. 1995. Vladimir Vapnik. Support vector networks. Machine Learning 20(3): 273–295.

David Heckerman. 1997. Bayesian Networks for Data Mining. Data Mining and Knowledge Discover 1(1): 79–119.

Du Z. 2009. Research on Some Variants of Support Vector Machine.

Jian L. 2006. Applications of SVM to Predict Silicon Content in Hot Metal.

Lin Q., Xu Q. & Wu J. 2005. The short-term load forecast using c-means clustering and BP network. Journal of shanghai university of electric power 21(4): 321–324.

Liu X.Y. et al. 2005. Application of Bayesian Network to Predicting Silicon Content in Hot Metal. Iron and Steel 40(3):17–21.

Liu X.Y., Li P. & Hao C.H. 2011. Fast Leave-One-Out Cross-Validation Algorithm for Extreme Learning Machine. Journal of Shanghai jiaotong university 45(8):1140–1145.

Liu Y.G. & Wang W.H. 2005. Application of Wavelet Analysis to Prediction of Silicon Content in Hot Metal. Iron and Steel 40(8):15–17.

Marti Hearst. 1998. Support Vector Machines. IEEE Transaction On Intelligent Systems 13(4):18–28.

Miche Y. et al. 2010. OP-ELM: Optimally pruned extreme learning machine. IEEE Trans Neural Network 21(1):159–162.

Radhakrishnan V.R. & Mohamed A.R. 2000. Neural netwoks for the identification and control of blast furnace hot metal quality. Journal of Process Control 6(10):509–524.

Sun T.D. & Yang Z.Y. 1996. Application of artificial network to predict to predict the silicon content in molten iron. Iron and Steel 31(3):17–20.

Vapnik V.N. 1995. An overview of Statistical Learning Theory. IEEE Trans Neural Network 10(5):988–999.

Wang Y.K. & Liu Y.G. 2012. Application of multiple support vector machines model based on FCM for temperature prediction in blast furnace. Metallurgical industry automation 36(3):18–23.

Wu Q. & Liu S.Y. Fuzzy least square support vector machines for regression, Journal of Xidian university 34(5):773–778.

Yan W.W. & Shao H.H. 2003. Application of support vector machines and least squares support vector machines to heart disease diagnoses. Control and Decision 18(3):358–360.

Control Engineering and Information Systems – Liu (Ed)
© 2015 Taylor & Francis Group, London, ISBN 978-1-138-02685-8

Research of software multi-level rejuvenation method based on prediction

J. Guo & X.F. Shi
College of Information Science and Engineering, Northeastern University, Shenyang, Liaoning, China

J. Cao
College of Computer, Tianjin University, Tianjin, China

M.H. Jia
Philips and Neusoft Medical Systems, Shenyang, Liaoning, China

ABSTRACT: Software aging phenomenon comes with the software running, and has a terrible effect in application system performance. In order to solve this problem, software rejuvenation methods are introduced. Traditional software rejuvenation technologies are not adjusted to the web application system with complicated structure and running environment. This paper presents a software multi-level rejuvenation method based on prediction according to the complicated web application system. The main works contains the partition of system rejuvenation levels, the decline trend prediction of system performance, and the draw-up of component-level rejuvenation execution strategy. At last, we have a simulation experiment to investigate the effectiveness of our methods.

1 INTRODUCTION

Recently, amount of studies and data manifest that one of reasons for software failure is software aging. Software aging is a phenomenon that software or system is breakdown for the conflict or run-out of system resources as software is running for a long time and errors are accumulated [1]. System performance degradation is always accompanied by software aging, which is ubiquity and is difficult to be eradicated. Software aging seriously threatens software reliability [2].

Aiming at this phenomenon, recently the widely method is Software Rejuvenation Technology (SRT) proposed by Huang et al. [3]. SRT is a preliminary reactive maintenance technology in case of software accidental failure. The fundamental of software rejuvenation is ending the procedure and restart the procedure to clear internal state, which can free operating system resources and regain software performance. The basic assumption and premise of SRT are stopping and restarting the application programs on schedule, which bring more little losses than bugs in software running process. Appropriate SRT can greatly decrease stop-time and losses these software aging causes, while high frequent using of SRT will increase them and low frequent using of SRT don't get the deserved results. So, the central questions of SRT contain how to search the best rejuvenation time, and how to partition rejuvenation levels of software system.

As for software rejuvenation, more researches are inclined to fine-grained multi-level rejuvenation strategies, which are superior to two-level rejuvenation strategies (system level and service level) [4] in decreasing rejuvenating costs. We propose a software multi-level rejuvenation method based on prediction, which improves the methods based on threshold values [5], taking account of the correlations among the factors influencing software aging [6] and no need of monitoring software system running state continuously [7]. System resources performance effects are discussed for the whole system performance in the run-time of web applications. And the performance metrics set that can reflect the system performance decline are established for every rejuvenation level. The method concludes the system performance declining trend through the pre-built software aging trend prediction model based on an improved ID3 (Iterative Dicremiser3) algorism. And lastly, based on RBF (Radial-Basis Function) neural network [8], we make a component-level rejuvenation implementation strategy to improve the previous rejuvenation strategies according to the trend.

The rest of the paper is organized as follows. In Section 2, we will describe the overall framework about software multi-level rejuvenation mechanism based on prediction. The software aging trend prediction model based on decision tree is presented in Section 3. And then Section 4 is dedicated to get regeneration strategies of components based on RBF neural network. Experimental evaluation is given in Section 5, followed by conclusion in Section 6.

2 OVERALL FRAMEWORK

We mainly do two key problems: the software aging trend prediction model based on decision tree [9] and the component rejuvenation strategies based on RBF neural network. As shown in Figure 1, we first primarily collect the performance indicators these software aging system related, and use the indicators to establish software aging trend prediction model, which can predict the trend of system performance decline. According to the forecast results, we acquire the reasonable rejuvenation decision strategy. When the results show the coming software aging, the right level of rejuvenation can be chosen depending on the current system state. Rejuvenation levels include system level, middleware level, application level and component level. And then we have respective operations for different levels based on rejuvenation decision-making results. Especially, component-level rejuvenating is complicated and needs a better strategy.

Component-level rejuvenation strategies provide continuous high reliability services with low

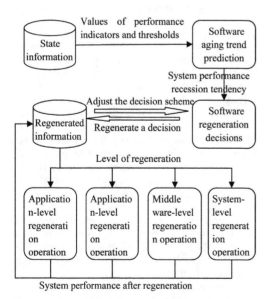

Figure 1. Process of software multi-level rejuvenation based on prediction.

costs, good effects and minimized key business interrupts.

3 SOFTWARE AGING TREND PREDICTION MODEL BASED ON DECISION TREE

3.1 *The foundation of decision tree T*

As is shown in (1), the attribute values number of candidate attribute A is v and the probability of each attribute value is P_j (j = 1, ..., v). The principle of minimum information entropy is used to extend A. B_j (j = 1, ..., v) is the selected attribute of j child node, and the corresponding entropy is Entropy (B_j). When choosing a new attribute, it considers information gain bringing by the next chosen attribute and deep nodes of the tree. The decision tree can be built by ID3 algorithm. Here we set the degree of user-interest U (0 < U < 1) to improved ID3 algorithm, which enhance the important attribute weights and reduce the reliance of the decision tree for more attribute values, and avoid the phenomenon of big data cover small data. As is shown in (2), user interest degree U is prior knowledge of influencing the decision tree rule generation and selection in the process of decision tree training. Gain(S, A) is information gain of attribute A in (3). S is the sample set.

$$\text{Entropy}^*(S,A) = \sum_{j=1}^{v} P_j \times \text{Entropy}(B_j) \quad (1)$$

$$\text{Entropy}_1^*(S,A) = \sum_{j=1}^{v} (P_j + U)\text{Entropy}(B_j) \quad (2)$$

$$\text{Gain}(S,A) = \text{Entropy}(S) - \text{Entropy}_1(S,A) \quad (3)$$

The process has 4 steps, as the following shows:

Step 1. All attributes are to be selected as the root node;

Step 2. Certain attributes are added with user interest degree U by prior knowledge, and one can be preferred test attribute A.

Step 3. We select A successor nodes $\{B_1, B_2, ..., B_v\}$ according to the principle of minimum information entropy. The formula (2) calculates the corresponding information entropy respectively, and then the formula (1) calculates Entropy*(S, A). When Entropy*(S, Anew) has the minimum value, Anew is regarded as the new selected attribute;

Step 4. If every B_j (j = 1, ..., v) is a leaf node, this node should be ceased to extend. Otherwise we repeat Step3 till finishing the whole tree.

3.2 The optimization of decision tree T

The decision tree is optimized by REP (Reduced Error Pruning) algorism [10]. From bottom to top, every sub-tree ST(except leaf-trees in T) is replaced by a leaf node to get a new tree T*. The leaf node is marked as the dominant class in ST. If the new tree with a smaller or equal to the number of test set classification errors having same performance, the ST is replaced by leaf nodes. We repeat the action to make all sub-trees be replaced by leaf nodes when not adding classification errors.

4 COMPONENT-LEVEL REJUVENATION EXECUTION STRATEGY BASED ON RBF NEURAL NETWORK

Executing component-level rejuvenation is more complex for determining the rejuvenation objects. Correlations usually exist among aging components, so it needs to determine component rejuvenation groups and regard one group as a rejuvenation object. Meanwhile, we also need a method that can determine group rejuvenation priority in order to avoid all components rejuvenation groups doing rejuvenation at the same time, which can deduce rejuvenation cost.

4.1 Foundation and priority evaluation of component rejuvenation group

The foundation of component rejuvenation group: We analyze their rejuvenation correlation based on the coupling relationship among components. The rejuvenation correlation is greater as the coupling relationship is greater. Coupling relationships contain control, content, data, common and indirect coupling. If the upper component invokes the lower to complete the specified function and the call is achieved through parameters exchange, it is known as control coupling. Content coupling is when a component directly enters into another component to access data or two-way call relationship. Data coupling is components exchange information through the interface parameter table, while common coupling is through common environments. Corresponding to the coupling relationships, rejuvenation correlation can be divided into four types, including functional correlation, two-way functional correlation, state correlation and mutual independence. The control coupling belongs to functional correlation, and content coupling belongs to two-way functional correlation while data coupling and common coupling belong to the state correlation.

The web application components are in set SET = {COM$_1$, COM$_2$, ..., COM$_n$}, and n is the number of components. Rejuvenation correlations among components are known. We use GROUP [COM$_i$] to represent the COM$_i$ group of rejuvenation correlated components. If component COM$_j$ (COM$_j$ ∈ SET) has the correlation with COM$_i$, we put COM$_j$ into GROUP [COM$_i$].

The priority of component rejuvenation group: The structure of RBF neutral network usually consists of input layer, hidden layer and output layer, and the numbers of units are N, L and M. We regard evaluating indicators (*rejuvenating duration, average invoking number, access latency, etc*) as the input and rejuvenating priority as the output (M = 1). X$_i$ (i = 1, ..., N) is the input of input layer while H$_j$ (j = 1, ..., L) is the output of hidden layer, and Y is the output of output layer. If the given input pattern is X$_i^m$ (m = 1, ..., P, P is the number of input pattern), the output of hidden layer is H$_j^m$ in (4). We can get the output Y^m of output layer.

$$H_j^m = R_j^m \Big/ \sum_{j=1}^{L} R_j^m \quad (j = 1, ..., L) \tag{4}$$

$$R_j^m = \exp\left[-\frac{1}{2} \sum \left(\frac{X_i^m - C_{ji}}{B_j} \right)^2 \right] \tag{5}$$

$$Y^m = \sum_{j=1}^{L} W_j H_j^m \tag{6}$$

C$_{ji}$ = {C$_{j1}$, C$_{j2}$, ..., C$_{jN}$}^{T} is the basis function center of hidden layer, and B$_j$ represents the basic function width of hidden layer. W$_j$ is the connection weight between a hidden layer element and the output layer element, and i and j are input element and output element.

4.2 Establishing a component rejuvenation chain

The execution strategy of Component rejuvenating is through a component rejuvenation chain (CHAIN). Firstly we determine the aging components set (SET) and its rejuvenation group (GROUP [COM$_i$]), and then simplify the group to exclude repetitive components in two or more

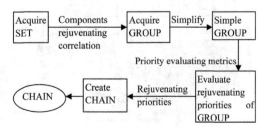

Figure 2. Creating process of CHAIN.

rejuvenation groups. Secondly we determine the rejuvenation priority of GROUP [COM$_i$] (PRIORITY {GROUP [COM$_i$]}), and finally we create CHAIN. When establishing CHAIN, we add GROUP [COM$_i$] to CHAIN according to the big-small order of the priorities.

5 EXPERIMENTS AND RESULTS ANALYSIS

This paper simulates a real WEB environment, which consists of an application server, a database server and a client. The hardware is Intel(R) core(TM) 2 Quad Q9500@2.83 GHz, 4 cores and 4 GB memory. The application server is sited up in LINUX, and client is in WINDOWS XP. We set up JDK1.5 in server side to run Tomcat and MySQL, which provide web services and database services respectively. We periodically collect web server performance information through performance monitoring tools HQ [11] (An open source monitoring system developed by SUN Microsystems), and the collection of time intervals and performance parameters is in its configuration file. On the client side, we use LOADRUNNER as a server load generator, simulating a large number of customer requests. An online shopping website [12] based on the MVC Struts framework is regarded as the research object, which is based on RUBIS.

1. Software aging process is so slow that we inject errors into the operating system, JVM and RUBIS in order to accelerate the aging of the system. In Figure 3, Figure 4 and Figure 5, we can see that memory usage; CPU usage and users visiting are all less at the beginning, so the system response time is shorter. However, the amount of users visiting is increasing continually in the 1200 sec, and reaches the top in the 2200 sec. Meanwhile, the memory usage speeds up, especially in [1800, 2600]. The amount of CPU usage fluctuates greatly in [1500, 2400], and system response time is increasing continually. At last, in [3000, 3600], users visiting

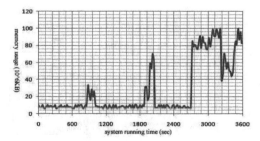

Figure 3. Memory usage.

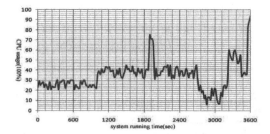

Figure 4. CPU usage.

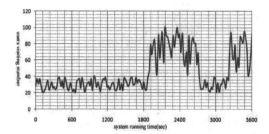

Figure 5. Users visiting.

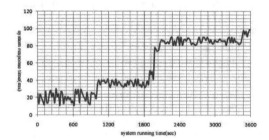

Figure 6. System response time.

and CPU usage go back to normal levels, while system response time is still increasing. The phenomenon shows the decline of system performance for the memory losses accumulating.

2. We evaluate the prediction accuracy and effectiveness of the software aging trend prediction model. Firstly, we find out the best attributes set {memory usage, CPU usage, users visiting} influencing the system response time. Memory usage A_MEN has 4 attribute values {M1, M2, M3, M4}, and CPU usage A_CPU has 6 attribute values {C1, C2, C3, C4, C5, C6}. Users visiting A_UV has 2 attribute values {U1, U2} and class attributes response time CLASS_R has 6 attribute values {CLASS1, CLASS2, CLASS3, CLASS4, CLASS5, CLASS6}. We can calculate the information entropies of A_CPU, A_MEM and A_UV. ENTROPY (A_CPU, S) is 0.5375, and ENTROPY (A_MEM, S) is 0.6763 while

ENTROPY (A_UV, S) is 1.1959. We choose the Min (A_CPU) to be the first layer node, and repeated the calculating process to finish the decision tree. REP algorism is adopted to optimize the tree. In Table 1, the improved method decreases the average relative error to 5.22%. And it increases the average predicting accuracy prediction accuracy from 83.87% to 85.69%, which is superior to traditional ID3 algorism.

3. We evaluate the effectiveness of the component level rejuvenation strategy through a contrast test. There are two servers running in the same environments with 100 online clients doing different actions. When the software is aging, we handle the servers with traditional method

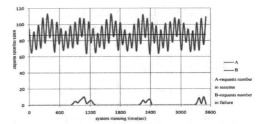

Figure 10.   Users visiting change of Server2.

Table 1.   Analysis of prediction errors.

	Improved ID3	Traditional ID3
Average relative error	5.22%	8.78%
Average predicting accuracy	85.69%	83.87%

and our method respectively. Each server runs for 60 minutes, and we take samples every 15 seconds. In the running period, Server1 has 4 rejuvenations for 120 s, while Server2 has 3 for 35 s. Server1 has 11752 invalid requests and Server2 has only 233. In Figure 7, the performance restoring of Server1 is not obvious after rejuvenating, and rejuvenating operations are too frequent, while the Server2 has moderate frequent rejuvenations and obvious results. And Server2 has less invalid requests because our rejuvenating strategy for component-level will stop when performance recovers to a certain degree. In conclusion, our strategy is more effective with obvious performance recovery after rejuvenating.

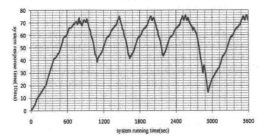

Figure 7.   Response time change of Server1.

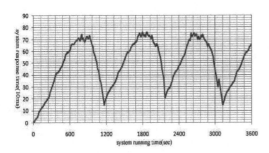

Figure 8.   Response time change of Server2.

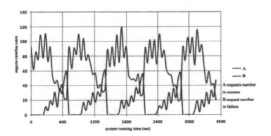

Figure 9.   Users visiting change of Server1.

6   CONCLUSION

In order to solve the software aging problems, software rejuvenation strategies are introduced. The researches are focus on determining rejuvenation executing time and rejuvenation priorities. The former directly affects the effectiveness of rejuvenating execution, while the later directly embodies the effect of rejuvenation. This paper proposes the software multi-level rejuvenation mechanism based on prediction according to the complicacy of a web application system. The mechanism eliminates the defects of the methods based on threshold value. A decision tree is built with several performance metrics these software aging related, and finally the software aging trend prediction model is determined. As for component level, we create the component rejuvenating chain based on the rejuvenating correlations among components, which can reduce the economic losses, increase availability and reliability of the system.

ACKNOWLEDGMENTS

This work was supported by the Key Technologies R&D Program of Shenyang City (F12-029-2-00), and the Fundamental Research Funds for the Central Universities (No. N120804001).

REFERENCES

Cassidy K.J., Gross K.C., Malekpour A. Advanced pattern recognition for detection of complex software aging phenomena in online transaction processing servers[C]// Dependable Systems and Networks, 2002. DSN 2002. Proceedings. International Conference on. IEEE, 2002: 478–482.

Castelli V., Harper R.E., Heidelberger P., et al. Proactive management of software aging[J]. IBM Journal of Research and Development, 2001, 45(2): 311–332.

Cun-hua Z.X.Q., Toshio N. Optimal Preventive Software Rejuvenation Policy with Periodic Testing [J]. Computer Science, 2009, 8: 039.

Garg S., van Moorsel A., Vaidyanathan K., et al. A methodology for detection and estimation of software aging[C]// Software Reliability Engineering, 1998. Proceedings. The Ninth International Symposium on. IEEE, 1998: 283–292.

Gray J., Siewiorek D.P. High-availability computer systems [J]. Computer, 1991, 24(9): 39–48.

Huang Y., Kintala C., Kolettis N., et al. Software rejuvenation: Analysis, module and applications[C]// Fault-Tolerant Computing, 1995. FTCS-25. Digest of Papers. Twenty-Fifth International Symposium on. IEEE, 1995: 381–390.

Hummel J., Strehmel N., Selbig J., et al. Decision tree supported substructure prediction of metabolites from GC-MS profiles [J]. Metabolomics, 2010, 6(2): 322–333.

Hyperic H.Q. Open Source Web Infrastructure Management Software. http://www.hyperic.com/.

Jiang L., Xu G., Zhou L. Abstraction of Software Aging Trend Based on Robust Locally Weighted Regression Algorithm [J]. Journal-Shanghai Jiaotong University-Chinese Edition-, 2006, 40(11): 1951.

Liu Y., Zhao S.L., Yi C. The forecast for corrosion of reinforcing steel based on RBF neural network[C]// Wavelet Analysis and Pattern Recognition, 2009. ICWAPR 2009. International Conference on. IEEE, 2009: 195–199.

Mu R. Design of Shopping Site Based on Struts Framework [M]//Informatics and Management Science VI. Springer London, 2013: 29–37.

Quinlan J.R. Simplifying decision trees [J]. International journal of man-machine studies, 1987, 27(3): 221–234.

Control Engineering and Information Systems – Liu (Ed)
© 2015 Taylor & Francis Group, London, ISBN 978-1-138-02685-8

Analysis of utility big data and its application

H.F. Yang, P. Zhang, X. Zhang, D.H. Li & M. Cheng
NARI Accenture Information Technology, Beijing, China

J. Ding
China Electric Power Research Institute, Beijing, China

ABSTRACT: Big Data is the hotspot of IT Development nowadays. Featured by advanced IM technologies and Analytical Methods, it defines research trends of the future. Power Big data embraces big data concepts and technologies with power industry characteristics, network operations and business performance management, etc. In this paper, we analyses the characteristics, research objects and specific methods of power big data, discusses the problems confronted, and depicts the application of big data technology in the future.

1 INTRODUCTION

The 3rd IT wave, featured by cloud computing and M2M, has turns ICT trends from pursuing higher computing power to deriving value from data minerals (WEISS 2007, Brantner et al. 2008, Buyya et al. 2008, Chen et al. 2009). The economic and scientific value of the data continually rises, makes it a new type of strategic resources equivalent to natural resources, human resources. Thus, big data has also become a hot topic for all the fields of industry and research. It utilizes and discerns the value of cloud computing and M2M, which will have a huge impact on federal service, enterprise management and individual lifestyle.

Big data finds its characteristics in the sheer volume and spanning variety of data (Gong et al. 2012, Meng et al. 2013). These features are concluded as 5V +1C: Volume, Variety, Velocity, Vitality, Value and Complexity: 1) Volume, The unit of PBs will become a new normal, tens, even hundreds of GBs of data will be generated day-to-day by smart devices. As estimated, the amount of data produced by the domestic internet companies per day is approaching TB level today; 2) Variety, All the structured (transactional data), semi-structured data (web page) and unstructured data (video and voice) will be processed and analyzed with different methods; 3) Velocity, Value of data decays with time, thus rapid processing speed become another must for big data, for instance, in e-commerce, real-time analytics is required to make replenishment both timely and effective; 4) Vitality, In the Internet age, business requirements updates frequently, big data analytical and processing models have to adapt accordingly; 5) Value, Big data

will reveals unprecedented business opportunities for the enterprise. Data mining and analytics techniques will incubates new business applications, as marketing analytics will also create tremendous value in the future. Complexity of Big data led to the retreat of traditional relational database, tools for data processing and analytics will be determined both by the data objects and business context (Zhao et al. 2010). In utility industry, the construction of strong smart grid and deployment of "three five" management system put forward higher requirements on the data management, data exchange and interoperability. Power big data is on its own way of maturity, manifested by the increase in the volume variety and velocity of data. In this regard, we need to reveal the potential demand in the big data context, to explore for theories and methods customized for electric power data science, to extend the frontier of information technology to further scope of data value mining.

2 A RESPONSE TO THE TECHNICAL CHALLENGES OF BIG DATA

The complexity of utility big data find it origin in the complexity of power grid, reflected in the "Quantity", "Classes", "Time" dimensions. "Quantity" refers to the amount of data: in Spatial dimension, widely used data-aware devices, monitoring terminals, smart devices and computer clusters together constitute a ubiquitous information-aware network; in Time dimension, the increase in density of monitoring terminal, and monitor frequency, lead to the increase in data amount acquired in unit time. "Classes" is the data type: data heterogeneity, including text, graphics, images, video, audio and other interrelated different forms; variety of data sources, including electricity primary equipment, secondary equipment and the all types of mobile IT equipment. "Time" reflects the timeliness of the data used, the balance of power supply and demand has become a real-time process, thus the electricity plans and other business must adapt accordingly. The current operation model of information systems is calculation-centered, the data, on the other side, is just the objects of calculation (Grossman & Gu 2008). This "all-purposed" computing model can no longer satisfy the complexity of utility data processing, considering the multi-sources and multi-modal characteristics of smart grid information. In this regard, we need to take the reconstruction of IT architecture into consideration, and turn to the data-centered operating mode. In this mode, calculation model will change according to data, thus the business rules, computing methods and data model will be connected in a flexible way (Deelman & Chervenak 2008, Deelman et al. 2004).

Among the various technical challenges confronted by utility big data processing, the following ones deserve particular attention.

2.1 The nature of utility big data is relationship web

Data within the utilities information system is often "silo" with discrete data connection. However, integration of these data and connection constitutes a vast relational web. For example, the Production Management System (PMS) and Information Management System (IMS) separates in the topology and data sources, but underlying, they share a common logical data model which lays an integration foundation for the two systems. The essence of Utility big data processing is to correlate data resources gathered from different time, different spots and even different physical space and cyberspace. The common characteristics of these data underlie in the data networks. We need to take business data of various types and in various forms of storage (unstructured, structured, history/quasi real-time, grid spatial data) as the object, using methodology of social relations, biological science (Ludanscher et al. 2005), from the perspective of graph theory to discern the basic parameters of the data network, such as the shortest path, to generate sub-tree, auditing, betweenness, etc. As the technology framework, we must break the existing barriers of current technology combinations, inheriting the technical characteristics of statistical science, fully utilize graph database, machine learning, classifier combination design and other types of information technology, to establish business oriented data scientific system.

2.2 The expansion of the data will continue to enhance the importance of information retrieval

Traditional information processing technologies are designed for specific data applications, thus the data size is usually under control. Therefore, the effort will often converge on the techs to gather and represent the data. However, the traditional techniques will find its deficiency in a big data context, since the data set applied, for most times. are unknown. The uncertainty of data will roar as the scale of data sets grow. Information retrieval is the technical field initiated specific for this uncertainty. We need to learn from the advanced Search-based Application, SBA, focus on the analysis of information retrieval technology in the information systems content and extension. From the service perspective, extend the original search engine to an optimized t integration data service system of heterogeneous data sources. To realize this objective, we need to design a stable, reliable, scalable and secure search architecture according to the typical business scenarios.

3 THE CONNOTATION AND EXTENSION OF POWER BIG DATA

In big data era, we need to handle larger data volumes, provide more rapid data processing speed, and analyze more various data types and mines deeper value of data. Comprehensive application of big data technologies will better support utilities development strategy. Big Data is extensively used in smart grid, three five system and two center constructions.

3.1 Smart grid

Smart grid devices that be observed, controlled, fully automated and integrated enabled by cloud computing and M2M will generate massive multi-type basic data; to provide technical support for smart grid development, and support information technology, automation, interactivity of smart grid, we will need techniques such as data quality and data mining; to improve system compatibility as well as the capabilities for defect analysis, risk reorganization and alerts, comprehensive application of big data technology in the decision-making, business management, customer service and other aspects.

3.2 "Three five system"

The deployment of the three five system constructs an integrated information system that contains data of the headquarter and all the multi-stage subsidiaries. As the business and technology develops and innovates, big data techniques (e.g. storage, computing, data mining) are required to support the integrated, professional, centralized and intensive management of the enterprise operation system.

3.3 *Two center*

Data for OMC will explode exponentially as the growth of business applications, thus in the future, OMC will be confronted with TB, PB data to govern and monitor, thus OMC should take into consideration of utilizing the big data techs such as stream data processing and distributed storage to support the comprehensive, real-time monitoring of enterprise management and operations performance in the future.

Customer service center in the future will generate various types of data as customer files data, voice data and history data. Big data techs as distributed computing and data mining will be required to reveal the value in this data to improve the user experience and service level.

4 APPLICATION ANALYSIS AND PROSPECTS

As conclusion to above-mentioned business situations and problems, big data will promote applications in various realms.

Data visualization: Visualization Technologies will be refined and enriched to satisfy the need to display 3D, virtual and interactive objects constructed by multi-type data from multi-source. Through the flexible combination of visualization components, visualization of files, videos, animations and GIS can be achieved simultaneously.

Data analytics: Extensive use of deep analysis techs (Moving average analysis, regression analysis and data correlation analysis, etc.) and optimization of large-scale parallel machine-learning and data mining algorithms will enable the high-efficiency analytics and queries of big data era.

Data integration: Improve data synchronization techs to support the real-time, multi-time and massive data transmission requirements.

IT Architecture: Invest in the research of advanced big data architecture and technologies such as MPP, stream data processing and construct a unified big data platform with the capabilities of providing standard data processing and service with rapid response, to accommodate the computing requirement of sheer volume, real-time data raise by business applications.

Data resource: Set up the common interface to access various type of data simultaneously

(structured, unstructured, real-time and GIS information). Find and manage the connections between these data to realize the centralized management of the data source, thus lay the foundation for comprehensive analytics of data from multi-business functions.

Hardware and software: Combine cloud computing and M2M infrastructure with the emerging DDN technologies to realize the unified control of hardware and software. Improve the flexibility and scalability of infrastructures to become more responsive, more quickly to the business requirements change.

5 CONCLUSION

Power big data features the combination of IM technology and computer science with power industry characteristics, it is the "Data Science" for the electric power information technologies. From the research perspective, it utilizes ongoing computing cloud and M2M infrastructure, focuses the research force on revealing the value of enterprise data assets rather than updating IT assets. From the methodology perspective, it replaces the traditional reductionism idea of "break and solve" with brand-new research paradigm adapted to the big data context. From the technology perspective, it matches best fuse point of various techs and methods as pattern reorganization, intelligent system, statistical methods etc. In brief, a clear and discern set of objective, subjects and methodologies for big data will contributes both to smart grid, "Three system and five platform", enterprise management and the construction of an environmental-friendly and efficient IT architecture.

REFERENCES

Brantner M., D. Florescuy, D. Graf, et al. Building a database on S3//ACM. Proceedings of the SIGMOD, Jun 9–12, 2008, Vancouver, BC. USA: ACM SIGMOD/PODS, 2008: 251–263.

Buyya R., C.S. Yeo, S. Venugopal. Market-oriented cloud computing: vision, hype, and reality for delivering IT service as computing utilities// Procedddings of the 10th IEEE International Conference on High Performance Computing and Communications, Sept 25–27, 2008, Dalian, China. Los Alamitos, CA, USA: IEEE CS Press, 2008.

Chen Kang, Zheng Weimin. Cloud computing: system instances and current research[J]. Journal of Software, 2009, 20(5): 1337–1348.

Deelman E., J. Blythe, Y. Gil, et al. Pegasus: Mapping scientific workflows onto the grid. Grid Computing, 2004, 3165: 131–140.

Deelman E. and A. Chervenak. Data management challenges of data-intensive scientific workflows//IEEE Computer Society. Proccedings of the IEEE International Symposium on Cluster Computing and the Grid, May 19–22, 2008, Lyon, France. USA: IEEE CS Press: 687–692.

Gong Xueqing, Jin Cheqing, Wang Xiaoling, et al. Data-intensive science and engineering: requirements and challenges[J]. Chinese Journal of Computers, 2012, 35(8): 1563–1578.

Grossman R., Y. Gu. Data mining using high performance data clouds: experients studies using sector and sphere//ACM. Proceedings of the 14th ACM SIGKDD International Conference on Knowledge Discovery and Data Mining, 2008, New York. USA: ACM SIGKDD, 2008: 920–927.

Ludanscher B., I. Altintas, C. Berkley, et al. Scienticfic workflow management and the Kepler system. Concurrency and Computation: Practice and Experience, 2005, 18(10): 1039–1065.

Meng Xiaofeng, Ci Xiang. Big data management: concepts, techniques and challenges[J]. Journal of Computer Research and Development, 2013, 50(1): 146–169.

Weiss A. Computing in the cloud[J]. ACM Networker, 2007, 11(4): 18–25.

Zhao Junhua, Wen Fushuan, Xue Yusheng, et al. Cloud computing: implementing an essential computing platform for future power system[J]. Automation of Electric Power System, 2010, 34(15): 1–8(in Chinese).

Control Engineering and Information Systems – Liu (Ed)
© 2015 Taylor & Francis Group, London, ISBN 978-1-138-02685-8

Implementation of psychology emotional stability tester

J.W. Ye
PLA Computer Technical Service Center, Guangzhou, China

J. Luan
PLA University of Science and Technology, Nanjin, China

J.F. Ye
College of Information Science and Technology, Dong Hua University, Shanghai, China

ABSTRACT: High psychology emotional stability with low measure problem is universal as to ordinary beings, and many specific jobs have special requirements to emotional stability, for example, the astronauts in orbit operation, soldiers in shooting, as well as doctors in operating all need relatively higher stable psychology emotion to keep regular work. However, the traditional evaluation method of measuring the emotional stability is a qualitative question while the quantitative analysis of stability is the main problem to figure out. The psychology emotional stability tester collects emotional stable data to conduct quantitative analysis, computation as well as comparison, and then quantitatively analyzes the result of emotional stable data with scientific algorithm to work out reasonable results. The system adopted in this paper has efficiently quantified several qualitative factors with scientific methods to get reasonable conclusions, thus effectively transforming the analysis method of psychology emotional stability from the traditional qualitative one to the quantitative one.

1 INTRODUCTION

The psychology emotional stability is an important branch in psychological study. The execution results are always qualitatively judged along with the stability of psychological stability while many specific jobs has special requirements, such as the astronauts in orbit operation, soldiers in shooting, as well as doctors in operating. The quantitative approach of evaluating personal psychological emotion with scientific instruments is the main focus of our research. However, the current researches are only confined to qualitative judgment without quantitative analysis. Then how to quantify several qualitative factors with scientific methods to get reasonable conclusions is the focus of this paper.

2 DESIGNED AND IMPLEMENTED FUNCTIONS

2.1 Template processor of control circuit

The template processor of control circuit is the control center of the system, the structure chart of which is shown in Figure 1. It can be seen that the single chip microcomputer is composed of the center control single chip microcomputer named 89C51 and a clock circuit, then emotional stability data are collected, stored and transferred to the main control PC for further processing.

2.2 Data collection of emotional stability tester

The psychology emotional stability test demands that one arm of the test subject get through three curvilinear pathes with different breadths and tortuosities, without touching the base plate or the curve wall. The measuring parameters such as touching frequency of different positions, operating time and speed, as well as touching positions are transferred to the main PC to get scientific conclusion using the emotional stability algorithm and the evaluation parameters.

2.3 USB 2.0 data communication interface

The communication interface adopts USB 2.0 connector which has nice commonality, strong

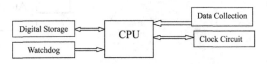

Figure 1. Diagram of microprocessor control circuit.

stability and high transmission speed. CP2101 is chosen as the interface which is a specific chip with relatively higher antijamming capability for USB data transmission.

Psychological Cognitive Ability *Test* Various kinds of cognitive ability test methods have been realized via computer software, including attention span test, landmark recognition, continuous computational test, visual discrimination test, graphics memory test and other related psychological cognitive ability test. A comprehensive scientific and accurate evaluation of the cognitive ability can be obtained through these tests.

2.4 *Intelligence scale test*

The computer software is combined with the popular Wechsler scales, norm standard score, as well as the deviation intelligence quotient algorithm. Then the conclusion can be drawn by the high integrated system with a wide range of sample and high reliability and validity, besides, the test results show higher stability.

2.5 *System testing, evaluation, result data output*

The parameter data generated from the series of system modules is gathered by the computer software to be compared with the data base for further comprehensive analysis. Moreover, the multiple-factor aggregation algorithm in psychological test (it will not be dealt with here because of space limitation) is also combined, and then the objective conclusion is obtained with form file output printed according to the requirement set.

2.6 *System testing, evaluation, result data output*

The parameter data generated from the series of system modules is gathered by the computer software to be compared with the data base for further comprehensive analysis. Moreover, the multiple-factor aggregation algorithm in psychological test (it will not be dealt with here because of space limitation) is also combined, and then the objective conclusion is obtained with form file output printed according to the requirement set.

3 SYSTEM COMPOSITION

The psychology emotional stability test system is composed of system test project and system auxiliary function. Modular design, real-time evaluation, and standardization analysis are adopted. Personal computer acts as the core for data processing and computation while single chip processor named AT89C51 is the center for data collection. The test items mainly include the following four functional test modules: emotional stability test, cognitive skill test, intelligence scale test and examination test. The auxiliary function test consists of user data input, profile view, data maintaining, data printing and expert consultation. The system functional diagram is shown in Figure 2.

4 HARDWARE SYSTEM STRUCTURE

The evaluation system adopts modularized design idea and advanced sensor technology for hardware design. The device hardware mainly consists of

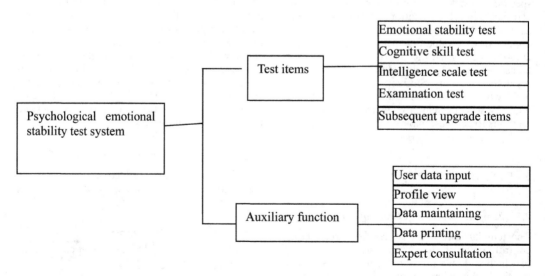

Figure 2. The system function frame.

Single chip processing circuit, data collection circuit, background sound generation system, and alarm indication circuit and USB communication interface circuit, as illustrated in Figure 3.

4.1 Data collection circuit design

The frequency of test subjects' hand shaking is recorded by the sensors, and the parameter data getting through the plate in Figure 4 is real-timely collected to get the raw data.

It is indeed hard to avoid any hand shaking when going through the emotional stability test curve. So the test woks like this: when the hand touches the side wall of the tester, the red square is recorded; when it touches the bottom board, the green one is recorded. The system records "E" when the test subjects' hand passes over and out of the curve panel, indicating that test errors occur because the hand gets out of the plate midway, as illustrated in Figure 4. When the test subjects'

hand goes through the curve from the beginning to the end, all the speed parameters of touching and "E" errors are scanned, collected, recorded by the system hardware and transferred to the main control computer. The operation process is transferred to the single chip processing circuit for preprocessing via scanning circuit, through which it is transformed into the required communication format, and then the formatted data is sent to the main control computer via USB interface. The system has initiated a unique emotional stability tester composed of three sine curves with different frequencies and breadths, as well as an emotional stability tester constituted of nine-hole meter. The parameters of the curve's width and depth, and the nine-hole meter's aperture as well as the testing pen's moving speed are adopted to evaluate the emotional stability through the multiple-factor fusion algorithm.

Besides, the operation trajectory tracking and restoring function is fulfilled along with the real-time evaluation of emotional stability. The operation locus is displayed graphically while the test subjects operate with the emotional stability tester, then the error numbers, error positions, moving speeds and background interferences, etc., are comprehensively analyzed to quantitatively study the emotional stability parameters of the test objects. It is an efficient measure for the psychological quality test to quantitatively analyze the emotional stability by extracting psychological features of different groups from various occupations and working environment and comparing those with the psychological characteristic database. The data with standard formats are finally analyzed using the specific software for comprehensive evaluation of psychological quality. The schematic diagram is shown in Figure 5.

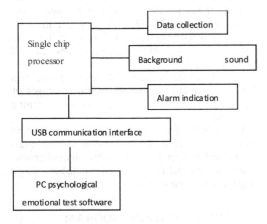

Figure 3. Hardware system structure diagram.

4.2 Background sound generation circuit

The background sound generation circuit is used to simulate working environment of the test subjects,

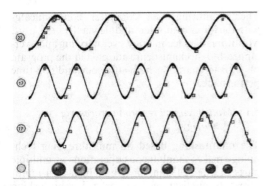

Figure 4. The emotional stability test curve and the data collection procedure chart.

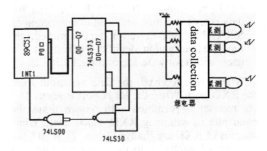

Figure 5. Schematic diagram of data-collecting electronic circuit.

then their sensory system would be stimulated by the simulative background noise, thus their psychological enduring capacity in the real scenarios would be tested.

4.3 Alarm indication circuit

The test subjects are reminded via lamplight, sound and other related stimulus methods when error occurs during the test procedure.

4.4 USB Communication interface

The data transmission interface adopts USB2.0 which is universal with high transmission speed. CP2101 is selected as the interface chip because of its strong anti-interference capability.

USB data transmission set: data bit 8, odd-even check 0 and baud rate 19200.

When QWY→PC:
10H—18H: Check the corresponding data transferred from the checking points via curve 1
20H—2EH: Check the corresponding data transferred from the checking points via curve 2
30H—3EH: Check the corresponding data transferred from the checking points via curve 3
40H—4EH: Check the corresponding data transferred from the checking points via nine-hole meter
51H—the test pen is grounding (error red ×)
52H—the test pen touched the bottom lines of the three curves (error green ×)
53H—the test is cancelled
5AH—standby signal. It is the communication signal with PC, which is consecutively sent per second. While the PC signal is not received, all the test lights died out. During the testing procedure, this signal ceases to be sent out, and the test lamps of the tested curves turn on and then die out when the test is over. When being sent out, the data are output from the TXD port. One frame is sent with 10 bits with one start bit and one stop bit, and the remaining eight are data bits. The data buffer instruction of SBUF is executed by CPU (for example, MOV SBUF, A). After the data bytes are written with read-in instruction of SBUF, the serial port transmitter is launched, and then the interruption identification is set as "1". (clear with a software, for example, CLR T1).

When being received, the data are input from the RXD port. When the REN port is set as "1", the receiver is permitted with sixteen times the baud rate of sampling RXD signal level. While the pins of RXD are changed from "1" to "0", the receiver is launched with the counter acting synchronously. The counter gathers the value of RXD for three times during each period, and exclusive

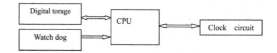

Figure 6. The processor control diagram

method with two of three vote is adopted to eliminate interruptions. Besides, the value of RI is set as "0" by the procedure when receiving signal, taking CLR RI for example. The real-time data collected by emotional stability tester is transferred to the host computer for further disposals via USB2.0 interface. CP2101 is selected as the chip for transmission interface of USB2.0 because of its anti-interference capability, as well as stable and reliable data transmission.

4.5 The processor control circuit module

The processor control circuit module is the control center of this system whose structure diagram is illustrated in Figure 6. CPU is composed of the device named AT89C52 with program memory of 8 K and data memory of 256 bytes. The reliability is extremely high because of few external components.

Digital storage is used to store ephemeral data during the treating processes, status messages of the selector switch and various kinds of warning information.

The watchdog is used to monitor the running state of AT89C52 CPU. It is reset at regular intervals if the CUP program runs normally, otherwise, a hard reset signal is generated to make the single chip back to normal running state.

5 COMMUNICATION PROGRAM DESIGN OF THE MONITORING HOST COMPUTER

The monitoring host computer communicates with the lower computer module via the USB port. Multithreading technology, event driving and overlapped I/O mechanism are adopted in the program of USB to guarantee high efficiency and real-time performance.

5.1 Multithreading real-time processing technology

The multitasking based on multithreading technology makes simultaneous operations of multiple parts during the same program. In fact, a multithreaded application has realized multitasking extension to give characteristics of parallel execution to the code. Thus, the use ratio of CUP has been improved to execute real-time and random operations.

The Microsoft Foundation Classes (MFC) is a general VC++ programming model, which supports multi-threading. Under VC++6.0, MFC application threads are presented by CWinThread objects. MFC divide threads into two classifications: one is UI threads related to user interface; the other is the worker threads which is irrelevant to that of user interface. Actually, the fundamental difference between the two models is that the former contains message loop while the later does not. In general, the main thread is automatically generated by MFC to receive and process various messages for UI threads while the worker threads are generated by the main thread for other time-consuming operations in the background, such as communication or operation.

A variety of synchronization methods exist among the threads, such as event objects, semaphores and mutexes. Event objects, the kernel objects with inspired and unexcited status, are adopted in this program.

The event objects are frequently applied to the situation where one thread waits for another thread to be executed. During the specific procedure, a objective function named CreateEvent () is generated, besides, the inspired status is set by function named SetEven () while the unexcited status set by ResetEvent(). The procedure is closed by the function of CloseHandle().

Multithreading real-time processing technology is adopted in the communication program, aimed at higher efficiency and better message response speed, as well as real-time communication.

The event on specific serial port is monitored by the event mask established by function SetCommMask(), and then its event mask is obtained by the function GetCommMask(). Besides, once the communication event is assigned by the function SetCommMask(), the application program may evoke the waiting function named WaitCommEvent() to wait for a communication event to happen.

Asynchronous mode is adopted in read-write operations of the serial port, as well as the waiting functions for communication events, in order to ensure the high-speed response time of the sub-thread to the main thread when receiving and writing, besides, the variation of event objects can be also effectively detected.

5.2 Event driven and overlapped I/O mechanism

Under the DOS system, interruption driving mechanism can be utilized to compile higher efficient program than that of inquiry mode. Meanwhile, a higher level of communication driving event is prepared under the WIN32 system, which

Table 1. The list of communication events for program processing.

Value	Description
EV_BREAK	Input termination is detected
EV_CTS	State of CTS is changed
EV_DSR	State of DSR is changed
EV_ERR	Line state error
EV_RING	Ring is detected
EV_RLSD	State of RLSD is changed
EV_RXCHAR	A character is received to be stored into input buffer
EV_RXFLAG	The event character is received to be stored into input buffer, such as EvtChar member of the architecture of DCB
EV_TXEMPTY	The last character if send out from the output buffer

encapsulates the underlying communication interruptions. The proposed method has provided a variety of intuitive communication events which have efficiently avoided direct manipulations of the complex underlying hardware. Due to the communication driving event, the transform can be checked without port checking, which is extremely useful especially for receiving bytes. Besides, the port status doesn't need to be continuously checked by application programs. Moreover, the CPU time is efficiently saved. The communication event applied by Windows is illustrated as Table 1.

6 CONCLUSION

The emotional stability tester is designed on basis of the famous emotional stability testing theory and technology. Besides, specific emotional stability testing module is designed for transferring the collected data to the host computer through sensor circuit and microprocessor. Thus, scientific evaluation results are obtained by database alignment, psychological multiple-factor aggregation algorithm and comprehensive analysis, effectively transforming the analysis method of psychology emotional stability from the traditional qualitative one to the quantitative one, and efficiently solving the problems of no testing equipment and no qualitative analysis in traditional psychology emotional stability test.

REFERENCES

Jianwei Ye & Tianxi Zhai. The design and implementation of targeted controller system [J]. Microcomputer information, 2010 23(26):84–86.

Jianwei Ye & Runhe Qiu. The principle and design of communication electronic circuit [M]. Beijing: electronic industry press, 2012.

Qi Zhang & Ningxi Zhu. The application system design of C51 single chip [M]. Beijing: Electronic industry press, 2009.

Xianyong Li. Visual C++ serial port communication technology and implementation [M]. Beijing: Posts and Telecom Press, 2004.

Xiaoke Wang, Jun Wang & Huidong Zhao. C# project development case records [M]. Beijing: Tsinghua university Press, 2012.

Control Engineering and Information Systems – Liu (Ed)
© 2015 Taylor & Francis Group, London, ISBN 978-1-138-02685-8

Design and implementation of the freight forwarding management information system based on ROR

J.H. Wan & Y.W. Zhang
Information Engineering Software Department, Anhui Xinhua University, Heifei, China

ABSTRACT: Industry Freight Forwarding is a industry that the agent accepts the entrustment of the consignee and consignor to transact cargo transportation or other relevant business in the name of principal for service remuneration. Industry Freight Forwarding Involves many aspects with trifles, such as booking, loading, customs clearance, reported insurance, container transport, bining and devanning, the issue of bills of lading and transport fees settlement which all make it not easy to do for human, so it is necessary to design and develop Freight Forwarding Management Information System for improving the freight forwarding service quality, enhancing agent efficiency and reducing operating costs.

Ruby on Rails is a new and developing Web application framework constructed on Ruby and it is single and efficient. This paper researchs advanced development mode and method on Ruby on Rails and applys them to Freight Forwarding Management Information System designed System management module, customer management module, warehouse management business management module based on the detailed demand analysis and business functions. As the same time, this paper carrys out a detailed analysis of the function of each module based on the role of system administrators, policymakers, business units, freight forwarding, import and export department, financial management department to accomplish the development of Freight Forwarding Management Information System by using the Ruby on Rails framework finally.

Keywords: freight forwarding management information system; ROR; MVC; decision subsystem

1 SUMMARY

After the accession to the WTO, with the rapid development of the national economy, China reduced barriers on import and export trade, making domestic freight forwarding enterprises mushroom. Foreign companies, state-owned enterprises and private enterprises are involving in the industry fast because freight forwarding market's potential is tremendous and would be a hot spot in investment in the future. It must inevitably lead to the freight forwarding industry be more competitive, and industry split and consolidation may be the trend [1]. In this context, all freight forwarding enterprises are expected on the premise of reasonable investment, by developing their own forwarder management information system, to constantly improve enterprise service ability and stable old customer team and capture new customers, making themselves always be invincible. Freight forwarding industry is service-oriented logistics industry and relys on the services and prices to complete market competition. the timeliness in the freight forwarding enterprise is demanding and important, so, through scien-

tific and reasonable arrangement to improve the work efficiency and service quality is very important [2]. In this case, the freight forwarding enterprise can use advanced management information system to further improve the enterprise's business process and strengthen enterprise management to enhance market service levels. So, how to combine the management information system with freight forwarding process reasonably has been regarded as necessary problem for modern freight forwarding enterprise, especially large and medium-sized freight forwarding companies.

2 RELEVANT TECHNICAL JANE INTRODUCED

2.1 RoR

RoR [3] is the abbreviation of Ruby on Rails. Ruby on Rails is a framework for writing web applications; it is based on the Ruby language and provides a powerful framework for application developer. Ruby on Rails includes two parts content: Ruby and Rails framework.

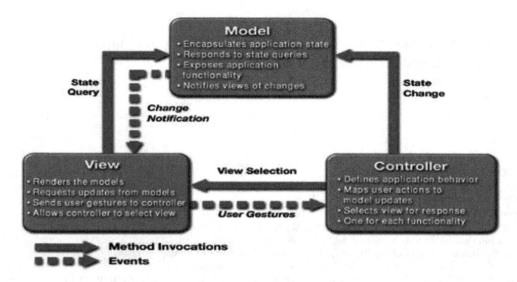

Figure 1. MVC component types and features.

2.2 *Ruby*

Ruby is regarded as an object-oriented and interpretative scripting language, it is fast and simple. Ruby can finish the work easily and quickly for it don't need variable declarations, no types of variables, and its grammar is relatively simple, be automatic memory management. At the same time, Ruby has adopted the very popular object-oriented programming ideas. Compared with orther programming languages, Ruby has advantages which is mainly embodied in: (1) Ruby can process the regular expression; (2) powerful string handling functions; (3) fully intelligent garbage collector; (4) Ruby has a good exception management mechanism; (5) to support an iterator.

2.3 *Rails*

The Rails framework is different from the existing complex Web development frameworks. Rails can be thought of as a more in line with the actual demand and high efficiency of Web development frameworks. Rails fully combines the advantages of PHP and Java systems, which make it from the birth has received extensive attention of the IT industry in a short time. Rails's main features are as follows: (1) the Full-Stack MVC; (2) less code; (3) good code generator; (4) zero turnover time; (5) with a support system.

2.4 *MVC*

The so-called MVC [4], is the abbreviation of Model—View—Controller in fact, Its Chinese

meaning is: Model View Controller. It is a new interactive application framework and was proposed by Trygve in the 1980s. In this framework, the application program mainly is divided into three components, namely the Model, View, and Controller. Their function and relationship between each other can be shown by Figure 1.

2.5 *The decision-making subsystem*

The major functions of the decision-making subsystem [5] is for a variety of data mining on the large relational databases and to discover the hidden law through a variety of advanced knowledge mining techniques, finally to provide decision-making information for managers. In Figure 2, the decision-making subsystem implementation flow chart is given.

3 DESIGN AND IMPLEMENTATION OF THE FREIGHT FORWARDING MANAGEMENT INFORMATION SYSTEM

3.1 *The system requirement analysis*

According to the actual situation of the freight forwarding company's business management, the users of the system mainly have three kinds of roles, namely, salesman, treasurer, system administrator. The Freight Forwarding Management Information System contains the main function is as follows: Business operation function: (1) business delegate

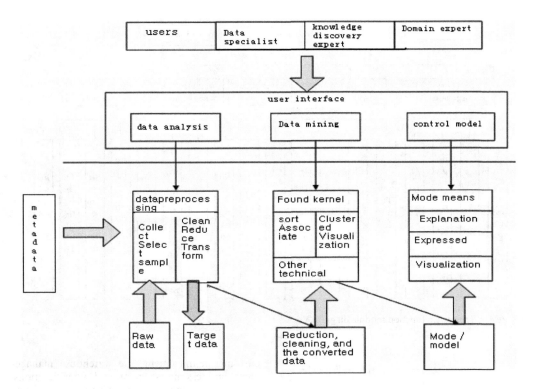

Figure 2. Decision subsystem flowchart.

of the naval and air import and export business;
(2) booking space; (3) FCL/LCL packing; (4) bill
of lading.

Business queries function: (1) the bill of lading
query; (2) order query; (3) the team query.

Financial management function: (1) the accrued
expenses management; (2) the cost of query;
(3) bill query; (4) the accounts receivable age
analysis; (5) the profit analysis.

Customer management function: (1) the customer management; (2) supplier management.

System maintenance function: for the system parameters setup and management, system
upgrade and repair.

3.2 The system architecture

System use B/S framework, as shown in Figure 3.

3.3 System function module design

According to the actual situation of the freight forwarding company, and the process and functional
requirements of the freight forwarding, The Freight
Forwarding Management Information System can
be divided into multiple modules, namely the system management module, supplier management

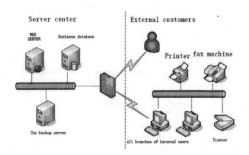

Figure 3. The system architecture diagram.

module, the customer management module, the
warehouse management module, business management module, etc. as shown in Figure 4.

3.4 The specific implementations

Before this system uses Rails for development, the
first step should be done is to build the complete
Rails development environment [6], at the same
time the development tools and database management system which are to used should be installed
in place, because of the paper length, here not to
expound one by one. A whole operation process of

921

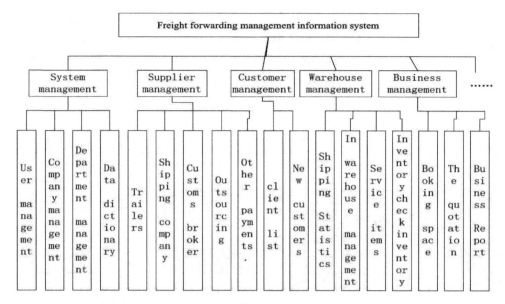

Figure 4.　The system function module diagram.

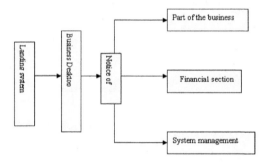

Figure 5.　System operation flow chart.

the system is: after the users' logins web site, they could process their business based on the notification and tasks received, as shown in Figure 5.

4　CONCLUSION

The Freight Forwarding Management Information System provides a web-based platform for the freight forwarding management and can realize the system management, supplier management, the customer management, the warehouse management, business management, and other functions. The system to some extent improves the efficiency of the freight forwarding management, have very strong practicability.

REFERENCES

[1] Jin Feng, Qin Jin. Operation Strategy Analysis and Development Expectations of the International Freight Forwarding Enterprise [J]. The Fortune Time, 2011(2):118–118.
[2] Lin Xing, Sun Jianjing. The Assessment of the Freight Forwarder Based on Fuzzy AHP, Logistics Sci-Tech, 2007(1):78–80.
[3] Nick Langley. Ruby on Rails takes Rest from Soap [J], Computer Weekly, 2008.7:34.
[4] Wang Fengling. Studies on Mainstream Web Framework Technology Based on MVC [J]. Journal of Nanning Polytechnic, 2011(3):94–97.
[5] Luo Lili, Cao Jihong, Zhang Dianye. Research on Intelligent Transportation Expert Decision-Making Subsystem [J]. Library Tribune, 2001, 21(5):37–39.
[6] Rui Lijie. Enterprise Application Framework, Ruby on Rails [J]. Electric Power Information Technology, 2010(2):89–92.

Control Engineering and Information Systems – Liu (Ed)
© 2015 Taylor & Francis Group, London, ISBN 978-1-138-02685-8

Research on target recognition technology in precision-guided

Q.H. Qu
Baicheng Normal University, Jilin, China

ABSTRACT: In terms of the development trend of the international modern national defense science and technology, it is obvious to get the following conclusion: precision guided weapon is one of the hot issues of the research on international weapons technology nowadays and also an important pillar of the future battlefield, in that way, how to effectively identify and track the target is the key of precision-guided. The main research object of this paper is the target recognition technology in the precision-guided. According to the characteristics of target recognition in missile system, the paper proposes an improved algorithm for target recognition. The algorithm improves the recognition quality and efficiency compared with the traditional target recognition methods.

1 INTRODUCTION

With the rapid development of science and technology and the change of operational mode, the new military and military technology revolution were born. The tide of revolution caused great attention of the world, and they formulated and revised their military technology development strategy in succession. This urged the key point of the weapon development to focus on two aspects of information technology and long-range precision strike technology. As the core technology of precision strike weapons, precision guided technology has become one of the hotspot of current international military technology and has been widely applied to various guided weapons such as ballistic missiles, air and missile defense, cruise missile and anti-armor missiles. It is the core weapon technology which can effectively improve the combat effectiveness and enhance conventional deterrence and air combat capability, and also is one of the weapons technology which the world today attached great importance to.

2 MOVING OBJECT RECOGNITION

Moving object recognition is an important part of target tracking and recognition system, and the accuracy of target recognition will influence tracking accuracy. Moving object detection involves many disciplines such as image processing, pattern recognition, artificial intelligence, computer graphics, artificial neural network, physiology, psychology, etc. Image of target recognition refers to a frame image in video sequence, moving object recognition divides the image into plural regions and finds the interesting region. In fact, it processes the video images by using digital signal, detects and extracts the foreground of the relative motion of image and background, and then further segments moving foreground into several independent goals according to the characteristics of the image such as gray, edge.

Difficulties in recognition of moving target: First, various factors make the target information incomplete, such as object occlusion phenomenon, deviation caused by the difference of target perspective, deviation of target recognition brought by the intensity of light. Second, under relatively complex circumstances, how to effectively extract the enemy target, ensure that the time complexity of the algorithm is as low as possible, and make the robustness of the algorithm as good as possible are very difficult at present. Again, due to the rich initial data of the image sequence, there is a strong correlation between adjacent frames. How to make good use of correlation in time domain and improve the detection and recognition of the moving target has the practical significance.

Moving target recognition in the image sequence is often based on the characteristics and gray. The detection based on the characteristics is mostly based on the image feature or its component model to detect moving targets. The method based on the gray is to realize motion segmentation on the basis of gray model of image. Commonly used methods are: optical flow method, frame difference method and background subtraction method. Optical flow calculation is complex, so not suitable for real-time processing; Inter-frame difference method compares the two continuous frames and then extracts information of the moving object. The integrity of moving object extraction is poor, but there is a strong environmental adaptability, background difference method is able to extract the target point, but is more sensitive to the dynamic change of scene.

2.1 The frame difference method

Frame difference method uses the difference between successive frame images in video sequence image to make the target detection and recognition, which is one of the most used methods, especially under the condition of constant background. After two or several frames or a few frames do subtraction, the part of gray which does not change is removed, including most of the background and a small part of the target. The advantages of the frame difference method are able to adapt to environmental changes, to more accurately detect moving target whose change is obvious, but effects on the part of target region which does not change obviously are not good.

The algorithm of the frame difference method is shown as the following:

Step 1, extract two successive frame images I_k, I_{k-1} and calculate the absolute gray difference image of two frames d_k:

$$d_k(x,y) = |I_k(x,y) - I_{k-1}(x,y)| \tag{1}$$

Step 2, set the threshold value TH, and make the difference image binaryzation. When the difference between a pixel is larger than the preset threshold, the pixel is considered to be the target pixel, on the contrary, considered to be the background pixels.

$$Ib_k(x,y) = \begin{cases} 0 & d_k(x,y) > TH \\ 1 & otherwise \end{cases} \tag{2}$$

where Ib_k is the binaryzation image, TH is the threshold value for binaryzation.

The purpose of binaryzation is to use image segmentation algorithm to separate foreground images from background pixels. The basic frame differencing method has very strong adaptability to dynamic environment without the need to obtain the background image, and the algorithm has low complexity, fast speed, and good real-time. But the basic frame difference method has its own shortcomings, when gray values of two adjacent frames images are close, this method cannot get the full contents of object and can only get the contour of object. This method may produce the same target superposition, and extracted target is larger than the actual target range.

2.2 The background subtraction method

The background subtraction method is an effective detection algorithm of moving object, and the basic idea is to approximate the background image pixels using a background parametric model, to make a difference comparison between the current frame and background image to detect moving region, the pixel area with the larger distinction is considered as the moving region, the pixel area with the smaller distinction is considered as the background region. Background subtraction method must have a background image, and the background image must be real-time updated with the change of the light or the external environment, so the key of background subtraction method is the background modeling and updating. The specific steps are as the following:

Step 1, read the background image B_k and the current image I_k, calculate the absolute gray image of two frames:

$$d_k(x,y) = |I_k(x,y) - B_k(x,y)| \tag{3}$$

Step 2, set the threshold value TH, make the difference image binaryzation, and extract moving object.

$$b_k(x,y) = \begin{cases} 0 & d_k(x,y) > TH \\ 1 & otherwise \end{cases} \tag{4}$$

Step 3, use mathematical morphology to make filter processing on the frame difference image b_k, then make the regional connectivity analysis of it, when the connected area is larger than a certain threshold, the region is considered as the target area.

The algorithm of the background subtraction is relatively simple, but more limited, requiring image background known. In order to contrast with image effects obtained by the frame difference method, selecting image and frame difference are correspondant. Compared with the frame difference method, background subtraction method can fully extract the target, but susceptible to the outside such as the weather, illumination and so on. in practical applications there are two basic problems to be solved: obtain the background model and update the background model.

2.3 Consecutive three-frame difference method

Because the above two algorithms have their own disadvantages, this paper takes a algorithm called as consecutive three-frame difference method to identify the target, consecutive three-frame difference method can effectively inhibit the phenomenon that the exposure of the background area is relatively larger due to larger displacement between two adjacent frames, can extract the moving target shape of the middle frame, namely motion object is obtained by calculating the continuous before and after the three frames of image sequences in the overlap between the frame difference. The steps of the algorithm are as the following:

Step 1, read consecutive three frames image $I_{k-1} I_k I_{k+1}$, and respectively calculate absolute gray value of continuous two frame images:

$$d_{k,k-1}(x,y) = |I_k(x,y) - I_{k-1}(x,y)| \qquad (5)$$

$$d_{k+1,k}(x,y) = |I_{k+1}(x,y) - I_k(x,y)| \qquad (6)$$

Step 2, set the threshold value TH, make the difference image binaryzation and extract moving object.

$$b_{k,k-1}(x,y) = \begin{cases} 0 & d_{k,k-1}(x,y) > TH \\ 1 & otherwise \end{cases} \qquad (7)$$

$$b_{k+1,k}(x,y) = \begin{cases} 0 & d_{k+1,k}(x,y) > TH \\ 1 & otherwise \end{cases} \qquad (8)$$

Step 3, extract the set of two frames difference and get moving target.

$$Ib_k(x,y) = \begin{cases} 0 & b_{k,k-1} \cup b_{k+1,k} = 0 \\ 1 & otherwise \end{cases} \qquad (9)$$

According to the above steps, the threshold value is used to extract foreground regions of difference image on the process of segmentation on the moving object in video sequences, thus the grayscale image is transformed into two values map. After operating the absolute value of the corresponding pixel difference on every two adjacent frames, the operation of making images binaryzationis is performed, and then the result of time frame difference for every two frames is smoothly processed and image noise is reduced.

3 RECOGNITION ALGORITHM MODEL

In the above analysis of the advantages and disadvantages of the algorithm, it is know that the interframe difference method is only slightly affected by the light changes. So the identification is affective and stable. But on the other hand, the precise location of the target cannot be identified, and the image information is not complete.

Background difference method can identify the location precisely and quickly, but it is vulnerable

to light, environmental changes. Besides, it is very hard to obtain the background image accurately. The frame difference mentioned in this paper is three frame difference methods.

This method can eliminate the tensile and cavity phenomenon of the moving object. By identifying of the difference of the detected frame and the image background, the moving target recognition of the background subtraction can be achieved. And by comparing the difference of the image background and the detected frame, the segmentation of the moving objects and the background images can be implemented.

However, background information and the moving target cannot be divided clearly due to the excessive redundant information introduced in the process of of image processing of the algorithm, thus leading to the low degree of readiness of the recognition.

Step 1, use the background subtraction method to identify the video sequence image.

$$M_b(x,y) = \begin{cases} 1 & f(x,y) > \xi(x,y) \\ 0 & otherwise \end{cases} \qquad (10)$$

In the above formula, 1 represents to be divided into motion target, 0 represents to be divided into background region, $M_b(x,y)$ represents pixels of the coordinates (x,y), $\xi(x,y)$ is the threshold of pixels (x,y).

Step 2, use the three frame difference algorithm to identify video sequence after the first step, and three frame difference method is only aimed at moving object region identified by background subtraction method in the first step to operate in order to simplify the algorithm.

$$M_f(x,y) = \begin{cases} 1 & D_k(x,y) = 1 \ and \ M_b(x,y) = 1 \\ 0 & otherwise \end{cases} \qquad (11)$$

In the above formula, $D_k(x,y)$ is the detected result of extracting the foreground region in the first step for the use of consecutive three frames subtraction method. When $D_k(x,y)$ is 1, divided into motion target. When $D_k(x,y)$ is 0, divided into the background region. That is to say, only when the background difference method and three frames difference method simultaneously identify the pixel as the moving object, can the pixel be divided into target.

Step 3, because of the influence of weather and temperature, the background of stance diagrams of the missile often appears as a field of motion. So it is possible that the motion of the background area may be recognized as the motion target. Due to this, the domain information should be fully considered impractical problems. Based on the above mentioned issues, the field of pixel difference method is applied to solve this problem.

$$D_n(x,y) = \min(|f(x,y) - \xi(x+i,y+i)|),$$
$$i,j \in \{-1,0,1\} \tag{12}$$

Generally speaking, there is the domain 5×5 or 3×3, and this paper mainly uses the domain 3×3.

Step 4, identify the motion pixels according to the above descriptions.

Step 5, make the morphological processing on $M_n(x,y)$ to get the motion image M.

4 THE EXPERIMENTAL RESULTS

The experimental result shows that the method that combines the three frame difference method and the background difference method together used in this paper is obviously more effective than the two frame difference method. As it is difficult to propose the target background accurately in the process of operation in real time, the background obtained is mostly inaccurate.

By utilizing the algorithm in this paper, the circumstance of the inaccurate image background can be adapted to. Under the condition of the inaccurate background, the error can be eliminated by twice subtraction of the three frames difference. So the method in this paper tries to combine the advantages of the background difference method and the three frame difference method. It has strong robustness.

REFERENCES

Ding Defeng. The moving target detection, recognition and tracking technology research [D]. Shanxi: Northwestern Polytechnical University, 2007.

He Guiming, Li Lingjuan, Jia Zhen-tang. A fast video segmentation algorithm based on symmetric difference. Mini-micro computer systems, 2003,24 (6): 966–968.

Hu Yusuo, Chen Huiyong, Chen Zonghai. Motion detection and object tracking strategy based on robust statistics [J]. Beijing: Journal of system simulation, 2006,18 (2): 439–304.

Lu Shengwei. Target recognition tracking research based on image processing [D]. Changchun: Changchun University of Science and Technology, 2008: 43–45.

Qin Xuan, Zhang Liujin. Infrared target tracking algorithm based on curve fitting prediction. Infrared technology, 2003,25 (4): 23–25.

Zheng Jiangbin, Li Xiuxiu, Zhang Yanning. Tracking algorithm of moving object in video surveillance [C]. Systems engineering and electronic technology, 2007, 11: 1991–1993.

Control Engineering and Information Systems – Liu (Ed)
© 2015 Taylor & Francis Group, London, ISBN 978-1-138-02685-8

The theory and practice of the digital campus construction

S. Ye & Q.G. Zhang
Training Department, Aviation University of Air Force, Changchun, China

ABSTRACT: Digital campus construction can make us realize highly sharing resources, changing traditional pattern and measure to teaching, science-research, administration and service, provide convenient, safe, efficient information applying environment and net information service. This paper introduces the conception, construction theory, overall design and practical application, and also some problems when in digital campus construction.

Keywords: digital campus; construction; theory; practice

1 INTRODUCTION

The notion of digital campus first originated from a large-scale research project, titled *Informationized Campus Program*, which was launched and directed by Professor Kenneth Green of Claremont Graduate University, U.S. in 1991. In terms of definition, digital campus uses computer, network and communication technology to integrate all information resources of teaching, research, administration and life service, for the sake of realizing uniform client management and authority control. And as a result, on the basis of a conventional campus will a digital space be constructed, where processes of campus activities, including teaching, research and administration service, etc., are digitized, and ultimately the information resources will get shared and efficiency of campus administration get elevated.

Digital campus is a major component of educational internationalization as its construction will further optimize teaching resources, enrich teaching tools, improve educational facilities and accelerate teaching reform. By integrating information technology and teaching process, an informationized education mode, featuring in multimedia instructional material, shared resources, personalized teaching, initiative study, cooperative activities, automatic management, virtual environment. And this new mode will gloriously give impetus to profound reforms in teaching notion, content, method and mode, provide favorable background for Essential-Quality-Oriented (EQO) education and innovative education and pave the way to comprehensively cultivate high-quality a new talented generation (Gao, 2004).

Digital campus could encompass most parts of teaching, research, administration and service and realize their uniform and standardized management, characterized by "definite responsibility and right for each individual". In this way, a wide scope of artificial delays and mistakes will be avoided at utmost. Therefore, not only will the efficiency of daily routine be increased and the management cost be decreased, but also will the relatively isolated tasks in different departments be closely linked and cooperatively dealt. Finally, the overall teaching, research and administrative level will be significantly elevated.

2 THE NOTION AND MODEL OF DIGITAL CAMPUS CONSTRUCTION

2.1 *The notion of digital campus construction*

The central idea of digital campus is digging out advanced management philosophy and using advanced computer network technology to integrate available resources, including teaching, research, administration, life and service, etc., for the sake of realizing uniform client management, resource management and authority control. And in this way, resources will be more effectively distributed and fully utilized, and management and service will be more optimized and coordinated. Ultimately, gearing for the need of social public users, as well as campus users, a virtual university, unconfined by time and space, will be constructed (Yin, 2004).

There are certain standards to measure whether a digital campus has been successfully constructed, including.

- Whether there is an advanced philosophy and a feasible plan for informationized construction.
- Whether a public service system has been built on the basis of digital campus.

- Whether the digital campus is accompanied with informationized management institutions and technological standards.

2.2 The model of digital campus construction

The model of digital campus is abbreviated to 1+2+3+x, namely a combination of one platform, two terms of uniformity, three core service systems and multiple application systems. More specifically speaking, this one platform refers to the supporting platform of digital campus applications and services. Two terms of uniformity indicate uniform data resource and uniform identity recognition. And the three core service systems include those of digital teaching management, digital research management and network teaching management. Finally, the multiple applications system refers to some informationized application systems constructed to cater the needs of service or managements from other departments.

3 PRESENT SITUATION OF DIGITAL CAMPUS CONSTRUCTION AMONG HIGHER LEARNING INSTITUTIONS

The informationized construction of domestic higher learning institutions has developed more than a decade, in which the platform of network hardware has gradually been built and the overall level of this construction has been effectively elevated. And this construction has positively given impetus to explore a new teaching management mode, improve teaching measures and elevate teaching and research administrative effectiveness and efficiency. However, some problems have been exposed in this constructional process, and they are dominantly confined to the following factors.

A *Lacking a uniform management mechanism, a comprehensive set of global plan and a set of uniform technical standards and normalizations in the process of digital campus construction. Consequently, redundant and blind construction, as well as resource wasting, is inevitable.*
B *For independently and dispersively constructed task management systems, due to the technical flow, framework and coding standards of each component could not effectively communicate with each other, "information islands" are built; with no guarantee on the consistency of data, uniform data management is seldomly realized. Also, this situation is adverse to the long-term development of each task management system.*
C *There is no uniform authorization management for information on public data, thus leads to indefinite authority distribution among a wide scope of departments. And this is evitable in a uniform*

data management environment, where "the constructor", "the maintainer" and the "user" are easily identified, and a uniform plan management is ensured.
D *There is no uniform authorization management for information on public data, thus leads to indefinite authority distribution among a wide scope of departments. And this is evitable in a uniform data management environment, where "the constructor", "the maintainer" and the "user" are easily identified, and an uniform plan management is ensured.*

Therefore, an independent user identification module has to be developed for each service management system. On one hand, it will lead to repeated restoration and concomitant redundancy. Moreover, the user management module in each application system will be inevitably repeatedly developed, which also causes many inconveniences in the management of users' identification information and users' application. One the other hand, repeated login is required when users switching to a different operation system.

The above problems are prevailing in the process of constructing domestic informationized higher learning institutions. And the primary task at present is to break down information islands, construct uniform platform of digital campus, integrate application systems and finally realize the uniformity of user management, resource management and authority control. This task is also a natural way for those higher learning institutions to develop leap-forwardly in the atmosphere of informationized construction. Domestic first-class universities, such as Peking University and Tsinghua University has chosen long-term technical cooperative partners to assist their digital campus construction by overall planning and the subsequent step-by-step realization, in which they have invested tens of million Yuan. Up to now, certain achievements have been obtained, including constructing a uniform platform, integrating applications and gathering resources, etc.; Their Embryos of digital campus have been basically constructed, which makes them set the pace for domestic higher learning institutions.

4 OVERALL DESIGN OF DIGITAL CAMPUS CONSTRUCTION

With the development of network technology and the flourishing of network application, conventional concepts of network design, in terms of their performance, scale, extension and service, are no longer satisfying the demands of contemporary large campus. Therefore, a "layered designed modularized model" has been developed to cater to new high-speed large campus network. To be more specific, on one hand, the so-called "layed designed" refer to a vertical division of the network into multiple layers, according to the differences in function and division of labor among department. The definitions of these layers have to be definite and professional; in other words, these layers are respectively responsible for a given specific function and duty. On the other hand, the "modularized design" is realized within a single layer, where it further divides the network into horizontally multiple modules, according to who and where the network get accessed. Here what worth to mention is that, although the access position and accessed object is different for each individual in charge, these responsible persons' functional role in the network are the same. Therefore, the layered modularization design has a good extensibility, enabling the extension of network with subsequent possible increases of network nodes (Li and Li, 2005).

The three vertical layers put forward are respectively foundations platform, application supporting system and application information service.

The foundations platform belongs to the bottom layer in digital campus. It includes infrastructural facilities of the bottom network, such as campus data network, CATV system and campus address system, etc., and corresponding fundamental network services of these facilities, such FTP, mass memory and integrated data base, etc. This whole module provides a series of basic services, enabling sharing data, visiting applications and offering users an uniform access interface, etc.

The application supporting system is the core of digital campus. It is constructed on the base of fundamental network facilities and services, and plays the role of informationized platform for various types of users in digital campus. The value-added service it providing include personalized portal site, integrated teaching management system, office automation system, digital library system, video of demand system, online teaching system, virtual practical teaching system and one-card-for-all system.

Application information service system directly interacts with the user, providing an uniform interface on which the user obtain a wide scope of services from multiple application systems. At an advanced stage, information service will be more personalized and initiative—according to the status of an individual user, it spontaneously offers him/her the very information and application services in need, thus the user could use these information at ease basing on their customized information and application systems which cater to their own demands in study and work and habits in leisure time.

5 DESIGN PROCEDURES OF DIGITAL CAMPUS

The general design scheme of digital campus is to make an overall plan which caters to campus development and students' and faculties' demands in informationized campus construction, implement project construction by stage on the condition of stable development, justly distribute finite campus resources for the sake of pacing the fundamental construction of hardware with development of application software as well as training of corresponding staff, and conform to the plan of project construction with awareness of combing independent development with purchased and commissioned development. There are four steps to realize this overall framework.

5.1 Step 1: optimalize the construction of fundamental network facilities and services

Infrastructural facilitiesis the material base in digital campus construction. In terms of military academies, with their development in the informationized progress, they proposed higher demands for network infrastructures and services for teaching, research and management, thus require further improvements and supplements of present facilities, including extending the bandwidths of the campus network output as well as the army comprehensive network output and upgrading the trunk speed of campus network from gigabit to 10 s gigabit, etc.

Moreover, although basic network services, such as e-mail, are sufficient to the basic operation in a campus, with the extension of the category of network applications, along with the influence by the Internet, advanced application systems have proposed more requirements for basic network services. In other words, more considerations may be given to new application services, such directory service and authentication service, etc. In this way, not only the user could benefit from added-value services, but also more basic service supports will be delivered to higher-layered network application.

Therefore, for those army academies which have not yet planned or are planning digital campus construction, in accordance with their own mode of campus development and staff training, they should macroscopically set a plan, vigorously verify the implementing scheme, justly allocate the fund and use the finite fund to reach the best effect.

5.2 Step 2: enhance constructions of data center and information resources

Due to the constraints from technical development and the dual status of the user and the administrator, at the early stage of campus network construction, many management information systems are stand-alone or LAN editions, in which the server, application system and information resource are managed and stored independently. As a result, many information islands have been built, leading to not only the situation featured by no linkages among individual systems and no cooperative construction and sharing, but also hidden dangers in data security. Therefore, for digital campus constructions among military academies, there are some things to be ensured at the first beginning, such as paying attention to the overall layout of system resource environment, establishing an uniform data center, adopting big capacity, reasonable performance-to-price ratio, and easy-to-manage storage facilities, processing the data with centralized storage and backup, developing all types of application systems on the same platform, and finally, on the basis of an uniform data base and on the condition of ensuring data security, elevating the utilization rate of data resource at the upmost limit and ensuring the smooth operations of application systems. Overall, the most vital aspect in digital campus construction is information resource, without which not only network applications and services bear no way to operate, but also the modernization of teaching and management becomes "an armchair strategist".

Generally, there are three types of information resources in the campus network—management, teaching and literature information resources. As the majors in military academies are characteristically designed, the first step is to unify the data standard according to the actual situations and development requirements of a given academy. And this is followed by a systematic, step-by-step gathering, reorganizing, converting and storing to the data base for these previous non-digitized information resources. And then the newly generated information resources should be suppled to the original data base in time. Finally, with long-term accumulations, online classified data bases will be built, forming intangible assets and creating beneficial conditions for the next step of network applications and services.

5.3 Step 3: constructing application supporting system and information service platform

The application supporting system, especially the construction of information platform, is the core of digital campus construction. After the network infrastructural facilities and services have been built up and are smoothly operated, the network information resources will be gradually enriched, and at this stage, the advanced development of digital campus will be the constructions of network application supporting system and information service platform.

The application system provides different user groups on campus with an informationized working platform, referring to all processes in teaching and management. And the information service platform is directly oriented to the user, providing the user with a uniform interface to access all application system.

Due to the actual situation in military academies with insufficient technical staffs, constructions of application supporting system and information service platform should rely more on purchase of mature application softwares or joint development (in which softwares are jointly researched and developed by technicals from both the academy and software companies). An enlightening example in the "Network Teaching Application System in Military Academies", researched and developed by multiple military academies and local companies following the assemblance by General-Staff-Department Ministry of Military Training and Arms. Up to now, it has been promoted among military academies and obtained satisfying instructional effect (Li and Li, 2005).

5.4 Step 4: construct a personalized portal site

The final product of digital campus construction is a virtual university, whose online form is a personalized portal site; A portal site could also be viewed as a gateway for a virtual university. Through the construction of the personalized portal site, the campus information and applicable resources get integrated. In another way, multiple applications in digital campus get integrated and a customized walk becomes possible, by which the user could visit the campus network resources corresponding to their authorities.

6 PRACTICE AND APPLICATION OF DIGITAL CAMPUS

Our university has initiated its digital campus construction since 2008. And up to now, relatively mature personalized portal site, teaching management system, office automation system and campus one-card-for-all system has been constructed. The following is a brief introduction for these products.

6.1 Personalized portal site

As a window of our digital campus, with the form of web browser, our personalized portal site presents resources and information in digital campus, seamlessly integrates all application systems in the whole campus, realizes real-time data update and communication among these systems and provides the user customized information services and supports with an easy-to-use web interface. In this way, different groups, such as students, teachers, administrators, etc., are provided with customized information, comprehensive services and convenient websites to access internal/external resources. As a portal site is a single access point in our whole digital campus, it provides the user with an uniform users' interface. After accessing this site, the user could retrieve all services matching with his/her status.

6.2 Comprehensive teaching management system

This system is a service-oriented comprehensive management system whose users include teaching management staffs, teachers and students. Our system has encompassed multiple modules, including registration management, academic status management, instructional plan, course selection management, student score management, automated course scheduling, user management, graduation assessment, toll management, College English Test (CET) 4 and 6 management, scientific research management, etc., and thus is a management informationized system which has integrated Client/Server and Browser/WebServer technologies, considered all components and links in teaching management and designed for the sake of all departments and different user groups. Within this system, teaching staff could get all daily teaching routines done; students could get all corresponding teaching-related data managed from their enrollment to graduation. Also, the student could select their courses and set study plans by accessing this teaching system.

6.3 Office automation system

Office automation system is one of the most significant projects in digital campus construction, whose construction will assist to accelerate the informationized construction among all departments, due to its effects of standardizing the workflow, motivating cooperative work and finally elevating working efficiency. Our office automation system has used e-mail as a platform for internal communication and realized the following primary functions.

1. Arranging personal office work, including E-mail, personal schedule, pending agenda and any arrangement in a whole day.
2. Processing public documents, including receiving management, sending management and electronic archive management.
3. Sharing public news, such as electronic bulletins, forums and journals.

4. Managing integrated Services, such as conference management and work presentations.

6.4 Digital library

Digital library employs the idea of distributed management, organizing and managing mass information and resources in the network environment. As a result, a convenient knowledge center is formed, unbounded by time and space. Therefore, digital library is not only a digitized library, but also a cultural mass media and a basis of organizing, developing and utilizing data resources. To be more specific, digital library virtually provides as a basic data base for digital information and resources of a given network application system. And the information stored includes voices, words and images. This information is rearranged to a set of well-organized digital information and operated in a high-speed broad-width network. Besides, several new technologies are applied, including data warehouse, data mining and data propulsion, providing the reader with more convenient, quick and comprehensive services.

6.5 Video on demand system

Video on Demand (VOD) system refers to a system in which a video is played following the orders from the audiences. And in a more broad sense, it indicate applications that providing interactive videos whenever the user need. In a more straightforward way, it supplies "whenever you want and whatever you order". This system is dominantly used for shared playings of multi-media files among multiple users. To realize this function, it adopts B/S structure, inserting a media player in the browser, thus enabling the playing of the ordered resources. So wherever a browser is available, VOD system works. Our VOD system at present is primarily concentrated in exquisite course construction, excellent courseware on demand, virtual multi-media classroom, practical training show and activities of campus life, such as departmental sport meetings, band events and theatrical festivals, etc. So it has actually merged with the campus study and life.

6.6 Campus one-card-for-all

Campus one-card-for-all centers on the idea of using IC card as a digital identity thus assist teachers and students to fulfill a series of status-related activities on campus, including registration, book borrowing, dining and shopping, etc. With campus network as its basis and network application systems as its agents, this card contributes to campus informationized management.

7 ISSUES CALLING FOR ATTENTION IN DIGITAL CAMPUS CONSTRUCTION AMONG HIGHER LEARNING INSTITUTIONS

7.1 Promoting realistically with clear constructional ideas

In the digital campus construction, not only excessive ambitious with its concomitant extravagancy should be avoided, but also the lacking of foresights. The layers of dital campus are required to be taken into consideration, when we clarify ideas, set plans and determine paces of campus informationized construction. Besides, the analysis will be made viewing the university as a whole. The relationships among campus information flow, cash flow and material flow, as well the links between different departments, have to be clarified. What's more, an uniform information standards and application supporting platform have to be built, on which campus application systems could be developed and integrated. And as a result, an organic integrated system will be formed (Wan et al., 2004).

7.2 Constructing synchronously soft- and hard-facilities

Construction of soft-facilities refers to the construction of resource base and the renewal of notions. If the construction of hard-facilities outpaces too far beyond the soft-facilities, assuming a constructed digital library with poor most-frequently used resource base of inadequate multi-media bank, courseware bank, item bank and case bank, the situation will be just an analog of "well-built road with no traffic flow". Besides, if the educational notions could not keep up with the upgrading hard-facilities, assuming many educators still view and handle educational problems in an old-fashioned way without consideration of the rich connotations of social educational notions and ideas regarding how to promote educational reforms with information technologies, then even the best overall plan will have no way to be realized.

7.3 Alerting for network security

Campus network is the base of digital campus, thus the smooth operation of campus activities is exposed to all threats jeopardizing the network security. Therefore, all possible safeguarding measures are required to be well operated. For instance, both hot backup and cold backup are demanded for key facilities, in case of data losing or damaging and for the sake of data security. Also, there are other helpful measures, including enhancing network security auditing, implementing invasion monitoring, adopting disaster tolerance system and assembling a group for security-related emergencies (Zhao, 2004).

7.4 Improving the construction of relevant institutions

A set of complete management institutions in digital campus construction will include intensified trainings of applying informationized technologies for teachers and students, encouraging policies of initiative course-related informationized reforms for teachers, managing measures of information and network security and all other institutions to ensure network resources being justly used and network operation conditions, as well as environment, being improved.

REFERENCES

Fei Li, Haixia Li. On Digital Compus Construction [R] Beijing, Tsinghua University Education Technology Institute 2005 (In Chinese).

Gelan Yin, Yunan Cao. The Inquiry to Some Problems on College Digital Campus Construction [J], China Higher Education Study 2004(7) (In Chinese).

Lipeng Wan, Ya Chen, Jianming Zheng. The Construction and Thinking of China College Digital Campus [J] Intelligence Science 2004.22 (3) (In Chinese).

Liuming Zhao. The Thinking of University Digital Campus Construction [J] Zhoukou Teachers College Journal 2004(5) (In Chinese).

Yinhe Gao. Digital Campus Construction Research [J] Jixi University Journal 2004(6) (In Chinese).

Control Engineering and Information Systems – Liu (Ed)
© 2015 Taylor & Francis Group, London, ISBN 978-1-138-02685-8

Analysis of military information system integration development

H.L. Zheng
Beijing Institute of Special Electromechanical Technology, Beijing, China

Z.W. Guo
New Star Research Institute of Applied Technology, Hefei, China

ABSTRACT: This paper focuses on the cognition and development of military information system integration. We investigate the characters and essence of various integration methods of military information system. Via summarization of military comprehensive integration practice, development stages of military information system integration are demonstrated.

1 INTRODUCTION

Military information system integration is an efficient method to form a military information system resource and improves the unit operation ability of information equipment. In a period time from now, integration would be the main development direction and key factor for unified operation of future army.

2 COMPREHENSION AND COGNITION ON MILITARY INFORMATION SYSTEM INTEGRATION

Since from 1980s, various countries all over the world develop military electronic information system integration in succession, taking the means of in large-scale construction or alteration of information equipment in active service. Though devoting a mass of manpower and material resources, the operation effect of information system can not reach the purpose of people (boquan 2009). We are initiated into thinking, claiming us comprehension and cognition on information system integration over again.

2.1 *Information system integration is not only technique problem, but also human integration*

Making a comprehensive survey on the development of military information system integration, the cognition of the integration experienced three periods. First period, information technique was considered as the impetus and key factor of information system integration. Information system integration was equipment integration based on communications, software, electronic and others.

Second period, system integration was integration, which means gathering administration, organizing, method, equipment, technique and facility together as integration, while the technique level is just the primary object of system integration (jifa et al. 2007). A series of problems with information system construction made us be convinced that "human" is the key factor of information system integration, also the determinant of information system's success. First of all, the ultimate purpose of information system integration is to shape for information system of uniform optimization, human priority, and man-machine integration. A integrated information system includes equipment, technique, administration, organizing and human integration, also consider over again and optimization design radically to operational flow and command process, taking "human" factor integrated into system and touching upon integration of a good many factors such as operational theory, equipment employment, etc. Secondly, the activity of information system integration carried out by human, and its fruits can be influenced by human cognition, organized system, coordination mechanism, etc. Therefore, Information system integration is not a pure technique problem, but it is a complex process from idea, cognition, method, technique to application management. Towards information system integration construction, we can use the idea of WSR to complete the analysis of integration method and model. By the way, information system integration must deal with the relationship of machine-machine, man-machine, man-man, mapping on methodology show physics, logic and human science respectively. Physics determines the actual effect of the integrated information system, logic determines the actual efficiency of the integrated information system, man-man integration determines the ultimate efficiency of information system integration.

2.2 *Information system integration require engineering way*

The assignment of military information system integration is hard and intricate. All product explored by human will leave brand of individual and organization, come under the historical restriction of cognition, method, technique then, as information system integration of the manpower intensive is still so. Thus information system integration should simplify complex problems, normalize simple problems, standardize normalization problems under the direction of scientific and explicit road map, making action information integration bear systematicness, scheduled, integrity, normative, inheritability, which means Information system integration need engineering.

Then we put forward the concept of "military information system integration". Military information system integration is the work which introduces and exerts correlated science theory, idea and technology, and actualizes activity systemic guidance and effective normalization into the domain of information system integration, making it operate in scientific, ordered, measurable, controlled way. In that way, it ameliorates the possible blindness, confusion, collision and other statement development in information system integration, so improve the development speed, manufacture efficiency and scientific, normalization construction level of information system.

2.3 *Information system integration pay attention to result, and process more*

With the progress of information technology, deepening in cognize and changing of the operational requirement, factors that constitute information system and corresponding circumstance are also developing. Integrated information system growth when accept new systems, replace older systems by improving. Military information system integration can not start over, instead integration construction based on developed achievements. Therefore information system integration is a long and dynamic process in developing. On the one hand, it must be paid attention to process management of integrated information system lifecycle, planning, controlling and improving the effect, efficiency and adaptability of process, including planning, implementation, checking, improving in the process of integration. On the other hand, the goals of information system integration is to achieve the best in the whole lifecycle of information system, claiming short-period fruits can be combined

with long developmental object. Aiming at certain developmental stage of information system integration, integration fruits must be important. In the long time of information system integration, infrastructure construction, development process like integration structure, method, organizing of all the information system may be more meaningful than specific result of single stage.

3 DIFFERENT STAGE OF MILITARY INFORMATION SYSTEM INTEGRATION

As mentioned above, military information system integration is long-term process. According to the further cognition and construction, technology development, considering the status of military information system integration, the development of information system integration can be divided into several development stages: integration based on communications network, platform, data, and servings. It's important to note that there is no strict phase's division among the phases, only different emphases in the integration of each stage.

3.1 *System integration based on communications network*

In order to achieve the exchange and connection between relatively independent services and arms, the foremost is with the communications web and other infrastructure construction as the center, set up the contact among the information system from material. There are three basic assignments in system integration based on communications web: firstly, complement and perfection of information equipment system rely on formation and operational mission of the army; secondly, constitute and perfect the connection among both each system inside and outside; thirdly, unify the equipment index, software version, communication protocols, receive and distribution information mechanism, achieving systems connection. At the end of last century, all military information system integration in the world is in the development stage, which achieves integrity of arms equipment system and communications web connection as the main construction content.

3.2 *System integration based on common platform*

System integration based on common platform aim for constitute and perfect integrated information system, achieving technology system consistency with storage, transmission, display and other basic process of each system. Currently multi-platforms exist in integrated information system, including information transmission platforms, information management platforms, information storage platforms, material platforms, loading platforms and so on, but differ in platforms system, technology, capability which confine efficiency of exchange and connection among systems. System integration based on common platform demand targeted integrated information system, change and reform each subsystem platform, extracting common part from various information systems, unified planning, unified system, unified construction, unified management, setting up common information operation circumstance supporting the operational system.

There are some problems in system integration based on common platform that should be regarded. First is the applicability of platform. The diversification of operation requirement determined the demand of each subsystem have difference, communications capability, information granularity, response speed, man-machine interface and others. Actually, it is not any system or technique meets the need of all levels of strategy, campaign, tactics and all types of artillery, anti-aircraft, and amour. Hence, how to assort the new resource with old, keep the scope and depth of common platform construction, deal with the relationship between flexibility and principle, which are problems need of important consideration with system integration based on common platform. Second is the reliability and security of platform (huazi 2010). When each subsystem based on common platform in co-operating, platform system or technical performance is immature, especially in the condition that technical performance of software and hardware can no satisfy with the operation requirement, we give priority to reliability and security of platform by weigh the pros and cons and household output-related.

3.3 System integration based on data

When system integration develops to a certain extent, supervisors and developers of system integration can not focus only on the realization of functional index. System integration based on data based on accomplishment of infrastructural construction and system criterion, focus on the problems of data interoperability and the use efficiency. The characteristics of integrated information system are data type's multiplicity, data structure intricacy, data source or output styles multiplicity, complex relationship between semantic data, different demand of data operation and others. It is described that integrated information system is "three tenth of technology, seven tenth of management, twelve tenth of data" (jianwei et al. 2010), which shows the importance of data integration. There are several problems need to be solved for system integration based on data. One is how to deal with a series of conflicts and problems arise from isomeric distributed data processing environment, achieving consistency and integrality of various operation data. Another is how to realize efficient information organization, scheduling and sharing for different combat troop's requirement.

The aim of system integration based on data is supplied with servings of information inquiry, data mining, decision support integrated, uniformly, securely, quickly. In order to meet the need, integrated data possess these characteristics:

1. Integration. Information system after integration is organic integration and associative storage of operation data in each system, not stacking simple, isolated, in the database system; achieving the convenience and quickness of inquiry and application with joint data space, not attaining from operation subsystem by inquiry or treatment.
2. Integrality. There are two aspects of data integrality and restriction integrality. Data integrality means integration pick-up data itself, and restriction integrality means integration between data relationship. Restriction integrality is premised on data issue and exchange, so it can be convenient in data processing and improve the body efficiency.
3. Consistency. Different information resource ensures the difference on semantics, from simple name semantic conflicts to complex structure semantic conflicts. It must ensure the consistency of system internal data applications after updating data source information.
4. Validity. The validity of data is the important factor in evaluation of data quality. With the information system complicates, corresponding data system become huge easily, bring the unnecessary redundancy. Data space of system integration based on data should be competent and efficient, avoiding the waste of information resource.
5. Security. System integration based on data need to set up complete consistent user rights management mechanism, information system interior according to the different levels, different business, different seats to reasonable access to access, edit, update; it must be able to offer security of sending and dispensing data on network, ensure the confidentiality of data interaction; it must be able to ensure the data loading and recycling effectively, with the ability of fault isolation and reconstruction, resisting network intrusion and damage etc.

3.4 Information system integration based on servings

Traditional information system integration, in generally, is integration from the angles of business process and data, such as interconnection integration based on point-to-point business process. Owing to lack of agility, it is far enough to adapt to the constant changing of operational environment and operational requirements despite of its ability to solve the information island problem. Under the condition of the integrated joint operation, the integration and application of information system are in face of a more profound confusion—how to complete the research and development of large systems by aggregating technological advantages and collaborating different units and manufacturers, how to improve the information system deployment and use, and interaction of flexibility, how to put complex comprehensive information system under easier maintenance, and how to further improve the operational effectiveness of comprehensive information system. It is information system integration based on service that will be the trend of meta-synthesis in our army.

1. Integration mechanism of information system based on service

The comprehensive of information system based on service attaches more importance to the service-oriented principle and realization of the transformation of information system's integration model from the conventional "system-oriented" principle to the "service-oriented" principle. The system integration based on service transforms various military information activities to service with unified interface standard, and realizes the organic combination and reusing of various services, thus achieving the goal of system integration, via providing flexible and convenient service integration mechanisms. Information system integration based on service is characterized by openness, cross-platform, coarse-grained and loose coupling (nicolai 2008).

The application of military electronic information system can adapt to the change and development of command or military business processing flexibly with slipping the leash of technical scheme, which benefits from the demand of paying attention to military business process and operational application without paying attention to inner system, internal service and layer realization. Thus, it is convenient to realize the agile and dynamic integration among existing military electronic information systems, between existing systems and new systems, and between future systems and existing systems, so as to transform the original relatively independent and closed information systems to open systems, and realize the interaction and matching of information and external environment.

2. Integration Process of information system based on service

Information system integration based on service mainly covers steps such as military information service resource planning, description and encapsulation of service resource, recombination and polymerization of information service and interaction and cooperation of information service.

a. Information service resource programming. We should construct the overall information service architecture involved in integrated information system and pay attention to hierarchy, multi-dimension and interaction of various information services, according to integrated operational requirements.

b. Information service description and encapsulation. We should adopt relevant information service technology mechanism to transform functions and relevant resources of various information system at all levels to military information service with unified interface standard.

c. Information service recombination and polymerization. We make recombination and polymerization of various information foundation service and information application service according to certain military business so as to construct new service flow and then support demanded operational application, through flexible and convenient service integration mechanism in face of information system integration demand.

d. Information service interaction and cooperation. After description, encapsulation, recombination and polymerization, various information services makes cross call and interaction by means of unified interface standard under the support of information service system, and realizes real-time and seamless exchange of operational information resources. Thus, cooperative work among several military information services in face of army forces under the particular operational environment and requirements.

3. Service Integration based on Human-computer Cooperation

As the growing improvement of information system integration construction, human will gradually evolved into people being a part of the system, owing to firstly, integration system ought to continue to meet the need of operational demands and secondly, operational personnel ought to also operate actually according to system development regulations. Therefore,

Table 1. Military information system integration developments.

Property	Interactive mode	Interactive content	Integration priorities	Efficiency focus	Interactive circumstance
Isolated information system	Mobile media, manual input	Disk files, voice info	None	None	Manual
System integration based on communications network	Isomorphism information interact directly	Text, email, image	Infrastructure construction on network	Network connectivity	Point to point
System integration based on common platform	Isomeric information communication	Formatted reports and signal	Circumstance construction of common platform	Integrated operation	Distributed
System integration based on data	Operational data interact; data separate from application	Operational information and data	Operational information resource construction	Operating efficiency	Integrated
System integration based on services	Information interact with related services	Data + application + circumstance	Operational service resource construction	Flexibility and opening	Common

integration based on service should be a development process in which "service of people" and "service of computer" are integrating and unifying by degrees, and finally an operational service resource system featured taking people as the foremost, human—computer cooperation and learning from other's strong points to offset one's weakness. Therefore, the trend of comprehensive integrated technology is to continuously improve of human-computer cooperation technology and interpersonal interaction mechanism and realize the coordination, integration and intelligence. Human, as a part of the comprehensive integration system, should exert merits of human and computer respectively and realize the leaps of operational effectiveness jointly. The characteristics of integrated information system in different stages are compared as Table 1.

4 CONCLUSION

The development of military information system integration is the process enhancing in depth and extent of information integration, also in efficiency.

Though information integration is complicated system engineering, the development follows certain rules and laws which are worth for us taking careful analysis, deepening cognition and exploring actively. Under the guidance of the roadmap in system development, we can improve the integration efficiency continually, achieving the great-leap forward in information system integration.

REFERENCES

Boquan, Xu. 2009. Strengthening theoretical bases and methods research of information system integration. *Journal of Institute of CECT*, 2009(4): 1–6.

Huazi, Huang. 2010. *Study on Synthetic Integration Management for Engineering Rick*. Journal of Water Resources and Architectural Engineering, 2010(6):157–160.

Jianwei, Dai. & Zhaolin, Wu. 2010. *The theory and technology of data engineering*. Beijing: National defence industry publishing.

Jifa, Gu. & Huanchen, Wang. 2007. *Meta-synthesis method system and system atolog Research*. Beijing: Science.

Nicolai M. 2008. *SOA in practice-the art of distributed system design*. Beijing: Publishing House of Electronics Industry.

Control Engineering and Information Systems – Liu (Ed)
© 2015 Taylor & Francis Group, London, ISBN 978-1-138-02685-8

Material requirements planning system FoxMRP based the ASP.NET MVC framework's research and implementation

W. Wang, Y.F. Liu, Y. Liu & R. Dang
Computer Engineering, Chengdu Technological University, Chengdu, Sichuan, China

ABSTRACT: MRP Material Requirements Planning is the final product from the production plan (independent requirements) to export-related materials (raw materials, parts, etc.)'s demand and needs of the time (related needs), according to the material needs of time and production (order) cycles to determine their begin production (order) of the time, parts of the production planning and preparation of procurement plans, reduce inventory, cost savings, Enterprise Resource Planning important (ERP) part. Traditional MRP is mainly based on experience, the lack of scientific method, the predicted function is poor. The proposed FoxMRP algorithm and the ASP.NET MVC framework material requirements planning implementation techniques, improved accuracy, its features, and the algorithm is proved technically feasible.

1 FoxMRP SYSTEM COMPONENTS, OPERATING ENVIRONMENT AND FEATURES

The MRP (Material Requirements Planning) for short. MRP's basic task is the final product from the production plan (independent requirements) to export-related materials (raw materials, parts, etc.)'s demand and needs of the time (related needs); according to the material needs of time and production (order) cycles to determine their begin production (order) of the time, MRP is the basic content of the preparation of parts production plans and procurement plans, reduce inventory and cost savings. It is the core of enterprise resource planning ERP; its use is currently at home and abroad is in the ascendant.

FoxERP enterprise resource planning system is the author's the research group for the Chengdu Little Fox Software Company after several years of development launched a new product, which is a corporate cash flow, logistics (production and marketing), personnel, information flow integration management system. It is about small and medium manufacturing enterprise resource planning system: the master production scheduling subsystem, sales management subsystem, WIP management subsystem, management subsystem equipment, inventory management subsystem, ledger management subsystem, personnel management subsystem, Quality management subsystem, JIT management subsystem and other 25 sub-systems. From the orders, manufacturing to shipping to the overall business process

of a comprehensive digital management, which FoxMRP (ie, MRP Material Requirements Planning) management subsystem is the FoxERP (Fox Enterprise Resources Planning) is an important part.

Our FoxMRP material requirements planning system consists of basic information management, MRP calculation management, query material planning management, traceability management, Action reports and exception reports, user management module. Proposed FoxMRP algorithm, and can provide regeneration, net reform (counting only changes in part) and action information. First in-transit order entry only after, ordering based on an ordering insufficient amount; computing fine to day, week, month; using the concept of retention; provide POQ, LFT, FOQ other bulk rule; provide information on traceability (Pegging Data) recorded gross demand and planned receipts data sources; support materials table alternative structures; provide safe storage quantity, minimum order quantity batch, multiple, fixed order quantity; provide a safe time, longer lead time than the more effective solution to the problem of late; provide Hedge Inventory; provide the projected inventory amount and future purchases\outsourcing amounts statements; support measurement units of characteristics, according to the way decimal\ decimal digits offs, revised gross requirements and so on.

It is based on the ASP.NET MVC framework, using ASP.NET 4.5 higher version, C # mainstream language and SQL Server database technology, good support for the B/S mode structure, and its interactive capability, friendly interface, speed fast, and also have a total system help and information page help of each page, to facilitate the majority of users. Set up a multi-level user permissions, to ensure system security.

This system not only for the actual material requirements planning enterprise management, and can be used for classroom teaching and laboratory practice, can also be used in the Internet environment for students in different regions as distance education software.

2 FoxMRP-MATERIAL REQUIREMENTS PLANNING SYSTEM KEY TECHNOLOGIES

MRP material requirements planning is based on certain material needs, and develop a production planning and procurement plan. But this production planning and procurement plans is how to develop it? This is the core of MRP. Briefly: According to the APICS (American Asset Management Association) a description of the dictionary, MRP is an automatically generated production plans and procurement plans for a set of processes. It is based on the material master file, master production scheduling, inventory and so on a series of data, referring to the standard APICS Dictionary, in turn calculated. Eventually, provides users with a feasibility strong production planning and procurement plans for users to choose.

Our FoxMRP system is based on international standards and that the ASP.NET MVC framework technology MRP system.

2.1 FoxMRP algorithm

To properly plan the preparation of parts, the first implementation of the product must be produced in the schedule, that is, the Master Production Schedule (MPS), which is the basis of MRP unfold. MRP also need to know the parts of the product structure, ie, BOM (Bill of Material), can the master production schedule expanded into part programs; meanwhile, we must know the inventory amount in order to accurately calculate the number of parts procurement.

Here in our FoxMRP's MRP implementation of key processes with the internal logic diagram, the principle of using such procedures will be described.

In Figure 1, inventory, order during fields are INVA010 from inventory system to obtain the selected table of two properties of a material. Safety Stock (SS), fixed order quantity, multiples of the minimum order quantity, the maximum order quantity, batch rule, these fields are planned from the material master file to obtain the class

Demand Detail   Week ⌄

Inventory quantity	8,940.00	Fixed order quantity	0.00	Batch rule	POQ
ss	5,000.00	Minimum order quantity	0.00	Order period	07
		Maximum order quantity	0.00	Multiple	0

Starting date of the current period	GR	SR	PORC	PAB
	0.00	0.00	0.00	8,948.00
2000/05/15	0.00	0.00	8,948.00	8,948.00
2000/05/21	8,100.00	0.00	22,500.00	22,500.00
2000/05/28	17,500.00	0.00	5,000.00	5,000.00

[Return]　　[Help]

Figure 1.　MRP material requirements report.

attribute. Starting date of the current period, Gross Requirements (GR), planned receipts (SR), but also from the master production schedule, in accordance with the number during the order (ie, number of days) to obtain.

Figure 1, the form of program order (PORC), projected inventory (PAB), is based on the above FoxMRP system formula data obtained out. MRP logic algorithm is as follows:

Which uses symbols such as WHILE, FOR, IF and other widely seen in the general computer program algorithm.

PROCEDURE MRP logic;

Single-stage project planning BOM Expand MPS orders issued PORt and joined his son piece GRt;

LLC $\leftarrow$ 1;

WHILE there is any item not handled DO

WHILE there is any item currently LLC is not handled DO

PAB0 $\leftarrow$ OH-AL + max (SR0, 0);

FOR t = 1 TO T DO

IF t = 1 THEN POH1 $\leftarrow$ PAB0 + SR1 - GR1-max (GR0, 0)

ELSE POHt $\leftarrow$ PABt-1 + SRt - GRt;

IF POHt <SS

THEN NRt $\leftarrow$ SS - POHt; PORCt bulk rule

(POH t, NR t, GR t);

ELSE NRt $\leftarrow$ 0; PORCt $\leftarrow$ 0;

ENDIF;

PABt $\leftarrow$ POHt + PORCt;

PORt-lt $\leftarrow$ PORCt;

Single stage of the project PORt BOM expand its components and add the GRt;

ENDFOR;

Print this item MRP reports;

ENDWHILE;

LLC $\leftarrow$ LLC +1;

ENDWHILE;

ENDPROCEDURE.

MVC structure is for those who need to provide multiple views of the same data applications and design, interface design for variability needs, MVC (Model-View-Controller) The composition of the interactive system into the model, view, controller three kinds of parts, it is well to achieve the data layer and the presentation layer separation, making it easier to maintain complex projects. MVC application's loose coupling between the three main components also facilitate parallel development. For example, a developer can work in view, the second controller logic developers can engage in the work, and the third model developers can focus on the business logic.

MVC architecture has won the hearts, Microsoft is not far behind, launched the ASP.NET MVC framework, which comes from the Castle's MonoRail, ASP.NET MVC framework for creating Web applications ASP.NET Web Forms model alternative models. Have been built in VS2010 MVC-related functions, if you use VS2008, you need to download and install Service Pack 1 ASP. NET MVC Framework, MVC is now 4.0 version. ASP.NET MVC 4.0 version includes the NuGet package runtime library (that is, our assemblies), as well as JavaScript libraries. In this way, the ASP. NET MVC framework and dynamic data technologies become eponymous ASP.NET 4 (2010) one of the two major new features. ASP.NET MVC framework (Web Forms-based applications, like) integrates with existing ASP.NET features, such as master pages and membership-based authentication. MVC framework defined in the System.Web.Mvc assembly. In FoxMRP system development, we use the following ASP.NET MVC framework technologies:

a. Separation tasks (input logic, business logic and display logic), use its testability, default test drive assembly and Test-Driven Development (TDD). In addition to using the MVC pattern to simplify, you can also make the application of the test work better than Web Forms-based ASP.NET Web application testing easier. For example, in the form of a Web-based ASP.NET Web application, a single class is used both for the display output and to respond to user input. ASP.NET Web Forms-based application to write automated tests may be a complex task, because to test a single page, you must instantiate the application page class, all its child controls and other related classes. Because you run the page and instantiate the class so much, it may be difficult to write applications specifically focused on individual parts of the test. Therefore, compared with the MVC application testing, ASP.NET Web Forms-based application testing more difficult to achieve. Moreover, based on an ASP.NET Web Forms application testing requires a Web server. MVC framework allows components to separate and extensive use interfaces, so that individual components can be separated from the rest with the framework for testing. ASP. NET MVC framework is a very high testability light frame, all MVC components are used and can be based on the interface mock objects to test, mock objects are objects to imitate the actual application behavior simulation object. Do not have to run the ASP.NET process controller can be used in the test. Makes testing faster and simple.

b. Using a scalable and can be inserted into the frame. Microsoft ASP.NET MVC framework designed for the purpose of the components can be easily replaced or customize them. We inserted FoxMRP system's own view engine, URL routing policy, action method overloading, and other components. ASP.NET MVC framework also supports the use of Dependency Injection (DI) and Inversion of Control (IOC) container model. DI object can be injected into the class, rather than relying class to create the object itself. IOC specify whether an object to other objects, the first object from the configuration file of the class should be an external source to get the second object. Thus, the test will be more relaxed.

c. Using the powerful routing url rewriting mechanism established more easily understandable and searchable url. After using the MVC than by link files to access the page, but through the controller for routing link.

When you create a new ASP.NET MVC application, the application has been configured to use ASP.NET routing. ASP.NET routing is set in two places. First, in your Web application configuration file (Web.config file) to enable ASP.NET routing. Second, in the application's Global.asax file creates a routing table. Routing tables are created during the Application Start event. When an MVC application is first run, it will call Application_Start () method. This method then calls the RegisterRoutes () method. RegisterRoutes () method creates a routing table. The default route table contains a route (named Default). Default routing the first part of the URL is mapped to the controller, and the second part is mapped to controller actions, the third part is mapped to a parameter called id.

Suppose in the browser's address bar, enter the following URL: / Home/Index/3

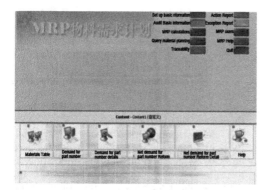

Figure 2. MRP material requirements master page layout.

The default route this URL mapping to the following parameters:

Controller=Home
Action=Index
id=3

When requested URL/Home/Index/3, it will execute this code: HomeController.Index (3).

Default routing contains all three parameters to their default values. If you do not provide the controller, the controller parameter defaults to Home. If you do not provide the action, then the action parameter defaults to the value Index. Finally, if you do not provide id, id parameter defaults to an empty string.

d. Since the ASP.NET page framework to support the existing (. Aspx files), user controls (. Ascx files), and master pages (. master files) markup file tag as a view template, so we will like nested master pages, inline expressions (<% = 02%>), declarative server controls, templates, data binding, localization and other existing ASP. NET features used in conjunction with the ASP. NET MVC framework.

e. Because of its existing ASP.NET program support, so you can use as we MVC forms authentication and Windows authentication, url certification, group management and rules, output, data caching, session, profile, health monitoring, configuration management system and other features.

2.3 *The use of master pages*

If you need to keep multiple Web pages look, you can use master page templates, and easily reuse. In addition, ASP.NET 4 theme can define the appearance of the control's standard features, in order to maintain the unity of the entire Web application.

In the menu, select Add New Item pop up the following window, select the master page, click Add, and then successfully entered the master page. Then the master page layout on the website menu, choose Add New Item, in the pop-up window, select Add Web Form tick in the check box and select the master page options, click Add and select the appropriate master pages. Finally the corresponding master page layout, as shown Figure 2.

First add DIV controls, and to run as a server control DIV control operation. Added code: DIV1. InnerHtml="<iframe width='766' height='306' noscolling=no src='set_bom.aspx'> </iframe>";

Then bind the control to use.

3 CONCLUSION

We develop FoxMRP system has been on Chengdu little fox software company successfully test run, which applies to small and medium manufacturing enterprises, will be further promotion for our enterprise information overtaken contributions.

ACKNOWLEDGEMENTS

This work was supported by Sichuan applied basic research project "based on multi-source information fusion and dynamic cloud computing intelligence service applied research" (project number 2013 JY0059).

REFERENCES

[America] Matthao MacDonald with, Booz studios translated. "ASP.NET4 Advanced Program Design (4th Edition)." Beijing: People's Posts and Telecommunications Press, 2011.

Liu Fu Ying, et al. "Web Cheng compiled practical technical tutorial (2nd Edition)." Beijing: Higher Education Press, 2008.

Liu Fu Ying, et al. "C # Programming Guide (3rd edition)." Beijing: Electronic Industry Press, 2012.

Ye Die. "Enterprise resource planning ERP". Beijing: Electronic Industry Press, 2002.

Control Engineering and Information Systems – Liu (Ed)
© 2015 Taylor & Francis Group, London, ISBN 978-1-138-02685-8

FoxOA with FoxSWF illegal and unhealthy information reporting system SOA-based integration calculation's research and application

J.Y. Li, Y.F. Liu & Y. Liu
Computer Engineering, Chengdu Technological University, Chengdu, Sichuan, China

L. Zhang
Computer Science, Sichuan Top IT Vocational Institute, Chengdu, Sichuan, China

ABSTRACT: Summary of current common OA and specific business systems integration computing applications mostly belonging to a tightly coupled system integration methods, its impact system flexibility scalability, the underlying business processes impede adjustment and optimization. This paper presents a new Service-Oriented Architecture (SOA) of OA (FoxOA) with specific business system (FoxSWF) integrated computing solutions. It can not only ensure that the original system data and logical security, but also to achieve loose coupling between systems to facilitate enterprises and business process reengineering and optimization. It implements the complementary advantages of the two systems, protecting existing IT infrastructure investments.

1 INTRODUCTION

Provincial network monitoring and management office is the provincial network culture construction and management of the organization and coordination of agencies to undertake the supervision and management of the province's cultural information network, the leading network publicity and education, coordination organization, qualification and training of personnel, etc., to develop its website management illegal and unhealthy Information Reporting and Information System is to maintain the normal order of network culture needs.

The author, etc. In recent years, has developed a applicable provincial network culture construction and a management "FoxSWF Site Management and illegal Information Reporting software System" and a "FoxOA office automation system". The former is based on Web technology "FoxSWF provincial network construction and management of information system", mainly consists of work dynamics, policies and regulations, network construction and management, website examined management, illegal Information Reporting Management (Report entrance, reported receiving, condemned exposure, informed recognition, etc.), the website informed management, web mail and other 10 modules, it is for Approval the Website, examined barcode distribution, receiving illegal information reported, to the Website notification (reward punishment), on the overall network culture monitoring and management processes

digitized management; latter FoxOA is to achieve Office Automation (OA) system, consisting of 13 sub-systems: enterprises subsystem, workflow management subsystem, document management subsystem, web conferencing subsystem, daily management subsystem, the administrative office automation conducted a comprehensive management. The two systems are a focused on network culture monitoring and management structures within the flow of data between resources, another focus on workflow approval. They generally operate independently of each other, providing separate functions. But in the Internet culture monitoring and management structures, often some business processes are run through the "Site Management and Information reporting" and OA among the two systems. If FoxSWF new website application process, approval process flow is accomplished by the FoxOA system, complete new website application form, payment, do file is FoxSWF system function, so users have to frequently switch between two systems to complete the new site applications. Internet culture monitoring and management structure in the use of OA workflow approval system, resulting in a lot of business data, and may be FoxSWF system data source, in order to avoid duplication of data and to ensure the uniqueness of the data source, will generate an OA system and FoxSWF system integration needs. In addition, OA system is the biggest characteristic workflow management, it has a powerful workflow customization features, can be adapted to various forms of enterprises and approval forms and process needs, and to meet the multi-level approval structure to support more complex Approval level; while FoxSWF workflow more to achieve the business logic on the data stream, it does not focus on the administrative structure of the approval, so for the inability to get rid of administrative examination and approval structure institutions, FoxSWF software in this regard pales. There, FoxSWF system mode is usually hard-coded into the business process application system overall structure, each business process changes are likely to cause significant changes in the structure of the program, this rigid architecture to increase system complexity and hinder system flexibility. By FoxSWF and OA integration, the use of OA's powerful workflow customization features can solve FoxSWF system problems, Our FoxSWF and FoxOA users are presented these integration requirements.

Current common OA and specific business systems integration computing applications, there are mainly two categories: First, the use of OA and specific business systems to provide access to the database each class of data between the two systems to achieve visits; secondly, through direct. The data for the two systems operating table way to achieve data access between different systems, and data consistency and real-time delivery. These methods are tightly coupled system integration approach. Will affect the flexibility of the system scalability, the underlying business processes hinder adjustment and optimization.

To this end, a new OA and specific business computing application systems integration solutions. It can not only ensure that the original system data and logical security, but also to achieve loose coupling between systems to facilitate enterprises and business process reengineering and optimization. This approach is a Service-Oriented Architecture (SOA) integration of computing methods. It implements the complementary advantages of the two systems, protecting existing IT infrastructure investments.

2 SOA-BASED SYSTEMS INTEGRATION FoxOA WITH FoxSWF THE KEY TECHNICAL COMPUTING APPLICATIONS

2.1 SOA nature and implementation techniques

2.1.1 SOA nature
In recent years, first proposed by the company BEA, derived from the Web Service Oriented Architecture SOA gradually rise into the mainstream of IT. In essence, SOA embodies a new system architecture. In a system based on SOA, application-specific functionality is provided by a number of loosely coupled, and a unified interface definition mode component (ie Service) combined to build up. As the name implies, the service infrastructure is set up to serve as the core infrastructure. It is a new type of enterprise software that helps users to deploy a service-oriented architecture that enables information to flow freely inside and outside the enterprise.

Enterprise management activities can be likened to the image of a social network of communication and coordination, SOA in the enterprise management software application value can be equated to ISO seven-layer of the OSI network model provides. Generally believed: SOA, service-oriented architecture is a component model, it's different functional units application packaged as services (Service), between different services through well-defined interfaces for communication. Interface definition of a neutral manner, independent of specific implementation services's hardware platforms, operating systems and programming languages. Such a unified and standard interface definition (not mandatory bind to specific implementations) feature called loose coupling between services. Suitable for this type of loosely coupled SOA integration computing applications.

2.1.2 SOA implementation techniques
The realization of SOA technologies, including Web Service and Enterprise Service Bus.

As one of the core elements of SOA, service aims to achieve with another telecommunications services, especially to achieve the interoperability of data sharing. The SOA architecture, the aim is to revolutionize the IT systems are built on the original building dedicated to establish a single application becomes more advanced and integrated applications, a distinctive feature of this application is to make full use of existing, can sharing and re-use functions, that is service. Web Services technology implemented using a series of standards and protocols related functions, service providers use WSDL (Web Services Description Language) describes the Web service, using UDDI (Universal Description, Discovery and Integration) to publish and registered agent services registered Web services, service requester via UDDI query to find the required services, the use of SOAP (Simple Object Protocol) to bind, invoke these services. Because given Web service WSDL address URL, in this paper the external WSDL URL provided directly through the corresponding Web Service calls without using the UDDI mechanism.

Enterprise Service Bus (ESB) is the identity of an intermediary service between requesters and service providers, so that the service requestor any service request, then sent to the service bus, the bus will request information from the service provider

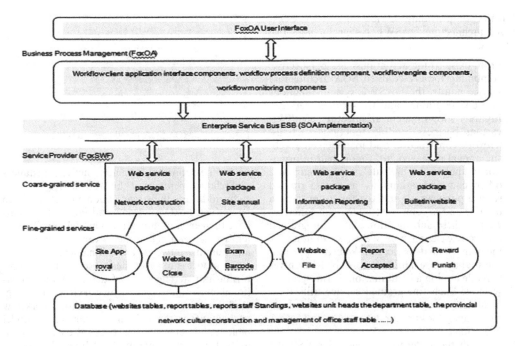

Figure 1. FoxOA SOA-based integration and application framework FoxSWF.

transferred who get the return information, service bus and then passed to the service requester.

2.2 Overall idea integrating computing applications

Enterprises and the flexibility of SOA will bring huge benefits. If the enterprises and IT infrastructure abstraction, its features to form coarse-grained services that come out, each service are clearly expressed its business value, the user can choose these services, regardless of their background to achieve specific technologies. Based on this, Fox-SWF with FoxOA systems integration computing applications can be considered as follows:

FoxOA workflow management subsystem, subsystem includes components of workflow process definition, workflow engine components, workflow client and application interface components and other parts. Workflow process definition is created in the process, the company's actual business process into a computer can process workflow model. The workflow engine is responsible for instantiating the workflow, implementation and management. Monitoring component is responsible for the execution of the workflow management, analysis and control. Workflow client is responsible for human-computer interaction, interface provides workflow execution to help complete the business process execution. The relevant application is

responsible for the application program interface, provides workflow execution software needed to assist in the proper execution of the workflow.

Through the analysis of their actual business processes, extract atomic-level corporate business activities. First through the workflow process definition component of these activities as well as activities related to information, people and events corresponding to integrate Web services unified, and then through the workflow engine components according to the defined business process model for the execution of the business, at the appropriate time activate the corresponding Web service, the parameters passed Web service, access to Web service processing results in order to achieve FoxOA systems and FoxSWF comprehensive system integration. SOA-based integration of applications FoxOA with FoxSWF system architecture is shown in Figure 1.

3 SOA-BASED SYSTEMS INTEGRATION FoxOA WITH FoxSWF COMPUTING APPLICATIONS TO ACHIEVE

3.1 Building SOA-based system loosely coupled FoxSWF

In order to achieve FoxOA integration between the systems with FoxSWF applications must be built with SOA architecture platform allows FoxSWF function of providing services to FoxOA. SOA-based architecture FoxSWF system includes two aspects:

a. From close to the actual business perspective, combining loosely coupled SOA architecture service points, the FoxSWF system functions into coarse-grained and fine-grained services. Figure 1 shows system architecture, FoxSWF various business module functions such as: network construction and management, web-site annual management, report bad information management, website management, and other communications as a coarse-grained service publication, and the function of each module and is composed by the sub-function, we put these sub-functions as a fine-grained service release. Such as Websites examined by the management services on Website approval, site shut down, examined the Website, archives barcode issuing a combination of fine-grained services. Each fine-grained service utilization data access logic components of the data-base table lookup, update, save and other operations.

b. Through the Enterprise Service Bus ESB the scattered centralized management of Web services. When the service requester sends a request message to the service bus, the first agent is sent to the agent service of the service bus, the agent is sent to the business services after receiving the service, business services transferred by it to fur-ther transferred to an external service provider.

3.2 Creating FoxOA integrated Web service workflow environment

Workflow environment to user-defined business processes and FoxSWF associate a Web Service, and to manage and control the business processes running, is run through the OA and FoxSWF system business process logic implementation. It includes process definition and process execution, monitoring two parts.

Process definition previously described. Imple-mentation and monitoring of the process: the workflow process definition deployed to the data-base, the workflow engine components in accord-ance with the process definition document to promote the process flow, and found certain activi-ties need to call the service, it is through the URL to the service request message sent on the bus, Web service bus according to the manager to deal with matters dealt notification workflow engine, and workflow engine to the next process or task; work-flow engine components, Service Bus, Web service needs articulated in workflow management and monitoring services in order to monitor the flow

Figure 2. FoxOA SOA-based user interface.

of business process instances, activity instances and related operation of the Web Service.

In addition, the organization model needs to be unified. FoxOA systems and FoxSWF systems have their own organizational model. FoxOA sys-tem's organizational model is to serve enterprises and administrative organizational level, FoxSWF tissue model is to serve the Internet culture moni-toring and management structure at the opera-tional level. In the process of workflow modeling, workflow process activities executor (workflow participant) is the reference organization model building. Therefore, the two systems must be unified organizational model. We use the organi-zational model for OA system redefined method to increase level number of the organizational structure of OA system, the FoxSWF users and OA users are in the new organizational structure reflected.

3.3 Accessing control mechanism called

Workflow engine called FoxSWF's Web service, the need for authentication; authenticated users can call the Web service interface methods.

This article by FoxSWF authentication Web service interface method implementation calls the Web service when access control. In the process the form input FoxSWF system's user name and password, via SOAP request message is passed to the authentication Web service, as an input parameter.

In the first visit to the Web service is required for authentication, then you can pass from the Session way to obtain continuous access to user informa-tion, until you exit the system or Session Timeout.

4 CONCLUSION

SOA-based systems integration FoxOA with Fox-SWF computing applications, facilitate network monitoring and management structures, cultural reorganization and optimization of business proc-esses, improve the ability to monitor the site, with a universal value.

ACKNOWLEDGEMENTS

This work was supported by Sichuan applied basic research project "based on multi-source information fusion and dynamic cloud computing intelligence service applied research" (project number 2013JY0059).

REFERENCES

Internet culture construction and management offices. Illegal and unhealthy information platform system development requirements. 2011.

Liu Fu Ying, et al. "Java EE Web programming tutorials." Beijing: Electronic Industry Press, 2010.

Liu Fu Ying, et al. FoxSWF illegal and unhealthy information platform for research. Chengdu Technological University Department of Computer Science, 2012.

Liu Yan, Wu Jian. SOA-based integration of OA and ERP applications of Computer Applications, 2008 (28) 3.

Control Engineering and Information Systems – Liu (Ed)
© 2015 Taylor & Francis Group, London, ISBN 978-1-138-02685-8

Research of parallel program performance analysis and optimization on multicore platform

Y.Y. Teng & Z.G. He
Department of Computer Science and Technology, Dalian Neusoft University of Information, Dalian, Liaoning Province, China

ABSTRACT: Parallel program can give full play to the computing power of multicore processors to meet user demand for high performance. With the number of cores increases, the currency of program can be increasingly higher and higher. Efficient parallel programs have become a focus of the industry. The paper focuses on the key issues in parallel program optimization, proposed cycles and methods of the parallel program performance analysis and optimization. Finally, gave the optimization process and optimize the results in case of Monte Carlo calculation of π.

1 INTRODUCTION

More independent execution core embedded into a single package on multi-core processor, each execution core has its own execution units and architecture resources, which is the material basis to achieve true parallelism. In the multicore era, how to design and develop parallel programs in order to give full play to the resource advantages of multi-core architecture? This is the problem to be solved by software developers.

Parallel program performance-related issues will focus around three aspects:

1. Quantitative indicators of measuring performance;
2. Measure program and obtain performance indicators;
3. Locate the program bottlenecks and improve program performance.

2 MULTI-CORE PARALLEL TECHNIQUES TYPES PARALLEL PROGRAM OPTIMIZATION

The task is divided into many different sub-tasks, and then deal with these sub-tasks simultaneously, can significantly improve program performance. In the sub-tasks completely unrelated case, improving performance is very evident. But most is not the case, quantifying the performance of parallel programs is an intuitive way.

Therefore running time (speedup) and parallel efficiency is very important performance indicators for parallel program. These two indicators complement each other and mutual restraint.

From the point of run-time optimization of performance, mainly consider reducing the critical path time, calculating portion is concern. From the perspective of improving the efficiency of the parallel optimize performance, the main consideration is to optimize the parallel algorithm and the associated implementation. Reducing the computational part can improve overall performance, fewer running time, but it may affect the scalability of the program, reducing the parallel efficiency. When we get higher performance by increasing speedup and efficiency, it is need to reduce overhead of solving data conflicts and synchronization in addition to improving the algorithm parallelism and data communications.

The following aspects will be considered to improve parallel program performance.

2.1 Parallel algorithm

For parallel application, consider the problem of parallel algorithms from the outset. Decomposition mode is to choose the data decomposition, task decomposition or data flow decomposition.

2.2 Parallel level

Determine the parallel level to optimize performance. As far as possible parallelize code on top, according to parallelism principle of using coarse-grained whenever possible.

2.3 FAQ of parallel application

Developers should pay attention to FAQ parallel application. These issues include whether the program scalability are still far away from the upper limit of Amdahl's law, whether parallel through the small, the program is to achieve load balancing, particle size is moderate, parallelization overhead and benefits issues.

2.4 *Special performance problems*

These issues include synchronization, data competition, how to lock, etc., but also include consideration and use of the performance of different API and different threads library (OpenMP, Win-Thread or TBB, etc.).

3 PARALLEL PROGRAM DEVELOPMENT AND OPTIMIZATION CYCLE

And traditional application development process, the parallel programs follow the same software development life cycle. Program optimization is the process of iterative process. Each iteration cycle is divided into the analysis, design, implementation, correctness checking and optimization.

3.1 *Analysis phase*

The main objective of the analysis phase is to find the part of the code, which part of the code from multi-threaded parallel gain most.

Analysis program bottlenecks and find intensive computing code regions. Analysis and determine whether the performance of the region can be improved by multi-threaded. This is the breakthrough to improve performance by multi-thread.

Amdahl's law states that reducing the proportion of the serial code can improve program performance. While reducing the proportion of the serial code, it is also necessary to determine which enhance the performance should be much greater than the multi-threaded system overhead.

From the perspective of load balance and granularity, large data blocks less dependent on the situation are suitable to optimize by multi-threading.

3.2 *Design and implementation*

The main task of the design phase is to modify the computationally intensive part of program to accommodate the multi-threaded mode. These sections are effectively broken down into several sub-problems and then designed into multiple threads. Multiple threads perform their duties, solve different problems, complete different tasks and improve performance. Common decomposition model includes task decomposition, data decomposition and data stream decomposition. Task decomposition is a decomposition method that the problem is divided into multiple functional modules which executed by different threads. Data decomposition is a decomposition method that the data to be processed is divided into a number of different data blocks which executed by different threads. Data stream decomposition is a decomposition method that a complex operation is divided into several simpler sub-operations which executed according to a certain sequence.

A lot of problems are needed to be considered in the design phase, such as whether several tasks decomposed can be performed independently, whether resources race and sharing can be effectively solved, whether the program paralleled meets the performance requirements and so on.

Model elements will be designed into real code after selected the appropriate threading model in implementation phase. Common programming model includes the Win32 API, Pthreads, OpenMP, TBB, etc. At the stage a series of parallel program-specific issues are need to face such as resource sharing and mutual exclusion, inter-thread synchronization and mutual exclusion, deadlock, which are inevitable in the implementation process.

3.3 *Check and debug correctness*

Using multiple threads to complete the application increases the risk of error. It is necessary to determine the correctness of programs.

Thread errors and memory errors are common mistakes for multithreaded programs. Threading errors and memory errors are common mistakes for multithreaded programs. Thread errors include thread deadlock and thread competition. These errors are caused by resources competition during threads operation. For most parallel software developers threads-based or OpenMP-based, these errors of simple code could easily be avoided. But for large-scale parallel tasks, some small inconspicuous variables will very easily lead to threads errors as resources, the algorithm complexity and uncertainty during real-time computing. And these errors are also very easy to be ignored by software developers in parallel, which requires special attention.

Memory errors include memory leaks, illegal memory access, read and write errors, etc. These errors mostly due to different threads unintended operation on the same address segment during thread running.

The exceptions will be found by performed many time because of parallelism of multithreaded programs on multi-core platform. Therefore developers need a static analysis capability, and can use software tools to find the error reasons.

3.4 *Performance optimization*

Any application performance optimization work is best completed with non-invasive tools, so it can provide more accurate facts. Multi-core parallel program should also follow this principle. Only more accurately know the actual behavior of the program, application developers can locate inefficient code and improved the code in the program.

By monitoring multithreaded programs and capturing the performance data, relevant information of program optimization will be obtained. First, locate hotspot functions and codes. Hotspot is the most time-consuming code area which is the most effective area to be paralleled. Second, analysis concurrency and find codes no effective use of processor resources. Third, the lock and wait for analysis. Find out where the program waits and determine the reason for waiting is I/O operation or synchronization. By analysis of locks and waits, it can be quantified for each synchronization operation to get the performance impact. Finally, compare analysis. There are many compared ways, such as comparing efficiency and performance of different parallel algorithms, comparing efficiency and performance of different the number of threads.

4 ANALYSIS AND OPTIMIZATION OF TYPICAL CASES

Monte Carlo method, also known as stochastic simulation method, is a calculation method based on probability and statistics theory. The basic idea of Monte Carlo is to estimate probability of random events by the frequency of occurrence.

4.1 *Algorithm description*

Shown in Figure 1, the unit circle inscribed in a square of side length 1, the circular area = $\pi/4$. Assuming randomly cast point to square area, the probability of point falls within the circle p = area of a circle/square area = $\pi/4$.

Let point coordinates (x, y), x and y is random number from 0 to 1. If the distance from point (x, y) to the center is less than the radius of the circle, the point falls inside the circle. Let n experiments, the number of random point falls within the circle is m, the frequency is m/n. When n is large, π is approximately equal to 4 * m/n.

Using C language to achieve the above algorithm, core code is shown below.

Code 1:
```
for(i=0;i<N;i++)
```

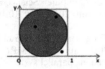

Figure 1.   Monte Carlo algorithm description chart.

```
{
    x=(rand()/(RAND_MAX+1.0));
//X Coordinate
    y=(rand()/(RAND_MAX+1.0));
//Y Coordinate
    if((pow((x–0.5),2.0)+pow((y-0.5),2.0))<=0.25){
        m=m+1;
}
pi=m/(double) N* 4;
```

Randomly cast 2*108 points, π is approximately equal to 3.141368380. Estimated time is 34.66 seconds.

In the Code 1 formula (x-0.5)2+(y-0.5)2≤0.25 is optimized to x*x + y*y <= 1.0. The core code is shown below.

Code 2:
```
for(i=0;i<N;i++)
{
    x=(rand()/(RAND_MAX+1.0));
//X Coordinate
    y=(rand()/(RAND_MAX+1.0));
//Y Coordinate
if (x*x + y*y <= 1.0){
    m=m+1;
}
pi=m/(double) N* 4;
```

Randomly cast 2*108 points, π is approximately equal to 3.141634740. Estimated time is 14.27 seconds.

4.2 *Parallel algorithm*

Through observation and analysis, hotspots of program concentrate in a loop. Insert one OpenMP pragmas before the loop starts to a multi-threaded execution and run program in a multithreaded manner. In parallel threads, pay attention to private and shared data. Use reduction to calculate summation. Parallel code is shown below.

Code 3:
```
#pragma omp parallel for private(x,y)
reduction(+:m)
for(i=0;i<N;i++)
{
    x=(rand()/(RAND_MAX+1.0));
//X Coordinate
    y=(rand()/(RAND_MAX+1.0));
//Y Coordinate
    if((pow((x-0.5),2.0)+pow((y-0.5),2.0))<= 0.25){
        m=m+1;
}
pi=m/(double) N* 4;   }
```

Randomly cast 2*108 points, π is approximately equal to 3.141548240. Estimated time is 5.5242 seconds.

4.3 Algorithm optimization

In the Monte Carlo parallel algorithm design, the use of Intel Math Kernel Library (MKL) for the sample immediately. MKL supports random number generator. Parallel code is shown below.

Code 4:
```
#pragma omp parallel private(x,y,i) reduction(+:m)
{
double r[BLOCK_SIZE*2];
    VSLStreamStatePtr stream;
vslNewStream(&stream, BRNG, (int)clock());
#pragma omp for
    for(j=0;j<iter/BLOCK_SIZE;j++) {
            vdRngUniform(METHOD, stream,
BLOCK_SIZE*2, r, 0.0, 1.0);
    for (i=0;i< BLOCK_SIZE;i++) {
            x=r[i]; //X Coordinate
y=r[i+BLOCK_SIZE]; //Y Coordinate
            if (x*x + y*y <= 1.0) {
                m++;
        }
    }
}
    vslDeleteStream(&stream);
}
 pi=m/(double) N * 4;
```

Randomly cast 2*108 points, π is approximately equal to 3.141507280. Estimated time is 2.549 seconds.

4.4 Analysis

The code comparison result is as shown in Table 1. From the data in the table, it can be seen that the running time is shorter and shorter and π values higher and higher precision after optimized.

The comparison result of serial code and parallel code is as shown in Table 2. Speedup has achieved 2.647. With the increasing number of loop iterations, π deviation is smaller and smaller.

Table 1. Test data of Monte Carlo.

Code	Runtime (second)	π value	Deviation
Code 1	34.66	3.141368380	0.000224274
Code 2	14.27	3.141634740	0.000042086
Code 3	5.5242	3.141548240	0.000054514
Code 4	0.963	3.141507280	0.000029816

Table 2. Test data of Monte Carlo.

Iterations (million)	π value	Serial time	Parallel time	Speed up
200	3.141507280	2.549	0.963	2.647
2000	3.141622460	21.915	9.354	2.342

5 CONCLUSION

With the traditional single-core processors, multi-core processor brings more parallel processing capabilities. How to write efficient parallel programs to take advantage of parallel computing advantage of multicore processors, which is to be solved by parallel program developers.

The process and method of parallel program performance analysis and optimization was proposed in the paper. OpenMP is used for Monte Carlo algorithm to calculate π and achieve better optimization results.

The experimental results showed that:

• With the increase in the number of loop iterations, the precision of π is increased highly. It fully demonstrates the effectiveness of the optimization process.
• In the optimization process, apart from relying on the experience of the developer, using a certain degree of parallel programming tools will make parallel program optimization process easier and more efficient.

REFERENCES

Cao Zhebo, Li Qing, "Research and design of parallel programming model on multi-core", Computer Engineering and Design, 2010, 31(13): 2999–3002.

Huang Hualin, Zhong Cheng, "Parallel program performance optimization for data-intensive applications on multi-core clusters". Computer Engineering and Applications, 2012, 48(30): 73–77.

Shengwei Peng, "Parallel Genetic Algorithm on Multi-Objective Scheduling Problem Subjected to Special Process Constraint", Proceedings of 2012 IEEE International Conference on Engineering Technology and Economic Management (ICETEM2012), IEEE Press, May 2012.

U E., Tian X., Girkar M., et al, Compiler Support for the Workqueuing Execution Model for Intel SMP Architectures, In Proceeding of the Fourth European Workshop on OpenMP, Roma, Sep 2002.

Weiming Zhou, "Multi-core Computing and Programming", Huazhong University Press, March 2009, pp. 19–124.

Wu Chao-Chin, Lai Lien-Fu, Yang Chao-Tung, et al. "Using hybrid MPI and OpenMP programming to optimize communications in parallel loop self-scheduling schemes formulticore PC clusters", Journal of Supercomputing, 2012, 60(1): 31–61.

Wu Y., Kumar A. "A parallel interval computation model for global optimization with automatic load balancing", Journal of Computer Science and Technology 27(4): 744–753, July 2012, doi: 10.1007/s11390-012-1260-x.

Yang Xuejun, Tang Tao, Wang Guibin, Jia Jia, Xu XinHai, "MPtostream: an OpenMP compiler for CPU-GPU heterogeneous parallel systems", Science China (Information Sciences), Vol. 55 No. 9: 1961–1971, Sep 2012, doi: 10.1007/s11432-011-4342-4.

Processes for delivering cloud computing resources

C.L. Li

College of Computer Science and Information Technology, Guangxi Normal University, Guilin, China

L. Li & M.M. Tao

Luzhai Vocational Education Center, Liuzhou, China

ABSTRACT: Cloud Computing is considered as a new delivering way for the IT services. It has its supply chain to deliver resources to customers. The paper analyses three resources delivering processes for the Cloud Computing services and how they ensure their quality of services. The direct delivering process is the provider to the end customers. The provider is always a big enterprise. It builds the centralized data centre to offer resources to customers. It signs service level agreements to ensure customers' quality of services. The other two are indirect delivering processes among customers or among providers, such as a customer offers resources to the other customers, or the enterprise first buys the resources from other enterprises and then offers to its objective users.

1 INTRODUCTION

In the Cloud Computing environment, there must be several providers provide kinds of resources to customers, such as computing resources, storage resources and applications resources, and so on. Just as any other resource has its supply chain, the Cloud Computing resources also have their supply chains, shown in Figure 1 [1].

In Figure 1, at the bottom, it's the massive, distributed data center infrastructure connected by IP networks and is called IaaS (Infrastructure-as-a-Service), which is a flexible infrastructure of distributed data center services connected via Internet-style networking. At the middle, it's Internet-scale software development platform and runtime environment and is considered as PaaS (Platform–as-a-Service), which provides the runtime environment for Cloud applications. At the top, it's IT as a Service (ITaaS) and provides on-demand services to users in the form of SaaS (Software-as-a-service). The infrastructure resources, platform resources, and software resources are delivered over the Internet from centralized data centres in the ways of on-demand and per-pay-per-use. It means that users can expand and downsize those services whenever they want and how much they pay [2].

In the supply chains, actors are the providers, the developers and the end customers. If we abstract the developers as providers, the relationships between the actors are providers to the end customers, either direct or indirect. The direct is a provider offer resources to an end customer. The indirect may first interchange internally between providers or customers, and then to the end customers. For example, the platform provider offers the platform to the developer to write application programs for the end customers. Platform resources are interchanging between the platform provider and the developer, and change into the application software for the end customers at the end. The other example is customers may offer their unused resources to the Cloud Control Centre in order that the other customers can use. At the moment, the Cloud Control Centre is an intermediate; the customers offered unused resources are the real providers. So we also see them as providers and the relationship is between the provided customers to the end customer.

Based on the relationships and their processes, this paper describes the resources delivering

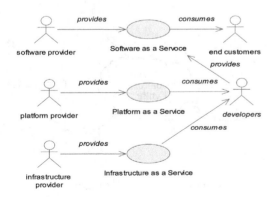

Figure 1. The supply chains of the cloud computing.

processes for Cloud Computing, including the provider to customer process, the provider to provider process and the customer to customer process. They not only can help providers pertinently provide resources to meet the needs of customers, but also help customers search the needed resources quickly and accurately.

The rest of this paper is organized as follows: section 2 analyses the provider to customer process, the customer to customer process is in section 3 and provider to provider process is in section 4 Section 5 are the conclusions.

2 PROVIDER TO CUSTOMER PROCESS

The provider to customer delivering process (shorted as P2C process) is which resources are provided by providers to different customers. Cloud Computing 's three types involving Private Cloud, Public Cloud and Hybrid Cloud belong to this process. Each Cloud is built by a large enterprise to provide a large resources pool for customers, shown in Figure 2.

But the customers for different Cloud are different. The customer for Private Cloud is enterprise itself. That means the enterprise built the Cloud only for it internal employees. Public Cloud is offered to public customers. Hybrid Cloud synthesizes the characters of the Private Cloud and Public Cloud and partial for interior and the other for public customers.

Delivering resources are always relating to the conditions because the resources or services are cost. There are some pricing models for providers to sell their resources or services in the P2C process, such as tiered pricing model, pay-per-use model, based-on-subscription model and dynamic pricing strategy, and so on. On one hand, providers get the money when they deliver resources or services to customers. On the other hand, customers can be ensured the quality of resources and service through SLAs (Service Level Agreements). An SLA is an agreement about the quality of services which is negotiated to sign between providers and customers [3]. It sets the expectations of the two parties and helps define the relationship between them. It is the cornerstone of how the service provider sets and maintains commitments to the service consumer, in which it defines the service's type, content, quality and payment, such as the bandwidth, memory, and CPU and others related to the performance [4]. An SLA quantifies the service quality measures and makes clearly some levels of service quality in order to avoid effectively misunderstandings and help to build the reputation mechanism of providers. While the resources and services offered by providers do not reach level maintained in the SLA, customers can ask to appropriate compensation. Furthermore, an SLA reflects the on-demand computing concepts when customers choose their own level terms.

For examples, Microsoft Windows Azure sets its series Cloud SLAs, including Microsoft Windows Azure storage SLA, Microsoft Windows Azure compute SAL, Microsoft Windows Azure platform App Fabric Access Control SLA, Microsoft SQL Azure SLA and Microsoft Windows Azure CDN SLA. Amazon also has SLAs for its Cloud services. For the services of Amazon S3, there is Amazon S3 SLA. For the Amazon Web Services, there is Amazon Web Services Customer Agreement. And Google Apps uses Google Apps SLA to make clearly the level for the offers and the exclusions.

3 CUSTOMER TO CUSTOMER PROCESS

In the delivering process of customer to customer (shorted as C2C process), customers as providers offer their unused resources to the Cloud Control Centre, the process is shown in Figure 3.

Figure 3 shows that customers from different places offer their idle resources to the Cloud Control Centre. The Centre integrates the distributed

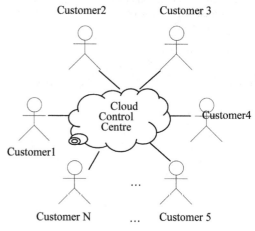

Figure 2. Resources delivering in P2C process.

Figure 3. Resources delivering in C2C process.

resources and builds a virtual and community-style data centre as a resources pool by the virtualized technology. It acts as an intermediary among customers. Other customers can apply those resources accessing the pool through the internet if they need. A customer may be a provider because they submit some resources. At the same time, he may be a customer because he use some resources which is offered by the other customers. In fact, a customer may be a person, or a company, or an organization.

It is a line in Figure 3 rather than an unidirectional arrow in Figure 2 indicating that the interactions are bidirectional. This is to say while a customer provides their resources they can also use the Cloud resource provided by other customers at the same time. In the C2C interactive pattern, resources and services are transferred among different customers groups.

Compared to the centralized data pool in the P2C process, the resources are central controlled and managed in a centralized data pool by enterprise providers. They will be ceased while the providers withdraw or close those services. The C2C process overcomes this disadvantage. It not only facilitates the resources sharing among customers, but also avoids from the risks of building large-scale centralized data centres which increase the resources coefficient of utilization. There is a great contribution on energy saving and environment protection using the existed resources in the C2C process. This Cloud Computing environment which builds by virtualized distributed clients' resources in the network is considered to be the Community Cloud Computing, shorted as Community Cloud. The virtual data centre formed by Community Cloud is a community [5]. So the C2C process in delivering Cloud resources can be named as Community Cloud.

The Community Cloud takes the free resources in the distributed personal computers place of the providers' centralized resources. There are kinds of roles in numbers of distributed computers and resources are set in and out the community at the same time. Therefore, some characters for the Community Cloud can be attributed as follows:

3.1 Open

The Community Cloud is not only independence on providers, but also liberates the parties for offering resources, which kinds of customers can upload their free resources to the Community. No mention to the customers' identity, culture background, education level, location, and wealthy or not, the resources' provision and use are open for all. You just need to register in the Community and get an account. If you login in, there are lots of resources or services ready to you.

3.2 Reliable

The whole Community owns the infrastructure of Community Cloud and no risk in competition weakened and innovation constrained. Because Community Cloud has numbers of nodes for offering resources, it is rust enough to free from fault avoiding the impact of someone's serving time and unusual circumstance. Even though a node does not enable to work, there are thousands of others to make up for the failure. In addition, the idle resources in customers' computers reach the sustainability of the environment.

3.3 Voluntary

Every one has its own actions according to its will in the Community. This is to say it is voluntary to offer or use resources because of pleasures or interests. When some one wants to offer his resources to the Community, he can upload them. When he does not want to offer resources again, he can stop offering. But those Thousands of others can also access them regardless of the locations. This is from the resources offering perspective. From the resources utilization perspective, some one who has an account in the Community, he can access the resources that he is affordable.

3.4 Affordable

The Community organizes those idle resources in order to use them repeatedly. There is no need to spend lots money in building the large-scale data centre and it makes the use of resources cheaper and more affordable. The Community makes use of the community currency to stimulate its members to share and interchange the resources. The currency may be virtual or real money. Who makes the greater resources contributions, the more community currencies he has, and the more resources he can affordable.

For example, YouTube (http://www.youtube.com) is one of the instances of Community Cloud, on which people can search, watch, upload, download or share videos. It is said that it has created huge of benefits and has greatly improved the scalable performance with a flexible capital funding model. The members are individuals, organizations, companies, and so on. Many international companies like CBS, BBC, Universal Music Group, Sony Music Group, Warner Music Group and NBA offer their videos contents in YouTube. YouTube provides a set of APIs development tools for them to integrate YouTube in their web so it conversely promotes YouTube's publicity and prevalence. There are other instances of Community Cloud such as Wikipedia and Facebook. All benefit from the community delivering process which apportions the cost of services to numbers of users in communities, reducing the pressures for new companies to enter the market.

4 PROVIDER TO PROVIDER PROCESS

The C2C process improves the efficiency of resources utilized among customers. It is the same to providers. When providers exchange their resources in the Cloud environment, Clouds are integrated into a cloud network and form a provider to provider delivering process (shorted as P2P process). The cloud network is considered to be the InterCloud. Its components are considered to be the Autonomous Clouds. Therefore, the P2P process can also be called InterCloud Computing (also as InterCloud) [6]. It means that the InterCloud is composed of several Autonomous Clouds, which are taken as the nodes of the network. A Autonomous Cloud is as a Cloud provided by an enterprise who owns the right for controlling the resources, such as computing, storage, network, application, and so on. But the delivery orders are sent by the InterCloud's manager.

As previously mentioned, three important components make up the InterCloud: Autonomous Clouds, Cloud Resource Manager (shorted as CRM) of Autonomous Clouds and Inter-Cloud Resource Manager (shorted as IRM). The resources delivering process is shown in Figure 4.

The CRM A and the CRM B are considered to be a resource manager of Cloud A and CRM B respectively. The CRM is responsible for the resources allocation and provisioning at the Autonomous Cloud level. Result A and Result B show the request resources results from Cloud A and Cloud B respectively. When CRM A monitors that Cloud A can not meet the customer's requirement, it will ask for help to IRM. And the Inter-Cloud delivering resources happen and the process is as below.

A customer firstly submits his resources allocation request to his local CRM A. When CRM A receives the request it searches resources in Cloud A to meet the customer's request. If Cloud A has those resources to meet the needs, it will pack the appropriate resources for Result A and return to the customer and ask for a fee. If there is on needed resources in Cloud A, CRM A will send the request to ask for help from the IRM. According to the content IRM finds relevant resources in cloud network (for instances, the right resources are found in Cloud B), then send the request to CRM B which

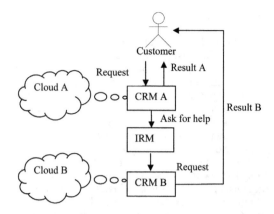

Figure 4. Resources delivering in P2P process.

returns the Result B to the customer. The customer uses the resources and pays for them.

For the perspective of the customer, he sends a request to a Cloud, and does obtain the right result from the Cloud. He takes for granted that the result is from the Cloud, but it is not the case. The source of the result is not visible for him, and it is no significance for him to notice this. The IRM is the tie of Autonomous Clouds and transfers information among them. Its responsibilities can be attributed to the followings [7]:

a. Responsibility to allocate resources at the Autonomous Clouds level

When a given CRM asks for help, IRM must be timely and accurately to allocate Cloud B to provide the right resources. It also provides Clouds selection capabilities to choose the trustworthy Cloud by the pre-defined prioritized mechanisms when there are multiple Clouds have the right results at the same time.

b. Responsibility to mediate the resources exchange peeing InterClouds

If IRM A can not allocate the suitable resources in InterCloud A, it must turn to IRM B to ask for help in InterCloud B. And there must be some terms of the peering agreements with InterClouds and some types of resources that the peering InterClouds can acquire from one another. So that IRM is able to manage requests of other IRM, otherwise it is unable to ask for help to meet the customers' requests.

c. Responsibility to monitor the execution of applications across multiple Autonomous Clouds

d. Responsibility to coordinate charges of InterCloud.

The IRM interacts with other entities including accounting systems that provide information on shares consumed by the peering InterClouds.

Although the InterCloud shows so good, their so many issues need to be solve in practice, building InterCloud through the P2P delivering process to share resources is just feasibility in theory. It is no instance for InterCloud so far. Integrating the Autonomous Clouds is so hard. They are independence for each other. It is not easy to standardize new technologies and business concepts. Different hardware, infrastructures and applications among Autonomous Clouds enterprises make standardization difficult. But the standardization is necessary for ensuring interoperability. Interoperability plays a crucial role for many enterprises. It shares different resources and reduces costs of the workforce and operation management. It also simplifies the exchange and re-use of resources both internally and externally. Luckily, the Internet is one of the successful examples which are some protocols and standards for the interfaces in order to realize the interaction. It is same to InterCloud. Building standardized interfaces is necessary for the Inter-Cloud interoperation, including the general policies, regulations and the ask-for-help services.

5 CONCLUSIONS

In the resources delivering processes, the emphasis is on the resources flowing and transferring between the providers and the customers. This article introduced three resources delivering processes which are the P2C process, the C2C process and the P2P process. Those delivering processes can help us to understand the Cloud Computing services. However, it will confront with many problems in practical. For example, in the P2P process, when Autonomous Cloud A asks to interchange resources with Autonomous Cloud B, there will be a fee problem. If the standard for the price is different between the two Autonomous Clouds, customers will doubt the fees when they pay for the services. And there will be some misunderstandings and contradictions. So we must do more research to deliver resources seamlessly in the Cloud Computing environment and solve them complementary with the form of policy and standards abided by providers and customers in order to maximize the share of Cloud resources.

ACKNOWLEDGMENTS

This work was partially supported by the general project in humanities and social sciences research base on the Ministry of Education of China under Grant 13YJC870012, and the general project in social science university humanity of Guangxi Province under Grant SK13YB015.

REFERENCES

Abawajy, J. 2009. Determining Service Trustworthiness in Intercloud Computing Environments. *IEEE computer society:* 784–788.
Briscoe, G. & Marinos, A. 2009. Digital Ecosystems in the Clouds: Towards Community Cloud Computing. *2009 3rd IEEE International Conference on Digital Ecosystems and Technologies:* 103–108.
Buyya, R. & Yeo, C.S. et al. 2008. Market-Oriented Cloud Computing: Vision, Hype, and Reality for Delivering IT Services as Computing Utilities. *IEEE Computer Society:* 5–13.
David, B. & Ludvigson, E. et al. 2009. Blueprint for the Intercloud-Protocols and Formats for Cloud Computing Interoperability. *ICIW/IEEE:* 328–336.
Lin, G. & Fu, D. et al. 2009. Cloud Computing: IT as a Service. *IEEE Computer Society:* 10–13.
Marinos, A. & Briscoe, G. 2009. Community Cloud Computing. *CloudCom:* 472–484.
Vivek, N. & Rami, B. et al. 2009. Self-Optimizing Architecture for Ensuring Quality Attributes in the Cloud. *Joint Working IEEE/IFIP Conference on Software Architecture/European Conference on Software Architecture:* 281–284.

A new model based on the LINGO solving for the data rectification

L.P. Wang, Q.X. Zhang & L. Yu
Shenyang Aerospace University, Shenyang, China

ABSTRACT: A new data correction model is presented based on the analysis of classical models. This mode need not identify the significant error before solving. The problem has been solved that the results of the classical solution model is away from the true values when significant error is not completely removed, and the solving steps is reduced to improve the solution accuracy. In this paper, the solution process is given based on the optimization model solver (LINGO), and the feasibility of the model is verified using the industrial data.

1 INTRODUCTION

In the process of industrial production process, because the process measuring data are easily affected by instrument deviation and even instrument damage, the measurement data contain a variety of errors. Process measuring data, however, as a reflection of the characteristics of the device running status information, are to realize the computer process control, simulation, optimization, and the basic foundation of production and management. Inaccurate and incomplete data will lead to many process optimizations, simulation and control can not effectively play a role, even a deviation for the decision. Therefore, the data must be calibrated in order to get real reaction industrial production conditions. Data correction aims to use plant all kinds of redundant information to identify and eliminate the error of the measurement data, eventually get closer to the real value of the calibration data.

2 CLASSICAL MODEL FOR DATA CORRECTION TECHNIQUES

In 1961 Kuehn and Davidson firstly proposed the concept of data coordination in the process of adjusting pipe flow and measuring of the temperature of the crude oil distillation in order to meet the material and energy balance. The rule is to make the sum of the squares of the deviation value between coordination degree and its corresponding measurement to be minimal under the condition of meeting the material balance and heat balance.

Measured value can be represented by the mathematical model:

$$\bar{x} = x - e \tag{1}$$

where $x(n \times 1)$ is the measured value of the variable, $\bar{x}(n \times 1)$ is the real value of the variable measured, and $e(n \times 1)$ is the measured values error.

The vector $u(m \times 1)$ is introduced to represent unmeasured variables (such as due to the instrument damage caused the loss of data or measurement cost is too high and did not design a measuring pipeline, its production data classed in the unmeasured variables), then the constraint conditions can be expressed as:

$$F(x,u) = 0 \tag{2}$$

For linear equality constraints, (2) can be expressed as:

$$Ax + Bu + c = 0 \tag{3}$$

where, A and B are the coefficient matrices of linear equality constraint equation, c is a constant vector of linear equality constraint equations.

Linear data coordination model of the system can be represented as:

$$\min(\bar{x} - x)^T Q^{-1}(\bar{x} - x) \tag{4}$$

$$S.t. \ A\bar{x} + Bu + c = 0 \tag{5}$$

where, the Q is covariance matrix of the system.

Yuan Yonggen shows the optimal solution of the problem:

$$\bar{u} = -[B^T(AQA^T)^{-1}B]^{-1}B^T(AQA^T)^{-1}(Ax+c) \tag{6}$$

$$\bar{x} = x - QA^T(AQA^T)^{-1}(Ax + B\bar{u} + c) \tag{7}$$

The model using the constraint relations of production makes the measurement value effectively move closer to the true value. But the classical calibration data model supposes measured variables only containing random errors. Therefore, gross error detection problem must be considered in the process of data correction. Although after gross error recognition theory is put forward, more than ten kinds of methods has already been used for the identification and detection of gross errors. The effect is remarkable, but it cannot guarantee to identify all significant error every time. Once the phenomenon of significant unrecognized error occurred, these significant errors will be passed to other variables through the data model, thus the correction result will deviate from the true value.

3 NEW MODEL DATA CORRECTION BASED ON LINGO METHOD

3.1 The presenting of model

The whole process of data correction, such as determination of data reconciliation purpose, the judgment of measuring data redundancy, solution of the correction and so on are all based on a reasonable mathematical model. Reasonable and strong fault tolerance mathematical model plays a key role in data reconciliation, gross error recognition and detection, calibration statistics estimation. Material balance model is based on the conservation of mass, chemical equilibrium and other production relations. Under normal circumstances, the mathematical model is built by the several methods:

1. Mechanism analysis: A series of equations can be got using the existing theory knowledge and can simulate the characteristics and changing trend of the object obtained.
2. Experience method: The appropriate equation is selected according to the characteristics of the research object, and the equation coefficients are obtained using actual measurement data.
3. The combination of above two kinds: The key features of the equation are determined through the theoretical study, and the equation coefficients are determined using actual measurement data.

Data correction on material balance model is based on the classical calibration model. The reliability of calibration data is improved by the variable divided into different classes, scale reduction, improving the gross error recognition rate, but it does not consider the process of establishing model. Establish proper model plays a decisive role not only in the theoretical study, but also in real application process.

The various algorithms about classical data coordination model are based on a hypothesis, which assumes that all measurement data only obey

the normal distribution of random errors, or the measured values with significant error have been identified and removed. Using classical model, however, the bad value (the measured variable with significant error) is not very complete, as a result, the bad value did not been detected, infect other measurements, and produce the deviation from the real value. Therefore, this paper puts forward a new objective function to overcome the shortcomings of classical model.

3.2 The establishment of model

According to the definition of data correction, combined with the classical model of data correction techniques, the absolute value of the difference between the measured data and the calibration data is the minimum in the condition of the measured data in the steady state and the correct data satisfying the premise of material balance. The absolute value sum replaces the sum of squares of the classical model, to prevent bad values affecting other data of measured data, the model is presented as follows:

$$\min \sum_{i=1}^{n} \left| (\bar{x}_i - x_i) \right| \cdot w_i \tag{8}$$

$$S.t. \ F(\bar{x}) = 0 \tag{9}$$

where, n is the number of industrial data corrected, is the ith industrial data corrected x_i is the ith industry data measured, ω_i is weights of influence coefficient between the industrial data changes and the target.

It has been verified by the experimental data, this method is effective and feasible. Due to least squares solution is no longer applies to new model of data correction in changing objective function of the classic model, a new solution method need be produced to got the optimal solution in this paper.

3.3 The learning algorithm of weight w

The choice of weights plays a decisive role on the model calibration results mentioned above, through (8), it can be seen that weight deviation function is to adjust the corresponding variables impact on target: the lower the accuracy of measurement variables are, the smaller weights are. On the other hand, the measured variable precision is higher, the corresponding weights is the bigger. Therefore the weight has a direct connection to the variable precision of measurement. For the system model corrected firstly, assume the weight ω is given as followed:

$$\omega_i^* = 1/\Delta x_i \tag{10}$$

The weight is normalized:

$$\omega_i = \frac{\omega_i^*}{\sum\limits_{j=1}^{N} \omega_j^*} \tag{11}$$

where $\triangle x_i$ is the measure precision of the measure variable x_i.

In the process of operation, instrument will be damaged with time, its precision is also affected, therefore, the weight ω also should been compensated, weight ω is learned through the formula (12):

$$\omega_i' = \frac{1}{2}\left(\omega_i + \overline{\omega}_i\right) \tag{12}$$

where:

$$\overline{\omega}_i = \frac{\overline{\omega}_i^*}{\sum\limits_{j=1}^{N} \overline{\omega}_j^*} \tag{13}$$

$$\overline{\omega}_i^* = \frac{\left|\overline{x}_i - x_i\right|}{\overline{x}_i} \tag{14}$$

ω_i' is the revised weight, ω_i is the recurrent weights, $\overline{\omega}_i^*$ Is calculated weights according to the last correction result, $\overline{\omega}_i$ is the normalization of $\overline{\omega}_i^*$, $\overline{\omega}_i$ is the correction value of measure variable I, x_i Is the measured value of the measured variable i. In this way, the weight data are corrected after each correction, to ensure the system model applied the actual production process, thus the system can operate reliably.

3.4 The solution based on LINGO

LINGO solver is a powerful optimization model, it is introduced by American LINDO Systems Company, the V13.0nis the latest version, there are six LINGO versions, they are demo version, the improved version, edition, professional edition, business edition and extended edition. Only one demo version is free, but the total number of variables, the condition number and so on have strict restrictions, the professional version should been used. LINGO's main features are:

1. It can not only solve the linear constraint problem, but also solve nonlinear constraint problem;
2. The model construction method is simple;
3. It has high solving speed and operation ability;
4. The description method of constraint equations is greatly simplified because of its modeling language and internal optimization functions;

5. New collective concept improve the modeling speed compared with other programming languages;
6. LINGO is fully compatible with c #, database and other software.

The data correction model based on LINGO is given below:
Mode:
...
min=@sum(value(i):@abs(measured-corrected)*w(i)); ! The objective function
@for(value:
@free(corrected);
);
@for(node(i):
@sum(value(j):flag(i,j)*corrected(j))=0;
);!Lists all the constraint conditions
...
End;

This model not only can get the optimal solution of the linear constraint condition, but also can directly get the optimal solution of nonlinear constraint conditions by changing the constraint conditions. Compared with the classical model, the obtained method of the optimal solution is more simple and quick.

4 SOLUTION EFFECT COMPARISON OF TWO MODELS

The simplify model of petrochemical enterprise production process is shown in Figure 1.

The experimental effect of the two models is compared by the steps as followed:

1. Drawing a set of harmonised data xs (1 × n) as reference data compared (that is, the real data);
2. A set of random data (1 × n) (the range of 0.2~0.2) is produced as the random error is attached to the real value xs;
3. Five data randomly are selected From 1~n (in this case, they are 26, 28, 22, 14, 7), some significant error are attached according to the

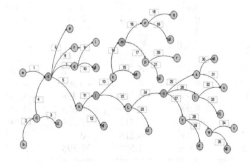

Figure 1. Part logistics diagram in a chemical company.

sample order, the additional significant error value is respectively + 1, + 2, + 3, + 4, + 5;

4. The result data x of Step 3) is as the measurement data, corrected by the new model, and are marked x in Table 1;

5. The result data x of Step 3) is as the measurement data, some significant errors are recognized, and eliminated by the MT – NT method, finally calibrated by the classical model, calibration results are described $\overline{x}^*$ in Table 1.

Two kinds of model calibration results are shown in Table 1.

Results analysis: It can be seen from Table 1, the value 7 measurement error is not identified by the classic models, thus the calibration result of

Table 1. Results comparison of new classical model and the calibration model.

No.	x_s	x	$\overline{x}$	$\overline{x}^*$
1	0.634	0.720	0.720	0.720
2	172.000	172.238	172.250	176.375
3	15.356	15.408	15.383	15.375
4	156.644	156.885	156.867	160.999
5	135.703	135.949	135.780	135.766
6	6.695	6.789	6.789	6.775
7	12.345	17.174	12.605	16.767
8	2.536	2.413	2.413	2.411
9	51.000	50.851	50.859	53.623
10	−38.655	−38.440	−38.253	−36.856
11	135.000	135.186	135.195	135.190
12	0.703	0.576	0.585	0.576
13	10.104	10.085	10.136	9.911
14	11.623	15.600	11.767	11.445
15	−1.519	−1.537	−1.631	−1.534
16	2.152	2.265	2.197	1.889
17	9.472	9.640	9.570	9.556
18	1.000	0.855	1.006	0.856
19	1.152	1.031	1.191	1.033
20	0.000	0.075	0.215	0.075
21	9.472	9.281	9.355	9.481
22	124.896	127.803	125.059	125.279
23	4.443	4.352	4.394	4.349
24	120.453	120.204	120.665	120.930
25	0.148	0.364	0.306	0.196
26	23.156	24.018	23.243	23.549
27	97.150	97.067	97.116	97.185
28	31.429	33.336	31.384	32.200
29	65.721	65.876	65.732	64.985
30	0.148	−0.063	−0.002	−0.058
31	0.000	0.202	0.308	0.254
32	0.000	−0.099	−0.073	−0.099
33	23.156	23.312	23.316	23.649
34	0.000	−0.215	−0.215	−0.215
35	31.429	31.598	31.598	32.414

Note: $\overline{x}$ is the new data model calibration results, is the classic model calibration result.

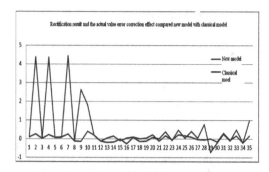

Figure 2. Error contrast diagram of two kinds of model.

measurement value 2,4,9 are deviated from the true value. However the calibration results of new model are not affected by the measured value 7, and are close to the real value. In order to more intuitively show the difference of two kinds of model, error contrast diagram is drawn in Figure 2.

It can be seen from Figure 2, the correction results of using the new model compared with the real value, its errors are near to zero, correction result are closer to the real values. The significant error of classic model is not all recognized, partial results are affected by the significant error, and have very big difference from real value, the calibration result is not very ideal.

5 CONCLUSION

A new data correction model is proposed in this paper, which provides a new way for the data correction technology development. Its correction result is consistent with real data, which achieves the goal of the data correction. The model compared with the classical data correction model, its advantage lies in that its correction result will not be influenced by the error of measurement data containing significant error, solve the problem that classical calibration model has not been fully identified in significant error, thereby increases the credibility of correction data in the production process, more conducive to industrial production. On the other hand, for the lower redundancy measurement data, the correction results can still be revised by historical production data, but the specific method need to be further studied.

REFERENCES

Amir Vasebi, Eric Poulin & Daniel Hodouin. 2012. Dynamic data reconciliation on the node imbalance autocovariance functions provides. The Computers and Chemical Engineering 43 (2012): 81–90.

Binder T., et al. Forward Multiscale Dynamic Data Reconciliation, Berber In R, Kravans C (Eds): Nonlinear Model -based Process Control. NATO ASI Series, Netherlands: Kluwer Academic Publishers.

Huang Chunpeng & Xia Maosen. 2004. Revised timing method application in the petrochemical plant data correction research. Machine integrated manufacturing system 10(7): 858–866.

Jiang Yuguang, et al. 2011. Based on the gross error detection and data GLR_NT coordination. Journal of east China university of science and technology 37(4): 502–508.

Kelly J.D. 2004. Techniques for solving industrial nonlinear data reconciliation the problems. Journal of Computers. Chem. Engng. 28(12): 2837–2843.

Kuelin R. 1961. Computer control 2, mathematics of control. Chem. Eng. Prog. J 57(6): 44–47.

Nie Qiuping, et al. 2010. Based on the consumption forecast of gas balance and data correction methods of iron and steel corporation, Chemical industry and the automation instrument 37(2): 14–18.

Song Kun & Li Lijuan, et al. 2011. Based on weighted MMMD soft measurement data of gross error detection. Journal of chemical engineering in colleges and universities 32(5): 1766–1769.

Yan Xuefeng, et al. 2007. Nodes coordinate with measurement data combination detection of data and application. Journal of chemical engineering 58(11): 2828.

Yuan Yonggen. 1985. Chemical process data correction and estimation method. Computer and applied chemistry 2(4): 259.

Zhang Zhengjiang, et al. 2011. A solving method based on experience to enhance data correction problem. Journal of college of chemical engineering 25(3): 482–488.

Control Engineering and Information Systems – Liu (Ed)
© 2015 Taylor & Francis Group, London, ISBN 978-1-138-02685-8

Study on daily runoff forecasting model

Z.S. Li
Fujian College of Water Conservancy and Electric Power, Yong'an, China

R.J. Zhang
Fujian College of Water Conservancy and Electric Power, Yong'an, China
Hohai University, Nanjing, China

ABSTRACT: Many predicting methods, such as ANN, etc. are not very precisely. Wavelet analysis is used to decompose daily runoff series, then ANFIS is used to modeling the decomposed series, in this paper; in the end it combined these series and predicted Poyang Lake's water level, it is the basic of influence research. The result shows that, the prediction accuracy rises a lot and is fit to use in daily runoff predict and water level predict.

1 INTRODUCTION

This paper predicts Poyang Lake's water level, it is the basic of influence research. Once there were many predicting methods, such as ANN [1,2], etc. But these methods are not very precisely. This paper uses wavelet analysis to decompose water level series, then it uses ANFIS to modeling the decomposed series, in the end it combined these series.

Wavelet analysis is a good multi-resolution frequency method, by the multi-scale sequence analysis it can effectively recognize the main frequency components and local information. ANFIS has strong nonlinear approximation functions and self-learning, adaptive characteristics. So they can be combined, and give full play to the advantage of both [3].

Poyang Lake is China's largest freshwater lake. It accepts Ganjing, Fu River, Xin River, Rao River and Xiu River's water, it is rich in water resources. There is about 145 billion m³ water injecting into Yangtze River through Poyang Lake annually. There is more than one-ninth of the middle and lower Yangtze River's water coming from the regulation of Poyang Lake [2].

Poyang Lake is rich in water resources and has special climate and geography. This two conditions gestate abundant and variety biology. It also preserves rare natural resources for our country, and provides superior natural condition for the development of region's economic and society.

2 ANFIS

Artificial Fuzzy Neural Network (ANFIS) [4,5] is combined by Artificial Neural Network and fuzzy theory. It uses ANN to construct the fuzzy system. According to input and output sample, it can automatically design and adjust the design parameters of the fuzzy system, and then it can achieve fuzzy system's self-learning and adaptive function. It can fit to complex input and output's linear and non-linear mapping relations. So it is especially applicable to complex non-linear hydrological system [6].

The network structure of Artificial Fuzzy Neural Network (ANFIS) is shown as Figure 1.

In Figure 1, the connecting line between the nodes only shows the flow of signal, it doesn't associate with weight. Square nodes represent nodes which have adjustable parameters; circular nodes represent nodes which don't have adjustable parameters. ANFIS's structure can be divided into 5 levels:

Level 1: Membership $z_{ij} = \mu_i(x_j, \theta_i)$ $i = 1,2,L,$ M_{num}, $j = 1,2,L,I_{num}$, M_{num} is the number of membership functions, I_{num} is the number of input variable, $\mu(\cdot)$ is generalized membership function, the commonly used membership functions are triangle membership function, trapezoidal membership function, Gaussian membership function and bell-shaped membership function. The form of

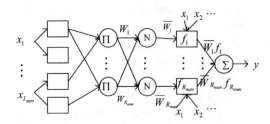

Figure 1. ANFIS's network structure.

Triangle and trapezoidal membership functions is simple, and their computing efficient is high. However, because their membership function is constructed by linear line, the corner points of some specified parameters are not smooth enough. Gaussian function and bell-shaped membership function have smooth and simple representation, so they are the most commonly used form to definite fuzzy sets. Formula 1 is bell-shaped membership function:

$$\mu_i(x, \theta_i) = 1 \left/ \left[1 + \left[\frac{x - c_i}{a_i} \right]^{2b_i} \right] \right.$$

x and $\theta_i = [a_i \ b_i \ c_i]$ are respectively ith membership function's input and original reasoning parameter set.

Level 2: kith incentive intensity, in which R_{num} is the number of fuzzy rules.

$$W_k = \prod^{I_{num}} Z_{ij}, \quad i \in [1, 2, ..., M_{num}], \quad i = 1, 2, ..., R_{num}$$

Level 3: normalized incentive intensity is the ratio of the rule's incentive intensity and the sum of all of the rule's incentive intensity.

$$\bar{W}_k = \frac{W_k}{\sum\limits_{j=1}^{R_{num}} W_j}, k = 1, 2, ..., R_{num}$$

Its vector form is:

$$\bar{W} = \left[\bar{W}_1, \bar{W}_2, ..., \bar{W}_{R_{num}} \right]^T$$

Level 4: fuzzy rule's conclusion, which is accurate output.

$$f_i = p_{i1} x_1 + p_{i2} x_2 + ... + p_{ij} x_i + r_i$$
$$i = 1, 2, ..., R_{num}, \quad j = I_{num}$$

The parameter set, which is composed by all of $\{p_{ij}, r_i\}$, is called consequent parameter set.

Level 5: after weight-average, the overall output of the net can get:

$$y = \sum_{i=1}^{R_{num}} \bar{W}^T F$$

In which $F = [f_1, f_2, L, f_{R_{num}}]$.

This paper uses hybrid learning algorithm to optimally select the parameter of ANFIS, this gets the smallest sum of square error between the finally output result and the goal. The core idea of this algorithm is: in the forward calculation, it keeps the value of all of the original reasoning parameter unchanged, and improves the value of consequent parameter by recursive least squares; then, it keeps the value of all of the improved consequent parameter unchanged, and improves the value of original reasoning parameter by error back-propagation.

3 WAVELET ANALYSIS

Wavelet analysis is a window fixed but the shape can variable (variable bandwidth and when wide) changing in the time-frequency analysis method. It has adaptive time-frequency window: high frequency, frequency-domain window increased, the time window reduced, the time window expands, and the frequency domain window reduced. Wavelet analysis is the key to satisfying certain conditions, the introduction of the basic wavelet function $\Psi(t)$ to replace the Fourier transform of basis functions $e^{-i\omega t}$. The next is stretching and translation functions:

$$\Psi_{a,b}(t) = |a|^{-1/2} \Psi\left(\frac{t-b}{a}\right) a, b \in R, a \neq 0$$

In Which: $\Psi_{a,b}$ is called analysis wavelet or continuous wavelet;

a is measurements (telescopic) factor, in a sense, is corresponding to frequency ω;

b is the time (translation) factor, it reacts time's translation.

Text is set in two columns of 9 cm (3.54") width each with 7 mm (0.28") spacing between the columns. All text should be typed in Times New Roman, 12 pt on 13 pt line spacing except for the paper title (18 pt on 20 pt), author(s) (14 pt on 16 pt), and the small text in tables, captions and references (10 pt on 11 pt). All line spacing is exact. Never add any space between lines or paragraphs. When a column has blank lines at the bottom of the page, add space above and below headings (see opposite column).

First lines of paragraphs are indented 5 mm (0.2") except for paragraphs after a heading or a blank line (First paragraph tag).

4 CALCULATION EXAMPLE

This paper uses wavelet analysis to find the suitable level of the water level series, by some trying, it find 2 level is the best level. The decomposed result is delayed as Figure 2.

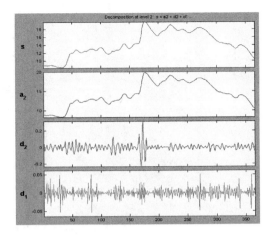

Figure 2. The decomposed result.

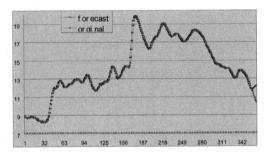

Figure 3. Comparison of original sequence and the final result.

In the figure we can clearly see that wavelet analysis can separate the tend item and the wave item. Next we use ANFIS to separately predict the tend item and the wave item.

We uses 1to 350 day as the modeling series and 351 to 365 day as the predict series. In order to combine them in the end, we select the same function for these series. This paper adopts anfis function of MATLAB's tool box to predict these series; the type of them is bell-shaped membership function. After 20th iterations, we can get a good simulating result. Then this paper combined the result of this series to make the final result. The contrast result of the forecast result and the original sequence is delayed as Figure 3.

Table 1. Comaperd result.

Relative error	5%	%5–20%	>20%
Proportion	93.68%	6.32%	0.00%

The evaluation of this model is delayed as Table 1.

From Figure 3 and Table 1 we can see that the result is very good. The relative error is nearly all less than 5%. So we can use wavelet-ANFIS model to forecast Poyang lakes water level. The precision will be very high. Next we can use the forecasting water level to evaluate the influence of the Three Georges to Poyang Lake. This is also the basic research of other water resources' researches.

5 CONCLUSION

This paper used wavelet analysis to decompose the water level series into three simple series, then it used ANFIS to model these series, in the end, it combined these series to make the final result. The result shows that Wavelet-ANFIS model is very fit to simulate water level. For water level's forecasting is the basic of other research, its precision is very important. This paper does some good to this. I think in the future the Wavelet-ANFIS model can be widely used in water level's forecast.

REFERENCES

Ben Zhang, et al. Research on Poyang Lake [M]. Shanghai: Shanghai technology press, 2008.

Jang J.S.R. ANFIS: Adaptive-network-based fuzzy inference system [J]. IEEE Transactions on System, Man and Cybernetics, 2011, 23(3):665–685.

Wensheng Wang, Jing Ding, Yueqing Li. Hydrological wavelet analysis [M]. Beijing: Chemical Industry Publishing House, 2005.

Xixia Ma, Haoze Mu, Huifang Guo. Reservoir Monthly Runoff Forecast Model Based on Wavelet-ANFIS Analysis [J]. Hydroelectric Energy, 2008,(1).

Xi-Xia Ma, Xiaoju H.E., Daochuan Zhao. BP Network hidden layer on the water quality impact analysis results of the evaluation [J]. Hydroelectric Energy, 2012,20 (3).

Zhixing Zhang, Chunzai Sun, Sunguyinger, et al. neuron fuzzy and soft computing [M]. xi'an:Xi'an Communication University press, 2010.

A novel multitask ensemble learning based on neural networks

Z.X. Li & W.H. Li
College of Computer Science, Northwestern Polytechnical University, Xi'an, China

ABSTRACT: Ensemble is widely used in pattern recognition, data mining, etc. Ensemble can divide a complex distribution into several simple distributions; then, constructs a particular classifier for every small and simple distribution. This is as simplifying a complex problem into several small and simple problems. Multitask learning can help to improve single task' accuracy by trained several related tasks in same classifier. Ensemble can divide distributions of related tasks into some same subspaces, which constitute the whole distribution. In this paper, we utilize this point and propose a novel multitask ensemble—multitask ensemble based on neural networks. Multitask ensemble can use same subspaces of related tasks to improve prediction. Experiments on real datasets prove multitask ensemble can help to improve prediction.

1 INTRODUCTION

An ensemble of classifiers consist of combination of different classifiers to jointly perform classification task. Ensemble construction is one of problems in artificial intelligence that receive most attentions. Ensemble uses combination of classifiers to classify a task. It is easy to get higher prediction than single classifier.

A classification problem of K classes and n observations whose classification membership have be known. Let $S = \{(\mathbf{x}_1, y_1), (\mathbf{x}_2, y_2), ..., (\mathbf{x}_n, y_n)\}$ be a set of n training samples where each instance $\mathbf{x}_i$ belongs to a domain X. Each label is an integer from the set $Y = \{1, ..., K\}$. A multiclass classifier is a function f: $X \rightarrow Y$ that maps an instance $\mathbf{x} \in X \subset R^D$ onto an element of Y.

The task is to find discriminant function $f(\mathbf{x})$. In a classifier ensemble framework we have a set of classifiers $C = \{C_1, C_2, ..., C_m\}$, each classifier performing a mapping of an instance vector $\mathbf{x} \subset R^D$ onto the labels $Y = \{1, ..., K\}$. Ensemble classification function is $f(\mathbf{x}) = \sum_{i=1}^{m} \alpha_i \cdot C_i(\mathbf{x})$. Training of ensemble contains two parts: training each classifier C_i; find optimal coefficients α_i to combine classifiers C_i into a uniform classifier function $f(\mathbf{x})$.

Ensemble can combine several weak classifiers to do classification task. This combination can construct a strong classifier in complex condition. In Hilbert space, data space would be linear, nonlinear or mixed. When meet a complex condition, ensemble can divide data space into some small and simple subspaces. These small subspaces are either linear or nonlinear. Ensemble can easily construct single classifier in such a small data

subspace, linear or nonlinear; then, merge these classifiers into a uniform and strong classifier.

This is different with SVM and neural network, which is constructing strong single classifier to do classification task. Due to each capability is limited, ensemble can construct a strong classifier in complex condition.

Boosting is popular method to construct ensemble of classifier. Its popularity is mainly due to ADABOOST. However, ADABOOST tends to perform very well for some problems but can also perform very poorly on other problems. One of the sources of the bad behavior of ADABOOST is that although it is always able to construct diverse ensembles, in some problems individual classifiers tend to have large training errors. Moreover, ADABOOST usually performs poorly on noisy problems.

In ensemble, SVM and neural network are widely used to construct base classifiers. In binary classification, SVM constructs a classifier which uses an optimal decision line or surface to classify in dimension reduced space. SVM, as a classifier or regression, is widely used in linear regression.

Neural network trains nodes in hidden layers. These nodes reconstruct features in hidden layers. So, neural network can construct classifier to resolve in nonlinear regression.

No matter SVM or neural network, it needs to construct a strong linear or nonlinear classifier. Recently, multitask learning has gotten people's attention. Multitask learning *R. Caruana 1997* is first proposed by Rich Caruna in 1997. Multitask learning is not a simple linear or nonlinear solution. It is either be linear or nonlinear, even be both. Before, we see tasks are noise to each other when they are trained in same learning machine, due to their differences. Rich Caruna proposed tasks, having related representations, can help to identify each other in same learning machine. In multitask learning, Rich Caruna showed related tasks trained in same learning machine can help to improve prediction.

Briefly, multitask learning is like comparison. Through training several related tasks in same classifier, each task can get a better prediction. Multitask learning utilizes related tasks have different attributes to improve prediction.

Now lots of work are devoted to regularization or linear resolution [*Argyriou 2007, Lozano 2012, Chapelle 2011, Liu 2009, Weinberger 2009, Argyriou 2008, Pillonetto 2010, Pillonetto 2010, Archambeau 2011*]. In 2007 Andreas Argyriou and Theodoros Evgeniou proposed a sparse multitask representation using 1 norm. This form combines the tasks and ensures that common features will be selected them. Aurelie C. Lozano and Grzegorz Swirszcz used a group of shared representations to weight vector. It used a group of fixed based weights to instead of random weights. Regularization can convert extremism problem into convex problem, and use linear regression to resolve nonlinear regression. Its deficiencies are: needs many observations to prevent ill-condition; performs badly in complex condition.

Except regularization, researchers are also devoted in probability region. Olivier Chapelle et al. *Chapelle 2011* proposed a multitask learning based on decision trees. They used gradient method to update each parameter. Qiuhua Liu et al. *Qiuhua 2009* proposed multitask learning based on Markov. In linear or nonlinear regression, probability methods perform badly.

In this paper, we propose a novel multitask ensemble based on neural networks. Multitask ensemble combines specialties of ensemble, neural network and multitask learning. Through ensemble, many related tasks can be trained in same learning machine and help each other to identify observations. Using different neural networks in each C_i, ensemble can resolve nonlinear or linear regression better.

This paper is organized as below: Section 2 explains deeply the proposed algorithm; Section 3 shows experiment setup, result and discussion; at last Section, we conclude our work.

2 MULTITASK ENSEMBLE BASED ON NEURAL NETWORKS

One layer neural network also can solve linear regression. However, due to one layer neural networks cannot combine different attributes to construct new abstract attributes, nonlinear regression needs neural networks has hidden nodes. Hidden nodes can recombine input attributes into new abstract attributes. For example, in computer vision, people cannot distinguish a 16×16 pixel picture represents what a picture is. But neural networks with hidden nodes can distinguish what an object is by many such 16×16 pixel pictures.

In order to apply nonlinear regression, we use hidden nodes in neural networks.

Briefly, we propose a novel multitask ensemble algorithm based on neural networks. Its discriminant function is (1) as below.

$$f(\mathbf{x}) = \frac{1}{m} \sum_{i=1}^{m} C_i(\mathbf{x}) \qquad (1)$$

Each $C_i(\mathbf{x})$ represents a base classifier neural networks. Each $C_i(\mathbf{x})$ use a subset to train. m represent the number of base classifier $C_i(\mathbf{x})$. In this paper, we use Basic Ensemble Method (BEM) in *Chen 2010* to calculate mixing coefficient α_i.

Although hidden nodes can resolve nonlinear regression, it is difficult to know how to combine with different attributes in hidden nodes and control how to combine. Add adjustment and combination from inside seems impossible. Image one scene, neural network with hidden nodes is a black box, we don't know how neural network runs in such a black box. All attributes combinations in hidden nodes can't be explained. This is a complete automatic process. As we know are input-attributes and output-class. Input-attributes are we can't change, it is foundation of all work. So, we consider whether we could add adjustment in output layer.

In regression, output layer transmits error signal $y_j - w_{ij} \cdot \mathbf{x}_i$ backward. Through backward mechanism, $y_j - w_{ij} \cdot \mathbf{x}_i$ is transmitted to previous layers nodes. Weights of previous nodes adjust. $y_j - w_{ij} \cdot \mathbf{x}_i$ is difference between sample's class label and estimated label. If it is small, weights vary small; otherwise, vary big. This method can reflect difference between sample's class label and estimated label. However, single $y_j - w_{ij} \cdot \mathbf{x}_i$ only can reflect sample $\mathbf{x}_i$ and one class difference. In multitask condition, this method can't reflect sample $\mathbf{x}_i$ with other classes. In order to reflect how close sample $\mathbf{x}_i$ is to other classifiers, we add parameter λ to represent.

$$\lambda_{ij} = 1 - \frac{\left| y_j - w_{ij} \cdot \mathbf{x}_i \right|}{\sqrt{\sum (y_k - w_{ik} \cdot \mathbf{x}_i)^2}} \qquad (2)$$

λ_{ij} represents how close sample $\mathbf{x}_i$ is to output class C_j. Bigger λ_{ij} shows sample $\mathbf{x}_i$ is closer to class

Data : A training set $S = \{(x_1, y_1), \ldots, (x_n, y_n)\}$, a base learning algorithm, L, and the number of iterations T.

Result: The final classifier:

$$C^*(\mathbf{x}) = \arg\max_{y \in Y} \sum_{t:C(\mathbf{x})=y} 1$$

Initialize m base classifiers
$C_0 = L(S)$
for t=1 to T-1 **do**

$\quad S' \subset S, \quad S' = \{x_i \in S : C_{t-1}(x_i) \neq y_i\}$.

$\quad$ Train network H with S' and get projection $\mathbf{P(x)}$ implemented by the hidden layer of H.

$\qquad C_t = L(\mathbf{P(S)})$

End

Algorithm 1. Multitask ensemble learning.

Data : A data point $(\mathbf{x}_i, y_i)$.
Result: The weight of classifier
for t=1 to T-1 **do**

Use (1), (2) to get λ_{ij} and output layer error.

Output layer delta function

$delta_i = error_i(\mathbf{x}) \cdot (b\_\tan/a) \cdot (a^2 - activation_i^2(\mathbf{x}))$

Hidden node delta function

$delta_{i'} = (b\_\tan/a) \cdot (a^2 - activation_i^2(\mathbf{x}))$

Use delta function to update weight and momentum.

$\Delta w_i + = \eta \cdot delta_{k+1} \cdot activation_k + \alpha \cdot momentum_k - \lambda \cdot w_k$

$momentum_k = \eta \cdot delta_{k+1} \cdot activation_k + \alpha \cdot momentum_k$

$\qquad\qquad -\lambda \cdot w_k$

end

Algorithm 2. Construction of base classifier.

C_j. Due to sample $\mathbf{x}_i$ is closer to class C_j, transmit error signal $y_j - w_{ij} \cdot \mathbf{x}_i$ bigger backward; or small. Then, whole error signal is

$$error_{ij} = \lambda_{ij} \cdot \left(y_j - w_{ij} \cdot \mathbf{x}_i \right) \qquad (3)$$

Error signal as shown in (3) does not only estimate error signal and measure in different tasks. Error signal can help to discriminate sample $\mathbf{x}_i$ belongs to which classifier.

In this paper, we manipulate input samples to train every classifier. First, divide samples into different subsets; then, use these subsets to train every classifier. Algorithm 1 shows framework of Multitask Ensemble Learning. Back-propagation

is a successful instance in neural networks. In this paper, we select back-propagation as base classifier. In Algorithm 2, we give construction of base classifier.

3 EXPERIMENTS AND DISCUSSION

3.1 *Experimental setup*

In order to validate multitask ensemble, we use four datasets to validate. Table 1 gives details of the four datasets. In practice, experiments often meet over-fitting, so we select 10 folds cross-validation to do experiments. No matter neural networks' hidden node or number, it needs to be predefined. Experiments approve 25 hidden nodes in hidden layer and 50 networks can effectively accuracy, so we use this as default set.

3.2 *Experimental results and discussions*

To validate function of multitask ensemble, we select AdaBoost, AdaBoost based on J48, Ada-Boost based on REPTree, Bagging, LMT, Logit-Boost, RandomCommittee, RandomForest and RandomSubSpace to compare in Table 2.

From Table 2, we could find that in super-market, multitask ensemble has highest accuracy in all algorithms, even exceeds ensemble algorithm—AdaBoost. Most ensemble algorithms perform normally in this dataset. In weather, we use its numeric features to run. AdaBoost based on REPTree and RandomSubSpace both get highest accuracy. Multitask ensemble and AdaBoost based on J48 have same accuracy. In pima indian datasets and iris, Bagging and LMT performs better than others. In all, performance of multitask ensemble does exceed LogitBoost, RandomCommittee and AdaBoost based on J48. Multitask ensemble performs close with AdaBoost.

Experiments above prove neural networks could improve accuracy of ensemble. Neural networks, as a method of nonlinear regression, can solve non-linear regression effectively. Ensemble, as combination of classifiers, can better discriminate complex distribution than single classifier. With multi-task trained simultaneously, we can improve accuracy

Table 1. Summary of data sets.

Data set	Cases	Classes	Features
Supermarket	4627	2	216
Weather	14	2	4
Pima indian datasets	768	2	8
Iris	150	3	4

Table 2. Summary of prediction for ensemble on neural networks as base learner (%).

Method	Supermarket	Weather	Pima[a]	Iris
AdaBoost	75.12	50.00	73.96	93.33
AdaBoost+j48	63.71	57.14	71.61	92.67
AdaBoost REPTree	63.71	64.29	75.65	94.00
Bagging	63.71	50.00	77.60	94.00
LMT	63.71	50.00	76.82	94.67
LogitBoost	76.70	50.00	74.74	92.67
RandomCommittee	63.71	50.00	72.79	93.33
RandomForest	63.71	50.00	73.96	94.00
RandomSubSpace	63.71	64.29	75.26	93.33
Mulitask Ensemble	79.86	57.14	73.18	93.33

[a]Pima is pima indian dataset.

of each task. This would help to resolve in nonlinear regression.

4 CONCLUSION

Combing with neural networks and ensemble, we proposed a novel ensemble algorithm—multitask ensemble. Facing nonlinear and complex data distribution, we can use multitask ensemble to resolve. This thinking is from ensemble—divide a complex distribution into several simple distributions. Ensemble algorithm utilizes related tasks trained simultaneously can help to improve each task's accuracy. Experiments prove multitask ensemble could improve accuracy.

In future, we would pay attention on how to explore performance of neural networks and apply in variety of regression.

ACKNOWLEDGMENT

Thanks for help from Ph.D candidate Rui Yao and Chenzhang Peng. Their help greatly improve our work.

The comments and suggestions from the anonymous reviewers greatly improve this paper. Minister foundation of R.P. China supported this research.

REFERENCES

Andreas Argyriou, Theodoros Evgeniou, Massimiliano Pontil, "Convex multi-task feature learning," Machine Learning, vol. 73, December 2008, pp 243–272, doi: 10.1007/s10994–007–5040–8.

Argyriou, A., Evgeniou, T. & Pontil, M., "Multi-task feature learning," In Advances in neural information processing systems (NIPS 2006), MIT Press, 2007, pp 41–48.

Caruana R., "Multi-Task Learning," Machine Learning, vol. 28, 1997, pp. 41–75, doi: 10.1023/A:1007379606734.

Cédric Archambeau, Shengbo Guo, Onno Zoeter, "Sparse Bayesian multi-task learning," Advances in Neural Information Processing Systems 24: 25th Annual Conference on Neural Information Processing Systems 2011 (NIPS 2011), Curran Associates Inc., 2011.

Gianluigi Pillonetto, Francesco Dinuzzo, Giuseppe De Nicolao, "Bayesian online multitask learning of gaussian processes," IEEE Transactions on Pattern Analysis and Machine Intelligence, vol. 32, February 2010, pp 193–205, doi: 10.1109/TPAMI.2008.297.

Gianluigi Pillonetto, Francesco Dinuzzo, Giuseppe De Nicolao, "Bayesian Online Multitask Learning of Gaussian Processes," IEEE Transactions on Pattern Analysis and Machine Intelligence, vol. 32, Feb 2010, pp 193–205, doi: 10.1109/TPAMI.2008.297.

Kilian Weinberger, Anirban Dasgupta, John Langford, Alex Smola, Josh Attenberg, "Feature hashing for large scale multitask learning," Proceedings of the 26th International Conference On Machine Learning, (ICML 2009), Omnipress, 2009, pp 1113–1120.

Lozano, Aurelie C., Swirszcz, Grzegorz, "Multi-level Lasso for Sparse Multi-task Regression," Proceedings of the 29th International Conference on Machine Learning (ICML 2012), Omnipress, 2012, pp 361–368.

Olivier Chapelle, Pannagadatta Shivaswamy, Srinivas Vadrevu, Kilian Weinberger, Ya Zhang, Belle Tseng, "Boosted multi-task learning," Machine Learning, vol. 85, October 2011, pp 149–173, doi: 10.1007/s10994–010–5231–6.

Qiuhua Liu, Xuejun Liao, Hui Li Carin, Jason R. Stack, Lawrence Carin, "Semisupervised multitask learning," IEEE Transactions on Pattern Analysis and Machine Intelligence, vol. 31, June 2009, pp 1074–1086, doi: 10.1109/TPAMI.2008.296.

Yuehui Chen, Ajith Abraham, Tree-Structure Based Hybrid Computational Intelligence, vol 2. Berlin: Springer, 2010, pp 90.

Control Engineering and Information Systems – Liu (Ed)
© *2015 Taylor & Francis Group, London, ISBN 978-1-138-02685-8*

The research of three dimensional network modeling based on boundary representation using Java 3D

L.L. Zhao, S. Liu, P. Yao & T.X. Wei
School of Engineering, Honghe University, Mengzi, P.R. China

L. Ma
Changjiang Institute of Survey, Planning, Design and Research, Wuhan, Hubei, P.R. China

ABSTRACT: Three-dimensional modeling is one of the core issues in three-dimensional Geographic Information System (3D GIS). The paper is focused on 3D modeling based on boundary representation based on Java 3D. Using primitive elements such as nodes, edges and faces, we present in this paper a new set of algorithms to model primitive objects by the combination of these primitive elements based on java 3D including quadrilateral faces, walls, pillars, etc and something useful is obtained.

1 INTRODUCTION

Three-dimensional modeling is one of the core issues in three-dimensional Geographic Information System (3D GIS). In general, the need of city 3D data is raising more and more, especially people involved in urban and landscape planning, cadastre, real estate, utility management, geology, tourism, army, etc. are keen on taking advantages of the third dimension [1].

The traditional two-dimensional GIS has failed to meet the information needs of city 3D modeling, it is gradually to 3DGIS Development. In fact, with the increase in the use of the Internet as a medium for information exchange, there has also arisen the need to develop applications that exchange 3D data seamlessly. So, we not only manipulate 3D data, but also share these data on the Internet.

In summary, 3D modeling methods can be generally summed up in two categories: the ones is that digital elevation models combined with images to build on 2.5-dimensional digital terrain model. The other is that the use of the block model combined with texture to build multiple levels of detail models (Levels of Detail, LOD) of the three-dimensional model (such as CityGML). CityGML is an international standard for the representation and exchange of semantic 3D city and landscape models adopted by the Open Geospatial Consortium (OGC) [2,3,4,5], which can construct and display not only the urban model and surface appearance, also involve semantics, thematic attribute expression, classification and collection. CityGML is utilized for use in many applications, including but not limited to urban planning, facility management, disaster management, homeland

security, personal navigation, etc. For example, how an emergency operator can find a suitable location and building for a field hospital after the explosion of a dirty bomb [6]. CityGML can provide important information for disaster management [7], which includes emergency route planning and indoor navigation [8,9].

In the context of GIS and of city modeling, 3D objects are often represented with boundary representation models, using primitive elements such as nodes, edges and faces. So we present in this paper a new set of algorithms to model primitive objects by the combination of these primitive elements based on java 3D including quadrilateral faces, walls, pillars, etc and something useful is obtained.

2 THREE-DIMENSIONAL MODELING BASED ON JAVA 3D

The basic elements of Java 3D are points, lines and surfaces. A plurality of surface can be closed to form a simple entity, and three-dimensional object can be constituted by a plurality of entities. Construction Object Model is generally constituted by a limited number of surface entities such as quadrilateral faces, walls, pillars, etc.

2.1 *Quadrilateral face*

Quadrilateral face is essentially quadrilateral with texture (QuadArray), the basic elements of which are four points, quadrilateral object and texture object. Its data structure definition is shown in Figure 1.

The sequence of the four points in quadrilateral face array affects the quadrilateral face generation and the texture visibility. Generally in accordance with the principle of the right hand, only the normal direction of the texture is visible. The points sequence in an array is distributed shown in Figure 2. The map texture is visible above the plane, and the quadrilateral face affixed texture is seen in Figure 3.

2.2 Wall

Wall object is closed by the six faces. Each face is a QuadSurface object. The basic elements are the two reference points, thickness and height of the wall, and two sets of points derived from both top and bottom. Its data structure definition is shown in Figure 4, where Point 1 and Point 2 are two reference points to build walls, the length of the wall can be obtained by subtracting two points. Width is the thickness of the wall. Height is the height of the wall. BottomPoints is composed of an array of four points on the bottom of wall, while AbovePoints is composed of an array of four points on the top of wall. Figure 5 identifies the two reference points (two points hollow) and

Class QuadSurface
{

    Point3f[ ] points;   // An array of four points

    QuadArray qa;   // Quadrilateral object

    Appearance app;   // Texture object

}

Figure 1.   Quadrilateral face data structure.

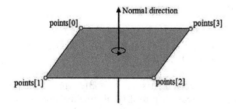

Figure 2.   The quadrilateral surface normal direction.

Figure 3.   The quadrilateral face affixed texture diagram.

Class WallSurface

{

    Point3f Point1;   // the reference point one of the wall

    Point3f Point2;   // the reference point two of the wall

    float Width;   // the wall length

    float Height;   // the wall height

    Point3f[] BottomPoints; // An array of four points on the bottom

    Point3f[] AbovePoints; // An array of four points on the top

}

Figure 4.   A data structure diagram of wall.

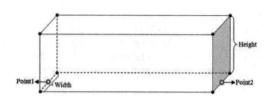

Figure 5.   The wall construction illustrating diagram.

parameters such as height and thickness, two sets of the four points on the bottom and on the top respectively can be calculated based on these parameters. According to the two reference points passed by the wall it can determine the structure of the wall is parallel to the X-axis, Y-axis or general sloping wall, the algorithm is slightly different in both cases. The wall affixed texture is showed in Figure 6.

2.3 Pillar

Pillar object is also closed by the six faces, and each object face is a QuadSurface, the basic elements of which are pillar central reference point, width, thickness, height and two sets of points derived from both top and bottom. Its data structure definition is shown in Figure 7, where Point is central reference point, Width is the width, Thick is the thickness, height is the height of the pillar. BottomPoints is composed of an array of four points on the bottom of wall, while AbovePoints is composed of an array of four points on the top of wall. Figure 8 identifies the pillar central reference point and parameters such as height and thickness, two sets of the four points on the bottom and on the top respectively can be calculated based on these parameters.

It can be set to 16 different states according to the case of the wall through the pillar in order to make pillar distinguish in walls. Each state loads different types of texture so as to achieve the effect of a ray. As showed in Figure 9 squares represent the pillars, lines represent the direction of the wall

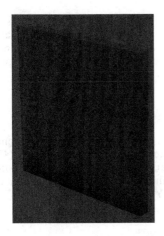

Figure 6.　The wall affixed texture diagram.

```
Class PillarSurface
{
    Point3f  Point;           // the pillar central reference point
    float   Width;            // the pillar width
    float   Thick;            // the pillar thickness
    float   Height;           // the pillar height
    Point3f[]  BottomPoints;  // an array of four points on the bottom of pillar
    Point3f[]  AbovePoints;   // an array of four points on the top of pillar
    int   Flag;               // pillar state value
}
```

Figure 7.　The data structure diagram of pillar.

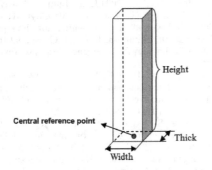

Figure 8.　The pillar construction illustrating diagram.

through; figures represent the state's number. State 0 is an independent pillar, no walls through. States 1, 2, 4 and 8 denote different pillars contacting with the wall surface, only one side connected to the wall. States 3 and 12 indicate wall across from the horizontal or vertical direction, respectively, left and right side or up and down the side in contact with the wall. States 7, 11, 13 and 14 indicate pillar in the butt at the two walls, there are three sides in contact with the wall. State 15 represents the pillar

in contact with the four wall surfaces. According to state case, surfaces in contact with the wall loaded with dark texture processing, thus makes sense to produce a shadow, results shown in Figure 10.

2.4　Stairs

Stairs are complex shape, which is constituted by the multi-stage ladders and corners; each section ladder has three sides and a number of steps, shown in Figure 11. Each step is a hexahedron, the two sides do not rule. According to the coordinates of starting point StartPoint and end point EndPoint, The total height of steps can be obtained, Each level step height can be obtained by dividing the number of steps, the direction of the step (Broken line arrows in Fig. 11) and reference point (Black solid point in Fig. 11) can also be obtained.

In the known width, height and reference point, the eight points' coordinates of hexahedral on each step can be obtained to construct the steps. According to the data structure diagram to define the step structure shown in Figure 12, where Point is central reference point, Width is the width, Thick is the thickness, height is the height of the stair. BottomPoints is composed of an array of four points on the bottom of stair, while AbovePoints is composed of an array of four points on the top of stair.

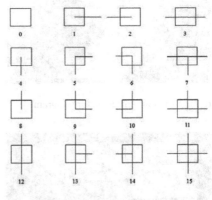

Figure 9.　Pillar state diagram.

a) No contact with the wall　b) one surface in contact with the wall　c) many surfaces contact with the wall

Figure 10.　The different states of texture renderings for pillars.

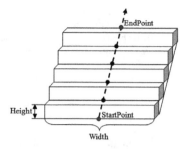

Figure 11.   Schematic diagram of the stair.

```
Class StairSurface

{

    Point3f  StartPoint;        //  the starting point of stair

    Point3f  EndPoint;          //  the ending point of stair

    float   Width;              //  the width of stair

    int    Count;               //  the number of stair

    Point3f[] RefPoints;        //  an array of reference points

}
```

Figure 12.   The data structure diagram of stair.

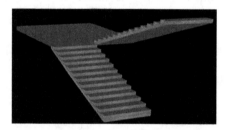

Figure 13.   The stair affixed texture diagram.

Figure 14.   The building construction.

3   EXPERIMENT

In this paper, we adopt a building construction as an example; its original is CAD data. The building consists of five floors, the first and second floor of the same structure, the third, fourth and fifth floors are the same structure, the building exterior is curved.

The constructed building floor can be constituted of walls, pillars, staircases and other basic body composition, these basic shapes of data structures and generation algorithm has already been introduced above, through the combination of transformation, the building can be constructed showed as Figure 14.

ACKNOWLEDGMENTS

This research is supported by the National Natural Science Foundation of China (41201418, 41301442), by Yunnan University Innovation Fund and by the Department of Education Research Fund of Yunnan Province (2012C198).

REFERENCES

CityGML Homepage: http://www.citygml.org Featherstone W.E, Sproule D.M (2006). Fitting AUSGeoid98 to the Australian Height Datum using GPS data and least squares collocation: application of a cross-validation technique,Survey Review 38(301):573–582.

Gröger, G., Kolbe, T.H., Czerwinski, A., Nagel, C., 2008. OpenGIS City Geography Markup Language (CityGML) Encoding Standard, Version 1.0.0, OGC Doc. No. 08-007r1, Open Geospatial Consortium.

Hamilton, A., Wang, H., Tanyer, A.M., Arayici, Y., Zhang, X., Song, Y., 2005. Urban information model for city planning, ITcon Vol. 10, Special Is-sue From 3D to nD modelling.

Held, G., Rahman, A.A. and Zlatanova, S. Web 3D GIS for urban environments[C]. Proceedings of the International Symposium and Exhibition on Geoinformation 2004 (ISG2004), 2004 Kuala Lumpur, Malaysia.

Kolbe, T.H., Gröger, G., Plümer, L., 2008. CityGML—3D City Models and their Potential for Emergency Response. In: Zlatanova, Li (eds.), Geospatial Information Technology for Emergency Response, Taylor & Francis.

Lapierre, A., Cote, P., 2007. Using Open Web Services for urban data management: A testbed resulting from an OGC initiative for offering standard CAD/GIS/BIM services. In: Coors, V., Rumor, M., Fendel, E.M., Zlatanova S. (eds): Urban and Regional Data Management. Proceedings of the 26th UDMS, October 10–12, 2007, Stuttgart, Taylor & Francis.

Lee, J., Zlatanova, S., 2008. A 3D data model and topological analyses for emergency response in urban areas. In: Zlatanova, Li (eds.), Geospatial Information Technology for Emergency Response, Taylor & Francis.

Mäs, S., Reinhardt, W., Wang, F., 2006. Conception of a 3D Geodata Web Service for the Support of Indoor Navigation with GNSS. In: Proc. of 3D GeoInfo 2006 in Kuala Lumpur, LNG&C, Springer.

Teller J., Keita A.K., Roussey C., Laurini R., 2005. Urban Ontologies for an improved communication in urban civil engineering projects. In: Proc. of the Int. Conference on Spatial Analysis and GEOmatics, Research & Developments, SAGEO 2005 Avignon, France, June, 20th–23rd.

Control Engineering and Information Systems – Liu (Ed)
© 2015 Taylor & Francis Group, London, ISBN 978-1-138-02685-8

A new strategy of statistical analysis based on graphic analysis

H.L. Yan & W.G. Duan
Key Laboratory of Molecular Biology for Sinomedicine, Yunnan University of Traditional Chinese Medicine, Kunming, China

Y. Yun
The Department of Pharmacology, Kunming Medical University, Kunming, China

ABSTRACT: The main aim of the study was to establish a new strategy in solving statistical analysis problem based on graphic analysis. By using two examples, we transformed values of different group factors or different level factors to distribution curves. The distribution curves were set in one coordinate system. The comparison was made by comparing the peak of different curves and by calculating certain areas enclosed by the curves and horizontal ordinate. The result of two examples based on graphic analysis was consistent with that of based on classical analysis, and suggested the strategy was feasible in statistical analysis though it needs some technical betterment.

1 INTRODUCTION

There are two tasks for statistics: one is statistical description, and the other is statistical inference (Zhenqiu Sun, 2005). The task of statistical description is mainly to describe the central tendency and dispersion tendency of a group data, while that of statistical inference is mainly to deduce whether the two or more samples are derived from one totality. The present statistical analysis depends heavily on its distribution pattern. However, there are multiple distribution patterns known or unknown, and the present statistical analysis depends especially on the known distribution pattern like normal distribution (Gaussian distribution), Poisson distribution, and Bernoulli distribution. Even further, the distribution pattern of different data should be the same or similar in one event of statistical analysis. If the distribution patterns are unknown or are different from each other, only rank sum test can be performed to make a statistical inference (V. Bansal, et al., 2010; H. Ito, et al, 2010).

By ranking the original data, rank sum test is the poorest method of statistical inference, because the statistical calculation uses the ranked data rather than the origin data; and which causes a lot of information loss. Actually, rank sum test is only a reluctantly candidate method used in statistical analysis (Table 1) (Jiqian, 2008; D. Lord and F. Mannering, 2010).

So far, optimal analysis method is needed in order to perform a good analysis inference, since the present statistical principles are mainly based on normal distribution or on the idea of normal distribution (D. Lord and F. Mannering, 2010). And the optimal selection of statistical method becomes a sophistical task according to the present statistical knowledge. In order to simplify the task,

Table 1. Methods of statistical analysis and their feature.

Methods	Data	Distribution pattern	Variance	Sig.
t-test	Measurement data	Normal distribution or similar	If homo-scedasticity uneven, need adjusting	U-test is a special t-test. T-test is only used for no more than two groups
Analysis of variance	Measurement data	Normal distribution or similar	If homo-scedasticity uneven, need adjusting	Suitable for multiple groups
chi-square test	Enumeration data	Same and known distribution		
Rank sum test	Both	Unknown or different distribution		A poorest test

here we put forward a strategy based on graphic calculation by using two examples, although there was graphic analysis that once used to solve crossing problems (H. Wen, et al., 2012).

2 PRINCIPLE OF TWO INDEPENDENT GROUPS BASED ON ONE FACTOR

There were two groups in Table 2 as an example, and we learnt that the data was based on one factor (Group factor). According to the data, we obtained their distribution curve by calculation through a probability function (Fig. 1). In Figure 1C, the Area Under the Curve (AUC) of group A and group B is defined as "1" or 100%, and the peak of the curve is something like mean, average, or median though we do not exactly know their distribution pattern, because we needn't to know their distribution pattern. We compared the peaks of the two groups by setting the two curves in one coordinate system. Then, if the peak of a curve was not merged by the other curve, there could be of significance. To quantitatively explore further, we calculated the area of A or B. If area A was more than a certain value, or if area B was less than a certain value, the two groups were more probably deriving from different totalities.

In Figure 1C, if more values of the data were close near to certain point in the horizontal ordinate, the peak of the curve could be higher, because the AUC is defined as 1, and the slope could be more steepen. Also, the sample size could affect the curve. A large sample size is prone to increasing the height of the curve peak and causes a more steepen slope. Based on the above description, we can tell the differences in Figure 2. In Figure 2, curve A is the distribution based on the data of a large sample size, curve B is that based on the data

Table 2. Data of two independent groups based on one factor (group factor).

Group A	Group B
1.57	1.63
1.65	1.68
1.66	1.7
1.51	1.59
1.74	1.8
1.59	1.69
1.63	1.72
1.65	1.77
1.7	1.64
1.66	1.75
1.64	1.69

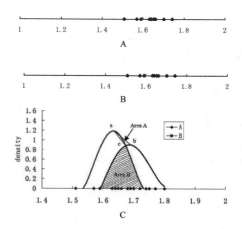

Figure 1. The principle of statistical analysis of two independent groups of one factor. The data in Table 1 was dotted in a number axis (see A and B). The density of the dots reflected the frequency of the group. The density was transformed to frequency curves (C). In C, there were two peaks in the two curves that did not merged, and caused two areas. Area A was a triangle based on dot a, b and c (a is the peak of curve A, b is the peak of curve B, and c is the cross dot of curve A and Curve C). Area B is an area enclosed by curve A, curve B and the horizontal ordinate.

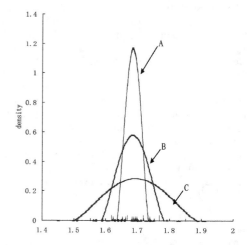

Figure 2. Different curves constructed base on the data of different sample sizes. Curve A is the distribution based on the data of a large sample size, curve B is that based on the data of a normal sample size, and curve C could be that based on the data of a small sample size.

of a normal sample size, and curve C could be that based on the data of a small sample size.

In order to compare two groups, we can put two distribution curves, which derived from original data, together in one coordinate system, and

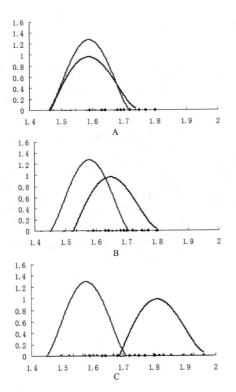

Figure 3. Three main types of graph statistical analysis. A, the two distribution curves almost merge completely, which suggests the two groups are derived from the same totality. B, the two distribution curves fairly separate because the two peaks are obvious, which suggests the two groups may be derived from different totalities but the decision should be made based on statistical calculation. C, the two distribution curves almost separate completely, which suggests the two groups are derived from different totalities even without statistical calculation.

there could be three main types (Fig. 3). From Figure 3A to Figure 3C, we found the distribution curves were separating each other more and more. Figure 3C suggested the data was most probably derived from different totalities.

If there are 3 or more independent groups based on one factor (Group factor), we can use similar method to compare them in one coordinate system.

3 PRINCIPLE OF TWO INDEPENDENT GROUPS BASED ON TWO FACTORS

There could be some situations on two or more factors. We obtained the data like Table 3 as an example (showed Group factor and Level factor). We put the data of group 1 in a coordinate system, and so did group 2. Then, we constructed the distribution

Table 3. Referred data of two independent groups based on two factors (group factor and level factor).

Height (m)	Group	Factor	Height (m)	Group	Factor
1.68	1	1	1.59	2	1
1.57	1	1	1.63	2	1
1.59	1	1	1.64	2	1
1.63	1	1	1.68	2	1
1.64	1	1	1.69	2	1
1.62	1	1	1.69	2	1
1.58	1	1	1.58	2	1
1.60	1	1	1.60	2	1
1.55	1	1	1.55	2	1
1.59	1	1	1.59	2	1
1.65	1	2	1.65	2	2
1.65	1	2	1.65	2	2
1.60	1	2	1.75	2	2
1.68	1	2	1.68	2	2
1.67	1	2	1.70	2	2
1.70	1	2	1.72	2	2
1.66	1	2	1.75	2	2
1.66	1	2	1.77	2	2
1.7	1	2	1.80	2	2
1.74	1	2	1.74	2	2

curves showed in Figure 4. The curve A and B were showed in red presenting group 1, and curve C and D showed in blue presenting group 2. Each curve has a peak, and each group has two, because the factor that the data of each group based on has two levels.

Therefore, we put the distribution curves in one three-dimensional coordinate system, and found the peak of curve A and curve C almost merged, while the peak of curve B and curve D separated fairly well. According to Figure 4C, the results groups 1 and groups 2 should be derived from different totalities; the results were confirmed by two way analysis of variance (two-way ANOVA). However, if only comparing the data between reciprocal levels of factor, we found there was no significance between curve A and curve C, and so between curve B and curve D. The results were also confirmed by classical statistical methods of one way Analysis of variance (one-way ANOVA). The results suggested that, we can obtain more meaningful information from a more carefully designed experiment.

Also, in the above example, if the Group factor has three or more, and if the level factor has three or more, we can obtain other graphs similar to Figure 4 with more distribution curves.

4 SUMMARY

There are two benefits manifest in the present statistical analysis. One is the fact that, all the original data is used in the analysis calculation; the other is the fact that we need not exactly know the distribution patterns.

Also there are technical problems that should be solved in the near future in the present statistical analysis. As showed above, the key step of the present method based on graphic analysis is transforming data distribution to distribution curve, and the most important problem is how to apply the best algorithm. The other problems may be associated the threshold of area A and area B, and how to defined the AUC as "1" in the distribution curve. Besides, graphic calculation could cause a heavy computation which could only be performed by a well-equipped computer.

ACKNOWLEDGEMENTS

The study is financially supported by National Natural Science Foundation of China (No. 81160495), and Yunnan Provincial Department of Science & Technology (No. 2010C1043).

REFERENCES

Bansal V., et al., 2010. Statistical analysis strategies for association studies involving rare variants, Nature Reviews Genetics, vol. 11, pp. 773–785.

Ito H., et al., 2010. A new graphic plot analysis for determination of neuroreceptor binding in positron emission tomography studies, Neuroimage, vol. 49, pp. 578–586.

Jiqian Fang, 2008. Health Statisticas. People's Medical Publishing House, pp. 96–108.

Lord D. and Mannering F., 2010. The statistical analysis of crash-frequency data: A review and assessment of methodological alternatives, Transportation Research Part A: Policy and Practice, vol. 44, pp. 291–305.

Wen H., et al., 2012. Graphic Analysis of the Crossing Problem, Dianzi Keji- Electronic Science and Technology, vol. 25, pp. 33–36.

Zhenqiu Sun, 2005. Medical Statistics. People's Medical Publishing House, pp. 113–128.

Control Engineering and Information Systems – Liu (Ed)
© 2015 Taylor & Francis Group, London, ISBN 978-1-138-02685-8

Analysis on key factors of community micro-blog influence[*]

H.M. Yi, Q. Wang, W. Cheng & C.S. Liu
Research Department, Beijing City University, Beijing, China

ABSTRACT: Virtual social management innovation is indispensable to that of social management. This paper dynamically tracks some community micro-blog sites in Beijing city; the micro-blog influence factors are summarized. Data mining tools such as parametric test, regression analysis are used to analyze the correlation between the above-mentioned factors and community micro-blog influence. With multi-party inspection of the correlation coefficient, key factors to improve the community micro-blog influence are selected. Some suggestions are proposed to strengthen the community micro-blog influence.

1 INTRODUCTION

Micro-blog has the cohesion and integration of information (Yu 2010). Chinese micro-blog is a hot spot in the domestic online social network (Wen 2012). Social management innovation can not be separated from the virtual social management innovation. The community micro-blog, which is used as an effective tool by social managers in virtual community, has developed rapidly in recent years. Take the nation's largest micro-blog site— Sina micro-blog for example, the latest statistics in March 2013 show that the number of Beijing Community micro-blog has reached 904. These micro-blogs covers 828 communities, 135 streets and all 16 districts and counties in Beijing. Compared with the community micro-blogs in the same period of last year, these micro-blogs has been significantly improved in terms of quantity and quality, and has a considerable influence in network. How to effectively improve the influence of community micro-blog have become increasingly important issues.

2 THE KEY FACTORS ANALYSIS OF COMMUNITY MICRO-BLOG INFLUENCE

This paper made a 2 years dynamic tracking research for Beijing Community Micro-blog in Sina Micro-blog site, from which, we summed and extracted forty kinds of factors that may affect the

community micro-blog influence. All these factors are divided into five categories and 22 small classes according to their origin.

2.1 The overall factors

There are four overall major factors—the quantity of fans, Micro-blog attention, and micro-Blog and Micro-blog reviews. These three factors affect each other. The annual data and the average amount of daily dynamic changes are the main research target as shown in Table 1.

2.2 Information factors

Information factors include: Introduction, labels, vocation information, and educational information. In addition, the bloggers' VIP certification is also considered as one of the elements as shown in Table 2.

2.3 Content factors

The frequency of releasing and the status of forwarding, collection and review are shown in Table 3.

Table 1. The overall factors and sub-factors.

The quantity of Micro-blog attention	The quantity in March, 2013
	The quantity in March, 2012
	Average dynamic increment
The quantity of fans	The quantity in March, 2013
	The quantity in March, 2012
	Average dynamic increment
The quantity of Micro-blog	The quantity in March, 2013
	The quantity in March, 2012
	Average dynamic increment
	Maximum as favorite
The quantity of reviews	The average quantity of review
	the Maximum quantity of review

*This paper is supported by project on the government purchasing services from social organization, "research on network interactive mode of Public opinion guiding (2013)", from the social committee of Beijing municipal committee of the communist party of China.

Table 2. The information factor and sub-factors.

The information factor	Whether is authenticated users
	Whether has introduction
	Whether has labels
	Whether has vocation information
	Whether has education information
	the age of the Micro-blog

Table 3. Content factors and sub-factors.

The frequency of releasing	The releasing date of the last Micro-blog
	The Micro-blog average interval
The quantity of forwarding	Average quantity was reproduced
	Largest quantity was reproduced
The number of collections	Average quantity being collected
	Maximum quantity being collected
The number of reviews	Average quantity of reviews
	Maximum quantity of reviews

Table 4. Interactive factors and sub-factors.

Name citation with "@"	The average quantity of citations
	The maximum quantity of citations
Participate in the discussion topic with "#"	The average quantity of each participation
	The maximum quantity of participation
Add pictures	The mean of each Micro-Bo
	The maximum of each Micro-Bo
Add links	The mean of each Micro-Bo
	The maximum of each Micro-Bo
Forwarding others' articles	The total quantity
	The rate of forwarding

2.4 Interactive factors

The owner of micro-blog always interacts with friends. These factors include the situations of forwarding and citation as shown in Table 4.

2.5 Street factors

The construction of neighborhoods and communities micro-blog is also an important factor as shown in Table 5.

The dynamic increase of fans is used as the index to evaluate the micro-blog influence. Compared with the traditional static amount of fans, it has better performance to describe the influence change and activity level.

Micro-blog network influence = f (The dynamic increase of fans) = f (the quantity of fans in March, 2013—that in March, 2012).

By data mining tools such as parametric test and regression analysis and others, the correlation

Table 5. Street factors and sub-factors.

The covering rate of street of communities micro-blog	The quantity
	The covering rate
The quantity of street communities micro-blog attention	The maximum quantity
	The average quantity
	Average incremental
The quantity of street communities micro-blog fans	The maximum
	The mean
	Average incremental
The quantity of street communities micro-blog	The maximum
	The mean
	Average incremental

between the above-mentioned factors and community micro-blog influence is analyzed. With multi-party inspection of the correlation coefficient, key factors to improve the community micro-blog influence are selected.

3 THE KEY FACTORS

3.1 The quantity of Micro-blog concern

The quantity of fans, Micro-blog attention, micro-Bos are important overall factors for influence evaluation and there is a strong correlation between them. The study found that the most influential factor to the dynamic increase of community micro-Blogging fans is "Average dynamic increment of the quantity of Micro-blog attention" followed by "The quantity of Micro-blog attention in March, 2013", "The quantity of micro-Bo in March, 2013" and "the Average dynamic increment of the quantity of micro-Bo in March 2013". While the "The quantity of micro-blog attention in March, 2012" and "the quantity of micro-blog in March 2012" is relatively small.

3.2 Show identification

For the information factors, "whether you are authenticated users" has great influence to the dynamic increase of fans. In addition, the micro-blog Introduction and the label information also have some influence. For non-authenticated users, the influence of the label information is greater than the micro-blog Introduction. While the vocation and education information has no relation to the dynamic increment of Micro-blog fans. In addition, the age of micro-blog also has little influence to the dynamic increase of fans.

3.3 Micro-blog quality

The forwarding times and review times reflected the quality of each micro-blog. Overall, The forwarding times and review times of the community micro-blog are not high. Take those micro-blog in the last three months for example, the average forwarding times is 1.21, the maximum is 72 times; the average review times is 0.39 times and the maximum number is 36. While at the same period, the average forwarding times of the Beijing government micro-blog is 8.89 and the maximum is 919; the average review times is 3.06 times and the maximum number is 218.

Review times, micro-blog releasing frequency and forwarding times have a considerable influence on the dynamic increment of the fans, and influence gradually weakened. While the influence of the total quantity of forwarding and review times are much larger than those average value. This shows that the high-quality micro-Bo is very important to its influence.

3.4 *Continued to release micro-blog*

In terms of micro-blog releasing interval, the fans of micro-blog with high releasing frequency increased significantly, but for those with an interval greater than 3 months and 1 year, things are same. In other words, the community micro-blog release only on some certain days or never release for a whole year can not be effective in attracting fans. The only way to really expand its influence is to continue releasing micro-blog.

3.5 *Direct interaction*

Community micro-blog adopt the following direct interaction methods, such as citation name with "@", add "#" to participate topics, add pictures, add links and forwarding the articles of others. Their relations with the fans dynamic incremental descended with an order as adding pictures, forwarding the articles of others, citation name with "@", add "#" to participate topics and adding link.

3.6 *Model establishment*

The Analysis of relationship between the fans and the street where the Community micro-blog belongs to indicated that the number and penetration of Community micro-blog have no significant influence to the fans dynamic increase; while the average number of communities micro-blog concerns and micro-Bo have significant influence to the fans dynamic increase, and the impact of the former is more stronger than the latter. This shows that streets community has a leading function to the community micro-blog, and this function is not enhanced by the quantity, but the quality of the community micro-blog. Running several influential

community micro-blog has good leading role in the development of the communities' micro-blog.

4 CONCLUSIONS

It can be learned from the above key factors analysis that the following issues should be paid attention to enhance community micro-blog influence.

A. Continue to focus on other micro-blog, especially enthusiastic users, "opinion leaders" and those community residents. In the formation process of public issues, opinion leader community played a very important role (Zeng 2012).

It is not enough to rely on a large number of micro-Bo releasing to attract Internet users. Measures such as network interaction, focusing on users, should be taken to effectively enhance influence.

B. It is very important to show its identity in the virtual society full of uncertainty.

Internet users are more inclined to trust the official views. Authenticated community micro-blog or those with clear identify introduction are more attractive to the fan. Therefore, we should carefully fill in the micro-blog information table, disclose the community Identity, and make the community micro-blog position and service goals clear. In addition, the precise positioning of the micro-blog content, clearly stating topics of interest in the label will also help to enhance the influence of micro-blog (Lin 2010).

C. We should improve the content and quality of micro-blog. More contents concerned by internet users should be released, and the content and language should be interesting and attractive; more Hot Topic Discussion should be actively participate in. all these will enhance forwarding rate and the quantity of reviews.

D. Direct interaction with Internet users should be enhanced, the management of fans should be strengthened; communications with the fans and other Internet users should be made frequently through technical means. In the micro-blog, adding more pictures and increasing content interesting are the direct way to attract the attention of users, followed by forwarding articles and citation user name with "@".

E. Bring up community micro-blog "opinion leaders", and depend on them to promote and enhance the community micro-blog influence as a whole.

Government affairs micro blogs ought to abandon the copycat thinking, and strengthen its core competitiveness in content and form level; its influence can be established and consolidated (Li 2012). Finally, there is a need to focus on the details of the community micro-blog management and construction, so that we can fully take advantage of these "micro-power", increase community management methods, establish new dialogue channels between managers and residents, and provide a bridge and platform for residents to monitor and participate in community management effectively, and resolve the contradiction between community residents.

REFERENCES

Bai Lin, Look from the "sina micro-blog" the spread of micro blog, young writer, July 2010 pp: 242–243.

Fanxu Zeng, Guangsheng Huang, Public Opinion Leader's Online Community: Structure, Interaction and Policy Influence, Harvard-Yenching Institute Panel, Beijing Forum 2012.

Guoming Yu, Micro-blog: the influence of the generation mechanism and function space, Zhongguancun, April 2010, pp: 89–91.

Hui Li, Three kinds of micro-blog, micro-blog digestion government influence, new media front, Jan. 2012, pp: 55–58.

Kunmei Wen, Review: Micro-blog, and Chinese micro-blog information processing researc, Journal of Chinese information processing, Vol. 26, No. 6, Nov. 2012, pp: 27–37.

Session 8: Engineering management

Control Engineering and Information Systems – Liu (Ed)
© 2015 Taylor & Francis Group, London, ISBN 978-1-138-02685-8

One result about ruin probability in discrete time risk model with constant interest rate

X. Luo, G.Z. Cui & C.Y. Wang
PLA Information Engineering University, Zhengzhou, China

ABSTRACT: In this paper we consider the discrete time insurance risk model with interest rate. First we prove the surplus is Markov chain. Second we use the Markov chain get series expansion and the surplus distribution at ruin of the instantaneous.

1 INTRODUCTION

Ruin probability problem is the insurance company business and risk management will face important problem, not only such, this problem is also risk theory research priorities. Risk model the way according to charge the premium can be divided into continuous model and discrete model two kinds. Because insurance companies operating process and the time closely related, in recent years about risk model research has concentrated continuous model, and got a series of conclusion. Discrete model studies have relatively lags behind some. Discrete model is divided insurance company business into uniform period, the amount of the claim for each time period as independent and identically distributed random variables. Comparative continuous model to discrete the model considering the problems with insurance company more integrity. In this paper, we consider a class of introducing interest rate risk model, the introduction of the interest rate of reality, to strengthen in the current model under the condition of the central bank set interest rates more practical significance.

2 MODEL DESCRIPTION

Consider premium income as random variables and join the discrete risk model interest rates. Hypothesis in time to consider with constant interest rate δ, u is initial reserve for insurance company, then in the moment n, the insurance company for the accumulation of surplus:

$$U_n = u(1+\delta)^n + \sum_{i=1}^{n} X_i(1+\delta)^{n-i+1} - \sum_{i=1}^{n} Y_i(1+\delta)^{n-i}, \tag{1}$$

Among them, $\{X_i, i \geq 1\}$ and $\{Y_i, i \geq 1\}$ are independent and identically distributed sequence of the nonnegative random variables. They mean respectively $[i-1,i)$ time interval insurance company premium income and claims spending.

We assume X_i and Y_i expectations are limited, make

$$Z_i = Y_i - (1+\delta)X_i$$

Then Z_i is also a net loss of independent and identically distributed random variable sequence.

So the surplus process can be rewritten for:

$$U_n = u(1+\delta)^n - \sum_{i=1}^{n}(1+\delta)^{n-i}Z_i. \tag{2}$$

type arrangement (2)

$$U_n = u(1+\delta)^n - \sum_{i=1}^{n}(1+\delta)^{n-i}Z_i$$

$$= (u(1+\delta)^{n-1} - \sum_{i=1}^{n-1}(1+\delta)^{n-1-i}Z_i)(1+\delta) - Z_n$$

$$= U_{n-1}(1+\delta) - Z_n.$$

In order to guarantee the normal operation of insurance companies, it must add certain risk load $E[Z_i] < 0$. Set respectively $F_X(x)$, $F_Y(y)$ are distribution function of X_1, Y_1, then the then distribution function of Z_1:

$$H(z) = P\{Z_1 \leq z\}$$

$$= \int_0^{+\infty} P\{Y_1 - (1+\delta)X_1 \leq z \mid X_1 = x\}dF_X(x)$$

$$= \int_0^{+\infty} F_Y((1+\delta)x + z)dF_X(x).$$

Can be seen from type, the distribution of net loss determine jointly by the distribution of premium and the distribution of claims.

Define $\Psi_\delta(u)$ is the ruin probability of risk process U_n, i.e.

$$\Psi_\delta(u) = P\{\bigcup_{n\geq 0}(U_n < 0)\,|\,U_0 = u\}.$$

Let $T_\delta = \inf\{n \geq 0\,|\,U_n < 0\}$ be the time of ruin, obviously T_δ is stopping time, then

$$\Psi_\delta(u) = P\{T_\delta < \infty\,|\,U_0 = u\}.$$

Because of difficulty, the risk model with interest rate risk analysis model is currently one of the hot researches; the class of problems research is still not fully developed. (Yang 1998) said to bring interest rate risk model discrete time were studied by martingale method have the ruin probability index is given. (Sun 2002) for a class of discrete time constant interest rate risk model was studied, and the obtained several important risk issues, using mainly update distribution equation method. This paper discussed more system of discrete constant interest rate risk model, using the markov chain, with state transition probability giving ruins probability, etc, and then gives the show the surplus distribution at ruin moment and the surplus distribution at ruin of the instantaneous.

3 HOMOGENEOUS MARKOV CHAIN

The following assumption $\delta > 0$. First proof $\{U_n\}_{n\geq 0}$ is a homogeneous markov chain.

Lemma 1 $\{U_n\}_{n\geq 0}$ is a homogeneous markov chain and state transition probability is $K(x, \Gamma) = P\{x(1 + \delta) - Z_1 \in \Gamma\}$. Γ. is any interval on real sets.

Proof Let $\mathcal{F}_n$ be the σ-algebra generated by $\{U_k, k \leq n\}$.

$$P\{U_n \in \Gamma|\mathcal{F}_{n-1}\} = P\{U_{n-1}(1 + \delta) - Z_n \in \Gamma|\mathcal{F}_{n-1}\}$$

Obviously $U_{n-1} \in \mathcal{F}_{n-1}$, Z_n and $\mathcal{F}_{n-1}$ are independent, thus

$$P\{U_n \in \Gamma|\mathcal{F}_{n-1}\} = P\{U_{n-1}(1 + \delta) - Z_n \in \Gamma|U_{n-1}\}$$

Then

$$P\{U_n \in \Gamma|\mathcal{F}_{n-1}\} = P\{U_n \in \Gamma|U_{n-1}\}$$

state transition probability $K(n-1, x; n, \Gamma)$ is

$$P\{U_{n-1}(1 + \delta) - Z_n \in \Gamma|U_{n-1} = x\}$$
$$= P\{x(1 + \delta) - Z_1 \in \Gamma\}.$$

Can be seen from the type of state transition probability $K(n-1, x; n, \Gamma)$ Has nothing to do with the time n, i.e. $K(x, \Gamma)$.

4 THE SURPLUS DISTRIBUTION AT RUIN OF THE INSTANTANEOUS

Number Insurance company is policy-holder are very concerned about the solvency of insurance company, the problems of the research at surplus at ruin can make the insurant and policy-holder have a more comprehensive understanding the solvency of insurance company. To find out whether there will be insurance company have too many difficulties to cope with, we then discuss the surplus distribution of insurance company at ruin of the instantaneous.

On the basis of Gerber and Kass functions, we define the surplus distribution of insurance company at ruin of the instantaneous $Q_\delta(u, y)$, i.e.

$$Q_\delta(u, y) = P\{T_\delta < \infty, 0 < U_{T_{\delta-1}} \leq y\,|\,U_0 = u\}$$

Theorem 6 The surplus distribution of insurance company at ruin of the instantaneous $Q_\delta(u, y)$ have the following expansion

$$Q_\delta(u, y) = I(u < y)\bar{H}(u(1+\delta)) + \sum_{n=2}^{\infty}\int_0^\infty K(u, dx_1)\cdots$$
$$\int_0^y K(x_{n-2}, dx_{n-1})\int_{-\infty}^0 K(x_{n-1}, dx_n)$$

Proof

$$Q_\delta(u, y) = \sum_{n=1}^{\infty} P\{T_\delta = n, 0 < U_{n-1} \leq y\,|\,U_0 = u\}$$

$$= I(u < y)P\{T_\delta = 1, 0 < U_0 \leq y\,|\,U_0 = u\}$$

$$+ \sum_{n=2}^{\infty} P\{T_\delta = n, 0 < U_{n-1} \leq y\,|\,U_0 = u\})$$

$$= I(u < y)P\{U_1 < 0, 0 < U_0 \leq y\,|\,U_0 = u\}$$

$$+ \sum_{n=2}^{\infty} P\{U_1 > 0, U_2 > 0, ..., 0 < U_{n-1} \leq y,$$

$$U_n < 0\,|\,U_0 = u\}$$

$$= I(u < y)P\{u(1+\delta) - Z_1 < 0\}$$

$$+ \sum_{n=2}^{\infty} \int_0^\infty K(u, dx_1)\cdots\int_0^y K(x_{n-2}, dx_{n-1})$$

$$\int_{-\infty}^0 K(x_{n-1}, dx_n)$$

Theorem 7 The surplus distribution of insurance company at ruin of the instantaneous $Q_\delta(u, y)$ satisfy the following integral equation

$$Q_\delta(u,y) = I(u < y)\bar{H}(u(1+\delta))$$
$$+ \int_{-\infty}^{u(1+\delta)} Q_\delta(u(1+\delta) - z, y)dH(z)$$

Proof When $u < y, n \geq 2$, we have

$$P\{T_\delta = n, 0 < U_{n-1} \leq y \mid U_0 = u\}$$
$$= \int_0^\infty P\{T_\delta = n-1, 0 < U_{n-2} \leq y \mid U_0 = z\}K(u,dz)$$

Sum to type on both sides by n, we get following integral equation

$$Q_\delta(u,y) - \bar{H}(u(1+\delta)) = \int_0^\infty Q_\delta(z,y)K(u,dz).$$

When $z < 0$, we have $Q_\delta(z, y) = 0$. From lemma 1, we get

$$Q_\delta(u,y) = \bar{H}(u(1+\delta)) + E[Q_\delta(u(1+\delta)) - Z_1]$$
$$= \bar{H}(u(1+\delta)) + \int_{-\infty}^{u(1+\delta)} Q_\delta(u(1+\delta) - z, y)dH(z).$$

When $u \geq y$, we get following integral equation similarly

$$Q_\delta(u,y) = \int_0^\infty Q_\delta(z,y)K(u,dz)$$

i.e.

$$Q_\delta(u,y) = \int_{-\infty}^{u(1+\delta)} Q_\delta(u(1+\delta) - z, y)dH(z).$$

REFERENCES

Bower N.L., H.U. Gerber, J.C. Hickman, D.A. Jones, C.J. Nes2bitt. Actuarial Mathematics [M]. Society of Actuaries, Itasca, IL, 1986.

Sun Li Juan, Gu Lan. Ruin problems for the discrete time in surance risk mode [J]. Chinese Journal of Applied Probability and Statistics, 2002, (3): 293–294.

Yang H., Non-exponential bounds for ruin probability with interest Effect included, Scandinavian Actuarial Journal, 1(1998), 66–79.

Control Engineering and Information Systems – Liu (Ed)
© 2015 Taylor & Francis Group, London, ISBN 978-1-138-02685-8

Optimization model of vehicle scheduling in logistics enterprises

Q.S. Jiang
College of Information Engineering, Tarim University, Alar, Xinjiang, China

W. Xing
College of Computer Science and Technology, Zhejiang University, Hangzhou, Zhejiang, China

ABSTRACT: By analyzing common types of vehicle scheduling problem and combining the practical situation of logistics enterprises in south Xinjiang corps, this paper adopted the step-down research method by dividing route branches. Firstly, this study established relatively simple optimization model of vehicle scheduling by simplifying assumptions and research objects. Moreover, empirical analysis was also presented according to the real situation of South Xinjiang corps. And with lingo software programming, model results were in great agreement with reality.

1 INTRODUCTION

The ultimate aim of the paper was to investigate strategic problem of sustainable development of modern logistics enterprises in south Xinjiang corps and achieve optimal allocation of limited resources in logistics enterprises. To achieve this aim, basing on the wide collection of policy documents, statistical data and research results related to modern logistics development, this research found out optimized decision method for the development of logistics enterprises in south Xinjiang by analyzing the data obtained in the investigation.

Transportation is capable of creating place utility and time utility of products. Hence, transportation decision is directly related to logistics cost and customer service level of enterprises. Transportation decision includes a wide range of contents, mainly: choice of transportation modes, choice of transportation routes and networks, choice of the number of deliveries and so on. Due to the special geographical location of south Xinjiang, there are limited transportation modes. Besides, as the most important problem in transportation decision, choice of transportation routes and networks is directly related to distribution cost as well as the time and accuracy of products delivered to customers. Moreover, the number of deliveries is directly related to distribution cost. Therefore, choice of transportation routes, that of transportation networks and that of the number of deliveries are the main problems in vehicle scheduling of logistics enterprises.

Choice of transportation routes can be divided into four types.

Starting point is different from terminal point, via other nodes on the way.

Starting point is the same with terminal point, but via different nodes on the way.

There are several starting points and several terminal points, without intermediate nodes.

There are several starting points and several terminal points, with intermediate nodes.

There are main traditional operations research methods for solving logistics scheduling problem such as linear programming, integer programming, unconstrained optimization, nonlinear programming, dynamic programming (Zhao & Dan, 2008) and so on. Among the methods, linear programming model can be used to deal with transportation decision problem in logistics, while storage theory can deal with the storage decision problem in logistics decision. In addition, assignment problem, loading problem, location problem etc. can be solved by integer programming.

To solve logistics scheduling problem, besides the traditional optimization algorithms of operations research, there are also modern optimization algorithms such as tabu search (Wang et al. 2010) simulated annealing (Yang et al. 2010), genetic algorithm (Gong et al. 2004), ant colony algorithm (Zhang & Liang, 2007) and so on.

South Xinjiang production and construction corps were established basing on the special geographical environment in Xinjiang in specific historical period. Now, south Xinjiang corps contain four divisions: the first division, the second division, the third division and the fourteenth division. There are fewer research objects in the corps. And the first division governs 16 regiments and its agricultural development is the most competitive and representative in the four divisions. Hence, the first division was taken as main research object of this study. Logistics service product is mainly the transportation of agricultural materials like cotton, chemical fertilizer, seeds and so on. Owing to fewer scheduling objects, scheduling problem in logistics service becomes relatively simple. Therefore, traditional linear programming model was adopted to deal with relevant problem.

Typical form of linear programming model is presented as follows:

$$\min \quad z = f(x) \tag{1}$$
$$s.t. \ g_i(x) \le 0 \quad (i = 1, 2, ..., m) \tag{2}$$

Model consisting of both (1) and (2) belongs to constrained optimization, while unconstrained optimization is obtained if the model only consisting of formula (1). In the model, $f(x)$ is defined as objective function and $g_i(x)$ denotes constraint conditions.

2 CONSTRUCTION OF MODEL FOR LOGISTICS SCHEDULING PROBLEM IN CORPS

2.1 Model assumptions

1. There are no severe natural disasters in south Xinjiang in recent years.
2. Dispatching center is provided with adequate resources for scheduling and adequate transportation capability.
3. There is only a type of transport vehicles in the company in simple model, while there are two types in complex scheduling model.
4. No time quantum is limited in scheduling process.

2.2 The construction of a simplified scheduling model

A simplified vehicle scheduling problem is presented first: there are p freights with volume q for each in resource point s. And c denotes freight rate per kilometer per ton, while w denotes total supply. Now there are m demand points, k routes and altogether a_i nodes on the route i. Moreover, the distance between logistics center s and the node j on the route i is $d_{ij}(i = 1, 2, ..., k; j = 1, 2, ..., a_i)$. It is presumed that altogether n vehicles are dispatched. And the number of vehicles dispatched to the route i is $n_i(i = 1, 2, ..., k)$, while that dispatched to the node j on the route i is

$$n_{ij}(i = 1, 2, ..., k; j = 1, 2, ..., a_i).$$

Besides, the demand of demand point i is

$$g_{ij}(i = 1, 2, ..., k; j = 1, 2, ..., a_i).$$

Basing on the data above mentioned, under constraint conditions that demand of each node is satisfied and cost is minimized, the optimal scheme with minimum total scheduling vehicles and shortest total journey is expected.

The number of dispatched vehicles n must be less than that of total vehicles, that is,

$$n \le p$$

The sum of the vehicles dispatched to all routes is equal to the number of dispatched vehicles. Hence,

$$\sum_{i=1}^{k} n_i = n(i = 1, 2, ..., k)$$

Supply of the node j on the route i is not less than demand. So,

$$n_{ij}q \ge g_{ij}(i = 1, 2, ..., k; \ j = 1, 2, ..., a_i)$$

All dispatched vehicle numbers of n, n_i and n_{ij} must be integers.

The minimum of costs Z was taken as objective function and the costs includes fixed royalty U and transportation cost T. Among the costs, workers' wages u_1 and maintenance charge of each vehicle u_2 and so on are contained in the fixed royalty. Hence, $U = u_1 + u_2 p$ is thus obtained. Transportation cost T depends on volume, freight rate and transport mileage.

Therefore, the objective function can be written as follows:

$$T = \sum_{i=1}^{k} \sum_{j=1}^{a_i} n_{ij}q d_{ij} c(i = 1, 2, ..., k; j = 1, 2, ..., a_i)$$

Therefore, the objective function can be written as follows:

$$Z = U + T$$
$$Z = u_1 + u_2 p$$
$$+ \sum_{i=1}^{k} \sum_{j=1}^{a_i} n_{ij}q d_{ij} c(i = 1, 2, ..., k; j = 1, 2, ..., a_i)$$

In conclusion, optimization model was established as follows:

$$\min Z = u_1 + u_2 p$$
$$+ \sum_{i=1}^{k} \sum_{j=1}^{a_i} n_{ij}q d_{ij} c(i = 1, 2, ..., k; \ j = 1, 2, ..., a_i)$$
$$s.t. \quad n \le p$$

$$\sum_{i=1}^{k} n_i = n(i=1,2,\dots,k)$$

$$n_{ij}q \ge g_{ij}(i=1,2,\dots,k;\; j=1,2,\dots,a_i)$$

n, n_i and $n_{ij}(i=1,2,\dots,k;\; j=1,2,\dots,a_i)$ are all integers.

2.3 Taking the first division in corps as an example

The station of the first division is in Aksu, south Xinjiang, China. With its fame dating back to the ancient Silk Road, Aksu is an important hinge on Eurasian Continental Bridge and provided with frequent logistics transportation. As mentioned above, the first division governs 16 regiments. And Aksu was taken as logistics center, while all regiments were taken as logistics nodes for the research. To investigate the simplified scheduling model, this paper took the transportation scheduling of chemical fertilizer in the first division as an example. The following are corresponding assumptions.

Now, with the headquarter of a chemical fertilizer company in Aksu and its fertilizer supplied for 16 surrounding regiment farms, the company expects to achieve cost-optimal vehicle scheduling scheme. Presently, there are 120 transport vehicles of the same type in the company and the volume of each vehicle is 2 tons. Besides, average maintenance charge of vehicles is 50 yuan, while monthly total wages of the workers are 12000 yuan. And cost-optimal vehicle scheduling table was obtained basing on the simplified model established above.

Basing on the known conditions and the simplified model mentioned above, each regiment farm in the first division was numbered first according to route distribution and there were 5 branches.

The appendix illustrates the route number of each regiment farm, the demand and the distance between each regiment farm and city center.

The simplified model above established was processed by lingo11 programming. The results were revealed in Table 1, and the detailed procedures were shown as follows:

```
model:
sets:
rows/1..5/:n1, a;
cows/1..11/:;
distn(rows, cows):d, n, g;
endsets
min = u1+u2*p+@sum(distn(i, j):n(i, j)*q*d(i, j));
n2 < p;
@sum(rows(i):n1(i)) = n2;
@for(rows(i):@for(cows(j):n(i, j)*q > g(i, j)));
@gin(n2);
@for(rows(i):@gin(n1(i)));
@for(distn:@gin(n));
@for(rows(i):@sum(cows(j):n(i, j)) = n1(i));

data:
u1 = 12000;
u2 = 50;
p = 120;
q = 2;
c = 2;
w = 250;
m = 16;
k = 5;
a = 1, 11, 1, 2, 1;

d =
1 0 0 0 0 0 0 0 0 0 0
30 40 60 90 120 130 128 140 150 160 165
70 0 0 0 0 0 0 0 0 0 0
20 30 0 0 0 0 0 0 0 0 0
100 0 0 0 0 0 0 0 0 0 0;

g =
16 0 0 0 0 0 0 0 0 0 0
15 16 12 16 20 14 12 10 12 10 10

16 0 0 0 0 0 0 0 0 0 0
10 12 0 0 0 0 0 0 0 0 0
16 0 0 0 0 0 0 0 0 0;
@ole('E:\LINGO11\work\eszydiaodu.xls',n,b, e) = n
n1, n2;
enddata
end
```

3 CONCLUSION

The results indicate that the number of vehicles could satisfy the demand of each regiment farm. Moreover, with the processing of lingo software, the demand was not only a scheduling scheme, but achieved maximum economic benefit. And the model is quite stable. In addition, the model is of great significance for wider application because only with suitable data change, more relatively complex problems can be solved.

ACKNOWLEDGMENTS

This research is supported by Youth special funds for scientific and technological innovation of the Xinjiang production and Construction Corps (2011CB001), the doctor foundation of the Xinjiang production and Construction Corps (2013BB013), the doctor Foundation of Tarim University (TDZKBS201205) and the national Spark Program Project (2011GA891010).

Here, thanks to the relevant company and organizations.

REFERENCES

Gong Y.C., X.F. Guo and X.L. You et al, "Vehicle Scheduling Problem Research of Logistics Distribution Basing on Genetic Algorithm", Mathematics in Practice and Theory, Vol. 34, pp. 93–97, 2004.

Wang X.B., H.J. Lu and H.M. Zhang, "Hybrid Tabu Search of Dynamic Scheduling Problem in Logistics", Computer Engineering and Applications, Vol. 46, pp. 228–231, 2010.

Yang S.L., H.W. Ma and T.J. Gu, "Simulated Annealing of Time-Varying Vehicle Scheduling Problem with Time Window", Operations Research Transactions, Vol.14, pp. 83, 2010.

Zhang Y., Y.C. Liang, "Improved Ant Colony Algorithm Basing on Routing Optimization", Computer Engineering and Applications, Vol. 43, pp. 60–63, 2007.

Zhao J., Q. Dan, Mathematical Modeling and Mathematical Experiments, 3rd edition, Beijing: Higher Education Press, 2008.

APPENDIX

Table 1. Scheduling table of the simplified scheduling model (Unit: Car).

| R | N | | | | | | | | | | |
| | NV | | | | | | | | | | |
	1	2	3	4	5	6	7	8	9	10	11
Route 1	8	0	0	0	0	0	0	0	0	0	0
Route 2	8	8	6	8	10	7	6	5	6	5	5
Route 3	8	0	0	11	0	0	0	0	0	0	0
Route 4	5	6	0	0	0	0	0	0	0	0	0
Route 5	8	0	0	0	0	0	0	0	0	0	0

(N: number; NV: the number of vehicles; R: route).

Table 2. Regiment farms distribution on each route.

| R | N | | | | | | | | | | |
	1	2	3	4	5	6	7	8	9	10	11
Route 1	5										
Route 2	6	7	8	16	9	10	12	11	13	14	15
Route 3	3										
Route 4	1	2									
Route 5	4										

(N: number; R: route).

Table 3. Distance between regiment farms of each route and headquarter (Unit: km).

| R | N | | | | | | | | | | |
	1	2	3	4	5	6	7	8	9	10	11
Route 1	10										
Route 2	30	40	60	90	120	130	128	140	150	160	165
Route 3	70										
Route 4	20	30									
Route 5	100										

(N: number; R: route).

Control Engineering and Information Systems – Liu (Ed)
© *2015 Taylor & Francis Group, London, ISBN 978-1-138-02685-8*

A study on DEA model with the optimal virtual DMU

Y.S. Xiong & Z.W. Wang

College of Transportation and Logistics, Central South University of Forestry and Technology, Changsha, China

ABSTRACT: The main problem of the traditional DEA model is that the distribution of index weight is not reasonable and cannot sort the decisionmaking units effectively. Many scholars constantly try to use various methods to improve the traditional DEA model, but the current research results either require value anticipation or they don't have the unique solution. The optimal virtual decision-making units are introduced to the DEA model in this paper, and by comparing the optimal virtual decision-making units with the actual decision-making units to determine the public weights, the decision-making units can be sorted effectively. Through example verification, this method can enhance the effectiveness of the decisionmaking units sorting. Comparing with other methods, this method can assign the unique weight reasonably without value anticipation, and further expand the scope of the decision making units' efficiency and improve their sorting effectiveness.

1 LITERATURE REVIEW

There are many common efficiency evaluation methods and DEA is just one of the commonest. However, as the research continues, more and more scholars question traditional DEA model (such as CCR). When DEA (CCR) is used to evaluate the efficiency, all decision-making units choose the weights which are the most beneficial to themselves, but the actual difference in the importance among these indexes is not taken into consideration. During the actual calculation process, different decision-making units are chosen and though the same input index and output index are adopted, the final weights are very inconsistent, which is contrary to actual situation. The results are that the evaluation discrimination degree is not high and the relatively effective sorting among these decision-making units is difficult. In order to resolve these problems, scholars at home and abroad propose many improvement methods, which are generally divided into three classes. The first class is cross evaluation method, with the cross efficiency evaluation model proposed by Sexton and others (1986) as the representative. Then, Doyle and Green (1994) proposed the two-step approach to resolve the non-uniqueness of cross efficiency. Wu Jie (2008) combined both of them and put forward DEA cross evaluation model which considered the interval mean cross efficiency of all weight information. Though these evaluation models resolve the sorting problem of decision-making units, the weight allocation problem is not resolved. The second class is the preference evaluation method, with the cone ratio C^2 WH model proposed by A. Charnes and others

(1989) which can adjust the index weight reflection preference as the representative. And on this basis, R.G. Thompson and others (1992) proposed the wider range of assurance region analysis method (DEA/AR). Then, Wu Yuhua and others (1992) constructed DEA model based on AHP constraint cone, Guan Zhongcheng (2007) proposed DEA model with fuzzy preference, and Yang Feng and others (2011) advanced DEA model which considered its own technical efficiency and weight preference simultaneously. Dong Weiwei (2013) proposes DEA model with independent subsystems. These improvements resolve the problems of decision-making unit sorting and weight allocation, but an important weakness is also brought, that is, value anticipation is required, which is not a big problem when the information is sufficient. However, such ideal condition is very rare and what is more is the condition of insufficient information, so the value anticipation will quite probably have a certain prejudice or inconsistency with the actual and thus can't be applied widely. The third class is the introduction of virtual DMU evaluation method, the DEA model with an introduced optimal DMU put forward by Li Guo (2000), Yang Yinsheng (2003) and others as the representative, but the weight is not unique and thus the evaluation results possess much high uncertainty. On this basis, Liu Yingping (2006), Sun Kai (2008) and others proposed DEA model introducing the optimal and worst virtual decision-making units that is to utilize the worst virtual decision-making unit as the constraint to solve the common weight after working out the infinite number of solutions which accord with the optimal virtual decision-making unit. Pan Yuhong and others (2011) applied this idea in the study on public rental housing allocation and made certain achievements. This model doesn't require value anticipation, which is one step closer to the resolution of weight allocation and decision-making unit sorting, but the possibility of non-unique solution still exists theoretically after introducing two virtual decision-making units.

This paper suggests that ideal DEA model shall meet two requirements, namely reasonable weight allocation under the premise of no value anticipation and effective sorting. Obviously, the above-mentioned three improvement methods can't meet the two requirements simultaneously, but the third class of introducing virtual DMU evaluation is apparently closest to ideal DEA model.

2 BASIC MODEL OF DEA-CCR MODEL

Suppose there are n decision-making units DMU_j (j = 1,2, ..., n) and each DMU has m input indexes and s output indexes. $X_j = (x_{1j}, x_{2j}, ..., x_{mj})^T > 0$, $Y_j = (y_{1j}, y_{2j}, ..., y_{sj})^T > 0$ and j = 1, ..., n are respectively used to represent the input and output vector of DMU_j, in which x_{ij} and y_{rj} represent the ith input quantity and the rth output quantity in DMU_j. $\omega = (\omega_1, \omega_2,..., \omega_m)^T$ is used to indicate the input weight vector and $\mu = (\mu_1, \mu_2 ..., \mu_s)^T$ represents the output weight vector. Then DMU_j efficiency of the evaluation unit can be represented as:

$$E_j = \frac{\mu^T Y_j}{\omega^T X_j}(j = 1,2, ..., n) \quad (1)$$

CCR model which solves the efficiency value could be represented by the following linear programming form:

$$\max \mu^T Y_0$$
$$\text{s.t. } \omega^T X_j - \mu^T Y_j \geq 0, j = 1, ..., n$$
$$\omega^T X_0 = 1 \quad (2)$$
$$\omega = (\omega_1, ..., \omega_m)^T \geq 0$$
$$\mu = (\mu_1, ..., \mu_s)^T \geq 0$$

Formula (2) or its dual model is usually used to evaluate DMU. If the solutions of Formula (2) ω^* and μ^* meet $\mu^{*T}Y_0 = 1$, then DMU_0 is weakly DEA efficient; if the solutions of Formula (2) meet the conditions of $\omega^* > 0$, $\mu^* > 0$ and $\mu^{*T}Y_0 = 1$, then DMU_0 is DEA efficient; and if the solutions of Formula (2) meet $\mu^{*T}Y_0 < 1$, DMU_0 is non-DEA efficient.

3 CONSTRUCT DEA MODEL WITH THE OPTIMAL VIRTUAL DECISION-MAKING UNIT

The original idea of the third improvement method is to utilize Formula (2) to introduce an ideal DMU with the minimum input and maximum output into a group of DMU to be evaluated, the ideal DMU is undoubtedly DEA efficient to other DMU, the weight determined with the efficiency of ideal DMU being maximum as the objective should be said to be reasonable and other DMU all adopt the weight of the ideal DMU as the common weight, thus forming an uniform evaluating standard. On this basis, Formula (1) is utilized to solve the efficiency of all DMU and carry out sorting. However, as the weight is solved through the linear programming, there are probably an infinite number of solutions most favorable for ideal DMU under the condition of only one group of ideal DMU for constraint. In order to resolve the problem of weight non-uniqueness, scholars later introduce one worst virtual decision-making unit on this basis, namely select a group of solutions which can make the worst DMU efficiency value the minimum among the infinite groups of solutions most favorable for the optimal decision-making unit. The common weight solved through the improvement and meeting the requirements is reduced from an infinite number to finite number and the probability of non-uniqueness still exists theoretically.

3.1 Introduction of the optimal virtual decision-making unit

As the third class of improvement method can't determine the unique weight, the sorting results based on this also have the probability of non-unique solutions, which is a great weakness. The idea of the paper is to adopt nonlinear programming to solve the weight, which is primarily in accordance with the weight calculation method in the grey correlation analysis and only introduces an optimal virtual decision-making unit to solve and determine the unique common weight. The optimal decision-making unit has the highest efficiency and should possess the minimum input index value and maximum output index value among all the decision-making units. Therefore, in practice, the minimum input index value and maximum output index value could be screened among a group of actual decision-making units to be evaluated to construct the optimal virtual decision-making unit DMU_0 and then compare the optimal virtual decision-making unit with the actual decision-making unit. The more similar the two are, the greater the correlation exists among the corresponding sequences and the bigger the weight and vice versa.

Thus, it can be seen that the input index and output index of other actual decision-making units are recorded as:

$$X_i = (X_{1i}, X_{2i}, ..., X_{mi})^T,$$
$$Y_i = (Y_{1i}, Y_{2i}, ..., Y_{si})^T$$

Then the input index and output index of the optimal virtual decision-making unit DMU_0 are:

$$X_0 = (X_{10}, X_{20}, ..., X_{m0})^T,$$
$$Y_0 = (Y_{10}, Y_{20}, ..., Y_{s0})^T$$

which must meet the following conditions:

$$X_0 = \min(X_{1i}, X_{2i}, ..., X_{mi}),$$
$$Y_0 = \max(Y_{1i}, Y_{2i}, ..., Y_{mi})$$

3.2 Determination of common weight

1. Calculation of the input common weight
   Correlation between X_0 and X_i

$$r_i = \frac{1}{n} \sum_{k=1}^{n} \frac{\Delta\min + \xi\Delta\max}{|X_0(k) - X_i(k)| + \xi\Delta\max}$$

In the equation,

$$\Delta\min = \min_i \min_k |X_0(k) - X_i(k)|,$$
$$\Delta\max = \max_i \max_k |X_0(k) - X_i(k)|$$

and ξ are the resolution ratio, which is between 0 and 1, usually 0.5.

After normalization processing of r_i, the input common weight of various factors X could be acquired

$$\omega_i^* = \frac{r_i}{\sum_{i=1}^{m} r_i}$$

2. Calculation of the input common weight
   Correlation degree between Y_0 and Y_i

$$r_i = \frac{1}{n} \sum_{k=1}^{n} \frac{\Delta\min + \xi\Delta\max}{|Y_0(k) - Y_i(k)| + \xi\Delta\max}$$

Wherein,

$$\Delta\min = \min_i \min_k |Y_0(k) - Y_i(k)|,$$
$$\Delta\max = \max_i \max_k |Y_0(k) - Y_i(k)|$$

and the value of ξ is the same as above.

After normalization processing of r_i, the output common weight of various factors Y could be acquired

$$\mu_i^* = \frac{r_i}{\sum_{i=1}^{s} r_i}$$

3.3 Efficiency evaluation

Substitute the common weights $\omega^{*T} \geq 0$ and $\mu^{*T} \geq 0$ into Formula (1) and get

$$E_j^* = \frac{\mu^{*T} Y_j}{\omega^{*T} X_j} (j = 1, 2, ..., n) \tag{3}$$

Find the efficiency of various decision-making units and then sort the decision-making units according to the size of $E_j^* (j = 1, 2, ..., n)$.

4 CASE STUDY

In order to compare with the third class of improvement methods mentioned above and verify the effectiveness of the models, take the data in Literature (Liu et al. 2006) as the example and conduct the contrast calculation, with the results as shown in Table 1.

Through comparison, the following two points could be seen:

First, both Product 1 and Product 3 identified by CCR model are optimal, which are against the fact; the paper identifies Product 1 as optimal, followed by Product 3, and Product 2 is the worst, which conforms to actual situations and is consistent with the results in Literature (Li et al. 2000) and Literature (Liu et al. 2006), which proves that the method resolves the problem of effective sorting. Besides, from the efficiency value range of the decision-making unit, the value is

Table 1. Comparison on the results based on different evaluation methods.

Product	CCR model (sorting)	Literature (Li et al. 2000) method (sorting)	Literature (Liu et al. 2006) method (sorting)	The method in this paper (sorting)
1	1.0000(1)	0.9131(1)	0.9231(1)	1.0000(1)
2	0.3556(3)	0.2108(3)	0.1998(3)	0.2166(3)
3	1.0000(1)	0.4455(2)	0.4143(2)	0.4488(2)

0.9131–0.2108 = 0.7023 in Literature (Li et al. 2000), 0.9231 – 0.1998 = 0.7233 in Literature (Liu et al. 2006) and 1.0000 – 0.2166 = 0.7834 in the present method. It shows that compared with Literature (Li et al. 2000) and Literature (Liu et al. 2006), the present method is more effective in improving the efficiency value range of the decision-making unit and the comparability of the decision-making unit.

Second, concerning the reasonable allocation of weight, Literature (Li et al. 2000) introduces one optimal decision-making unit in the evaluation, but the satisfactory weight value has an infinite number of solutions theoretically, which is difficult to screen. Literature (Liu et al. 2006) introduces one worst decision-making unit again on the basis of Literature (Li et al. 2000), trying to resolve the weight screening problem by adding constraint conditions. However, just as Literature (Liu et al. 2006) acknowledges, after introducing two extreme decision-making units, only the uncertainty of weight selection could be reduced and the unique weight still can't be determined theoretically. While the present method only introduces one optimal decision-making unit, which reduces the amount of calculation, and equation calculating is adopted, so the unique weight could be determined and the problem of reasonable weight allocation on the premise of no value anticipation is effectively resolved.

5 CONCLUSION

When CCR model sorts the decision-making units, it adopts different weight allocation on indexes according to the advantages of the decision-making units, which could probably cause that different decision-making units have different evaluation standards and often several decision-making units are effective simultaneously and can't be sorted accurately. Later scholars have tried to resolve the problem, but they haven't resolved the problem of reasonable weight allocation and simultaneously effective sorting under the premise of no value anticipation. The paper constructs a DEA model with the optimal decision-making unit to resolve the weight and sorting problems. The model introduces an optimal decision-making unit and works out the unique weight. From the modeling process, it can be seen that no value anticipation is required to make different decision-making units and it has the same evaluation standards and finally the weight is substituted into CCR model to

work out the efficiency of these decision-making units. Through the comparative analysis on the studies made by predecessors, the present method can improve the efficiency value range of the decision-making units and more effectively sort the decision-making units.

REFERENCES

Charnes A, Cooper W, Wei Q L, Huang Z M. Cone ratio data envelopment analysis and multi-objective programming [J]. International Journal of System Science, 1989,20(7):1099–1118.

Dong Weiwei. A New Method of Decision-making Unit DEA Sorting with Independent Subsystems [J]. Journal of Shandong University (Natural Science), 2013, 48(1):89–99.

Guan Zhongcheng, Xu Hui and Xiong Huiqin. Application of DEA Based on Fuzzy Preference in the Scientific Research Institution Evaluation [J]. Research Management, 2007, 28(2):9–14.

Li Guo, Shen Xiaoyong and Wang Yingming. A New Method of Sorting Decision-making Units [J]. Forecast, 2000, 19(4):51–53.

Liu Yingping, Lin Zhigui and Shen Zuyi. Data Envelopment Analysis Method for Effectively Distinguishing Decision-making Units [J]. Systems Engineering Theory and Practice. 2006(3):112–116.

Pan Yuhong, Xiong Yuanjun and Meng Weijun. Study on Public Housing Rental Method Based on Introduced Ideal Decision-making Unit DEA [J]. Urban Development Research, 2011,18(8):14–26.

Sun Kai and Ju Xiaofeng. Industrial Enterprises Technological Innovation Ability Evaluation Based on DEA Model [J]. Systems Engineering-Theory Methodology Application, 2008, 17(2):134–145.

Thompson R G, Lee E, Trall R M. DEA/AR efficiency of U.S. independent oil/gas producers over time[J]. Computers and Operations Research, 1992, 19:377–391.

Wu Jie and Liang Liang. An Interval Cross Efficiency Sorting Method Considering All Weight Information [J]. Journal of Systems Engineering and Electronics. 2008, 30(10):1891–1894.

Wu Yuhua, Zeng Xiangyun and Song Jiwang. DEA Model with AHP Constraint Cone [J]. Journal of Systems Engineering. 1999, 14(4):330–333.

Xu Xiangpeng. Application of Improvement-based DEA Model in the Efficiency Evaluation [J]. Research of Finance and Accounting, 2011, (6):60–62.

Yang Feng, Yang Shenshen, Liang Liang and Xu Chuanyong. Study on Decision-making Unit Sorting Based on Common Weight DEA Model [J]. Journal of Systems Engineering. 2011, 26(4):551–557.

Yang Yinsheng and Li Hongwei. Non-uniform Evaluation of Green Products Based on Data Envelopment Analysis (DEA) Model [J]. Chinese Journal of Mechanical Engineering, 2003, 14(11):964–966.

Research and application of project teaching method in teaching of "Management Information System"

J. Liu

School of Finance, Qilu University of Technology, Shandong, Jinan, China

ABSTRACT: The project teaching can help students to deeply understand the teaching objective. It also can help the students to can establish courage of innovation, grasp the ideas and methods to analyze and solve problems. Then they can achieve the requirements of professional knowledge and skills finally. The application of project teaching method in the course of "Management Information System" obtained successful teaching results. It proves the practicability of the project teaching method. This paper explores the project teaching system and its realization mechanism.

1 WHY TO APPLY PROJECT TEACHING METHOD

The major of Computer information management established for junior college students should focus on "specialists" rather than "generalists". Namely, the course setting should focus on professional knowledge and application, attaches great importance to professional theories especially professional skills. So that the students will be equipped with good occupation morals and proficient operation skills. So they will be competent for informationization construction's related work in all kinds of enterprises after graduation.

"Management Information System", as the core course of major of computer information management, relates to the basic knowledge of many industries, covers many knowledge points, thus the teachers are required holding strong comprehensive ability. However, classroom teaching is taken as the current major teaching mode, and the teachers spend the majority of time in preparing and explaining theory in textbooks, and ignore practical content; in addition they often cram in the class, thus they are tired but with no sense of achievement. In this imperfect teaching mode, students have no chance to access real project training, and they have no faith to apply what they learn into practice, thus it is un-doubtful that they will lose motivation to learn the course for they believe the content is boring. Hat's worse, they may sleep, read novels, play mobile phone or chat in the class, followed by poor teaching effect. At the end of the semester, most students only have been accustomed to cramming for the final test. How can they really equip themselves with the theories and practical ability required by this course.

Although many universities have realized this problem and adopted advanced teaching methods. But this has not thoroughly changed the traditional teaching mode that is featured in classroom theory cram. In order to resolve the problems mentioned above, teachers must reform the teaching mode and method according to the needs of curriculum and students. Project teaching has certainly becomes the best choice.

At present, easy-to-difficult step-by-step teaching rule is followed in my class, and project teaching system is adopted in teaching, including four hierarchies, namely basic knowledge and skills, professional knowledge development, comprehendsive practice ability and innovation ability. The system also includes three links, namely experiment, training design and extracurricular practice. Students' ability to analyze and solve problems is increasingly fostered, They are provided teaching environment rich in subjective consciousness.

2 INTRODUCTION OF PROJECT TEACHING METHOD

2.1 Definition

In many specific professions, project is the process to create valuable products. For example, information management can involve website development and maintenance, development and application of information management system, LAN construction and other projects.

The project teaching method is firstly put forward in early twentieth Century by American education scholars. The following two classical definitions are widely adopted in literature:

Frey defines it that, "With the project teaching method, learners, work on some field in groups and implement a project. Team members self-plan and execute their work and usually end with a visible product (such as equipment, instruments, documents, performance and so on). In the project teaching method, the key does not lie in its final product but the independent process to manufacture the products by students themselves." [1]

The definition in Baidu Encyclopedia is that, "cross-curriculum subject is also known as project teaching method which is adopted in teaching via "project". In order to enable the students to follow a complete way to solve problems, the "project" covers many courses. With project teaching method, teachers will tasks the students who will independently finish the project themselves. Namely, they will independently complete information collection, project design, project implementation and final evaluation. In the whole process they will well understand the basic requirements of procedures".[2]

The above definition shows that project teaching method is open and student-centered teaching form. It emphasizes the formulation of plan, process to solve the problems, setting of corresponding teaching environment and other important factors. With the help of project teaching method, teachers design tasks which are given to the students who are divided into groups and make plans according to the requirements of project, then jointly complete the task. The teacher is the guider. This means the learning process will transform from passive accepting into active practice to well complete the project.

2.2 *Characteristics and advantages of project teaching method*

This method is most featured in the innovation that the teachers complete the teaching and guidance according to the actual requirement of project rather than just cram the students with theory and static knowledge on textbooks. The project was born from actual needs of society; its theme is closely related with the real world while the students would like independent thinking and practice to indeed apply theory into practice in active learning process.

These characteristics determine the great advantages of project teaching method. That is to say, the traditional imperfect way that teachers are the dominant of the class and take explaining theory as center is transformed into new pattern to be students-centered and practice-focused.

3 THE CONCRETE IMPLEMENTATION METHOD OF PROJECT TEACHING METHOD IN "MANAGEMENT INFORMATION SYSTEM"

3.1 *To design effective project*

Before implementing the project teaching, teachers should firstly survey students' current learning status and characteristics of the teaching content and study the feasibility of project to be taught. In the implementation process, it is critical to design and develop specific project tasks, maintain the integrity of the project as far as possible. Teachers must focus on the integration of theory and practice in teaching. So that the students will enjoy the process to participate into the mission, and their professional skills will be greatly strengthened. Through the exploration of practice, some points are summarized about how to design and develop project:

Firstly, typical and practical project tasks should be selected according to the needs of the society, especially the small and medium-sized enterprises' recruitment needs, so as to arouse students' enthusiasm.

Secondly, the project must be designed based on students' actual theoretical knowledge and professional skills taken full account, and the difficulty should be in line with their current level. Teachers can't be anxious for success in order to avoid the students' fear of difficulties.

Thirdly, the project selected must embrace fun factors to drive students' enthusiasm.

Fourthly, many projects should be designed, so that the students can select one according to their interests and their own professional level.

3.2 *To make requirements for project implementation clear*

In order to help students to develop project plan, specific requirements shall be clearly raised to students before the implementation of project:

The first, students should collect data extensively and review the theories learned before.

The second, each team shall submit the report of demand analysis and description of project design. At the end of the project, each of the students shall make a submission of project implementation.

The third, never be late or leave early, never be absent from school. If there is no special reasons, any student who is absent from class for 4 times will be scored 0.

The fourth, they should mutually respect, with solidarity and cooperation. They should firmly believe to successfully complete the project tasks as is required. When any problem arises, they should actively look for solutions.

The Fifth, they should never be satisfied with the existing achievements, but re-examine their own process with thought of innovation and development and work out better solutions.

3.3 To help students to well understand the project and divide groups

Students should well know the project objectives and requirements, which is a basic prerequisite to smoothly finish project task. When the project list is given, teachers should help students to interpret the content of project and select appropriate topics.

Since the students are weak in theory and practice of project management, teachers should make corresponding guidance. The student at high grade who successfully complete the project can be asked to introduce the preparation of project (including how to choose the topic, how to prepare relevant data, how to review theories and others' experience), specific methods adopted in the process of project implementation (including how to complete the report of demand analysis, outline design and detailed design, as well as how to select the programming tool and other points needing attentions In the coding process), and project submission and evaluation (including program test and writing of curriculum design report). In the process of theirs introduction, the teacher should take some necessary explanation to ensure that the students can understand better. In addition, teachers should give answers and guidances on the questions raised by students.

Theoretically students should consider their characteristics as well as their strengths and weaknesses when they independently form project team. However, teachers should try to stop students with good results forming team, so that the students with poor performance or popularity will be ignored.

3.4 To determine the project division and planning

After determining the project, the teams should hold project meeting to elect leader and deputy leader. The team leader is directly responsible for the project. The deputy leader organizes the preparation and improvement of all documents except the curriculum design report. Then the leader make group members' respective responsibilities clear. Three days later, the team leaders organize members to discusst and develop the implementation plan of the project.

3.5 Steps in implementation

After determining the project implementation plan and division of responsibilities, the project will be formally put into effect, according to the following five steps.

Step 1: students are guided to establish projects. The methods are shown in section 3.3.

Step 2: students are required finishing demand analysis. The team leaders should be trained to well complete arrangements and procedures.

Step 3: outline design and detailed design should be finished. The teacher may also guide leaders to do the specific work.

Step 4: program code should be completed. This stage determines the successful implementation of the project task. In order to guarantee the complete of program code on schedule, teachers shall play guiding role to support students' in technology, so as to eliminate the incorrect operation of the codes.

step 5: project shall experience acceptance and evaluation. The project code shall be mainly checked in correct operation; students' achievement shall be assessed; the students shall be guided in modification of final curriculum design report, and filing various document and archive.

3.6 Project evaluation

Assessment of student achievement consists of teacher evaluation, team internal evaluation and

Table 1. Table of evaluation rules and parameter.

Assessment rules	Assessment requirements
Preparation	If the actual situation is fully taken into account in demand analysis, and if the analysis is overall-round Yes Medium level No
Function realization	If the function module is reasonably divided Yes Medium level No If the required functions achieves the desired requirements Yes Medium level No
Interface	If it is friendly Yes Medium level No If it is beautiful Yes Medium level No
Maintenance and repair	Whether to have functions in maintenance and repair Yes No Whether the maintenance is easy to be carried out Yes Medium level No
Other opinions	

self-evaluation for comprehensive measure. The teachers will evaluate whether the main program design can be correctly operated, whether the requirements of project can be well followed, and whether individual innovation is integrated. With the way of team internal evaluation, all group members evaluate each other on the basis of attitude and contribution to the project during the process. Each student shall have self-assessment, to find out whether his own level of knowledge and skills have been improved. The ratio is 6:3:1.

In the specific operation process, each team leader will report the completed project in the class to list the problems and solutions encountered in the project development, and answer questions raised by teachers and other students on site. In this way, students can realize mutual communication. Finally, the teacher select the best project works, and make comments. Specific assessment rules and parameters are shown in Table 1.

4 REALISTIC SIGNIFICANCE OF THE PROJECT TEACHING METHOD

The significant effect of project teaching method in the teaching practice shows that, when the teachers are never theoretical database in activity but truly become the bridge of theory and practice, they will lead the students to explore and master new knowledge, and help them to forge ahead in the study for innovation. The students shall improve theoretical level and professional skills and cultivate their sense of cooperation and team spirit. In the guiding process, teachers shall not only improve students' professional level, but also further their own teaching ability. To sum up, the project teaching method can greatly improve the teaching level, and promote the quantity and quality of talent training.

REFERENCES

Baidu encyclopedia: encyclopedia card. http://baike.baidu.com/ view/2062970.htm.2013–2-20. *(references)*.

Frey, Karl. Die Projekt methode. Behz-Verlag. Weinheim, Basel 2002.*(references)*.

Ling Wei, Xiao-yu Sun, De-wei Dai. Research on practical teaching reform of information management and information system. Education Exploration. 2012(6).

Marcus Deininger, Kurt Schneider. Teaching Software Project Management by Simulation-Experiences with a Comprehensive Model. Software Engineering Education. 1994.

Shu-dong Jing. Study on the characteristics and adaptability of the project teaching. Education and Vocation, 2012(18).

Shuo Xu. Connotation, Education Pursuit and Teaching Characteristics of Project Teaching Method. Vocational and Technical Education. 2008(28).

Control Engineering and Information Systems – Liu (Ed)
© 2015 Taylor & Francis Group, London, ISBN 978-1-138-02685-8

City emergency capacity evaluation based on grey cluster method

H.Y. Wang
Hebei Vocational College of Rail Transportation, Shanxi, China

Y.N. Wang
College of Economic and Management, Hebei University of Science and Technology, Hebei, China

ABSTRACT: Based on the existing evaluation theory and model, a more scientific assessment index system of emergency capacity and assess model are structured. Then using this methods, this paper evaluates emergency capacity of SJZ city, especially Longitudinal comparing emergency capacity in 2009 with 2007 and 1995, the conclusion shows that 2009 belongs to the middle level, 2007 belongs to the middle level, and 2005 belongs to the weak level. But 2009 is also close to middle level. Finally some measure is given to improve emergency capacity.

Keywords: city emergency capacity; evaluation; grey cluster

1 INTRODUCTION

Now the research on evaluation of emergency capacity is mainly about index system construction and evaluation method. This paper designed indexes of emergency capacity, including: Pre-disaster early-warning capacity, Disaster response capacity and Disaster recovery capacity. These indexes coordinate with emergency capacity. In the end, emergency capacity is assessed by grey cluster[1–2].

2 PRINCIPLE OF INDEX SYSTEM DESIGNING

Indexes system assessing emergency capacity should be based on the above concepts and features. They are indicators collection which can comprehensively, systematically and briefly reflect the conditions of emergency capacity. We should comply with the following five principles when we design them[3].

2.1 *Scientificity*

Evaluation indexes of emergency capacity should fully reflect and embody the meaning of emergency capacity. From a scientific point, users can systematically and accurately analyze and grasp the essence of emergency capacity.

2.2 *Generality*

Evaluation indexes system of emergency capacity should be relatively complete; it means that, indexes system as a whole should basically reflect the main aspects or features of emergency capacity, rather than unrelated branches.

2.3 *Feasibility*

Indexes of emergency capacity can be quantified. These indicators should not be too complicated to calculate and the required data are not only easier to gain but also basically reliable.

2.4 *Guidance*

Indexes system of emergency capacity can not only guide but also lead and drive emergency capacity in a certain degree.

2.5 *Independence*

It refers that all indicators included in the indexes system of emergency capacity should be unrelated and independent with each other. Then on one hand indexes system has clearer structure, on the other hand, indicators indexes system can be ensured to be analyzed separately.

3 CONSTRUCTION OF INDEXES SYSTEM

According to the idea of basic principle of construct indexes system, indicator system follows "system-targets-indexes" three-level framework structure[4] the model of emergency capacity can consist of 3 hierarchy, 3 sub-system and 10 indicators (Table 1).

Table 1. Evaluation system of emergency capacity.

System	Targets	Indexes
Evaluation system of emergency capacity (A)	Pre-disaster early-warning capacity (B1)	Disaster monitoring capability %—X11
		Disaster reduction capacity %—X12
		Disaster forecasting capability %—X13
	Disaster response capacity	Ability to identify disaster %—X21
		Disaster rescue capability %—X22
		Government emergency response capacity %—X23
		Residents of the emergency response capacity %—X24
	Disaster recovery capacity	Capacity of social security—X31
		Capacity of post-disaster reconstruction %—X32
		Capacity of disaster damage evaluation %—X33

The first is system-level, namely evaluation system of emergency capacity. Second is target-level, namely three dimensions including Pre-disaster early-warning capacity, Disaster response capacity and Disaster recovery capacity. Third is concrete evaluation index of every target-level.

These indicators were got by literature polymerization[5–17]. We divide these indicators into three categories: Pre-disaster early-warning capacity (B1), Disaster response capacity (B2) and Disaster recovery capacity (B3). They together express emergency capacity (A). Pre-disaster early-warning capacity includes Disaster monitoring capability (X11), Disaster reduction capacity (X12) and Disaster forecasting capability (X13). Disaster response capacity includes Ability to identify disaster (X21), Disaster rescue capability (X22), Government emergency response capacity (X23) and Residents of the emergency response capacity (X24). Disaster recovery capacity includes Capacity of social security (C31), Capacity of post-disaster reconstruction (X32) and Capacity of disaster damage evaluation (X33) (Table 1).

4 EVALUATION MODEL

To fully assess the development of the system, multi-indicator comprehensive evaluation methods are adopted. Author makes use of grey cluster in this paper. Fix grey cluster is very applicable to assess this system[18].

Suppose n objects, m indicators and k grey classes, the sample observation x_{ij} means that object i is about the indicator $j(i = 1,2, ..., n; j = 1,2, ..., m)$. We must assess object i in the light of x_{ij}, and concrete steps are as follows:

Step 1: divided grey classes k evaluation requirements, and range of indicator is divided into s grey classes.

Step 2: define indicator j and k subclass whitenization weight function: $f_j^k(\cdot)$. If the sample observation of object j is x_{ij}, then membership function $f_j^k(x_{ij})$ can be got by whitenization weight function. Specific forms are as follows:

1. Lower measure whitenization weight function:

$$f_j^k(x_{ij}) = \begin{cases} 0, & x_{ij} \notin [0, x_j^k(4)] \\ 1, & x_{ij} \in [0, x_j^k(3)] \\ \dfrac{x_j^k(4) - x_{ij}}{x_j^k(4) - x_j^k(3)} & x_{ij} \in [x_j^k(3), x_j^k(4)] \end{cases}$$

2. Upper measure whitenization weight function:

$$f_j^k(x_{ij}) = \begin{cases} 0, & x_{ij} < [x_j^k(1)] \\ \dfrac{x_j^k(4) - x_{ij}}{x_j^k(4) - x_j^k(3)} & x_{ij} \in [x_j^k(1), x_j^k(2)] \\ 1, & x_{ij} \geq x_j^k(2) \end{cases}$$

3. Moderate measure whitenization weight function:

$$f_j^k(x_{ij}) = \begin{cases} 0, & x_{ij} \notin [x_j^k(1), x_j^k(4)] \\ \dfrac{x_{ij} - x_j^k(1)}{x_j^k(2) - x_j^k(1)} & x_{ij} \in [x_j^k(1), x_j^k(2)] \\ \dfrac{x_j^k(4) - x_{ij}}{x_j^k(4) - x_j^k(2)} & x_{ij} \in [x_j^k(2), x_j^k(4)] \end{cases}$$

In the function, $x_j^k(1), x_j^k(2), x_j^k(3), x_j^k(4)$ is turning point of $f_j^k(\cdot)$

Step 3: determine each indictor cluster weight η_j and then by membership function $f_j^k(x_{ij})$ and cluster weight η_j, we can calculate comprehensive cluster coefficient:

$$\sigma_i^k = \sum_{i=1}^m f_j^k(x_{ij})\eta_j$$

In the end, we think object i is belong to grey classes k^* because of $\max_{1\leq k\leq s}\{\sigma_i^k\} = \sigma_i^{k^*}$.

5 THE EVALUATION OF EMERGENCY CAPACITY

According to the index which established in this paper and making use of grey cluster, we assess SJZ city emergency capacity by comparing 2005, 2007 and 2009.

5.1 Determine weight and range of grey classes

By Delphi, we get weight of each indicator as follow. Then we divide level of emergency capacity into three classes: weak status, middle status and strong status, and likewise range of indicator is divided into three grey classes (Table 2).

5.2 Evaluation

We get the data of emergency capacity related indicators about SJZ city as follows (Table 3). The data in this table come from experts scoring. We invited 10 experts, and asked them give score of every indicator about 2009, 2007 and 2005, and then we respectively sum and average them.

Using formula given in IV, we can get σ^k as follow (Table 4).

By $\max_{1\leq k\leq s}\{\sigma_i^k\} = \sigma_i^{k^*}$, we know the level of emergency capacity is belong to weak level in 2005, and in this year σ^1 is far more larger than σ^3 but is close to σ^2, which means the level is similar to middle level, but have more deference with strong level.

Table 2. Indicator system of emergency capacity.

Indicators	Weights %	Level of emergency capacity		
		Weak	Middle	Strong
X11	8	<4	4–7	>7
X12	15	<5	5–8	>8
X13	5	<3	3–6	>6
X21	12	<5	5–8	>8
X22	7	<5	5–8	>8
X23	13	<5	5–7	>7
X24	6	<6	6–8	>8
X31	12	<6	6–9	>9
X32	6	<3	3–9	>9
X33	14	<5	5–7	>7

Table 3. Data of emergency capacity.

Indicators	2009	2007	2005
X11	9.13	6.27	7.12
X12	6.74	5.14	4.97
X13	6.97	7.32	5.43
X21	8.19	6.87	4.52
X22	6.62	5.48	6.13
X23	5.43	6.19	5.5
X24	6.17	5.53	5.67
X31	8.35	8.16	7.87
X32	6.4	7.14	5.3
X33	9.45	6.18	5.25

Table 4. Three years cluster coefficient.

Indicators	σ^1	σ^2	σ^3
2009	0.4375	0.597	0.6539
2007	0.5814	0.6465	0.39658
2005	0.5594	0.4879	0.392

Table 5. Three years cluster coefficient in 2009.

Indicators	σ_j^1	σ_j^2	σ_j^3
X11	0.000	0.543	0.809
X12	0.098	0.765	0.431
X13	0.314	0.981	0.353
X21	0.000	0.304	0.957
X22	0.696	0.370	0.104
X23	0.814	0.540	0.035
X24	0.800	0.500	0.000
X31	0.144	0.757	0.413
X32	0.371	0.933	0.547
X33	0.000	0.571	0.750

The level of emergency capacity belongs to middle leaval in 2007. Further comparing σ^1 with σ^2 in 2007, we know that σ^2 is close to σ^1 and larger than σ^3, which means the level in 2007 is just into middle level from weak level, but has huge gap to strong level.

The level of emergency capacity belongs to strong level in 2009. In this year σ^3 is larger than σ^1 or σ^2, which means middle level in 2009, is more obvious than other two levels. By comparing, σ^3 is more close to σ^2, which means the level in 2009 has more characteristic of middle level.

From Table 5, we can get more details of 2006.

Further analyzing the level in 2009, we know that in order to reach strong level of emergency capacity, SJZ city should make better in Disaster rescue capability, Government emergency response capacity and Residents of the emergency response capacity.

ACKNOWLEDGMENTS

The Education Department of Hebei Province Natural Science Youth Fund Project (2011140).

REFERENCES

Adam Rostis. Make no mistake:the effectiveness of the lessons-learned approach to emergency management in Canada[J]. International Journal of Emergency Management, 2007, 4(2): 197–210.

Ajinder Walia, Sujata Satapathy. Review of the Kumbakonam school fire in India: Lessons learned[J]. Journal of Emergency Management, 2007, 5(1): 58–62.

Alain Normand. Translating emergency management concepts into reality: where the rubber meets the road[J]. International Journal of Emergency Management, 2006, 3(2/3):238–246.

Cozman, F.G. Graphical models for imprecise probabilities. International Journal of Approximate Reasoning, 2005, 39(2): 167–184.

Fagiuoli, E., Zaffalon, M. 2U: An exact interval propagationalgorithm for polytrees with binary variables[J]. Artificial Intelligence, 1998, 106(1): 77–107.

Li Mingsheng, He Tianxiang. Comprehensive evaluation of sustainable development about regional foreign trade. Explorations, 2005(2), pp.8–11.

Liuwei. Study on the capmus emergency management capacity. China university of mining and technology, 2009.

Lui Sifeng. Entropy-effectiveness grey system theroy and application. Bejing science press.1999(3). pp.95–102.

Pi Zu-Xun, Liu He-Qing. Grey Incidence Assessment Model for Campus Safety and Its Application. China Safety Science Journal, 2008(6): 134–140.

Tian Yilin, Yang Qing. Study of the Evaluation index system model of the Emergency capability on Emergency. journal of basic science and engeineering. 2008(4): 200–207.

Wang Ruilan. Research on the performance Evaluation index system for government's emergency management. journal of Anhui university. 2009(1):35–39.

Yang Jin. Evaluation of emergency management executive capability of the unexpected accidents in the universities. Science-Technology and Management. 2008(5): 116–119.

Yoon, K. & Hwang, C.L. Multiple attribute decision making: An introduction. California: Sage, 1995.

Yu Jian-Xin, Liu Huan-Chun Wang. Campus Security Assessment Based on Hierarchy-grey Theory. Safety and Environmental Engineering, 2011(3): 43–46.

Zhihong Yu, Guoyou Ren. Study on evaluation index of campus security based on analytical hierarchy process. Journal of Safety Science and Technology, 2009(10): 50–54.

Control Engineering and Information Systems – Liu (Ed)
© 2015 Taylor & Francis Group, London, ISBN 978-1-138-02685-8

Construction of the evaluation index system of enterprise informatization

S. Lu & Y.P. Yang

School of Information, Capital University of Economics and Business, Beijing, China

ABSTRACT: This article describes the meaning of enterprise informatization index system building. It proposes the enterprise informationization index system construction principle and design ideas. The enterprise informatization index is calculated by the mathematical model of fuzzy comprehensive evaluation.

1 SIGNIFICANCE OF ESTABLISHMENT OF INDEX SYSTEM OF ENTERPRISE INFORMATIZATION

Enterprise informatization is the basis of the national economy, is an important symbol of enterprise modernization. At present, the overall situation is not so satisfactory enterprise information in China, most enterprise information weak low level, compared with the developed countries have a larger gap. Under the wave of global information technology, enterprises are faced with a more wide space, more anxious time, competition on the strength of better living environment, we only have the opportunity for development, play advantages, actively and steadily promote the informationization of enterprises. Therefore, as soon as possible to establish a scientific and objective reflect the status quo of enterprise informatization in China and the guiding roles of enterprise informatization of enterprise informatization evaluation index system of important theoretical and practical significance.

In theory, makes up for in Enterprise informatization evaluation index system of enterprise information insufficient research to further strengthen enterprise Informationization evaluation content. Governments around the world are all on Enterprise Informationization construction of great concern, and have commissioned a special study, developed a number of sets of evaluation index system, but there are some shortcomings. Formulated a comprehensive, scientific and practical Enterprise informationization evaluation index system on the promotion of rapid economic growth is of great strategic significance and practical significance.

In practice, enterprises informatization construction, if there is no set of evaluation index

system of scientific in reasonable, situation of informatization construction of enterprises would not be able to judge their own condition, unable to determine further information construction of Center of gravity, may cause abnormal development of enterprise informatization construction. An evaluation index system, to a certain extent, a certain frame of reference, in the construction of enterprise Informationization, a certain reference, provides information for enterprise application assessment tools, so as to further raise the level of informatization of enterprises to improve their own capacity and market competitiveness, further optimization of enterprise information systems, will help advance the process of enterprise information. At the same time, establishment of evaluation index system of enterprise Informationization to objective evaluation of enterprise informatization development, compare the differences and characteristics of enterprise Informationization level between, for the Government to develop economic development strategies, improvement of information industry development policy, optimizing information construction environment, proper guidance is of important significance to the development of enterprise information.

2 PRINCIPLES OF INDEX SYSTEM OF ENTERPRISE INFORMATIZATION

Index system of enterprise informatization in scope, contains informative. Its primary focus is on content to set up a scientific, standardized, systematic, standardized set of indicators system, to make a full and objective picture of the enterprise Informationization construction in its entirety. According to characteristics and evaluation of index system for evaluating Enterprise Informationization and its own characteristics, the establishment of index system of enterprise information must adhere to the following principles.

2.1 Reasonable structure index layer and index number

To any object, you need to build a number of quantitative targets and indicators, and evaluation to achieve the desired effect. However, the number of indicators and targets are not the more, the better, because it according to the actual situation. If the number of index levels and indicators are too more, it will result in the evaluation process is too complex, reduced the evaluation accuracy. If the indicator levels and indicators are too less, it will not be able to evaluate the scientific and effective. Therefore, the number structure indicator levels and indicators must be reasonable.

2.2 Principle of dynamic continuity

Enterprise informatization is a dynamic process of development and constantly. Therefore, it must reflect the status quo of enterprise informatization indicator system, while reflecting the trends and potential and reveal its inherent law. Principle of dynamic continuity is embodied in two aspects, the use of independent assessment and evaluation of the various modules to reflect the effect of enterprise informationization; B is the index that took into account the static and dynamic indicators of integration, using static indexes reflect the status and development trend of enterprise information, using dynamic indicators reflect the enterprise informationization development prospects with predictive function.

2.3 Principle of quantitative and qualitative combined

Enterprise informatization evaluation is a very complex engineering system in identification and comprehensive evaluation of enterprise informatization should integrate quantitative and qualitative indicators of the effects of enterprise information. Enterprise informatization evaluation system should be based on quantitative indicators, and moderate use of qualitative indicators, quantitative analysis is the key to an accurate evaluation of enterprise informatization. Because some cannot direct statistical evaluation of enterprise informatization of qualitative indicators. Such as: intangible economic benefits of enterprise information, business information, social benefits, the enterprise

information environment indicators only as qualitative judgments.

2.4 Principle of objective

The establishment of enterprise informatonization index system, aiming to make enterprises realize their informatization level, conduct effective information system construction. By measuring the status of the informatization, the enterprises will find out the shortcoming and solve it, sum up the experience of it and improve it.

2.5 Principle of overall

Evaluation index system of enterprise informationization level must be accurately and fully reflected the status of enterprise informatization. As informatization technology of enterprise not only involved material things but also involved contains such as the quality of personnel, business culture, spiritual factors. The aim is improving the economic efficiency of enterprises, including raising social benefits.

2.6 Principle of science

Establishment of index system of enterprise informatization, must be based on the theory of economic theory and statistical information, combined with its own characteristics, drawing on published national information on indicators and international comparison of general theories and methods of informatization, as Paula methods and methods of information index, raised the enterprise informationization evaluation indicator system and evaluation methods.

2.7 Principle of operability

The meanings are clear, in formations are concentration, data are easy to get in the process of constructing evaluation index. In addition, selected indicators to be simplified, straightforward calculation method, enables the selected indicators to objectively reflect the problems, reduce or remove some minor indicator of effects on the evaluation results. If there is no maneuverability will lose its significance, establishment of indicator system always around the target application can actually receive.

2.8 Principle of comparability

Comparability principle is referring to the comparison between different periods and different objects; this requires designing evaluation index system, not only to continue in time, but also to expand on the content. It is easy for the enterprise compare with others and also with different time of itself.

2.9 *Principle of independence*

At the time of designing evaluation index, contains relationships should be avoided as far as possible. It is evident that among the indicators, the implicit correlation handle as far as possible be weakened, elimination, seeks to target independent, minimizing the overlap of indicators, to make evaluation more credible and efficient.

3 DESIGN OF EVALUATION INDEX SYSTEM OF ENTERPRISE INFORMATIZATION

Currently, scholars have made has some evaluation programme on enterprise information evaluation, but due to on enterprise information meaning of awareness exists differences, effect enterprise information level of factors comparison complex, and information construction of effectiveness in many area mutual cross reflected, causes, makes these programme more will focus placed in information technology in enterprise in the of application area, or too stressed soft, and hardware of technology conditions, does not too attention information technology and enterprise management business of combined; Or index system of qualitative, subjective too strong, operability and quantify the degree is not enough.

The evaluation index system is mainly from the information environment, information support system construction, information technology application, information talent four aspects evaluation. Enterprise informatization construction should pay attention to benefits based on the information, success of the construction of

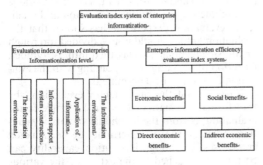

Figure 1. Composition of evaluation index system of enterprise informatization.

information can benefit from information construction reflected, one thing can be determined not revenue-generating information construction is definitely fails. Enterprise informatization efficiency evaluation index system is used for testing the success of enterprise information. Evaluation index system is evaluate from economic benefits and social benefits. Economic is divided into direct and indirect economic benefits.

4 ENTERPRISE INFORMATION INDEX CALCULATION

Fuzzy Comprehensive Evaluation method (called FCE) is a very broad and very effective method of fuzzy mathematics. FCE was introduced, whose mathematical model has expanded from the initial model for multilevel model and operator models. FCE are described below apply to the general process of assessment of enterprise informatization.

Fuzzy comprehensive evaluation method is using an evaluation system based on fuzzy sets theory for comprehensive evaluation methods. This programme test object can be a system, product, or categories of persons. Fuzzy evaluation based on fuzzy mathematics, mathematical method of processing "fuzziness" issue, mainly refers to the objective of so-called fuzzy thing of differential intermediate "fuzzy". For example, "good" and "bad", such opposition there is no distinct boundaries between concepts, to express the fuzzy concept to address practical issues of vagueness, general concepts of collection must be promoted, this is fuzzy subset. Fuzzy subset entirely by its membership function characterizations, in this sense, fuzzy subsets is the membership function. Given a fuzzy subset a, was to mean that given its membership function μA, when U A = 1, represents an element μ fully belong to A, when u A = 0, μ does not belong to A. Closer μ A to 1, μ is belongs to A in larger degree. In addition, depending on the size of membership and the weighting of factors and considering the multiple experts this is fuzzy.

Evaluation of enterprise informatization indexes are very many, some parameter is a "good" degree membership, some parameter is a "poor" or "fair" membership, make overall assessment of evaluation is not clear. It always has the error with subjective judgment for the parameter belonging to which membership. Fuzzy mathematics is applied in the decision of enterprise information, distinguishing subjective judgment into the unified objective of quantitative calculation of orbits.

4.1 Mathematical model of fuzzy comprehensive evaluation

The initial model. Factors set: U = (U1,U2,U3, ..., UN), type in the Uj (j = 1,2,3,...,n) representative enterprise information evaluation indicators system of indicators, comments set: V = (v1,v2,v3,v4, V5,V6,v7) = (very good, better, best, general, poor, poorer, poorest), indicators weights by levels analysis law determines, again integrated the expert of weight, through familiar degrees found, eventually statistics for integrated average weights vector A = (A1,A2,A3, ..., An) and to meet: $\sum_{i=1}^{n} Ai = 1$.

From U to V of a fuzzy mapping R like (vector) R (U) = (ri1,ri2, ..., rim) is called a single-factor evaluation, it is a fuzzy subset of V on which riy () from UI to consider that things can be named as vi membership. Mapping the fuzzy R = (riy) is called the comprehensive assessment of the transformation matrix. In this way, power distribution and a transformation matrix R is known, application of fuzzy matrix comprehensive evaluation of complex operations can be carried out, resulting in initial fuzzy comprehensive evaluation model:

$$AoR = B(b1b2 \ldots bm) \qquad (1)$$

$$bj = V(ai^{rij}) \cdot bj = V(ai^{rij}) \ (0<b<1)$$

Multilevel models. Through the division of tiered factors set, above model can be extended to multi-level mode through the division of hierarchical factor set, above model can be extended to multi-level fuzzy comprehensive evaluation model. It is the original model used in each layer on multilayer evaluation on the results and evaluation of input, until such time as the top. Factor set is a partition p in U = {u1,u2 ..., un}, can be two-level fuzzy comprehensive evaluation model. The formula is:

$$AoR = Ao \begin{bmatrix} A1 & o & R1 \\ A2 & o & R2 \\ & | & \\ An & o & Rn \end{bmatrix} \qquad (2)$$

A is weight distribution Ui of n-factor in U/P = {U1,U2, ..., Un}; Ai is weigh distribution uiy of k-factor in Ui = {ui1,ui2, ..., uik}; R and Ri, respectively to U/P and comprehensive evaluation of the Ui of the transformation matrix.

Following the most simple example of a two-level comprehensive evaluation model for (Figure 2).

If further separated from U/P, you can get three levels as well as more levels comprehensive assessment model.

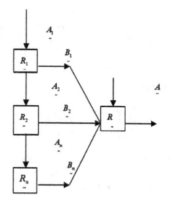

Figure 2. Two layer synthetic evaluation model.

4.2 General procedures for evaluation of enterprise informatization

Determine the participated enterprises, X = {xk}.

According to the needs and the actual situation, select some typical enterprises as a participating enterprise, to evaluate the degree of information.

Establishment of evaluation index system of enterprise informationization, give the factors set U = {ui}. Evaluation index system of evaluation object is defined by the characterization of multiple indexes in each aspect of information characteristics and their relationship (factors) organic whole composed of. To establish a set of evaluation index system of enterprise informatization and accordingly determine the evaluation factor set U = {ui} (I = 1,2, ..., n).

V = {vj} (j = 1,2, ..., m) Determine the evaluation rating and the corresponding standard, gives comments set. Ultimately, the purpose of establishing indicator system is used to rate the participated enterprises information system and metric. Indicators determining the evaluation criterion and the corresponding level are the basis for such an evaluation and measurement, as well as a bridge to combine the quantitative evaluation and qualitative evaluation. Experience has shown that the evaluation grades should not be divided too rough or too small, should be classified into 5~7 grades. Criterion of meaning corresponding with the evaluation division to be determined.

Carrying out the single factor evaluation, the transformation matrix R = {rij}. Single-factor evaluation, is from the individual factors (evaluation indicators) UI to determine the participated enterprises technology innovation capability can be rated as attached to the extent of the various judges rank VJ rij. There are basically two ways to determine: the experience table look-up method, that is, through experience-based reasoning, from arbitrary comment the found fuzzy quantization tables; The second is peer review of statistical law, through peer review and summarize evaluation results, drawn from statistical rate comments.

Comprehensive evaluation and strike a final comprehensive evaluation results. Due to the multi-level evaluation index system of enterprise information pack structure, it must be used in multi-level fuzzy comprehensive evaluation model for comprehensive evaluation. For a comprehensive evaluation of s determines, by principle of maximum proximity to determine level, comprehensive decision provides as follows:

- Set $sR = \max\{si\}$, calculate $\sum_{i=1}^{k-1} si$ and $\sum_{i=K+1}^{m} si$. If $\sum_{i=1}^{k-1} si. \leq 1/2 \sum_{i=1}^{m} si.$ according to its assessment grade. If $\sum_{i=1}^{k-1} si. \geq 1/2 \sum_{i=1}^{m} si.$ according to grade which SK−1 (or SK+1) belongs to.
- If $S = (s1,s2,s3, ..., sN)$, q $(q \leq m)$ equivalent to the maximum number, respectively, as required by 1 first shift calculations, the shift after a rating is discrete, then after the transfer is taken by grading Center.

  If centre has two grades, take the weight ratings, such as: $S = (s1,s2,s3,s4) = (0.25, 0.25, 0.25, 0.25)$. Shift grading after calculation, press s2 or s3 belongs.

5 CONCLUSIONS

Enterprise informatization is continuously to advance in depth and breadth. With the establishment of a set of scientific, operational enterprise informatization evaluation index system, is the inevitable choice for enterprises to deal with the global economic integration challenges. To improve the objectivity and reliability of enterprise informationization evaluation results, enterprise information index calculation should be used fuzzy comprehensive evaluation method. Establishment of evaluation system of enterprise informatization, will enable the evaluation of the enterprise more reasonable, scientific, objective and fair.

ACKNOWLEDGEMENTS

The paper would express sincere appreciation to the support from Beijing Higher Education Young Elite Teacher Project; improve scientific research funding of Capital University of economics and business (No. 0791254430107); the National Natural Science Foundation (No.71240002,71371128); Ministry of Education humanities social sciences research project (No. 13YJC630012); Beijing Philosophical Social Science Project (No. 13SHB015,11 JGB077); Beijing Natural Science Foundation Project (No. 9122003, 9123025); Beijing education science Project (No. AAA13003, KM2012100038001).

REFERENCES

[1] Benaroch M, Kauffman R. Justifying electronic banking network expansion using real options pricing analysis [J]. MIS quarterly, 2000(24):197–225.

[2] Bernroider EWN, Stix V. Profile distance method-a multi-attribute decision making approach for information system investments [J]. Decision Support Systems, 2006(42): 988–998.

[3] Brian L. Information technology investments: Characteristics, choices, market risk and value [J]. Information systems Frontiers, 2003, 5(3):289–301.

[4] Chen W, Sheng YH. A Possibilistic Adjusting Model for Portfolio Selection with Transaction Costs [J]. Information: An International Interdisciplinary Journal, 2012, in press.

[5] Chen W, Yang Y-P, Ma H. Fuzzy Portfolio Selection Problem with Different Borrowing and Lending Rates [J]Mathematical Problems in Engineering, 2011: 1–15.

[6] Hui Ma, Yiping Yang. Research on Quality Appraisal and the Body of Knowledge for Software Quality. Posts & Telecommunications Press, 2009 (In Chinese).

[7] Keng Siau, YuanWang. Cognitive evaluation of information modeling methods. Information And Software Technology. 2007.49(5): 455–474.

[8] Koh A. Differential Evolution Based Bi-Level Programming Algorithm for Computing Normalized Nash Equilibrium [J]. Advances in Intelligent and Soft Computing, 2011, 96: 97–106.

[9] Xu L, Li Q-S, Tang W-S. Hybrid Intelligent Algorithm for Solving the Bilevel Programming Models with Fuzzy Variables [J]. Systems Engineering—Theory & Practice, 2008, 28(7): 100–104.

[10] Yiping Yang. The fusion of modern software engineering techniques and CMM, People's posts and telecommunications press, 2002 (In Chinese).

[11] Zadeh LA. Fuzzy Sets as a basis for a theory of possibility [J]. Fuzzy Sets and Systems, 1978, 1: 3–28.

[12] Zadeh LA. Fuzzy sets [J]. Information and Control, 1965, 8: 338–353.

Control Engineering and Information Systems – Liu (Ed)
© 2015 Taylor & Francis Group, London, ISBN 978-1-138-02685-8

Social network in accounting information systems analysis

H.F. Yang

School of Business, Shandong University of Technology, Zibo, China

ABSTRACT: This paper introduces social network analysis as an alternative research method for conducting accounting information systems related research. With advances in information and communication technologies, transaction data are being recorded in electronic form, resulting in a variety of research opportunities to examine dyadic interactions. Social network research focuses on how outcomes are influenced not just by the attributes of the nodes, but also by the ties connecting nodes to each other. The nodes are typically conceptualized as actors, such as individuals, teams, or organizations. A unique network structure is created to reflect each different type of tie. Social network analysis can be used for research examining individual, dyadic or network levels of analyses, and is a powerful tool for conducting multi-method research. Given the vast amounts of trace electronic data collected via accounting information systems, this paper reviews how social network analysis not only opens new research avenues for accounting information systems researchers, but identifies opportunities for the field of accounting information systems to inform social network research by identifying new network structures and dynamics leveraging transactional data.

1 INTRODUCTION

There has been a general shift in management research over the past decade towards more relational theories of organizations that view actions and actors not as independent, autonomous agents, but as embedded within socio-technical systems. In contrast to theories that examine individuals based on their attributes, such as gender, age, education, or occupation, social network perspectives focus on how the relationships between entities, such as individuals, functional units, or organizations, influence interactions and outcomes. The concept of a "network" is broad and can be applied to a variety of phenomena where a set of relations is ascribed to an identified set of actors. What unites social network perspectives is the focus on the patterns and implications of the ties within a collective. For example, at the individual level ties facilitate the spread of information among network participants; enable the flow of both tangible and intangible resources among network members, and place constraints on each member's behavior.

Social network research focuses on the significance of relationships as essential for understanding social action, but varies widely in the attributes that are studied. A network is defined as a set of nodes connected by ties. Nodes are typically "actors", and can be people, teams, organizations or information systems. Relations, or ties, connect the actors and can vary in content, direction, and relational strength, all of which influence the dynamics of the network. The content of ties refers to the resource exchanged or common bond, such as information, money, advice, or kinship. The direction of ties indicates an "ego" who gives the resource, and the "alter" who receives it, although ties in some networks are undirected, such as a shared attribute (e.g. gender), or joint membership on a team. The relational strength of ties pertains to the level of activity, such as quantity of communications, or the intensity, such as the social influence exerted by the tie, indicating that ties can be valued or weighted. For instance, the relational strength of ties could indicate the amount of energy, emotional intensity, intimacy, commitment or trust connecting the actors. Relational ties are often studied in management research as important aspects of social influence that can exert control, such as social punishments or ostracism. Other aspects of social influence foster cohesion and pro social behavior in the network, such as trust, identification, the diffusion of information and commitment.

Each tie defines a different network, and while some ties are often related, ties are often assumed to function differently. Not all ties are considered to have positive outcomes; for instance, network research is often used to map the flow of disease or terrorist networks. Therefore, some network research focuses on how to improve the flow of the resource through the network, such as adoption of a new accounting information system, or how to disrupt the flow of resources in the network, such as taking out key nodes in a fraud network. Depending upon the theory being applied, some studies examine network variables as independent variables causing consequences, such as adoption of a technology or improved performance, while other studies examine network variables as dependent variables, identifying the causes underlying the pattern of network connections, how networks come to be, and how networks change over time.

2 SOCIAL NETWORK ANALYSIS IN RESEARCH

While a rich stream of research utilizing social network analysis exists within the sociology, anthropology, management and information systems disciplines, accounting researchers have been slow to add this method to their toolkits. Traditionally, social network analysis research has investigated networks in three manners: network connections which facilitate flows of resources between nodes, ties between nodes that influence behavior and enforce social norms, and networks themselves as either independent or dependent variables. In the following section, we provide a brief overview of Accounting Information Systems (AIS) research to date that has used social network analysis. This is followed by potential research questions where social network analysis might inform existing AIS phenomena.

2.1 Research on network enterprises

Accounting researchers have become increasingly interested in examining the dynamics of inter organizational relations. In response to changing organizational dynamics and calls for research that investigate accounting and management control across organizational boundaries, researchers have expanded their view of the organization beyond traditional hierarchical perspectives to include network enterprises. A network enterprise is comprised of members with independent goals that cooperate and coordinate in an effort to share costs, gain access to resources, and affect a common end. Network enterprises vary in size and scope, from smaller consultancies that cooperate and coordinate to perform complex engagements, to a network of regulatory agencies that collectively set accounting and auditing standards. Social network analysis expands the scholar's view of the network enterprise as a collection of traditional organizations that band together to achieve a common goal by enabling a more granular examination of how position within, composition of, and links among the network participants influence organizational outcomes.

Coordination and exchange within network enterprises often differs from traditional forms of governance. Rather than managing economic activities as a series of arms-length transactions, exchanges often occur within the fabric of embedded social relationships. Network governance suggests that economic exchanges are characterized by higher levels of trust, fine-grained information transfer between trading partners, and joint problem-solving arrangements. Network enterprises utilize social controls to safeguard against opportunism and malfeasance, instead of relying on contractual obligation, formal control mechanisms, or hierarchical structures.

2.2 Research where accounting information systems are nodes

Traditionally, actors in a network have been viewed as human entities. However, this need not always be the case. Networks may include nonhuman actors such as software, hardware, information systems and infrastructure standards. When viewed in this light, networks of human and nonhuman elements represent stable social structures comprised of actors whose interests have been aligned. The stability of the network and any variability in outcomes is theorized to be a result of how goals are translated, resources are enrolled, and irreversibility is established in proposed courses of action. Traditionally, qualitative methods such as case studies have been used to investigate and explain the relationships between human and non-human actors in a network.

For AIS researchers who subscribe to the view that information systems represent significant actors within a socially constructed network, social network analysis provides an additional tool in investigating the effects of information systems on organizational outcomes and performance by examining the manner in which information systems interact with other nodes in a network. This approach moves information systems beyond simply being viewed and studied as electronic ties between nodes in a network, but rather as autonomous and powerful nodes within a network.

Given this view of information systems as nodes, studies aimed at addressing the following research questions could contribute to the AIS literature:

- What is the optimal nature and strength of ties between information systems linking nodes to facilitate the flow of information or to foster innovation?
- How does the position of the information system in the network affect competitiveness and strategic advantage?

How can information systems that link nodes help manage risks associated with transactions, improve innovation, and increase collaboration?

Can we predict changes in the usage of transaction processing systems and related internal controls that have occurred, signaling adjustments to business processes and/or practices which have not yet been identified nor documented by either the firm's internal or external auditors?

2.3 *Research on connection between human actors*

By far, the most common use of social network analysis in the accounting literature has been as a tool to identify the relationships among and between individuals and groups. Social network analysis has been used to investigate Board memberships across organizations, accounting scholars' connections and productivity, composition and connections of advice networks, socialization of new staff auditors, and the role of accounting and financial experts in organizational decision making under conditions of uncertainty.

While scholarly inquiry in this vein has certainly been the most prolific, future studies could inform the following questions:

- How can communication and coordination structures within work units, divisions, firms and/or professional associations be more efficiently and effectively organized?
- How effective are ties and relationships at encouraging/discouraging ethical business practices as compared to more formal and traditional means?
- How can accounting and audit personnel better position themselves within their organizations to increase their effectiveness, perceived value, access to resources, and/or access to critical information?

3 SOCIAL NETWORK ANALYSIS

Recently, scholars have turned to social network analysis as a tool to investigate one of the oldest accounting crimes in human history: fraud. The intersection of social network analysis' relative maturity, electronic data capture, and public interest created a perfect storm of sorts. The e-mail records subpoenaed by the Department of Justice, and subsequently released as the "Enron corpus", have allowed researchers to investigate the pattern of communication and coordination among the various actors in this fraud. Early efforts have proven insightful as they suggest how communication and coordination patterns changed as the fraud progressed.

While early work using social network analysis to investigate fraud has been largely exploratory in nature, there is the potential to combine communication and exchange theories with social network analysis to investigate fraud in ways previously unexplored. Rather than simply examining the relationships between members of a network, it may be possible to examine how fraud is facilitated by different types of ties among members of a social network and to identify characteristics of the social structures within those networks that may have an influence on mitigating or prohibiting opportunities for fraud.

AIS and audit researchers can look to several applicable theories from the social psychology and criminology domains to better understand how frauds are conceived and executed through the social networks that exist within an organization, or at the boundary of the organization between employees and clients. For instance, social capital theory has been applied to reveal a negative relationship between social capital and criminal activities within a community. Building upon this understanding and applying it to the context of criminal networks may allow us to understand how community members engage in communication and other activities that enhance the social capital available within the community and, subsequently, establish obstacles for behaviors that encourage fraud opportunities, or are otherwise harmful to the community.

Perhaps an even more intriguing theoretical lens from which to examine a fraud network is provided by social disorganization theory, which attributes criminal behavior to the breakdown of communal relationships that traditionally foster mutual benefits among community members. First, social disorganization theory suggests that as interactions among members of a community become less frequent, less structured, and less beneficial to community members, social controls which serve to influence member actions are weakened, thereby resulting in opportunities for negligent or criminal behaviors. In terms of its direct application to fraud activities, social disorganization theory suggests that social networks that are limited in terms of strong ties between members are more likely to be infested with crime. Members of this weak tie network, if given the opportunity and pressure, are positioned to exploit the lack of social controls present within the network and rationalize their behaviors accordingly.

Another potentially rewarding application of social disorganization theory to the examination of fraud related criminal networks concerns the identification and understanding of collusion structures among social network members. Social exchange theory posits that all exchanges, including social exchanges, are a subjective cost–benefit analysis comparing alternatives. The main assumptions underlying why individuals engage in social exchange include anticipated reciprocity and expected gain in reputation, influence, or other rewards. While dense network structures facilitate the flow of information about the reputations and actions of actors, these strong network ties are also high in expectations of obligation and reciprocity, possibly including obligations to engage in and support criminal activity. Although only indirectly supported by evidence from the Enron email corpus, the potential for a curvilinear relationship among social activity and criminal activity structures exists, whereby a lack of social ties results in individual criminal activities within a network, but an over-abundance of social ties leads to collusion among the criminally-inclined. Exploratory research in this area may prove beneficial.

While the extant literature involving social disorganization theory is primarily concentrated on explaining criminal behaviors within neighborhoods and social networks that occur in physical spaces, we believe that the application of this theory to fraud investigation and prevention within digital social networks is justifiable. An area ripe for future research is the investigation of fraud and crime in digital social networks, such as face book, twitter and other online communities, and how tie strength differs between electronic and physical community relationships. Another area of research that social network analysis invites is the inclusion of characteristics shared among actors in a dyad, not simply the pattern of transactions, to help identify fraud. For instance, using electronic trace data from accounting information systems, social network analysis can uncover areas of high density transactions, low density transactions and transactions among nodes that are closely tied versus sparsely connected or share some common attributes that may have the potential to identify patterns of fraudulent activities.

4 CONCLUSION

We provided an overview of current accounting research utilizing social network analysis. From this examination, three distinct literature streams emerged: transactions and control mechanisms in the context of inter-organizational dynamics, networks where accounting systems are viewed as nodes, and connections between human actors in an accounting context. From these streams, we presented a series of research questions which could serve to expand our knowledge in these areas and which social network analysis is uniquely positioned to inform in the near-term. Subsequently

we discussed one of these examples in depth, to provide deeper insights into how social network analysis could be applied in the context of fraud detection and investigation.

While social network analysis offers opportunities to expand and contribute to the research streams mentioned above, we would be remiss if we didn't expand our discussion of how social network analysis can be leveraged to inform other areas of interest to AIS researchers. To this end, we looked to the topics typically addressed in the International Journal of Accounting Information Systems for areas where social network analysis might provide distinct benefits over other methods and foster a better understanding of these. Although there are numerous topics where we feel this method might prove fruitful, we focused on three that we feel have the most promise.

4.1 Control and auditability of information systems

Since the Sarbanes Oxley Act of 2002, accounting researchers have focused anew on investigating issues around control and accountability of information systems. Of particular interest has been evaluating the nature and effectiveness of the control environment. The control environment represents the foundation for an organization's system of internal controls and includes management's ethical values and integrity, organizational structure, and authority and reporting arrangements.

Social network analysis could prove useful for research aimed at better understanding the nature and effectiveness of the control environment as well as the overall system of internal controls. For example, it could be used

- as a method for examining communication and coordination between key stakeholders important to the identification and control of risk,
- as a method for examining the effectiveness of coordination mechanisms inscribed in organizational structures as well as informal coordination mechanisms,
- as a method for examining the effectiveness of traditional controls compared to social controls embedded in the fabric of relationships.

4.2 Electronic dissemination of accounting information

Information from accounting and transaction processing systems is a necessary component of planning, budgeting and management control. For this information to be useful, it must be accessible to decision makers in a proper format and timely fashion. However, it is often difficult to identify which actors require specific information, especially in globally dispersed organizations or networked enterprises with diverse stakeholder groups. As we have previously discussed, actors can be individuals, work units, organizations, regulatory bodies or even other information systems.

Social network analysis provides a useful methodological approach to examine the inter connections between diverse actors and how these linkages can inform questions around which actors need accounting information and in what format. For example, it could be used

- as a method to examine flows of information between linkages in inter-organizational systems connecting networked enterprises, to identify whether key information is kept confidential or is transferred beyond organizational boundaries,
- as a method to identify key sources of accounting information as well as consumers of this information.

4.3 Organizational perspectives on accounting

Prior research has demonstrated that informal networks augment or supersede formal hierarchies with respect to knowledge sharing. An extension of research aimed at understanding the dissemination of accounting information would be to more fully explore where and to whom people look for advice and expertise.

Social network analysis provides a new approach to more fully explore how accounting and business process knowledge is dispersed across organizational or professional boundaries, as well as understand the manner in which it is accessed. For example, it could be used

- as a method to identify the location of expertise within and outside traditional hierarchies and organizational structures,
- as a method to identify informal expertise networks and provide insight on how people perform their work,
- as a method to aid in the design of systems aimed at connecting people with expertise,
- as a method to explain attributes of individuals and groups critical to internal control, dissemination and creation of accounting-related information, and policy and standards.

To conclude, the purpose of this article was to present social network analysis as an alternative method that has high potential for expanding AIS research. Social network analysis focuses

on the pattern of ties connecting nodes within a network, and emphasizes that both the pattern of relationships and characteristics of the dyads influence the actions of network members. While most prior research using social network analysis has focused on network dynamics in face to face networks, advances in information and communication technologies, especially accounting information systems, open new possibilities for the types of network structures and research questions that can be investigated.

REFERENCES

Baker F., Hubert L. "The analysis of social interaction data". Sociological Methods and Research 1981; 9:339–931.

Bart C., Turel O. "IT and the board of directors: an empirical investigation into the "governance questions" Canadian board members ask about IT". J Inf Syst 2010, 24:147.

Borgatti S.P., Cross R. "A relational view of information seeking and learning in social networks". Manage Sci 2003, 49:432–45.

Bowen P., Cheung M.-Y.D. "Rohde F. Enhancing IT governance practices: a model and case study of an organization's efforts". Int J Acc Inf Syst 2007, 8:191–221.

Conyon M., Muldoon M. "The small world of corporate boards". J Bus Finance Acc 2006, 33:1321–43.

Kane G., Alavi M. "Casting the net: a multimodal network perspective on user–system interactions". Inf Syst Res 2008, 19:253–72.

Katz R. "Re-examining the integrative social capital theory of crime". West Criminol Rev 2002, 4:30–54.

Klamm B., Watson M.W. "SOX 404 reported internal control weaknesses: a test of COSO framework components and information technology". J Inf Syst 2009, 23:1.

Masquefa B. "Top management adoption of a locally driven performance measurement and evaluation system: a social network perspective". Manage Acc Res 2008, 19:182–207.

Mouritsen J., Thrane S. "Accounting, network complimentarities and the development of inter-organisational relations". Acc Organ Soc 2006, 31:241–75.

Murthy U., Taylor E. "Knowledge sharing among accounting academics in an electronic network of practice". Acc Horiz 2009, 23:151.

Richardson A. "Regulatory networks for accounting and auditing standards: a social network analysis of Canadian and international standard-setting". Acc Organ Soc 2009, 34:571–88.

Wakefield R. "Networks of accounting research: a citation-based structural and network analysis". Br Acc Rev 2008, 40:228–44.

Control Engineering and Information Systems – Liu (Ed)
© *2015 Taylor & Francis Group, London, ISBN 978-1-138-02685-8*

Recent advances of e-government in the public sector

A. Abuduaini
Industry and Commerce Administration Institute, Xinjiang Economic University, Urumqi, China

ABSTRACT: This paper aims to thoroughly review the research literature concerning e-government in the public sector (2001–2013) for the purpose of summarising and synthesising the arguments and ideas of the main contributors to the development of e-government research and explore the different perspectives. In addition, the paper attempts to identify the key characteristics of e-government; and to gather conceptual perspectives on the nature, scope, and transformation to e-government. Most of the literature has focused on the underlying perspectives of approaches to e-government. It clearly acknowledges that contextual issues and factors influence e-government. However, there is still no standard definition of the concept and vagueness about what exactly e-government is, creating confusion and comparability issues, and making it difficult for researchers to build on each others'work. In addition, the key underlying theme throughout the literature is that e-government in the public sector necessitates closer working relationships between government stakeholders.

1 INTRODUCTION

Electronic government is one of the most interesting concepts in the field of public administration in recent years as it greatly improves the convenience and accessibility of government services and information to citizens. The term e-government is also used to describe the "use of information technology to support government operations, engage citizens, and provide government services". There is still no universally accepted definition of e-government. These concepts noted above are generally centred on the fundamental premise roles that information and communication technologies play in making government services more accessible, more relevant to citizens, more customer-focused, and more responsive to citizens' needs. e-government implementation promotes major innovations in the utilization of Information communication technology to the government operation, management, and its activities organization.

Electronic government (e-government) suggests the use of Information Technology (IT) and systems to provide efficient and quality governmental services to citizens, employees, businesses and agencies. Moreover, it increases the convenience and accessibility of government services and information to citizens (Carter and Belanger, 2005). The multiplicity of anticipated benefits that may stem from the im plementation of e-government has led governments to invest heavily in technologies and systems. The aim of the governments to provide not only improved and computerised but also innovative services in e-government has spanned services innovation literature in the public sector and boosted the study of New Service Development (NSD). However, a major portion of the literature on NSD has concentrated on the financial-service sector and hospitality industry (Kitsios et al., 2009), and there has been relatively no significant research on NSD in e-government and public sector. Moreover, e-government, sometimes perceived as buzzword in public administration (Yildiz, 2007), implies different things to different stakeholder groups (Grant and Chau, 2005; Halchin, 2004). Despite its numerous benefits—such as greater public access to information and a more efficient, cost-effective government—e-government is contingent upon the willingness of the citizens to adapt it. Although implementing NSD remains a challenge for researchers and practitioners alike, there has been relatively little research exploring the implementation of NSD in e-government. To address this gap and under the Critical Success Factors (CSFs) prism (Shah and Siddiqui, 2006), The main argument developed in this paper is that the implementation of NSD in e-government is multifaceted, and since the benefits of e-government are much anticipated by governments but the financial investments involve high risk, it is necessary to suggest a model based both on previous literature in the field and research, which will take under consideration the majority of the factors that secure the successful outcome of future investments and implementations of NSD in e-government.

The paper aims to present a comprehensive review of the e-government literature in the public sector, focusing on human, technological, and other types of challenges related to activities that produce an e-government that meet the needs of all stakeholders. From a managerial perspective, those activities represent processes that are critical to any organization and strategic assets that must be understood, managed, and improved to provide value-added services, and enable the organization to be more capable of change. That is how the review of the e-government literature is linked to Business Process Management (BPM), which focuses on aligning all aspects of an organization to promote effectiveness, efficiency, flexibility, and integration with technology, and is considered an approach to integrate a "change capability" to an organization—both human and technological.

2 MAJOR E-GOVERNMENT INTERACTIONS, TRANSFORMATION, AND RELEVANT STAKE HOLDERS

Most of the existing literature refers to four types of e-government interactions in public organizations:

1. Government to Government (G2G);
2. Government to Citizen (G2C);
3. Government to Business (G2B); and
4. Government to Employee (G2E).

Governmental organizations interact with its citizens in the downstream channel and with the business community in the upstream channel. These interactions can include information-based interactive exchanges, negotiation, promotion flows, title exchanges, and service flows. In the downstream channel, it includes concepts such as Citizen Relationship Management (CRM and e-CRM). In addition, the e-government concept can be applied in an intra-agency context. the agency or department is a customer of another agency or department. and the e-government concept subsumes such intra-organizational interactions. For example, those involved in e-government processes could be individuals, constituents, businesses, employees or perhaps government organization. In literature, e-government stakeholders in the public sector context are not clearly determined or defined. The bulk of e-government literature tends to deal with e-public service users as citizens (Alford, 2002; Bollettino, 2002; Jaeger, 2002; Lenk and Traunmüller, 2002; Means and Schneider, 2000; Brown and Brudney, 2003; Bowers and Martin, 2007), although sometimes they may not be citizens (e.g. issues regarding immigration), ignoring other groups involved either in providing or using the service. The relevant stakeholders are not just those who make use of the service. They can be involved in any stage of e-government development and delivery, from those who are responsible for organizing and supervising services, those who possess and provide the necessary background knowledge for designing and implementing services, those who provide the necessary technological knowledge or the development of the service, adopters who contributes to getting other ready for the e-government (Heeks and Santos, 2009), to the end-users that make use of the service. Even the end-users being served by government can be grouped according to the extent to which the service is designed to help them as primary, secondary, compliers or volunteers.

3 SELF-SERVICE TECHNOLOGIES AND E-GOVERNMENT

Text Most authors have described e-government, for all practical purposes, as "self-service", relying on users to navigate through options and help themselves (Colby, 2002; Dobholkar, 2000; Meuter et al., 2000; Zhu et al., 2002; Salomann et al., 2006; Ostrom et al., 2002; Surjadaja et al., 2003). Those authors differentiate between self-service and e-government in that self-services include using all technological interfaces available for customers while the e-government is limited to using the internet. So, it can be concluded e-government is one of the Self-Service Technologies (SSTs). In addition, with self-service operations, a customer has to go to the technology (such as an automated teller machine) to receive a service, whereas in e-government, a customer can receive the service through the internet at home or in other places.

Recently, academic research has explored the SSTs topic, with the focus specially directed to the online services (Zeithaml et al., 2000; Szymanski and Hise, 2000; Ruyter et al., 2001; Montoya-Weiss et al., 2001). These authors do not refer to it as e-government, preferring to use self-service as a synonym, although they focus their research only on the online services.

A second theme in the literature is the focus on technology assimilation into the citizen as a part of the self-service format. The shift from high touch to high tech as a feature of the self-service approach has specifically occupied scholars' attention (Lu et al., 2007; Kahraman et al., 2007; Georgolios et al., 2007; Chang et al., 2007).

According to their studies, technology is considered an enabler of self-service by providing interfaces for interaction with the citizen. In general, the main focus of this technology-centric view is the examination of how to successfully make use of technology in forging stronger citizen relationships. A criticism of the studies mentioned above is that they focus on automating and digitizing existing processes rather than toward transformation of public services. Making service electronic is not simply a matter of putting existing services on the internet. What should and will be happening are permanent changes in the public processes themselves and possibly the oncept of the service itself. e-government initiatives should be accompanied by re-conceptualization of the public service itself. Only then, the full benefit of providing will be realized only when organizational changes accompany technological changes.

4 CONCLUSIONS AND IMPLICATIONS FOR FUTURE RESEARCH

In seeking to take a wider perspective than that implicit in the business e-government literature, this paper has reviewed research and gathered conceptual perspectives on e-government in the public sector context within developing countries environment. The growing interest of the e-government topic is evident as well as the speedy growth in the e-government research along with e-government research over the past few years. In conclusion, this review of research indicates: Most of reviewed papers have focused on the underlying perspectives, definitions, and approaches to e-government. There is a significant understanding of the human and contextual factors that influence and affects the e-government introduction and development. However, most of these perspectives are based on private sector assumptions, and there confused positions about e-government in the public sector and about underlying philosophies supporting it.

The research uses a wide range of research methods, including interviews, document analysis, questionnaires, etc. although much of it draws mainly from a positivism philosophical paradigm.

There is still lack of consensus on the impact of e-government on both individuals and businesses. However, the general underlying theme throughout the literature is that the nature of e-government in the public sector necessitates closer working relationships between government stakeholders. The development of meaningful and effective relationships between central government, individual government agencies and users of public e-government are critical to the success of e-government. Some models which described the stages of e-government are generally descriptive in nature. They ignored some important dimensions such as barriers of e-government adoption and focused mainly on integration issues and supportive functions of formal government primarily provided by technology.

Also, they did not offer any guidance to howthe process for implementation could actually take place in real life. Further, the majority of these models propose the evolvement of e-government as linear. This does not have to be the situation in all e-government initiatives.

ACKNOWLEDGMENT

This work was financially supported by the Social Science Foundation of Xinjiang Uyghur Autonomous Region (No.11CSH064).

REFERENCES

Abanumy, A., Al-Badi, A. and Mayhew, P. (2009), "E-government website accessibility: in-depth evaluation of Saudi Arabia and Oman", The Electronic Journal of e-Government, Vol. 3.

Affisco, J.F. and Soliman, K.S. (2009), "E-government: a strategic operations management framework for service delivery", Business Process Management Journal, Vol. 12 No. 1, pp. 13–21. No. 3, pp. 99–106.

Agarwal, P. "Economic growth and poverty reduction: evidence from Kazakhstan", Asian Development Review, (2009) Vol. 24 No. 2, pp. 90–115.

Andersen, K.V. (2010), "E-government: five key challenges for management", Electronic Journal of e-Government, Vol. 4 No. 1, pp. 1–8.

Borghi, V. and Berkel, R. Individualized service provision in an era of activation and new governance, International Journal of Sociology and Social Policy, Vol. 27 Nos 9/10, pp. 413–24.

Eason, G., Noble, B. and Sneddon, I.N. "On certain integrals of Lipschitz-Hankel type involving products of Bessel functions," Phil.

Control Engineering and Information Systems – Liu (Ed)
© 2015 Taylor & Francis Group, London, ISBN 978-1-138-02685-8

The influence of online interaction on consumers buying intention based on flow theory

H.L. Su
School of Business, Jiaxing University, Jiaxing, China

ABSTRACT: Due to the rising of the network society, consumption structure has consequently changed. It is significance that studying in-depth on the influence of network interaction on buying intention. This article, from the perspective of flow theory, constructs a structural model of interaction, experience and buying intention, and proofs it through empirical research. The study found that people interaction affect partly the online experience, machine interaction positively affect the online experience, online experience positively affect the buying intention.

1   INTRODUCTION

Due to the rising of the network society, consumption structure has consequently changed. The network consumption as a new consumption pattern became popular in the world. China Internet Survey data shows that the online shopping users reached 210 million, increased 8.2% compared with the end of 2011 users at the end of June 2012. Starting in 2011, the growth of online shopping users is gradually stable [1]. In the future, the development of online shopping market relies on not only the growth of the subscribers, but also the consumption depth. According to the Internet media and economic analysis, the particularity of the Internet is the "interaction" and "personal channels". Focusing on two features, there is only experience economy in accordance with internet industry. Network interaction is new forms of interaction of network users based on IT technology. Network interaction is communication and dialogue among network users in the network, is an extension of the real interaction in the virtual space. Because of online experience is different from the physical store experience, it is significance that studying in-depth on the influence of network interaction on buying intention, making marketing strategies for attracting more consumers to online shop.

2   LITERATURE REVIEW

2.1   *Flow theory*

Csikszentmihalyi (1975) first proposed the "Flow" concept. He believes that "Flow" is a state that body operating at maximum potential state. People can feel the flow in entertainment and daily activities [2].

Hoffman and Novak (1996) introduced flow experience into the network environment [3]. Subsequent studies defined the flow in accordance with Hoffman and Novak (1996) which are most about human-computer interaction and network activity. Although the names of the flow are difference, the essence is the same. The flow is a person state. During interacting with the action objects, people devote themselves wholeheartedly, forget the surrounding environment, and generate sense of losing self. This experience makes people feel their best, their own happiness, the most efficient moment.

The flow theory is one of the main theories of the experience theory. It is also part of consumer psychology. Flow theory can not only provide theoretical basis for online experience dimension of online shopping, online games and other virtual environment, but also opens up new ideas of studying on online experience and online consumer behavior based on psychology. Flow experience can attract consumers; reduce the consumers' price sensitivity; positively impact on the subsequent consumer attitudes and behavior.

2.2   *Online interaction*

Previous studies have shown that the interaction is an important feature of modern media. Consumers and the sales staffs are no longer directly face-to-face in the network. On the contrary, the consumer experience through the network virtual screen designed and synthesized by the computer instead of physical environment. People participate in computer activities promote interaction generated [4]. Base on communication among people, interaction can be defined that information in turn related; especially the information can be used for later description. The interaction can be also defined as the need of mutual communications between communicators [5].

In shopping, there will be positive consumer experience if the shopping website integrates use of rich content, positive understanding and cooperation. This interaction can not only be the interaction between people, but also the interaction between people and technology. The interaction can affect a person's knowledge or behavior at least. The communication of hypermedia environment is different from the traditional environment. Based on the above studies, the interaction of online shopping will be divided into two kinds, one is people interaction, and another one is machine interaction.

3 RESEARCH HYPOTHESIS

In the past few years, the flow theory in the information technology and computer controlled environmental context has been analyzed, and is recommended as the one of most accepted theory in online consumer experience theories.

The flow experience is difficult to define, since its operation, test, and application have many methods. However in the study of flow experience, we get some useful information from the emotional and cognitive factors which are intrinsic enjoyment, conscious control and concentration. The first two variables also responded to those delightful factors as well as controlled over the physical environment [6]. This structure can be applied to the measurement of the dimension of the online consumer experience. Therefore, this article chooses intrinsic enjoyment, conscious control and concentration as the composition of the online experience dimension.

People interaction in online consumption is the consumers communicate with the sales staffs through website media. In the offline shopping environment, the people interaction is one of the important parts of the offline shopping environment. In the offline shopping; the people include not only the consumers and waiters, but also the other consumers at that time. Clearly, the people interaction between people is an important factor influencing consumers' perceived services. With the continuous development of network economy, the shopping websites achieve the interaction between consumers and sales staffs, including the interaction among consumers. The more people interaction the more the social existence consumers feeling, so people interaction influences the consumers' cognitive behavioral control [7]. On the basis of the above scholars' studies, this article hypothesizes that:

H1: People interaction will positively affect the online consumers' experience

H1a: People interaction will positively affect the online consumers' intrinsic enjoyment

H1b: People interaction will positively affect the online consumers' conscious control

H1c: People interaction will positively affect the online consumers' concentration

Machine interaction in online consumption is consumers participate with the website system. The information provided by the website and the degree of machine interaction affect the customer experience of shopping websites. With the increasing of the level of machine interaction in online shopping, the consumers' experience will be corresponding increased. The more relevant information for each product consumers get, the higher the experience consumers feel. In online media environment, the high level of machine interaction makes consumers feel more pleasure and excitement. On the basis of the above scholars' studies, this article hypothesizes that:

H2: Machine interaction will positively affect online consumers' experience

H2a: Machine interaction will positively affect the online consumers' intrinsic pleasure

H2b: Machine interaction will positively affect the online consumers' conscious control

H2c: Machine interaction will positively affect the online consumers' concentration

Consumers usually evaluate activities according to inherent fun which is inner emotional response. The fact is this fun is measured by consumers with appropriate psychological scales [8]. In the case of online shopping, the inner happiness is shopping enjoyment. Past research has shown that the online shopping enjoyment is the decisive factor of consumer loyalty [9] [10] [11]. Similarly, the flow experience studies have shown that inner enjoyment can generate positive affects on the application of e-mail, computer softwares [12] and the network [8]. Therefore, we assume that shopping enjoyment will positively affect the online consumers' buying intention.

In the flow theory, the level of control had been defined as the control of a person and his corresponding behavior. For online consumers, the information environment will be very different between online consumption and the entity consumption. There will be many utilitarian consumers because of the trivial time for shopping, and the explosion of online information and variety of products. The utilitarian consumers want more control, less effort and greater efficiency in the online shopping [9] [10]. The characteristics of the network can solve these challenges and make it easier for consumers to find what they need through engine website and recommended agents. The more understanding of the product, the more quickly consumer find it online. All of these network characteristics make network consumers have high level of control and convenience [3]. Online consumers will have higher loyalty on the shopping website than others if the shopping website provides them a series of control. Therefore, we assume that control will positively affect the online consumers' buying intention.

Individuals in flow experience must focus on their activities. Therefore, the concentration and flow experience are closely related. Because of the limited time and information processing, the online consumers have short concentration. However, the degree of concentration is very important to complete the purchase effectively. The research shows that the concentration as a measurement of network experience have positive impact on the online consumers' experience and the intention of repeated using the shopping system. Similarly, we know that interference will affect the online consumers' online shopping satisfaction. Therefore, we assume that concentration will positively affect the online consumers' buying intention.

On the basis of the above scholars' studies, this article hypothesizes that:

Hypothesis 3: Online experience will positively affect online consumers' buying intention.

H3a: Intrinsic enjoyment will positively affect online consumers' buying intention.

H3b: Conscious control will positively affect online consumers' buying intention.

H3c: Concentration will positively affect online consumers' buying intention.

4 EMPIRICAL ANALYSIS

4.1 Scale design and data collection

This paper formulates the people interaction and machine interaction dimension referring to Hoffman and Novak (1996) [3], Jyh-Jeng et al. (2005) [13] research, the experience including intrinsic enjoyment and control and concentration dimension referring to Ghani et al. (1991) [14] research, the buying intention dimension referring to Zeithaml, Berry & Parasuraman (2002) [15] research.

This paper analyses the data with SPSS 13.0 and AMOS5.0. There are two kinds of internet users who have representation and development potential including college students and company employees. So this paper chooses these two kinds people as research object. The formal survey is administered by the experimental center of a Zhejiang university which has a complete computer network facilities and an office building in Zhejiang region. All the survey objects are either university students or company employees. There were 500 questionnaires, 421 valid questionnaires, questionnaires. The questionnaire total effective rate was 84.2%. Sample Description.

4.2 Analysis of reliability and validity

This study uses software SPSS13.0 and adopts Cronbach's α value to measure the reliability of each variable; taking Cronbach's α value greater than 0.7 as standard. The running results show: the test index of people interaction, machine interaction, enjoyment, control, concentration and buying intention all are between 0.708 to 0.848, so we can say the research questions of each variable, to some degree, have internal consistency and stability.

The study tests and constructs validity by using Confirmatory factor analysis. According to the result of SPSS analysis, all the factors' loading are above 0.55, reaching statistical significant or very significant level.

4.3 Structural equation analysis

This study uses software Amos 5.0 to construct the structural model of observed variables and latent variables, latent variables and latent variable, determine the path coefficients between variables and test hypotheses.

Each fitting indicator of models mentioned above has reached the standard. The path coefficients obtained through the structural equation model show in Figure 1.

The path coefficient of people interaction to control is 0.07. It means people interaction positively affect control. The path coefficient of people interaction to enjoyment is 0. It means people interaction has no influence on enjoyment. The path coefficient of people interaction to concentration is –0.04. It means people interaction negative affect concentration. So research hypothesis H1b is supported, H1a and H1 are not supported.

Machine interaction positively affect the enjoyment, control and concentration, the path coefficient are 0.29, 0.35 and 0.37. Machine interaction affects the online experience factors, the most obvious of which is concentration. So research hypothesis H2a, H2b, H2c are supported.

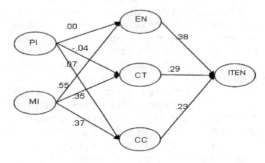

Figure 1. Path relationship and path coefficients of the structural model.

The online experience factors including enjoyment, control and concentration positively affect the online buying intention. The path coefficient is 0.38, 0.29 and 0.23. So research hypothesis H3a, H3b, H3c are supported.

5 CONCLUSION

Shopping website should ensure the advantages of information resources through richness, effectiveness, usefulness and accuracy. At the same time, the website administrators should arrange the website outward appearances reasonable, the menu functions more convenient, the web interface beautifully vivid with multimedia technology, and the commodity pictures more clear and detailed. These means can make intangible services online tangible, allow consumers to fully impressed and experience the pleasure, so that the website can be tangible in the minds of the consumers.

Interaction has significant positive impact on online experience. Website administrators can provide consumers with the tools and space to communicate with each other create interactive community, encourage information exchange between consumers and sellers. All of these can enhance the interaction quality, making consumers shopping on the website an immersive sense of pleasure, and optimal experience, the trading opportunities increased. This article shows that interactions between consumers and staffs have no significant effect online experience, as well as the interactions between consumers and machines have positive influence effect on the online experience.

In China, interaction between staffs and consumers is still an important part of online shopping in the many shopping websites. People living in different places communicate and online transactions conveniently because of the characteristics that network without borders and unlimited space. In the whole transaction process, consumers and staffs are rarely able to face-to-face contact which affects the consumer website experience. The website should reduce the service interaction in consumer shopping process appropriately, improve shopping website page, improve network consumer self-service. Online shopping website should add self-service functions. There are many ways for consumer self-service, for example ATM, online banking, and remote education. Shopping website should provide consumers with shopping platform, as well as various types of modules. Consumers feel experience through self-design, self-production, self-consumption. When websites interact with consumers, the approach must be more targeted and personalized using information technology to provide consumers with advertises which have interactive effect and more machine interactive services. The shopping website should provide innovative online services such as analog situations with kind technology, so that the consumers are immersed among online shopping and feel a sense of pleasure which is unforgettable experience that accelerates consumers buying decisions.

REFERENCES

CNNIC. The Survey of Chinese Internet Network Hot topic. [R]. 2011, 10–21.
Csikszentmihalyi, M. Beyond Boredom and Anxiety: The Experience of Play in Work and Games. Jossey-Bass Publishers, San Francisco, CA, 1975.
Eighmey John, Lola McCord. Adding value in the information age: Uses and gratifications of sites on the World Wide Web. Journal of Business Reasearch. 1998, 41(3), 187–19.
Ghani, Jawaid A., Satish P. Deshpande. Task characteristics and the experience of optimal flow in human-computer interaction. J. Psychology, 1994, 128(4), 381–391.
Ha, L. and James, E.L. Interactivity re-examined: a baseline analysis of early business web site. Journal of Broadcasting and Electronic Media, 1998, Vol. 42. No.4, 457–474.
Hoffman, Donna L. & Novak, Thomas P. Marketing in hypermedia computer-mediated environments: Conceptual foundations [J]. Journal of Marketing, 1996, July, 60, 50–68.
Hoffman, Donna L. & Novak, Thomas P. How to acquire customers on the Web [J]. Harvard Business Review, 2000, 78(3), 179–188.
Javenpaa, Sirkaka L., Peter A. Todd. Is there a future for retailing on the internet? R.A. Peterson, ed. Electronic Marketing and the consumer. Sage, Thousand Oaks, CA, 1997a.
Javenpaa, Sirkaka L., Peter A. Todd. Consumer reactions to electronic shopping on the world wide web. Internet Journal of Electronic Commerce. 1997b, 1(2), 59–88.
Jyt-Jeng Wu and Yong-Sheng Chang. Towards understanding members' interactivity, trust, and flow in online travel community [J]. Industrial Management, 2005,105, 7935–954.
Nelson, T.H. The right way to think about software design. The Art of Human-Computer Interface Design, Addison-Wesley, Reading, MA, 1990.
Russell J.A. "Affective Space Is Bipolar", Journal of Personality and Social Psychology, 1979, 37, 345–356.
Sukpanich N. Machine interactivity and person interactivity: The driving forces behind influences on consumers' willingness to purchase online. Doctor, the University of Memphis, 2004.
Webster, Jane, Trevino, Linda K. & Ryan, Lisa. The dimensionality and correlates of flow in human computer interactions [J]. Computers in Human Behavior, 1993, 9(4), 411–426.
Zeithaml, Valarie A., A. Parasuraman and Arivind Malhotra. Service Quality Delivery through Web sites: a critical review of extant knowledge [J]. Journal of the Academy of Marketing Science, 2002, 30(4), 362–75.

Research on intelligent E-maintenance platform for electronic commerce

Q.S. Jiao, J.C. Lei & J.Y. Zhang
School of Economics and Commerce, South China University of Technology, Guangzhou, China

ABSTRACT: In the fierce market competition environment, it is crucial to build an intelligent E-maintenance platform with the characteristics of agility, intelligence, and dynamic, network and socialization for the enterprise. It can reduce cost, improve efficiency and keep its competitiveness for the enterprise in the global market. Based on the analysis of the intelligent E-maintenance and the platform type electronic commerce, a solution of intelligent E-maintenance platform for electronic commerce was present in this paper. First of all, the platform structure characteristics and function composition were discussed. Then, the detailed design of the platform is put forward, and the working process of the platform scheme was designed. In the meantime, the paper expounds some key technologies of the realization of the platform.

1 INTRODUCTION

Along with the rapid development of science and technology, the machine equipment now is developing to large-scale, quick running, electromechanical integration, as well as complicated structure. The traditional later maintenance, periodic preventive maintenance and the isolated closed maintenance system already cannot satisfy the requirement of modern production. In order to guarantee the normal operation of equipment, enterprises not only have to take conservative strategy and reduce equipment failure rate in extremely low status, but also need to arrange with sufficient technical personnel for maintenance. In order to solve timely and flexible maintenance requirements of the device, intelligent E-maintenance arises at the historic moment. Intelligent E-maintenance refers to the fusion of remote monitoring technology, intelligent diagnosis technology, remote maintenance support system architecture and network technology, decision—making optimization technology[1] in maintenance system, real-time monitoring of equipment, making maintenance strategy according to the monitoring data, and makes the equipment in the whole life cycle to achieve almost zero fault and reach the optimal production status of maintenance mode.

On the other hand, electronic commerce platform[2] has become an inevitable trend, and it is the inevitable stage of e-commerce mode after the practical application. Compared with the traditional business mode, low cost and high information content of e-commerce have been widely recognized. Enterprises carry out the e-commerce

to improve competitive ability of the enterprise itself constantly. But when the enterprises develop e-commerce alone, it will become the information isolated island. E-commerce platform can solve such a problem, and gradually account for the important position in the electronic commerce.

Under such opportunities, in order to solve the technical bottlenecks and challenges of intelligent E-maintenance, the combination of intelligent E-maintenance with e-commerce and E-maintenance are necessary, which can strengthen cooperation between enterprises, promoting enterprise to regard product as the core, to serve as the means. Intelligent E-maintenance platform can realize the sharing of maintenance technology and resources on the basis of intelligent E-equipment, and fast monitoring and maintenance of equipment to the greatest degree, and improve the speed and quality of equipment maintenance services.

2 INTELLIGENT E-MAINTENANCE PLATFORM FOR ELECTRONIC COMMERCE

2.1 Characteristic of intelligent E-maintenance platform for e-commerce

As an innovation of equipment maintenance mode[3], Intelligent E-maintenance orienting e-commerce is combination and epitaxial of intelligence E-maintenance and e-commerce under the prescient maintenance strategy, and is the realization of digital, intelligent of prescient maintenance strategy. This platform can support equipment manufacturers, equipment maintenance outsourcing service providers and users to carry out equipment maintenance work together intelligently based on device state in the framework of predicted maintenance strategy, and to realize the maximization of maintenance benefits through the maintenance resource optimization configuration and maintenance of process workflow management. Compared with the traditional pattern of equipment maintenance, intelligent E maintenance orienting e-commerce mode has the following characteristics:

1. Information sharing. To ensure the seamless coordination on the maintenance and management work between enterprise and suppliers, effectively play the professional knowledge of equipment suppliers, you must realize all the based data sharing through the equipment life cycle. To ensure that enterprises and suppliers can get all kinds of information rapidly, accurately and comprehensively, electronic data interchange system, the database Shared technology, knowledge base management system and Internet data transmission technology to should be fully used to realize the integration of the maintenance information and guarantee the reliability of the equipment parameters and fault recording, predictable equipment solutions, and other important information sharing.
2. The union. Intelligent E maintenance platform orienting e-commerce emphasizes the cooperation and coordination between equipment users, equipment manufacturers and equipment maintenance outsourcing service provider, so as to establish a win-win relationship. The platform adopts collaborative commerce model, realizing the integration with other information management system among the device users, and realizing the combination of equipment users, equipment manufacturers, equipment maintenance outsourcing service providers and relevant experts on the intelligent E maintenance platform, all of which form as the dynamic alliance of diagnosis and maintenance on equipment, and to achieve overall target under multilateral participation in collaborative working process during the maintenance of business processes.
3. The dynamicity. Intelligent E maintenance orienting e-commerce is dynamic, real-time and online, its maintenance system is mainly carried out in the form of information storage, transmission and processing through the Internet/ Intranet platform. Real-time data collection and analysis, timely maintenance and update of database should be used on equipment with monitoring condition, so that when you can take the most appropriate maintenance program as soon as possible.

4. Intelligent. Intelligent E-maintenance emphasizes equipment maintenance work in the form of intelligent. On the basis of the fault model, fault diagnosis technology, maintenance experience, equipment engineering technology and maintenance decision-making, E-maintenance supports intelligent monitoring and evaluation of device status, intelligent diagnosis of fault, aid decision making and intelligent planning of maintenance plan.

Intelligent E maintenance platform system for e-commerce includes equipment, users, and intelligent E maintenance platform for e-commerce. It is showed in Figure 1. These three components contact each other and none can be dispensed with, which embodied in:

1. Equipment is the main part of the platform, the platform provides maintenance for equipment, and equipment real-time state is monitored by the platform. Equipment's running information in real time after the acquisition, storage, and processing is transferred through remote network to E maintenance platform server. The platform proceeds status analysis, fault diagnosis, at the same time generates and returns the equipment maintenance strategy and program.
2. Users are accepted to the service, which include the equipment users, equipment manufacturers and equipment maintenance outsourcing service providers. The platform provides customers with information, technology, knowledge, supply and maintenance decision, etc., responding to the questions from users. The users provide the necessary equipment initial information, previous fault information and predictable solutions to equipment failure to the platform.
3. Users by adding platform for access to gain the use of the platform and equipment and the intelligent maintenance on the scene, at the same time they provide engineering technical data and equipment customization requirements, etc.

2.2 *Detailed design of intelligent E-maintenance platform for e-commerce*

Intelligent E maintenance platform[4] is showed in Figure 2. In the intelligent E-maintenance

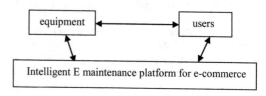

Figure 1. Intelligent E-maintenance platform system for e-commerce.

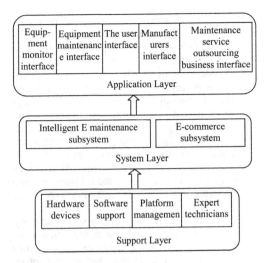

Equip-ment monitor interface	Equipment maintenance interface	The user interface	Manufacturers interface	Maintenance service outsourcing business interface
Application Layer				

Intelligent E maintenance subsystem	E-commerce subsystem
System Layer	

Hardware devices	Software support	Platform managemen	Expert technicians
Support Layer			

Figure 2. Intelligent E-maintenance platform framework for e-commerce.

platform for e-commerce, core business is the platform provider. Through the process of equipment maintenance, equipment manufacturers, equipment maintenance outsourcing service providers and equipment using enterprises are closely linked together around the core enterprise. Intelligent E-platform for e-commerce adopts three-layer architecture: application layer, system layer and support layer.

1. Support layer includes hardware devices, software support which including data support and security support, platform management and expert technicians. Support layer is the basis level and core resource of the system, ensuring that the system can be realized.

2. System layer contains E maintenance subsystem and subsystem of e-commerce. System layer provides intelligent E maintenance content through the electronic commerce. The content of the system is the main part of the platform to process information and is the method and process for the transformation of the external and internal information resources.

   Intelligent E maintenance subsystem includes local diagnosis and remote diagnosis. When equipment failure is tested, the information collection device installed at the bottom of the equipment will collect and process the fault information, and send to a local fault diagnosis expert system which will use the diagnosis reasoning machine diagnosis to determine the cause of the problem and make maintenance decision and fault decision in advance through the search network diagnostic knowledge base, method base and case store. When the local

diagnosis doesn't solve the problem, the remote diagnosis can be applied. Remote diagnosis means that the expert analyses and judges on the basis of the information transmission from field test system, discussing with the field monitoring person, proceeding real-time detection according to needs and providing diagnosis and strategy.

E-commerce subsystem includes smart E maintenance and equipment market. Maintenance market is a marketing and trading market for intelligence E maintenance services. The users use intelligence E maintenance market to do e-commerce transactions of intelligence E maintenance.

3. Application layer includes the equipment monitor interface, equipment maintenance interface, user interfaces, manufacturer and maintenance service provider interface. The application layer provides a platform of information input and output function, which is the external interface of the platform. The layer is the implement of function for the platform, and it is the interface and way for users, enterprises and others.

Equipment manufacturers can more quickly find the customer demand for equipment and make corresponding plan through the platform. Also manufacturers can improve manufacturing enterprise's innovation capability and core competitiveness through cooperation and competition with mutual information sharing between manufacturers. Equipment users receive a perfect after-sale service, which can reduce the rate of equipment failure and improve the quality and speed of the equipment maintenance, so that it can not only improve equipment production capacity, but also get new device information and deepen the cooperation with equipment manufacturing enterprises. Through information sharing of E maintenance content between Platform and equipment manufacturers, equipment service providers and equipment using enterprises, the whole industry chain of information can be collected in the platform, making the equipment manufacturers and equipment users effectively solve the problem of asymmetric information between businesses.

3 INTELLIGENT E-MAINTENANCE PLATFORM WORKFLOW FOR E-COMMERCE

This platform scheme is mainly objected to intelligent e maintenance of equipment. Its working process can be divided into two parallel processes, which are maintenance transaction process and equipment transaction process respectively. The process is shown in Figure 3.

Figure 3. Intelligent E-maintenance platform workflow for e-commerce.

During the maintenance transaction process, firstly device manufacturers initialize the data of the support layer, input device information into database, initialize to the device maintenance basic knowledge in the knowledge base, and add equipment predictable maintenance methods into the method base, thus accomplish the initialization data support. Equipment using enterprises buy the equipment, then register and start using it. They use the platform for intelligent E maintenance on equipment. Maintenance of equipment through remote monitoring for decision-making, which includes degenerative prediction, making maintenance decision and fault decisions in advance, applying for using local diagnosis or remote diagnosis according to different types of equipment maintenance and accomplishing maintenance work. After that, the platform adds maintenance instance into instance library. Data support updates information through self-learning on distributed database and complete the maintenance decision-making. During the maintenance decision-making process, the users, intelligent E maintenance e-commerce platform and equipment control the interaction by the way of trading.

Equipment trading process is the business process trading on the equipment. The difference between equipment transaction process and general equipment is that it adds the information update capacity of equipment, which makes the expert process equipment improvement and new equipment research through the queries for knowledge base and case database. In the meantime, equipment manufacturing enterprise can do new product marketing and trading through the device information updates, and the implementation of the intelligent E maintenance process.

4 KEY TECHNOLOGY OF E-MAINTENANCE PLATFORM FOR E-COMMERCE

Key Technologies of intelligent E-maintenance platform[5] for e-commerce are:

1. Remote monitoring technology and IP sensor technology. Research network data support communication interface, rapid monitoring system environment, define the data transmission and processing of graphical programming software tools, improve the portability of detection software, can rapid build of monitoring diagnosis software platform according to different monitoring objects. IP sensor technology is the material basis for the maintenance system and the implementation of comprehensive real-time understanding of equipment operation, which can provide multidimensional information and parameters for timely forecast and diagnose of the condition of product performance.

2. Intelligent forecast and diagnosis technology. The model of the effective degradation state of product operation, product performance prediction need establish. Before the product performance degradation or failure, it can make maintenance decision in advance and take appropriate measures to maintain it. Intelligent E maintenance platform is based on intelligent information processing methods and genetic algorithm, such as neural network, fuzzy theory and research on the equipment fault intelligent diagnosis technology and different intelligent diagnosis method of fusion technology.

3. System architecture and network technologies that support for remote maintenance. The basis of realizing remote maintenance is the use of network technology to achieve information for smooth transmission, building a network of remote maintenance to improve each node's mobility and portability of maintenance network. According to the remote diagnosis, it maintains each node (diagnosis maintenance center, users, manufacturers, service providers and diagnosis expert) of the network for information transmission requirements, taking into account network equipment price, equipment technical difficulties, researching the realize of the system and application solutions from the aspects such as hardware, software and integration.

4. The embedded network access technology. Research and develop the built-in 10 m/100 m Ethernet interface with high-performance embedded microcontroller processor and Embedded Operating System (EOS) as the core, realizing the equipment condition monitoring unit of the network access and network transmission of equipment condition monitoring data and improving the transmission efficiency. It realizes the embedded Web Server, system maintenance function based on the network, and Web publishing for equipment state data which is used to meet the demand for amount of data of system.

5 CONCLUSION

Nowadays, as the explosive development of network communication technology and information technology, e-commerce emerged at the end of the 20th century. It puts forward a kind of new business opportunities, challenges, requirements and rules. It represents the future development direction of information industry. Meanwhile great changes in equipment maintenance system of equipment manufacturing have taken place. Intelligent E maintenance platform for e-commerce provides a feasible way for intelligent E maintenance commercialization. On the basis of literature research, this paper puts forward intelligent E maintenance platform for e-commerce solution, and the features of platform, the platform architecture, detailed design, workflow, and some key technologies to realize the platform are expounded.

ACKNOWLEDGMENT

The authors gratefully acknowledge financial support for this research by the project named "Research on e-commerce oriented plant E-maintenance mode" (x2jmD2118150) supported by "the Fundamental Research Funds for the Central Universities".

REFERENCES

Liu, J., Yu, D.J. & Li, D.G. 2004. System Framework and key Enabling Technology of E-Maintenance. China Mechanical Engineering 15(13): 1171–1175.

Sheng, T.W., Chen, X.H. & Yi, S.P. 2008. E Maintenance Oriented Network Group Decision Support System Research. Journal of Southwest University 30(3): 146–151.

Sheng, T.W., Chen, X.H. & Yi, S.P. 2008. Multi-agent Collaborative Commerce System of Embedded Intelligence E Maintenance. Journal of Chongqing University 31(4): 371–375.

Zhang, H., Chen, H. & Yang, J.G. 2011. Intelligent Maintenance Application in Rail Transit Environment and Equipment Monitoring and Control System. Journal of southeast university 37(4): 457–461.

Zhao, T., Han, J.T. & Qi, H.Y. 2004. The Research on Overall Scheme and Implementation Technology on Manufacturing E Maintenance Platform. School of Management, Tianjin University 26(3): 12–15.

Author index